Heterocyclic Chemistry in Drug Discovery

Heterocyclic Chemistry in Drug Discovery

Edited by

Jie Jack Li
Bristol-Myers Squibb Company

WILEY

A JOHN WILEY & SONS, INC., PUBLICATION

Published by John Wiley & Sons, Inc., Hoboken, New Jersey.

Published simultaneously in Canada.

Library of Congress Cataloging-in-Publication Data:

Heterocyclic chemistry in drug discovery / edited by Jie Jack Li.
p. ; cm.
Includes bibliographical references and index.
ISBN 978-1-118-14890-7 (cloth)
I. Li, Jie Jack.
[DNLM: 1. Drug Discovery—methods. 2. Heterocyclic Compounds—chemistry. 3. Heterocyclic Compounds—pharmacology. QD 400]

615.1'9—dc23 2012030054

10 9 8 7 6 5 4 3 2 1

Dedicated To

Li Jing Ya, Li (Zhen) Cheng-Cheng, Li Chun, and Li Lei

Contents

Detailed Table of Contents

Preface

There is a disconnection in our education of organic chemists whose inspiration is to work on drug discovery in either industry or academia. The traditional textbooks are no longer adequate in preparing our undergraduate and graduate students in entering the pharmaceutical industry. The original philosophy was that one could learn medicinal chemistry "on the job" after a strong synthetic chemistry background.

In this book, attempts have been made to fuse the two fields: heterocyclic chemistry and drug discovery. I hope it will give our undergraduate and graduate students a "jump-start" in this competitive employment market. As a matter of fact, there is no sacrificing of a solid education in "authentic" heterocyclic chemistry here. All aspects of reactions, reactivity, and mechanisms are still intact, except they are discussed in the context of medicinal chemistry and drug discovery.

I welcome your critique! Please send your comments to me directly: lijiejackli@hotmail.com.

Jack Li
June 1, 2012

Contributing Authors:

Dr. Nadia M. Ahmad
Eli Lilly and Company
Erl Wood Manor
Windlesham
Surrey, GU20 6PH
United Kingdom

Dr. Narendra B. Ambhaikar
Dr. Reddy's Laboratories
CPS
Bollaram Road, Miyapur
Hyderabad-500 049
A. P., India

Professor Adam M. Azman
Department of Chemistry
Butler University
4600 Sunset Avenue
Indianapolis, IN 46208, United States

Connor W. Brown
Hamilton College
198 College Hill Road
Clinton, NY 13323, United States

Dr. Timothy T. Curran
Chemical Development
Vertex Pharmaceuticals
130 Waverly Street
Cambridge, MA 02139, United States

Professor Amy B. Dounay
Department of Chemistry and Biochemistry
Colorado College
14 East Cache La Poudre Street
Colorado Springs, CO 80903, United States

Tyler W. Farnsworth
Division of Natural Sciences and Engineering
University of South Carolina Upstate
800 University Way
Spartanburg, SC 29303, United States

Professor Micheal Fultz
Department of Chemistry
West Virginia State University
Institute, WV 25112, United States

Professor Timothy J. Hagen
Department of Chemistry and Biochemistry
Northern Illinois University
DeKalb, IL 60115, United States

Dr. Jie Jack Li
Medicinal Chemistry
Bristol-Myers Squibb Company
Route 206 and Province Line Road
Princeton, NJ 08540, United States

Dr. Sha Lou
Process Research and Development
Bristol-Myers Squibb Company
New Brunswick, NJ 08901, United States

Professor Richard J. Mullins
Department of Chemistry
Xavier University
3800 Victory Parkway
Cincinnati, OH 45207-4221, United States

Dr. Jennifer Xiaoxin Qiao
Medicinal Chemistry
Bristol-Myers Squibb Company
P.O. Box 5400
Princeton, NJ 08543-5400, United States

William Rollyson
Department of Chemistry
West Virginia State University
Institute, WV 25112, United States

Professor Joshua V. Ruppel
Division of Natural Sciences and Engineering
University of South Carolina Upstate
800 University Way
Spartanburg, SC 29303, United States

Professor Nicole L. Snyder
Department of Chemistry
Davidson College
Box 7120
Davidson, NC 28036, United States

Alexander D. Thompson
Hamilton College
198 College Hill Road
Clinton, NY 13323, United States

Dr. Ji Zhang
HEC R&D Center
Pharmaceutical Science
Process Research and Development
HEC-Hi-Tech Park, Dongguan
Guang Zhou, Guang-Dong Province
P. R. China

Dr. Zheng Zhang
Department of Chemistry and
Biochemistry
Northern Illinois University
DeKalb, IL 60115, United States

Professor Alexandros L. Zografos
Department of Chemistry
Aristotle University of Thessaloniki
Thessaloniki 54124, Greece

Chapter 1 Introduction

Jie Jack Li

1.1 Nomenclature of Heterocycles

What's in a name? That which we call rose by any other name would smell as sweet. [William Shakespeare, *Romeo and Juliet* (II, ii, 1–2)].

Contrary to Shakespeare's exclamation, *naming heterocycles* is an integral part of our learning of heterocyclic chemistry. They are the professional jargon that we routinely use to communicate with our peers.

Heterocycles, as the name suggests, are cyclic compounds containing one or more heteroatoms such as N, O, S, P, Si, B, Se, and Se. They may be further divided into aromatic heterocycles and saturated heterocycles. This book will focus largely on aromatic heterocycles. Saturated heterocycles represent a smaller portion of drugs. Another way of naming heterocycles is using the size of the heterocyclic rings. Therefore, they may be classified as three-, four-, five-, six-, and seven-membered heterocycles, and so on.

Three-membered heterocycles are important reaction intermediates in organic chemistry and in preparing medicines. But they usually do not exist in final drugs because they are reactive in physiological environments. Exceptions are found in cancer drugs such as epothilone A and mitomycin C (see Section 1.4, page 9), where their reactivities are taken advantage of for therapeutic purposes.

The most frequently encountered three-membered heterocycles are oxirane, thiirane, aziridine, and azirine.

oxirane thiirane aziridine azirine

Four-membered heterocycles include oxetane, 2*H*-oxete, thietane, 2*H*-thiete, azetidine, and azete. In the field of drug discovery, oxetanes and azetidines are more and more incorporated into drugs for modulating biological and physical properties as well as for expanding intellectual properties space.

oxetane 2*H*-oxete thietane 2*H*-thiete azetidine azete

Five- and six-membered heterocycles are of utmost importance to both life and drug discovery. The most common five-membered heterocycles with one heteroatom are pyrrole, furan, and thiophene.

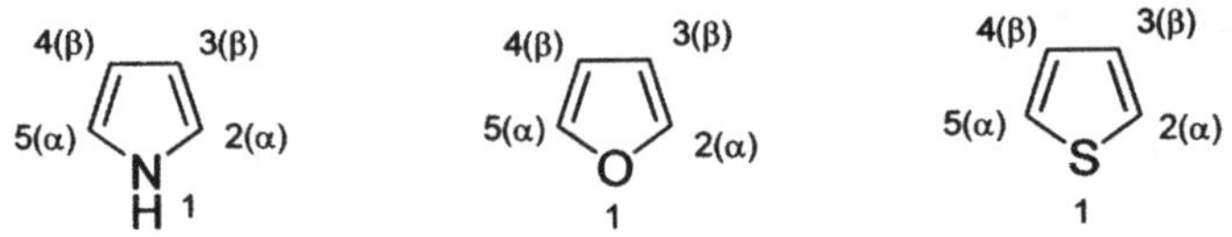

pyrrole furan thiophene

Popular five-membered heterocycles with two heteroatoms include pyrazole, imidazole, oxazole, isoxazole, thiazole, and isothiazole.

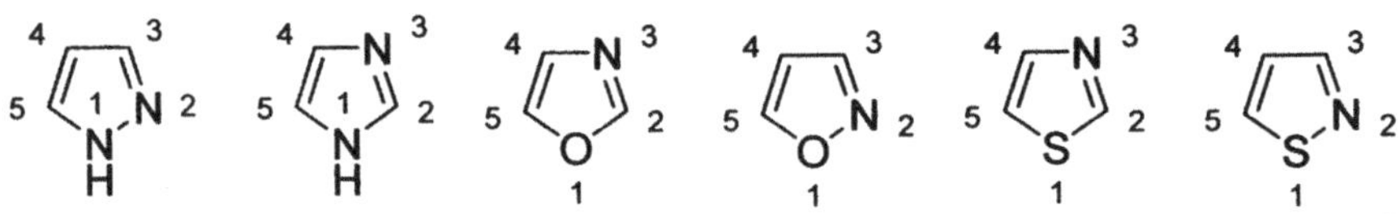

pyrazole imidazole oxazole isoxazole thiazole isothiazole

All these aromatic heterocycles have their counterparts in the corresponding saturated heterocycles. Among those, pyrrolidines, tetrahydrofurans, and oxazolidines are more frequently encountered in drug discovery.

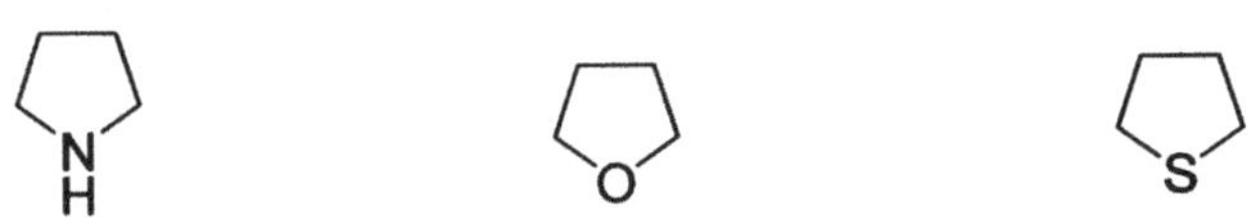

pyrrolidine tetrahydrofuran tetrahydrothiophene

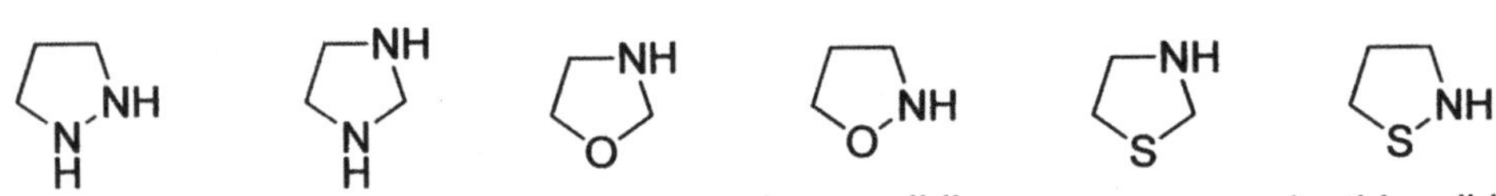

pyrazolidine imidazolidine oxazolidine isoxazolidine thiazolidine isothiazolidine

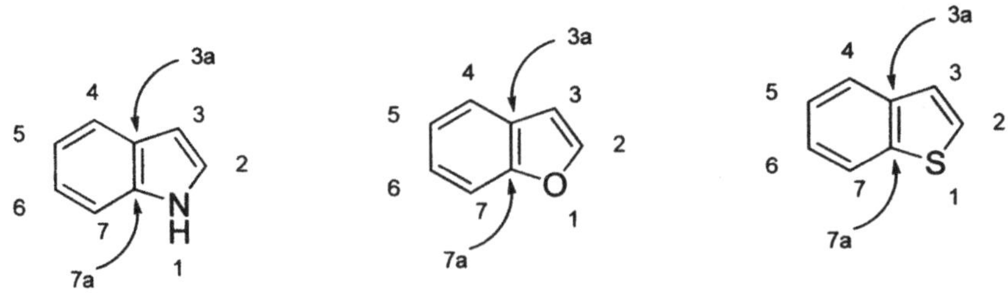

indole benzofuran benzothiophene

Some of the important benzene-fused five-membered heterocycles are indole, benzofuran, benzothiophene, benzimidazole, benzoxazole, and benzothiazole. The numbering of these heterocycles is shown below:

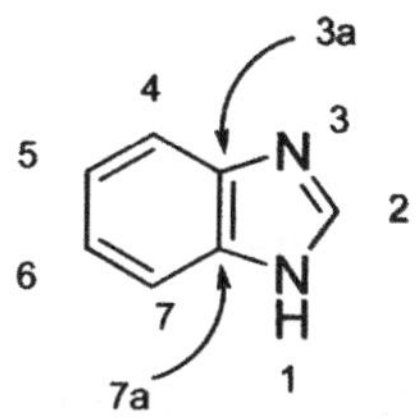
benzoimidazole

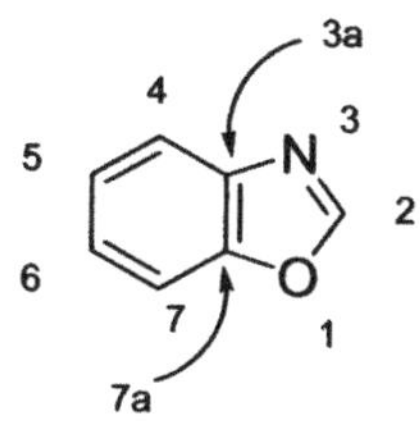
benzooxazole

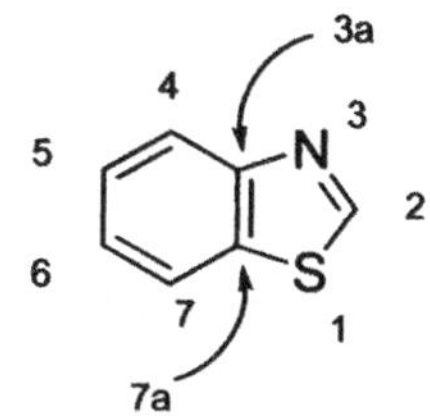
benzothiozole

Chief among the six-membered heterocycles, pyridine and its benzene-fused derivative quinoline are most ubiquitous. Pyrazine and its benzene-fused analogue, quinoxaline, also play an important role in heterocyclic chemistry.

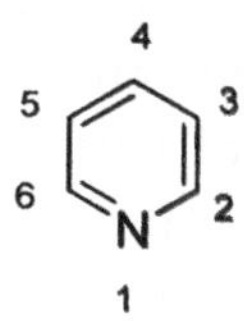
pyridine

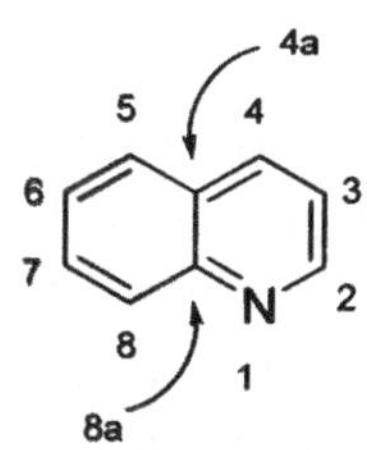
quinoline

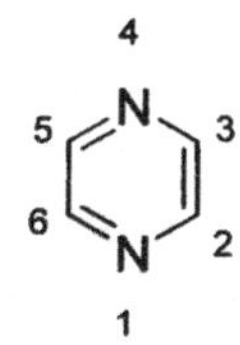
pyrazine

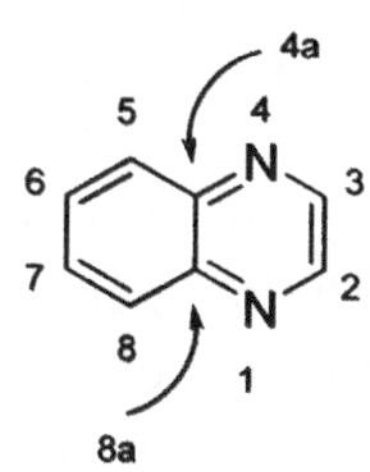
quinoxaline

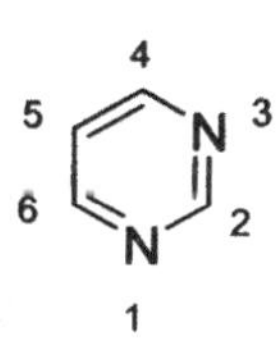
pyrimidine

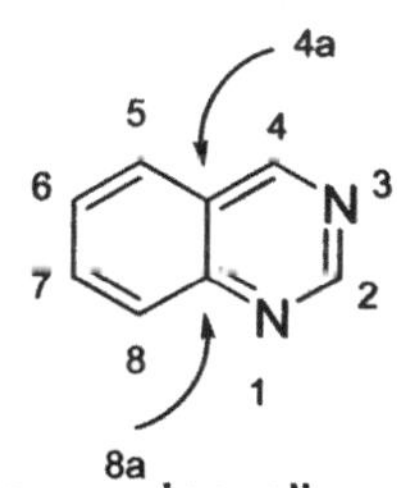
quinazoline

Their corresponding saturated derivatives often encountered in drug discovery are piperidine and piperazine.

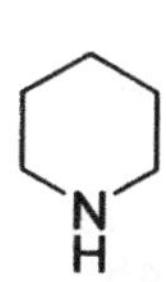
piperidine

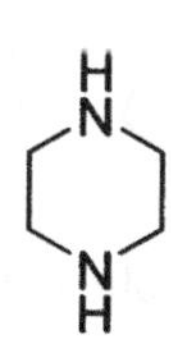
piperazine

1.2 *Aromaticity of Heterocycles*

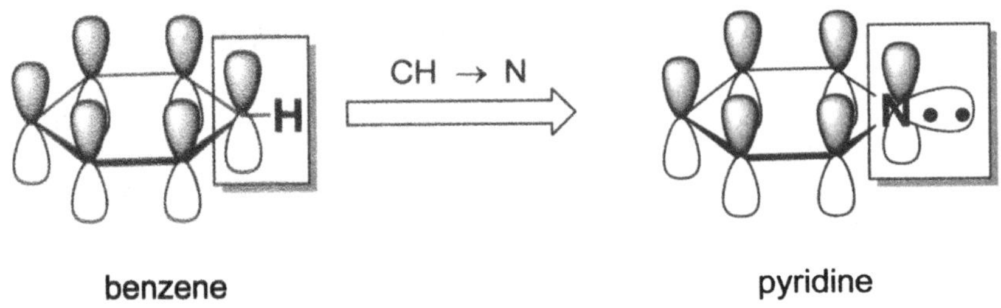

The major thrust of this book is aromatic heterocycles. According to Hückel's rule of aromaticity, a cyclic ring molecule is aromatic when the number of its π-electrons equals 4n + 2, where n is zero or any positive integer. The most common aromatic compound is benzene, which has 4 + 2 = 6 π-electrons. Pyridine, an electron-deficient aromatic heterocycle, also has 6 π-electrons. In comparison with benzene, pyridine has an additional lone pair of electrons at the nitrogen atom after it contributes a pair of two electrons to make up the 6 π-electrons for aromaticity. These lone pair electrons are responsible for much of pyridine's unique physical and chemical properties. On the other hand, furan, an electron-excessive aromatic heterocycle also with 6 π-electrons, is different from both benzene and pyridine. The oxygen atom has two lone pairs of electrons, one of which contributes to the 6 π-electrons to achieve the aromaticity. The second pair of electrons is located in an sp^2 hybrid orbital in the plane of the furan ring. Thiophene is similar to furan in its aromaticity although thiophene is more "aromatic" because the S atom is larger than the O atom.

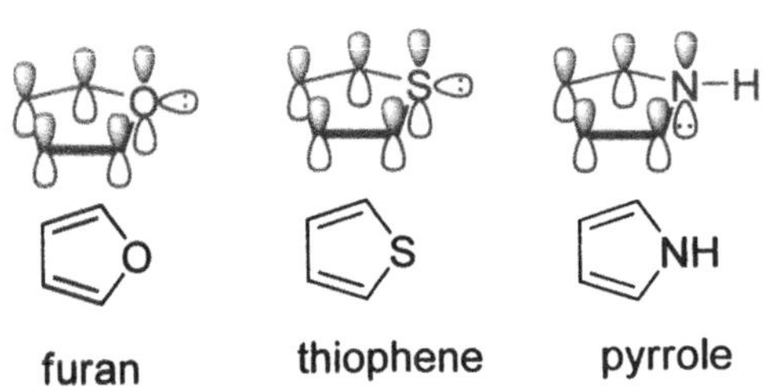

The relative aromaticity of common heterocycles is shown below:

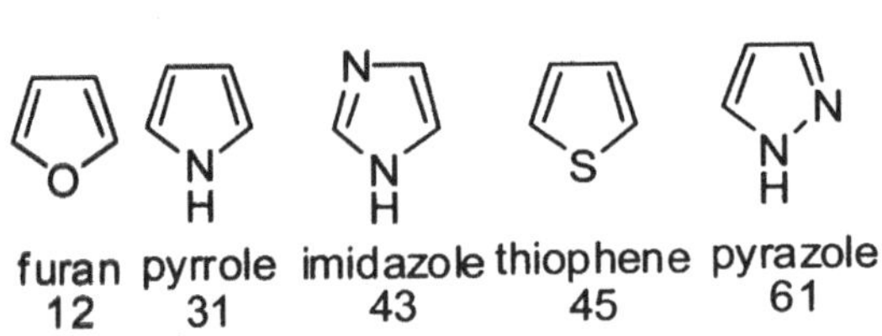

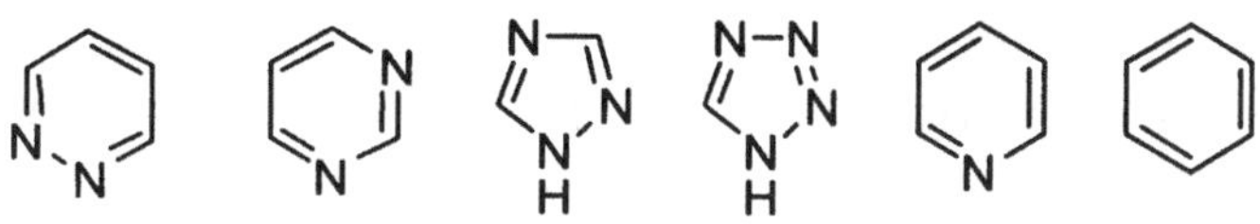

pyridazine 65 | pyrimidine 67 | triazole 71 | tetrazole 80 | pyridine 82 | benzene 100

Pyrrole, also an aromatic heterocycle with 6 π-electrons, is probably the most unique of all among the aromatic heterocycles. Different from furan and thiophene, the nitrogen atom on the pyrrole ring only has one lone pair of electrons, which both contributed to the 6 π-electrons to achieve the aromaticity. As a consequence, although pyrrole is also an electron-excessive aromatic heterocycle, just like furan and thiophene, pyrrole has many of its own characteristics. For instance, it is probably the most reactive as a nucleophile among all aromatic heterocycles (see Chapter 2). In addition, pyrrolc's conjugation effect outweighs the nitrogen's inductuve effect in the contributing dipole moment, with the partial positive charge resting at the nitrogen atom.

1.3 Importance of Heterocycles in Life

The importance of heterocycles in life was recognized as the nascent stage of organic chemistry two centuries ago with isolation of alkaloids such as morphine from poppy seeds, quinine from cinchona barks, and camptothecin from the Chinese joy tree. Today, heterocycles are found in numerous fields of biochemical and physiological such as photosynthesis, amino acids, DNA bases, vitamins, endogenous neurotransmitters, and so on.

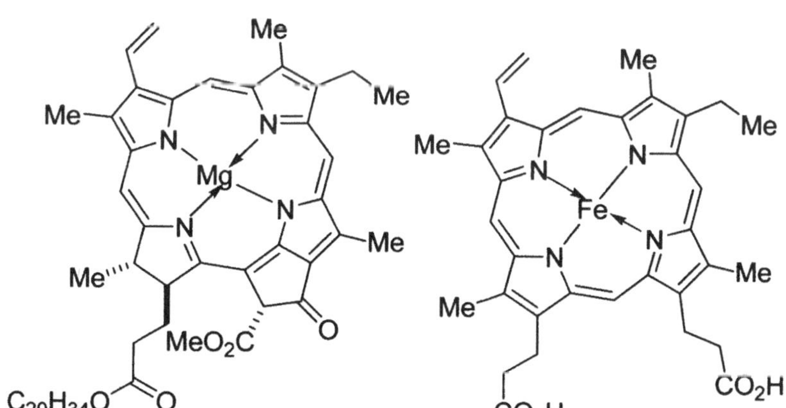

chlorophyll-a | heme

To begin with, chlorophyll is porphyrin (a tetramer of pyrrole) surrounding a magnesium atom. It is the molecule that absorbs sunlight and uses its energy to synthesize carbohydrates from CO_2 and water. This process, known as photosynthesis, is the basis for sustaining the life processes of all plants.

On the other hand, the heme consists of a porphyrin ring surrounding an *iron* atom. The ring contains a large number of conjugated double bonds, which allows the molecule to absorb light in the visible part of the spectrum. The iron atom and the attached protein chain modify the wavelength of the absorption and give hemoglobin its characteristic color.

Several amino acids, the building block of life, are made of heterocycles. Histidine has an imidazole; tryptophan has an indole; yet proline has a pyrrolidine.

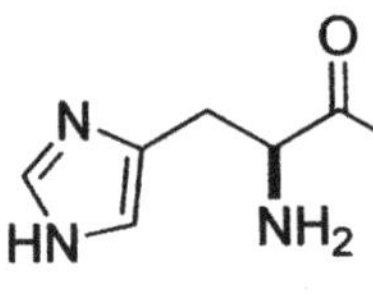

histidine

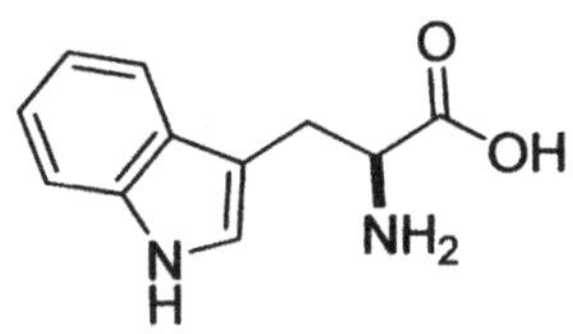

tryptophan

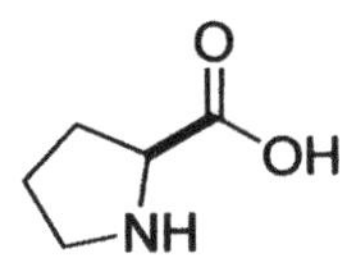

proline

Heterocycles also play an important role as endogenous neuro-transmitters. Chief among them are serotonin and histamine, which are of paramount importance in modulating the body's physiological and biochemical processes.

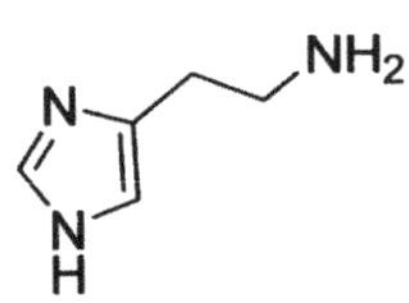

histamine

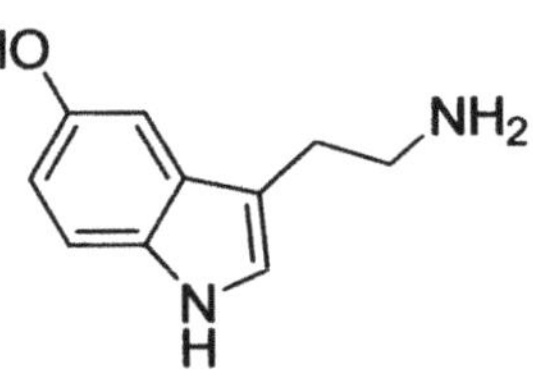

serotonin

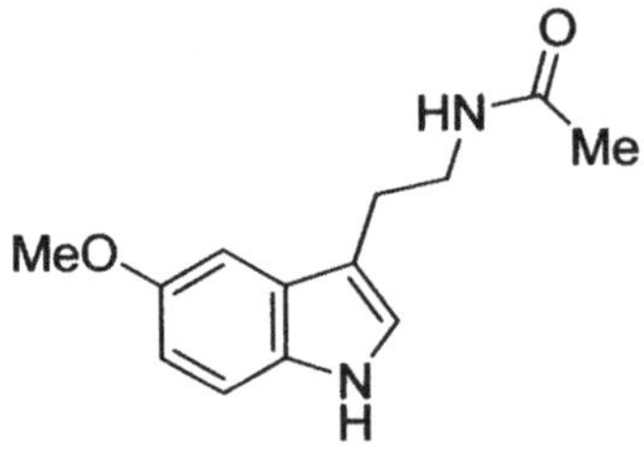

melatonin

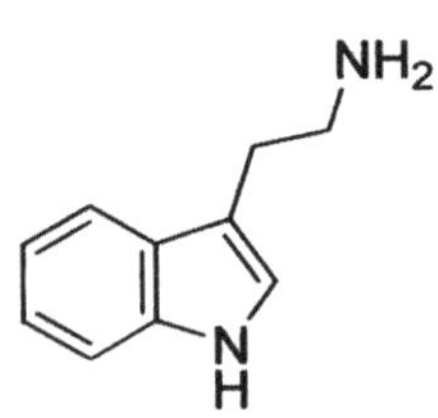

tryptamine

Melatonin regulates circadian rhythms, most noticeably sleep, whereas tryptamine is closely related to melatonin and the amino acid tryptophan.

The double helix of DNA, the code of life, comprises two base pairs: adenine/thymine (A/T) and cytosine/guanine (C/G).

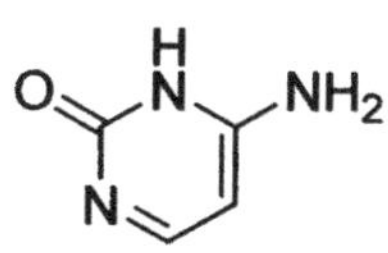
cytosine

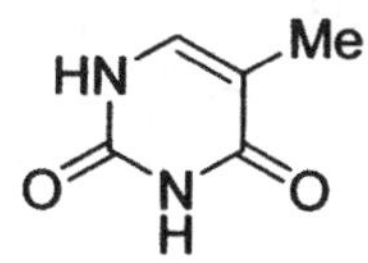
thymine

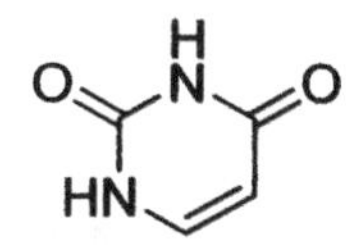
uracil

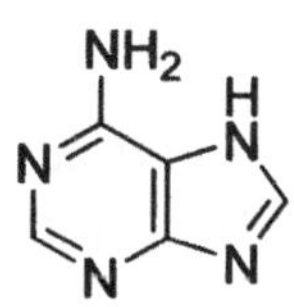
adenine

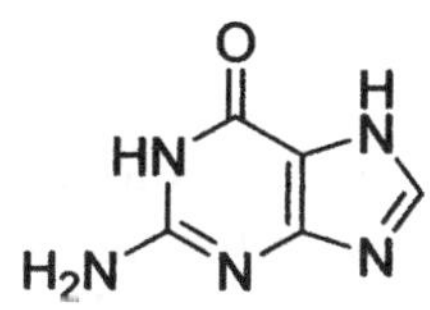
guanine

By adding the ubiquitous sugar fragments, we are left with nucleic acids contaning pyrimidine bases, including cytosine, thymine, and uracil and purine bases such as adenine and guanine.

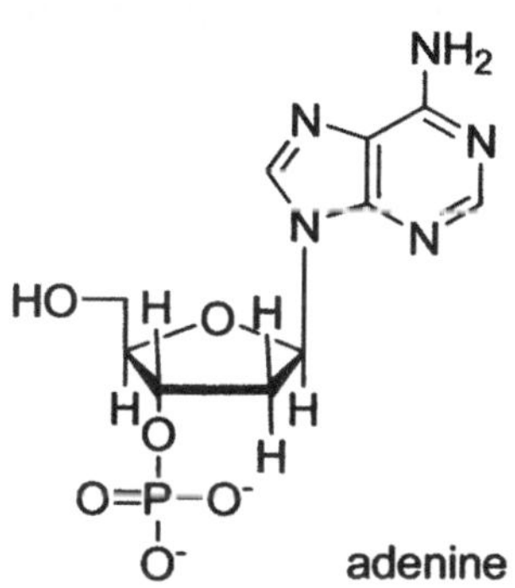
adenine

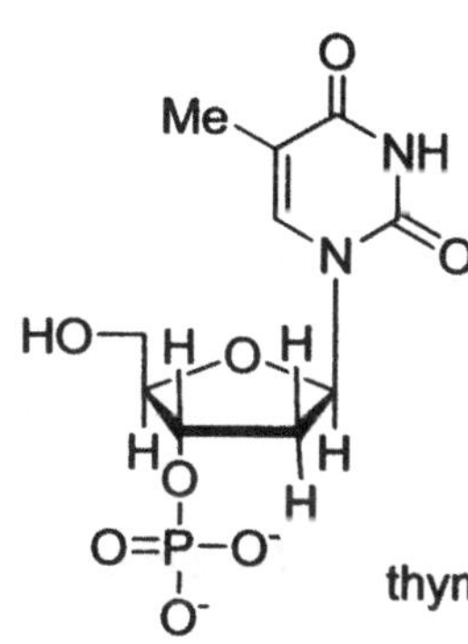
thymine

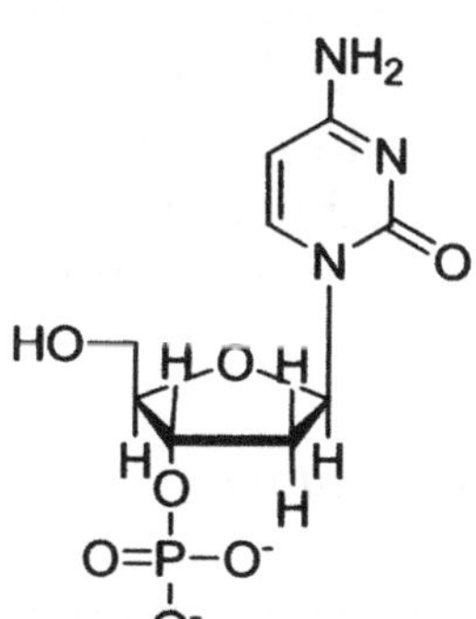
cytosine

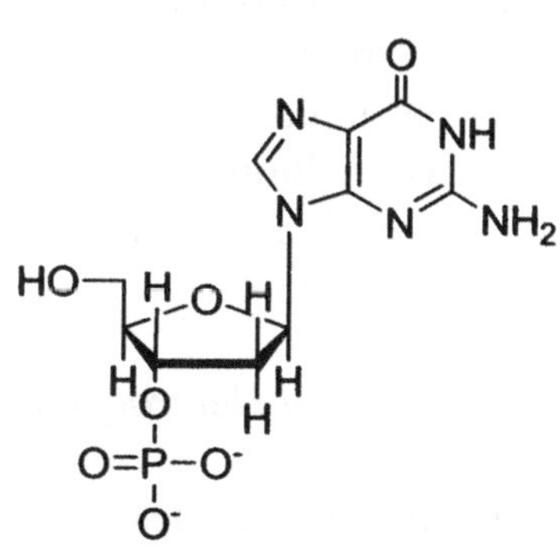
guanine

Thiazoles also play a prominent role in nature. For example, the thiazolium ring present in vitamin B_1 serves as an electron sink and its coenzyme form is important for the decarboxylation of α-keto-acids. The left-hand fragment of vitamin B_1 is an aminopyrimidine.

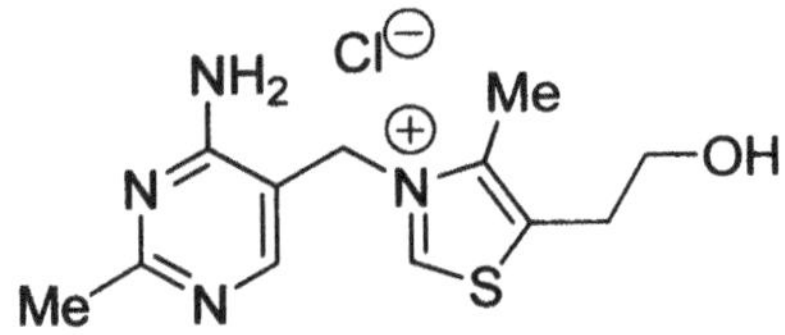

Vitamin B_1

Vitamin B_5 (nicotinic acid amide) and vitamin B_6 (pyridoxal) are pyridine-based molecules, whereas vitamin B_7 (biotin) is a bi-heterocycle fusing reduced imidazole and thiophene.

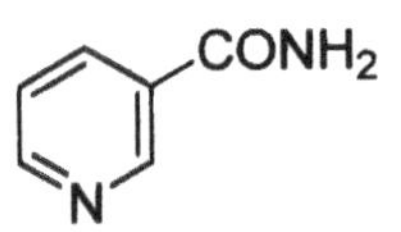

vitamin B_5 (nicotinic acid amide)

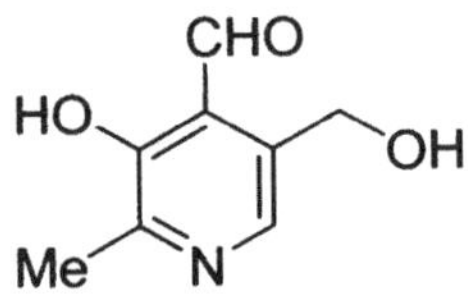

vitamin B_6 (pyridoxal)

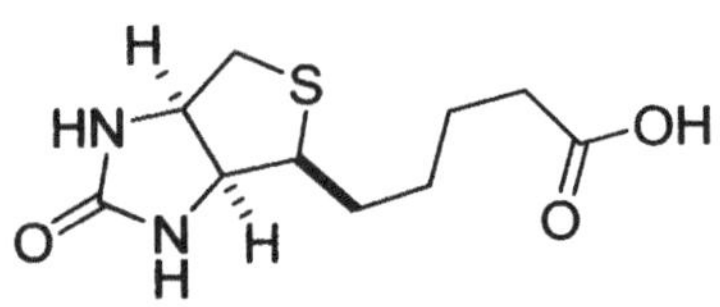

vitamin B_7 (biotin)

1.4 Importance of Heterocycles in Drug Discovery

It will be evident from the ensuing chapters that heterocycles play an extremely important role in drug discovery, in general, and in medicinal chemistry, in particular. Heterocycle-containing drugs are found in all therapeutic areas including cardiovascular and metabolic diseases, central nervous system (CNS), anti-cancer, anti-inflammatory, anti-ulcer, anti-infective drugs, and so on.

1.4.1 Five-Membered Heterocycles with One Heteroatom

Three-membered heterocycles are usually not fragments of drugs because they are reactive toward nucleophiles in physiological environments. Cancer drugs such as epothilone and mitomycin are exceptions rather than the rules. The epothilones have shown their eminent cytotoxic activity against tumor cells, taxol-like mitose inhibition and toxicity against multiple drug-resistant tumor cell lines. On the other hand, mitomycin C is isolated from a strain of bacteria called *Streptomyces lavendulae*. It is a chemotherapy agent because of its anti-tumor properties. It is indicated as a useful therapeutic agent in combination with other anticancer drugs for the treatment of disseminated adenocarcinoma of the pancreas and the stomach.

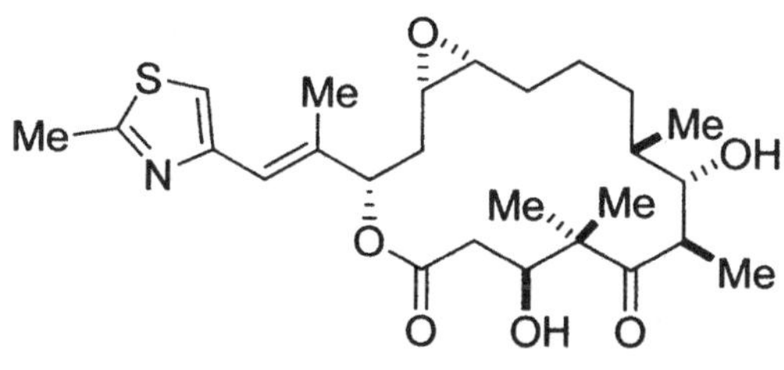

epothilone A

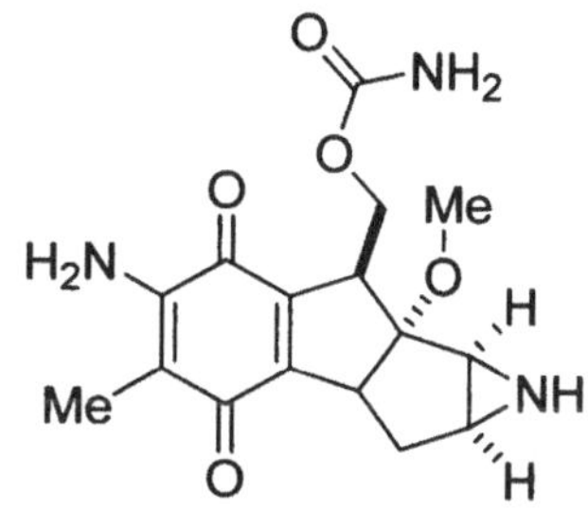

mitomycin C

Not many drugs contain four-membered heterocycles either. The best-known drug containing an azetidine-ring is Schering-Plough's ezetimibe (Zetia). Launched in 2002 as a cholesterol absorption inhibitor, its mechanism of action is the inhibition of the Nieman–Pick C1-like 1 (NPC1L1) protein.

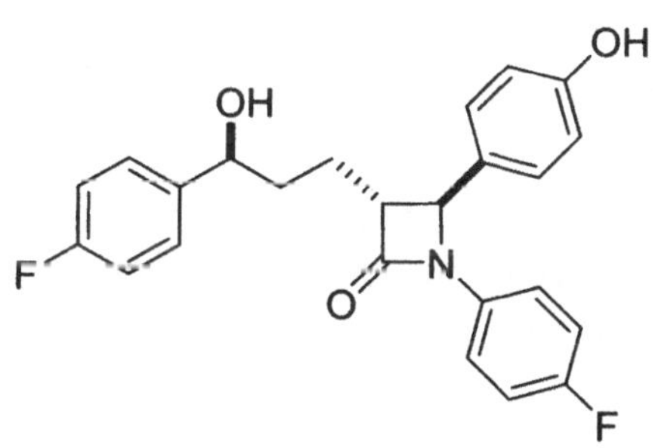

ezetimibe (Zetia), NPC1L1 inhibitor

Just as in life, five-membered heterocycles are of utmost importance to drug discovery. The most conspicuous of all is probably atorvastatin (Lipitor), an HMG-CoA inhibitor. Another bioactive pyrrole shown below is an antipsychotic agent.

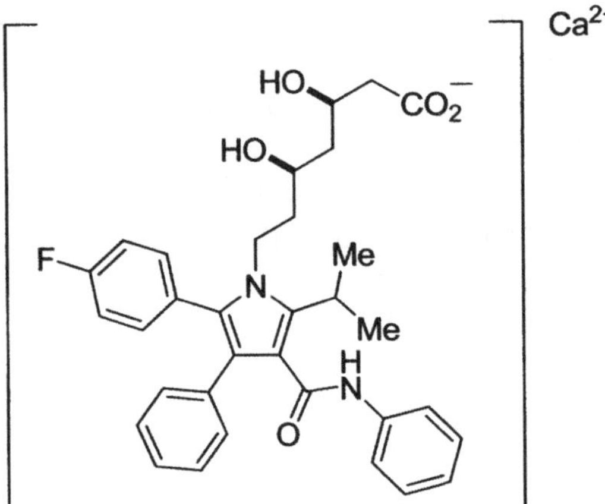

atorvastatin (Lipitor)

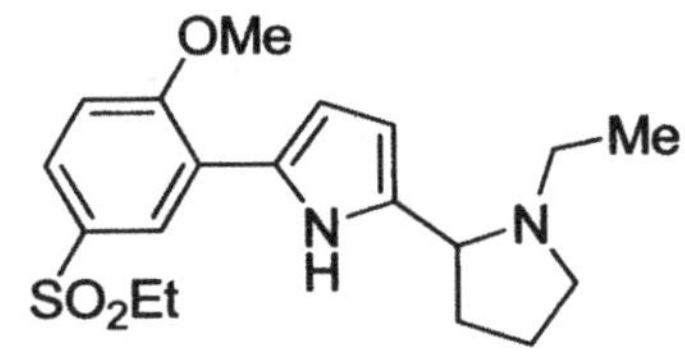

antipsychotic

Many drugs contain the indole-ring as their core structures. Fluvastatin sodium (Lescol) is an HMG-CoA reductase inhibitor.

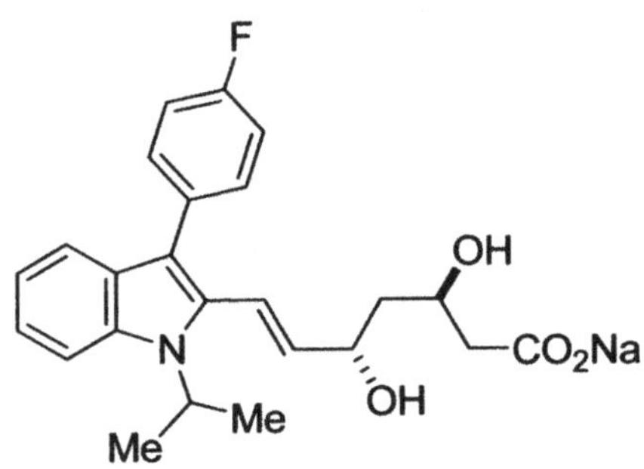

fluvastatin sodium (Lescol)

In addition, sumatriptan succinate (Imitrex), a serotonin receptor (5-HT_{1B1D}) agonist, is used to treat migraines. And naratriptan (Naramig) is a "me-too," indole-containing anti-migraine drug on the market.

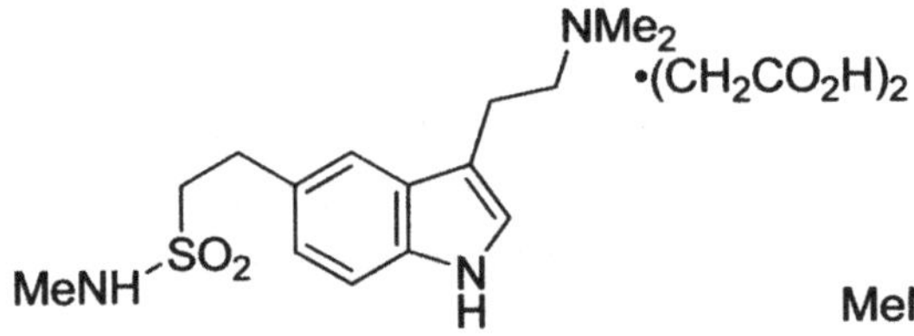

sumastriptan (Imitrex)

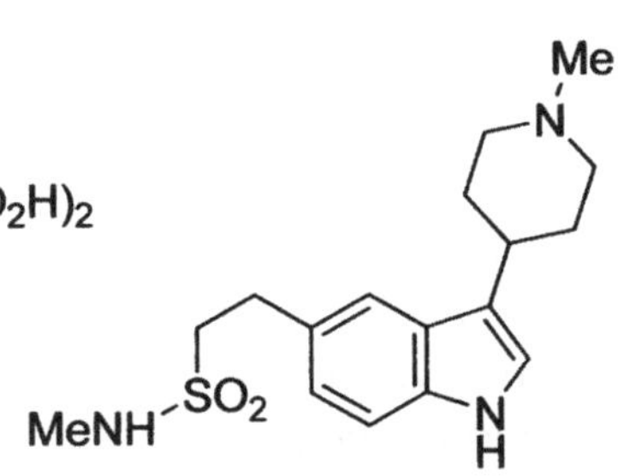

naratriptan (Naramig)

Furthermore, delavirdine (Rescriptor) is a novel HIV-1 reverse transcriptase inhibitor for HIV-positive individuals and zafirlukast (Accolate) is an antiasthma drug.

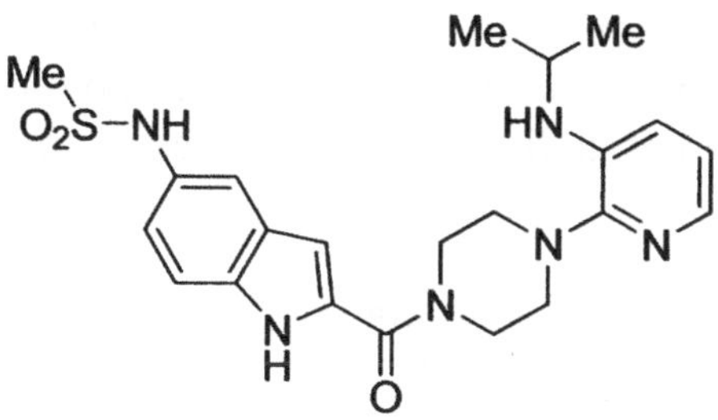

delavirdine (Rescriptor)

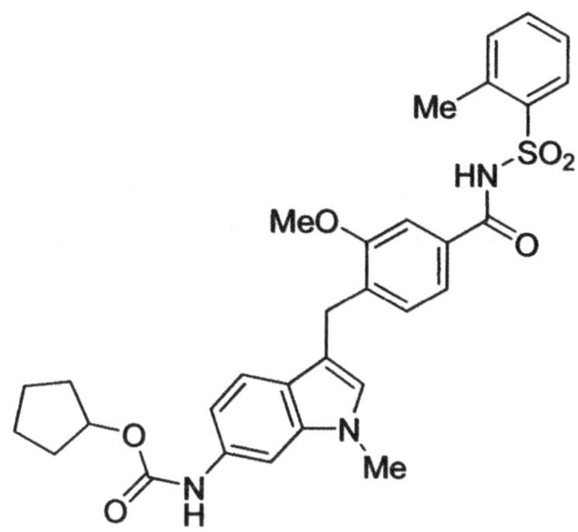

zafirlukast (Accolate)

Bristol-Myers Squibb's thiophene-containing clopidogrel (Plavix) inhibits platelet aggregation induced by adenosine diphosphate (ADP), a platelet activator that is released from red blood cells, activated platelets, and damaged endothelial cells. Clopidogrel, launched in 1993, achieved great commercial success. But its mechanism of action (MOA) was not elucidated until 1999: through the antagonism of the P2Y12 purinergic receptor and prevention of binding of ADP to the P2Y12 receptor by its active metabolite. Therefore, clopidogrel is a *bona fide* pro-drug.

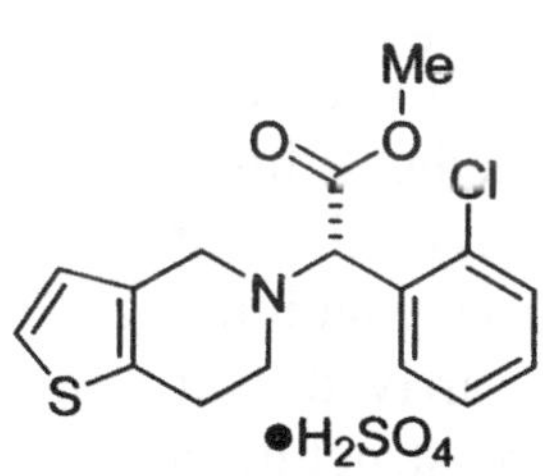

clopidogrel (Plavix)

Eli Lilly's raloxifene (Evista) is a selective estrogen receptor modulator (SERM) indicated for osteoporosis and breast cancer. Its core structure is a benzothiophene.

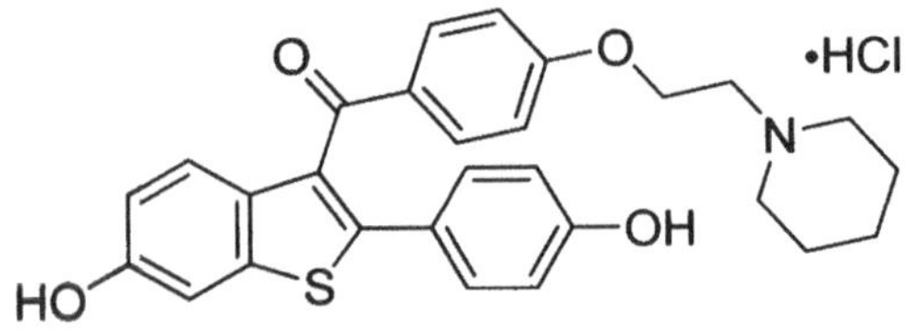

raloxifene (Evista)

Thiophene seems to be very popular in Li Lilly drugs. Its dual selective serotonin and norepinephrine reuptake inhibitor (SSNRI) for depression, duloxetine (Cymbalta), contains a thiophene. And its atypical antipsychotic drug olanzapine (Zyprexa) has a fused thiophene as its core structure.

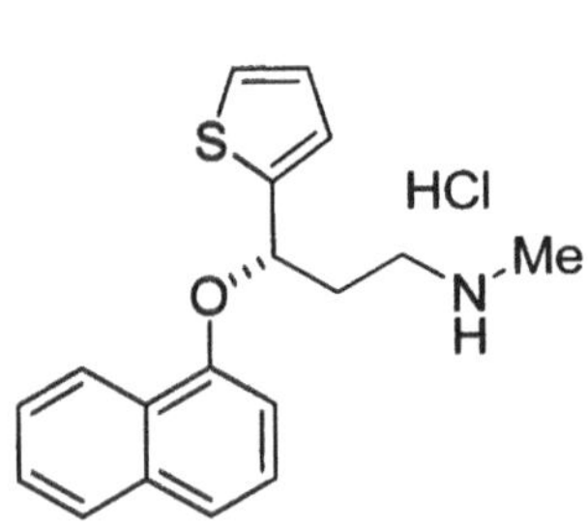

duloxetine (Cymbalta)

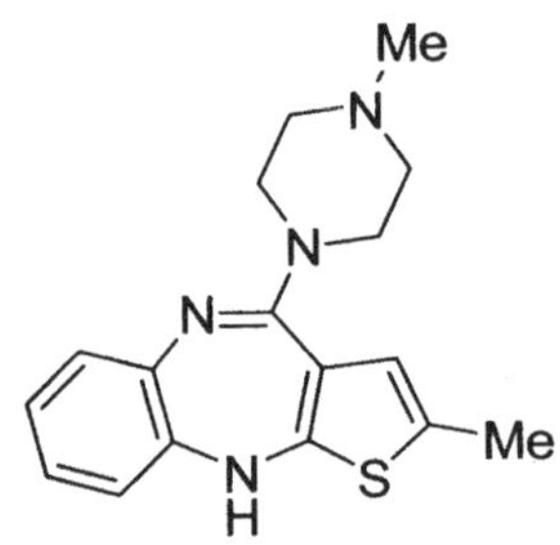

olanzapine (Zyprexa)

1.4.2 Five-Membered Heterocycles with Two Heteroatoms

Histamine-2 (H_2) receptor antagonists as anti-ulcer drugs best showcased the versatility of heterocycles in drug discovery. Marketed in the United States in 1977, SmithKline & French's cimetidine (Tagamet) became the first blockbuster drug ever in medical history in 1985. Transforming the imidazole ring into the dimethylamino-furan in combination with replacing cyanoguanidine with nitrovinyl guanidine gave rise to ranitidine (Zantac). Later on, Yamanouchi arrived at famotidine (Pepcid) using guanidinothiazople as its core structure and sulfamoyl-amidine as its side chain. All of these drugs went on to become blockbuster drugs.

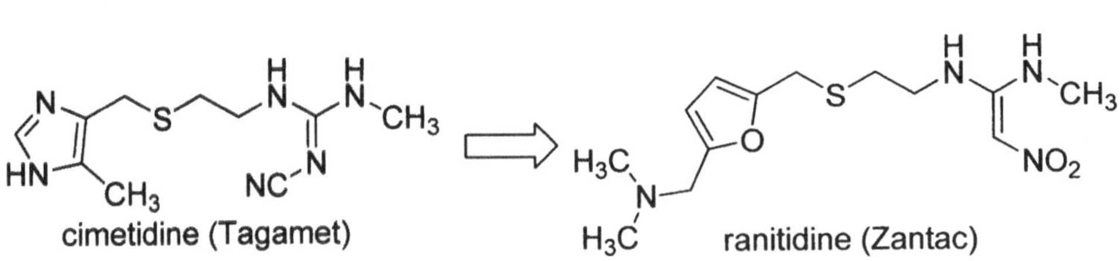

cimetidine (Tagamet) ranitidine (Zantac)

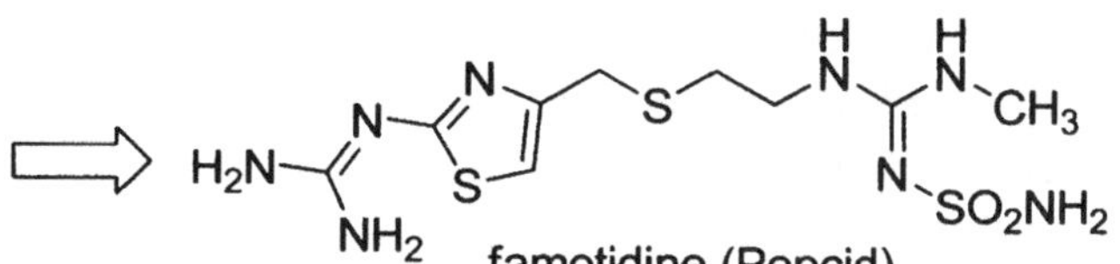

famotidine (Pepcid)

Anti-inflammatory cyclooxygenase-2 (COX-2) selective inhibitor celecoxib (Celebrex) has the tri-substituted pyrazole as its core structure.

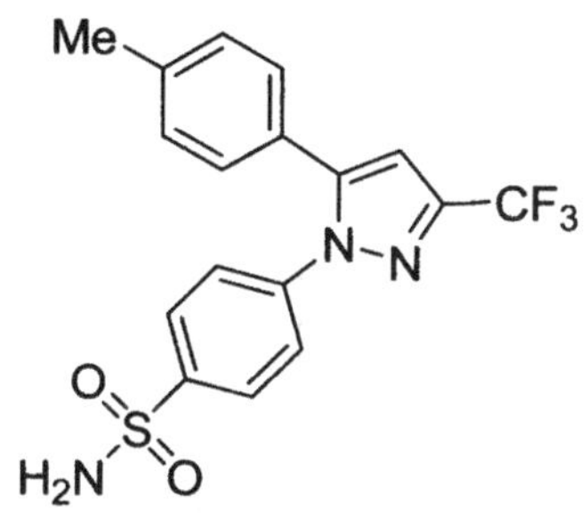

celecoxib (Celebrex)

1.4.3 Six-Membered Heterocycles with One Heteroatom

As we mentioned in Section 1.2, pyridine has an additional lone pair of electrons at the nitrogen atom after it contributes a pair of two electrons to make up the 6 π-electrons for aromaticity. These lone pair electrons are responsible for much of pyridine's unique physical and chemical properties.

One prominent example is AstraZeneca's H^+/K^+-ATPase inhibitor, pyridine-containing omeprazole (Prilosec) and its enantiomerically pure follow-up esomeprazole (Nexium).

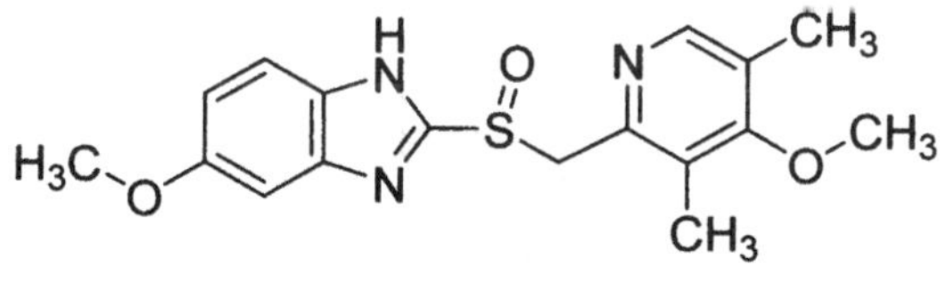

omeprazole (Prilosec)

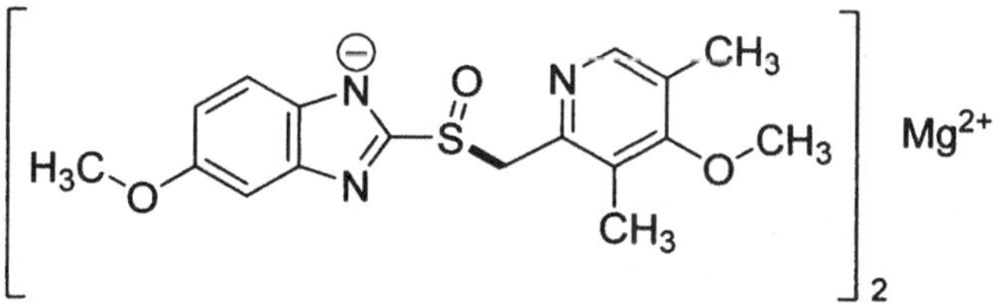

esomeprazole (Nexium)

They are both pro-drugs and their MOA is through the "omeprazole cycle," initiated by pyridine's lone pair of electrons. In fact, pyridine's lone pair of electrons could be viewed as the engine that propels the "omeprazole cycle." The pyridinium sulfydryl intermediate is the actual inhibitory species.

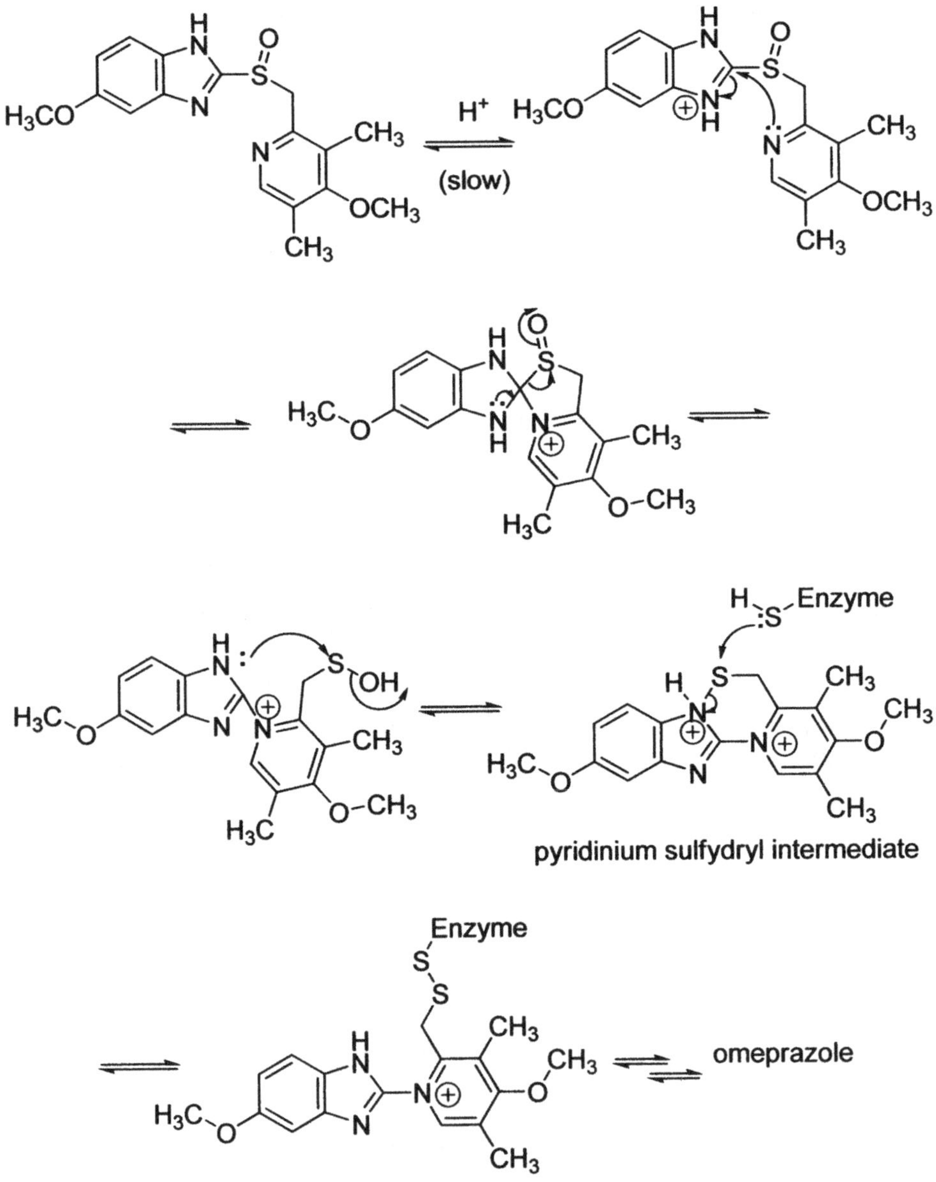

There are quinoline-containing drugs from both nature and synthesis. Natural product drugs may be exemplified by quinine, an anti-malarial drug used for three centuries. Synthetic quinoline-containing drugs are represented by pitavastatin calcium (Livalo), Sankyo's HMG-CoA inhibitor for lowering cholesterol.

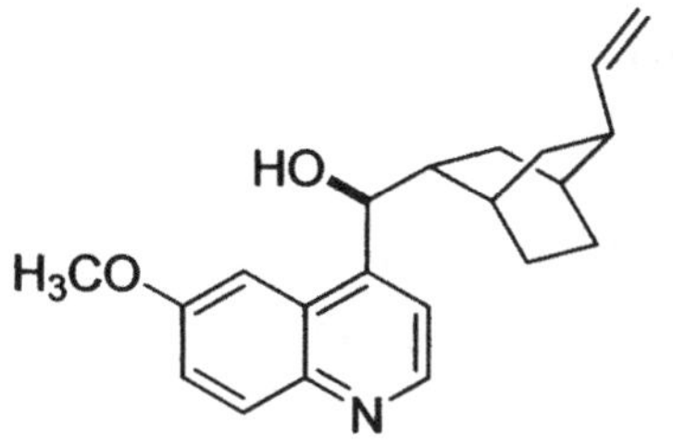

quinine

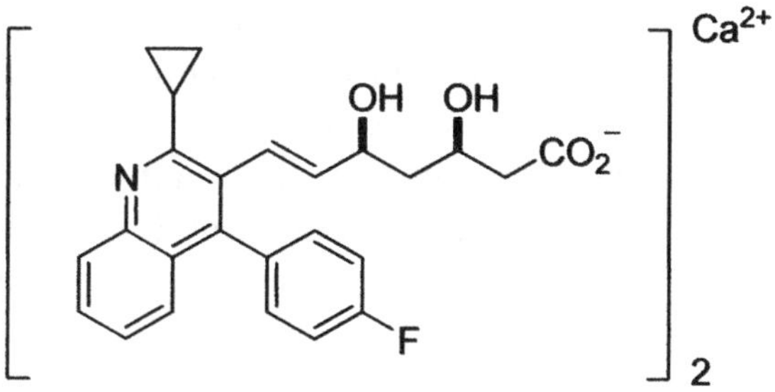

pitavastatin calcium (Livalo)

1.4.4 Six-Membered Heterocycles with Two Heteroatoms

The best-known pyrimidine-containing drug of today is probably AstraZeneca's rosuvastatin (Crestor) as an HMG-CoA redutase inhibitor for lowering cholesterol. By choosing a sulfonamide substituent, a unique intellectual property position was achieved.

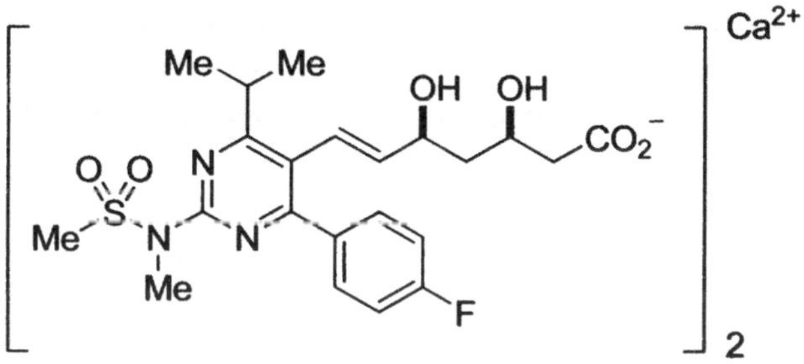

rosuvastatin calcium (Crestor)

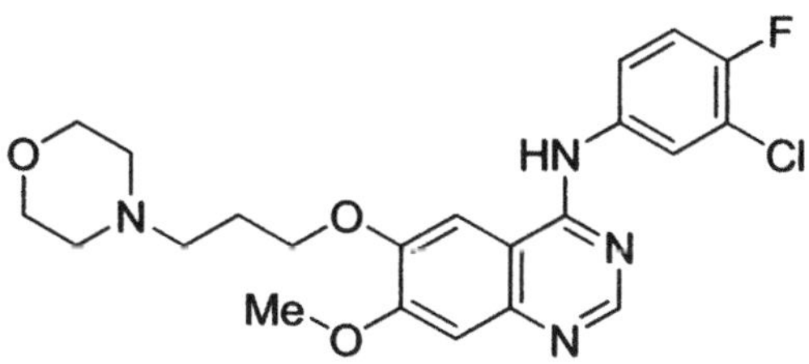

gefitinib (Iressa)

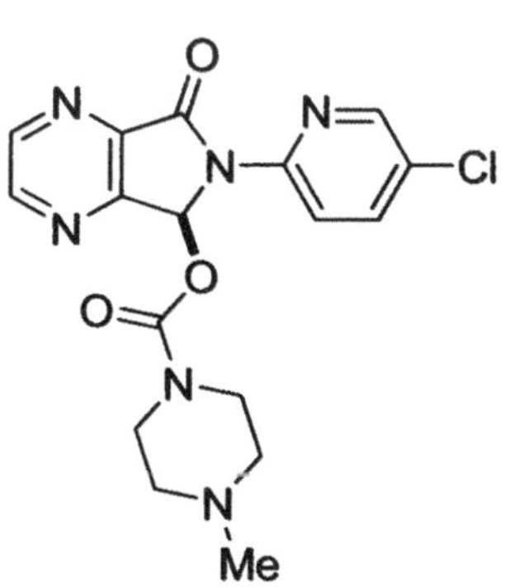

eszopiclone (Lunesta)

AstraZeneca's gefitinib (Iressa)'s core structure is a quinazoline. It is an epidermal growth factor receptor (EGFR) tyrosine kinase inhibitor (TKI) indicated for the treatment of cancers. Several other protein kinase inhibitors also used the quinazoline ring as their core structure. They include OSI's erlotinib (Tarceva) and GSK's lapatinib (Tykerb).

Sepracor's eszopiclone (Lunesta) contains a pyrazine ring. It is a $GABA_A$ receptor agonist for the treatment of insomia.

Finally, Pfizer's varenicline (Chantix) used a fused quinoxaline ring. It is an α4β2 nicotinic receptor partial agonist for smoking cessation.

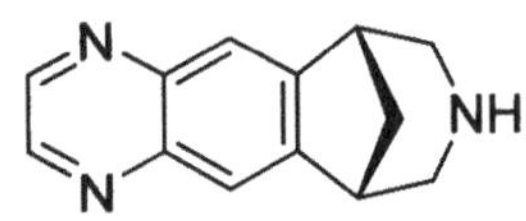

varenicline (Chantix)

In this section, only a small portion of marketed drugs are shown to illustrate the importance of hetereocyclic chemistry in drug discovery. Many drugs containing saturated heterocycles, heterocycles with more than two heteroatoms, and non-heterocycles. In the ensuing chapters, the most popular types of heterocycles in drug discovery are reviewed for their physical and chemical properties, their constructions in the context of medicinal chemistry, and their potential liabilities as drugs when applicable.

PART 1 FIVE-MEMBERED HETEROCYCLES WITH ONE HETEROATOM

Chapter 2 Pyrroles

Jie Jack Li

2.1 Introduction

The parent compound pyrrole is a colorless, flammable liquid with a boiling point of 131 °C. It turns to a light-amber color upon exposure to air and/or light. Pyrrole has a mild aniline-like odor.

Pyrrole, with 6 π-electrons, is an electron-*excessive* (also known as electron-*rich*) aromatic heterocycle because the electron density on each ring atoms is greater than one. Its lone pair electrons take part in the delocalization thus essential to pyrrole's aromaticity. Pyrrole's aromaticity is between furan and thiophene, which is in accordance with Pauling's electronegativity for O (3.5), N (3.0), and S (2.5):

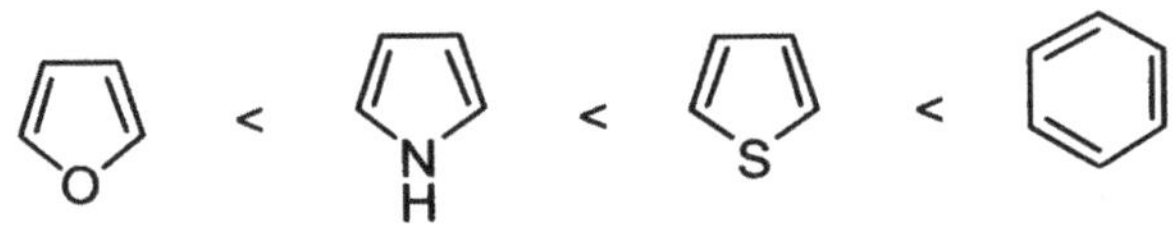

Aromaticities

The reason why pyrrole is an electron-*excessive* aromatic heterocycle is because the electron density on each ring atom is greater than one. Pyrrole has a dipole moment of 1.55 D, similar to that of pyrrolidine in number although with *opposite direction.* (Here, the direction of the dipole moment vector is represented by an arrow and is properly defined so that the arrow is directed from the positive fractional charge to the negative fractional charge).

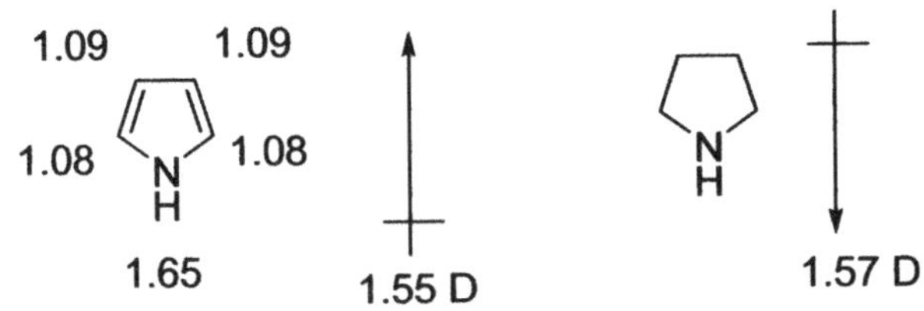

Electron Densities of the Atoms on the Pyrrole Ring and Dipole Moments

Although the direction of pyrrolidine's dipole moment is easily rationale by the nitrogen atom's inductive effect, that of pyrrole's is more nuanced. As shown in the following resonance structures:

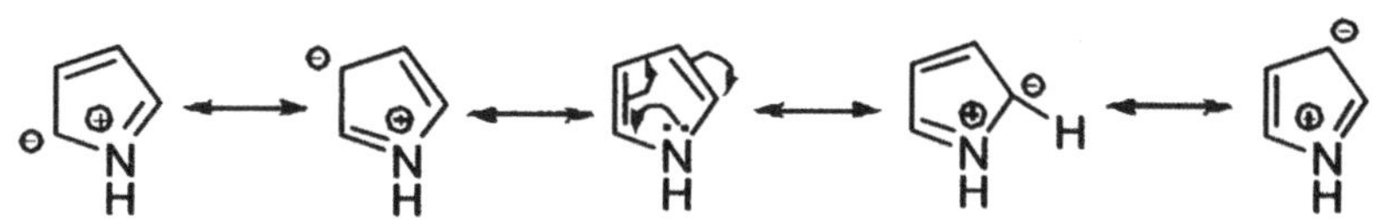

As a consequence, pyrrole has a resonance hybrid that places the partial positive charge on the nitrogen atom and the partial negative charges on the four carbon atoms. For pyrrole, the resonance effect overpowers the inductive effect exerted by the nitrogen atom, whereas the inductive effect for furan and thiophene was a stronger force than their resonance effect.

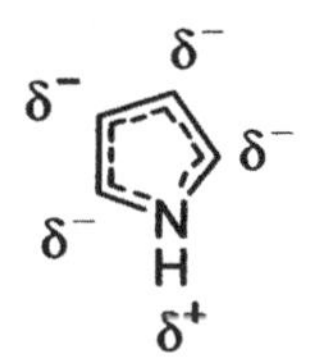

As far as furan and thiophene are concerned, their resonance effect is not as strong as the inductive effect. Therefore, their dipole moments are in the same direction of pyrrolidine.

Geographically, the pyrrole ring is a plane pentagon, with bond angles and bond lengths almost the same as a regular pentagon:

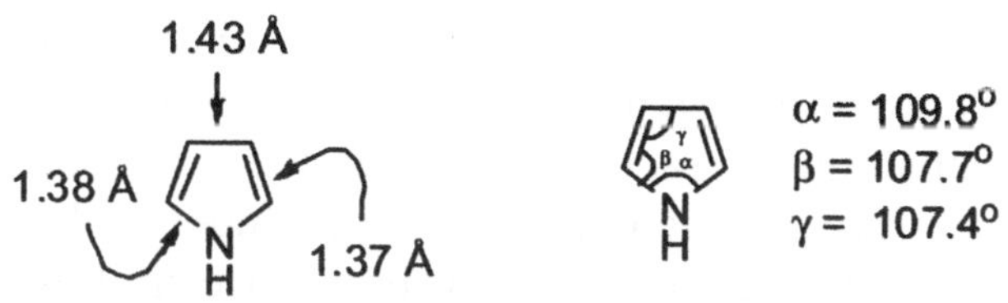

Pyrrole's Bond Lengths and Angles

For its ^{1}H-NMR (Nuclear Magnetic Resonance Spectroscopy), the two β-protons (Hβ) show up at 6.32 ppm, whereas the two α-protons (Hα) show up at 6.87 ppm, further down field from the two β-protons, because of the inductive effect from the nitrogen atom. As far as the NH is concerned, its chemical shift often is affected by solvents and concentrations for the NMR samples. The coupling constant between Hα and Hβ is 2.6 Hz, whereas the coupling constant between Hβ and Hβ is 3.4 Hz.

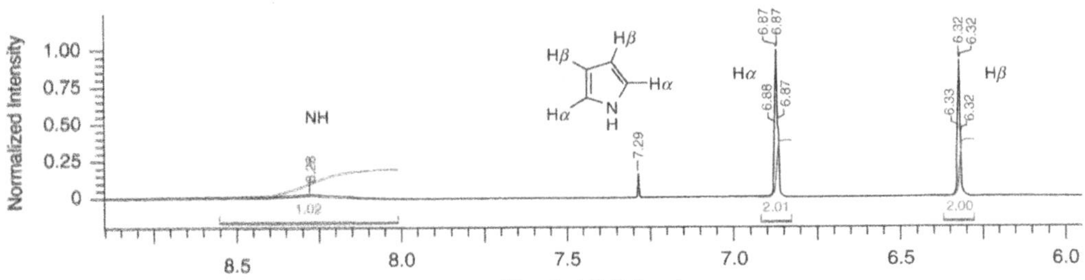

^{1}H-NMR of Pyrrole

For the ^{13}C-NMR spectrum, Cα is further down field with a chemical shift of 117.9 ppm again thanks to the NH's group's inductive effect. Cβ has a chemical shift of 108.1 ppm.

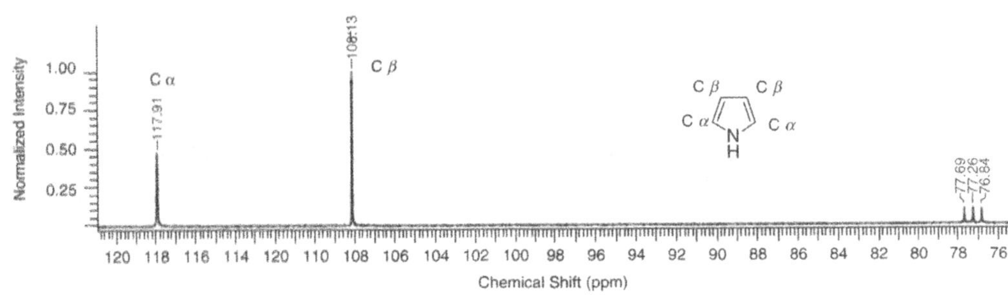

^{13}C-NMR of Pyrrole

Pyrrole is possibly one of the most reactive heterocycles thanks to its lonepair electrons at the nitrogen atom. The enormous reactivity of pyrrole in electrophilic substitution reactions explains the occurrence of more than 100 naturally occurring halogenated pyrroles. Indeed, the pyrrole ring is widely distributed in nature. It occurs in both terrestrial and marine plants and animals. An illustration of the abundant complex natural pyrroles is konbu'acidin A, a sponge metabolite that inhibits cyclin-dependent kinase 4 (CDK4).

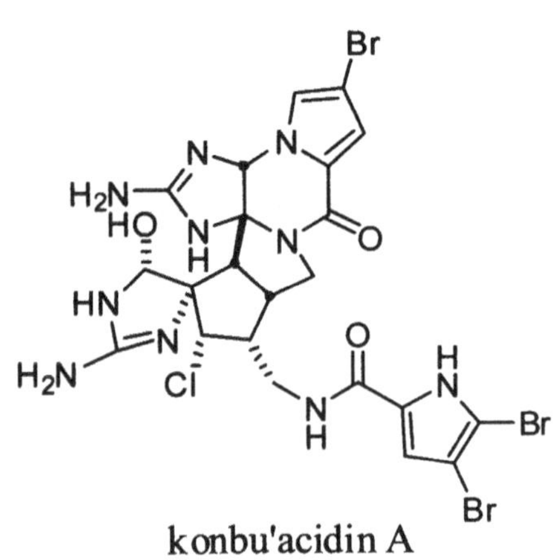

konbu'acidin A

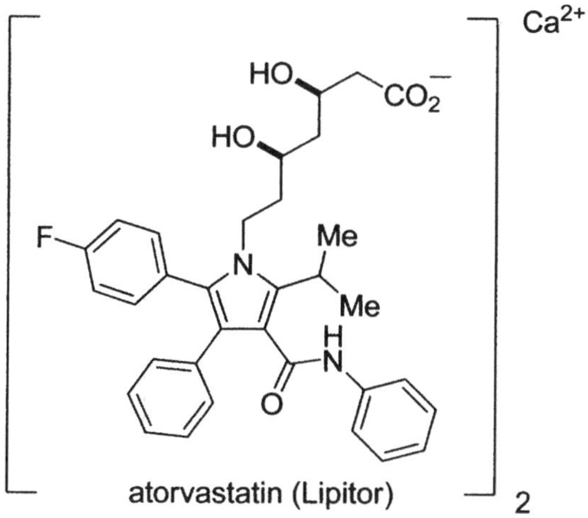

atorvastatin (Lipitor)

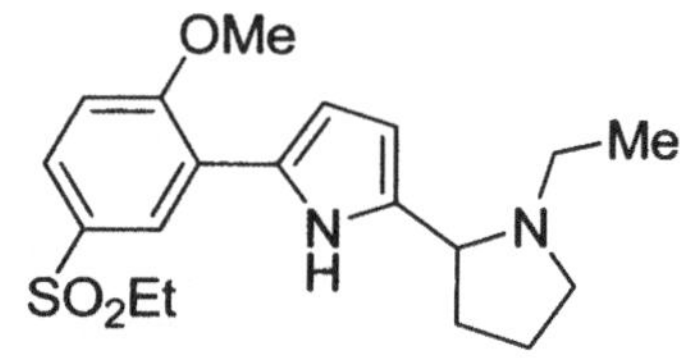

antipsychotic

The pyrrole ring has found great use in the design and development of pharmaceuticals. Atorvastatin (Lipitor), a HMG CoA inhibitor, is the best selling drug ever. Other bioactive pyrroles shown as follows include an antipsychotic agent, a sodium-independent dopamine receptor antagonist, and a DNA cross-linking agent.

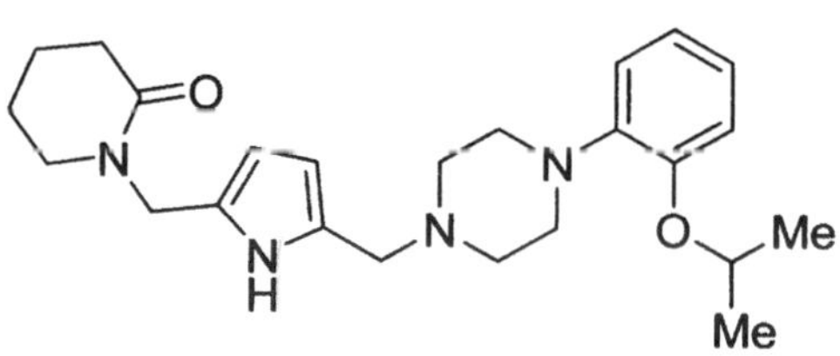

dopamine receptor antagonist

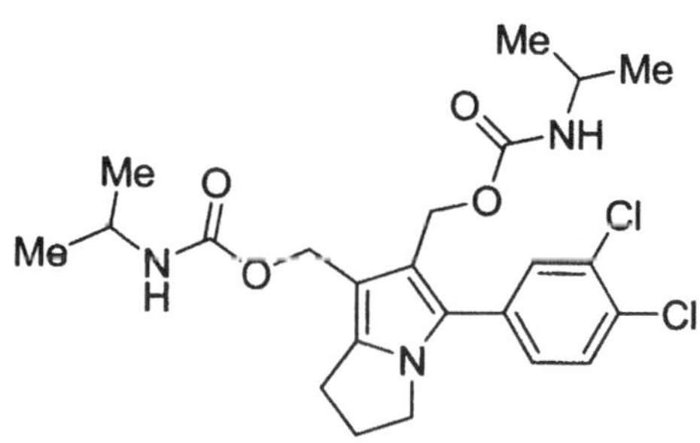

DNA cross-linking agent

However, pyrroles are most frequently found in analgesic and anti-inflammatory drugs. Tolmetin and zomepirac were the first pyrrole-acetic acids to be used as nonsteroidal anti-inflammatory drugs (NSAIDs). Ketorolac (Toradol for inflammatory indications and Acular for ophthalmic indications), also an analgesic and anti-inflammatory, was discovered by Syntex. It was one of the top 100 selling drugs during its heydays during the 1980s. Pyrrolnitrin is an antifungal and an antibiotic.

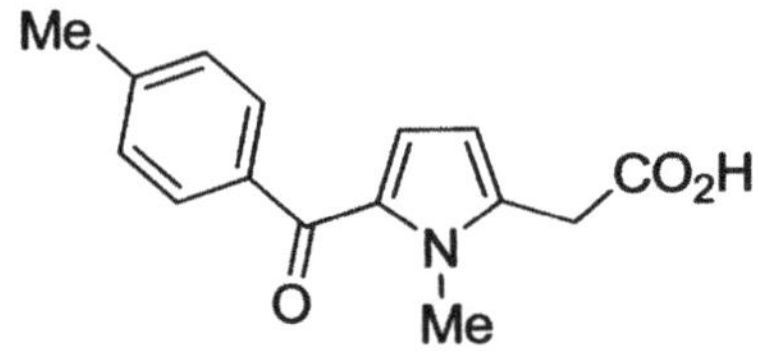

tolmetin
analgesic and
anti-inflammatory

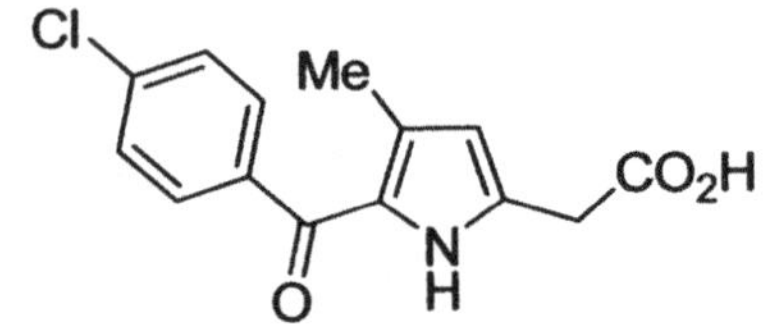

zomepirac
analgesic and
anti-inflammatory

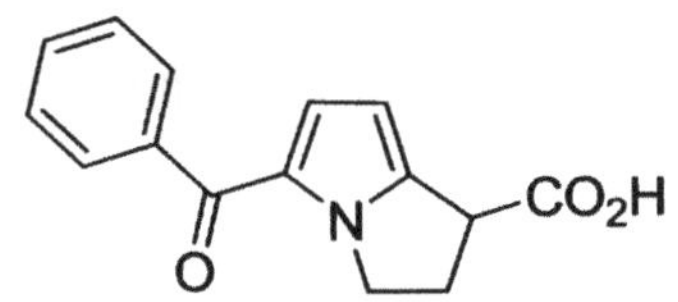

Ketorolac
analgesic and
anti-inflammatory

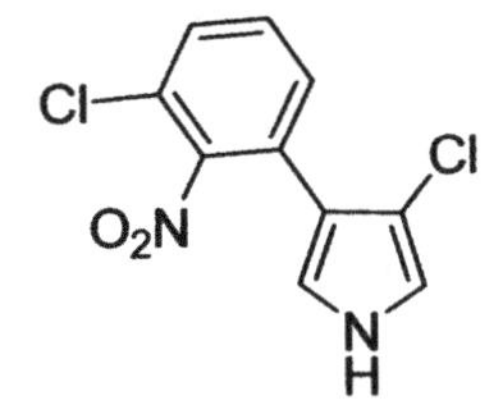

pyrrolnitrin
antifungal
and antibiotic

2.2 *Reactivity of the Pyrrole Ring*

2.2.1 *Protonation*

The pyrrole nitrogen atom (p*K*a = –3.80)[1] is only weakly basic, pyrrole loses its aromaticity when protonated because the lone pair electrons are involved in the aromatic sextet. When treated with strong acid such as sulfuric acid, the protonation does not occur at the nitrogen atom, but most of the protonation takes place at C2 (80%) and 20% of the protonation takes place at C3 (20%).

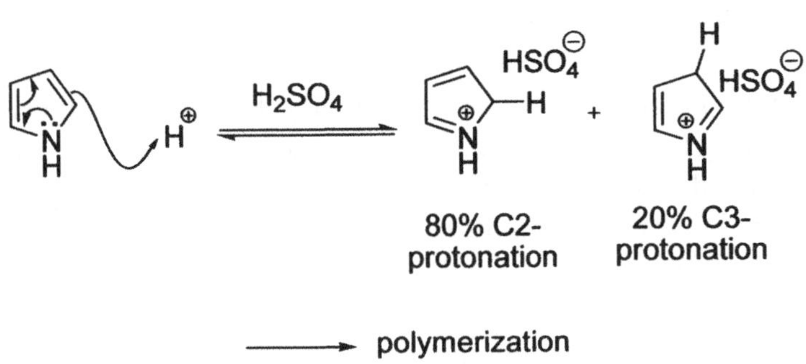

2.2.2 *C2 Electrophilic Substitution*

Pyrrole's lone pair of electrons is the engine that propels many of its unique reactivities. Contrary to the indole where C3 electrophilic substitution takes place predominantly, when treated with an electrophile ($E^{\oplus}$), the lone pair of electrons pushes the pyrrole ring to attack the electrophile at its *C2* position. Therefore, *C2*-electrophilic substitution is the most fertile ground with regard to pyrrole's reactivities. Nonetheless, C3 electrophilic substitution and poly-substitution still take place from time to time, often with concurrent *C2*-electrophilic substitution.

Here is why the pyrrole ring prefers *C2*-electrophilic substitution. First of all, N1-electrophilic substitution is not favored because the positive charge would be localized on the nitrogen atom. On the other hand, *C2*-electrophilic substitution is favored over the C3-electrophilic substitution because the intermediate for the *C2* substitution is more delocalized than that of the C3 substitution. This preference of more delocalization is reflected by the protonation as well. As shown in the previous section, 80% of the protonation occurs on the *C2* position.

C2-electrophilic substitution predominates:

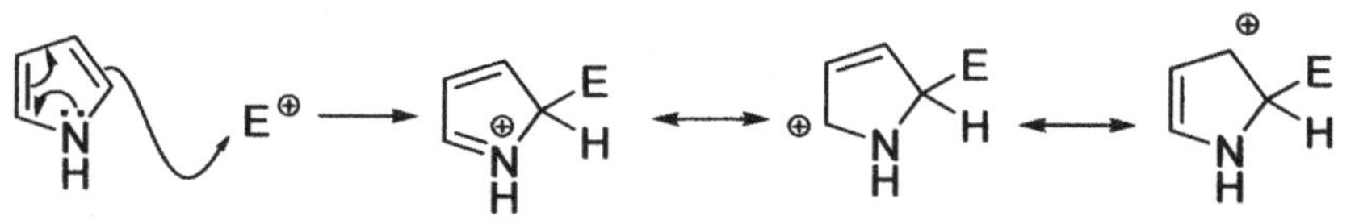

C3-electrophilic substitution is less preferred:

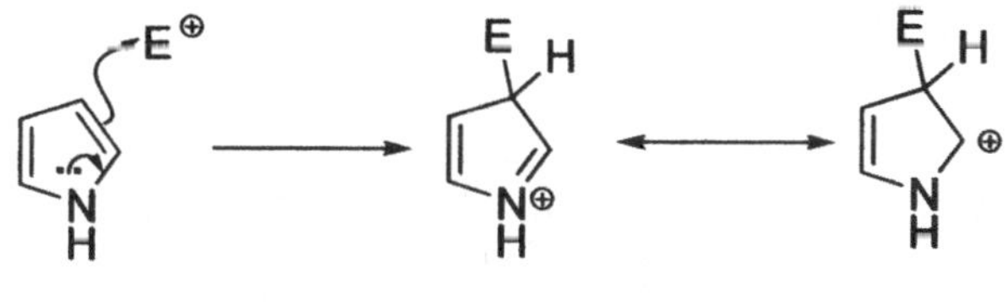

Halogenation

The aforementioned regioisomer formation is showcased by bromination of pyrroles. The *C2*(α)-bromination is prevalent for bromination of the 1*N*-Boc protected pyrrole.[2]

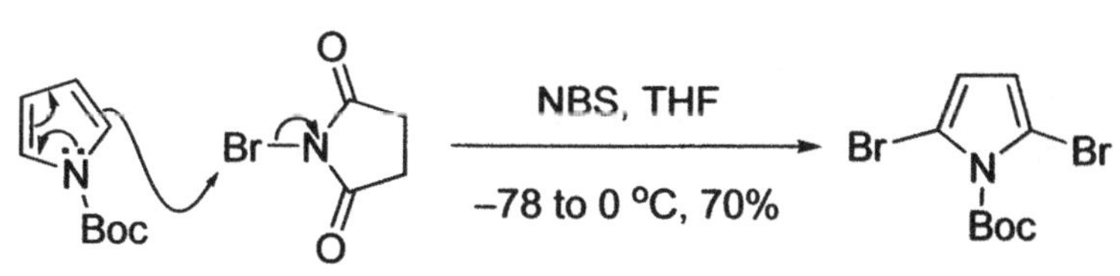

There are exceptions to the rule. For instance, when there is a bulky group on the nitrogen such as triisopropylsilyl group (TIPS), bromination occurs predominantly on the C3 (β) position instead of on the C2 (α) position.[3]

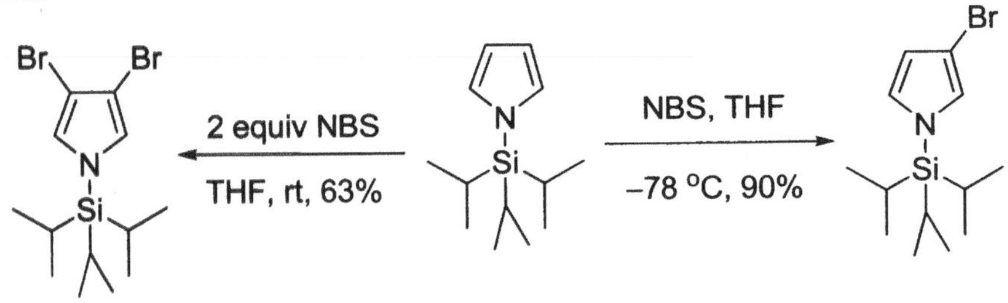

Bromination on both α- and β-positions is also possible[4]:

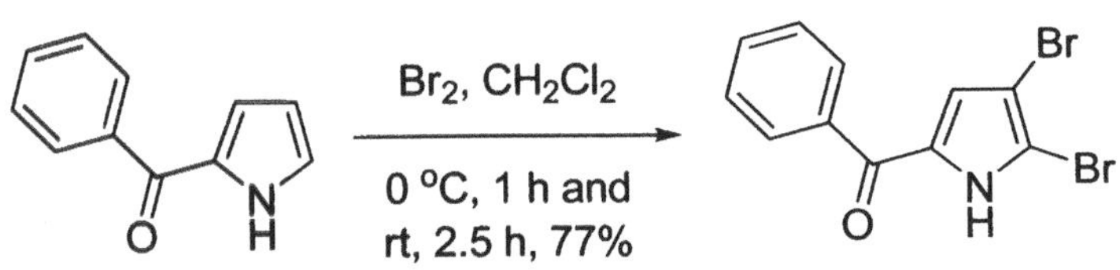

Depending on the reaction conditions, 1-methylpyrrole can be brominated at C2 with NBS (*N*-bromosuccinate) to give 2-bromo-1-methylpyrrole or at C3 with NBS and catalytic PBr_3 to give 3-bromo-1-methylpyrrole. Both reactions are essentially quantitative, but both bromides decompose on silica gel.[5]

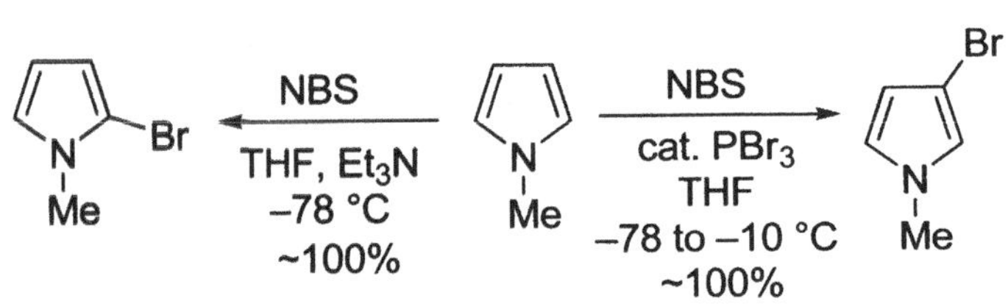

Methyl-protected bipyrrole was chlorinated at both C2 positions on treatment with NCS (*N*-chlorosuccinate) in excellent yield.[6]

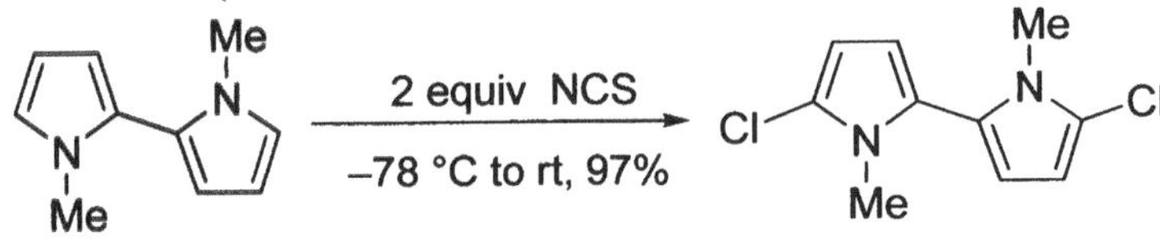

Mannich reaction

For the Mannich reaction with pyrrole, the substitution occurs predominantly at the C2 position as well.[7]

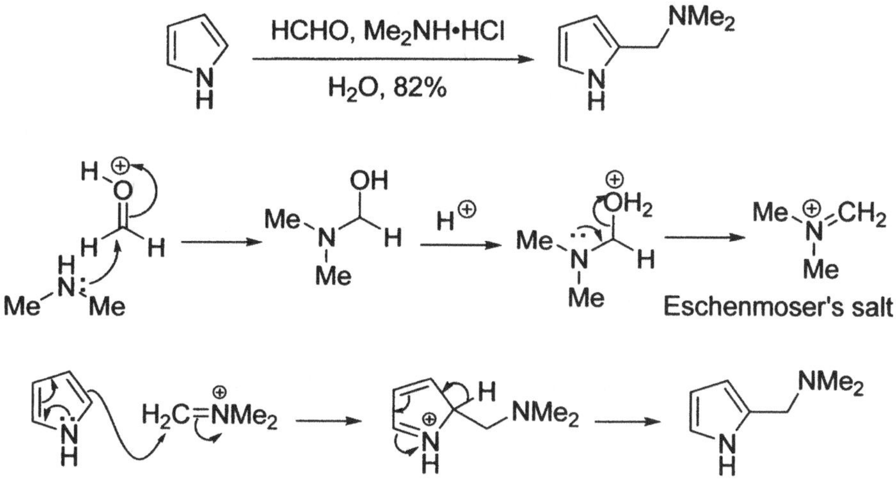

Pyrrole Mannich bases have been transformed into the tertiary ammonium salt as a good leaving group. Therefore, treatment of the quaternary ammonium salt with sodium sulfinate to give the corresponding sulfonyl pyrrole, which in turn, could undergo another Mannich reaction to synthesize sulfonyl pyrrole Mannich bases as germicides.[8]

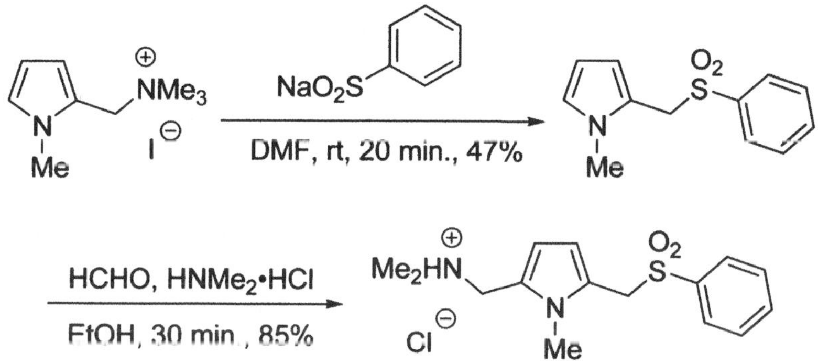

germicide

A similar strategy was employed in preparing a pilot scale of FR143187, a novel nonpeptide angiotensin II receptor antagonist for the treatment of hypertension.[9] At first, acid-catalyzed cyclization of ethyl 4-aminobenzoate with dialkoxytetrahydrofuran formed the *N*-aryl pyrrole. The reaction is sometimes known as the Clauson–Kaas pyrrole synthesis. Next, the Mannich base was prepared quantitatively by the treatment of the *N*-phenylpyrrole with paraformaldehyde and dimethylamine hydrochloride. The procedure was superior to the original standard Mannich conditions using 37% aqueous formaldehyde (47% yield) in EtOH. The quaternary ammonium salt was formed in 93% yield by mixing the Mannich base with methyl iodide in EtOAc. The reaction was carried out in a 125-kg scale.

Reduction of the quaternary ammonium salt was achieved using borane•pyridine complex in 1,3-dimethylimidazolinone (DMI). The resulting methylpyrrole was then transformed into FR143187 in additional 7 steps.

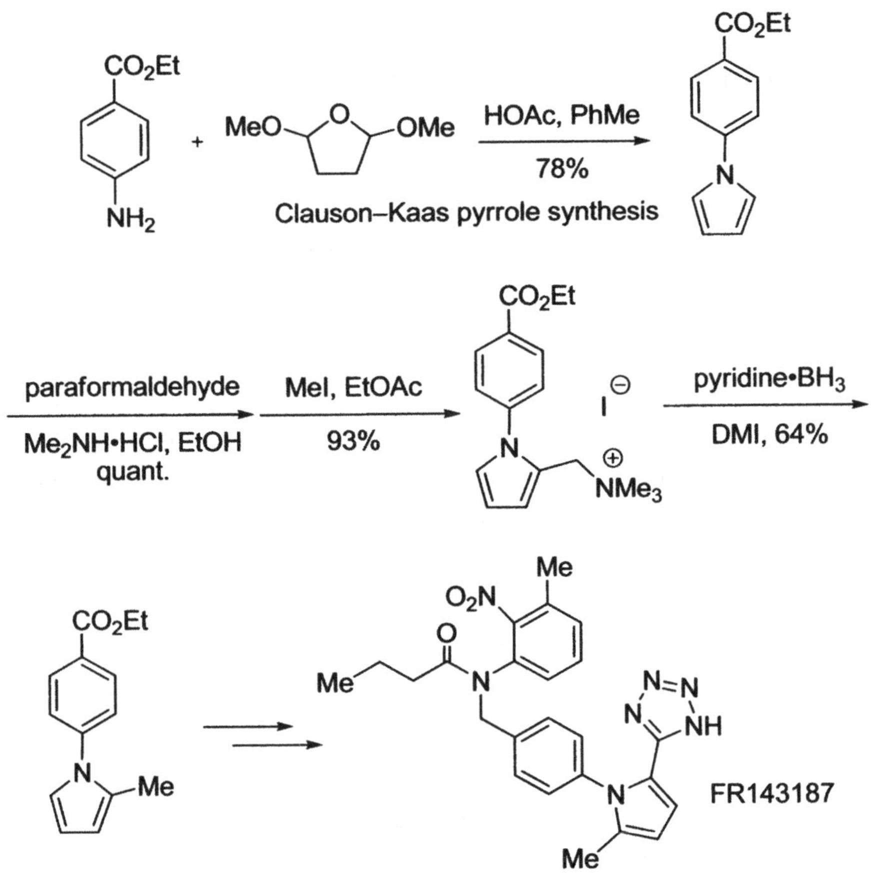

The Mannich reaction between 2-phenylpyrroles and phenylpiperazines provided Mannich bases as a new class of potential antipsychotics. They served as conformationally restricted benzamide analogues.[10]

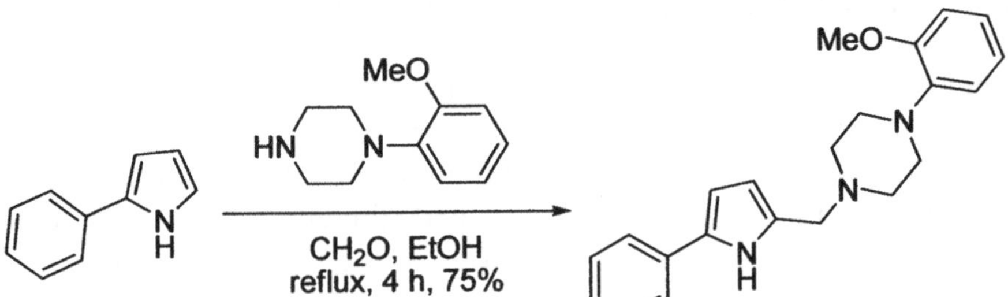

antipsychotic

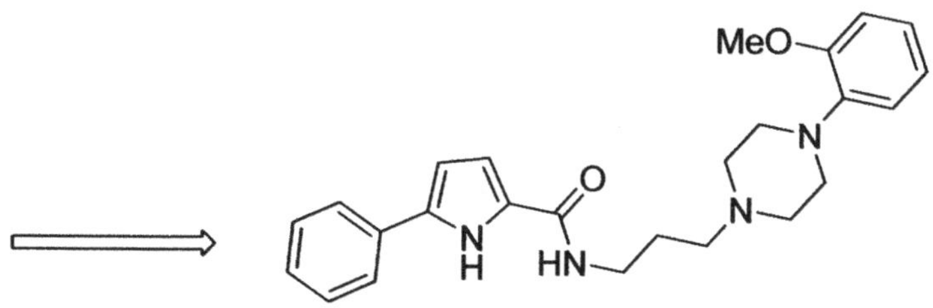

Pyrrole Mannich bases have been prepared as potential antipsychotic agents that do not have the extrapyramidal side effects (EPS). In one case, *N*-methylpyrrole was amidomethylated with 1-(hydroxymethyl)azepan-2-one, which was assembled by condensation between the seven-membered lactam and formaldehyde.[11,12] The amidomethylated *N*-methylpyrrole then underwent the Mannich reaction with arylpiperazine and formaldehyde in the presence of trifluoroacetic acid (TFA). The pyrrole Mannich bases synthesized in this manner exhibited a high affinity for the serotonin 5-HT-1A and 5-HT-1B binding sites. Although these arylpiperazines interact weakly with dopamine D-1 and D-2 receptors, they were reasonably potent in an *in vivo* model in the rat CAR (conditioned avoidance responding), an indication of potent antipsychotic activity.

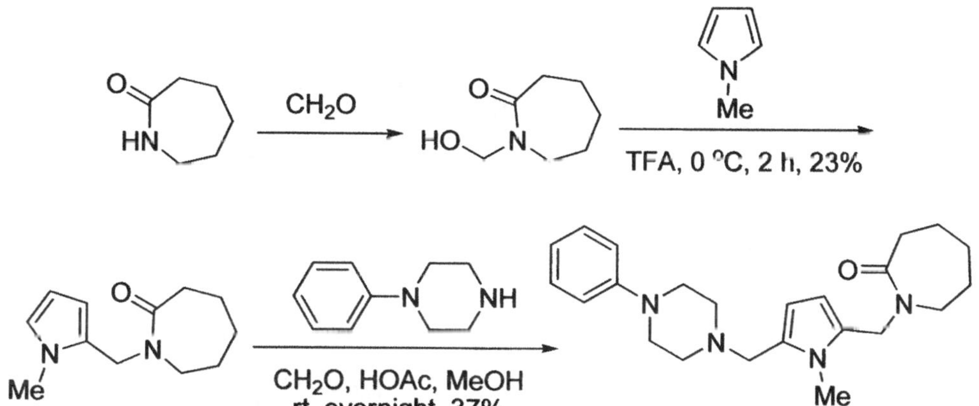

antipsychotic

Outside the CNS (central nervous system), pyrrole Mannich bases have found utility in other therapeutic areas as well. The Mannich reaction between iminoibitol and 9-deazahypoxanthine took place at the C3 position to provide an *N*-pyrrolylmethyl substituted iminoribitol as an inhibitor of a purine-specific nucleoside hydrolase.[13] In terms of regiochemistry, this particular Mannich reaction of 9-deazahypoxanthine behaved similarly to indole rather than to pyrrole. The resulting Mannich bases are potential treatment for parasitic infections.

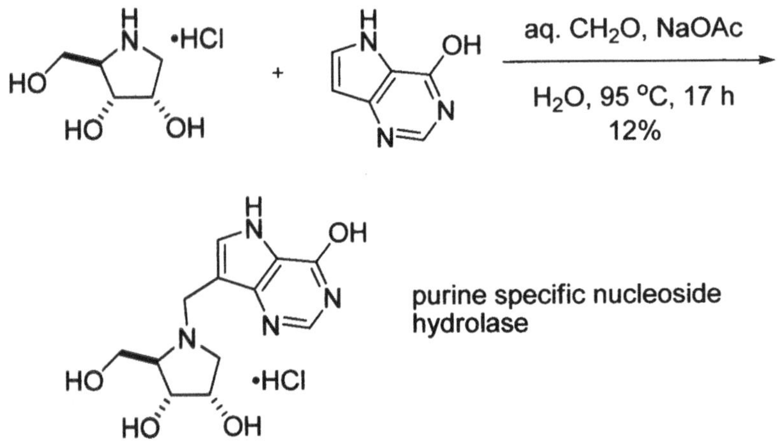

C-10 pyrrole Mannich bases of artemisinin have been accessed as potential anti-malarial agents.[14] Treatment of dihydroartemisinin with *N*-methylpyrrole in the presence of a Lewis acid gave rise to the C-10 pyrrole analogue. The subsequent Mannich reaction used preformed Eschenmoser's salt to afford the dimethylaminopyrrole, which was an anti-malarial with enhanced water solubility.

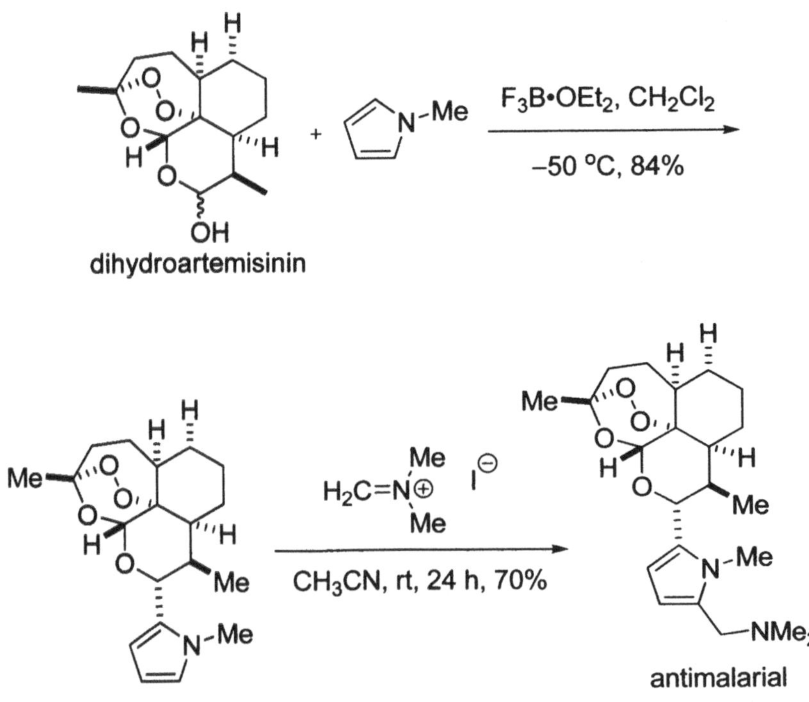

Vilsmeier–Haack reaction

The Vilsmeier–Haack reagent, a chloroiminium salt, is a weak electrophile. Therefore, the Vilsmeier–Haack reaction works better with electron-rich carbocycles and heterocycles. Since pyrrole is very electron-rich, the Vilsmeier–Haack reaction readily takes place. Formylation of methyl pyrrole-2-carboxylate was achieved using the Vilsmeier–Haack reaction.[15] The mechanism is shown below. The resulting methyl 5-formylpyrrole-2-carboxylate, in turn, was converted into nonpeptidic analogues of neurotesin(8–13), which are potential treatment for neuropsychiatric diseases such as schizophrenia and Parkinson's disease.

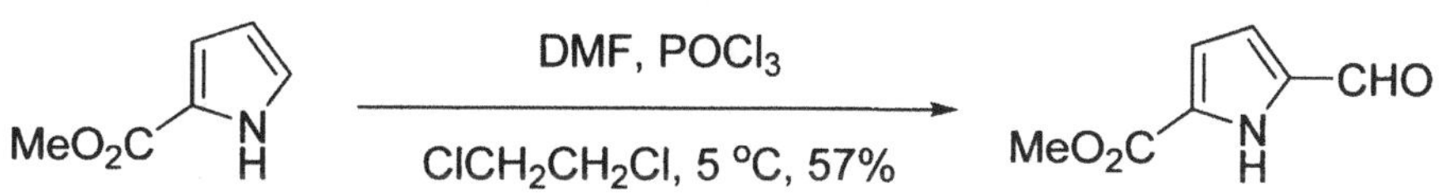

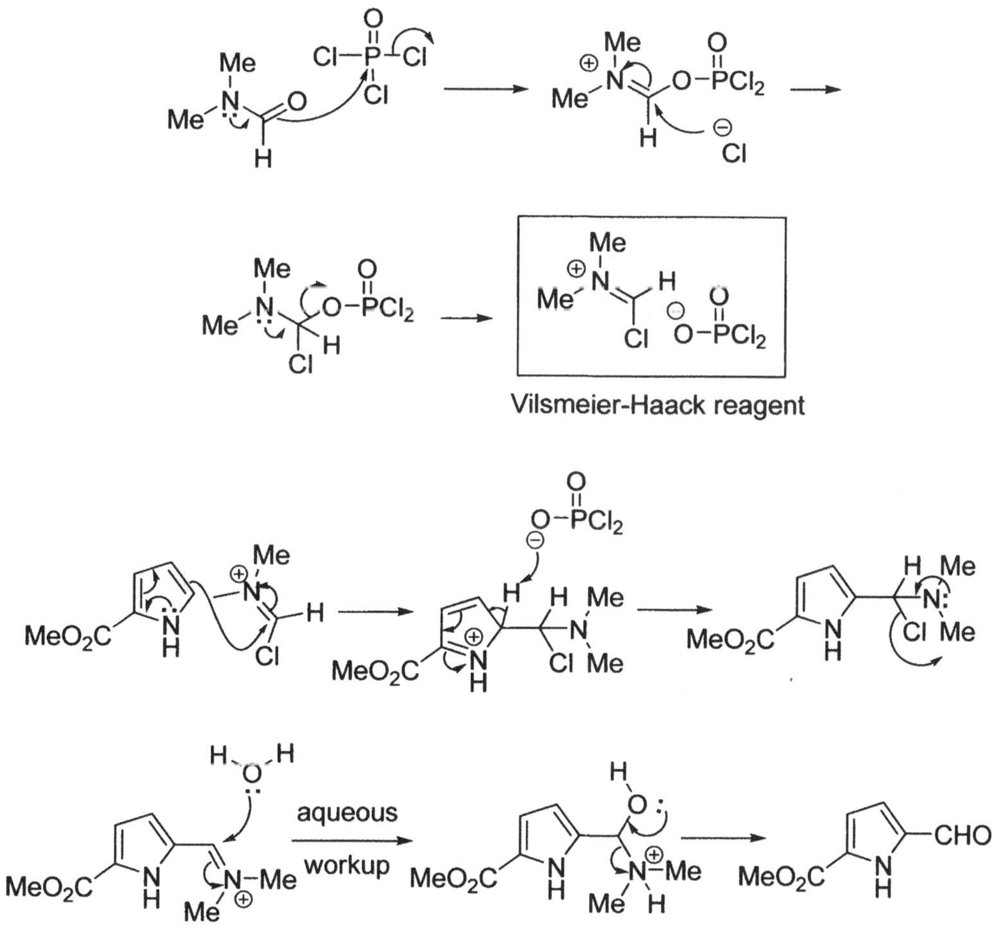

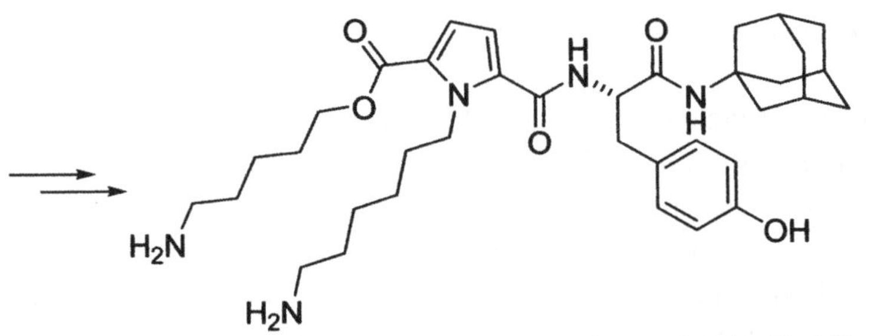

nonpeptidic analogue of neurotesin (8–13)

Oxidative coupling

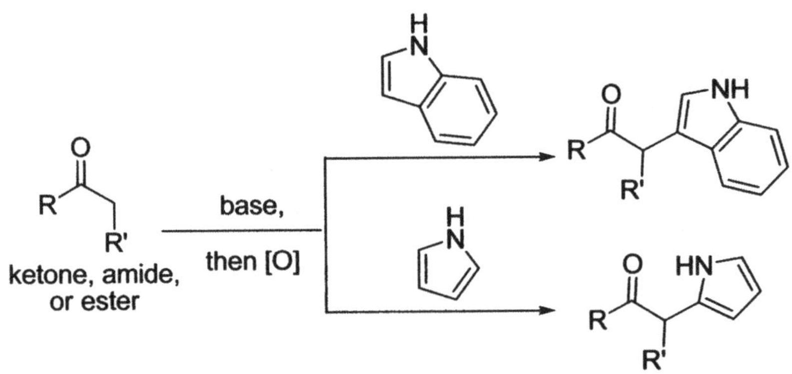

An exciting new development of oxidative cyclization of pyrroles emerged from Baran's Laboratories at Scripps Research Institute. Building on their success with the oxidative cyclization of indoles,[16] Baran *et al.* expanded the methodology to pyrroles.[17]

Applying the strategy of direct coupling of pyrroles with carbonyl compounds, Baran *et al.* developed a short enantioselective synthesis of (*S*)-ketorolac, Syntex's analgesic and anti-inflammatory agent.[17] The antipode of ketorolac is more potent and causes fewer side effects.

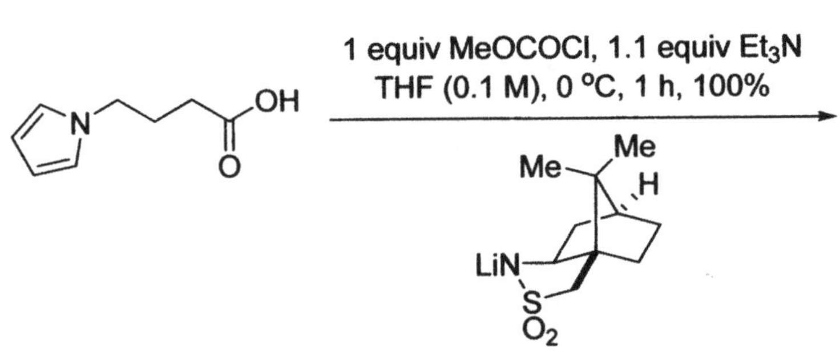

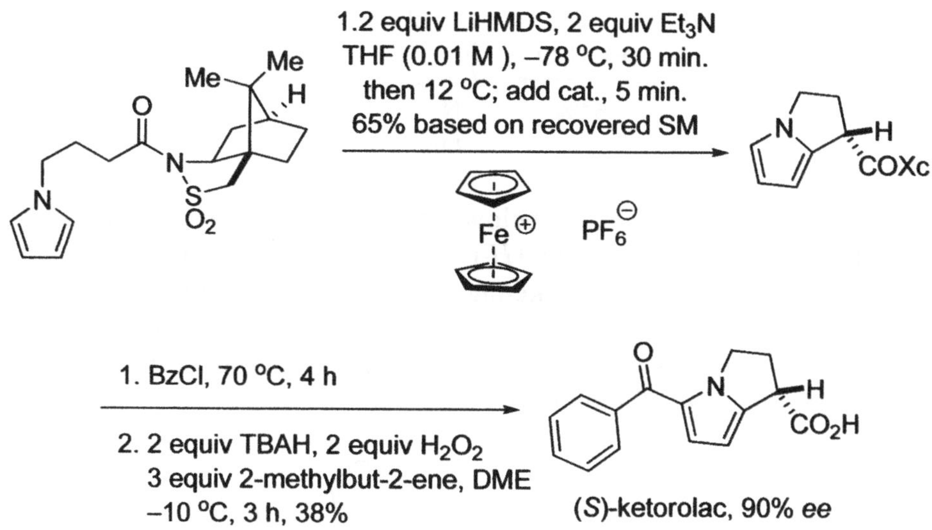

2.2.3 *C3 Electrophilic Substitution*

In the last section, we saw many C2 halogenations. Depending on the substrates, C3 electrophilic substitutions do occur although often accompanying the C2 electrophilic substitutions. The C3 electrophilic substitutions generally take place more sluggishly and often at lower yields, although this is not always the case.

Various iodinated pyrroles have been prepared by direct iodination or via initial thallation. For example, 3-iodo-*N*-TIPS-pyrrole is prepared in 83% yield from *N*-TIPS-pyrrole.[18] And 3,4-diiodo-2-formyl-1-methylpyrrole is available in 54% yield via a bis-thallation reaction.[19] Although *N*-protected 2-lithiopyrroles are readily generated and many types are known, these intermediates have not generally been employed to synthesize halogenated pyrroles.

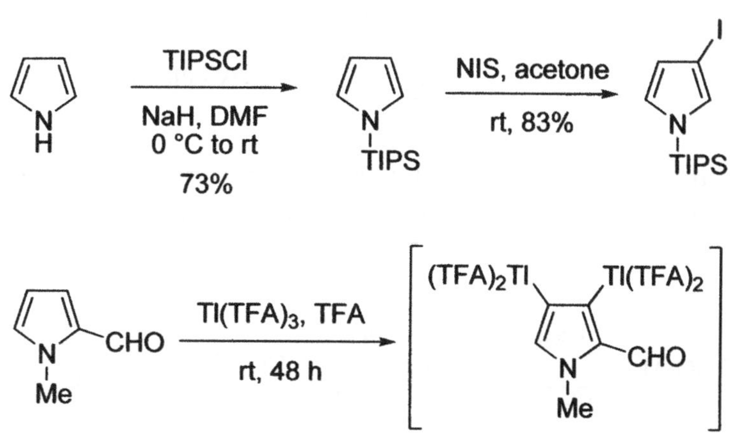

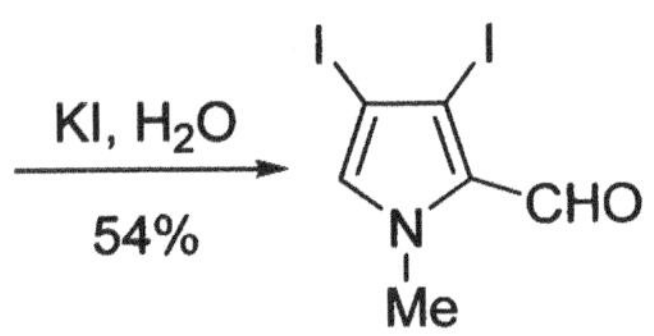

As in the example shown below, the Vilsmeier–Haack reaction was performed on 2-benzoyl-1-methyl-1*H*-pyrrole to afford the C2 formylation product in 44% yield and the C3 ketone affords the C2 formylation product in 44% yield in 56% yield.[20] The two pyrrolyl aldehydes, in turn, were converted into the corresponding hydroxamates, which are a new class of histone deacetylase (HDAC) inhibitors.

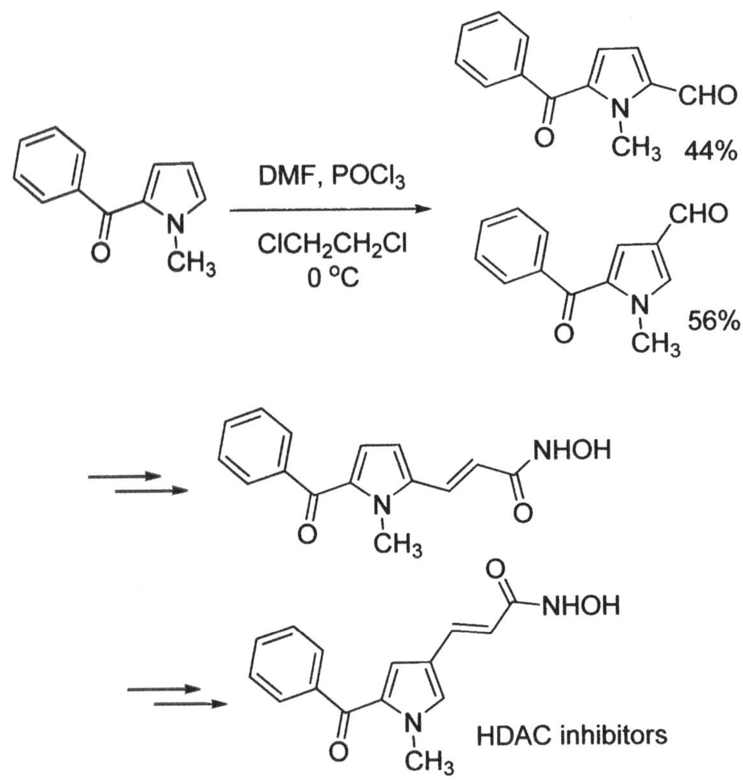

A similar tactic was employed in transforming methyl pyrrole-2-carboxylate into hepatitis C virus (HCV) helicase inhibitors.[21]

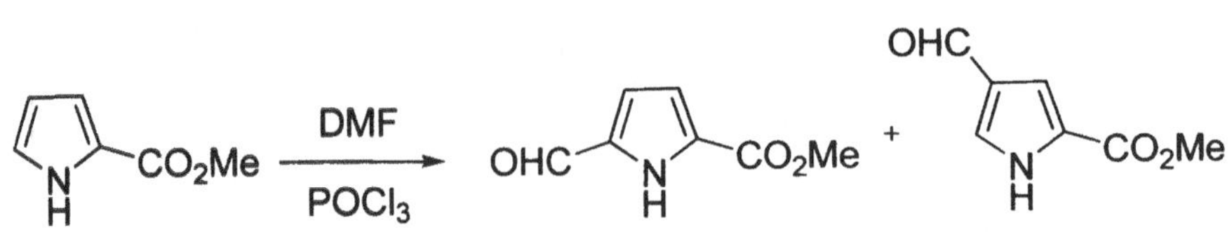

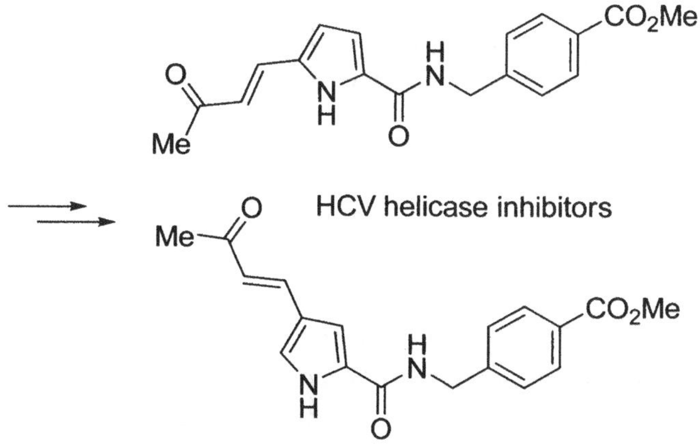

When C3 is the only position open, the C3 electrophilic substitutions obviously occur exclusively. The following Vilsmeier–Haack reaction was applied to the synthesis of a novel class of glycine site antagonists of the ionotropic *N*-methyl-D-aspartate (NMDA) glutamatergic receptor.[22]

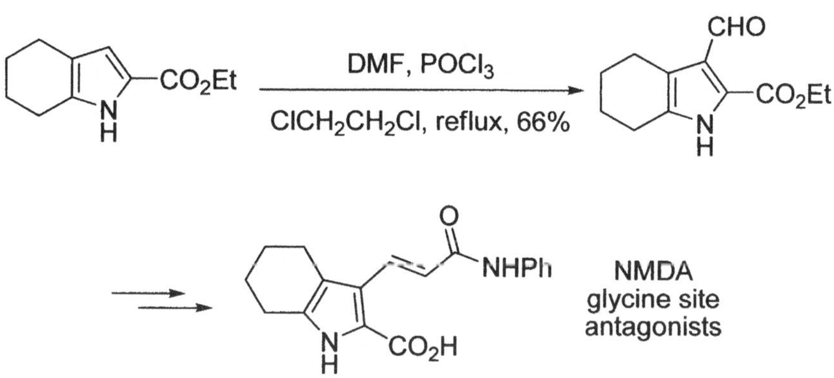

A Friedel–Crafts acylation of 2-(1-methyl-1*H*-pyrrol-2-yl)acetonitrile with benzoyl chloride gave the C3 substitution product in 21% yield, whereas the C2 substitution product was obtained in 25% yield.[23] The C2 adduct, in turn, was hydrolyzed to the corresponding acid, which is a potent anti-inflammatory agent and was active in the *in vivo* animal models.

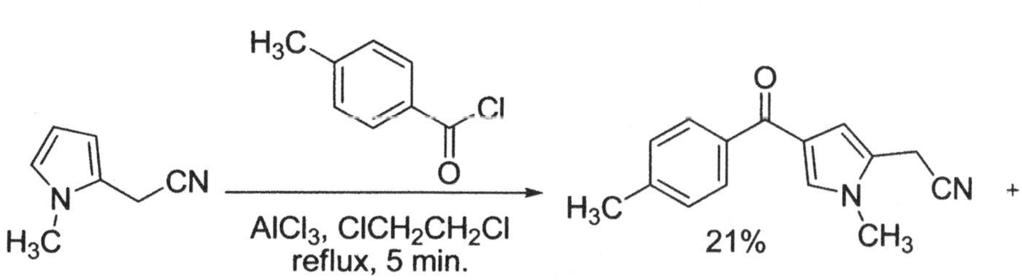

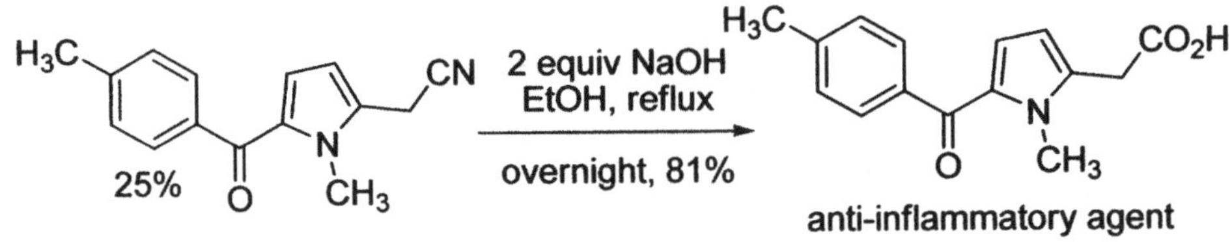

2.2.4 *Metalation*

The NH pyrrole proton is acidic with a p*Ka* of 17.5. As a consequence, bases such as NaH, Grignard reagents, *n*-BuLi, and NaOEt readily deprotonate pyrrole. For instance, alkylation of 4-nitro-pyrrole-2-ester was achieved by treatment of the pyrrole with sodium ethoxide in the presence of cyclopropylmethyl bromide.[24] The resulting alkylated pyrrole was converted to pyrrole tetraamides, which are minor groove DNA binders and serve as potent antibacterial agents.

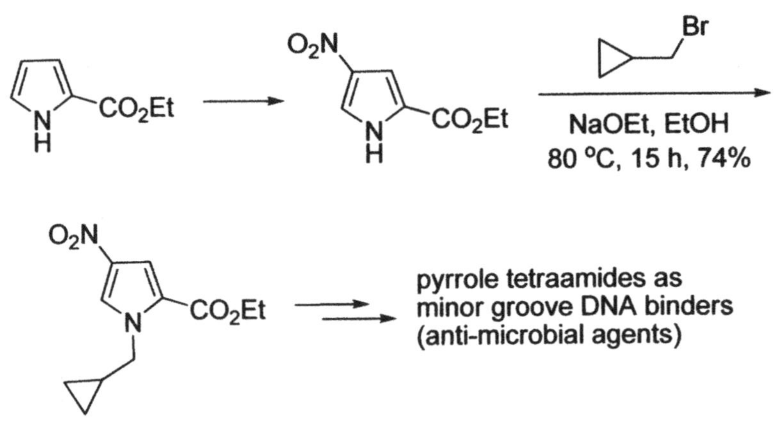

2.3 *Construction of the Pyrrole Ring*

2.3.1 *Knorr Pyrrole Synthesis*

The Knorr pyrrole synthesis involves the reaction between an α-amino ketone and a second carbonyl compound, having a reactive α-methylene group, to give a pyrrole.[25] The amine is often generated *in situ* by reduction of an oximino group.

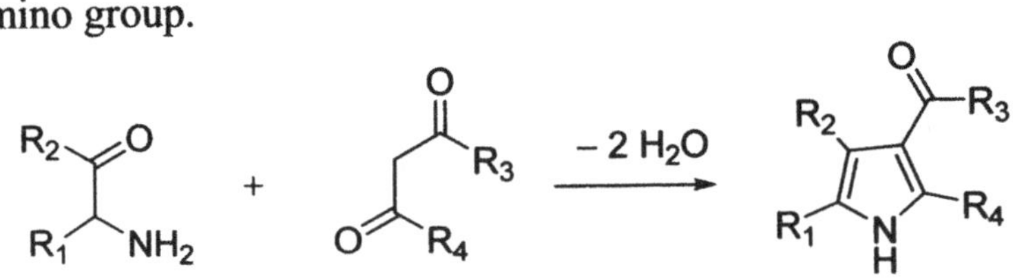

R_1–R_4 = various alkyl, acyl, aryl groups

The mechanism of the original Knorr pyrrole synthesis entails *in situ* reduction of the oxime moiety to an amine, condensation with the second carbonyl compound, and cyclization with loss of a second molecule of water to give a pyrrole.[25] Several studies have demonstrated that different pathways and pyrrole products obtain depending on the substrates.

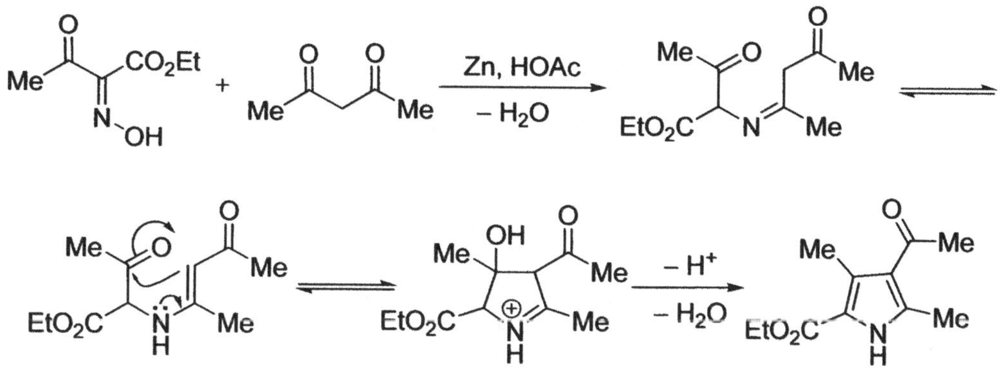

The Knorr pyrrole synthesis was applied to make butyrophenone analogues of molindone, a typical anti-psychotic first marketed in the United States in 1974. The Knorr condensation of 2-hydroxyimino-3-pentanone with 1,3-cyclohexadione in 70% acetic acid in the presence of zinc powder at reflux afforded the dihydroindolone,[26] which was transformed into QF-0400B in six additional steps. QF-0400B had similar affinities for D_1, D_2, and 5-HT_{2A} receptors to those of molindone.

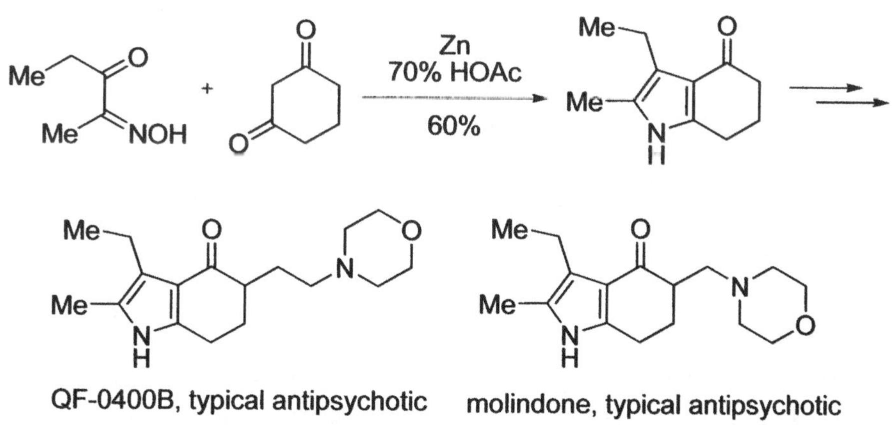

The same tactic was employed to prepare new cyclic butyrophenone derivatives in the indole series as potential atypical anti-psychotics. The Knorr pyrrole synthesis provided a simple and practical access of 6-aminomethyltetrahydroindol-4-ones and their affinities for D_2, and 5-HT_{2A} receptors were evaluated for their potential as atypical anti-psychotics. As shown below, the Knorr condensation between 2-isonitroso-3-pentanone with hydroxylmethyl alcohol, as a masked cyclohexadione, in 70% acetic acid in the presence of zinc powder at reflux gave a mixture of two tetrahydroindole-4-ones.[27] The acetate was easily converted into the corresponding alcohol on treatment with 10% ethanolic potassium hydroxide.

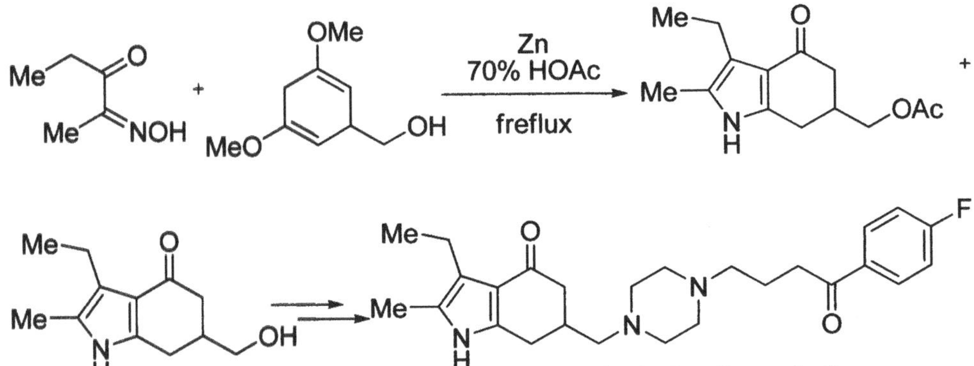

atypical anti-psychotic

A modified Knorr pyrrole synthesis was key to a practical synthesis of the potent δ-opioid agonist SB-342219 by GSK.[28] SB-342219 is a selective δ-opioid agonist undergoing preclinical evaluation for the potential treatment of neuropathetic pain. The medicinal chemistry route used the conventional Knorr pyrrole synthesis. Condensation of the ketone with α-aminoketone, which was generated *in situ* by reduction of the requisite phenylhydrazone using zinc, gave the desired pyrrole in 63% yield.

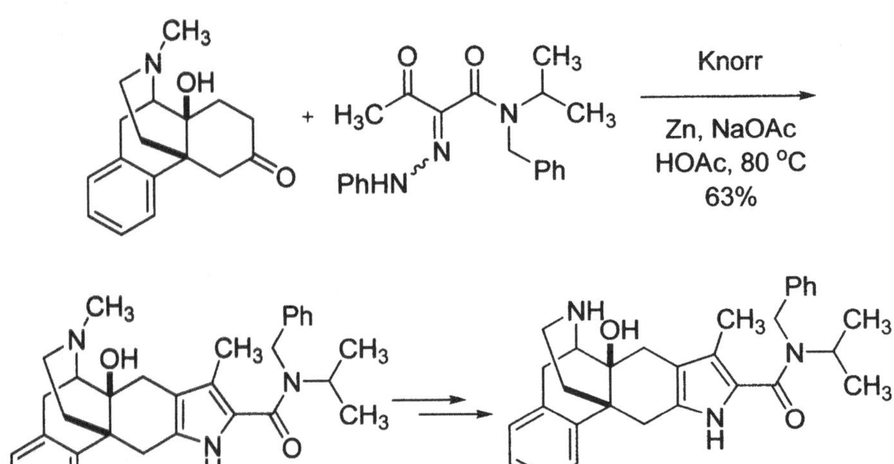

SB-342219, opioid agonist

The conventional Knorr pyrrole synthesis delivered SB-342219 on a small scale for medicinal chemistry. Nonetheless, use of metallic zinc could be problematic for scale-up. Therefore, Carey et al. devised a modified Knorr pyrrole synthesis where they use an amine instead of the phenylhydrazone, thus avoiding the use of zinc metal.[28] Therefore, the condensation between the ketone and the aminoketone in the presence of NaOAc and HOAc gave the desired pyrrole in 68% yield, which was easily converted into SB-342219 in two additional steps.

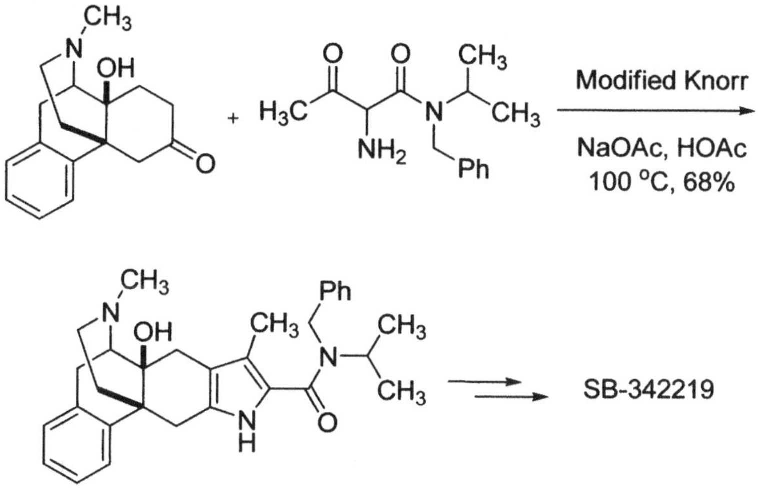

2.3.2 *Paal–Knorr Pyrrole Synthesis*

Discovered more than a century ago, the Paal–Knorr pyrrole synthesis is similar to the Knorr synthesis. It is the intermolecular condensation of a primary amine (or ammonia) with a 1,4-diketone (or 1,4-dialdehyde) to give pyrroles.[25]

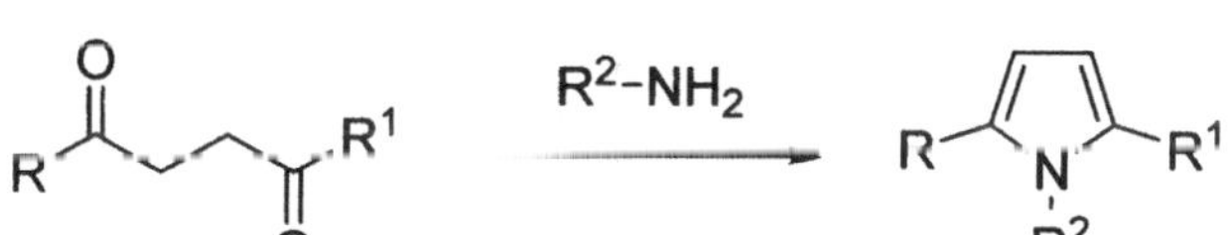

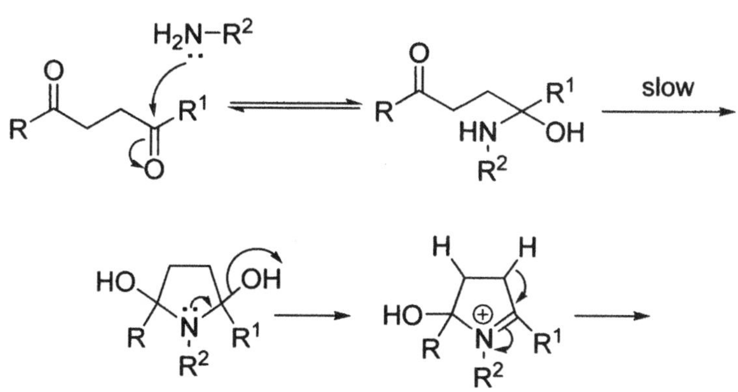

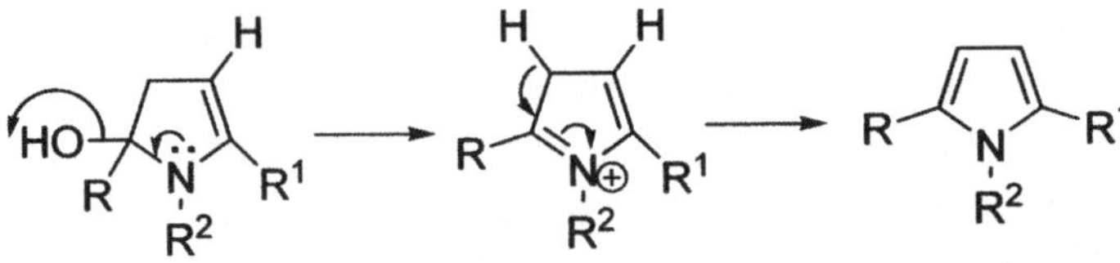

L-167307 is an orally bioavailable inhibitor of p38 kinase. *In vivo*, it reduces secondary paw swelling in the rat adjuvant arthritis model with an ID_{50} of 7.4 mg/kg/day. Triarylpyrrole L-167307 was assembled using the Paal–Knorr pyrrole synthesis of a 1,4-diketone and ammonium acetate.[29]

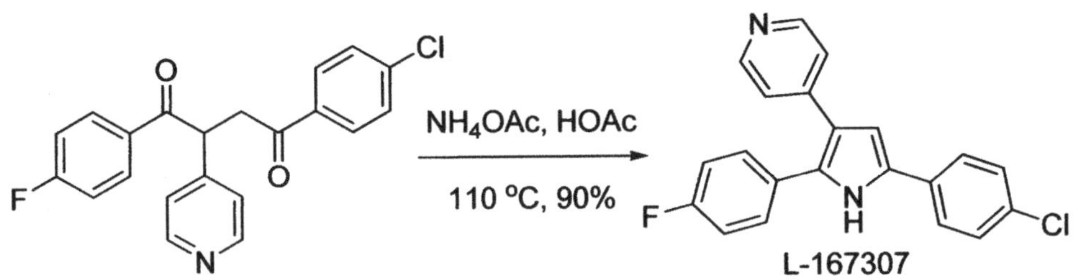

Celecoxib (Celebrex) is a selective cyclooxygenase-2 (COX-2) inhibitor prescribed as a nonsteroidal anti-inflammatory drug (NSAID). The Paal–Knorr cyclization was the crucial step in preparing tri-substituted keto-pyrroles as COX-2 inhibitors. Here, the tri-ketone substrates were prepared *in situ* from phenacyl bromide and 1,3-diketone.[30]

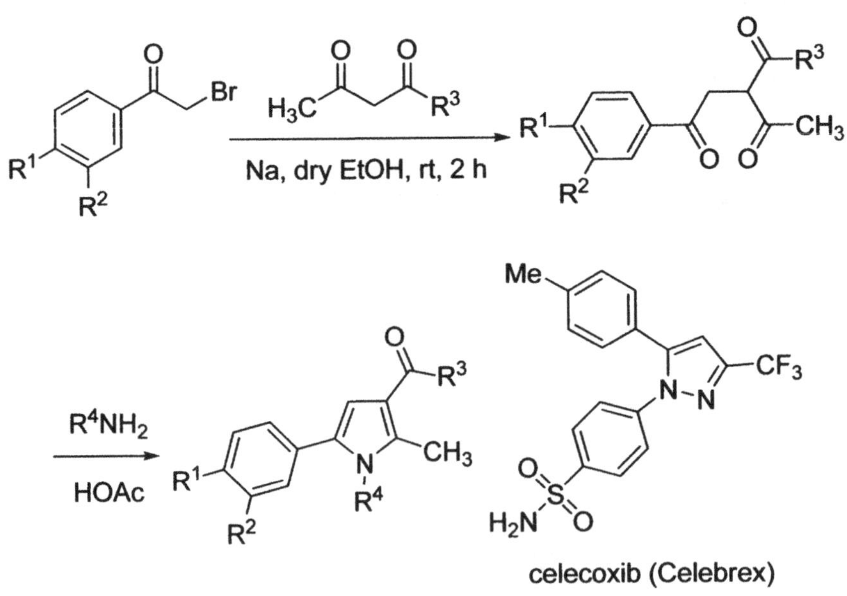

The Paal–Knorr cyclization was employed to produce highly aryl-substituted pyrrole carboxylates as useful medicinal chemistry leads.[31] Therefore, 1,4-diketone-2,3-diester was assembled from an S_N2 displacement of ethyl 2-bromoacetoacetate with the anion of the ketoester. Condensation with an aniline then provided a library of fully substituted pyrroles.

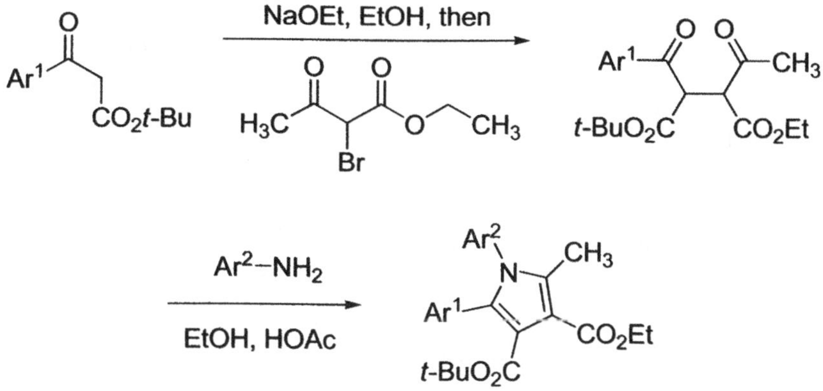

When Paal and Knorr discovered the pyrrole synthesis more than a century ago, they had no idea that the reaction bearing their names would have contributed greatly to the manufacture of atorvastatin (Lipitor). Indeed, synthesis of Lipitor is probably the *tour de force* for the Paal–Knorr pyrrole synthesis.

After more than one year's exploration, the carefully controlled conditions were worked out to prepare penta-substituted pyrrole in 43% yield.[32] The conditions entailed the condensation of the diketone with the diethyl acetal of 3-amino-propanal in the presence of 1 equivalent of pivalic acid. It was also significant that it was demonstrated that a totally convergent synthesis was possible. With this result in hand, it became possible to envision a route in which a fully elaborated side chain could be combined with the appropriate 1,4-diketone to assemble the entire molecule into one operation.

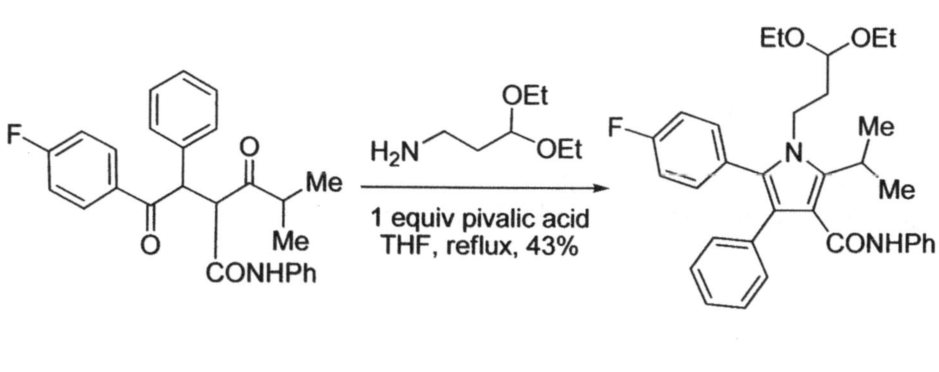

When the fully functionalized, stereochemically pure side chain and the fully substituted diketone were treated under very carefully defined conditions (1 equiv. pivalic acid, 1:4:1 toluene/heptanes/THF), a 75% yield of the penta-substitued pyrrole was obtained.[33] Deprotection and formation of the hemi-calcium salt produced stereochemically pure atorvastatin calcium in a convergent, high-yielding, and commercially viable manner.

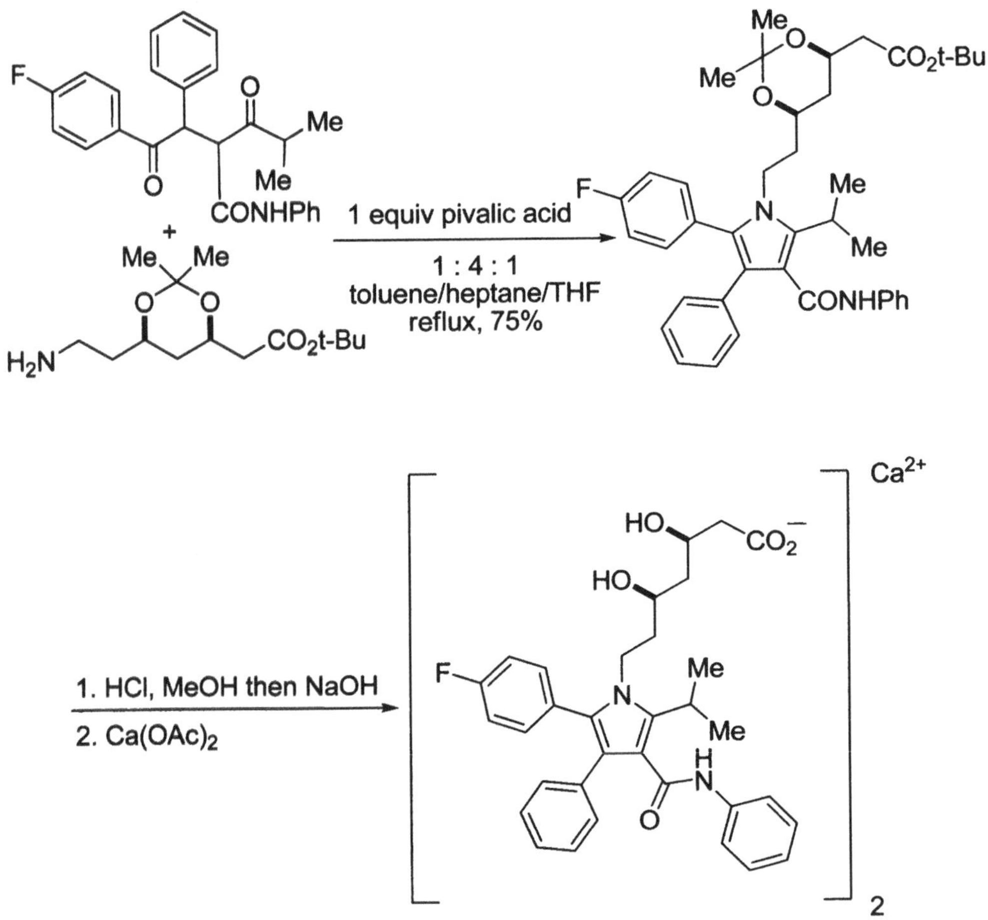

atorvastatin calcium (Lipitor)

2.3.3 *Hantzsch Pyrrole Synthesis*

The Hantzsch pyrrole synthesis is the condensation of β-ketoesters with primary amines (or ammonia) and α-haloketones to give substituted pyrroles.

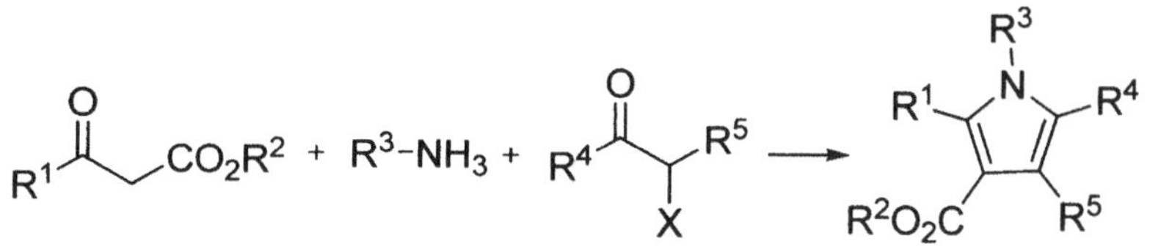

It is possible that the mechanism of the Hantzsch pyrrole synthesis commenced with the condensation between the amine and the ketoester. The resulting imine then undergoes an S_N2 replacement reaction with the α-haloketone via the intermediacy of an enamine. The adduct as an enamine ketone then undergoes an intramolecular C–N bond formation to deliver the final pyrrole after extrusion of a molecule of water.

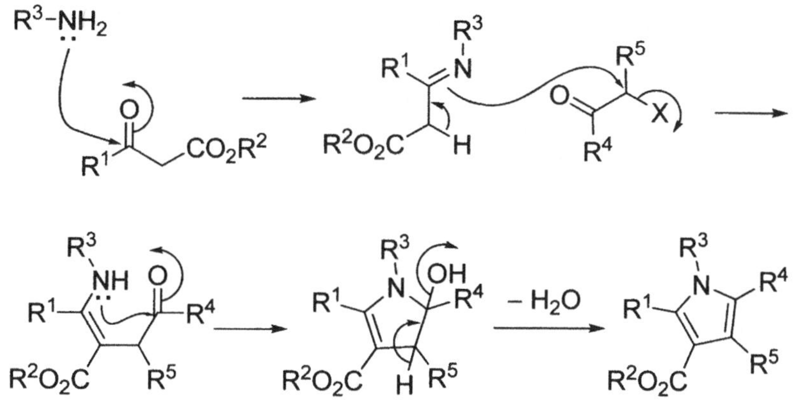

The Hantzsch pyrrole synthesis was employed to prepare pyrrole-2-acetic acids as anti-inflammatory agents.[23b] A transient precipitate of a white crystalline solid was formed when diethyl acetone-dicarboxylate was mixed with aqueous methylamine. After chloroacetone was added rapidly with cooling before the disappearance of the precipitate, a good yield of ethyl 1,4-dimethyl-3-ethoxycarbonylpyrrole-2-acetate was produced. Further functional group transformations then produced pyrrole-2-acetic acids as anti-inflammatory agents.

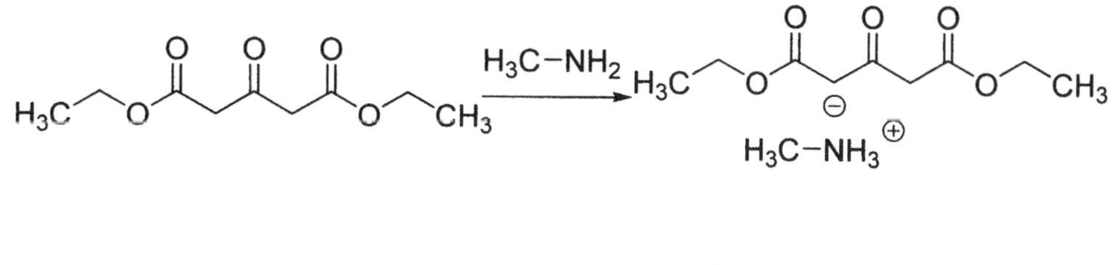

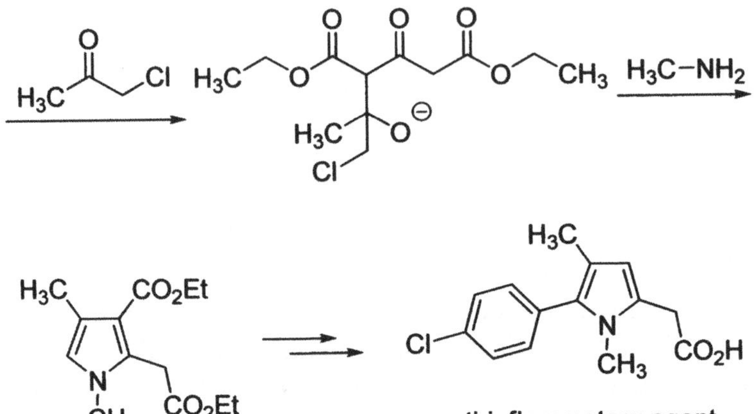

anti-inflammatory agent

A variant of the Hantzsch pyrrole synthesis was applied to prepare LY231514 (Alimta), an antifolate as a potential anti-cancer drug.[33] As is typical of the Hantzsch pyrrole synthesis, the Feist–Benary reaction is frequently the competing pathway to afford the corresponding furan. As shown below, condensation of 2,6-diaminopyrimidin-4(3*H*)-one with α-chloro-ketone gave rise to a 1:1 mixture of the corresponding furan and pyrrole in 40% yield each. It is interesting that when α-bromo-ketone was used in place of the α-chloro-ketone, no Feist–Benary reaction was observed to give the furan, and only the pyrrole was isolated.[34]

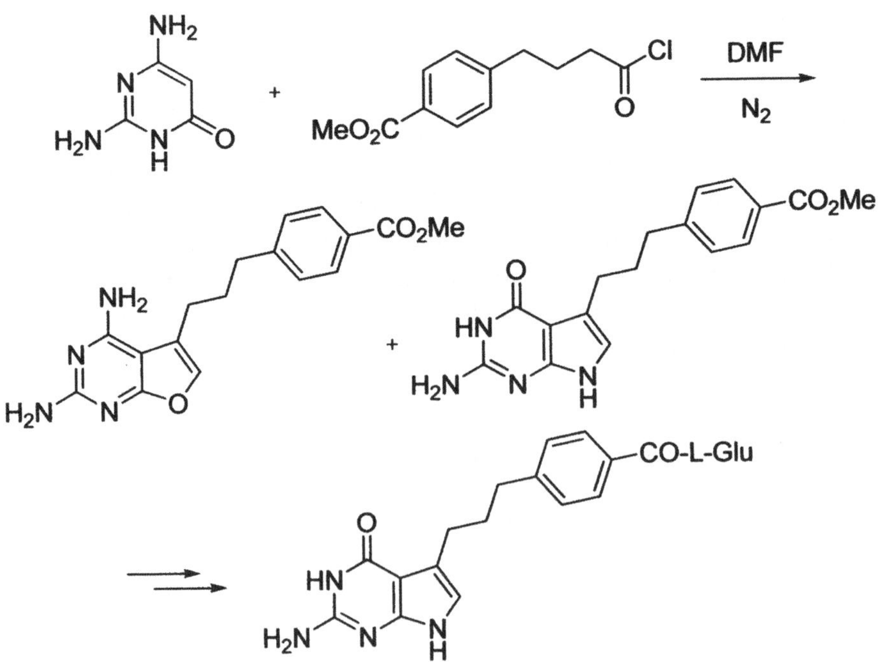

LY231514 (alimta, anti-cancer agent)

2.3.4 Barton–Zard Reaction

The Barton–Zard reaction refers to the base-induced reaction of nitroalkenes with alkyl α-isocyanoacetates to afford pyrroles.[35] Solvents used are THF or alcohols (or mixtures), and the reaction often proceeds at room temperature. The Barton–Zard pyrrole synthesis is similar both to the van Leusen pyrrole synthesis that uses Michael acceptors and TosMIC and to the Montforts pyrrole synthesis that uses α,β-unsaturated sulfones and alkyl α-isocyanoacetates. An alternative to the use of the reactive nitroalkenes is their *in situ* generation from β-acetoxy nitroalkanes, which are readily prepared via the Henry reaction between an aldehyde and a nitroalkane followed by acetylation.

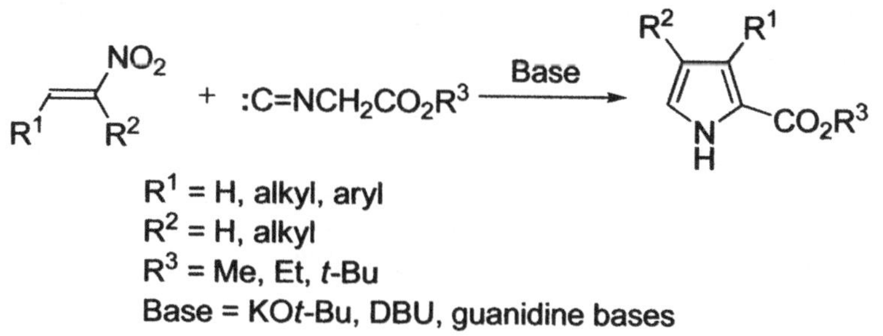

The mechanism is presumed to involve a pathway related to those proposed for other base-catalyzed reactions of isocyanoacetates with Michael acceptors. Thus, base-induced formation of enolate is followed by Michael addition to the nitroalkene and cyclization of nitronate to furnish the nitro-isocyanide after protonation. Loss of nitrous acid (HNO_2) and aromatization then affords the pyrrole ester.

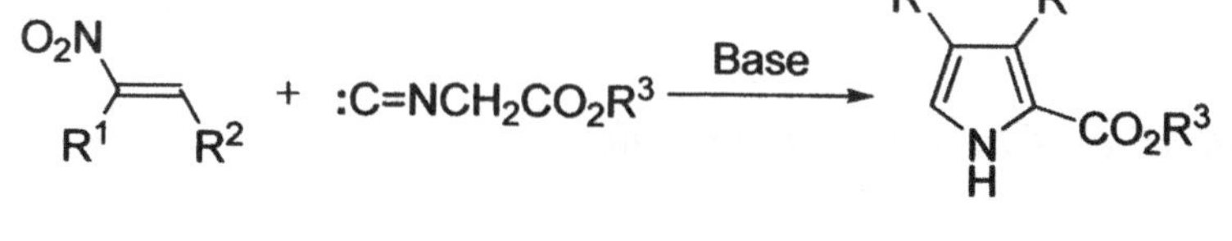

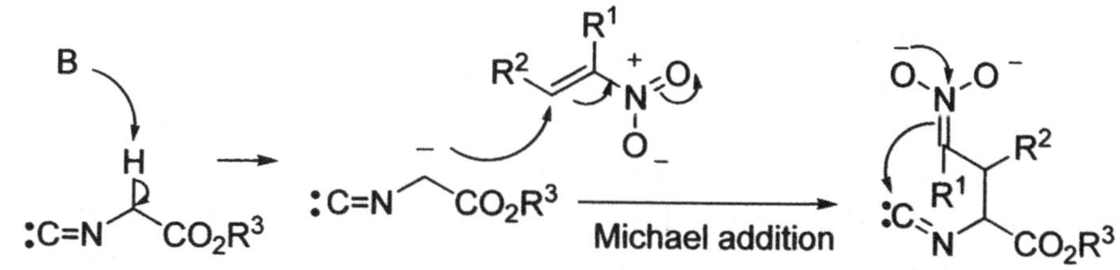

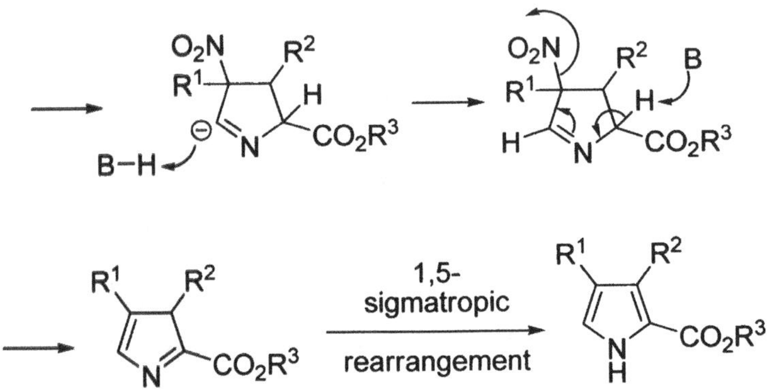

The Barton–Zard reaction found its application in the synthesis of LY2059346, a positive allosteric modulator of the α-amino-3-hydroxyl-L-aspartate (AMPA) receptor as a potential treatment of neurological and psychiatric disorders.[36]

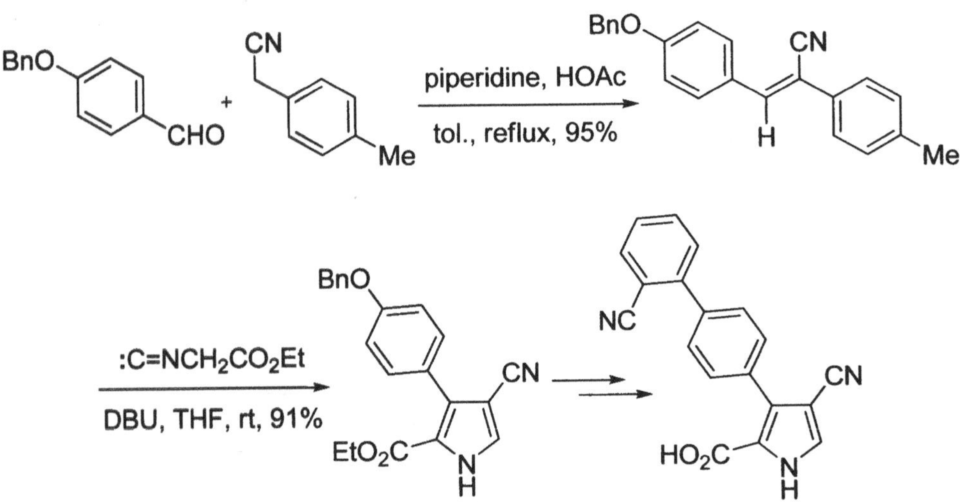

LY2059346

2.4 *Palladium Chemistry of Pyrroles*

A series of novel 2,5-bis(guanidino-aryl)-1-methyl-1*H*-pyrroles was synthesized starting from 1-methyl-1*H*-pyrrole employing the Stille coupling as the key operation.[37] The bis-stannylpyrrole was obtained from 1-methylpyrrole via 2,5-dilithiation. Subsequent Stille coupling afforded di(nitrophenyl)-pyrrole, which was used to craft novel diguanidine anti-fungal agents. The derivative shown below was found to be equipotent or more potent than fluconazole against most of the tested fungus strains.

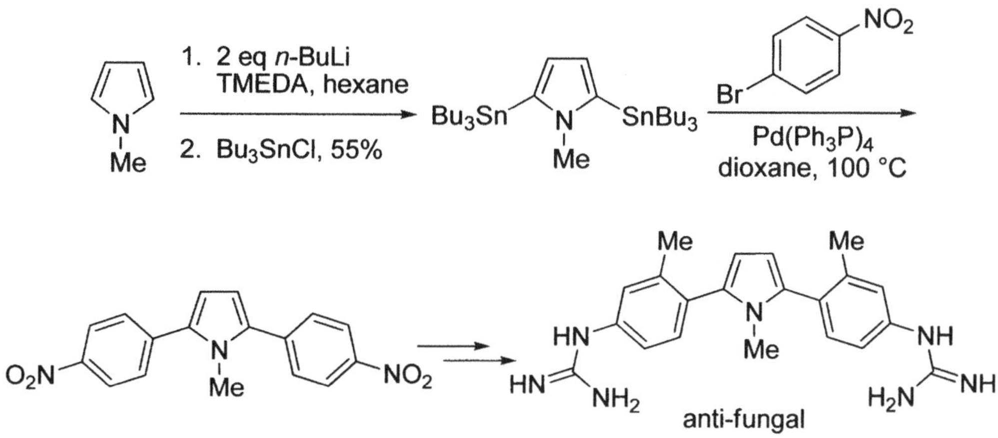

The stannylpyrrole aldehyde shown below underwent the Stille coupling with the aldehyde group unmolested. The Stille coupling between stannylpyrrole aldehyde and acid chloride was used to synthesize the sponge metabolite mycalazol 11 and related compounds, which have activity against the P388 murine leukemia cell line.[38]

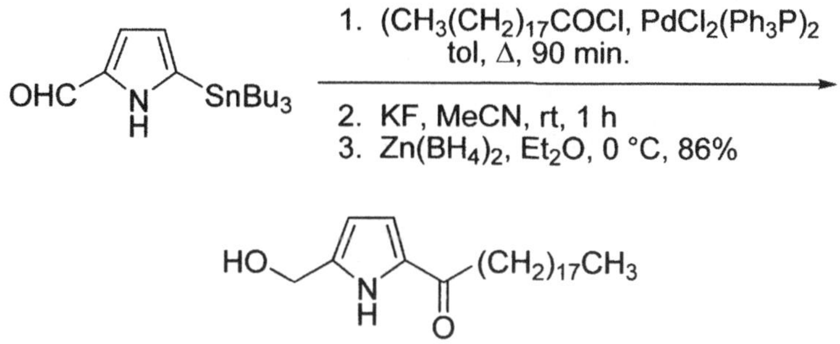

The Suzuki coupling of a bis-pyrrolo-2-triflate with pyrrolo-2-boronic acid afforded a triple pyrrole such as prodigiosin. Although the mechanism is unknown at the moment, prodigiosins and copper prodigiosins cleave double-strand DNA and show promise in cancer treatment.[39]

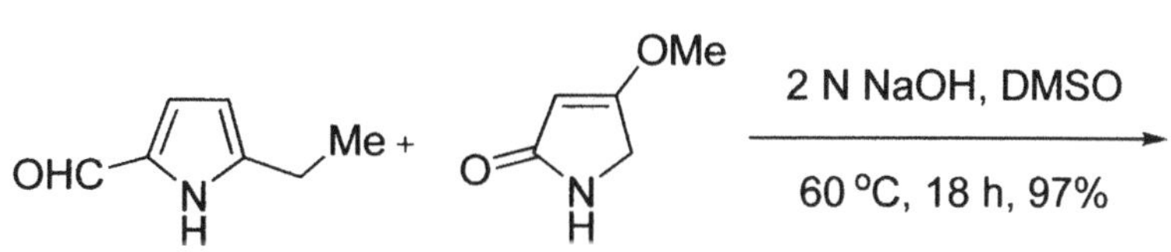

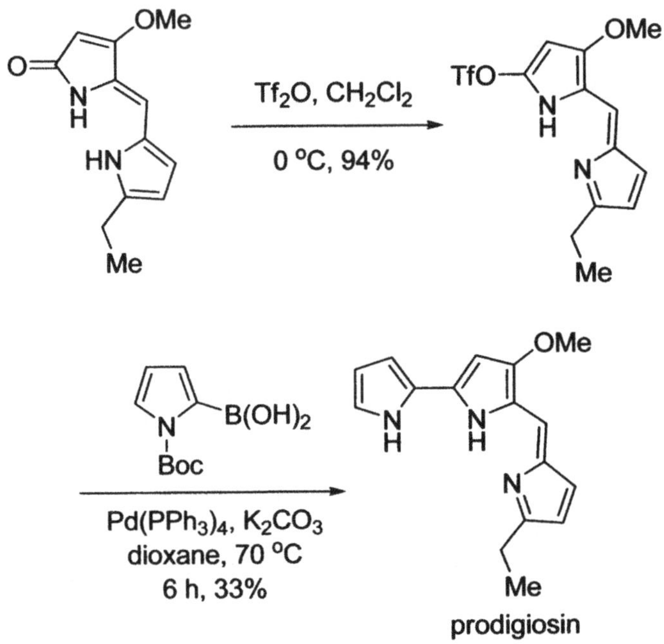

2.5 *Possible Liabilities of Pyrrole-Containing Drugs*

Because the pyrrole ring is extremely electron-rich, pyrrole-containing drugs are easily oxidized by cytochrome P-450 (CYP-450) enzymes in the liver. The resulting metabolic oxidation products are prone to nucleophilic replacement by physiological nucleophiles such as the thiol group. The consequence is toxicities such as agranulocytosis, hepatotoxicity, and so on. An anti-hypertensive agent, mopidralazine (MDL-899), was extensively investigated with regard to the metabolic oxidation of its pyrrole ring in rats and dogs.[40–42] Isolation and characterization of mopidralazine's metabolites led to the hypothesis that the biotransformations of pyrrole may involve the introduction of molecular oxygen into the pyrrole ring. The intermediacy of 1,2-dioxetane explains that an oxidative cleavage of the pyrrole ring could provide all metabolites identified.

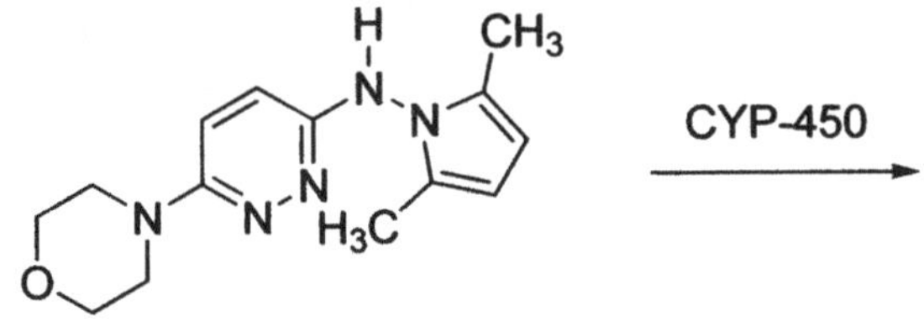

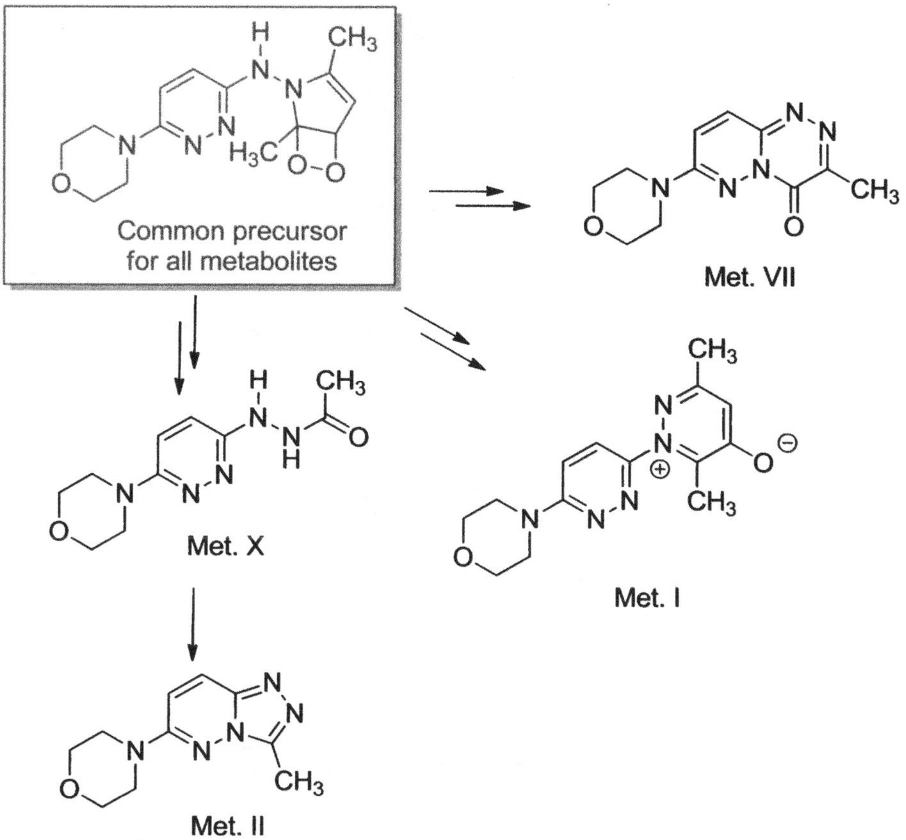

As one could imagine, the highly reactive intermediates from pyrrole could wreak havoc in the physiological system. Nucleophiles such as the thiol group could induce toxicities. As a result, the development of mopidralazine was subsequently discontinued.

Similar oxidative metabolism was observed for premazepam, an anti-anxiety drug, in the rat and the dog[43]; prinomide, an anti-inflammatory agent, in six species of laboratory animals[44]; and pyrrolnitrin, an anti-fungal agent, in rats.[45]

Nonetheless, if one assumes that all pyrrole-containing drugs are toxic, we would have missed atorvastatin (Lipitor). Atorvastatin is remarkably safe barring the mechanism-based safety concerns such as rhabdomyolysis that are associated with all HMG-CoA inhibitors.[46,47] Although atorvastatin contains the pyrrole ring, it has at least three factors that strongly attenuate its nucleophilicity. First, it is fully substituted at all

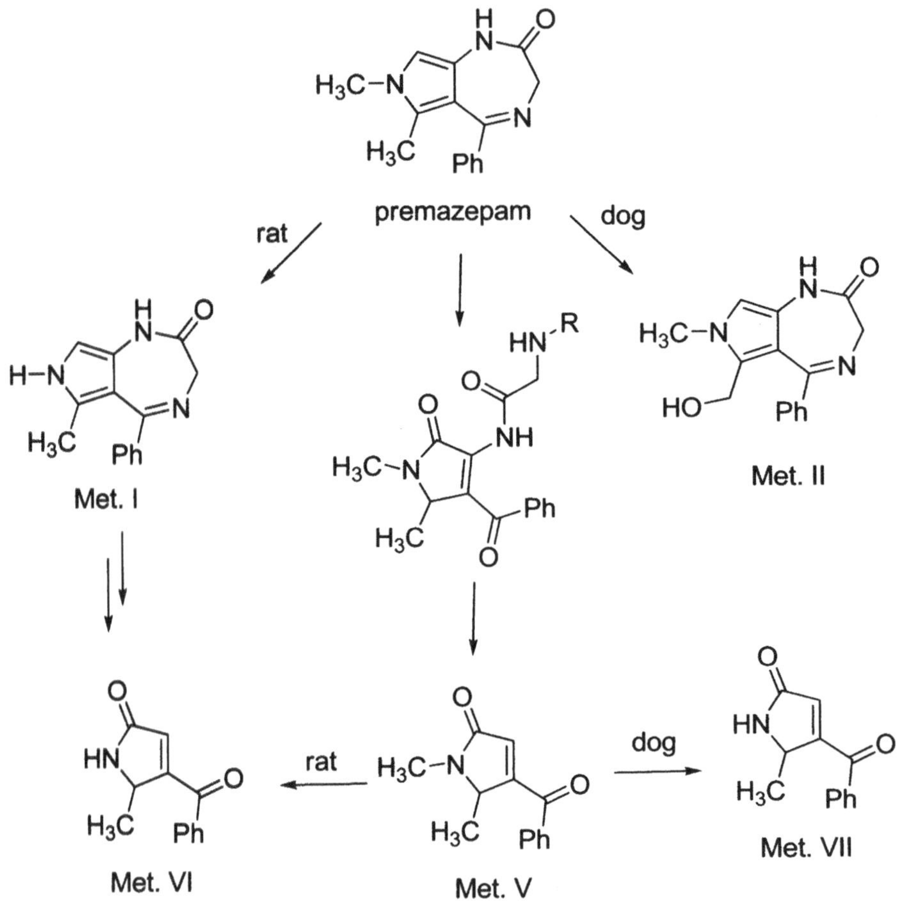

Reproduced from Reference 43 with permission

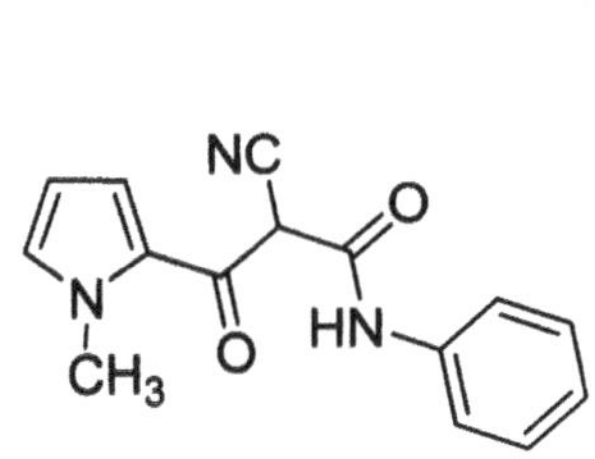

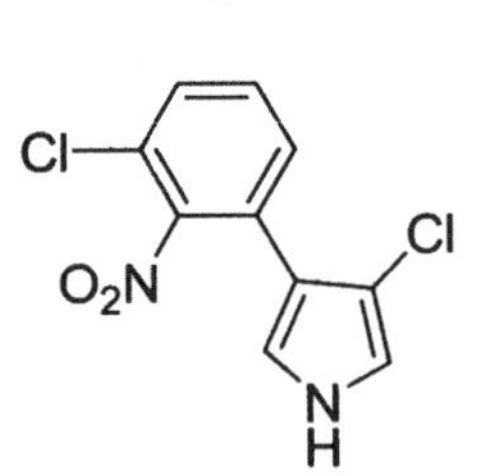

possible positions—it is a penta-substituted pyrrole, thus, the steric hindrance would block most substitutions. Second, two phenyl and one amide substitution form large delocalization to disperse the electronic density of the pyrrole ring. Third, *para*-fluorophenyl and amide are both electron-withdrawing, further diminishing the electronic density of the pyrrole ring.

In drug discovery, as in many other things in life, there are always exceptions to the rules. Frequently, the safety and efficacy of a drug can only be determined by clinical trials as the touchstone.

2.6 *Problems*

2.6.1 Explain why the dipole moments for pyrrole and pyrrolidine have the opposite directions?

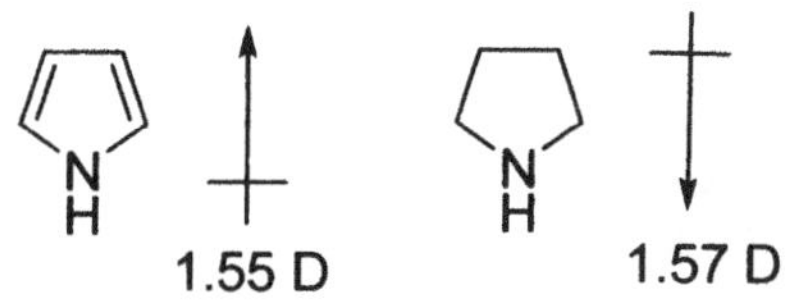

2.6.2 Why is pyrrole-2-aldehyde less reactive than benzaldehyde as an electrophile?

2.6.3 Predict the structures of adducts **A**.[48]

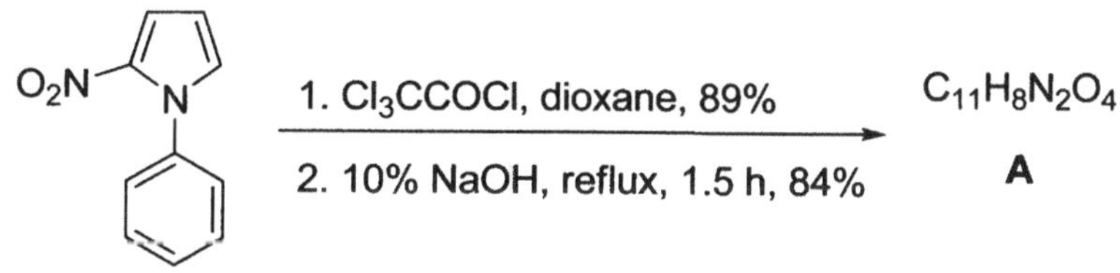

2.6.4 Predict the structure of product **B**.[49]

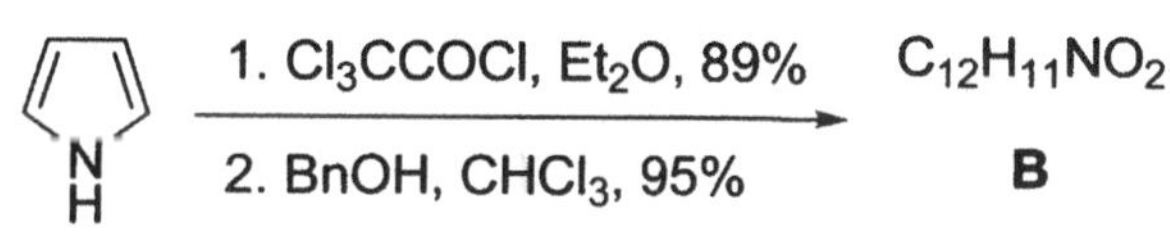

2.6.5 Predict the structures of adducts **C** and **D** with special attention on the regiochemistry[50]:

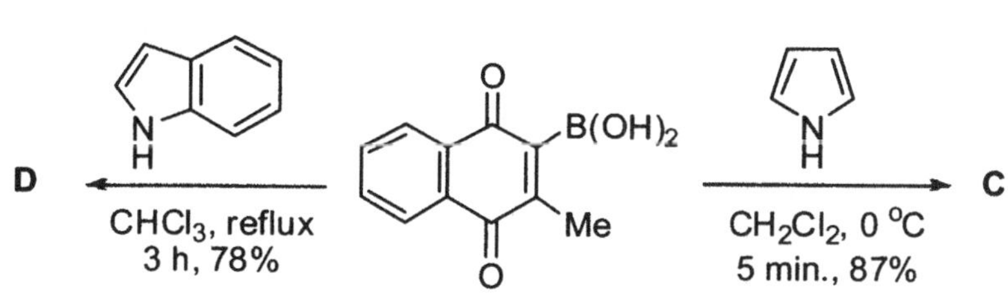

2.6.6 Propose the mechanism of the following transformations[51]:

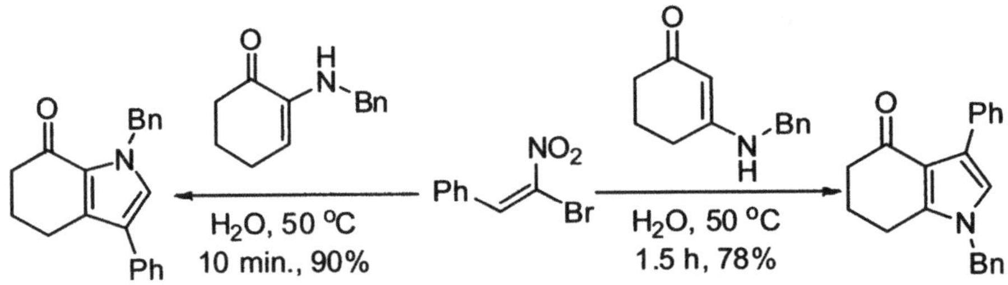

2.6.7 What is the structure of product **E**?[52]

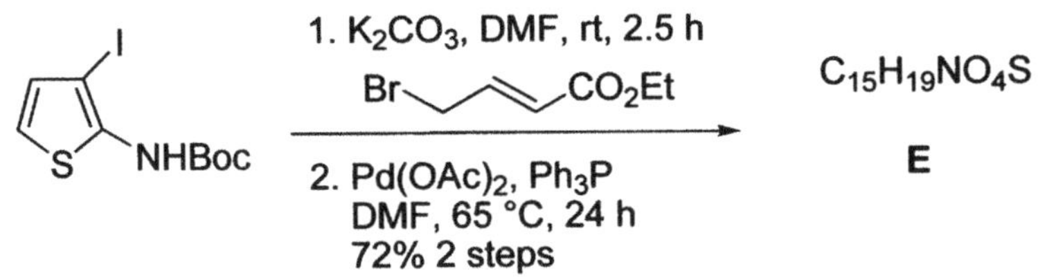

2.6.8 Provide the mechanims of the Ciamician–Dennstedt Rearrangement[53]:

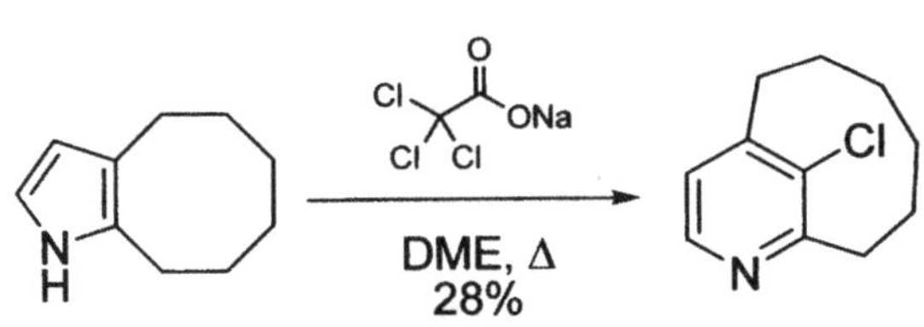

2.6.9 Provide the mechanims of the Clauson–Kaas pyrrole synthesis.[54]

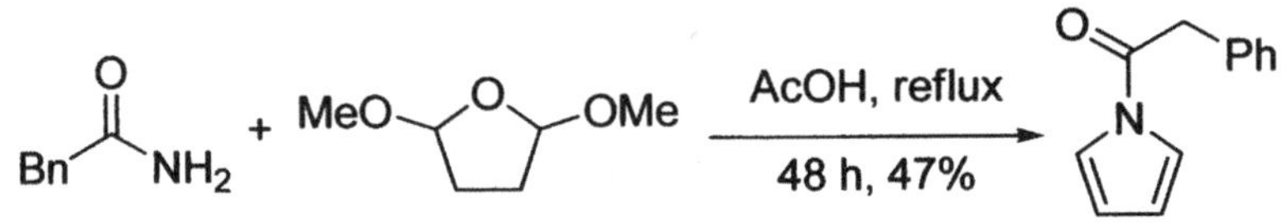

2.6.10 Propose the mechanism of the following pyrrole formation[55]:

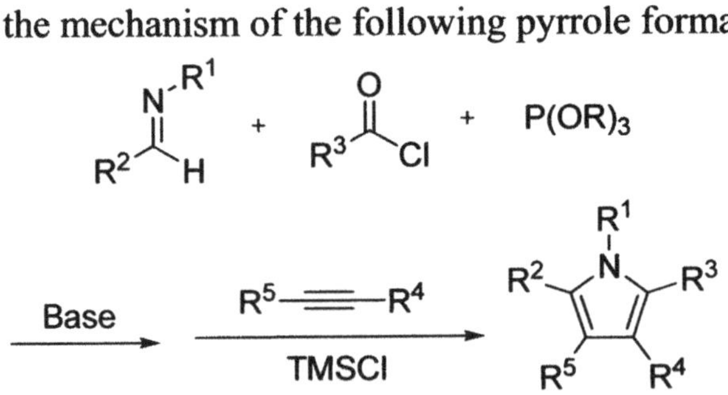

2.7 References

1. Chiang, Y.; Whipple, E. B. *J. Am. Chem. Soc.* **1963,** *85,* 2763–2767.
2. Balón, M.; Carmona, M. C.; Munoz, M. A.; Hidalgo, J. *Tetrahedron* **1989,** *45,* 7501–7504.
3. Fürstner, A.; Krause, H.; Thiel, O. R. *Tetrahedron* **2002,** *58,* 6373–6380.
4. Franco, F.; Greenhouse, R.; Muchowski, J. *J. Org. Chem.* **1982,** *47,* 1682–1688.
5. Dvornikova, E.; Kamiefska-Trela, K. *Synlett* **2002,** 1152–1154.
6. Gribble, G. W.; Blank, D. H.; Jasinski, J. P. *Chem. Commun.* **1999,** 2195–2196.
7. Kim, I. T.; Elsenbaumer, R. L. *Tetrahedron Lett.* **1998,** *39,* 1087–1090.
8. Messinger, P.; Gompertz, J. *Arch. Pharm.* **1977,** *310,* 249–255.
9. Zanka, A.; Nishiwaki, M.; Morinaga, Y.; Inoue, T. *Org. Process Res. Dev.* **1998,** *2,* 230–237.
10. van Wijngaarden, I.; Kruse, C. G.; van der Heyden, J. A.; Tulp, M. T. *J. Med. Chem.* **1988,** *31,* 1934–1940.
11. Scott, M. K.; Martin, G. E.; DiStefano, D.; Fedde, C. L.; Kukla, M. J.; Barrett, D. L.; Baldy, W. J.; Elgin, R. J., Jr.; Kesslick, J. M.; Mathiasen, J. R.; Sahnk, R. P.; Vaught, J. L. *J. Med. Chem.* **1992,** *35,* 552–558.
12. Scott, M. K.; Baxter, E. W.; Bennett, D. J.; Boyd, R. E.; Blum, P. S.; Codd, E. E.; Kukla, M. J.; Malloy, E.; Maryanoff, B. E.; Maryanoff, C. A.; et al. *J. Med. Chem.* **1995,** *38,* 4198–4219.
13. Goeminne, A.; Berg, M.; McNaughton, M.; Bal, G.; Surpateanu, Ge.; Van der Veken, P.; De Prol, S.; Versees, W.; Steyaert, J.; Haemers, A.; Augustyns, K. *Bioorg. Med. Chem.* **2008,** *16,* 6752–6763.
14. Pacorel, Be.; Leung, S. C.; Stachulski, A. V.; Davies, J.; Vivas, Li.; Lander, H.; Ward, S. A.; Kaiser, M.; Brun, R.; O'Neill, P. M. *J. Med. Chem.* **2010,** *53,* 633–640.
15. Hong, F.; Zaidi, J.; Pang, Y.-P.; Cusack, B.; Richelson, E. *J. Chem. Soc., Perikin Trans. 1* **1997,** 2997–3003.
16. Baran, P. S.; Richter, J. M. *J. Am. Chem. Soc.* **2004,** *126,* 7450–7451.
17. Baran, P. S.; Richter, J. M.; Lin, D. W. *Angew. Chem., Int. Ed.* **2005,** *44,* 609–612.
18. Morrison, M. D.; Hanthorn, J. J.; Pratt, D. A. *Org. Lett.* **2009,** *11,* 1051–1054.
19. (a) Hollins, R. A.; Colnago, L. A.; Salim, V. M.; Seidl, M. C. *J. Heterocyclic Chem.* **1979,** *16,* 993–996. (b) Gribble, G. W.; Blank, D. H.; Jasinski, J. P. *Chem. Commun.* **1999,** 2195–2196.
20. Ragno, R.; Mai, A.; Massa, S.; Cerbara, I.; Valente, S.; Bottoni, P.; Scatena, R.; Jesacher, F.; Brosch, G. *J. Med. Chem.* **2004,** *47,* 1351–1359.
21. Kandil, S.; Biondaro, S.; Vlachakis, D.; Cummins, A.-C.; Coluccia, A.; Berry, C.; Leyssen, P.; Neyts, J.; Brancle, A. *Bioorg. Med. Chem. Lett.* **2009,** *19,* 2935–2937.
22. Balsamini, C.; Bedini, A.; Diamantini, G.; Spadoni, G.; Tontini, A.; Tarzia, Di Fabio, R.; Feriani, A.; Reggiani, A.; Tedesco, G.; Valigi, R. *J. Med. Chem.* **1998,** *41,* 808–820.
23. (a) Carson, J. R.; McKinstry, D.; Wong, S. *J. Med. Chem.* **1971,** *14,* 646–647. (b) Carson, J. R.; Wong, S. *J. Med. Chem.* **1973,** *16,* 172–174.
24. Dyatkina, N. B.; Roberts, C. D.; Keicher, J. D.; Dai, Y.; Nadherny, J. P.; Zhang, W.; Schmitz, U.; Kongpachith, A.; Novikov, A. A.; Lou, L.; Velligan, M.; Khorlin, A. A.; Chen, M. S. *J. Med. Chem.* **2002,** *45,* 805–817.
25. Gribble, G. W. in *Name Reactions in Heterocyclic Chemistry*, Li, J. J.; Corey, E. J. Eds., Wiley: Hoboken, NJ, **2005,** 79–89.

26. Raviña, E.; Masaguer, C. F.; Cid, J.; Casariego, I.; Fontenla, J. A.; Ferreiro, T. G.; Cadavid, M. I.; Loza, M. I.; de Ceballos, M. L. *Bioorg. Med. Chem. Lett.* **1995,** *5*, 579–584.
27. Masaguer, C. F.; Raviña, E.; Loza, I.; Fontenla, J. A. *Bioorg. Med. Chem. Lett.* **1997,** *7*, 913–918.
28. Bellingham, R. K.; Carey, J. S.; Hussain, N.; Morgan, D. O.; Oxley, P.; Powling, L. C. *Org. Process Res. Dev.* **2004,** *8*, 279–282.
29. de Laszlo, S. E.; Visco, D.; Agarwal, L.; et al. *Bioorg. Med. Chem. Lett.* **1998,** *8*, 2689–2694.
30. Danchev, N.; Bijev, A.; Yaneva, D.; Vladimirova, S.; Nikolova, I. *Archiv. Pharm.* **2006,** *339*, 670–674.
31. Brower, P. L.; Butler, D. E.; Deering, C. F.; Le, T. V.; Millar, A.; Nanninga, T. N.; Roth, B. D. *Tetrahedron Lett.* **1992,** *33*, 2279-2282.
32. Baumann, K. L.; Butler, D. E.; Deering, C. F.; Mennen, K. E.; Millar, A.; Nanninga, T. N.; Palmer, C. W.; Roth, B. D. *Tetrahedron Lett.* **1992,** *33*, 2283–2284.
33. Gangjee, A.; Zeng, Y.; McGuire, J. J.; Mehraein, F.; Kisliuk, R. L. *J. Med. Chem.* **2004,** *47,* 6893–6901.
34. Deng, Y.; Wang, Y.; Cherian, C.; Hou, Z.; Buck, S. A.; Matherly, L. H.; Gangjee, A. *J. Med. Chem.* **2008,** 5052–5063.
35. Gribble, G. W. in *Name Reactions in Heterocyclic Chemistry*, Li, J. J.; Corey, E. J. Eds.; Wiley: Hoboken, NJ, **2005,** pp. 70–78.
36. Zarrinmayeh, H.; Tromiczak, E.; Zimmerman, D. M.; Ramkl, N.; Ho, K. H.; Diminguez, E.; Castaño, A.; Escriano, A.; Fernandez, C.; Jimenenz, A.; Hornback, W. J.; Nisenbaum, E. S. *Bioorg. Med. Chem. Lett.* **2006,** *16*, 5203–5206.
37. Jana, G. H.; Jain, S.; Arora, S. K.; Sinha, N. *Bioorg. Med. Chem. Lett.* **2005,** *15*, 3592–3595.
38. Nabbs, B. K.; Abell, A. D. *Bioorg. Med. Chem. Lett.* **1999,** *9*, 505–508.
39. Melvin, R. A.; Tomlinson, J. T.; Saluta, G. R.; Kucera, G. L.; Lindquist, N.; Manderville, R. A. *J. Am. Chem. Soc.* **2000,** *122,* 6333–6334.
40. Assandri, A.; Perazzi, A.; Baldoli, E.; Ferrari, P.; Ripamonti, A.; Bellasio, E.; Tuan, G.; Zerilli, L. F.; Tarzia, G. *Xenobiotica* **1985,** *15*, 1069–1087.
41. Assandri, A.; Perazzi, A.; Bellasio, E.; Ciabatti, R.; Tarzia, G.; Ferrari, P.; Ripamonti, A.; Tuan, G.; Zerilli, L. F. *Xenobiotica* **1985,** *15*, 1089–1102.
42. Assandri, A.; Tarzia, G.; Bellasio, E.; Ciabatti, R.; Tuan, G.; Ferrari, P.; Zerilli, L.; Lanfranchi, M.; Pelizzi, G. *Xenobiotica* **1987,** *17*, 559–573.
43. Assandri, A.; Barone, D.; Ferrari, P.; Perazzi, A.; Ripamonti, A.; Tuan, G.; Zerilli, L. *Drug Metab. Dispos.* **1984,** *12*, 257–263.
44. Egger, H.; Itterly, W.; John, V.; Shimanskas, C.; Stancato, F.; Kapoor, A. *Drug Metab. Dispos.* **1988,** *16*, 568–575.
45. Murphy, P. J.; Williams, T. L. *J. Med. Chem.* **1972,** *15,* 137–139.
46. Walsh, K. M.; Albassam, M. A.; Clarke, D. E. *Toxicol. Pathol.* **1996,** *24,* 468–476.
47. Li, J. J. *Triumph of the Heart—the Story of Statins,* Oxford University Press: New York, NY, **2009**.
48. Rotas, G.; Kimbaris, A.; Varvounis, G. *Tetrahedron* **2004,** *60,* 10825–10832.
49. Schmuck, C.; Bickert, B.; Merschky, M.; Geiger, L.; Rupprecht, D.; Dudaczek, J.; Wich, P.; Rehm, T.; Machon, U. *Eur. J. Org. Chem.* **2008,** *15,* 324–329.
50. Veguillas, M.; Ribagorda, M.; Carreño, M. C. *Org. Lett.* **2011,** *13*, 656–659.
51. Rueping, M; Parra, A. *Org. Lett.* **2010,** *11*, 5281–5283.

52. Wensbo, D.; Annby, U.; Gronowitz, S. *Tetrahedron* **1995**, *51*, 10323–10342.
53. Dhanak, D.; Reese, C. B. *J. Chem. Soc., Perkin Trans. 1* **1987,** 2829–2832.
54. Menger, F. M.; Donohue, J. A. *J. Am. Chem. Soc.* **1973,** *95*, 432–437.
55. Morin, M. S. T.; St-Cyr, D. J.; Arndtsen, B. A. *Org. Lett.* **2010,** *12*, 4916–4919.

Chapter 3 Indoles, Oxindoles, and Azaindoles

Jie Jack Li

3.1 Introduction

The parent compound, indole itself, is a white crystalline solid at room temperature with a melting point of 52–54 °C. It has a pungent, naphthalene-like odor. Indole's bond lengths are shown below:

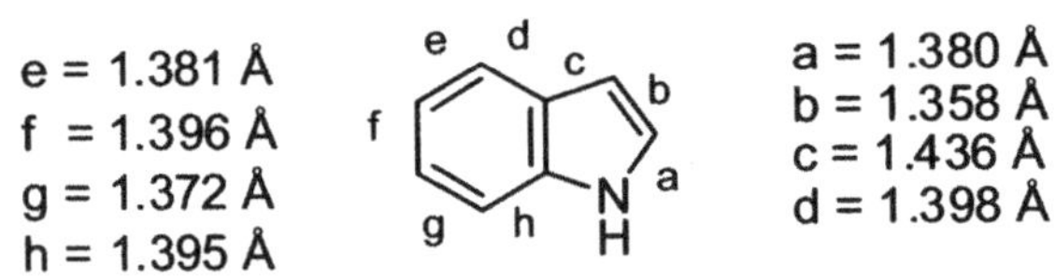

Indole's Bond Lengths

In its ^{1}H-NMR (Nuclear Magnetic Resonance Spectroscopy), the chemical shifts of H_2 and H_3 follow the trend of those of pyrrole. H_2 (~7.2 ppm) is much further down filed than H_2 (~6.6 ppm), again thanks to the inductive effect exerted by the N atom. The chemical shifts for the benzene ring are more nuanced, with H_4 (~7.7 ppm) showing up most down field and H_6 showing up at ~7.4 ppm. Similar to that of pyrrole, the NH's chemical shift often changes in different solvents and concentrations for the NMR samples.

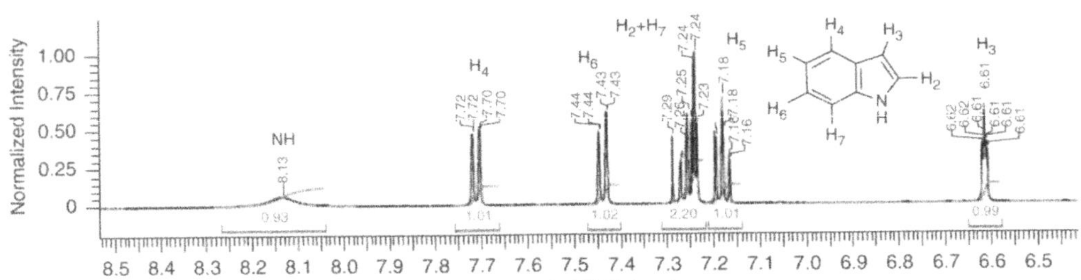

^{1}H-NMR Spectrum of Indole

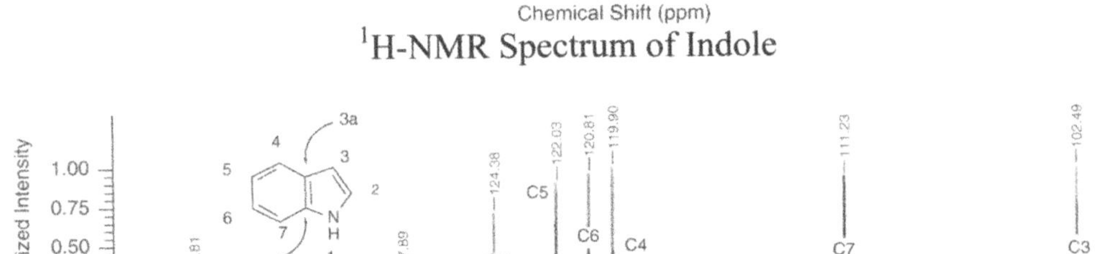

^{13}C-NMR Spectrum of Indole

As far as the ^{13}C-NMR spectrum of indole is concerned, the chemical shifts of C_2 and C_3 follow the trend of those in pyrrole. C_2 (124.4 ppm) is much more down field in comparison with that of C_2 (102.5 ppm), again thanks to the inductive effect of the N atom. The two quaternary carbon atoms C_{7a} and C_{3a} show up at 135.8 and 127.9 ppm, respectively. In addition, the heights for these two quaternary carbon atoms are significantly short in comparison with the carbon atoms with a hydrogen atom adjacent to them. The reason is that the two quaternary carbon atoms lack the nuclear Overhauser effect (nOe) that made the carbon peak taller with more protons ($–CH_3 > –CH_2 > –CH > –C$), also known as proton-enhancement.

Indole is perhaps the most visible heterocycle in all of chemistry. Since Adolf von Baeyer proposed the structure of indole as a heteroaromatic compound 140 years ago, indole has embodied a myriad of natural products, pharmaceutical agents, and a growing list of polymers. In the human body, serotonin modulates 5-hydroxytryptamine (5-HT), a monoamine neurotransmitter primarily found in the gastrointestinal (GI) tract and central nervous system (CNS), and modulates vasoconstriction and many brain activities. Melatonin regulates circadian rhythms, most noticeably, sleep. Tryptamine is closely related to melatonin and the amino acid tryptophan.

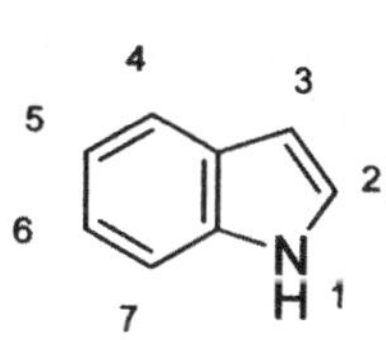

indole

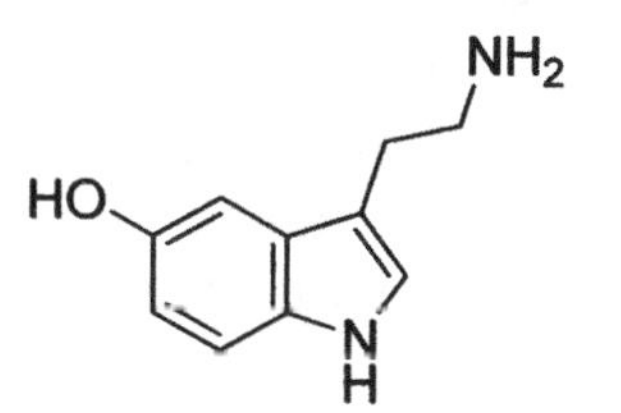

5-hydroxyindole (serotonin)

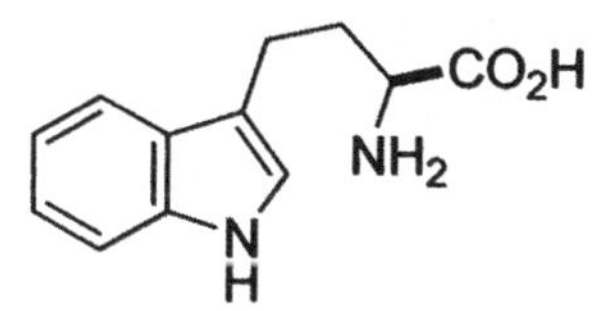

tryptophan

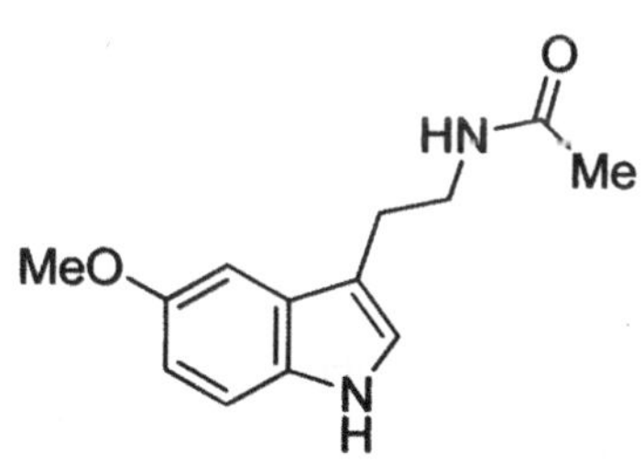

melatonin

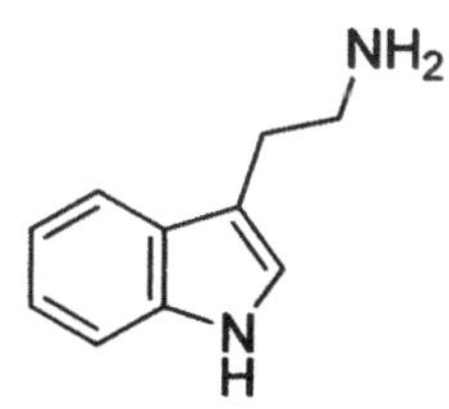

tryptamine

In addition to the hundreds of well-known indole plant alkaloids (*e.g.*, yohimbine, reserpine, strychnine, ellipticine, lysergic acid, and physostigmine), the indole ring is present in an array of other organisms. The indigo analogue Tyrian purple is the ancient Egyptian dye produced by Mediterranean mollusks. It was so precious that it was only used to dye the robes of Roman zemperors. Indigo, the dye used to dye jeans, was initially extracted from the indigo plant. In 1882, Baeyer developed the Baeyer–Drewson indigo synthesis using inexpensive 2-nitrobenzaldehyde and acetone in the presence of sodium hydroxide. Baeyer's revolutionary synthesis made indigo an easily accessed dye that all commoners could afford. Indigo, in no small way, contributed to the rise of the German chemical industry although William Perkin in England was the first to synthesize Mauve, a purple dye from the coal tar ingredients aniline and toluidine.[1]

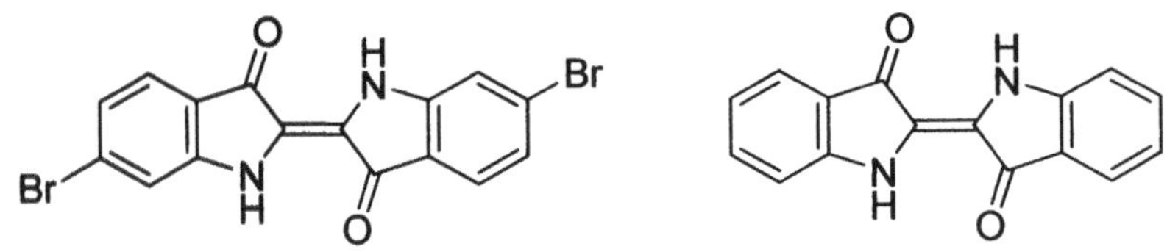

Tyrian purple indigo

The central importance of indole derivatives such as serotonin and tryptophan in living organisms has inspired medicinal chemists to design and synthesize thousands of indole-containing pharmaceuticals. Chief among them is fluvastatin sodium (Lescol), an HMG-CoA reductase inhibitor. Although fluvastatin is very potent *in vitro*, its *in vivo* potency is lower than many other statins. The combination of an electron-rich indole core and an allylic alcohol might contribute to its instability in peptic acid. Rosuvastatin calcium (Crestor), with its allylic alcohol attached to an electron-deficient heterocycle pyrimidine, is much more potent *in vivo* than fluvastatin. With peak sales of $734 million in 2003, fluvastatin was not among the top-selling statins.[2]

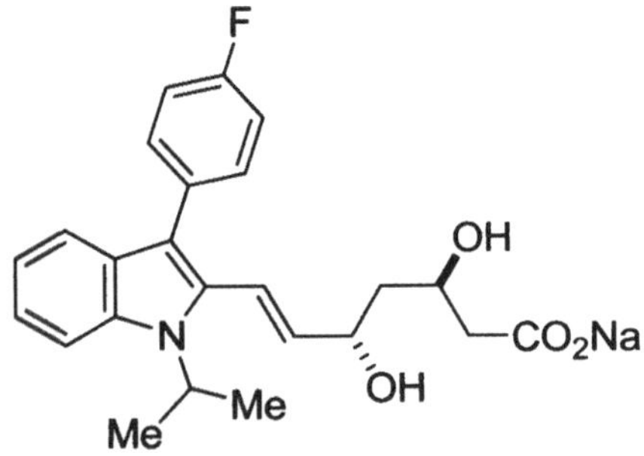

fluvastatin sodium (Lescol)

In addition, sumatriptan succinate (Imitrex), a serotonin receptor (5-HT_{1B1D}) agonist used to treat migraines had annual sales in the United States of $970 million in 2008. Following the highly effective and commercially successful sumatriptan, three "me-too" indole-containing anti-migraine drugs have been put on the market. They are naratriptan (Amerge), zolmitriptan (Zomig), and rizatriptan (Maxalt).[3]

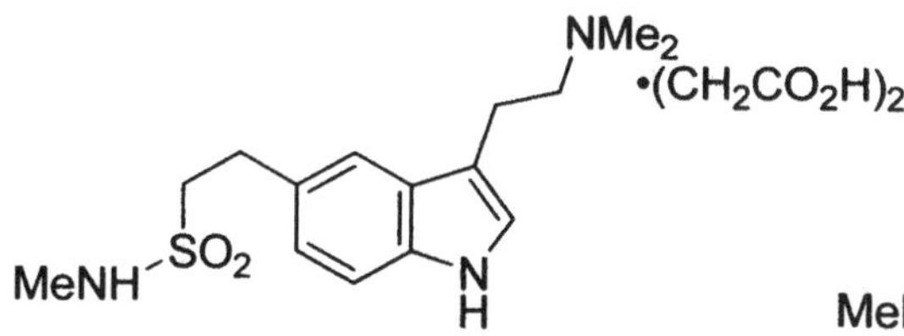

sumastriptan (Imitrex)

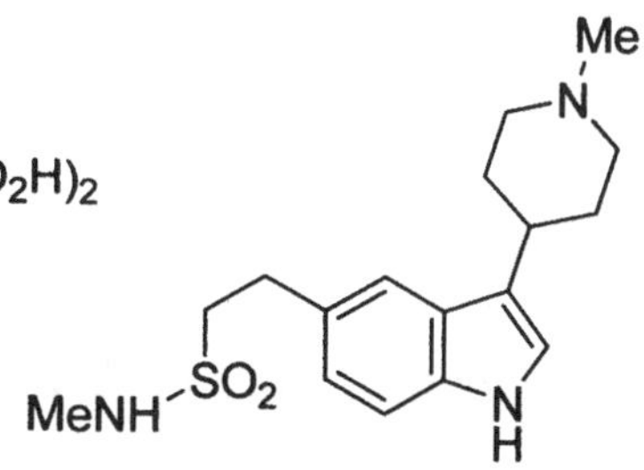

naratriptan (Naramig)

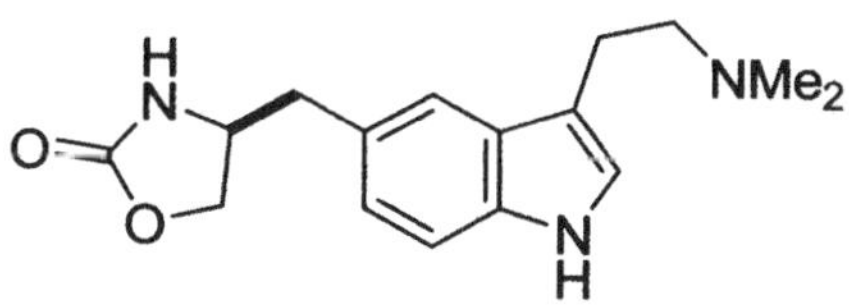

zolmitriptan (Zomig)

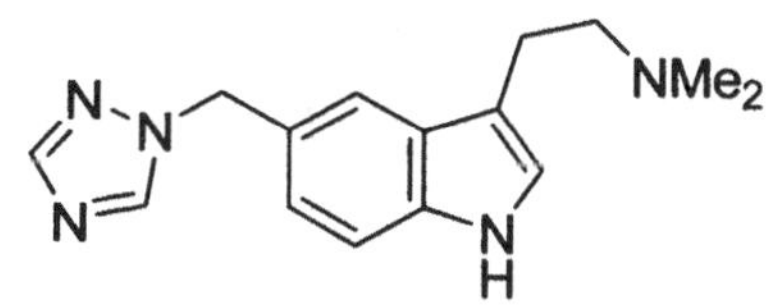

rizatriptan (Maxalt)

Furthermore, delavirdine (Rescriptor) is a novel HIV-1 reverse transcriptase inhibitor for HIV-positive individuals and zafirlukast (Accolate) is an anti-asthma drug. The anti-emetics ramosetron (Nasea) and dolasetron (Anzemet) are potent and highly selective 5-HT_3 receptor antagonists for the treatment of chemotherapy-induced nausea and vomiting.

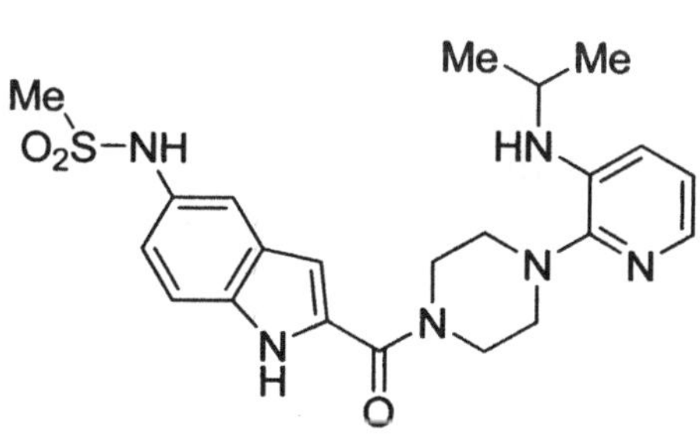

delavirdine (Rescriptor)

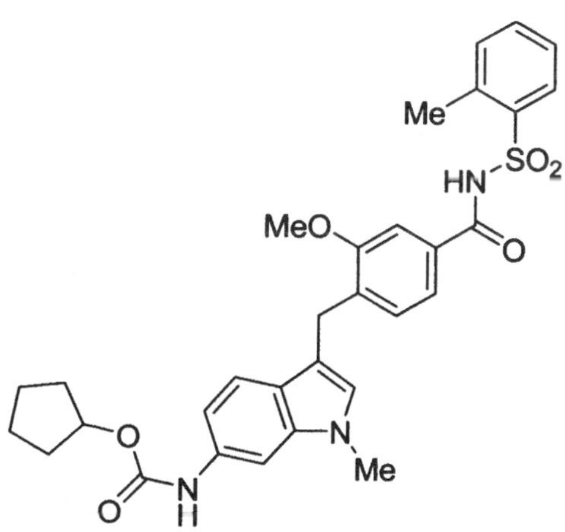

zafirlukast (Accolate)

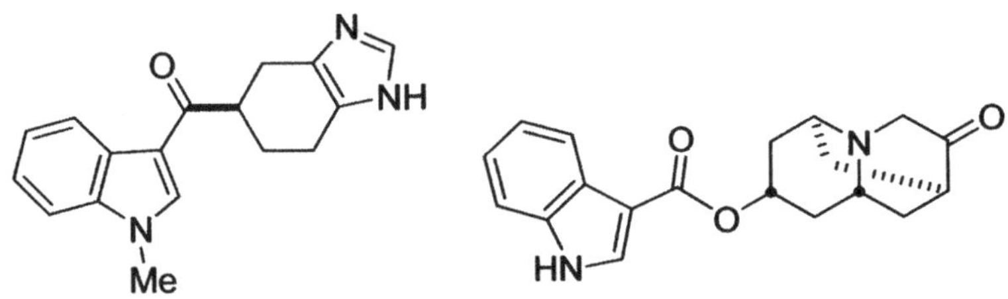

ramosetron (Nasea) dolasetron (Anzemet)

3.2 *Reactivity of the Indole Ring*

3.2.1 Protonation

While pyrrole has 6 π-electrons, indole, with 10 π-electrons, is also an *electron-rich* aromatic heterocycles. Indole is not as reactive as pyrrole because its electrons are more delocalized. An indole lone pair of electrons takes part in the delocalization essential to indole's aromaticity. Therefore, the indole nitrogen atom (p*K*a = –3.5)[4] is not basic: Indole loses its aromaticity when protonated. The protonation takes place predominantly at the C3 position to form the 3*H*-indolium ion, which combines with another molecule of indole to give oligomers.

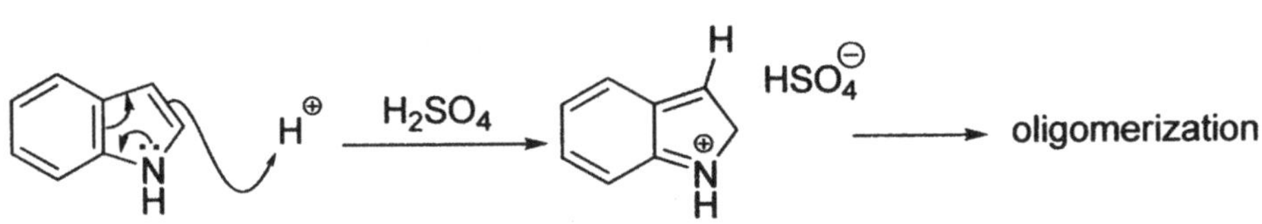

3.2.2 C3 Electrophilic Substitution

The indole lone pair of electrons is the engine that propels many of its unique reactivities. C3 electrophilic substitution is the most fertile ground with regard to indole's reactivities. When treated with an electrophile ($E^{\oplus}$), the lone pair of electrons stabilizes the transition state leading to attack at the C3 position.

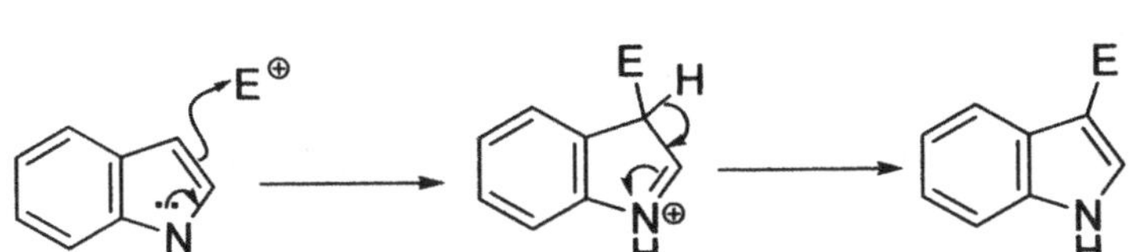

For instance, treatment of indole with bromine gives exclusively 3-bromoindole. In the same vein, Michael addition with nitroethene, ethyl acrylate, and the Vilsmeier reagent all take place at C3. The adduct between indole and the Vilsmeier reagent can be hydrolyzed under basic conditions to give 1*H*-indole-3-carbaldehyde. Meanwhile, nitration and treatment with oxalyl chloride all give rise to the C3 electrophilic substitution products.

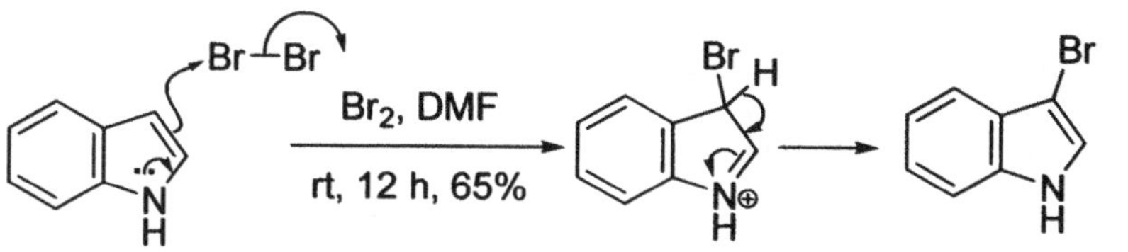

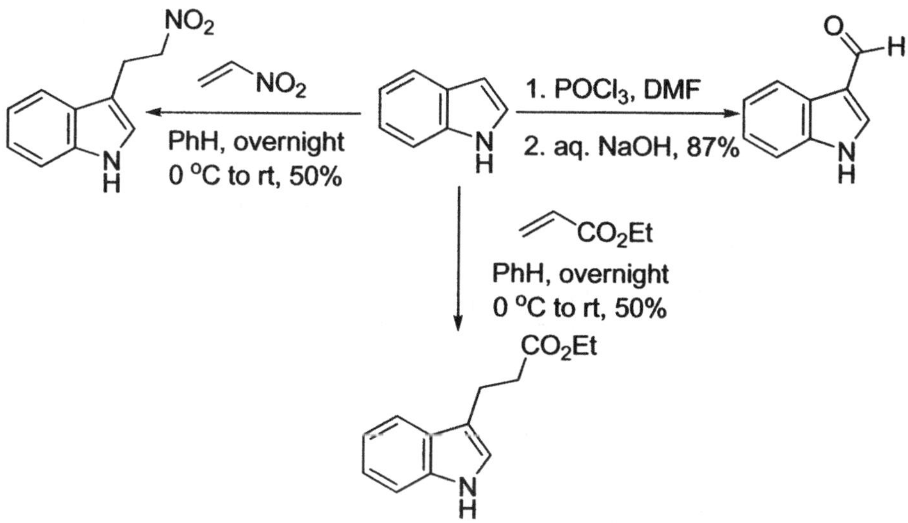

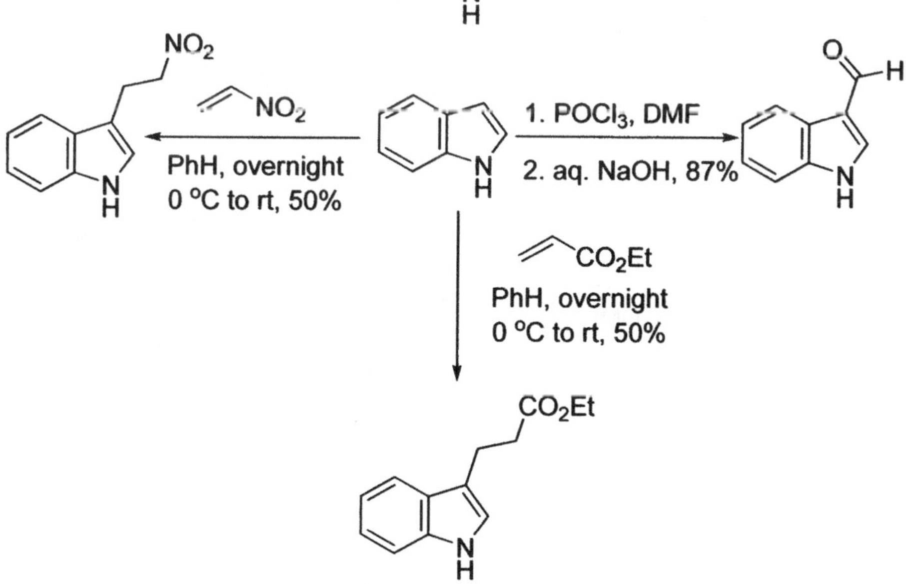

An olefin attached to an electron-deficient heterocycle such as pyridine may be viewed as a Michael acceptor as well. When indole was heated in acetic acid with vinyl pyridine or vinyl pyrimidine, C3 electrophilic substitutions readily take place, giving rise to the adducts.[5] Interestingly, when indole and 4-vinylpyridine were treated with sodium metal in the presence of $CuSO_4$, Michael addition at the indole nitrogen took place.

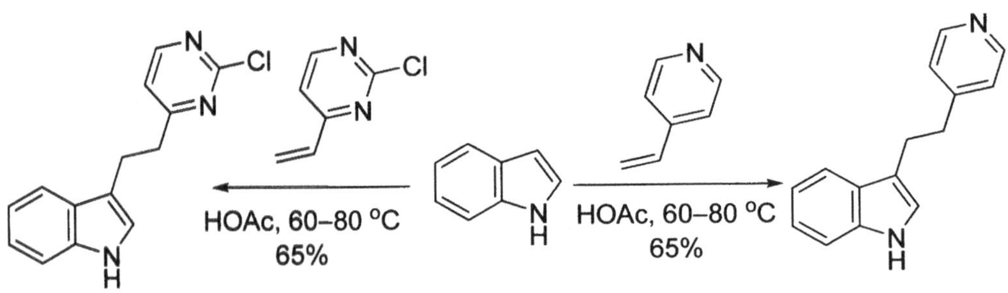

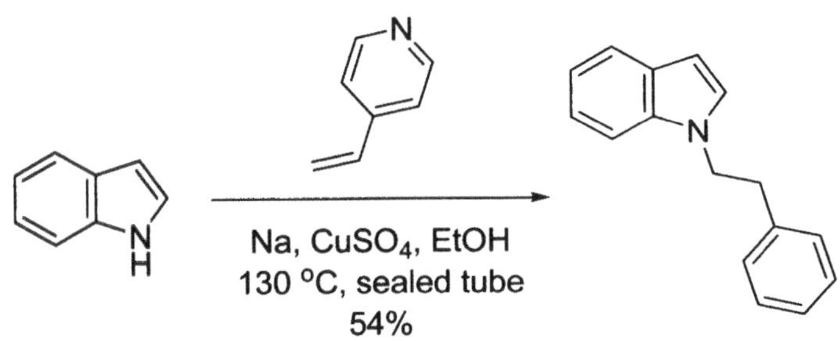

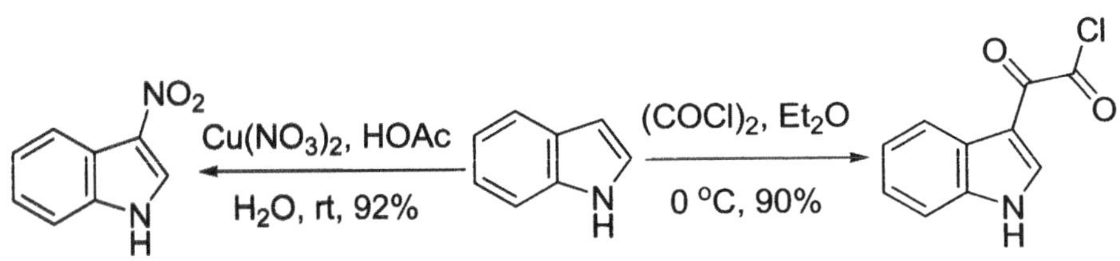

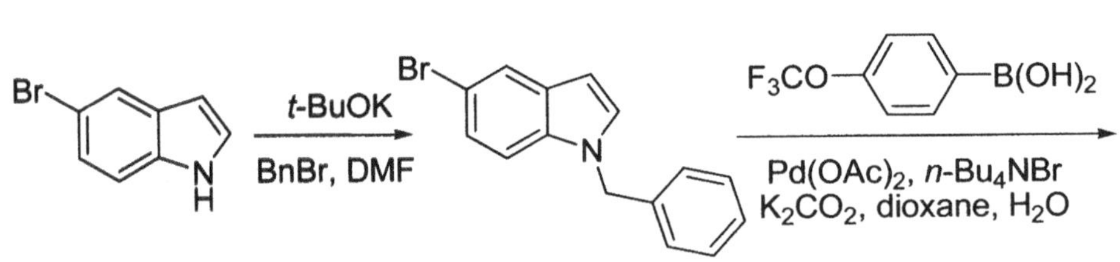

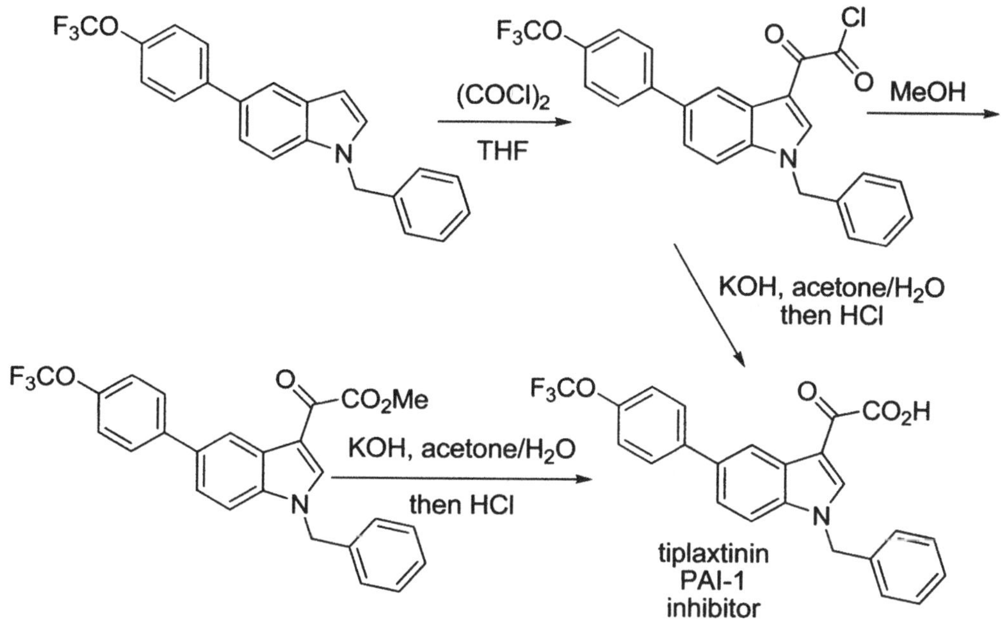

Incidentally, the synthesis of tiplaxtinin took advantage of the C3 electrophilic substitution of oxalyl chloride. Tiplaxtinin is a potent and selective inhibitor of plasminogen activator inhibitor 1 (PAI-1), and it demonstrated oral efficacy in multiple models of acute arterial thrombosis. It was investigated in phase I clinical trials but is not in active development at this time. In one of its synthesis shown below,[6] preparation of 1-benzyl-5-bromoindole was carried out by alkylation of 5-bromoindole with benzyl bromide in THF using potassium *tert*-butoxide as a base (sodium hydride was not used because of the inherent hazards associated with it on a large scale). The Suzuki coupling of the benzylated bromoindole with 4-trifluoromethoxyphenylboronic acid installed the trifluoromethoxyphenyl substitution. Reaction of the indole derivative with oxalyl chloride in tetrahydrofuran yielded the oxo acid chloride derivative, which was found to be stable and crystalline. Quenching of the oxo acid chloride with methanol produced the methyl ester, which was hydrolyzed under basic conditions, followed by acidification and crystallization to furnish tiplaxtinin. Alternatively, the oxo acid chloride could be hydrolyzed under basic conditions to give tiplaxtinin directly. Multi-kilogram batches of tiplaxtinin were produced using the later synthetic route with an overall yield of greater than 65%.

The strong reactivity of the lone pair of electrons cannot be diminished even by a strong electron-withdrawing group. For example, 1-tosylindole was readily nitrated in cold concentrated nitric acid.

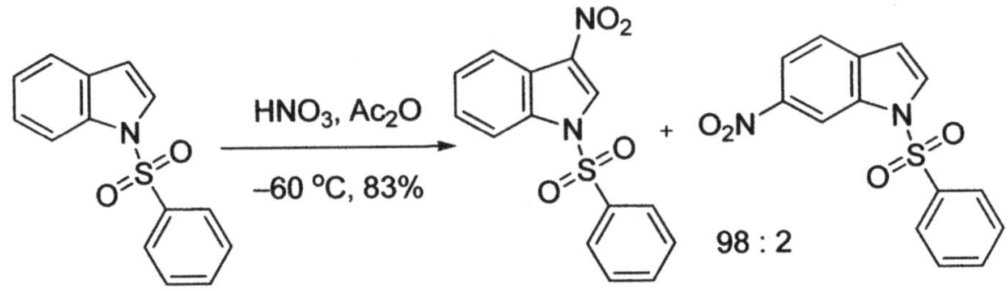

The Mannich condensation of indole results in the C-3 substitution in good yield.

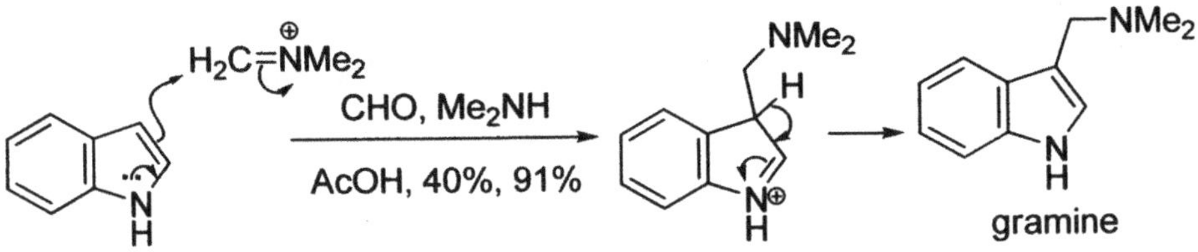

The dimethylaminomethyl group in gramine resulting from the Mannich reaction is labile and readily undergoes elimination when a nucleophile is present to take part in a Michael addition. For example:

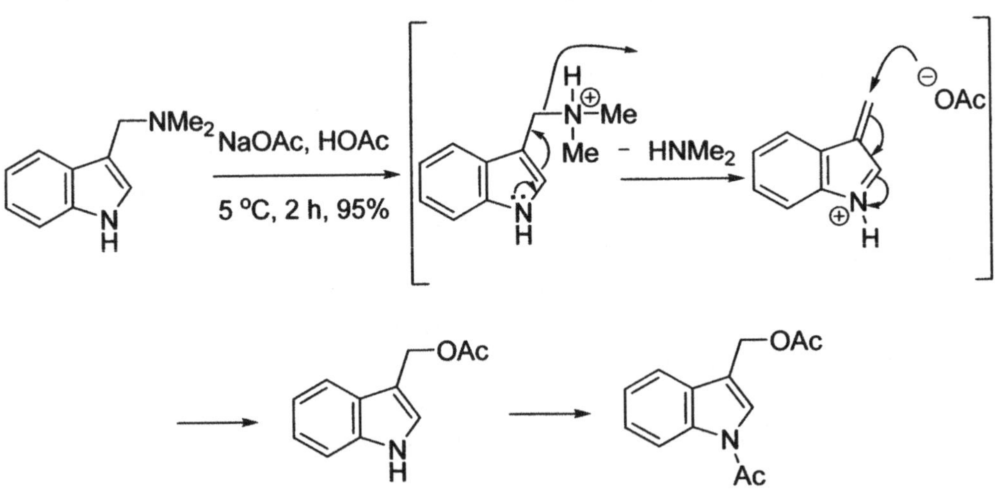

The preference for indole to undergo C3 electrophilic substitution has been applied to the synthesis of many drugs. One example is found in the synthesis of zafirlukast (Accolate), which is a potential drug used in the treatment of pulmonary disorders such as asthma. Zafirlukast acts by antagonizing the pharmacological actions of one or more of the arachidonic acid metabolites known as leukotrienes. In a process synthesis, a C3 electrophilic substitution was carried out between 5-nitroindole and the benzyl bromide in the presence of silver oxide and dioxane yielded the corresponding adduct.[7]

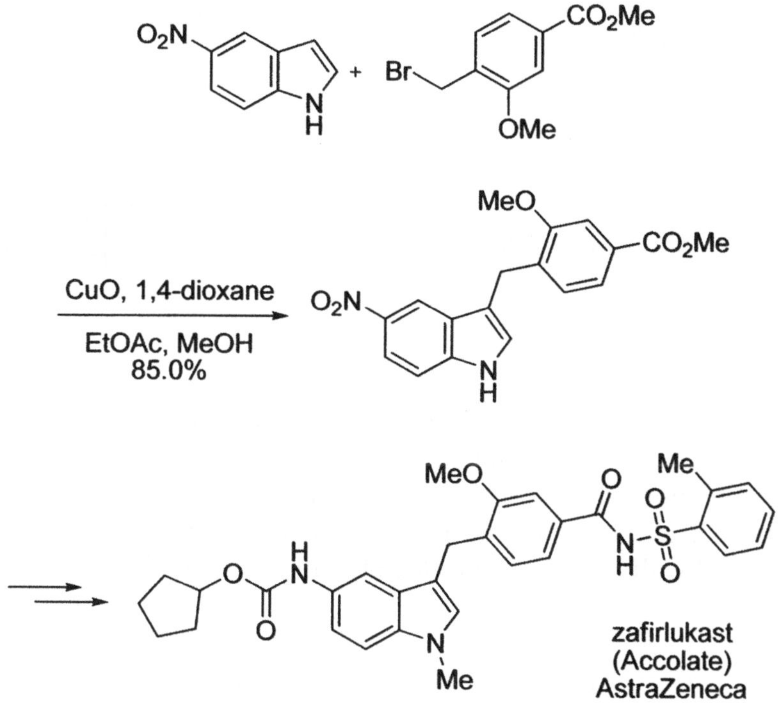

zafirlukast
(Accolate)
AstraZeneca

3.2.3 *C2 Electrophilic Substitution*

Electrophilic substitutions on the indole ring overwhelmingly favor the C3 position. However, when the energy outcome is favorable, C2 electrophilic substitutions do occur, especially for intramolecular substitutions. For instance, in the synthesis of tadalafil (Cialis), the Pictet–Spengler reaction is used to prepare the *cis*-β-carboline where the C2 electrophilic substitution takes place. In the presence of trifluoroacetic acid (TFA), *D*-tryptophan methyl ester is condensed with piperonal. The C2-carbon of the indole adds to the resulting iminium ion to give a mixture of the *cis*-β-carboline and the *trans*-isomer.[8,9]

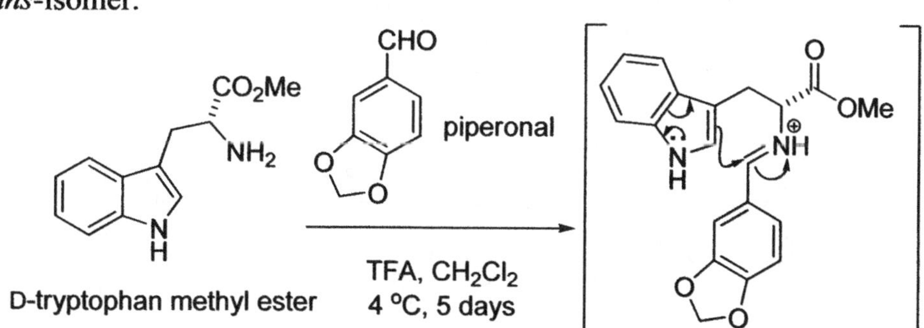

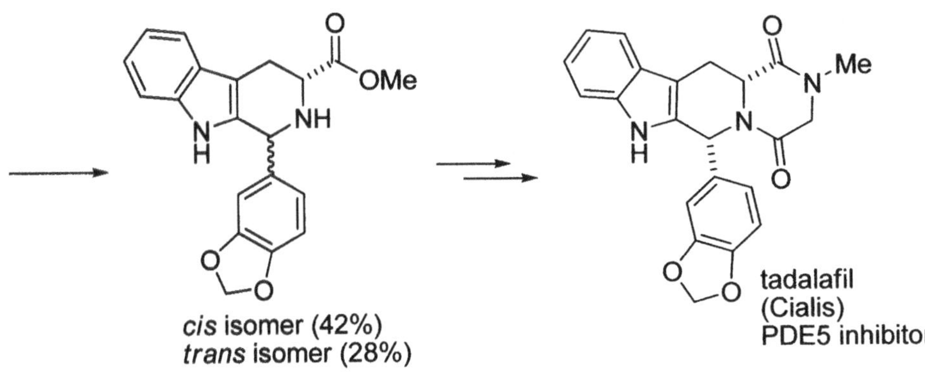

3.2.4 Metallation

The outcome of metallation of indoles depends on many factors, but the base is a crucial player. For instance, indole treated with NaH followed by addition of MeI undergoes an S_N2 reaction at N1 to give 1-methyl-1*H*-indole exclusively. On the other hand, indole treated with a Grignard reagent and then quenched by allyl bromide produced 3-allyl-1*H*-indole predominantly.

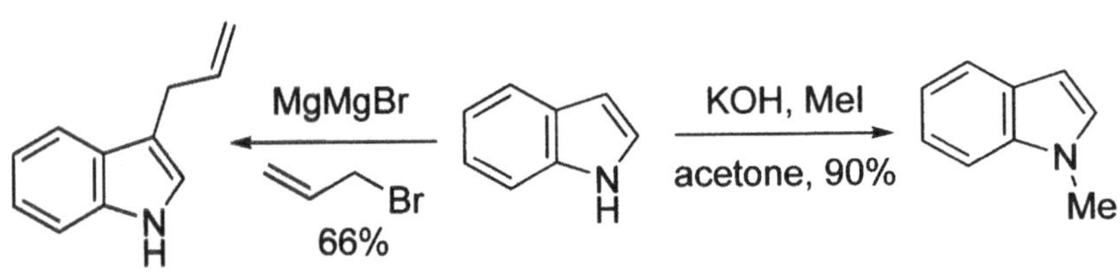

As far as N1 protected indoles are concerned, metallation primarily takes place at the C2 position. As a consequence, a "dummy protective group" could be engineered to steer the metallation to the C2 position. Indole is treated with *n*-BuLi and then carbon dioxide is added. The *N*-carboxylated intermediate is then treated with another equivalent of *t*-BuLi and quenched with an electrophile.

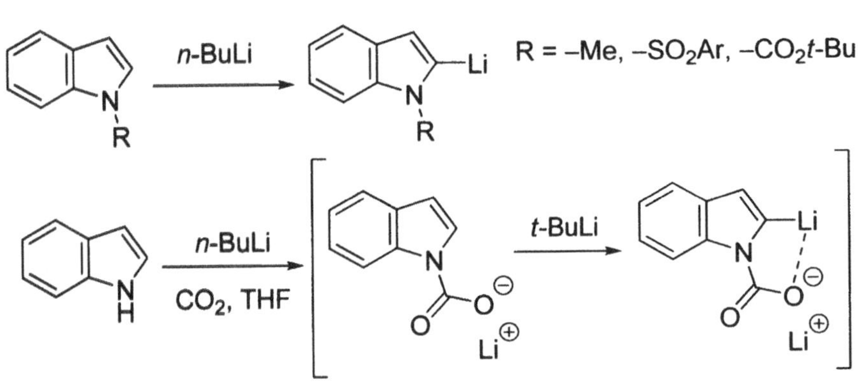

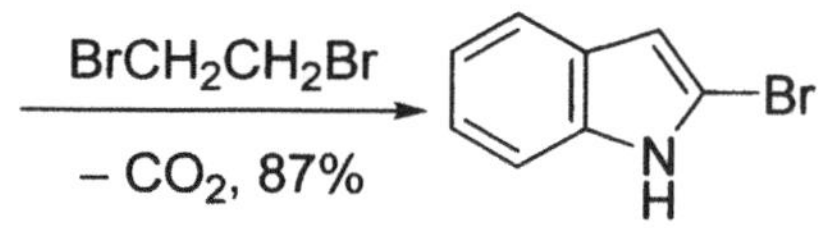

3.3 *Construction of Indole Rings*

Indole synthesis has provided an ample playground for both organic chemists and medicinal chemists. Each year, thousands of indole derivatives are made in pursuit of the life-saving medicines.

3.3.1 Fischer Indole Synthesis

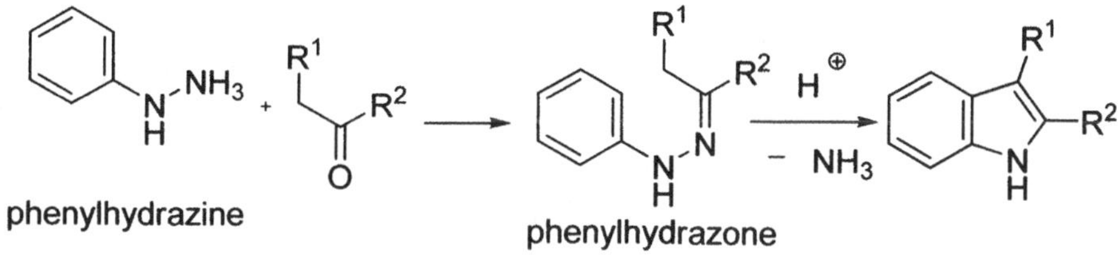

The Fischer indole synthesis has become the most popular method to prepare indole rings since its discovery in 1883 by Emil Fischer.[10] In essence, the Fischer indole synthesis can be regarded as the cyclization of an arylhydrazone, prepared from an arylhydrazine, an aldehyde, or a ketone, by treatment with an acid catalyst or effected thermally to form the indole nucleus. The mechanism has been the subject of intensive investigations for over a century, and many intermediates have been isolated and characterized. By consensus from both theoretical and experimental evidence, the Fischer indole synthesis proceeds as shown below[11]:

The arylhydrazone, prepared from condensation of an arylhydrazine and a carbonyl compound, undergoes protonation and isomerization to the enamine tautomer (ene–hydrazine). The protonated enamine tautomer then undergoes an *irreversible* electrocyclic rearrangement (*i.e.*, [3,3]-sigmatropic rearrangement) where the N–N bond is broken. The resulting double imine then re-aromatizes the benzene ring to provide an anilino-imine, whose nucleophilic amine group attacks the imine intramolecularly to afford the amino-indoline. Loss of a molecule of ammonia and aromatization then deliver the indole.

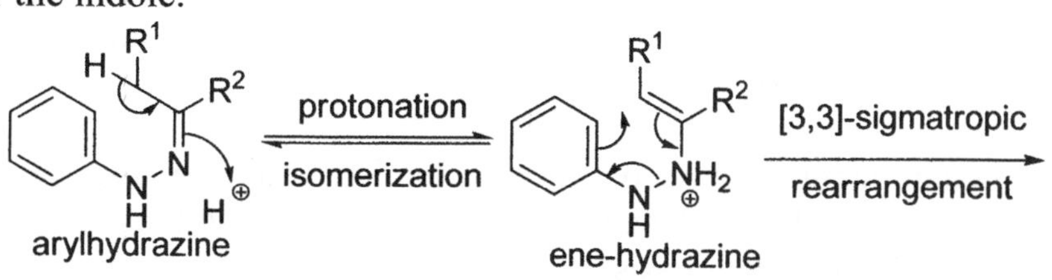

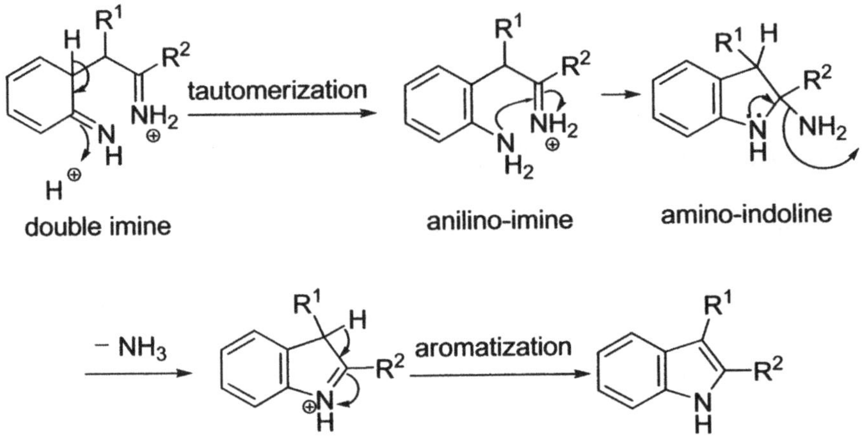

Not surprisingly, the Fischer indole synthesis is the workhorse in medicinal chemistry. For instance, frovatriptan (Frova), a migraine treatment by SmithKline/Elan, was assembled using the Fischer indole synthesis.[12,13] Condensation of HCl salt of *p*-hydrazinylphenylamide with the phthalimide cyclohexanone in refluxing acetic acid yielded the indole in 51% yield. The indole was then manipulated to afford frovatriptan in three additional steps. Frovatriptan belongs to a group of medicine called triptans, which are a group of indole-containing serotonin receptor (5-HT_{1B1D}) agonists used to treat migraines.

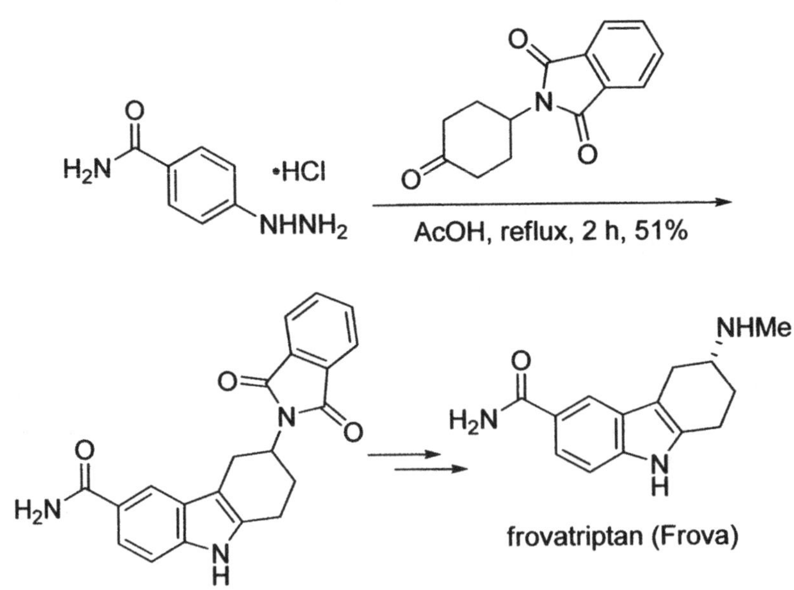

frovatriptan (Frova)

A better understanding of this mechanism aids our appreciation of how the reaction works, and helps us design novel reactions. The Japp–Klingemann hydrazone synthesis is such an example. A hydrazone is an important intermediate for the Fischer indole synthesis. The Japp–Klingemann hydrazone synthesis is a way of preparing the hydrazones by treating a compound with an active methinyl carbon and a diazonium salt involving the formation of an unstable azo compound that undergoes hydrolytic cleavage.[14] A synthesis of sumatriptan (Imitrex) utilized a Japp–Klingemann reaction/decarboxylation strategy.[15] The sulfonamide aniline was diazotized with sodium nitrite to give the diazonium salt, which underwent a Japp–Klingemann reaction with ethyl 2-(3-dimethylamino-propyl)-3-oxobutanoate in the presence of sodium acetate to give the corresponding hydrazone. Fischer cyclization of the hydrazone proceeded smoothly when treated with acetic acid and HCl gas at ambient temperature. One of the advantages of this strategy is that the 2-carboethoxy substituted indole is not subject to electrophilic attack and by-products can be avoided. The ethyl ester was hydrolyzed and the resulting acid was decarboxylated by heating at 200 °C in quinoline with copper powder to give sumatriptan.

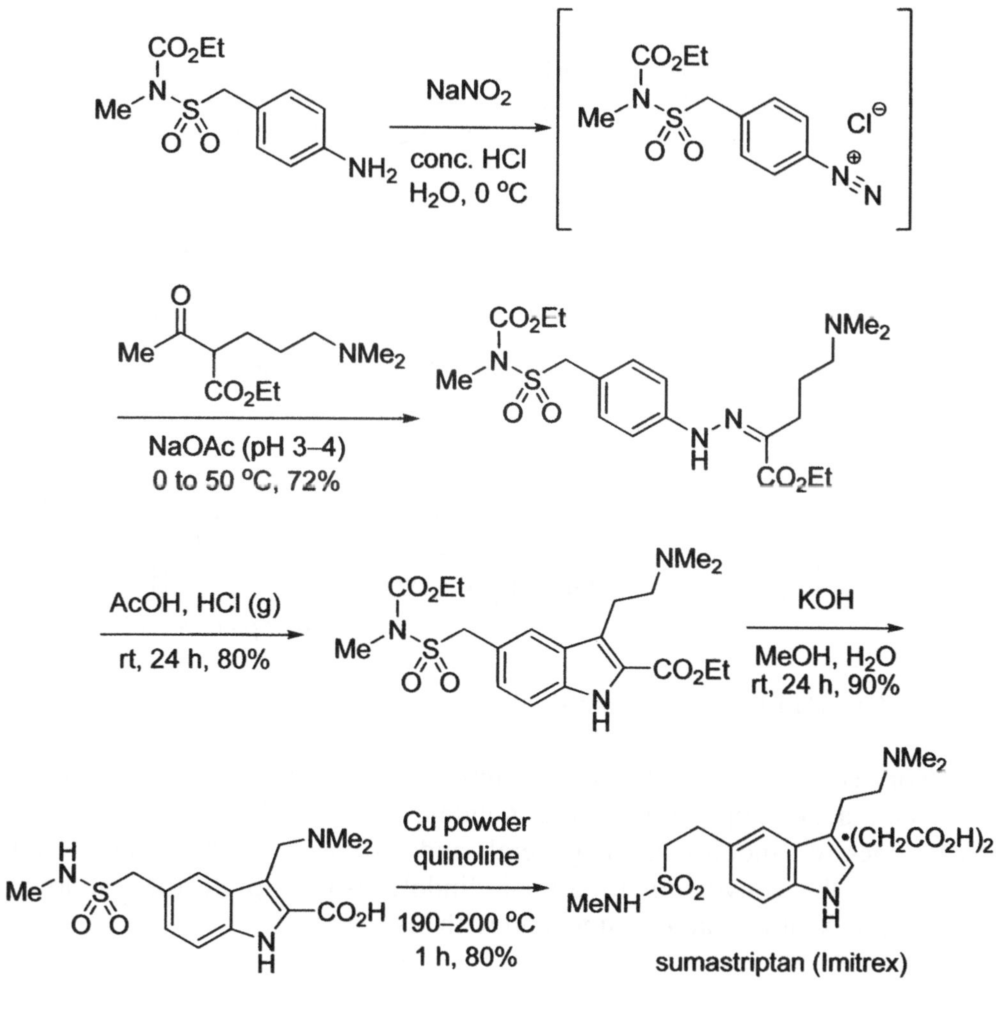

The Grandberg modification of the Fischer indole reaction is another example of understanding the reaction mechanism of the Fischer indole synthesis.[16] It takes advantage of the fact that one molecule of ammonia is released in the Fischer indole synthesis. When an alkyl halide is present, ammonia attacks it in an S_N2 fashion to give a primary amine. For instance, the Grandberg modification was applied in a synthesis of almotriptan (Axert), Almirall's triptan for the treatment of migraine.[17] As shown below, the hydrazine was treated with 4-chlorobutanal diethyl acetal in aqueous HCl and the resulting hydrazone precipitated out and was isolated by filtration. Treatment of the crude hydrazone with aqueous HCl buffered with Na_2HPO_4 to pH 5 promoted the Grandberg modification of the Fischer indole synthesis whereby the alkyl chloride underwent aminolysis under the reaction conditions to afford the tryptamine in 58% yield. Reductive alkylation of the indoloethylamine with formaldehyde in the presence of $NaBH_4$ then provided almotriptan.

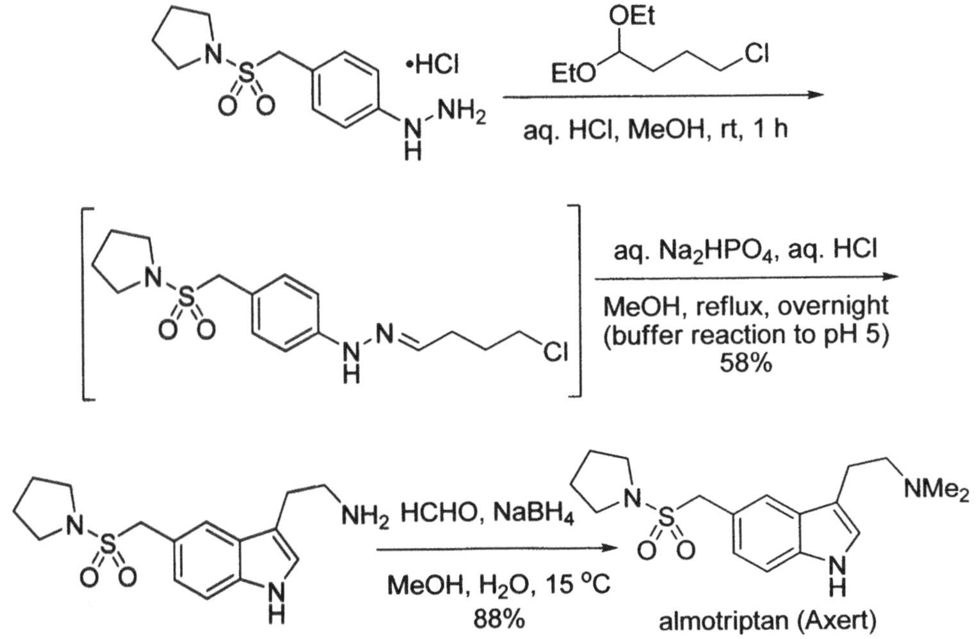

3.3.2 *Mori–Ban Indole Synthesis*

The Mori–Ban indole synthesis[18] is referred to the intramolecular version of the Heck reaction applied to synthesis indoles. The cyclization of *o*-halo-*N*-allylanilines to indoles is a general and efficient methodology. For example,[18a] the conversion to 3-methylindole can be performed at lower temperature, shorter reaction time, and with less catalyst.

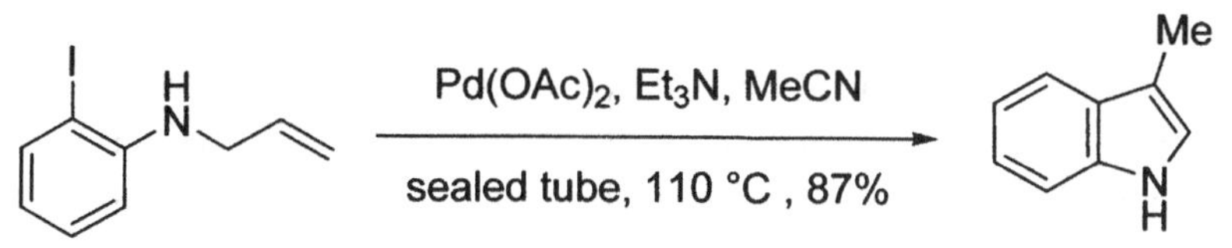

Larock's improved conditions,[19] which have been widely adopted, are catalytic 2% $Pd(OAc)_2$, *n*-Bu_4NCl, DMF, base (usually Na_2CO_3), 25 °C, 24 h (also known as the Jeffrey's conditions). Larock extended his work in several ways, particularly with regard to Pd-catalyzed cross-coupling of *o*-allylic and *o*-vinylic anilides with vinyl halides and triflates to produce 2-vinylindoles. In a program to synthesize CC-1065, an antibacterial, analogs, Sundberg prepared indole from *o*-bromo-*N*-allylaniline in excellent yield[20] using Jeffrey's conditions. Silver carbonate and sodium carbonate were less effective than triethylamine.

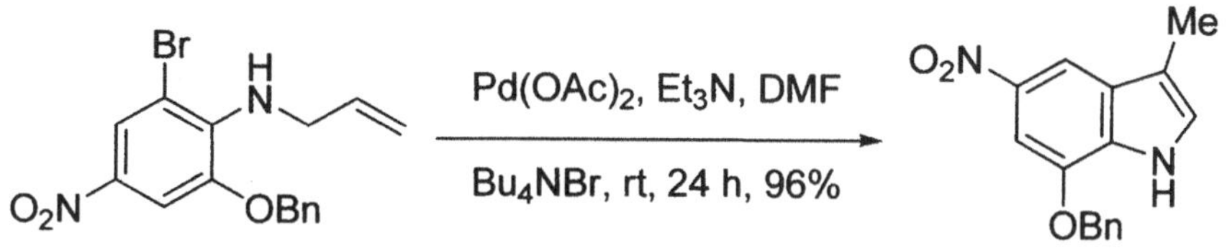

Macor also exploited the Mori–Ban indole synthesis to synthesize several anti-migraine analogues of sumatriptan and homo-tryptamines as potent and selective serotonin reuptake inhibitors (SSRIs).[21] Noticeably, the presence of the second bromine (the bromine "passenger") on the substrate was not significantly deleterious to the reaction although a small amount of the 7-bromoindole might be sacrificed at the end of the reaction to consume the active palladium catalyst. The approach to 7-bromoindole could provide a general method to access 7-bromoindoles (a rare class of indole derivatives), which then could be further manipulated for the synthesis of more complex 7-substituted indoles.

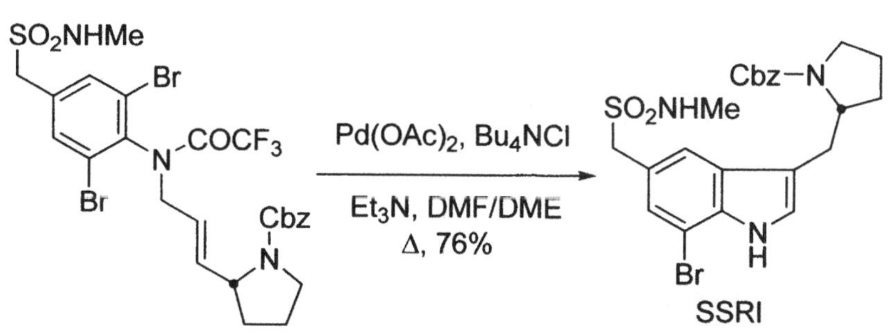

In synthesis of eletriptan (Relpax), Pfizer's triptan for the treatment of migraine, Macor again used the Mori–Ban indole synthesis as the key operation to assemble the indole ring.[22] Jeffrey's conditions were once again employed to transform the *o*-bromo-*N*-allylaniline into the indole core.

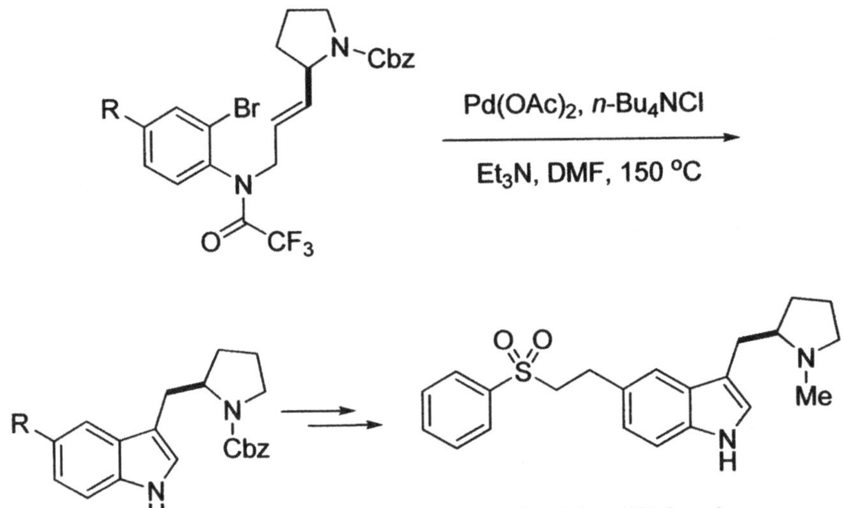

eletriptan (Relpax)

In a synthesis of another anti-migraine agent almotriptan (Axert), the Mori–Ban indole synthesis was the key operation to construct the indole core.[17] Under Jeffrey's conditions, the *o*-iodo-*N*-allylaniline cyclized with concomitant deprotection of the trifluoroacetyl group to form the indole-acetic methyl ester.

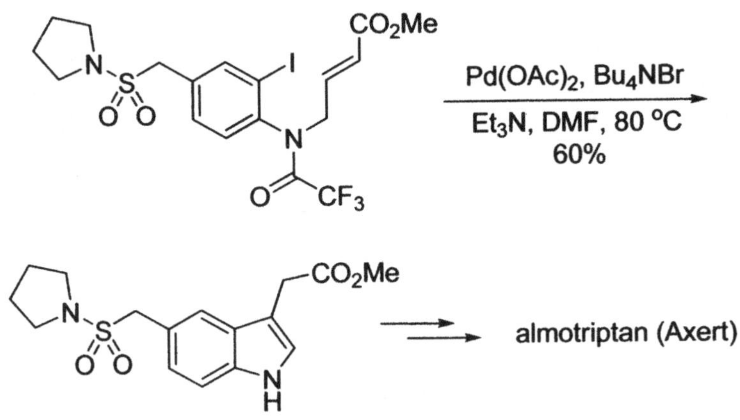

3.3.3 *Larock Indole Synthesis*

Larock and co-workers described the one-step Pd-catalyzed reaction of *o*-haloanilines with internal alkynes to give indoles.[23] This reaction involves oxidative addition of the aryl halide (usually iodide) to Pd(0), *syn*-insertion of the alkyne into the ArPd bond, nitrogen displacement of the Pd in the resulting vinyl-Pd intermediate, and final reductive elimination of Pd(0).

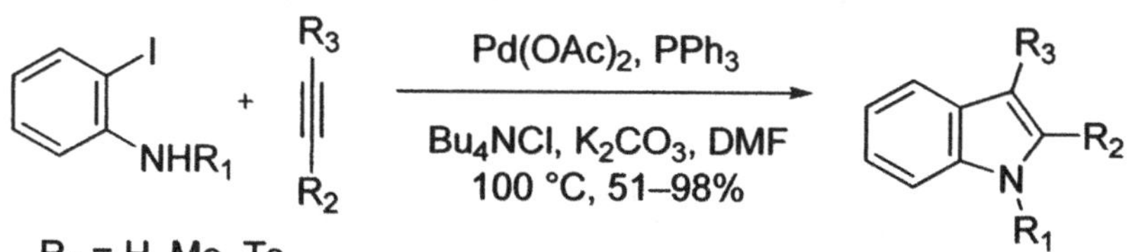

R_1 = H, Me, Ts
R_2, R_3 = *n*-Pr, *t*-Bu, cyclohexyl, TMS, Ph, CH_2OH, $C(Me){=}CH_2$, $(CH_2)_2OH$, CMe_2OH

The reaction can be regioselective with unsymmetrical alkynes, and this is particularly true with silylated alkynes wherein the silyl group always resides at the C-2 indole position in the product. This is noteworthy because silyl-substituted indoles are valuable substrates for other chemistry such as halogenation and Heck coupling. 7-Azaindoles were also prepared using a Larock indole synthesis.[24]

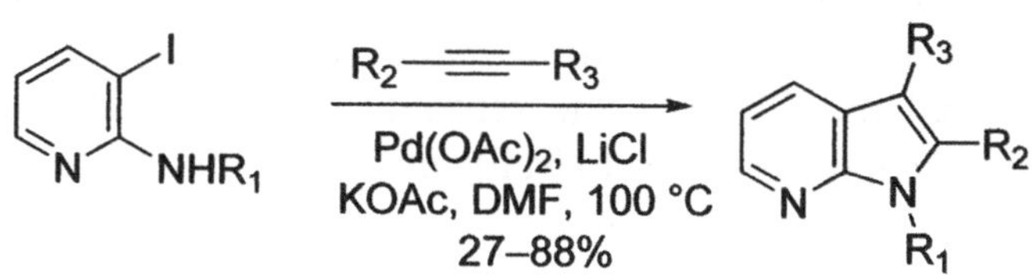

R_1 = Me, Bn, H, *p*-MeOBn
R_2 = TMS, Ph, Pr
R_3 = CH_2OH, CH_2CH_2OH, Me, Ph, Pr

The Larock synthesis was used to synthesize Merck's rizatriptan (Maxalt, 5-(triazolylmethyl)tryptamine, MK-0462), a potent 5-HT_{1D} receptor agonist for the treatment of migraine.[25–28] The reaction was carried out on a 25 kg scale.

As shown in the next scheme, palladium-catalyzed coupling of iodoaniline with bis-triethylsilyl protected butynol in the presence of Na_2CO_3 provided a mixture of indoles. This mixture was desilylated with aqueous HCl in MeOH to furnish the tryptophol in two steps. Protection of the alkyne prevented coupling at the terminal carbon of the alkyne and triethylsilyl (TES) was found to be optimal because it offered the correct balance between reactivity (rate of coupling) and stability. Notably, this palladium-catalyzed process does not require the use of triphenyl phospine, tetrabutyl ammonium chloride, or lithium chloride. The alcohol was converted into the mesylate and then treated directly with 40% dimethylamine to afford rizatriptan, which was purified via recrystallization after forming the benzoate salt.

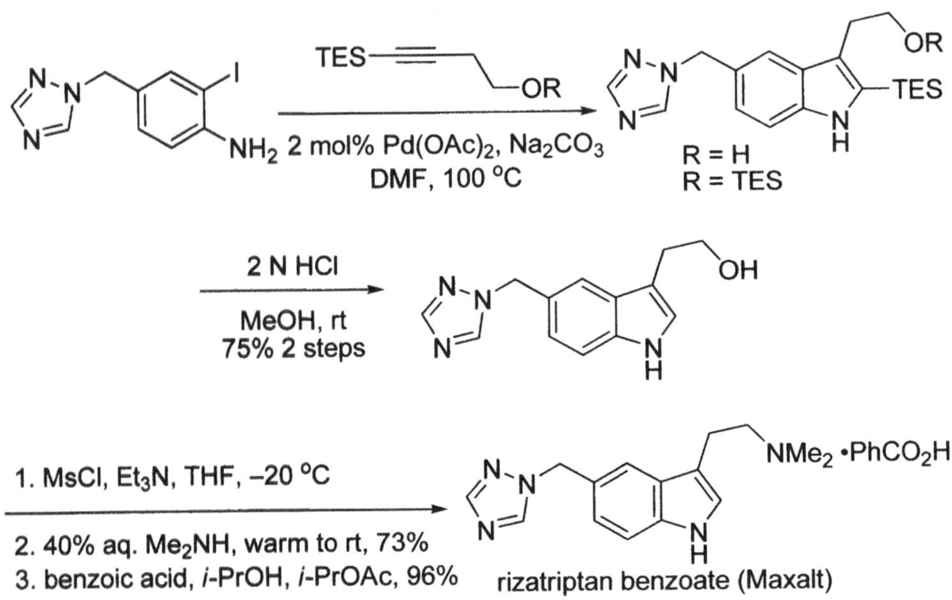

Rosso and co-workers have employed this method for the industrial-scale synthesis of an anti-migraine drug candidate. The triethylsilyl "masked protective group" was removed under strong acidic conditions to reveal the C2–H. The removal of spent palladium was best effected by trimercapto-triazine (TMT) although many techniques were explored.[29]

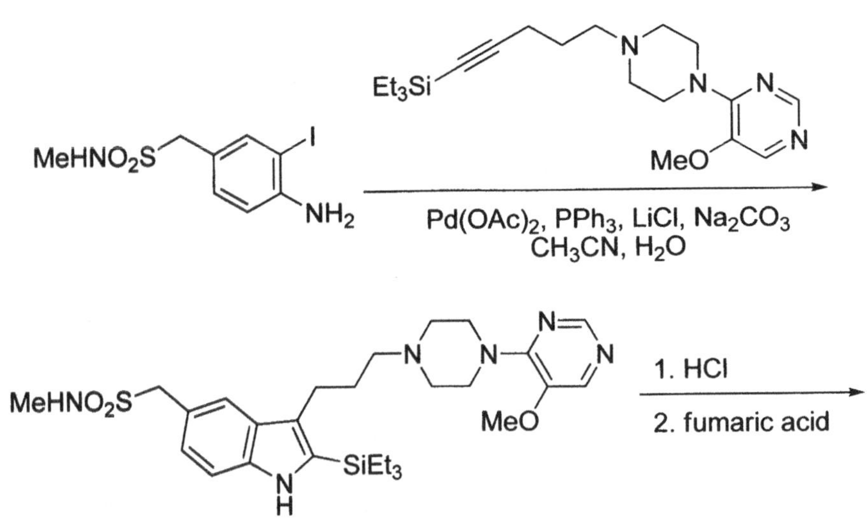

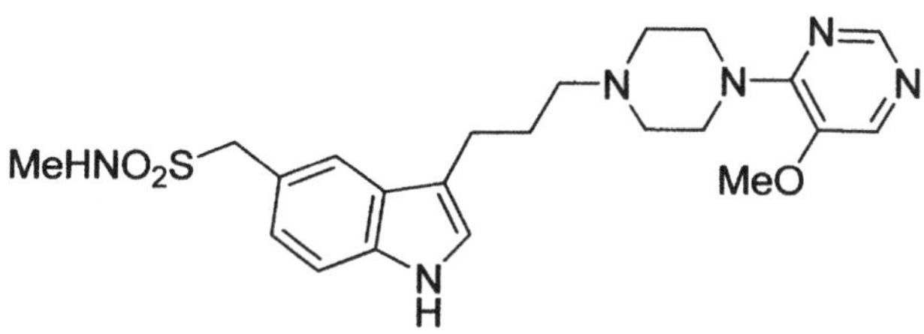

anti-migraine drug candidate

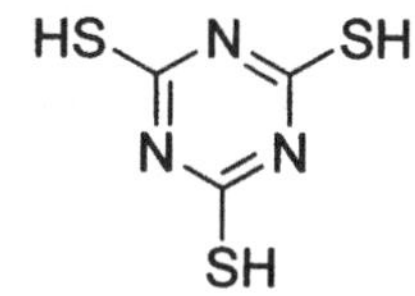

trimercaptotriazine (TMT)

3.3.4 Bischler–Möhlau Indole Synthesis

The Bischler–Möhlau indole synthesis, also known as the Bischler indole synthesis, is the formation of a 2-arylindole from an α-bromoacetophenone and excess aniline.[30] Although not widely used in making indoles, one of the syntheses of fluvastatin sodium (Lescol) took advantage of the Bischler–Möhlau indole synthesis to assemble its indole core. As shown below, reaction of α-chloroketone with *N*-*i*-Pr-aniline at elevated temperature generated the tertiary amine. The resulting *N*-*i*-Pr-aniline underwent a $ZnCl_2$-mediated Bischler–Möhlau indole synthesis also at an elevated temperature to afford the indole core structure of fluvastatin.[31]

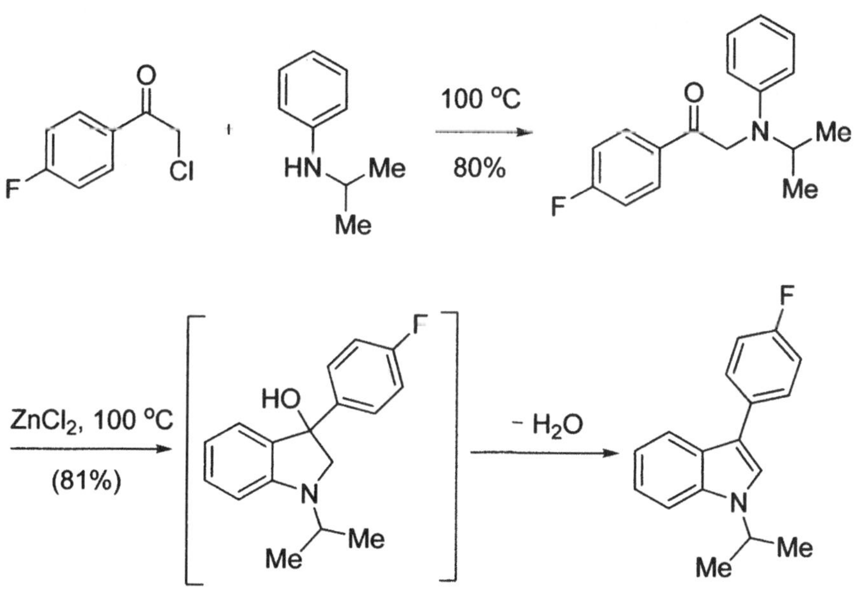

fluvastatin's indole core

A Bischler–Möhlau indole synthesis was also pivotal in constructing the indole ring of bazedoxifene acetate (Viviant), Wyeth's novel and highly selective indole estrogen. It is a selective estrogen receptor modulator (SERM) for the treatment of and prevention of osteoporosis. It was found that the yields and reproducibility of this reaction could be increased by conducting the reaction in two steps but in one pot. Therefore, condensation between α-bromopropiophenone and 4-(benzyloxy)-aniline hydrochloride yielded the 3-methylindole core.[32,33] Installation of the side chain was followed by additional functional group manipulations to deliver bazedoxifene as its acetic acid salt.

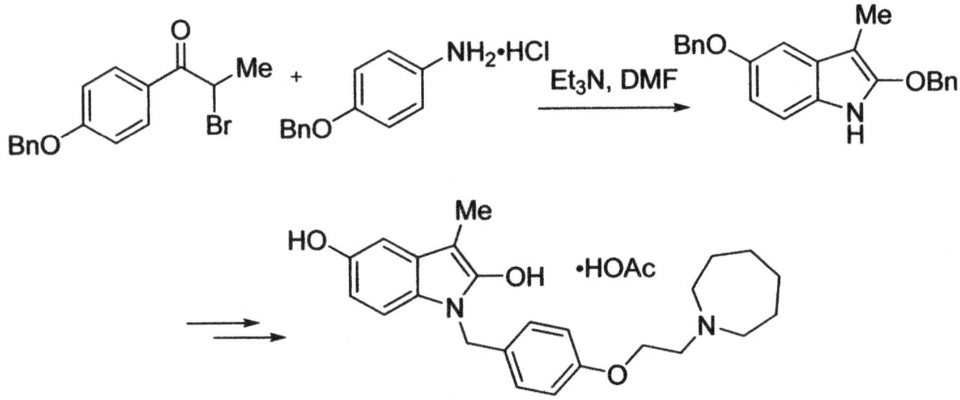

bazedoxifene (Viviant)

Instead of using α-haloketones, an unsymmetrical benzoin can be utilized as starting material for the Bischler–Möhlau indole synthesis. RWJ 68354, a potent inhibitor of the p38 MAP kinase, was prepared via a variation of the Bischler–Möhlau indole synthesis under mild conditions with 55% isolated yield; only 2–3% regioisomer could be isolated from the mother liquor.[34]

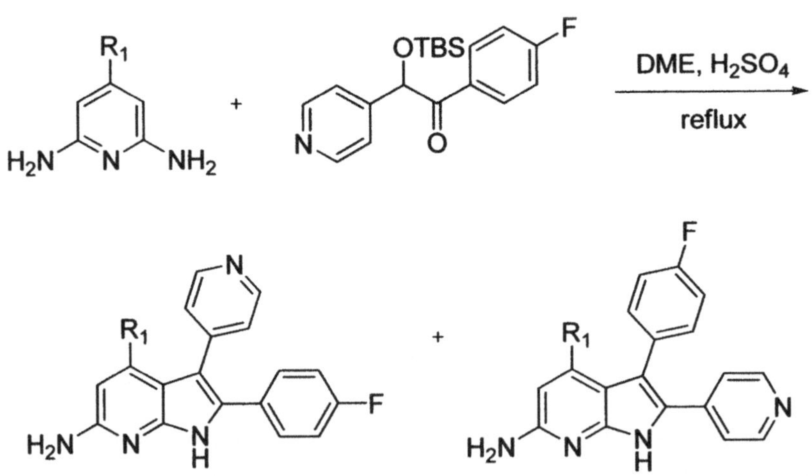

49–58%
p38 MAP kinase inhibitor

2–3%

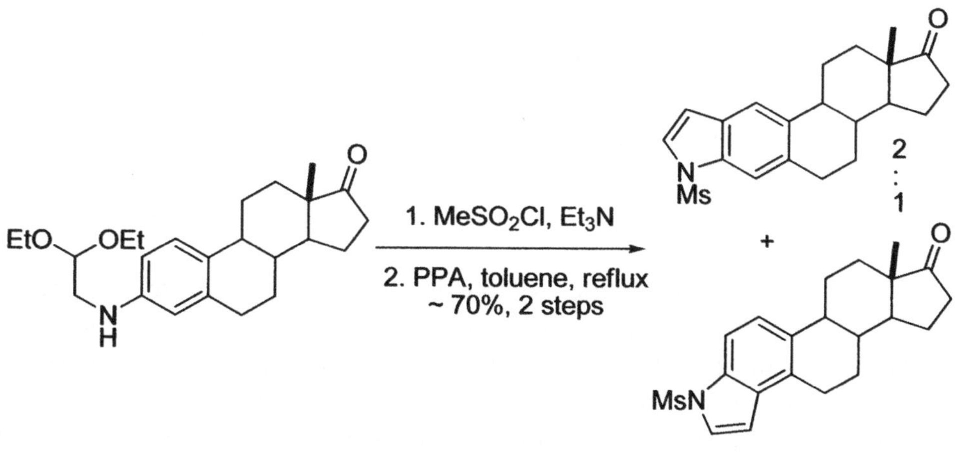

Steroids bearing heterocycles fused to the A-ring of the steroid nucleus have been of pharmaceutical interest. During the design and synthesis of novel estrogen receptor ligands for the treatment or prevention of disorders and diseases, two novel indolyl steroids were prepared using a modified Bischler–Möhlau method.[35] The estrieno[2,3-*b*] and [3,4-*c*]pyrroles were isolated in ~70% yield with a 2:1 ratio of regioselectivity.

3.3.5 Nenitzescu Indole Synthesis

The Nenitzescu indole synthesis[36] is the formation of a 5-hydroxylindole from the condensation of *p*-benzoquinone and β-aminocrotonate.

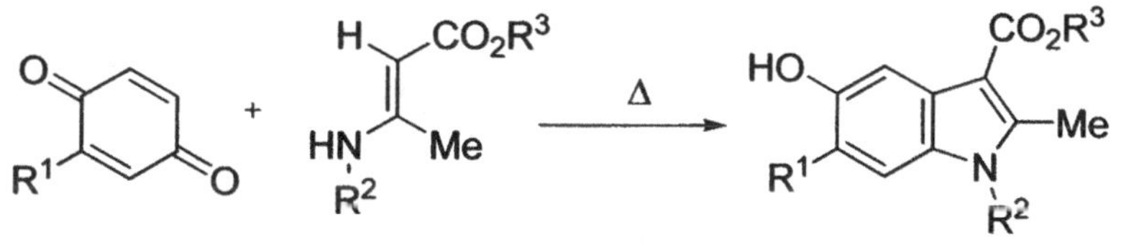

The *bona fide* mechanism of the Nenitzescu indole synthesis involves complicated, multiple redox cycles. A simplified version is shown below:

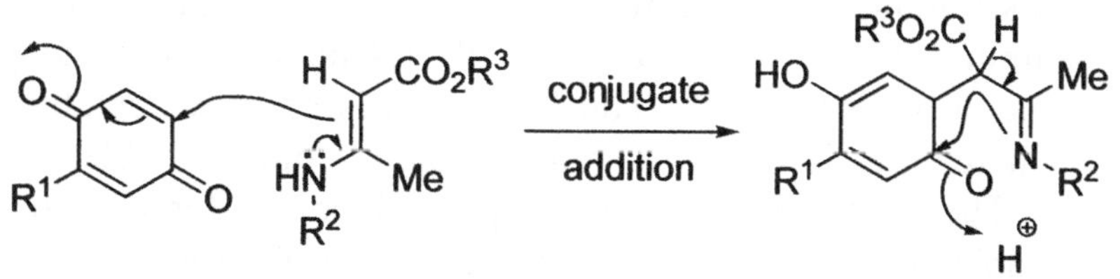

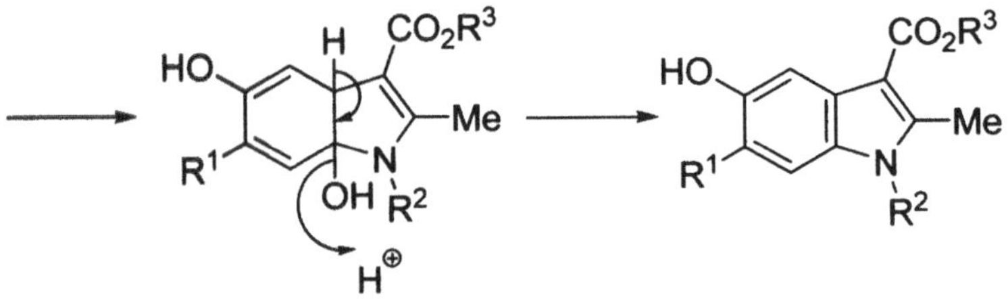

The Nenitzescu indole synthesis was employed to synthesize LY311727, an indole acetic acid-based selective inhibitor of human non-pancreatic secretory phospholipase A2 (hnpsPLA2) by Eli Lilly as a potential treatment for sepsis.[37] The Nenitzescu condensation of benzoquinone with the β-aminoacrylate was carried out in CH_3NO_2 to provide the desired 5-hydroxylindole. An additional seven steps then afforded LY311727.

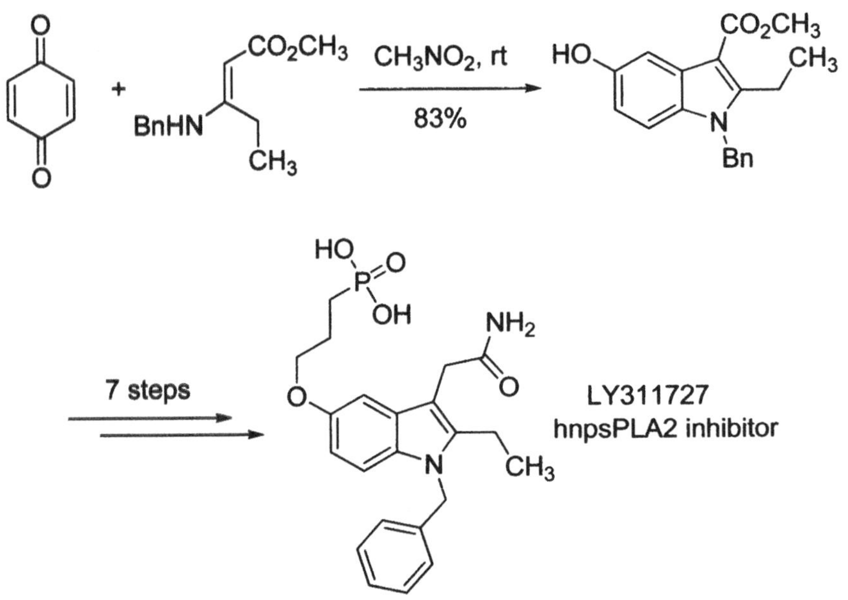

A Nenitzescu indole synthesis was applied to synthesize benzoxazines as highly selective antagonists at M_4 muscarinic receptors.[38] Ammonia was bubbled into the methanolic solution of the β-keto-ester to give the enamine, which was coupled with 1,4-benzoquinone to give the Nenitzescu indole in 6% yield! The low yield is often acceptable in drug discovery where the key is to make potent, bio-available, and safe compounds. The synthesis could be improved only after the compounds showed desirable profiles. The Nenitzescu indole underwent a Vilsmerier reaction en route to deliver a benzoxazine as a potent and selective M_4 muscarinic antagonist. These selective M_4 muscarinic antagonists may present a potential target for Parkinson's research by showing beneficial therapeutic effects without cholinergic effects.

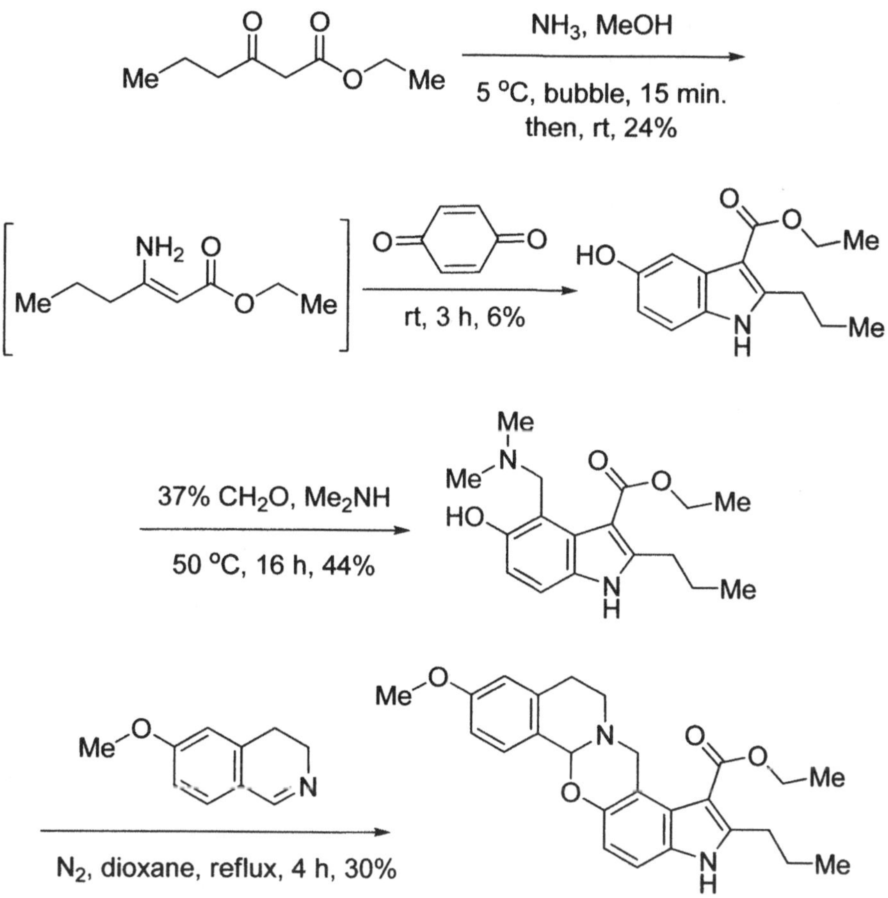

M4 muscarinic antagonist

3.3.6 Bartoli Indole Synthesis

Invented merely 20 years ago in 1989, the Bartoli indole synthesis refers to the formation of 7-substituted indoles from the reaction of *ortho*-substituted nitroarenes and excess of vinyl Grignard reagents.[39,40]

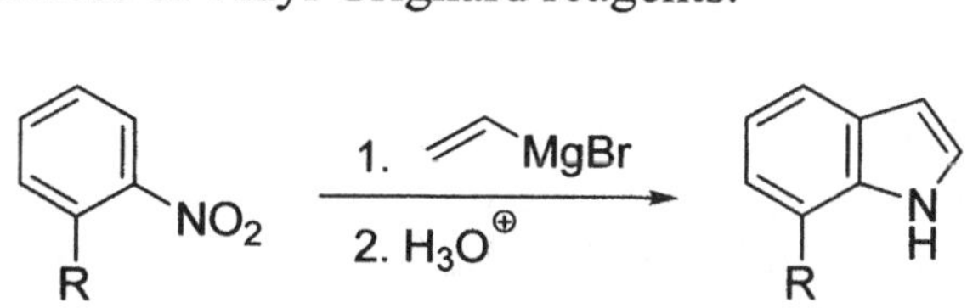

The mechanism for the Bartoli indole synthesis, postulated by Bartoli et al. based on experimental evidence,[41] is rather unique. As shown below, the first equivalent of the Grignard reagent attacks the N=O double bond on the nitro group at the oxygen atom. And the nitroarene is reduced to nitrosotoluene via vinyl enolate elimination to give the nitroso intermediate. The second equivalent of the Grignard reagent adds to the N=O double bond of the nitroso group also at the oxygen atom to provide the *N*-aryl-*O*-vinylhydroxylamino magnesium salt, which then undergoes a [3,3]-sigmatropic rearrangement, followed by a rapid ring closure to form the bicyclic intermediate. The third equivalent of the Grignard reagent behaves as a base on this bicyclic intermediate to re-aromatize the six-membered ring to provide the intermediate, which undergoes an elimination of water upon workup to deliver the indole.

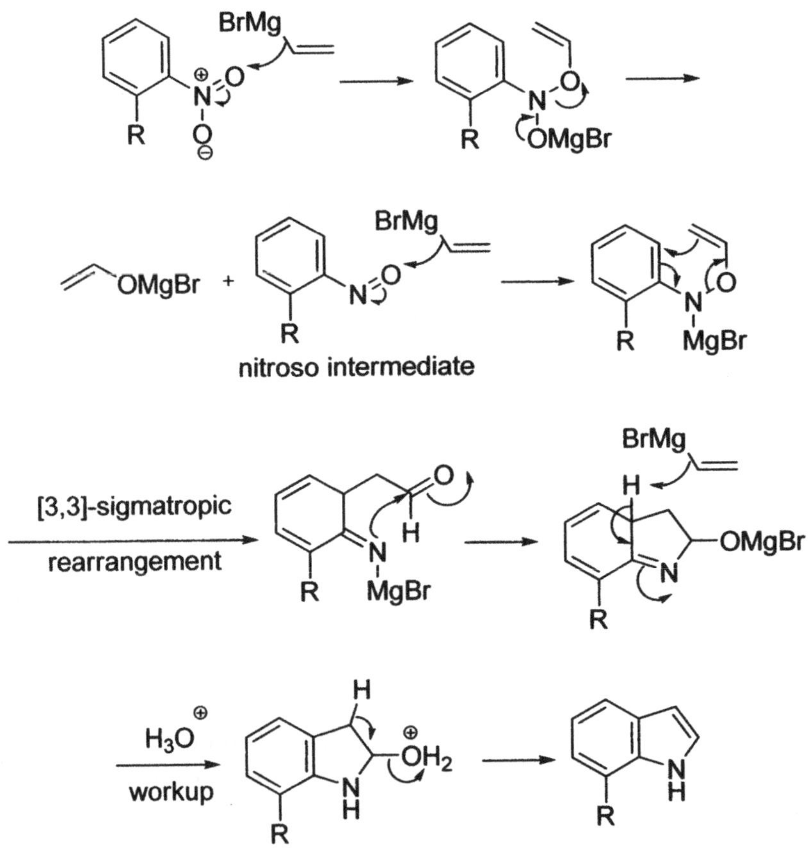

The Bartoli indole synthesis has found utility in medicinal chemistry despite its infancy. For instance, the nitro-diarylether ketone was protected as a ketal and treated with vinylmagnesium bromide to afford the indole derivative. The resulting phenyl-indolyl ether is an intermediate for preparing serotonin (5HT) receptor ligands for treatment of nervous system disorders.[42] Specifically, the diphenyl ether ligands exhibited activity as $5HT_2$ antagonists.

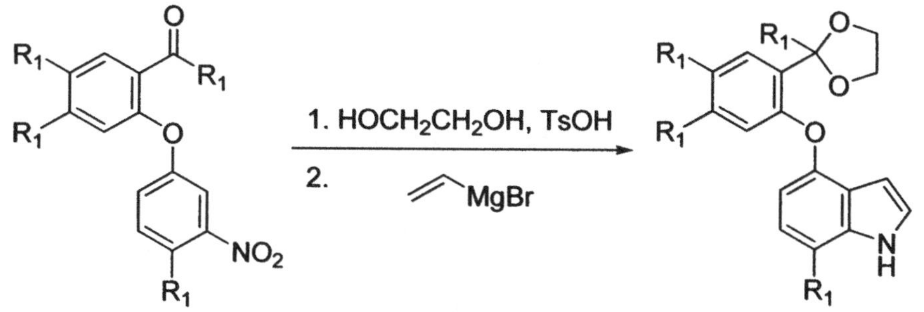

On the other hand, the benzhydryl protected 7-hydroxyindole is also the key intermediate in the synthesis of a 7-carboxyindolylglycine derivative, proposed as a putative mGluR1 receptor antagonist.[43] Using the Gilmore's modification[44] of the Bartoli indole synthesis, the protected nitro-phenol was converted into 7-hydroxyindole in 44% yield.

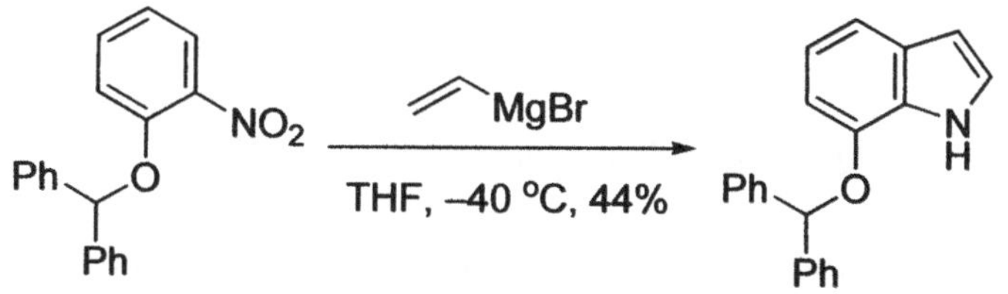

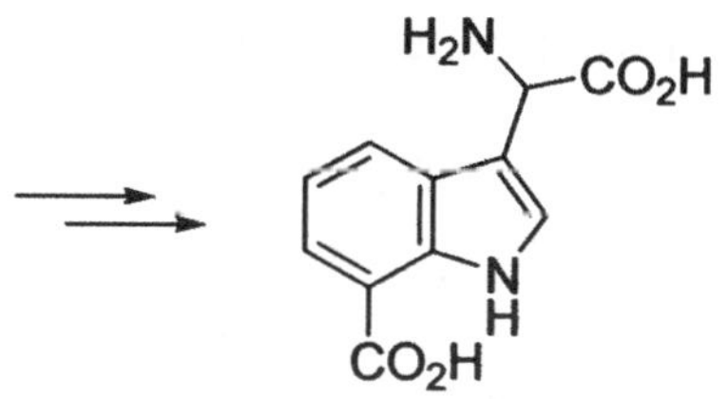

The Bartoli indole synthesis was used to prepare DG-041,[45] a small molecule antagonist of the EP3 receptor for prostaglandin E2 that is in clinical development for treatment of peripheral artery disease (PAD). The interplay between the EP3 receptor, a GPCR (G-Protein-Coupled Receptor), and a related receptor, the IP (Iloprost) receptor, regulates the homeostatic response to platelet activation by prostanoids. In deCODE's drug discovery route, treatment of 2-bromo-4-fluorobenzene with 3 equivalents of 1-allylmagnesium bromide gave the indole in 42% yield. Although adequate for small scale, the requirement of a low temperature (–50 °C) was cumbersome and purification of the indole product required silica gel chromatography. The process synthesis was later accomplished by a tandem Heck reaction (see Section 3.5.5 on the Heck reaction).

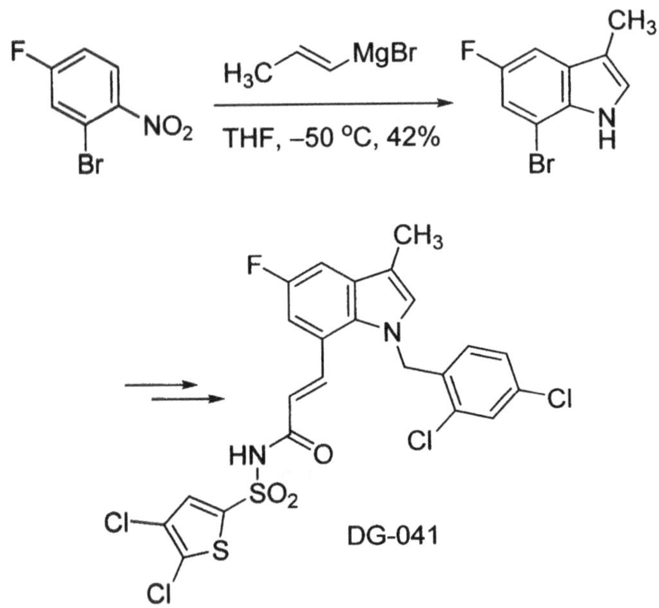

3.3.7 *Batcho–Leimgruber Indole Synthesis*

The Batcho–Leimgruber indole synthesis is referred to condensation of *o*-nitrotoluene derivatives with formamide acetals, followed by reduction of the *trans*-β-dimethylamino-2-nitrostyrene to furnish indole derivatives.[46]

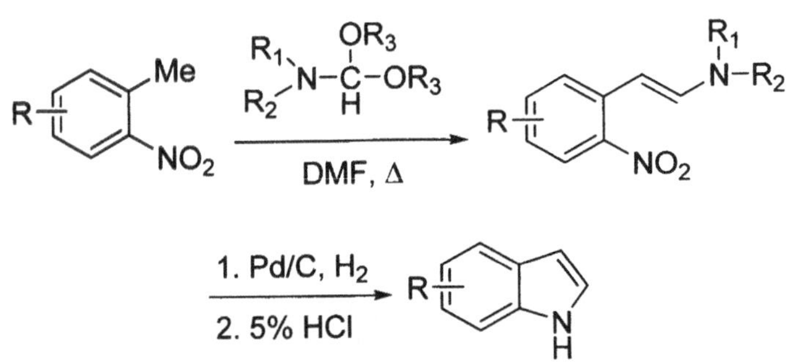

There are no detailed reports on the mechanism of the Batcho–Leimgruber process, but it is proposed below. In the presence of the base, *o*-nitrotoluene is coupled with *N,N*-dimethylformamide dimethyl acetal [DMFDMA, $Me_2NCH(OMe)_2$] to offer the dimethylamino imine. Hydrogenation then reduces the nitro group to aniline. When workup with an acid, cyclization and re-aromatization then delivered the indole.

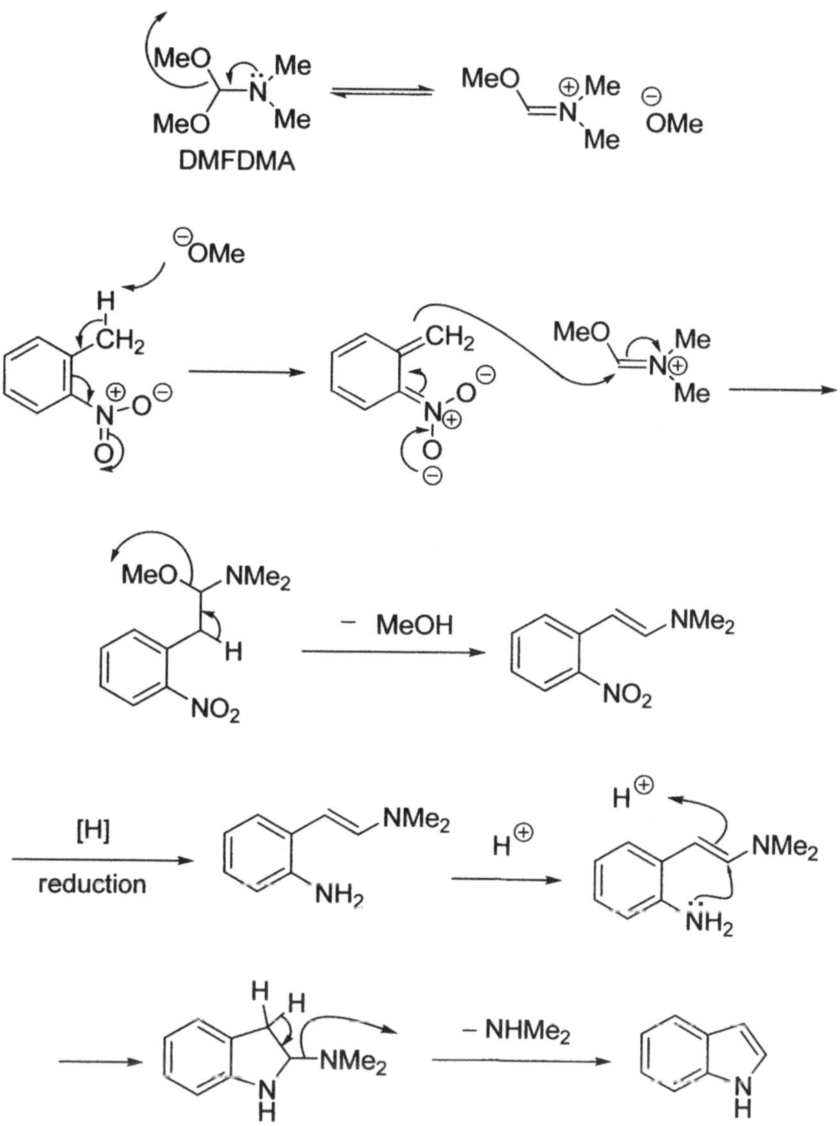

Two new series of benzonitrile derivatives on position 6 or 4 of the indole ring were successfully synthesized via a Batcho–Leimgruber reaction.[47] The benzophenone aniline was condensed with DMFDMA at 110 °C to provide the crude enamine, which was directly used in catalytic hydrogenation using Raney nickel in ethanol to afford the key intermediate in 15% yield. After further manipulations, the resulting racemic 4-substituted indole was evaluated *in vitro* on the inhibition of aromatase (CYP19) and 17α-hydroxylase-C17,20-lyase (CYP17). It showed a high level of inhibitory activity toward CYP19.

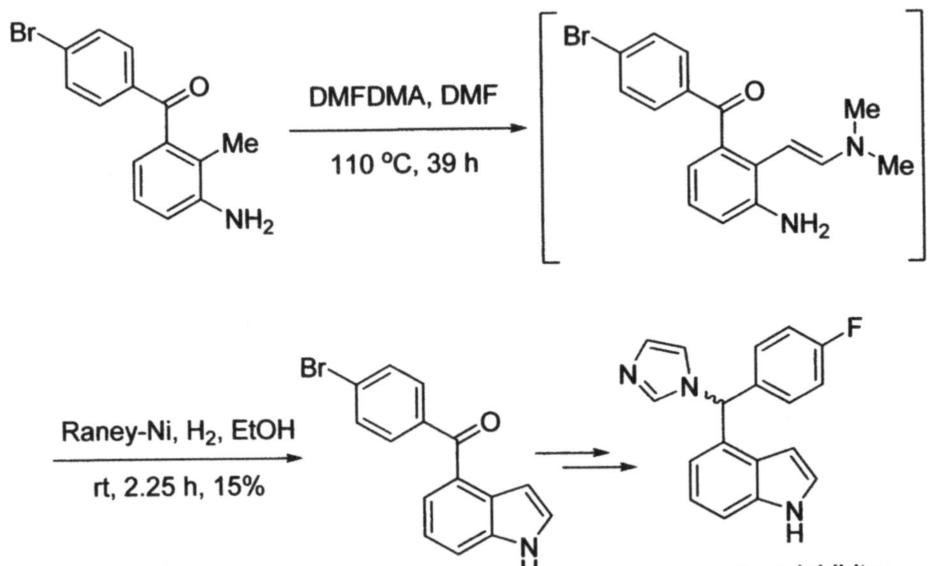

aromatase inhibitor

Process chemists at Eli Lilly developed a manufacturing process for a 1-piperidin-4-yl)-1H-indole as a key intermediate for protein kinase C (PKC) inhibitor LY317615,[48] an investigatory new drug for the treatment of glioblastoma. The mode of action of LY317615 is to prevent angiogenesis by cancer cells during tumor growth.

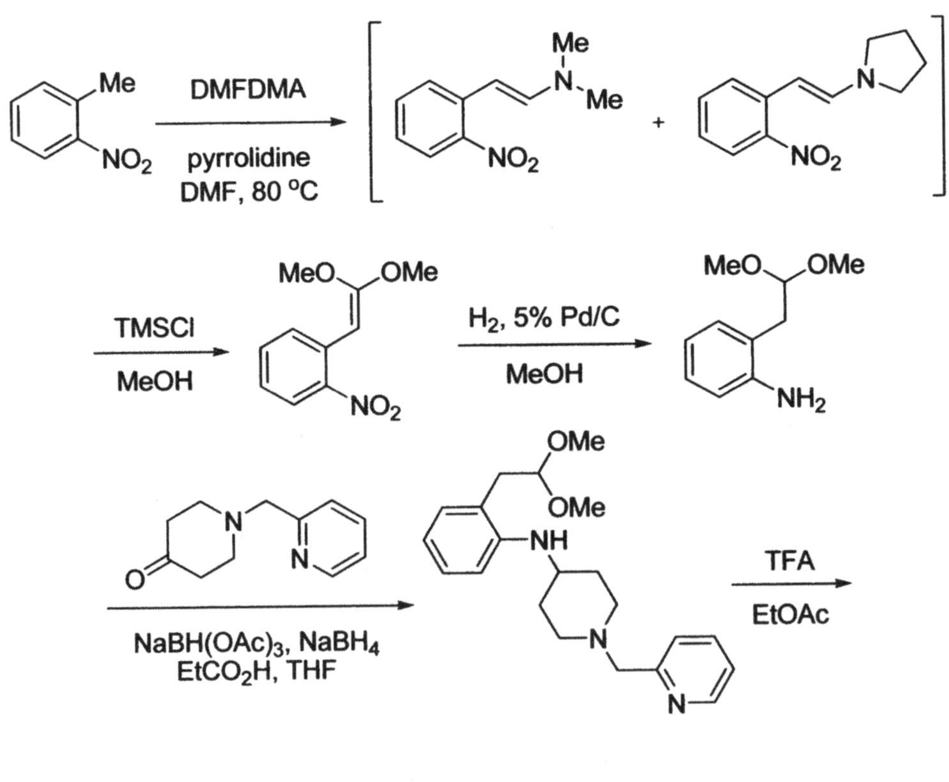

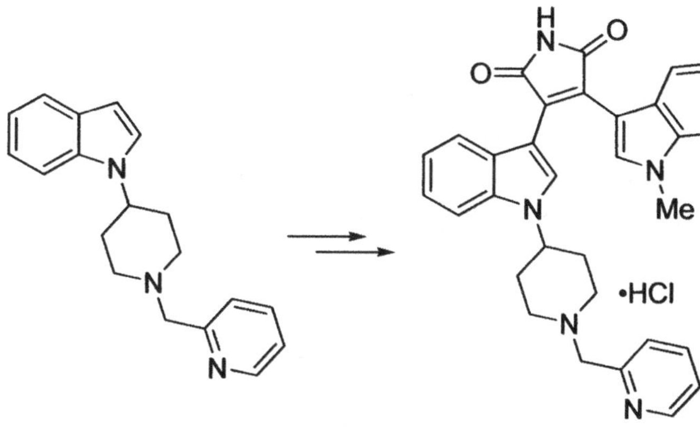

LY317615
PKC inhibitor

When benzyl ether is present during reduction of the nitro group for the Batcho–Leimgruber reaction, debenzylation could occur. For instance, hydrogenation with 5% Rh/C resulted in 55% of the desired indole and 34% of the phenol.[49] However, with the addition of 10 mol% of either $Ni(NO_3)_2{\bullet}6H_2O$, $Fe(OAc)_2$, or $Co(acac)_2$, debenzylation was greatly suppressed. The 6-(benzyloxy)-1*H*-indole was an intermediate for the synthesis of Merck's indolocarbazole as a DNA topoisomerase I inhibitor. The drug candidate has emerged as a potent cancer chemotherapy agent because of its potent cytotoxic activity against human cancer cells and its wide safety margin and is currently in clinical trials.

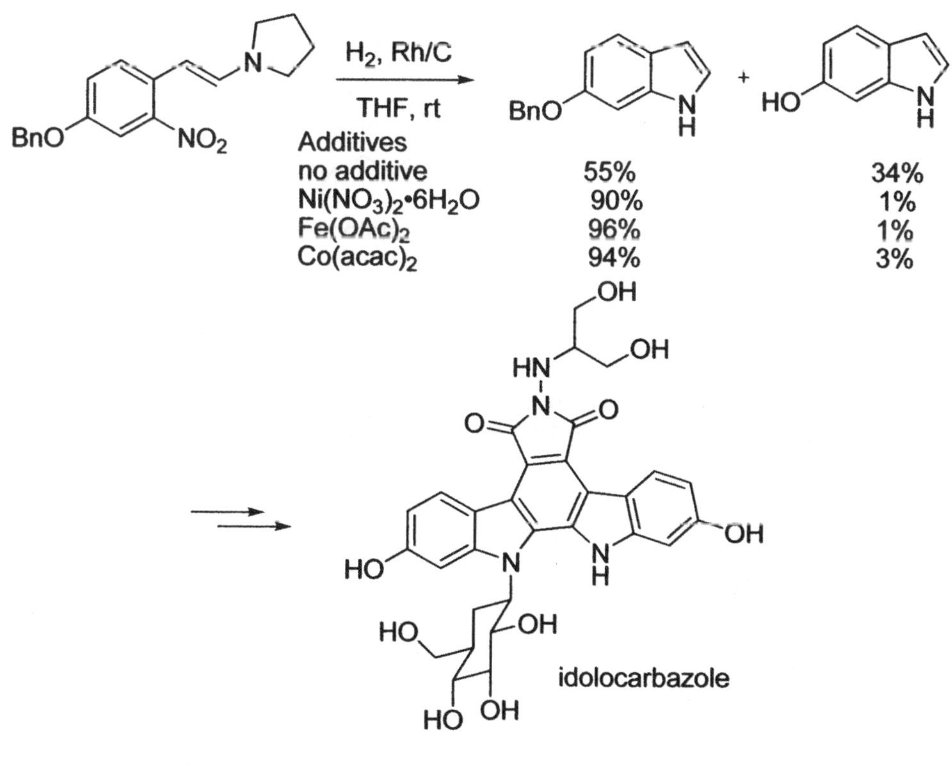

3.3.8 Gassman Indole Synthesis

The Gassman indole synthesis involves a one-pot process in which a hypohalite, a β-carbonyl sulfide derivative, and a base are added sequentially to an aniline or a substituted aniline to provide 3-thioalkoxyindoles.[50] The sulfur can be readily removed by hydrogenolysis or Raney nickel.

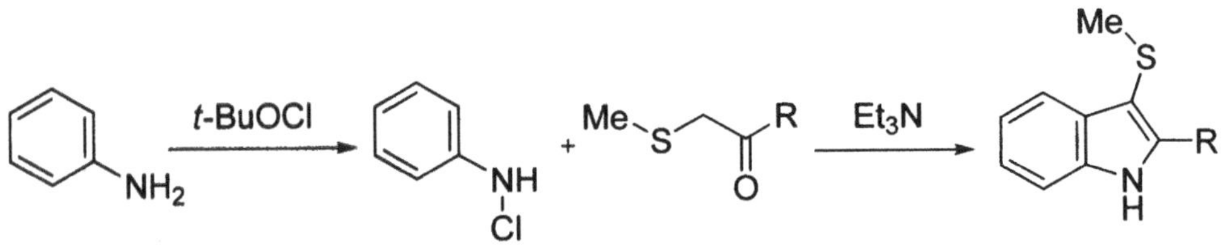

The mechanism of the Gassman indole synthesis bears some resemblance to that of the Fischer indole reaction. An S_N2 displacement of the *N*-chlorinated aniline with the α-keto sulfide gives rise to a sulfonium ion. After deprotonation, a [2,3]-sigmatropic rearrangement (Sommelet–Hauser rearrangement) takes place where the weak N–S bond is broken and a strong C–C bond is formed. Similar to the Fischer indole reaction, the resulting imine was transformed into indole after re-aromatization and cyclization.

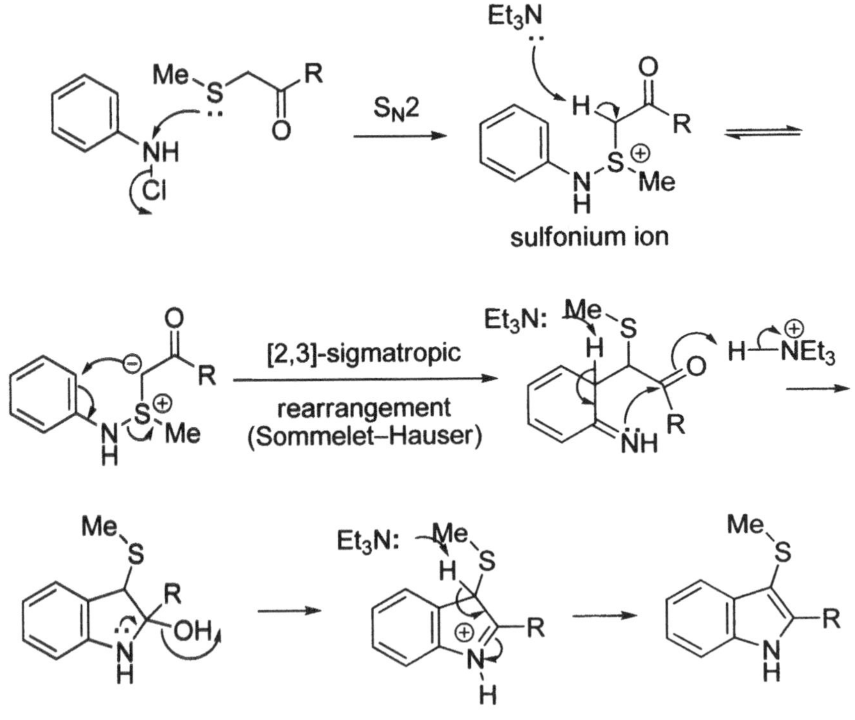

A total synthesis of (+)-paspalicine and (+)-paspalinine took advantage of the Gassman indole synthesis.[51] The thiomethyl ketone was treated with *N*-chloroaniline (prepared by mixing aniline with *t*-butyl hypochlorite) to afford the azasulfonium salt, which was transformed into anilinylsulfide upon treatment with triethylamine. Desulfurization with Raney nickel was followed by the acid-catalyzed cyclization to give the hexacyclic indole intermediate.

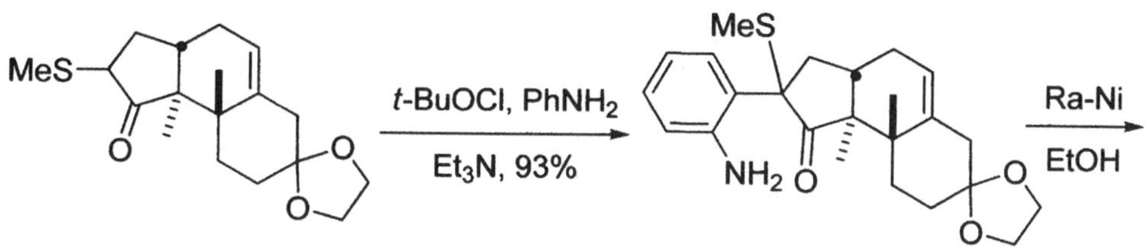

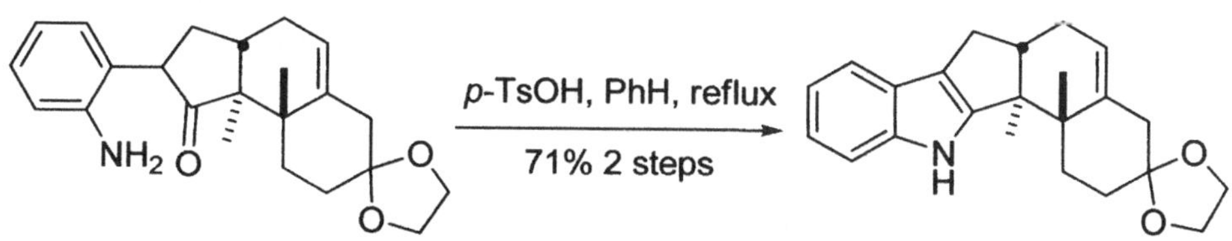

In the synthesis of an indoloquinolone antibacterial, the Gassman indole synthesis was explored to serve as the central operation.[52] Therefore, difluoroaniline was selectively monobrominated. Although the Fischer indole cyclization for the resulting 2-bromo-4,5-difluoroaniline only gave the indole in poor yield, the Gassman indole synthesis worked well in one pot. Thus, treatment of the aniline with *t*-butyl hypochlorite was followed by addition of ethylthio-2-propanone and treatment with triethylamine to produce the indole in excellent yield.

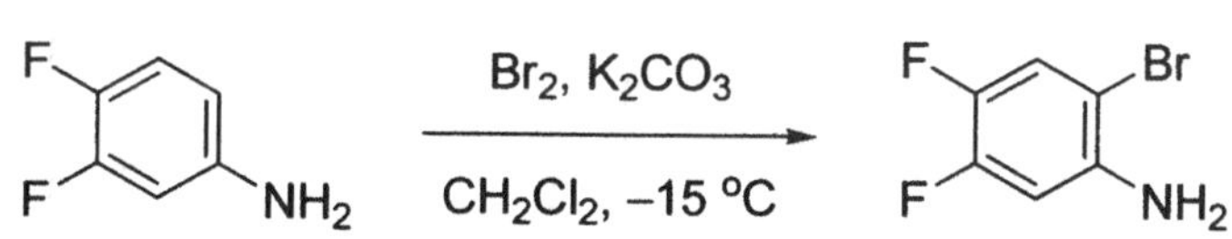

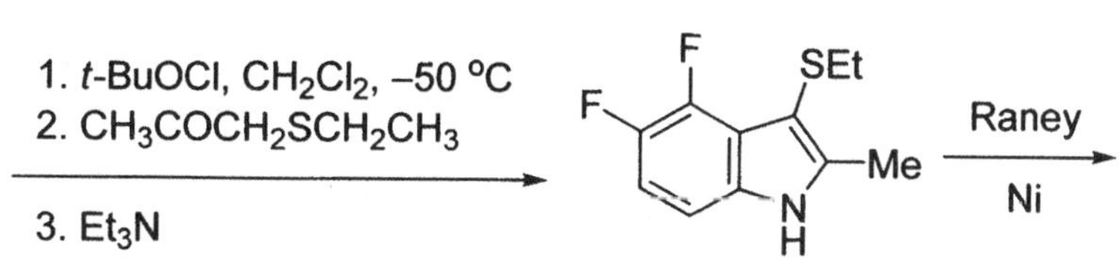

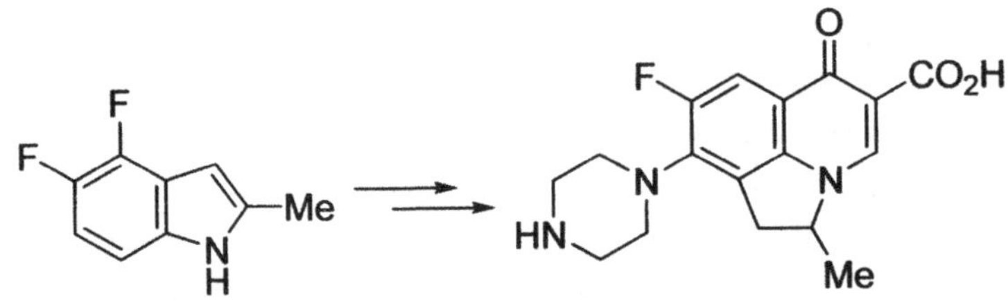

Although a few PPARγ (peroxisome proliferator-activated receptor-γ) inhibitors including rosiglitazone (Avandia) and pioglitazone (Actos) are already on the market for the treatment of type-2 diabetes, the pursuit is still on-going for finding better PPARγ inhibitors. In the discovery synthesis of a selective PPARγ modulator by Merck,[53] the Gassman indole synthesis between *m*-trifluoro-methoxy-aniline and (ethylthio)acetone provided the desired indole regioisomer in a 2:1 ratio. Nonetheless, the major isomer of indole was converted to a potent and selective PPARγ modulator after functional group manipulations.

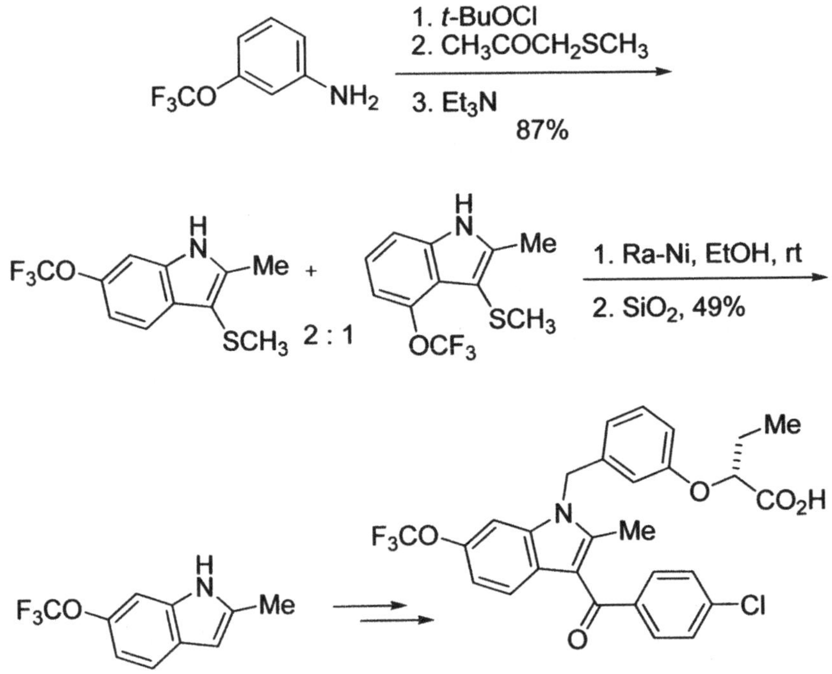

3.3.9 *Cadogan–Sundberg Indole Synthesis*

The Cadogan–Sundberg indole synthesis, often known as the Cadogan reaction, refers to the deoxygenation of *o*-nitrostyrenes or *o*-nitrostilbenes with trialkyl phosphite and subsequent cyclization of the resulting nitrene to form indoles.[54,55] The mechanism of the Cadogan–Sundberg indole synthesis is similar to that of the Sundberg indole synthesis, which involves a nitrene as the key intermediate.

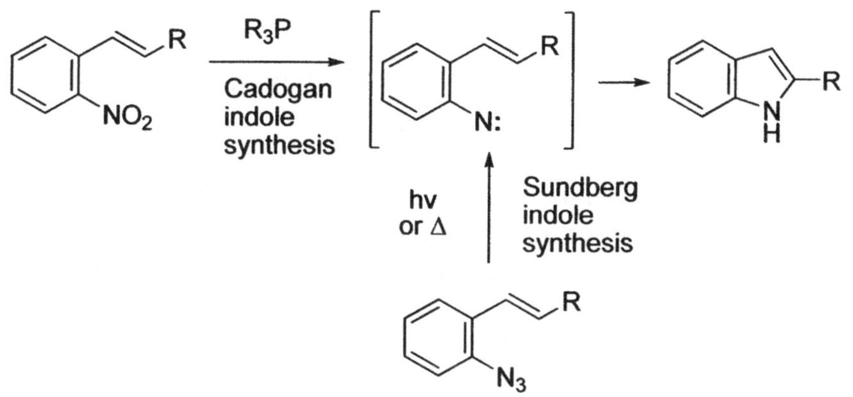

The Cadogan–Sundberg indole synthesis has recently found utility in medicinal chemistry. It was employed to prepare a novel class of kinase domain receptor (KDR) inhibitors.[56] The KDR kinase is a tyrosine kinase that has a high affinity for vascular endothelial growth factor (VEGF) and is believed to be a primary mediator of tumor induced angiogenesis. Compounds that inhibit, modulate, or regulate the KDR receptor are useful for the prevention and treatment of tumor induced angiogenesis. Reductive cyclization of the nitrostyrene with $P(OEt)_3$ afforded the indole intermediate. The deprotection of the masked quinolin-2-one moiety of chloroquinoline was accomplished by hydrolysis of chloroquinoline in a 1:1 mixture of $AcOH/H_2O$ to give the freebase of the final product.

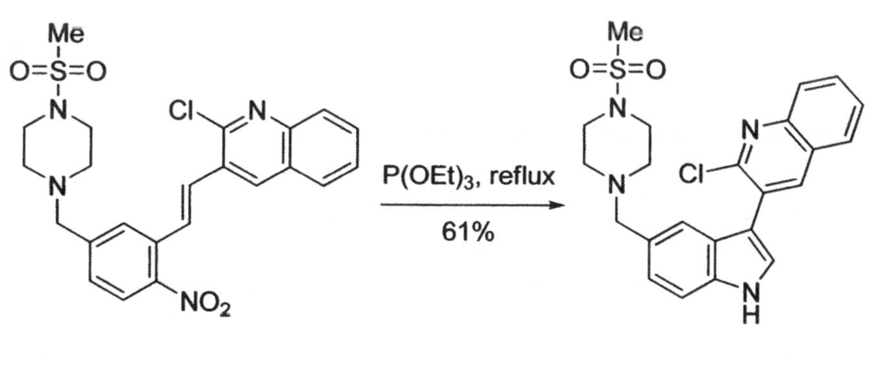

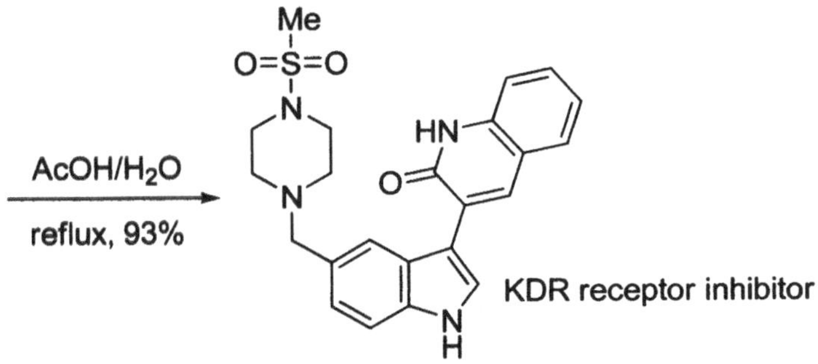

The Cadogan–Sundberg indole synthesis was also employed to prepare botulinum neurotoxin A light chain (BoNT/A LC) endopeptidase inhibitors.[57] Botulinum neurotoxins secreted by strains of the anaerobic spore forming bacterial species *Clostridium botulinum* are the most potent neurotoxins known and are categorized as category A (highest priority) bioterrorist agents by the Centers for Disease Control and Prevention (CDC). In the synthesis, the stilbene was refluxed in neat triethyl phosphite for 1 h to deliver the key indole intermediate, which was converted into botulinum neurotoxin A inhibitors in several additional steps.

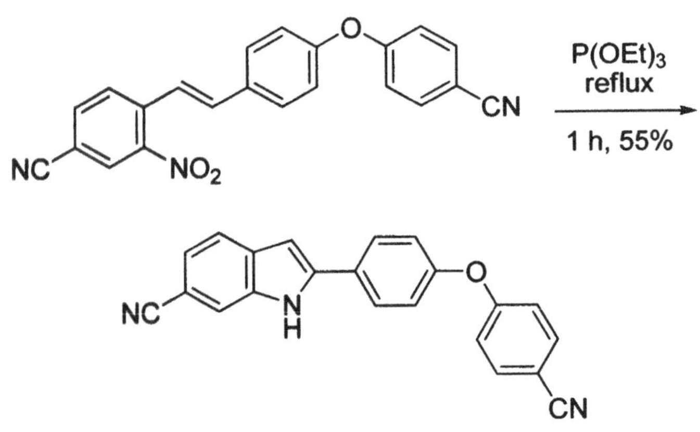

3.4 Oxindole-containing Drug Synthesis

Although not as popular as indoles, oxindoles exist in some drugs. One example is Pfizer's atypical anti-psychotic ziprasidone (Geodon). Its right-hand fragment oxindole started with a Wolff–Kishner reduction of isatin to give the oxindole.[58–60] Friedel–Crafts acylation with chloroacetyl chloride afforded the aryl ketone, which was reduced with triethylsilane in trifluoroacetic acid to the phenethyl chloride. The alkyl chloride fragment was joined by alkylation with the piperazine fragment in the presence of NaI and Na_2CO_3 to give ziprasidone in 20% yield in isoamyl alcohol. The yield of the coupling step was improved dramatically (86%) when the reaction was conducted in water.

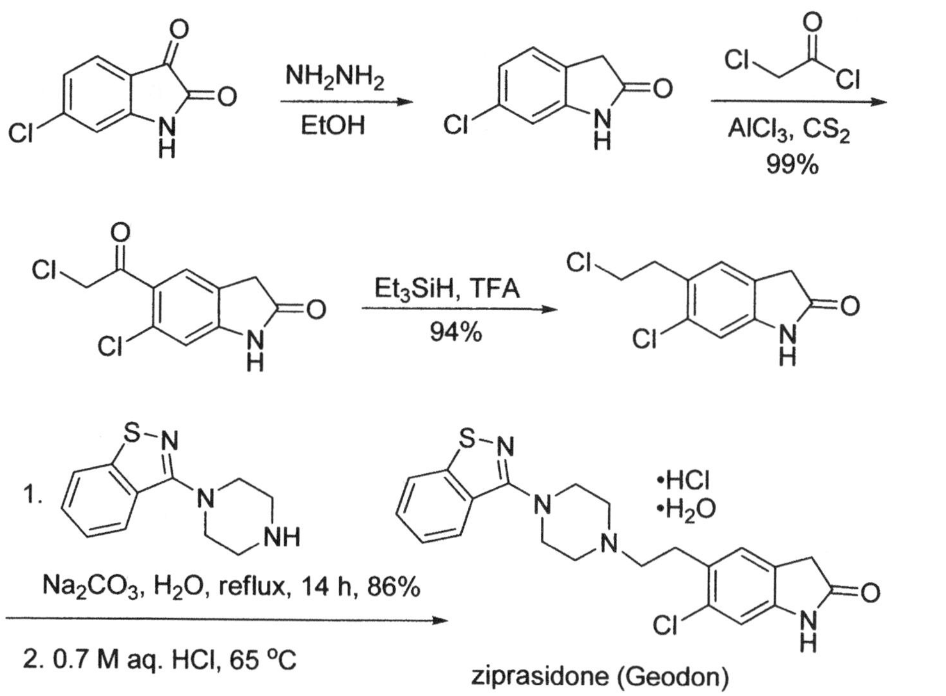

Another oxindole-containing drug is sunitinib (Sutent), discovered by Sugen. Sunitinib, an angiogenesis inhibitor, inhibits a panel of kinases including VEGFR1, VEGFR2, VEGFR3, PDGFRα, PDGFRβ, Flt-3, and c-Kit. Its core structure 5-fluorooxindole was constructed from 5-fluoroisatin in two steps.[61] Thus, heating a neat mixture of 5-fluoroisatin and hydrazine hydrate to 110 °C effected a Wolff–Kishner reduction and ring-opening to give an acyl hydrazide. The crude material was subjected to an intramolecular acylation reaction by exposure to aqueous HCl at room temperature to afford the oxindole in 73% yield over two steps.

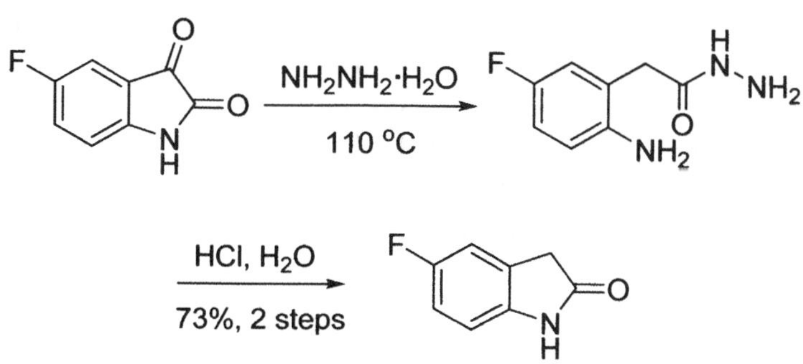

The process chemists developed an ingenious route to couple 5-fluorooxindole with the pyrrole-containing right-hand piece.[62] As shown below, pyrrole was added to a solution of the Vilsmeier reagent in acetonitrile at room temperature. Following the completion of the formylation reaction, the resulting intermediate was treated with 5-fluorooxindole and pulverized KOH, which led to precipitation of the desired product. Filtration of the reaction mixture afforded the sunitinib free base in 74% yield.

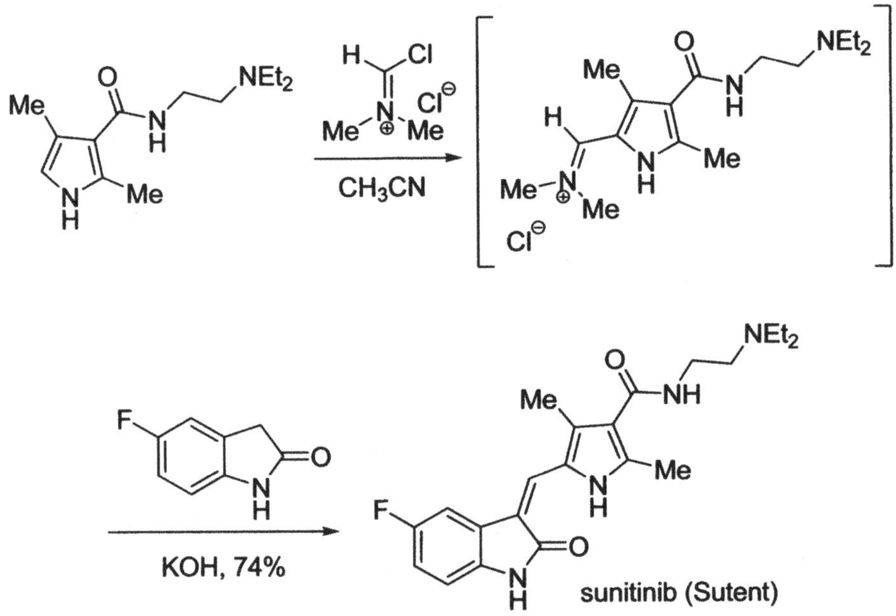

Ropinirole (Requip), a prejunctional dopamine receptor agonist, is used for the treatment of Parkinson's disease and restless legs syndrome (RLS). It acts as a dopamine D_2, D_3, and D_4 receptor agonist with highest affinity for D_3. Ropinirole also exhibits weak activity at the 5-HT_2 and α2-adrenergic receptors and shows little affinity for the 5-HT_1, benzodiazepine, GABA, muscarinic, and α1- and β-adrenergic receptors. The syntheses of ropinirole provide ample examples for oxindole chemistry. In SmithKline & French's original synthesis,[63] the *o*-methyl-nitrobenzene compound was deprotonated and quenched by diethyl oxalate. The resulting α-ketoacid was decarboxylated under basic oxidative conditions. The nitro group was reduced using hydrogenation and the resulting amino acid subsequently cyclized *in situ* to deliver ropinirole.

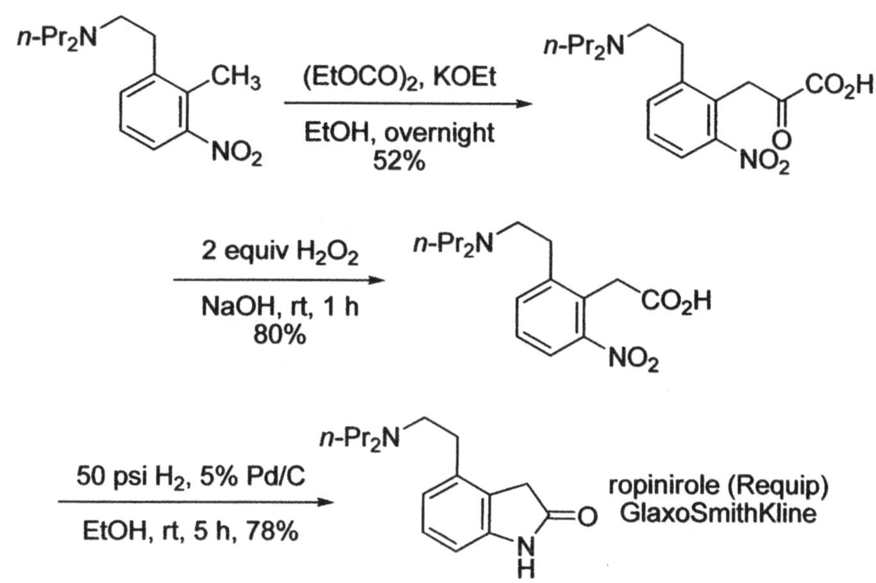

In a later synthesis of the regioisomers of ropinirole, the isatin was formed from cyclization of a hydrazone under strong acidic conditions.[64] After removal of the ketone functionality *via* hydrogenolysis and removal of the protective group, reductive amination afforded the regioisomer of ropinirole.

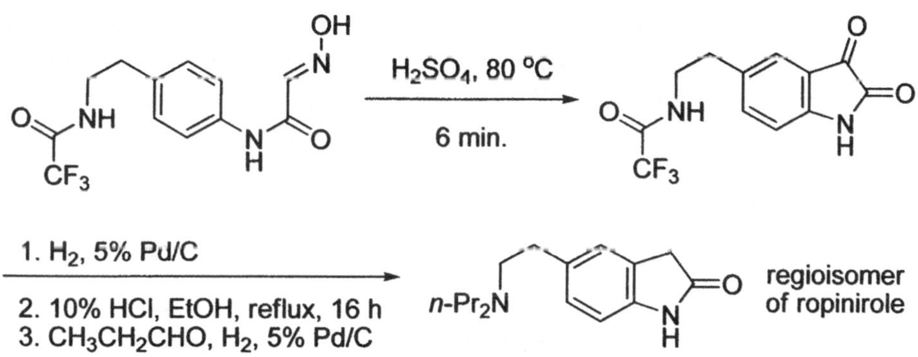

3.5 *Cross-coupling Reactions for Indoles*

Metal-catalyzed cross-coupling reactions have emerged as an important advancement in organic chemistry during the last few decades. Meanwhile, due to indole's importance in medicinal chemistry and many other fields, metal-catalyzed cross-coupling reactions have been extensively applied in the field of indole synthesis.[65]

3.5.1 Palladium-Catalyzed Oxidative Coupling

The oxidative coupling/cyclization reaction is the intramolecular or inter-molecular union of two arenes with formal loss of H_2 promoted by a Pd(II) species, typically $Pd(OAc)_2$. Thanks to the entropy benefit, intramolecular oxidative coupling, also known as oxidative cyclization, is much more facile than the inter-molecular oxidative coupling. Hill described the oxidative coupling: $Pd(OAc)_2$-mediated oxidative cyclization of bisindolylmaleimides to give indolo[2,3-*a*]pyrrolo[3,4-*c*]carbazoles,[66] which is the core ring system in numerous natural products, many of which have potent protein kinase activity.

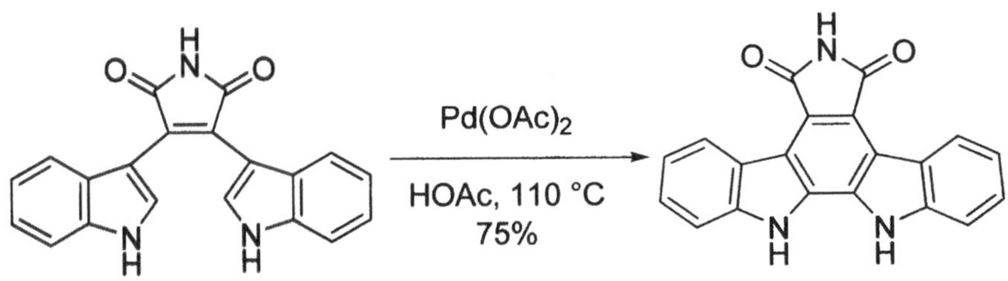

Intermolecular palladium(II)-mediated oxidative couplings with indoles are well established, although initial results were unpromising. In the total syntheses of clavicipitic acid and costaclavine, one key step is the oxidative coupling of 4-bromo-1-tosylindole with methyl 2-Boc-amino-acrylate to give a dehydrotryptophan derivative.[67] The use of chloranil as a reoxidant to recycle Pd(0) to Pd(II) greatly improves the coupling over earlier conditions. For example, chloranil was more effective than DDQ, MnO_2, Ag_2CO_3, Co(salen)$_2$/O_2, and $Cu(OAc)_2$. In the absence of chloranil, the yield was 31%.

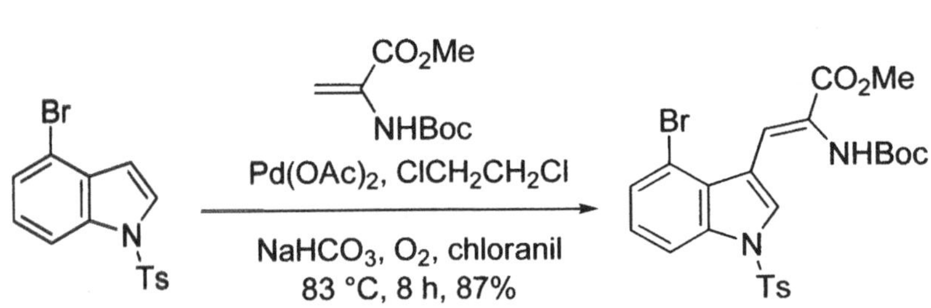

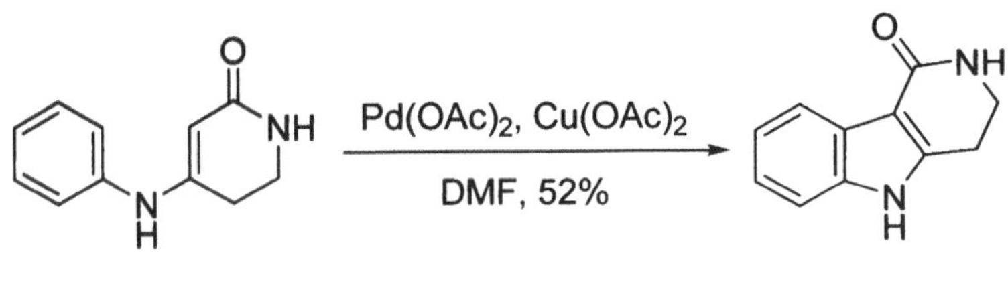

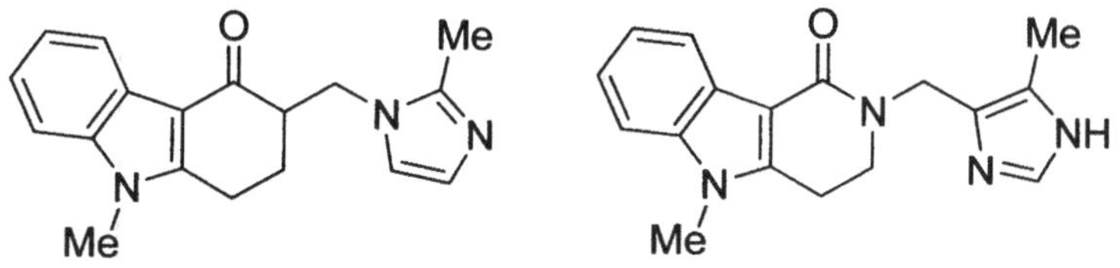

ondansetron (Zofran)
GlaxoSmithKline

alosetron (Lotronex)
GlaxoSmithKline/Prometheus Lab

3.5.2 Negishi Coupling

The Negishi reaction is the palladium-catalyzed cross-coupling of organozinc reagents with organohalides or triflates. It is compatible with many functional groups including ketones, esters, amines and nitriles. The organozinc reagents are usually generated and used *in situ* by transmetalation of Grignard or organolithium reagents with $ZnCl_2$. In addition, some halides may oxidatively add to Zn(0) to give the corresponding organozinc reagents.

Although the Negishi coupling has been less frequently used in indole synthetic manipulations than either Suzuki or Stille couplings, we will see in this chapter that Negishi chemistry is often far superior to other Pd-catalyzed cross-coupling reactions involving indoles. One of the first such examples is the coupling of 1-methyl-2-indolylzinc chloride with iodopyrimidine to give the indolyl-pyrimidine-dione.[68]

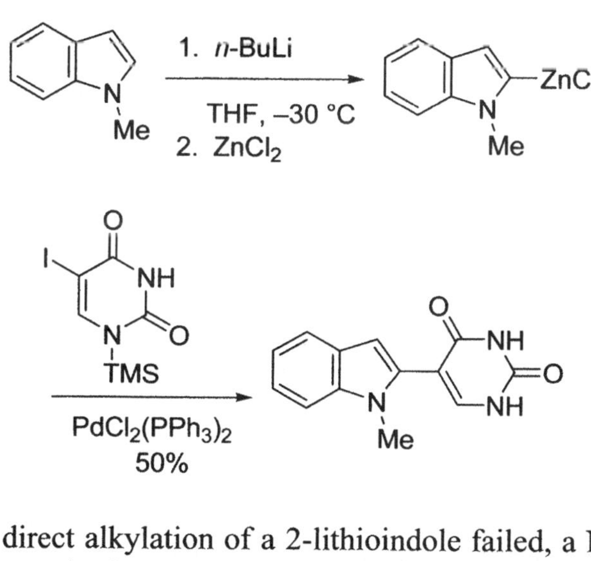

Since direct alkylation of a 2-lithioindole failed, a Negishi protocol was utilized to synthesize a 2-benzylindole, an intermediate for the novel CNS agent.[69]

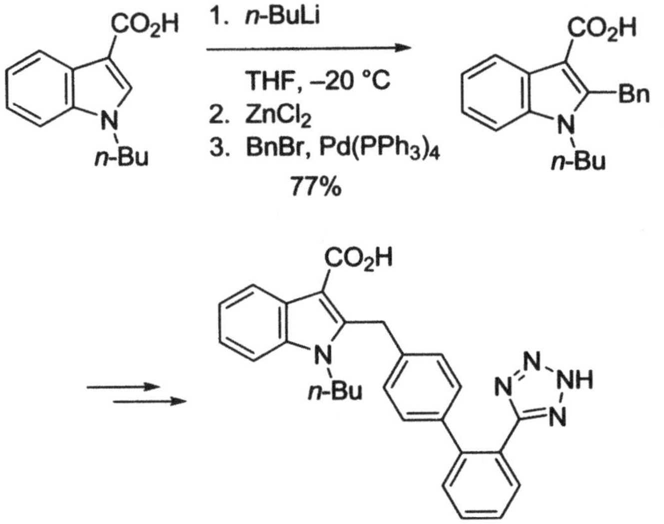

Negishi methodology can also be used to achieve the 3-acylation of indoles. Thus, a series of 3-acylindoles was prepared from 1,2-substituted indole.[70] The 1,2-substituted indole could also be iodinated cleanly at C-3 with *N*-iodosuccinimide.

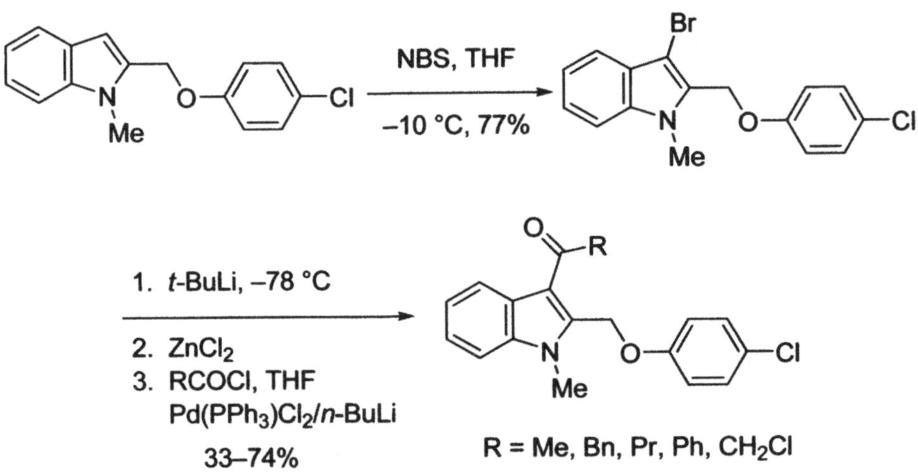

An asymmetric synthesis of prolino-homotryptophane was accomplished *via* amino-zinc-ene-enolate cyclization followed by transmetalation of the cyclic zinc intermediate with indolyliodide.[71] The use of palladium catalyst derived from [*t*-Bu_3PH]·BF_4 was required to avoid the undesired β-hydride elimination. The resulting proline chimeras are useful tools for medicinal chemistry and/or biological applications.

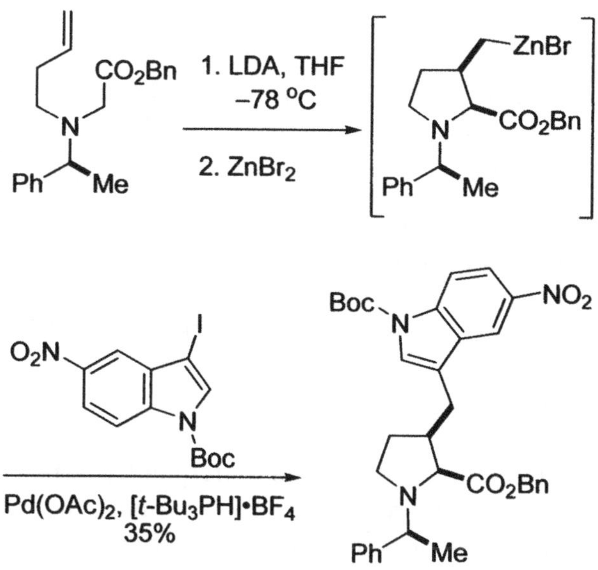

The 1-(4-fluorophenyl)indole moiety is present in many CNS drugs such as the anti-psychotic drug serindole, serotonin 5-HT_{2A} and dopamine D_2 receptor antagonists. In Lundbeck's synthesis of central adrenergic α1 receptor modulator, a Negishi coupling took place on the indole ring at the C5 position.[72]

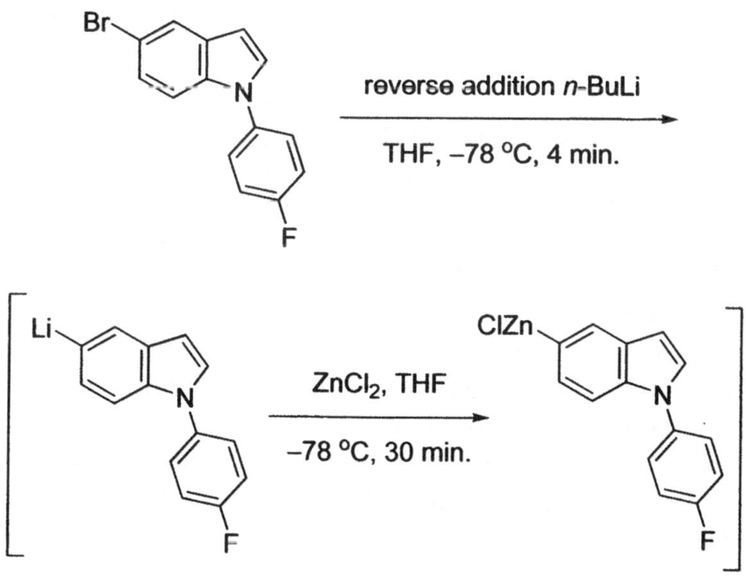

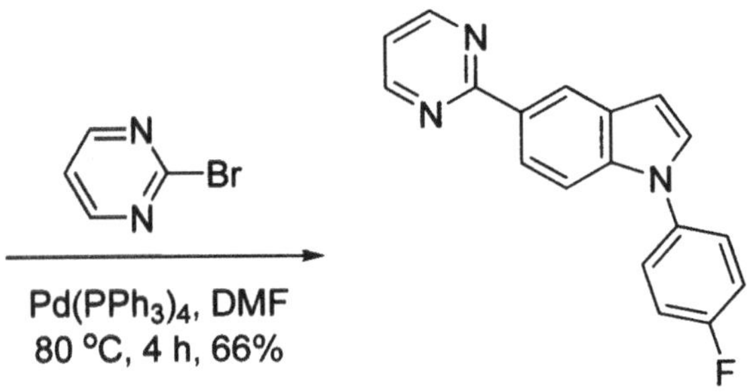

α1 receptor modulator

3.5.3 *Suzuki Coupling*

The Suzuki reaction is one of the most frequently employed palladium-catalyzed C–C bond-forming reactions, and involves the cross-coupling of organoboron reagents with organohalides (or triflates). This transformation has two significant advantages over related Stille couplings of organostannanes. First of all, organoboron reagents have very low toxicity, especially compared with aryl(trialkyl)tin derivatives. In addition, the boron-containing side products of these transformations are usually easily removed through a simple alkali workup, whereas the trialkyltin halide by-products of Stille coupling reactions are much more difficult to remove. For these reasons, Suzuki coupling is a more attractive choice than Stille coupling for syntheses conducted on a large scale. It has been widely applied to medicinal chemistry.

Indoxole [2,3-bis(4-methoxyphenyl)indole] is a potent anti-inflammatory, anti-arthritic, and anti-pyretic agent in animals, with activity comparable with, or greater than, aspirin and indomethacin in certain assays. Indoxole's important anti-inflammatory properties opened the door for the development of potent COX-2 inhibitors for arthritic pain relief. Older synthesis of Indoxole involved *de novo* indole construction. Gribble devised a clever scheme to prepare 2,3-dihaloindoles and applied a bis-Suzuki cross-coupling reaction with arylboronic acids to prepare 2,3-diarylindoles including Indoxole.[73] Treatment of indole with *n*-BuLi and quenched by iodine gave the expected 3-iodoindole, which was then treated with LDA and phenylsulfonyl chloride to provide the protected 3-iodoindole. Treatment of the intermediate with LDA selectively deprotonated the C2 position on indole, which was quenched with iodine to afford 2,3-di-iodoindole. The subsequent Suzuki coupling worked well, whose protection was removed to deliver Indoxole.

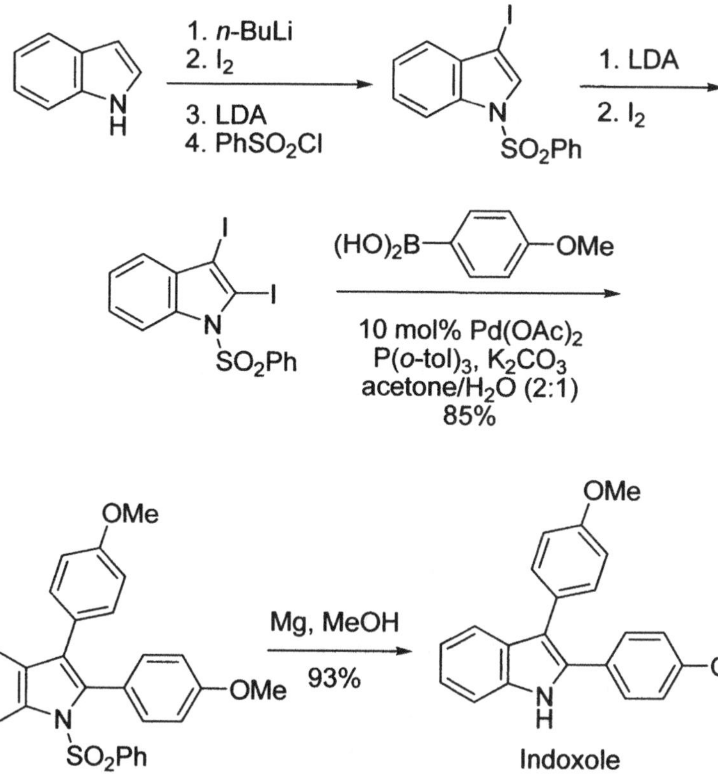

MDL 103371 is an *N*-methyl-D-aspartate (NMDA)-type glycine receptor antagonist for the potential treatment of stroke. Aventis's Process Group explored a synthesis of MDL 103371 involving the Suzuki coupling.[74] As shown in the scheme below, Horner–Emmons coupling of 3-indolylaldehyde with α-bromo-phosphonate ester provided the desired (*Z*)-vinyl bromide substrate in 65% yield after fractional crystallization from cyclohexane. The Suzuki coupling between the vinyl bromide with arylboronic acid then assembled the main frame structure, which could be manipulated to MDL 103371 and its derivatives.

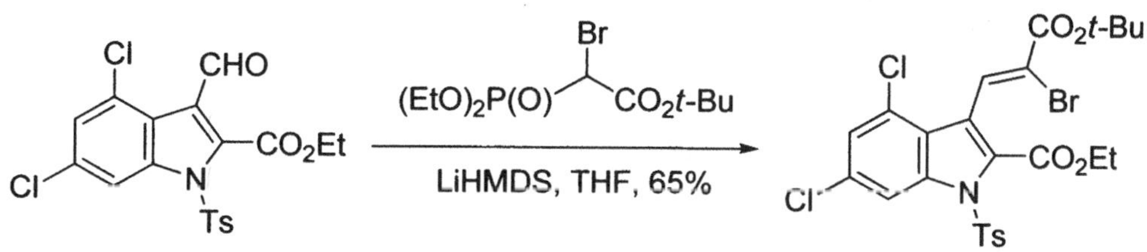

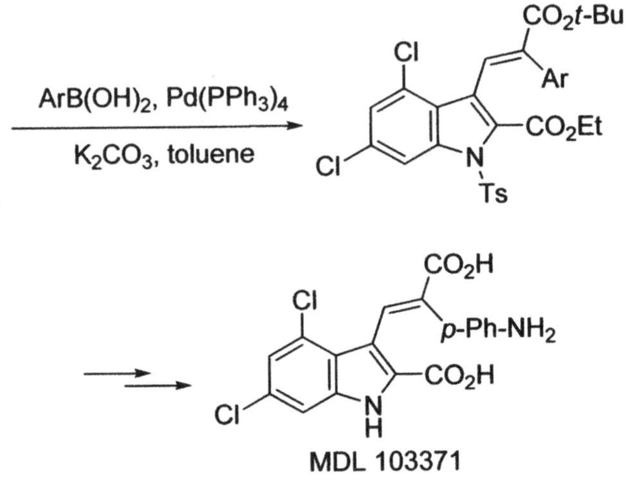

A Suzuki coupling was also employed during the synthesis of obatoclax. Obatoclax's novel indolylprodigiosin derivative demonstrated its ability to antagonize multiple members of the B-cell lymphoma (Bcl) family of antiapoptotic proteins. The compound has shown potent anticancer activity in several animal tumor models. Obatoclax is now in Phase II clinical trials directed against multiple hematologic and solid tumor malignancies. In a scalable process for the synthesis of obatoclax,[75] the Suzuki coupling between the bromo enamine and 1-(*N*-methoxycarbonyl)-indole-2-boronic acid gave a coupled intermediate, followed by hydrolysis of both the enamine and *N*-methoxycarbonyl groups to afford the indolylpyrrole aldehyde. This transformation, whereby a methyl carbamate on indole was prepared, proved to be a cost-effective alternative to using the commercially available 1-[*N*-(*tert*-butoxycarbonyl)]-indole-2-boronic acid. Finally, the acid-mediated condensation of the indolylpyrrole aldehyde with 2,4-dimethyl-1*H*-pyrrole provided the HCl salt of obatoclax.

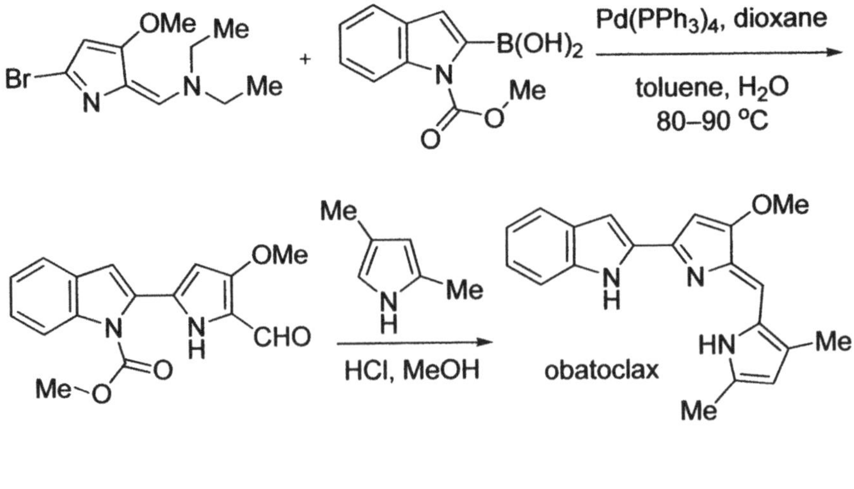

3.5.4 Sonogashira Coupling

The Sonogashira reaction allows for the direct cross-coupling of terminal alkynes with aryl halides under mild conditions through use of catalytic bis(triphenylphosphine)-palladium dichloride and CuI as the co-catalyst in the presence of an aliphatic amine. The formation of diphenylacetylene exemplifies the original Sonogashira reaction conditions[76]:

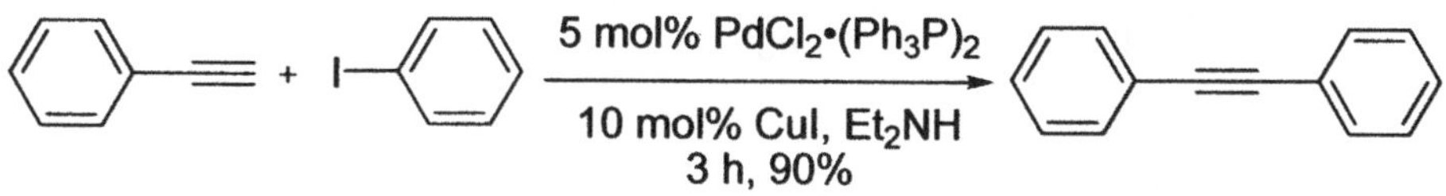

The greatest application of the Sonogashira coupling is of course the Larock indole synthesis, which is reviewed in Section 3.3 in this chapter. A Sonogashira coupling was carried out between protected phenyliodide and alkyne.[77] With *N*-methanesulfonyl protection, the coupling product spontaneously cyclized to the indole, which was converted into an indole-based insulin mimic.

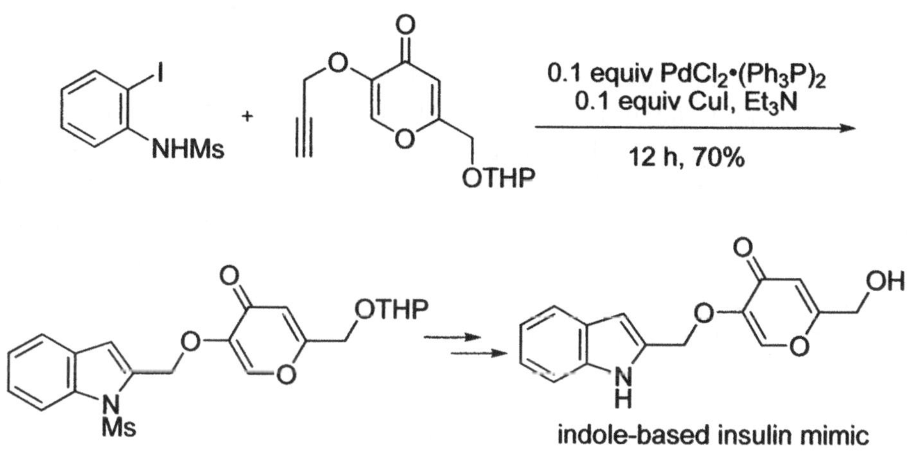

The Sonogashira coupling was applied to the synthesis of an indole-containing KDR kinase inhibitor by tandem Sonogashira coupling/5-*endo*-*dig*-cyclization as a key step.[78]

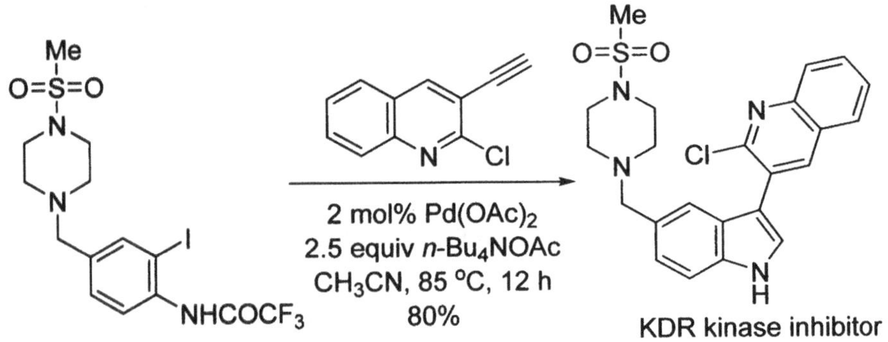

3.5.5 Heck Reaction

The Heck reaction, first disclosed independently by Mizoroki and Heck in the early 1970s, is the Pd-catalyzed coupling reaction of organohalides (or triflates) with olefins. In recent years, this transformation has become an indispensable tool for organic chemists. Inevitably, many applications to heterocyclic chemistry have been pursued and successfully executed. The greatest utility of the intramolecular Heck reaction was captured in Section 3.2 in this chapter. Intermolecular Heck reactions will be focused on this section.

Triptans have provided a futile ground for indole synthesis employing both the Heck and the intramolecular Heck reaction. In Glaxo's process synthesis of naratriptan,[79] the reaction of 5-bromoindole with *N*-methyl-4-piperidone gave the *N*-methyl-tetrahydropyridinyl substituted indole. The Heck reaction between the 5-bromoindole and the *N*-methylvinylsulfonamide was carried out on 700-g scales. Both double bonds of the resulting diene were hydrogenated to give naratriptan hydrochloride as white crystals in 71% yield after recrystallization from hot water.

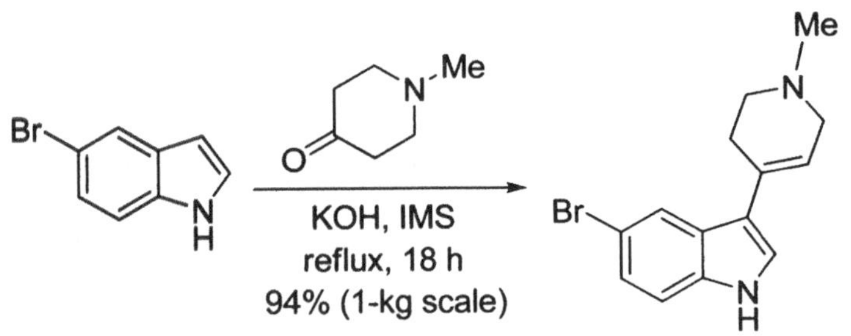

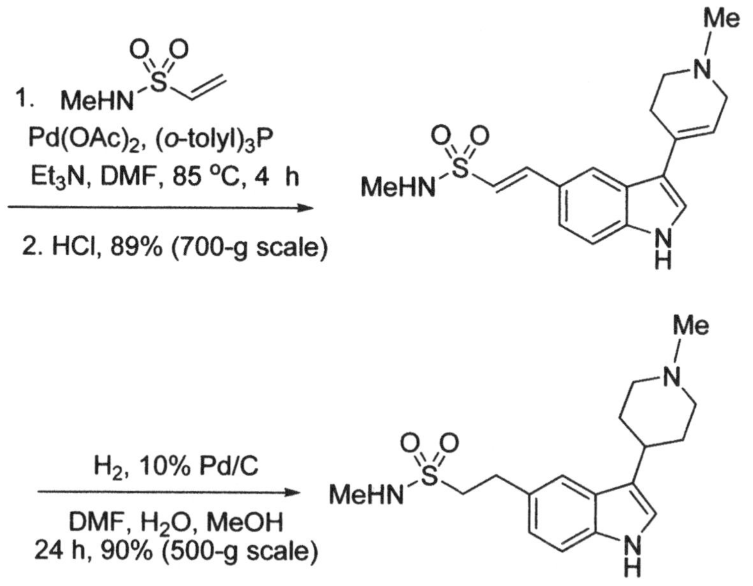

naratriptan (Naramig)

Eletriptan (Relpax) is a conformationally restricted analog of sumatriptan. In Pfizer's original synthesis, the Heck reaction between 5-bromo-*N*-methylpyrrolidinylmethyl indole with phenylvinyl sulfone actually gave a decent yield (61%).[80] However, the unprotected Heck adduct is prone to dimerization, where the indole nitrogen adds to the vinyl sulfone of another molecule. The formation of this impurity reduces the yield and complicates the purification. Therefore, indole is protected with an acetyl group and then subjected to the Heck coupling with phenylvinyl sulfone to give the adduct, which is no longer prone to dimerization. The reactive double bond is reduced by catalytic hydrogenation, and the acetyl group is removed by hydrolysis to give eletriptan of high purity in good yield.

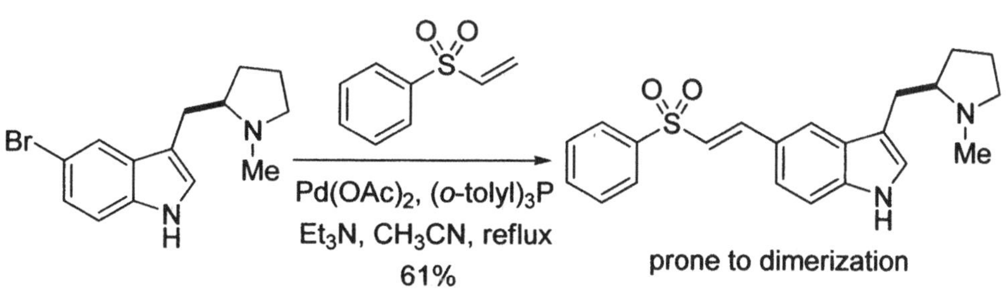

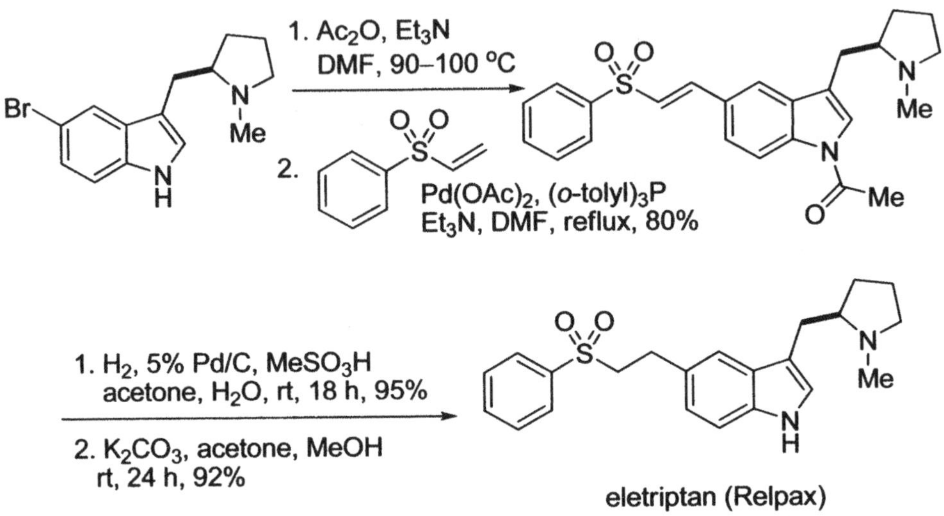

A Heck reaction was used to homologate an indole in the synthesis of DG-041, a small molecule antagonist of the EP3 receptor for prostaglandin E2. In deCODE's final optimized four-step route to DG-041,[45] 2,6-dibromo-4-fluoroaniline was alkylated with allyl bromide in a solution of potassium *tert*-butoxide in THF at room temperature. The resulting 2,6-dibromo-*N*-allylaniline was converted into the 3-methylindole via a Mori–Ban indole synthesis, followed by an intermolecular Heck coupling with acrylic acid. After the tandem Heck reactions, deprotonation of the indolyl acrylic acid using 2.2 equiv of potassium *tert*-butoxide in THF followed by quenching with 2,4-dichlorobenzyl chloride was found to efficiently give the desired penultimate compound in excellent yield. Finally, coupling of the acrylic acid with 2,3-dichlorothiophene-5-sulfonamide catalyzed with EDCI and HOBt afforded DG-041 in good yield.

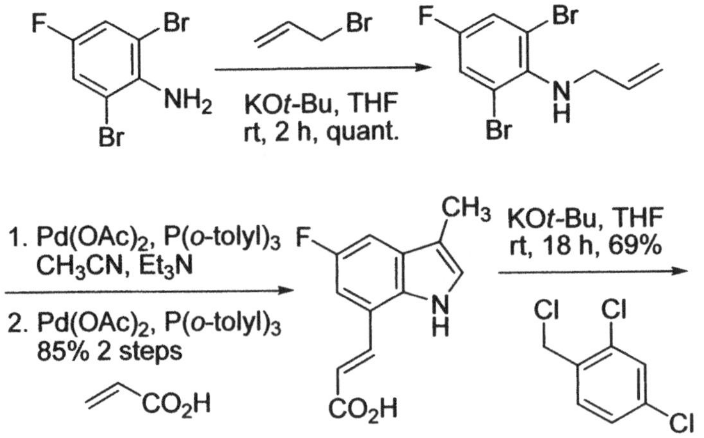

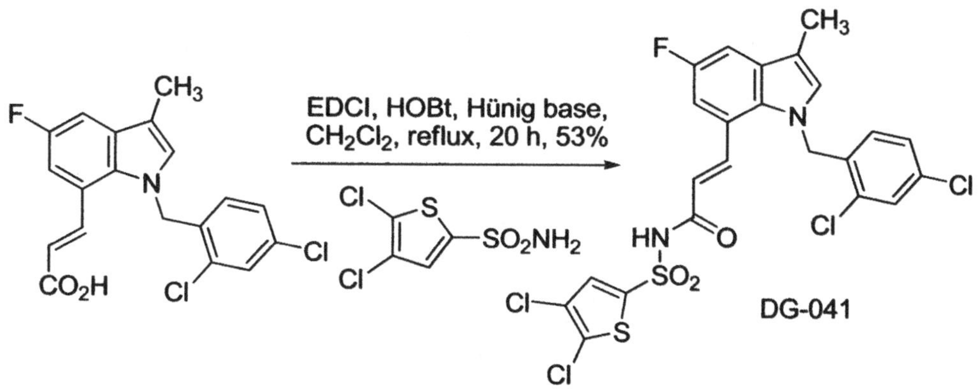

In another example of intermolecular Heck reaction shown below, 4-bromoindole was coupled with a BOC-protected 7-vinyl-1,2,3,4-tetrahydronaph-thyridine olefin in 63% yield.[81] The resulting alkene was manipulated to a potent $\alpha_v\beta_3$ antagonist, a vitronectin receptor that has been identified as a promising potential target for the treatment of osteoporosis, diabetic retinopathy, and cancer.

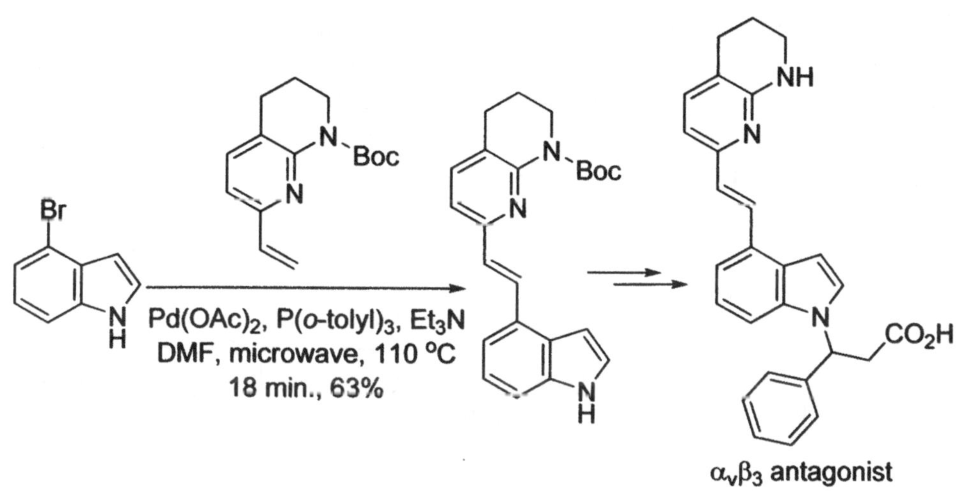

Three representative examples of intramolecualar Heck reaction are shown below. The first one is the Mori–Ban reaction; the second is an intramolecualar Heck reaction taking place on the indole ring; and the third one is an intramolecualar heteroaryl Heck reaction connecting two aromatic rings.

The Mori–Ban reaction has been widely used in drug discovery and development. In the process synthesis of a selective PPARγ modulator by Merck, the enamine was cyclized to afford the indole core.[53]

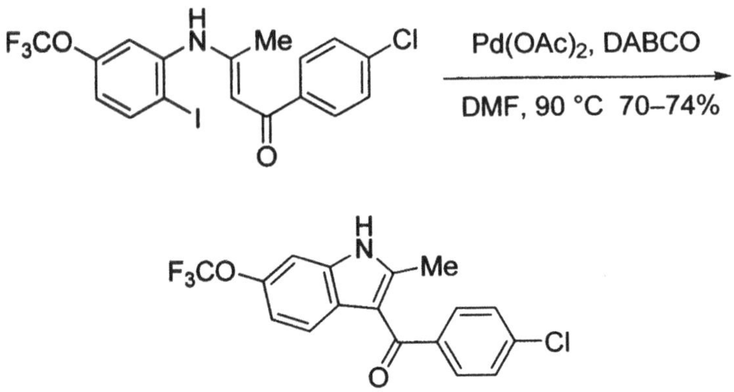

An intramolecular Heck reaction was used for the synthesis of tricyclic indole-2-caboxylic acids as potent NMDA-glycine antagonists. In one example, 3-iodo-indole with pendent α,β-unsaturated ester was cyclized to afford the core tricyclic indole scaffold.[82] Although radical cyclization of the same substrate could conveniently afford the desired cyclized product, use of tributyltin hydride should be avoided in a large-scale synthesis because of its toxicity. For the intramolecular Heck reaction, the addition of both silver(I) phosphate and an appropriate amount of triethylamine cooperatively to the palladium(II) acetate-triphenylphosphine system in DMF, improved the yield. On the other hand, NMDA-glycine antagonists are potential treatments of stroke and neurodegenerative disorders such as Alzheimer's and Huntington's diseases.

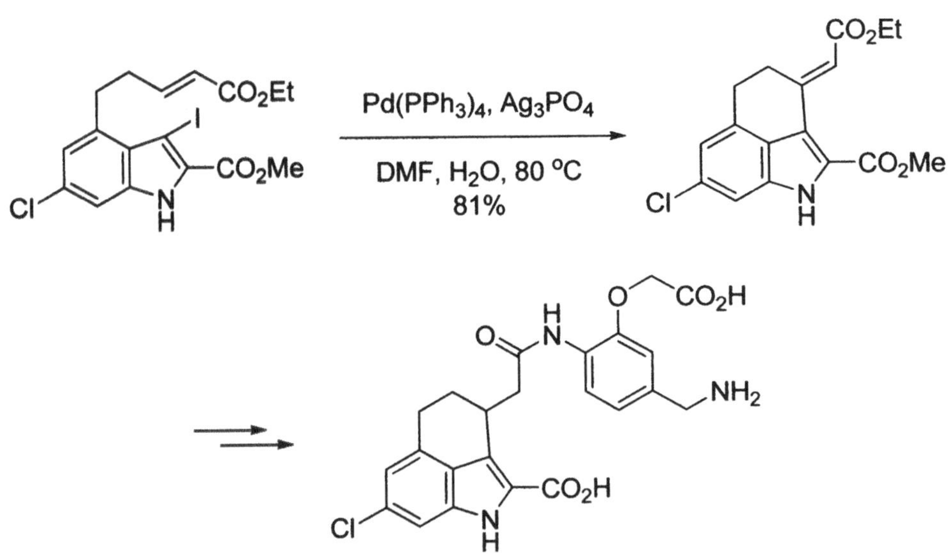

NMDA-Glycine antagonist

Another intramolecular Heck reaction could be called a heteroaryl Heck reaction because the C–C bond formation involves a benzene and an indole ring.[83] The resulting phenylcarbazoles are potential anti-cancer agents functioning as DNA topoisomerase I inhibitors.

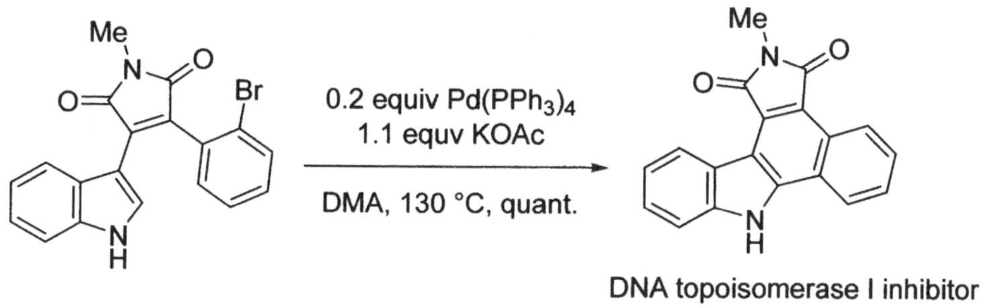

3.6 *Azaindoles*

Azaindoles, also known as pyrrolopyridines, are biosteres for indole. Depending on the position of the nitrogen atom, they are named 4-azaindole, 5-azaindole, 6-azaindole, and 7-azaindole, respectively, as shown below.

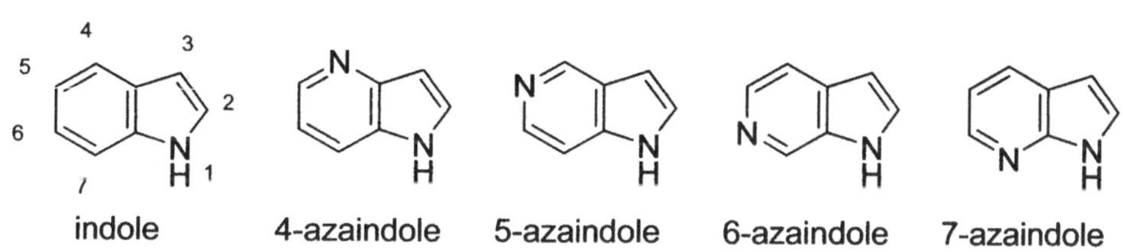

The addition of a nitrogen atom to the indole ring in a prospective drug could potentially modulate its potency, solubility, and intellectual property position. Because of the pyridine ring is electron-deficient, many classic indole synthesis methods do not work as well. For instance, the Fischer indole synthesis generally gives poor yields using pyridyl hydrazines and it also requires harsh conditions. There are tactics to circumvent pyridine's electron-deficiency by addition of the electron-pushing group such as methoxyl and methylsulfide groups. For example, 4- and 6-azaindoles are prepared using the Fischer reaction.[84] 7-Azaindole formation is unfavorable electronically via this "aza-Fischer" synthesis.

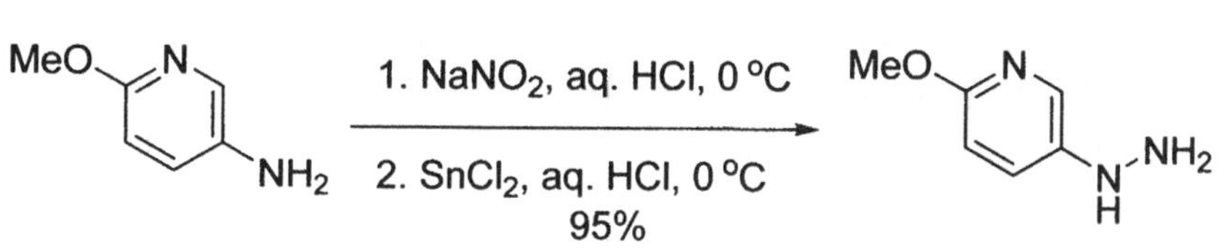

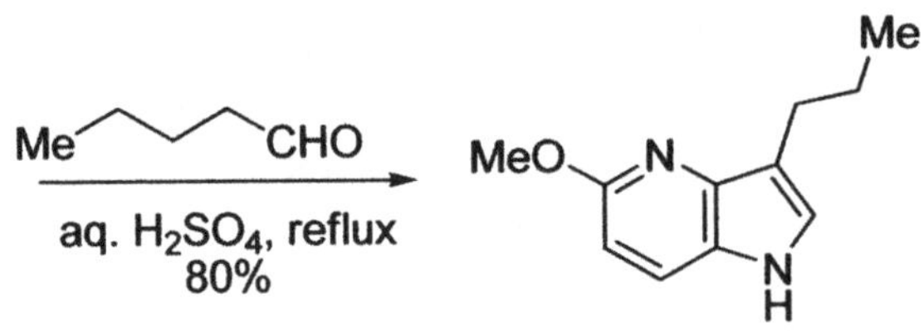

4-azaindole

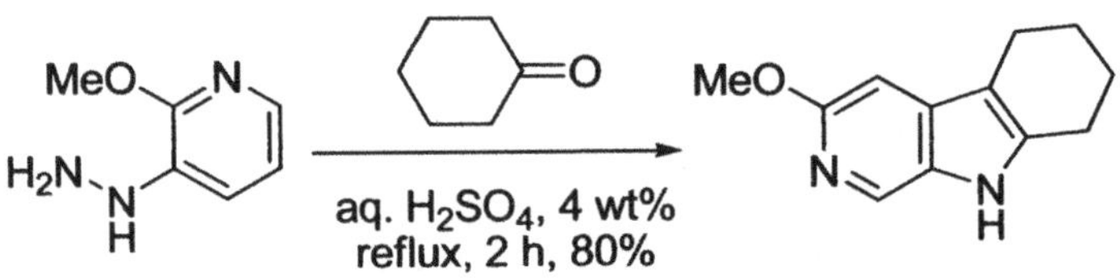

6-azaindole

Thankfully, some of the conventional and modern indole synthesis methods do apply to make azaindoles. They include the Larock reaction, Mori–Ban indole synthesis, Bartoli indole synthesis, Batcho–Leimgruber indole synthesis, and Cadogan–Sundberg indole synthesis.

3.6.1 Larock Reaction

The Larock reaction is applicable to make all azaindoles. In one case, 4-azaindoles were synthesized as novel inhibitors of C-Met kinase.[85] Interestingly, 4,7-diazaindole was assembled via a Larock reaction. The resulting product as the C-Met kinase inhibitor is potentially a cancer therapy by disruption of signal pathways that mediate tumor formation and growth.

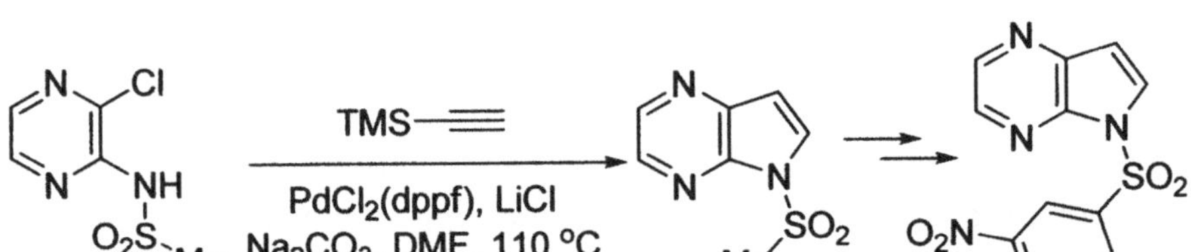

3.6.2 Bartoli Reaction

Although having been discovered only 20 years ago, the Bartoli indole synthesis has found more and more applications in indole and azaindoles synthesis. Applying the classic Bartoli conditions, nitropyridyl-piperazines were converted into 4-azaindole and 6-azaindole in 17 and 51% yield, respectively.[86] Both azaindoles were prepared as potential 5-HT_6 inhibitors for the indication of schizophrenia.

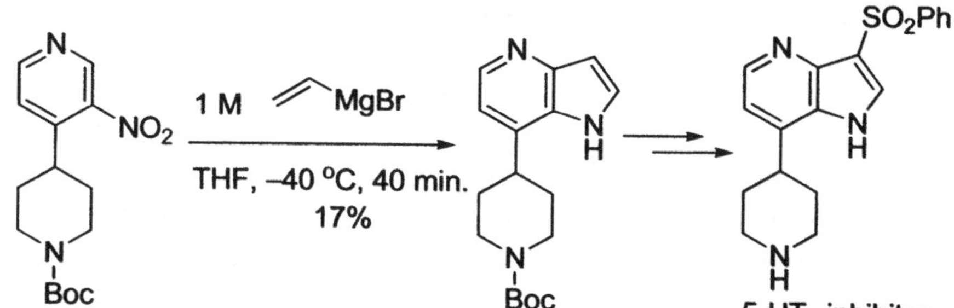

5-HT_6 inhibitor

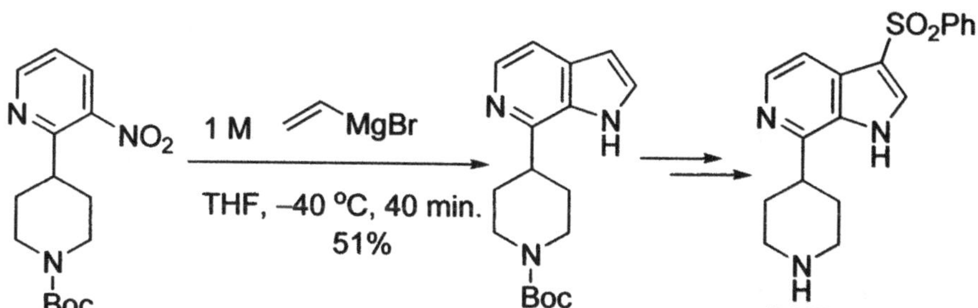

5-HT_6 inhibitor

Δ^9-Tetrahydrocannabinol, the principal pharmacologically active component of *Cannabis sativa,* acts on G-protein-coupled receptors known as $CB_{1,2}$, $CB_{2,3}$, and GPR55. Since agonism of CB_1 receptors or CB_2 receptors can produce analgesic effects in animal models of pain, consequently targeting one, or both, of these receptors is an attractive strategy for drug discovery in pain. GSK554418A, a brain penetrant 5-azaindole CB_2 agonist for the treatment of chronic pain, was prepared using the Bartoli reaction.[87] GSK employed the Bartoli conditions and converted the azaindole intermediate into CB_2 inhibitors.

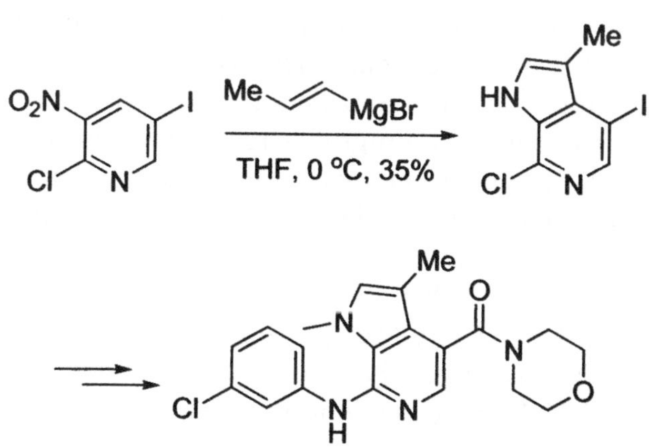

CB_2 inhibitor

For the BMS antiviral program targeting human immunodeficiency virus type 1 (HIV-1) attachment, all four azaindoles proved to be superior to the indole analogue.[88] The screening lead indole was found to be a potential inhibitor of HIV-1 attachment. Systematic replacement of each of the unfused carbon atoms in the phenyl ring of the indole moiety by a nitrogen atom provided four different azaindole derivatives that displayed a clear SAR for antiviral activity and all of which displayed marked improvements in pharmaceutical properties. In one case, the 6-azaindole scaffold was arrived at by taking advantage of the Bartoli reaction.

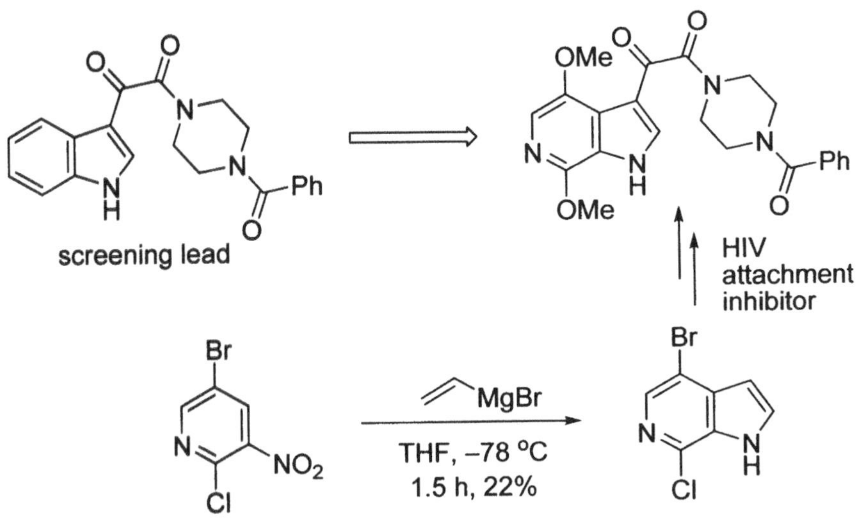

3.6.3 *Batcho–Leimgruber Reaction*

Human immunodeficiency virus type 1 (HIV-1), the causative epathogen of AIDS, replicates utilizing three essential enzymes encoded in the HIV pol gene: reverse transcriptase (RT), protease (PR), and integrase (IN). In an anti-HIV program, Pfizer found azaindoles hydroxamic acids as potent HIV-1 integrase inhibitors.[89] The 6-azaindole core structure, in turn, was constructed using the Batcho–Leimgruber reaction. Starting from 4-methyl-5-nitro-pyridin-2-ol, bromination was followed by palladium-catalyzed cyanization. Acidic hydrolysis under acidic condition converted the nitrile into the ester, which became the substrate of the Batcho–Leimgruber reaction. Treatment of the substrate with *N,N*-dimethylformamide dimethyl acetal [DMFDMA, $Me_2NCH(OMe)_2$] afforded the enamine intermediate, which was reduced via hydrogenolysis to give 6-azaindole. The scaffold was transformed into HIV-1 integrase inhibitors after further functional group manipulations.

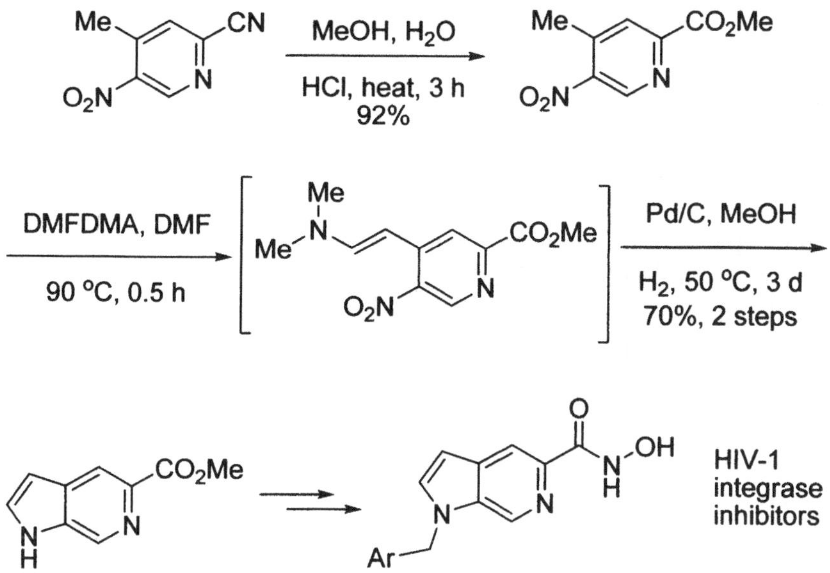

3.6.4 Cadogan–Sundberg Indole Synthesis

In 1972, the Cadogan–Sundberg indole synthesis was the key operation in preparing azaindoles as anthelmintic agents.[90] Thermal decomposition of the azido-pyridine *N*-oxide provided the corresponding azaindole *N*-oxide. On the other hand, refluxing the nitro-pyridine *N*-oxide with triethyl phosphite in dry benzene for seven days gave rise to the corresponding azaindoles because the *N*-oxide was concurrently reduced by triethyl phosphite.

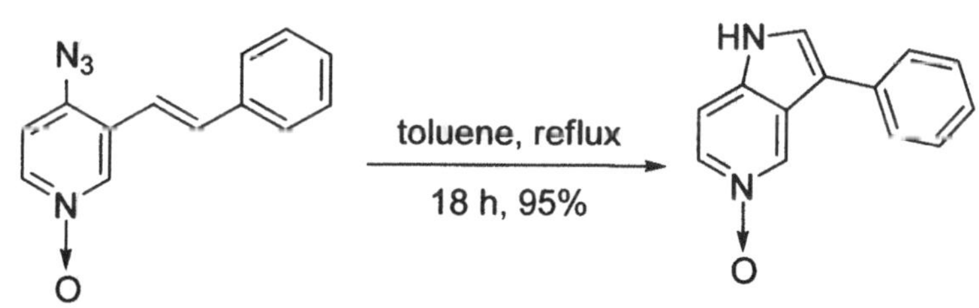

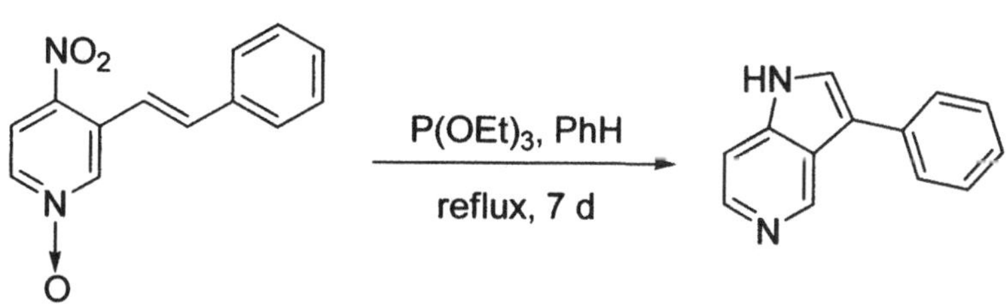

3.7 *Possible Liabilities of Drugs Containing 3-Methylindole*

The indole-ring system exists in a plethora of endogenous amino acids, neurotransmitters, and drugs. The metabolic 2,3-oxidation of the indole ring by CYP450 takes place from time to time, but its correlation to *in vivo* toxicity is not often observed.

One particular indole 3-methylindole, unfortunately, has been associated with higher risk of adverse outcomes, namely, pneumotoxin in animals. Evidence was found to support the formation of 2,3-epoxy-3-methylindoline as a reactive intermediate of the pneumotoxin 3-methylindole.[91] 3-Methylindole has been shown to form adducts with glutathione, proteins, and DNA using *in vitro* preparations.[92]

The P450-mediated bioactivation of 3-methylindole may be summarized below:

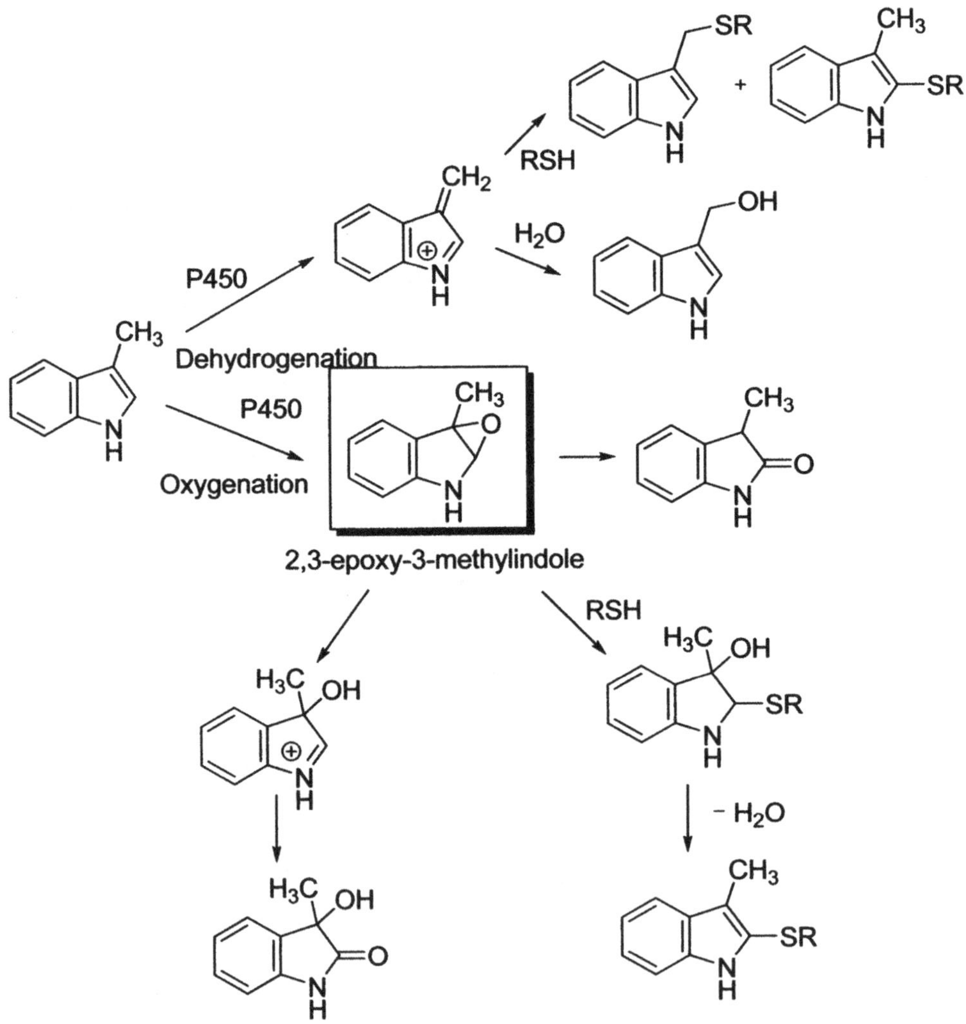

Oxidation of the 3-methyl group occurs either directly via deoxygenation or via epoxidation of the 2,3-double bond leading to 2,3-epoxy-3-methylindole, the reactive intermediate that can be trapped by endogenous nucleophiles, such as glutathione.

The presence of a leaving group on the C3-methyl increases the likelihood of formation of electrophilic reactive intermediates.

Zafirlukast (Accolate) is a leukotriene antagonist indicated for the treatment of mild-to-moderate asthma, but the drug has been associated with occasional idiosyncratic hepatotoxicity. Structurally, zafirlukast is similar to 3-methylindole because it contains an *N*-methylindole moiety that has a 3-alkyl substituent on the indole ring. The results presented here describe the metabolic activation of zafirlukast via a similar mechanism to that described for 3-methylindole. NADP(H)-dependent biotransformation of zafirlukast by hepatic microsomes from rats and humans afforded a reactive metabolite, which was detected as its GSH adduct.[93] The formation of this reactive metabolite in human liver microsomes was shown to be exclusively catalyzed by CYP3A enzymes. Evidence for *in vivo* metabolic activation of zafirlukast was obtained when the same GSH adduct was detected in bile of rats given an *i.v.* or oral dose of the drug.

The observation of *in vitro* metabolic activation of the 3-benzylindole moiety in zafirlukast, an anti-asthma drug, to give the glutathione adduct is an indication that the 3-methyl-indole activation pathway applies to other activated 3-alkyl indoles as well.[94]

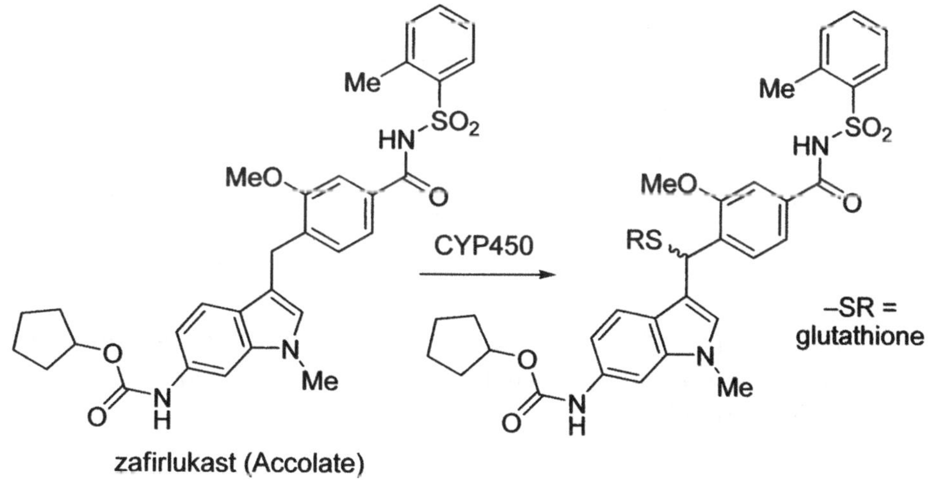

3.8 *Problems*

3.8.1 Propose the mechanism of the Baeyer–Drewson indigo synthesis.[95]

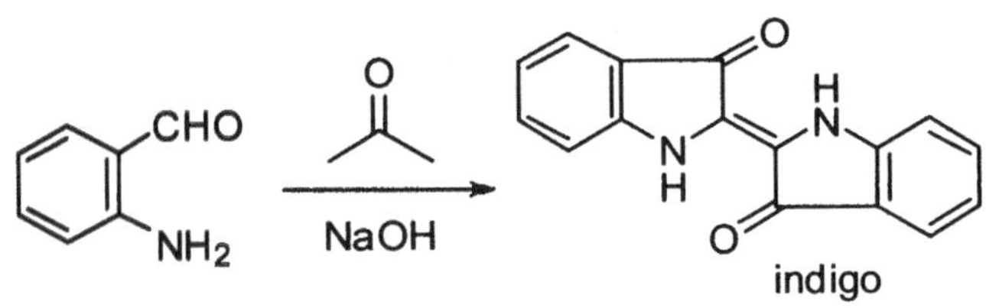

3.8.2 Propose the mechanism of the following indole formation[29]:

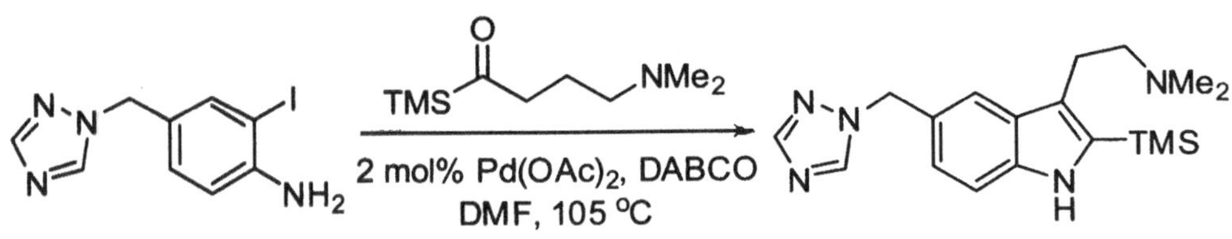

3.8.3 Explain the regiochemical outcome for the oxidative coupling of indole vs. pyrrole.[96,97]

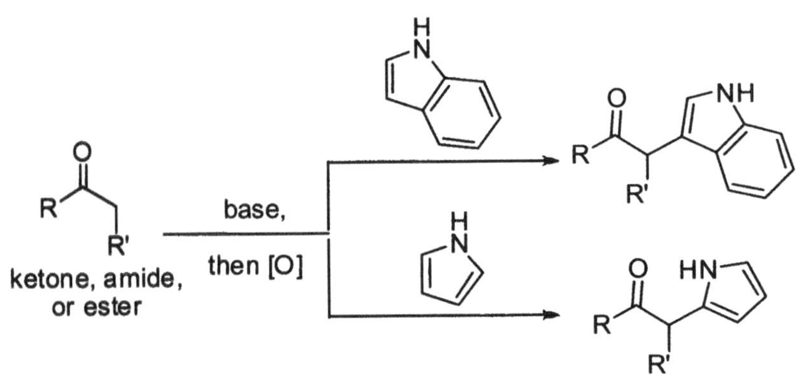

3.8.4 When 3-propylindole was treated with sodium hydride followed by dimethyl sulfate, a mixture of 2-methyl-3-propylindole and of 2-propyl-3-methylindole was found. Why?

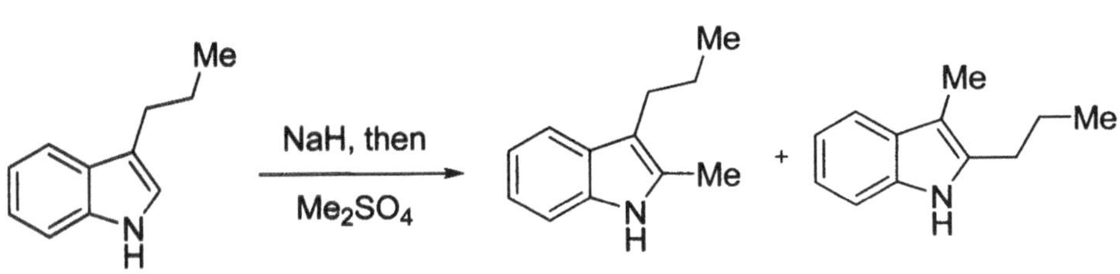

3.8.5 The optically pure (*R*)-indolylamine **A** was treated with sodium ethylmalonate in refluxing toluene to afford racemic adduct B. Propose a reasonable mechanism for the transformation.[98]

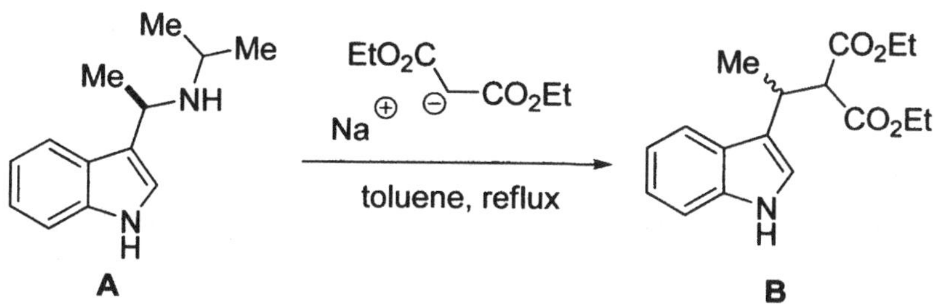

3.8.6 Delineate the mechanism of the second sulfenylation of indole.[99,100]

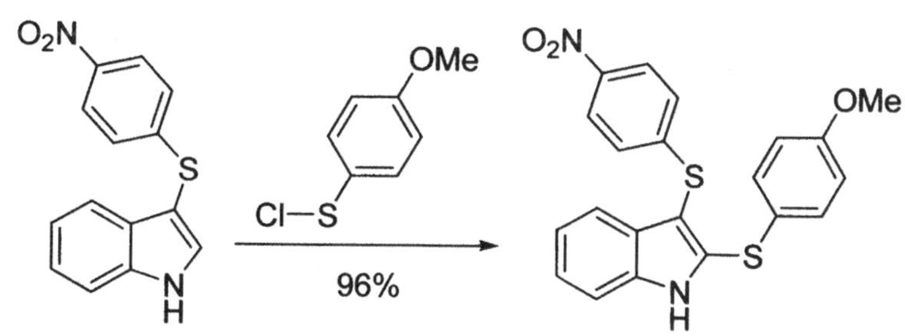

3.8.7 Propose a reasonable mechanism for the following indole formation.[88]

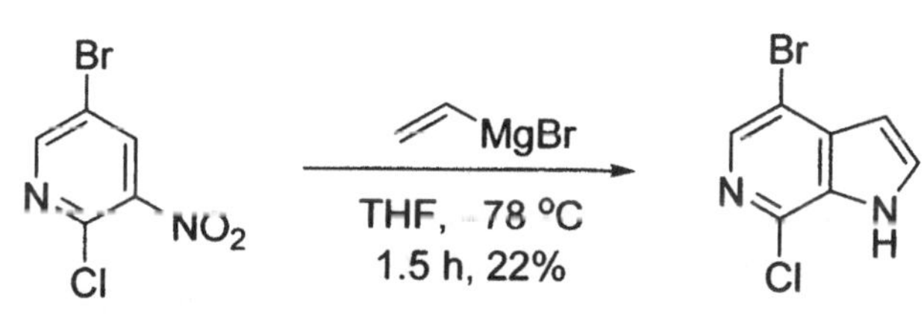

3.8.8 The Fischer indole synthesis has been the staple of assembling indoles for over a century. However, it does not work for ALL substrates. Propose a reasonable mechanism explaining why some Fischer indolizations fail.[101]

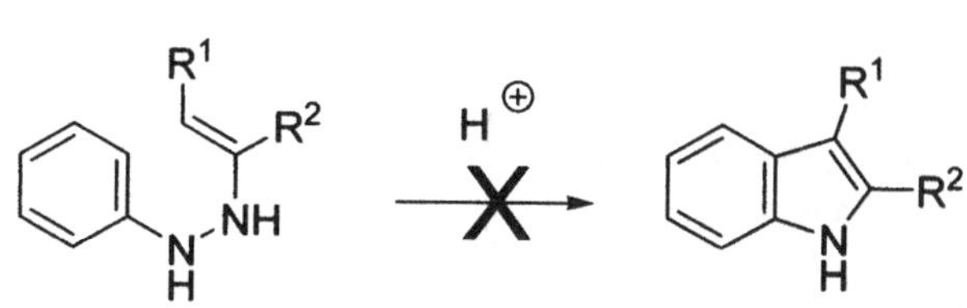

3.8.9 Propose a reasonable mechanism for the following indole formation.[102]

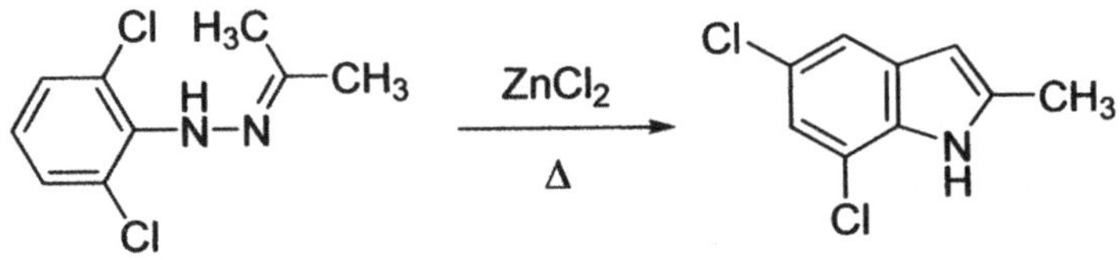

3.8.10 Propose a reasonable mechanism for the following indole formation.[103]

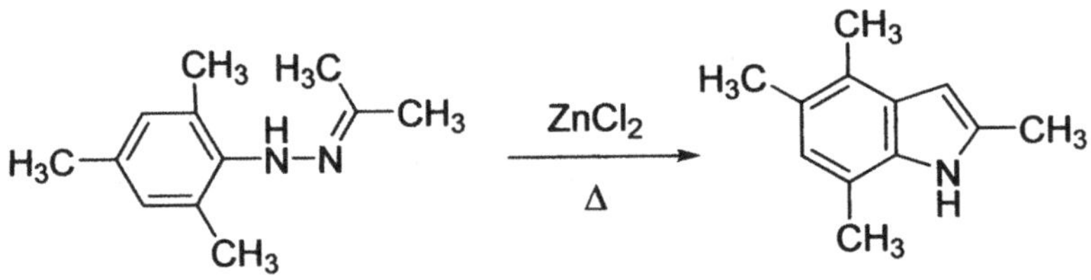

3.8.11 Propose a reasonable mechanism for the following transformation.[104]

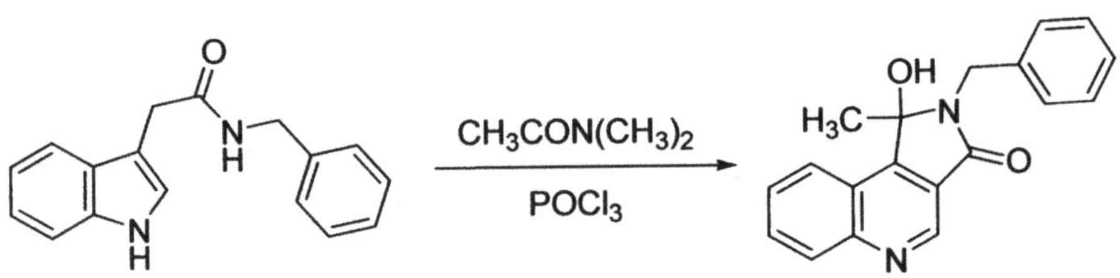

3.9 References

1. Garfield, S. *Mauve, How One Man Invented A Color That Changed the World,* W. W. Norton & Company: New York, **2000**.
2. Li, J. J. *Triumph of the Heart—The Story of Statins,* Oxford University Press: New York, NY, **2009**.
3. (a) Li, J. J.; Johnson, D. S.; Sliskovic, D. D.; Roth, B. D. *Contemporary Drug Synthesis,* Wiley: Hoboken, NJ, **2004**; pp 161–187. (b) Wu, Y.-J. *New Indole-Containing Medicinal Compounds,* in *Indoles*, Gribble, G. W., Ed.; Springer: Helderberg, Germany, **2010**, pp. 1–29.
4. Hinman, R. L.; Lang, J. *J. Am. Chem. Soc.* **1964,** *86*, 3796–3806.
5. Elokdah, H.; Abou-Gharbia, M.; Hennan, J. K.; McFarlane, G.; Mugford, C. P.; Krishnamurthy, G.; Crandall, D. L. *J. Med. Chem.* **2004,** *47*, 3491–3494.
6. Gill, A. L.; Frederickson, M.; Cleasby, A.; Woodhead, S. J.; Carr, M. G.; Woodhead, A. J.; Walker, M. T.; Congreve, M. S.; Devine, L. A.; Tisi, D.; O'Reilly, M.; Seavers, L. C. A.; Davis, D. J.; Curry, J.; Anthony, R.; Padova, A.; Murray, C. W.; Carr, R. A. E.; Jhoti, H. *J. Med. Chem.* **2005,** *48*, 414–426.
7. Goverdhan, G.; Reddy, A. R.; Sampath, A.; Srinivas, K.; Himabindu, V.; Reddy, G. M. *Org. Process Res. Dev.* **2009,** *13*, 67–72.

8. Daugan, A. C.-M. US 5859006 (1999); US 6025494 (2000).
9. Maw, G. N.; Allerton, C. M. N.; Gbekor, E.; Million, W. A. *Bioorg. Med. Chem. Lett.* **2003,** *13*, 1425–1428.
10. Li, J.; Cook, J. M. *Fischer Indole Synthesis* in *Name Reactions in Heterocyclic Chemistry*, Li, J. J., Ed., Wiley: Hoboken, NJ, **2004**, pp 116–127.
11. Li, J. J. *Name Reactions, A Collection of Detailed Reaction Mechanisms, Fourth Edition,* Springer: Heidelberg, Germany, **2009**, pp 227–228.
12. King, F. D.; Gaster, L. M.; Kaumann, A. J.; Young, R. C. US 5464864 (1995).
13. Borrett, G. T.; Kitteringham, J.; Porter, R. A.; Shipton, M. R.; Vimal, M.; Young, R. C. US 5962501 (1999).
14. Li, J. *Japp-Klingemann Hydrazone Synthesis* in *Name Reactions for Functional Group Transformations*, Li, J. J.; Corey, E. J., Eds., Wiley: Hoboken, NJ, 2007, pp 630–634.
15. Pete, B.; Bitter, I.; Harsanyi, K.; Toke, L. *Heterocycles* **2000,** *53*, 665–673.
16. The Grandberg modification of the Fischer indole reaction: Grandberg, I. I. *Zh. Org. Khim.* **1983,** *19*, 2439–2452.
17. Bosch, J.; Roca, T.; Armengol, M.; Fernandez-Forner, D. *Tetrahedron* **2001,** *57*, 1041–1048.
18. (a) Mori, M.; Chiba, K.; Ban, Y. *Tetrahedron Lett.* **1977,** *18*, 1037–1040; (b) Ban, Y.; Wakamatsu, T.; Mori, M. *Heterocycles* **1977,** *6,* 1711–1715. (c) Terpko, M. O.; Heck, R. F. *J. Am. Chem. Soc.* **1979,** *101*, 5281–5283.
19. (a) Larock, R. C.; Yang, H.; Pace, P.; Cacchi, S.; Fabrizi, G. *Tetrahedron Lett.* **1998,** *39,* 1885–1888. (b) Larock, R. C.; Pace, P.; Yang, H. *Tetrahedron Lett.* **1998,** *39,* 2515–2518. (c) Larock, R. C.; Pace, P.; Yang, H.; Russell, C. E.; Cacchi, S.; Fabrizi, G. *Tetrahedron* **1998,** *54*, 9961–9980. (d) Larock, R. C.; Babu, S. *Tetrahedron Lett.* **1987,** *28,* 5291–5294. (e) Larock, R. C.; Hightower, T. R.; Hasvold, L. A.; Peterson, K. P. *J. Org. Chem.* **1996,** *61*, 3584–3585.
20. Sundberg, R. J.; Pitts, W. J. *J. Org. Chem.* **1991,** *56*, 3048–3054.
21. (a) Macor, J. E.; Blank, D. H.; Post, R. J.; Ryan, K. *Tetrahedron Lett.* **1992,** *33*, 8011–8014. (b) Macor, J. E.; Ogilvie, R. J.; Wythes, M. J. *Tetrahedron Lett.* **1996,** *37*, 4289–4292.
22. Macor, J. E.; Wythes, M. J. US 5545644 (1996).
23. (a) Larock, R. C.; Yum, E. K. *J. Am. Chem. Soc.* **1991,** *113*, 6689–6690. (b) Larock, R. C.; Yum, E. K.; Refvik, M. D. *J. Org. Chem.* **1998,** *63*, 7652–7562.
24. Park, S. S.; Choi, J.-K.; Yum, E. K.; Ha, D.-C. *Tetrahedron Lett.* **1998,** *39*, 627 630.
25. Chen, C.; Lieberman, D. R.; Larsen, R. D.; Reamer, R. A.; Verhoeven, T. R.; Reider, P. J.; Cottrell, I. F.; Houghton, P. G. *Tetrahedron Lett.* **1994,** *35*, 6981–6984.
26. Chen, C.; Lieberman, D. R.; Street, L. J.; Guiblin, A. R.; Larsen, R. D.; Verhoeven, T. R. *Synth. Commun.* **1996,** *26*, 1977–1984.
27. Chen, C. Y.; Larsen, R. D.; Verhoeven, T. R. WO 95/32197 (1995).
28. Chen, C. Y.; Larsen, R. D. WO 98/06725.
29. Rosso, V. W.; Lust, D. A.; Bernot, P. J.; Grosso, J. A.; Modi, S. P.; Rusowicz, A.; Sedergran, T. C.; Simpson, J. H.; Srivastava, S. K.; Humora, M. J.; Anderson, N. G. *Org. Process Res. Dev.* **1997,** *1*, 311–314.
30. Zhang, J. *Bischler–Möhlau Indole Synthesis* in *Name Reactions in Heterocyclic Chemistry- II*, Li, J. J., Ed., Wiley: Hoboken, NJ, **2011**, pp. 84–90.
31. Walkup, R.E.; Linder, J. *Tetrahedron Lett.* **1985,** *26*, 2155–2158.
32. Miller, C.P.; Collini, M. D.; Tran, B. D.; Harris, H. A.; Kharode, Y. P.; Marzolf, J. T.; Moran, R. A.; Henderson, R. A.; Bender, R. H. W.; Unwalla, R. J.; Greenberger,

L. M.; Yardley, J. P.; Abou-Gharbia, M. A.; Lyttle, C. R.; Komm, B. S. *J. Med. Chem.* **2001,** *44,* 1654–1657.
33. Miller, C. P.; Harris, H. A.; Komm, B. S. *Drug Fut.* **2002,** *27,* 117–119.
34. Henry, J. R.; Dodd, J. H. *Tetrahedron Lett.* **1998,** *48,* 8763–8764.
35. Zhang, X.; Sui, Z. *Tetrahedron Lett.* **2003,** *44,* 3071–3073.
36. Li, J.; Cook, J. M. *Nenitzescu Indole Synthesis*, in *Name Reactions in Heterocyclic Chemistry;* Li, J. J., Corey, E. J., Eds.; Wiley: Hoboken, NJ, **2005**, pp 145–153.
37. Pawlak, J. M.; Khau, V. V.; Hutchison, D. R.; Martinelli, M. J. *J. Org. Chem.* **1996,** *61,* 9055–9059.
38. Boehme, T. M.; Augelli-Szafran, C. E.; Hallak, H.; Pugsley, T.; Serpa, K.; Schwarz, R. D. *J. Med. Chem.* **2002,** *45,* 3094–3102.
39. Dalpozzo, R.; Bartoli, G. *Current Org. Chem.* **2005,** *9,* 163–178.
40. Li, J.; Cook, J. M. *Bartoli Indole Synthesis*, in *Name Reactions in Heterocyclic Chemistry*; Li, J. J.; Corey, E. J. Eds.; Wiley & Sons: Hoboken, NJ, **2005**, pp 100–103.
41. Bosco, M.; Dalpozzo, R.; Bartoli, G.; Palmieri, G.; Petrini, M. *J. Chem. Soc. Perkin Trans. 2* **1991,** 657–663.
42. Claffey, M. M.; Fliri, A. F. J.; Gallaschun, R. J.; O'Donnell, C. J. WO 2006/027684 (2006).
43. Clark, B. P.; Harris, J. R *Synth. Commun.* **1997,** *27,* 4223–4234.
44. Dobson, D. R.; Todd, A.; Gilmore, J. *Synth. Commun.* **1991,** *21,* 611–617.
45. Zegar, S.; Tokar, C.; Enache, L. A.; Rajagopol, V.; Zeller, W.; O'Connell, M.; Singh, J.; Muellner, F. W.; Zembower, D. E. *Org. Process Res. Dev.* **2007,** *11,* 747–753.
46. Li, J.; Cook, J. M. *Batcho–Leimgruber Indole Synthesis*. In *Name Reactions in Heterocyclic Chemistry*; Li, J. J.; Corey, E. J., Eds.; Wiley: Hoboken, NJ, **2005**, pp 104–109.
47. Lézé, M.-P.; Palusczak, A.; Hartmann, R. W.; Le Borgne, M. *Bioorg. Med. Chem. Lett.* **2008,** *18,* 4713–4715.
48. Boini, S.; Moder, K. P.; Vaid, R. K.; Kopach, M.; Kobierski, M. *Org. Process Res. Dev.* **2006,** *10,* 1205–1211.
49. Akao, A.; Nonoyama, N.; Mase, T.; Yasuda, N. *Org. Process Res. Dev.* **2006,** *10,* 1178–1183.
50. Li, J.; Cook, J. M. *Gassman Indole Synthesis*. In *Name Reactions in Heterocyclic Chemistry*; Li, J. J., Corey, E. J., Eds.; Wiley: Hoboken, NJ, **2005**, pp. 128–131.
51. Smith, A. B., III; Kingery-Wood, J.; Leenay, T. L.; Nolen, E. G.; Sunazuka, T. *J. Am. Chem. Soc.* **1992,** *114,* 1438–1459.
52. Ishikawa, H.; Uno, T.; Miyamoto, H.; Ueda, H.; Tamaoka, H.; Tominaga, M.; Nakagawa, K. *Chem. Pharm. Bull.* **1990,** *38,* 2459–2462.
53. Maligres, P. E.; Humphrey, G. R.; Marcoux, J.-F.; Hillier, M. C.; Zhao, D.; Krska, S.; Grabowski, E. J. J. *Org. Process Res. Dev.*, **2009,** *13,* 525–534.
54. Li, J. J. *Cadogan–Sundberg Indole Synthesis* in *Name Reactions in Heterocyclic Chemistry- II,* Li, J. J., Ed., Wiley: Hoboken, NJ, 2011, pp. 112–124.
55. Gribble, G. W. *Perkin 1* **2000,** 1045–1075.
56. Kuethe, J. T.; Wong, A.; Qu, C.; Smitrovich, J.; Davies, I. W.; Hughes, D. L. *J. Org. Chem.* **2005,** *70,* 2555–2567.
57. Li, B.; Pai, R.; Cardinale, S. C.; Butler, M. M.; Peet, N. P.; Moir, D. T.; Bavari, S.; Bowlin, T. L. *J. Med. Chem.* **2010,** *53,* 2264–2276.
58. Howard, H. R.; Lowe, J. A.; Seeger, T. F.; Seymour, P. A.; Zorn, S. H.; Maloney, P. R.; Ewing, F. E.; Newman, M. E.; Schmidt, A. W.; Furman, J. S.; Robinson, G. L.; Jackson, E.; Johnson, C.; Morrone, J. *J. Med. Chem.* **1996,** *39,* 143–148.

59. Yevich, J. P.; New, J. S.; Smith, D. W.; Lobeck, W. G.; Catt, J. D.; Minielli, J. L.; Eison, M. S.; Taylor, D. P.; Riblet, L. A.; Temple, D. L. *J. Med. Chem.* **1986,** *29*, 359–369.
60. Walinsky, S. W.; Fox, D. E.; Lambert, J. F.; Sinay, T. G. *Org. Process Res. Dev.* **1999,** *3*, 126–130.
61. Sun, L.; Liang, C.; Shirazian, S.; Zhou, Y.; Miller, T.; Cui, J.; Fukuda, J. Y.; Chu, J.-Y.; Nematalla, A.; Wang, X.; Chen, H.; Sistla, A.; Luu, T. C.; Tang, F.; Wei, J.; Tang, C. *J. Med. Chem.* **2003,** *46*, 1116–1119.
62. Vaidyanathan, R. in *Process Chemistry in the Pharmaceutical Industry*, Gadamasetti, K.; Braish, T., Eds.; CRC Press: Boca Raton, FL, **2008**, pp. 49–63.
63. Gallagher, G. Jr.; Lavanchy, P. G.; Wilson, J. W.; Hieble, J. P.; DeMarinis, R. M. *J. Med. Chem.* **1985,** *28*, 1533–1536.
64. DeMarinis, R. M.; Gallagher, G. Jr.; Hall, R. F.; Franz, R. G.; Webster, C.; Huffman, W. F.; Schwartz, M. S.; Ross, S. T.; Wilson, J. W.; Hieble, J. P. *J. Med. Chem.* **1986,** *29*, 939–947.
65. Li. J. J. *Chapter 6. Metal-catalyzed Cross-coupling Reactions for Indoles*, in *Indoles*, Gribble, G. W., Ed.; Springer: Helderberg, Germany, 2010, pp. 193–234.
66. Harris, W.; Hill, C. H.; Keech, E.; Malsher, P. *Tetrahedron Lett.* **1993,** *34*, 8361–8364.
67. Yokoyama, Y.; Matsumoto, T.; Murakami, Y. *J. Org. Chem.* **1995,** *60*, 1486–1487.
68. Vincent, P.; Beaucourt, J. P.; Pichart, L. *Tetrahedron Lett.* **1984,** *25*, 201–202.
69. Fisher, L. E.; Labadie, S. S.; Reuter, D. C.; Clark, R. D. *J. Org. Chem.* **1995,** *60*, 6224–6225.
70. Faul, M. M.; Winneroski, L. L. *Tetrahedron Lett.* **1997,** *38*, 4749–4752.
71. Mothes, C.; Lavielle, S.; Karoyan, P. *J. Org. Chem.* **2008,** *73*, 6706–6710.
72. Balle, T.; Andersen, K.; Vedsø, P. *Synthesis* **2002,** 1509–1512.
73. Liu, Y.; Gribble, G. W. *Tetrahedron Lett.* **2000,** *41*, 8717–8721.
74. Watson, T. J. N.; Horgan, S. W.; Shah, R. S.; Farr, R. A.; Schnettler, R. A.; Neville, C. R.; *etc. Org. Process Res. Dev.* **2000,** *4*, 477–487.
75. Daïri, K.; Yao, Y.; Faley, M.; Tripathy, S.; Rioux, E.; Billot, X.; Rabouin, D.; Gonzalez, G.; Lavallee, J.-F.; Attardo, G. *Org. Process Res. Dev.* **2007,** *11*, 1051–1054.
76. (a) Sonogashira K.; Tohda, Y.; Hagihara, N. *Tetrahedron Lett.* **1975,** 4467–4470. (b) Rossi, R.; Carpita, A.; Bellina, F. *Org. Prep. Proc. Int.* **1995,** *27*, 129–160. (c) Campbell, I. B. in *Organocopper Reagents*, **1994,** Taylor, R. J. K., Ed.: IRL Press, Oxford, UK, 217–235. (d) Sonogashira, K. *J. Organomet. Chem.* **2002,** *653*, 46–49.
77. Xiong, X.; Pirrung, M. *Org. Lett.* **2008,** *10*, 1151–1154.
78. Palimkar, S. S.; More, V. S.; Kumar, P. H.; Srinivasan, K. V. *Tetrahedron* **2007,** *63*, 12786–12790.
79. Naratriptan: Blatcher, P.; Carter, M.; Hornby, R.; Owen, M. R. WO 95/09166 (1995).
80. Perkins, J. F. EP 1088817 (2001).
81. Ogilvie, R. J. WO 02/50063 (2202).
82. Raboisson, P.; Desjarlais, R. L; Reed, R.; Lattanze, J.; Chaikin, M.; Manthey, C. L; Tomczuk, B. E; Marugan J. J. *Eur. J. Med. Chem.* **2007,** *42*, 334–343.
83. Routier, S.; Merour, J.-Y.; Dias, N.; Lansiaux, A.; Bailly, C.; Lozach, O.; Meijer, L. *J. Med. Chem.* **2006,** *49*, 789–799.
84. Jeanty, M.; Blu, J.; Suzenet, F.; Guillaumet, G. *Org. Lett.* **2009,** *11*, 5142–5145.
85. Porter, J.; Lumb, S.; Franklin, R. J.; Gascon-Simorte, J. M.; Calmiano, M.; Le Riche, K.; Lallemand, B.; Keyaerts, J.; Edwards, H.; Maloney, .; Delgado, J.; King,

L.; Foley, A.; Lecomte, F.; Reuberson, J.; Meier, C.; Batchelor, M. *Bioorg. Med. Chem. Lett.* **2009**, *19*, 2780–2784.

86. Ahmed, M.; Briggs, M. A.; Bromidge, S. M.; Buck, T.; Campbell, L.; Deeks, N. J.; Garner, A.; Gordon, L.; Hamprecht, D. W.; Holland, V.; Johnson, C. N.; Medhurst, A. D.; Mitchell, D. J.; Moss, S. F.; Powles, J.; Seal, J. T.; Stean, T. O.; Stemp, G.; Thompson, M.; Trail, B.; Upton, N.; Winborn, K.; Witty, D. R. *Bioorg. Med. Chem. Lett.* **2005,** *15*, 4867–4871.
87. Giblin, G. M. P.; Billinton, A.; Briggs, M.; Brown, A. J.; Chessell, I. P.; Clayton, N. M.; Eatherton, A. J.; Goldsmith, P.; Haslam, C.; Johnson, M. R.; Mitchell, W. L.; Naylor, A.; Perboni, A.; Slingsby, B. P.; Wilson, A. W. *J. Med. Chem.* **2009,** *52*, 5785–5788.
88. Wang, T.; Yin, Z.; Zhang, Z.; Bender, J. A.; Yang, Z.; Johnson, G.; Yang, Z.; Zadjura, L. M.; D'Arienzo, C. J.; DiGiugno Parker, D.; Gesenberg, C.; Yamanaka, G. A.; Gong, Y.-F.; Ho, H.-T.; Fang, H.; Zhou, N.; McAuliffe, B. V.; Eggers, B. J.; Fan, L.; Nowicka-Sans, B.; Dicker, I. B.; Gao, Q.; Colonno, R. J.; Lin, P.-F.; Meanwell, N. A.; Kadow, J. F. *J. Med. Chem.* **2009,** *52*, 7778–7787.
89. Plewe, M. B.; Butler, S. L.; R. Dress, K.; Hu, Q.; Johnson, T. W.; Kuehler, J. E.; Kuki, A.; Lam, H.; Liu, W.; Nowlin, D.; *et al. J. Med. Chem.* **2009,** *52*, 7211–7219.
90. Fisher, M. H.; Schwartzkopf, G., Jr.; Hoff, D. R. *J. Med. Chem.* **1972,** *15*, 1168–1171.
91. Skordos, K.; Skiles, G. L.; Laycock, J. D.; Lanza, D. L.; Yost, G. S. *Chem. Res. Toxicol.* **1998,** *11*, 741–749.
92. Regal, K. A.; Laws, G. M.; Yuan C, Yost, G. S.; Skiles, G. L. *Chem. Res. Toxicol.* **2001,** *14*, 1014–1024.
93. Kassahun, K.; Skordos, K.; McIntosh, I.; Slaughter, D.; Doss, G. A.; Baillie, T. A.; Yost, G. S. *Chem. Res. Toxicol.* **2005,** *18*, 1427.
94. Blagg, J. *Structural Alerts* in *Burger's Medicinal Chemistry, Drug Discovery and Development, 7th ed.*, Abraham, D. J.; Rotella, D. P., eds.; Wiley: Hoboken, NJ, **2010**, Vol. 3, pp. 301–334.
95. McKee, J. R.; Zanger, M. *J. Chem. Educ.* **1991,** *68,* A242.
96. Baran, P. S.; Richter, J. M. *J. Am. Chem. Soc.* **2004,** *126*, 7450–7451.
97. Baran, P. S.; Richter, J. M.; Lin, D. W. *Angew. Chem., Int. Ed.* **2005,** *44*, 609–612.
98. Snyder, H. R.; Matteson, Donald S. *J. Am. Chem. Soc.* **1957,** *79*, 2217–2221.
99. Hamel, P.; Preville, P. *J. Org. Chem.* **1996,** *61*, 1573–1577.
100. Hamel, P. *J. Org. Chem.* **2002,** *67*, 2854–2858.
101. Çelebi-Ölçüm, N.; Boal, B. W.; Huters, A. D.; Garg, N. K.; Houk, K. N. *J. Am. Chem. Soc.* **2011,** *79*, 2217–2221.
102. Carlin, R. B.; Wallace, J. G.; Fisher, E. E. *J. Am. Chem. Soc.* **1952,** *74,* 990–994.
103. Carlin, R. B.; Harrison, J. W. *J. Org. Chem.* **1965,** *30,* 563–566.
104. Dolby, L. J.; Furukawa, S. *J. Org. Chem.* **1963,** *28*, 2512–2514.

Chapter 4 Furans, Benzofurans, Thiophenes, and Benzothiophenes

Joshua V. Ruppel, Nicole L. Snyder, Alexander D. Thompson, and Tyler W. Farnsworth

4.1 Introduction

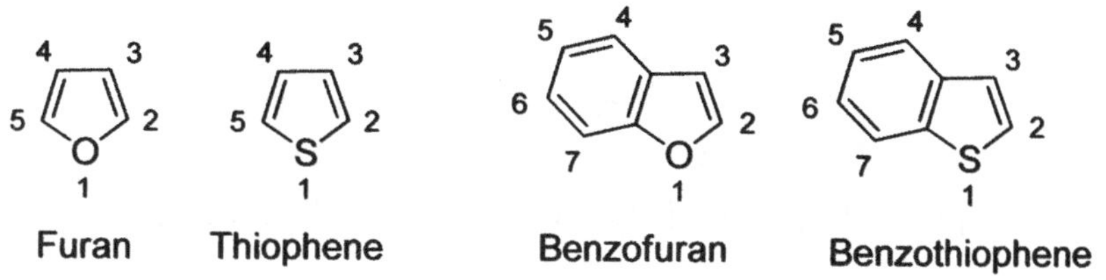

Furan is a colorless, water-insoluble liquid with a boiling point of 31 °C and is described as having a pleasant odor. Polymerization of furan occurs readily at room temperature but can be stabilized by the addition of hydroquinone or phenol. Thiophene, the sulphur congener of furan, is a stable, colorless liquid with a boiling point of 84 °C and an odor similar to benzene.

Benzofuran is a colorless oil with a boiling point of 173 °C. It occurs naturally in coal tar and is believed to form during carbonization by cyclodehydration of 2-ethylphenol. In contrast, benzothiophene is a white solid with a melting point of 32 °C. It is considered an aquatic toxin with acute and long-lasting effects.

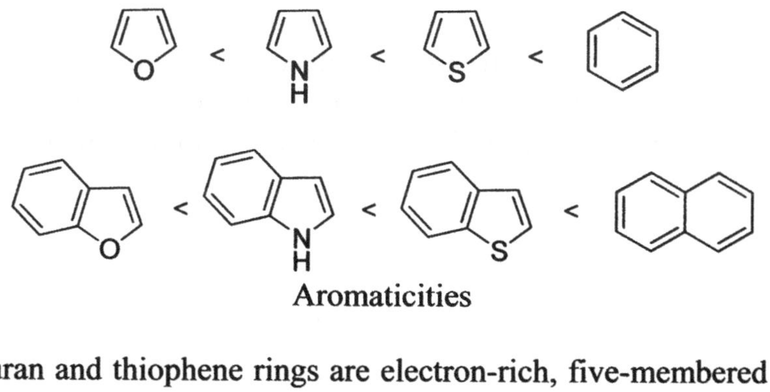

Aromaticities

Furan and thiophene rings are electron-rich, five-membered aromatic systems. The donation of a pair of electrons by the heteroatom into the π-system is essential to the aromaticity of these compounds. Due to the highly electronegative oxygen heteroatom, the aromaticity of furan is comparable to pyrrole.[1] The aromaticity of benzofuran is comparable to indole for similar reasons. On the other hand, the aromaticity of thiophene resembles benzene due to the decreased electronegativity of the sulfur heteroatom. For similar reasons, the aromaticity of benzothiophene is comparable to naphthalene.[1]

Furan has a dipole moment of 0.70 D, while thiophene has a dipole moment of 0.51 D. The dipole moments of furan and thiophene are in the opposite direction of pyrrole due largely to the relatively strong inductive effect caused by the oxygen and sulfur in relation to weaker resonance effects. In the case of pyrrole, as described in Chapter 2, the resonance hybrids of the molecule result in the inversion of the dipole.

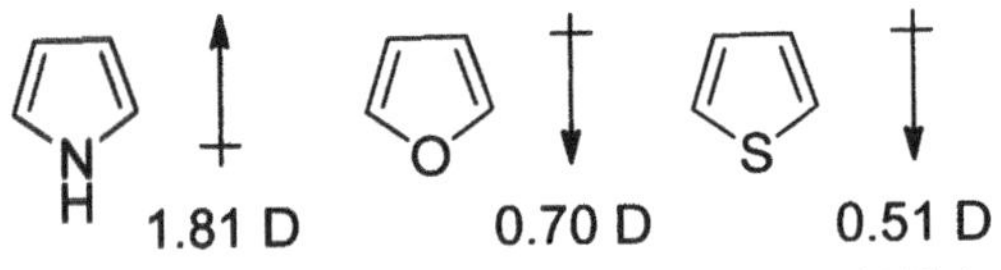

Dipole Moments of Pyrrole, Furan, and Thiophene

Furan is a planar aromatic compound with bond angles and lengths that are affected by the atomic radius of the oxygen atom. This is evidenced by observing the length of the C–O bonds and the slightly exaggerated bond angle.

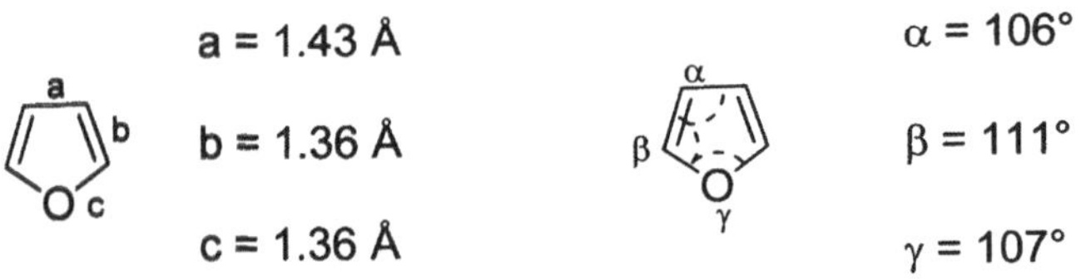

Bond Lengths and Angles of Furan

The ^{1}H-NMR of furan has two distinct peaks corresponding to the two separate sets of nonequivalent hydrogens in the ring system. The two α-protons (adjacent to the oxygen) are deshielded by the inductive effect of the oxygen atom and can be found at 7.54 ppm. The two β-protons are further upfield (effectively less deshielded than their α-proton neighbors) at 6.49 ppm.

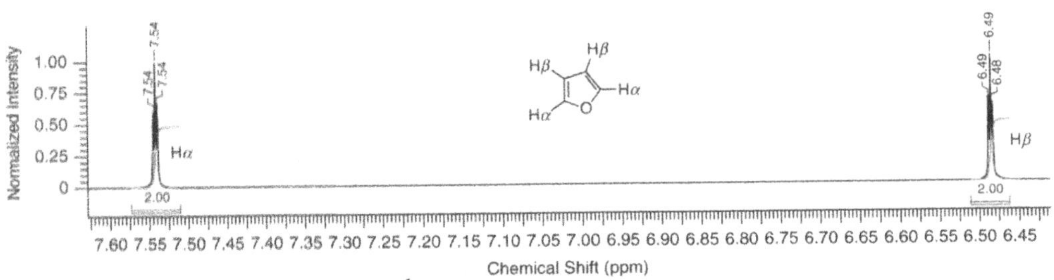

^{1}H-NMR of Furan

Predictably, the ^{13}C-NMR of furan shows two signals. As expected, the α-carbons are located further downfield (142.5 ppm) due to the inductive effects of the oxygen, while the β-carbons (109.4 ppm) are located further upfield due to their distance from the oxygen atom.

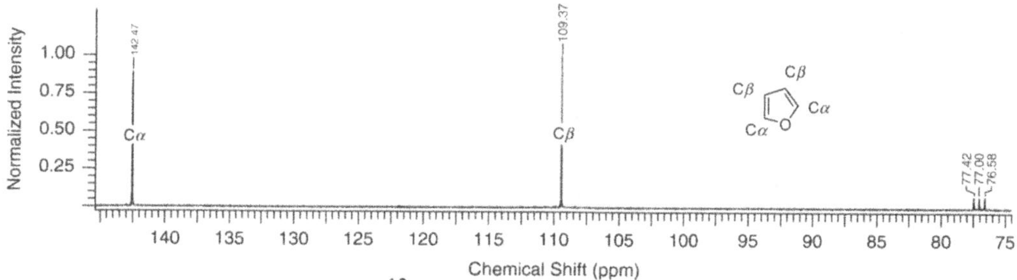

^{13}C-NMR of Furan

The ^{1}H-NMR spectra of benzofuran is more complex than furan due to the fused aromatic ring and six nonequivalent protons. *H2, H4,* and *H7*, which are deshielded by the oxygen in the furan moiety, are located downfield between 7.75 ppm and 7.63 ppm with *H2* being the most deshielded due to its proximity to the oxygen atom. Correspondingly, *H3, H5* and *H6*, are less shielded, with *H3* being the least shielded (6.85 ppm) based on the electron delocalization of the heteroatom. Although the effect is not as pronounced, *H6* is also slightly downfield of *H5* for the same reason.

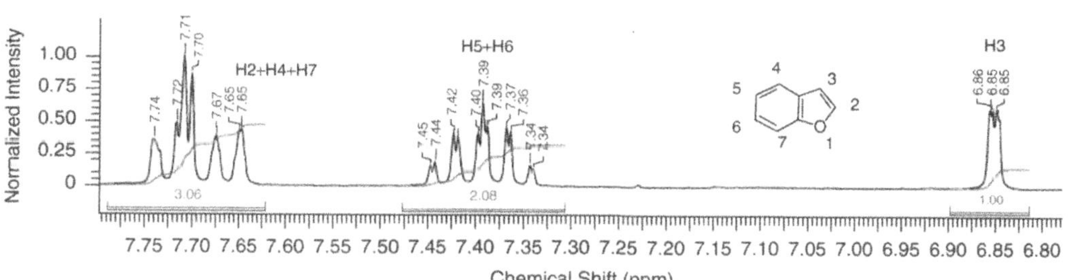

^{1}H-NMR of Benzofuran

The ^{13}C-NMR spectrum of benzofuran presents eight distinct signals. The quaternary carbons *C7a* and *C3a* are found downfield at 155.1 ppm and 127.5 ppm, respectively. Of the remaining carbon signals *C2* (144.9 ppm) is located the furthest downfield due to proximity of oxygen atom followed by *C6, C5, C4, C7*, and *C4*.

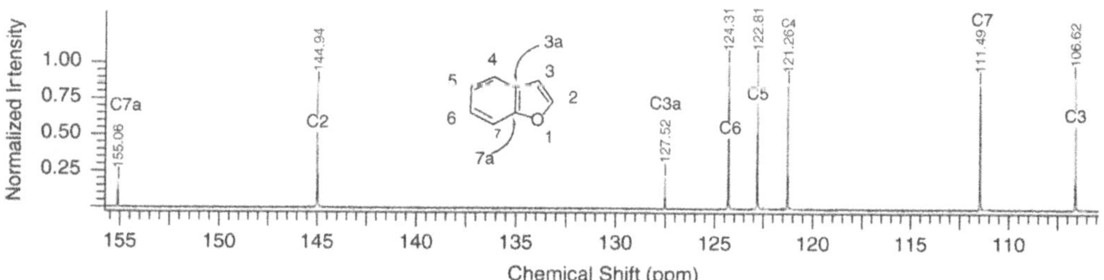

^{13}C-NMR of Benzofuran

Thiophene, a planar aromatic compound, has bond angles and lengths that are affected even more so than furan by the large atomic radius of the sulfur atom. This is evidenced by observing the length of the C–S bonds and the exaggerated bond angle.

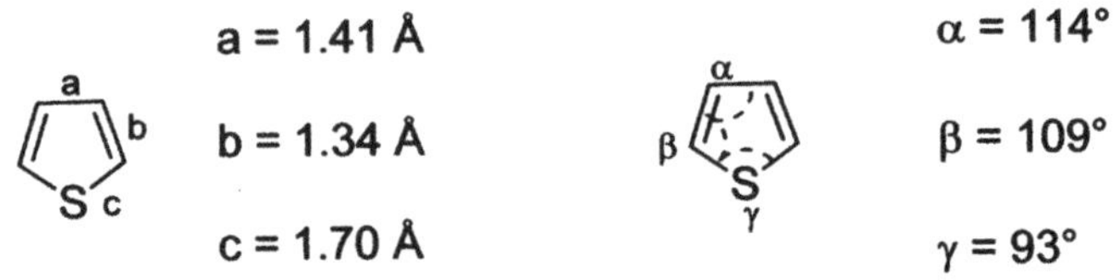

Bond Lengths and Angles of Thiophene

The ^{1}H-NMR of thiophene has two distinct peaks corresponding to the two separate sets of nonequivalent hydrogen atoms in the ring system. The two α-protons (adjacent to the sulfur) are deshielded by the inductive effect of the sulfur atom and can be found at 7.39 ppm. The two β-protons are further upfield (effectively less deshielded than their α-protons neighbors) at 7.17 ppm.

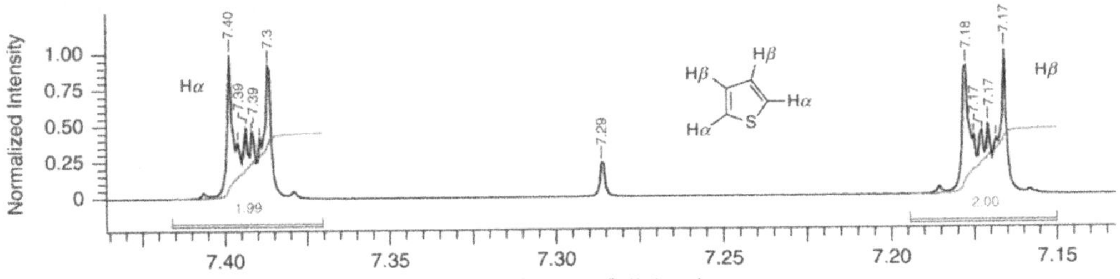

^{1}H-NMR of Thiophene

The ^{13}C-NMR of thiophene provides a predictable spectrum with the α-carbons found further downfield (126.9 ppm) than the β-carbons (125.1 ppm) in the aromatic region; again, these patterns are largely the result of the inductive effect of the sulfur.

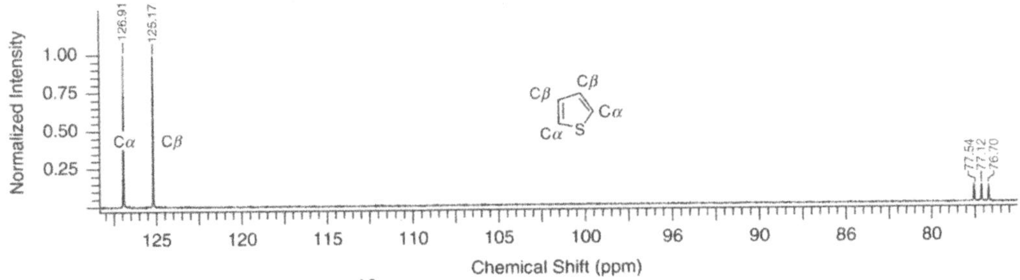

^{13}C-NMR of Thiophene

The ^{1}H-NMR spectra of benzothiophene is much more complex than thiophene due to the fused aromatic ring and six nonequivalent protons. Protons *H7* and *H4* are located farthest downfield at 8.04 ppm and 7.98 ppm because they are deshielded by the adjacent thiophene moiety. As expected *H2* is deshielded and is detected further downfield of *H3* due to its proximity to the sulfur. Although the effect is not as pronounced, *H6* (7.53 ppm) is also slightly downfield of *H5* (7.50 ppm) for the same reason.

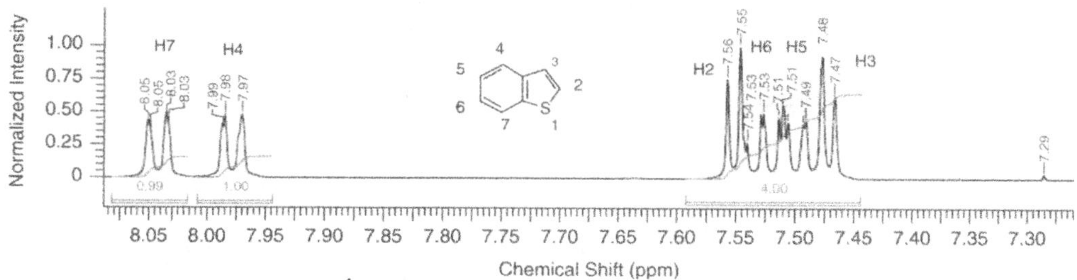

^{1}H-NMR of Benzothiophene

The ^{13}C-NMR spectrum of benzothiophene presents eight distinct signals. The quaternary carbons *C7a* and *C3a* are found downfield at 139.8 ppm and 139.7 ppm, respectively. Of the remaining carbon signals, *C2* (126.3 ppm) is downfield due to the proximity of the sulfur atom followed by *C6*, *C5*, *C3*, *C4*, and *C7*.

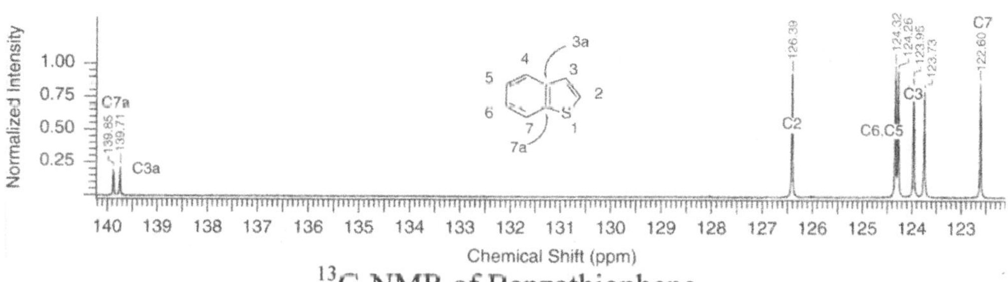

^{13}C-NMR of Benzothiophene

Furan and benzofuran ring systems are found in a variety of natural products and pharmaceutical compounds. One of the most successful furan-based pharmaceutical compounds is ranitidine (Zantac), an important H_2-histamine receptor antagonist for the treatment of acid reflux disease. In 1986, just three years after receiving FDA approval, sales of Zantac reached a record $1 billion, making it the largest selling prescription drug in the world at the time. In addition, sales from furan-based antimicrobials such as nitrofurazone (Furin), furazolidone (Furoxone), nitrofurantion (Niftran), cefuroxine (Ceftin), and their generic derivatives, generate over $50 million in revenue per annum in the United States combined. More recently, Lapatinib (Tyverb) has become an increasingly important therapeutic for the treatment of breast cancer and other solid tumors, earning nearly $1 billion since its approval by the FDA in 2007. Other important furan-based therapeutics include prazosin (Minipress) and furosemide (Lasix), which are used to treat hypertension and hyperplasia, and dantrolene (Dantrium), an effective skeletal muscle relaxant.

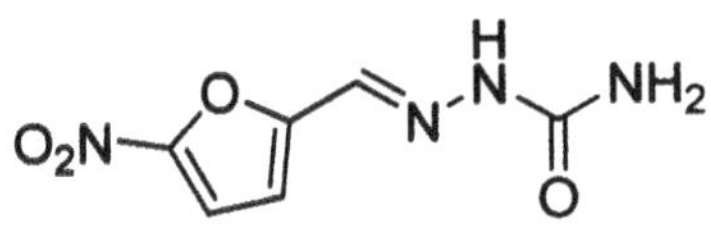

nitrofurazone (Furin)

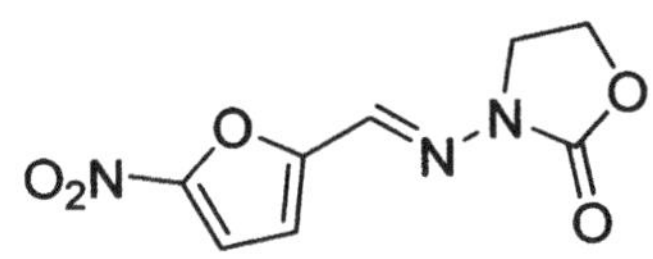

furazolidone (Furoxone)

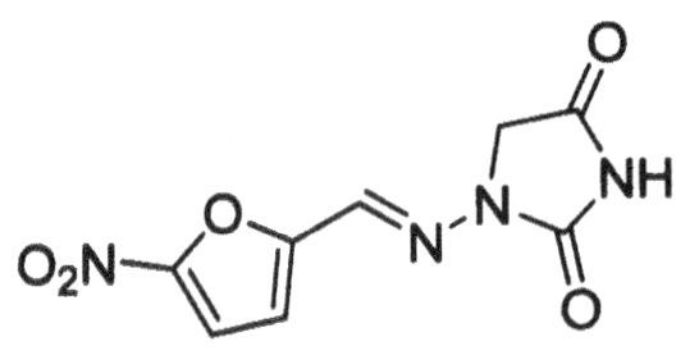

nitrofurantoin (Niftran)

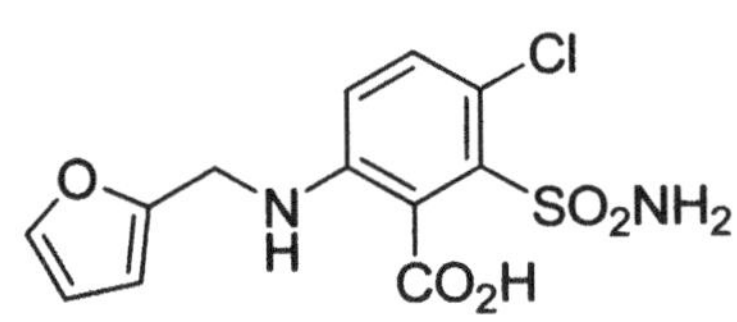

furosemide (Lasix)

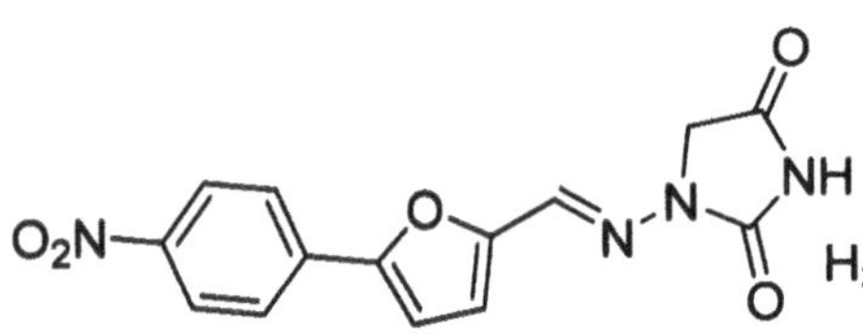

dantrolene (Dantrium)

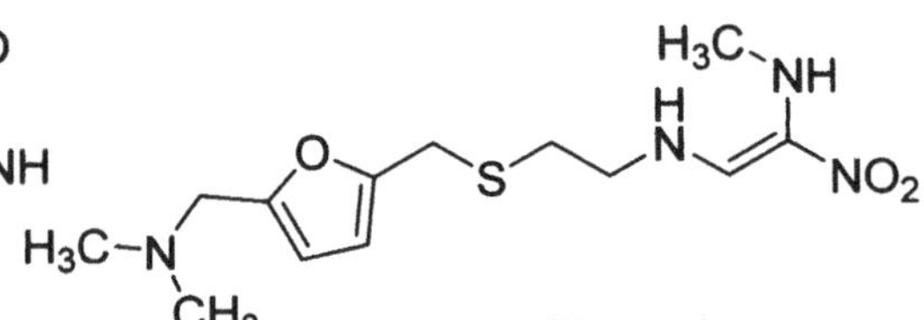

ranitidine (Zantac)

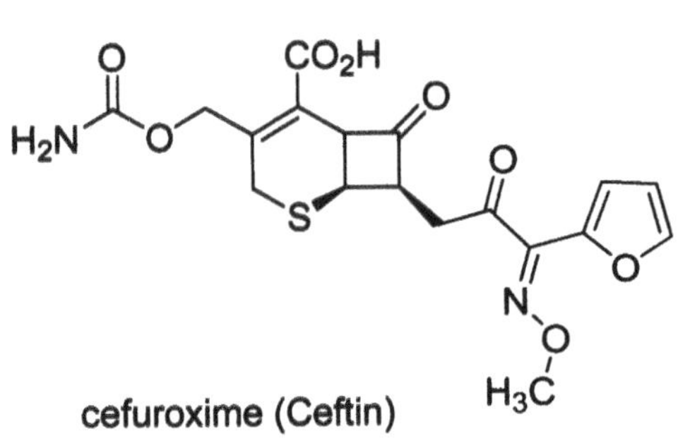

cefuroxime (Ceftin)

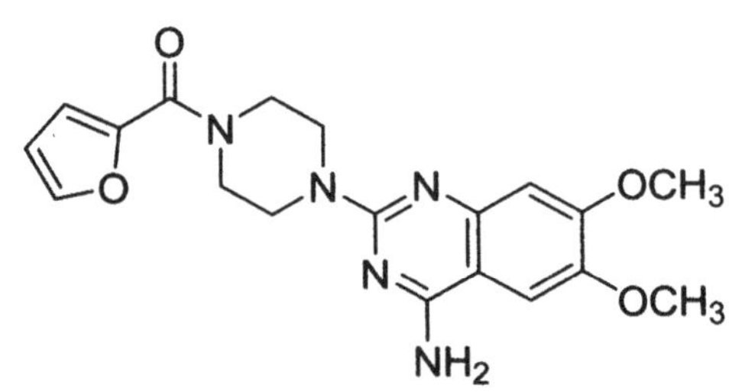

prazosin (Minipress)

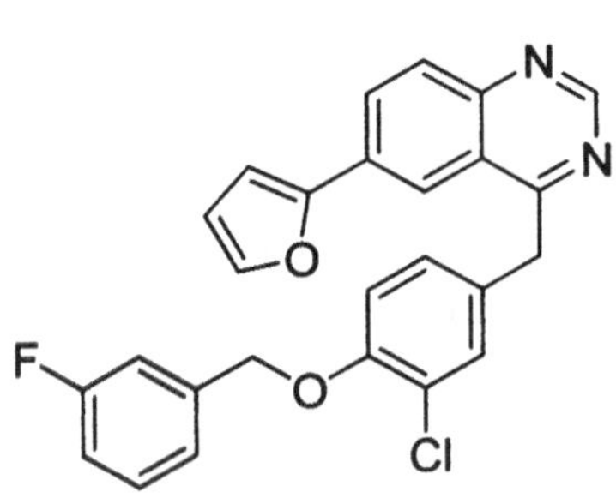

lapatinib (Tyverb)

Benzofuran rings are also important functional moieties in commercial pharmaceuticals. Saprisartan is a useful compound for the treatment of hypertension and heart failure, while amiodarone (Cordarone) is used as an antiarrhythmic agent for the treatment of tachyarrhythmias. Finally, methoxsalen (Oxsoralen) and its derivatives are important compounds used to treat psoriasis, eczema, and certain skin cancers.

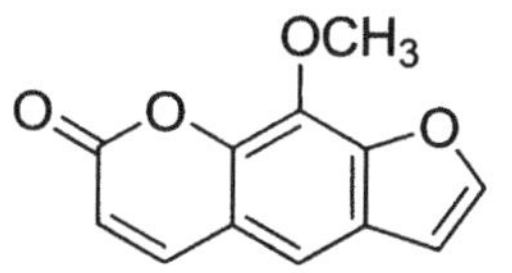

methoxsalen (Oxsoralen)

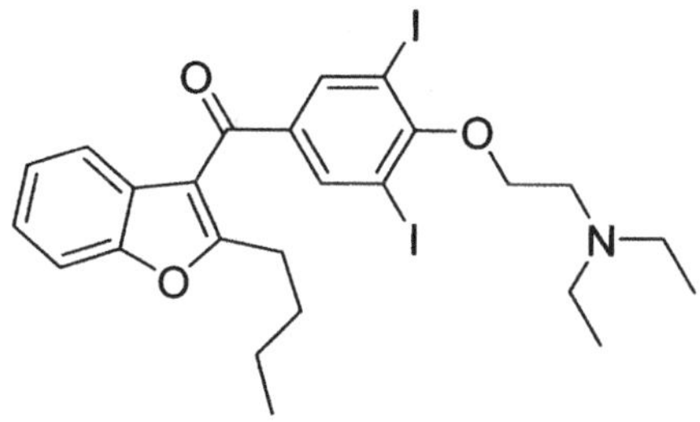

amiodarone (Cordarone)

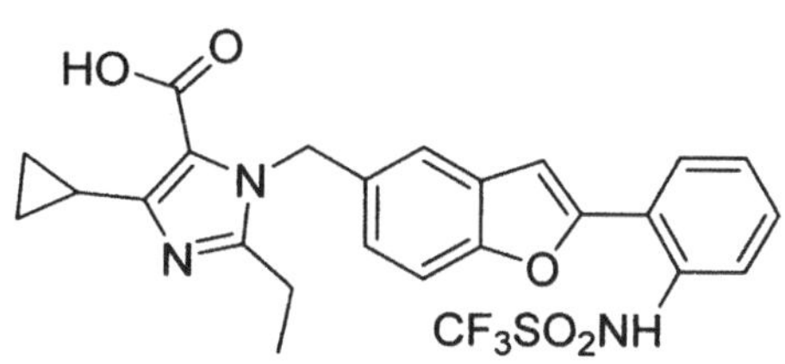

saprisartan

The electron-rich thiophene and benzothiophenes are widely used as isosteres of their phenyl cousins in medicinal chemistry. These ring systems can be found in many pharmaceuticals with varied therapeutic applications such as the inhibition of platelet aggregation, treatment of asthma, chronic obstructive pulmonary disorder (COPD), bipolar disorder, psychosis, and prevention of osteoporosis, among many others.[2]

The importance of thiophenes in medicinal chemistry can be highlighted by the enormous economic success of thiophene-containing drugs. Pharmaceuticals containing thiophenes such as duloxetine (Cymbalta), olanzapine (Zyprexa), tiotropium (Spiriva), raloxifene (Evista), and zileuton (Zyflo) had combined sales of more than $6 billion in the U.S. in 2010, while the blockbuster drug clopidogrel (Plavix) alone obtained sales in excess of $6 billion in the same year.

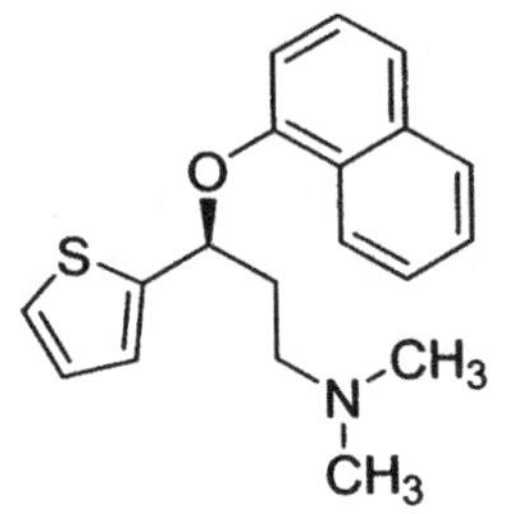

duloxetine (Cymbalta)

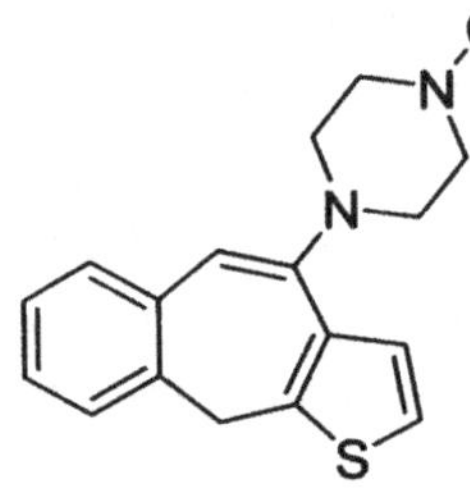

olanzapine (Zyprexa)

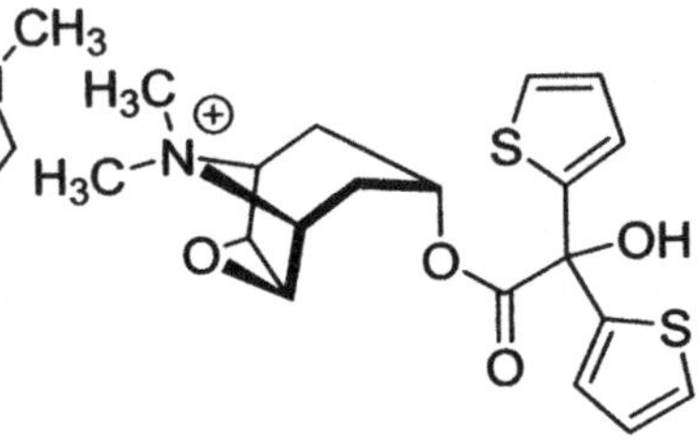

tiotropium (Spiriva)

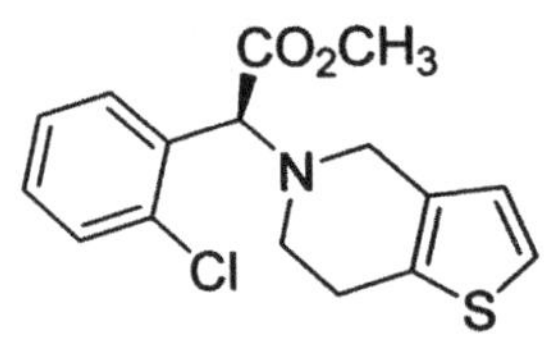

clopidogrel (Plavix)

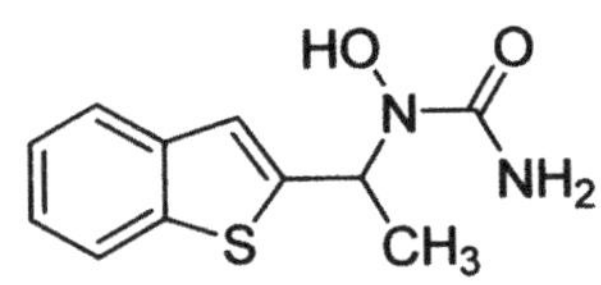

zileuton (Zyflo)

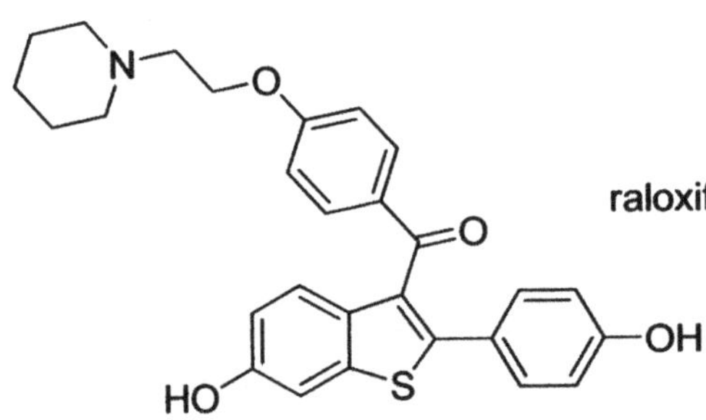

raloxifene (Evista)

4.2 Furans and Benzofuran

4.2.1 Reactions of Furans and Benzofurans

4.2.1.1 Reactions at C2

There are few general methods for the functionalization of existing furan rings due to their low stability under acidic and aerobic conditions. For this reason, most functionalized furan derivatives are prepared by the cyclization of acyclic precursors using acid, base, or transition metal catalysis.

Despite the sensitivity of furans, several methods have been developed for the direct functionalization of these ring systems. The general reaction is shown below with *C2* substitution preferred over *C3*. This is due to in part to the greater HOMO coefficient of the *C2*-carbon atoms as well as greater delocalization of the positively charged intermediate. Substitution at the *C3* position can occur when the *C2* position is occupied. Several examples, including halogenation, nitration, metal–halogen exchange,

Friedel–Crafts reactions, Wittig rearrangements, Michael additions, and intramolecular cyclizations are highlighted below.

Preferred *C2*-electrophilic substitution:

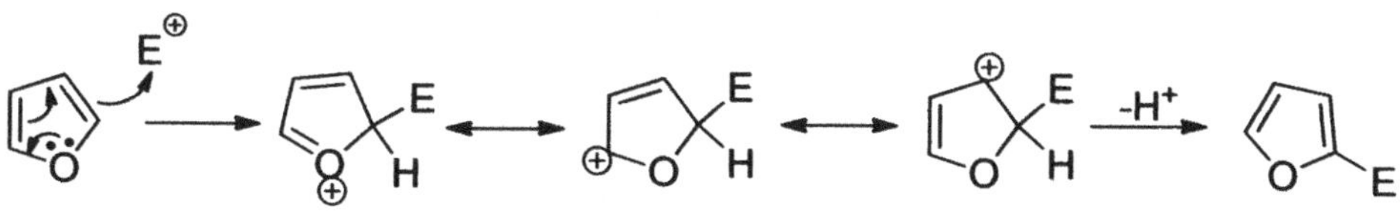

Less-favored *C3*-electrophilic substitution:

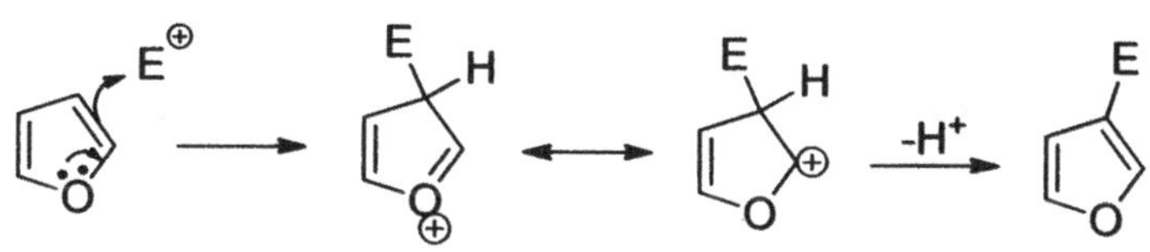

Halogenation (Bromination)

Monobromination of furan to produce 2-bromofuran can be accomplished by addition of one equivalent of bromine in DMF in 70% yield.[3] Two equivalents of bromine under the same reaction conditions provides the disubstituted product, 2,5-dibromofuran, in 48% yield.

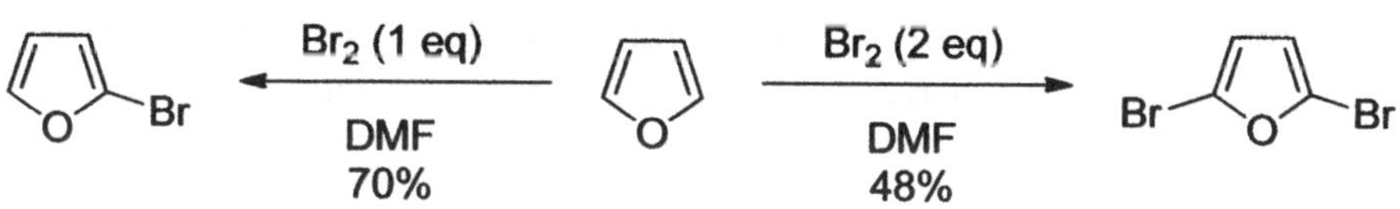

Halogenated furans are important intermediates for the synthesis of a number of functionalized furan derivatives through metal–halogen exchange and transition metal-catalyzed cross-coupling reactions, as will be demonstrated throughout this chapter.

Nitration

Nitration of furan rings is accomplished by treatment of furan with concentrated nitric acid and sulfuric acid. This protocol was employed by Fathali and co-workers in their concise synthesis of the antibacterial nitrofurazone.[4] Treatment of 2-furaldehyde with nitric acid, followed by treatment with acetic anhydride, gave the corresponding 5-nitrofurfuraldiacetate. Reaction of 5-nitrofurfuraldiacetate with semicarbazide hydrochloride gave the desired product, nitrofurazone, in 44% over three steps.

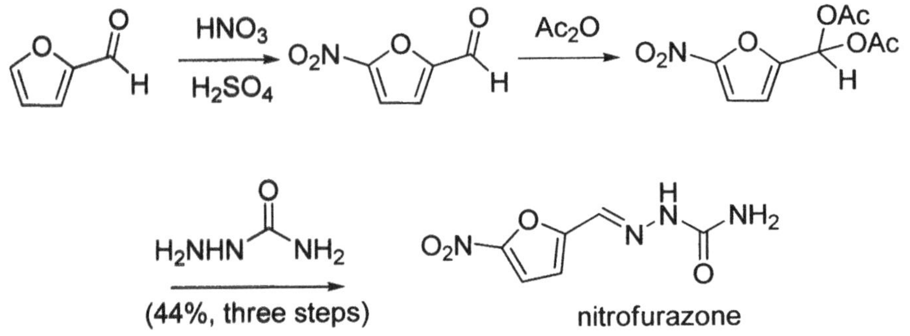

Metal–Halogen Exchange

Metal–halogen exchange has become an effective method for the formation of carbon–carbon bonds between furans and other functional moieties. In particular, furylmagnesium compounds, which can be generated under mild conditions and are tolerant of a number of functional groups including esters, nitriles, or amides, have been especially important in the synthesis of highly substituted furans. Recently, Knochel and co-workers[5] performed a magnesium–halogen exchange with a 2-bromofuran derivative bearing an electron-withdrawing ethyl ester at *C5* to produce the corresponding furylmagnesium derivative. Subsequent addition of an iminium salt produced the corresponding 2,5-disubstituted furan bearing a propargyl amine at *C2* in 65% yield over two steps. This protocol was shown to be effective and selective with other halogenated furans.

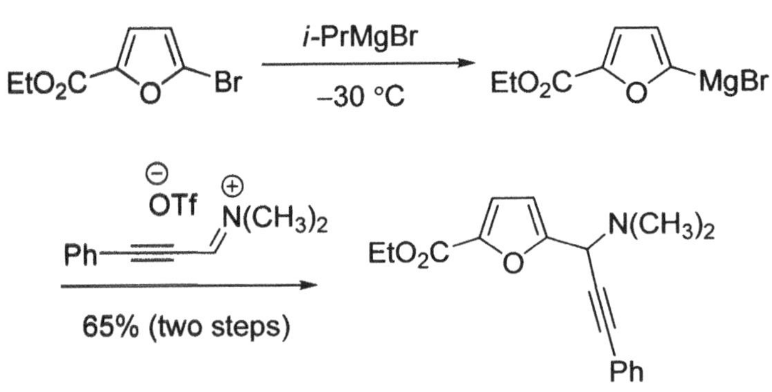

Friedel–Crafts Reactions

The Friedel–Crafts alkylation and acylation reactions have become important methods for the synthesis of substituted furans. The most useful reactions employ chiral catalysts to generate products in good yield and with high diastereoselectivities and enantioselectivities. In turn, these products serve as important intermediates in the synthesis of furan-containing therapeutics and natural products.

Jurczak and co-workers prepared chiral furfuryl alcohols via the Friedel–Crafts reaction.[6] Treatment of 2-methylfuran with alkyl glyoxylates in the presence of a salen-cobalt(II) complex under high pressure provided the desired product in 47% yield with modest enantioselectivity.

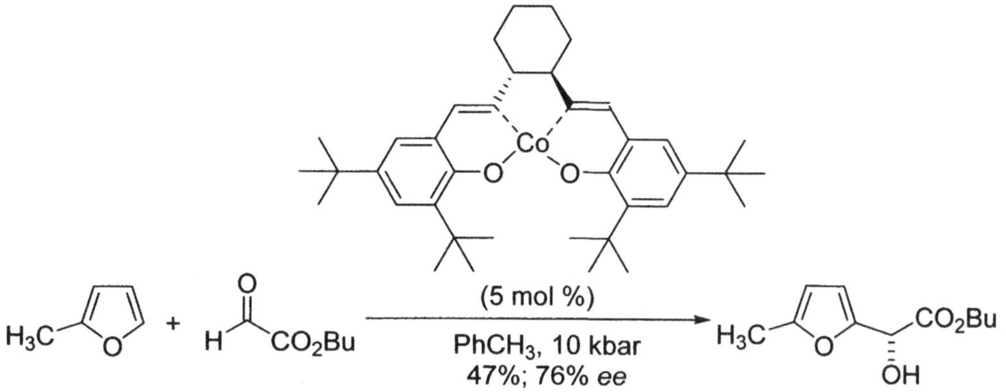

Friedel–Crafts reactions have also been employed in the synthesis of benzofurans. Ohishi and co-workers used a Friedel–Crafts acylation protocol to produce a highly substituted benzofuran with inhibitory activity for leukotriene B-4.[7] Reaction of 3-(4-chlorophenyl)benzofuran with 2-chloroacetyl chloride provided the corresponding benzofuran derivative in good yield with a handle for further functionalization. A Hantzsch thiazole synthesis was then used to produce the final amino-thiazole inhibitor.

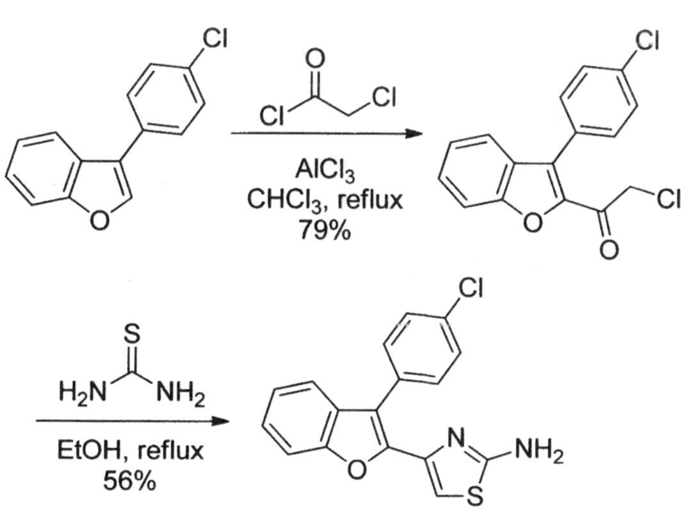

Wittig Rearrangement

The Wittig rearrangement has been employed as a mild method for the synthesis of 2,3-substituted furans. Treatment of 3-furylmethyl ether with LDA in THF gave the corresponding 2,3-furan in 60% yield as a racemic mixture.[8]

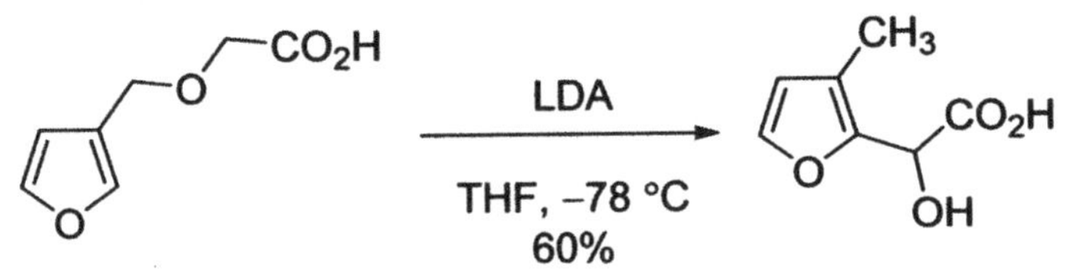

Michael Addition

Substituted furans have also been prepared by Michael addition of furans to α,β-unsaturated enones. Harada and co-workers used this approach to produce optically active furans in high yield and with high enantioselectivity from 2,3-dimethylfuran and hex-4-en-3-one using a chiral oxazaborolindinone.[9]

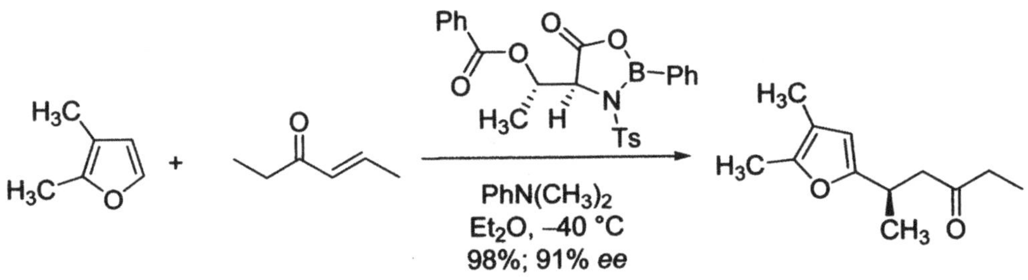

Intramolecular Cyclization

Mukai and co-workers used a reduction/cyclization cascade to prepare (–)-nakadomarin, a manzamine-related alkaloid with cytotoxic activity, inhibitor activity against cyclin-dependent kinase 4 (CDK-4), and anti-fungal and antibacterial activity.[10] Treatment of the tetracyclic precursor with DIBAL-H followed by reaction with HCl led to the desired pentacyclic system in 41% yield.

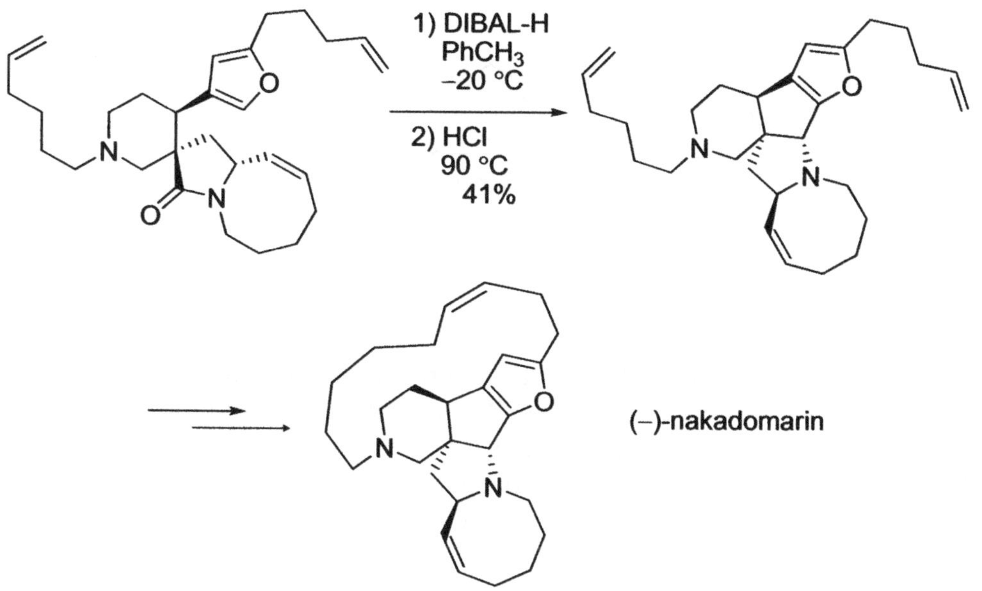

4.2.1.2 Reactions at C3

Electrophilic substitution reactions of furans take place predominately at the *C2* (*C5*) position. In many cases reaction at the *C3* position is only favored when the *C2* and *C5* positions are already occupied or severe sterics are involved.

Halogenation (Iodination)

Chen and Lu used a mercuric acetate-mediated annulation of homopropargylic alcohols to generate iodinated furans for use in cross-coupling reactions.[11] Treatment of 2-(3-methylbut-1-yn-1-yl)-2-phenyl-1,3-dithiolane with *n*BuLi, followed by reaction with *n*-butylaldehyde, produced a mercuric acetate intermediate that was subsequently trapped with iodine to give the corresponding 3-iodofuan. Further cross-coupling with a series of compounds (in this example naphthalene) led to the desired tetrasubstituted furan in 87% yield over three steps.

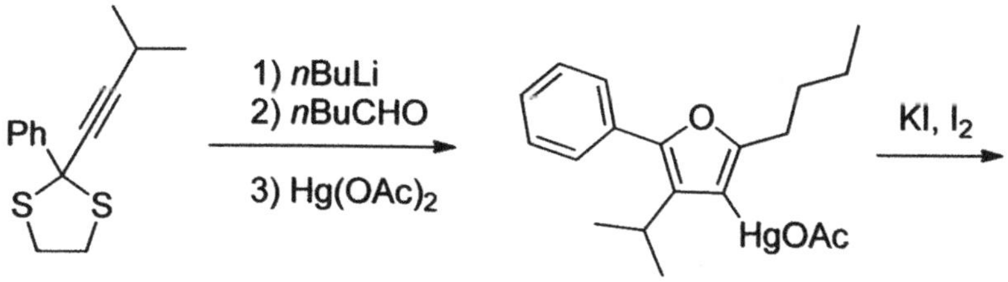

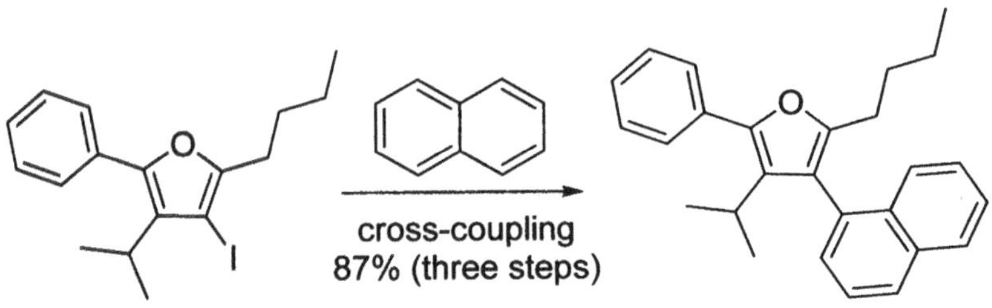

Metal–Halogen Exchange

Knochel and co-workers used a magnesium–halogen exchange reaction with TMPMgCl·LiCl to generate 3-arylsulfinylfurans.[12] The authors went on to illustrate the utility of this method by demonstrating that an iterative magnesium–halogen exchange approach could be used to functionalize all four positions on the furan ring in a sequential and selective fashion.

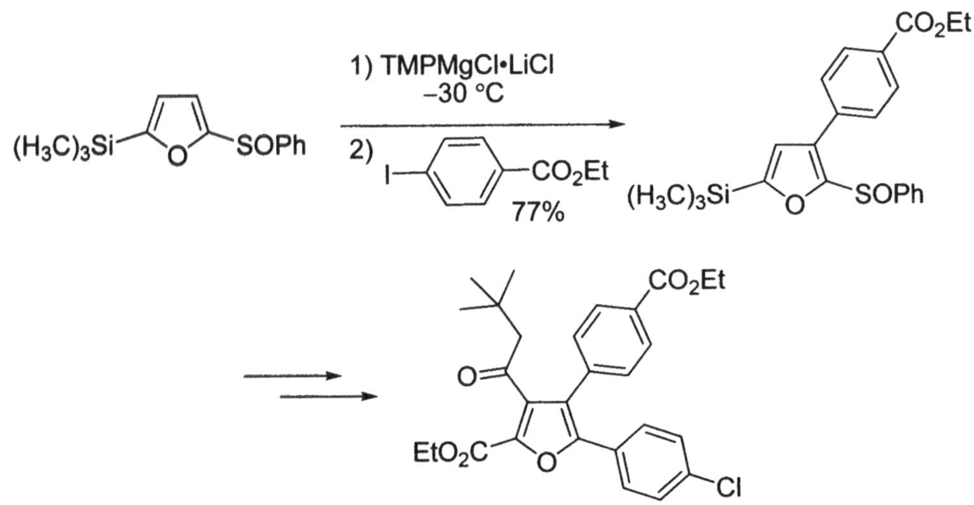

Friedel–Crafts Reaction

A regioselective Friedel–Crafts acylation was used to synthesize a novel derivative of amiodarone, a pharmaceutical compound used to treat ventricular and supraventricular arrhythmias.[13] Treatment of 2-methylbenzofuran with tin(IV) chloride in the presence of *p*-methoxybenzoyl chloride gave the desired product, 3-*p*-methoxybenzoyl-2-methylbenzofuran, in 63% yield.

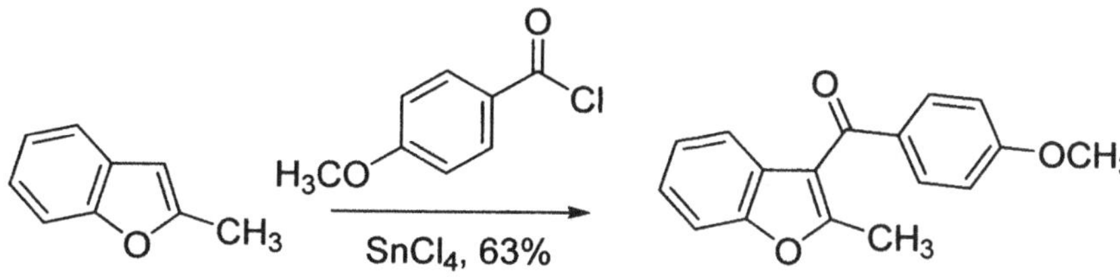

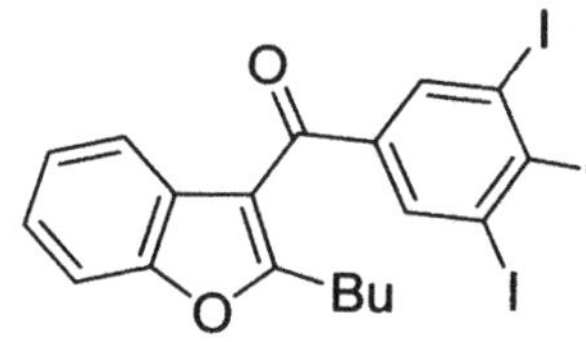

amiodarone

Intramolecular Cyclization

Padwa and co-workers, *en route* to (±)-selaginoidine, showed the *C3* position of a 2,5-disubstituted furan could react under protic conditions to form a highly substituted tetracyclic intermediate incorporating a trisubstituted furan ring.[14] Trifluoroacetic acid-induced Pictet–Spengler reaction of a furan-tethered tetrahydroindolinone provided the desired tetracyclic system in 96% yield.

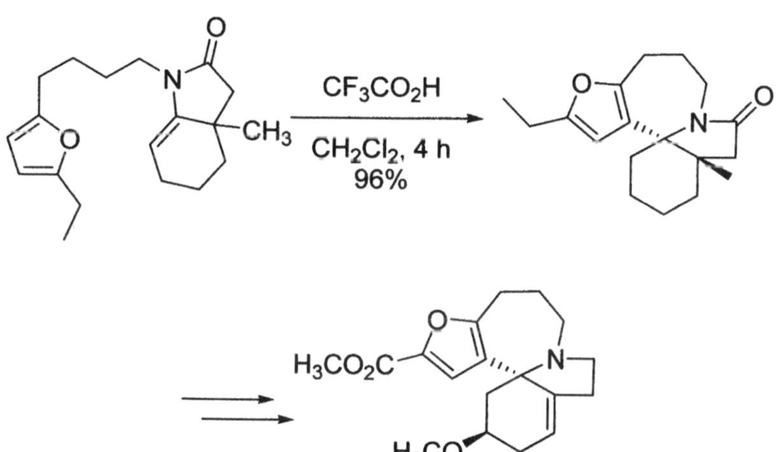

(±)-selaginoidine

Andersen and co-workers used an acid-catalyzed polyene cyclization to form the fused ring system of the oncolytic PI3Kα inhibitor liphagal.[15] Treatment of a terpine-based benzofuran with chlorosulfonic acid gave the desired tetracyclic system in 43% yield.

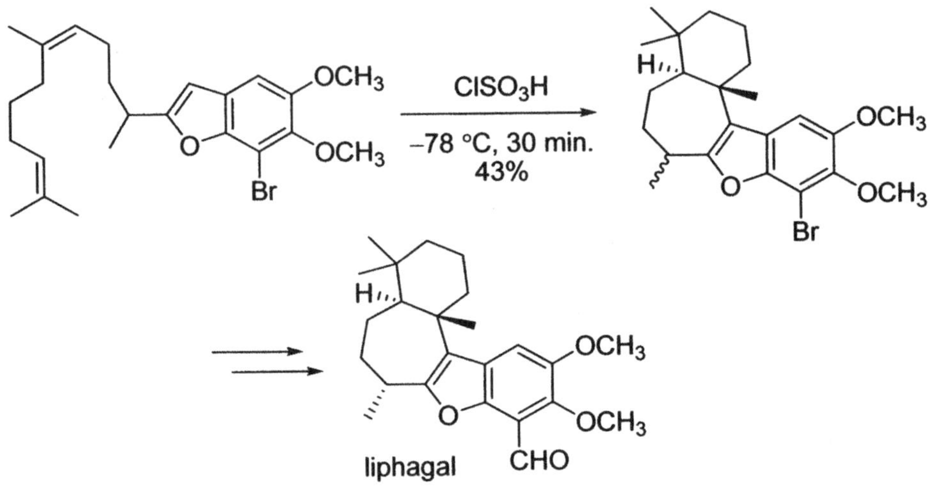

Shanmugham and White used an intramolecular base-catalyzed cyclization reaction to construct the tricyclic framework of 6-β-hydroxyeuryopsin, a compound with strong antifeedant activity against insects of the genera *Leptinotarsa* and *Myzus*.[16] Treatment of a chiral aldehyde precursor with TMSOTf in the presence of 2,6-lutidine gave the desired tetrasubstituted furan in quantitative yield. Two additional steps were required to access the target compound, 6β-hydroxyeuryopsin.

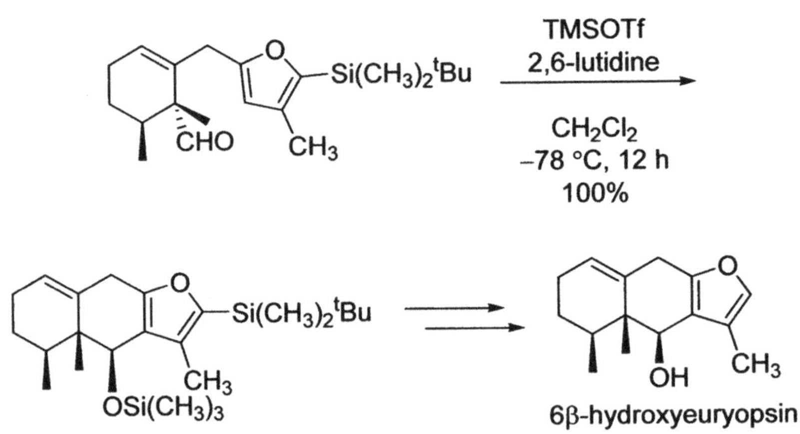

4.2.2.3 *Transition Metal-Catalyzed Cross-Coupling Reactions*

Transition metal-catalyzed cross-coupling and C–H activation have become an important method for the formation of carbon–carbon bonds in the synthesis of furan and benzofuran containing pharmaceutical compounds. The reaction conditions required for these transformations are generally mild and tolerant of a number of different functional groups including ethers, esters, and amines to name a few.

Stephenson and co-workers recently demonstrated a photo-redox-promoted direct intermolecular *C2*–H functionalization using diethyl-bromomalonate in the presence of a $Ru(bpy)_3Cl_2$ catalyst to produce 2-furyldiethylmalonate in 67% yield.[17]

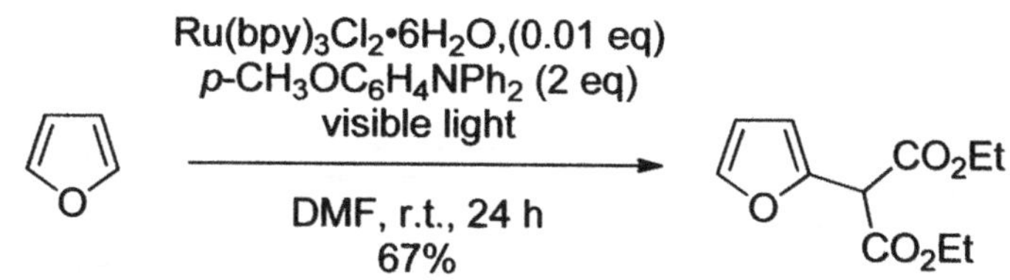

Takai and co-workers used C–H insertion to generate several 2,3-disubstituted furans. Reaction of 3-*t*-butyliminofuran with phenyl isocyanate in the presence of a rhenium catalyst gave the desired disubstituted furan in good yield.[18]

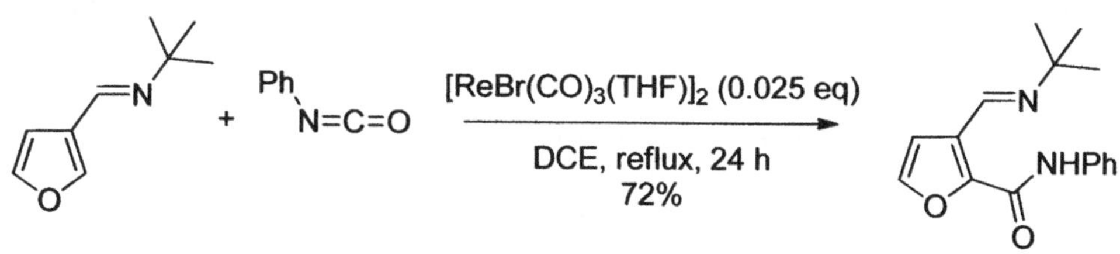

The Suzuki coupling of a 2,4-dibromofuran with (2,4-di-*tert*-butoxypyrimidin-5-yl)boronic acid gave the corresponding furanylpyrimidine derivative in 86% yield.[19]

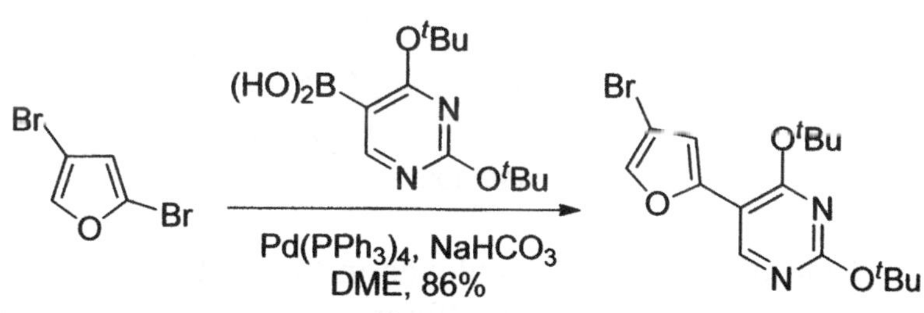

Bach and Krueger used a Stille coupling to synthesize rosefuran, a minor constituent of rose aroma. Cross-coupling of methyl 4,5-dibromofuran-2-carboxylate with tributyl(3-methylbut-2-en-1-yl)stannane gave the 2-substituted product selectively in 73% yield.[20,21] Two additional steps were required to access rosefuran.

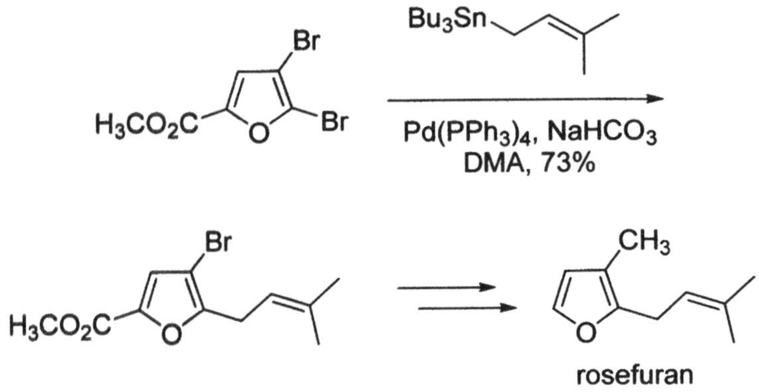

Bach and Bartels used a Negishi cross-coupling reaction as a key step in their synthesis of eupomatenoid.[22] Members of this class of compound have been shown to exhibit anti-parasitic activity against *Leishmania* and *Trypanasoma cruzi*. Reaction of 2,3,5-tribromobenzofuran with (3,4-dimethoxyphenyl)zinc(II) chloride gave the desired product selectively in 75% yield.

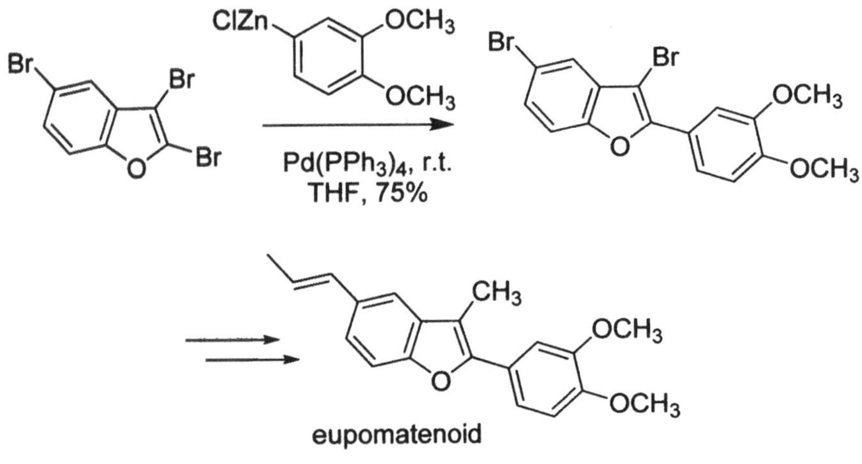

A C–H insertion was also employed in the synthesis of 3-substituted furans. Coupling of 2,5-dimethylfuran with 4-methystyrene using palladium(II) acetate gave the *trans*-alkenyl furan selectively and in modest yield.[23]

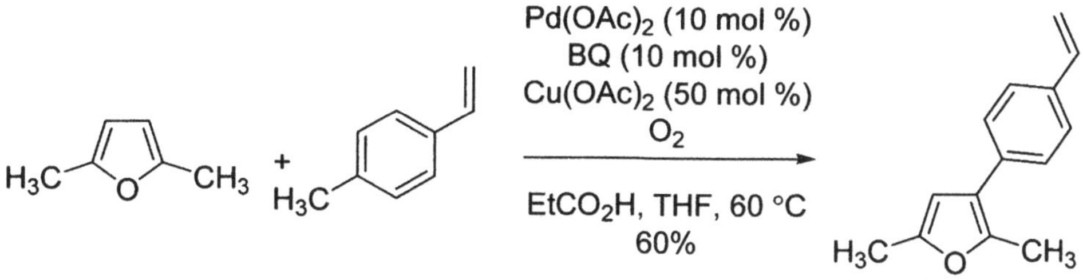

Bach and Krueger used a Negishi coupling to synthesize a derivative of the naturally occurring furanoid fatty acid F_5.[24,25] Treatment of 3-bromofuran with methylzinc chloride gave the corresponding trisubstituted furan in 66% yield (over two steps). Two additional steps were required to generate the desired fatty acid derivative.

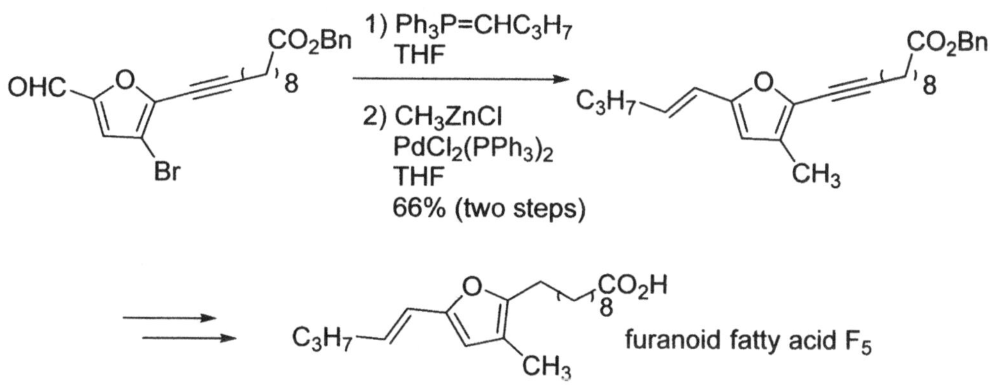

4.2.2 *Furan and Benzofuran Synthesis*

4.2.2.1 *Furan Synthesis*

The Feist–Bénary and Paal–Knorr syntheses are commonly employed in the preparation of furan ring systems. In special cases where furan derivatives are difficult to prepare by other methods, Diels–Alder and retro-Diels–Alder reactions have become important methods for their synthesis. Finally, transition metal-catalyzed cyclization and cycloisomerization reactions have recently gained significant attention for their utility in the synthesis of highly functionalized furans. Key examples of these syntheses are highlighted in the sections below.

Feist–Bénary Furan Synthesis

The Feist–Bénary furan synthesis, first described in 1902,[26] is especially useful for the synthesis of substituted furan rings. This reaction occurs between an α-halocarbonyl, such as chloroacetaldehyde or chloroacetone, and a β-dicarbonyl, such as ethyl acetoacetate or a derivative of ethyl acetoacetate, in the presence of a base (usually pyridine, triethylamine, or sodium hydroxide) under thermal conditions. The resulting product, a 3-furoate, is produced in modest to good yield.

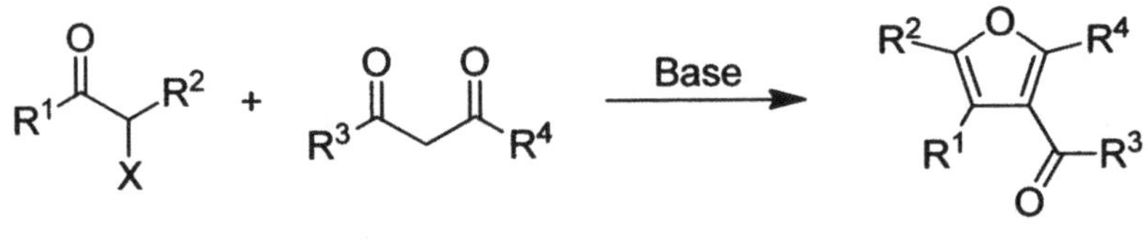

The Feist–Bénary synthesis proceeds via an aldol reaction followed by intramolecular *O*-alkylation and dehydration to yield the furan product as illustrated below.[27] The reaction can be modified such that a substituted 1,4-dicarbonyl is produced.[28] This can be then used to synthesize furan derivatives under Paal–Knorr conditions.

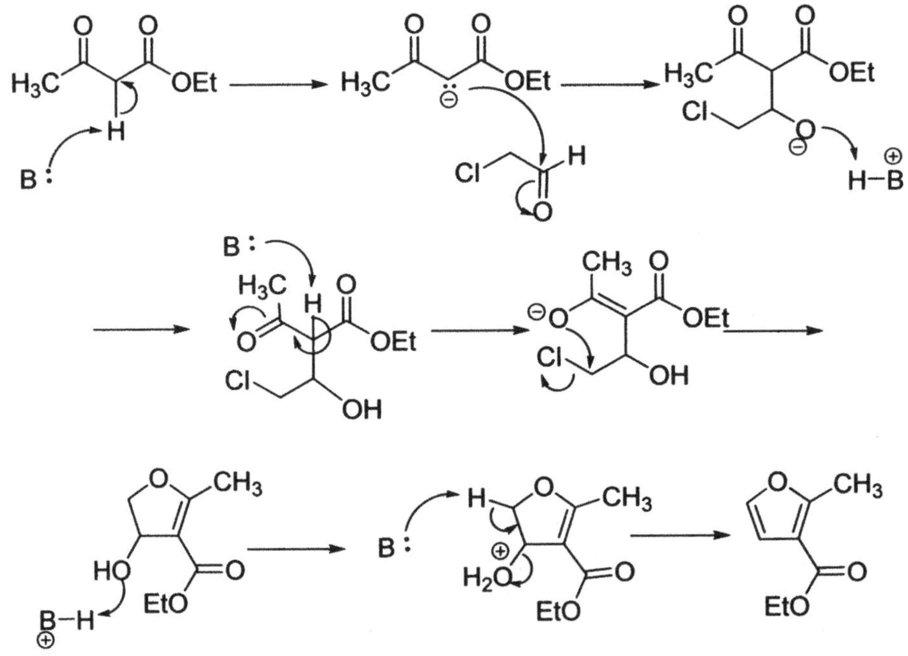

The Feist–Bénary furan synthesis is most commonly used for the preparation of 2-substituted 3-furoates, including ethyl 2-methyl-3-furoate, the original compound prepared by Bénary.[29] The resulting ester is generally converted into a carboxylic acid for use in a variety of transformations, including decarboxylation to produce the corresponding 2-substituted furan. For example, reaction of ethyl 7-methyl-3-oxooct-6-enoate with chloroacetaldehyde in pyridine led to the desired furan.[30] Saponification of the ester yielded the corresponding acid, which readily underwent decarboxylation to produce the final 2-alkylfuran, a derivative of rosefuran, in 33% over three steps.

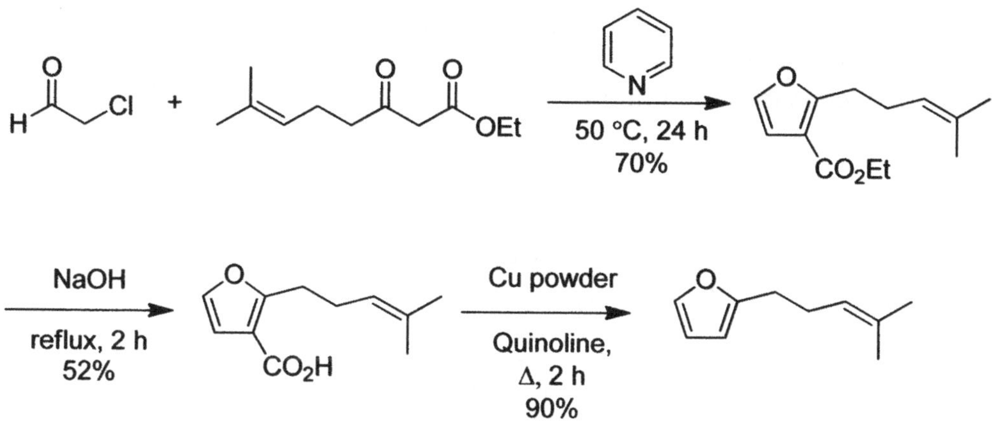

The Feist–Bénary reaction has also been used to prepare trisubstituted furanoates as well. At present, only one 2,4-disubstituted 3-furoate has been prepared using the Feist–Bénary reaction. Reaction of chloroacetone with ethyl acetoacetate in cold hydrochloric acid followed by exposure to cold triethylamine provided ethyl 2,4-dimethyl-3-furoate in 54–57% yield over multiple trials.[31–33]

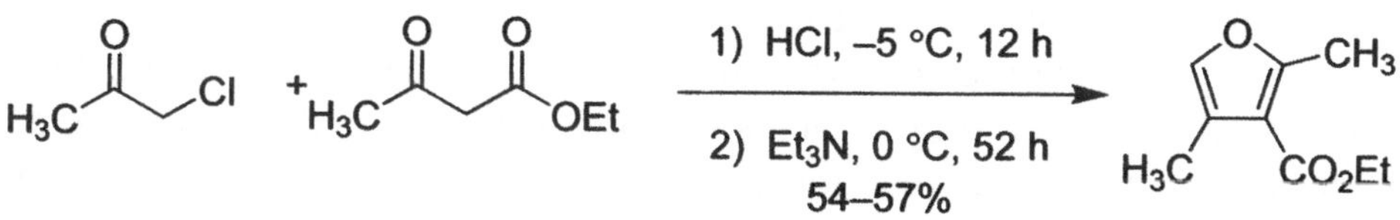

Several tetrasubstituted furan derivatives have been prepared by the Feist–Bénary reaction. For example, Stetter and Lauterbach demonstrated that 1,3-cyclohexanedione could serve as a β-dicarbonyl in combination with ethyl 2-chloroacetoacetate in the presence of potassium hydroxide to yield the corresponding fused tetrasubstituted furan derivative in good yield.[34]

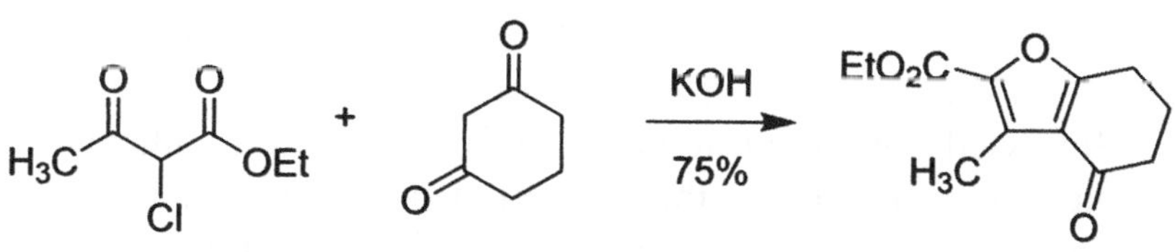

Magnus and co-workers[35,36] extended on work by Stetter and Lauterbach[34] to produce an intermediate common to the natural products linderalactone, isolinderalactone and neolinderalactone. Compounds of this class have been shown to exhibit a host of biological properties including analgesic, anti-inflammatory, antiemetic, antibacterial and antiviral. The Feist–Bénary reaction of ethyl 2-chloroacetoacetate and 5-(methoxymethyl)-5-methylcyclohexane-1,3-dione in the presence of potassium hydroxide gave the desired tetrasubstituted furan in 57% yield.

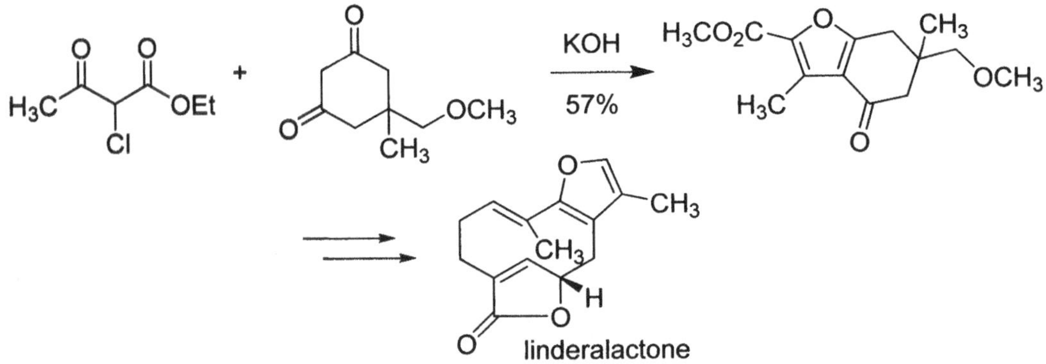

Paal–Knorr Furan Synthesis

The Paal–Knorr furan synthesis involves the treatment of a 1,4-dicarbonyl with catalytic acid to generate the corresponding furan.[37] A variety of differentially substituted 1,4-dicarbonyls have been used in the Paal–Knorr reaction to synthesize the corresponding mono-, di-, tri-, and tetrasubstituted furans. Commonly employed acids include sulfuric, hydrochloric, and *p*-toluenesulfonic acid. The reaction generally takes place at room temperature or under thermal conditions.

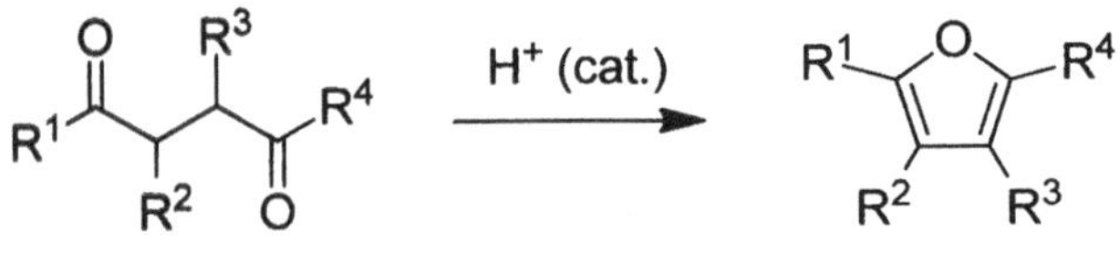

The mechanism of the Paal–Knorr furan synthesis was established by Arnath and co-workers using the *d,l-* and *meso*-diastereomers of 2,3-disubstituted 1,4-diketones.[38] Protonation of one of the carbonyls of the β-diketone, followed by deprotonation of the proton adjacent to the unprotonated ketone, results in the formation of an enolate that attacks the protonated ketone to form the corresponding dihydrofuran. Elimination of water then generates the furan.

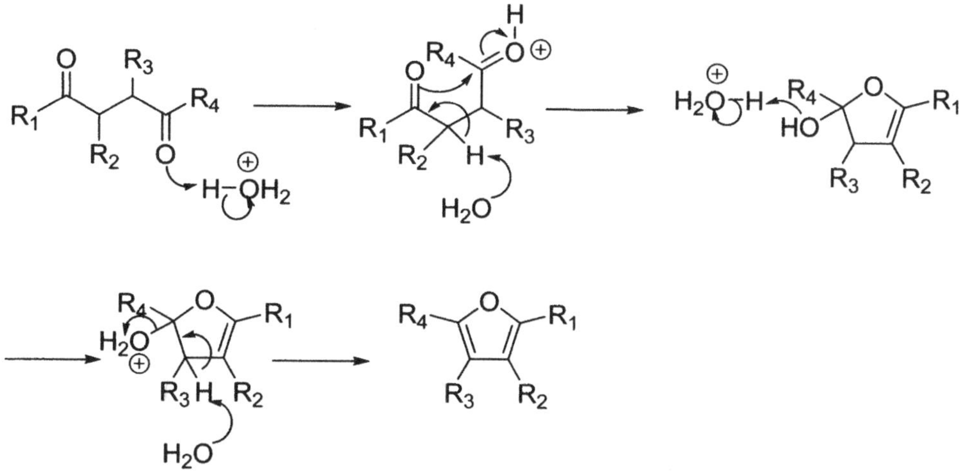

Only a single example employing the Paal–Knorr synthesis for the generation of mono-substituted furans has been reported to date. Molander demonstrated that 2-(methyldiphenylsilyl)furan could be prepared in good yield via treatment of 4-(methyldiphenylsilyl)-4-oxobutanal with hydrochloric acid in THF.[39]

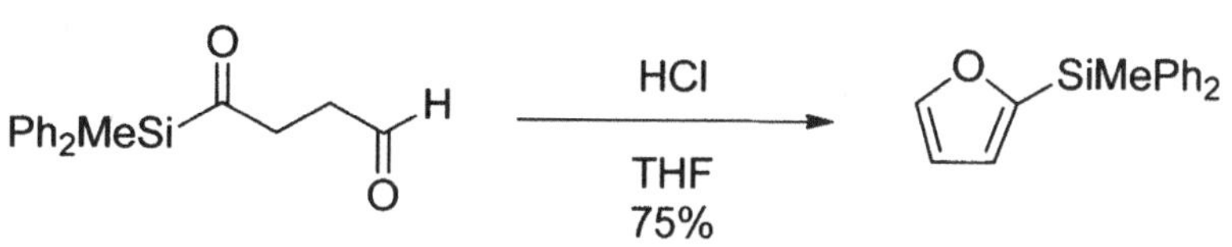

The Paal–Knorr synthesis is more commonly used for the synthesis of 2,5-disubstituted furans, especially 2,5-diaryl furans, which have shown significant antiviral activity against RNA viruses. For example, Wilson and co-workers demonstrated that diaryl-substituted diones treated with hydrochloric acid could be used to generate the corresponding 2,5-disubstituted furans in good yields.[40]

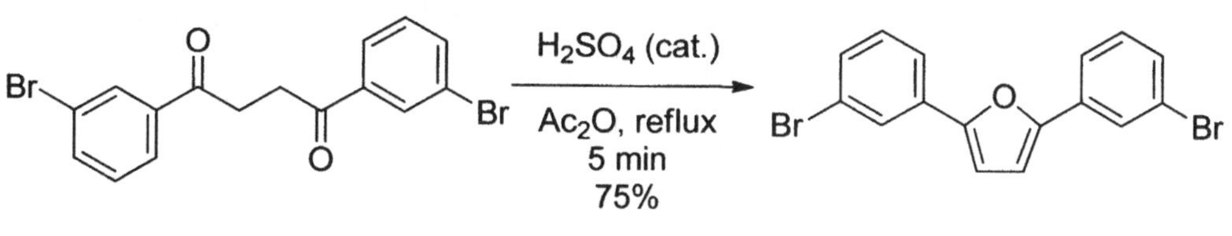

The Paal–Knorr synthesis was also used to prepare 2,5-disubstituted furans that were investigated as potential enzyme inhibitors. Reaction of a highly substituted 1,4-dione with catalytic sulfuric acid produced the corresponding furan, which was evaluated for its ability to bind to a retenoic acid receptor.[41]

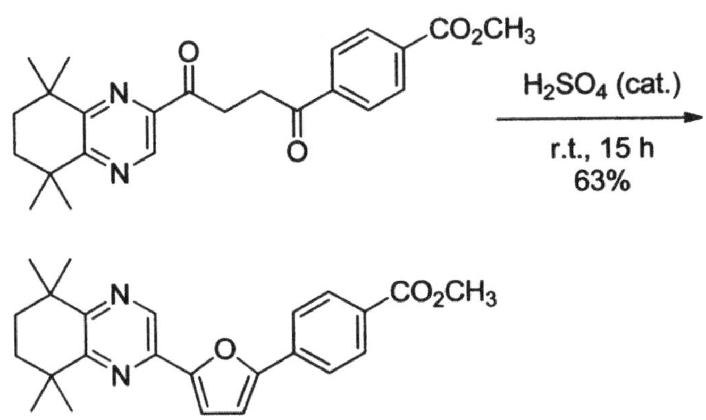

Denisenko and co-workers employed the Paal–Knorr synthesis in their preparation of a steroid-based furan with potential anti-inflammatory properties.[42] Treatment of the corresponding chiral 1,4-dione precursor with catalytic *p*-toluenesulfonic acid generated the desired furan in 35% yield.

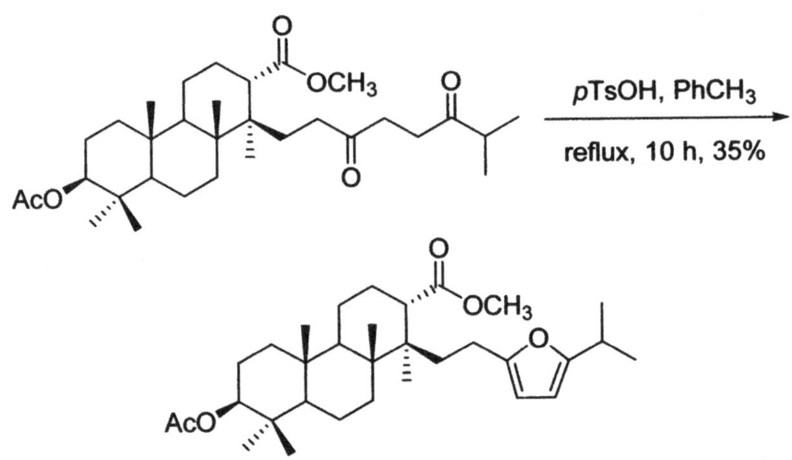

The Paal–Knorr synthesis has also been employed in the synthesis of triaryl furans. For example, de Laszlo and co-workers prepared several 2,3,5-triarylfurans as potential inhibitors of P38 kinase by treating the 1,4 diketo-precursors with *p*-toluenesulfonic acid as illustrated below.[43]

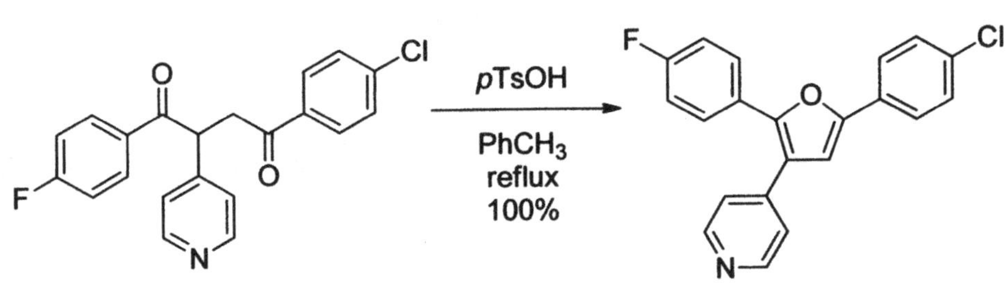

Another interesting example of the use of the Paal–Knorr synthesis is in the preparation of trialkylfurans. General compounds of this class have been shown to play important roles in biological signalling. Weirsum and co-workers used *p*-toluenesulfonic acid in refluxing ether to produce a sterically congested tri-*t*-butylfuran from the corresponding dione.[44]

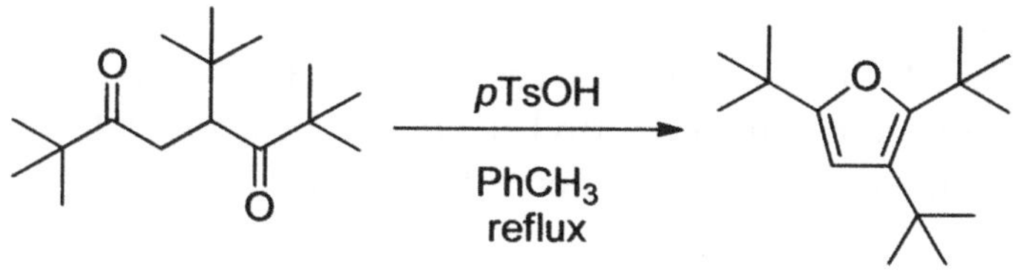

Tetrasubstituted furan derivatives have also been synthesized via the Paal–Knorr synthesis. In one example, Katzenellenbogen and co-workers prepared numerous alkyl triarylfurans in modest to high yield by treating the corresponding 1,4-dione precursors with *p*-toluenesulfonic acid.[45,46] These compounds, which were designed to mimic estradiol, were evaluated for activity toward the estrogen receptor (ER) α, and it was shown that analogues with basic side chains on the *C*4 phenol were high-affinity antagonists.

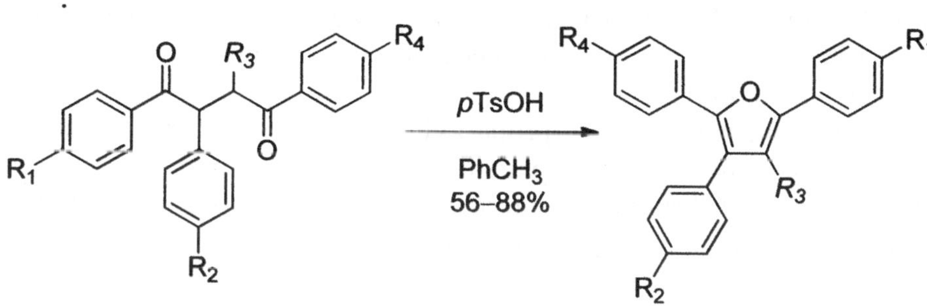

Diels–Alder/retro-Diels–Alder Reactions

Combination Diels–Alder/retro-Diels–Alder reactions have been used to prepare substituted furan derivatives from furans and oxazolidinones. Reaction of a furan or oxazolidinone derivative with a disubstituted alkyne (usual dimethyl acetylenedicarboxylate or diethyl acetylenedicarboxylate) produces the Diels–Alder adduct, which can undergo a retro-Diels–Alder reaction to give the desired furan derivative.

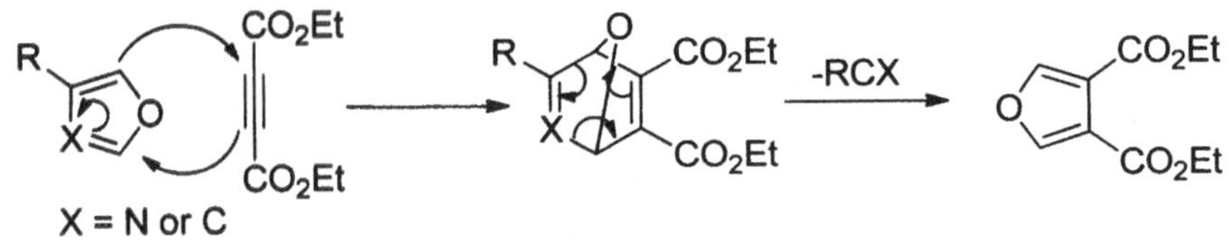

Aoyama and co-workers used this approach in their synthesis of 2,3,4-trisubstituted furan derivatives.[47] Reaction of 2-cyclohexyl-4-(trimethylsilyl)oxazole with dimethyl acetylenedicarboxylate under thermal conditions gave the corresponding derivative in good yield.

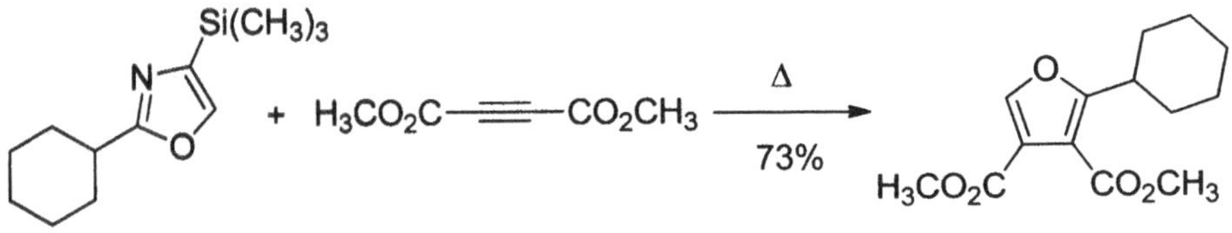

Transition Metal-Promoted Cyclization Reactions

A powerful method for generating furans is via transition metal-catalyzed isomerization reactions of unsaturated acyclic precursor including allenyl and propargyl ketones. In recent years, propargyl ketones have largely replaced the use of allenes in these reactions because they are much more stable than their allene counterparts.

Cylcoisomerization of propargyl ketones is often assisted by transition metals such as gold, palladium and copper, and the reaction is believed to proceed through an intermediary allene. Di- and trisubstituted furans have been prepared in good yields via this method.

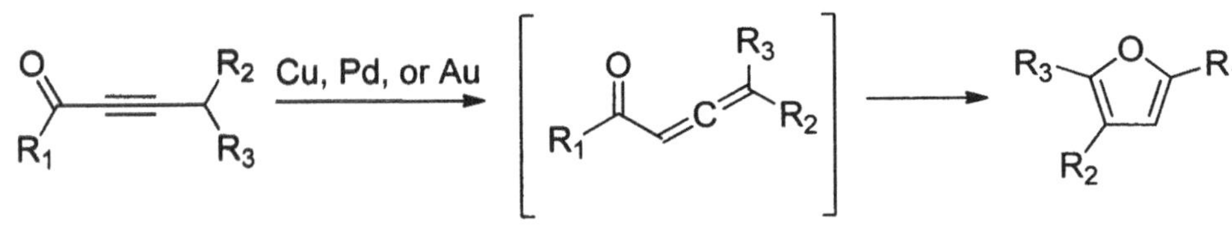

Kel'in and Gevorgyan illustrated the utility of this process with the copper(I)-catalyzed cyclization of a series of propargyl ketones to form the corresponding 2,5-disubstituted furans.[48] Treatment of 8-methylnon-7-en-5-yn-4-one with copper(I) iodide under basic conditions provided the corresponding furan in 88% yield.

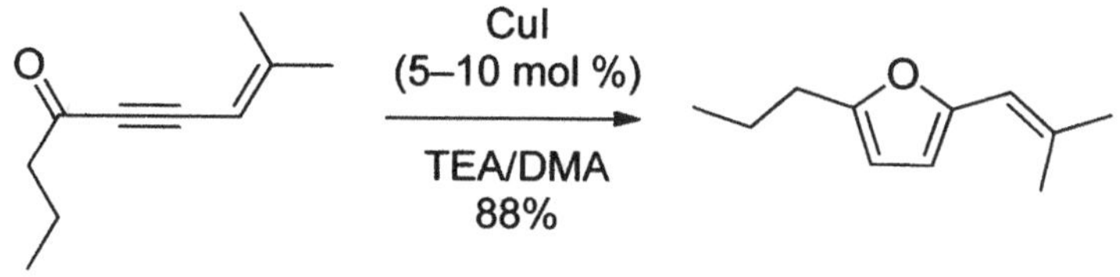

A number of cycloisomerization reactions involving α,β-unsaturated propargyl ketones have also been employed in the synthesis of substituted furans. These reactions are generally catalyzed by either gold(III) or copper(I), and are believed to proceed through an oxonium ion intermediate, which is subsequently trapped by a nucleophile as illustrated below.

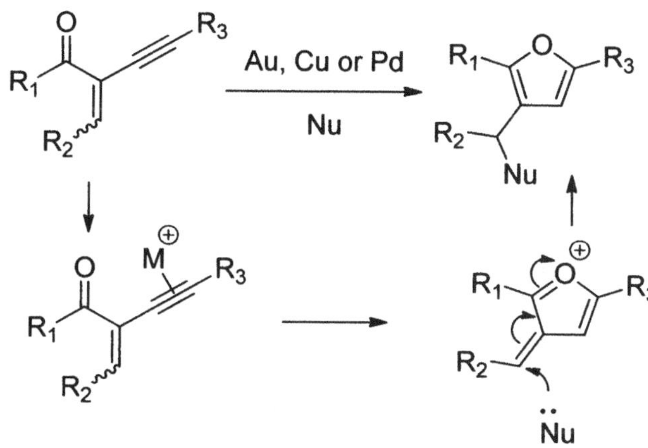

Monteiro and co-workers recently used this approach in their synthesis of 2,3-disubstituted furocoumarins.[49] These compounds have proven to be important functional components with a host of interesting biological properties. Cycloisomeratization of 4-methoxy-3-(phenylethynyl)-2*H*-chromen-2-one with methyl iodide in the presence of palladium tetraphenylphosphine in DMF gave the corresponding 2-aryl-3-alkyl furocomarin in modest yield. This approach was also employed by Cheng and Hu in their synthesis of substituted furocoumarins.[50]

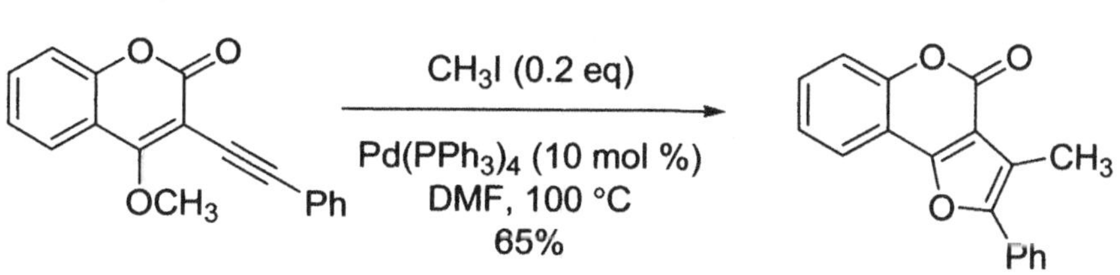

Metal-catalyzed cyclization/conjugate addition sequences involving substituted propargyl epoxides have also been employed in the synthesis of highly functionalized furans. These reactions are generally catalyzed by either gold(III) or palladium, proceed through an allene intermediate, and involve the addition of a nucleophile as illustrated below.

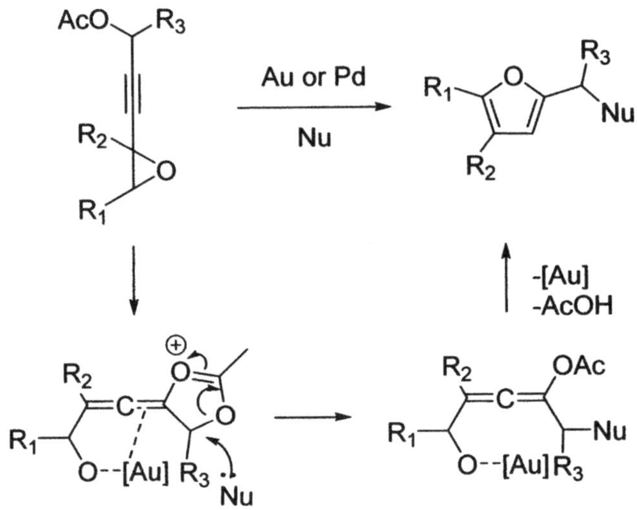

A recent example by Pale and co-workers used the cationic gold complex PPh_3AuSbF_6 as a catalyst and methanol as a nucleophile to generate the desired 2,3-substituted furan in 92% yield.[51] 2,3,5-Trisubstituted furans, such as 2-(methoxy(2-nitrophenyl)methyl)-4,5,6,7-tetrahydrobenzofuran (shown below) were also prepared in modest yields using this methodology.

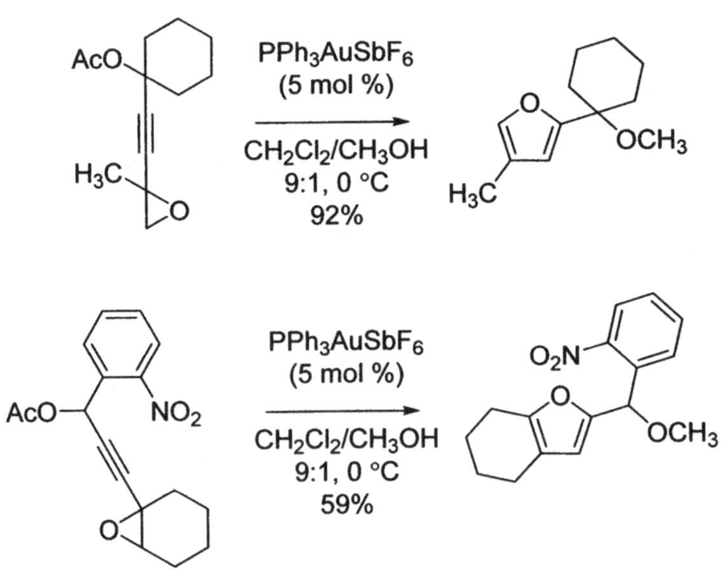

Aurrecoechea and co-workers used a palladium(II) catalyst to perform a palladium-catalyzed cyclization-coupling reaction to prepare highly functionalized tetrasubstituted furans.[52] Treatment of a functionalized epoxy-propargyl ester with samarium(II) iodide, followed by palladium-catalyzed cylcoisomerization/conjugate addition yielded the desired tetrasubstituted furan in high yield. Several additional examples were reported with yields between 20% and 80%.

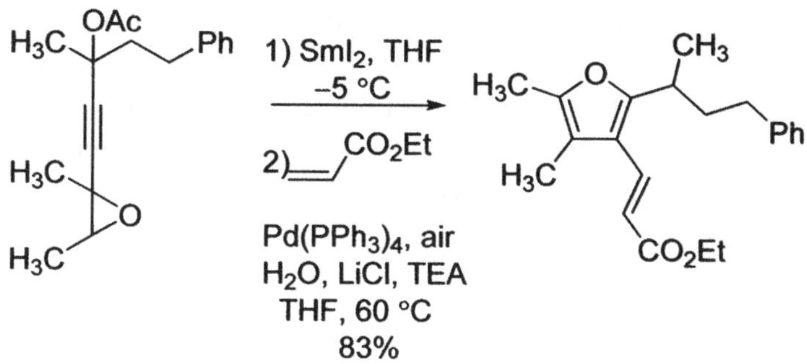

Ring-closing metathesis (RCM) has also been employed in the synthesis of substituted furans. Donohoe and co-workers used a ring-closing metathesis strategy to form the furan moiety of (–)-(*Z*)-deoxypukalide,[53] a compound from the family cembranolides, which have been shown to exhibit neurotoxicity and anti-inflammatory properties. Treatment of a mixed acetal with Grubbs' II catalyst, followed by *in situ* aromatization with PPTS, yielded the disubstituted furan in 85% yield over two steps. Seven additional steps were required to reach of (–)-(*Z*)-deoxypukalide.

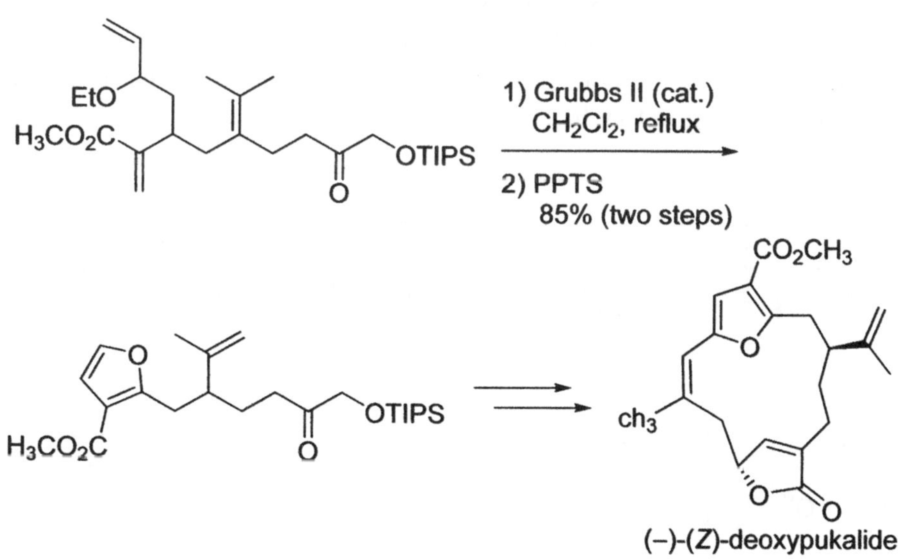

Acid- and Base-Catalyzed Cyclization

Brønsted and Lewis acid-catalyzed cyclizations of enones have become important methods for the synthesis of fused-furans and benzofurans. Mukai and co-workers[54] recently applied this strategy to the synthesis of the tetracylic ring system of (+)-nakadomarin A, a compound with cyctotoxic, inhibitory and antimicrobial activity. Dihydroxylation of the isolated double bond of the α-pentenyl-α,β-cyclopentenone framework, followed by acid-catalyzed ring closure, gave the desired tetracyclic product in 80% yield. A similar acid-catalyzed strategy was employed by Dixon and co-workers in their synthesis of the furan moiety of (–)-nakadomarin A.[55]

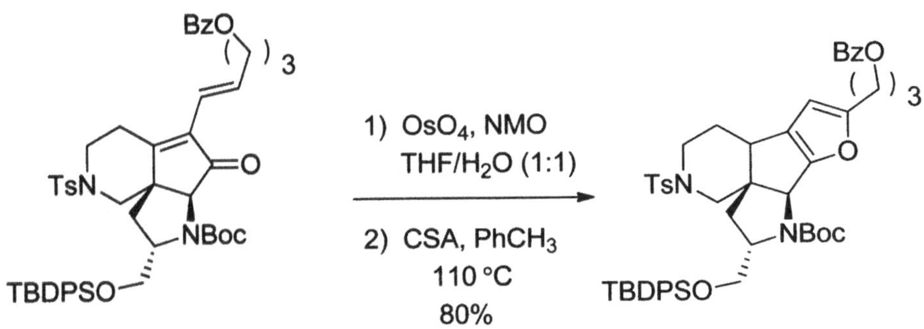

Liao and co-workers used an acid-catalyzed ring-closing reaction to form a polyfuctionalized *cis*-decalin ring system as a key intermediate for the synthesis of (±)-3β-angeloyloxyfuranoeremophilane and (±)-3β-methacryloyloxyfuranoeremophilane.[56] Treatment of a selectively protected bicyclic ketone with *p*-toluenesulfonic acid in aqueous THF gave the furan-fused *cis*-decalin in good yield.

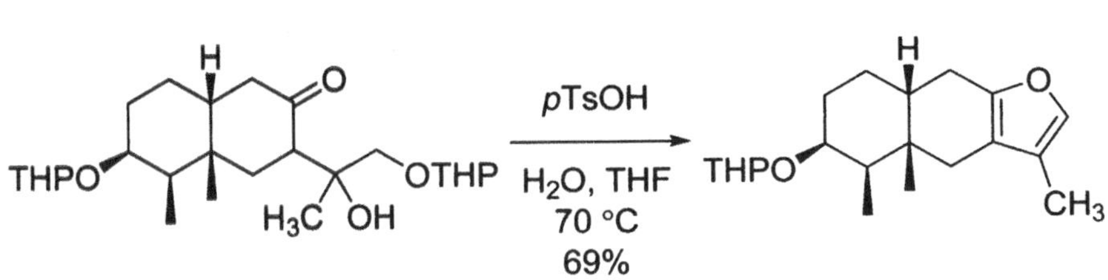

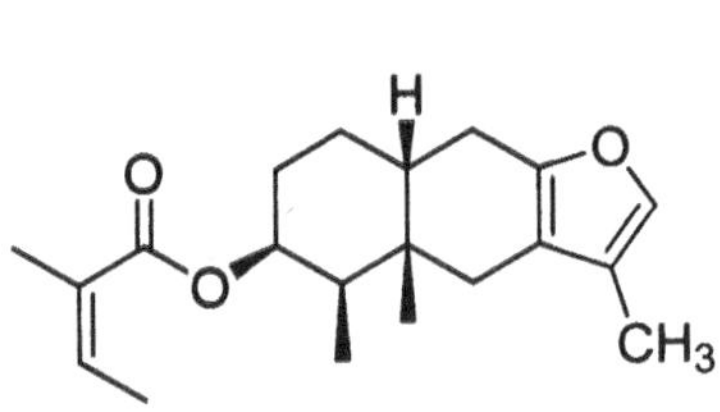

(±)-3-β-angeloyoxy furanoeremophilane

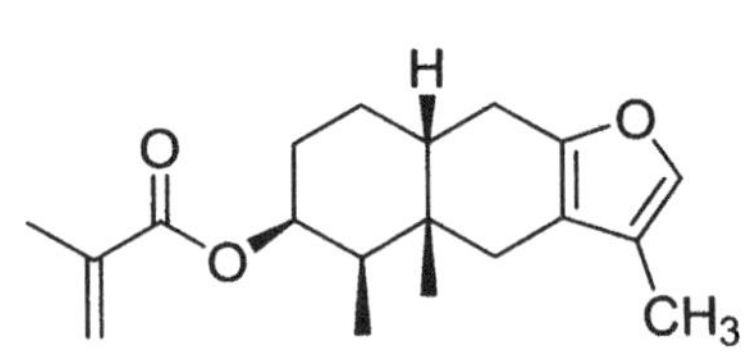

(±)-3-β-methacryloyloxy furanoeremophilane

Lewis acids have also been used to synthesize highly functionalized furan derivatives with biological signalling capabilities. For example, Sello and co-workers used scandium(III) triflate in their biomimetic synthesis of several signalling molecules of the bacterium *Streptomyces*.[57] Reaction of methyl 5-methyl-3-oxohexanoate and 2,2-dimethyl-1,3-dioxan-5-one in the presence of scandium (III) triflate in methanol gave the corresponding furan in 60% yield.

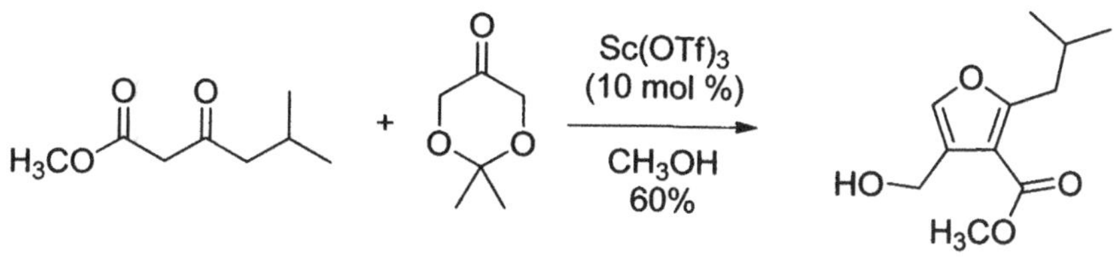

Synthesis of Furans from Carbohydrates

Carbohydrates have been used as chirons to generate a number of interesting furan derivatives from industrial commodities such as 2-furaldehyde, to more complex, biologically significant compounds with therapeutic potential. Several interesting examples of the use of carbohydrates as starting materials for the synthesis of furans are highlighted below.

2-Furaldehyde, an important intermediate in the synthesis of a number of natural products and pharmaceutical reagents, is prepared industrially by treating plant residues rich in pentoses (pentosan) with dilute sulphuric acid. Steam distillation is then used to obtain 2-furaldchydc as a pure substance for further functionalization.

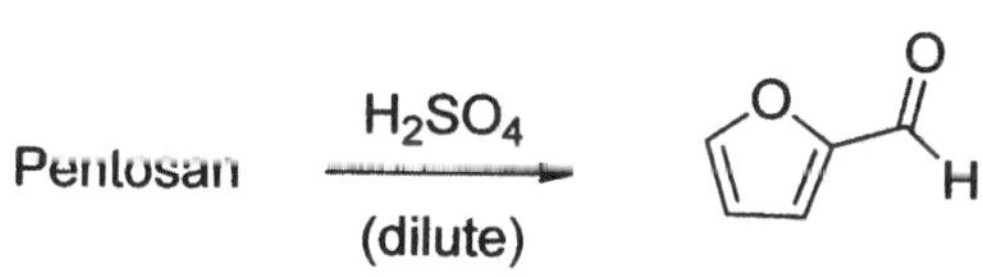

Marcantoni and co-workers[58] used glucose as a chiron in the silica-supported Knoevenagel condensation of aldose sugars with β-dicarbonyl compounds (the Garcia–Gonzalez reaction) to produce 2,3,5-trisubstituted furan derivatives for use as scaffolds in the synthesis of biologically active compounds. Treatment of a mixture of α- and β-glucose with cerium(III) chloride hepta-hydrate and sodium iodide in the presence of silica produced the desired furan in high yield.

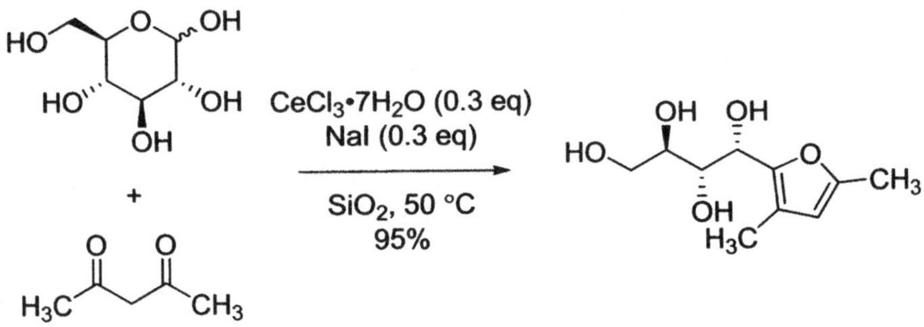

Shaw and co-workers used a mixed Lewis acid approach to synthesize chiral furanmethanol derivatives from glycals.[59] These compounds are important intermediates in the synthesis of many biologically significant compounds and serve as precursors for furanmethanes and 3-acylfurans. Reaction of a selectively protected 6-azido glycal with zirconium(IV) chloride and zinc(II) iodide gave the desired chiral furan selectively in 74% yield.

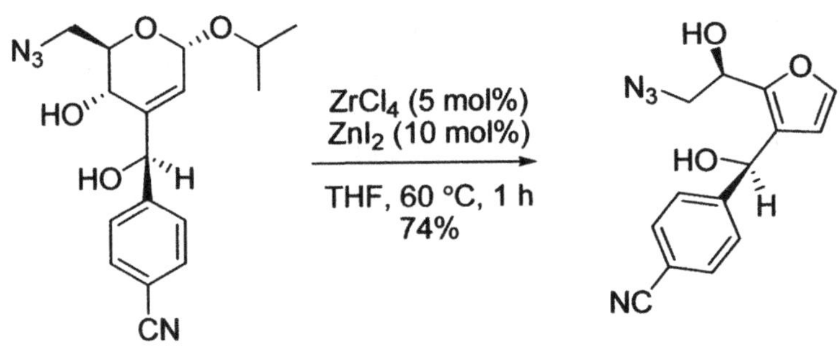

In another example, Postel and co-workers used functionalized carbohydrates to produce β-substituted polyoxygenated furans.[60] Treatment of functionalized benzylidene precursors with TMS-N_3 in the presence of dibutyltin oxide gave the corresponding furan in 10% yield. The authors used a computational approach to rationalize furan formation via a series of cascade fragmentation reactions obtained during 1,5-C–H insertion from alkylidenecarbenes.

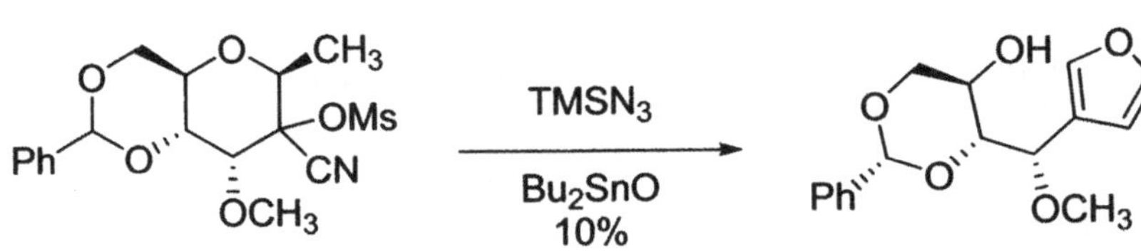

4.2.2.2 Benzofuran Synthesis

Perkins Reaction

Benzofurans can be prepared from α-halo-coumarin compounds.[1] Base-catalyzed ring-contraction followed by dehydrative decarboxylation gives rise to the desired benzofuran.

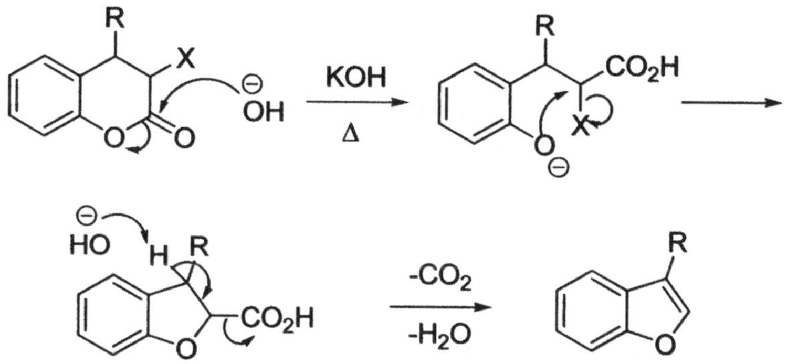

The synthesis of benzofuran is also readily accomplished by base-catalyzed Perkins rearrangement of 2,3-dibromo-coumarin.[1]

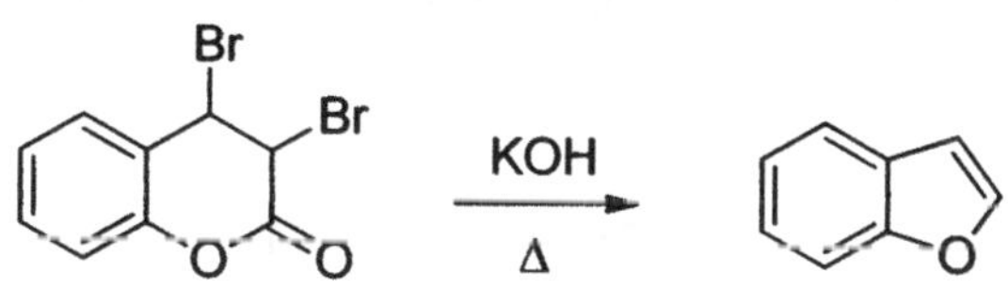

Acid/Base-Catalyzed Cyclization Reactions

Erhardt and Khupse used an acid-catalyzed cyclization as a key step in their synthesis of the benzofuran-containing lespedezol A_1.[61] Members of this family have been shown to posses significant anti-oxidant, anti-allergic, anti-cancer, and anti-fungal activities. Treatment of the isoflavanone precursor with a catalytic amount of hydrochloric acid in a methanol/trimethyl-orthoformate mixture gave the target compound in 71% yield.

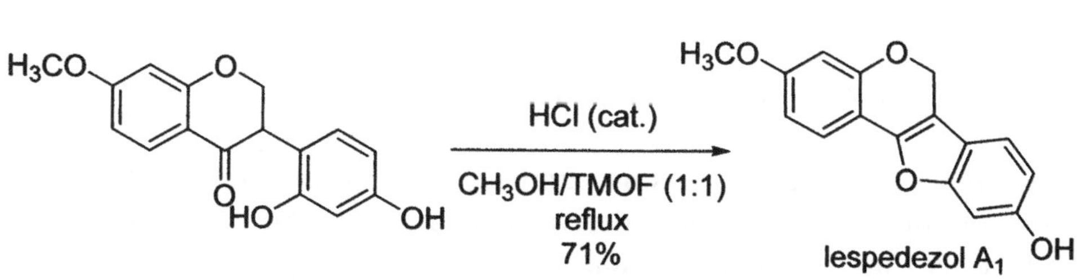

Mehta and Likhite[62] used a Lewis acid-catalyzed cyclization route to construct the benzofuran ring system of frondosin B, an IL-8 antagonist, and an inhibitor of protein kinase C (PKC). Treatment of a highly functionalized chiral ketone precursor with boron trifluoride-diethyl etherate gave the corresponding furan in 95% yield over two steps. This strategy was also employed by Ovaska and Li in their synthesis of the same compound.[63]

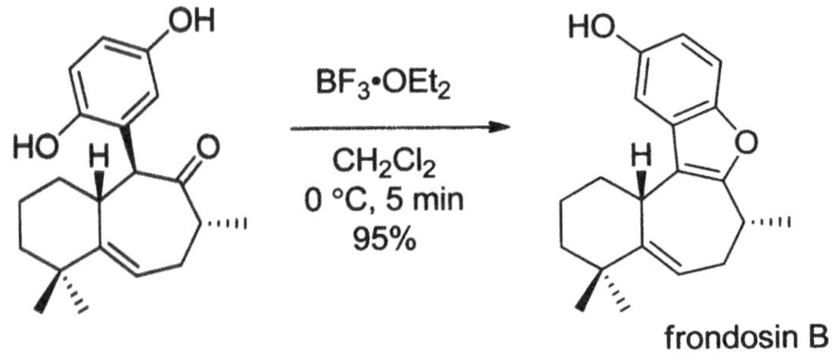

Phosphazene and its derivatives have been used as bases to catalyze the synthesis of highly substituted benzofurans. Krause and co-workers used phosphazenes in a key step of their total synthesis of amurensin H, an anti-inflammatory compound.[64] Treatment of a highly substituted benzophenone precursor with P4-^{t}Bu in benzene gave amuresin H after deprotection of the methyl esters with boron tribromide.

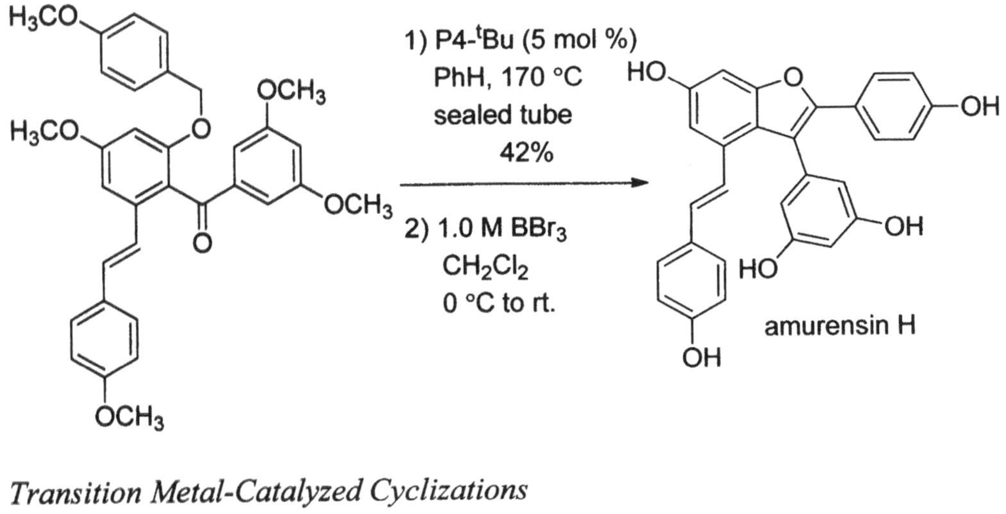

Transition Metal-Catalyzed Cyclizations

There are a few reports using transition metal-catalyzed ring-closure reactions to produce benzofuran derivatives. One example, reported by Chen and Dormer, used a copper-catalyzed protocol to prepare several 2,3-disubstituted benzofuran derivatives.[65] Treatment of the β-ketoester, ethyl 2-

(2-bromophenyl)-3-oxo-3-phenylpropanoate, with copper(I) iodide in the presence of potassium phosphate gave the corresponding benzofuran in 91% yield.

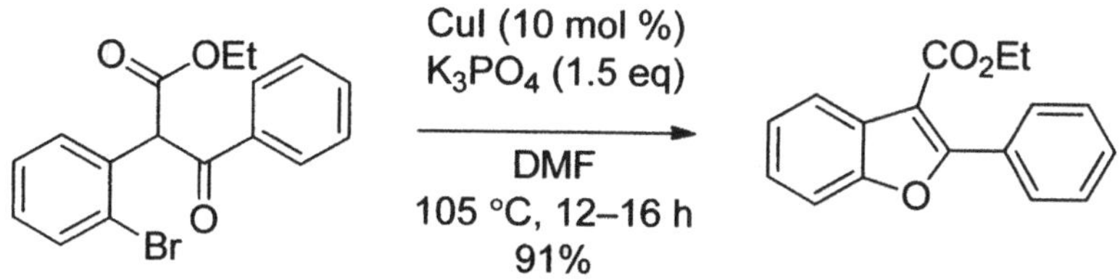

4.2.3 Synthesis of Furan- and Benzofuran-Containing Drugs

The syntheses of several commercially available furan and benzofuran based pharamaceutics have been published. The examples below are meant to illustrate some of the more general methods employed in the synthesis of these important compounds. Note that the majority of the syntheses highlighted below take advantage of preformed or readily available furan or benzofuran derivatives.

Dantrolene is a muscle relaxant that is currently the only effective treatment for malignant hyperthermia, a rare and life-threatening disorder that can be triggered by general anesthesia in some patients. Dantrolene has been synthesized by a number of different routes starting from readily available furaldehyde and its derivatives.

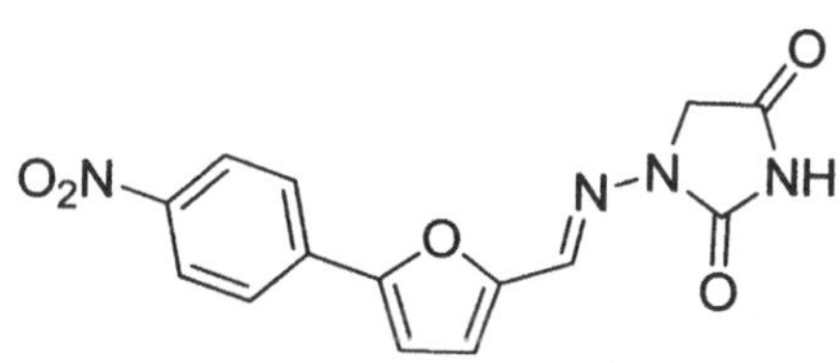

dantrolene

The original synthesis of dantrolene by Snyder and co-workers (shown below) employed a Meerwein arylation to produce the corresponding 5-aryl-2-furaldehyde intermediate.[66,67] Condensation of this intermediate with 1-aminohydantoin under acidic conditions produced the target compound dantrolene.

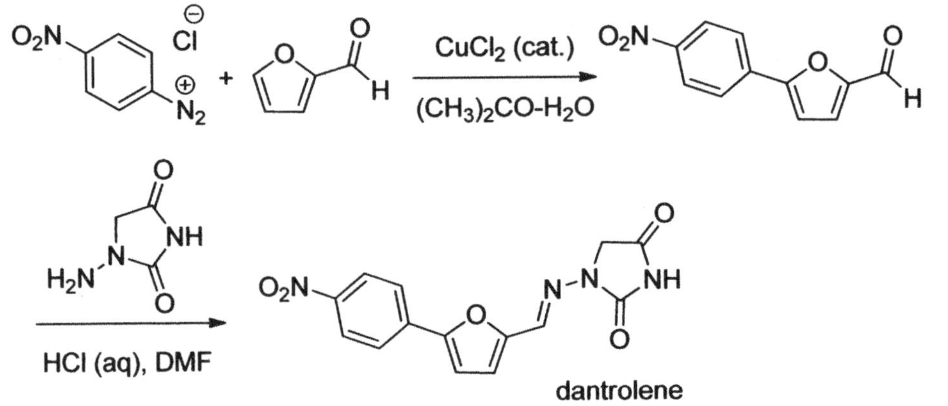

Suzuki and co-workers recently applied a Stille coupling reaction to synthesize dantrolene and several of its congeners.[68] Coupling of 4-nitro iodide with 5-(tri-butylstannyl)-2-furaldehyde in the presence of dichloro-palladium(II) triphenylphosphine gave the corresponding 5-*p*-nitrophenyl-2-furaldehyde in 75% yield. Condensation with 1-aminohydantoin produced the desired dantrolene in 94% yield (71% yield over two steps).

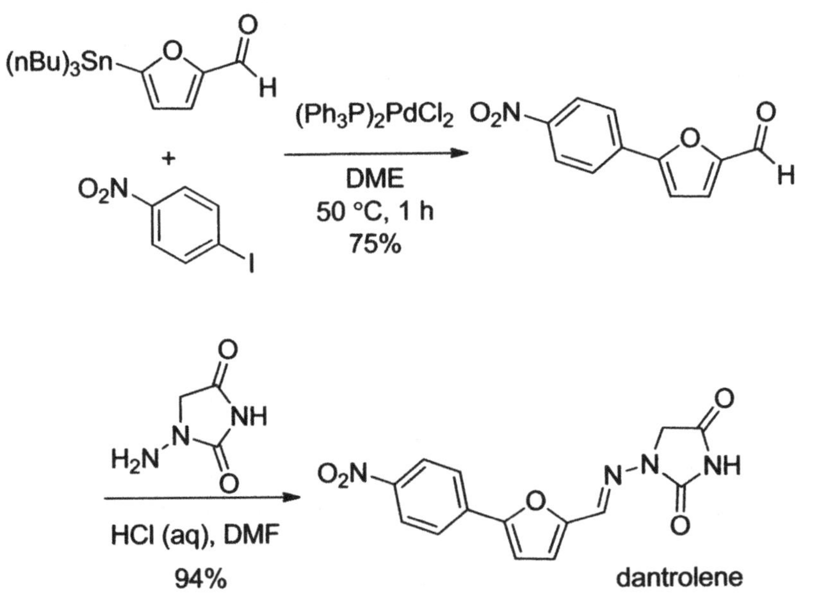

Ceftin is a second-generation broad-spectrum cephalosporin antibiotic used to treat bacterial infections resulting in Lyme's disease, bronchitis, sinusitis, tonsillitis, otitis, skin infections, gonorrhea, and urinary tract infections, and it is considered the antibiotic of choice for otitis media. Ceftin functions by inhibiting bacteria cell wall biosynthesis.

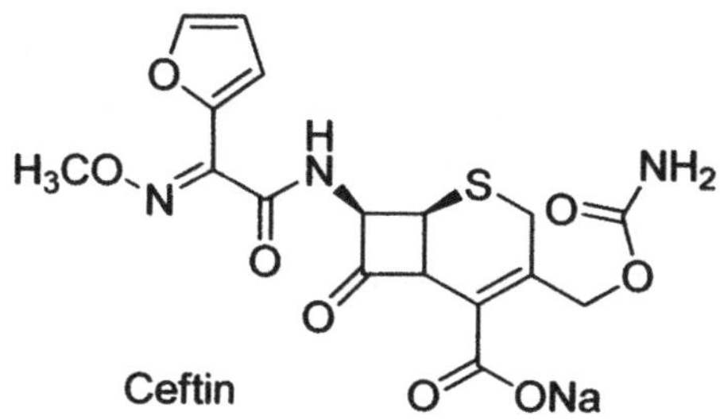

Ceftin

Jain and co-workers reported a concise total synthesis of cefuroxin (Ceftin) from readily accessible 7-amino-3-acetoxymethyl-3-cephem-4-carboxylic acid.[69] Hydrolysis of 7-amino-3-acetoxymethyl-3-cephem-4-carboxylic acid in aqueous methanol followed by acidification with tetrabutylammonium hydroxide provided the corresponding alcohol in 88% yield. Condensation of this compound with the ammonium salt of 2-furanyl-(*Z*)-2-methoxyiminoacetyl chloride provided the core structure in 77% yield. Protection of the free alcohol as a carbamate with chlorosulfonyl isocyanate, followed by conversion to the sodium salt, gave the desired compound in 83% yield (56% overall yield).

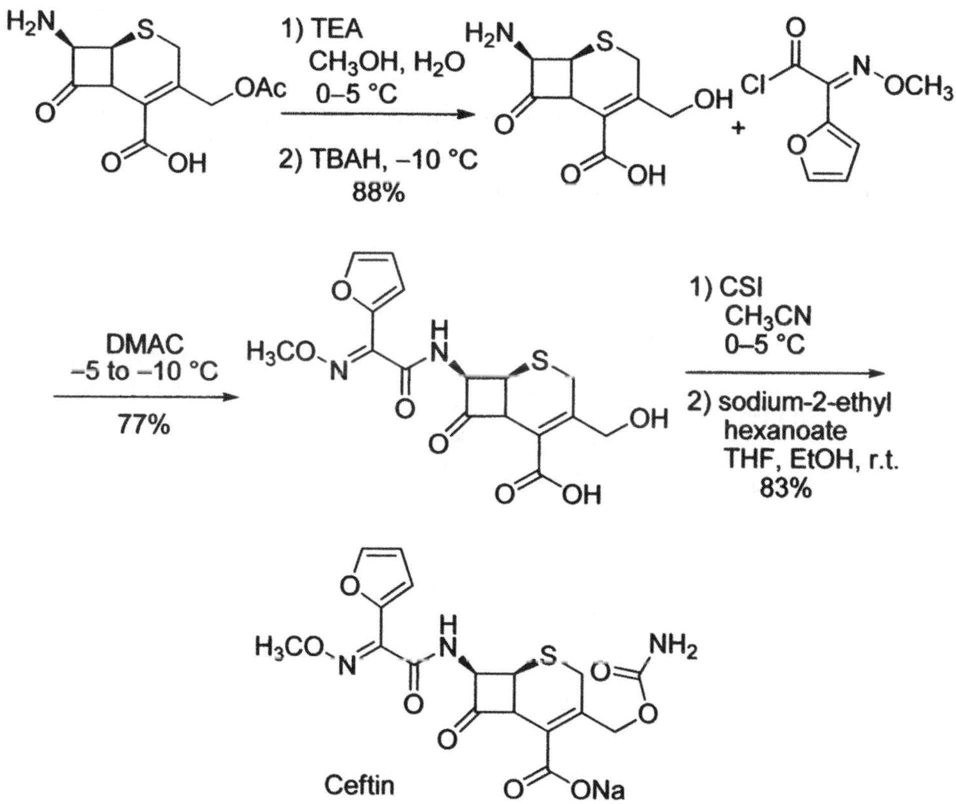

Methoxsalen and trioxsalen are derived from a family of benzofuran-containing plant natural products known as psoralens. Both methoxsalen and trioxsalen act as photosensitizers that decompose in the presence of UV light into active metabolites that crosslink DNA. These compounds have found significant use in the treatment psoriasis, eczema, and vitiligo.[70–72]

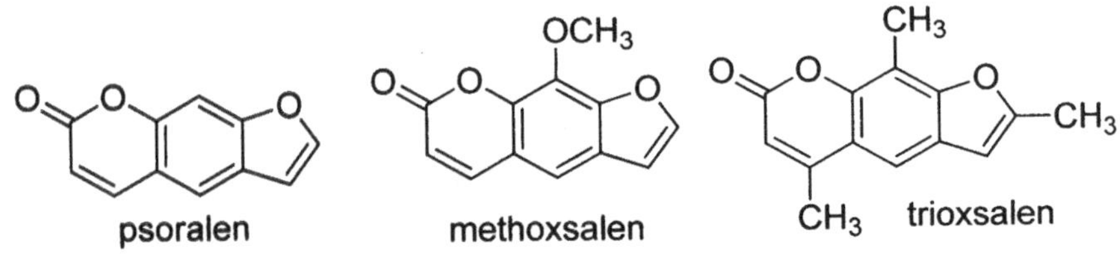

Spath and Pailer reported the first synthesis of methoxsalen in 1936.[73] Their original report involved the condensation of a dihydroxy coumarin derivative with malic acid to afford trace amounts of methoxsalen.

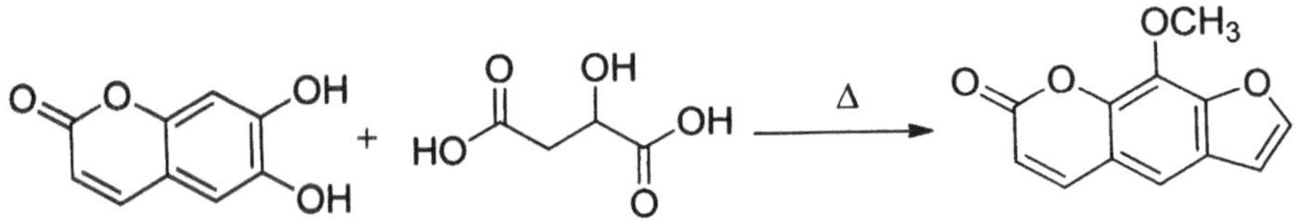

Mai and co-workers later published an improved synthesis of methoxsalen.[70] Anionic [4 + 2]-cycloaddition followed by a retro-Diels–Alder reaction and esterification formed the key tricyclic intermediate in low yield. Two additional steps were required to reach the target compound.

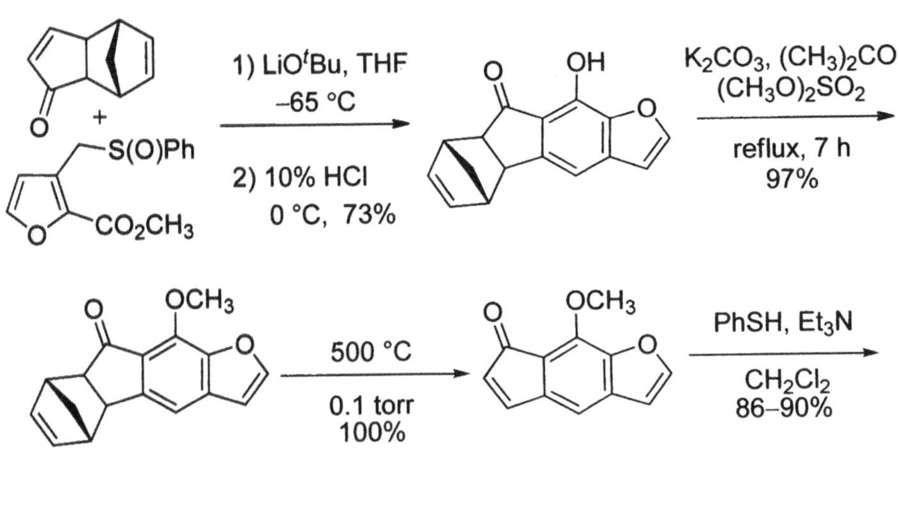

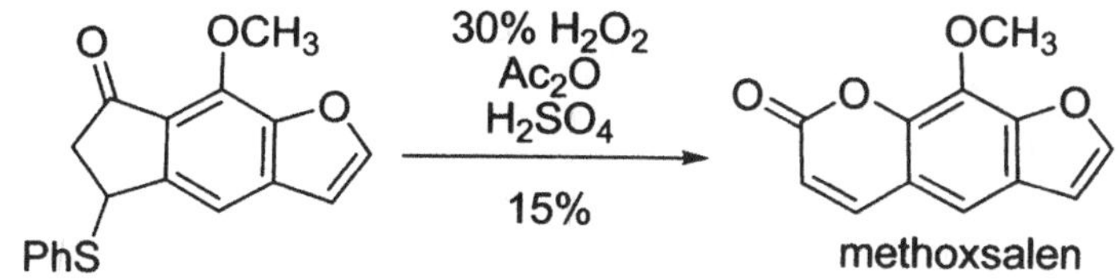

The benzofuran triflamide compound saprisartan has been approved as an angiotension II (AT-II) antagonist for the treatment of hypertension and heart failure. Saprisartan's triflamide group is credited with its high bioavailability.

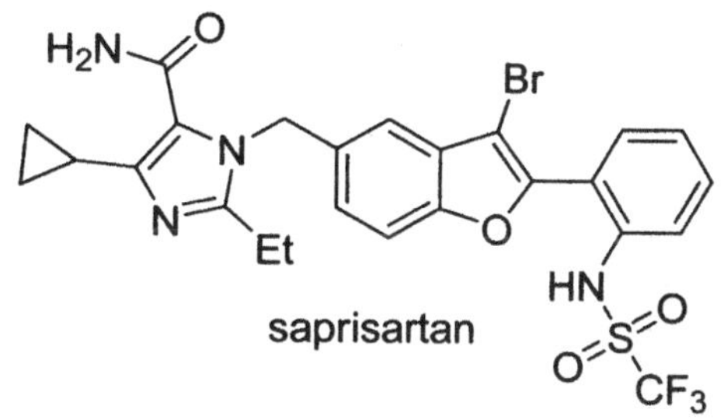

The synthesis of saprisartan, reported by Judd and co-workers,[74] used a Suzuki reaction as a key step to couple a benzofuran-based boronic acid with methyl 2-bromobenzoate in 30% yield. The resulting compound was then brominated and the methyl ester was hydrolyzed to produce the corresponding acid in 92% yield. Conversion of the acid to the Boc protected amine occurred with 18% yield. A second bromination, followed by substitution with a highly functionalized imidazole, gave the desired intermediate in 60% over two steps. Three additional steps were required to generate saprisartan in 1% overall yield.

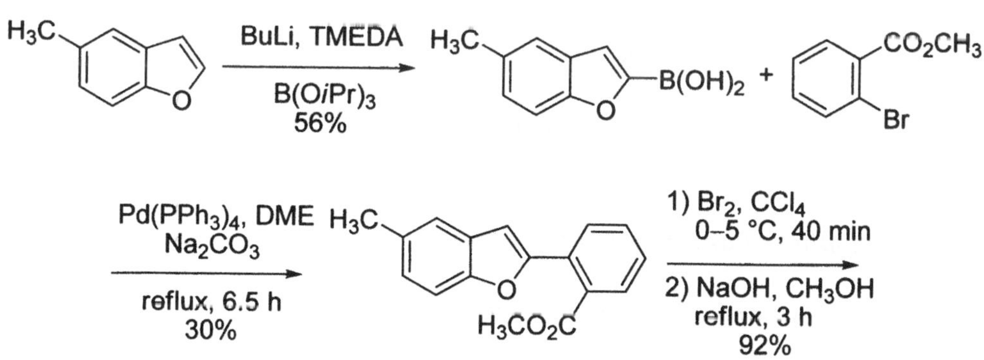

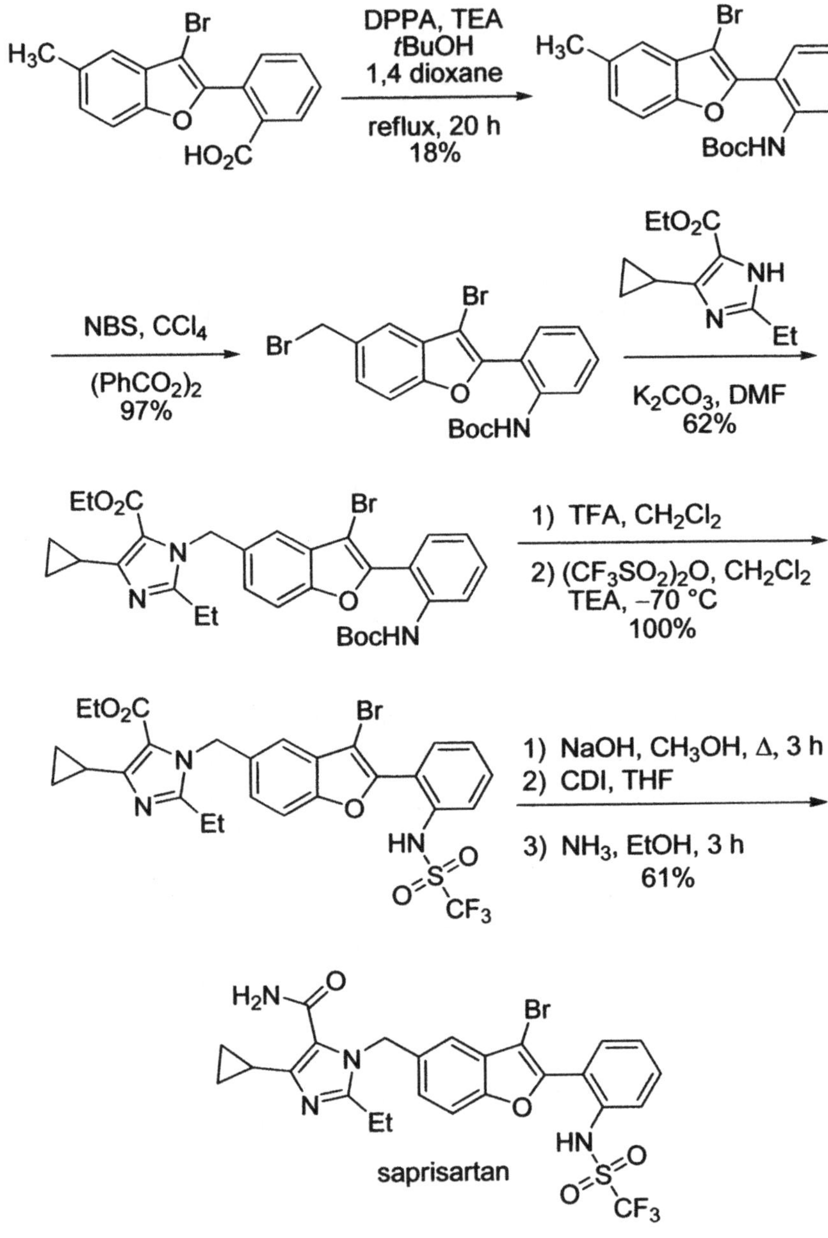

4.3 *Thiophenes and Benzothiophenes*

4.3.1 *Reactions of Thiophene and Benzothiophene*

4.3.2.1 *Reactions at C2*

Electrophilic Substitution

Electrophilic substitution of the thiophene ring occurs predominately at the *C2* position. As seen below, this is facilitated by the lone pairs on the sulfur atom.

Preferred *C2*-electrophilic substitution:

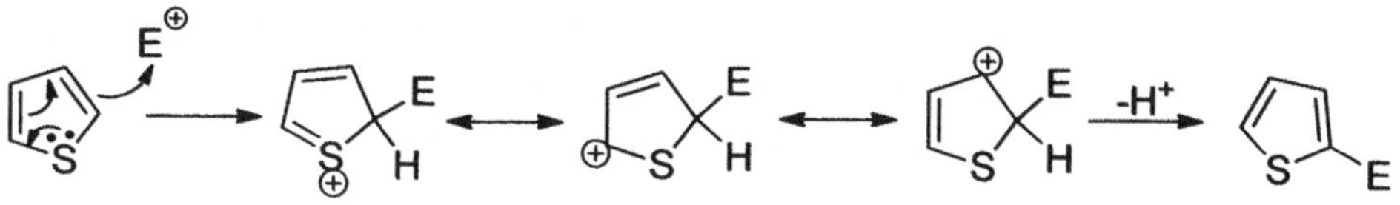

Addition at the *C3* position of the thiophene is also possible, although unlikely unless the *C2* position is already substituted. The electron-rich thiophene ring prefers electrophilic substitution at the *C2* position because the intermediate has greater charge delocalization and, therefore, more stabilization in comparison to the less-favored *C3* position.

Less-favored *C3*-electrophilic substitution:

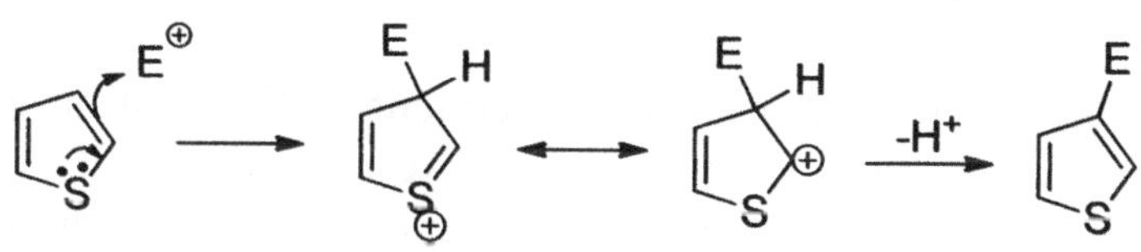

Halogenation (Bromination)

Monobromination of thiophene at the *C2* position can be achieved in excellent yields by the slow addition of hydrobromic acid in the presence of hydrogen peroxide under reduced temperatures.[75]

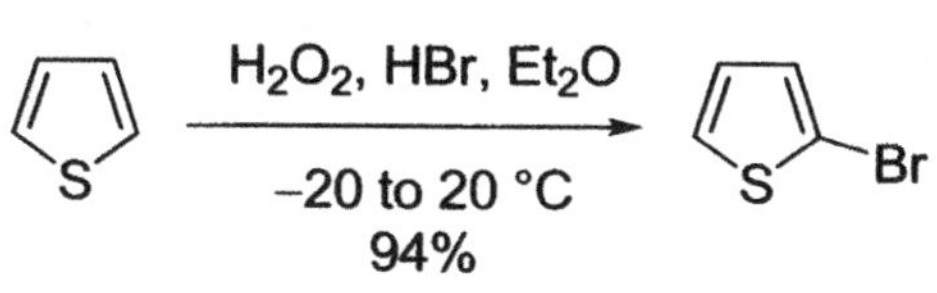

Bromination at both *C2* and *C5* is also possible with the addition of excess bromine.[75]

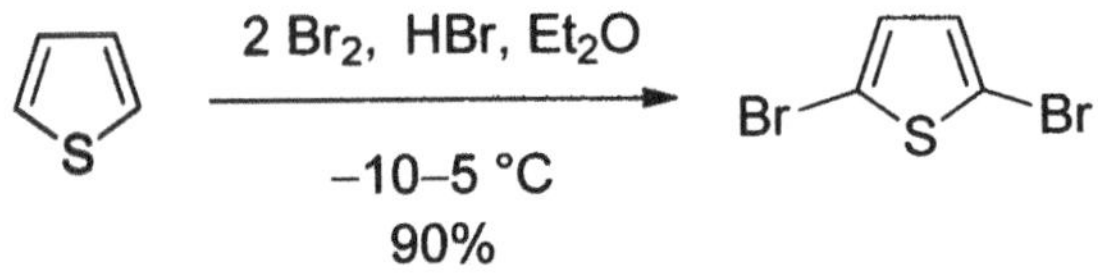

As seen throughout this chapter, halogenation produces functionalized thiophenes that are valuable intermediates in metal–halogen exchange reactions, perfluoroalkylation, and cross-coupling reactions, among others. One such example is the solid-phase bromination of thiophene at the *C2* position in greater than 95% yield. The bromination product can then be exploited further through various cross-coupling reactions.[76]

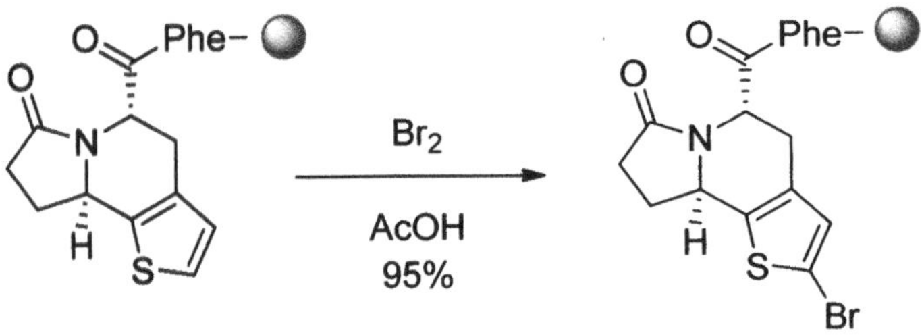

Lithiation/Methylation

Lithiation of 3-methylthiophene and subsequent addition of an electrophile, in this case methyl iodide, produces a mixture of products at the *C2* or *C5* positions, with the 2,4-disubstituted thiophene predominating. However, a strong electron-withdrawing or sterically hindered *C3*-substituent will direct the formation of the *C5* methylated thiophene to the more sterically favorable position.[77]

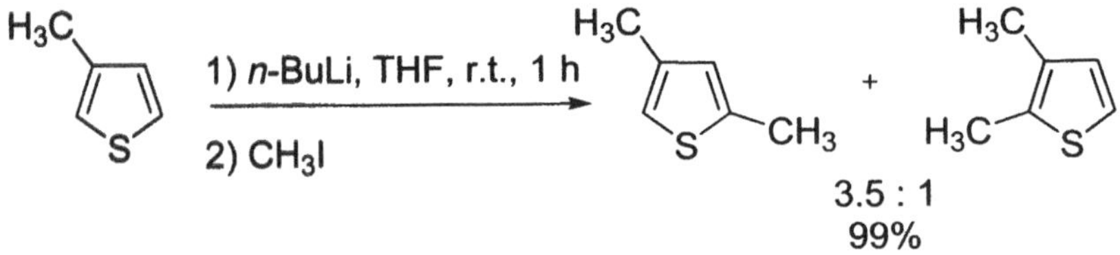

The selection of the lithiation reagent also has a strong influence over the regioselectivity of the electrophilic substitution; the bulkier the base, the more favorable the reaction at the *C5* position. For example, lithium 2,2,6,6-tetramethylpiperidine (LiTMP) directs the formation of the *C5*-substituted product at a 79:1 ratio to the *C2*-substituted thiophene. A variety of electrophiles can be substituted after lithiation by LiTMP.[77]

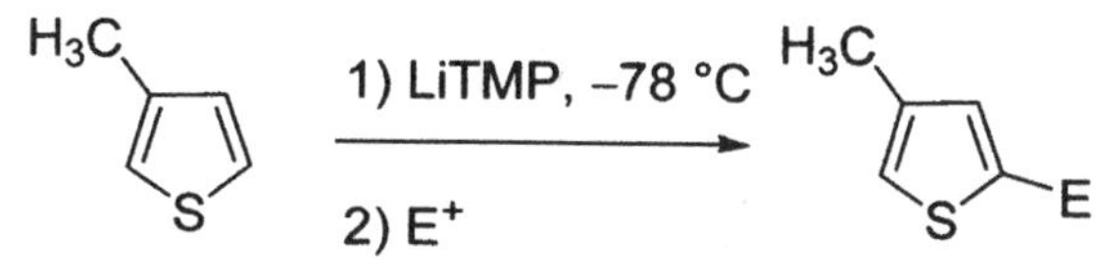

Metal–Halogen Exchange

The Grignard reaction is an important carbon–carbon bond-forming reaction that is widely used in organic synthesis. The utility of the Grignard reaction in synthesis of functionalized thiophenes can be seen in the reported synthesis of zileuton (Zyflo), an inhibitor of 5-lipoxygenase. The benzothiophene-based Grignard reagent displaces the sulfone of tert-butyl hydroxy(1-(phenylsulfonyl)ethyl)carbamate to form the zileuton precursor in 76% yield.[78]

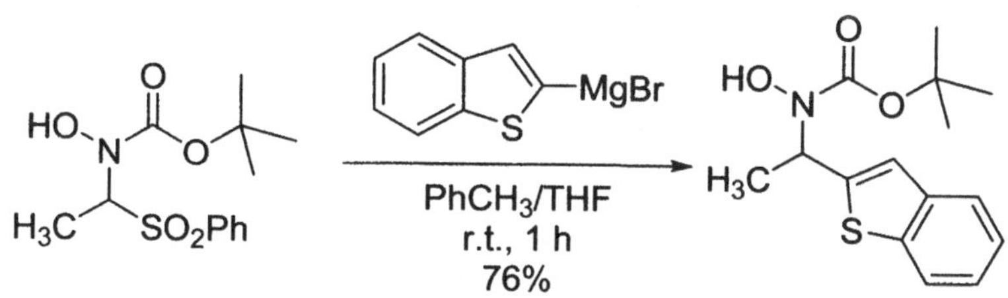

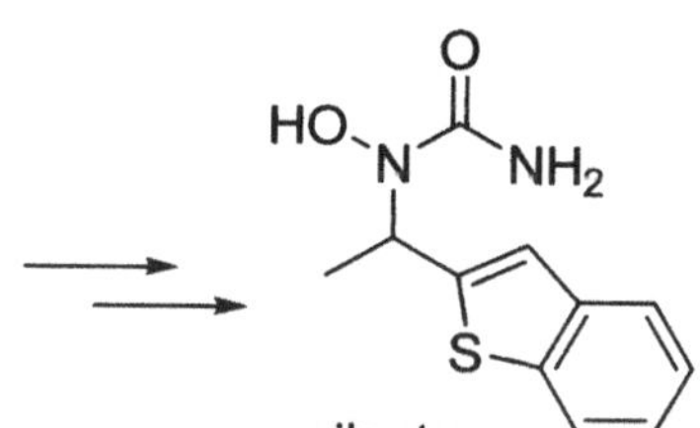

zileuton

Friedel–Crafts Reaction

The importance and versatility of the Friedel–Crafts reaction in thiophene chemistry can be seen in the following examples of duloxetine synthesis involving Friedel–Crafts acylations. Duloxetine (Cymbalta) is a serotonin–norepinephrine reuptake inhibitor (SNRI) known for its use as an anti-depressant.

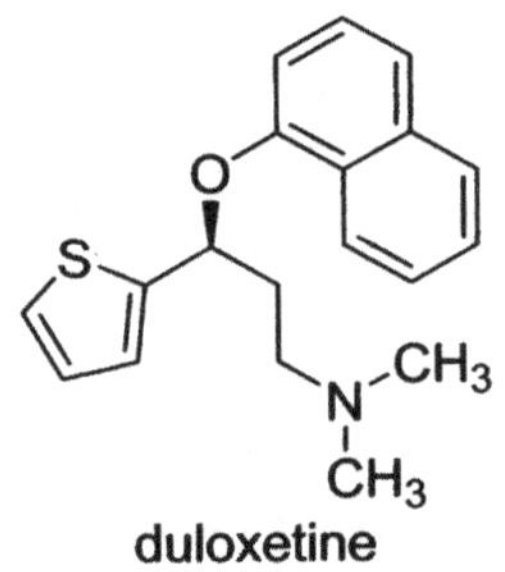

duloxetine

The following acylation reactions of thiophenes are regioselective (*C2*) and high yielding. Both reactions are key C–C bond forming reactions that have been used in the synthesis of duloxetine.[79,80]

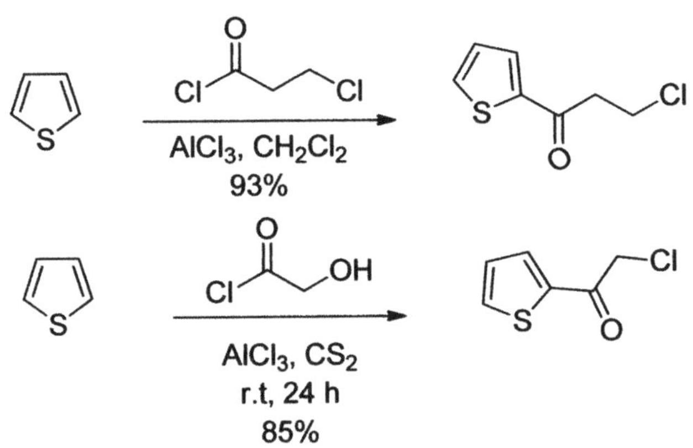

Jurczak and co-workers have developed an enantioselective variation of the Friedel–Crafts reaction to produce hydroxyl(thiophene-2-yl)acetates from the reaction of thiophenes with glyoxylates in the presence of a chiral BINOL–titanium catalyst.[81] The desired thiophenes can be produced in high enantiomeric excess and can be utilized as a key intermediates in the synthesis of duloxetine.

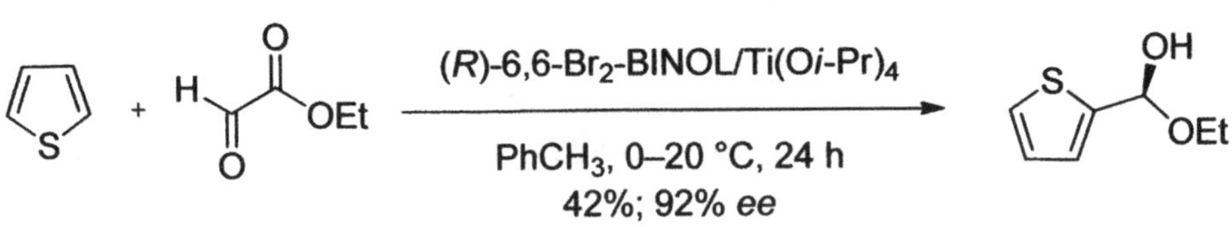

Vilsmeier–Haack Reaction

The Vilsmeier–Haack reaction has been used in the synthesis of thiophenes with potential anti-inflammatory properties. The electron-rich ring of the thiophene system makes it an ideal nucleophile for the Vilsmeier–Haack formylation reaction. Treatment of thiophene or a thiophene derivative with DMF and phosphorus oxychloride generates the corresponding aldehyde which can be further functionalized. For example, additional synthetic transformations on the Vilsmeier–Haack product of a 2-substituted thiophene produced an anti-inflammatory agent that displayed limited desired biological activity.[82]

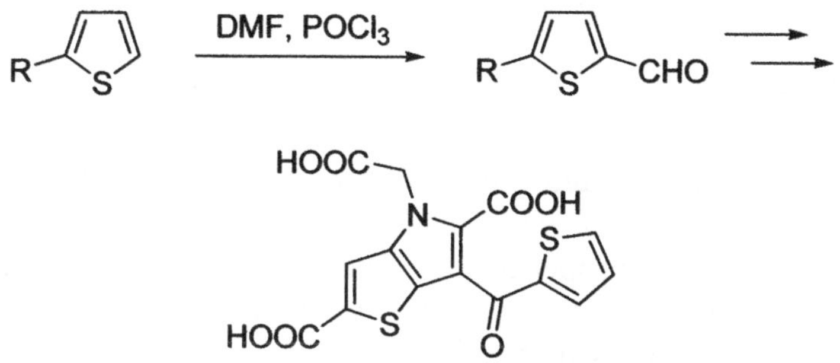

Aromatic nitriles, including thiophenes, can be accessed through a Vilsmeier–Haack protocol followed by the addition of iodine and aqueous ammonia. Benzothiophene (not shown) was found to be unreactive in this system. The authors theorized that this was possibly due to insufficient electron density.[83]

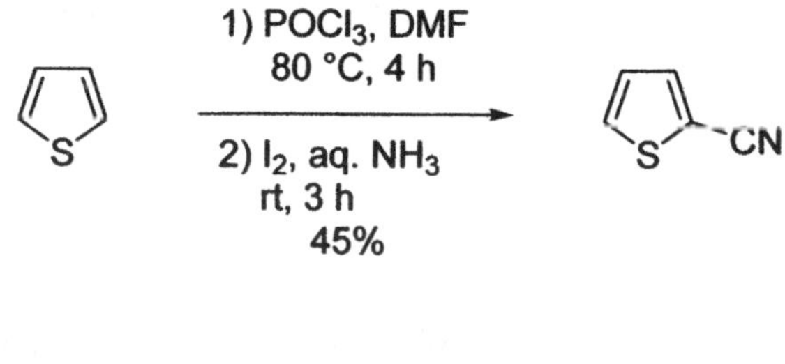

Carboxylation

Carboxylation of the thiophene ring may occur in a variety of ways. A classic method involves a Friedel–Crafts acylation followed by oxidation.[82]

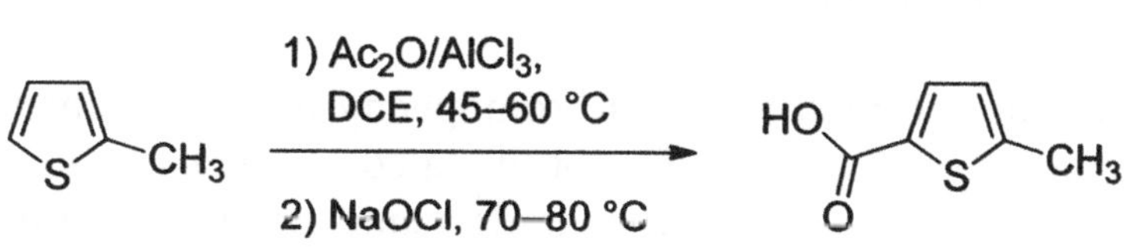

Other methods of carboxylation include the use of palladium(II) acetate in the presence of carbon dioxide. Carboxylation is favored predominately at the *C2* position (or *C5* if *C2* is substituted). Interestingly, when a thiophene containing a *C3* methyl substituent was subjected to carboxylation under these conditions, it was found that the *C3* positions had little effect on the selectivity of the *C2* or *C5* carboxylation.[84]

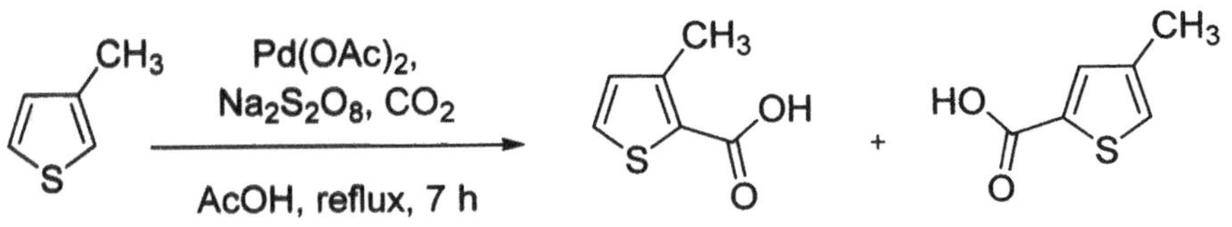

Perfluoroalkylation

Tsuji and co-workers reported the perfluoroalkylation of bromo-thiophenes to produce a compound that was found to have excellent anti-inflammatory and immune-regulatory properties.[85] The reaction yielded only a small quantity of the desired product (~15%), with the reduced (–H) compound (not shown) being the major by-product.

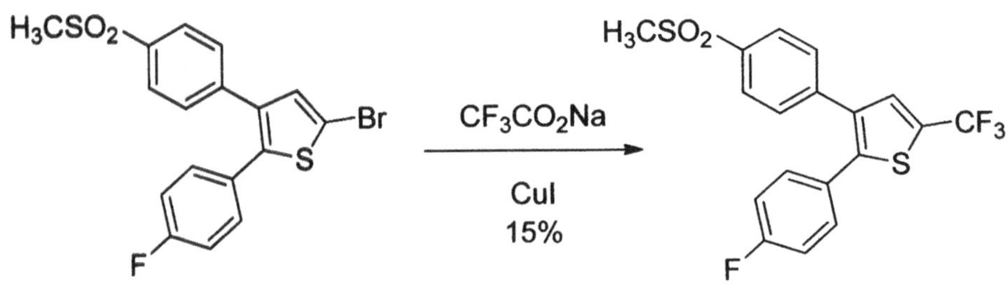

4.3.1.2 Reactions at C3

Electrophilic Substitution

Electrophilic substitution reactions of thiophenes predominate at the *C2* (*C5*) position. In many cases, reaction at the *C3* position is only favored when the *C2* and *C5* positions are already substituted or severe sterics are involved.

Halogenation (Bromination)

Bromination of 6-methoxy-2-(4-bromophenyl)benzothiophene occurs smoothly at the *C3* position.[86] The brominated product is able to undergo further reactions to form raloxifene, trade name Evista. Marketed by Eli Lilly and Company, this drug is a selective estrogen receptor modulator (SERM) used to prevent osteoporosis in postmenopausal women.

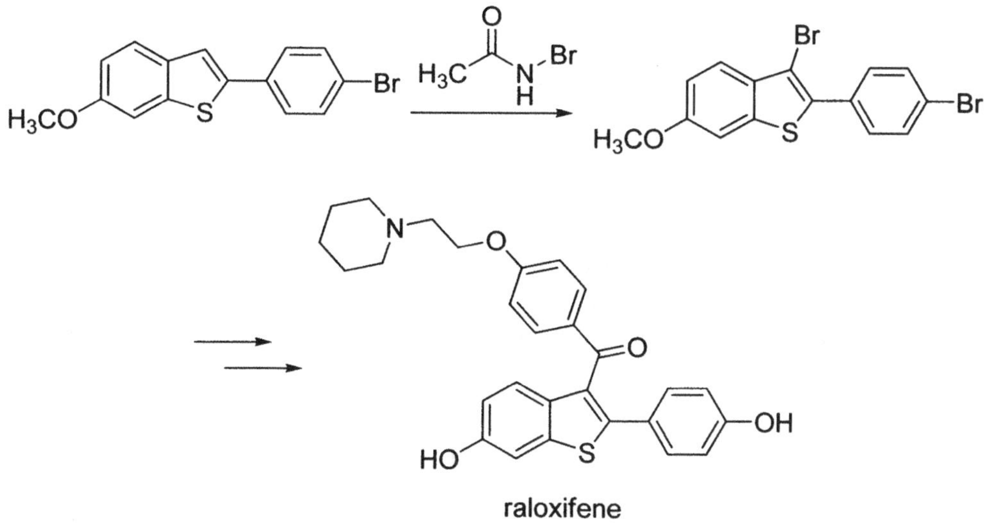

Friedel–Crafts Reaction

Friedel–Crafts acylations of benzothiophene are known to give mixtures of both the 2- and 3-isomers; however, the 3-isomer predominates by 80%. Steric hindrance by the *C5*-substituent was shown to play a substantial role in the product formation, generally producing higher yields of acetylation compared to acylation. The reactions below illustrate several different reagents and *C5*-benzothiophene substituents that are capable of undergoing *C3*-substituted Friedel–Crafts reactions.[87]

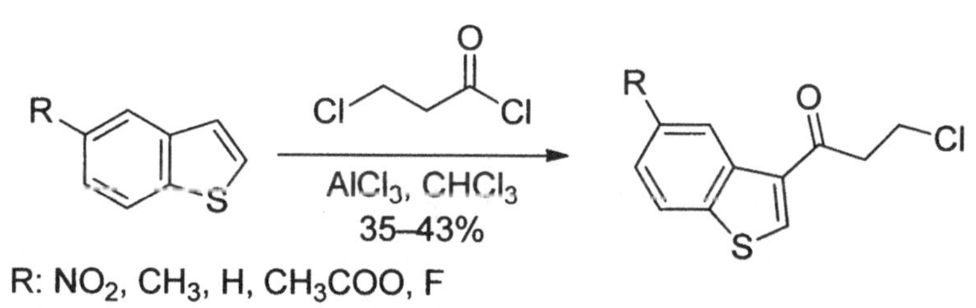

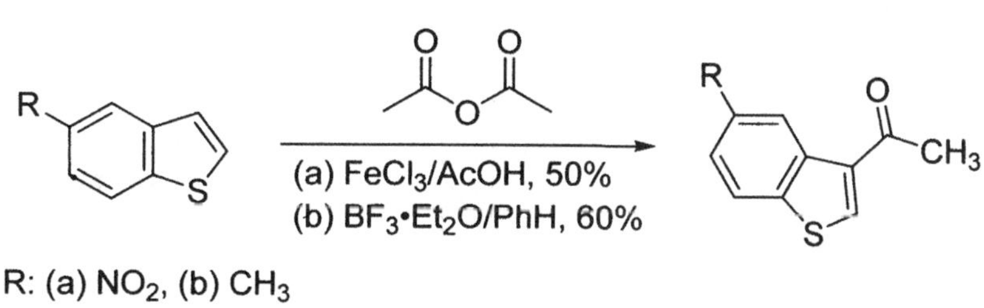

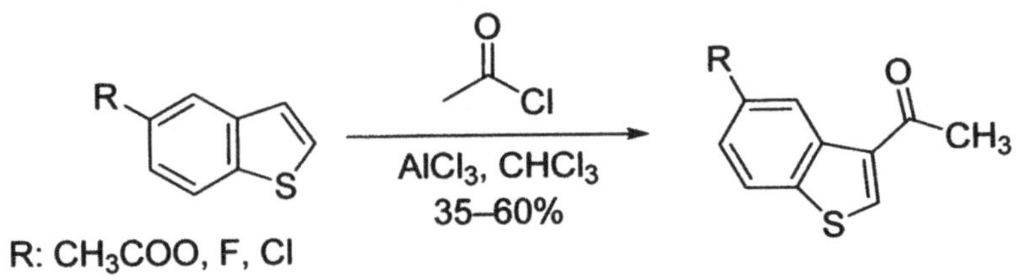

The Friedel–Crafts acylation reaction at the *C3* position of a substituted benzothiophene was shown to be a key C–C bond forming step in the synthesis of raloxifene, as first reported in 1984 and illustrated below.[88]

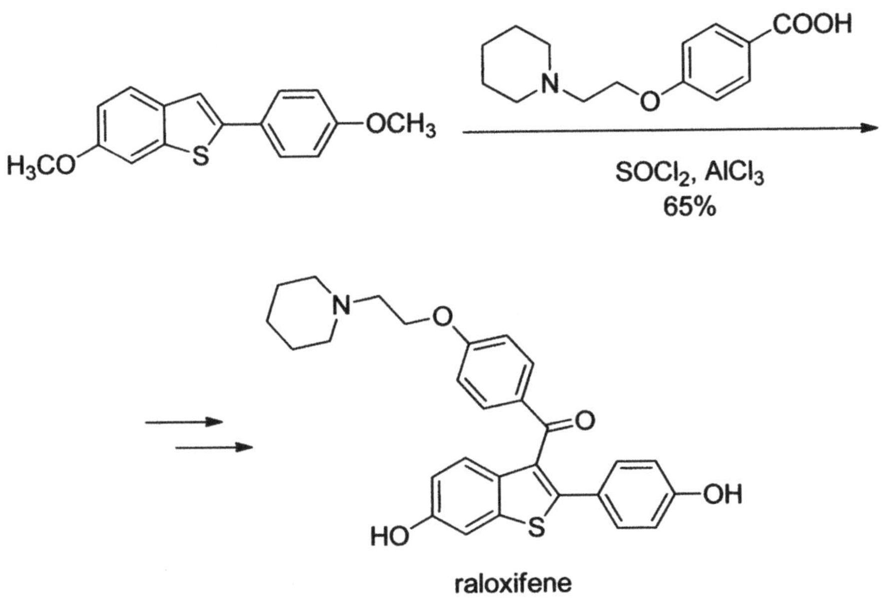

Carboxylation

The *C3* position understandably undergoes carboxylation when both the *C2* and *C5* positions are substituted, albeit with diminished yields in comparison to the reaction at the *C2* position. For example, the palladium(II)-catalyzed reaction of the 2,5-dichlorothiophene under thermal conditions gave the corresponding 2,3,5-trisubstituted thiophene with 1.6% yield.[84]

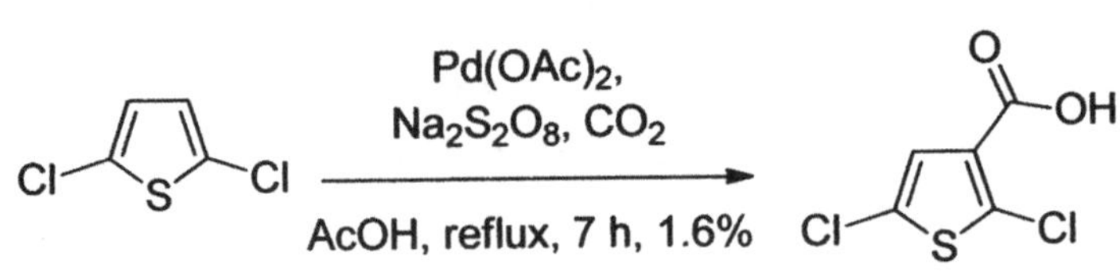

Intramolecular Cyclization

Katritzky and co-workers illustrated the reactivity of the *C3* position with the formation of a thiophene dimer via an intramolecular cyclization.[89] The cyclization was performed first by lithiation followed by the addition of zinc bromide under refluxing conditions to produce the dimerized product.

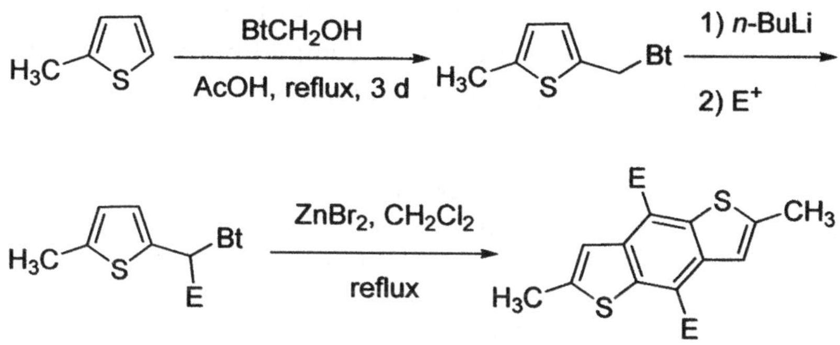

The synthesis of thieno-[3,2-*b*]-pyrroles reported by Reddy and co-workers is another interesting example of an intramolecular cyclization at the *C3* position.[82] When the product of a Vilsmeier–Haack reaction is condensed with ethyl azidoacetate, the resulting azido compound undergoes an intramolecular cyclization under refluxing conditions to yield a thieno-[3,2-*b*]-pyrrole ring system. The thieno-[3,2-*b*]-pyrrole ring system can also be accessed through condensation of a enol ester and a nitro group at the *C3* position.

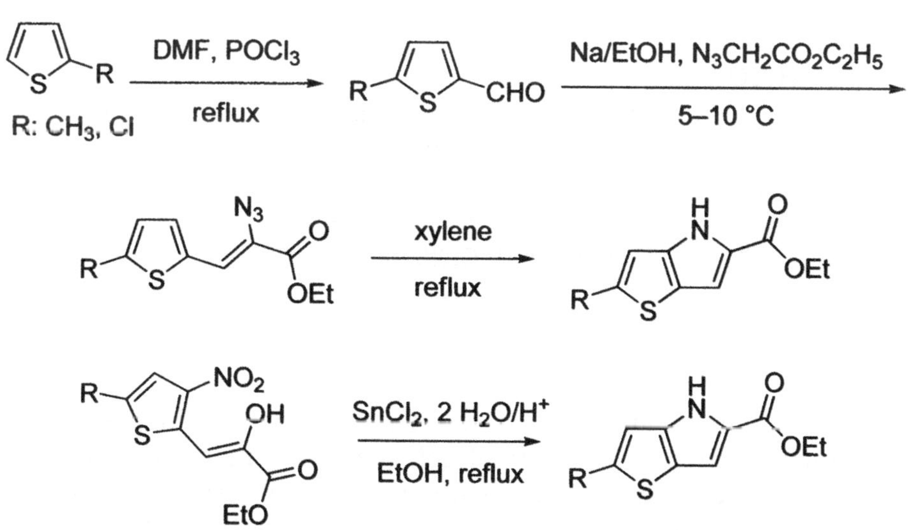

4.3.1.3 *Transition Metal-Catalyzed Cross-Coupling Reactions*

Cross-coupling reactions have proven to be a powerful tool in organic synthesis; this is no different for the manipulations of thiophenes and benzothiophenes.[90] The following examples, reported by Larock and co-workers, highlight the versatility of halo-benzothiophenes (synthesized via electrophilic cyclization) as substrates for various cross-coupling reactions such as the Heck, Sonogashira, and Suzuki–Miyaura reactions.[91] These compounds were synthesized to create a screening library of benzothiophene-based analogues as biologically active compounds.

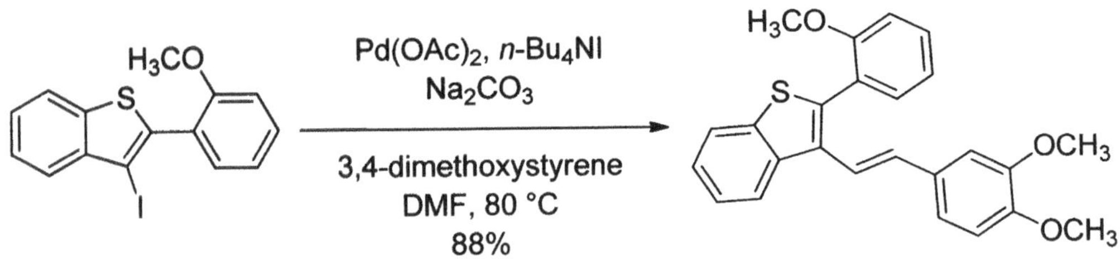

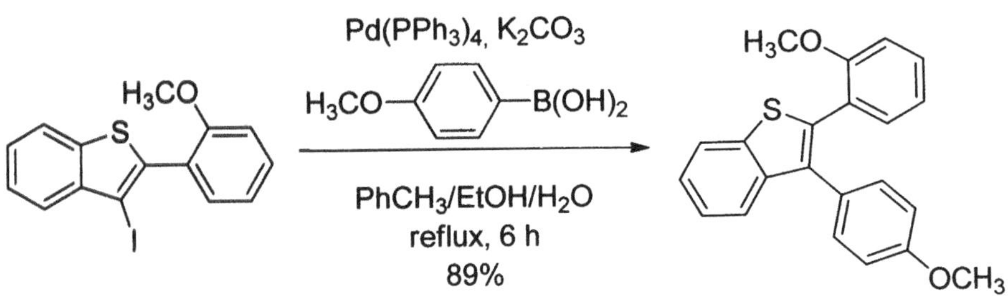

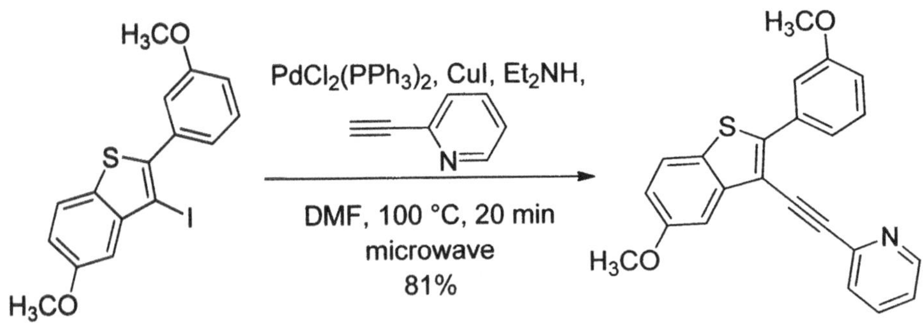

Direct functionalization of the *C2* position of thiophene can be accomplished through an oxidative-Heck reaction employing palladium(II) acetate in the presence of silver carbonate. The resulting products are obtained in good to high yields with the *trans*-alkene isomer being favored. Typically less than 10% branched coupling products were observed (not shown).[92]

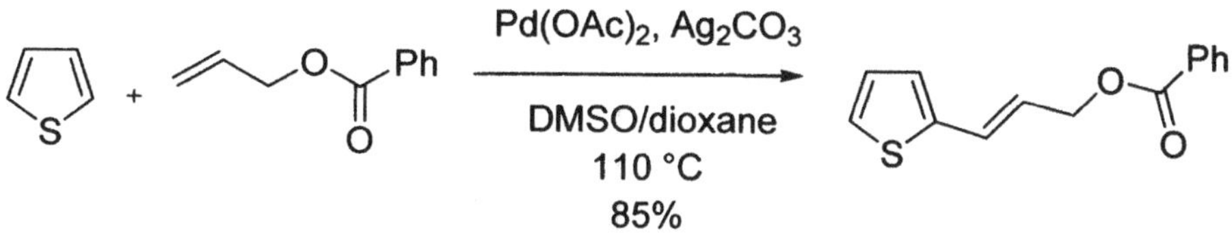

The Suzuki–Miyaura reaction has proven useful for synthetic chemists in the formation of aryl–aryl bonds. Benefits include generally mild reaction conditions, compatibility with most functional groups, and the use of readily available boronic acids known for their stability. In the reaction shown below, Meldal and co-workers reported a palladium-catalyzed Suzuki cross-coupling reaction of a bromothiophene using a solid-phase synthesis protocol.[93]

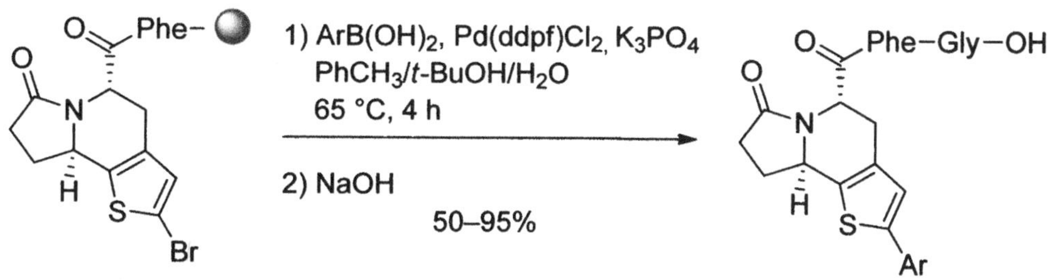

Like Larock, Flynn and co-workers utilized various electrophilic cyclization reactions to produce desirable halo-benzothiophenes for use in cross-coupling reactions. The following benzothiophene synthesis involved a Suzuki cross-coupling reaction to form new aryl–aryl bonds at the *C3* position. The product has been investigated in the search for potent tubulin-binding agents.[94]

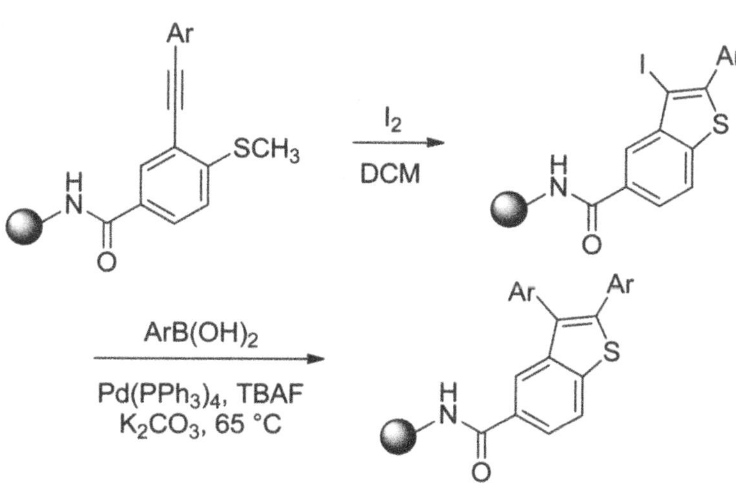

The Suzuki–Miyarua reaction at the *C3* position of benzothiophenes were shown to be useful in producing raloxifene derivatives such as desketo-raloxifene.[95]

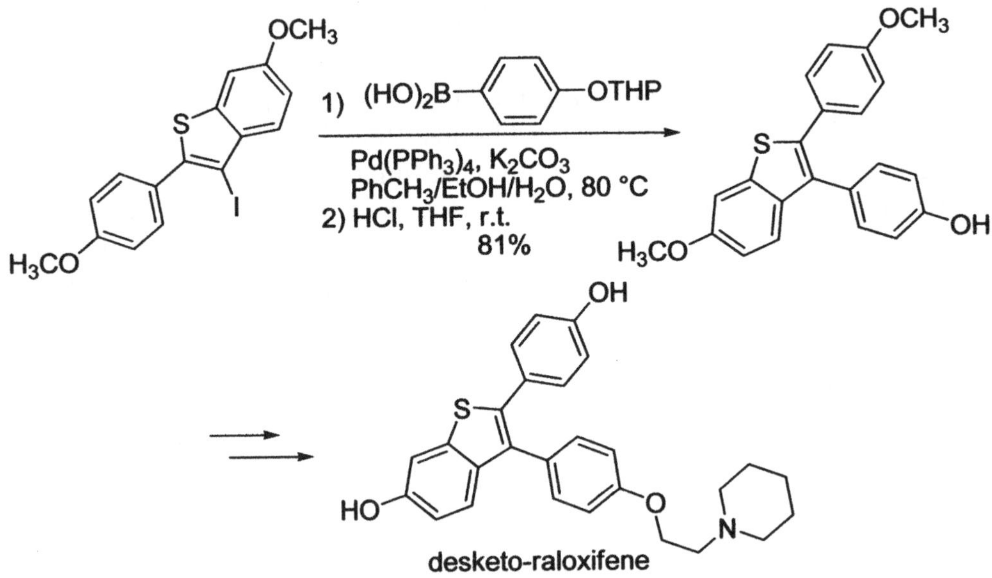

The Sonogashira coupling of bromothiophenes led to the synthesis of promising anti-tumor agents. Coupling of the *C5* bromothiophene with alkynyl alcohols using copper iodide, palladium(II) chloride, and triphenylphosphine in the presence of triethylamine, gave the corresponding propargylic thiophene in 74% yield. Several additional steps were required to synthesize a highly potent antifolate inhibitor shown below.[96]

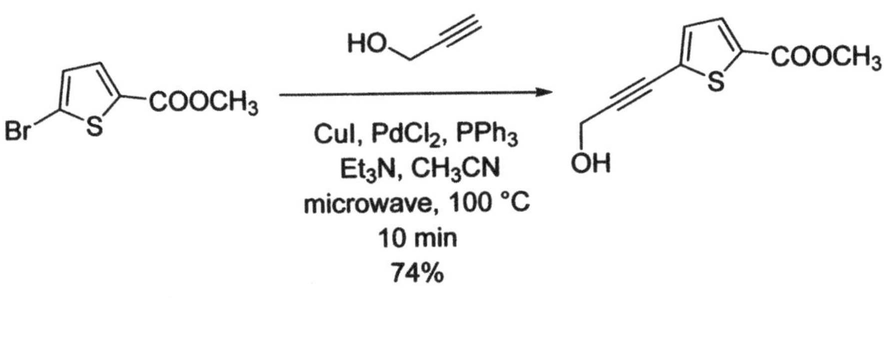

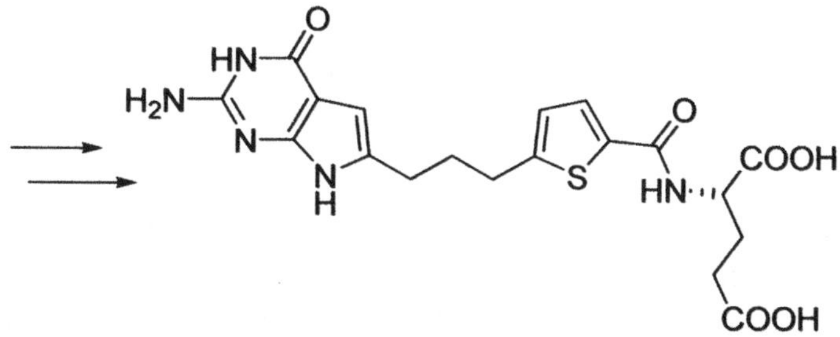

The Buchwald–Hartwig reaction has been widely used in organic chemistry for the palladium-catalyzed cross-coupling of amines with aryl halides. In particular, it has been used in conjunction with benzothiophenes for arylamination of the thiophene. The final diarylamine product below demonstrated anti-fungal activity against dermatophytes, yeasts, and the *Aspergillus* species.[97]

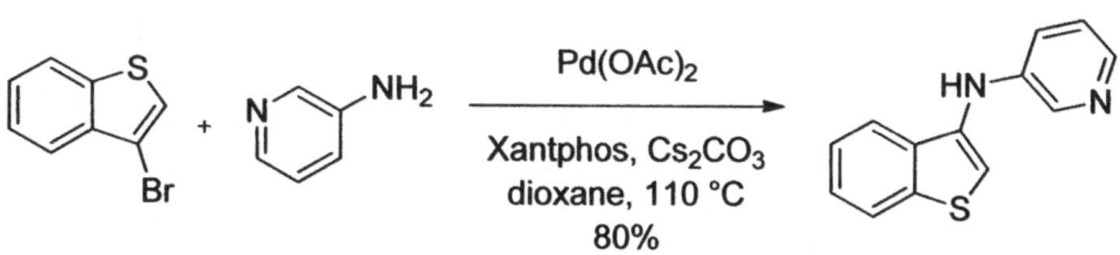

4.3.2 Synthesis of Thiophene and Benzothiophene

The Fiesselman and Gewald thiophene syntheses arc among the most important methods for constructing thiophenes.[98] These methods are useful for producing functionalized thiophenes from readily accessible starting materials. Both methods allow for thc synthesis of varied substitutions and functionalities on the ring system. This has proven to be advantageous to medicinal chemists in performing structure-activity relationships (SAR) and lead optimization studies.

4.3.2.1 Thiophene Synthesis

Fiesselman Thiophene Synthesis

The classic example of a Fiesselman thiophene synthesis is the reaction involving a methylthioglycolate and an α,β-acetylenic ester in the presence of a base.

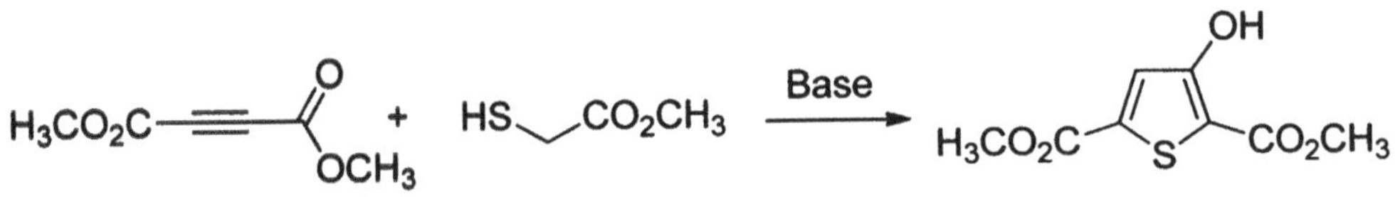

The mechanism involves the formation of a thioacetal followed by a Dieckman condensation to produce the requisite five-membered ring. The reaction is completed with elimination and tautomerization to form the desired product.[98,99]

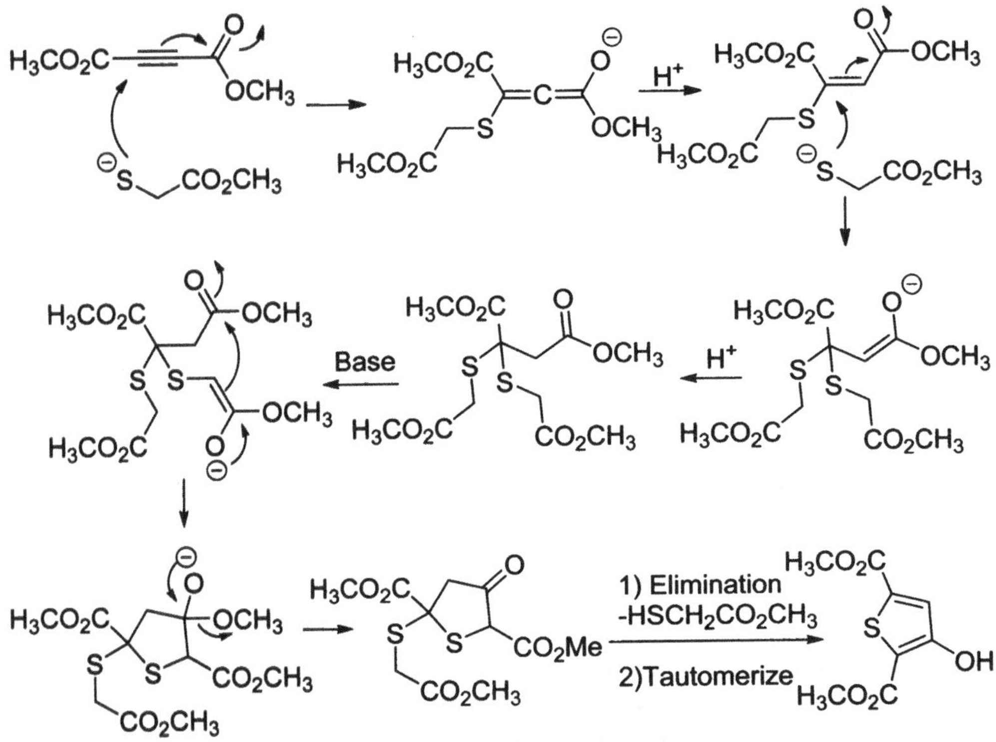

One widely used variation of the Fiesselman reaction is the use of cyanoalkynes in the presence of a thioglycolate under basic conditions to produce 3-aminothiophenes. This strategy was used in the production of 2-aminothiopehene-based p38 kinase inhibitors.[100]

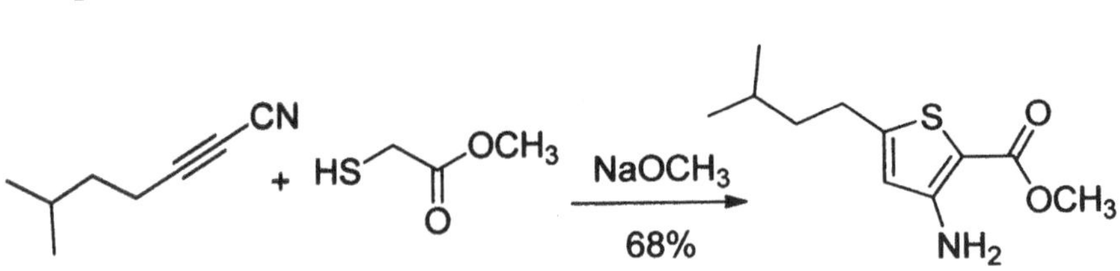

An interesting example of the Fiesselman thiophene synthesis was reported during the investigation of the susceptibility of golfomycin A toward nucleophilic attack. Treatment of TBS-protected golfomycin A with methyl thioglycolate in the presence of base produced the strained thiophene product in 20% yield.[101]

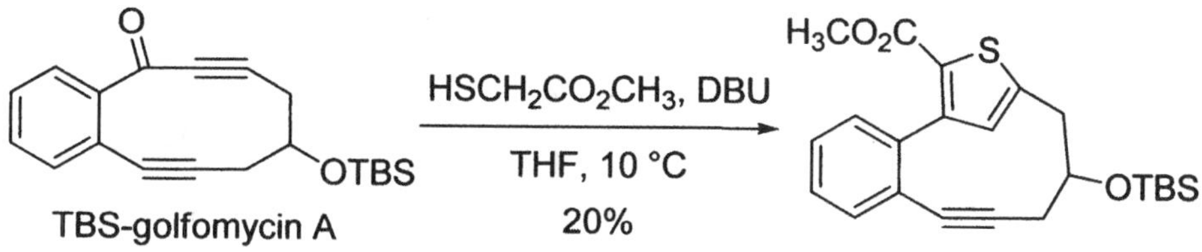

Ronzoni and co-workers sought to synthesize analogues of pyrrolomorphinans to investigate the role of the pyrrole ring in the binding of opioid receptors in an effort to increase the lipophilicty of this compound and to impove its ability to cross the blood–brain barrier.[102] Their aim was to replace the pyrrole unit with a thiophene, which they achieved through a modified Fiesselman synthesis as shown below. The key thio-intermediate[103] was cyclized in the presence of base to create a single regioisomer. The substitution of the thiophene moiety resulted in nano-molar affinity for the γ-opioid receptor; additionally, the lipophilicity was found to be markedly improved.[102]

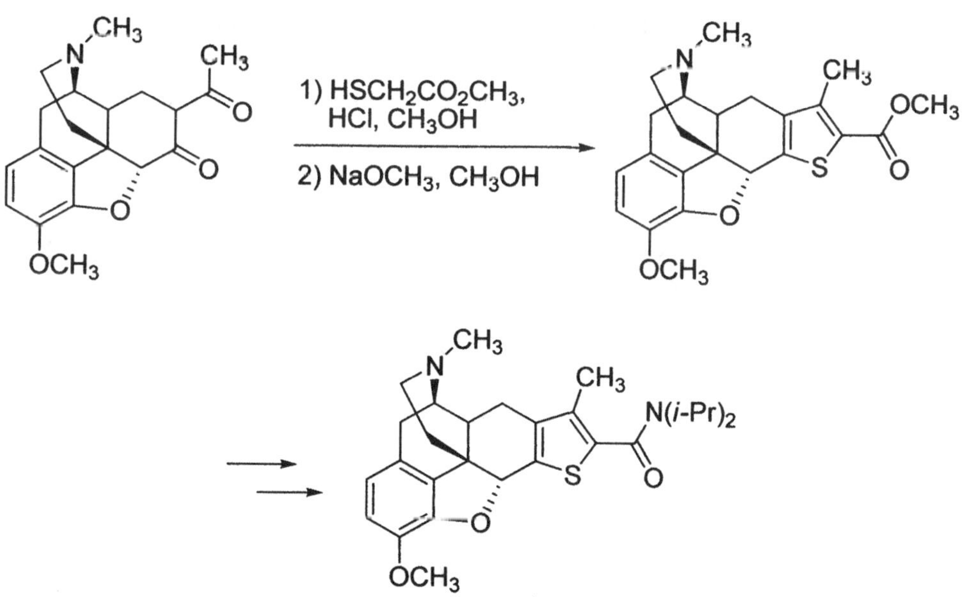

The Fiesselmann thiophene synthesis has also been used to produce various 2,3-diarylthiophenes to investigate their potential anti-inflammatory properties. The thiophene intermediate synthesized via the Fiesselman reaction was further modified through a series of reactions to produce a compound that displayed broad-spectrum anti-inflammatory and immune-regulatory activities.[85]

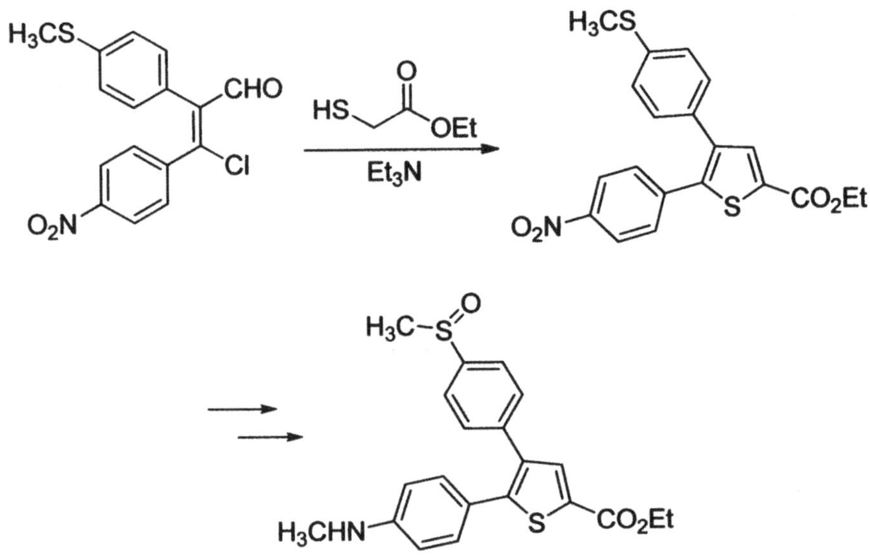

The synthesis of several methylsulfonyl steroid derivatives was undertaken to investigate the structural requirements necessary to obtain high androgen receptor (AR) affinity.[104] Thiophene analogues were produced in good yield through the reaction of β-chloro-α,β-unsaturated aldehydes under the Fiesselman conditions. Unfortunately, the authors found that the thiophene compound showed no significant binding; however, the furanyl derivative (not shown) produced results similar to the active control compound.

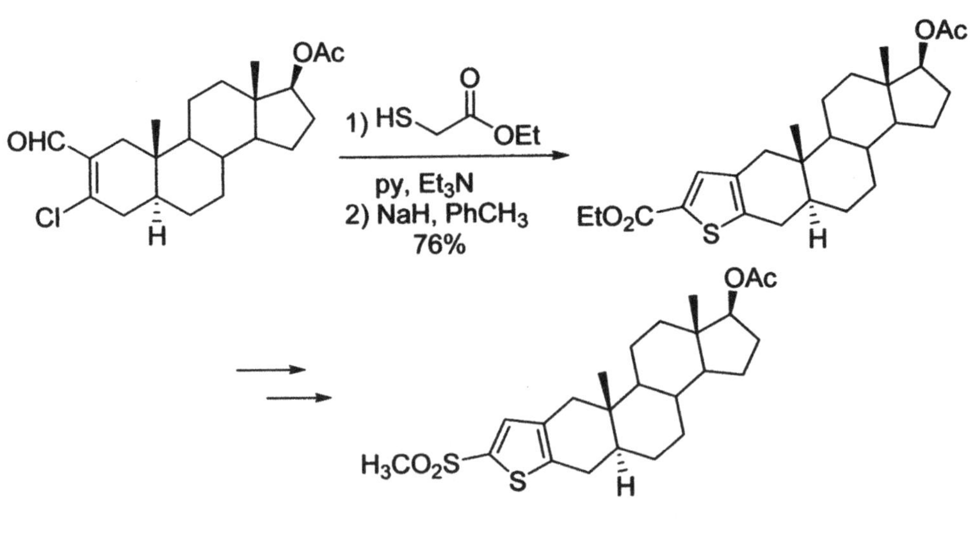

Gewald Aminothiophene Synthesis

The Gewald aminothiophene synthesis involves the condensation of carbonyl compounds (containing an α-proton) with activated nitriles in the presence of a base and elemental sulfur.

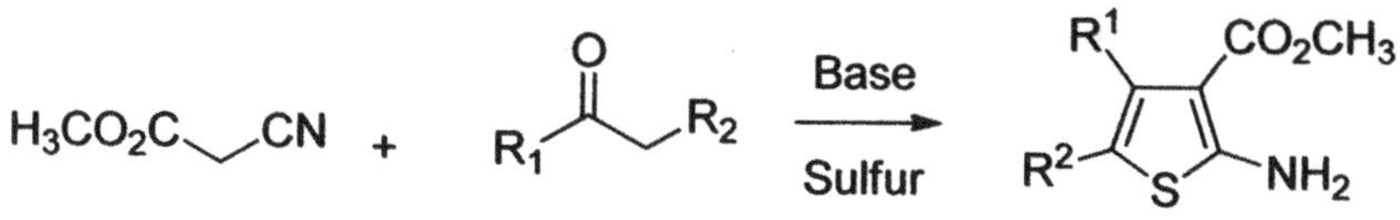

The mechanism involves a Knoevenagel condensation of the nitrile and carbonyl compound to produce the acrylonitrile intermediate. Thiolation by elemental sulfur and subsequent cyclization followed by tautomerization produces the 3-aminothiophene.[98,99]

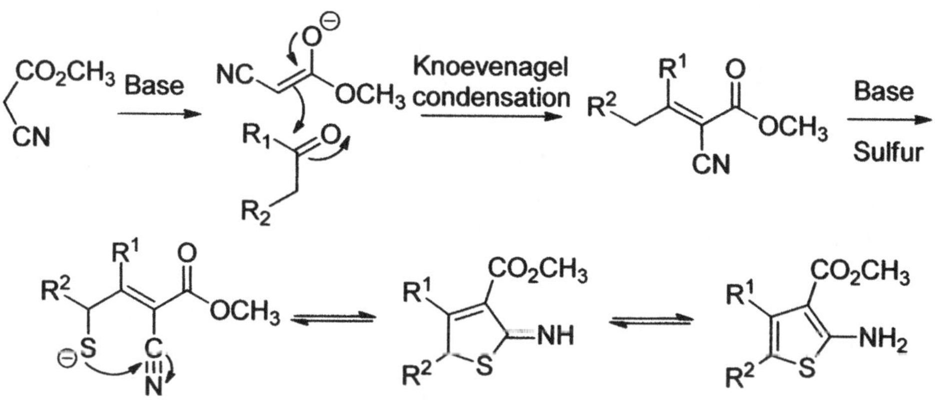

The Gewald aminothiophene synthesis was employed in the production of thienotriazine-4-ones as potential antimicrobial agents. The most potent derivative, illustrated below, exhibited inhibition for both Gram-positive and Gram-negative bacteria in the micro-molar range.[105]

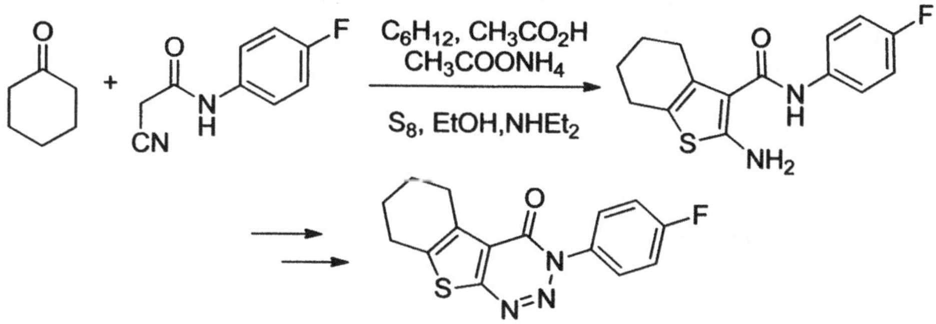

In another example of the Gewald aminothiophene synthesis, a series of thiophene-containing compounds were synthesized for use as allosteric enhancers of A_1 adenosine agonist binding. The utility of the Gewald reaction is highlighted in this example by the ability to access various substituted thiophenes depending upon the nature of the starting material. This proved to be a versatile method for the synthesis of fully substituted aminothiophenes containing various functional groups.[106]

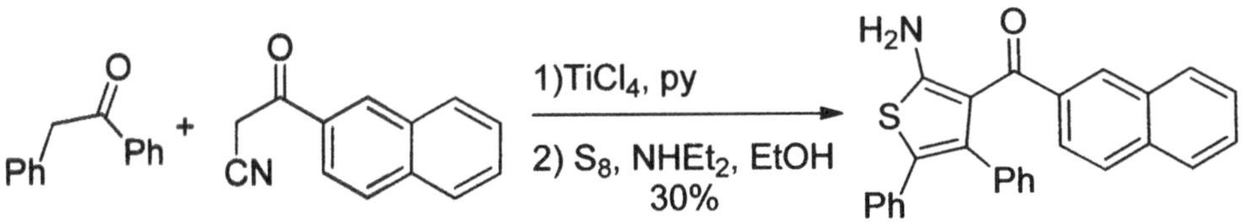

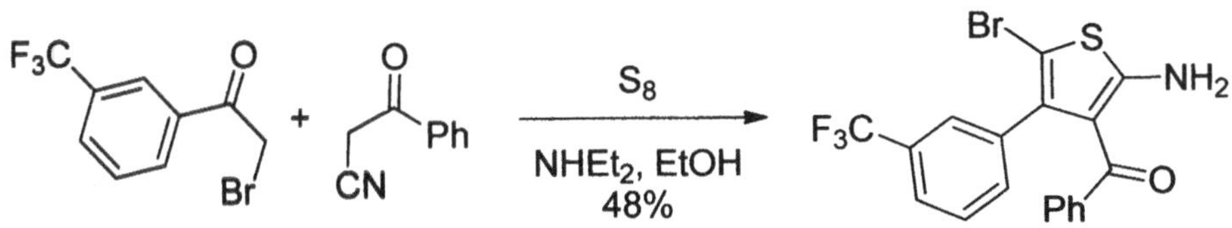

The Gewald aminothiophene synthesis was also performed under ultrasonic aqueous conditions to give high yields of the corresponding substituted aminothiophenes in several minutes.[107]

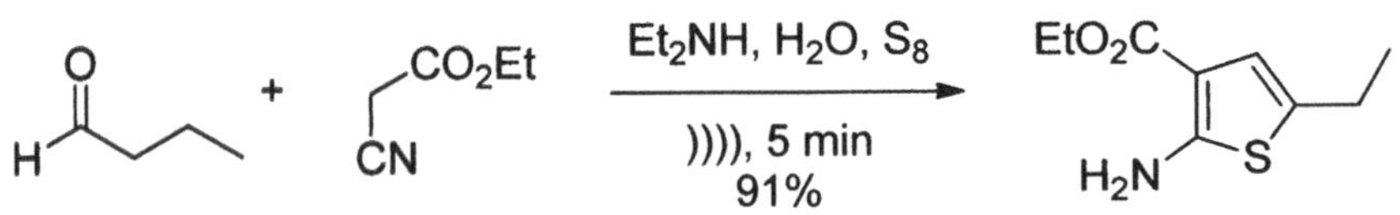

Peseke and co-workers investigated the effect of incorporating a thienopyrimidine moiety into *C*-nucleoside analogues to increase cell membrane permeability and antimicrobial and anti-tumor activity.[108] The *C*-nucleoside derivatives were constructed through the Gewald amino thiophene synthesis in 55% yield under standard conditions. The thiophene intermediate was further converted into a thienopyrimidine to produce the desired compound.

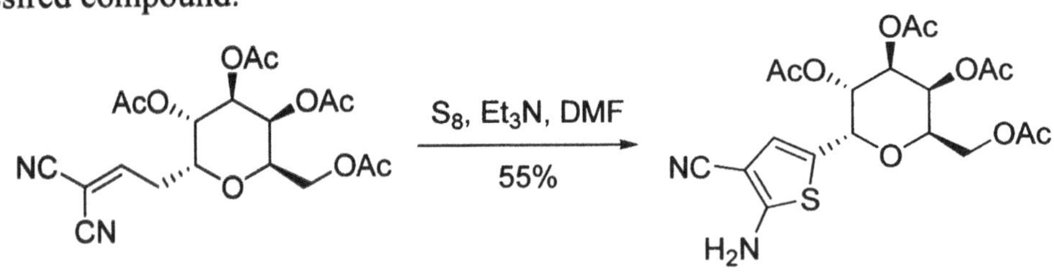

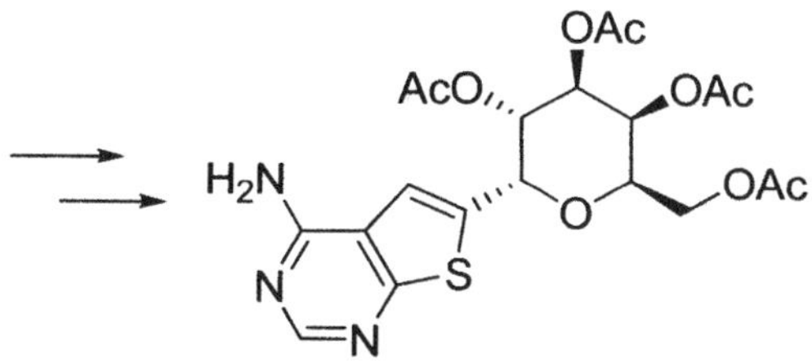

Rosowsky and co-workers sought to synthesize thiophene-containing derivatives of trimetrexate and piritrexim in their search for potential inhibitors of *Pneumocystis carinii* and *Toxoplasma gondii* dihydrofolate reductase.[109] Their goal was to improve selectivity and limit the potential side effects found in trimetrexate and piritrexim. The thiophene moieties were installed by means of a Gewald aminothiophene reaction similar to those used in the creation of the *C*-nucleosides shown previously. Desired substitution patterns were accessed by the use of appropriate substrates as shown below.

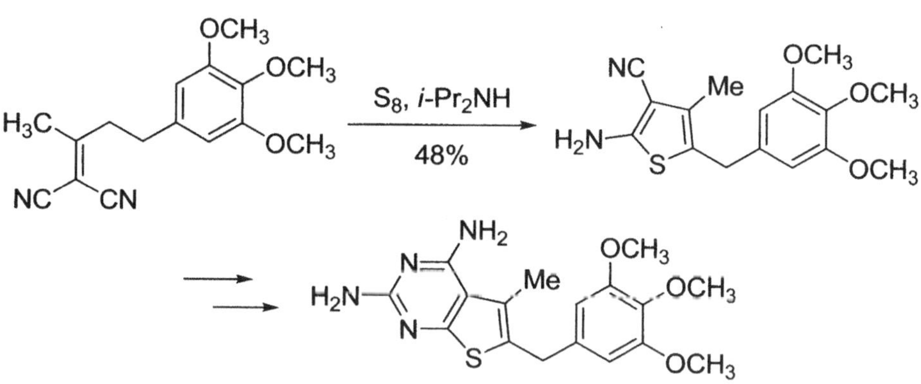

The synthetic versatility of the Fiesselman and Gewald reaction is further highlighted in the synthesis of both 2-amino- and 3-aminothiophene derivatives by Romagnoli and co-workers.[110,111]

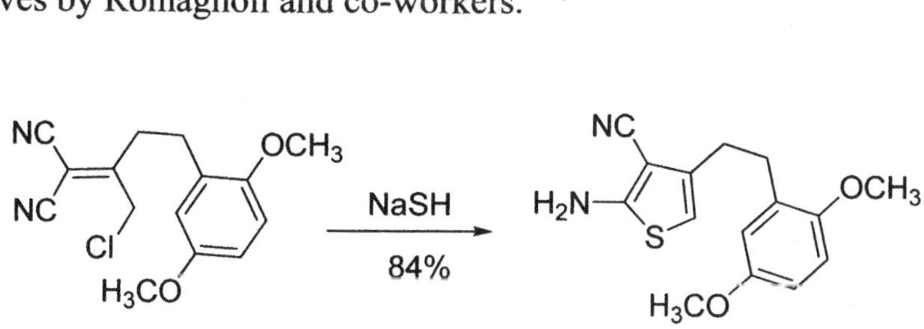

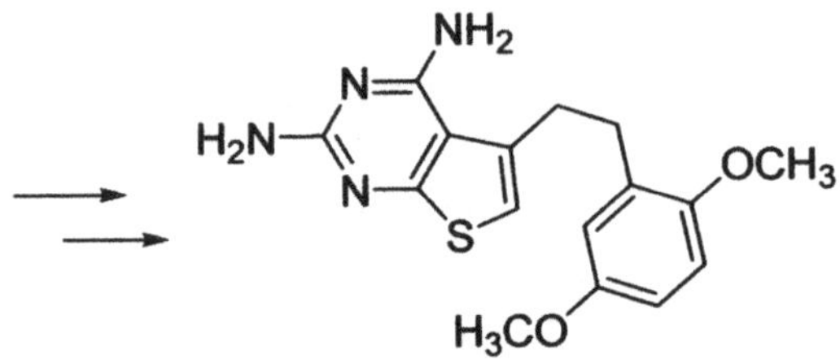

The Gewald aminothiophene synthesis was employed in the generation of 2-amino-3-(3,4,5-trimethoxybenzoyl)-5-aryl thiophenes as derivatives of the anti-tubulin agent combretastatin A.[110] Condensation of 2,5-dihydroxy-1,4-dithiane with 3-oxo-3-(3,4,5-trimethoxyphenyl)-propionitrile and elemental sulfur in the presence of triethylamine afforded the thiophene product in 74% yield. Further synthetic steps yielded an anti-tubulin agent that not only inhibited tubulin polymerization but also showed anti-proliferation activity against both L1210 and K562 cancer cell lines.

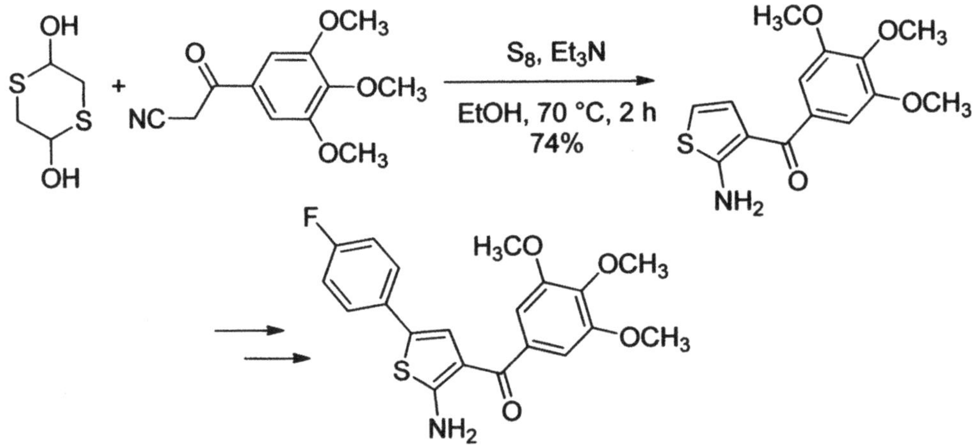

Further investigation of anti-tubulin agents by Romagnoli *et al.* led to the development of a modified Fiesselmann thiophene synthesis to produce 3-amino-2-(3,4,5-trimethoxybenzoyl)-5-aryl-thiophenes.[111] The α-mercapto-ketone anion was generated *in situ* by treating *O*-ethyl-*S*-[2-oxo-2-(3,4,5-trimethoxyphenyl)-ethyl]dithiocarbonate with piperidine. Reaction of the α-mercapto anion with β-chloro-arylcinnamonitriles under thermal conditions afforded the desired thiophene. Much like the 2-amino derivatives, these 3-amino-2-(3,4,5-trimethoxy-benzoyl)-5-aryl-thiophenes were shown to be potent anti-proliferation and anti-tubulin agents as well.

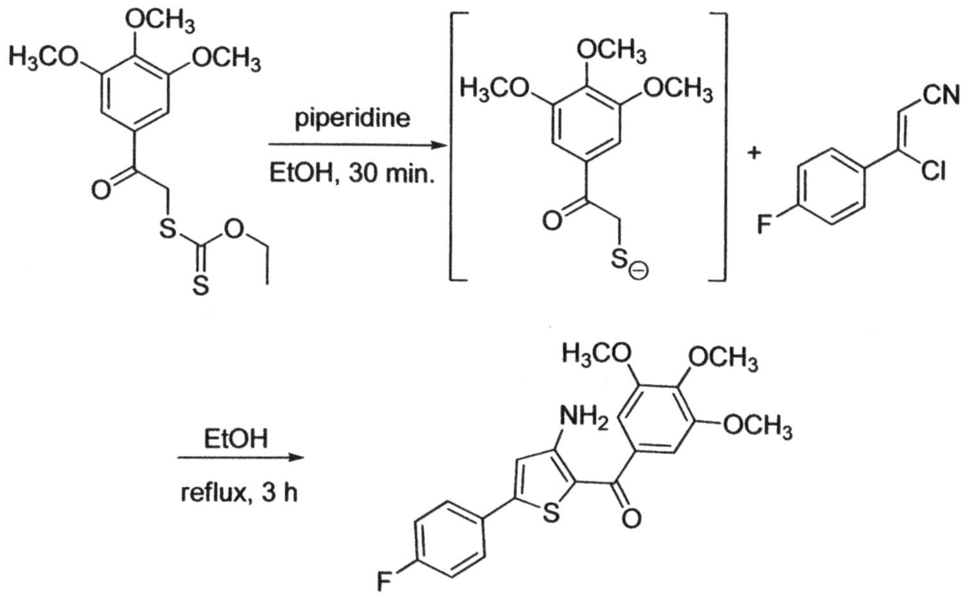

4.3.2.2 *Benzothiophene Synthesis*

Benzothiophenes may be accessed by a variety of methods including variations of the Fiesselmann and Gewald thiophenes synthesis. In addition, acid-catalyzed and electrophilic (iodo) cyclizations have been used to create valuable synthetic thiophene and benzothiophene intermediates with varied substitution patterns.

Fiesselmann Benzothiophene Synthesis

In the search for new atypical antipsychotic agents for the treatment of schizophrenia, Hrib and co-workers synthesized P-9236 by means of a Fiesselmann reaction with lithium hydroxide.[112] Several additional synthetic steps were required to produce P-9236, which displayed an activity profile similar to that of clozapine.

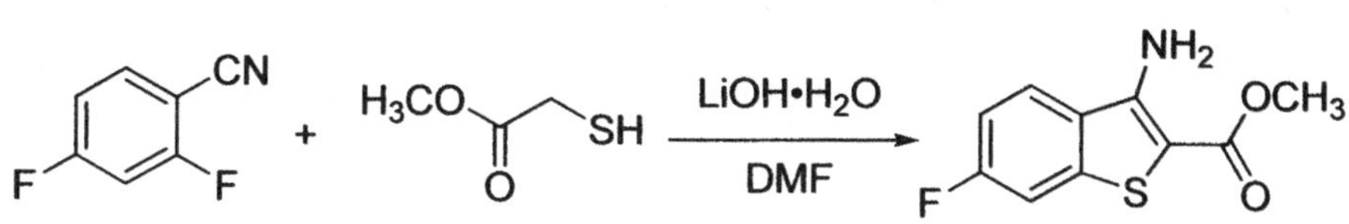

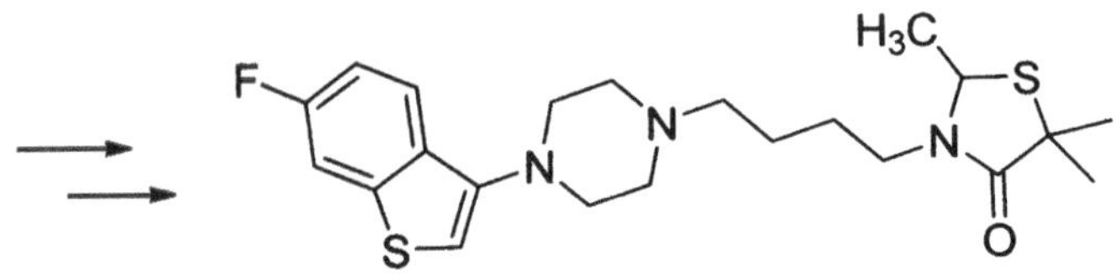

P-9236

A modification of the Fiesselman synthesis was used to produce a 5-substituted benzothiophene from the corresponding benzaldehyde derivative as illustrated below.[113] This methodology was also employed in the synthesis and study of zileuton analogues as anti-inflammatory and anti-nocieptive agents.[114]

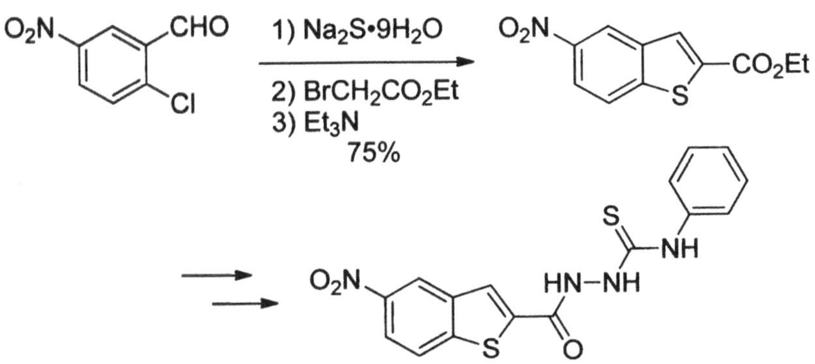

Gewald Aminobenzothiophene Synthesis

Synthesis of 2-aminobenzothiophene derivatives were achieved via a modified Gewald aminobenzothiophene synthesis under the Willgerodt–Kindler conditions.[115] This procedure produced the desired 2-aminobenzothiophenes, such as *N*-methyl-5-nitrobenzo[b]thiophen-2-amine shown below, in moderate to good yields.

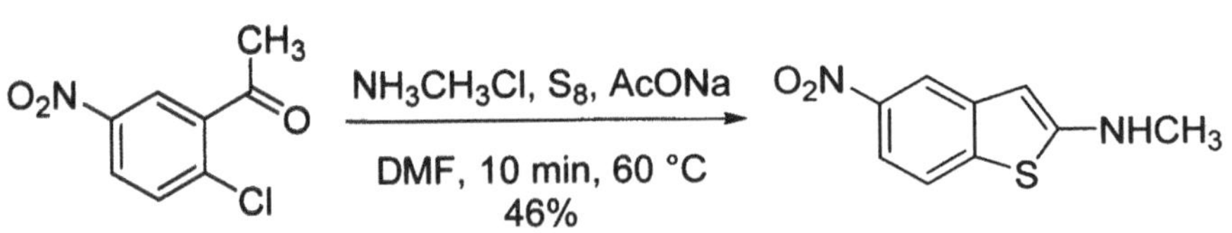

Acid-Promoted Cyclization

Synthesis of benzothiophenes can also be accomplished by an acid-promoted intramolecular cyclization reaction. The 3-substituted thiophene undergoes a rearrangement under acidic conditions to produce 2-substituted benzothiophenes.[116]

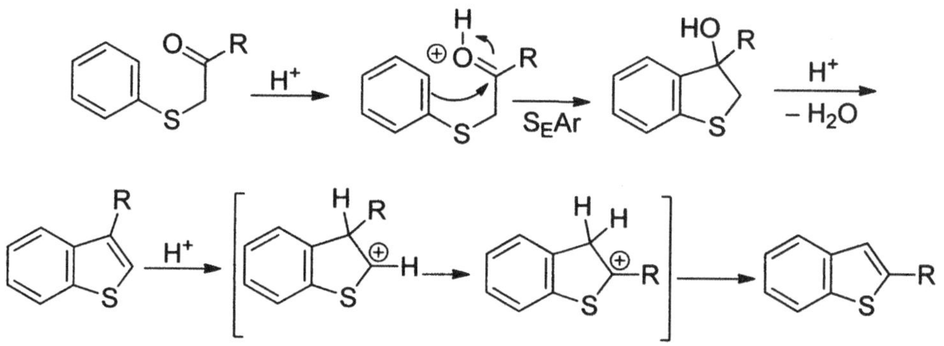

This methodology was used by researchers at Eli Lilly to generate benzothiophene derivatives of tamoxifen.[88] The cyclization and rearrangement reaction to produce the desired 2-aryl benzothiophene derivatives were promoted by polyphosphoric acid (PPA). These 2-aryl benzothiophene were then subjected to several additional transformations to produce LY156758 (raloxifene).

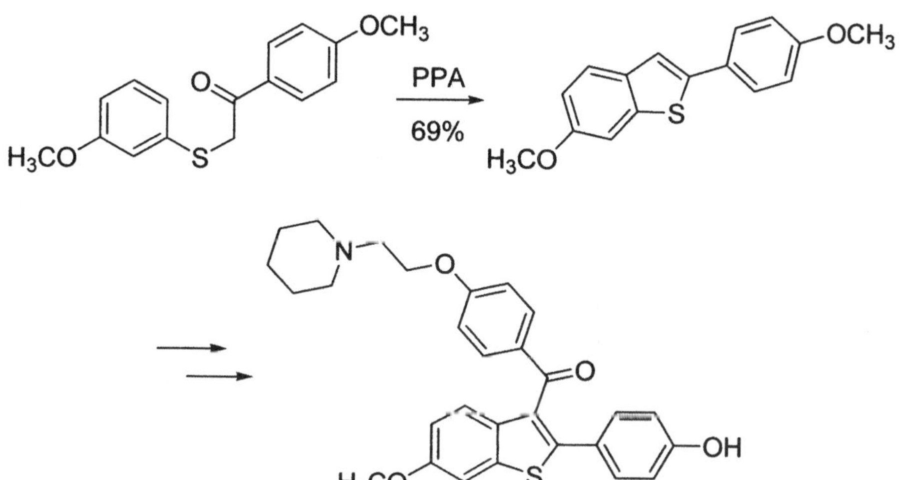

LY156758 (raloxifene)

In pursuit of additional derivatives, Eli Lilly scientists sought to synthesize constrained analogues of raloxifene.[117] The synthesis of these analogues used a unique method of constructing the benzothiophene ring system through an acid-promoted "dehydrative carbocationic cyclization" of the hydroxythioacetamide using methanesulfonic acid as illustrated below.

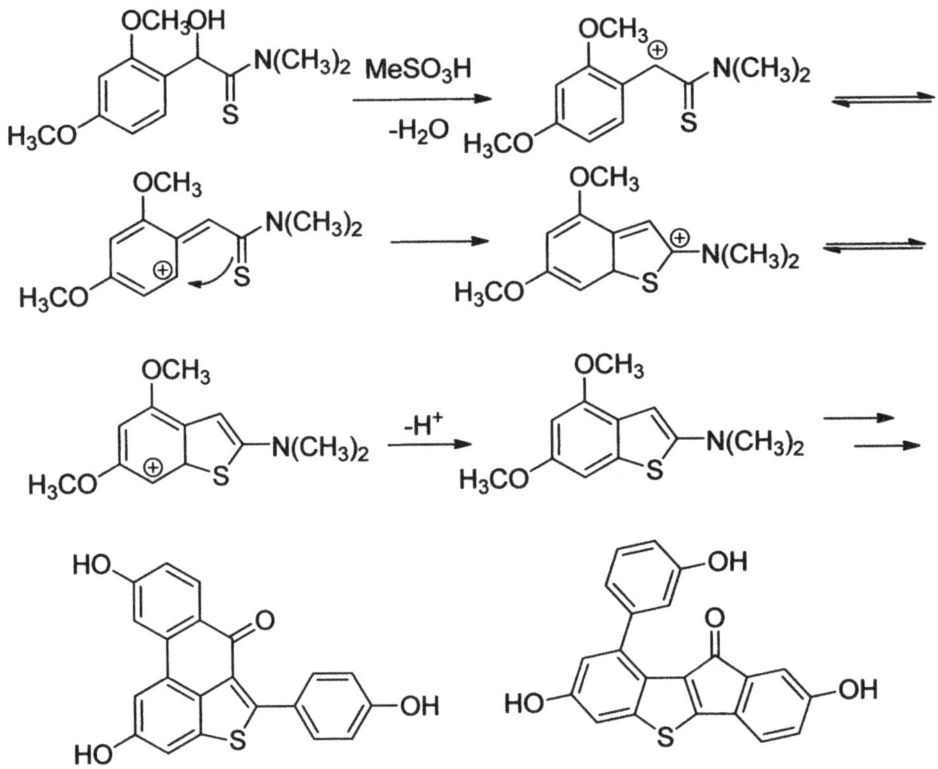

Electrophillic (Iodo-) Cyclization

Methods to synthesize benzothiophenes via electrophilic (iodo) cyclization were separately published by Flynn and Larock in 2001.[118,119] The reaction typically involves electrophilic addition to an alkynyl substituent of a phenyl sulfide or thiophenol in the presence of iodine. Subsequent nucleophilic attack of the sulfur establishes the 5-membered ring system.[120] Loss of a proton produces the corresponding benzothiophene functionalized at the *C3* position with iodine. The resulting 3-iodothiophene can then undergo further transformation as described in previous sections.

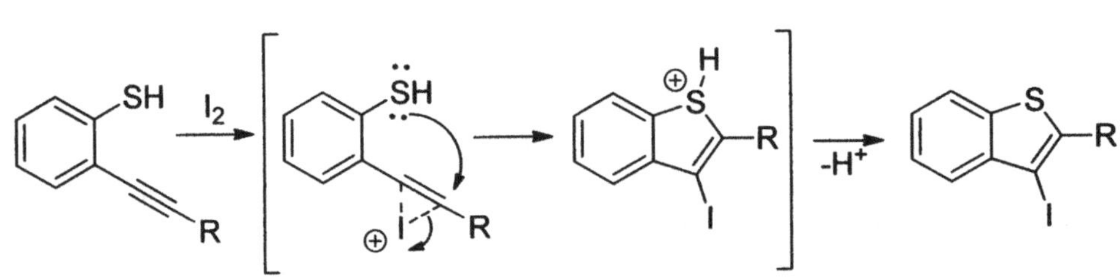

4.3.3 Synthesis of Thiophene- and Benzothiophene-Containing Drugs

As biosteres of benzene, thiophene and benzothiophene have enjoyed much success in drug discovery. In fact, pharmaceuticals containing thiophene or benzothiophene moieties accounted for several billion dollars a year in sales through much of the 2000's. Syntheses of many of these therapeutics are well known and provide excellent case studies in the chemistry of thiophenes and benzothiophenes and their utility.

The synthesis of raloxifene (Evista) is one such example. Raloxifene is currently approved for use as a therapy to treat and prevent osteoporosis. Researchers at Eli Lilly published the first synthesis and biological evaluation of LY156758, which would later be named raloxifene, in 1984.[88]

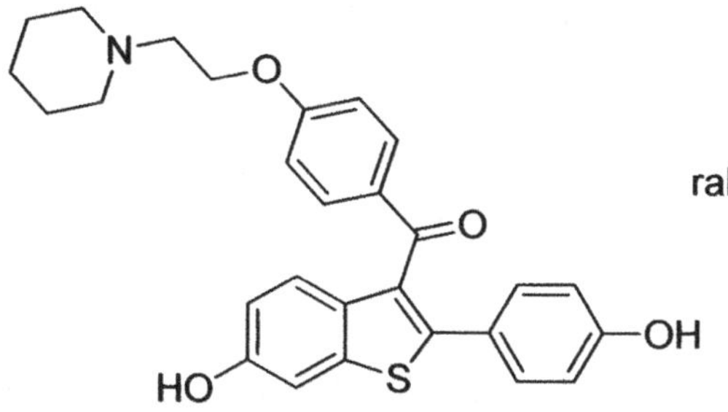

raloxifene

The synthesis of raloxifene began with the construction of the benzothiophene core structure by acid-promoted cyclization and rearrangement as discussed previously.[116]

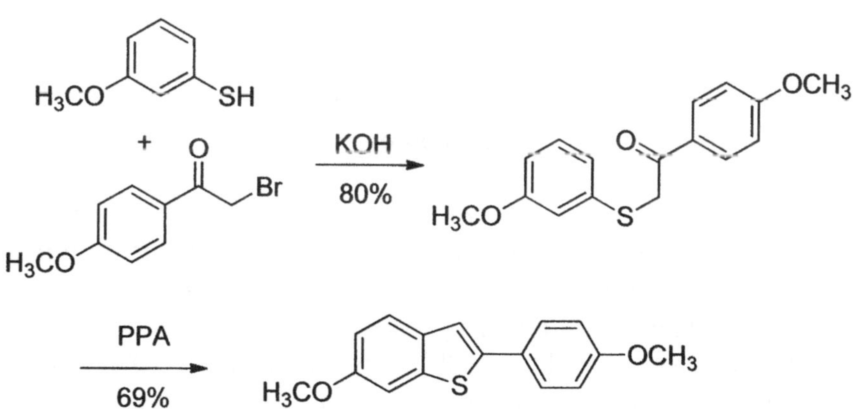

Benzothiophene synthesis was followed by a Friedel–Crafts acylation at the *C3* position. Subsequent deprotection resulted in the synthesis of the desired compound. A one-pot Friedel–Crafts reaction and deprotection protocol improved the overall yield of the final product. The lack of stereocenters results in a relatively simple synthesis with an overall yield of 50% starting with the Friedel–Crafts transformation.

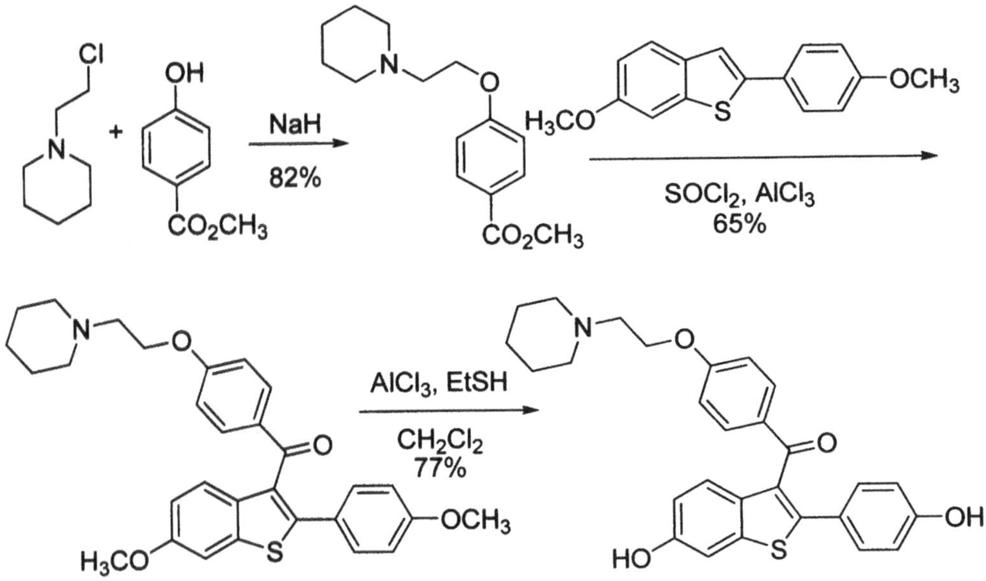

The blockbuster drug, clopidogrel (Plavix), is a thiophene-containing inhibitor of platelet aggregation and is used to treat a variety of conditions caused or aggravated by blood clotting.

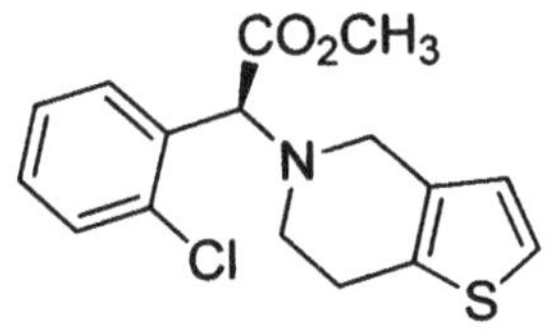

clopidogrel

One possible synthesis of clopidogrel begins with benzylic halogenation followed by nucleophilic substitution of the commercially available 4,5,6,7-tetrahydrothieno[3,2-*c*]pyridine.[121] The resulting compound is subjected to reduction followed by esterification to produce the racemic clopidogrel in an overall yield of 60% stepwise or up to 70% in a one-pot protocol. Kinetic resolution of clopidogrel can be performed by L-CSA (L-camphorsulfonic acid) in toluene to give 88% yield and 98% *ee* of the desired enantiomer. Higher optical purity (99.5% *ee*) can be achieved by washing the purified product with cold isopropanol.

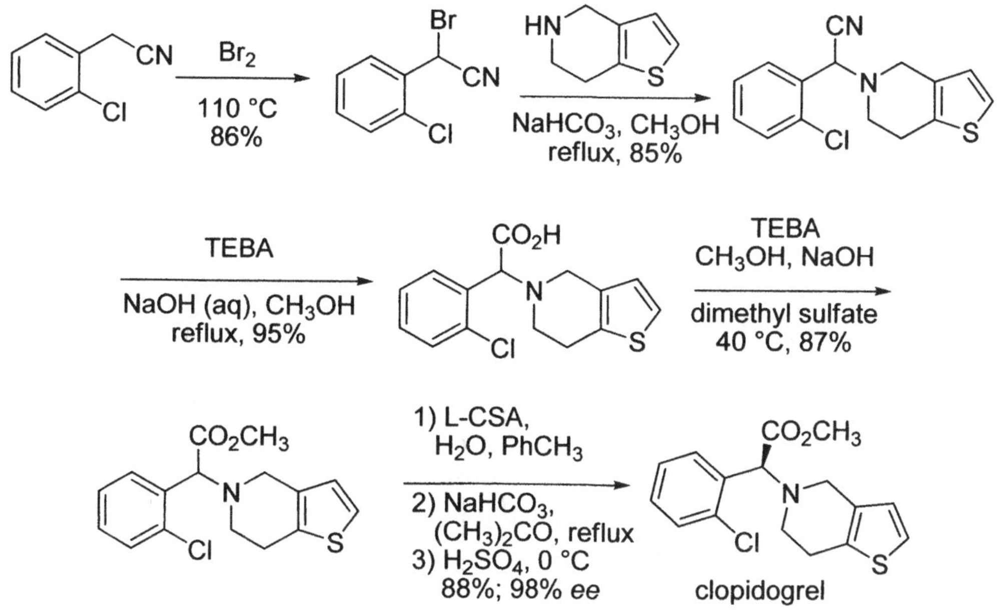

Recently, two methodologies have been published that use very unique approaches to synthesize 4,5,6,7-tetrahydrothieno[3,2-*c*]pyridine, a key thiophene-containing intermediate *en route* to clopidogrel. One such approach is the copper-catalyzed ring expansion of vinyl thiiranes followed by oxidation and subsequent removal of the amine protecting group.[122]

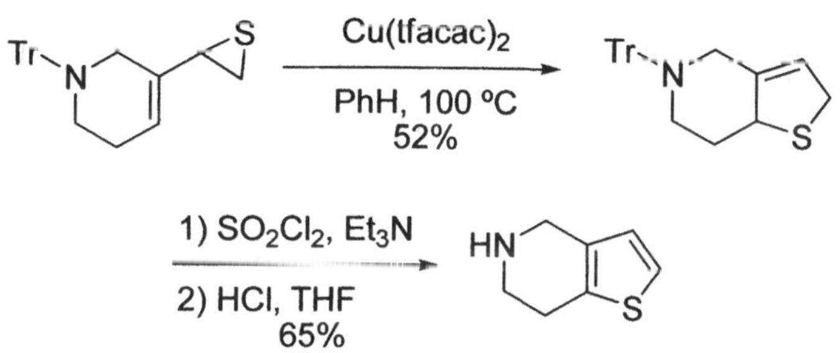

Another approach utilizes sulfur oxide (SO) transfer conditions.[123] Under the conditions reported, dienes are converted into thiophenes directly by the use of an SO transfer reagent. Excess SO transfer reagent was required for high yield.

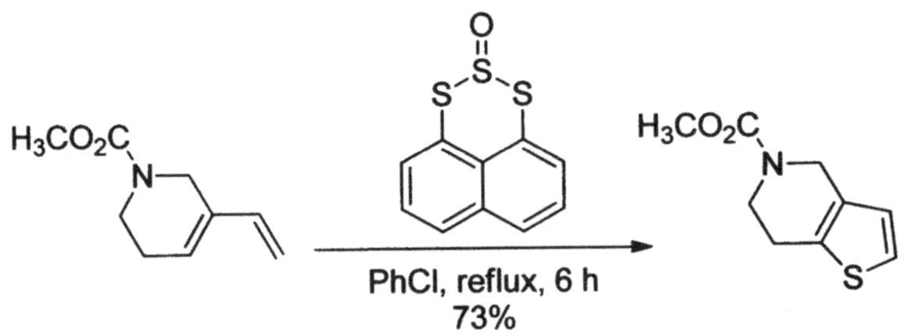

4.4 *Possible Liabilities of Furan- and Thiophene-Containing Drugs*

Electron-rich furan and thiophene ring systems are susceptible to oxidation by cytochrome P-450 (CYP450) enzymes. The oxidized products are then capable of reacting with various biological nucleophiles; the resulting metabolites can lead to toxicity, typically hepatotoxicity.[124–128]

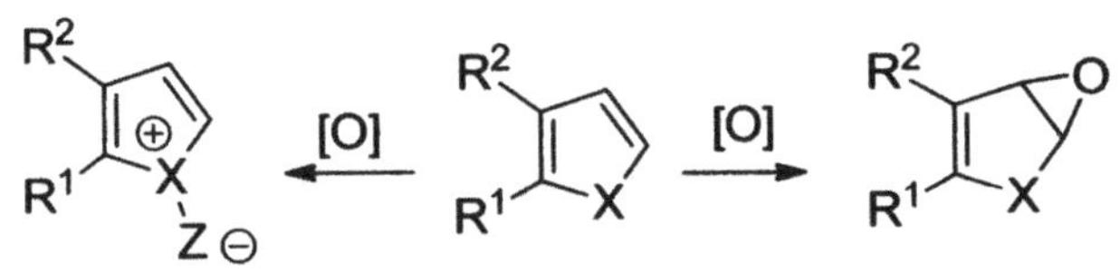

In general, the furan ring system appears to far less reactive than the thiophene system, and it is therefore much less toxic. This is most likely due to the higher electronegativity of the oxygen atom, which reduces the reactivity of the ring toward oxidation.

Of the furan-containing drugs on the market, furosemide, a diuretic, has been shown to cause hepatic necrosis in mice.[128] The mechanism involves metabolic activation of furosemide by oxidation of the furan ring by CYP450 followed by conjugation to glutathione to produce a furosemide-glutathione conjugate. Despite these results, furosemide has not been shown to present significant toxicity to humans.

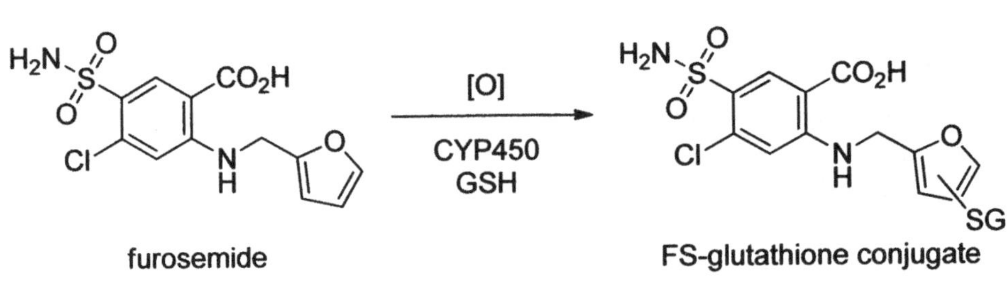

Tienilic acid, a thiophene-based diuretic used to treat hypertension, has been shown to cause hepatotoxicity. Oxidation of the thiophene and subsequent reactions of the activated product with nucleophilic proteins is responsible for the observed pathology.[129] Tienilic acid was withdrawn from the market shortly after there was evidence of drug induced hepatitis. Additionally, tienilic acid was found to be a "suicide" inhibitor for the cyctochrome P-450 enzyme (CPY2C9) to which it became covalently linked.[130]

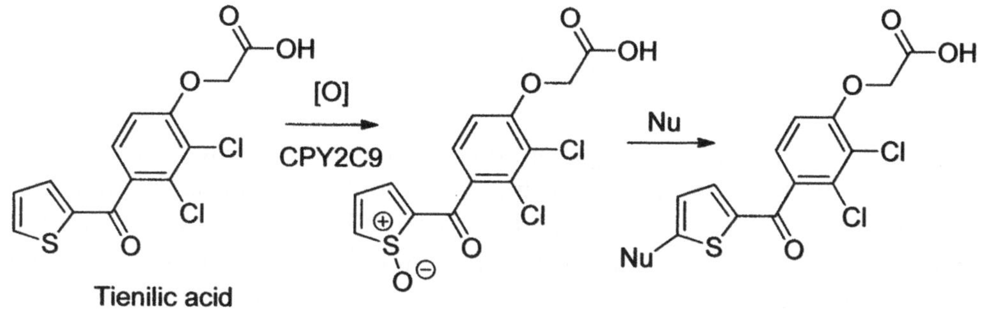

Although the electron-rich thiophene may lead to toxicity, the metabolic chemistry of thiophene can also lead to desirable therapeutic effects as in the case of clopidogrel (Plavix). The parent compound is oxidized by cytochrome P-450, and further oxidation in the presence of water opens the thiophene ring to produce an electrophilic sulfenic acid.[129] This electrophillic intermediate is then susceptible to a nucleophilic thiol found on the $P2Y_{12}$ receptor.[131] Creation of this disulfide bond modifies the receptor and inhibits platelet aggregation leading to the desired therapeutic effect.

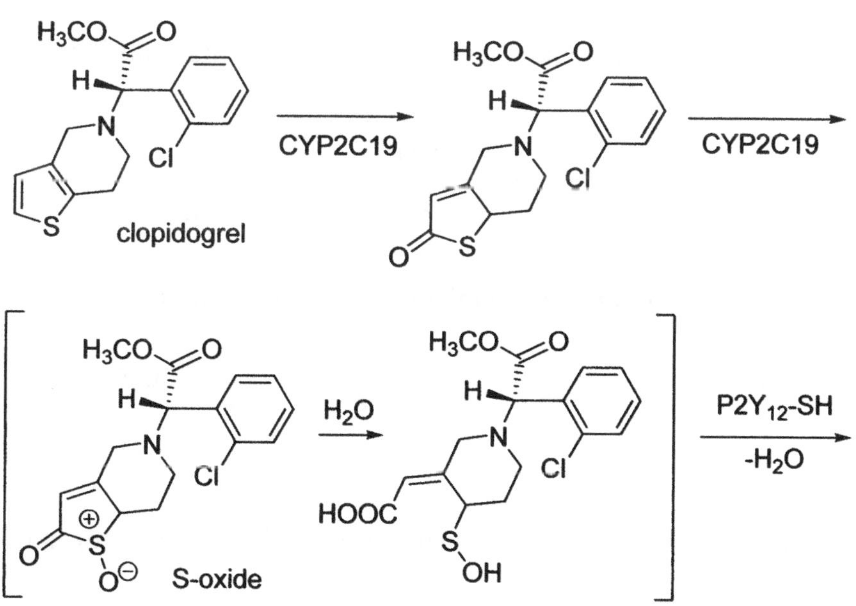

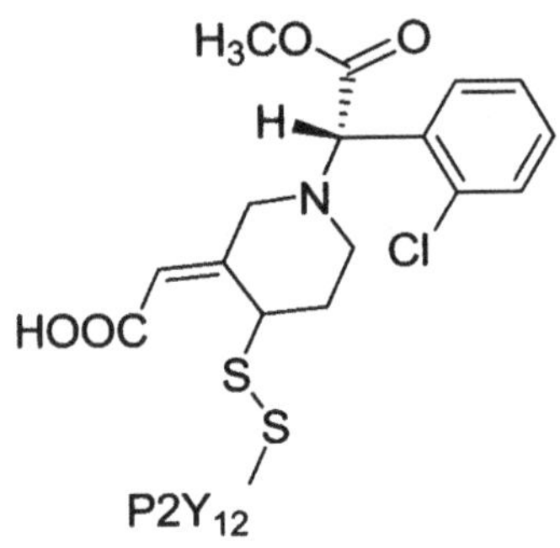

Potential toxicity arising from the furan and thiophene ring structures is a concern in designing potential drug leads, but potential leads should not be eliminated on those grounds alone. The success of current pharmaceuticals that contain furan and thiophene moieties is a clear indication that furan- and thiophene-containing therapies can be made safe, and even exploited as in the case of clopidogrel.

4.5 Problems

4.5.1 Propose a reasonable mechanism for the following transformation:

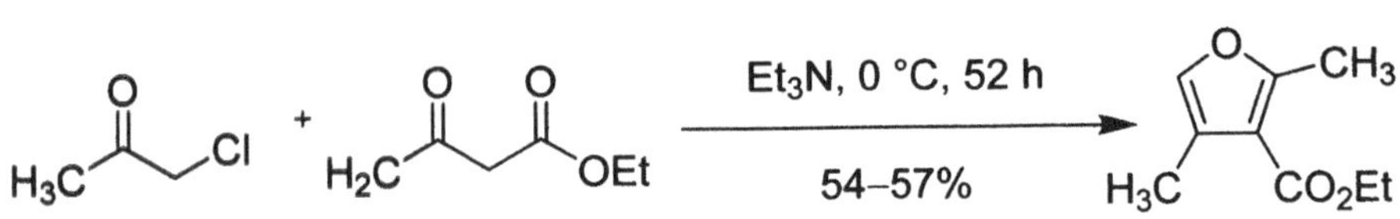

4.5.2 Propose a reasonable mechanism for the following transformation:

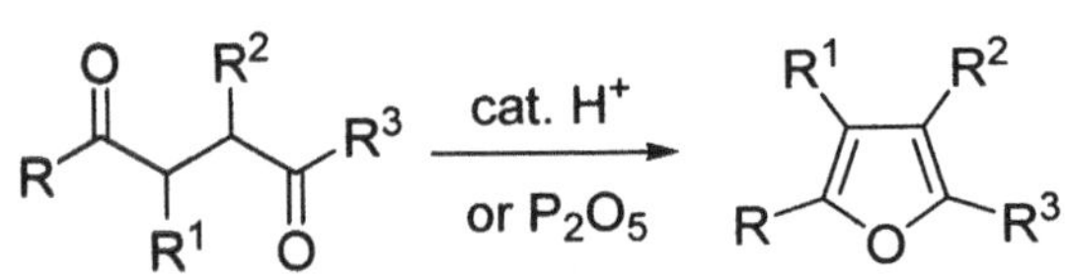

4.5.3 Propose a reasonable mechanism for the following transformation:

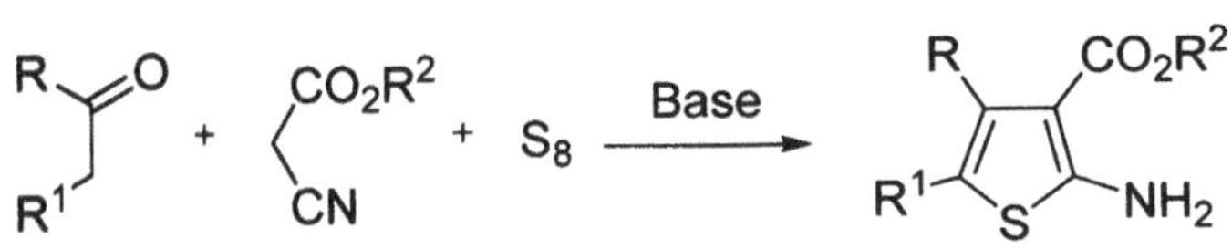

4.5.4 Propose a reasonable mechanism for the following transformation:

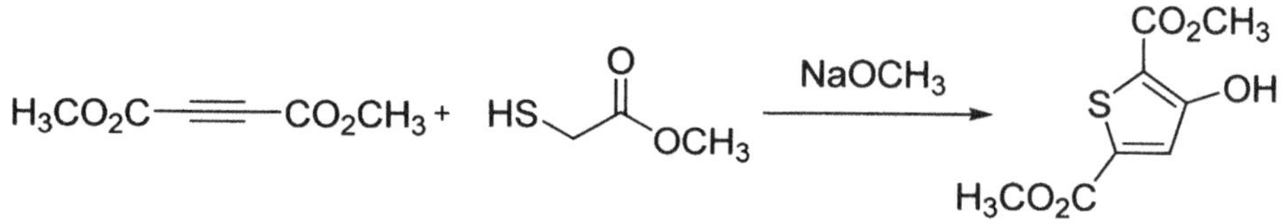

4.5.5 Propose a reasonable mechanism of the following transformation[132]:

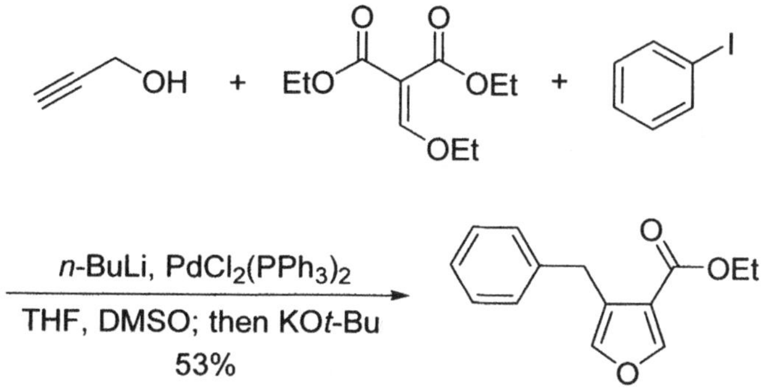

4.5.6 Predict the structures of products **A**, **B**, and **C**[22]:

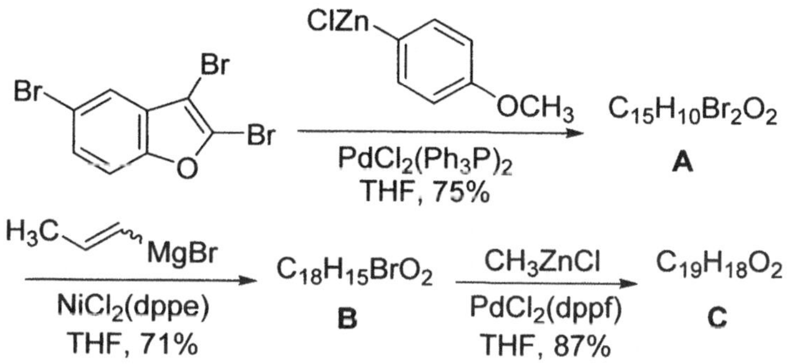

4.5.7 Predict the structures of products **D**.[133]

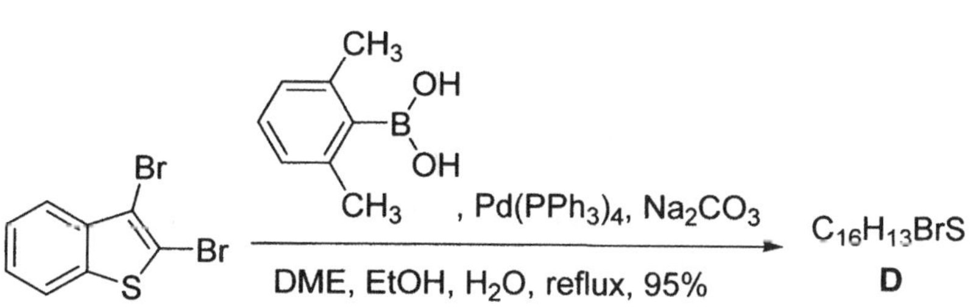

4.5.8 Predict the structures of products **E** and **F**.[134]

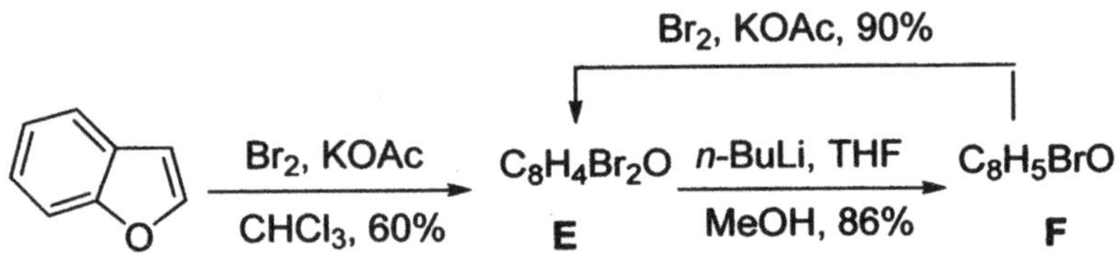

4.5.9 Suggest a plausible mechanism for the following two-step, one-pot, iron-catalyzed synthesis of α-carbonyl furans.[135]

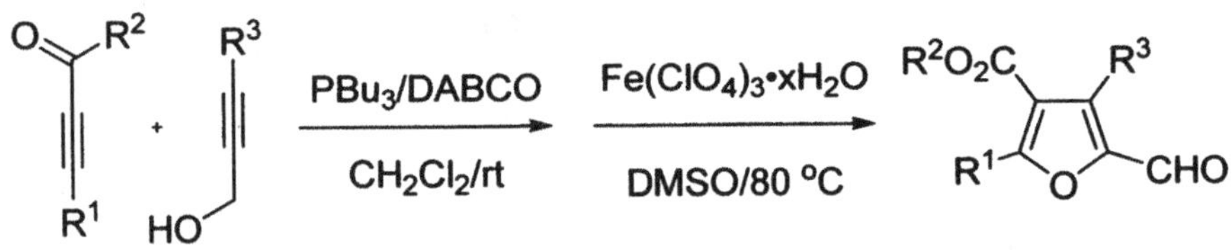

4.5.10 Propose a mechanism for the synthesis of benzofurans using the Grignard reagent as shown below.[136]

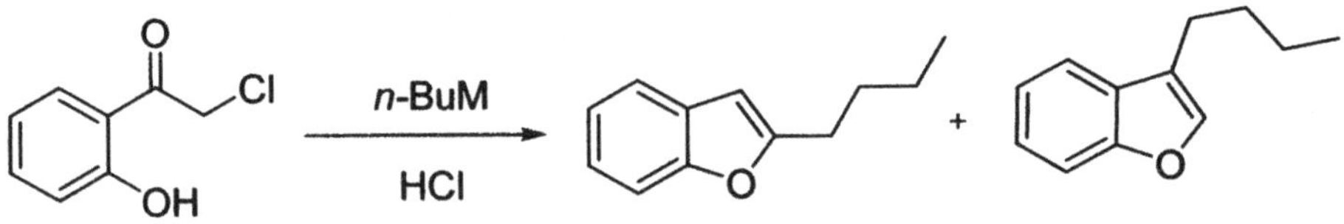

4.5.11 Propose a mechanism for the following copper-catalyzed furan synthesis involving a tandem propargylation/cycloisomerization step.[137]

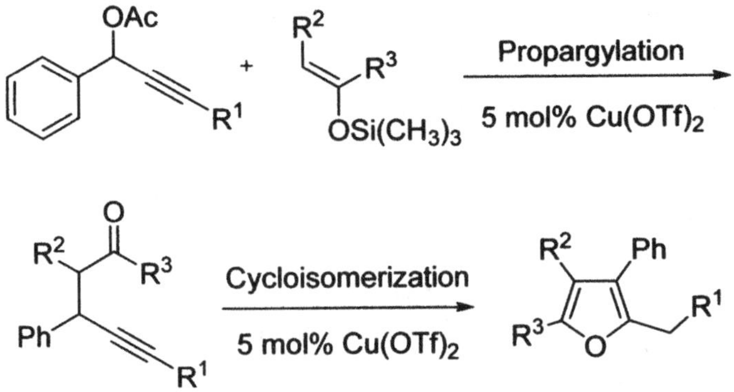

4.5.12 Propose a mechanism for the following ruthenium-catalyzed cyclization of epoxyalkyne derivative.[138]

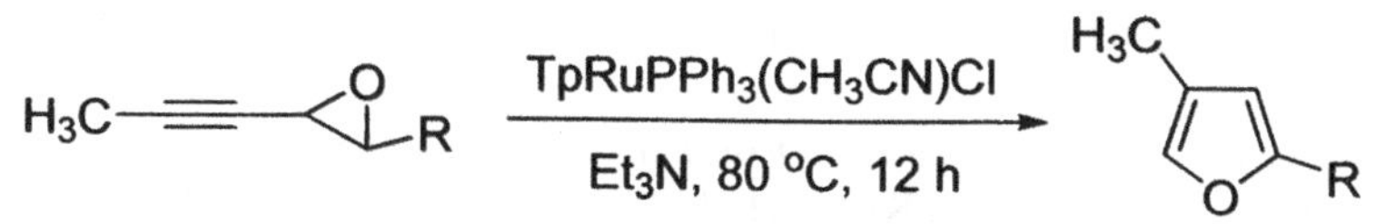

Tp = trispyrazolylborate

4.5.13 Propose a reasonable mechanism for the synthesis of benzo[*b*]thiophene-2-carboxylates and benzo[*b*]thiophene-2-carboximides.[139]

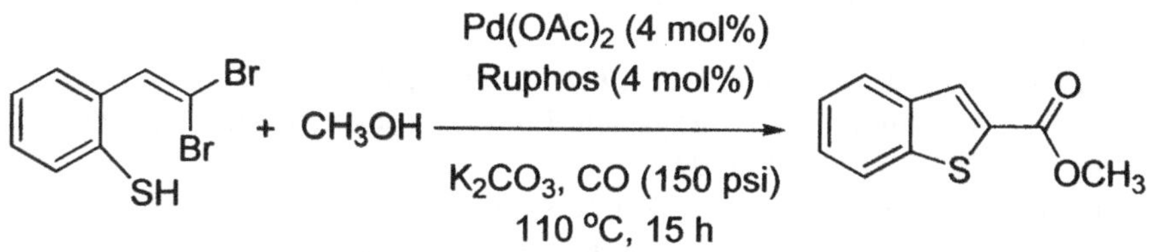

4.5.14 Propose a mechanism for the three-step synthesis of 3-nitro-2-substituted thiophene from 1,4-dithiane-2,5-diol.[140]

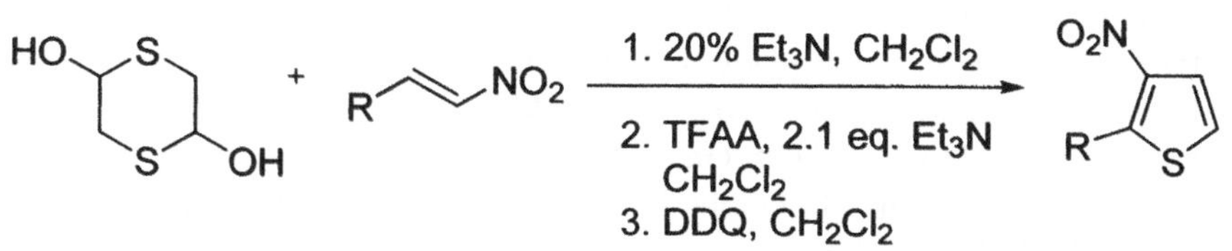

4.6 *References*

1. Eicher, T.; Hauptmann, S. *The Chemistry of Heterocycles*; Wiley-VCH: New York, NY, 1995.
2. Press, J. B. *Biologically Active Thiophene Derivatives Revisited: 1983–1988*. In *The Chemistry of Heterocyclic Compounds*; Gronowitz, S., Ed.; Wiley-Interscience: New York, NY, 1991; *Vol. 44*, pp 397–502.
3. Keegstra, M. A.; Klomp, A. J. A.; Brandsma, L. *Synth. Comm.* **1990,** *20*, 3371–3374.
4. Tehrani, M. H. H.; Zarghi, A.; Fathali, S. *Iranian J. Pharm. Res.* **2003,** 67–69.
5. Ila, H.; Baron, O.; Wagner, A. J.; Knochel, P. *Chem. Comm.* **2006,** 583–593.
6. Kwaitkowski, P.; Wojaczynskya, E.; Jurczak, J. *Tetrahedron: Asymmetry* **2003,** *14*, 3643–3645.
7. Sakata, Y.; Kuramoto, M.; Ando, K.; Yamaguchi, M.; Kawasaki, I.; Kunimoto, J.; Yokimizo, T.; Ohishi, Y. *Org. Biomol. Chem.* **2007,** *5*, 3083–3086.
8. Devarie, N. O.; Kim, W. S.; Smith, A. B. III; Xian, M. *Org. Lett.* **2009,** *11*, 1861–1864.
9. Adachi, S.; Tanaka, F.; Watanabe, K.; Harada, T. *Org. Lett.* **2009,** *11*, 5206–5209.

10. Inagaki, F.; Kinebuchi, M.; Miyakoshi, N.; Mukai, C. *Org. Lett.* **2010,** *12*, 1800–1803.
11. Chen, C. W.; Luh, T. Y. *Tetrahedron Lett.* **2009,** *50*, 3263–3265.
12. Melzig, L.; Rauhut, C. B.; Knochel, P. *Chem. Comm.* **2009,** 3536–3538.
13. Carlsson, B.; Singh, B. N.; Temciuc, M.; Nilsson, S.; Li, Y.-L.; Mellin, C.; Malm, J. *J. Med. Chem.* **2002,** *45*, 623–630.
14. Rose, M. D.; Cassidy, M. P.; Rashatasakhon, P.; Padwa, A. *J. Org. Chem.* **2007,** *72*, 538–549.
15. Marion, G.; Williams, D. E.; Patrick, B. O.; Hollander, I.; Mallon, R.; Kim, S. C.; Roll, D. M.; Feldberg, L.; von Soest, R.; Andersen, R. J. *Org. Lett.* **2006,** *8*, 321–324.
16. Shanmugham, M. S.; White, J. D. *Chem. Comm.* **2004,** 44–45.
17. Furst, L.; Matsuura, B. S.; Narayanam, J. M. R.; Tucker, J. W.; Stephenson, C. R. *J. Org. Lett.* **2010,** *12*, 3104–3107.
18. Kuninobu, Y.; Tokunaga, Y.; Takai, K. *Chem. Lett.* **2007,** *36*, 872–873.
19. Wellmar, U.; Hoernfeldt, A.-B.; Gronowitz, S. *J. Heterocyl. Chem.* **1995,** *32*, 1159–1163.
20. Bach, T.; Krueger, L. *Eur. J. Org. Chem.* **1999,** 2045–2057.
21. Bach, T.; Krueger, L. *Synlett.* **1998,** 1185–1186.
22. Bach, T.; Bartels, M. *Tetrahedron Lett.* **2002,** *43*, 9125–9127.
23. Aout, C.; Thiery, E.; Le Bras, J.; Muzart, J. *Org. Lett.* **2009,** *11*, 4096–4099.
24. Bach, T.; Krueger, L. *Tetrahedron Lett.* **1998,** *39*, 1729–1732.
25. Bach, T.; Krueger, L. *Eur. J. Org. Chem.* **1999,** 2045–2057.
26. Feist, F. *Ber. Dtsch. Chem. Ges.* **1902,** *35*, 1545.
27. Gupta, R. R.; Kumar, M.; Gupta, V. *Heterocyclic Chemistry*, Springer: New York, NY, 1999; *Vol. 2*, 87–88.
28. Joule, J. A.; Mills, K. *Heterocyclic Chemistry*, 4_{th} ed., Blackwell Science: Cambridge, 2000, 310–311.
29. Bénary, E. *Ber. Dtsch. Chem. Ges.* **1911,** *44*, 493.
30. Ho, T.-L.; Ho, M.-F. *J. Chem. Soc., Perkin Trans. 1* **1999,** 1823.
31. Bisagni, É.; Marquet, J.-P.; Bourzat, J.-D.; Pepin, J.-J.; André-Louisfert, J. *Bull. Soc. Chim. Fr.* **1971,** 4041–4047.
32. Alexander, E. R.; Baldwin, S. *J. Am. Chem. Soc.* **1951,** *73*, 356–358.
33. Wenkert, E.; Khatiya, H.; Klein, P. S. *Tetrahedron Lett.* **1999,** *40*, 5171–5174.
34. Stetter, H.; Lauterbach, R. *Chem. Ber.* **1960,** *93*, 603.
35. Lavoisier-Gallo, T.; Rodriguez, J. *Synth. Commun.* **1998,** *28*, 2259–2269.
36. Courtheyn, D.; Verhé, R.; De Kimpe, N.; De Buyck, L.; Schamp, N. *J. Org. Chem.* **1981,** *46*, 3226–3229.
37. Paal, C. *Ber. Dtsch. Chem. Ges.* **1884,** *17*, 2756.
38. Amarnath, V.; Amarnath, K. *J. Org. Chem.* **1995**, *60*, 301–307.
39. Siedem, C. S.; Molander, G. A. *J. Org. Chem.* **1996,** *61*, 1140–1142.
40. Xiao, G.; Kumar, A.; Li, K.; Rigl, C. T.; Bajic, M.; Davis, T. M.; Boykin, D. W.; Wilson, W. D. *Bioorg. Med. Chem.* **2001,** *9*, 1097–1113.
41. Kikuchi, K.; Hibi, S.; Yoshimura, H.; Tokuhara, N.; Tai, K.; Hida, T.; Yamauchi, T.; Nagai, M. *J. Med. Chem.* **2000**, *43*, 409–423.
42. Denisenko, M. V.; Pokhilo, N. D.; Odinokova, L. E.; Denisenko, V. A.; Uvarova, N. I. *Tetrahedron Lett.* **1996,** *37*, 5187–5190.
43. de Laszlo, S. E.; Visco, D.; Agarwal, L.; Chang, L.; Chin, J.; Croft, G.; Forsyth, A.; Fletcher, D.; Frantz, B.; Hacker, C.; Hanlon, W.; Harper, C.; Kostura, M.; Li, B.;

Luell, S.; MacCoss, M.; Mantlo, N.; O'Neill, E. A.; Orevillo, C.; Pang, M.; Parsons, J.; Rolando, A.; Sahly, Y.; Sidler, K.; Widmer, W. R.; O'Keefe, S. J. *Bioorg. Med. Chem. Lett.* **1998,** *8*, 2689.

44. Wynberg, H.; Wiersum, U. E. *J. Chem. Soc., Chem. Commun.* **1990,** 460.
45. Mortensen, D. S.; Rodriguez, A. L.; Carlson, K. E.; Sun, J.; Katzenellenbogen, B. S.; Katzenellenbogen, J. A. *J. Med. Chem.* **2001,** *44*, 3838–3848.
46. Mortensen, D. S.; Rodriguez, A. L.; Sun, J.; Katzenellenbogen, B. S.; Katzenellenbogen, J. A. *Bioorg. Med. Chem. Lett.* **2001,** *11*, 2521–2524.
47. Hari, Y.; Iguchi, T.; Aoyama, T. *Synthesis* **2004,** 1359–1362.
48. Kel'in, A. V.; Gevorgyan, V. *J. Org. Chem.* **2002,** *67*, 95–98.
49. Rafa, G.; Rusch, M.; Balme, G.; Monteiro, N. *Org. Lett.* **2009,** *22*, 5254–5357.
50. Cheng, G.; Hu, Y. *J. Org. Chem.* **2008,** *73*, 4732–4735.
51. Blanc, A.; Aliz, A.; Weibel, J.-M.; Pale, P. *Eur. J. Org. Chem.* **2010,** 1644–1647.
52. Aurrecoechea, J. M.; Durana, A.; Perez, E. *J. Org. Chem.* **2008,** *73*, 3650–3653.
53. Donohoe, T. J.; Ironmonger, A.; Kershaw, N. M. *Angew. Chem. Int. Ed.* **2008,** *47*, 7314–7316.
54. Inagaki, F.; Kinebuchi, M.; Miyakoshi, N.; Mukai, C. *Org. Lett.* **2010,** *12*, 1800–1803.
55. Jakubec, P.; Cockfield, D. M.; Dixon, D. J. *J. Am. Chem. Soc.* **2009,** *131*, 16632–16633.
56. Hsu, D.-S.; Hsu, P.-Y.; Lee, Y.-C.; Liao, C.-C. *J. Org. Chem.* **2008,** *73*, 2554–2563.
57. Davis, J. B.; Bailey, J. D.; Sello, J. K. *Org. Lett.* **2009,** *11*, 2984–2987.
58. Bartoli, G.; Fernandez-Bolanos, J. G.; Di Antonio, G.; Foglia, G.; Giuli, S.; Gunnella, R.; Mancinelli, M.; Marcantoni, E.; Paoletti, M. *J. Org. Chem.* **2007,** *72*, 6029–6036.
59. Saquib, M.; Husain, I.; Kumar, B.; Shaw, A. K. *Chem. Eur. J.* **2009,** *15*, 6041–6049.
60. Cordonnier, R.; Van Nhien, A. N.; Soriano, E.; Marco-Contelles, J. *Tetrahedron*, **2010,** *66*, 736–742.
61. Khupse, R. S.; Erhardt, P. W. *J. Nat. Prod.* **2008,** *71*, 275–277.
62. Mehta, G.; Likhite, N. S. *Tetrahedron Lett.* **2008,** *49*, 7113–7116.
63. Li, X.; Ovaska, T. *Org. Lett.* **2007,** *9*, 3837–3840.
64. Kraus, G. A.; Gupta, V. *Tetrahedron. Lett.* **2009,** 7180–7183.
65. Chen, C.-Y.; Dormer, P. G. *J. Org. Chem.* **2005,** *70*, 6964–6967.
66. Snyder, H. R., Jr.; Davis, C. S.; Bickerton, R. K.; Halliday, R. P. *J. Med. Chem.* **1967,** *10*, 807.
67. Davis, C. S.; Snyder, H. R., Jr. US3,415,812 (1968).
68. Hosoya, T.; Aoyama, H.; Ikemoto, T.; Kihara, Y.; Hiramatsu, T.; Endo, M.; Suzuki, M. *Bioorg. Med. Chem.* **2003,** *11*, 663–673.
69. Singh, S. K.; Tiwari, R. P.; Jain, P. *Synth. Comm.* **2003,** *33*, 2475–2482.
70. Mai, D.; Bandhyopadhyay, M.; Datta, K.; Marry K. V. S. N. *Tetrahedron.* **1998,** *54,* 7525–7538.
71. Kaufman, K. D. US3,201,421 (1965).
72. Bender D. R. US4,398,031 (1983).
73. Spath, E.; Pailer, M. *Ber.* **1936,** *69*, 767.
74. Judd, D. B.; Dowle, M. D.; Middlemiss, D.; Scopes, D. I. C.; Ross, B. C.; Jack, T. I.; Pass, M.; Tranquillini, E.; Hobson, J.E.; Panchal, T.A.; Stuart, P.G.; Paton, J. M. S.; Hubbard, T.; Hilditch, A.; Drew, G. M.; Robertson, M. J.; Clark, K. L.; Travers, A.; Hunt, A. A. E.; Polley, J.; Eddershaw, P. J.; Bayliss, M. K.; Manchee, G. R.;

Donnelly, M. D.; Walker, D. G.; Richards, S. A. *J. Med. Chem.* **1994,** *37*, 3108–3120.
75. Keegstra, M. A.; Brandsma, L. *Synthesis* **1988,** 890–891.
76. Le Quement, S. T.; Nielsen, T. E.; Meldal, M. J. *J. Comb. Chem.* **2008,** *10*, 447–455.
77. Smith, K; Barratt, M. L. *J. Org. Chem.* **2007,** *72*, 1031–1034.
78. Guinchard, X.; Denis, J.-N. *J. Org. Chem.* **2008,** *73*, 2028–2031.
79. Kamal, A.; Malik, S. H.; Shaik, A. A.; Azeeza, S. *J. Molec. Catalysis B: Enzymatic* **2009,** *58*, 132–137.
80. Kamal, A.; Ramesh Khanna, G. B.; Ramu, R.; Krishnaji, T. *Tetrahedron Lett.* **2003,** *44*, 4783–4787.
81. Majer, J; Kwiatkowski, P; Jurczak, J. *Org. Lett.* **2009,** *11*, 4636–4639.
82. Kumar, P. R.; Raju, S.; Goud, P. S.; Sailaja, M.; Sarma, M. R.; Reddy, G. O.; Kumar, M. P.; Reddy, V. V. R. M. K.; Suresh, T.; Hegde, P. *Bioorg. Med. Chem.* **2004,** *12*, 1221–1230.
83. Ushijima, S.; Togo, H. *Synlett.* **2010**, 1067–1070.
84. Itahara, T. *Chem. Lett.* **1983,** 127–128.
85. Tsuji, K.; Nakamura, N.; Ogino, T.; Konishi, N.; Tojo, T.; Ochi, T.; Seki, N.; Matsuo, M. *Chem. Pharm. Bull.* **1998,** *46*, 279–286.
86. Qin, Z.; Kastrati, I.; Chandrasena, R. E. P.; Liu, H.; Yao, P.; Petukhov, P. A.; Bolton, J. L.; Thatcher, G. R. J. *J. Med. Chem.* **2007,** *50*, 2682–2692.
87. Oficialdegui, A. M.; Villanueva, H.; Orus, L.; Monge, A. *J. Heterocyclic Chem.* **2001,** *38*, 1025–1030.
88. Jones, C. D.; Jevnikar, M. G.; Pike, A. J.; Peters, M. K.; Black, L. J.; Thompson, A. R.; Falcone, J. F.; Clemens, J. A. *J. Med. Chem.* **1984,** *27*, 1057–1066.
89. Katritzky, A. R.; Serdyuk, L.; Kie, L.; Ghiviriga, I. *J. Org. Chem.* **1997,** *62*, 6215–6221.
90. Nicolaou, K. C.; Bulger, P. G.; Sarlah, D. *Angew. Chem. Int. Ed.* **2005,** *44*, 4442–4489.
91. Cho, C; Neuenswander, B.; Lushington, G. H.; Larock, R. C. *J. Comb. Chem.* **2009,** *11*, 900–906.
92. Zhang, Y.; Li, Z.; Liu, Z.-Q. *Org. Lett.* **2012,** *14*, 226–229.
93. Le Quement, S. T.; Nielsen, T. E.; Meldal, M. *J. Comb. Chem.* **2008,** *10*, 447–455.
94. Bui, C. T.; Flynn, B. L. *J. Comb. Chem.* **2006,** *8*, 163–167.
95. Cho, C.-H.; Jung, D.-I.; Larock, R. C. *Tetrahedron Lett.* **2010,** *51*, 6485–6488.
96. Wang, L.; Desmoulin, S. K.; Cherian, C.; Polin, L.; White, K.; Kushner, J.; Fulterer, A.; Chang, M.-H.; Mitchell-Ryan, S.; Stout, M.; Romero, M. F.; Hou, Z.; Matherly, L. H.; Gangjee, A. *J. Med. Chem.* **2011,** 54, 7150–7164.
97. Pinto, E.; Querioz, M. R. P.; Luis, A. V.; Oliveria, J. F.; Begouin, A.; Begouin, J.; Kirsch, G. *Bioorg. Med. Chem.* **2008,** *16*, 8172–8177.
98. Li, J. J. *Named Reactions in Heterocyclic Chemistry*; Wiley: Hoboken, NJ, 2005; pp 183–217.
99. Li, J. J. *Name Reactions: A collection of Detailed Reaction Mechanisms;* Springer-Verlag: Heidelberg, Germany, 2003.
100. Redman, A. M.; Johnson, J. J.; Dally, R.; Swartz, S.; Wild, H.; Paulsen, H.; Caringal, Y.; Gunn, D.; Renick, J.; Osterhout, M.; Kingery-Wood, J.; Smith, R. A.; Lee, W.; Dumas, J.; Wilhelm, S. M.; Housley, T. J.; Bhargava, A.; ranges, G. E.; Shrikhande, A.; Young, D.; Bombara, M.; Scott, W. J. *Bioorg. Med. Chem. Lett.* **2001,** *11*, 9–12.

101. Nicolaou, K. C.; Skokotas, G.; Furuya, S.; Suemune, H.; Nicolaou, D. C. *Angew. Chem. Int. Ed. Engl.* **1990,** *29*, 1064–1067.
102. Ronzoni, S.; Cerro, A.; Dondio, G.; Fronza, G.; Petrillo, P.; Raveglia. L. F.; Gatti, P. A. *Org. Lett.* **1999,** *1*, 513–515.
103. Donoso, R.; Jordan de Urries, P.; Lissavertzky, J. *Synthesis* **1992,** 526–528.
104. Kumar, V. ; Daum, S. J.; Bell, M. R.; Alexander, M. A.; Christiansen, R. G.; Ackerman, J. H.; Krolshi, M. E.; Pilling, G. M.; Herrmann, J. L.; Winneker, R. C.; Wagner, M. M. *Tetrahedron*, **1991,** *47*, 5099–5110.
105. Saravanan, J.; Mohan, S.; Roy, J. J. *Eur. J. Med. Chem.* **2010,** *45*, 4365–4369.
106. Lutjens, H.; Zickgraf, A.; Figler, H.; Linden, J.; Olsson, R. A.; Scammells, P. J. *J. Med. Chem.* **2003,** *46*, 1870–1877.
107. Mojtahedi, M. M.; Abaee, M. S.; Mahmoodi, P.; Adib, M. *Synth. Comm.* **2010,** *40,* 2067–2074.
108. Garcia, I.; Feist, H.; Cao, R.; Michalik, M.; Peseke, K. *J. Carbohyr. Chem.* **2001,** *20,* 681–687.
109. Rosowsky, A.; Mota, C. E.; Wright, J. E.; Freisheim, J. H.; Heusner, J. J.; McCormack, J. J.; Queener, S.F. *J. Med. Chem.* **1993,** *36*, 3103–3112.
110. Romagnoli, R.; Baraldi, P. G.; Pavani, M. G.; Tabrizi, M. A.; Preti, D.; Fruttarolo, F.; Piccagli, L.; Jung, M. K.; Hamel, E.; Borgatti, M.; Gambari, R. *J. Med. Chem.* **2006,** *49*, 3906–3915.
111. Romagnoli, R.; Baraldi, P. G.; Remusat, V.; Carrion, M. D.; Cara, C. L.; Preti, D.; Fruttarolo, F.; Pavani, M. G.; Tabrizi, M. A.; Tolomeo, M.; Grimaudo, S.; Balzarini, J.; Jordan, M. A.; Hamel, E. *J. Med. Chem.* **2006,** *49*, 6425–6428.
112. Hrib, N. J.; Jurcak, J. J.; Bregna, D. E.; Dunn, R. W.; Geyer, H. M.; Hartman, H. B.; Roehr, J. E.; Rogers, K. L.; Rush, D. K.; Szczepanik, A. M.; Szewczak, M. R.; Wilmot, C. A.; Conway, P. G. *J. Med. Chem.* **1992,** *35*, 2712–2715.
113. Oficialdegui, A. M.; Villanueva, H.; Orus, L.; Monge, A. *J. Heterocyclic Chem.* **2001,** *38*, 1025–1030.
114. Fakhr, I. M. I.; Radwan, M. A. A.; El-Batran, S.; Abd El-Salam, O. M. E.; El-Shenawy, S. M. *Eur. J. Med. Chem.* **2009,** *44*, 1718–1725.
115. Solovyev, A. Y.; Androsov, D. A.; Neckers, D. C. *J. Org. Chem.* **2007,** *72*, 3122–3124.
116. Kost, A. N.; Dubylin, V. A.; Matveeva, E. D.; Sterligov, D. O. *Zh. Org. Khim.* **1970,** *6*, 1503–1505.
117. Kalinin, A. V.; Redd, M. A.; Norman, B. H.; Snieckus, V. *J. Org. Chem.* **2003,** *68*, 5992–5999.
118. Flynn, B. L.; Verdier-Pinard, P.; Hamel, E. *Org. Lett.* **2001,** *3*, 651–654.
119. Larock, R. C.; Yue, D. *Tetrahedron Lett.* **2001,** *42*, 6011–6013.
120. Godoi, B.; Schumacher, R. F.; Zeni, G. *Chem. Rev.* **2011,** *111*, 2937–2980.
121. Wang, L.; Shen, J.; Tang, Y.; Chen, Y.; Wnag, W.; Cai, Z.; Du, Z. *Org. Process. Res. Dev.* **2007,** *11*, 487–489.
122. Rodgers, E.; Araki, H.; Batory, L. A.; McInnis, C. E.; Njardarson, J. T. J. *Am. Chem. Soc.* **2007,** *129*, 2768–2769.
123. Grainger, R. S.; Patel, B.; Kariuki, B. M.; Male, L. Spencer, N. *J. Am. Chem. Soc.* **2011,** *133*, 5843–5852.
124. Mansuy, D.; Valadom, P.; Erdelmeier, I.; Lopez-Garcia, P.; Amar, C.; Girault, J.-P.; Dansette, P. M. *J. Am. Chem. Soc.* **1991,** *113*, 7825–7826.
125. Valadon, P.; Dansette, P. M.; Girault, J.-P.; Amar, C.; Mansuy, D. *Chem. Res. Toxicol.* **1996,** *9*, 1403–1413.

126. Treiber, A.; Dansette, P. M.; Amri, H. E.; Girault, J.-P.; Ginderow, D.; Mornon, J.-P.; Mansuy, D. *J. Am. Chem. Soc.* **1997,** *119*, 1565–1571.
127. Blagg, J. Structural Alerts For Toxicity. In *Burger's Medicinal Chemistry, Drug Discovery, and Development, Seventh Edition*; Abraham, D. J.; Rotella, D. P., Eds.; Wiley: Hoboken, NJ, 2010; pp 301–334.
128. Williams, D. P.; Antoine, D. J.; Butler, P. J.; Jones, R.; Randle, L.; Payne, A.; Hoard, M.; Gardner, I.; Blagg, J.; Park, B. K. *J. Pharm. Exp. Ther.* **2007,** *322*, 1208–1220.
129. Stepan, A. F.; Walker, D. P.; Bauman, J.; Proce, D. A.; Baillie, T. A.; Kalgutkar, A. S.; Aleo, M. D. *Chem. Res. Toxicol.* **2001,** *24*, 1345–1410.
130. Bonierbale, E.; Valadon, P.; Pons, C.; Desfosses, B.; Dansette, P. M.; Mansuy, D. *Chem. Res. Toxicol.* **1999,** *12*, 286–296.
131. Dansette, P. M.; Libraire, J.; Bertho, G.; Mansuy, D. *Chem. Res. Toxicol.* **2009,** *22*, 369–373.
132. Garcon, S.; Vassiliou, S.; Cavicchioli, M.; Hartmann, B.; Monteiro, N.; Balme, G. *J. Org. Chem.* **2001**, *66*, 4069–4073.
133. Heynderickx, A.; Samat, A.; Guglielmetti, R. *Synthesis* **2002**, 213–216.
134. (a) Bach, T.; Bartels, M. *Synthesis* **2003**, 925–939. (b) Malamas, M. S.; Sredy, J.; Moxham, C.; Katz, A.; Xu, W.; McDevitt, R.; Adebayo, F. O.; Sawicki, D. R.; Seestaller, L.; Sullivan, D.; Taylor, J. R. *J. Med. Chem.* **2000**, *43*, 1293–1310.
135. Jiang, H.; Yao, W.; Cao, H.; Huang, H.; Cao, D. H. *J. Org. Chem.* **2010,** *75*, 5347–5350.
136. Pei, T.; Chen, C.; DiMichele, L.; Davies, I. W. *Org. Lett.* **2010**, *12*, 4972–4974.
137. Pan, Y.-M.; Zhao, S.-Y.; Ji, W.-H.; Zhan, Z.-P. *J. Comb. Chem.* **2009**, *11*, 103–109.
138. Lo, C.-Y.; Guo, H.; Lian, J.-J.; Shen, F.-M.; Liu, R.-S. *J. Org. Chem.* **2002**, *67*, 3930–3932.
139. Zeng, F.; Alper, H. *Org. Lett.* **2011,** *13*, 1425–2871.
140. O'Conner, C. J.; Roydhouse, M. D.; Przybyl, A. M.; Wall, M. D.; Southern, J. M. *J. Org. Chem.* **2010,** *75*, 2534–2538.

PART II FIVE-MEMBERED HETEROCYCLES WITH TWO HETEROATOMS 197

Chapter 5 Pyrazoles, Pyrazolones, and Indazoles

Jie Jack Li

5.1 Introduction

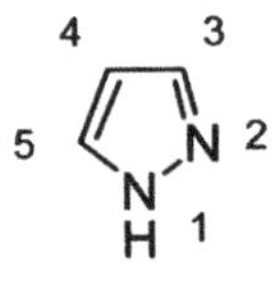

pyrazole

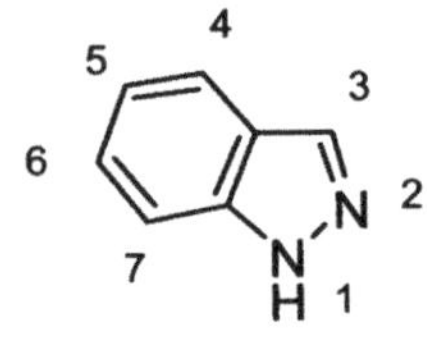

indazole

The parent compound pyrazole itself is a colorless crystalline solid with a melting point of 66–70 °C. It is corrosive and is an irritant to both skin and eyes. Pyrazole is a five-membered aromatic heterocycles with two N heteroatoms. Its N-1 is similar to the NH of pyrrole, and its N2 behaves similarly to that pyridine. Pyrazole's aromaticity lies somewhere in the middle of the scale in comparison with other heterocycles:

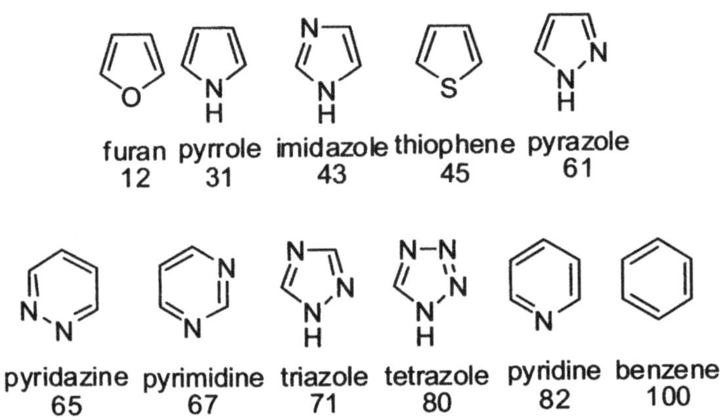

Relative Aromaticity of Heterocycles

The bond lengths are shown below, which impact the coupling constants in their ^{1}H-NMR spectra.

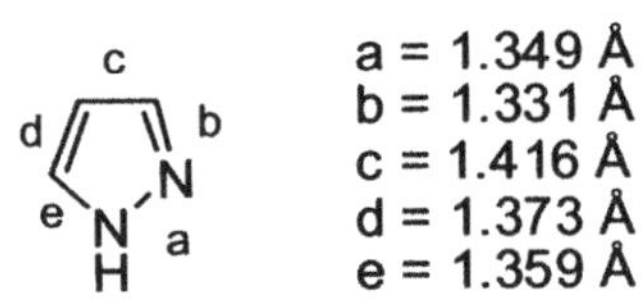

Pyrazole's Bond Lengths

In pyrazole's ^{1}H-NMR spectrum, the chemical shift for both H3 and H5 is the same (thanks to the tautomerization between the two nitrogen atoms) at 7.61 ppm, further down field that that of C4 (6.32 ppm). The phenomenon is readily explained using the two nitrogen atoms' inductive effect. Like other NH-containing heterocycles, the chemical shift of pyrazole's NH largely depends on the solvent used to do the NMR spectrum.

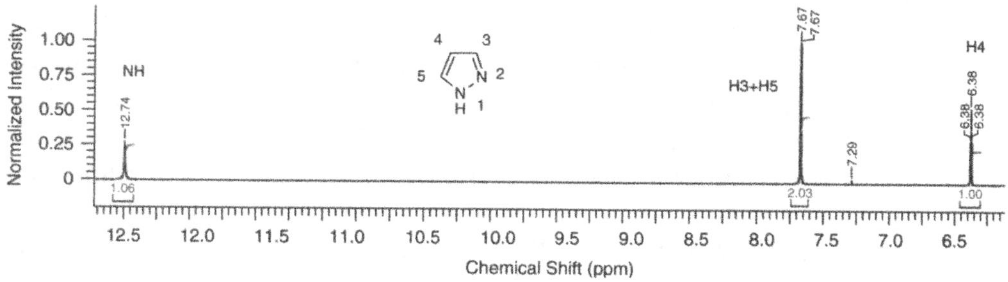

In pyrazole's ^{13}C-NMR spectrum, the chemical shift for both C3 and C5 is the same again thanks to the tautomerization between the two nitrogen atoms) at 133.5 ppm, further down field than that of C4 (104.8 ppm). In the same vein, the phenomenon is readily explained using the two nitrogen atoms' inductive effect.

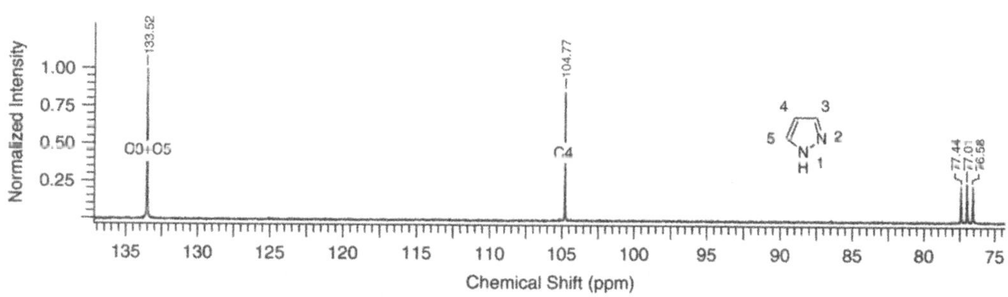

The other parent compound indazole is a white solid at room temperature with a melting point of 147–149 °C. It does not have the tautomerization effect due to the restrain of the benzene ring.

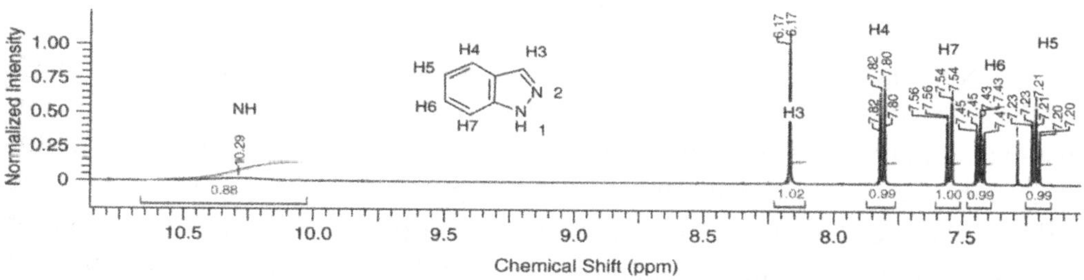

Indazole's ^{1}H-NMR spectrum is much more complicated than those of pyrazole and pyrazolone. With a chemical shift of 8.11 ppm, H3 is the most down field due to N2's inductive effect. With a chemical shift of 7.14 ppm, H3 is the most up field because it is the least affected by two nitrogen atoms' inductive effect. H4, H6, and H7 have chemical shifts of 7.81, 7.44, and 7.55 ppm, respectively. Like other NH-containing heterocycles, the chemical shift of indazole's NH largely depends on the solvent used to do the NMR spectrum.

In indazole's ^{13}C-NMR spectrum, the chemical shifts for C7a and C3, at 140.0 and 134.6 ppm, respectively, are the most downfield due to the two nitrogen atoms' inductive effect. The other quaternary carbon C3a shows up at 123.1 ppm. Both quaternary carbons C3a and C7a are shorter in height because their lack of the nuclear Overhauser effect (nOe) that makes the other carbon atoms directly attached to a proton atom taller. The remainder of the carbon atoms on the benzenes have the chemical shifts as follows: C6, 126.8 ppm; C5, 120.9 ppm; C4, 120.8 ppm; and C7, 109.7 ppm. C7 is the most up-field of all carbon atoms on the indazole molecule.

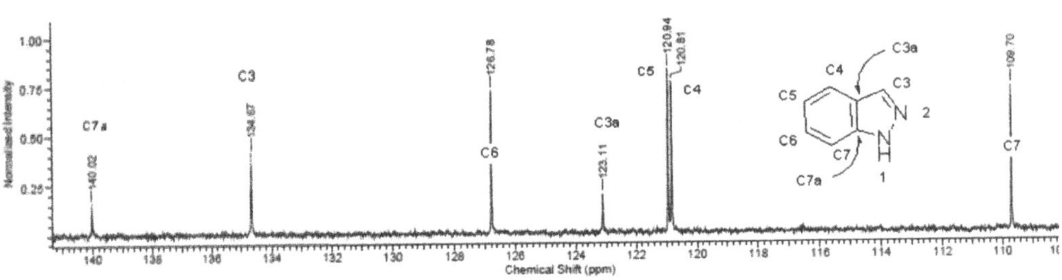

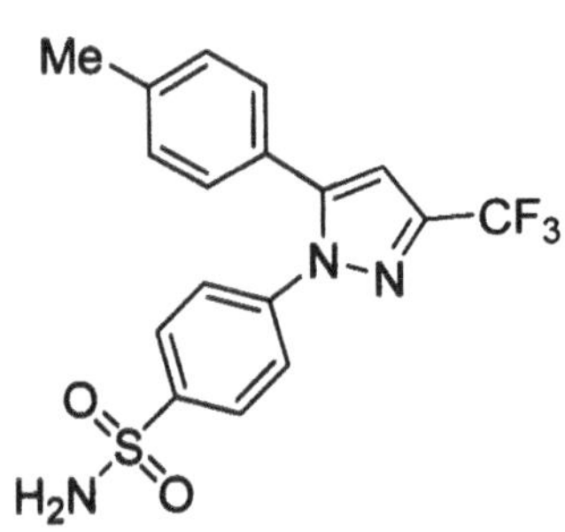

celecoxib (Celebrex)

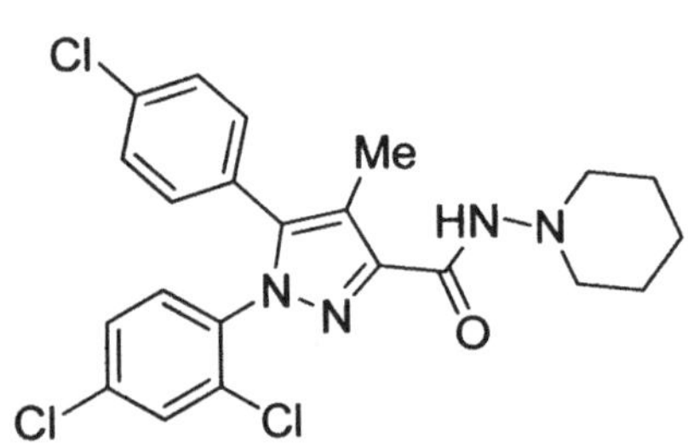

rimonabant (Acomplia)

Not many pyrazoles and indazoles exist in nature. However, many synthetic medicines do contain pyrazoles, pyrazolones, and indazoles. For instance, an anti-inflammatory cyclooxygenase-2 (COX-2) selective inhibitor celecoxib (Celebrex) has the tri-substituted pyrazole as its core structure. A fully substituted pyrazole, rimonabant (Acomplia), is a selective inverse agonist for the cannabinoid receptor type 1 (CB_1). Acomplia was marketed in 56 countries for the treatment of obesity by Sanofi-Aventis starting in

2006, but it was withdrawn in 2008 due to an unfavorable benefit/toxicity profile including the risk of serious psychiatric problems.

Pyrazoles as drugs were reviewed in 2002.[1]

5.1.1 *Basicity and Acidity*

With a p*K*a of 2.5, pyrazole is significantly less basic than imidazole, whose p*K*a is 7.1. In fact, having an adjacent heteroatom near the N atom always has the effect of lowering the basicity of the N because of their inductive effect. Therefore, the N atoms on isothiazole (p*K*a, –3.0) and isoxazoles (p*K*a, –0.5) are less basic than those of thiazole (p*K*a, 2.5) and oxazole (p*K*a, 0.8), respectively. Nonetheless, pyrazole is basic enough to be protonated with most strong inorganic acids.[2,3]

Pyrazole's *N*-hydrogen can be deprotonated by many bases such as NaH, Cs_2CO_3, and K_2CO_3. When the pyrazole is symmetrically substituted, alkylation is devoid of regioselectivity issues. For instance, dimethyl 1*H*-pyrazole-3,5-dicarboxylate is alkylated with 1,2-dibromoethane to give the adduct in 71% without the regioselectivity issues. The alkylated pyrazole was then converted into an orally active, long-acting non-peptide fibrinogen receptor antagonist.[4]

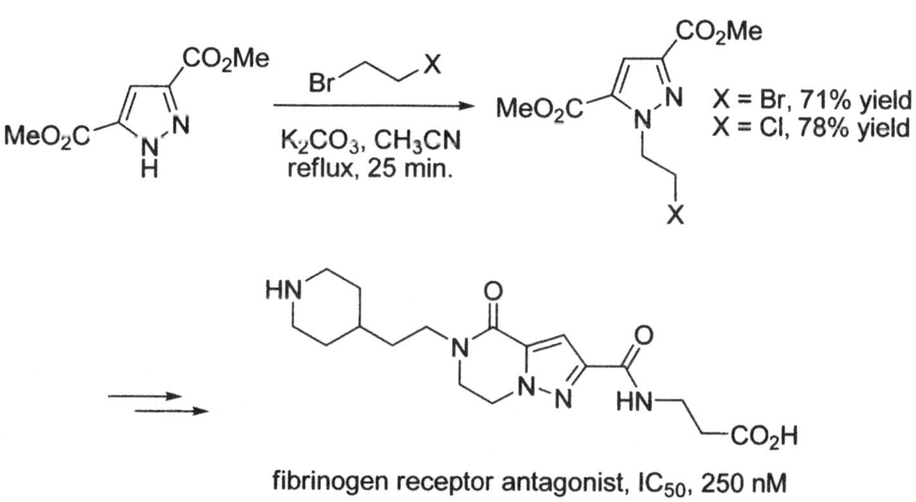

fibrinogen receptor antagonist, IC_{50}, 250 nM

5.1.2 *Tautomerization*

When the pyrazole is unsymmetrically substituted, it may exist as a mixture of two tautomers. For instance, 5-methylpyrazole and 3-methylpyrazole coexist in a solution.

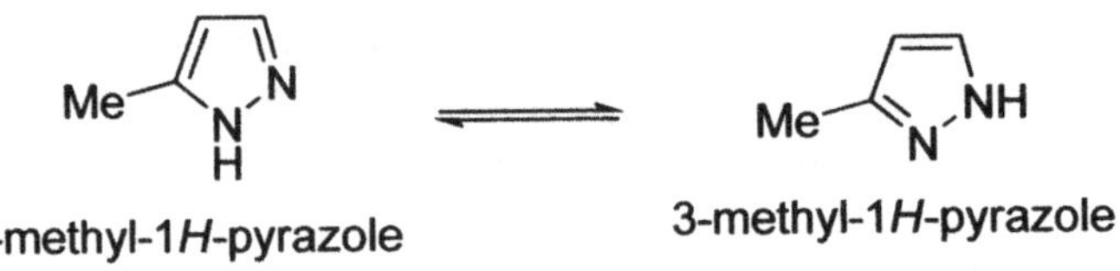

The ramification of the tautomerization is that alkylation of unsymmetrically substituted pyrazoles often gives rise to a mixture of two isomers, one is the *N*-1 alkylation and the other is the *N*-2 alkylation. The ratio depends on the nature of the substrate and the electrophile as well as on the solvent and base (*vide infra*).

5.2 *Reactivities of the Pyrazole Ring*

5.2.1 Alkylation of the Pyrazole Ring

An example delineating the ramification of unsymmetrical pyrazole's tautomerization is shown below:[5]

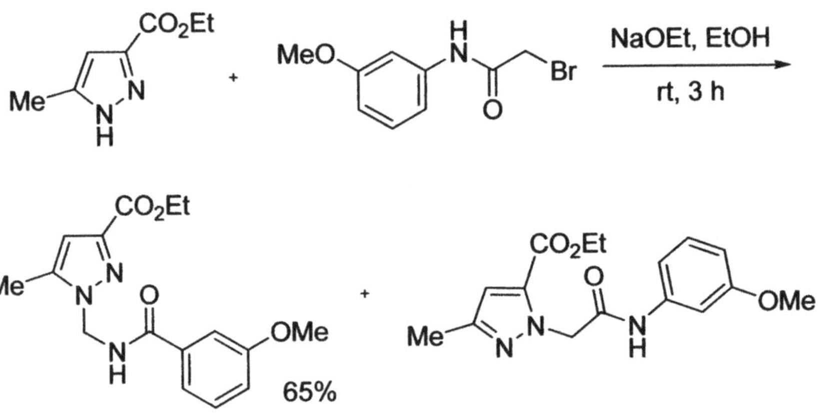

Similarly, alkylation of unsymmetrically substituted 4,5-diphenylpyrazole led to two regioisomers in 69% and 31% yield, respectively.[6] Saponification of the methyl esters to the corresponding carboxylic acids gave rise to nonprostanoid prostacyclin (PGI2) mimetics that inhibited ADP-induced human platelet aggregation.

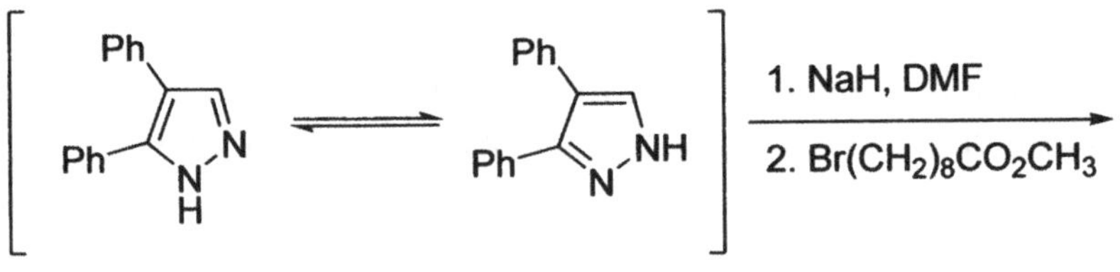

In the same vein, Michael additions of unsymmetrically substituted pyrazoles also result in two regioisomers. 3,4-Diphenylpyrazole was treated with acrylonitrile in ethanolic KOH to afford the adduct in 69% yield, which contained 92% of the 3,4-diphenyl isomer, along with 8% of the 4,5-diphenyl isomer.[7] The major isomer was then converted into the corresponding 3,4-1*H*-diphenylpyrazole-1-propanamines as anti-depressants.

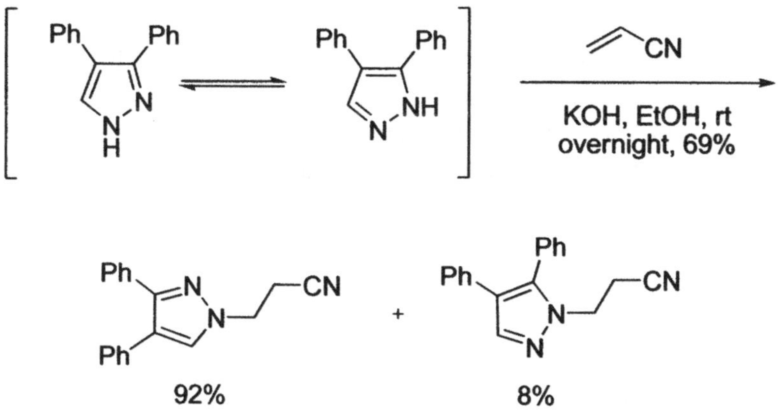

5.2.2 C4 Electrophilic Substitution

Electrophilic substitution always takes place at the 4-position for the neutral pyrazole and the cation.

En route to the assembly of COX-2 inhibitors, when 1,3,5-tri-substituted pyrazole was treated with chlorine gas in glacial acetic acid, the C4 electrophilic substitution product was isolated in 75% yield.[8]

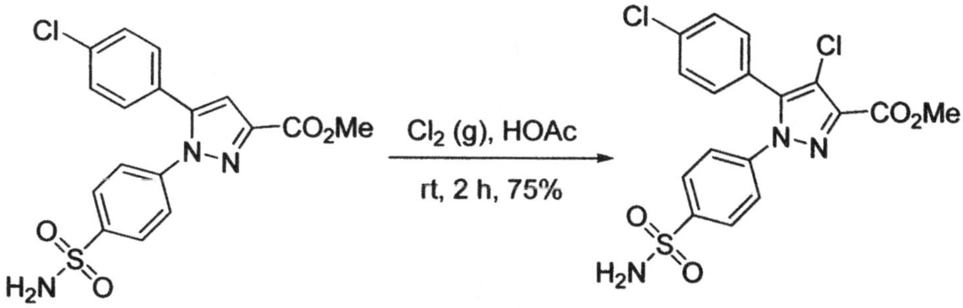

Similarly, when 1,5-di-substituted pyrazole was treated with NBS, the C4 electrophilic substitution product, 4-bromopyrazole, was the sole product isolated from the reaction mixture.[9] The resulting 4-bromopyrazole served as an intermediate for the synthesis of nelotanserin and related 5-HT_{2A} inverse agonists for the treatment of insomnia.

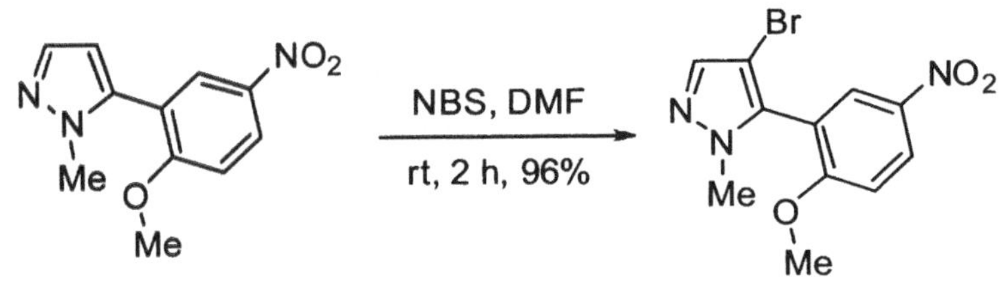

In addition to NBS, bromine may serve as the electrophile in the bromination reaction as well. As shown below, bromination of the pyrazole-ester with bromine as –5 °C in CH_2Cl_2 produced the corresponding 4-bromopyrazole, with concurrent *o*-bromination of the *p*-methoxyphenyl group at the C5 position.[10]

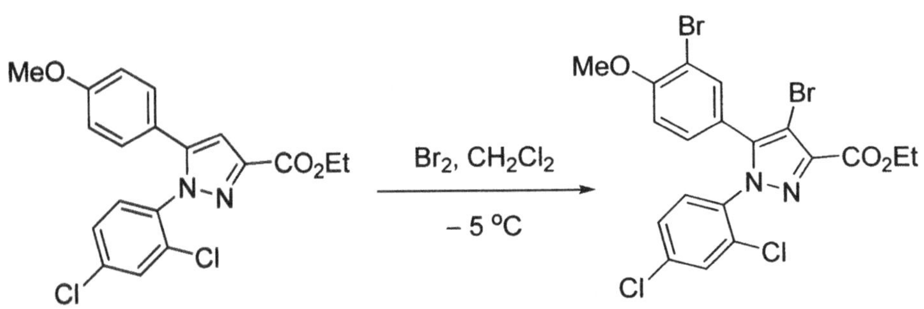

Nitration of pyrazole is a high energetic reaction. Applying a continuous processing approach using a micro-reactor, 4-nitro-pyrazole was prepared from pyrazole-5-carboxylic acid in 73% yield.[11]

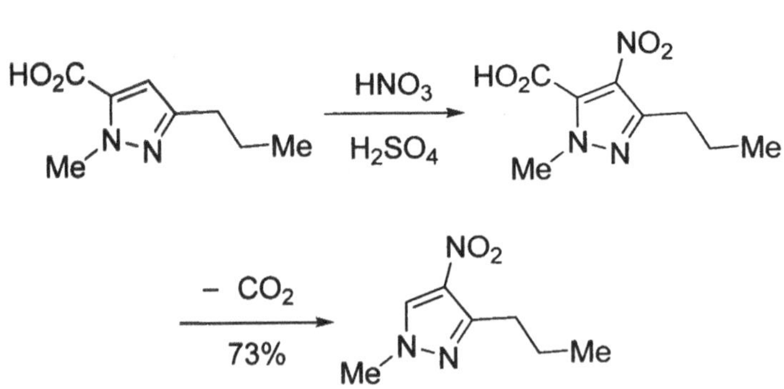

Some pyrazole rings are electron-rich enough to undergo the Friedel–Crafts reaction. With the aid of $AlCl_3$ as the Lewis acid, the pyrazole below underwent the Friedel–Crafts aroylation with *o*-fluorobenzoyl chloride to afford the ketone, an advanced intermediate for the synthesis of novel potential antipsychotic agents.[12]

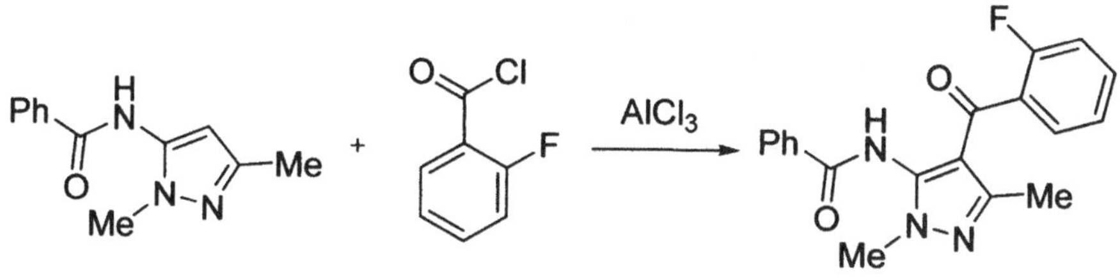

5.2.3 C5-Metallation

C-Metallation of pyrazoles generally requires protection of the NH group. This could be achieved using the dialkylaminomethyl group, easily installed using the Mannich reaction conditions. Refluxing pyrazole with pyrrolidine and formaldehyde in ethanol provided 1-(1-pyrrolidinomethyl)-pyrazole in 56% yield.[13] C5-Lithiation was then achieved using *n*-BuLi. The resulting anion could be quenched with a variety of electrophiles such as diphenylketone. The protective group was then removed under acidic conditions.

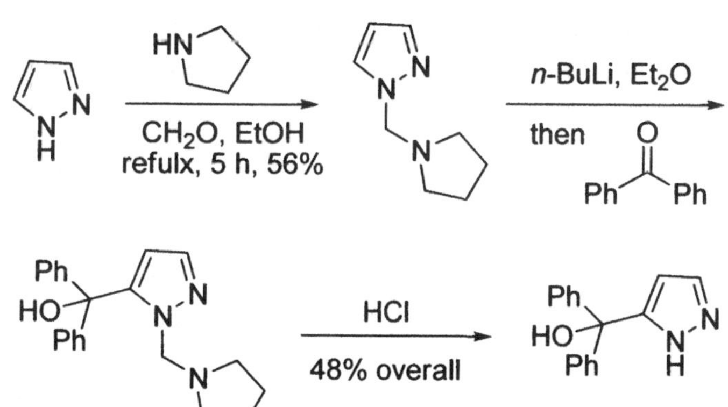

Pyrazole could also be protected as its corresponding tetrahydropyranyl (THP) derivative by trcating pyrazole with 3,4-dihydro-2*H*-pyran (DHP) with the aid of a catalytic amount of TFA.[14] C5-Lithiation of THP-pyrazole could also be achieved using *n*-BuLi. Pyrazole-5-boronic acid was prepared in this fashion.

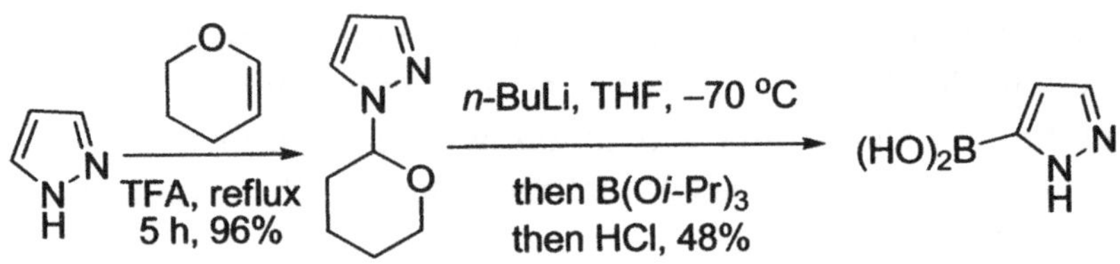

However, C5-lithiation of tosylpyrazole required the use of *t*-BuLi. For example, lithiation of 1-tosylpyrazole at –78 °C in THF took place almost instantaneously.[15]

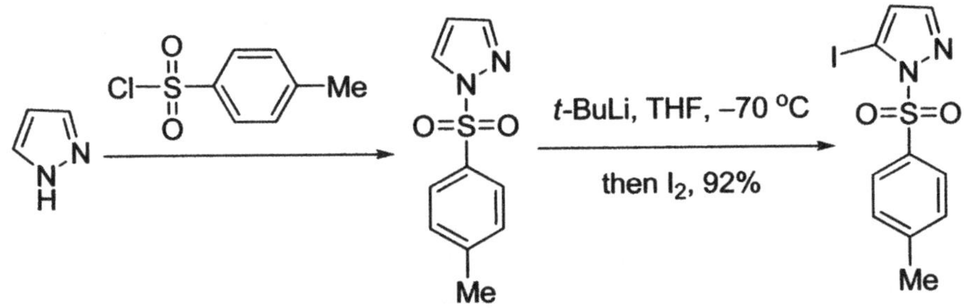

5.3 *Construction of the Pyrazole Ring*

5.3.1 *Knorr Pyrazole Synthesis*

Similar to the Paal–Knorr pyrrole synthesis, the Knorr pyrazole synthesis entails the reaction of hydrazine, or substituted hydrazine, with 1,3-dicarbonyl compounds to provide the pyrazole ring systems.[16]

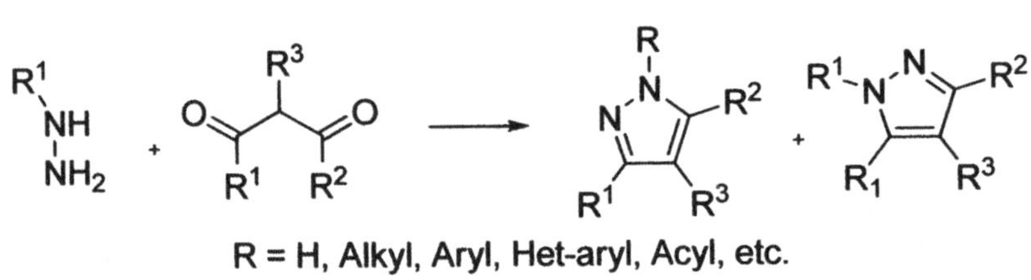

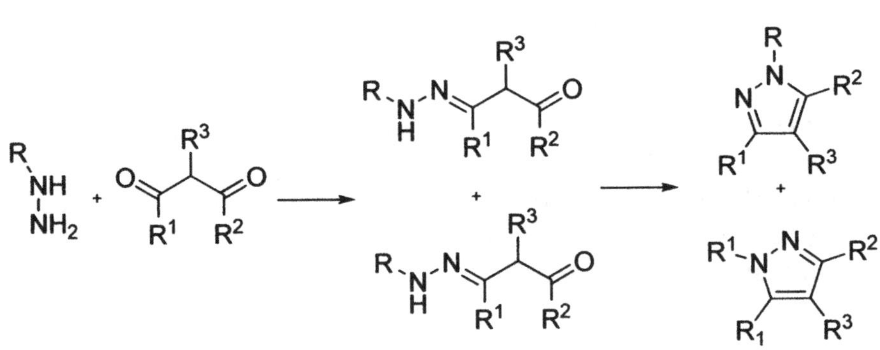

Presumably, the Knorr pyrazole synthesis begins with formation of the hydrazone with the more active one of the two carbonyls. In reality, a mixture of the two regioisomers would arise. The hydrazones then cyclize to afford a mixture of regioismeric pyrazoles.

The Knorr pyrazole synthesis has bestowed a futile ground for medicinal chemists to practice their craft in drug discovery. The most conspicuous example is the synthesis of celecoxib (Celebrex), a selective COX-2 inhibitor prescribed as an analgesic. As shown below, the substrate dione was prepared by the Claisen condensation of 4-methylacetophenone with ethyl trifluoroacetate in the presence of NaOMe in methanol under reflux. Subsequent diarylpyrazole formation from the condensation of the dione and 4-sulfonamidophenyl-hydrazine hydrochloride then delivered celecoxib.[8]

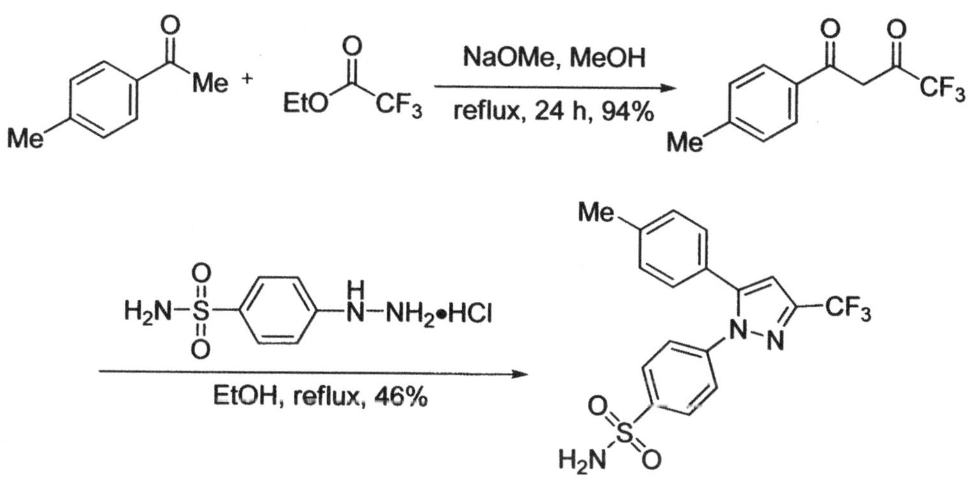

celecoxib (Celebrex)

Tepoxalin is a potent inhibitor of both the cyclcooxygenase and lipoxygenase pathways of the arachidonic acid cascade. In a facile and more environmentally friendly synthesis,[17] condensation between methoxyphenyl-hydrazine and the bis-ketone with a pendant hydroxamic acid group at the end of the molecule delivered tepoxalin in 70% yield.

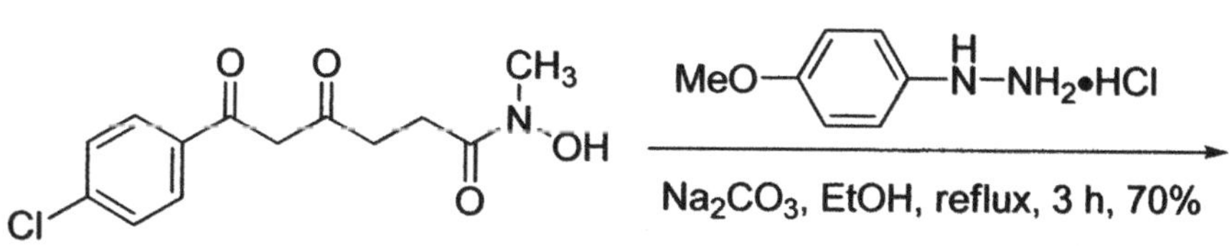

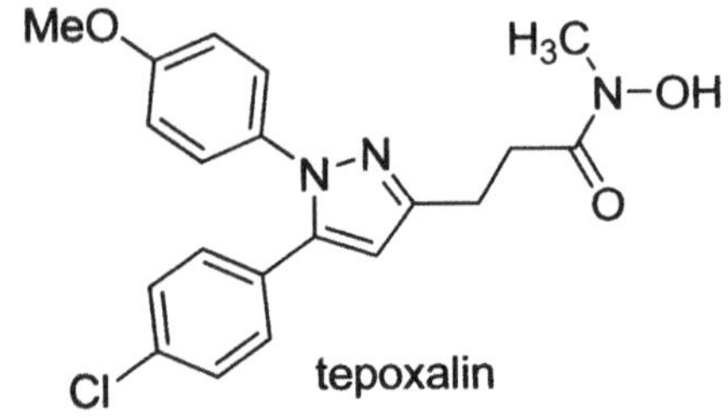

Another important application of the Knorr pyrazole synthesis is the preparation of rimonabant.

The synthesis of rimonabant in Sanofi's 1997 patent by Barth et al.[18] commenced with the Claisen condensation of 4-chloropropiophenone and ethyl oxalate in the presence of LHMDS to give the enolate of the corresponding bis-keto-ester in 37% yield. However, simply switching the solvent from ether to methyl cyclohexane[18] or THF,[19] the yield for the condensation was improved to 59% and 70%, respectively. Pyrazole formation was promoted by hot acetic acid. The resulting pyrazole ester was then converted to rimonabant by hydrolysis and hydrazide formation.

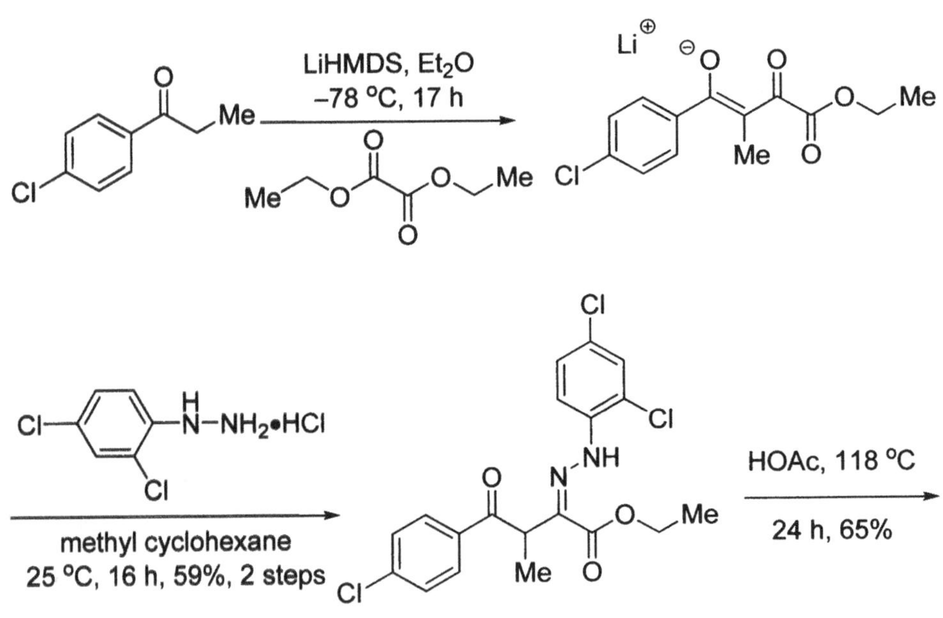

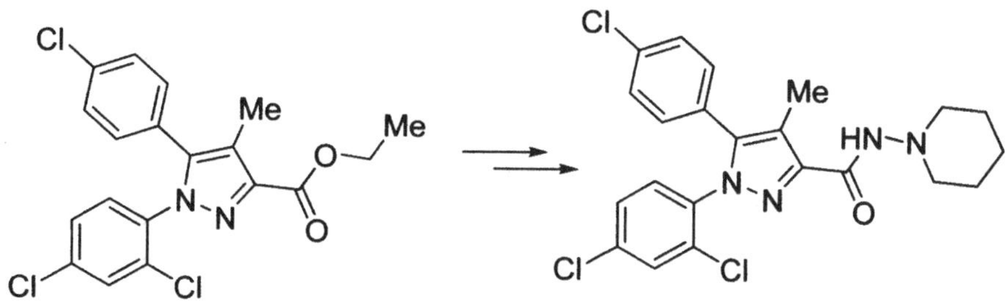

rimonabant (Acomplia)

Generic companies attempted to develop process routes to make rimonabant without infringing on Sanofi's patents. For example, an improved process was developed by taking advantage of a trivial fact that sulfuric acid could promote pyrazole formation and hydrolysis of the ethyl ester simultaneously.[20]

Remarkably, fine-tuning the chemical environment around the two carbonyl groups can completely alter the regiochemical outcome for the hydrazone formation.[21] Triethylamine-mediated amide formation gave the corresponding hydrazide, which deactivated the neighboring carbonyl so that the hydrazone was formed at the distal carbonyl. Condensation with 2,4-dichlorophenylhydrazine HCl salt then delivered the regioisomer of rimonabant in 30% yield.

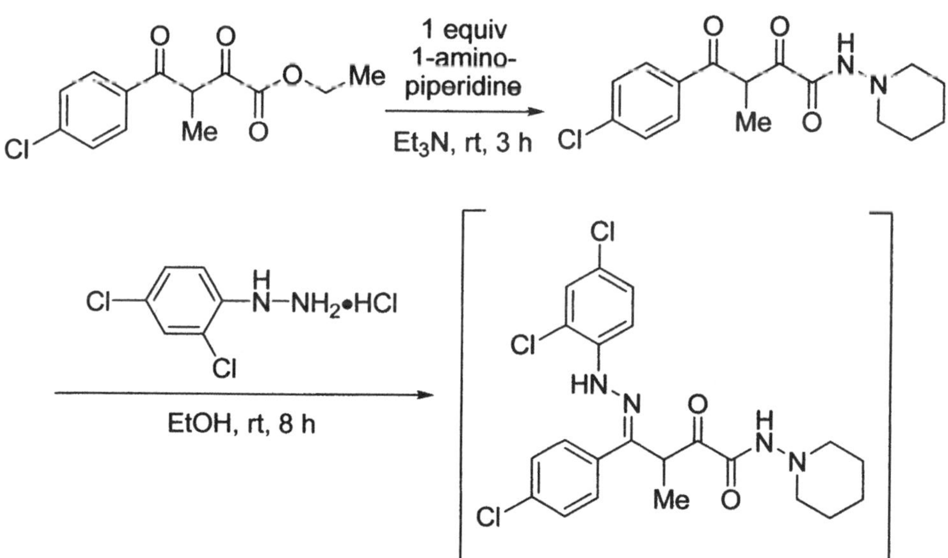

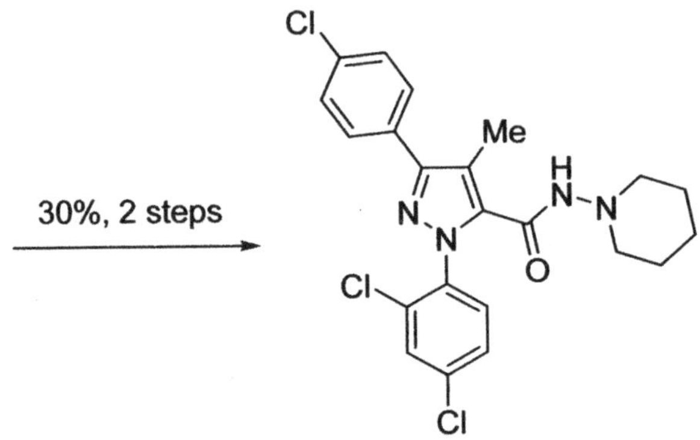

regioisomer of rimonabant

5.3.2 Variations of the Knorr Pyrazole Synthesis

In addition to 1,3-dicarbonyl as a reactant for the Knorr pyrazole synthesis, several variants exist as the 1,3-dicarbonyl group surrogate. Enaminone is one example. Reaction between ethyl-4-iodobenzylacetate and *N,N*-dimethylformamide dimethylacetal (DMFDMA) provided the enaminone.[22] Subsequent condensation between the enaminone and 2,4-dichlorophenyl-hydrazine afforded the pyrazole as an advanced intermediate for 1,5-diarylpyrazole derivatives as CB_1 receptor antagonists.

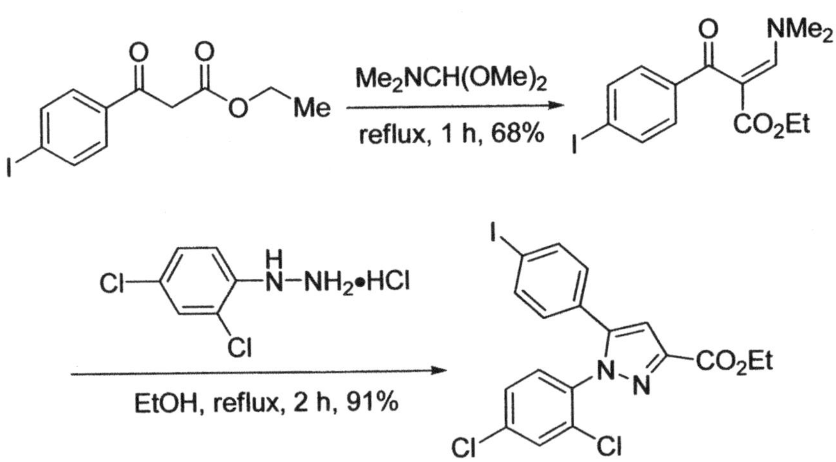

The tactic of using the enaminone as the 1,3-dicarbonyl variant was also successfully applied to the synthesis of a new series of 5-(biphenyl-4-ylmethyl)pyrazoles as AT_1 selective angiotensin II receptor antagonists.[23]

Another 1,3-dicarbonyl equivalent is keto-enol. Treatment of the ketone with ethyl formate in the presence of NaH afforded the keto-enol as the 2-hydroxy-methylene ketone.[24] Condensation of the keto-enol with hydrazines provided the corresponding pyrazole fused bile acid methyl ester as advanced intermediates for substrates for bile acid transporters. Meanwhile, the keto-enol tactic has been applied to synthesize pyrazolo[3,4-*c*]pyridines as novel inhibitors of human eosinophil phosphodiesterase.[25]

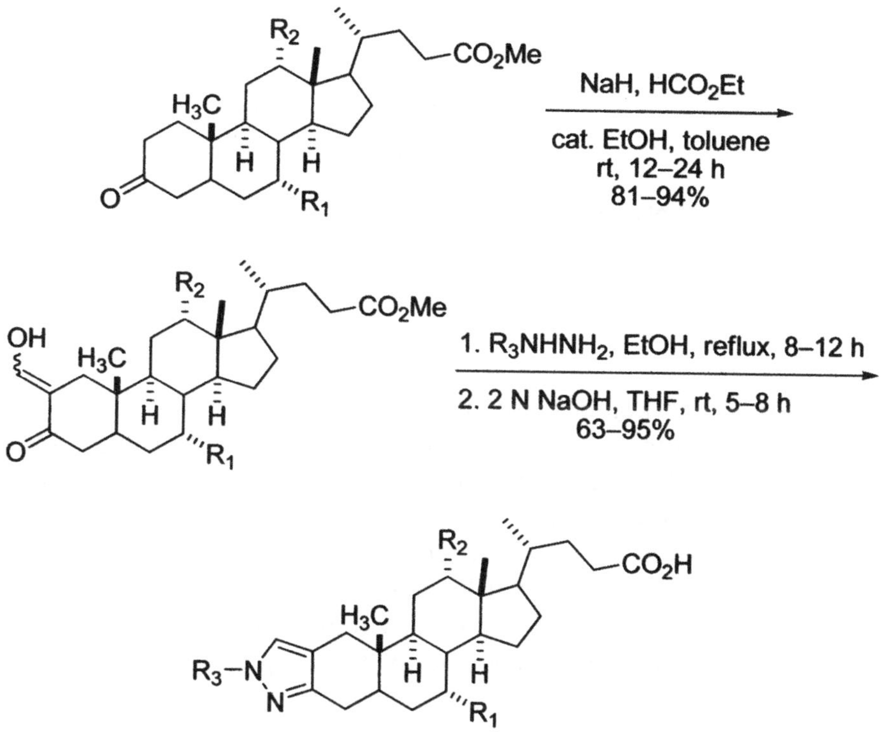

Enone may also serve as a substrate for preparing pyrazolines, which epoxy ketone has been used to make hydroxyl-pyrazolines.[26] For instance, the enone-ester substrate was used to make diaryl dihydropyrazole-3-carboxamides as CB_1 receptor antagonists.[27]

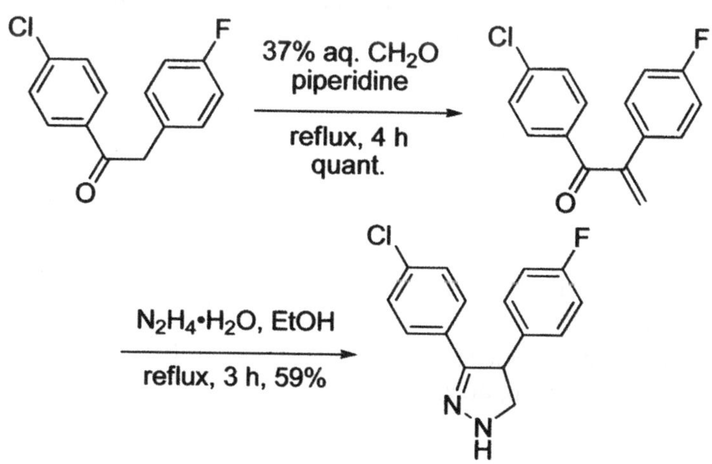

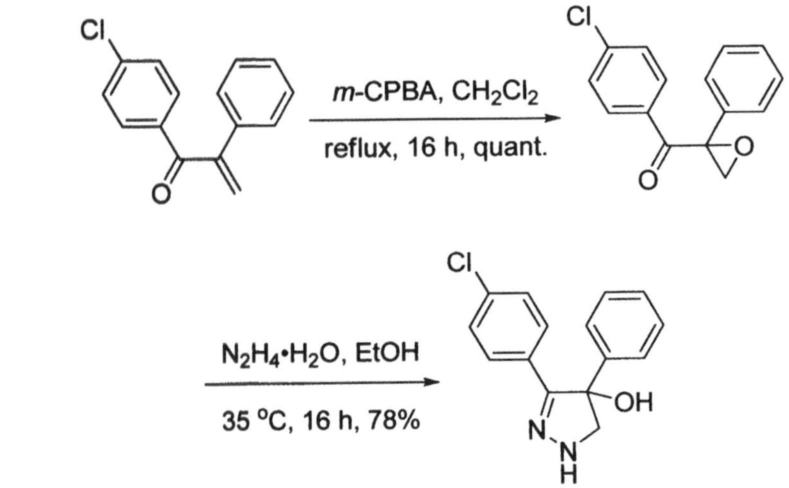

Methoxyimino-ketones are not only variants of the 1,3-dicarbonyl group, but also they have the advantage of affording excellent regioselectivity for the pyrazole formation. Chemoselective condensation between the diketoester and methoxyamine hydrochloride produced methoxyimino-ketoester.[28] After the alkylation with 4′-(bromomethyl)biphenyl-2-carbonitrile, the highly regioselective pyrazole synthesis was achieved by treating methoxyimino-ketoester with a hydrazine. The intermediate was used to prepare nonpeptide angiotensin II antagonists.

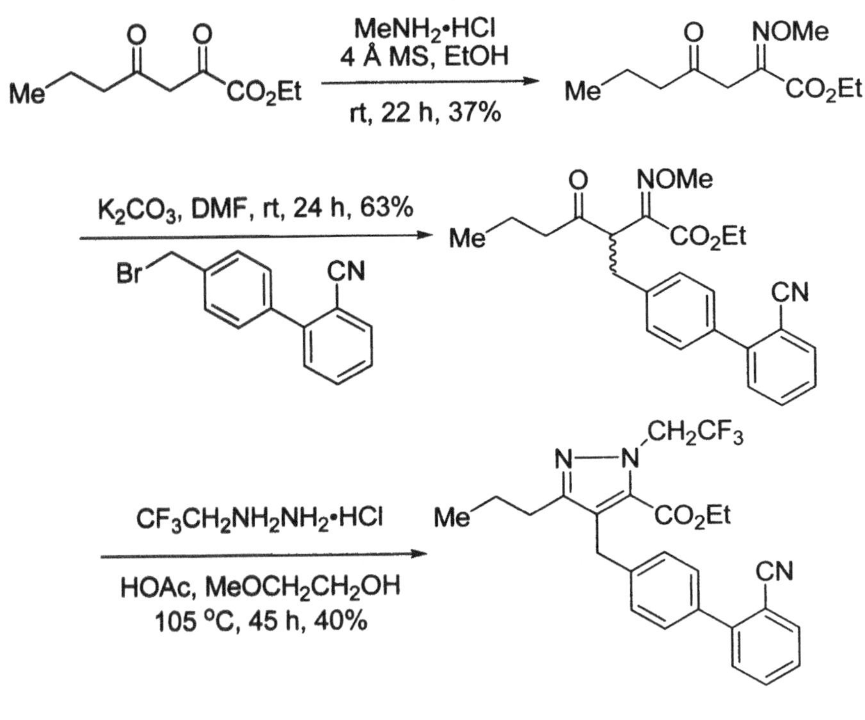

Methoxyimino-ketones have been taken advantage of to synthesize a series of novel non-amidine factor Xa inhibitors.[29]

Aminopyrazoles may be assembled with the 3-chloropropenenitrile intermediates, easily prepared by sequential treatment of the corresponding ketone with $POCl_3$/DMF and hydroxylamine. Condensation of the intermediate with hydrazine then provided the aminopyrazole as the penultimate intermediate to the COX-2 inhibitors.[23]

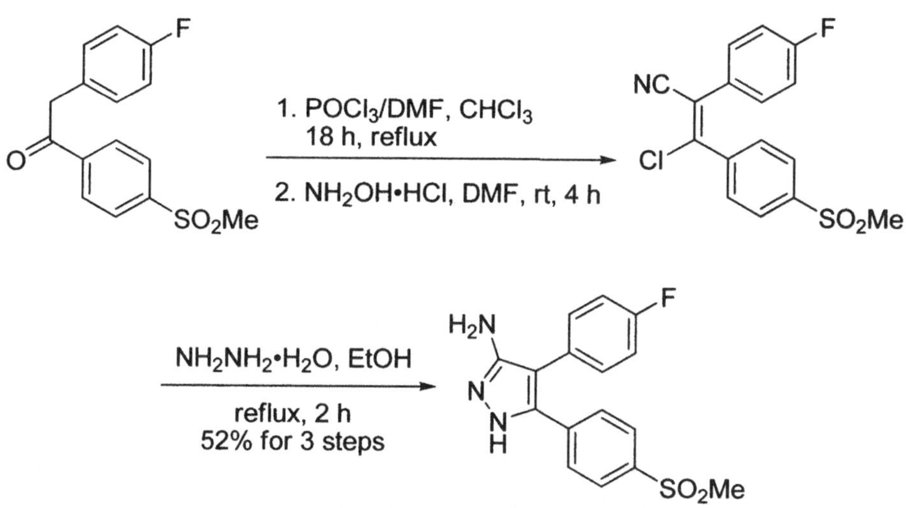

(2*E*)-Ethyl 2-chloro-3-cyano-3-ethoxyacrylate is a useful building block for constructing aminopyrazoles. Its condensation with phenylhydrazine provided a series of aminopyrazoles as intermediates of preparing antitubercular agents.[30]

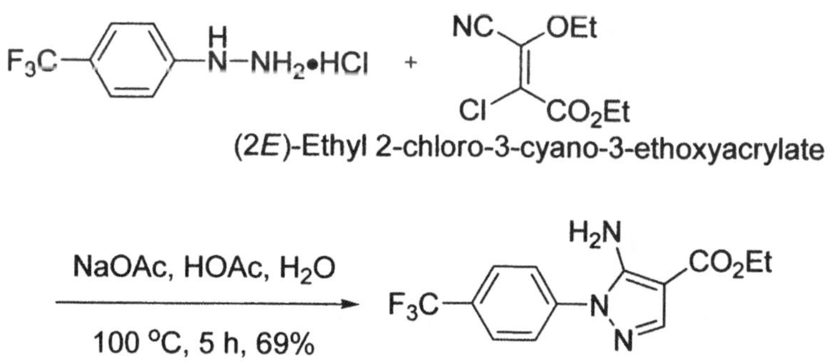

One reactant to make aminopyrazoles may be prepared from ketones as well. As shown below, cyanoketone was obtained by treating the lithium anion derived from acetonitrile with methylbenzoate. Condensation between the cyanoketone and phenylhydrazine afforded 5-aminopyrazole to serve as an advanced intermediate for the synthesis of RO3201195, an orally bioavailable and highly selective inhibitor of p38 MAP kinase.[31]

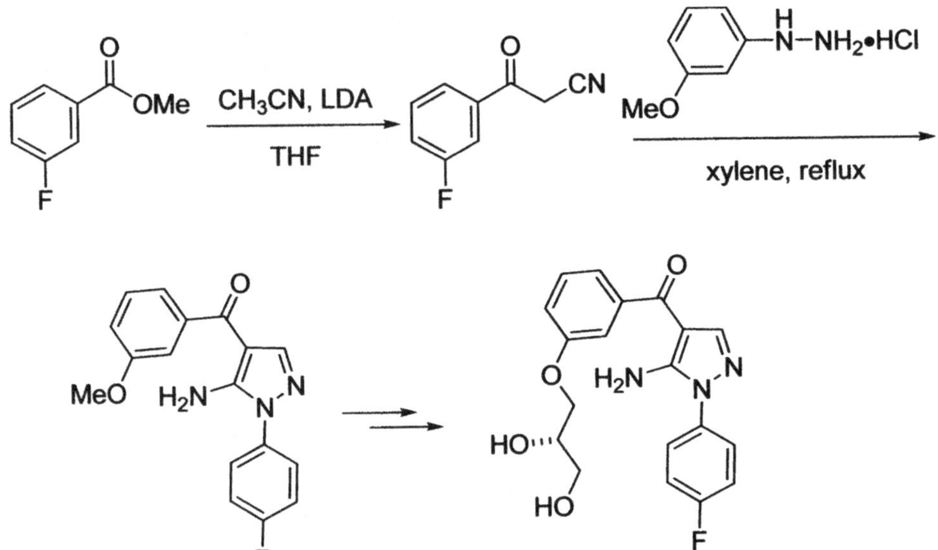

RO3201195, a p38 MAP kinase inhibitor

Due to their importance in drug discovery, pyrazolone-containing drugs will be discussed in a separate section, Section 5.4.

5.3.3 Pechmann Pyrazole Synthesis

The 1,3-dipolar cycloaddition between diazoalkanes and alkynes resulting in pyrazole formation is known as the Pechmann pyrazole synthesis.[32]

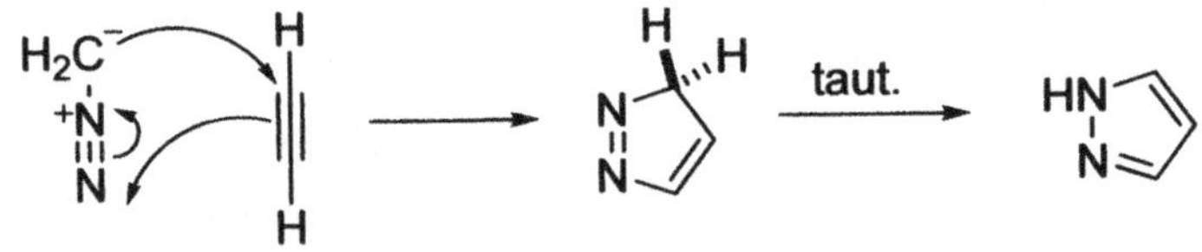

The original reaction discovered by Pechmann involved the cycloaddition of diazomethane and acetylene. Although a better understanding of the reaction has led to the common use of more electron-deficient alkynes, diazomethane continues to be synthetically useful.[33] A recent elegant example of the use of diazomethane as the 1,3-dipole was demonstrated in the preparation of 2,3-benzodiazepine derivatives as potential non-competitive AMPA antagonists.[34] Beginning with the alkyne, the pyrazole moiety could be incorporated into the benzodiazepine structure, using the Pechmann pyrazole synthesis, to produce the 2,3-benzodiazepines.

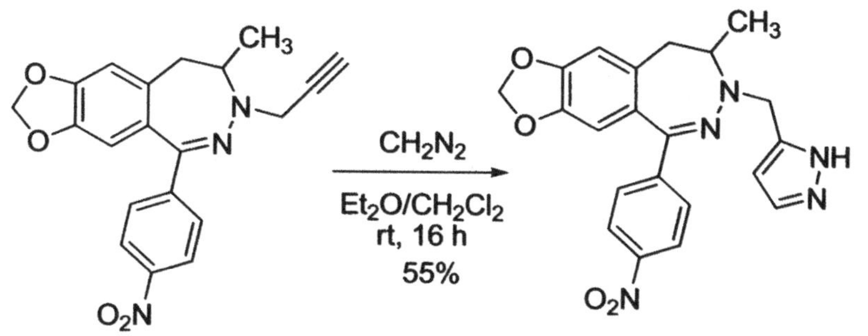

A 1,3-dipolar cycloaddition of ethyl propiolate and ethyl diazoacetate was catalyzed by $InCl_3$ in water to afford pyrazole bis-ester.[35] For secondary diazo compounds, the $InCl_3$-catalyzed 1,3-dipolar cycloaddition often gives a mixture of two regioisomers.

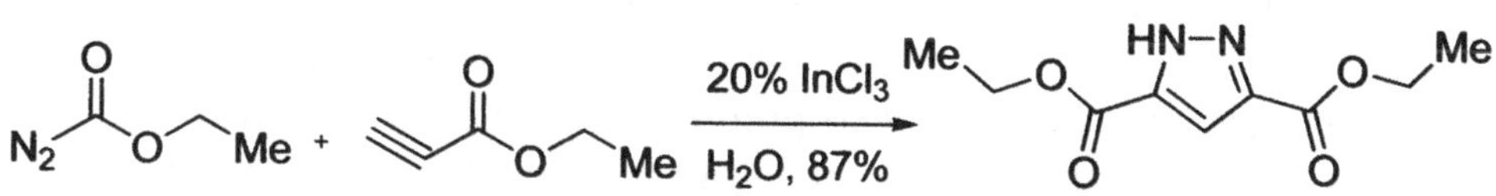

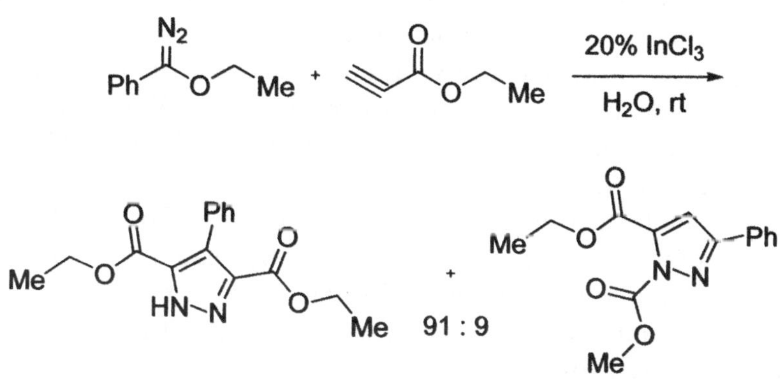

Ethyldiazoacetate (EDAC) has become a commonly utilized 1,3-dipole in the Pechmann pyrazole synthesis. In efforts toward the rational design of growth inhibitors of *Mycobacterium tuberculosis*, Kozikowski and co-workers utilized this strategy for the synthesis of the pyrazole ester.[36] Treatment of a mixture of alkyne and EDAC under microwave conditions resulted in the preparation of the pyrazole ester as a mixture of 1*H* and 2*H* tautomers. The pyrazole ester proved inactive in the anti-TB assay, thereby proving the importance of an isoxazole moiety for anti-TB activity, as previous work by these authors suggested.[37,38]

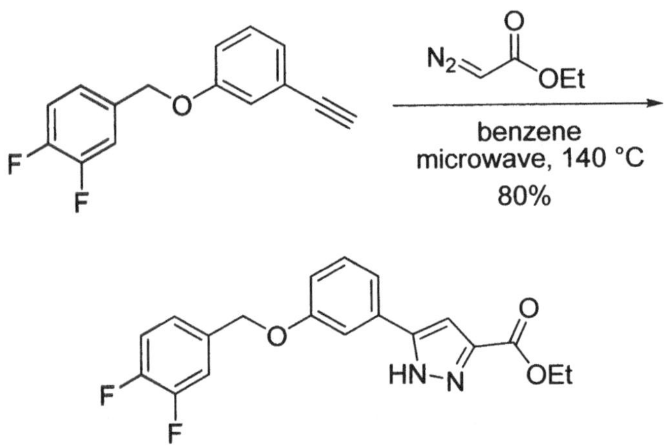

Although the Pechmann pyrazole synthesis routinely features the use of simple diazo compounds such as diazomethane and ethyldiazoacetate, several complex diazo compounds have also found utility in the title reaction. One such reagent has been used to produce an analogue of pyrazofurin[31] in order to evaluate the importance of intramolecular hydrogen bonding in its intracellular conversion to the 5′-monophosphate, a process that contributes to its antitumor and antiviral properties. The diazo compound was produced in a two-step process from the acetyl amide. Reaction of the acetyl amide with nitrogen dioxide and acetic acid, to produce the *N*-nitrosamide, was followed by treatment with aqueous potassium hydroxide to yield the diazo compound. The reaction of the diazo compound and methylpropiolate resulted in efficient formation of pyrazole-5-ester, in a 76% yield over the three steps. Completion of the synthesis eventually resulted in formation of the pyrazofurin analogue in high yield. Similar work in the same laboratory utilized the Pechmann pyrazole synthesis for preparation of an acyclic analogue of 4-deoxypyrazofurin.[39]

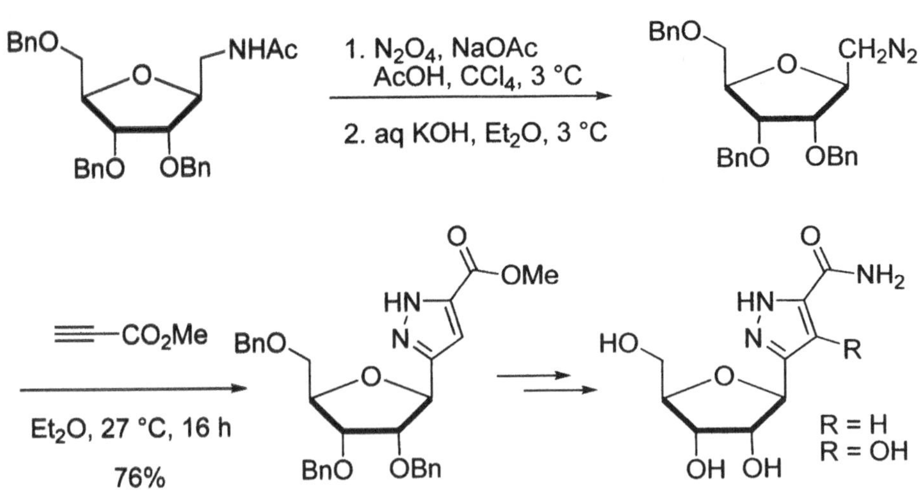

Finally, a facile and regioselective synthesis of rimonabant was accomplished through an enamine-directed 1,3-dipolar cycloaddition.[40] In the presence of triethylamine, hydrazonoyl iodide was converted into the nitrile imine *in situ*. The subsequent 1,3-dipolar cycloaddition with the morpholine enamine provided the 1,5-diarylpyrazole, which was transformed into rimonabant.

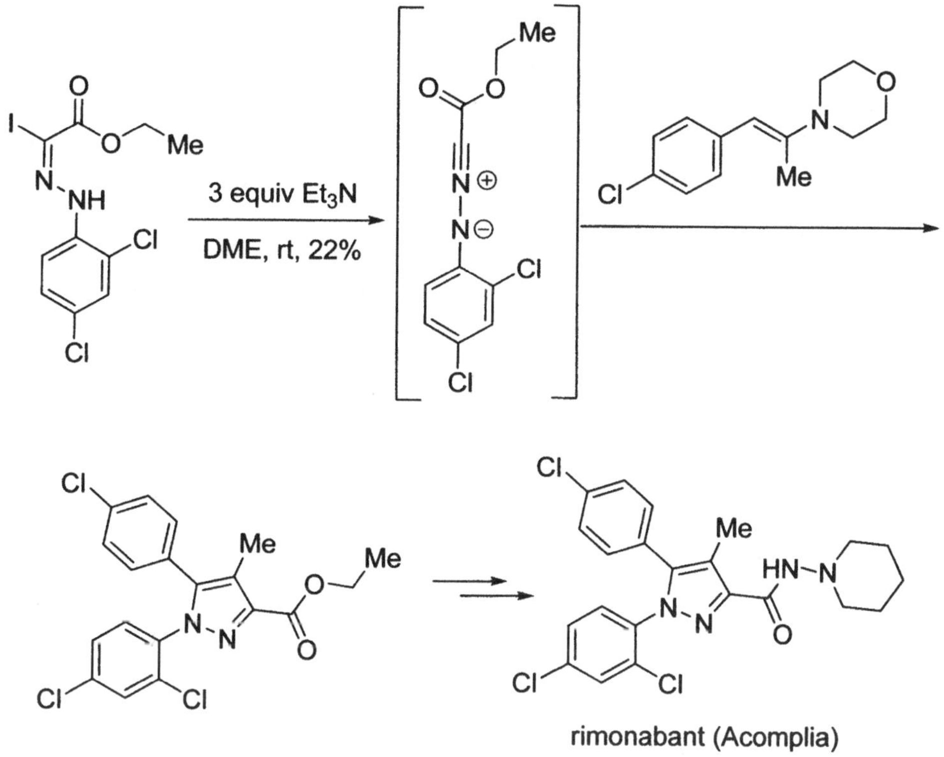

5.4 *Pyrazolone-containing Drugs*

For ketoester the substrate, the hydrazone forms exclusively between the hydrazine and with more active ketone. The corresponding hydrazone-ester then cyclizes to produce the pyazolone. Mechanistically, the more basic and less substituted hydrazine amine condenses with the ketone. The resultant hydrazine–ester then cyclizes to the pyrazolol, which could also tautomerize to the corresponding pyrazolone.

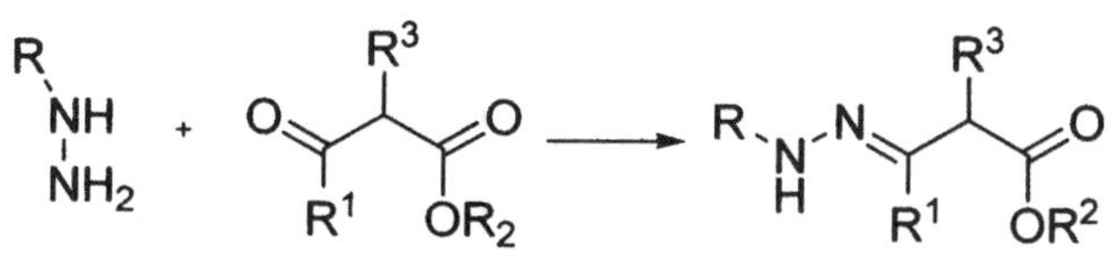

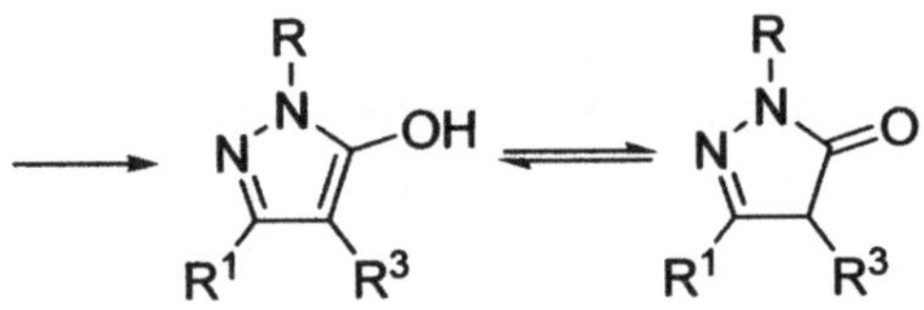

The preparation of nifemazone was a good example for making pyrazolone-containing drugs.[41] Condensation between phenylhydrazine and ethyl acetoacetate produced the pyrazolone, which was subsequently protected as its N-1*H* position with the methyl group to afford an anti-inflammatory agent, antipyrine. The 4-nitroso-substitution was readily achieved by treating antipyrine with sodium nitrite in the presence of HCl. Reduction of the 4-nitroso group was followed by acylation with a nicotinic acid derivative to give nifemazone.[42]

Nifemazone had good antiphlogistic activity, showing prompt onset and sustained action especially in albumin-induced edema. It had antipyretic and analgesic effects similar to aminophenazone and phenylbutazone (*vide infra*) but much less toxic than the latter.

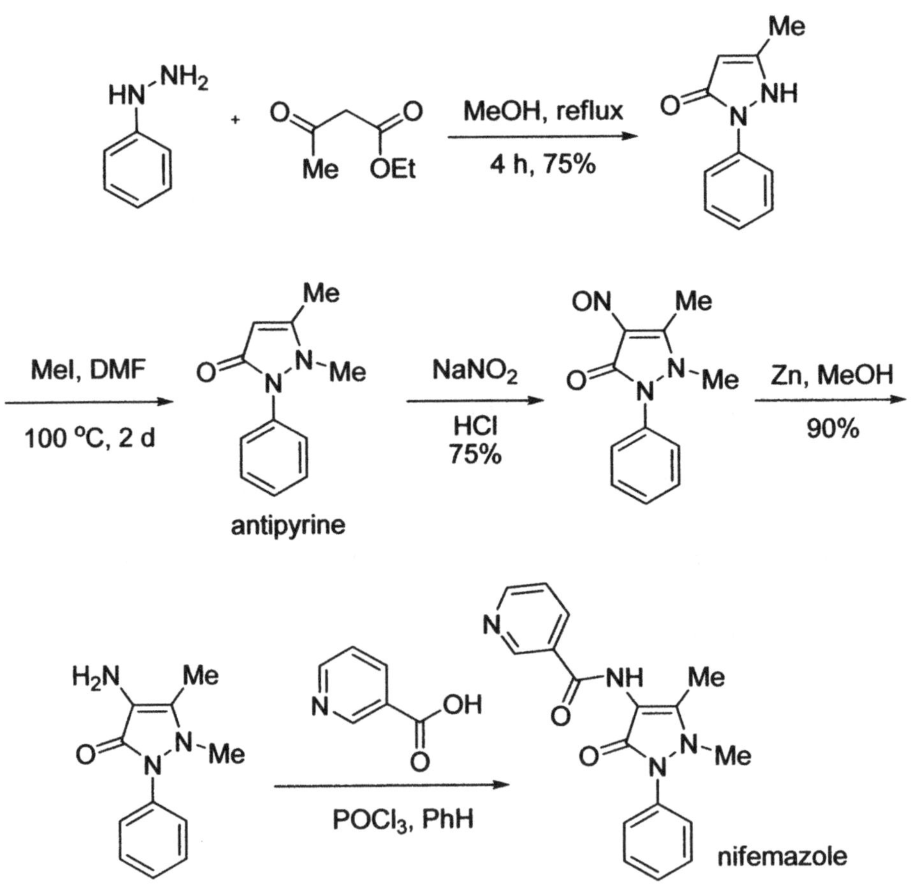

Phenylbutazone, on the other hand, was the first and most widely used anti-inflammatory agent. Its synthesis began with preparation of (*Z*)-1,2-diphenyldiazene from phenylaniline via the diazonium salt intermediate.[43] Reduction of the diphenyldiazene using hydrazine afforded the diphenyldiazine, which was subsequently condensed with diethyl 2-butylmalonate to deliver phenylbutazone.

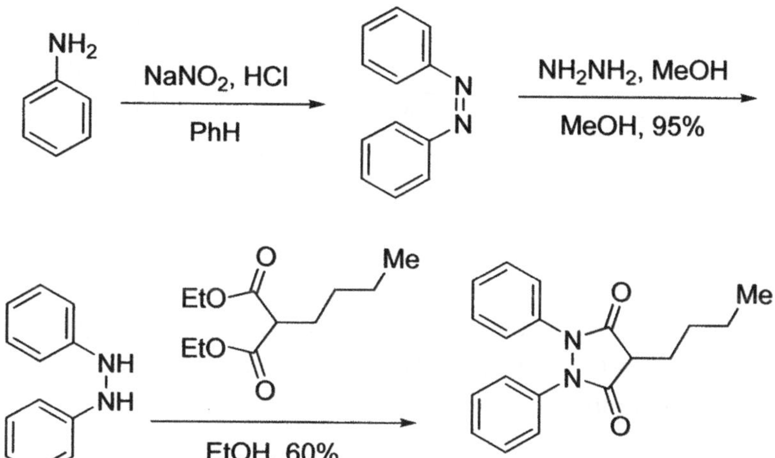

phenylbutazone

Two phenylbutazone-like anti-inflammatory agents, oxyphenbutazole[44] and sulfipyrazole,[45] were also prepared using similar chemistry shown above.

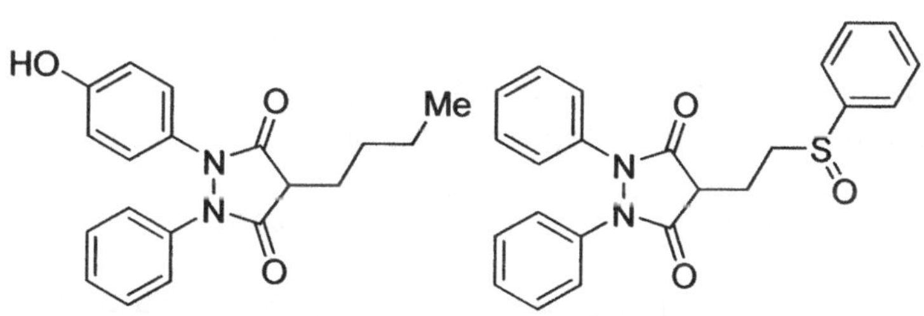

oxyphenbutazole sulfipyrazole

In addition to the conventional method of making pyrazolones using keto-ester, diethyl 2-(ethoxymethylene)malonate could be used as a surrogate for the keto-esters. As shown below, treatment of phenylhydrazine with diethyl 2-(ethoxymethylene)malonate in ether gave rise to pyrazolols, which was acylated with acyl chloride. The pyrazolols/pyrazolones served as intermediates for the preparation of a potent multi-drug resistance modulator.[46]

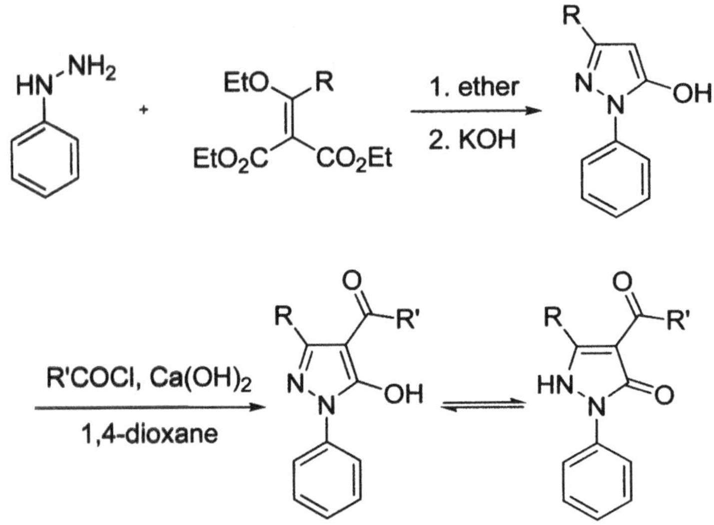

Finally, pyrazolol/pyrazolone derivatives have recently been synthesized as inhibitors of *Mycobacterium tuberculosis*[47] and as potent antihyperglycemic agents.[48]

5.5 *Indazole-Containing Drugs*

Indazoles have been used as the biosteres of indoles in drug discovery. In the synthesis of the nonsteroidal anti-inflammatory drug (NSAID) bendazac, benzylaniline was used as its starting material. Nitrosolation was accomplished using *nitrous acid.* The resulting *N*-nitroso intermediate was reduced with sodium thiosulfate to the corresponding hydrazine, which cyclized to the indazolone. Subsequent alkylation with methyl chloroacetate was followed by hydrolysis to deliver bendazac.

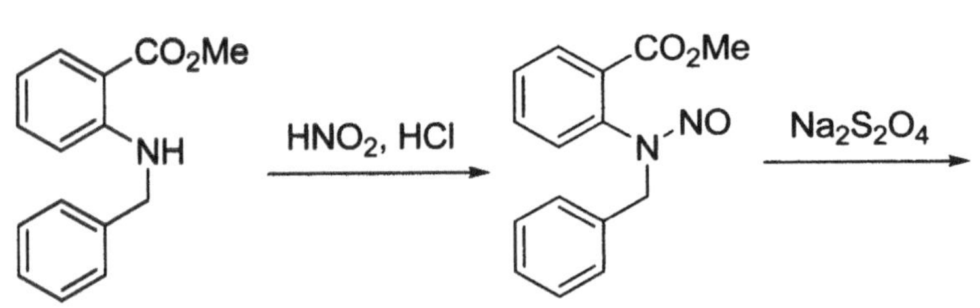

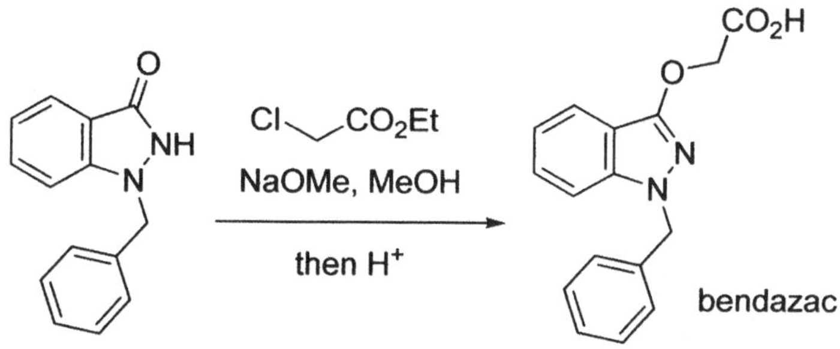

Interestingly, bendazac is one of the agents that have been introduced for the treatment of cataracts.[49] Bendazac and its main metabolite, the 5-hydroxy derivative, provide antioxidant effects as scavengers of oxygen-derived free radicals.

Beecham's granisetron (trade name Kytril) is a serotonin 5-HT_3 receptor antagonist used as an antiemetic to treat nausea and vomiting following chemotherapy.[50] In expanding the structure-activity relationship (SAR) around granisetron, interesting indazole chemistry has been developed. For instance, 5-methoxyindazole was protected as its SEM derivative, which was then installed in the ethyl ester on the 3-position. Switching the SEM group with a methyl group was followed by hydrolysis to provide the 5-methoxy-1-methyl-indazole-3-carboxylic acid. Coupling with the bicyclic amine produced the amide and the methyl ether was converted into the benzyl ether. The end result was the benzyloxy derivative of granisetron.[51]

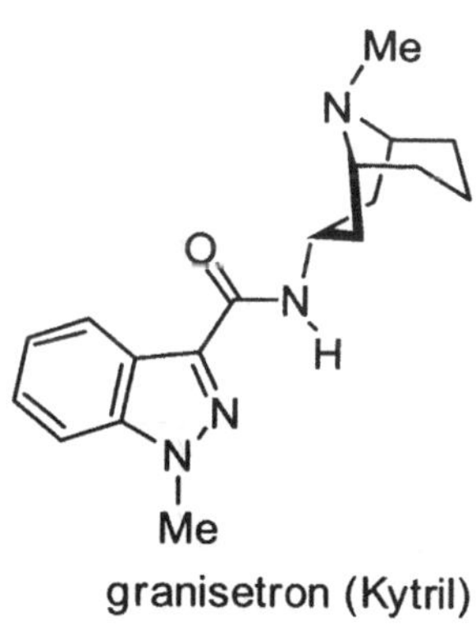

granisetron (Kytril)

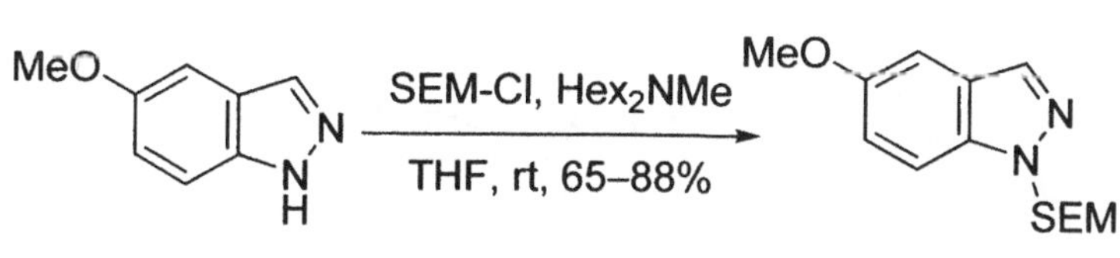

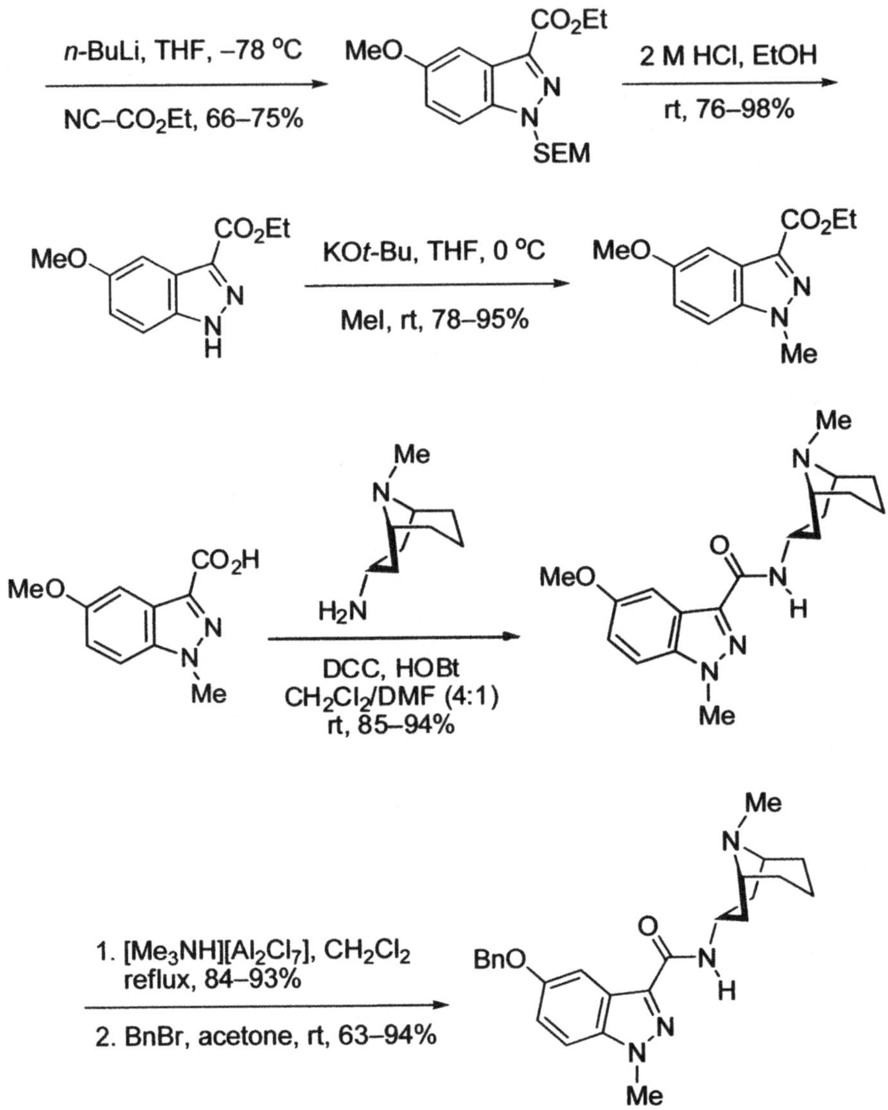

Recently, 1-aryl-1*H*-indazoles have been prepared via palladium-catalyzed intramolecular amination of aryl halides.[52] For example, the bromo-aryl-hydrazone cyclized to give the corresponding indazole under the influence of a palladium catalyst in the presence of a base.

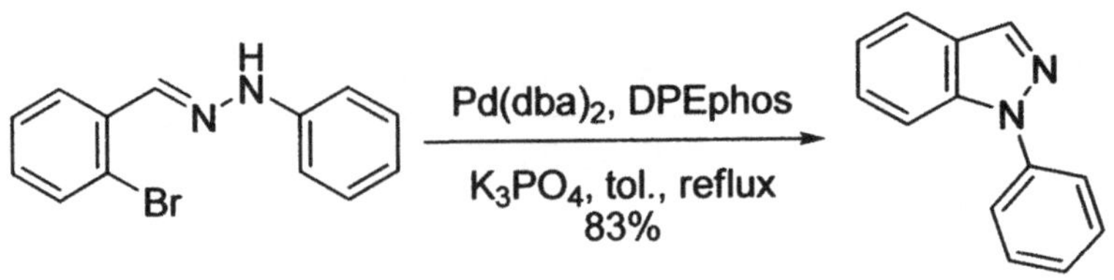

Indazoles may be assembled via thermolysis as well. Treatment of diphenylcarbamoyl chloride with sodium azide gives rise to diphenylcarbamoyl azide. Thermolysis of the diphenyl-carbamoyl azide then afforded indazolol. Subsequent alkylation of the indazolol then produced a series of indazole derivatives as activators of the nitric oxide receptor, soluble guanylate cyclise.[53]

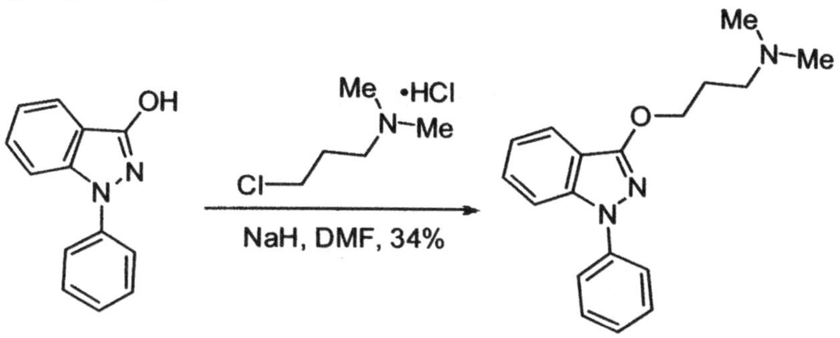

The synthesis of indazols and indazolones was reviewed in 1978.[54]

5.6 *Problems*

5.6.1 Predict the bromination products **A** and **B**.[55,56]

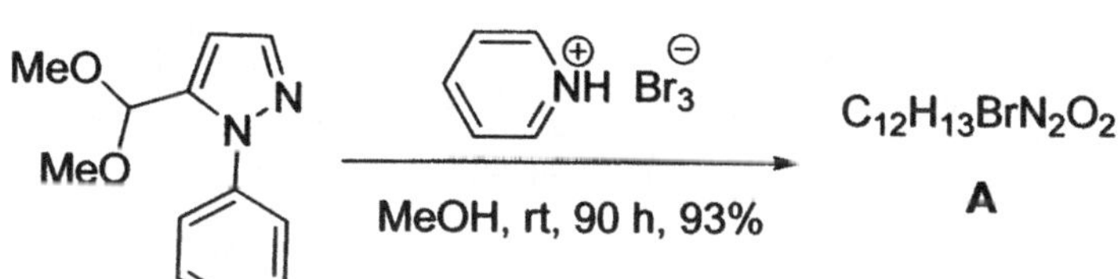

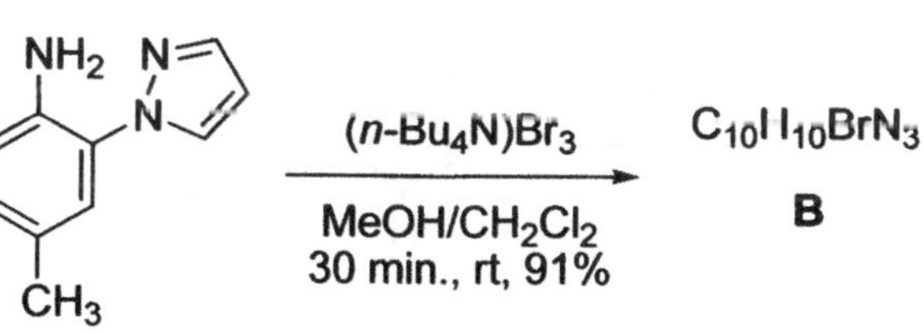

5.6.2 Predict the bromination product **C**.[57]

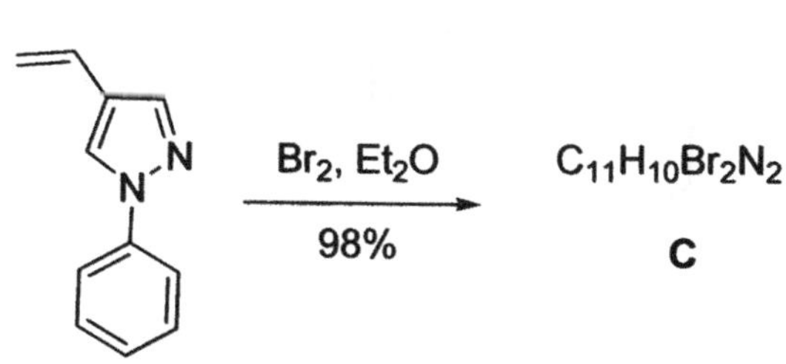

5.6.3 Propose the mechanism of the transformation of *indole* into its corresponding *indazole*-aldehyde[58]:

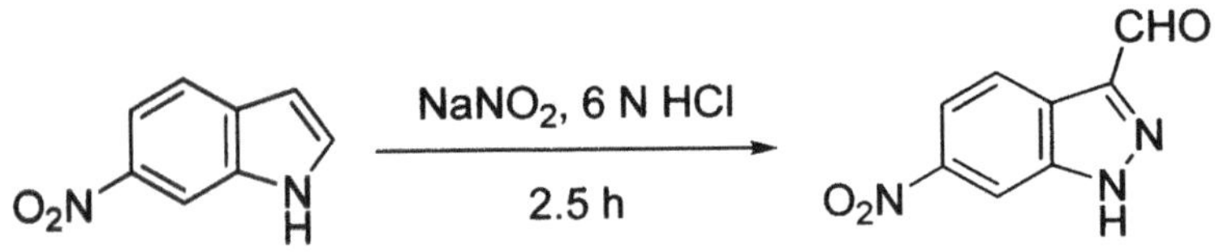

5.6.4 Propose a mechanism for the following transformations[59–61]:

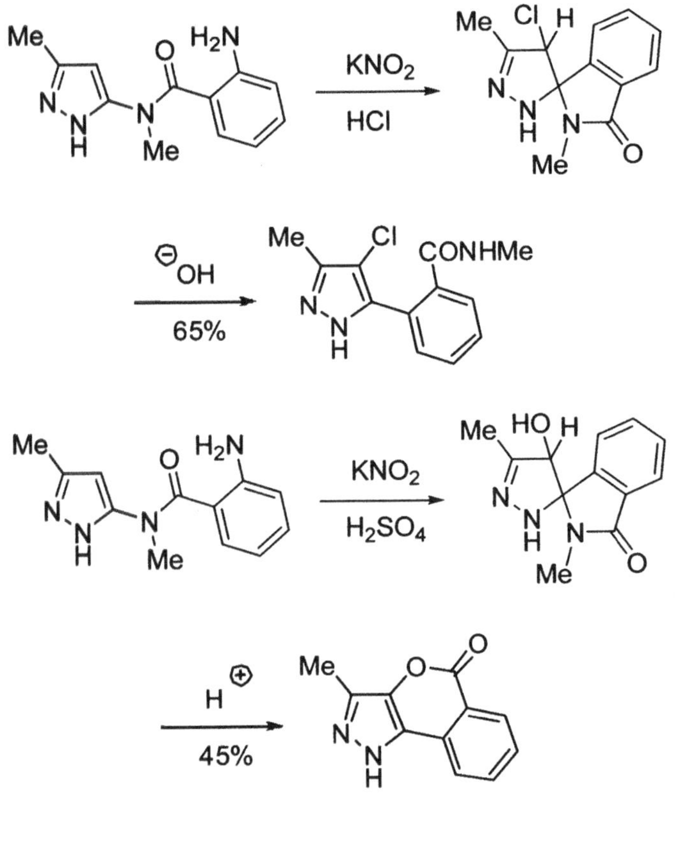

5.6.5 Predict the major products **D**, **E**, and **F**[62]:

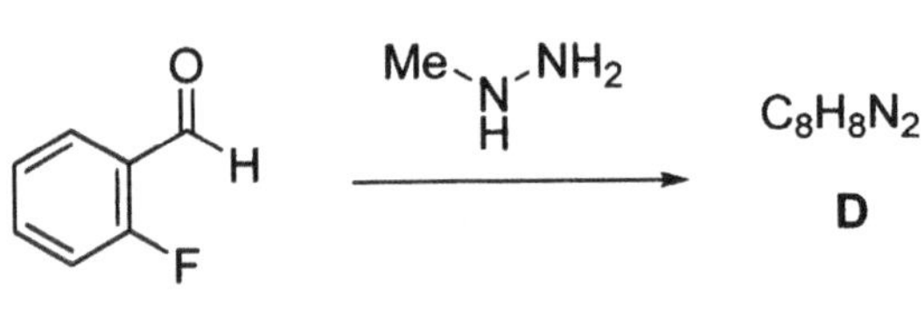

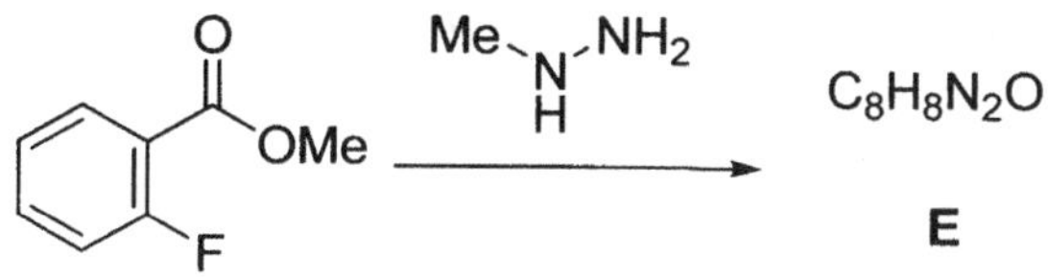

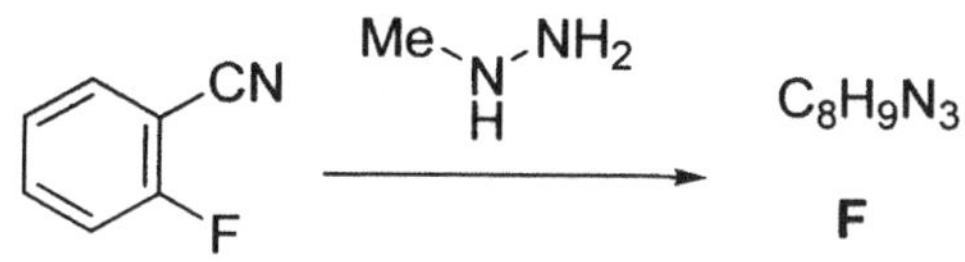

5.6.6 Predict the major product **G**:

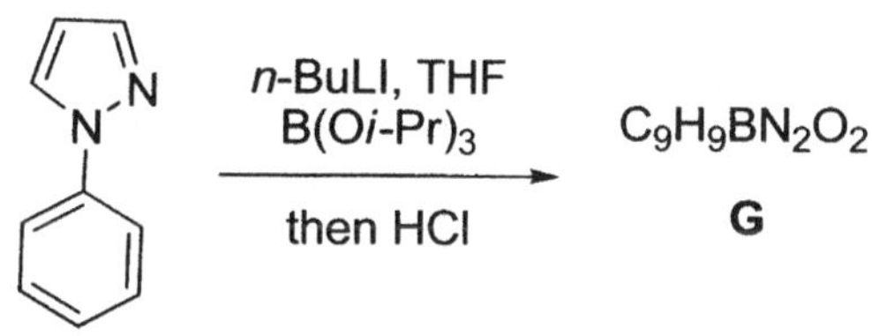

5.6.7 Deduce the outcome of the following reactions and the identities of compounds **H** and **I**[63,64]:

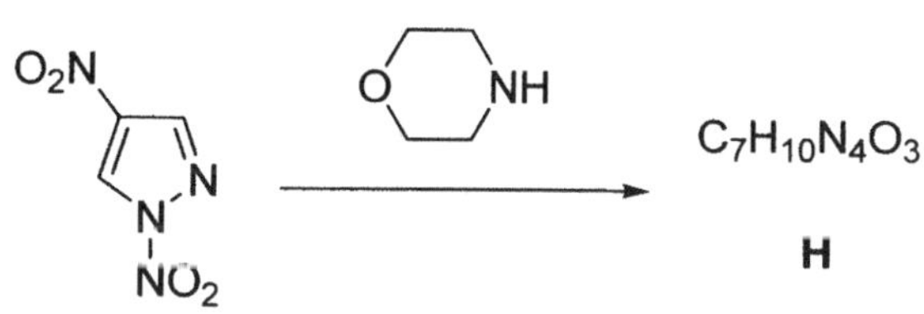

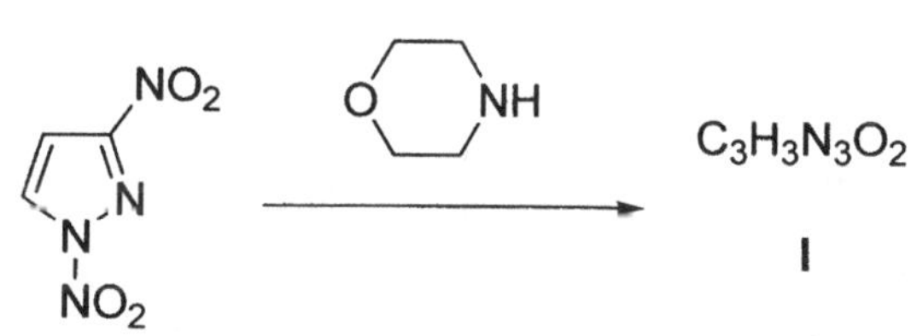

5.6.8 Propose the mechanism of the following transformation[65]:

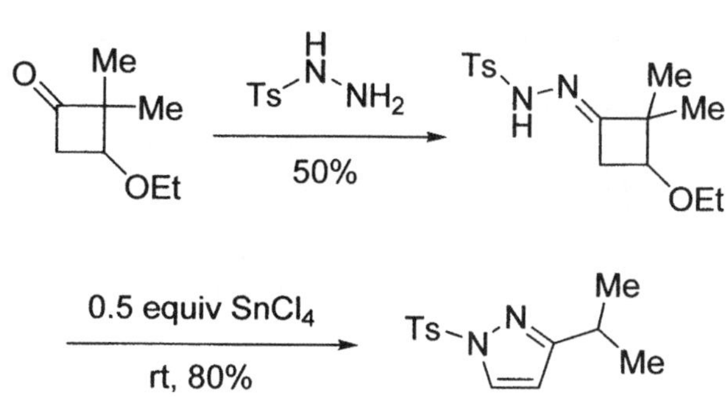

5.6.9 Predict the major product and propose a reasonable mechanism[66]:

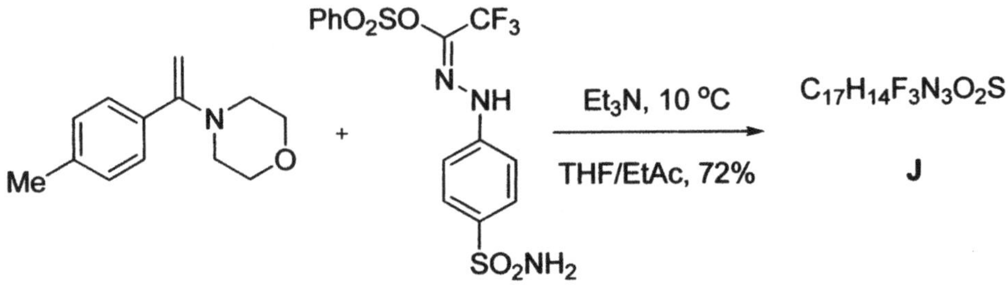

5.6.10 Predict the major products **K** and **L**[67]:

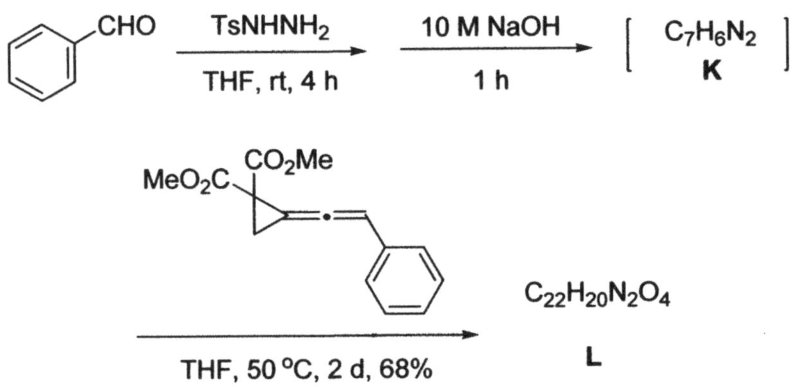

5.7 References

1. Elguero, J.; Goya, P.; Jagerovic, N.; Silva, A. M. S. *Pyrazoles as Drugs: Facts and Fantasies,* in *Targets in Heterocyclic Systems,* **2002,** *6,* 52–98.
2. Chiang, Y.; Whipple, E. B. *J. Am. Chem. Soc.* **1963,** *85,* 2763–2767.
3. Balón, M.; Carmona, M. C.; Munoz, M. A.; Hidalgo, J. *Tetrahedron* **1989,** *45,* 7501–7504.
4. Askew, B. C.; Bednar, R. A.; Bednar, B.; Claremon, D. A.; Cook, J. J.; McIntyre, C. J.; Hunt, C. A.; *et al. J. Med. Chem.* **1997,** *40,* 1779–1788.
5. Tabrizi, M. A.; Baraldi, P. G.; Preti, D.; Romagnoli, R.; Saponaro, G.; Baraldi, S.; Moorman, A. R.; Zaid, A. N.; Varani, K.; Borea, P. A. *Bioorg. Med. Chem. Lett.* **2008,** *16,* 2419–2430.
6. Meanwell, N. A.; Rosenfeld, M. J.; Wright, J. J. K.; Brassard, C. L.; Buchanan, J. O.; Federici, M. E.; Fleming, J. S.; Seiler, S. M. *J. Med. Chem.* **1992,** *35,* 389–397.
7. Bailey, D. M.; Hansen, P. E.; Hlavac, A. G.; Baizman, E. R.; Pearl, J.; DeFelice, A. F.; Feigenson, M. E. *J. Med. Chem.* **1985,** *28,* 256–260.
8. Penning, T. D.; Talley, J. J.; Bertenshaw, S. R.; Carter, J. S.; Collins, P. W.; Docter, S.; Graneto, M. J.; Lee, L. F.; Malecha, J. W.; Miyahiro, J. M.; Rogers, R. S.; Rogier, D. J.; Yu, S. S.; Anderson, G. D.; Koboldt, C. M.; Perkins, W. E.; Seibert, K.; Veenhuizen, A. W.; Zhang, Y. Y.; Isakson, P. C. *J. Med. Chem.* **1997,** *40,* 1347–1365.

9. Teegden, B. R.; Li, H.; Jayakumar, H.; Strah-Pleynet, S.; Dosa, P. I.; Selaya, S. D.; *et al.* *J. Med. Chem.* **2010,** *53,* 1923–1936.
10. Katoch-Rouse, R.; Pavlova, O. A.; Caulder, T.; Hoffman, A. F.; Mukhin, A. G.; Horti, A. G. *J. Med. Chem.* **2003,** *46,* 642–645.
11. Panke, G.; Schwalbe, T.; Stirner, W.; Taghavi-Moghadam, S.; Wille, G. *Synthesis* **2003,** 2827–2830.
12. Wise, L. D.; Butler, D. E.; DeWald, H. A.; Lustgarten, D.; Coughenour, L. L.; Downs, D. A.; Heffner, T. G.; Pugsley, T. A. *J. Med. Chem.* **1986,** *29,* 1628–1637.
13. Katritzky, A. R.; Rewcastle, G. W.; Fan, W. Q. *J. Org. Chem.* **1988,** *53,* 5685–5689.
14. Young, M. B.; Barrow, J. C.; Glass, K. L.; Lundell, G. F.; Newton, C. L.; Pellicore, J. M.; *et al.* *J. Med. Chem.* **2004,** *47,* 2995–3008.
15. Booker-Milburn, K. I. *Synlett* **1992,** 327–328.
16. Sakya, S. M. *Knorr Pyrazole Synthesis*, In *Name Reactions in Heterocyclic Chemistry*, Li, J. J.; Corey, E. J., eds, Wiley: Hoboken, NJ, **2005**, pp. 292–300.
17. Murray, W. V.; Hadden, S. K. *J. Org. Chem.* **1992,** *57,* 8662–86663.
18. Barth, F.; Casellas, P.; Congy, C.; Martinez, S.; Rinaldi, M.; Anne-Archard, G. US5624941 (1997).
19. Gustafsson, T.; Ponten, F.; Seeberger, P. H. *Chem. Commun.* **2008,** 1100–1102.
20. Kotagiri, V. K.; Suthrapu, S.; Reddy, J. M.; Rao, C. P.; Bollugoddu, V.; Bhattacharya, A.; Bandichhor, R. *Org. Process Res. Dev.* **2007,** *11,* 910–912.
21. Kotagiri, V. K.; Reddy, J. M.; Suthrapu, S. K.; Rao, C. P.; Reddy, P. P.; Bhattacharya, A.; Bandichhor, R. *Monatsh. Chem.* **2008,** *139,* 1091–1093.
22. Menozzi, G.; Fossa, P.; Cichero, E.; Spallarossa, A.; Ranise, A.; Mosti, L. *Eur. J. Med. Chem.* **2008,** *43,* 2627–2638.
23. Almansa, C.; de Arriba, A. F.; Cavalcanti, F. L.; Gómez, L. A.; Miralles, A.; Merlos, M. - GarcíaRafanell, J. G.; Forn, J. *J. Med. Chem.* **2001,** *44,* 350–361.
24. Bhat, L.; Jandeleit, B.; Dias, T. M.; Moors, T. L.; Gallop, M. A. *Bioorg. Med. Chem. Lett.* **2005,** *15,* 85–87.
25. Duplantier, A. J.; Andresen, C. J.; Cheng, J. B.; Cohan, V. L.; Decker, C.; DiCapua, F. M.; *et al.* *J. Med. Chem.* **1998,** *41,* 2268–2277.
26. Lange, J. H.; Coolen, H. K. A. C.; van Stuivenberg, H. H.; Dijksman, J. A. R.; Herremans, A. H. J.; Ronken, E.; *et al.* *J. Med. Chem.* **2004,** *47,* 627–643.
27. Srivastava, B. K.; Joharapurkar, A.; Raval, S.; Patel, J. Z.; Soni, Rina; Raval, P.; Gite, A.; Goswami, A.; Sadhwani, N.; Gandhi, N.; Patel, H.; Mishra, B.; Solanki, M.; Pandey, B.; Jain, M. R.; Patel, P. R. *J. Med. Chem.* **2007,** *50,* 5951–5966.
28. Ashton, W. T.; Hutchins, S. M.; Greenlee, W. J.; Doss, G. A.; Chang, R. S. L.; Lotti, V. J.; Faust, K. A.; Chen, B.-B.; Zingaro, G. J.; Kivlighn, S. D.; Siegl, P. K. S. *J. Med. Chem.* **1993,** *36,* 3595–3605.
29. Jia, Z. J.; Wu, Y.; Huang, W.; Goldman, E.; Zhang, P.; Woolfrey, J.; Wong, P.; Huang, B.; Sinha, U.; Park, G.; Reed, A.; Scarborough, R. M.; Zhu, B.-Y. *Bioorg. Med. Chem. Lett.* **2002,** *12,* 1651–1655.
30. Sutherland, H. S.; Blaser, A.; Kmentova, I.; Franzblau, S. G.; Wan, B.; Wang, Y.; Ma, Z.; Palmer, B. D.; Denny, W. A.; Thompson, A. M. *J. Med. Chem.* **2010,** *53,* 855–866.
31. Goldstein, D. M.; Alfredson, T.; Bertrand, J.; Browner, M. F.; Clifford, K.; Dalrymple, S. A.; Dunn, J.; Freire-Moar, J.; Harris, S.; Labadie, S. S.; La Fargue, J.; Lapierre, J. M.; Larrabee, S.; Li, F.; Papp, E.; McWeeney, D.; Ramesha, C.; Roberts, R.; Rotstein, D.; San Pablo, B.; Sjogren, E. B.; So, O.-Y.; Talamas, F. X.; Tao, W.; Trejo, A.; Villasenor, A.; Welch, M.; Welch, T.; Weller, P.; Whiteley, P. E.; Young, K.; Zipfel, S. *J. Med. Chem.* **2006,** 1562–1575.

32. Mullins, R. J. *Pechmann Pyrazole Synthesis,* in *Name Reaction in Heterocyclic Chemiatry-II,* Li, J. J., ed.; Wiley: Hoboken, NJ, 2005, pp. 327–336.
33. Padwa, A.; Pearson, W. H., Eds., *Synthetic Applications of 1,3-Dipolar Cycloaddition toward Heterocycles and Natural Products*; John: New York, NY, 2002.
34. Kertész, M.; Pallagi, I.; Sólyom, S. *J. Heterocyclic Chem.* **2007,** *44*, 431–435.
35. Jiang, N.; Li, C.-J. *Chem. Commun.* **2004**, 394–395.
36. Lilienkampf, A.; Pieroni, M.; Wan, B.; Wang, Y.; Franzblau, S. G.; Kozikowski, A. P. *J. Med. Chem.* **2010,** *53*, 678–688.
37. Mao, J.; Wan, B.; Wang, Y.; Franzblau, S. G.; Kozikowski, A. P. *ChemMedChem* **2007,** *2*, 811–813.
38. Lilienkampf, A.; Mao, J.; Wan, B.; Wang, Y.; Franzblau, S. G.; Kozikowski, A. P. *J. Med. Chem.* **2009,** *52*, 2109–2118.
39. Sauer, D. R.; Schneller, S. W. *J. Org. Chem.* **1990,** *55*, 5535–5538.
40. Donohue, S. R.; Halldin, C.; Pike, V. W. *Tetrahedron Lett.* **2008,** *49*, 2789–2791.
41. Varvounis, G. *Adv. Heterocycl. Chem.* **2009,** *98*, 143–224.
42. Ponngratz, A.; Zirm, K. L. *Monatsh.* **1957,** *88,* 330–335.
43. Stenzl, H. US2,562,830 (1957).
44. Pfister, R.; Häfliger, F. *Helv. Chim. Acta* **1957,** *40,* 395–401.
45. Pfister, R.; Häfliger, F. *Helv. Chim. Acta* **1961,** *44,* 232–237.
46. Singh, P.; Paul, K.; Holzer, W. *Bioorg. Med. Chem.* **2006,** *14*, 5061–5071.
47. Castagnolo, D.; De Logu, A.; Radi, M.; Bechi, B.; Manetti, F.; Magnani, M.; Supino, S.; Meleddu, R.; Chisu, L.; Botta, M. *Bioorg. Med. Chem.* **2008,** *16*, 8587–8591.
48. Kees, K.; Fitzgerald, J. J., Jr.; Steiner, K. E.; Mattes, J. F.; Mihan, B.; Tosi, T.; Mondoro, D. M.; McCaleb, M. L. *J. Med. Chem.* **1996,** *39,* 3920–3928.
49. Shen, H.; Gou, S.; Shen, J.; Zhu, Y.; Zhang, Y.; Chen, X. *Bioorg. Med. Chem. Lett.* **2010,** *20*, 2115–2118.
50. Bermudez, J.; Fake, C. S.; Joiner, G. F.; Joiner, K. A.; King, F. D.; Miner, W. D.; Sanger, G. J. *J. Med. Chem.* **1990,** *33,* 1924–1929.
51. Vernekar, S. K. V.; Hallaq, H. Y.; Clarkson, G.; Thompson, A. J.; Silvestri, L.; Lummis, S. C. R.; Lockner, M. *J. Med. Chem.* **2010,** *53,* 2324–2328.
52. Lebedev, A. Y.; Khartulyari, A. S.; Voskoboynikov, A. Z. *J. Org. Chem.* **2005,** *70,* 596–602.
53. Selwood, D. L.; Brummell, D. G.; Budworth, J.; Burtin, G. E.; Campbell, R. O.; Chana, S. S.; Charles, I. G.; *et al. J. Med. Chem.* **2001,** *44,* 78–93.
54. Baiocchi, L.; Corsi, G.; Palazzo, G. *Synthesis* **1978,** 633–648.
55. Schläger, T.; Schepmann, D.; Wuerthwein, E.-U.; Wuensch, B. *Bioorg. Med. Chem.* **2008,** *16*, 2992–3001.
56. Liddle, B. J.; Lindeman, S. V.; Reger, D. L.; Gardinier, J. R. *Inorg. Chem.* **2007,** *46*, 8484–8486.
57. epulveda-Arques, J.; Medio-Simon, M.; Piqueres-Vidal, L. *Monatsh. Chem.* **1989,** *120*, 1113–1118.
58. Buchi, G.; Lee, G. C. M.; Yang, D.; Tannenbaum, St. R. *J. Am. Chem. Soc.* **1986,** *108,* 4115–4119.
59. Daidone, G.; Plescia, S.; Maggio, B.; Sprio, V.; Benetollo, F.; Bombieri, G. *J. Chem. Soc., Perkin Trans. 1* **1993,** 285–291.
60. Daidone, G.; Maggio, B.; Raffa, D.; Plescia, S.; Benetollo, F.; Bombieri, G. *J. Chem. Soc., Perkin Trans. 1* **1998,** 2891–2898.
61. Maggio, B.; Raffa, D.; Raimondi, M. V.; Cascioferro, S.; Plescia, S.; Sabatino, M. A.; Bombieri, G.; Meneghetti, F.; Daidone, G. *ARKIVOC* **2008,** *(16)*, 130–143.

62. Wheeler, R. C.; Baxter, E.; Campbell, I. B.; Macdonald, S. J. F. *Org. Process Res. Dev.* **2011,** *15*, 565–569.
63. Habraken, C. L.; Poels, E. K. *J. Org. Chem.* **1977,** *42*, 2893–2895.
64. Cohen-Fernandes, P.; van Eendenburg, E. C. G. M.; Verhoeven, J. J.; Habraken, C. L. *J. Org. Chem.* **1979,** *44*, 4156–4160.
65. Shan, G.; Liu, P.; Rao, Y. *Org. Lett.* **2011,** *13,* 1746–1749.
66. Oh, L. M. *Tetrahedron Lett.* **2006,** *47,* 7943–7946.
67. Wu, L.; Shi, M. *J. Org. Chem.* **2010,** *75,* 2296–2301.

Chapter 6 Oxazoles, Benzoxazoles, and Isoxazoles

Adam M. Azman and Richard J. Mullins

6.1 Introduction

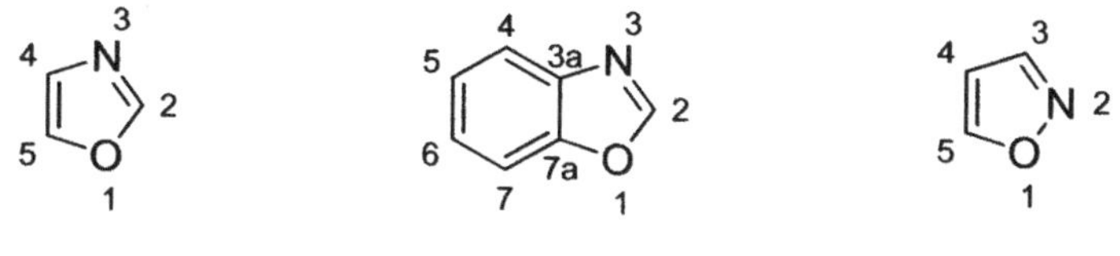

oxazole   benzoxazole   isoxazole

Oxazole is a liquid at room temperature, with a boiling point of 69 °C. It is weakly basic and miscible with water and many other organic solvents.[1] Spectroscopically, the ^{1}H NMR of oxazole is relatively simple, featuring three distinct signals for H2, H4, and H5. Consistent with the reactivity described below, the signal for H2 is the most deshielded, appearing at 7.72 ppm as a broad singlet. Similarly, the signal for C2 is the most deshielded of the three signals present in the ^{13}C spectrum, appearing at 150 ppm. The site of electrophilic substitution on the oxazole ring, C5, however, gives rise to the more shielded signal at 125 ppm.

The ^{1}H NMR spectrum of benzoxazole is substantially more complex. The most deshielded proton is still H2 and appears as a singlet. H4 and H7, because of the electronic similarity of the two positions, appear as overlapping multiplets between 7.28 and 7.37 ppm. In contrast, H5 appears as a complex multiplet slightly upfield (7.72–7.78 ppm) as compared to H6 (7.52–7.58 ppm). The ^{13}C NMR of benzoxazoles displays similar patterns, with C2 being most de-shielded at 151 ppm. The bridgehead carbons, C3a and C7a, appear around 140 ppm and 150 ppm, respectively, with the difference being due to the electron donating ability of the oxygen of the oxazole ring and its ability to delocalize electron density onto C7a. Finally, the carbons of the benzo-ring (C4–C7) appear between 110 and 126 ppm.

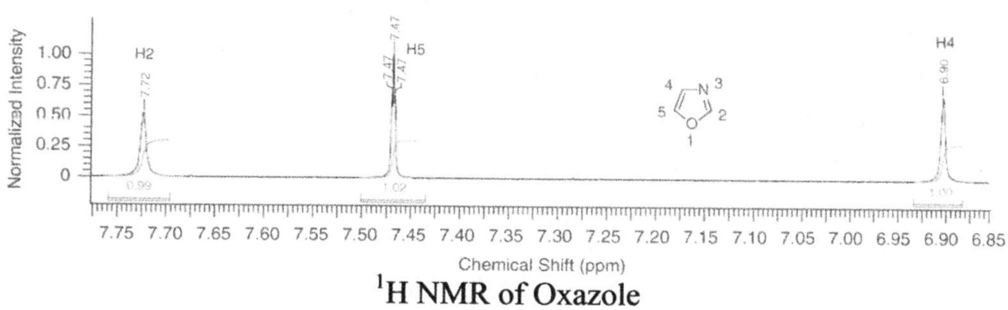

^{1}H NMR of Oxazole

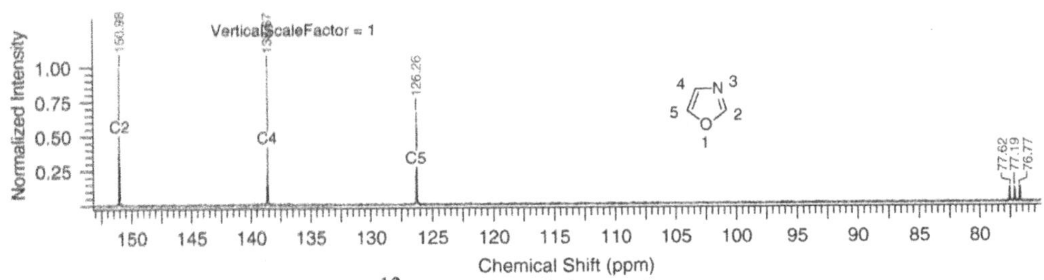

^{13}C NMR of Oxazole

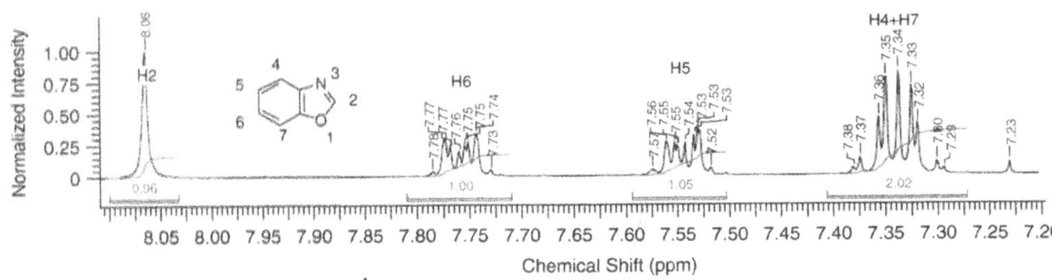

^{1}H NMR of Benzoxazole

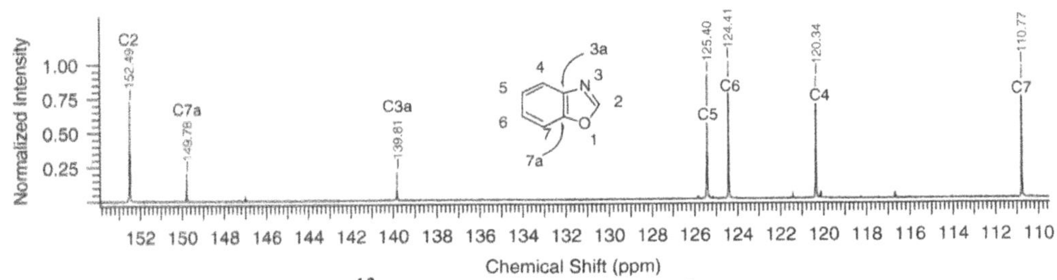

^{13}C NMR of Benzoxazole

The oxazole is a π-electron-excessive heterocycle. While it resembles an aromatic molecule and has the requirements laid out by Hückel for aromaticity, its properties and reactivity suggest the delocalization to be incomplete. Implications of this partial aromatic character are that, similar to furans, the oxazole nucleus readily participates in the Diels–Alder reaction, while rarely undergoing the electrophilic substitution characteristic of other aromatic ring systems.[1] The C2 position is partially electropositive due to the electron-withdrawing influence of the flanking electronegative heteroatoms. As such, oxazoles are susceptible to nucleophilic attack at C2. In contrast, the C5 position is partially electron-rich, and although rare, electrophilic substitution preferentially occurs at C5. Benzoxazoles show similar electronic distribution; however, electrophilic substitution occurs with different regioselectivity than oxazoles. Electrophilic substitution of benzoxazoles occurs preferentially at C6 and sometimes C5. Isoxazoles swap the C2 carbon atom and the nitrogen atom. Now, C3 and C5 are electrophilic and C4 is nucleophilic.

The C2 proton of oxazole is mildly acidic with a theoretical pK_a value of ~27.[2] The C2 benzoxazole proton is slightly more acidic than oxazole with an experimental pK_a value of 24.8.[3] The most acidic proton in isoxazole is the C5 proton, with a theoretical pK_a value close to the pK_a value of oxazole at ~27.[2] Oxazole, benzoxazole, and isoxazole are weak bases at nitrogen with a conjugate acid pK_a value for oxazole and benzoxazole of ~1, and a conjugate acid pKa value for isoxazole of ~−3.

Oxazoles are known in natural products and also find wide incorporation into synthetic medicinal compounds. Oxazoles can be found in the natural products (−)-hennoxazole A (antiviral)[4] and leucascandrolide A (cytotoxic and antifungal).[5] Benzoxazoles and isoxazoles are less common in natural products. Pseudo-pteroxazole (benzoxazole-containing anti-tubercular agent)[6] and ibotenic acid (isoxazole-containing neurotoxin) are two examples.

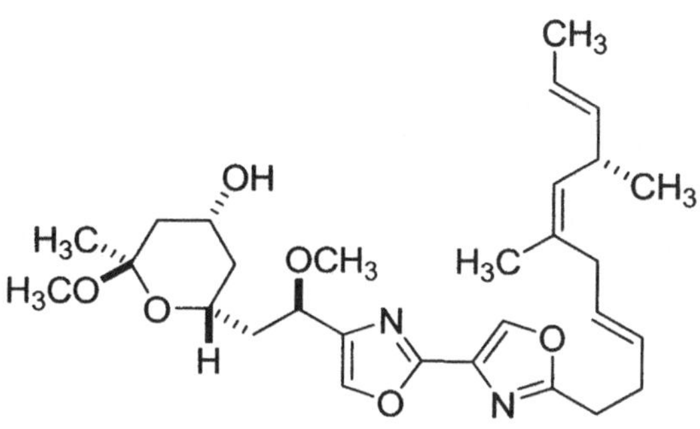

(−)-hennoxazole A

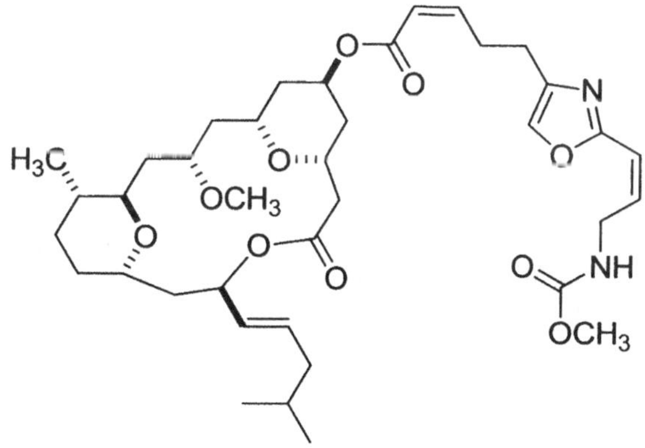

leucascandrolide A

Oxazoles and benzoxazoles with an aryl substituent at C2 are strongly fluorescent and can act as an optical brightener. Some washing agents include 4,4′-bis(benzoxazol-2-yl)stilbene as an additive. The additive is absorbed by clothes fibers, and the blue fluorescence of the benzoxazole makes clothes appear "whiter than white."

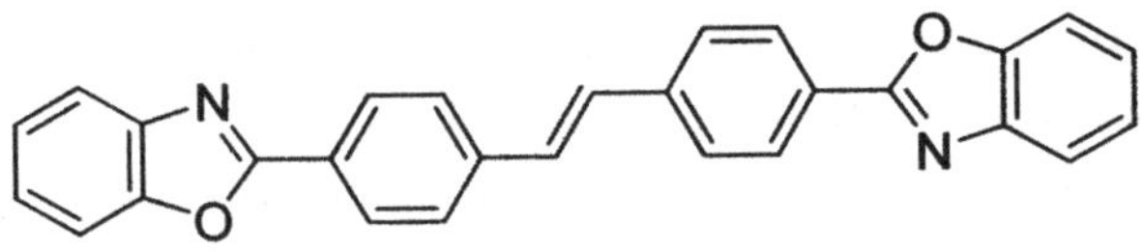

4,4'-bis(benzoxazol-2-yl)stilbene

Tilmacoxib is a COX-2 inhibiting, oxazole-containing, nonsteroidal anti-inflammatory agent,[7] and valdecoxib is an isoxazole NSAID.[8] Formation of the sodium salt of an *N*-acylated derivative of valdecoxib affords parecoxib,[9] a water-soluble valdecoxib prodrug amenable to injectable formulations. Valdecoxib was withdrawn from the US market in 2005 due to concern of increased risk of heart attack or stroke. While parecoxib is approved for use in the EU, the US FDA issued a letter of non-approval for the drug, also in 2005. While no documentation for the non-approval was made public, proximity to the 2004 withdrawal of rofecoxib (Vioxx) cannot be overlooked.

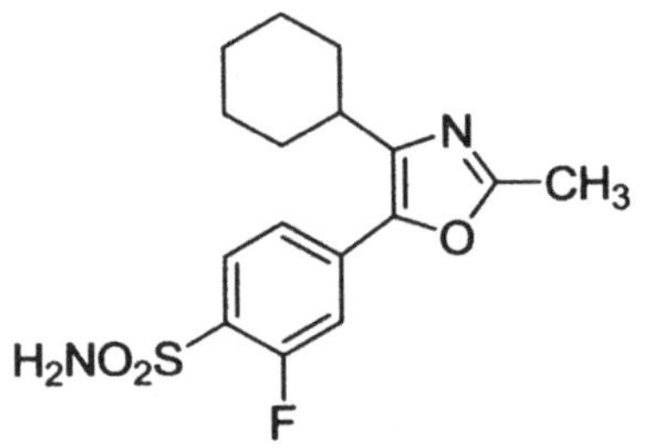

tilmacoxib

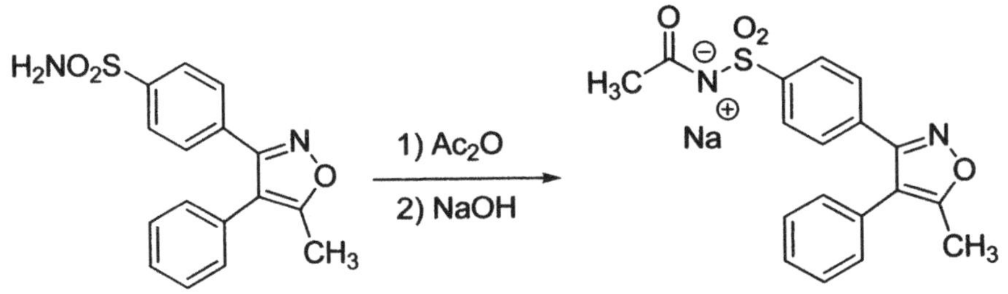

valdecoxib
(Bextra)

parecoxib
(Dynastat)

6.2 *Construction of the Heterocyclic Ring*

6.2.1 *Construction of the Oxazole Ring*

The oxazole ring can be dissected retrosynthetically in a variety of places to yield oxazole precursors. The synthetic routes to oxazoles can loosely be classified by the location of the C2, C4, and C5 substituents prior to heterocycle formation. In a linear synthesis, all substituents would be contained in the same precursor. In a more convergent synthesis, one or two substituents are contained in one reagent, and the remaining substituents are contained in the other. A number of disconnections in all of these categories is known; several of them will be highlighted here.

Robinson–Gabriel synthesis

The Robinson–Gabriel oxazole synthesis is a versatile, although linear, preparation of oxazoles first described in 1909.[10] It is the cyclodehydration of 2-acylamidoketones, where the oxazole precursor contains all three oxazole substituents. 2,5-Disubstituted oxazoles and 2,4,5-trisubstituted oxazoles can be prepared, and the substituents can be alkyl, aryl, or heteroaryl groups. Labeling studies have shown that the amide oxygen atom is the most Lewis basic and is the oxygen atom incorporated into the oxazole ring.[11] A variety of dehydrating agents has been reported including polyphosphoric acid,[12] H_3PO_4/Ac_2O,[13] Ph_3PI_2,[14] $Et_3NSO_2NCO_2Me$ (Burgess reagent, microwave conditions),[15] TFAA (solid-supported synthesis),[16] and $POCl_3$.[17] Godfrey and co-workers at Eli Lilly took advantage of the Robinson–Gabriel synthesis in the synthesis of a dual PPAR α/γ agonist.[18] Other bioactive molecules prepared utilizing a Robinson–Gabriel synthesis include the NSAID oxaprozin,[19] and the muscle relaxant azomolene.[20]

The Robinson-Gabriel Oxazole Synthesis:

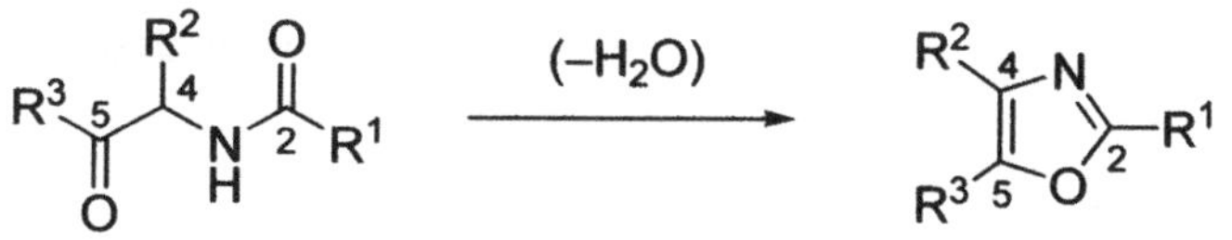

Godfrey's Application:

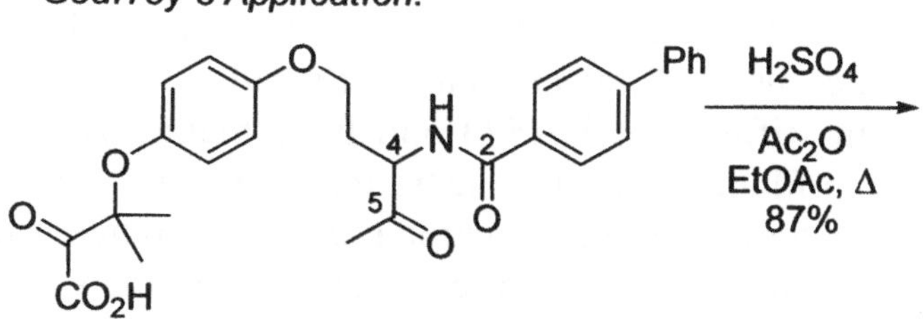

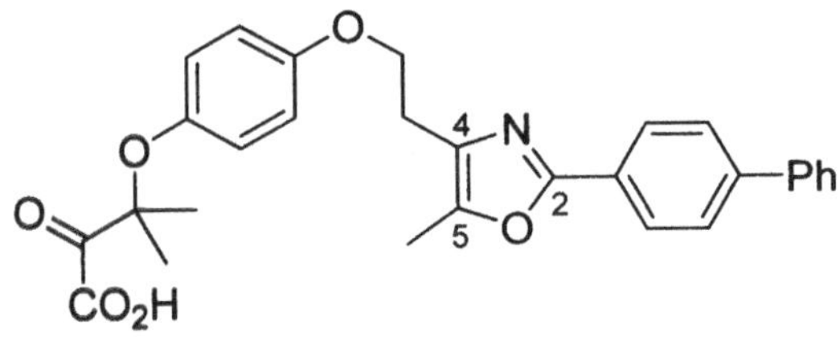

Davidson oxazole synthesis

The Davidson oxazole synthesis is the reaction of an acylated α-hydroxy carbonyl with an ammonium cation, typically ammonium acetate to yield oxazoles.[21] The Davidson synthesis is also linear, as all three substituents are contained in the precursor. The reaction is most efficient in preparing 2,4,5-trisubstituted oxazoles, where C5 is aryl substituted. Yields suffer when preparing a 2,4-disubstituted or a mono-substituted oxazole. The COX-2 inhibitor tilmacoxib was prepared by Haruta in this manner.[22]

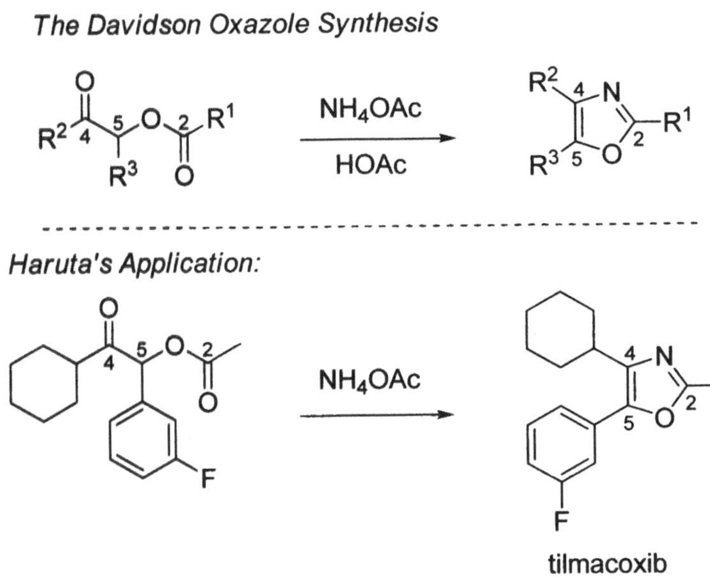

Wipf–Williams oxazole synthesis

Wipf and Williams recently reported a one-pot synthesis of oxazoles from β-hydroxy amides.[23] Similar to other dehydrative methods, these groups expanded on the use of electrophilic fluoride reagents,[24,25] namely diethylaminosulfur trifluoride (DAST) and bis(2-methoxyethyl)aminosulfur trifluoride (Deoxo-Fluor), by demonstrating the high functional group tolerance of this system. The high functional group tolerance is demonstrated by the oxazole synthesis in the presence of a chemically labile epoxide.

Following cyclization with DAST, *in situ* oxidation with DBU and $BrCCl_3$ provides the 2,4-substituted oxazole in decent yield. Notably, if desired, the intermediate oxazoline can be isolated in stereochemically pure form.

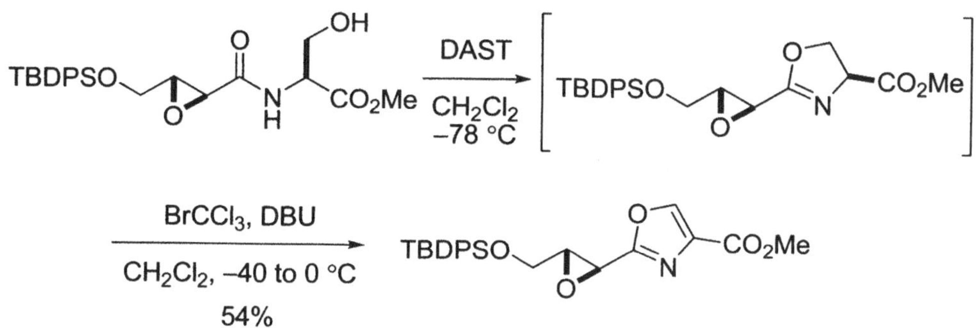

Fisher oxazole synthesis

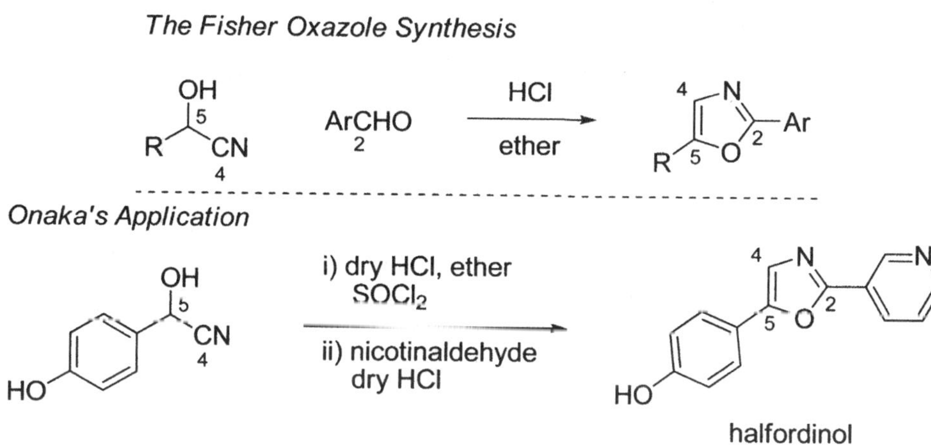

The condensation of an aromatic aldehyde with an aldehyde cyanohydrin under anhydrous acidic conditions provides oxazoles. The convergent synthesis couples the C2 substituent contained in the aromatic aldehyde and the C5 substituent in the cyanohydrin (C4 must necessarily remain unsubstituted). Through a modified procedure, Onaka prepared the alkaloid halfordinol in one step,[26] utilizing the Fisher oxazole synthesis.

Japp oxazole synthesis

The Japp oxazole synthesis is the coupling of 1,2-diketones with ammonia and an aromatic aldehyde. The diketone contributes the C4 and C5 substituents, and the aromatic aldehyde contributes the C2 substituent. The synthesis also proceeds when reacting an α-ketooxime with a benzylic halide under mildly basic conditions. Biftu utilized the Japp oxazole synthesis as a key step in their preparation of an anticoccidial agent.[27]

The Japp Oxazole Synthesis:

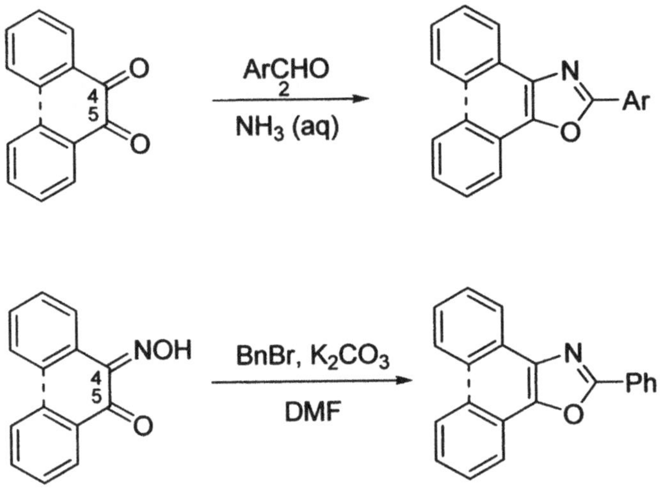

Biftu's Application:

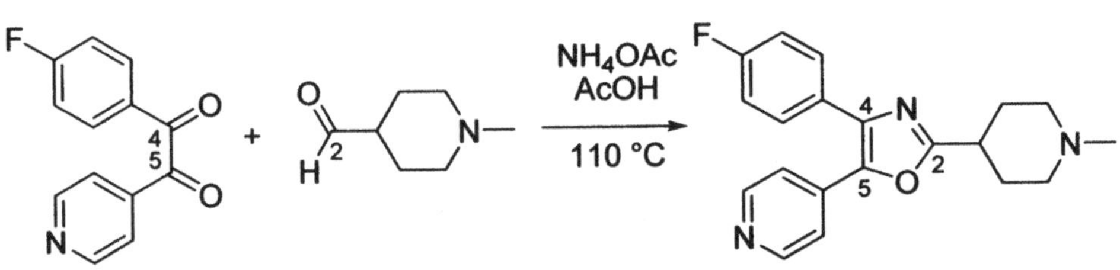

Schöllkopf oxazole synthesis

A widely used oxazole synthesis, the Schöllkopf reaction,[28] couples an isocyanide bearing the C4 substituent with an acylating agent bearing the C5 substituent. C2 must necessarily remain unsubstituted. The reaction allows for the versatile synthesis of a number of synthetically useful 4,5-disubstituted oxazoles. The isocyanide is typically substituted with electron-withdrawing groups, typically esters. The acylating agent can be substituted with a variety of alkyl groups, aromatic groups, and it can be derived from chiral amino acids. Acid chlorides, esters, amides, anhydrides, acyl azides, and lactones can function as acylating reagents. This protocol has been incorporated into a number of syntheses of bioactive molecules and natural products. Brescia and co-workers[29] utilized the Schöllkopf reaction in the evaluation of a series of oxazoles as prostacyclin receptor agonists.

The Schöllkopf Reaction

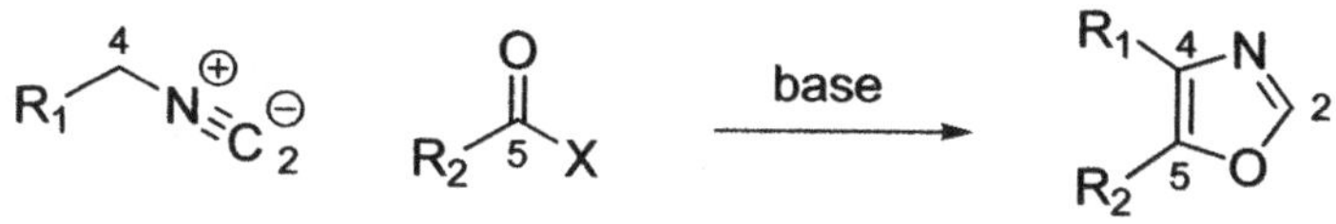

Brescia's Application

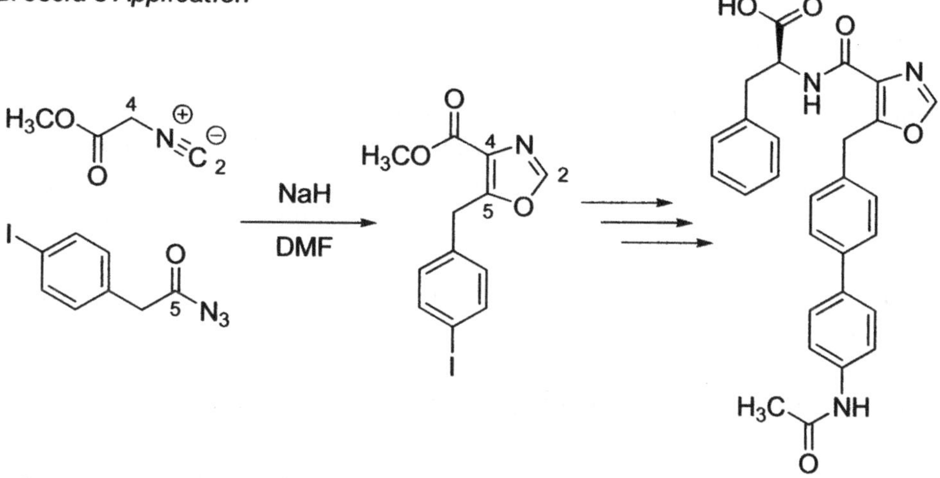

van Leusen oxazole synthesis

p-Tolylsulfonylmethyl isocyanide (TosMIC) is an active methylene compound that reacts with aldehydes under basic conditions to yield 5-substituted oxazoles.[30] The reaction is quite similar in appearance to the Schöllkopf reaction. Here, the C5 substituent is introduced through the aldehyde, and the C4 substituent arises from the TosMIC reagent (C2 must necessarily remain unsubstituted). Aromatic and heteroaromatic aldehydes are tolerated in this synthesis. The isocyanide can be substituted with an alkyl, aryl, allyl, or benzyl group to prepare 4,5-disubstituted oxazoles.[31] The key intermediate in the synthesis of VX-497, a hepatitis C drug candidate, incorporates the van Leusen synthesis in 96% yield.[32]

The van Leusen Oxazole Synthesis

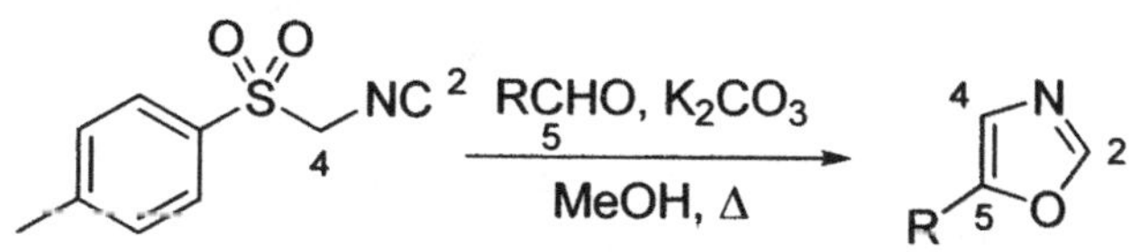

Herr's Application:

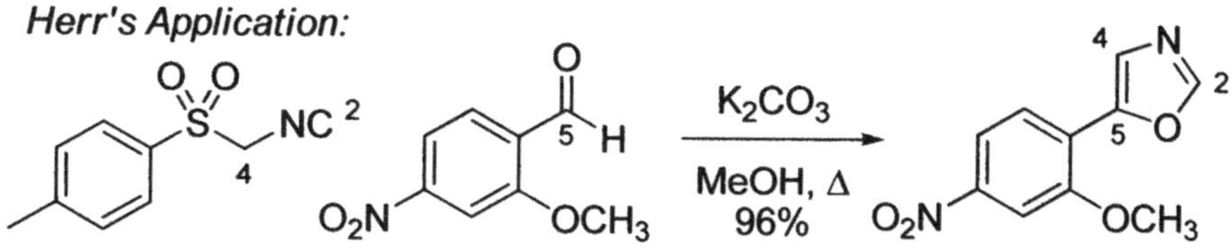

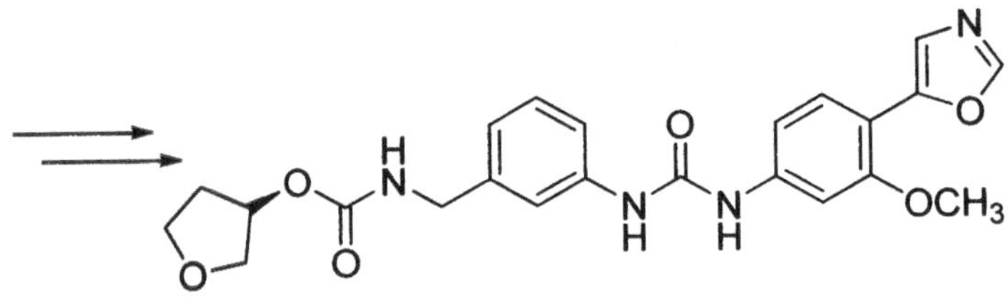

VX-497

Other oxazole syntheses

Recently, the direct synthesis of oxazoles from aldehydes has been enabled by conducting the cyclization reaction in the presence of an oxidant, such as $BrCCl_3$.[33] Similarly, the Wang group has developed a practical, metal-free, iodine-catalyzed tandem oxidative cyclization providing for the synthesis of 2,5-disubstituted oxazoles.[34] As demonstrated in the reaction below, the reaction, proceeding through a sequential condensation/cyclization/oxidation process, results in the preparation of a wide variety of substituted oxazoles. A large number of aromatic and heteroaromatic aldehydes has been shown as viable substrates in this very general process.

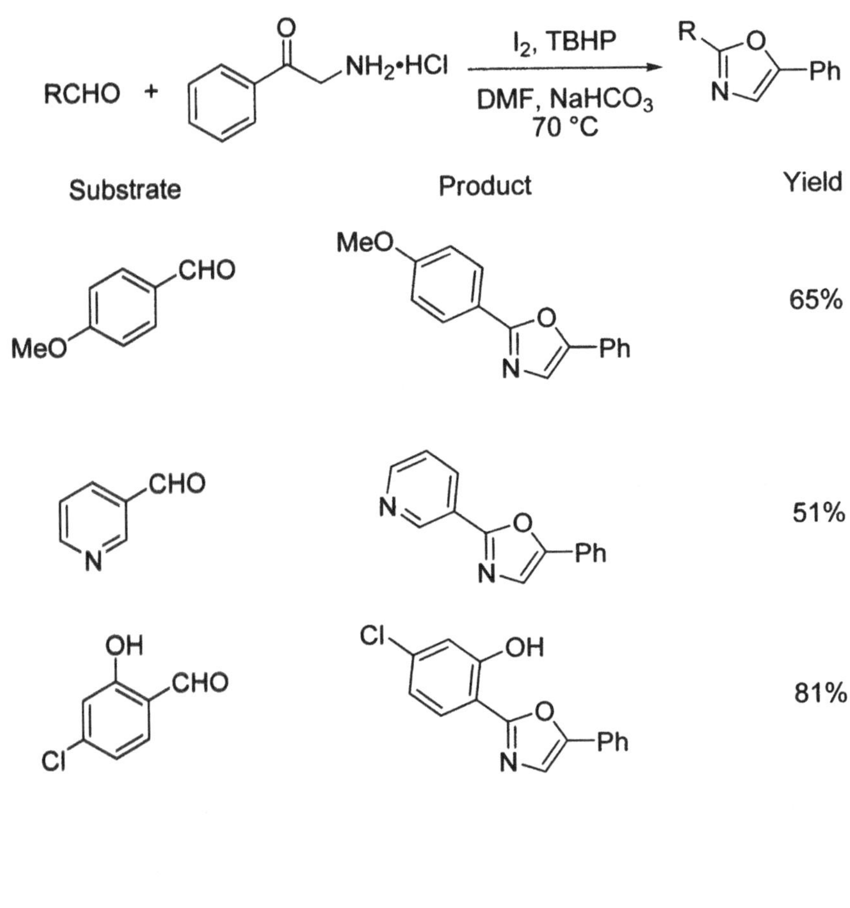

Similar in approach to linear methods discussed previously, the cyclodehydration of α-amido ketones is an alternative strategy for preparation of the oxazole nucleus. Treatment of the ketone with PPh_3 and I_2 results in cyclization to provide the fully substituted oxazole nucleus in excellent yield. This strategy was recently used in the synthesis of oxazole-modified glycopeptides, designed to target two classes of arthritis-associated proteins.[35] The oxazole was probed as a conformationally locked dipeptide mimic.

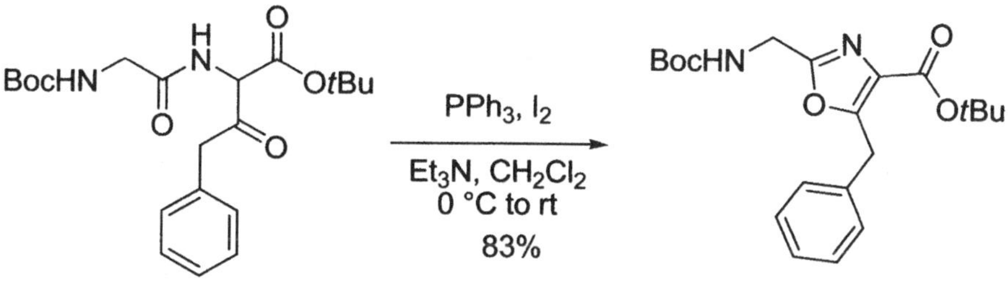

An alternative method for synthesis of 2,5-disubtituted oxazoles involves the cycloisomerization of propargyl amides. One such strategy, developed by Wipf and co-workers, relies on a mild silica-mediated process that produces substituted oxazoles with wide functional group compatibility.[36] As a representative example, the fully substituted oxazolyl ketone below was produced in good yield from the starting propargylic alcohol in two steps. The propargylic alcohol was produced via alkyne addition to the corresponding aldehyde, demonstrating the convergent nature of the approach.

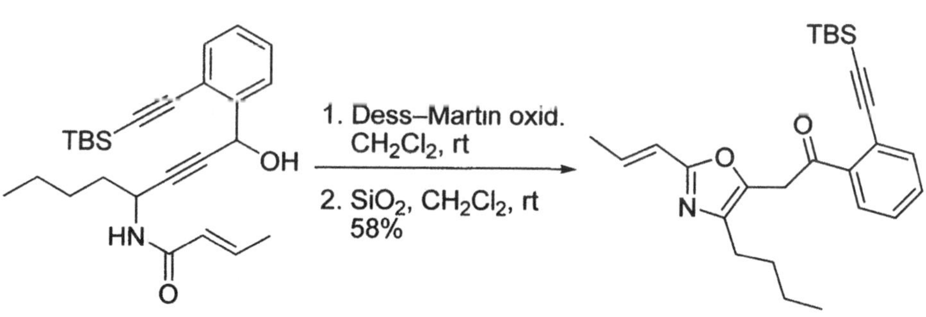

6.2.2 *Construction of the Benzoxazole Ring*

Benzoxazoles are classically created through the cyclization of a 2-aminophenol with a carboxylic acid under highly acidic conditions at elevated temperatures. Alternatively, the amine of a 2-aminophenol is acylated and subsequently subjected to a number of cyclodehydration conditions to form benzoxazoles. The main drawback is the harsh reaction conditions when highly acidic or high-temperature conditions are employed.

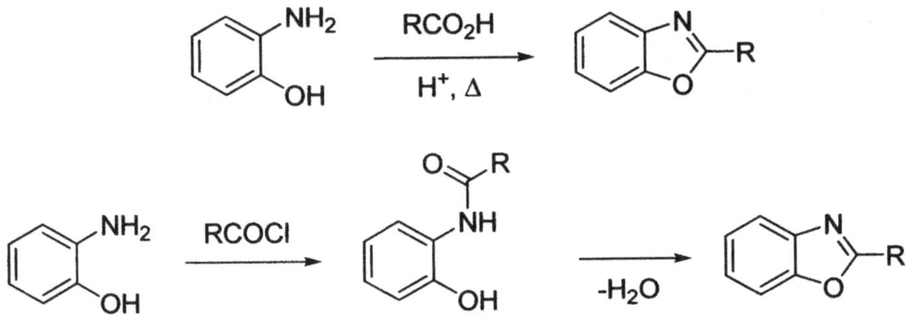

Several reagents have been shown to convert 2-hydroxyaryloximes into benzoxazoles through a Beckmann rearrangement. Zeolite catalysis in a continuous flow reactor has been shown to be successful in benzoxazole formation,[37] as has diethyl chlorophosphate[38] and $POCl_3$. Rindhe and co-workers utilized $POCl_3$ in the formation of pyrazolyl benzoxazoles for use as antimicrobial agents.[39]

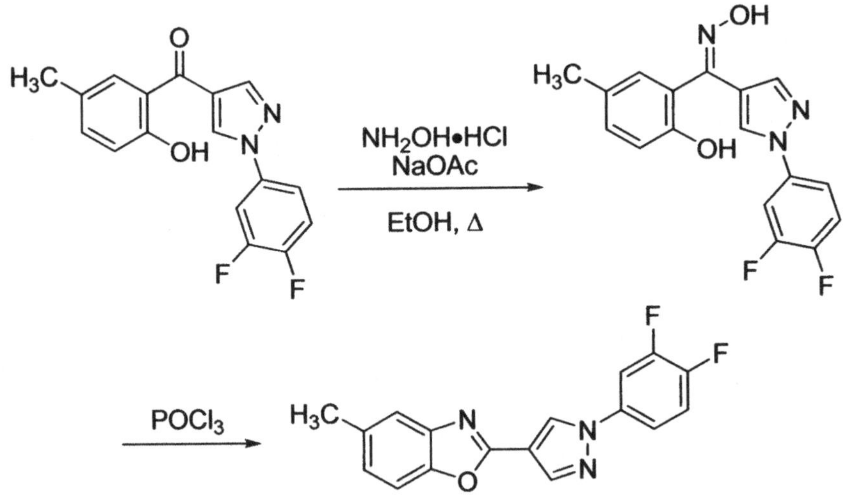

Under solid-phase reaction conditions, benzoxazoles can be prepared through a Mitsunobu reaction.[40] The mild reaction conditions of the Mitsunobu alleviate the hazards associated with other strongly acidic conditions.

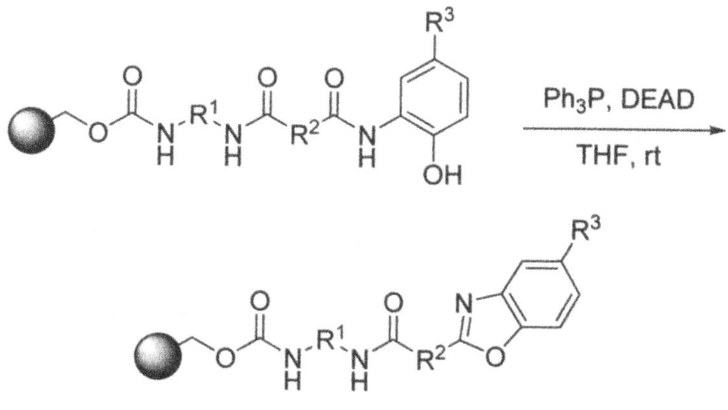

6.2.3 *Construction of the Isoxazole Ring*

Claisen isoxazole synthesis

In 1888, Claisen reported that condensation of β-ketoesters with hydroxylamine provided a mixture of 3-hydroxyisoxazoles and 5-oxazolones. The product distribution was largely dependent on the substitution of the initial β-ketoester and the pH of the reaction mixture.[41] The differentiation occurs due to the initial attack of the hydroxylamine at either the ester carbonyl or the ketone carbonyl. Krogsgaard-Larsen has prepared isoxazoles selectively starting with Meldrum's acid in a three-step procedure.[42] When acetyl chloride (R = CH_3) is used in this procedure, the soil fungicide 5-methyl-3-oxazazolol is prepared in 57% overall yield.

Original Claisen Method:

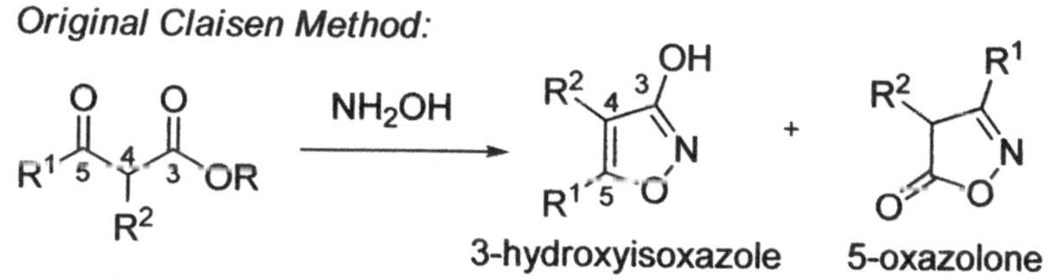

Meldrum's Acid Method:

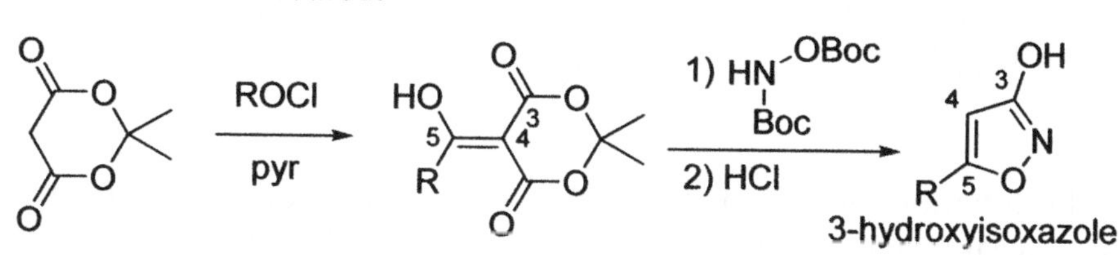

6.3 *Reactivity*

6.3.1 *Acid/Base Reactivity*

Oxazoles are acidic at C2 with a theoretical pK_a value of ~ 27.[2] Metalation is facile at C2 with strong bases. C2-metalated oxazoles are in equilibrium with their ring-opened isonitrile. Due to this equilibrium, C2-metalated oxazoles are nucleophilic at either C2 or C4, and care must be taken when quenching the C2-metalated oxazole to ensure proper chemoselectivity (see Section 6.3.3). Benzoxazoles are acidic at C2 like oxazoles (experimental pK_a 24.8),[3] and the metalated benzoxazole is also in equilibrium with its ring opened isonitrile. Isoxazoles, without a C–H group flanked by two heteroatoms, are acidic at the C5 position with a theoretical pK_a value of ~ 27;[2] however, in practice, deprotonation with hydroxide occurs at C3 to give a ring-opened β-ketonitrile intermediate. All of these heterocycles can be deprotonated with metal amides or carbanion bases.

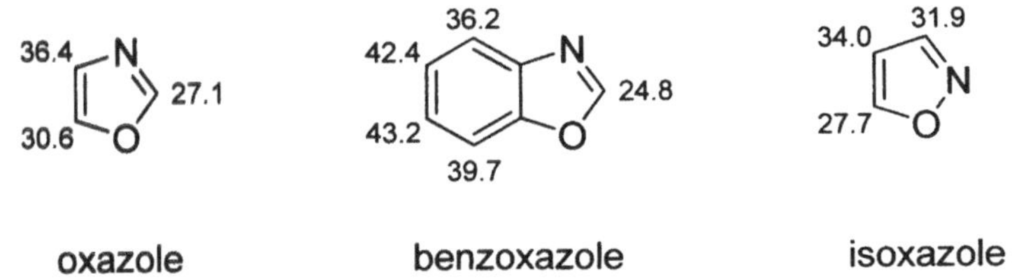

heterocycle pK_a values

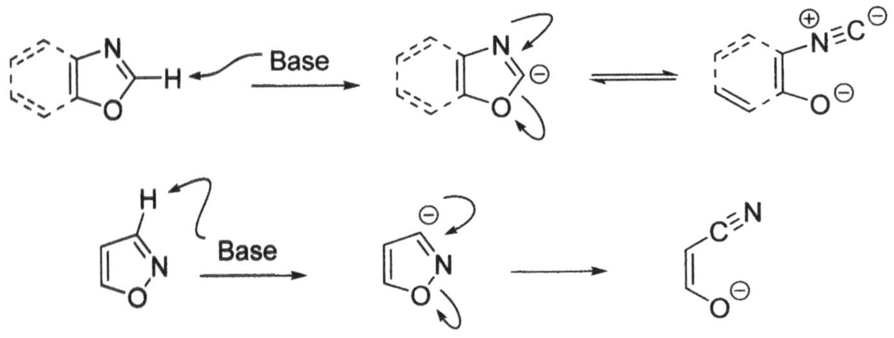

Protonation of all three heterocycles occurs at the nitrogen atom, which is weakly basic. The resulting oxazole and benzoxazole conjugate acids have a pK_a value of ~ −1, and the conjugate acid of isoxazole has a pK_a value of ~ −3.

6.3.2 *Electrophilic Substitution*

As mentioned previously, examples of electrophilic substitution to the oxazole nucleus are relatively rare, especially in the modern synthetic literature. Introduction of an electron-donating substituent substantially increases the rate of substitution.[1] An early example was demonstrated by Wiley, wherein 2-phenyloxazole was brominated with NBS to give 5-bromooxazole as expected.[43] More recently, efforts toward the synthesis of molecules with a high affinity for the GABA receptor made use of this reaction to give the fully substituted 5-bromooxazole.[44]

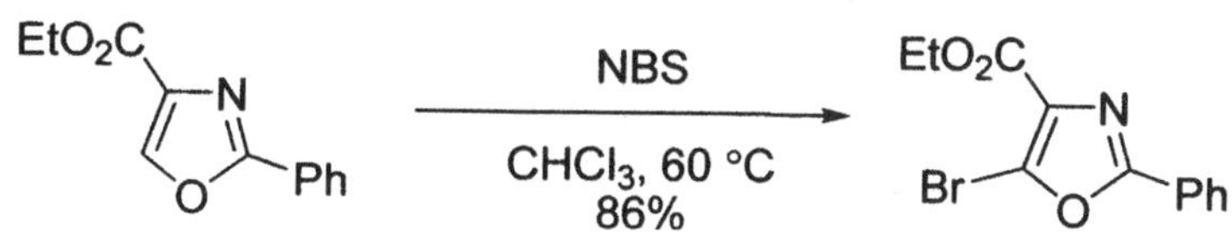

Electron-rich isoxazoles are also known to undergo electrophilic substitution under similar conditions. The bromination of 5-aminoisoxazole provided the 4-bromo derivative that was utilized as part of a synthesis of a library of compounds screened for their ability to act as bradykinin B_2 receptor antagonists.[45]

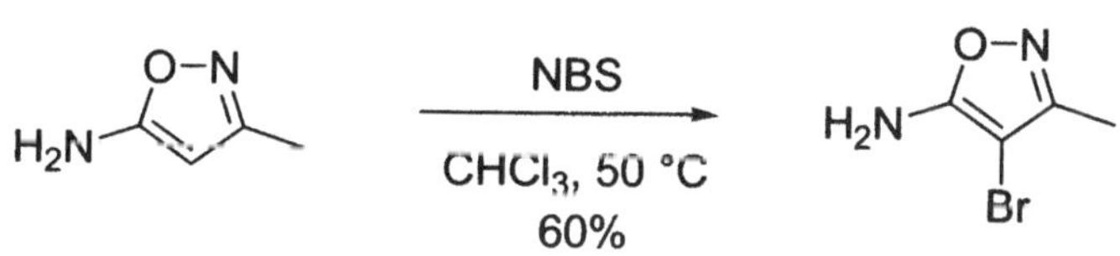

6.3.3 *Metalation and Nucleophilic Substitution*

As mentioned above, the heterocycles can be easily metalated with a variety of bases. The oxazole ring and the benzoxazole ring can be metalated at C2 with a variety of strong bases. The major limitation of metalated oxazoles and benzoxazoles is their equilibrium with an open chain β-alkoxynitrile (see Section 6.3.1). In the ring-closed form, the oxazole is nucleophilic at C2. In the open chain form, the anion is nucleophilic at C4. This dual reactivity of the oxazole anion can lead to undesired regiochemistry unless carefully monitored.

The reactivity pattern between the C2-lithiated, ring-closed oxazole and the *O*-lithiated, ring-opened oxazole appears to be temperature related, with the ring-opened form being reactive at low temperatures (–78 °C) and the ring-closed form being reactive at higher temperatures (≥ 0 °C).[46] The regioselective trapping of these nucleophiles, then, can sometimes be controlled by the reactivity of the electrophile; however, complex product mixtures are also common. Highly reactive electrophiles react with the ring-opened form at colder temperatures to give C4-substituted oxazoles, and less-reactive electrophiles react with the ring-closed form to yield C2-substituted oxazoles.

Hughes and co-workers performed extensive deuterium quenching experiments and determined preferential deuterium incorporation was correlated to the pK_a of the deuterium source.[47] Inverse quenching of lithiated oxazole with d_4-acetic acid showed preferential deuterium incorporation at C4, whereas quenching with D_2O showed nearly exclusive deuterium incorporation at C2 (perhaps through initial *O*-deuteration followed by subsequent valence-bond tautomerization). Hodges and co-workers demonstrated that more-reactive aldehydes regioselectively react to yield C4-substitued oxazoles, whereas less-reactive amides regioselectively react to yield C2-substituted oxazoles.

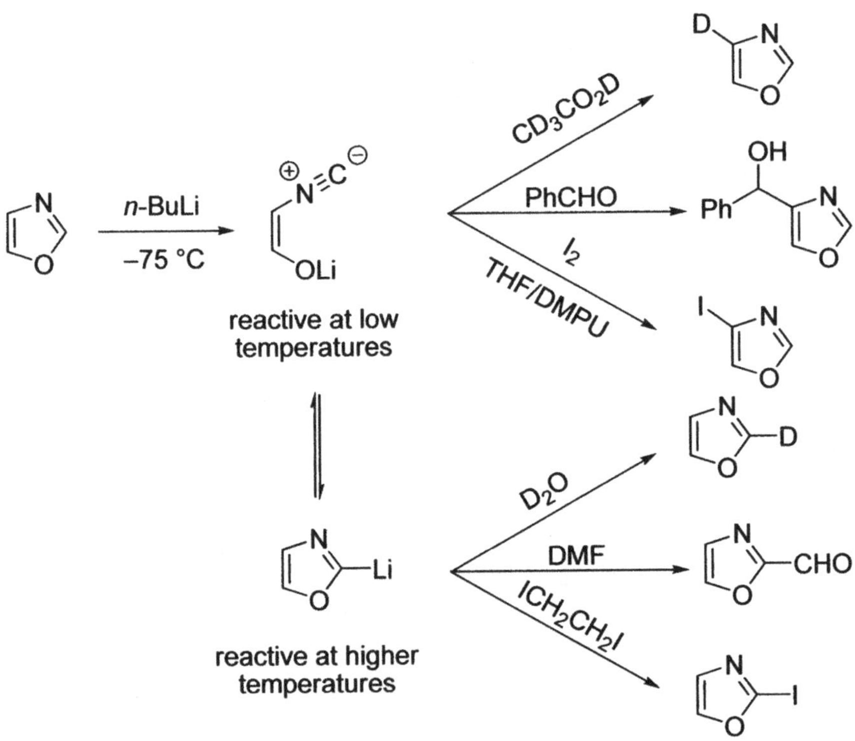

Vedejs and co-workers developed a regioselective iodination of lithiated oxazole.[48] Iodination at the C4-position can be affected by dissolution of the oxazole substrate in THF with DMPU co-solvent prior to deprotonation and iodination by I_2. Without DMPU as co-solvent, C2 iodination can be accomplished using less-reactive 1,2-diiodoethane as the electrophile.

In developing alternative ways to overcome regioselectivity issues, Vedejs postulated that deactivation of the lone pair of electrons on nitrogen prior to deprotonation might inhibit the ring-opening equilibrium.[49] When oxazole is initially complexed with borane, the Lewis acid/Lewis base complex can be deprotonated at C2 by *n*-BuLi or *s*-BuLi. Reaction with electrophiles proceeded rapidly and exclusively at C2, even with electrophiles previously shown to react at C4 under otherwise identical conditions. Boger utilized this methodology in the synthesis of a series of fatty acid amide hydrolase inhibitors.[50]

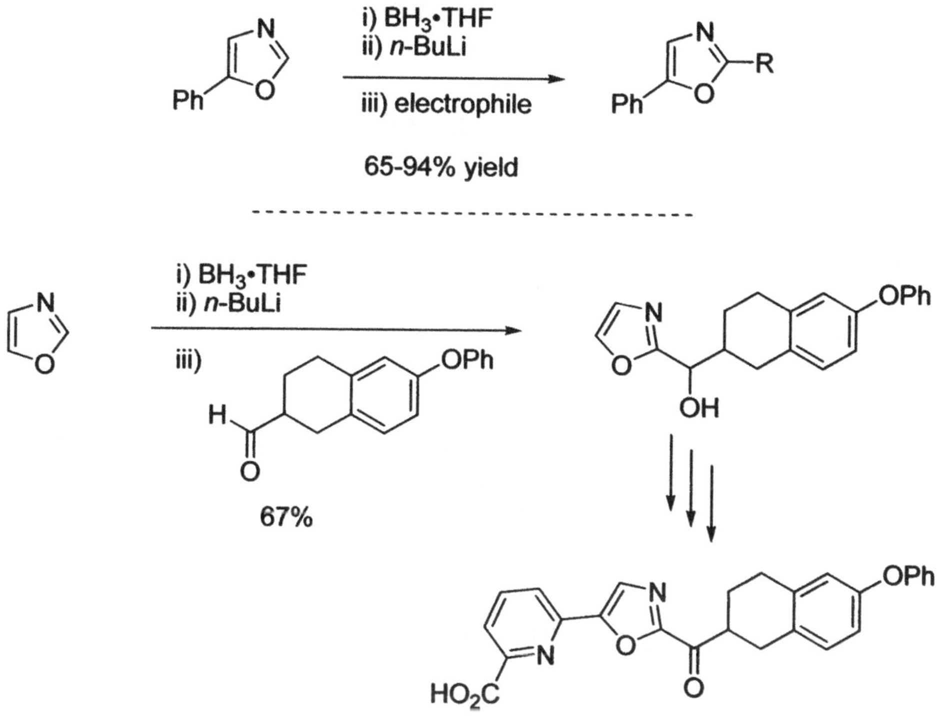

Of course, the issue of chemoselectivity in oxazole metalation can be alleviated if one or more positions are blocked in the starting oxazole. When the C2 position is blocked, the next most acidic site is C5. C5-metalated oxazoles do not suffer from ring opening equilibrium the way C2-metalated oxazoles do. If both C2 and C5 are blocked, metalation occurs at C4.

C2 silylation as a protecting group strategy for oxazoles has historically been difficult due to the preferential oxophilicity of standard silyl chlorides; protection of the ring opened oxygen atom typically predominates rather than protection of the C2 carbon atom. Miller has shown that treatment of unsubstituted oxazole with *n*-BuLi followed by triisopropylsilyl triflate yields C2-silylated oxazole in near quantitative yield.[51] Furthermore, this protected oxazole is stable to nonacidic aqueous workup and chromatography, an improvement over other C2-silylated oxazoles. The C2-silylated oxazole can be lithiated at C5 with *n*-BuLi at –10 °C and quenched with a variety of electrophiles. Various alkyl halides, aldehydes, silyl chlorides, and tributyltin chloride proved effective in this reaction. The C2-protected, C5-substituted product can be isolated, or acidic workup deprotects the C2-silylated oxazole to reveal the C5-substituted oxazole product. If C5 is already substituted, the reaction proceeds smoothly to substitute the C4 position.

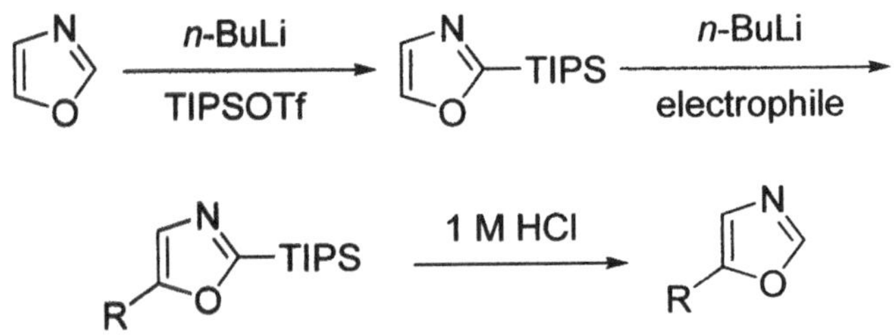

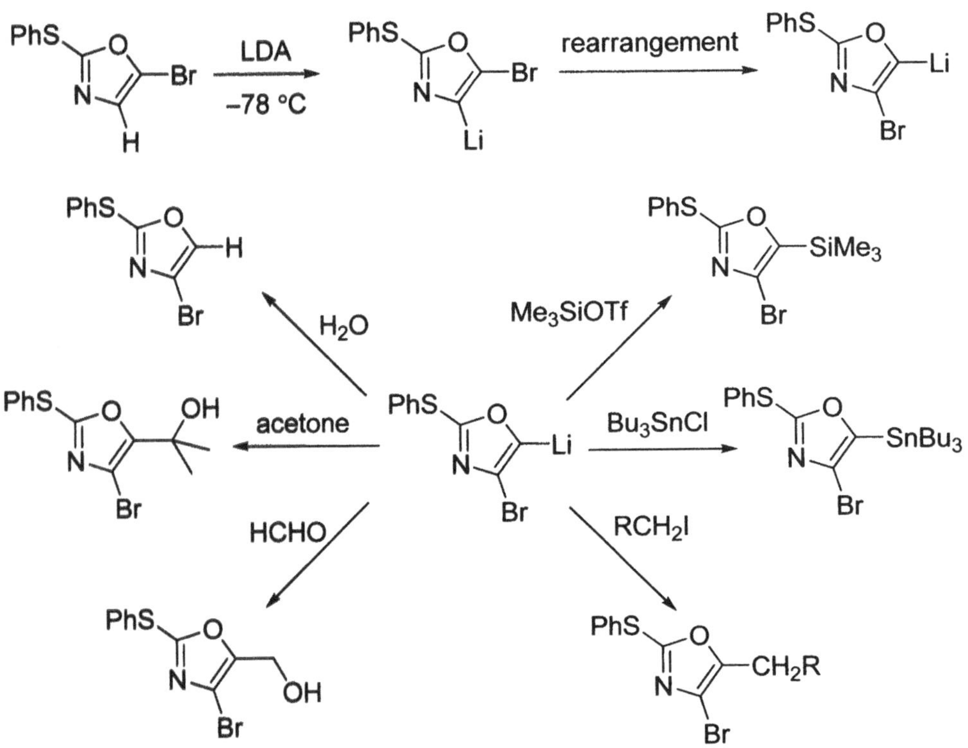

Williams and Fu[52] have recently demonstrated a novel method for production of fully substituted oxazoles. Following the kinetic deprotonation of the 4-position of 5-bromo-2-phenylthio-1,3-oxazole to give the 4-lithio species, a thermal rearrangement occurs resulting in the 5-lithio-4-bromo species. This anion can be hydrolyzed to yield the 4-bromo-2-phenylthio-1,3-

oxazole. On the other hand, the 4-lithio intermediate can be quenched by a variety of electrophiles, including alkyl halides, silyltriflates, stannylchlorides, aldehydes, and ketones to produce a diverse array of substituted oxazoles efficiently. Additionally, the lithio species can be transmetalated with $ZnCl_2$ and utilized directly in a Negishi coupling.

6.3.4 Pericyclic Reactions

Oxazoles are competent dienes for Diels–Alder reactions with alkenes, alkynes, and singlet oxygen[53]; however, the initial cycloadduct is unstable and decomposes to yield different products depending on the nature of the dienophile.

The Diels–Alder reaction of oxazoles with alkenes forms substituted pyridine rings after decomposition of the initial cycloadduct.[54] The regioselectivity can usually be predicted by assessing the partial charges of the dienophile. The C2 position of the oxazole typically carries a partial positive charge, and the C5 position typically carries a partial negative charge. Incoming dienophiles react in kind, with carbon atoms bearing electron-withdrawing groups bonding with C2.

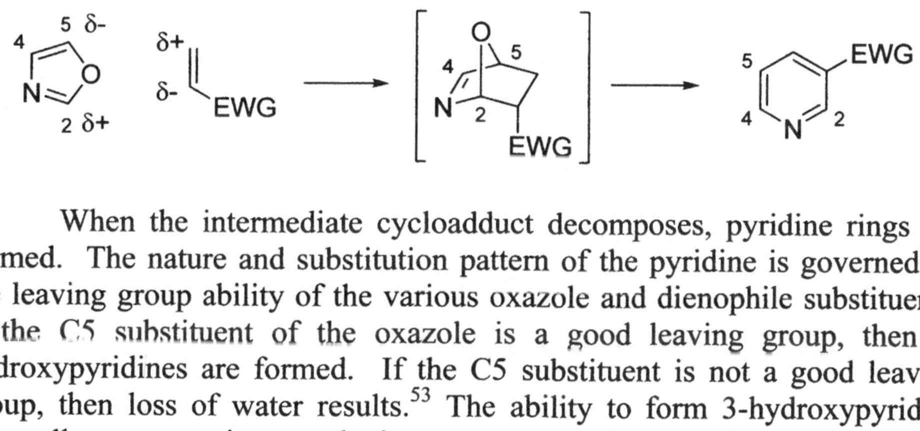

When the intermediate cycloadduct decomposes, pyridine rings are formed. The nature and substitution pattern of the pyridine is governed by the leaving group ability of the various oxazole and dienophile substituents. If the C5 substituent of the oxazole is a good leaving group, then 3-hydroxypyridines are formed. If the C5 substituent is not a good leaving group, then loss of water results.[53] The ability to form 3-hydroxypyridine rings allows convenient synthetic access to pyridoxine (vitamin B_6) and its derivatives.[55]

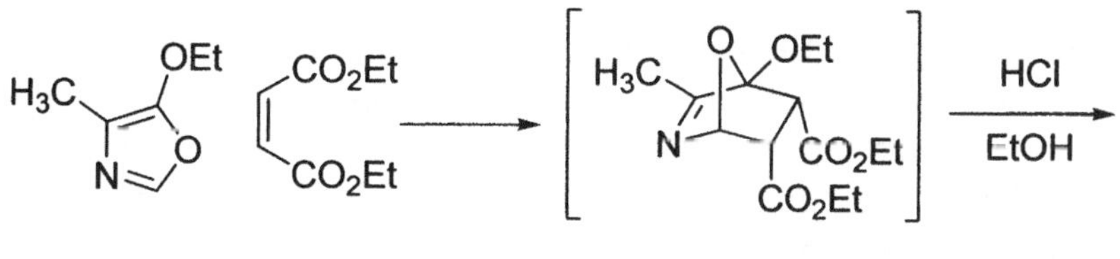

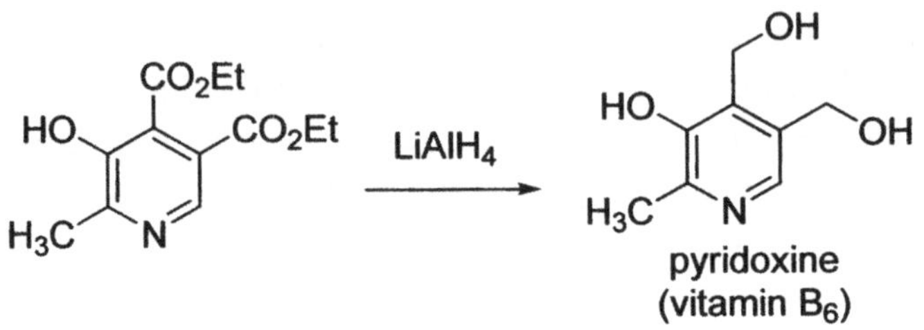

The bicyclic intermediate arising from Diels–Alder reaction of oxazoles with alkynes extrudes nitriles (comprised of the nitrogen atom and C4 of the oxazole) to form furans as the ultimate product of the cycloaddition.[56] The same regioselectivity seen in alkene Diels–Alder reactions is noted here.

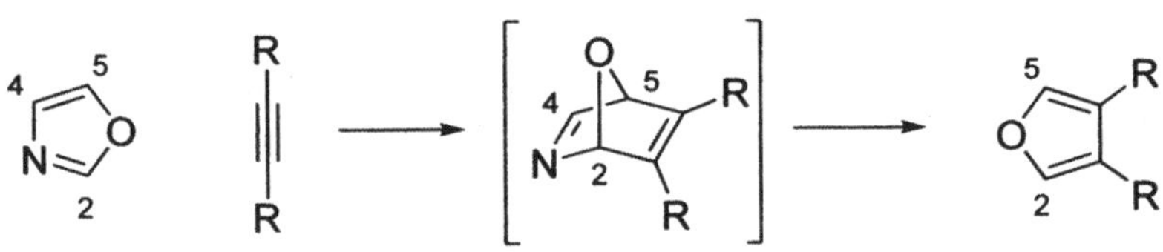

6.4 *Cross-Coupling Reactions*

The palladium-catalyzed cross-coupling reactions of oxazoles through the Negishi, Stille, Suzuki, and Sonogashira reactions are known, and have recently been reviewed.[57] Reactions utilizing the oxazole as the organometallic coupling partner or the electrophilic coupling partner are known, with examples in the C2, C4, and C5 position in most cases. Ease of precursor synthesis typically dictates whether the oxazole will be the organometallic or the electrophilic coupling partner.

6.4.1 *Preparation of Halo- and Trifloyloxazoles*

At the C2 position

As detailed previously, Vedejs showed that trapping of 2-lithiooxazole with 1,2-diiodoethane yields 2-iodooxazole.[48] Daugulis has demonstrated non-cryogenic conditions for the direct bromination of C2-unsubstituted oxazoles using dibromotetrafluoroethane as the electrophile.[58] The Sandmeyer reaction can be utilized to prepare 2-chlorooxazoles from 2-aminooxazoles.[59] Hexachloroethane can also be used as an electrophile for the direct chlorination of 2-lithiooxazole.[60] The conversion of 2-oxazolones to 2-trifloyloxazoles is also possible; however, 2-trifloyloxazoles decompose at high temperatures.[61]

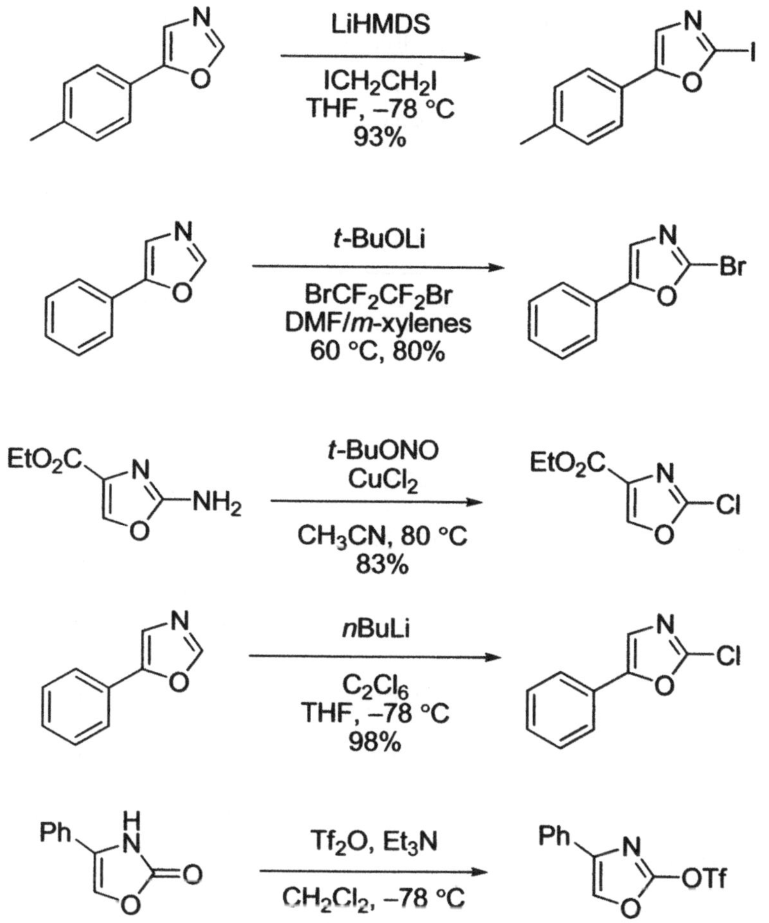

At the C4 position

To prepare 4-iodooxazoles, Vedejs' protocol with DMPU cosolvent allows for efficient synthesis. When DMF is used as the solvent, 2,4-unsubstituted oxazoles can be regioselectively brominated at the 4-position with NBS on kilogram scale.[62] Hunsdiecker reaction of a C4-carboxylic acid also provides the C4 bromide.[59] With a substituted C2 position, Nicolaou and co-workers achieved chlorination of the C4 position using NCS to prepare an intermediate in the partial synthesis of diazonamide A.[63] Similar to C2 trifloyloxazole synthesis, trapping of 4-oxazolones with triflic anhydride yields 4-trifloyloxazoles.[64] These triflates, in contrast to 2- or 5-trifloyloxazoles are stable and can generally be utilized in cross-coupling reactions.

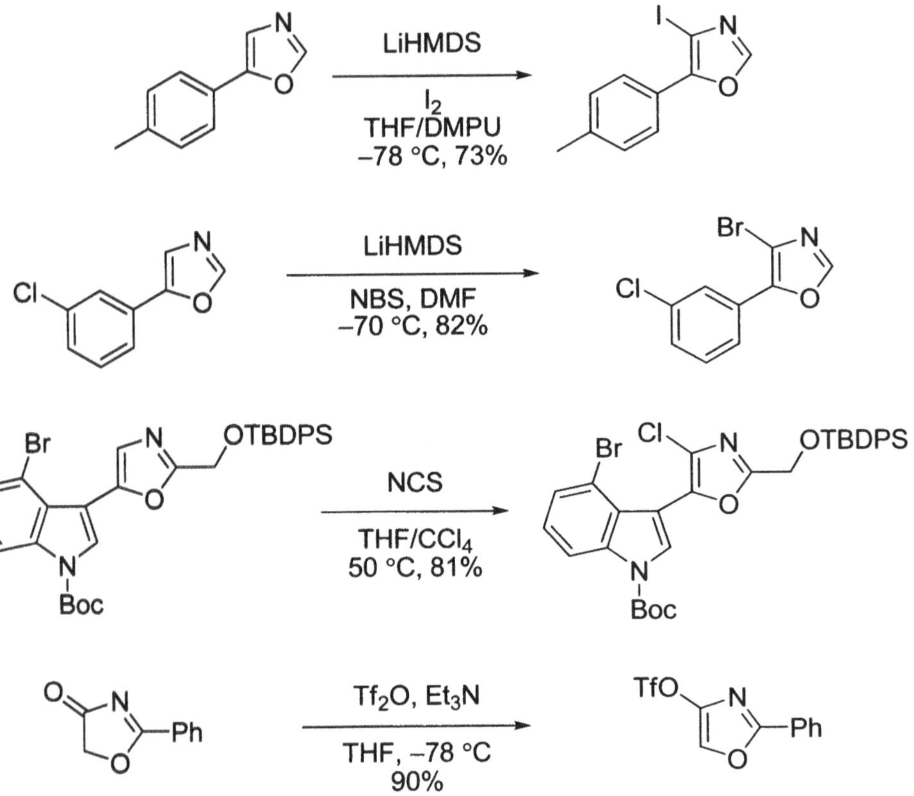

At the C5 position

To functionalize the C5 position, Williams and Fu developed a 2-phenylsulfonyl substituted oxazole.[65] The C5 position of this oxazole can be cleanly deprotonated with LDA and trapped with either NIS or NBS to form the 5-iodo- or 5-bromo-2-phenylsulfonyloxazole in good yield. The same report details that the 2-phenylsulfonyl group can subsequently be displaced with alkyl, alkenyl, or aryl lithium reagents to form 2,5-disubstituted oxazoles efficiently. A triflate at the C5 position can be prepared from the corresponding oxazolone; however, the oxazolone decomposes at room temperature, and Kelly reported that attempted Stille coupling with C5 triflates failed due to decomposition of the triflate.[66]

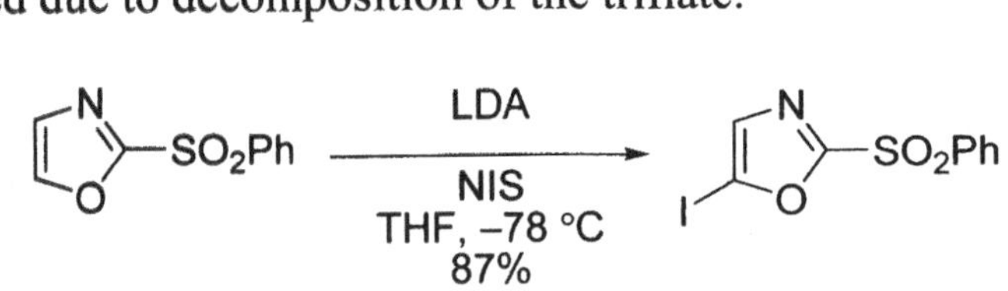

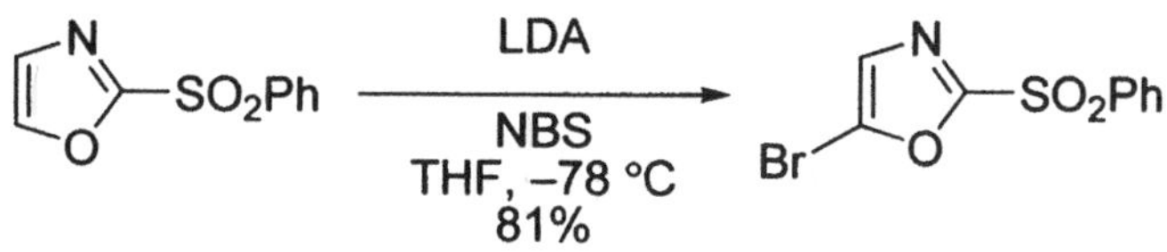

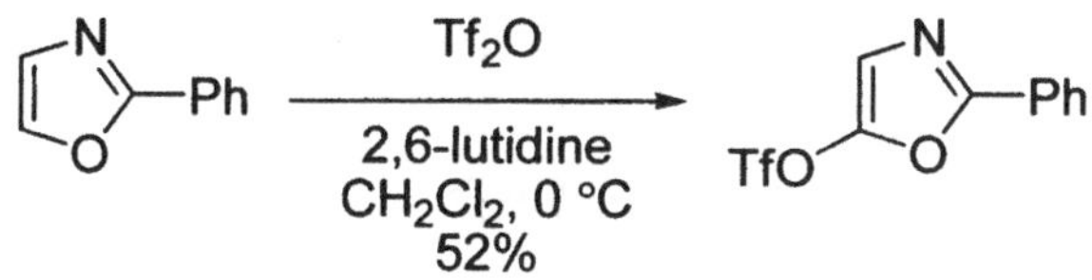

6.4.2 *Stille Coupling*

Preparation of stannyloxazoles

Oxazole can be lithiated with *n*-BuLi and quenched with either trimethyltin chloride or tributyltin chloride to prepare 2-stannyloxazoles.[67–69] Benzoxazole can be lithiated and trapped with trimethyltin chloride in a similar manner.[70] When the C2 and C5 positions of the oxazole ring are substituted, C4 can be lithiated and trapped with tributyltin chloride to provide the 4-stannyloxazole in good yield.[71] However, once prepared, the 4-stannyloxazole proved recalcitrant in the Stille coupling. The reaction can be coaxed into producing decent yields, however, when a stoichiometric amount of CuO is added. As long as the C2 position is substituted, lithiation of the oxazole ring and trapping with stannyl chlorides provides the 5-stannyloxazole. Both Williams and Miller showcase general methods for the preparation of C5-stannyloxazoles from their 2-phenylsulfonyloxazole[65] or 2-triisopropylsilyloxazole,[51] respectively.

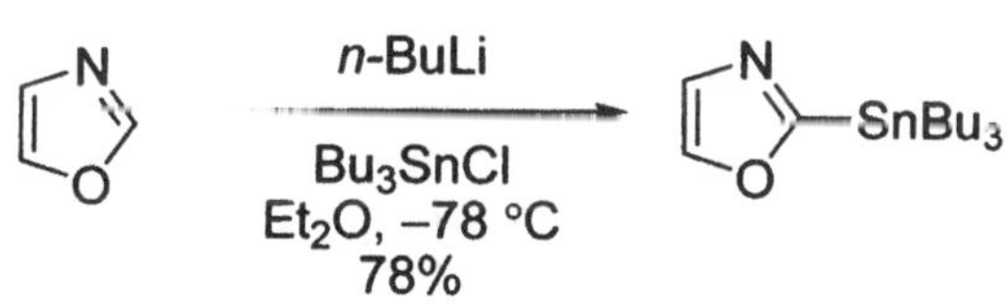

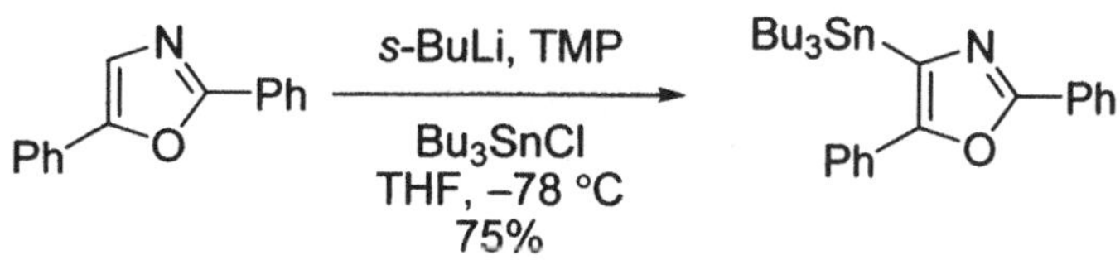

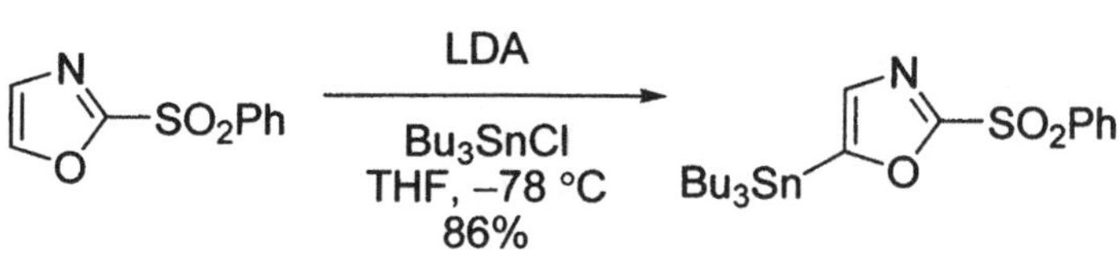

Scope of the reaction

In total synthesis papers, most reported Stille reactions involving oxazoles occur remotely to the oxazole ring. Most papers detailing direct Stille reaction of an oxazole ring are methodology papers. Williams showed that 2-phenylsulfonyl-5-stannyloxazole is a competent coupling partner with a vinyl iodide in a Stille reaction,[65] and Taylor utilized a series of 2-chlorooxazoles as coupling partners for Stille reaction with tributylvinylstannane.[72]

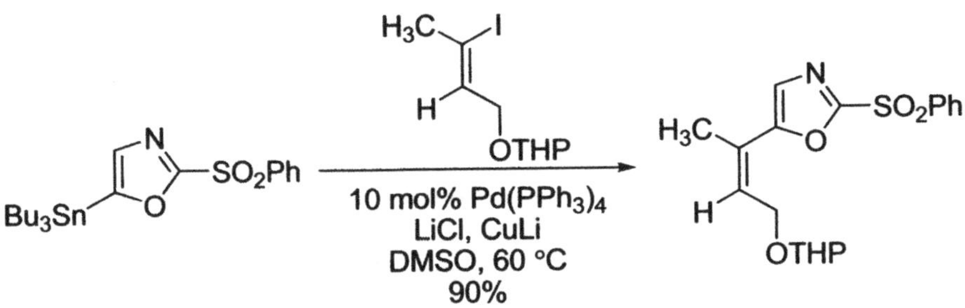

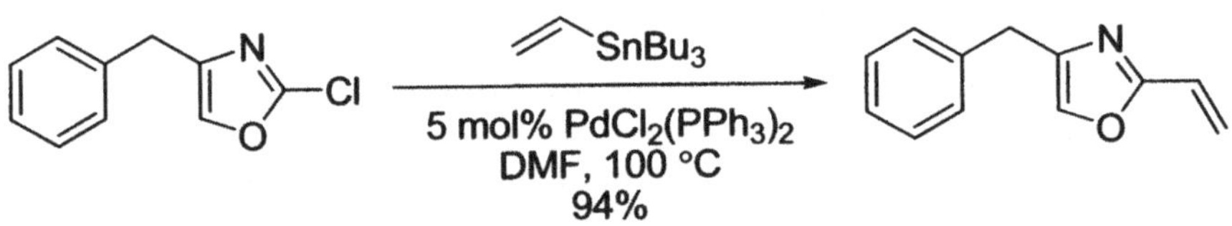

Hodgetts[59] and Stambuli[73] independently surveyed halooxazoles for their utility in a variety of cross-coupling reactions. With the Stille coupling, Hodgetts demonstrates 2-chlorooxazoles, and 4- or 5-bromooxazoles can be coupled with tributylvinylstannane. Stambuli couples 4- and 5-iodooxazoles with vinyl stannanes in a Stille coupling with LiCl as an additive. The thioether is used as a handle in subsequent palladium-catalyzed cross-coupling reactions with organozinc reagents to functionalize the C2 position.

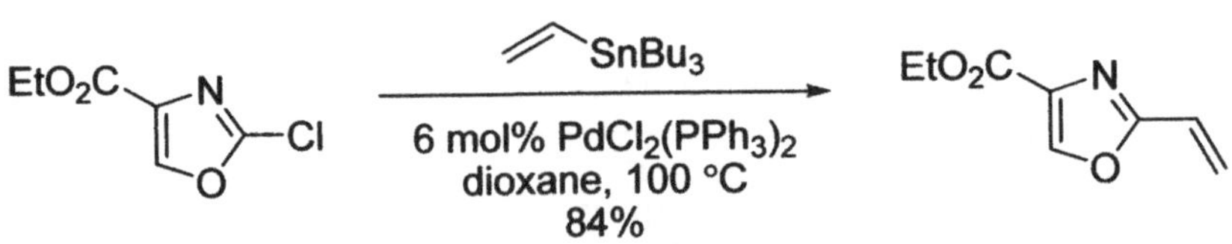

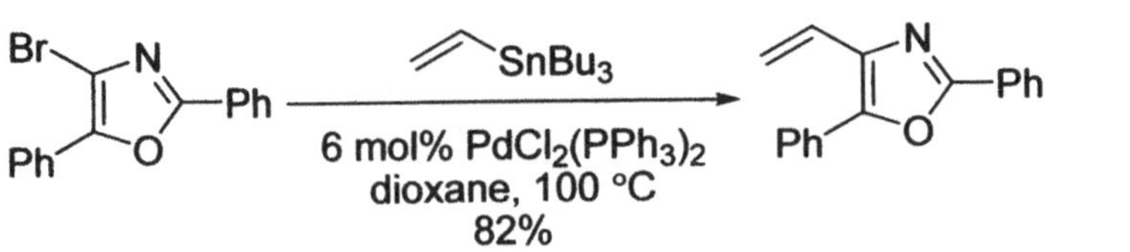

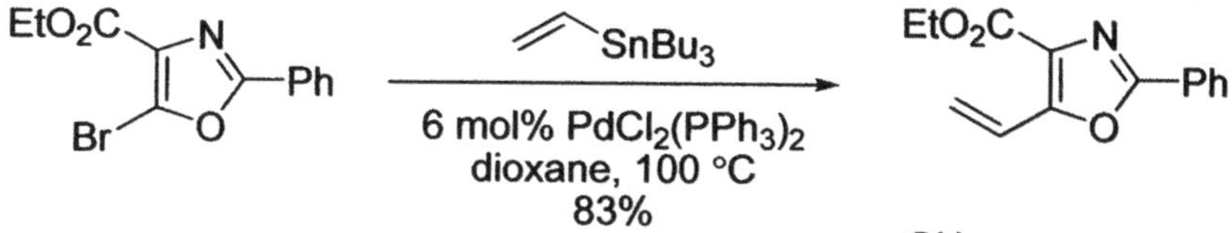

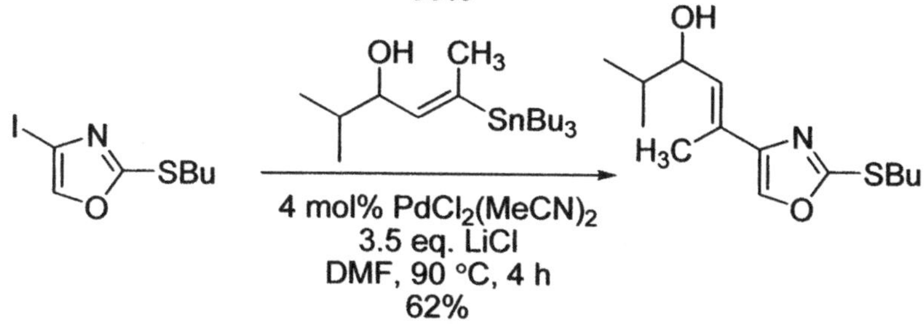

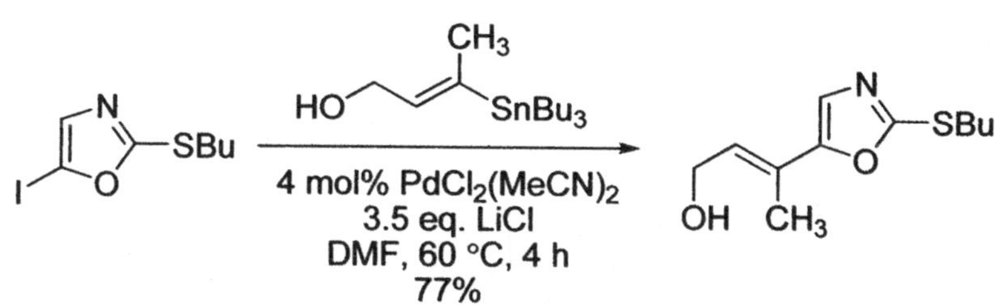

Smith prepared orthogonally functionalized 2-chloromethyl-4-trifloyloxazole and subjected the molecule to Stille reaction with tributylvinyltin chloride. The reaction occurred preferentially at the triflate to yield the 2-chloromethyl-4-vinyloxazole.[74] Kelly synthesized an intermediate along the way to dimethyl sulfomycinate, a methanolysis product of the naturally occurring antibiotic sulfomycin I, utilizing a Stille coupling with 2-phenyl-4-trifloyloxazole.[66] The Stille coupling of oxazole rings has also found utility in the syntheses of the natural products diazonamide,[75] and phorboxazole A.[76]

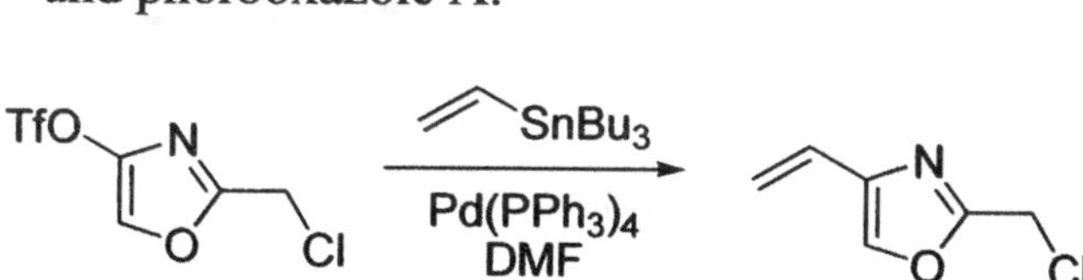

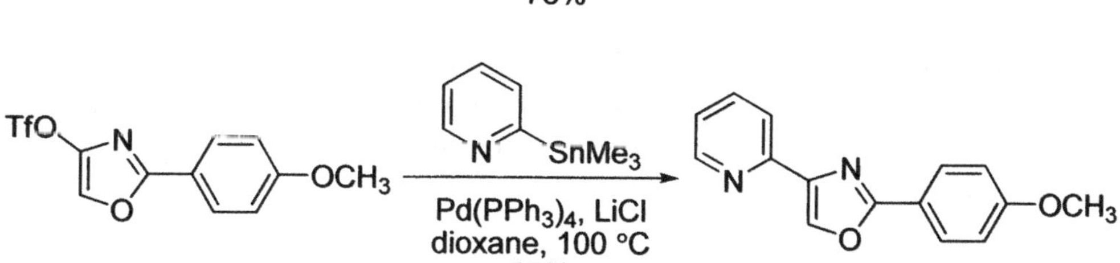

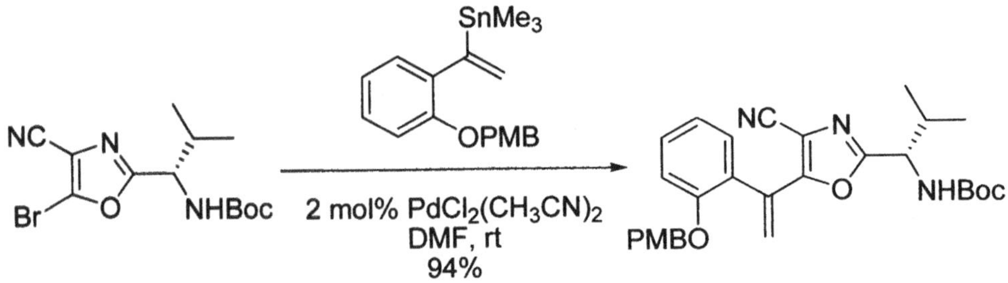

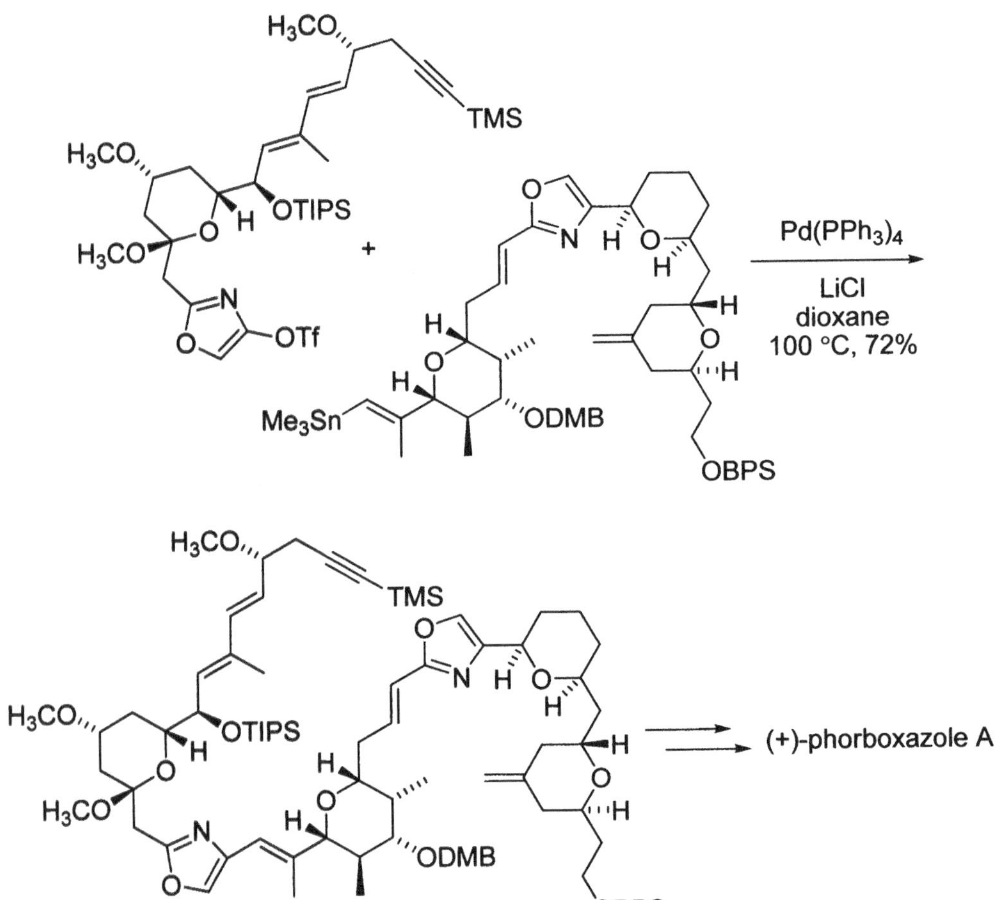

Medicinally relevant, Doi, Takahashi and co-workers derivatized the tris-oxazole subunit of the antiproliferative telomestatin with a Stille reaction (as well as Suzuki reaction and Pd-catalyzed amination reaction) utilizing a 5-bromooxazole.[77]

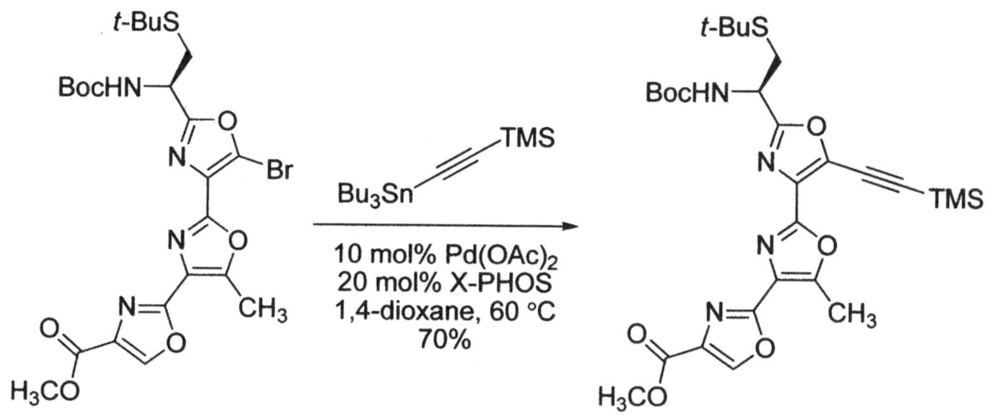

6.4.3 *Suzuki Coupling*

Preparation of oxazolylboronic acids

Preparation of C2 oxazolylboronic acids is troublesome, likely due to the ring-opening liability of 2-lithiooxazoles. The C4 position can be made into a oxazolylboronate by palladium-catalyzed reaction of a C4 triflate with bis(pinacolato)diboron.[78] The C4 boronate can also be prepared through lithium-halogen exchange of a C4 bromide and reaction with B(O*i*-Pr)$_3$ and pinacol.[79] C2-TIPS protected oxazole can be lithiated at the C5 position and trapped with B(O*i*Pr)$_3$ to form the C5 boronic acid.[80]

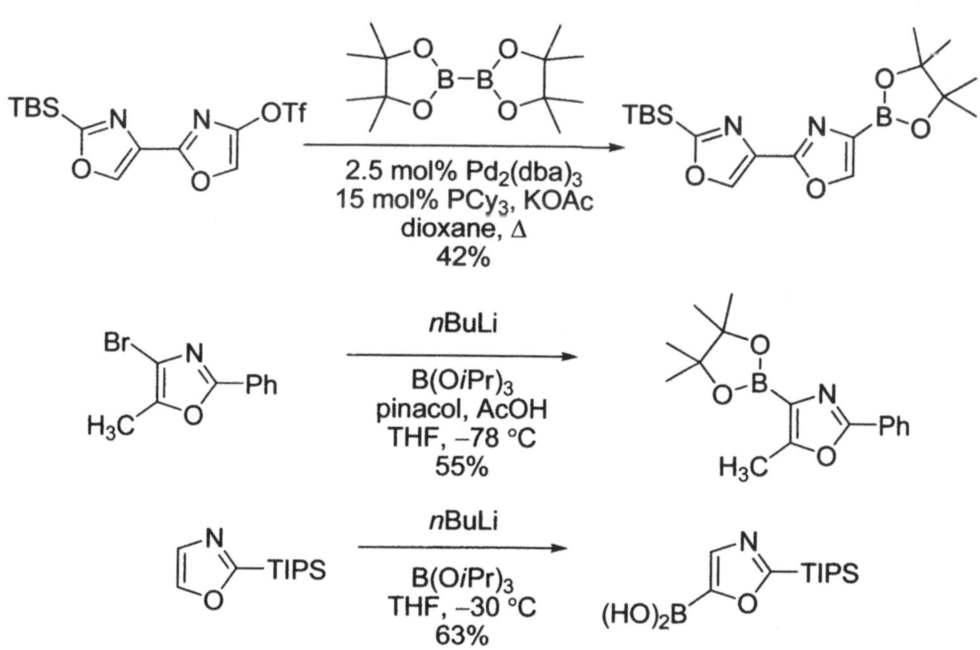

Scope of the reaction

By careful choice of ligand on palladium, 2,4-diiodooxazole can regioselectively engage in Suzuki coupling with a variety of aryl and heteroaryl boronic acids. Addition of 1,3,5-triaza-7-phosphaadamantane gave regioselective cross-coupling reaction at C2, and addition of the Xantphos ligand provided regioselective cross-coupling reaction at C4.[81]

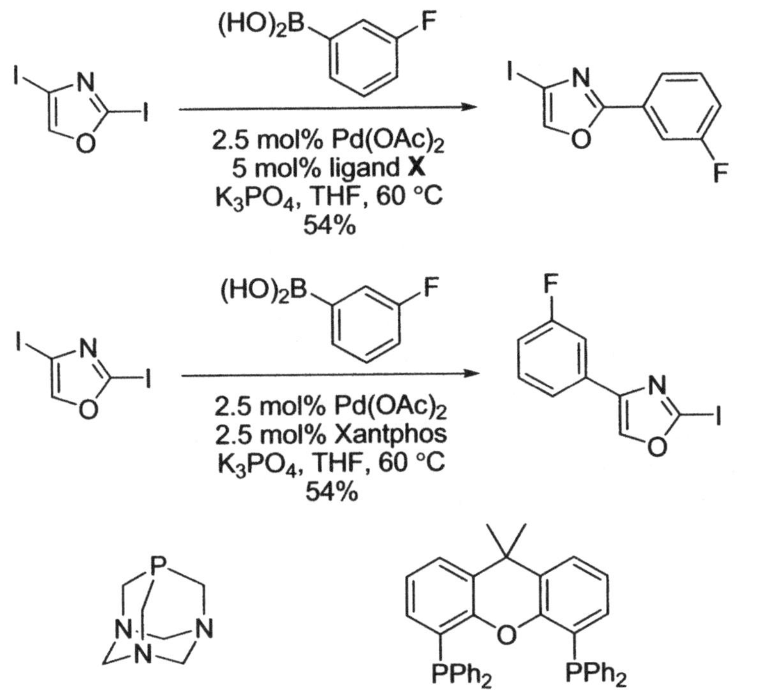

X Xantphos

Under noncompetitive conditions, Greaney functionalized C2-chlorides or C4-triflates with a variety of aryl and heteroaryl boronic acids. The authors attempted Suzuki reactions with the C2-sulfonates but found facile thermal decomposition with both triflates and nonaflates.[61]

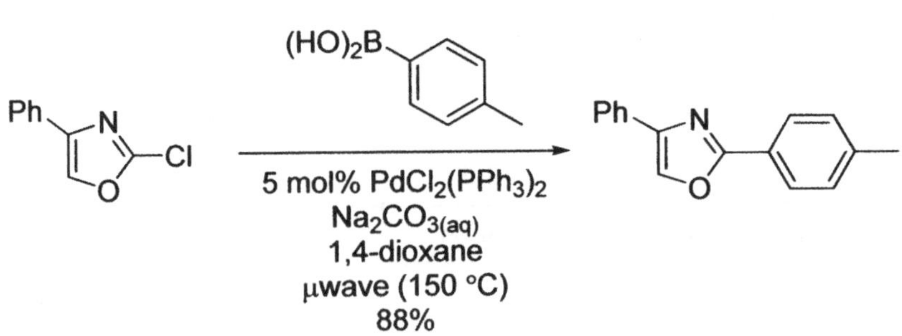

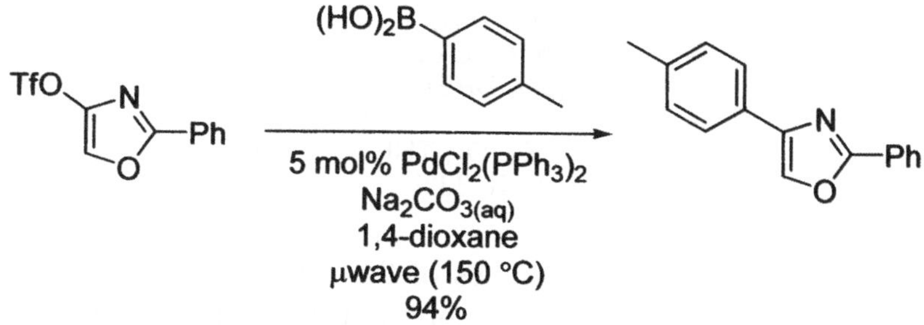

In their cross-coupling survey, Hodgetts showed C2-chlorides, C4-bromides or C5-bromides are competent partners in the Suzuki reaction.[59] Stambuli surveyed bromo- and iodo-oxazoles and found iodooxazoles more reactive in the Suzuki coupling at both the 4- and the 5-position.[73]

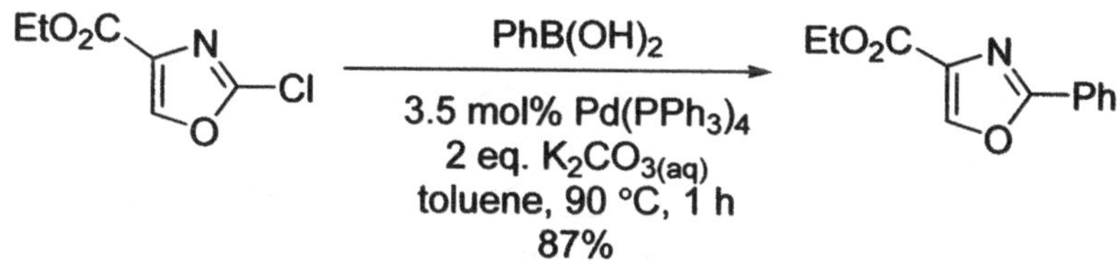

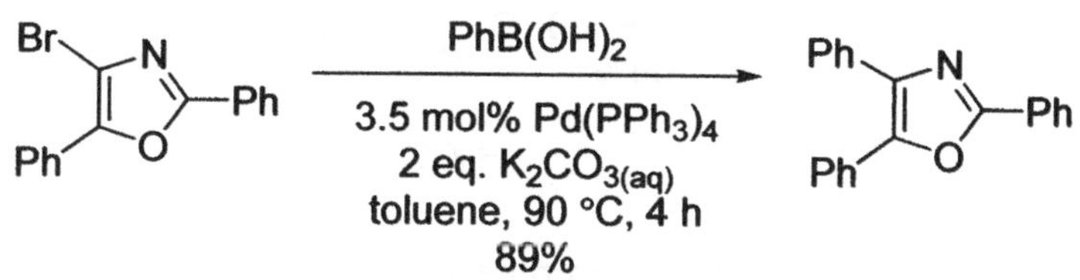

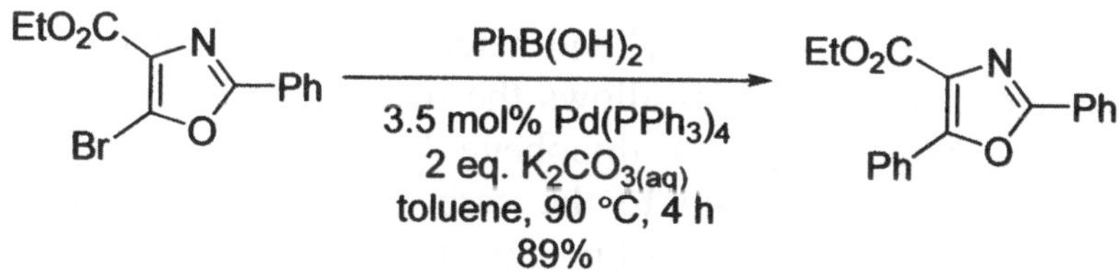

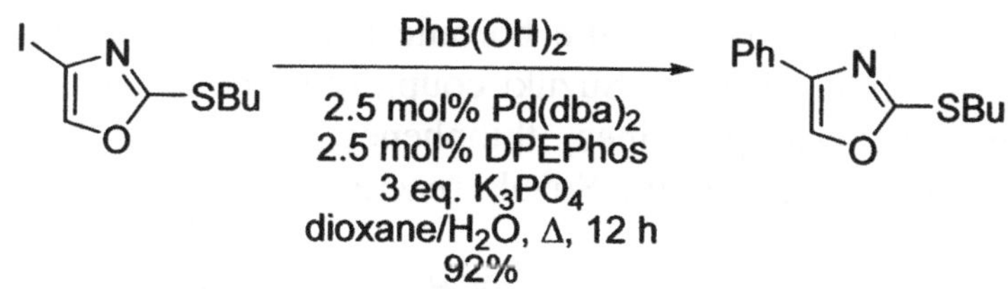

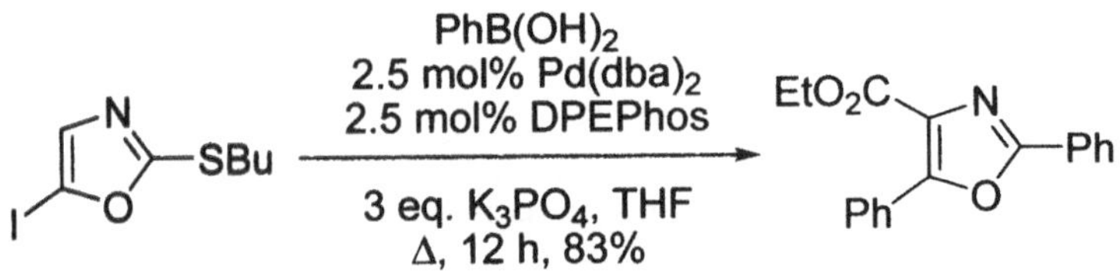

Aside from preparing analogues of the anti-proliferative telomestatin through Stille coupling as described previously, Doi, Takahashi and co-workers prepared a number of telomestatin derivatives through Suzuki coupling. Several aryl and heteroaryl substituents were incorporated in this manner.[77]

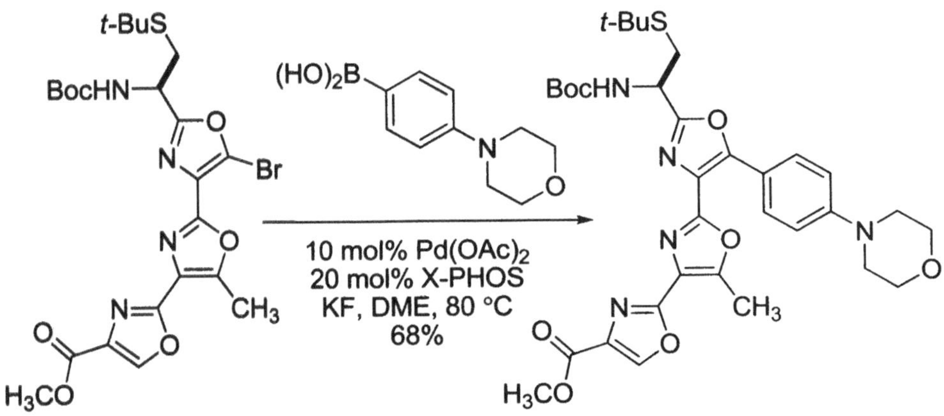

Liebeskind and Srogl have prepared a series of heteroaryl thioethers, which are useful starting materials in the Suzuki reaction with a variety of electron-rich and electron-poor aryl and heteroaryl boronic acids.[82] The use of the thioether starting material allows the reaction to be run base-free by adding stoichiometric copper(I) thiophene-2-carboxylate (CuTc). After oxidative addition of palladium to the carbon-sulfur bond, the copper is proposed to coordinate with sulfur and the carboxylate oxygen atom coordinates with boron. This facilitates a six-membered cyclic transmetallation step to transfer the boronic acid aryl group to palladium. Reductive elimination gives the Suzuki coupling product. Benzoxazoles, as well as benzothiazoles, pyridines, thiophenes, and other heteroaromatic thioethers, give good yields in the Suzuki reaction described.

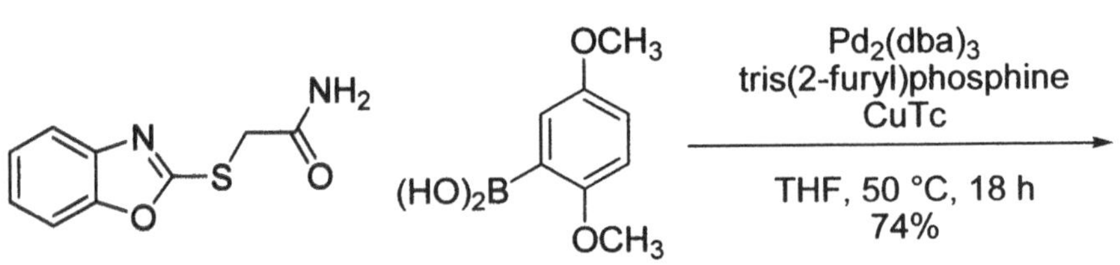

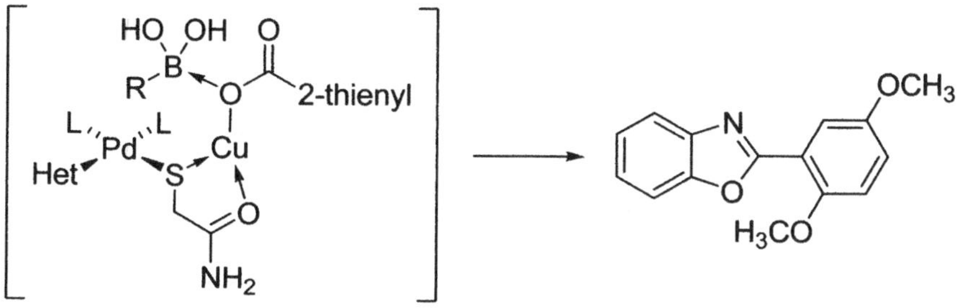

6.4.4 Negishi Coupling

Preparation of oxazolylzinc chlorides

Oxazolylzinc chlorides are best known in the C2 position. No generally practical examples of C4 oxazolylzinc chlorides are known in the literature, and few examples of C5 oxazolylzinc chlorides exist. The C2 position of oxazole or benzoxazole can be lithiated and transmetalated with zinc chloride to yield the oxazol-2-ylzinc chloride, which can be used directly in Negishi reactions. The use of solid zinc chloride rather than ethereal solutions allowed for convenient scale up of Negishi reaction of oxazoles and benzoxazoles to kilogram scale.[83] Zinc has a relatively low oxophilicity and forms a strong covalent bond with carbon,[84] allowing oxazolylzinc reagents to overcome the ring opening liability of 2-lithiooxazoles. In this way, C2 metalated oxazoles can be utilized in Negishi coupling reactions where similar Suzuki coupling reactions may fail.

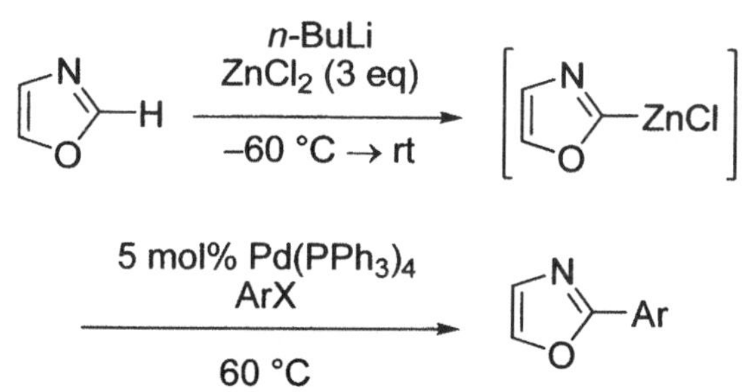

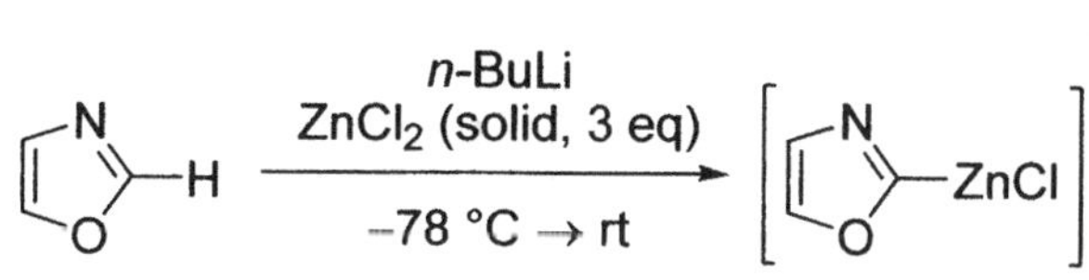

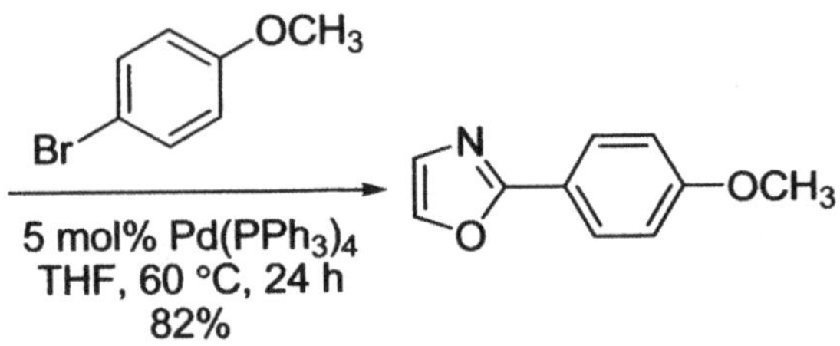

Scope of the reaction

Molinksi and co-workers had needed to prepare a C2 zinc chloride where the use of *n*-BuLi to deprotonate the C2 position was untenable. Instead, treatment of a 2-iodooxazole (prepared by Sandmeyer reaction of the corresponding 2-aminooxazole) with LiCl and Zn under Knochel conditions[85] to provide the 2-oxazolylzinc chloride necessary for Negishi coupling.[86] The oxazolylzinc chloride was stable under N_2 at 4 °C in the dark for more than one month. Negishi coupling to a vinyl iodide provided a key intermediate in the total synthesis of the marine natural product enigmazole A.

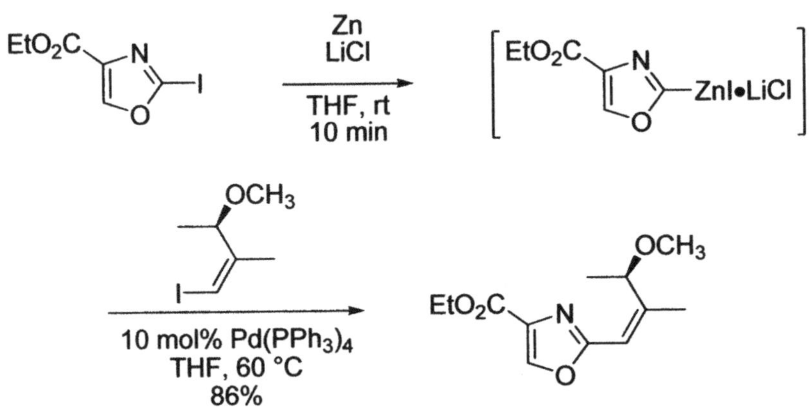

Vidal and co-workers used 2-oxazolylzinc chloride to enhance an SAR study of orally efficacious A_{2B} adenosine receptor antagonists. The oxazole was also appended through the C4 position by utilizing a Stille reaction of the C4 stannane.[29]

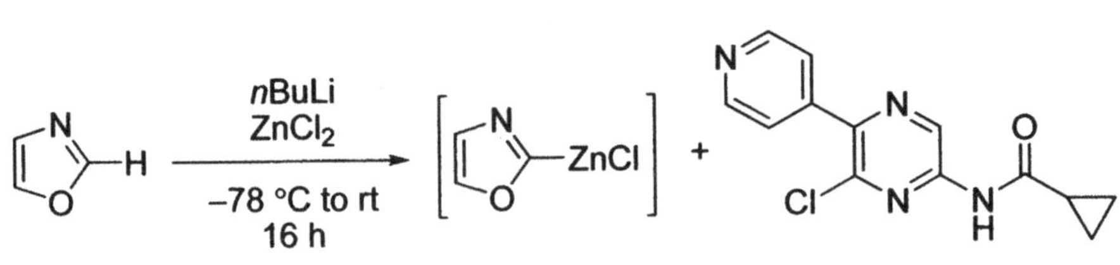

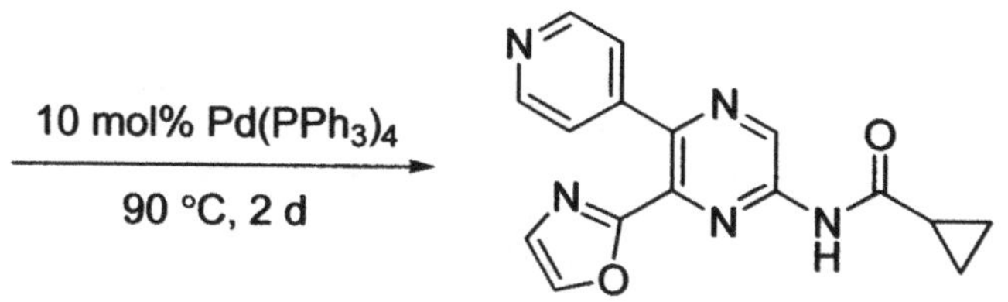

Vedejs utilized a Negishi reaction in synthesizing the core of aziridinomitosene A. The C5 position of a C2-substituted oxazole was lithiated and transmetalated with $ZnCl_2$. The 5-oxazolylzinc chloride was utilized in an intermolecular Negishi reaction with a vinyl triflate in 70% yield with no reported intramolecular reaction at the alkyl iodide.[87]

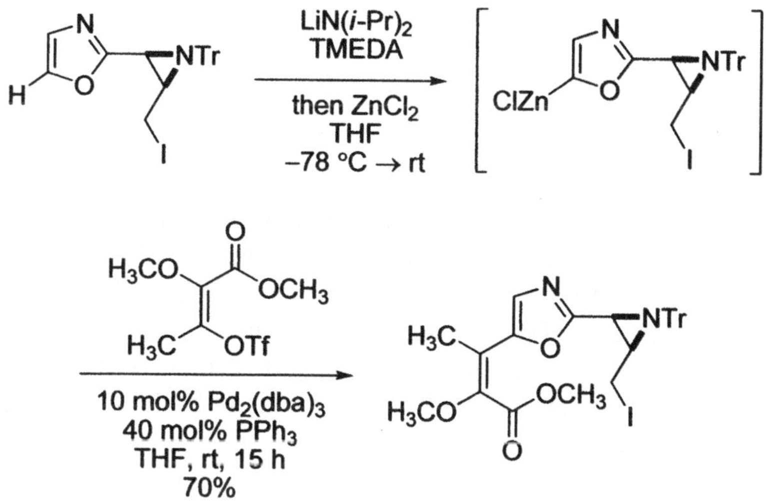

Hodgetts has shown that C2 chlorides and C4 and C5 bromides all participate in Negishi coupling with 2-pyridylzinc bromide in approximately 70% yield.[59] Stambuli did not attempt Negishi coupling of the halooxazoles, probably due to the lability of the thioether under Negishi-like conditions.[73]

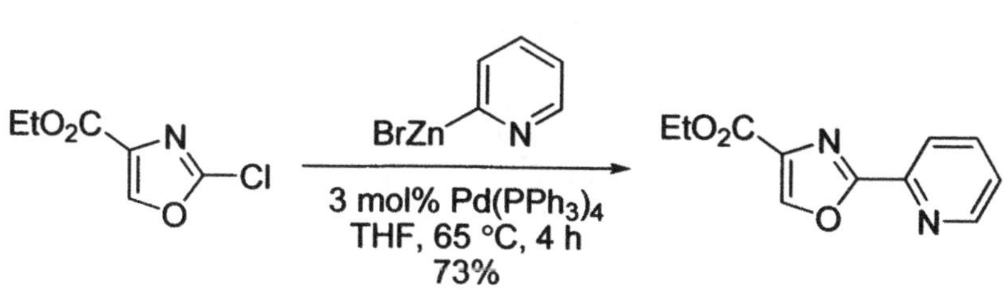

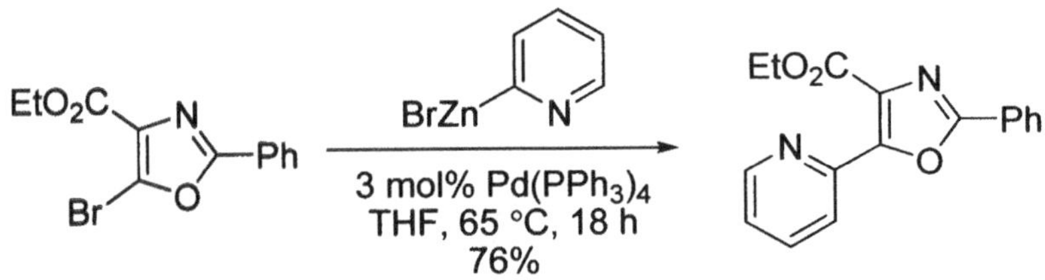

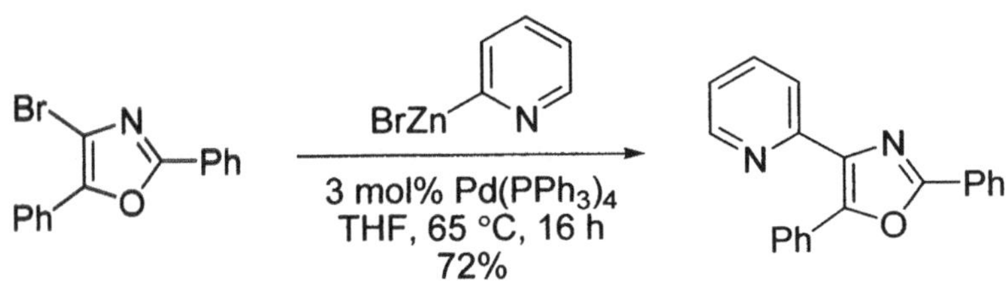

6.4.5 Sonogashira Coupling

Scope of the reaction

The first report of a Sonogashira coupling reaction of an oxazole was by Yamanaka in 1987.[88] Oxazoles substituted with bromine at the 4- or 5-positions were coupled with phenylacetylene yielded the alkyne in 83% and 89% yield, respectively. The Sonogashira reaction with 2-halooxazoles was not attempted; however, 2-halothiazoles and 2-halo-*N*-methylimidazoles were subjected to Sonogashira conditions. Yields in both cases were low and not synthetically useful.

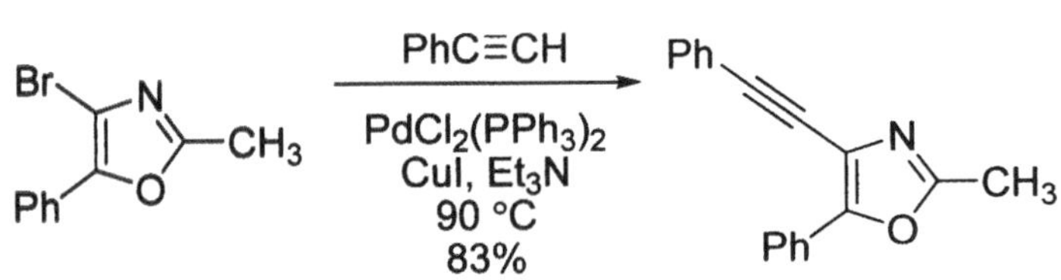

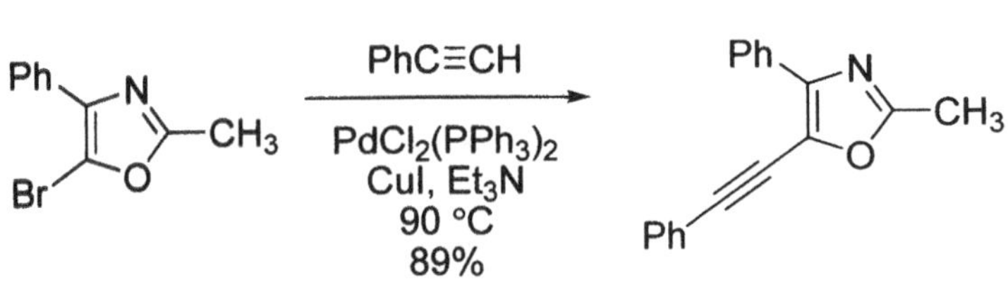

Panek and co-workers showed that trifloyloxazoles react quite cleanly in Sonogashira coupling reactions.[89] The reactions require heating to 65 °C and work best with copper(I) iodide (CuI > CuBr > $(MeCN)_4CuPF_4$ ≈ CuCl). DMF or 1,4-dioxane proved to be the optimal solvents for the reaction. Under optimal conditions (CuI, Et_3N, $Pd(PPh_3)_4$, DMF), 4- and 5-

trifloyloxazoles reacted cleanly. Gratifyingly, the Sonogashira reaction proceeded without any homocoupling of the acetylene, a common by-product in the Sonogashira reaction. Benzoxazoles and isoxazoles are also competent coupling partners in the Sonogashira reaction. If triethylamine is used with 2-trifloyloxazoles, then the oxazole decomposes back to the oxazolone. However, 2,6-lutidine suppresses this decomposition and allows the Sonogashira coupling to proceed in good yield. This method was utilized in the synthesis of leucascandrolide A.

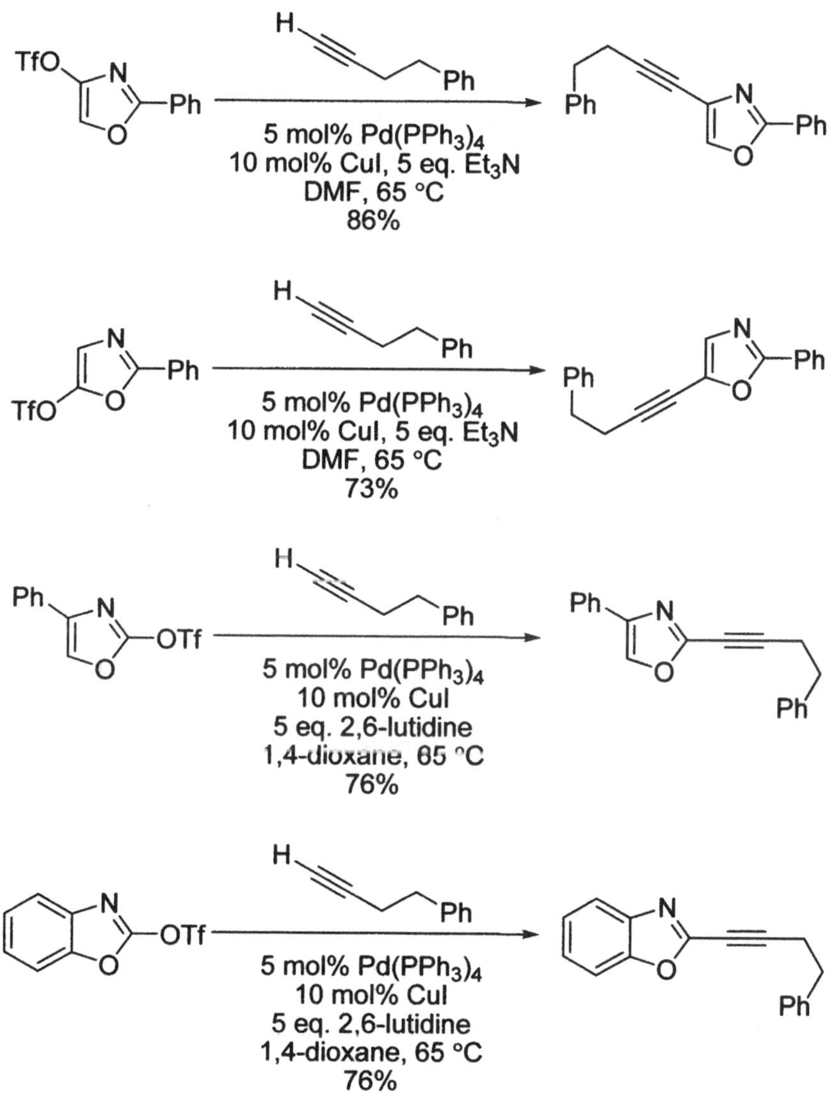

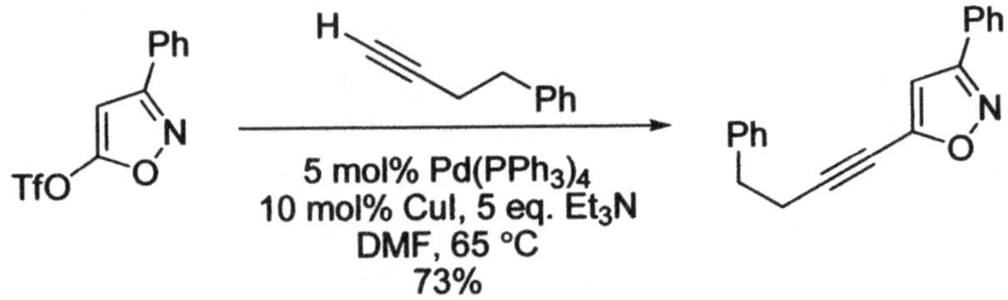

Both Hodgetts and Stambuli surveyed Sonogashira conditions for halooxazoles. Hodgetts has shown that while 4- and 5-bromooxazoles engage in Sonogashira coupling in just under 80% yield, 2-chlorooxazoles yield an intractable "resinous material" where the product could not be isolated.[59] Stambuli has demonstrated that ethers, esters, and nitriles are well tolerated in the Sonogashira coupling.[73]

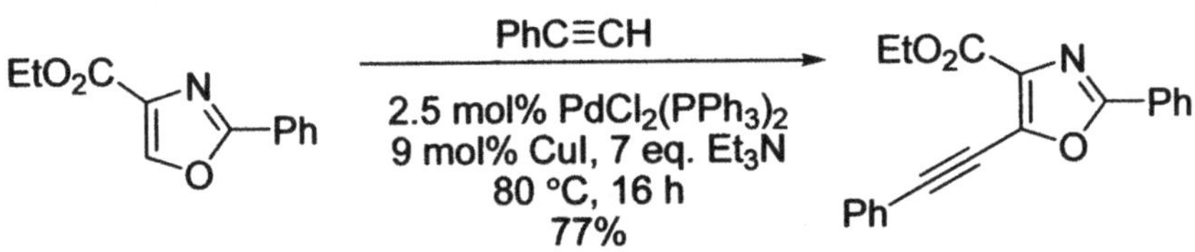

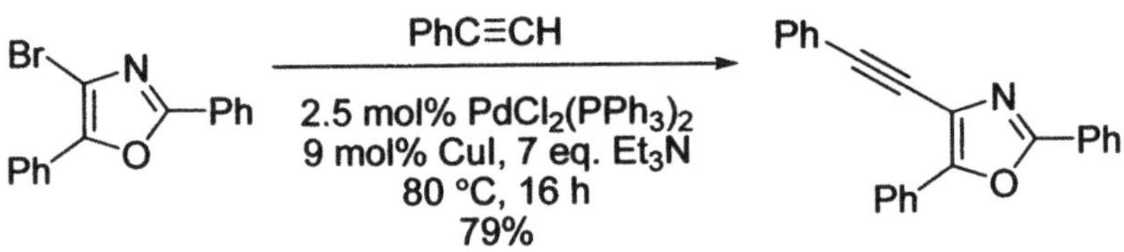

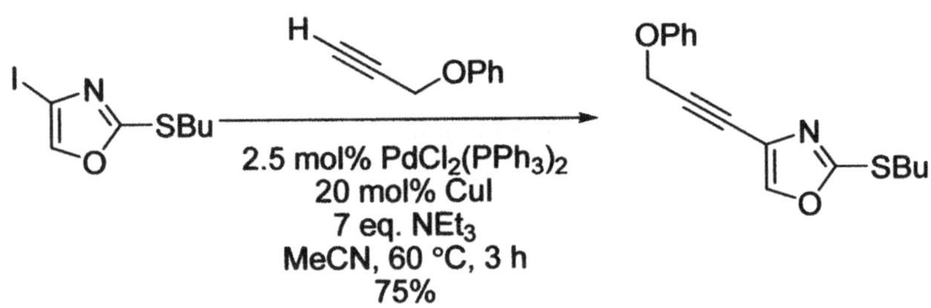

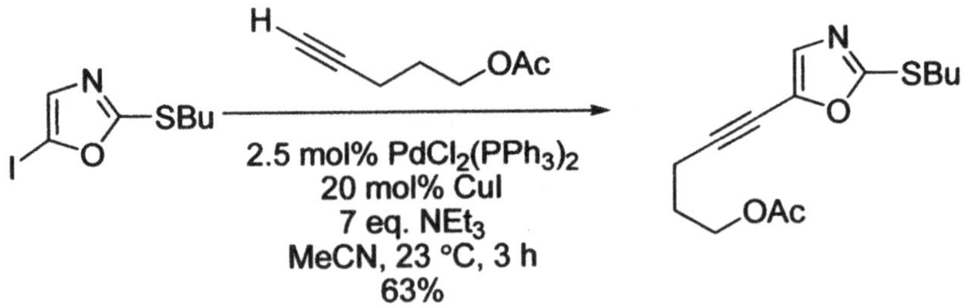

Yin and co-workers prepared a series of 3-dimethylaminopropyl-substituted heterocycles, including oxazoles, as inhibitors of the protein calcineurin through Sonogashira coupling.[90] The target compound is prepared after subsequent hydrogenation.

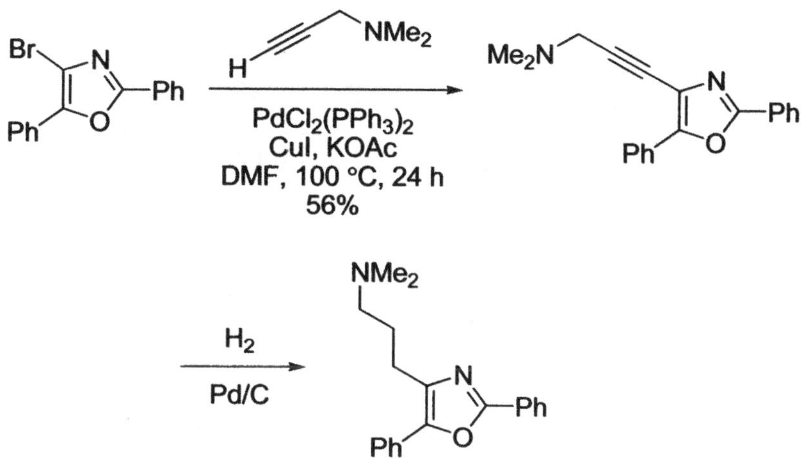

6.4.6 Heck Coupling

Scope of the reaction

Vinyl oxazoles are efficiently prepared through Stille reaction of a halooxazole with a vinyl stannane; therefore, there are relatively few reported Heck reactions of halooxazoles with olefins. Yamanaka's 1987 Sonogashira paper[88] included a section of Heck coupling reactions. The authors noted poor yields with 2-bromooxazoles, and yields that were variable at best for 4- and 5-bromooxazoles.

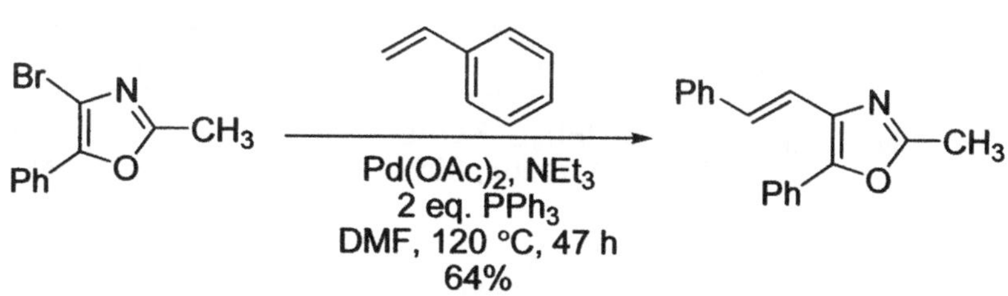

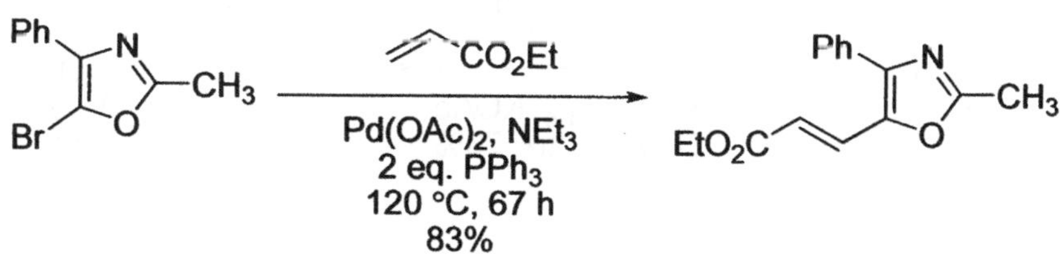

The Heck reaction has also found utility in the search for new nonpeptide glycoprotein GPIIb/GPIIIa antagonists. Specifically, the preparation of the oxazolepiperidine scaffold has been enabled by the Heck reaction between a bromooxazolopyridine with methyl acrylate in the presence of palladium acetate.[91]

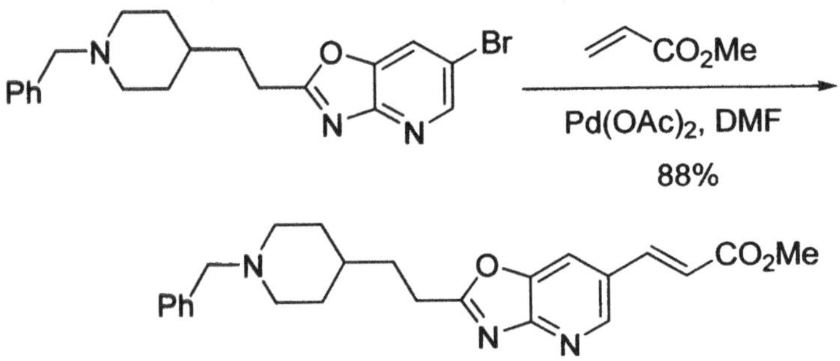

Oxazoles and benzoxazoles are viable participants in the heteroaryl Heck reaction, whereby the alkyl halide or aryl halide is coupled to the unfunctionalized oxazole. First developed by Ohta and colleagues, it was demonstrated that a diverse array of aromatic heterocycles can be substrates in the reaction with chloropyrazines.[92] While substitution was expected to occur at the 2-position, the reaction with chloropyrazine as the aryl halide resulted in substitution at the 5-position. When benzoxazole is used as the coupling partner, substitution is effected at the 2-position. More recently, a systematic study by Strotman and co-workers has demonstrated that slight modifications of the reaction conditions can allow completely regioselective coupling at either the 2- or 5-position.[93]

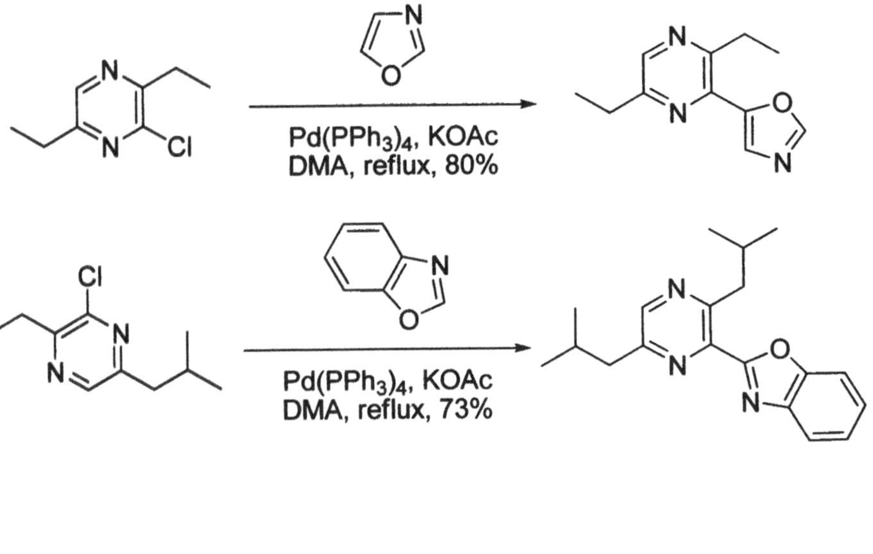

The heteroaryl Heck reaction has recently found utility in the preparation of vascular endothelial growth factor receptor-2 (VEGFR-2) inhibitors.[94] The authors proposed that inhibition of the kinase activity of VEGFR-2 would result in reduction of angiogenesis. Because tumors require angiogenesis to grow beyond a certain size, it follows that VEGFR-2 inhibition would also result in suppression of tumor growth. As such, the oxazole nucleus could be introduced onto the pyrazine scaffold under the conditions described above. The resulting molecule unfortunately displayed diminished potency in the biological assay.

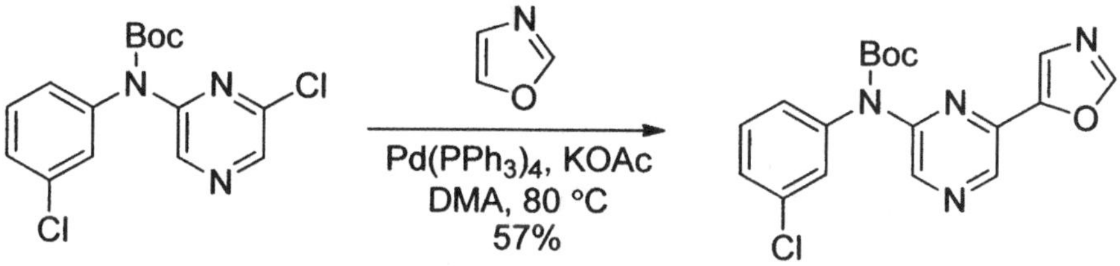

The heteroaryl Heck reaction was similarly used in the preparation of potential A_{2B} adenosine receptor agonists, compounds with the potential to become a new class of drugs for the treatment of human asthma.[95]

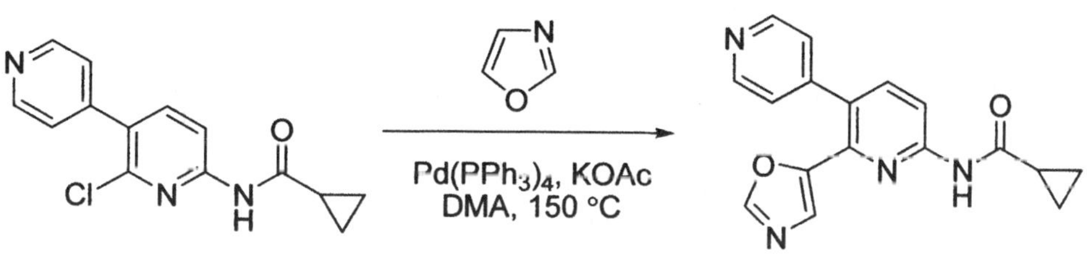

Finally, 2-iodonitrobenzene has also been effectively utilized in the heteroaryl Heck reaction to provide a fully substituted oxazole ring.[44]

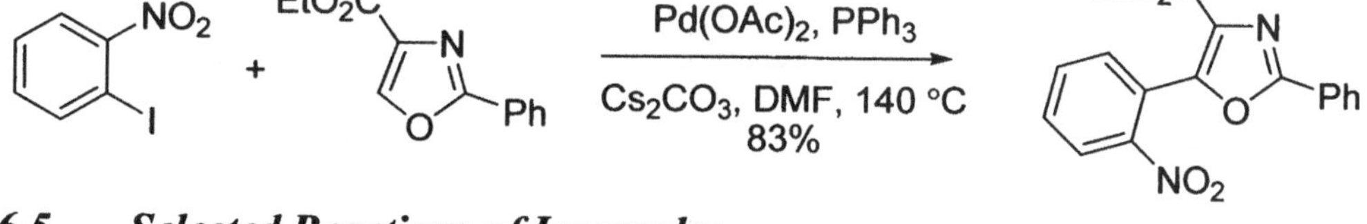

6.5 *Selected Reactions of Isoxazoles*

Reactions of isoxazoles are less prolific due to facile isoxazole ring cleavage. Isoxazoles can be treated with *N*-bromosuccinimide to brominate the C4-position of the isoxazole. In contrast to oxazoles, when an isoxazole engages in a [4 + 2] cycloaddition, the isoxazole acts as a dienophile, not a diene.[96]

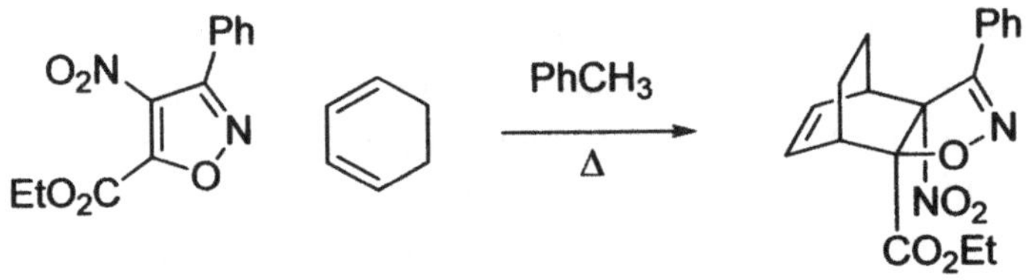

Other reagents such as reductants, strong bases, or thermal conditions cause cleavage of the isoxazole ring. For example, reaction with strong bases opens the ring to yield β-ketonitriles. Additionally, heating (or photochemical irradiation) of the isoxazole or benzisoxazole ring goes through diradical intermediates to yield oxazoles.[97]

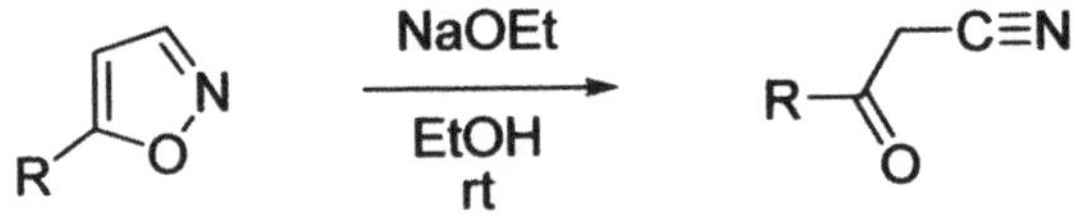

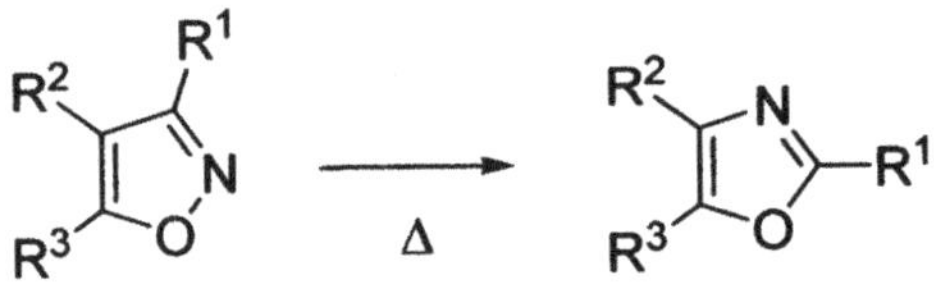

6.6 *Possible Liabilities of Oxazole-Containing Drugs*

As shown previously, C2-lithiated oxazoles are in equilibrium with their ring-opened counterpart. C2-unsusbtituted oxazoles are therefore potentially unstable to strongly basic conditions.

Ammonia can react with the C2 position of oxazoles resulting in ring cleavage and formation of an imidazole ring. Ring cleavage by this mechanism occurs more frequently than S_NAr substitution.

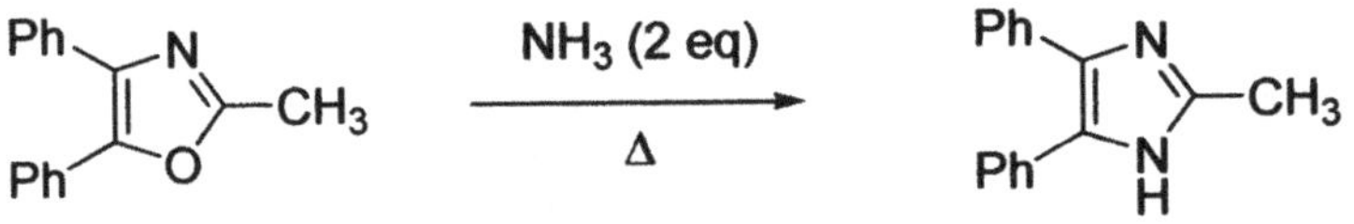

Oxazoles with an acyl substituent at C4 can undergo a thermal rearrangement essentially exchanging the C4 and C5 substituents.[98] The facility of this rearrangement is improved when the C5 substituent is a heteroatom (–OR, –SR, –Cl). Dewar and Turchi observed deuterium scrambling of a labelled oxazole ester under thermal conditions. They propose a nitrile ylide intermediate in the rearrangement mechanism.[99] This rearrangement is known as the Cornforth Rearrangement,[100] and could be exploited to the synthetic chemist's advantage if properly planned.

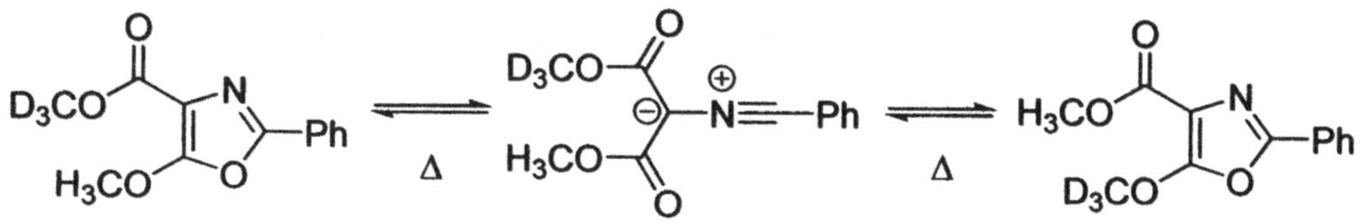

6.7 *Problems*

6.7.1 Vedejs and co-workers synthesized the following oxazole intermediate in their attempted synthesis of aziridinomitosene A. Upon treatment with a cyanide source, the molecule participates in a [3 + 2] cyclization to form tetracycle **A**. Propose a complete, arrow-pushing mechanism for this transformation.[87]

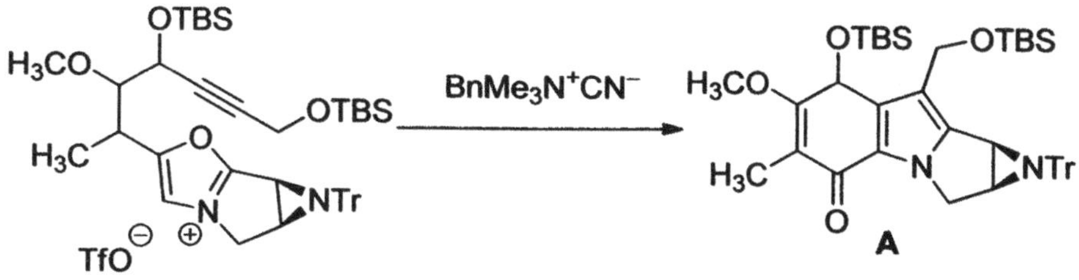

6.7.2 Kuo and co-workers recently synthesized a novel series of vascular endothelial growth factor receptor-2 inhibitors, similar to the following.[94] Suggest a reasonable retrosynthetic analysis for constructing the indicated bond.

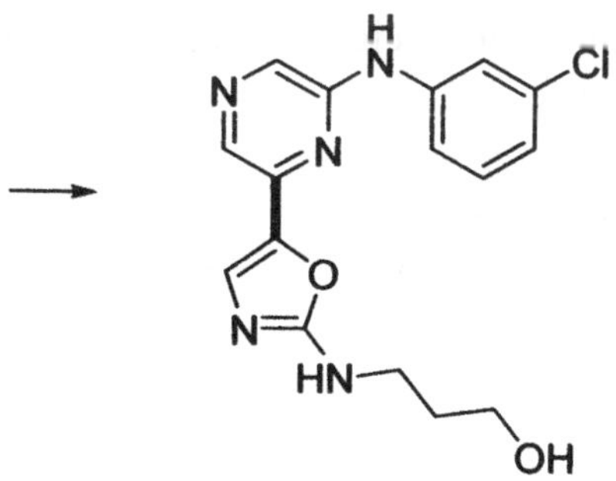

6.7.3 Provide a reasonable, arrow-pushing mechanism that justifies the outcome of the reaction below.[36]

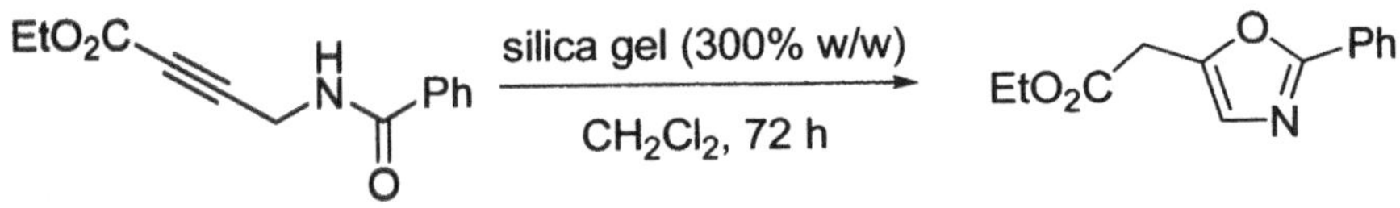

6.7.4 The below synthesis was utilized in efforts toward an oxazole analogue of the natural product tubulysin U.[101] Predict the product of each step.

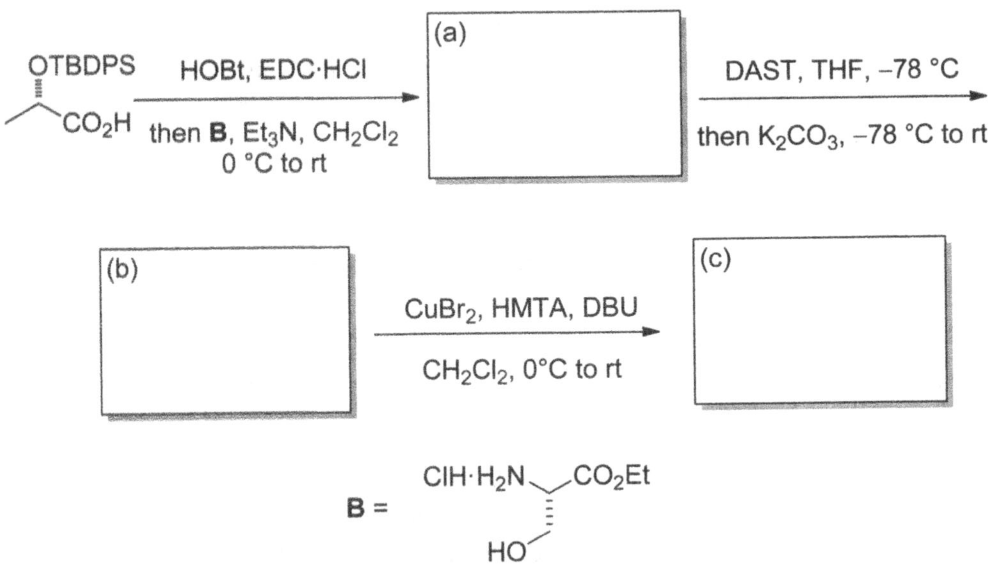

6.7.5 Beginning with 2-phenyloxazole, suggest three separate syntheses of **C**, making use of (a) a heteroaryl Heck reaction, (b) a Negishi coupling, and (c) a Stille coupling reaction.

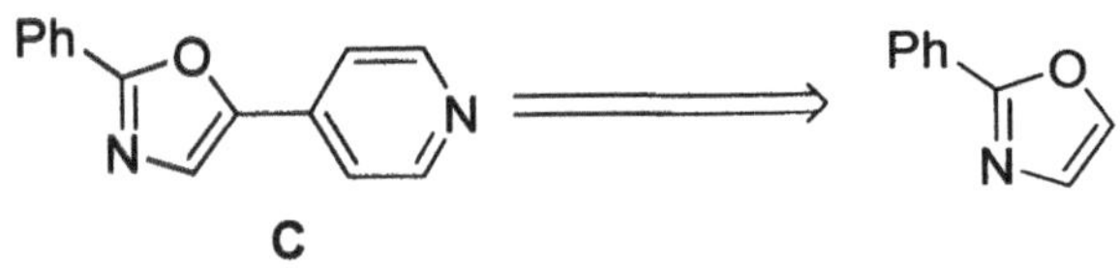

6.7.6 Provide a reasonable, arrow-pushing mechanism that justifies the outcome of the reaction below.[39]

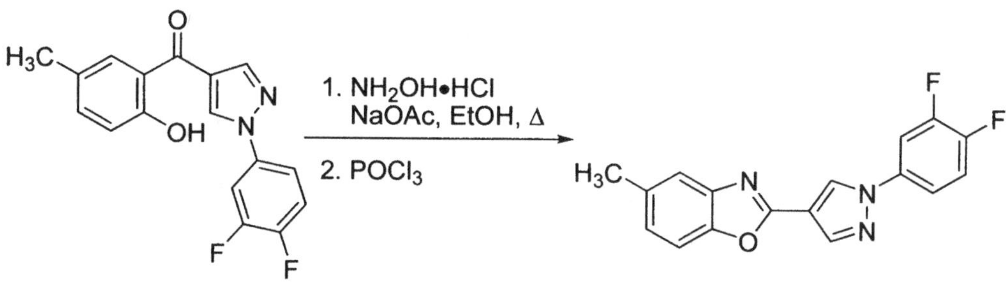

6.7.7 Provide a reasonable, arrow-pushing mechanism that justifies the outcome of the reaction below.[34]

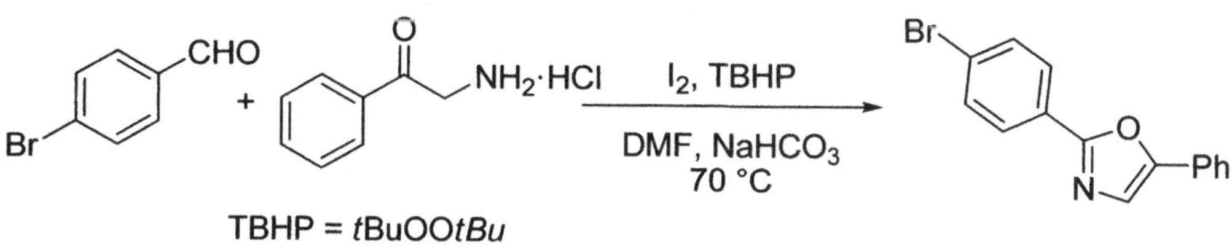

6.7.8 Provide a reasonable, arrow-pushing mechanism that justifies the outcome of the reaction below.[102]

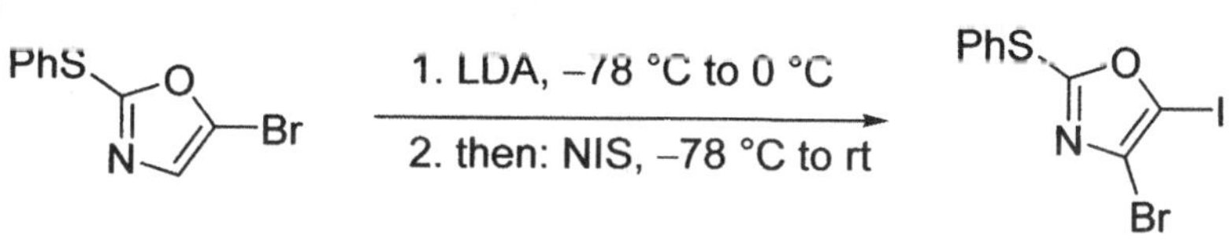

6.7.9 Provide a reasonable, arrow-pushing mechanism that justifies the outcome of the reaction below, which produces an intermediate in the synthesis of pyridoxine.[55]

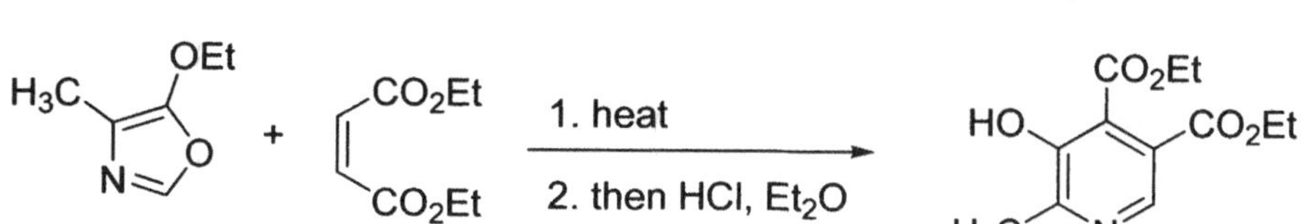

6.7.10 Suggest a synthesis of **E** from **D**.[65]

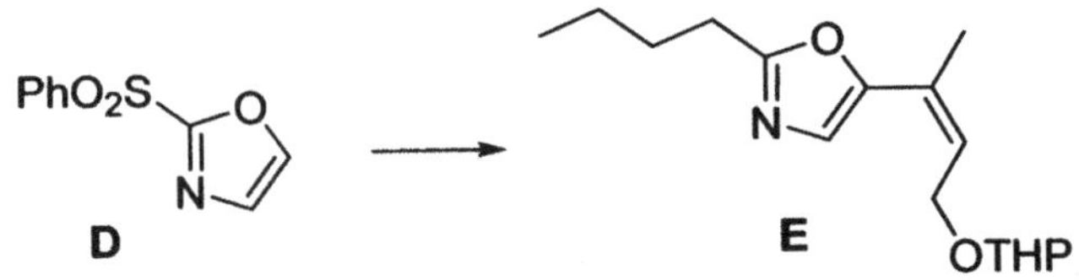

6.7.11 Propose a complete, arrow-pushing mechanism for the conversion of an oxazole to an imidazole upon treatment with ammonia.

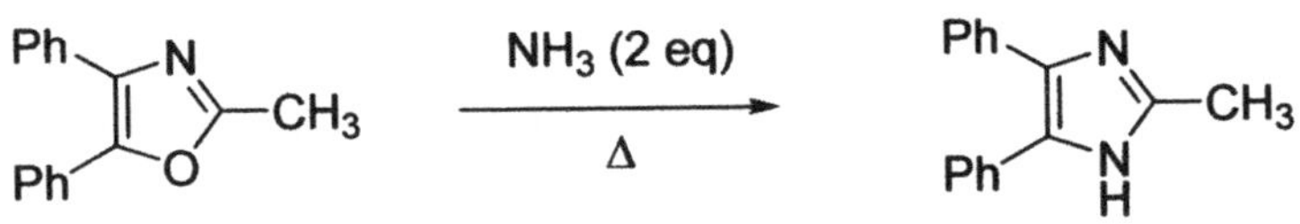

6.7.12 Propose a plausible arrow-pushing mechanism for the two degradation pathways of isoxazoles.

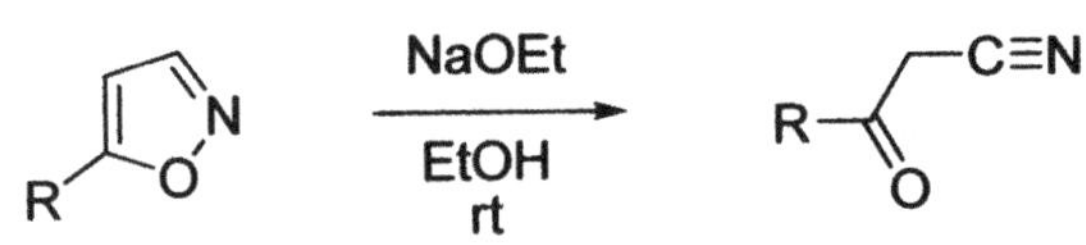

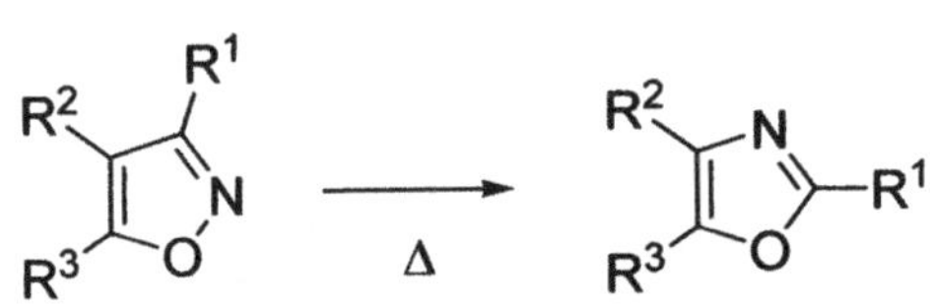

6.7.13 Propose a reasonable mechanism for the following transformations:

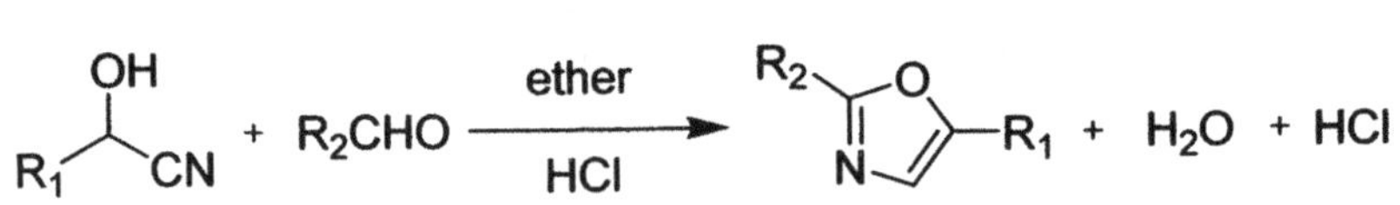

6.7.14 Propose a reasonable mechanism for the following transformations:

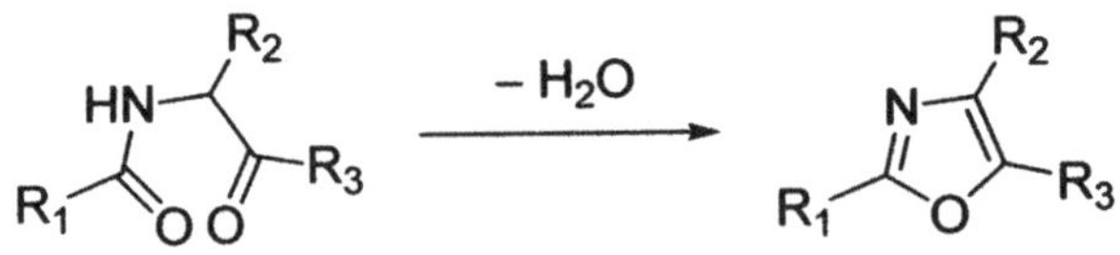

R_1, R_2, R_3 = alkyl, aryl, heteroaryl

6.7.15 Propose a reasonable mechanism for the following transformations:

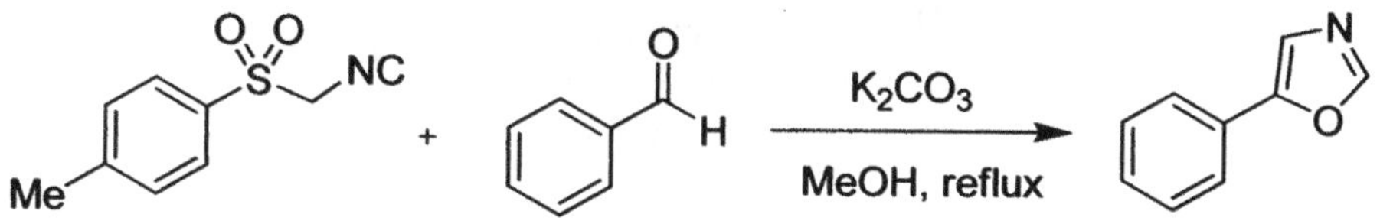

6.7.16 Propose a reasonable mechanism for the following transformations:

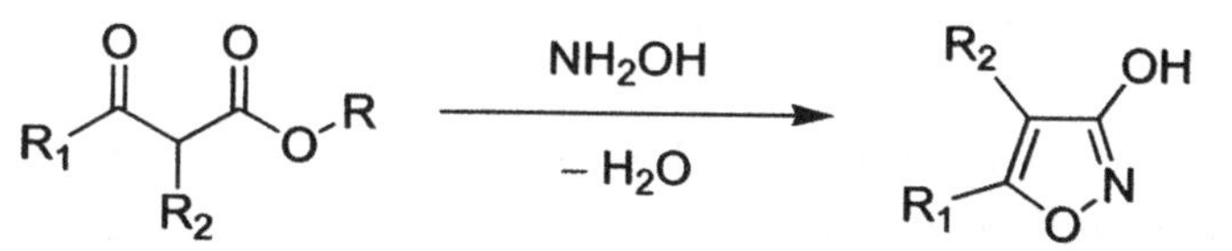

6.7.17 Propose a reasonable mechanism for the reactions of trifluoromethyl-acrylonitrile with 5-ethoxyoxazoles to assemble trifluoromethylated pyridine derivatives.[103]

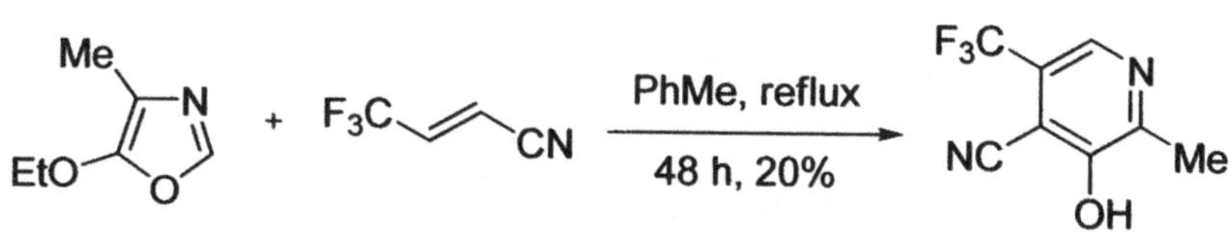

6.7.18 Propose a reasonable mechanism for the following transformations[103]:

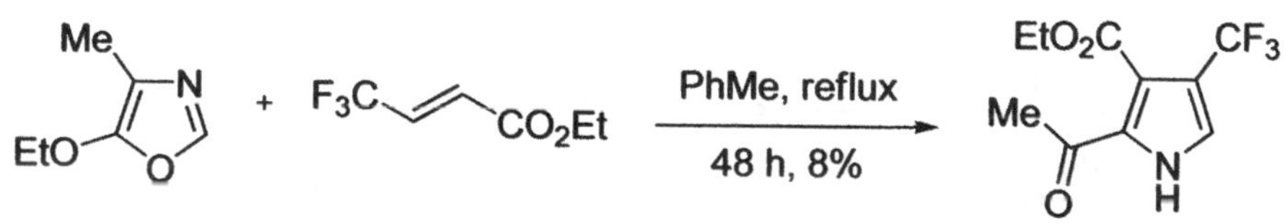

6.7.19 Propose a reasonable mechanism for the following transformation[104]:

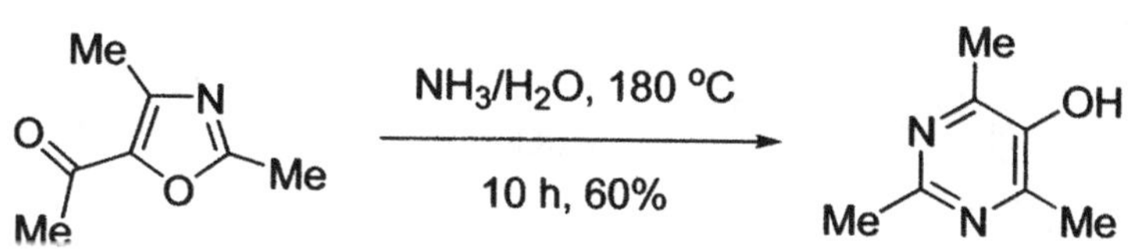

6.7.20 Predict the structure of product **F** from of the following reaction.[105]

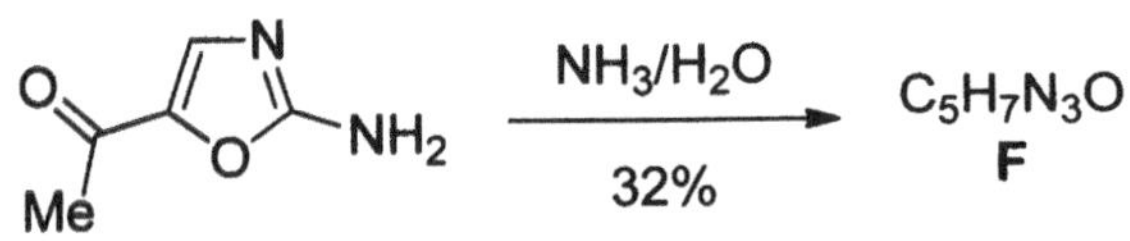

6.7.21 Predict the structure of product **G** from of the following reaction.[106]

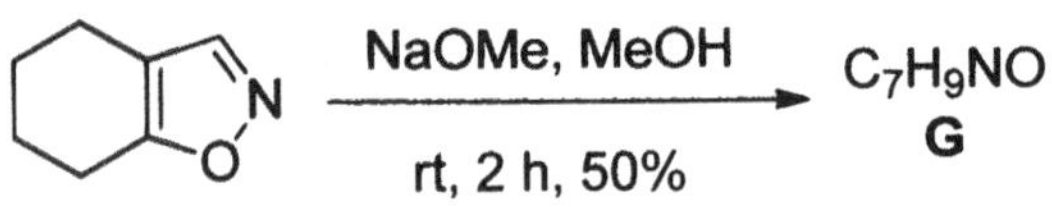

6.7.22 What was reagent **H** for the following transformation? State the identity of intermediate **I**. Propose a mechanism for the transformation of **I** to the oxazole-containing product.[107]

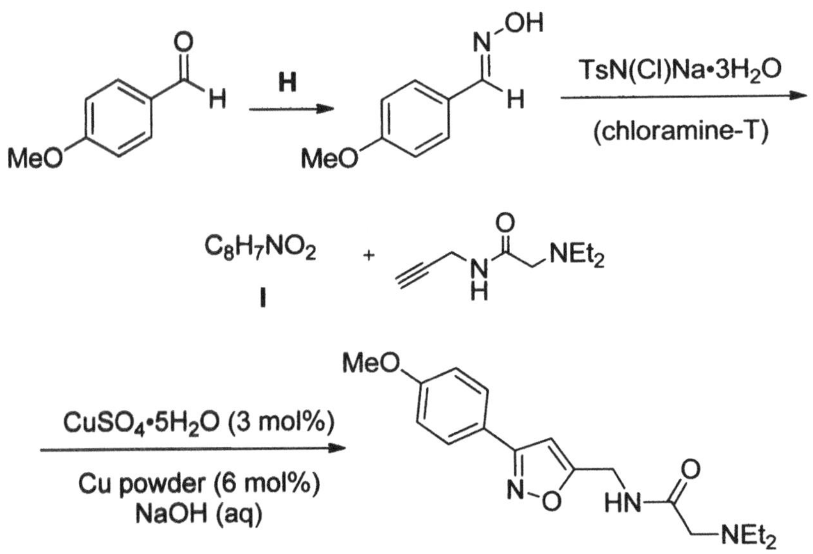

6.7.23 Suggest a mechanism for the following isoxazole synthesis using gold catalysis.[108]

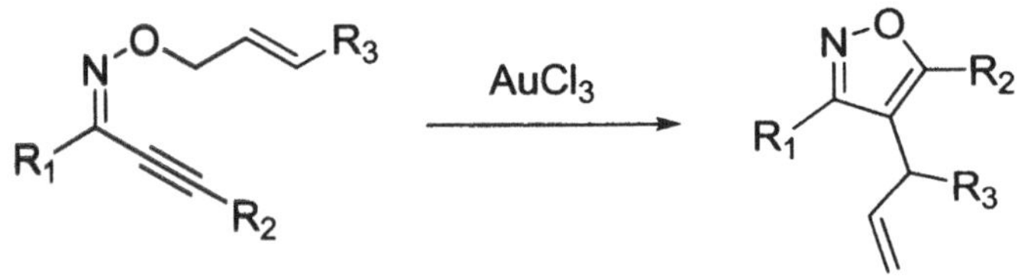

6.7.24 Propose a mechanism for the one-pot synthesis of isoxazole-4-carboxaldehyde through a cascade reaction.[109]

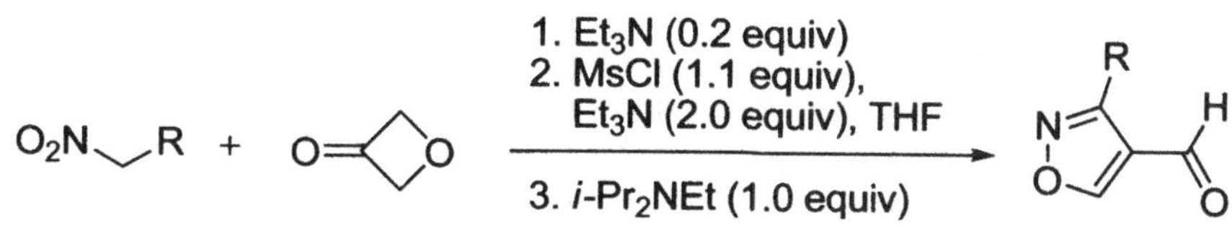

6.7.25 Propose the rearrangement mechanism from isoxazole to oxazole.[110]

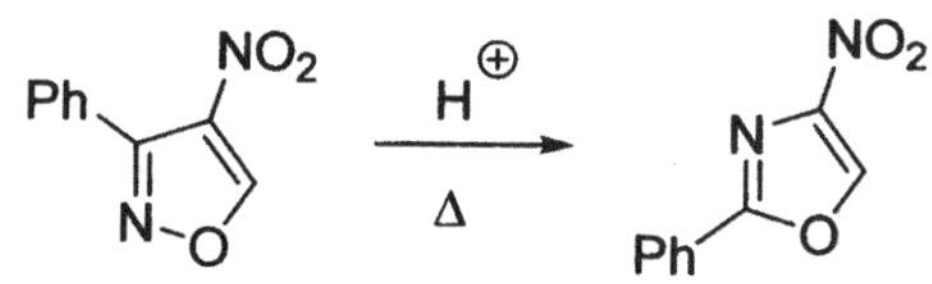

6.7.26 Propose a mechanism for the conversion of an epoxide into an oxazolidinone under basic conditions.[111]

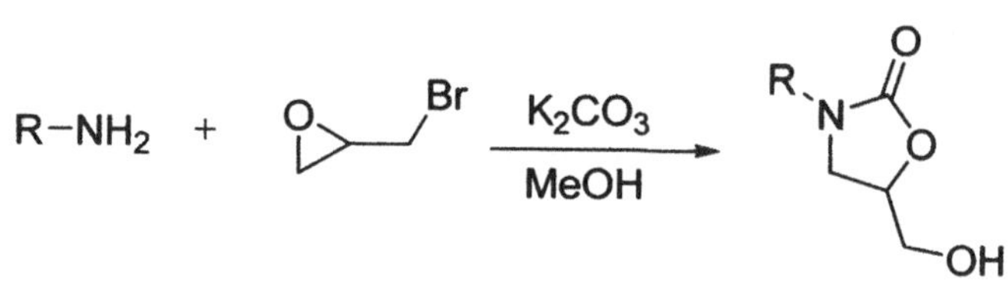

6.7.27 Propose the mechanism for the conversion of an enantiomerically pure aziridine-2-methanol to an oxazolidinone.[112]

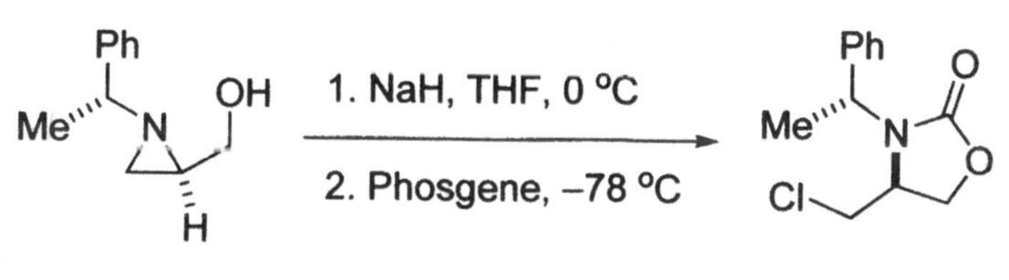

6.7.28 Propose a mechanism for the one-pot synthesis of oxazolidinones via allylation of α-dicarbonyl compounds.[113]

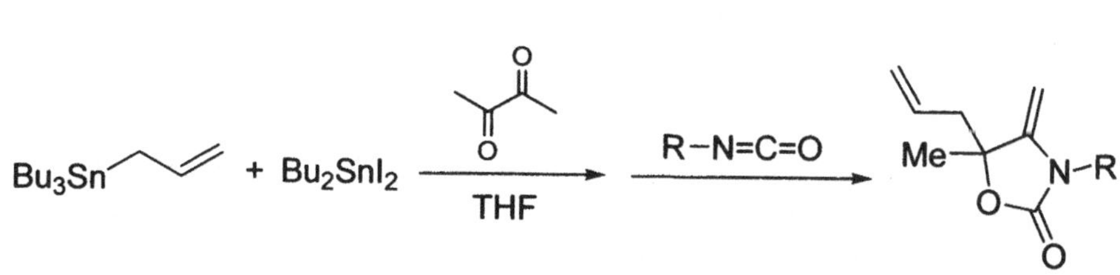

6.8. *References*

1. Bansal, R. K. *Heterocyclic Chemistry*; 3rd ed., New Age International: New Delhi, India, 1999; Chapter 10.
2. Shen, K.; Fu, Y.; Li, J.-N.; Liu, L.; Guo, Q.-X. *Tetrahedron* **2007,** *63*, 1568–1576.
3. Bordwell, F. G. *Acc. Chem. Res.* **1988,** *21*, 456–463.
4. Ichiba, T.; Yoshida, W. Y.; Scheuer, P. J.; Higa, T.; Gravalos, D. G. *J. Am. Chem. Soc.* **1991,** *113*, 3173–3174.
5. D'Ambrosio, M.; Guerriero, A.; Pietra, F.; Debitus, C. *Helv. Chim. Acta* **1996,** *79*, 51–60.
6. Rodríguez, A. D.; Ramírez, C.; Rodríguez, I. I.; González, E. *Org. Lett.* **1999,** *1*, 527–530.
7. Yamamoto, H.; Kondo, M.; Nakamori, S.; Nagano, H.; Wakasa, K.; Sugita, Y.; Chang-de, J.; Kobayashi, S.; Damdinsuren, B.; Dono, K.; Umeshita, K.; Sekimoto, M.; Sakon, M.; Matsuura, N.; Monden, M. *Gastroenterology* **2003,** *125*, 556–571.
8. Talley, J. J.; Brown, D. L.; Carter, J. S.; Graneto, M. J.; Koboldt, C. M.; Masferrer, J. L.; Perkins, W. E.; Rogers, R. S.; Shaffer, A. F.; Zhang, Y. Y.; Zweifel, B. S.; Seibert, K. *J. Med. Chem.* **2000,** *43*, 775–777.
9. Talley, J. J.; Bertenshaw, S. R.; Brown, D. L.; Carter, J. S.; Graneto, M. J.; Kellogg, M. S.; Koboldt, C. M.; Yuan, J.; Zhang, Y. Y.; Seibert, K. *J. Med. Chem.* **2000,** *43*, 1661–1663.
10. Robinson, R. *J. Chem. Soc., Trans.* **1909,** *95*, 2167–2174.
11. Wasserman, H. H.; Vinick, F. J. *J. Org. Chem.* **1973,** *38*, 2407–2408.
12. Wiegand, E. E.; Rathburn, D. W. *Synthesis* **1970,** 648–649.
13. Kerr, V. N.; Hayes, F. N.; Ott, D. G.; Lier, R.; Hansbury, E. *J. Org. Chem.* **1959,** *24*, 1864–1866.
14. Wipf, P.; Miller, C. P. *J. Org. Chem.* **1993,** *58*, 3604–3606.
15. Brain, C. T.; Paul, J. M. *Synlett* **1999,** 1642–1644.
16. Pulici, M.; Quartieri, F.; Felder, E. R. *J. Comb. Chem.* **2005,** *7*, 463–473.
17. Nicolaou, K. C.; Rao, P. B.; Hao, J.; Reddy, M. V.; Rassias, G.; Huang, X.; Chen, D. Y.-K.; Snyder, S. A. *Angew. Chem. Int. Ed.* **2003,** *42*, 1753–1758.
18. Godfrey, A. G.; Brooks, D. A.; Hay, L. A.; Peters, M.; McCarthy, J. R.; Mitchell, D. *J. Org. Chem.* **2003,** *68*, 2623–2632.
19. Brown, K. Oxazoles. U.S.3,578,671 (1971).
20. White, R. L.; Wessels, F. L.; Schwan, T. J.; Ellis, K. O. *J. Med. Chem.* **1987,** *30*, 263–266.
21. Davidson, D.; Weiss, M.; Jelling, M. *J. Org. Chem.* **1937,** *2*, 319–327.
22. Haruta, J.; Hashimoto, H.; Matsushita, M. Heterocyclic Aromatic Oxazole Compounds and Use Thereof. U.S. 5,994,381 (1999).
23. Phillips, A. J.; Uto, Y.; Wipf, P.; Reno, M. J.; Williams, D. R. *Org. Lett.* **2000,** *2*, 1165–1168.
24. Burrell, G.; Evans, J. M.; Jones, G. E.; Stemp, G. *Tetrahedron Lett.* **1990,** *31*, 3649–3652.
25. Lellouche, J.-P.; Lafargue, P.; Guenot, P. *Heterocycles* **1995,** *41*, 947–958.
26. Onaka, T. *Tetrahedron Lett.* **1971,** *12*, 4393–4394.
27. Biftu, T.; Feng, D.; Ponpipom, M.; Girotra, N.; Liang, G.-B.; Qian, X.; Bugianesi, R.; Simeone, J.; Chang, L.; Gurnett, A.; Liberator, P.; Dulski, P.; Leavitt, P. S.; Crumley, T.; Misura, A.; Murphy, T.; Rattray, S.; Samaras, S.; Tamas, T.; Mathew, J.; Brown, C.; Thompson, D.; Schmatz, D.; Fisher, M.; Wyvratt, M. *Bioorg. Med. Chem. Lett.* **2005,** *15*, 3296–3301.
28. Schöllkopf, U.; Schröder, R. *Angew. Chem. Int. Ed.* **1971,** *10*, 333.

29. Brescia, M.-R.; Rokosz, L. L.; Cole, A. G.; Stauffer, T. M.; Lehrach, J. M.; Auld, D. S.; Henderson, I.; Webb, M. L. *Bioorg. Med. Chem. Lett.* **2007,** *17*, 1211–1215.
30. van Leusen, A. M.; Hoogenboom, B. E.; Siderius, H. *Tetrahedron Lett.* **1972,** *13*, 2369–2372.
31. van Leusen, A. M.; Possel, O. *Heterocycles* **1977,** *7*, 77–80.
32. Herr, R. J.; Fairfax, D. J.; Meckler, H.; Wilson, J. D. *Org. Process Res. Dev.* **2002,** *6*, 677–681.
33. Graham, T. H. *Org. Lett.* **2010,** *12*, 3614–3617.
34. Wan, C.; Gao, L.; Wang, Q.; Zhang, J.; Wang, Z. *Org. Lett.* **2010,** *12*, 3902–3905.
35. Andersson, I. E.; Batsalova, T.; Dzhambazov, B.; Edvinsson, L.; Holmdahl, R.; Kihlberg, J.; Linusson, A. *Org. Biomol. Chem.* **2010,** *8*, 2931–2940.
36. Wipf, P.; Aoyama, Y.; Benedum, T. E. *Org. Lett.* **2004,** *6*, 3593–3595.
37. Thomas, B.; George, J.; Sugunan, S. *Ind. Eng. Chem. Res.* **2008,** *48*, 660–670.
38. Sardarian, A.; Shahsavari-Fard, Z. *Synlett* **2008,** 1391–1393.
39. Gadakh, A. V.; Pandit, C.; Rindhe, S. S.; Karale, B. K. *Bioorg. Med. Chem. Lett.* **2010,** *20*, 5572–5576.
40. Wang, F.; Hauske, J. R. *Tetrahedron Lett.* **1997,** *38*, 6529–6532.
41. Katritzky, A. R.; Barczynski, P.; Ostercamp, D. L.; Yousaf, T. I. *J. Org. Chem.* **1986,** *51*, 4037–4042.
42. Sørensen, U. S.; Falch, E.; Krogsgaard-Larsen, P. *J. Org. Chem.* **2000,** *65*, 1003–1007.
43. Wiley, R. H. *Chem. Rev.* **1945,** *37*, 401–442.
44. Hodgetts, K. J.; Kershaw, M. T. *Org. Lett.* **2003,** *5*, 2911–2914.
45. Gibson, C.; Schnatbaum, K.; Pfeifer, J. R.; Locardi, E.; Paschke, M.; Reimer, U.; Richter, U.; Scharn, D.; Faussner, A.; Tradler, T. *J. Med. Chem.* **2009,** *52*, 4370–4379.
46. Hodges, J. C.; Patt, W. C.; Connolly, C. J. *J. Org. Chem.* **1991,** *56*, 449–452.
47. Crowe, E.; Hossner, F.; Hughes, M. J. *Tetrahedron* **1995,** *51*, 8889–8900.
48. Vedejs, E.; Luchetta, L. M. *J. Org. Chem.* **1998,** *64*, 1011–1014.
49. Vedejs, E.; Monahan, S. D. *J. Org. Chem.* **1996,** *61*, 5192–5193.
50. Ezzili, C.; Mileni, M.; McGlinchey, N.; Long, J. Z.; Kinsey, S. G.; Hochstatter, D. G.; Stevens, R. C.; Lichtman, A. H.; Cravatt, B. F.; Bilsky, E. J.; Boger, D. L. *J. Med. Chem.* **2011,** *54*, 2805–2822.
51. Miller, R. A.; Smith, R. M.; Marcune, B. *J. Org. Chem.* **2005,** *70*, 9074–9076.
52. Williams, D.; Fu, L. *Synlett* **2010,** 591–594.
53. Turchi, I. J.; Dewar, M. J. S. *Chem. Rev.* **1975,** *75*, 389–437.
54. Karpeiskii, M. Y.; Florent'ev, V. L. *Russ. Chem. Rev.* **1969,** *38*, 540–546.
55. Harris, E. E.; Firestone, R. A.; Pfister, K. III; Boettcher, R. R.; Cross, F. J.; Currie, R. B.; Monaco, M.; Peterson, E. R.; Reuter, W. *J. Org. Chem.* **1962,** *27*, 2705–2706.
56. Grigg, R.; Jackson, J. L. *J. Chem. Soc. C* **1970,** 552–556.
57. Inoue, M. *Mini Rev. Org. Chem.* **2008,** *5*, 77–84.
58. Popov, I.; Do, H.-Q.; Daugulis, O. *J. Org. Chem.* **2009,** *74*, 8309–8313.
59. Hodgetts, K. J.; Kershaw, M. T. *Org. Lett.* **2002,** *4*, 2905–2907.
60. Atkins, J. M.; Vedejs, E. *Org. Lett.* **2005,** *7*, 3351–3354.
61. Flegeau, E. F.; Popkin, M. E.; Greaney, M. F. *Org. Lett.* **2006,** *8*, 2495–2498.
62. Li, B.; Buzon, R. A.; Zhang, Z. *Org. Process Res. Dev.* **2007,** *11*, 951–955.
63. Nicolaou, K. C.; Snyder, S. A.; Huang, X.; Simonsen, K. B.; Koumbis, A. E.; Bigot, A. *J. Am. Chem. Soc.* **2004,** *126*, 10162–10173.
64. Schaus, J. V.; Panek, J. S. *Org. Lett.* **2000,** *2*, 469–471.
65. Williams, D. R.; Fu, L. *Org. Lett.* **2010,** *12*, 808–811.

66. Kelly, T. R.; Lang, F. *J. Org. Chem.* **1996,** *61*, 4623–4633.
67. Dondoni, A.; Fantin, G.; Fogagnolo, M.; Medici, A.; Pedrini, P. *Synthesis* **1987,** 693–696.
68. Krebs, O.; Taylor, R. J. K. *Org. Lett.* **2005,** *7*, 1063–1066.
69. Flegeau, E. F.; Popkin, M. E.; Greaney, M. F. *J. Org. Chem.* **2008,** *73*, 3303–3306.
70. Jutzi, P.; Gilge, U. *J. Organomet. Chem.* **1983,** *246*, 159–162.
71. Clapham, B.; Sutherland, A. J. *J. Org. Chem.* **2001,** *66*, 9033–9037.
72. Young, G. L.; Smith, S. A.; Taylor, R. J. K. *Tetrahedron Lett.* **2004,** *45*, 3797–3801.
73. Counceller, C. M.; Eichman, C. C.; Proust, N.; Stambuli, J. P. *Adv. Synth. Catal.* **2011,** *353*, 79–83.
74. Smith, A. B. III; Minbiole, K. P.; Freeze, S. *Synlett* **2001,** 1739–1742.
75. Jeong, S.; Chen, X.; Harran, P. G. *J. Org. Chem.* **1998,** *63*, 8640–8641.
76. Smith, A. B. III; Verhoest, P. R.; Minbiole, K. P.; Schelhaas, M. *J. Am. Chem. Soc.* **2001,** *123*, 4834–4836.
77. Shibata, K.; Yoshida, M.; Doi, T.; Takahashi, T. *Tetrahedron Lett.* **2010,** *51*, 1674–1677.
78. Araki, H.; Katoh, T.; Inoue, M. *Tetrahedron Lett.* **2007,** *48*, 3713–3717.
79. Araki, H.; Katoh, T.; Inoue, M. *Synlett* **2006,** 555–558.
80. Primas, N.; Bouillon, A.; Lancelot, J.-C.; Rault, S. *Tetrahedron* **2009,** *65*, 6348–6353.
81. Strotman, N. A.; Chobanian, H. R.; He, J.; Guo, Y.; Dormer, P. G.; Jones, C. M.; Steves, J. E. *J. Org. Chem.* **2010,** *75*, 1733–1739.
82. Liebeskind, L. S.; Srogl, J. *Org. Lett.* **2002,** *4*, 979–981.
83. Reeder, M. R.; Gleaves, H. E.; Hoover, S. A.; Imbordino, R. J.; Pangborn, J. J. *Org. Process Res. Dev.* **2003,** *7*, 696–699.
84. Anderson, B. A.; Harn, N. K. *Synthesis* **1996,** 583–585.
85. Krasovskiy, A.; Malakhov, V.; Gavryushin, A.; Knochel, P. *Angew. Chem. Int. Ed.* **2006,** *45*, 6040–6044.
86. Skepper, C. K.; Quach, T.; Molinski, T. F. *J. Am. Chem. Soc.* **2010,** *132*, 10286–10292.
87. Bobeck, D. R.; Warner, D. L.; Vedejs, E. *J. Org. Chem.* **2007,** *72*, 8506–8518.
88. Sakamoto, T.; Nagata, H.; Kondo, Y.; Shiraiwa, M.; Yamanaka, H. *Chem. Pharm. Bull.* **1987,** *35*, 823–828.
89. Langille, N. F.; Dakin, L. A.; Panek, J. S. *Org. Lett.* **2002,** *4*, 2485–2488.
90. Yin, L.; Liebscher, J.; Erdmann, F. *J. Heterocycl. Chem.* **2005,** *42*, 1369–1379.
91. Mérour, J.-Y.; Grumel, V.; Guillaumet, G. *Heterocyles* **2001,** *55*, 1329–1345.
92. Ohta, A.; Aoyagi, Y.; Inoue, A.; Koizumi, I.; Hashimoto, R.; Tokunaga, K.; Gohma, K.; Komatsu, J.; Sekine, K.; Miyafuji, A.; Kunoh, J.; Honma, R.; Akita, Y. *Heterocycles* **1992,** *33*, 257–272.
93. Strotman, N. A.; Chobanian, H. R.; Guo, Y.; He, J.; Wilson, J. E. *Org. Lett.* **2010,** *12*, 3578–3581.
94. Kuo, G.-H.; Wang, A.; Emanuel, S.; DeAngelis, A.; Zhang, R.; Connolly, P. J.; Murray, W. V.; Gruninger, R. H.; Sechler, J.; Fuentes-Pesquera, A.; Johnson, D.; Middleton, S. A.; Jolliffe, L.; Chen, X. *J. Med. Chem.* **2005,** *48*, 1886–1900.
95. Eastwood, P.; Gonzalez, J.; Paredes, S.; Nueda, A.; Domenech, T.; Alberti, J.; Vidal, B. *Bioorg. Med. Chem. Lett.* **2010,** *20*, 1697–1700.
96. Nesi, R.; Giomi, D.; Papaleo, S.; Corti, M. *J. Org. Chem.* **1990,** *55*, 1227–1230.
97. Perez, J. D.; De Diaz, R. G.; Yranzo, G. I. *J. Org. Chem.* **1981,** *46*, 3505–3508.
98. Cornforth, J. W. In *The Chemistry of Penicillin*; Clarke, H. T.; Johnson, J. R.; Robinson, R., Eds., Princeton University Press: Princeton, N. J., 1949, 700.

99. Dewar, M. J. S.; Turchi, I. J. *J. Org. Chem.* **1975,** *40,* 1521–1523.
100. Dewar, M. J. S. *J. Am. Chem. Soc.* **1974,** *96,* 6148–6152.
101. Sreejith, S. P.; Sani, M.; Saunders, F. R.; Wallace, H. M.; Zanda, M. *Synlett* **2011,** 1673–1676.
102. Williams, D. R.; Fu, L. *Synlett* **2010,** 591–594.
103 Sandford, G.; Wilson, I.; Timperley, C. M. *J. Fluorine Chem.* **2004,** *125,* 1425–1430.
104 Valgimigli, L.; Brigati, Gi.; Pedulli, G. F.; DiLabio, G. A.; Mastragostino, M.; Arbizzani, C.; Pratt, D. A. *Chem. Eur. J.* **2003,** *9,* 4997–5010.
105 LaMattina, J. L.; Mularski, C. J. *Tetrahedron Lett.* **1984,** *25,* 2957–2960.
106 Ferris, J. P.; Trimmer, R. W. *J. Org. Chem.* **1976,** *141,* 13–19.
107 Yermolina, M.V.; Wang, J.; Caffrey, M.; Rong, L. L.; Wardrop, D. J. *J. Med. Chem.* **2011,** *54,* 765–781.
108 Ueda, M.; Sato, A.; Ikeda, Y.; Miyoshi, T.; Naito, T.; Miyata, O. *Org. Lett.* **2010,** *12,* 2594–2597.
109 Burkhard, J. A.; Tchitchanov, B. H.; Carreira, E. M. *Angew. Chem., Int. Eng.* **2011,** *50,* 5379–5382.
110 Nesi, R.; Turchi, S.; Giomi, D. *J. Org. Chem.* **1996,** *61,* 7933–7936.
111 Osa, Y.; Hikima, Y.; Sato, Y.; Takino, K.; Ida, Y.; Hirono, S.; Nagase, H. *J. Org. Chem.* **2005,** *70,* 5737–5740.
112 Park, C. S.; Ki, M. S.; Sim, T. B.; Pyun, D. K.; Cheol, H. L.; Choi, D.; Lee, W. K., *J. Org. Chem.* **2003,** *68,* 43–49.
113 Shibata, I.; Kojima, R.; Tsunoi, S.; Nozaki, T.; Watanabe, T.; Ninomiya, A.; Yasuda, M.; Baba, A., *Org. Biomol. Chem.,* **2010,** *8,* 2009–2011.

Chapter 7 Thiazoles and Benzothiazoles

Narendra B. Ambhaikar

7.1 Introduction

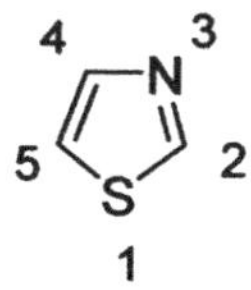

thiazole

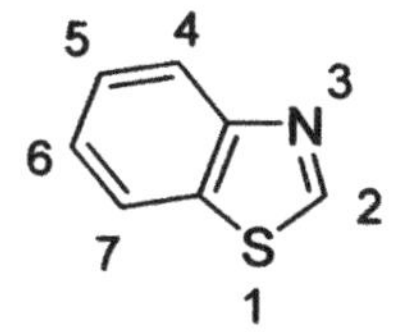

benzothiazole

1,3-Thiazole is the parent compound in this interesting class of heterocycles. It is a clear, colorless liquid with an odor resembling pyridine, a boiling point in the range of 116–118 °C, and a specific gravity of 1.2. Benzothiazole is a colorless liquid with a boiling point range of 227–228 °C. The thiazole ring is planar aromatic and characterized by more π-electron delocalization than the analogous oxazoles due to the empty *d*-orbitals in sulfur.[1] The aromaticity is brought about by delocalization of a lone pair of electrons from the sulfur atom satisfying the requirement of 6π electrons dictated by Hückel's rule. Bond lengths in thiazole have been determined with the aid of microwave spectroscopy and have been shown below.[1] The thiazole structure is very close to an average of the structures of thiophene and 1,3,4-thiadiazole.

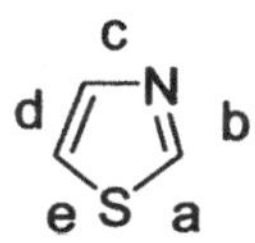

a = 1.713 Å
b = 1.304 Å
c = 1.372 Å
d = 1.367 Å
e = 1.713 Å

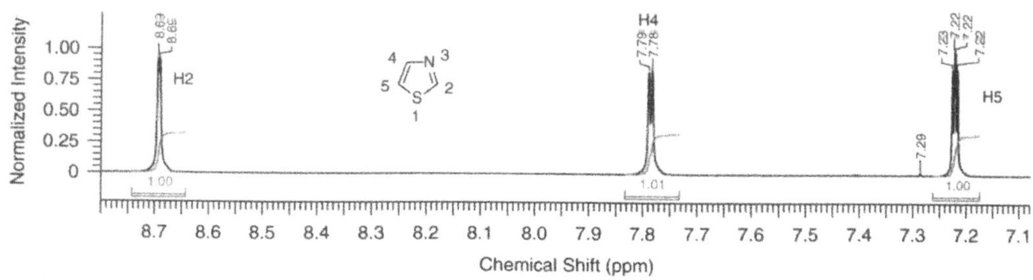

The ^{1}H-NMR spectrum of thiazole shown below is simple and first order. It shows the following chemical shifts: H-2 8.69 ppm, H-4 7.79 ppm, and H-5 7.22 ppm. It shows down-field values that are distinctive of aromatic hydrogen atoms. Deshielding of C2 is a result of the combined electron-withdrawing effects of the adjoining nitrogen and sulfur. Such a substantial deshielding leads to a further down-field shift for H-2 compared to the other hydrogens, and it can be correlated with the occurrence of a ring current associated with the aromatic sextet of thazole and the magnetic anisotropy of both the sulfur and nitrogen heteroatoms.

^{13}C NMR spectral data reveal three peaks: C-2 152.58 ppm, C-4 143.22 ppm, and C-5 118.52 ppm, all values reflecting aromatic carbons.

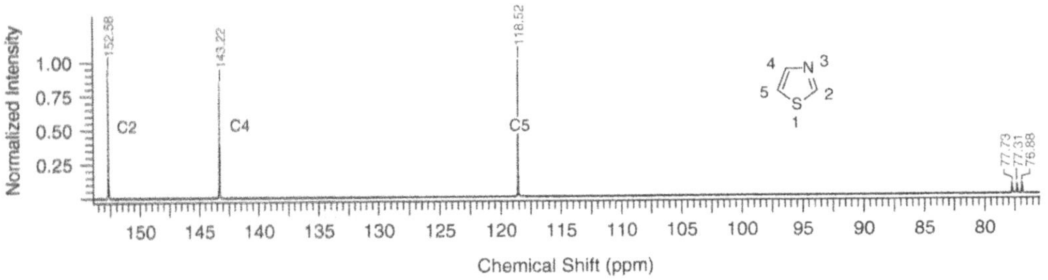

Benzothiazoles consist of a 5-membered, 1,3-thiazole ring fused to a benzene ring. The ^{1}H NMR of unsubstituted benzothiazole is shown below. As expected, all peaks fall in the aromatic region and appear more downfield than usual. It is noteworthy that H2 especially appears at 9.03 ppm as against ~ 7.46 shown by H2 in unsubstituted benzoxazole, clearly indicating the deshielding impact of the presence of sulfur instead of oxygen in the ring.

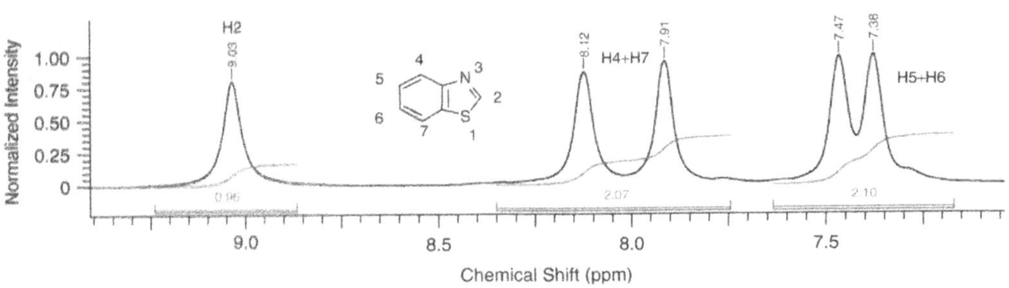

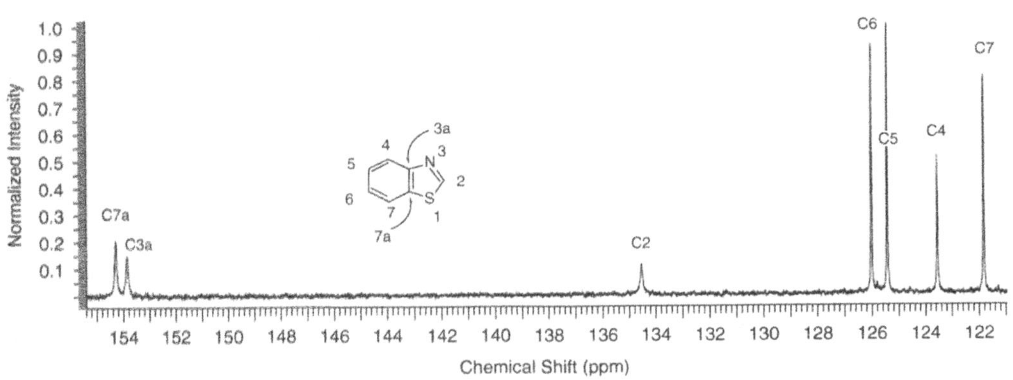

^{13}C NMR data of unsubstituted benzothiazole show the C3 and C7 peaks being deshielded. The C2 peak appears up-field compared to that in 1,3-thiazole.

The physical and spectral properties of 1,3-thiazole are similar to those of pyridine in some ways. The electronegative *N*-atom at the 3-position makes C(2) partially electropositive and consequently susceptible to nucleophilic attack.

Thiazole is known to occur naturally in various forms. The most widely recognized thiazole lies in the essential vitamin B_1 or thiamin in the form of its thiazolium salt. It is a water-soluble vitamin of the B complex. Thiamin pyrophosphate is the essential coenzyme in the enzymatic decarboxylation of pyruvate to aldehyde.

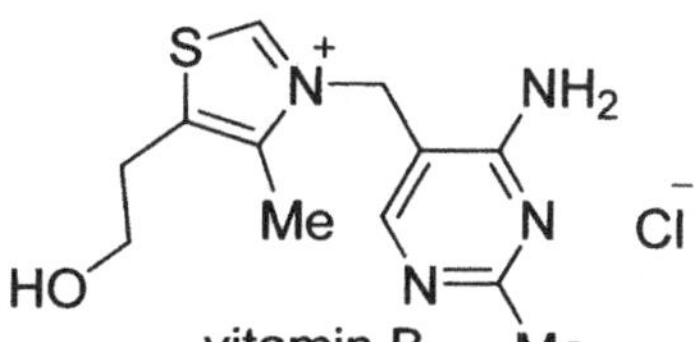

vitamin B_1

Several natural products including penicillins are thiazolidine derivatives. Other natural antibiotics such as althiomycine[2,3] and thiazole antibiotics like micrococcin[4a] include the thiazole heterocycle and they are shown on the next page.[5] Cystothiazole A in particular is a bithiazole metabolite isolated from the myxobacterium *Cystobacter fuscus* and shows activity against a broad range of fungi as well as *in vitro* cytotoxicity toward human colon carcinoma HCT-116 and leukemia K562 cells.[4b] Metabolic products of living organisms such as aeruginoic acid isolated from *Pseudomonas aeruginosa* contains a 1,3-thiazole ring.[6] Firefly luciferin is an interesting light-emitting natural product found in firefly. This compound is actually responsible for bioluminescence and chemiluminescence. It possesses a structure comprising both a benzothiazole and a Δ^2-thiazoline ring.[7]

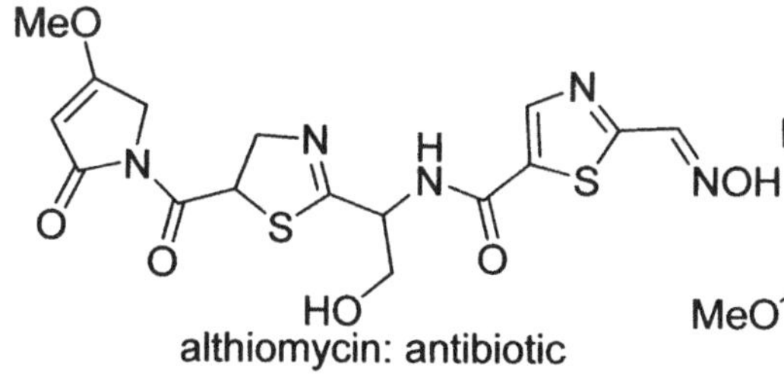

althiomycin: antibiotic

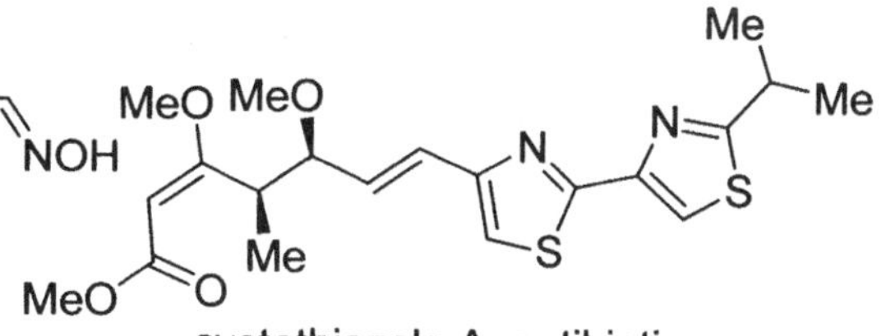

cystothiazole A: antibiotic

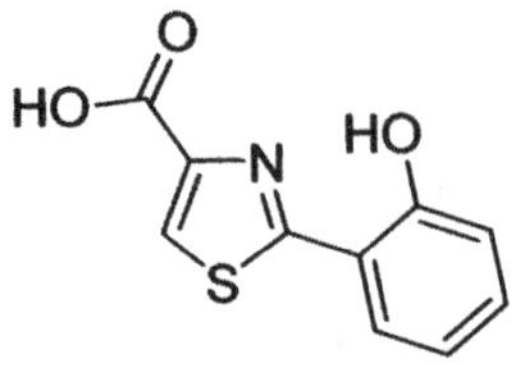

aeruginoic acid

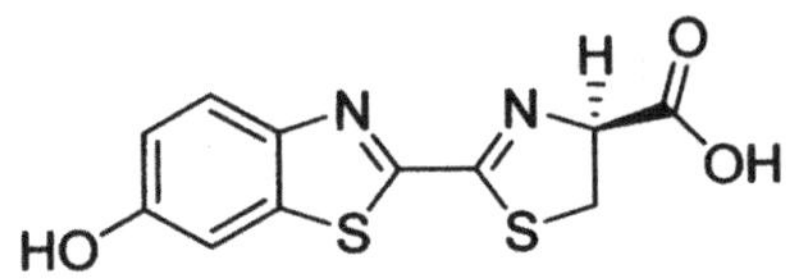

luciferin

Thiazoles have been often discovered as key components of novel and structurally diverse natural products exhibiting a range of biological activities. Their antitumor, antiviral, and antibiotic activities; their ubiquitous occurrence in peptides; and their ability to bind to proteins, DNA and RNA have all led to several synthetic studies and their inclusion in drug discovery. Perhaps the best-known example of such a complex natural product featuring a thiazole substructure is the epothilones, which have demonstrated activity against taxol resistant tumor cell lines.[8]

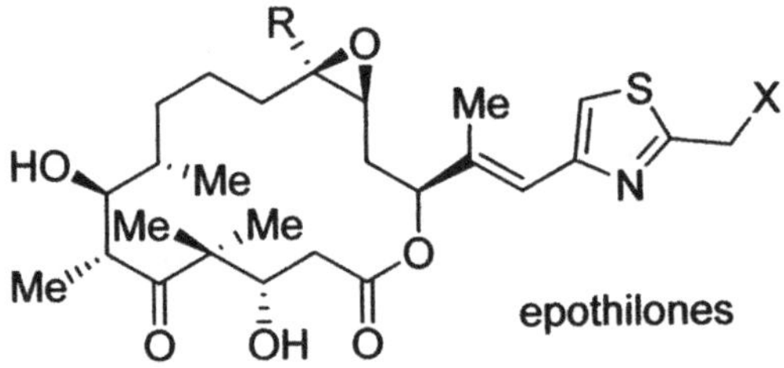

epothilone A: R = H; X = H
epothilone B: R = Me; X = H
epothilone E: R = H; X = OH
epothilone F; R = Me; X = OH

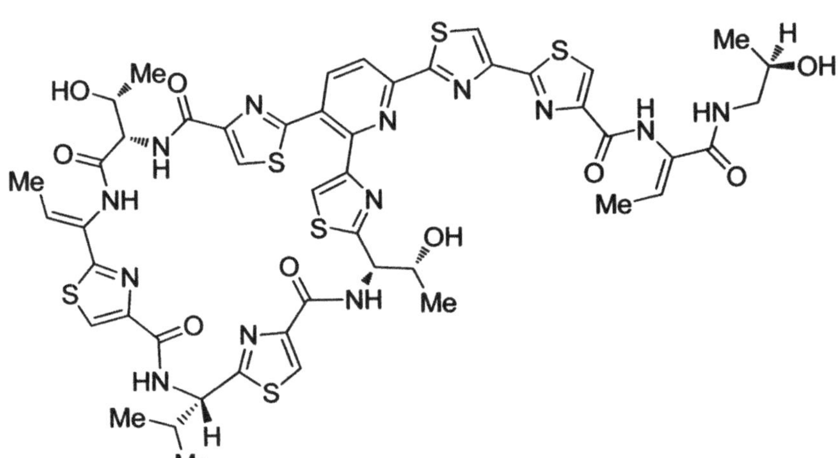

micrococcin P1: a thiopeptide antibiotic

Another example of a well-established class of natural products containing the thiazole heterocycle is the thiopeptide antibiotics. Often referred to as thiazolyl peptides, most of these highly modified, macrocyclic peptides inhibit protein synthesis in bacteria.[9] They display a high inhibition of protein synthesis against Gram-positive bacteria while showing little activity against Gram-negative bacteria. A case in point is micrococcin P_1, shown below. Structural ambiguities with respect to stereochemistry of this compound surrounded micrococcin P_1. Recently Ciufolini and co-workers settled them by confirming the structure.[10] Other thiopeptide antibiotics have also been subjects of landmark total synthetic endeavors.[11]

Thiazoles have made their presence felt in peptide research, an example of which is the pseudopeptide dolastatin 10, a powerful anti-neoplastic agent that has been in the human clinical trials for the treatment of cancer.[12a]

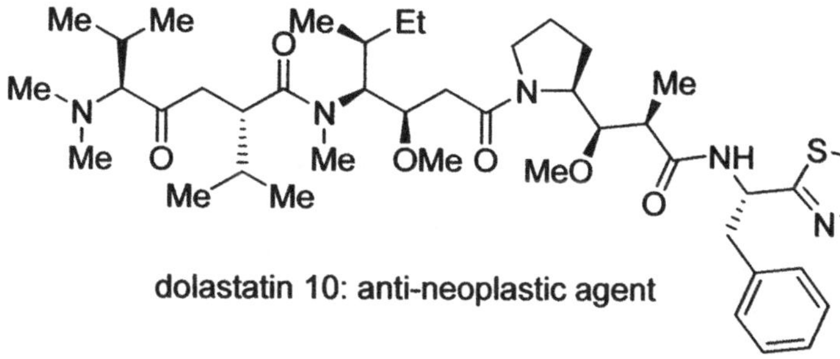

dolastatin 10: anti-neoplastic agent

In the early twentieth century, one of the first commercial synthetic drugs containing thiazole was sulfathiazole, a short-acting sulfa drug.[13]

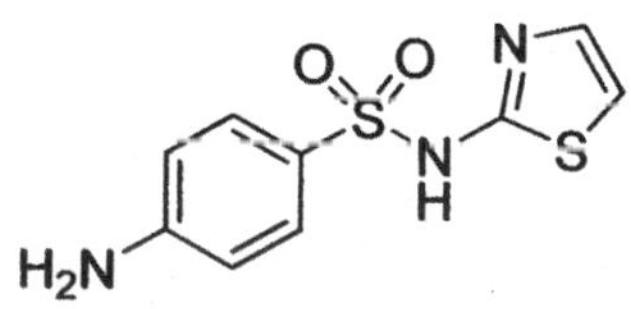

sulfathiazole: anti-microbial agent

Several drugs that are on the market incorporate the thiazole or benzothiazole substructure and a few examples are shown below. For instance, the orally administered, semisynthetic, and broad-spectrum third-generation cephalosporin cefdinir showing antibacterial activity against *Staphylocossus* species was the first thiazole drug to enter the list of Top 200 Drugs.[14] It was approved by the Food and Drug Administration (FDA) in 1997. Cefotaxime and ceftriaxone are also third-generation cephalosporins, showing broad-spectrum activity against Gram-positive and Gram-negative bacteria for the treatment of bacterial meningitis and pneumonia among other infections.[15,16] The thiazole structure here influences the pharmacodynamics and kinetics, resulting in a decrease in bioavailability.[17]

HIV-1 protease inhibitor ritonavir (Norvir), a peptidomimetic anti-viral compound approved in 1996 possesses two different thiazoles in the same molecule.[18a] Although approved as an HIV protease inhibitor, ritonavir has mainly been administered at subtherapeutic doses in combination with other HIV protease inhibitors to "enhance" their plasma concentrations.[18b] Another example of a thiazole drug is pramipexole (Mirapex), a dopamine D_2-agonist comprising a fused bicyclic tetrahydrobenzothiazole.[19] Famotidine (Pepcidine), an H_2-receptor antagonist prescribed for the treatment of peptic ulcer contains a guanidine substituted with thiazole.[20] Febuxostat, a nonpurine selective inhibitor of xanthine oxidase, is another thiazole drug prescribed in the treatment of gout to lessen uric acid levels.[21] Other drugs that include the thiazole heterocycle are bleomycin,[12c] a glycopeptides antibiotic also used for the treatment of cancer chemotherapy, and abafungin, a broad-spectrum anti-fungal agent.[12d]

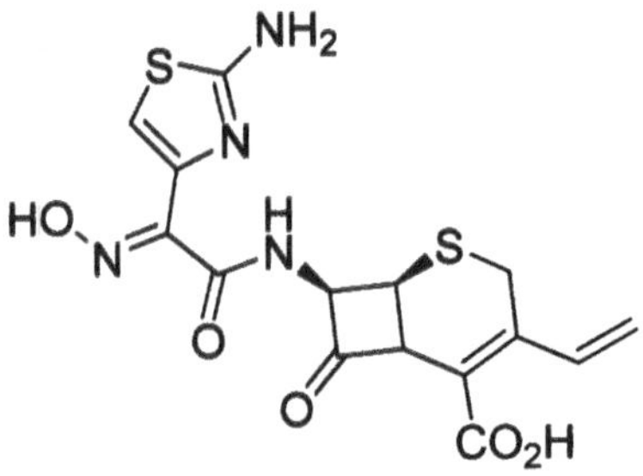

cefdinir: antibiotic

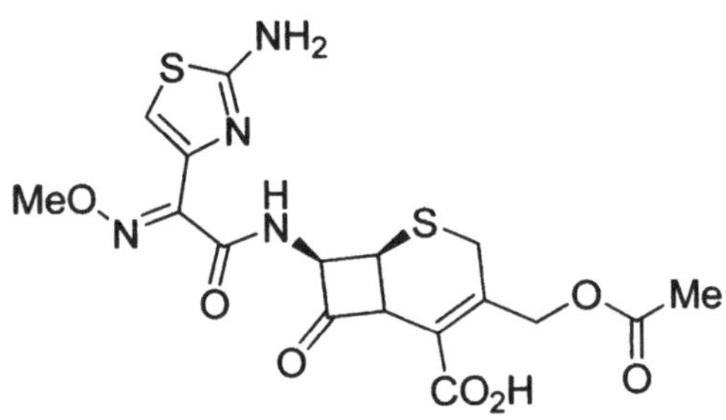

cefotaxime:antibiotic

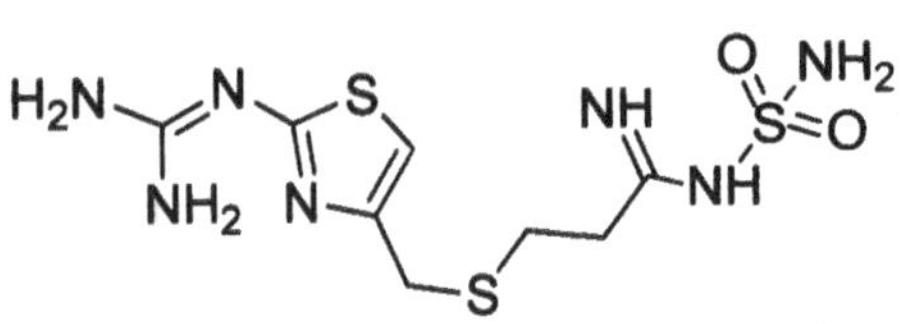

famotidine: H_2 receptor antagonist

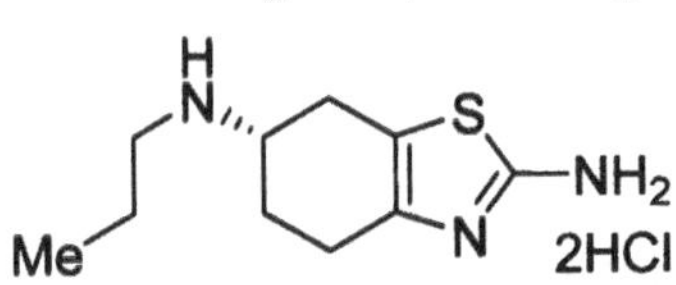

pramipexole: dopamine D_2 agonist

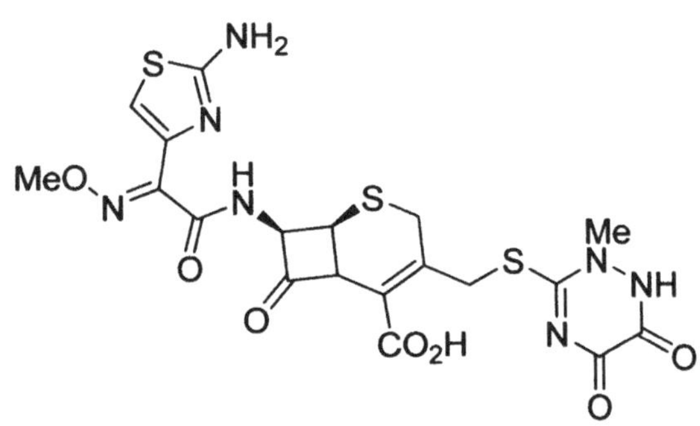

ceftriaxone:antibiotic

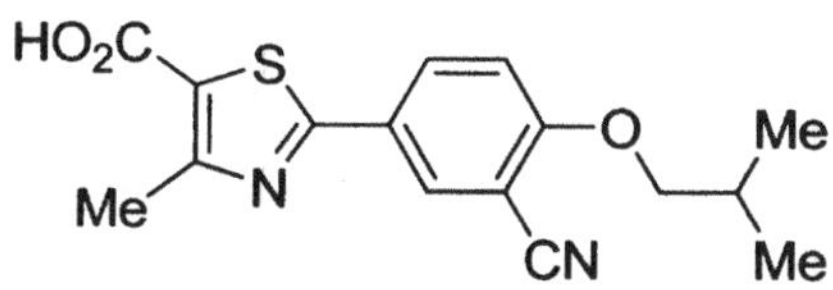

febuxostat: Xanthine oxidase inhibitor (gout)

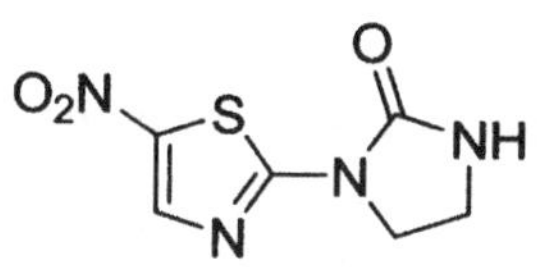

niridazole: schistosomicide

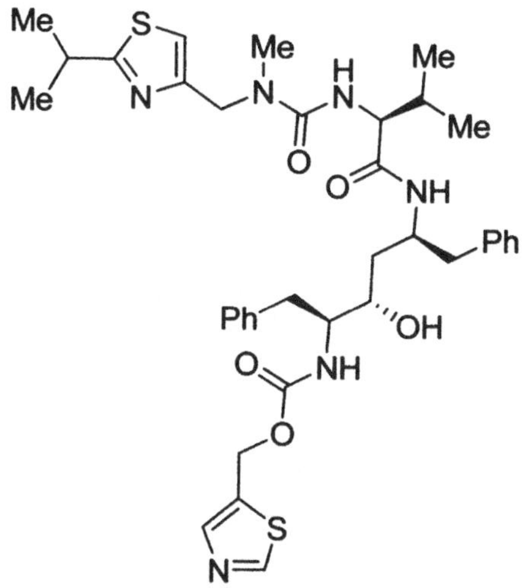

ritonavir: HIV-1 protease inhibitor

The benzothiazole heterocycle is also found in a variety of pharmaceutical agents. For example, anthelmintic activity is exhibited by tioxidazole[22] and frentizole, both possessing benzothiazole.[23] Fostedil is a calcium channel blocker that shows vasodilator activity.[24] Riluzole is another benzothiazole-containing molecule that has found use in the treatment of amyotrophic lateral sclerosis, where it delays progression of the disease thereby prolonging life.[25]

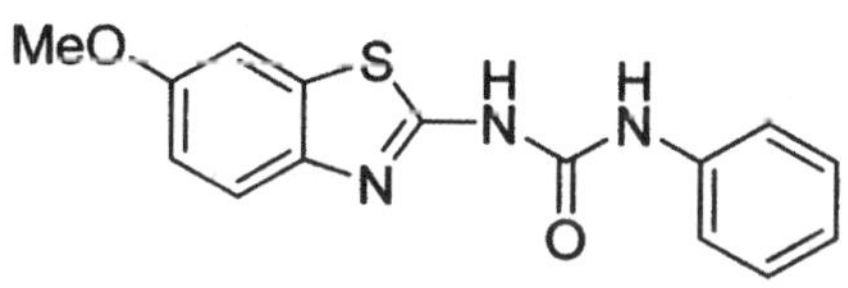

frentizole: anthelmintic

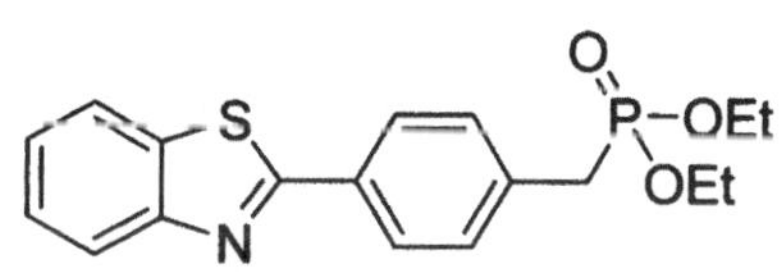

fostedil: calcium channel blocker

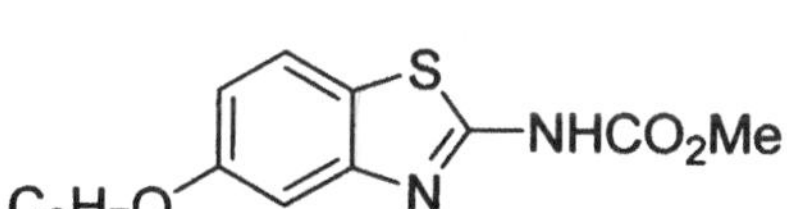

tioxidazole: anthelmintic

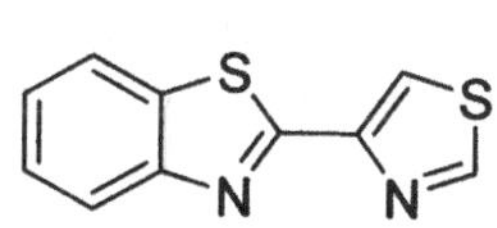

thiabendazole: fungicide

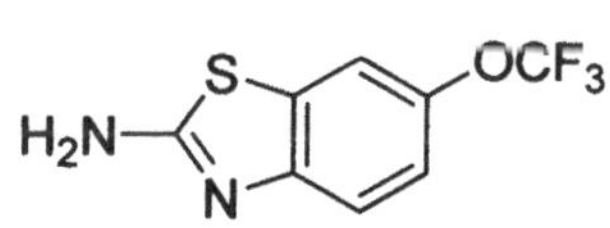

riluzole: amyotrophic lateral sclerosis

The above examples of thiazole- and benzothiazole-containing molecules demonstrate that these substructures are important heterocyclic components responsible for various types of biological activities, justifying the need to understand their reactivity and syntheses.

7.1.1 *Basicity of Thiazoles*

Unsubstituted 1,3-thiazole is more basic than the corresponding oxazole and less basic than pyridine. Nitrogen with its lone pair of electrons in 1,3-thiazoles is less reactive than in pyridine due to enhanced aromatic character. The lower reactivity is responsible for a more effective stabilization of the positive charge. Typically, alkyl groups add to basicity due to their +I effect, which decreases in the order 2-, 4-, and 5-alkyl in agreement with the net π-charge on nitrogen.[26] A conjugated +M2 amino group noticeably enhances basicity while the introduction of a strongly –M nitro group into the 5-position results in a decrease of the p*K*a. In case of a fused benzene ring such as benzothiazole, there is a reducing effect on the basicity of the nitrogen atom. Thus, benzothiazole with a p*K*a of 1.2 is a weaker base than thiazole, which has a p*K*a of 2.52.

7.2 *Reactions of the Thiazole Ring*

7.2.1 *Electrophilic Attack at Carbon*

Thiazole can be attacked at sulfur, nitrogen, or at the C-5 carbon atom, with alkylation typically occurring on the nitrogen atom. When it comes to electrophilic aromatic substitution reactions, 1,3-thiazole behaves much like pyridine. Therefore such reactions are difficult for thiazoles to undergo. The ring nitrogen atom deactivates the heterocycle toward an electrophile in spite of its π-excessive character. Electrophilic substitution occurs only when there is an electron donor group on the thiazole ring. For example, when there is a hydroxy or an amino group, substitution occurs at the 5-position if unsubstituted or at the 4-position if the 5-position is already substituted.[27] These reactions normally proceed in strongly acidic media and involve the protonated form of thiazole or quarternized derivative that displays an even lower reactivity than the free base.

Sulfonation

Thiazole sulfonation requires forced conditions such as oleum at 250 °C for 3 h in Hg(II) sulfate to yield 65% 5-thiazolesulfonic acid.[28] But when activated by an electron donor group, as in the case of 2-amino thiazole, sulfonation takes place at 0 °C to produce 2-sulfamic acid, which then thermally rearranges to 2-amino-5-sulfamic acid.[29]

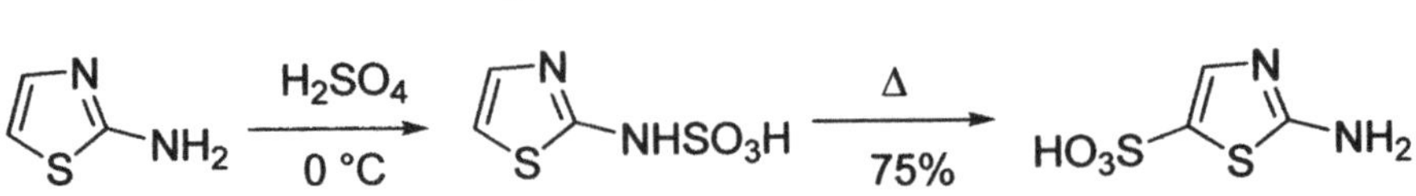

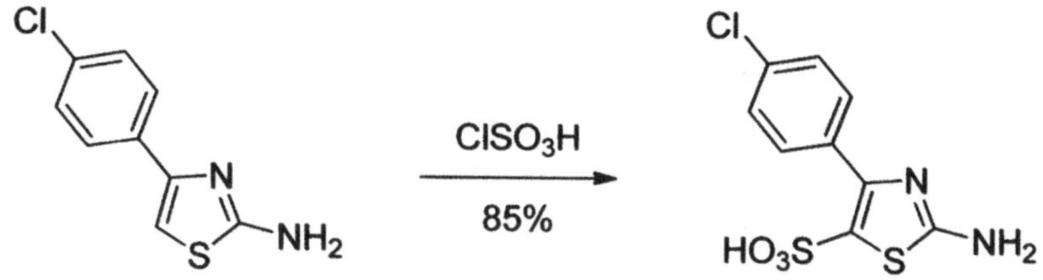

Nitration

Nitration of thiazoles has not been reported to take place under the classical nitration conditions even under extreme condition. Thiazole itself is untouched by nitric acid or oleum at 160 °C, but methyl thiazoles are sufficiently activated to undergo substitution and preferentially yield the 5-nitro over the 4-nitro regioisomer.[31]

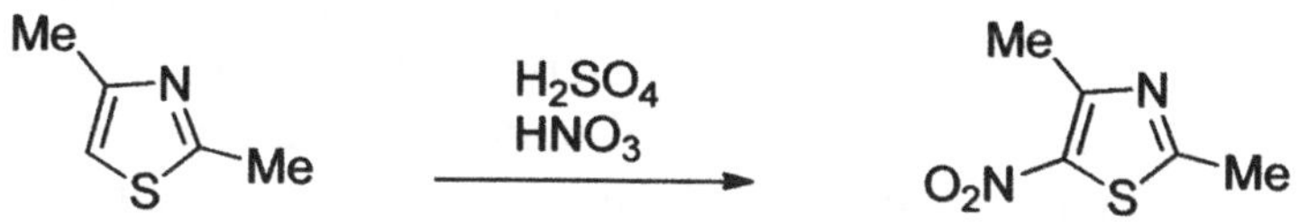

Halogenation

There is no reaction of unsubstituted 1,3-thiazole with chlorine or bromine in an inert solvent. However, when activated by a donor such as the amino group, bromination has been reported to take place. An example is shown below.[32]

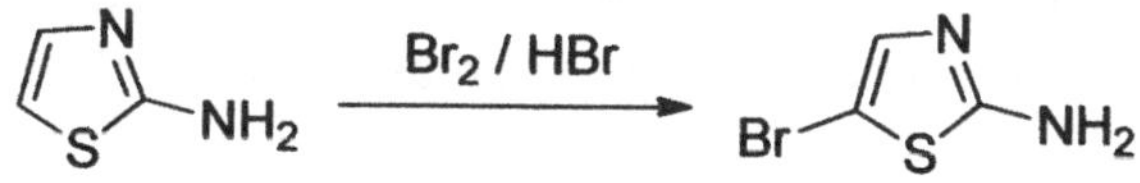

Similarly, when there are two substituents on the thiazole, the unsubstituted position gets halogenated. Activation of thiazoles with the aid of donor groups has been documented.[33] The reaction scheme shown below illustrates this point where electrophilic substitution of 2'-phenyl-*4R*-2,4'-bisthiazoles has been studied in order to establish relationships between structure and reactivity of the two thiazole rings.

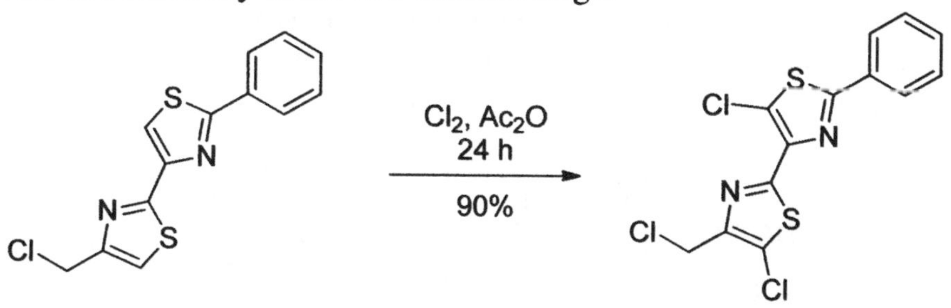

This reaction has also been used in the syntheses toward diamidino 2,5-bis(aryl)thiazoles exhibiting antiprotozoal activity against *Trypanosoma brucei rhodensiense* (T. b. r.) and *Plasmodium falciparum (P. f.),* shown below. Here, the synthesis involved conversion of derivatized thiazoles to bromothiazoles as substrates for Suzuki cross-couplings.[34]

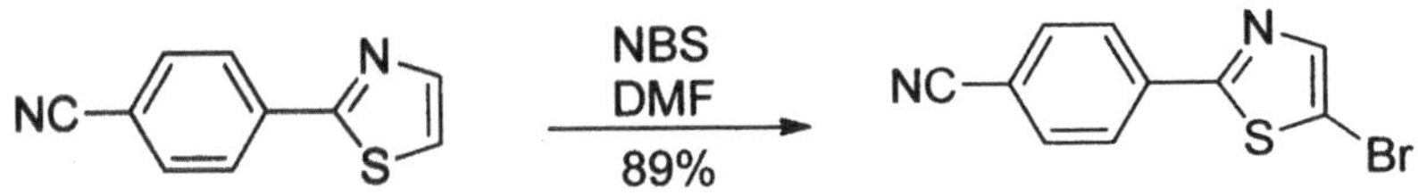

7.2.2 C-Metalation

The proton on the C2 position of the thiazole ring possesses higher kinetic and thermodynamic acidity than the ones on C4 and C5.[35] Organolithiates and Grignard reagents attack thiazoles and benzothiazoles thereby acting as strong bases *via* deprotonation. Deprotonation of thiazoles and benzothiazoles at the C2 position is well understood and has been employed in combination with various electrophiles. For example, thiazole deprotonation with *n*-butyl lithium at low temperatures and the reaction of its corresponding lithiate with an aldehyde is well known.[36]

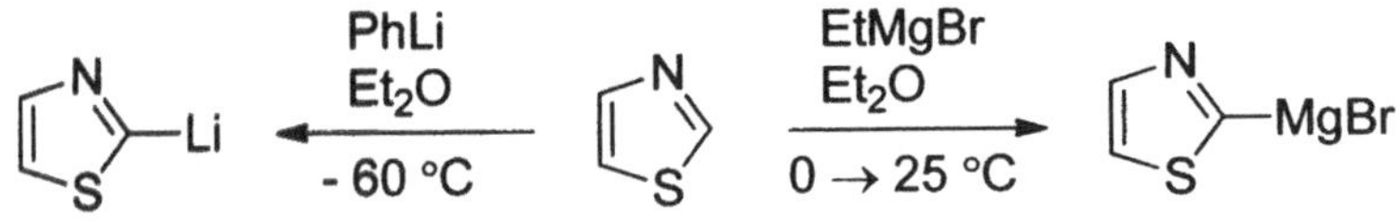

Benzothiazoles undergo reactions similarly at the 2-position through deprotonation in the presence of a base such as *n*-butyl lithium.[37]

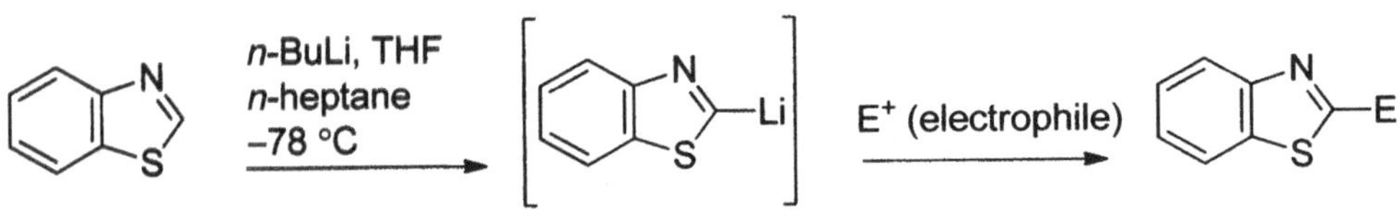

Florio et al. have found that 2-1ithio-4-methylthiazole can be conveniently generated almost quantitatively by the deprotonation of 4-methyl-thiazole with lithium diisopropylamide (LDA) in THF at −78 °C. Addition of α-chloroacetone to the solution yields the corresponding chlorohydrins.[38]

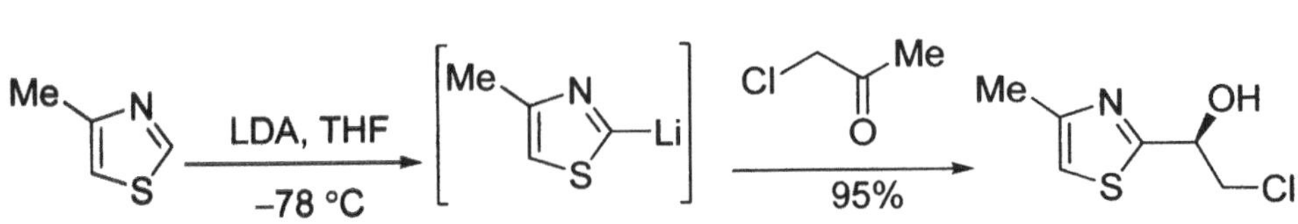

Similarly, Myllymäki et al. have employed this concept in the synthesis of aryl ketones *via* the reaction of the lithio benzothiazole with an aryl acid chloride to produce the desired ketone.[39] A series of such carbonyl compounds was synthesized and evaluated as potential fatty acid amide hydrolase (FAAH) inhibitors.

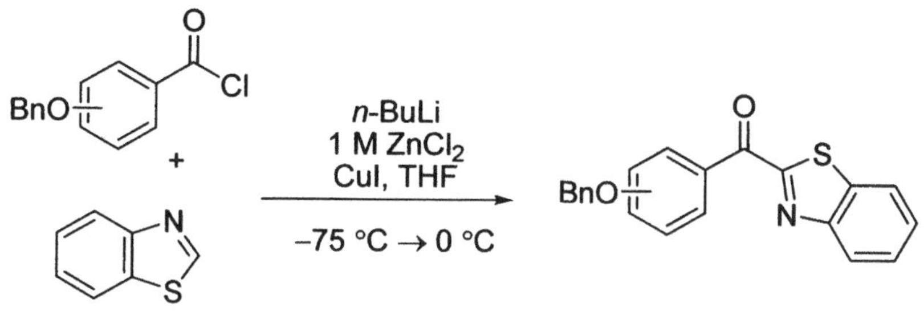

Other metalated compounds such as silanes and stannanes can be derivatized likewise. For example, the corresponding stannanes of lithio benzothiazole can also be synthesized with relative ease by treating the lithiated species with organotin chlorides.[40]

Grignard reagents of thiazoles and benzothiazoles have also been prepared and applied on scale. Kenney et al. have described an efficient process to produce kilogram quantities of argininylbenzo[*d*]thiazole intermediate in the synthesis of argynyl-tryptase inhibitor RWJ-56423, a compound demonstrating the potential to treat allergic and inflammatory diseases. The developments during the scale-up work on this compound led to the selection and optimization of the reaction between an argininyl imidazolide ester and benzothiazol-2-yl–MgCl nucleophile.[41,42]

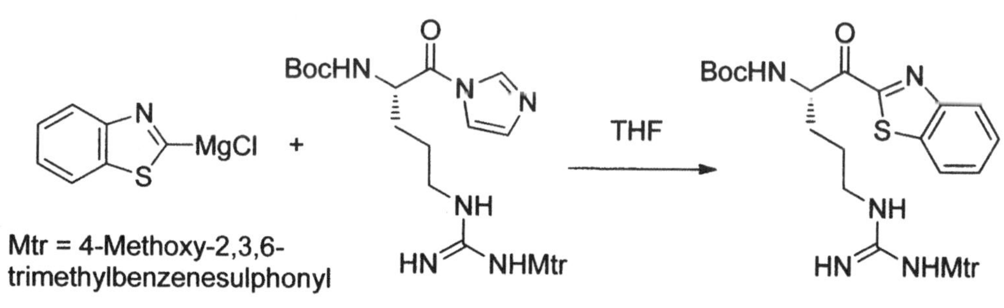

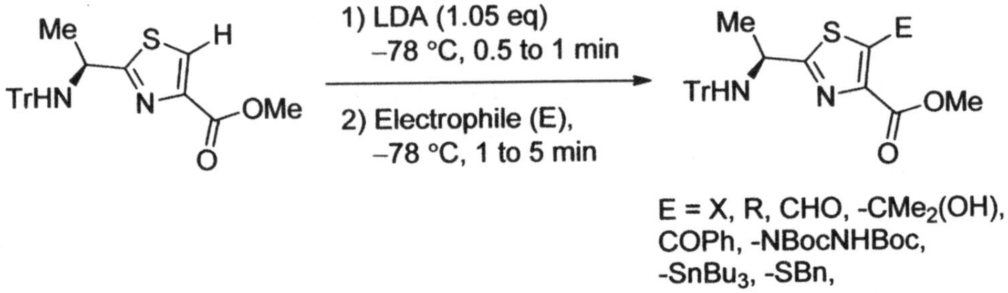

Taunton and co-workers have reported a one-step racemization free organolithium-mediated diversification of peptide thiazoles.[43] A mild procedure for the C5 lithiation of 2,4-disubstituted thiazoles has been described in this report, and an example is shown below. The method is compatible with *N*-Boc, *N*-trityl, carboxylic ester, and carboxamide protecting groups and has been used to functionalize directly the thiazole ring of cyclopeptide natural products.

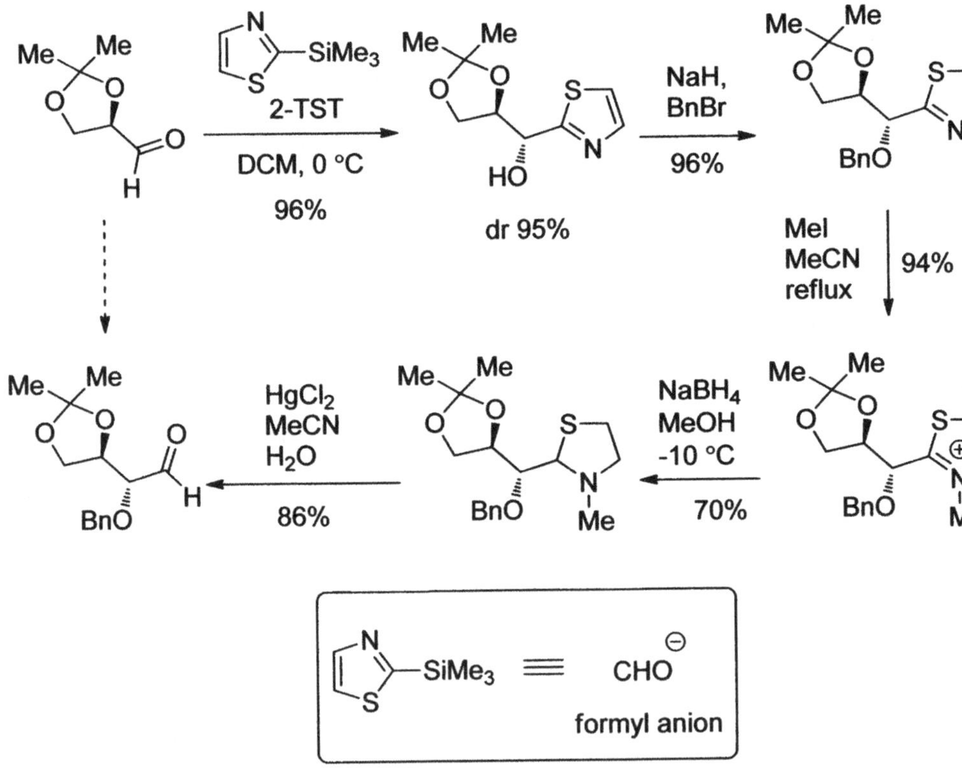

A key feature of the thiazole deprotonation chemistry is its potential as a latent formyl group in the assembly of tightly functionalized, chiral molecular architecture. The use of 2-(trimethylsilyl) thiazole (TST) as a formyl anion equivalent has been well investigated and popularized by

Dondoni.[44] 2-TST is a metalated heterocycle that was prepared in Dondoni's laboratory and employed as a synthetic auxiliary. 2-TST undergoes rapid and spontaneous carbodesilylation reactions with various *C*-electrophiles for instance ketenes, acyl chlorides, aldehydes, and heteroaryl cations producing the corresponding 2-substituted thiazoles in good yields. Reaction with chiral aldehydes has been shown to occur with good *anti*-diastereoselectivity to produce compounds, which upon conversion of thiazole to formyl group furnishing aldehydes, possess one extra carbon atom. There are two main properties of the thiazole ring that have render this methodology effective: i) its tolerance to a wide range of reaction conditions allowing for elaboration of the substrate in which it has been introduced and ii) easy transformation into the formyl group under almost neutral conditions, thus leaving stereocenters untouched as well as acid- or base-sensitive functionally groups present in the molecule. After the addition of 2-TST, a sequence of simple and high-yielding transformations, namely OH-protection, *N*-methylation, reduction and finally hydrolysis, can be conveniently carried out to yield the desired aldehyde, as shown below.

Chemists at Hoffman-La Roche applied this methodology toward the large scale synthesis of the HIV protease inhibitor saquinavir (Ro 31-8959), a drug that was later approved by the USFDA in 1995 for the treatment of AIDS. As part of this work, they employed 2-TST to produce a starting material comprising the chiral aldehyde for the hydroxyethylamine isosteric dipeptide intermediate, shown below. With the mono-protected nitrogen atom on the amino aldehyde, the desired *anti*-alcohol was formed as a minor product, the ratio of major:minor being 3:2. The target aldehyde was obtained in low overall yield, and as a result, the approach had to be abandoned in favor of another approach.[45] Yet the potential of this approach is well reflected due to its operational simplicity. Interestingly, Dondoni et al. later demonstrated that the *anti*-selectivity could be well realized by utilizing the equivalent bis-*N,N'*-protected amino aldehyde (phenylalaninal), however the process involved additional steps.[46] The homologation methodology has been described as practical, simple, highly stereoselective, and chemically effective.[47]

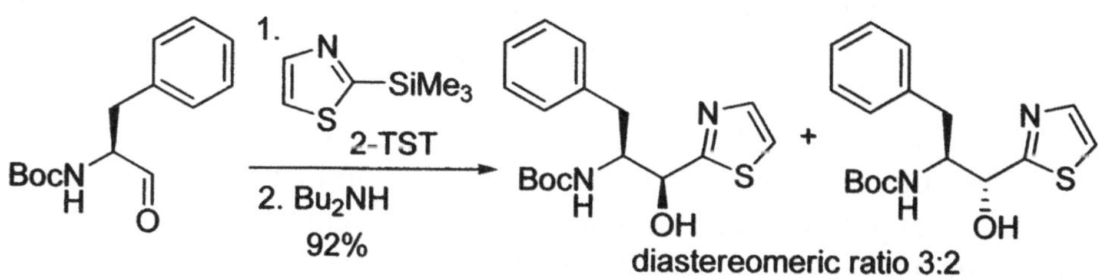

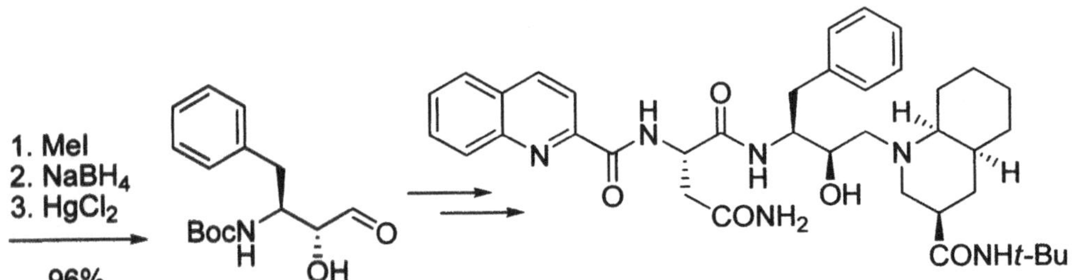

Ro 31-8959 (saquinavir)

Preparative, strong-base deprotonation takes place preferentially at C-2 position or at C-5 if C-2 is blocked. This has been demonstrated by Dondoni when unsubstituted 1,3-thiazole undergoes deprotonation with *n*-butyl lithium and subsequent quenching with trimethyl silyl chloride formed 2-TST.[48] Further lithiation and silylation gives exclusively 2,5-bis(trimethylsilyl)thiazole, thus confirming that preferential metalation occurs at C-5 with respect to C-4 in unsubstituted thiazoles.

7.2.3 Alkylation

Alkylation of 1,3-thiazole occurs on the nitrogen atom leading to its quarternization and generates the thiazolium salt, which can react readily with various nucleophiles.[49]

thiazolium salt

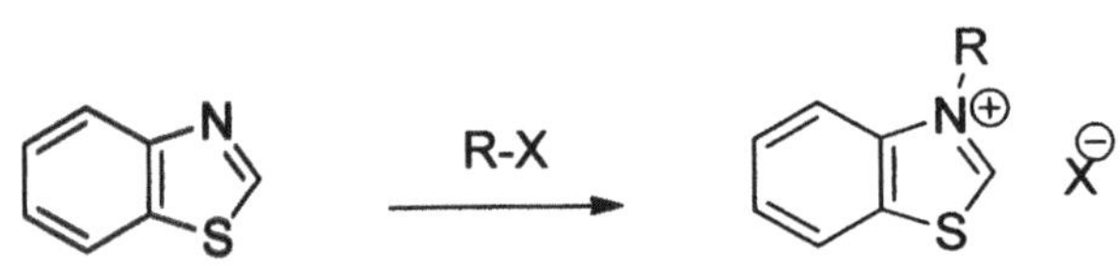

benzothiazolium salt

Benzothiazoles undergo alkylation similarly and can react further. With nucleophiles, these salts can undergo ring opening and form the corresponding products.

N-Alkylated thiazolium salts are often used to introduce complexity into the thiazole heterocycle. The synthesis of imidazothiazole–chalcone derivatives as anti-cancer and apoptosis inducing agents by Kamal et al. at the Indian Institute of Chemical Technology illustrates this point.[50] Here, a new class of imidaza[2,1-*b*]thiazole chalcone derivatives were synthesized from the corresponding imidazol[2,1-*b*]thiazole aldehydes, which in turn were derived by alkylation of aminothiazoles with α-bromoketones. The *N*-alkylated were directly subjected to cyclization to yield imidazol[2,1-*b*]thiazole which could then be used to prepare the chalcone derivatives.

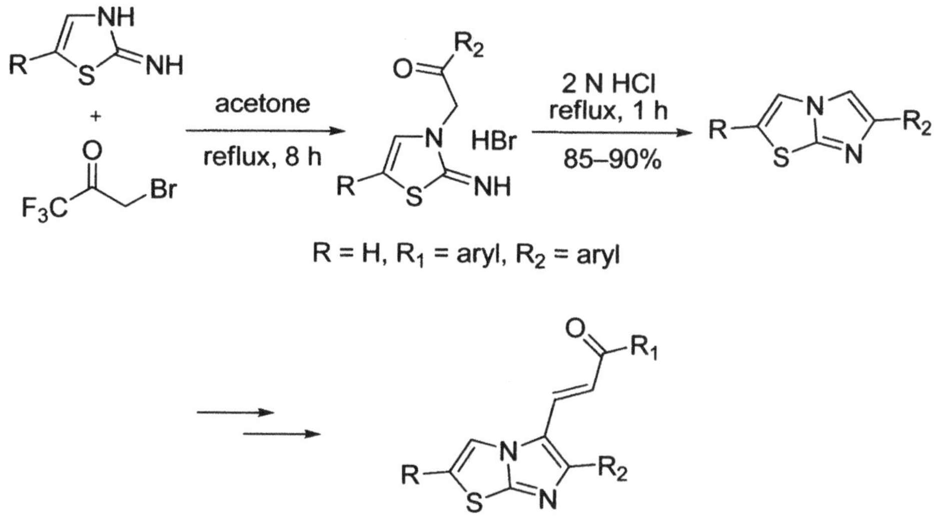

imidazothiazole-chalcone derivatives

N-Alkylated thiazolium and benzothiazolium salts also experience base-promoted deprotonation at the 2-position to form ylides. Such compounds, often referred to as *N*-heterocyclic carbene (NHC), are nucleophilic catalysts in benzoin condensation. In 1943, Ugai and co-workers reported that thiazolium salts catalyze self-condensation of benzaldehyde to generate benzoin *via* an umpoulong process.[51] Breslow at Columbia University in 1958 proposed thiazolium ylide as the actual catalyst for this transformation.[52] In this mechanism, the catalytically active species was represented as a thiazolium zwitterion, the resonance structure of an NHC, and the reaction was postulated to ensue *via* the enaminol or the "Breslow intermediate."[53]

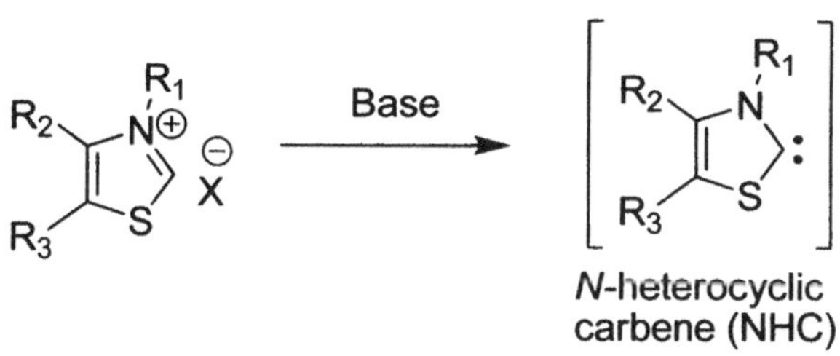

N-heterocyclic carbene (NHC)

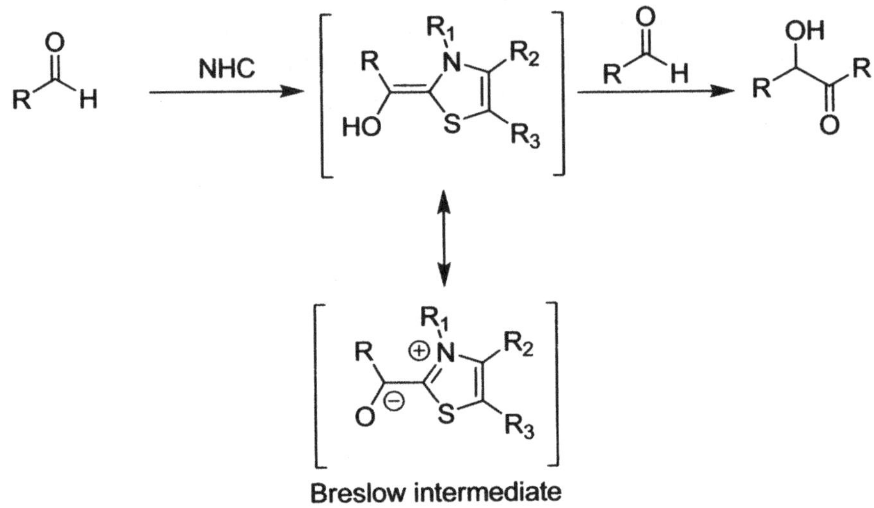

With Breslow subsequently citing the nucleophilic carbene to be responsible for catalysis in the process that utilizes a thiamine related NHC, this "umpoulong" chemistry has evolved into distinct area of research called NHC catalysis, and it has found many applications in the recent past. A detailed discussion on this topic is, however, beyond the scope of this book and has been reviewed elsewhere.[54]

Tosylates as well as alkyl halides can also cause *N*-alkylation of thiazoles and involve an S_N2 quarternization reaction. The thiazolium salt formation concept has been well utilized in medicinal chemistry. For example, Kerwin and co-workers at the University of Texas at Austin have reported the role of 2-alkynylbenzothiazolium component in DNA cleavage.[55] They prepared a series of "moderately potent" *N*-propargyl-2-alkylbenzothiazolium aza-enediynes DNA cleavage agents, an example of which is shown below. Propargyl triflate was employed as the alkylating agent.

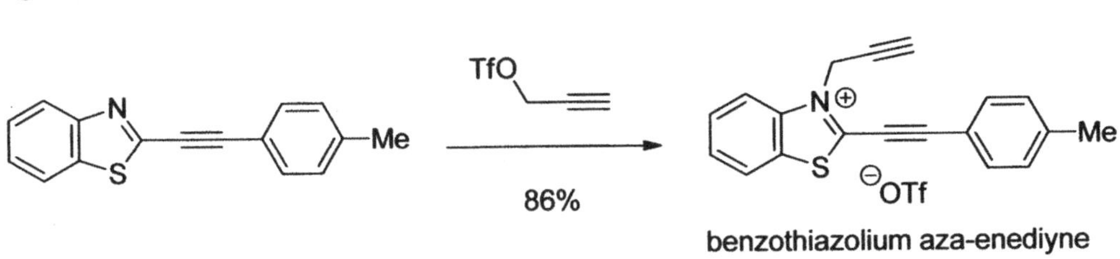

7.2.4 N-*Oxidation*

Thiazole *N*-oxides may result from the direct oxidation of thiazoles with hydrogen peroxide, tungstic acid, or peracetic acid.[56]

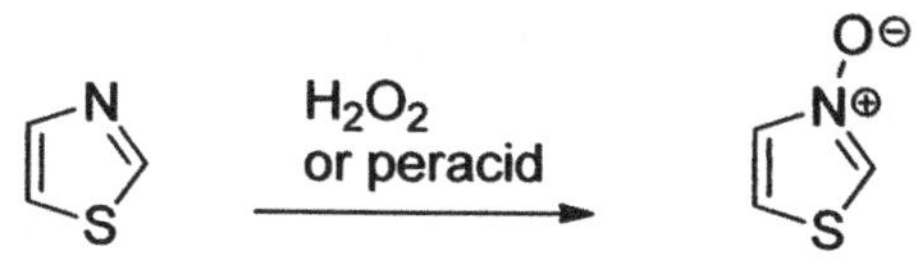

Although they have not found extensive utility in drug discovery, a few examples do exist where thiazole *N*-oxides were prepared as target molecules. One such case is shown below where these compounds were tested for their anthelmintic and antifungal activity in 1972.[57] Interestingly, the oxidation of thiazole took place selectively and required a peracid derived from a strong acid, with no oxidation taking place on the imidazo[1,2-*a*]pyridine ring. Peracetic acid did not bring about oxidation at all.

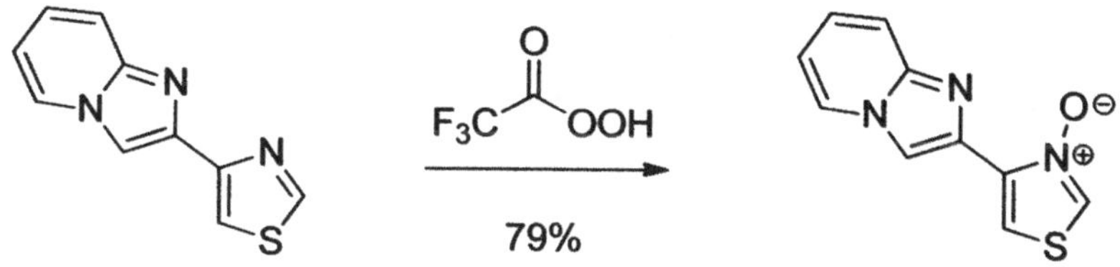

7.2.5 *Cycloaddition*

Thiazole does not have a relative diene character possessed by other heterocycles, unlike other heterocyclic structures such as 1,3-oxazole and furan. So 1,3-thiazole does not undergo Diels–Alder reactions very easily due to the its low reactivity on account of greater aromaticity than oxazoles as well as due to highly nucleophilic thiazole.[58,59] However, there are exceptions where an intramolecular Diels–Alder reaction of thiazole with alkyne activated by a methoxyl group is facilitated to produce a thiophene, albeit in a rather low yield of 48%, after three days at reflux in degassed mesitylene.[60] This happens through the extrusion of acetonitrile.

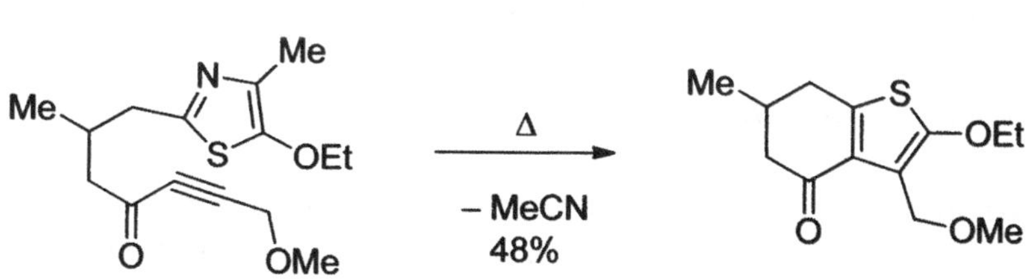

However, 1,3-dipolar cycloaddition of acetylenic dipolarophiles such as dimethylacetylenedicarboxylate (DMAD) have been reported. Thus, thiazole and its alkyl derivatives undergo condensation with dimethyl acetylenedicarboxylate, giving 1:2 adducts. For example, Potts et al. reported that thiazole when quarternized as an ylide with alkylators such as bromoacetophenone ($BrCH_2COPh$), undergoes 1,3-dipolar cycloaddition with dipolarophiles. The cycloadduct further rearranges to give a fused heterocycle shown below.[61]

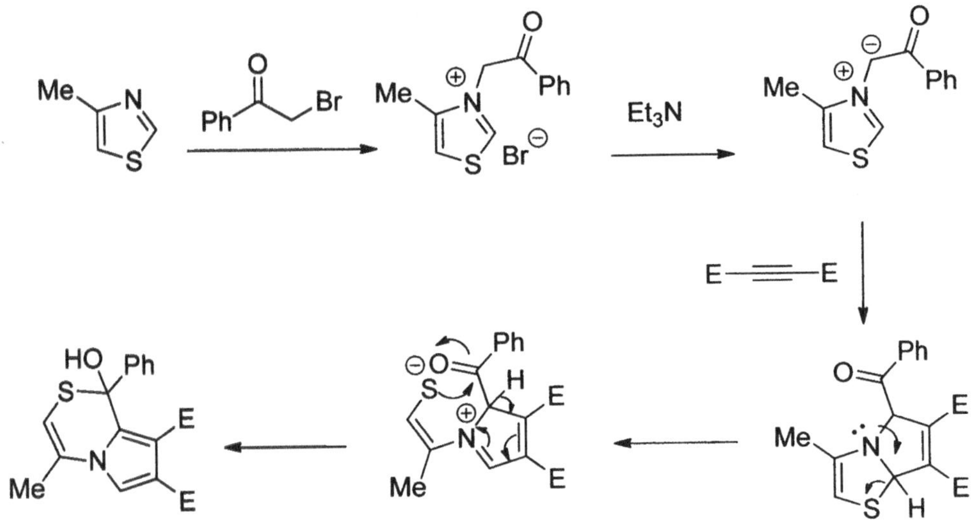

E = electron withdrawing group

7.3 *Palladium Chemistry Undergone by Thiazoles and Benzothiazoles*

In this section, standard reactions that have been encountered in the field of synthetic medicinal chemistry have been covered. These are typically the Suzuki, Negishi, Heck, Sonogishira, and Stille couplings among others. While a detailed review covering all the palladium catalyzed cross-coupling reactions in drug discovery may not be within the scope of this book, a few representative cases of these reactions have been included in this section to indicate the range of chemistry thiazole- and benzothiazole-containing substrates can create. Examples revealing handiness of these reactions on thiazole and benzothiazole have been presented.

7.3.1 *Suzuki–Miyaura Reaction*

Suzuki–Miyaura reactions are perhaps the most widely employed palladium catalyzed cross-couplings in the realm of thiazole medicinal chemistry. They typically take place only when the thiazole is an electrophile in the transformation. The nucleophilic thiazole boronic acid or ester, especially at the 2-position, is relatively unstable and therefore difficult to prepare.[62] The electrophiles namely the 2-, 4-, or 5-substituted halothiazoles are often readily accessible in terms of their synthetic ease or commercial availability. A remarkable application has been described by Jang et al. in the discovery of 7-*N*-piperazinyl[5,4-*d*]pyrimidine analogues as a novel class of immunosuppressive agents.[63]

The synthesis involved construction of a heteroaryl–aryl coupling, where the chloro-thiazole[5,4-*d*]pyrimidine was coupled with *p*-fluorophenyl boronic acid in a Suzuki coupling. The reaction proceeded well even in the presence of a free amino group on the pyrimidine.

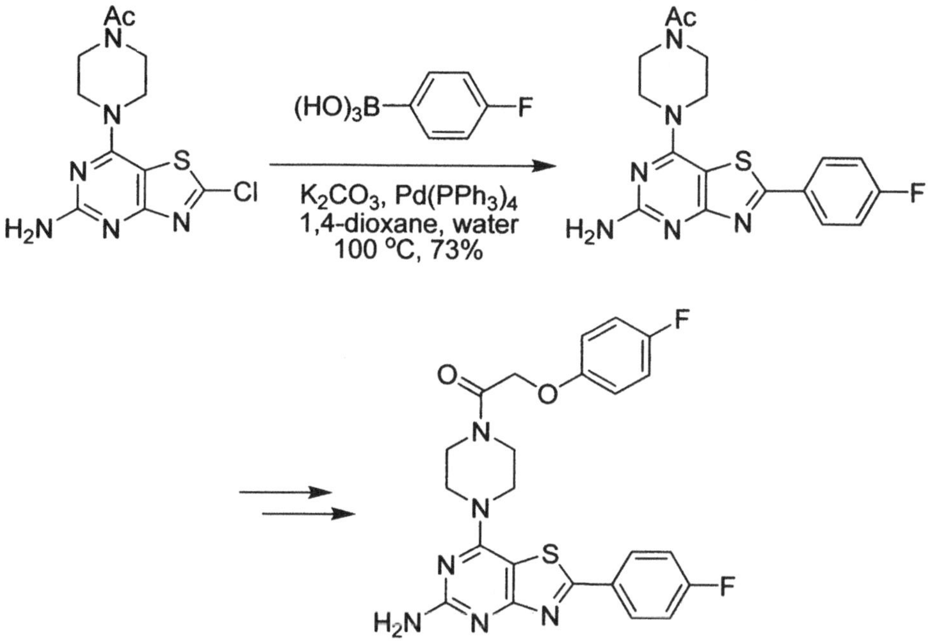

Novel immunosuppressive agent

Wang et al. at Abbott Laboratories have applied the Suzuki coupling in the syntheses of benzothiazole-based analogues as selective fatty acid amide hydrolase inhibitors (FAAH) inhibitors.[64] 2-Chlorobenzothiazole was the electrophilic partner in the coupling with the boronic ester in the presence of tetrakis(triphenylphosphine)palladium(0) and base. In this study, the piperidine and benzothiazole substructures were found to be the key components responsible for biological activity. The benzothiazole derivative shown below was observed to be the most potent analogue, an observation supported by modeling study, indicating that hydrophobic interactions of the benzothiazole ring with the enzyme contributed to its impressive potency.

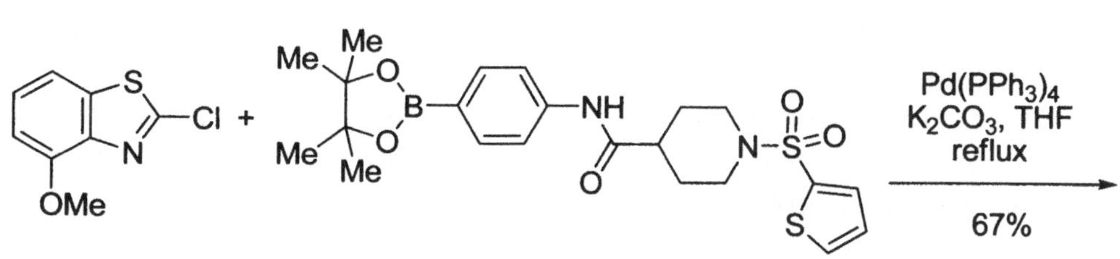

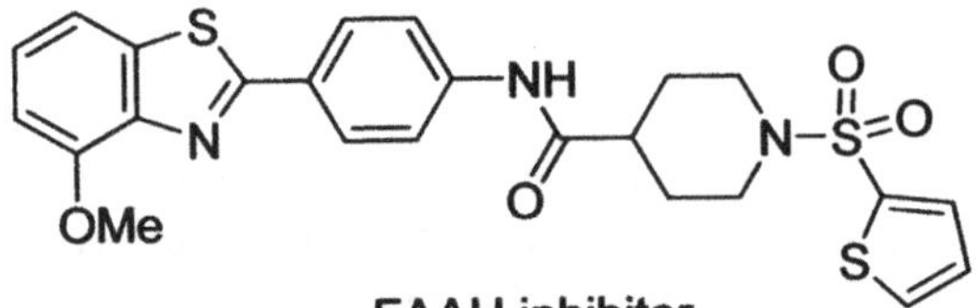

FAAH inhibitor

Romagnoli et al. have reported a convergent synthesis of a class of microtubule targeting agents where they applied the Suzuki–Miyaura reaction to highly substituted 5-bromothiazoles.[65] With various aryl boronic acid, highly substituted thiazole derivatives were prepared and evaluated for their anti-proliferative activity against a panel of human tumor cell lines.

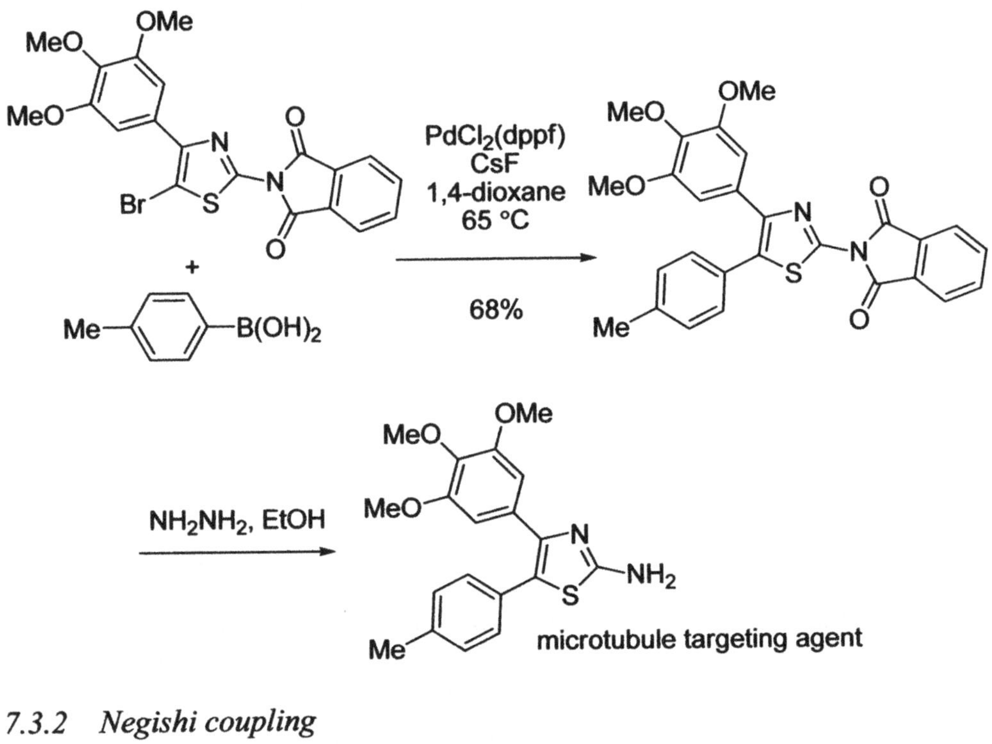

microtubule targeting agent

7.3.2 Negishi coupling

Thiazoles and benzothiazoles are suitable partners in the Negishi coupling where typically they have been used as the corresponding nuclophilic zinc halides. 2-Thiazolylzinc halides can be prepared by deprotonation with *n*-BuLi and treatment of the anion with zinc halide (ZnX_2). The examples below illustrate the convenience of such a C–C bonding forming transformation.

Slee and co-workers at Neurocrine Biosciences have synthesized 2-amino-*N*-pyrimidin-4-ylacetamides as adenosine A_{2A} receptor antagonists.[66] A_{2A} is a type of G-protein-coupled receptor (GPCR) known to modulate activity of medium spiny neurons. In an attempt to overcome metabolic liabilities associated with unsubstituted furans, a range of novel heterocyclic compounds was synthesized and their SAR was studied. Successful identification led to replacement of the unsubstituted furan moiety with either a methyl furan or thiazole while maintaining potency and selectivity. The thiazole-containing analogue was prepared by coupling thiazole with the chloropyrimidine starting material under Negishi conditions in 70% yield. Here, 1,3-thiazole was employed as a nucleophile in the form of its zinc chloride derivative as the coupling partner.

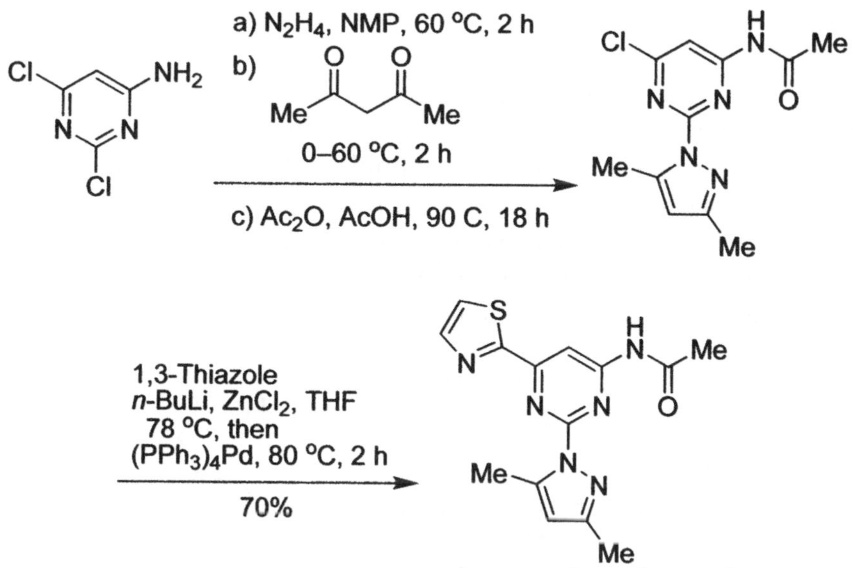

A_{2A} receptor antagonist

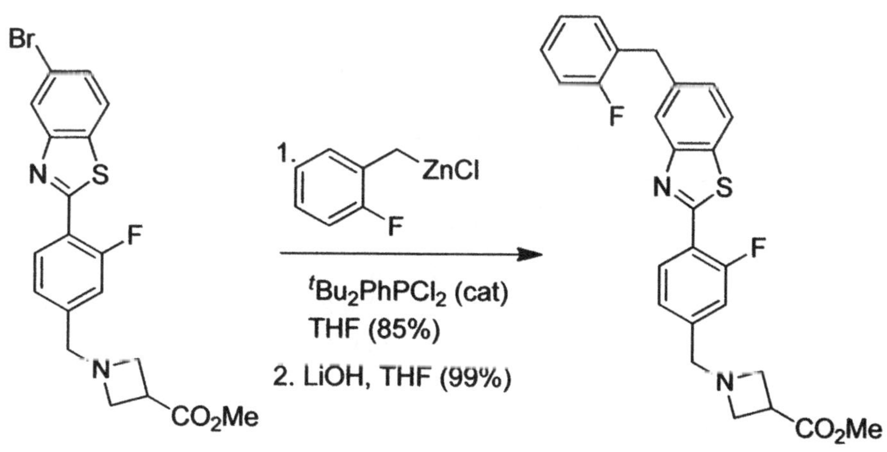

$S1P_1$ agonist

There are several examples where halothiazoles have been used as electrophiles in Negishi couplings in medicinal chemistry. An interesting example comes from Lanman and co-workers at Amgen on the discovery of a selective sphingosine-1-phosphate receptor 1 ($S1P_1$) agonist with minimal activity at the $S1P_3$ receptor leading to a benzothiazole. A Negishi coupling was carried out to form the benzyl–benzothiazole carbon–carbon bond.[67] The bromobenzothiazole was the electrophile, while (2-fluorobenzyl)zinc chloride was the nucleophile in this transformation. Hydrolysis of the product led to the isolation of a potent $S1P_1$ agonist with EC_{50} = 0.042 μM.

7.3.3 *Heck Reaction*

Relatively few cases of Heck reactions as applied to thiazoles or benzothiazoles have been reported.[68] For example, an intermediate for a benzothiazole-containing structurally simple inhibitor of lanosterol 14α-demethylase (L14DM) was prepared with the aid of Heck conditions. This inhibitor showed excellent pharmacokinetic properties and in a mouse model of acute Chagas disease and was found to show efficacy comparable with that of posaconazole, an anti-fungal compound.

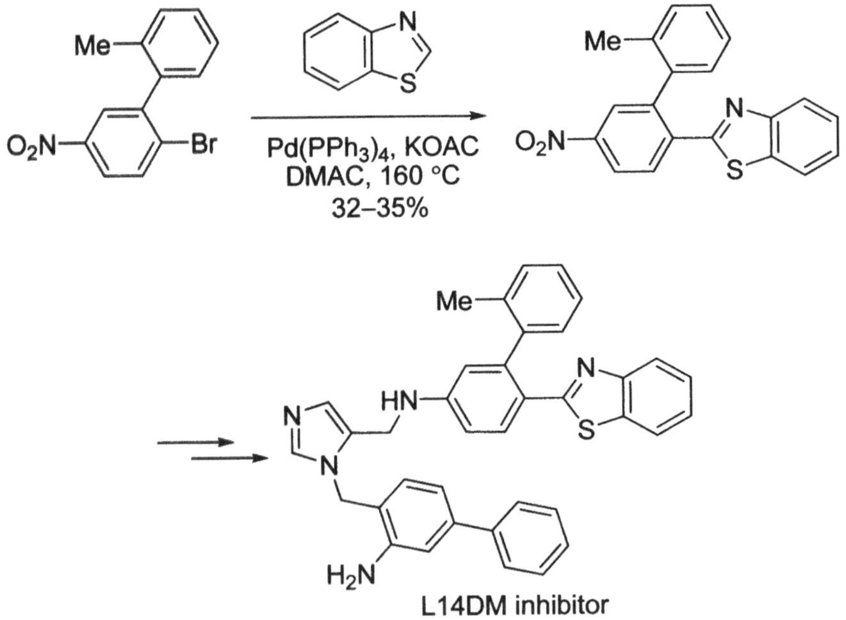

7.3.4 *Sonogashira Coupling*

Quite a few examples of this reaction on thiazoles in drug discovery have been described in the literature. An interesting application of this reaction has been illustrated by Gu et al. in a study of structure–toxicity relationship

(STR) that led to the identification of the alkyne linker as the key motif responsible for serious neurological and cardiovascular liabilities in a class of acetyl-CoA carboxylase 2 (ACC2) inhibitors.[69] To study this toxicity, the alkyne derivatives were prepared by first employing the Sonogashira reaction conditions shown below, followed by deprotection of the phthalimide group.

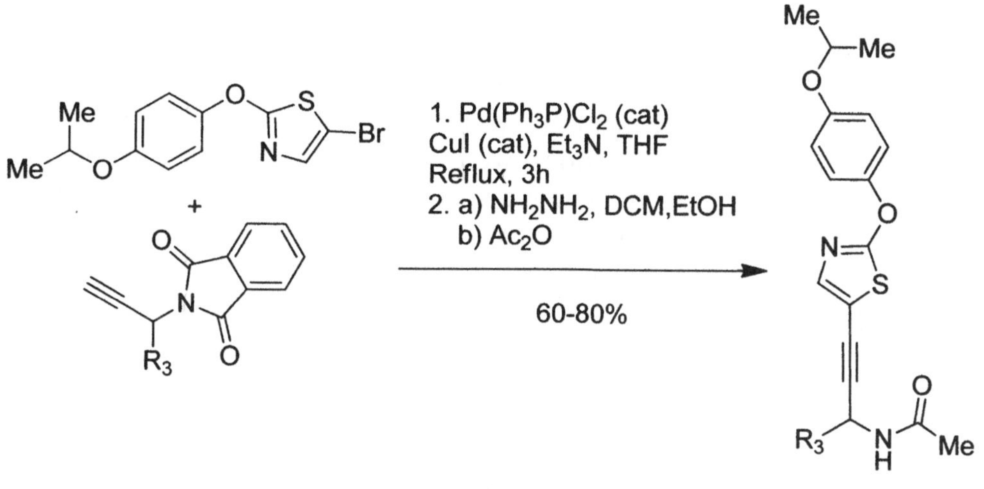

A variation of this transformation to functionalize thiazoles has been reported by Panek and co-workers and extended to triflates as the electrophilic coupling partners. Here 2-, 4-, and 5-trifloyl thiazoles were demonstrated to undergo the Sonogashira reaction with terminal alkynes as shown in the example below in yields ranging from 70–95%.[70] They developed the methodology with a variety of substrates and found that the 2-alkynyl-4-trifloylthiazole was a suitable substrate in a second palladium(0)-mediated cross-coupling such as a Stille or another Sonogashira reaction. This methodology provides rapid access to highly functionalized thiazoles despite the presence of multiple heteroatoms on the substrate.

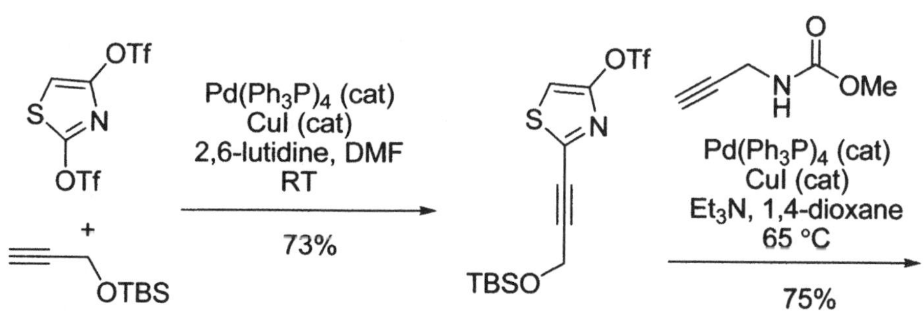

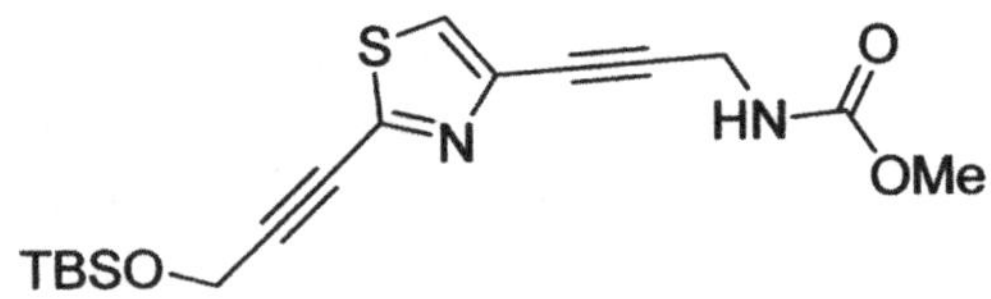

7.3.5 Stille Coupling

The Stille coupling has been found of great value in medicinal chemistry providing rapid access to carbon–carbon bonds. Since the account on the synthesis of micrococcinic acid by Kelly[71] in 1991 where multiple and sequential Stille coupling reactions were applied, the transformation has found several applications in analogue synthesis in discovery. Li (editor of this book) and co-workers for example, reported a synthesis of 2-amino-3-heteroaryl-quinoxalines as antagonists for interleukin-8 (IL-8) receptor to study the structure–activity relationship (SAR), one of which was a thiazole derivative.[72] The team devised a novel synthetic route to overcome the rather lengthy original synthesis by exploiting the Stille cross-coupling. The electrophilic coupling partner obtained from phenylene diamine in three steps was refluxed with the 2-thiazolyl stannane in tetrahydrofuran with palladium catalyst and copper iodide(I) to yield the desired Stille product in 81% yield.

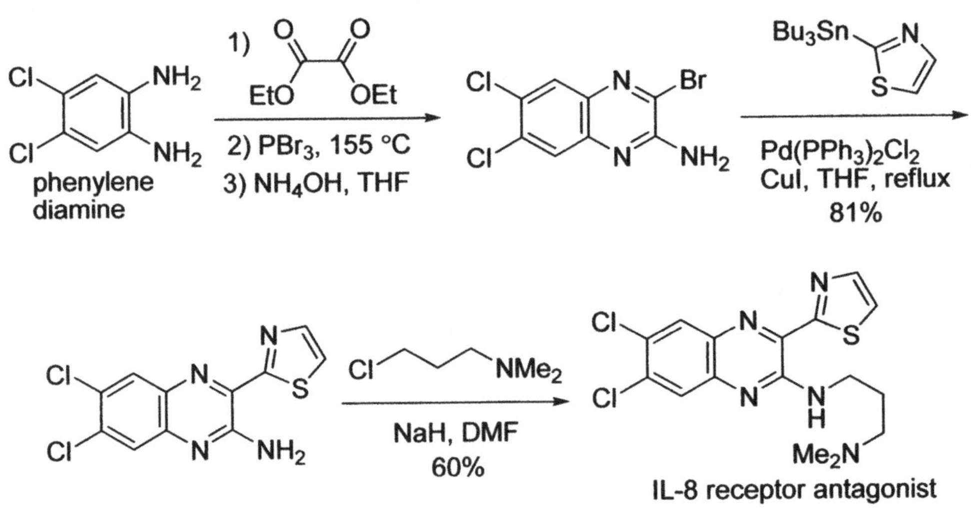

In the structure–toxicity relationship (STR) by Gu mentioned in the preceding section on the Sonogashira reaction, the toxicity-inducing alkyne-link group was replaced with a heterocycle such as thiazole. The replacement led to a dramatic reduction in the cardiovascular and neurological liabilities while maintaining ACC2 inhibitory activity. Such a bis-thiazole compound was prepared *via* a Stille coupling by converting the bromothiazoles substrate to its corresponding stannane and then allowing it to react with 2-

bromothiazole in the presence of palladium catalyst in dimethyl formamide (DMF) as the solvent to yield 95% product.

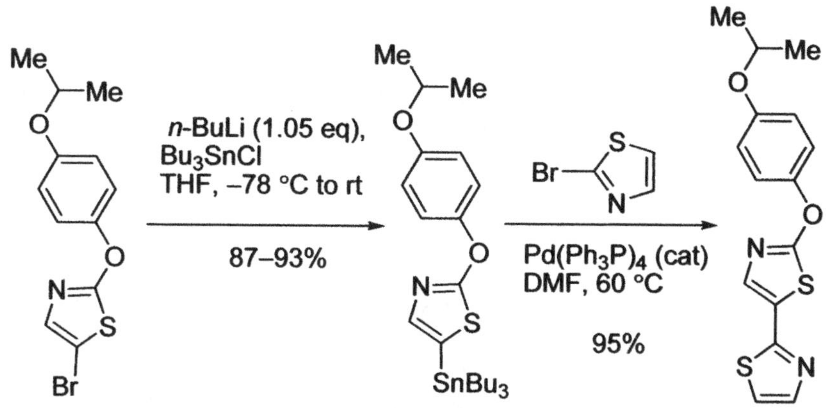

7.4 *Construction of the Thiazole Ring*

7.4.1 Hantzsch Method

The first synthesis of thiazole was described by Arthur Hantzsch in 1887.[73] This reaction in its most fundamental form as well as other forms with variations has been widely applied in the construction of a variety of molecules since then. In its most basic form, the reaction is illustrated in the scheme below and involves the use of an α-halocarbonyl compound with reactants comprising the N–C–S linkage.

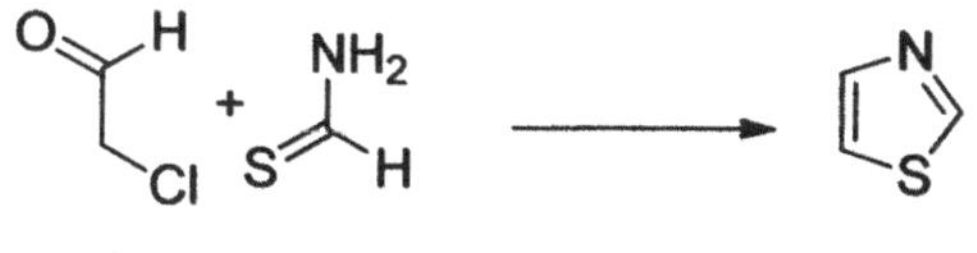

The reaction can also be carried out in the presence of all three carbon atoms being substituted with suitable alkyl or aryl groups. Moreover, it also works with thiourea instead of thioamide to yield the corresponding aminothiazole.

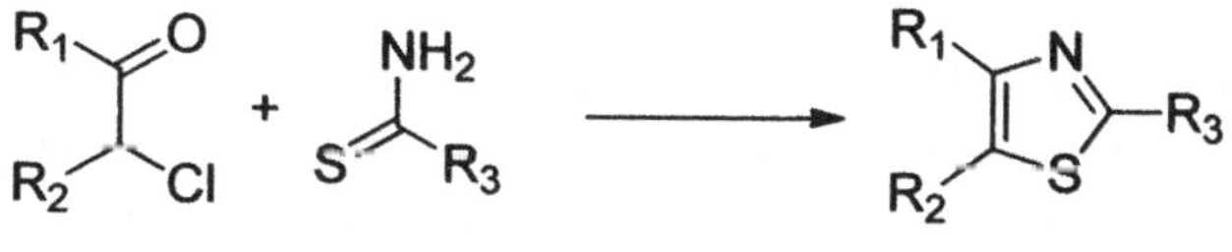

Traditionally, the mechanism of this cyclization has been postulated to be as shown below wherein an intermediate **i** is formed *via* displacement of the halide, after which the nucleophilic nitrogen in **ii** adds intramolecularly across the activated carbonyl to form hydroxy intermediate **iii**. Dehydration of **iv** yields the thiazole.[74]

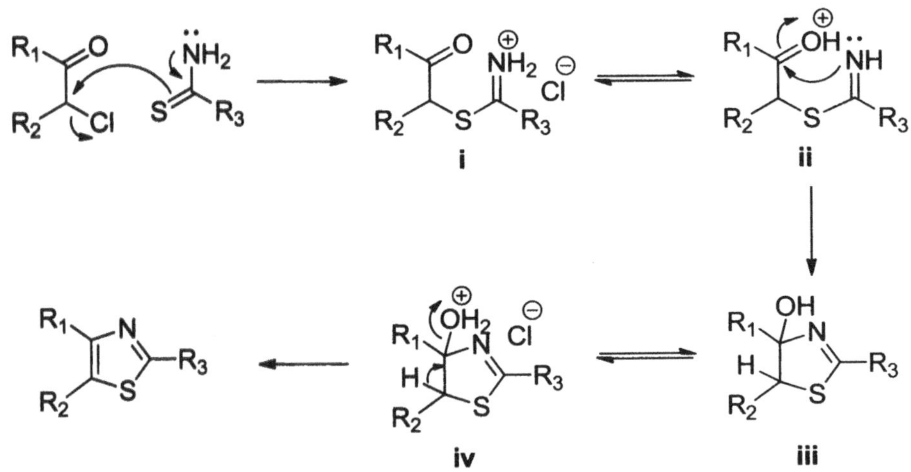

The reaction has been employed in several molecules in drug discovery and development. For example, Ikemoto et al. have developed a practical route to a thiazole as part of their efforts toward a kilogram-scale synthesis of a β-adrenergic receptor agonist as a candidate for the treatment of obesity. The thiazole ring was assembled by allowing the haloketone and thioamide to react.[75]

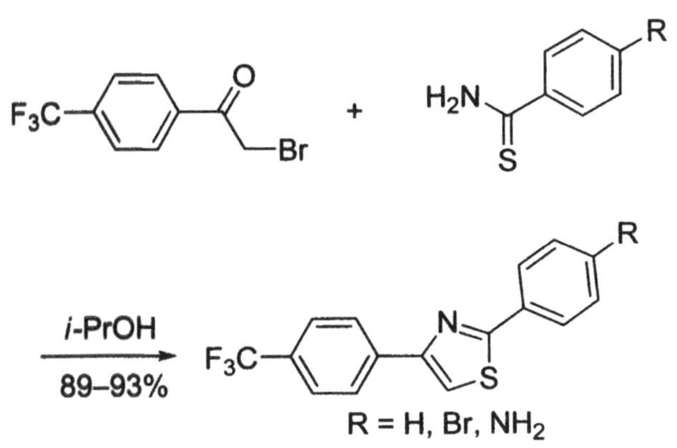

Sodelglitazar is a Phase II clinical candidate for the treatment of type 2 diabetes. It is a panagonist active toward all three peroxisome proliferators-activated receptors (PPAR). Scientists in the Chemical Development Laboratories at GlaxoSmithKline developed a robust, efficient process for its synthesis. The thiazole-containing starting material was built from the readily available benzonitrile *via* its reaction with thioacetic acid as the thionation agent chosen after considerable screening to yield the corresponding thioamide in greater than 95% conversion. The Hantzsch reaction of the crude thioamide with the chloroketo ester yielded the desired thiazole with which the structure of sodelglitazar could be put together.[76]

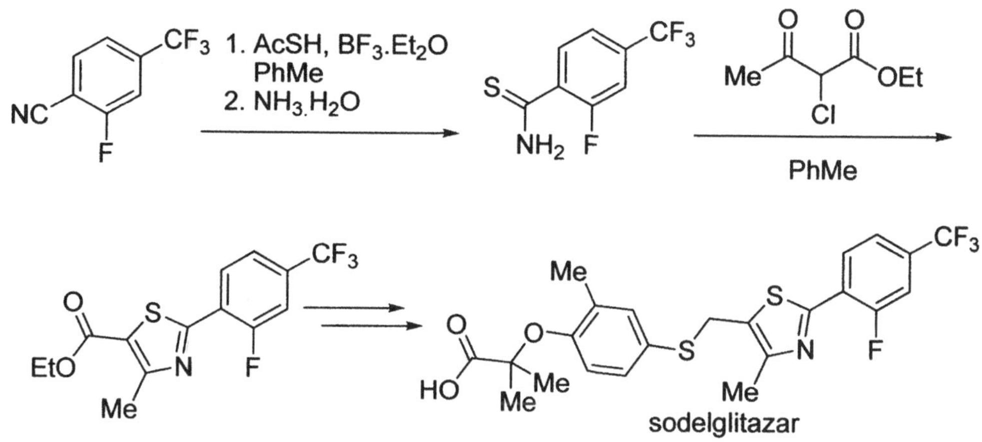

Febuxostat (Uloric), a novel xanthine oxidase inhibitor was approved by the FDA in 2009. It works through the noncompetitive blockage of xantine oxidase thereby reducing the amount of uric acid. It has been used in the treatment of hyperuricemia in gout. The synthesis of febuxostat requires the preparation of the thioamide intermediate for the Hantzsch cyclization with chloro ketoester. The thioamide intermediate is prepared from the dicyano building block prepared in two steps from *p*-nitrobenzonitrile. The less-hindered cyano group undergoes reaction with thioacetamide, forming the corresponding thioamide as the substrate for the thiazole synthesis.[77]

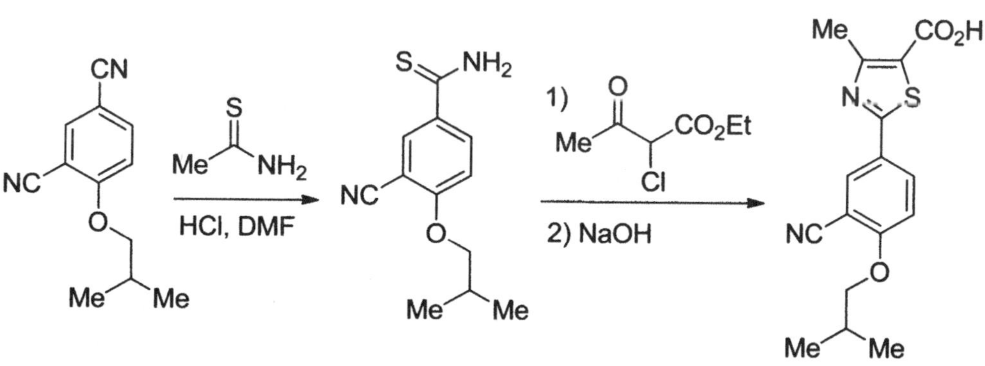

An impressive application of the Hantzsch on process scale is the kilogram-scale synthesis of ravucanozole, a novel thiazole containing antifungal development candidate at Bristol-Myers Squibb. Ravucanozole shows the longest half-life among all known azole antifungals and exhibits activity against *Cryptocossus neoformans, Candida alibcans,* and *Aspergillus* species. The safe, robust, and cost-effective process developed for this molecule began from commercially available raw materials. The cyano group in the advanced intermediate is converted to the thioamide by refluxing with diethyl dithiophosphate instead of the typically practiced hydrogen sulphide. The thioamide is then refluxed with bromoketone to yield the desired thiazole.[78]

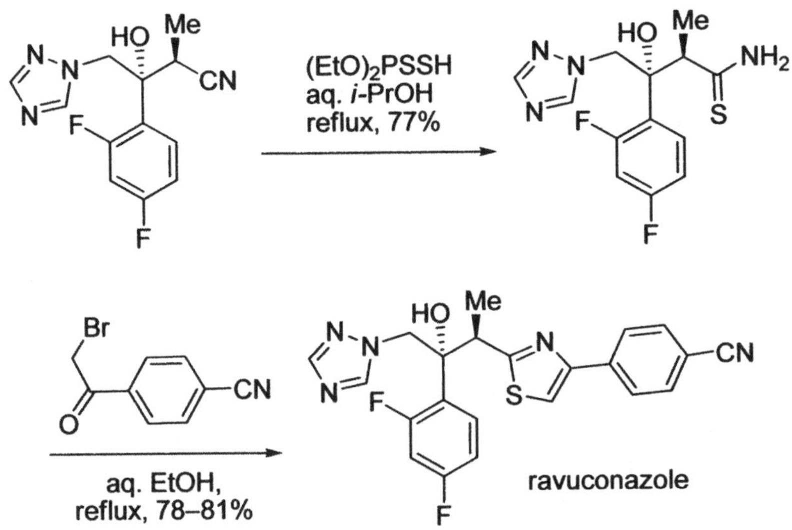

Aminothiazole can also be synthesized similarly by utilizing urea. A slight variation of the Hantzsch approach involves a one-pot procedure where the ketone with a reactive hydrogen atom, a halogen, and thiourea are mixed at a suitably elevated temperature.[79]

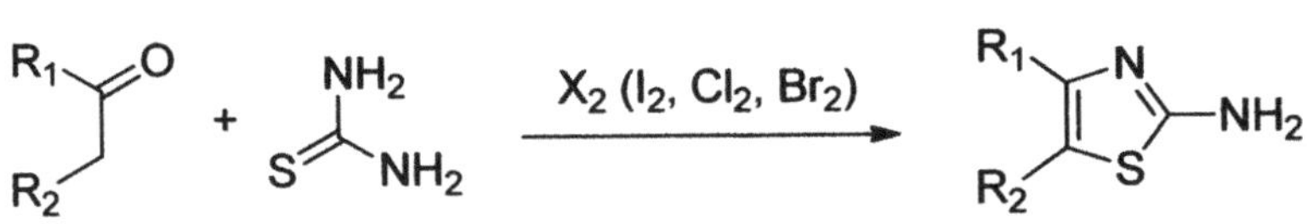

The Hantzsch strategy has been applied in the synthesis of the dopamine D_2-agonist pramipexole, a molecule comprising a fused bicyclic tetrahydrobenzothiazole substructure. The synthesis starts with α-bromination of protected 4-aminocyclohexanone followed by the Hantzsch condensation with thiourea. Deprotection and then chiral resolution yield the *S*-enantiomer. Reductive amination leads to pramipexole, which is isolated as the dihydrochloride.[80]

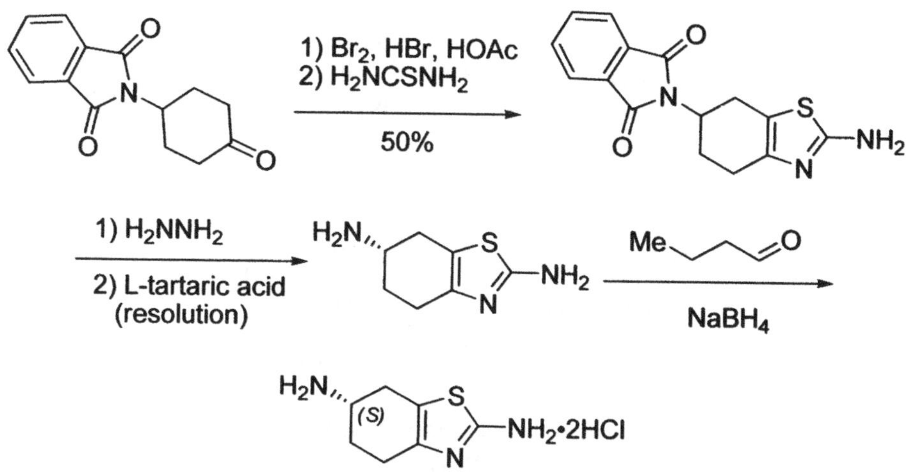

pramipexole

Another application of the Hantzsch method is in the synthesis of the aminothiazole coupling partner in the synthesis of the antibiotic cefdinir. The reactive methylene carbon of ethyl acetoacetate is first functionalized to the oxime after which α-chlorination is brought about to furnish the chloro-oxime substrate for Hantzsch synthesis with thiourea. *N,N'*-Dimethyl aniline is used as a base in the thiazole formation.[81]

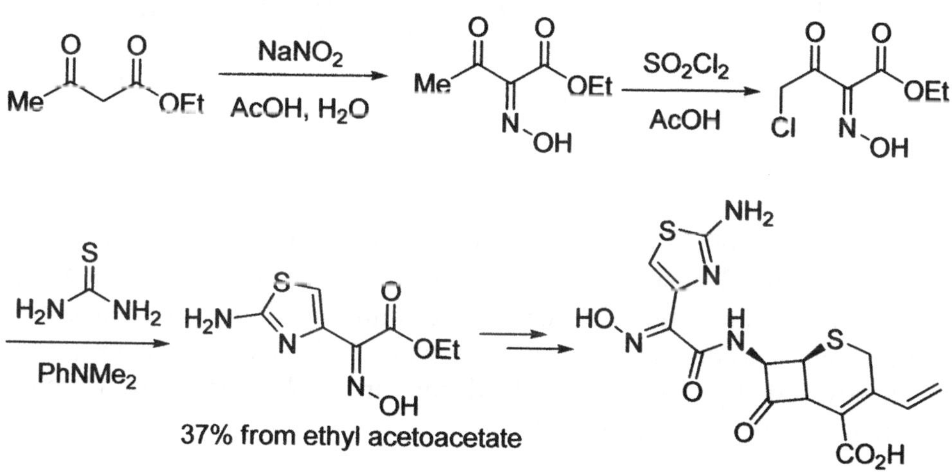

Famotidine, the orally administered H_2 receptor antihistamine drug for ulcer, possesses an interesting structure with a guanidine substituent on the 2-position of the thiazole ring. The thiazole is constructed in the early stages of the synthesis *via* a Hantzsch reaction in 84% yield wherein dichloroacetone and thiourea are condensed.

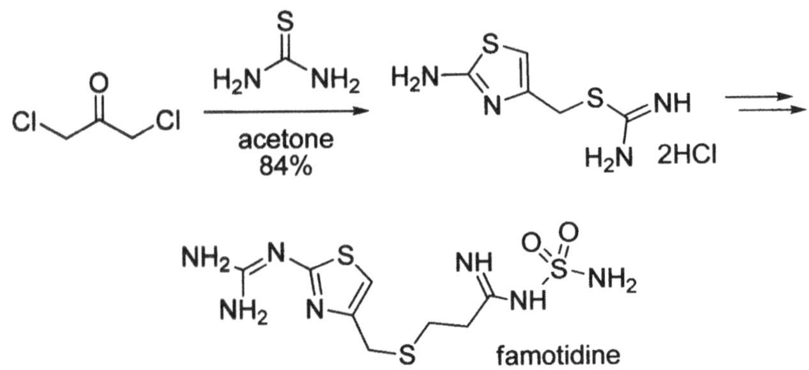

The Hantzsch reaction has also been incorporated in a multi-component reaction described by Rao et al. to produce 2-pyrazol-4-yl-substituted thiazole system in one step.[82] The reaction involves a one-pot cyclization to form thiazole, pyrazolone and cyclopropane rings *via* the reaction of aryl bromomethylketonewith, thiosemicarbazide, and α-acetyl-γ-butyro lactone in phosphorus oxychloride.

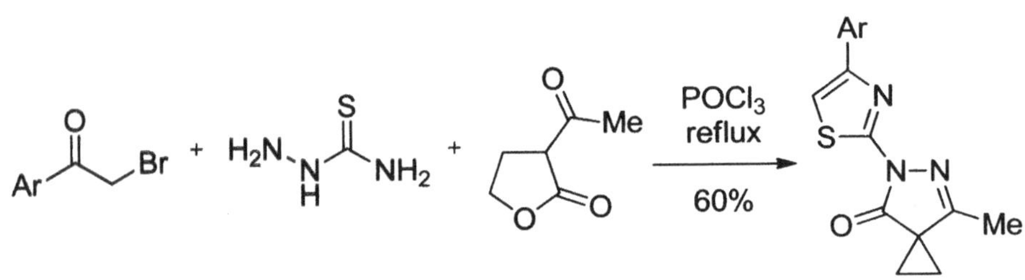

7.4.2 *Cook–Heilbron Synthesis of Thiazoles*

The Cook–Heilbron synthesis yields 5-amonothiazoles with various substituents on the 2-position of the thiazole.[83] Such a construction is brought about under mild conditions by reacting α-amino nitrile with salts and esters of dithioacids, carbon disulfide, carbon oxysulfide, isothiocyanates, *etc.* It was discovered by Cook and Heilbron in 1947 prior to which 5-aminothiazoles were less known.[84]

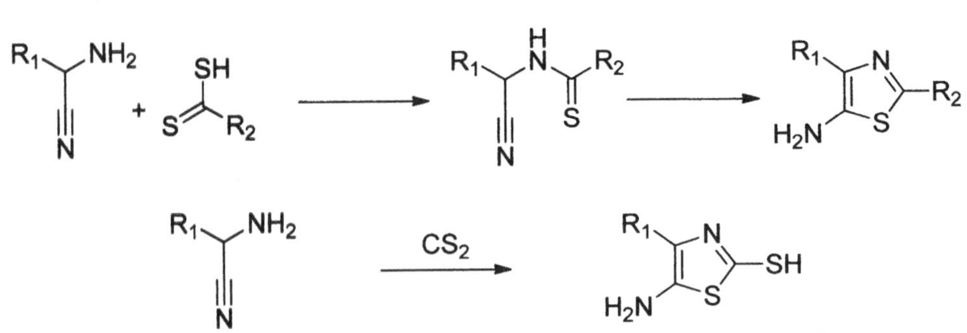

The reaction has not been extensively applied although a few examples in the literature exist. For example, Scott et al. have used this reaction in the synthesis of thiazolyl bisamide CSF-1R kinase inhibitors for the treatment of cancer.[85] The 2-methyl-5-aminothiazoles with different substituents were prepared *via* condensation of aminoacetonitrile and ethyl dithioacetate followed by cyclization and then coupled with carboxylic acid to yield bis-amides that were tested for *in vivo* activity.

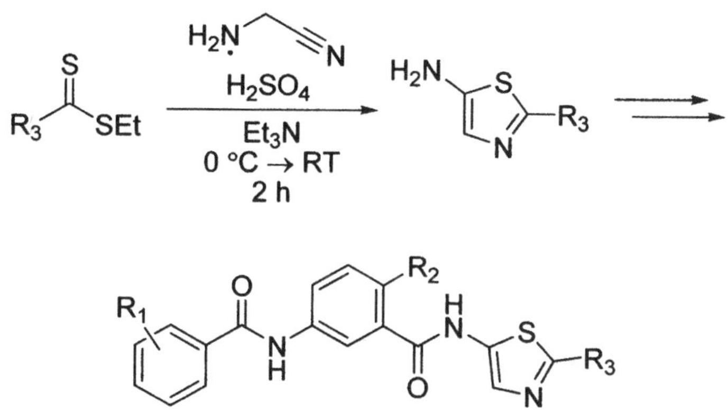

thiazolyl bisamides
R_1 = CF_3, Cl, Me
R_2 = Cl, Me,
R_3 = H, Me, ^{i}Pr, cyclopropyl

Another example involves an application of this reaction toward the synthesis of purines by Biagi and co-workers.[86] 5-Amino-2-phenylmethylaminothiazole-4-carboxamide was prepared as a substrate for a base-catalyzed thiazole rearrangement to yield a product that could be employed as a starting material for the synthesis of substituted purines.

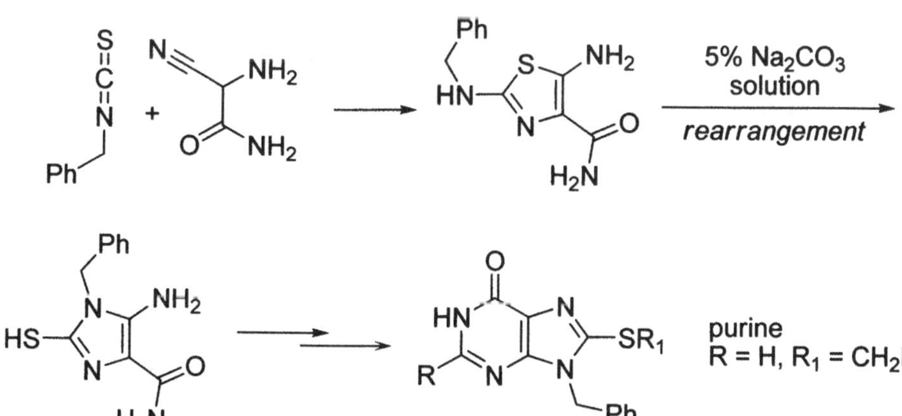

purine
R = H, R_1 = CH_2Ph

7.4.3 *Gabriel Synthesis of Thiazoles*

Warming an acylamino ketone with a stoichiometric amount phosphorus pentasulfide yields the corresponding thiazole product. The reaction was first discovered by Gabriel in 1910.[87]

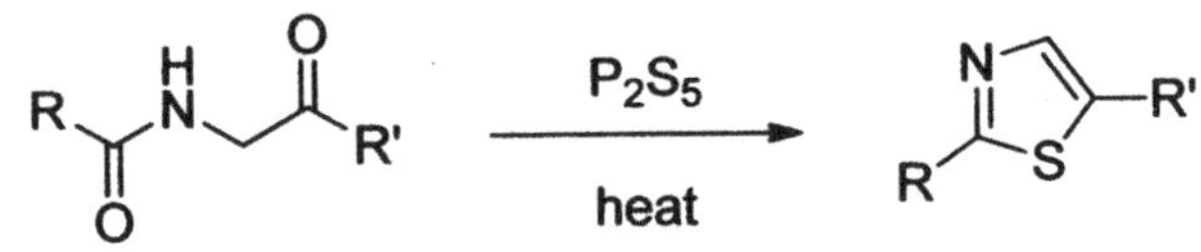

Several examples of this widely practiced name reaction have been applied in drug discovery abound. For example in a study describing the design and synthesis of inhibitors of hepatitis C virus (HCV) NS5B polymerase, thiazole intermediate was constructed *via* the Gabriel reaction.[88]

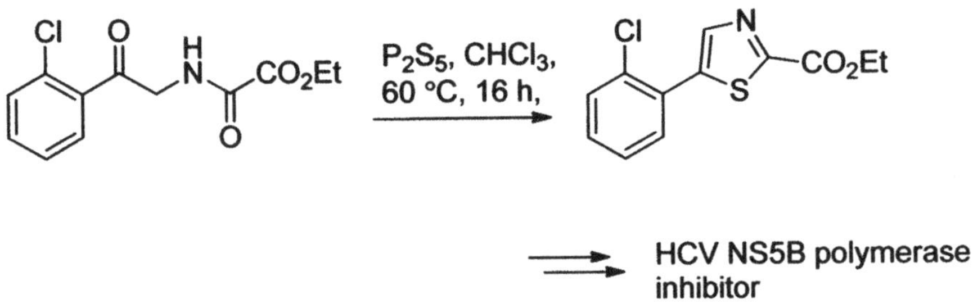

Similarly, Boyd and co-workers[89] at the R. W. Johnson Pharmaceutical Research Institute used this reaction for the synthesis of several imidazolyl thiazole derivatives to study significant R2A/D adrenoceptor binding affinity and *in vivo* anti-nociceptive activity in mice as part of a program to develop analgesic agents.

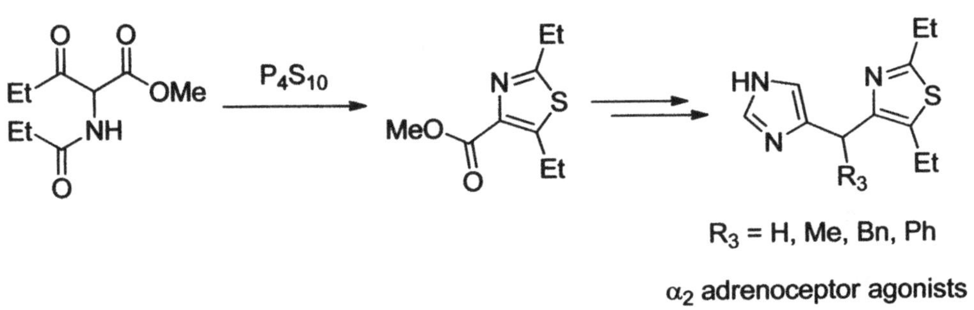

A variant of this transformation has been reported with Lawesson's reagent. Kiryanov *et al.* during their studies on cyclization of 1,4-dicarbonyl compounds mediated by Lawesson's reagent, found that 2-alkoxy-1,3-thiazole could be synthesized readily using a conventional microwave oven in high yields with little or no by-product being formed in short reaction times in the absence of a solvent.[90]

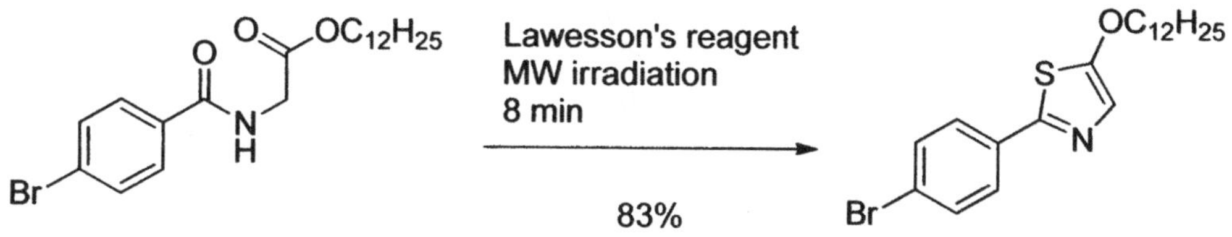

7.5 *Construction of the Benzothiazole Ring*

7.5.1 *From 2-Aminobenzenethiols*

In 1880, Hofmann reported the first preparation of benzothiazole from 2-aminobenzenethiol and formamide.[91] Hofmann stated that benzothiazoles were formed by the interaction of 2-aminobenzenethiol and aldehydes. Bogert et al. later confirmed the same as part of their own studies on the derivatives of thiazoles,[92] during which they synthesized 6-chloro-2-phenylbenzothiazole. Hofmann had noted only formation of 2- anilinobenzothiazole from the reaction of 2-aminothiophenol and phenyl isothiocyanate.

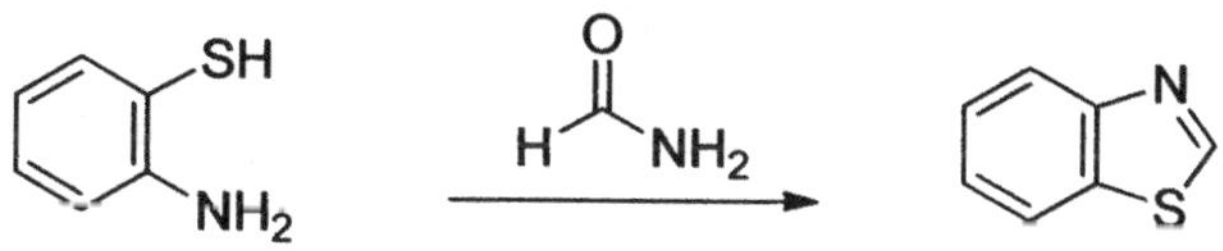

Variations of this reaction have been applied over the years, typically involving the condensation of 2-aminothiophenols with substituted carboxylic acids, acyl chlorides, aldehydes, and nitriles. Initially, the reaction involves the formation of an imine that cyclizes spontaneously and then oxidation to form benzothiazole. An application of this chemistry has been showcased in the synthesis of 2-(4-aminophenyl)benzothiazoles and the evaluation of their *in vitro* and *in vivo* activities against breast cancer cell lines, with compound **a** exhibiting the most potent growth inhibition.[93] Unfortunately, there are limitations due to the difficulties met during the syntheses of readily oxidizable *o*-aminothiophenol-bearing substituents.

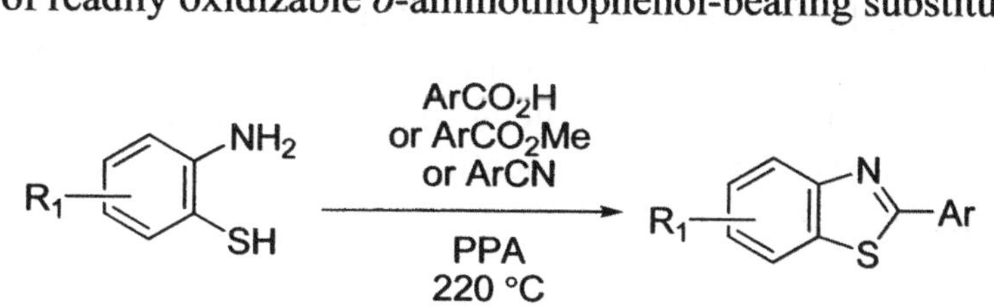

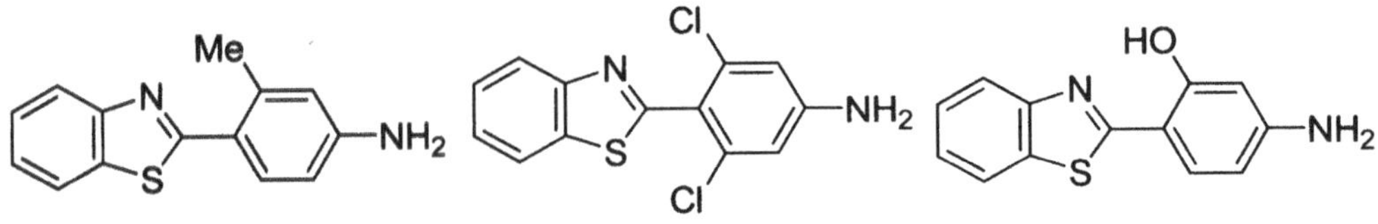

a **b** **c**

7.5.2 Hugerschoff Synthesis

The Hugerschoff reaction is a classical method to convert arylthioureas into aminobenzothiazoles under oxidative conditions. It was discovered by Hugerschoff in the early 1900s, who found that an arylthiourea can be cyclized with liquid bromine in chloroform to form a 2-aminobenzothiazole.[94] Solvents such as carbon tetrachloride and carbon disulfide can also be used. This transformation is aided by thiophilic bromine or its equivalent and requires an intramolecular aromatic electrophilic substitution reaction of the aryl ring to the thiocarbonyl group of thiourea. Attempts have been made to avoid the use of bromine due to the handling concerns it poses. For example, Jordan and coworkers have used benzyltrimethylammonium tribromide.[95,96]

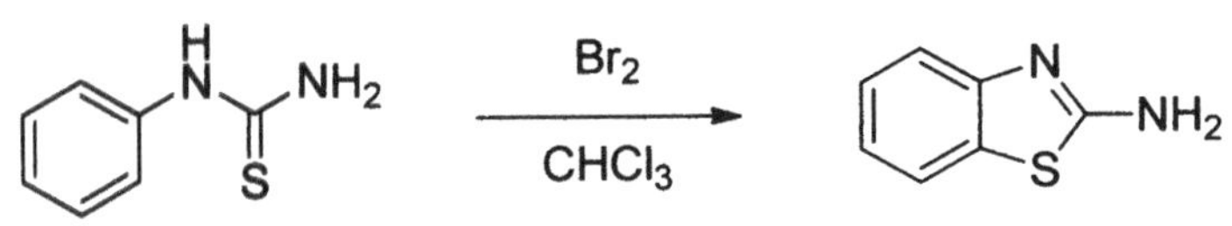

A side product possessing a thioamido guanidine moiety presumably forming *via* a disulfide intermediate is often observed along with the desired 2-aminobenzothiazole. Yella and co-workers have observed that for the Hugerschoff product to form, the aryl ring has to be sufficiently activated.[97] An application of this method is illustrated below where treatment of arylthiourea with bromine in dichloromethane provided 4-methoxybenzothiazol-2ylamine in > 95% yield.[98] The Medicinal Chemistry team at Amgen had identified AMG 628 as a highly efficacious vanilloid receptor-1 antagonist, the large-scale synthesis of which required the ready availability of the benzothiazole raw material. The Process Research team developed a scalable process for 4-methoxybenzothiazol-2ylamine, a commercially available yet expensive intermediate. The original cyclization conditions involved bromine in dichloromethane yielding 60% product. Optimization efforts led to the identification of bromine with lithium bromide in acetic acid.

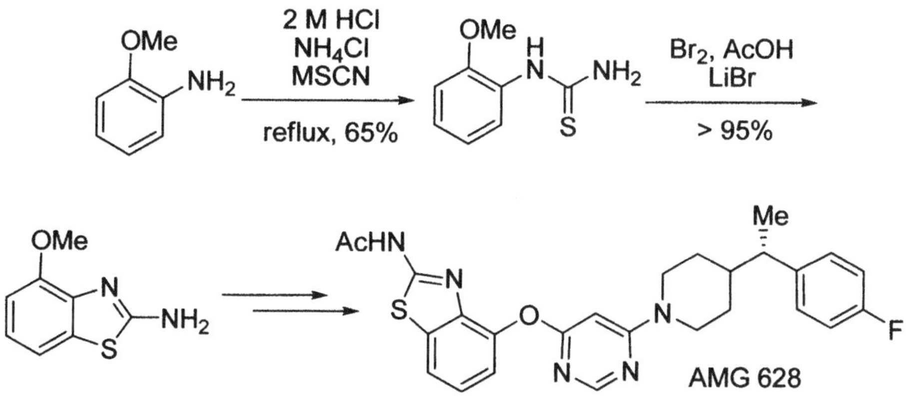

A variation of the oxidative cyclization of aryl thioureas has been reported to yield aminobenzothiazoles.[99] Based on this approach, Martinez et al. have described a facile and versatile approach to cyclize *N*-alkyl, *N′*-aryl, or heteroaryl thioureas *via* intramolecular oxidative cyclization in the presence of sulfuryl chloride in dichloromethane.[100]

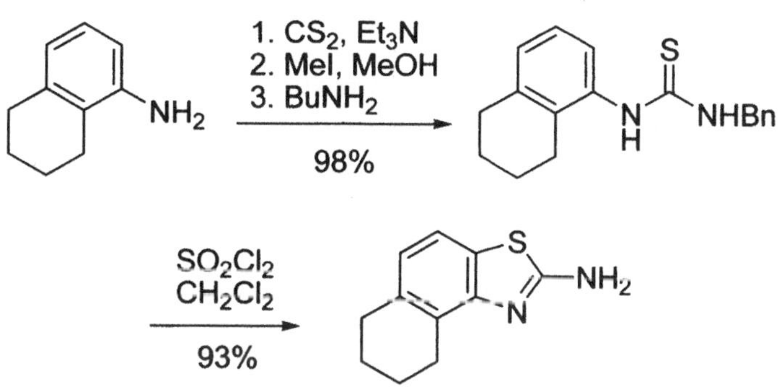

The Process Research and Development team at Janssen has applied this approach toward a facile and large-scale preparation of the anti-tumor agent R116010. It was demonstrated on a multi-kilogram scale. The intermediate *N*-phenyl-2-benzothiazolamine was synthesized by means of oxidative ring closure of the corresponding diphenylthiourea employing a mix of bromine and 48% aqueous hydrogen bromide. Subsequent Friedel–Crafts acylation with 2-chloropropionyl chloride led to α-chloroketone as the hydrochloride.[101]

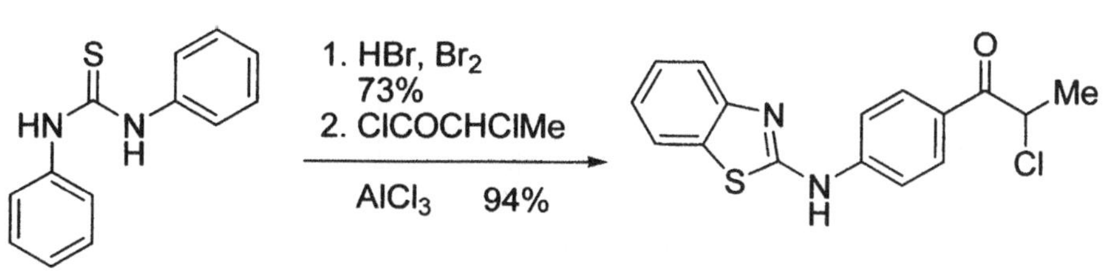

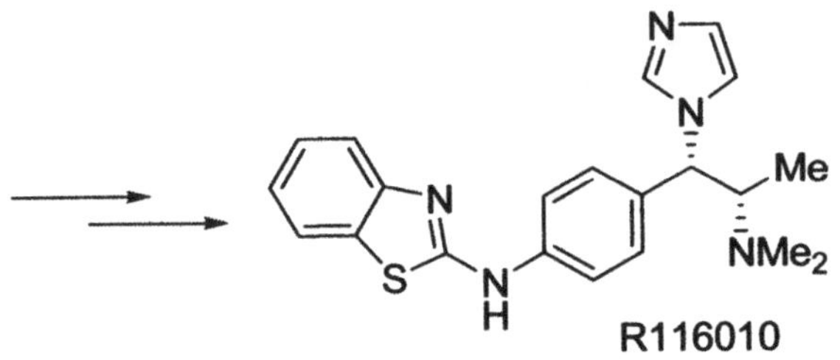

7.5.3 *Jacobson Cyclization*

In 1886, Jacobson reported the oxidative cyclisation of an aryl thioamide (or thiobenzanilide) to the corresponding 2-phenylbenzothiazole. Since then, the method has gained considerable acceptance and found applications in various syntheses of benzothiazole-containing molecules.[102]

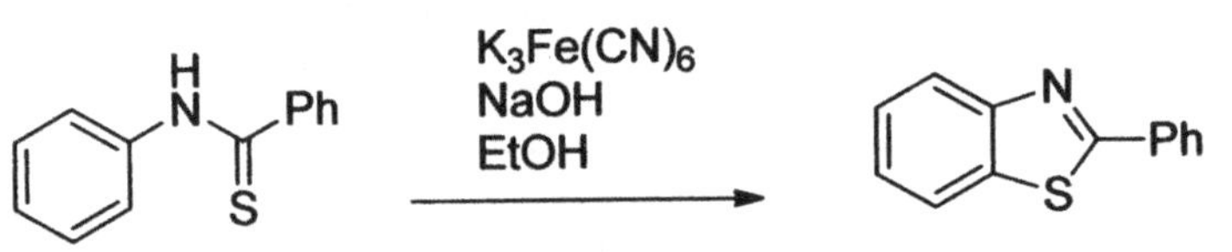

The mechanism of this transformation has been studied to some extent by Downer and Jackson[103] during their synthetic work on synthesis and structure verification of an analogue of kuanoniamine A, an alkaloid that has demonstrated *in vitro* cytotoxicity.[104]

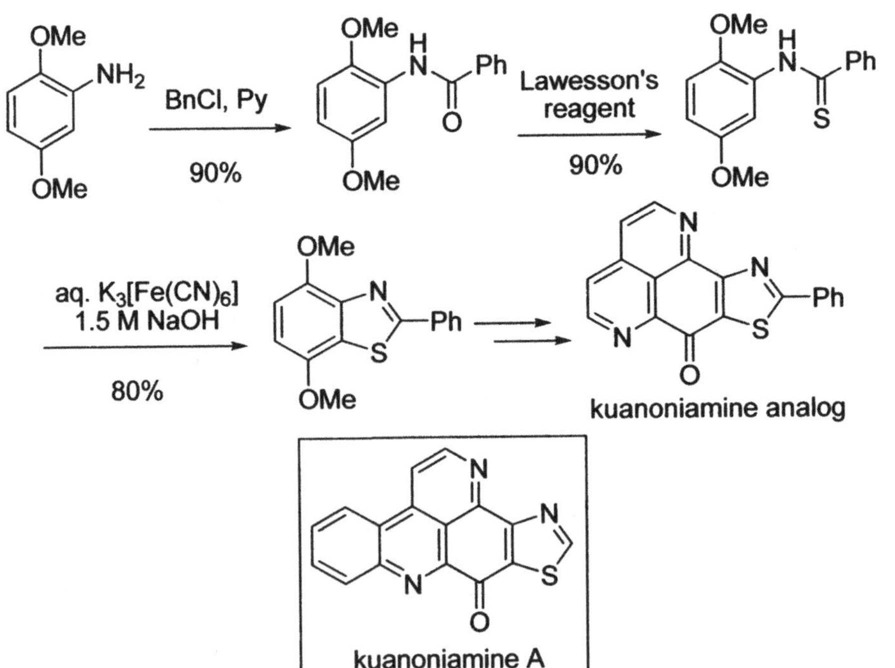

The benzothiazole synthesis in this case involved *ortho*-methoxythiobenzamides. Their studies indicated that the presence of electron-withdrawing groups on the primary ring caused a substantial decrease in the yield of this reaction. Jackson et al. studied this reaction and made some interesting observations *en route* to an analogue of the natural product. Two types of mechanism have been proposed for this transformation. The mechanistic proposal by Metzger and Planck is shown below and suggests the intermediacy of a thioimidic cation that attacks the benzene ring intramolecularly with the loss of a proton.[105]

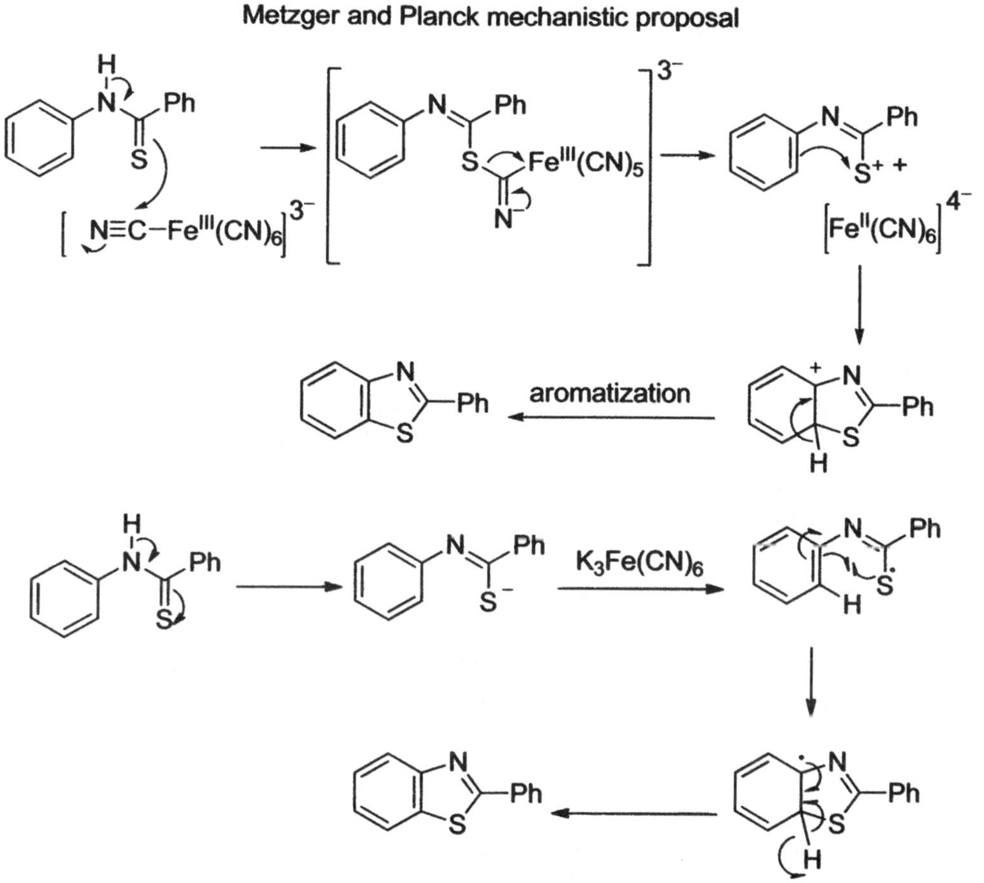

Stevens et al. have proposed another likely mechanism for the Jacobson synthesis where a single-electron transfer process appears to operate. Here thiobenzamide reacts with the base to generate thiolate ion that undergoes oxidation to form a thiol radical shown below.[106] The thiol radical can attack the unsubstituted *ortho* position and form a five-membered ring that aromatizes through the elimination of a hydrogen radical and form benzothiazole. Stevens et al. during this study also noted that the availability of N–H group was a significant factor to facilitate thione–thiolate conversion in thiobenzanilides and it was emphasized by the inability to cyclize *N*-alkylthiobenzanilides.

7.5.4 Miscellaneous Methods to Form Thiazole and Benzothiazole

Despite the use of well-established methods to construct thiazoles and benzothiazoles, several interesting methods employing varied strategies continue to emerge. For example, Zhan and co-workers at Xiamen University have reported a facile one-pot synthesis of three differently substituted thiazoles starting from propargylic alcohols.[107]

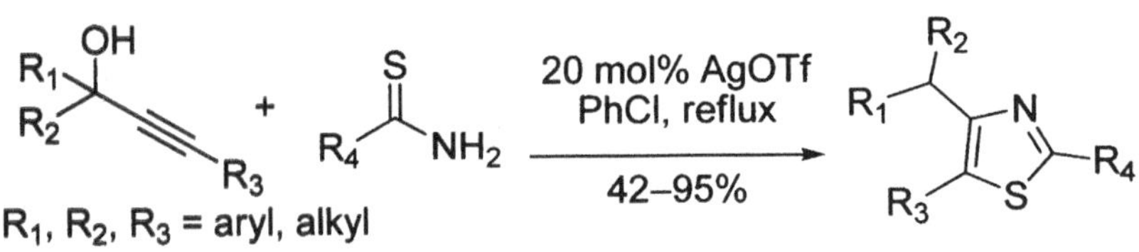

A range of thiazoles with various combinations of R_1, R_2, and R_3 substituents was synthesized using this method. The silver-catalyzed cycloaddition of propargylic alcohol with thioamide was proposed to proceed through the intermediacy of a propargylic cation or the corresponding allenyl cation and its subsequent reaction with the nucleophilic sulfur of thioamide, followed by a 5-*exo-dig* attack by nitrogen.

The formation of several 2-aminobenzothiazoles *via* palladium-catalyzed, direct intramolecular oxidative C–H functionalization was recently demonstrated by Batey at the University of Toronto.[108] The substrates used were *N*-aryl thioureas which in the presence of an interesting co-catalytic $Pd(PPh_3)_4/MnO_2$ system under oxygen atmosphere, yielded the desired products. In terms of the mechanism, this transformation proceeded presumably through electrophilic palladation, suggested by the higher reactivity of the more electron-deficient substrates.

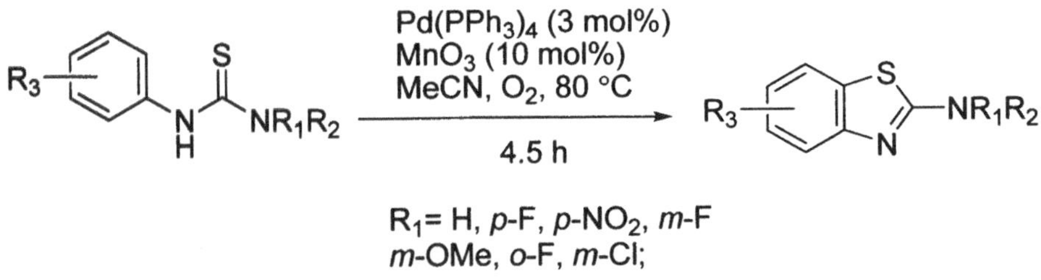

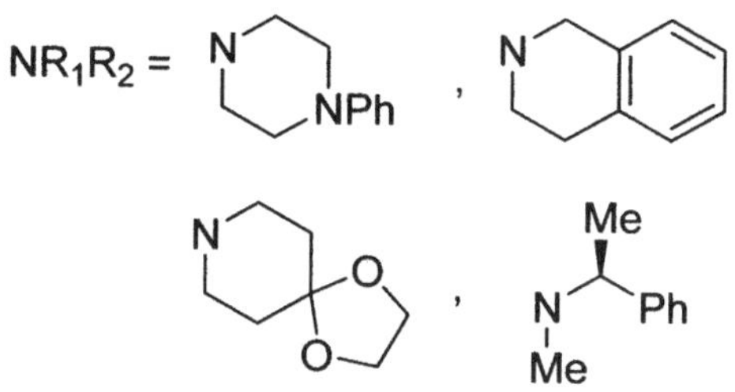

Photoisomerization of benzisothiazole to benzothiazole has been reported by Sharp et al. at Pfizer during their efforts to study photo-stability of ziprasidone, an anti-psychotic drug.[109] The transformation was proposed to proceed through an azirine intermediate followed by a nucleophilic displacement by sulphur either at nitrogen or carbon of the azirine ring, presumably assisted by an orthogonal aromatic π-bonding system in relation to the azirine nitrogen.

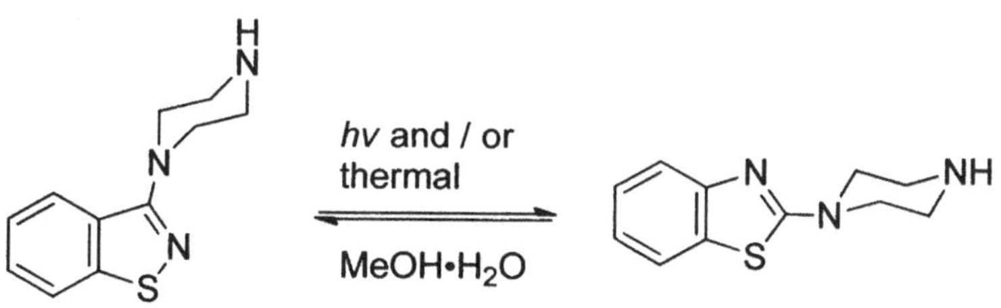

7.6 *Possible Liabilities of Drugs Containing Thiazoles and Benzothiazoles*

Commercially outsourced libraries as well as pharmaceuticals often have 1,3-thiazoles or benzothiazoles.[110] Yet thiazoles especially 2-aminothiazoles are considered as structural alerts and often excluded when considering the design of new drug candidates. The risk associated with structural alerts like these can be reduced by inducing an alternative metabolic pathway or simply low clinical exposure, but such an approach cannot accurately predict the human clinical response. An understanding of the "metabolic purpose" and the tendency to from reactive metabolites is therefore necessary. This is especially relevant from a position concerning hit triage and follow-up tactic strategies. The function of most cytochrome P450 (or CYP) enzymes, the major enzymes involved in drug metabolism, is to catalyze the oxidation of organic substances. 1,3-Thiazoles are prone to oxidative metabolism and typically undergo epoxidation. This happens at the 4,5-double bond and causes the formation of α-dicarbonyl metabolites and thioamide derivatives, such as thioamides, thioureas, or acylated thioureas. Both types of metabolites are capable of undergoing further metabolism to form reactive intermediates. For example, compounds with documented adverse reactions due to associated downstream reactive intermediates have been observed.[111]

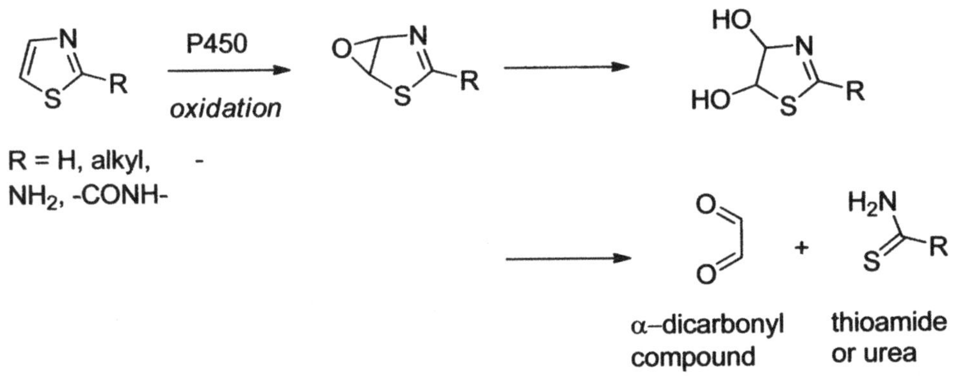

The existence of substituents at the 4- or 5-carbon can, however, delay this oxidative pathway as seen in some examples like meloxicam shown below. Meloxicam and sudoxicam are nonsteroidal anti-inflammatory drugs (NSAIDs) and belong to the enol–carboxamide category. They are structurally very similar; the only difference is the presence of an additional methyl group on the 5-carbon in the thiazole ring. Here, the 5-methyl group in meloxicam undergoes oxidative metabolism, which prevents the oxidative ring opening of the 2-amidothiazole. Sudoxicam, which has an unsubstituted 2-amidothiazole, has been observed to form oxidative ring opened products *in vivo.*[112] Thus, it has been associated with severe hepatotoxicity precluding its further use, while meloxicam has been on the market for more than a decade showing much less hepatotoxicity.[113]

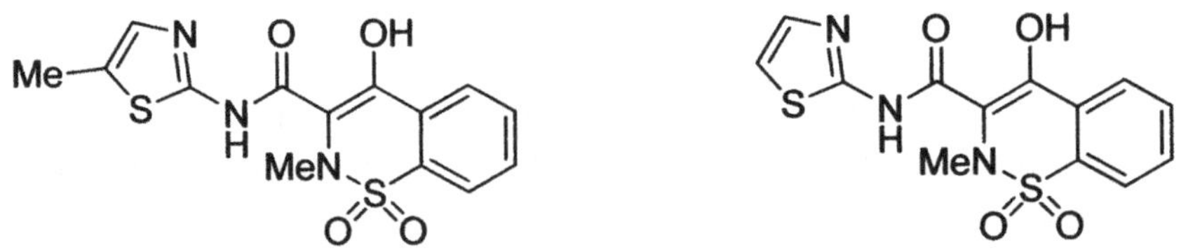

meloxicam sudoxicam

Another example is that of the HIV protease inhibitor ritonavir that contains two thiazole-rings and involves the oxidative ring-opening of thiazole to reactive intermediates. The oxidation appears to be a rate-limiting step in the mechanism based inactivation of the cytochrome P450 enzyme CYP3A4. Nonthiazole containing HIV protease inhibitors like indinavir, nelfinavir, or saquinavir are metabolized by CYP3A4. The inhibition of CYP3A4 by ritonavir has been found to result in a decreased metabolism of simultaneously administered protease inhibitors such as saquinavir or indinavir, thereby causing them to be cleared from the body more slowly.[114] Thus, ritonavir can boost the efficacy of other protease inhibitors, enabling the clinician to lower their dosing frequency.[115]

7.7 *Thiazoles and Benzothiazoles as Bioisosteres*

In the realm of medicinal chemistry, the concept of bioisosteres is commonly employed to find novel scaffolds starting from known, biologically active compounds. Bioisosteres are not exact structural mimetics but compounds with substituents or substructures that produce alike biological rather than physical properties in a molecule.[116] Ring-equivalent bioisosteres have often been used in drug discovery programs.[117] In several cases, thiazoles and benzothiazoles as bioisosteres have been utilized broadly toward improving potency, enhancing selectivity, altering physical properties, redirecting metabolism, as well as acquiring novel intellectual property. Some recent cases are discussed below.

Recently, Jang and co-workers discovered a novel class of compounds *in vitro* and *in vivo* immunosupressant activity, useful for preventing graft rejection after organ transplantation. Here, 7-*N*-piperazinyl[5,4-*d*]pyrimidine analogues were described as a novel class of immunosuppressive agents useful for preventing graft rejection following organ transplantation, with *in vivo* biological activity.[118] 1,3-Thiazole being a well-known isostere of pyridine, was introduced to replace the pyridine ring in an earlier analogue. A structure–activity relationship (SAR) was conducted. A comparative study of the thiazolo[5,4-*d*]pyrimidines with cyclosporine A, the marketed immunosuppressive drug suggested that the activity of the former was equal to that of the latter. Thus, 7-*N*-piperazinyl[5,4-*d*]pyrimidine was observed to be an excellent point toward the development of new generation of immunosuppressive drugs.

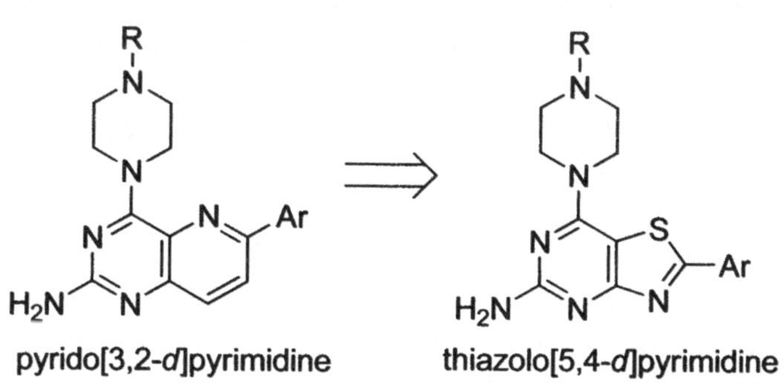

Another recent example where a thiazole was introduced as an isosteric replacement for a pyrazole was reported by medicinal chemists at AstraZeneca.[119] Compound AZ960 with IC50 $\leq$ 3 nM had been identified as a potent and selective Janus kinase 2 inhibitor (JAK2) acting as a competitive inhibitor of ATP. As part of their continued efforts, they demonstrated that AZ960 could be modified further by specifically replacing pyrazol-3-yl amine with thiazol-2yl amine, leading to potent JAK2 ATP-competitive inhibitors. Based on some earlier work, they hypothesized that the replacement was possible for two reasons: i) C–H hydrogen bond donor interaction between the pyrazole and pyrazine in AZ960 as well as A and ii) a favorable electrostatic interaction between the sulfur and nitrogen of the adjacent ring in the thiazole analogue B. The thiazole analogue A was found to show excellent JAK2 inhibition, thus demonstrating the bioisosteric role of thiazole in sustaining co-planarity with pyrazine ring.

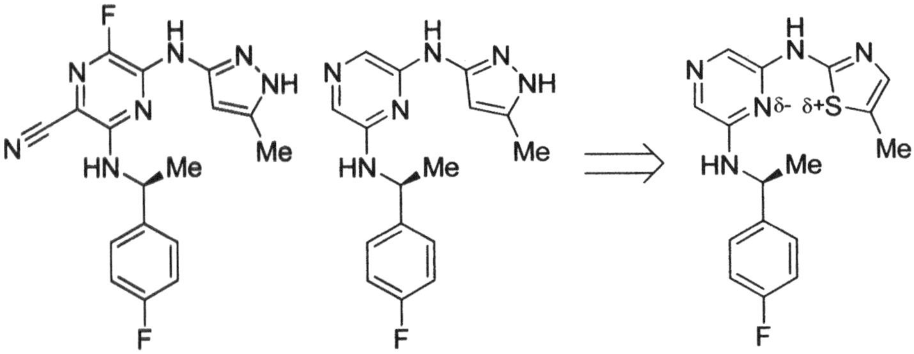

AZ960 A: JAK2 IC50 (μM) 0.004 B: JAK2 IC50 (μM) < 0.003

Another case is that of a benzothiazole as a bioisostere of benzofuran from Lanman et al.,[120] a program inspired by the impressive efficacy of the immunomodulator fingolimod (FTY-720) for multiple sclerosis. The team aimed at finding an $S1P_1$ receptor agonist with minimal activity at the $S1P_3$ receptor, since $S1P_3$ activity responsible for heart rate reduction. In their earlier work, the benzofuran-containing molecule was chosen as the lead molecule. A liability of pro-convulsive activity at oral doses, attributed to the benzofuran structure was observed. An investigation in pursuit of closely-related analogues displaying equivalent or improved pharmacological behavior without pro-convulsive activity was carried out. A variety of bioisosteric heterocycles were seen as possible replacements for benzofuran. On the whole, noteworthy *in vitro* as well as *in vivo* activity was demonstrated by the benzothiazole analogue with no pro-convulsive activity up to 100 mg/Kg, higher activity at $S1P_1$ receptor, and lower activity at $S1P_3$ than the molecule with benzofuranyl core.

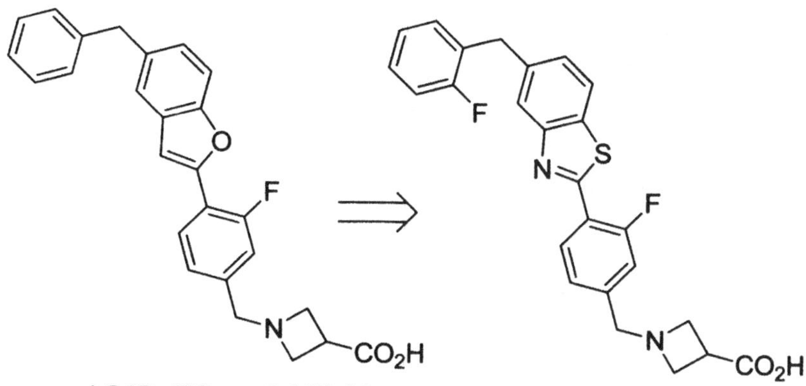

hS1P1 EC_{50} = 0.057μM
hS1P3 EC_{50} = 1.73 μM
Pro-convulsive in rats at
≥ 40 mg/Kg

hS1P1 EC_{50} = 0.042μM
hS1P3 EC_{50} = 3.47 μM
Non-convulsive in rats at
100 mg/Kg

7.8 *Problems*

7.8.1 Propose a reasonable mechanism for the following transformation[121]:

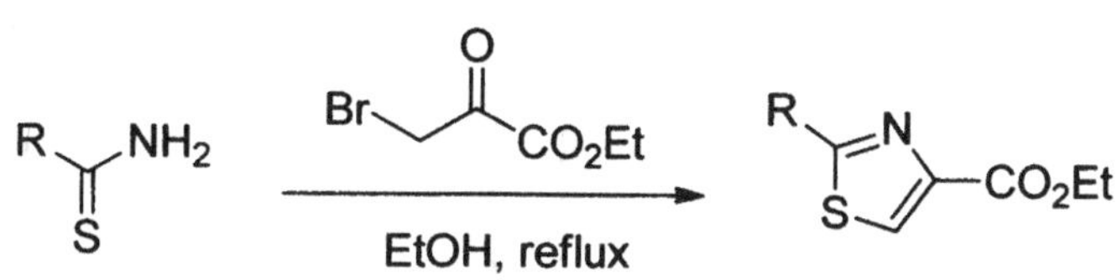

7.8.2 The compound 2-lithiothiazole like 2-(trimethylsilyl)thiazole (2-TST) has been used as a formyl anion equivalent used by Dondoni in the scheme below to prepare a chiral compound by a process called aminohomolgation of Garner aldehyde. Propose structures for products **A**, **B**, **C**, and **D** below.[122]

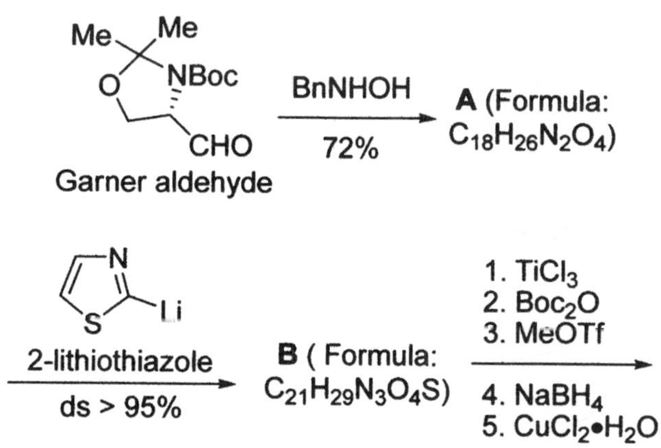

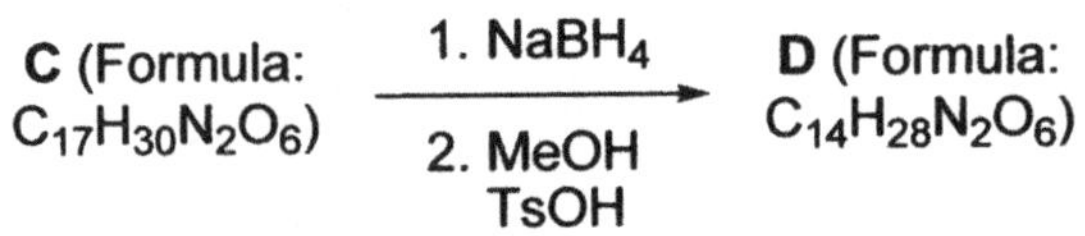

7.8.3 Propose a reasonable mechanism for the following transformation:

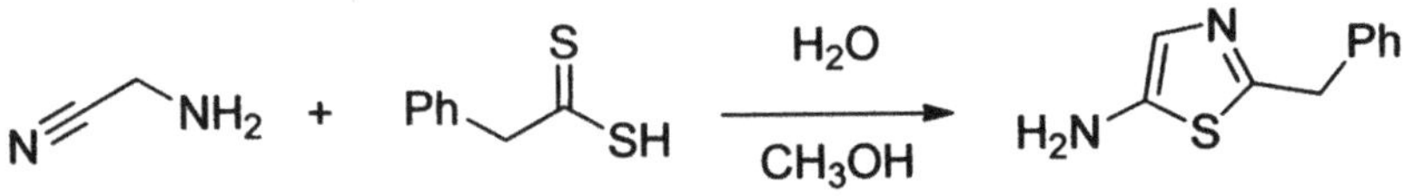

7.8.4 Predict the structure of product **E–H** from one of the following reactions.[123]

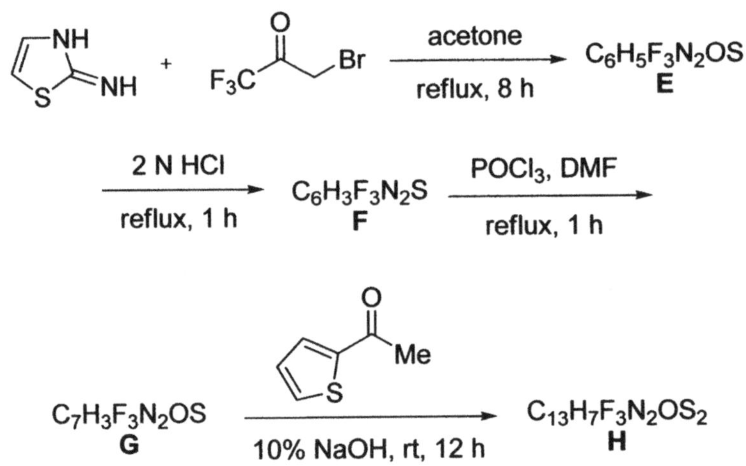

7.8.5 Predict the structure of product **I** from the following reaction.[124]

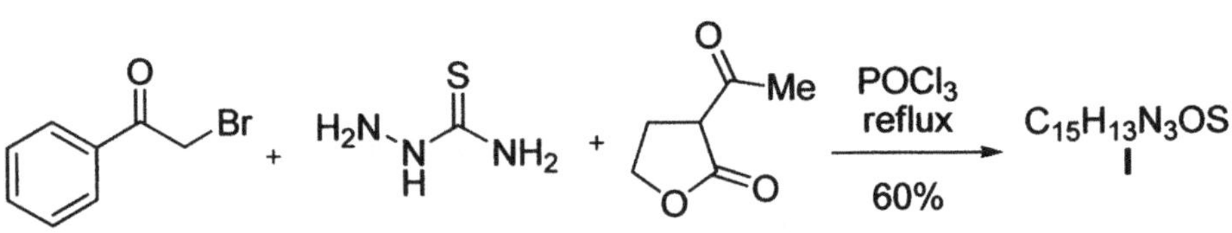

7.8.6 Predict the structure of product **J** from the following reaction.[125]

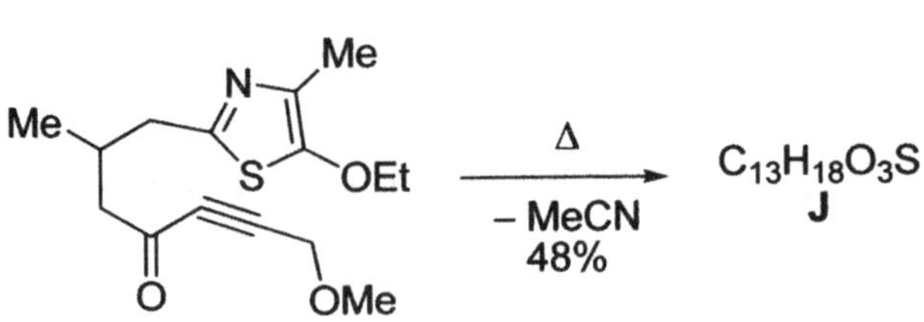

7.8.7 Propose a reasonable mechanism for the formation of the following 2-aminbenzothiazole.[126]

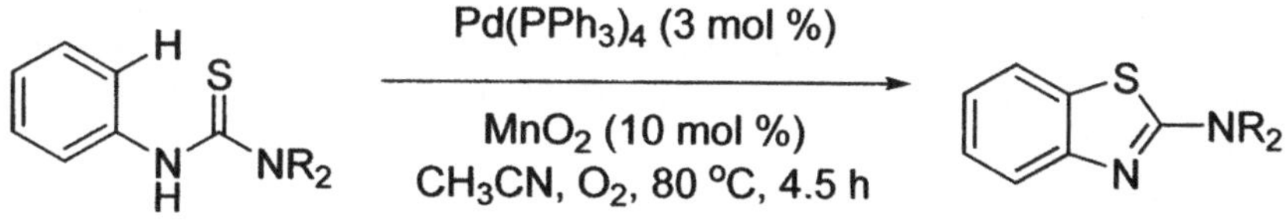

7.8.8 Suggest a reasonable mechanism for the following reaction to give the benzathiazole shown.[127]

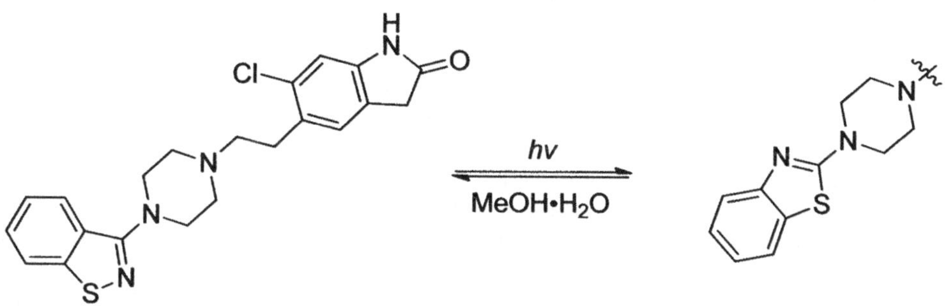

ziprasidone (Geodon)

7.8.9 Propose a reasonable mechanism for the conversion of an alkynyl alcohol and to the corresponding thiazole.[128]

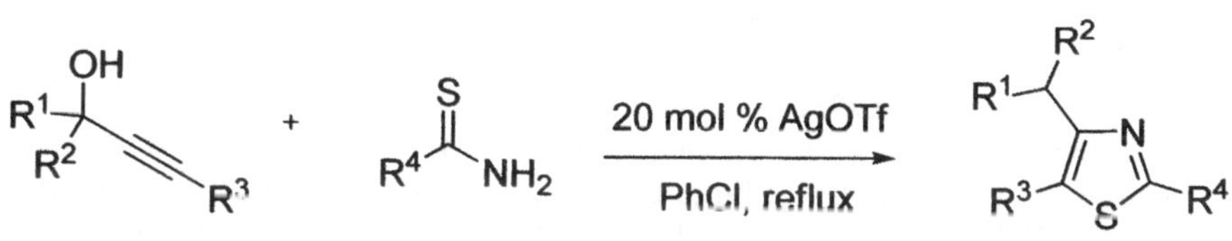

7.8.10 Bose and Idrees reported the following transformation in 2006 in yields ranging from 85% to 95%. Explain this transformation mechanistically.[129]

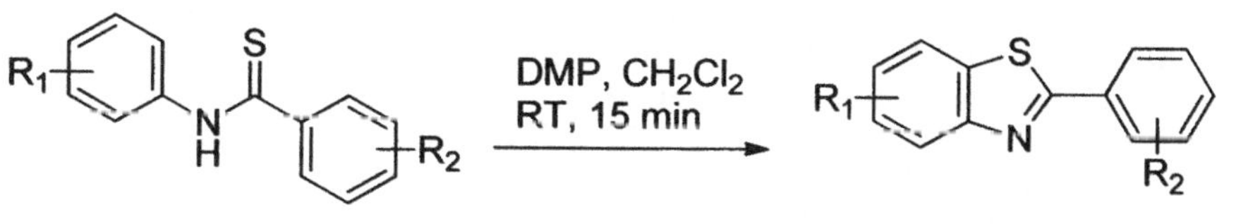

7.9 References

1. Nygaard, L.; Asmussen, E.; Høg, J. H.; Maheshwari, P. C.; Nielsen, C. H.; Petersen, I. B.; Rastrup-Andersen, J.; Sorensen, C. O. *J. Mol. Struct.* **1971,** *9,* 222.
2. Meyer, R.; Meyer, M.; Bares, D.; Vincent, E. J. *Thermochim. Acta*, **1975,** *11*, 211.
3. Toogood, P. L.; Hollenbeck, J. J.; Lava, H. M.; Li, L. *Bioorg. Med. Chem. Lett.* **1996,** *6*, 1543.
4. (a) Butler, M. S.; Buss, A. D. *Biochem. Pharmacol.* **2006**, *71*, 919. (b) Cystothiazole A.; Gebauer, J.; Arseniyadis, J.; Cossy, J. *Eur. J. Org. Chem.* **2008**, 2701.
5. James, M. N. G.; Watson, K. J. *J. Chem. Soc. (C)* **1966,** 1361.
6. Jadol, J.; Casimir, J.; Warin, R. *Bull. Soc. Chim. Belg.* **1969,** *78*, 299.
7. White, E. H.; Wörther, H.; Field, G. F.; McElroy, W. D. *J. Org. Chem.* **1965,** *30*, 2344.
8. Goodin, S.; Kane, M. P.; Rubin, E. H. *J. Clin. Oncol.* **2004,** *22*, 2015.
9. Bagley, M. C.; Dale, J. W.; Merritt, E. A.; Xiong, X. *Chem. Rev.* **2005,** *105*, 685.
10. LeFranc, D.; Ciufolini, M. A. *Angew. Chem. Int. Ed.* **2009,** *48*, 4198.
11. A substructure micrococcinic acid was synthesized for the first time by T. Ross Kelly, Boston College. Reference: Kelly, T. R.; Jagoe, C. T.; Zhengxiang, G. *Tetrahedron Lett.* **1991,** *32*, 4263. In 2005, Nicolaou *et al.* published an elegant total synthesis of thiostrepton. References: (a) Nicolaou, K. C.; Zak, M.; Safina, B. S.; Estrada, A. A.; Lee, S. H.; Nevalainen, M. *J. Am. Chem. Soc.* **2005,** *127*, 11159. (b) Nicolaou, K. C.; Safina, B. S.; Zak, M.; Lee, S. H.; Nevalainen, M.; Bella, M.; Estrada, A. A.; Funke, C.; Zecri, F. J.; Bulat, S. *J. Am. Chem. Soc.* **2005,** *127*, 11176.
13. Land, A. H.; Zeigler, C.; Sprague, J. M. *J. Org. Chem.* **1946,** *11*, 617.
14. (a) Inamoto, Y.; Chiba, T.; Kamimura, T.; Takaya, T. *J. Antibiot.* **1988,** *41*, 828. (b) Ternansky, R. J.; Christopher, L.; Jordan, F.; Bell, W.; Theresea, G.; Kasher, S. *J. Antibiot.* **1993,** *46*, 1897.
15. Brogden, R. N.; Spencer, C. M. *Drugs* **1997,** *53,* 483.
16. Guay, D. R. P. *Clin. Ther.* **2002,** *24*, 473.
17. Mooney, M. T.; Deguchi, S.: Tada, T.; Fujioka, M.; Okamoto, Y.; Yasuda, T. *Chem. Pharm. Bull.* **1995,** *92*, 374.
18. (a) Kempf, D. J.; Marsh, K. C.; Denissen, J. F.; McDonald, E.; Vasavanonda, S.; Flentge, C. A.; Green, B. E.; Fino, L.; Park, C. H.; Kong, X. P. *Proc. Natl. Acad. Sci. USA* **1995,** *92*, 2484. (b) Xu, L.; Liu, H.; Murray, B. P.; Callebaut, C.; Lee, M. S.; Hong, A.; Strickley, R. G.; Tsai, L. K.; Stray, K. M.; Wang, Y.; Rhodes, G. R.; Desai, M. C. *ACS Med. Chem. Lett.* **2010,** *1*, 209.
19. (a) Bennett, J. P.; Jr.; Piercey, M. F. J. *Neurol. Sci.* **1999,** *163*, 25. (b) Radad, K.; Gille, G.; Rausch, W.-D. *Pharmacol. Rep.* **2005,** *57*, 701. (c) Antonini, A.; Barone, P. *Neurol. Sci.* **2008,** *29*, S371.
20. Yanagisawa, I.; Hirata, Y.; Ishii, Y. *J. Med. Chem.* **1987,** *30*, 1787.
21. (a) Okamoto, K.; Eger, B. T.; Nishino, T.; Kondo, S.; Pai, E. F.; Nishino, T. *J. Biol. Chem.* **2003,** *278,* 1848. (b) Becker, M. A.; Schumacher, H. R., Jr.; Wortmann, R. L.; MacDonald, P. A.; Eustace, D.; Palo, W. A.; Streit, J.; Joseph-Ridge, J. *New. Engl. J. Med.* **2005,** *353,* 2450.
22. Paget, C. J.; Sands, J. L. DE2003841, (1970).
23. Paget, C. J.; Sands, J. L. U.S.4088768 (1978).
24. Yoshino, K.; Kohno, T.; Uno, T.; Morita, T.; Tsikamota, G. *J. Med. Chem.* **1986,** *29*, 820.
25. Bensimon, G.; Lacomblez, L.; Meininger, V. *N. Engl. J. Med.* **1994,** *330*, 585.

26. Metzger, J. V. *Chem. Heterocycl. Compd.* **1979,** *34-1*, 1.
27. Zahradnik, R.; Koutecky, J. *Collect. Czech. Chem. Commun.* **1961,** *26*, 156.
28. Erlenmeyer, K. *Helv. Chim. Acta.* **1945,** *28*, 985.
29. (a) Clark, J. H.; English, J. P.; Winnek, P. S.; Marson, H. W.; Cole, Q. P.; Clapp, J. W. *J. Am. Chem. Soc.* **1946,** *68*, 96. (b) Ochiai; Nagasawa, *Beritche* **1939,** *74*, 1470.
30. Lyashchuk, S. N.; Enya, V. I.; Doroshenko, T. F.; Skrypnik, Yu. G. *Russ. J. Org. Chem.* **2004,** *40,* 1647.
31. Ganapathi, V. *Proc. Ind. Acad. Sci., Section A* **1945,** *22*, 362.
32. English, J. P.; Clark, J. H.; Clapp, J. W.; Seeger, D.; Ebel, R. H. *J. Am. Chem. Soc.* **1946,** *58*, 453.
33. Oniga, O.; Grosu, I.; Mager, S.; Simiti, I. *Monatshefte fuer Chemie* **1998,** *12,* 661.
34. Branowska, D.; Farahat, A. A.; Kumar, A.; Wenzler, T.; Brun, R.; Liu, Y.; Wilson, W. D.; Boykin, D. W. *Bioorg. Med. Chem.* **2010,** *18,* 3551.
35. Metzer, J. V. In *The Chemistry of Heterocyclic Compounds. Thiazole and its Derivatives*; Wiley: New York, 1979; *Vol. 34, Part 1*, Chapter 1.
36. Dondoni, A.; Franco, S.; Merchan, F. L.; Merino, P.; Tejero, T. *Tetrahedron Lett.* **1993,** *34*, 5475.
37. (a) Tosa, M.; Pilbak, S.; Moldovan, P.; Paizs, C.; Szatzker, G.; Szakacs, G.; Novak, L.; Irimie, F.-D.; Poppe, L. *Tetrahedron: Asymmetry* **2008,** *19*, 1844. (b) Chikashita, H.; Ishibaba, M.; Ori, K.; Itoh, K. *Bull. Chem. Soc. Jap.* **1988,** *61*, 3637.
38. Florio, S.; Troisi, L.; Capriati, C. *Tetrahedron Lett.* **1998,** *39,* 7951.
39. Myllymäki, M. J.; Saario, S. M.; Kataja, A. O.; Castillo-Melendez, J. A.; Nevalainen, T.; Juvonen, R. O.; Järvinen, T.; Koskinen, A. M. P. *J. Med. Chem.* **2007,** *50*, 4236.
40. Molloy, K. C.; Waterfield, P. C.; Mahon, M. F. *J. Organomet. Chem.* **1989,** *365*, 61.
41. Kenney, B. D.; Breslav, M.; Chang, R.; Glaser, R.; Harris, B. D.; Maryanoff, C. A.; Mills, J.; Roessler, A.; Segmuller, B.; Villani, F. J., Jr. *J. Org. Chem.* **2007,** *72*, 9798.
42. Metzger, J. In *Comprehensive Heterocyclic Chemistry*; Katritzky, A. R.; Rees, C. W. Eds.; Pergamon Press: Oxford, 1984; *Vol. 6*, pp 235–331.
43. Deng, S.; Taunton, J. *Org. Lett.* **2005,** *7*, 299.
44. (a) Dondoni, A.; Marra, A. *Chem. Rev.* **2004,** *104,* 2557. (b) Dondoni, A.; Merino, P. *Org. Syn.* **1998,** *Coll. Vol.* **9**: 952.
45. Parkes, K. E. B.; Bushnell, D. J.; Crackett, P. H.; Dunsdon, S. J.; Freeman, A. C.; Gunn, M. P.; Hopkins, R. A.; Lambert, R. W.; Martin, J. A.; Merrett, J. H.; Redshaw, S., Spurden, W. C.; Thomas, G. J. *J. Org. Chem.* **1994,** *59*, 3656.
46. Dondoni, A.; Perron, D.; Merino, P. *J. Org. Chem.* **1995,** *60*, 8074.
47. Dondoni, A. *Pure Appl. Chem.* **1990,** *62*, 643.
48. Dondoni, A.; Fantin, G.; Fogagnolo, M.; Medici, A.; Pedrini, P. *J. Org. Chem.* **1988,** *58*, 1748.
49. *The Chemistry of Heterocycles: Organic Chemistry Monograph Series*. Editors Theophil Eicher, Siegfried Hauptmann. Translated by Hans Suschitsky 1995, Thieme.
50. Kamal, A.; Dastagiri, D.; Ramaiah, M. J.; Reddy, J. S.; Bharathi, E. V.; Srinivas, C.; Pushpavalli, S. N. C. V. L.; Pal, D.; Pal-Bhadra, M. *ChemMedChem* **2010,** *5*, 1937.
51. Ugai, T.; Tanaka, S.; Dokawa, S. *J. Pharm. Soc. Jpn. (Yakugaku Zasshi)* **1943,** *63*, 296.
52. Breslow, R. *J. Am. Chem. Soc.* **1958,** *80*, 3719.
53. Biju, A. T.; Kuhl, N.; Glorius, F. *Acc. Chem. Res.* **2011,** *44,* 1182.
54. For example: Phillips, E.; Chan, A.; Scheidt, K. A. *Aldrichimica Acta* **2009**, *42*, 55.
55. Kumar, D.; David, W. M.; Kerwin, S. M. *Biorg. Med. Chem. Lett.* **2001,** *11*, 2971.

56. Begtrup, M.; Hansen, L. B. L. *Acta Chem. Scand.* **1992,** *46*, 372.
57. Fisher, M. H.; Lusi, A. *J. Med. Chem.* **1972,** *15*, 982.
58. Acheson, R. M.; Foxton, M. W.; Miller, G. R. *J. Chem. Soc.* **1965,** 3200.
59. Reid, D. H.; Skelton, F. S.; Bonthron, W. *Tetrahedron Lett.* **1964**, *5*, 1797,
60. Jacobi, P. A.; Egbertson, M.; Frechette, R. F.; Miao, C. K.; Weiss, K. T. *Tetrahedron* **1988,** *44*, 3327.
61. Potts, K. T.; Choudhury, D. R.; Westby, T. R. *J. Org. Chem.* **1976,** *41*, 187.
62. Schnürch, M.; Hämmerle, J.; Mihovilovic, M. D.; Stanetty, P. *Synthesis* **2010,** *5*, 837.
63. Jang, M.-Y.; Lin, Y.; De Jonghe, S.; Gao, L.-J.; Vanderhoydonck, B.; Froeyen, M.; Rozenski, J. H.; Louat, T.; Van Belle, K.; Waer, M.; Herdewijn, P. *J. Med. Chem.* **2011,** *54*, 655.
64. Wang, X.; Sarris, K.; Kage, K.; Zhang, D.; Brown, S. P.; Kolasa, T.; Surowy, C.; El Kouhen, O. F.; Muchmore, S. W.; Brioni, J. D.; Stewart, A. O. *J. Med. Chem.* **2009,** *52*, 170.
65. Romagnoli, R.; Baraldi, P. G.; Brancale, A.; Ricci, A.; Hamel, E.; Bortolozzi, R.; Basso, G.; Viola, G. *J. Med. Chem.* **2011,** *54*, 5144.
66. Slee, D. H.; Chen, Y.; Zhang, X.; Moorjani, M.; Lanier, M. C.; Lin, E.; Rueter, J. K.; Williams, J. P.; Lechner, S. M.; Markison, S.; Malany, S.; Santos, M.; Gross, R. S.; Jalali, K.; Sai, Y.; Zuo, Z.; Yang, C.; Castro-Palomino, J. C.; Crespo, M. I.; Prat, M.; Gual, S.; Díaz, J.-L.; Saunders, J. *J. Med. Chem.* **2008,** *51*, 1719.
67. Lanman, B. A.; Cee, V. J.; Frohn, M.; Golden, J.; Muller, K. M.; Neira, S. C.; Pickrell, A. J.; Buys, J.; Fiorino, M.; Gore, A.; Horner, M.; Itano, A.; McElvain, M.; Middleton, S.; Schrag, M.; Vargas, H. M.; Xu, H.; Xu, Y.; Zhang, X.; Siu, J.; Burli, R. W.; Cheruku, S. R.; Lin, J.; Lobera, M.; Marantz, Y.; Rivenzon-Segal, D.; Schutz, N.; Sharadendu, A.; Yu, X.; Zhang, Z. *Med. Chem. Lett.* **2011,** *2*, 102.
68. Suryadevara, P. K.; Olepu, S.; Kraus, J. M.; Gelb, M. H.; Verlinde, C. L. M. J.; Schoepe, J.; Karimi, M.; Van Voorhis, W. C.; Buckner, F. S.; Lockman, J. W.; Ohkanda, J.; Hamilton, A. D. *J. Med. Chem.* **2009,** *52*, 3703.
69. Gu, Y. G.; Weitzberg, M.; Clark, R. F.; Xiangdong, X.; Li, Q.; Lubbers, N. L.; Yang, Y.; Beno, D. W. A.; Widomski, D. L.; Zhang, T.; Hansen, T. M.; Keyes, R. F.; Waring, J. F.; Carroll, S. L.; Wang, X.; Wang, R.; Healan-Greenberg, C. H.; Blomme, E. A.; Beutel, B. A.; Sham, H. L.; Camp, H. S. *J. Med. Chem.* **2007,** *50*, 1078.
70. Langille, N. F.; Dakin, L.; Panek, J. S. *Org. Lett.* **2002,** *4*, 2485.
71. Kelly, T. R.; Jagoe, C. T.; Gu, Z. *Tetrahedron Lett.* **1991,** *32*, 4263.
72. Li, J. J.; Carson, K. G.; Trivedi, B. K.; Yue, W. S.; Ye, Q.; Glynn, R. A.; Miller, S. R.; Connor, D. T.; Roth, B. D.; Luly, J. R.; Low, J. E.; Heilig, D. J.; Yang, W.; Qin, S.; Hunt, S. *Bioorg. Med. Chem.* **2003,** *11*, 3777.
73. Hantzsch, A.; Weber, J. H. *Ber.* **1887,** *20*, 3118.
74. Wiley, R. H.; England, D. C.; Behr, L. L. *Organic Reactions* **1951,** *6*, 367.
75. Ikemoto, N.; Liu, J.; Brands, K. M. J.; McNamara, J. M.; Reider, P. J. *Tetrahedron* **2003,** *59*, 1317.
76. Brown, A. D.; Davis, R. D.; Fitzgerald, R. N.; Glover, B. N.; Harvey, K. A.; Jones, L. A.; Liu, B.; Patterson, D. E.; Sharp, M. J. *Org. Proc. Res. Dev.* **2009,** *13*, 297.
77. Hasegawa, M. *Heterocycles* **1998,** *47*, 857.
78. Pesti, J.; Chen, C.-K.; Spangler, L.; Del Monte, A. J.; Benoit, S.; Berglund, D.; Bien, J.; Brodfuehrer, P.; Chan, Y.; Corbett, E.; Castello, C.; DeMena, P.; Discordia, R. P.; Doubleday, W.; Gao, Z.; Gingras, S.; Gross, J.; Haas, O.; Kacsur, D.; Lai, C.; Lueng, S.; Miller, M.; Muslehiddinoglu, J.; Nyugen, N.; Qui, J.; Olzog, M.; Reiff, E.; Thoraval, D.; Totleben, M.; Vanyo, D.; Vemishetti, P.; Wasylak, J.; Wei, C. *Org. Proc. Res. Dev.* **2009,** *13*, 716.

79. Dodson, R. M.; King, L. C. *J. Am. Chem. Soc.* **1945,** *67,* 2242.
80. Schneider, C.; Griss, G.; Hurnaus, R.; Kobinger, W.; Pichler, L.; Bauer, R.; Mierau, J.; Hinzen, D.; Schingnitz, G. Tetrahydrobenzothiazole, their Production and Use as Intermediates and Pharmaceuticals. Eur. Patent 186087B1, July 2, 1986.
81. González, M.; Rodriguez, Z.; Tolón, B.; Rodriguez, J. C.; Veléz, H.; Valdez, B.; Lopéz, M. A.; Fini, A. *Farmaco* **2003,** *58,* 409.
82. Rao, V. R.; Reddy, V. R. *Synth. Comm.* **2010,** *40,* 3186.
83. Cook, A.; Heilbron, I.; Levy, A. L. *J. Chem. Soc.* **1947,** 1594. Cook, A. H.; Heilbron, I.; MacDonald, S. F.; Mahadevan, A. P. *J. Chem. Soc.* **1949,** 1064.
84. Ahmad, N. M. *Cook-Heilbron 5-aminothiazole synthesis.* In *Name Reactions in Heterocyclic Chemistry,* Li, J. J., ed.; Wiley: Hoboken, NJ, 2005.
85. Scott, D. A.; Aquila, B. M.; Bebernitz, G. A.; Cook, D. J.; Dakin, L. A.; Deegan, T. L.; Hattersley, M. M.; Ioannidis, S.; Lyne, P. D.; Omer, C. A.; Ye, M.; Zheng, X. *Biorg. Med. Chem. Lett.* **2008,** *18,* 4794.
86. Biagi, G.; Giorgi, I.; Livi, O.; Pacchini, F.; Salerni, O. L.; Scartoni, V. *J. Het. Chem.* **2005,** *42,* 743.
87. (a) Gabriel, S. *Ber.* **1910,** *43,* 134. (b) Gabriel, S. *Ber.* **1910,** *43,* 1283.
88. Pfefferkorn, J. A.; Greene, M. L.; Nugent, R. A.; Gross, R. J.; Mitchell, M. A.; Finzel, B. C.; Harris, M. S.; Wells, P. A.; Shelly, J. A.; Anstadt, R. A.; Kilkuskie, R. E.; Koptab L. A.; Schwendea, F. J. *Bio. Med. Chem. Lett.* **2005,** *15,* 2481.
89. Boyd, R. E.; Press, J. B.; Rasmussen, C. R.; Raffa, R. B.; Codd, E. E.; Connelly, C, D.; Bennett, D. J.; Kirifides, A. L.; Gardocki, J. F.; Reynolds, B.; Hortenstein, J. T.; Reitz, A. B. *J. Med. Chem.* **1999,** *42,* 5064.
90. Kiryanov, A. A.; Sampson, P.; Seed, A. J. *J. Org. Chem.* **2001,** *66,* 7925.
91. Hofmann, A. *Ber.* **1880,** *13,* 1223. Hofmann, A. W. *Ber.* **1887,** *20,* 1788.
92. Bogert, M. T.; Corbitt, H. B. *J. Am. Chem. Soc.* **1926,** *48,* 783.
93. Shi, D.-F.; Bradshaw, T. D.; Wrigley, S.; McCall, C. J.; Lelieveld, P. Fichtner, I.; Stevens, M. F. G. *J. Med. Chem.* **1996,** *39,* 3375.
94. (a) Hugerschoff, A. *Chem. Ber.* **1901,** *34,* 3130. (b) Hugerschoff, A. *Chem. Ber.* **1903,** *36,* 3121.
95. Jordan, A. D.; Luo, C.; Reitz, A. B. *J. Org. Chem.* **2003,** *68,* 8693.
96. Le, Z.-G.; Xu, J.-P.; Rao, H.-Y.; Ying, M. *J. Heterocycl. Chem.* **2006,** *43,* 1123.
97. Yella, R.; Murru, S.; Ali, A. R.; Patel, B. K. *Org. Biomol. Chem.* **2010,** *8,* 3389.
98. Thiel, O. R.; Bernard, C.; King, T.; Dilmeghani-Seran, M.; Bostick, T.; Larsen, R. D.; Faul M. M. *J. Org. Chem.* **2008,** *73,* 3508.
99. Griffin, T., Woods, T.; Klayman, D. Adv. *Het. Chem.* **1975,** *18,* 99.
100. Castr, A.; Martinez, A. *J. Het. Chem.* **1999,** *36,* 991.
101. Aelterman, W.; Lang, Y.; Willemsens, B.; Vervest, I.; Leurs, S.; De Knaep, F. *OPRD,* **2001,** *5,* 467.
102. Jacobson, P. *Ber.* **1886,** *19,* 1068. Jacobson, P. *Ber. Dtsch. Chem. Ges.* **1886,** *19,* 1067.
103. Downer, N. K.; Jackson, Y. A. *Org. Biomol. Chem.* **2004,** *2,* 3039.
104. Lyon M. A.; Lawrence, S.; Williams, D. J.; Jackson, Y. A. *J. Chem. Soc., Perkin Trans. I* **1998,** 437.
105. Metzger, J.; Planck, H. *Chim. Ind.* **1956,** *75,* 930.
106. Stevens, M. F. G.; McCall, C. J.; Lelieveld, P.; Alexander, P.; Richter, A. *J. Med. Chem.* **1994,** *37,* 1689.
107. Gao, X.; Pan, Y.; Lin, M.; Chen, L.; Zhan, Z. *Org. Bio. Chem.* **2010,** *8,* 3259.
108. Joyce, L. L.; Batey, R. A. *Org. Lett.* **2009,** *11,* 2792.
109. Sharp, T. R.; Leeman, K. R.; Bryant, D. E.; Horan, G. J.; *Tetrahedron Lett.* **2003,** *44,* 1559.

110. *Mass Spectroscopy in Drug Metabolism and Disposition: Basic Principles and Applications*. Editor Lee, M. S.; Zhu, M. Chapter 3.
111. Burger's Medicinal Chemistry, Drug Discovery and Development, Seventh Edition. Editors Donald J. Abraham and David P. Rotella, 2010, Wiley: Hoboken, NJ, p301.
112. Hobbs, D. C.; Twomey, T. M. *Drug Met. Disp.* **1977,** *5*, 75.
113. Obach, R. S.; Kalgutkar, A. S.; Ryder, T. F.; Walker, G. S. *Chem. Res. Toxicol.* **2008,** *21*, 1890.
114. Zeldin, R. K.; Petruschke, R. A. *J. Antimicrob. Chemother.* **2004,** *53*, 4.
115. Merry, C.; Barry, M. G.; Mulcahy, F.; Ryan, M.; Heavey, J.; Tjia, J. F.; Gibbons, S. E.; Breckenridge, A. M.; Back, D. J. *AIDS* **1997,** *11,* F29.
116. Meanwell, N. A. *J. Med. Chem.* **2011,** *54*, 2529.
117. Burger, A. *Prog. Drug Res.* **1991,** *37*, 287.
118. Jang, M.-Y.; Lin, Y.; De Jonghe, S.; Gao, L.-J.; Vanderhoydonck, B.; Froeyen, M.; Rozenski, J. H.; Louat, T.; Van Belle, K.; Waer, M.; Herdewijn, P. *J. Med. Chem.* **2011,** *54*, 655–688.
119. Ioannidis, S.; Lamb, M. L.; Almeida, L.; Guan, H.; Peng, B.; Bebernitz, G.; Bell, K.; Alimzhanov, M.; Zinda, M. *Biorg. Med. Chem. Lett.* **2010,** *20*, 1669.
120. Lanman, B. A.; Cee, V. J.; Frohn, M.; Golden, J.; Muller, K. M.; Neira, S. C.; Pickrell, A. J.; Buys, J.; Fiorino, M.; Gore, A.; Horner, M.; Itano, A.; McElvain, M.; Middleton, S.; Schrag, M.; Vargas, H. M.; Xu, H.; Xu, Y.; Zhang, X.; Siu, J.; Burli, R. W.; Cheruku, S. R.; Lin, J.; Lobera, M.; Marantz, Y.; Rivenzon-Segal, D.; Schutz, N.; Sharadendu, A.; Yu, X.; Zhang, Z. *Med. Chem. Lett.* **2011,** *2*, 102.

121 Kempson, J. In *Name Reactions in Heterocyclic Chemistry-II*, Li, J. J., Ed.; Wiley: Hoboken, NJ, 2011, pp 299–309.

122 Dondoni, A.; Merchan, F. L.; Merino, P.; Tejero, T.; Bertolasi, V. *J. Chem. Soc., Chem. Commun.* **1994,** 1731.

123 Kamal, A.; Dastagiti, D.; Pal-Bhadra, M.; *et al. ChemMedChem* **2010,** *5,* 1937–1947.

124 Rao, V. R.; Reddy, V. R. *Synth. Commun.* **2010,** *40,* 3186–3195.

125 Jacobi, P. A.; Egbertson, M.; Frechette, R. F.; Miao, C.; Weiss, K. T. *Tetrahedron* **1988,** *44,* 3327–3338.

126 Joyce, L. L.; Batey, R. A. *Org Lett.* **2009,** *11*, 2792–2795.

127 Sharp,T. R.; Leeman, K. R.; Eugene Bryant, D.; Horan, G. J. *Tetrahedron Lett.* **2003,** *44*, 1559–1561.

128 Gao, X.; Pan, Y.-M.; Lin, M.; Chen, L.; Zhan, Z.-P. *Org. Bio. Chem.* **2010,** *8*, 3259–3266.

129 Bose, D. S.; Idrees, M. *J. Org. Chem.* **2006,** *71*, 8261–8263.

Chapter 8 Imidazoles and Benzimidazoles

Amy B. Dounay and Timothy T. Curran

8.1 Introduction to Imidazole

The parent compound 1*H*-imidazole[1] is a white to off-white solid at room temperature that has a melting point of 89–91 °C. This is due largely to the ability of the *N*-unsubstituted material to self-associate and not due to its polarity. Imidazole is a planar, 5-membered heterocyclic ring containing 2 nitrogen atoms at the 1- and 3-positions. The two nitrogen atoms have differing reactivities: one like pyridine and the other like pyrrole. Imidazole is a highly polar, water-soluble compound with a calculated dipole of 3.61 D. Imidazole is classified as an aromatic compound with 6π electrons and has a resonance energy of ~ 50 kJ/mol. The resonance energy of imidazole is lower than benzene yet is high enough to not readily abandon its aromaticity.

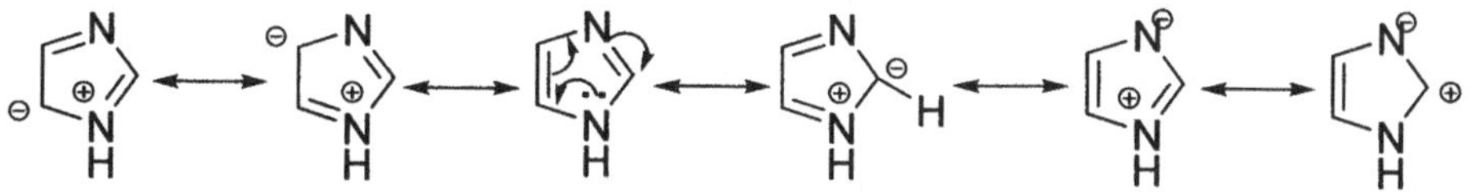

Resonance Structures of Imidazole

Imidazole is an amphoteric molecule, acting as both an acid and a base. As an acid, it has a p*K*a of about 14.9, making it less acidic than carboxylic acids or phenol but more acidic than alcohols. As a base, imidazole has a p*K*b of about 7, making imidazole much more basic than pyridine.

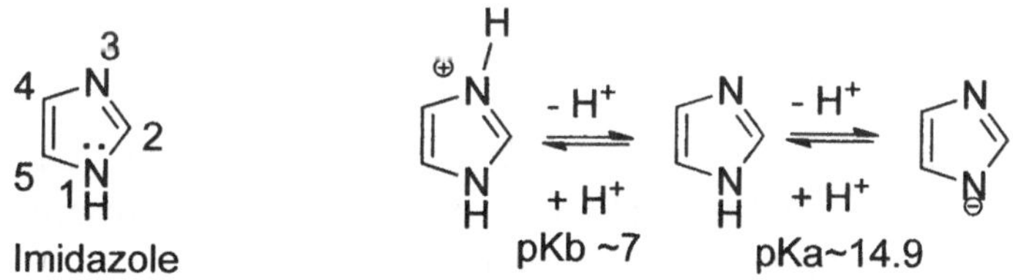

Imidazole Numbering and p*K*a's

In addition to its acid and base nature, 4-substituted-1*H*-imidazole can readily convert into 5-substituted materials. This tautomerization provides some synthetic challenges and also is manifested in peculiar nomenclature. For example, *N*-unsubstituted imidazoles are named as the 4(5)-substituted imidazole. Electron-withdrawing groups favor the 4-substituted tautomer.

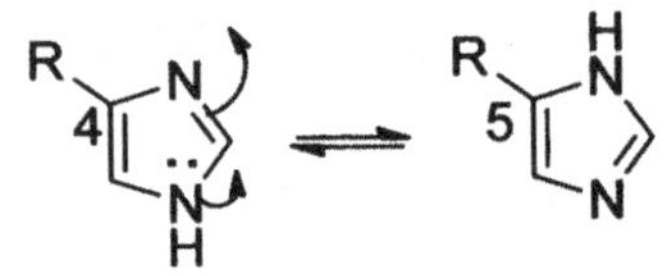

1*H*-Imidazole Conversion between 4- and 5-Substituted Products

Additionally, a hydroxyl or thiol substituent at the 2-position of imidazole may exhibit reactivity of the ketone or thione as illustrated in the resonance structures below.

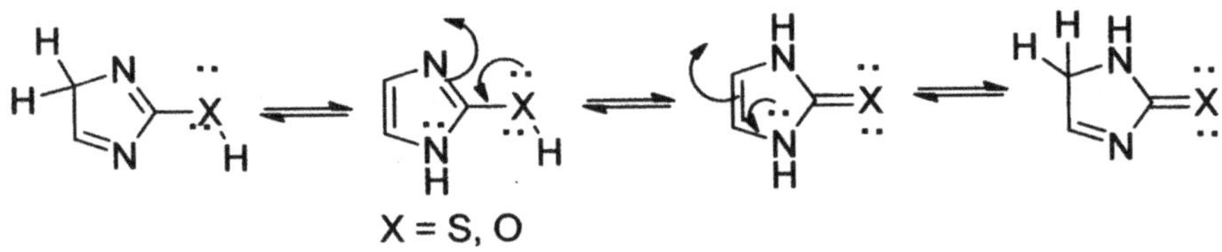

Tautomerism of 2-Hydroxy- or Thioxy-Imidazole

The proton NMR spectrum of 1H-imidazole in $CDCl_3$ shows two distinct aromatic signals. The H4 and H5 give a single peak due to tautomerization, and the H2 is at the higher field. The ^{1}H-NMR coupled with NOE experiments typically enables determination of substituents at the 4- or 5-positions.

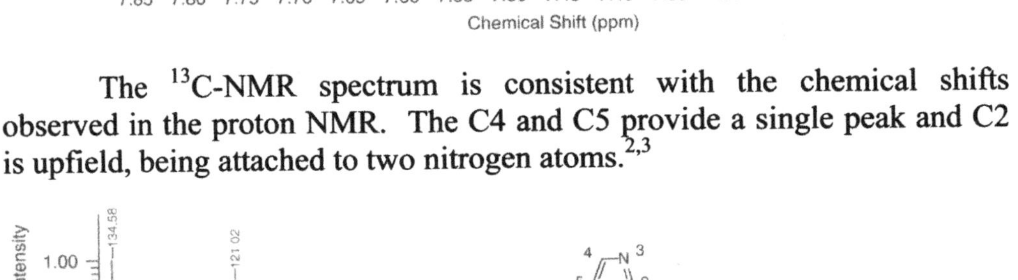

The ^{13}C-NMR spectrum is consistent with the chemical shifts observed in the proton NMR. The C4 and C5 provide a single peak and C2 is upfield, being attached to two nitrogen atoms.[2,3]

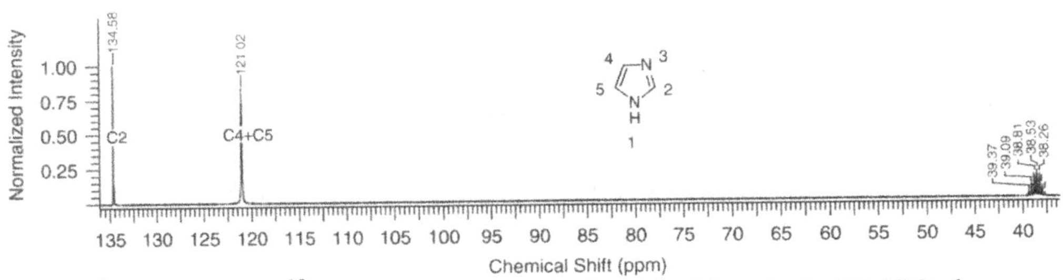

^{1}H-NMR and ^{13}C-NMR Spectrum of 1*H*-Imidazole in DMSO-d_6

The imidazole molecule is incorporated into many biologically relevant molecules. For example, the amino acid histidine is present in proteins and enzymes, while histamine (its decarboxylated product) is important in the body's immune response system. The imidazole ring has been found in a number of natural products, particularly the oroidin-derived alkaloids. One of the structurally simple compounds in this class of natural products is girolline, containing a 2-amino-imidazole attached to a chlorohydrin moiety.[4]

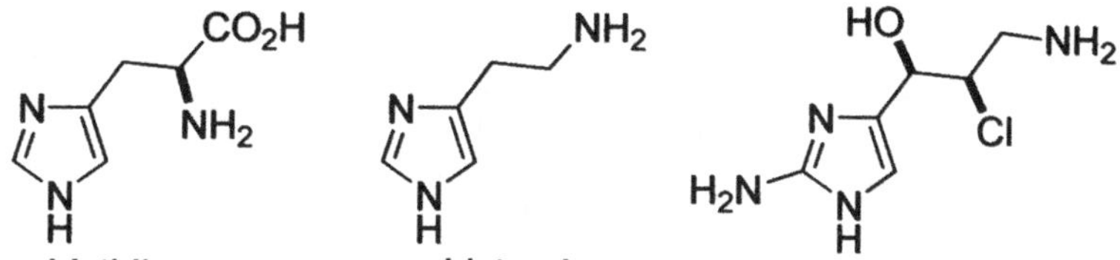

histidine
α-amino acid

histamine
immune response

girolline

Similarly, several medicines or potential medicines have been made that contain imidazole. Structurally, the molecules can be vastly different and can affect very unique disease states. For example, metronidazole is an antibiotic effective against anaerobic bacteria, and it also acts as an anti-parasitic agent. It is a relatively simple functionalized imidazole.[5] Cipralisant is a potent H3 receptor antagonist; H3 antagonists are believed to have potential therapy to aid in sleep, cognition, food intake, and memory.[6]

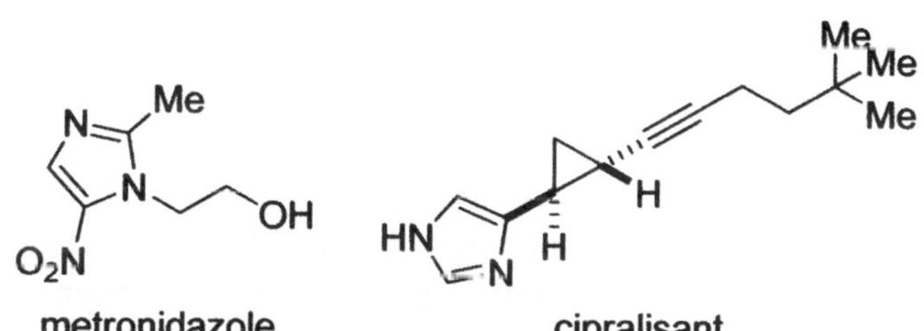

metronidazole
antiprotozoal/antibiotic

cipralisant
H3 receptor antagonist

8.2 *Reactivity of the Imidazole Ring*

8.2.1 *Nitrogen Alkylation*

Alkylation of the nitrogen occurs readily and is either via direct S_N2 or S_N2', depending on the basicity of the reaction mixture and the electrophile. Sterics of the *N*-alkylating group with other substituents will also play a role. For example, 4(5)-iodo-imidazole is reported to yield the *N*-tosyl compound in 69% yield. Fully deprotonated imidazole predominantly provides direct S_N2-type reactions and tends to provide less quaternization. Excellent yield of alkylation is obtained using NaH as base with several electrophiles (MOMCl, SEMCl, and BnBr). The mixture of resulting 4- and 5-iodo-products can be converted to 4-iodoimidazole selectively by warming with the electrophile in DMF at elevated temperature, thereby obtaining a high yield of a single regioisomer. This likely occurs via quaternization.[2–4]

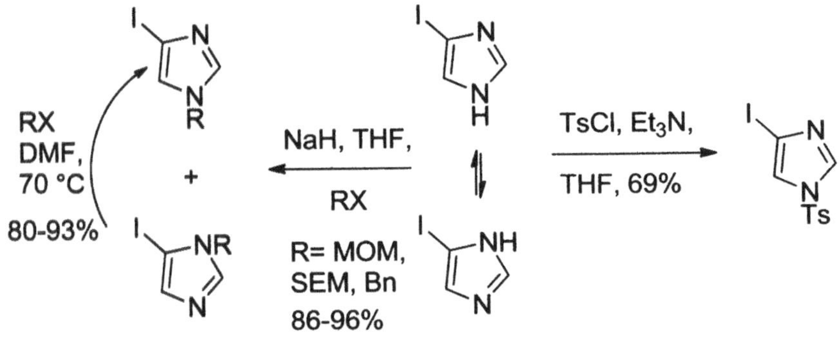

A similar thermal equilibration has been shown for the preparation of 4-nitro-*N*-methoxymethyl imidazole. The initial alkylation ratio was about 3:1 with a mixture of 4- and 5-nitro compounds. The addition of more base and MOMCl to the mixture with heating served to convert the compound into the 4-nitro imidazole.[7]

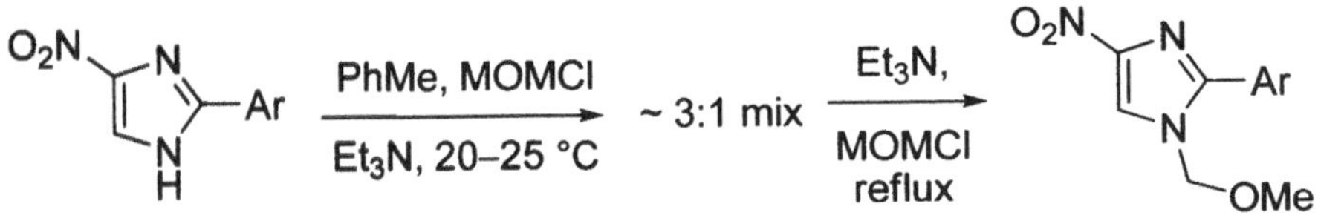

Quaternization is sometimes difficult to avoid but can be used positively. Quaternization can provide entry into 5-substituted *N*-alkylated imidazoles. For example, *N*-ethylation of *N*-benzoyl-4-phenylimidazole with a powerful alkylating agent like triethyloxonium tetrafluoroborate provides the salt that undergoes hydrolysis under mildly basic conditions to provide 5-phenyl-*N*-ethylimidazole.[2]

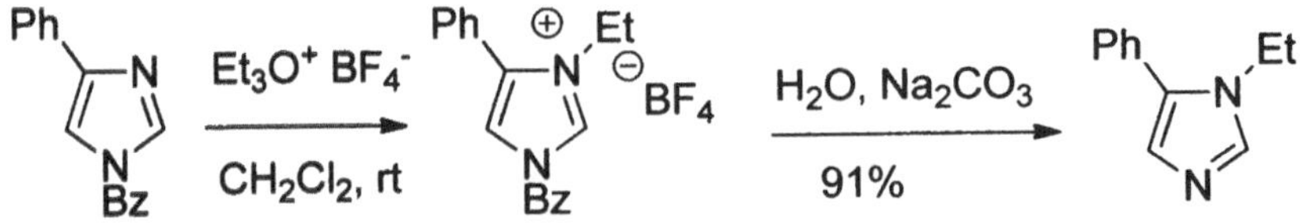

While Friedel–Crafts acylations do not occur readily with imidazoles, quaternization can lead to 2-acylimidazoles. The reaction of *N*-phenyl imidazole with benzoyl chloride and Et_3N in MeCN provides first the *N*-

benzoyl compound followed by ylide formation and rearrangement to the 2-acyl compound.[2,3]

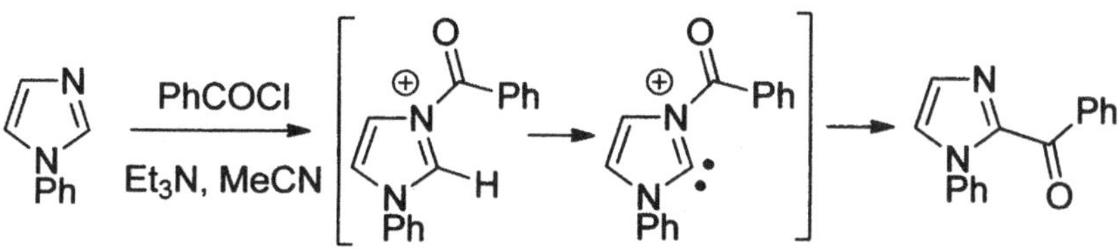

Addition of a trifluoroacetate moiety in high yield was recently reported using TFAA and PhMe. Bis-addition of imidazole to the trifluoroacetyl group was observed in MeCN.[8]

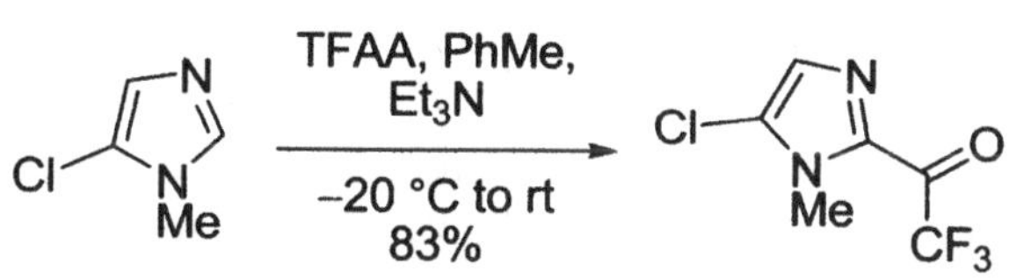

8.2.2 *Electrophilic C-Substitution*

Reactions of the imidazole carbon atoms occur easily under basic or neutral conditions; however, once protonated, electrophilic substitution is slowed. For example, Friedel–Crafts type alkylations and acylations do not readily occur under protic or Lewis acid conditions, which has led to the development of syntheses of imidazoles that more readily allow the desired carbon alkylated or acylated products. Nitration and halogenations of both *N*-un-substituted and *N*-substituted imidazoles take place with preferential addition to the 4- or 4- and 5-positions.[2]

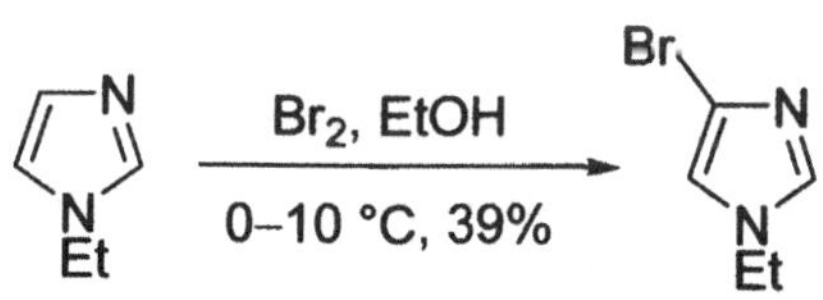

The addition of weak acids can facilitate the formation of polyhalogenated materials.

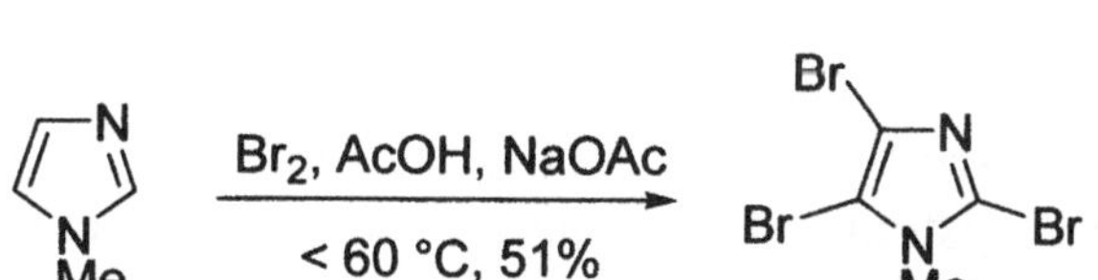

Nitration, like halogenations, occurs preferentially at the 4- or 5 position in *N*-unsubstituted and *N*-substituted imidazoles. As with all nitrations, poly-nitration can occur, and there is the potential for subsequent thermal hazards or runaway reactions. Despite its lack of reactivity, polynitration products have been reported.

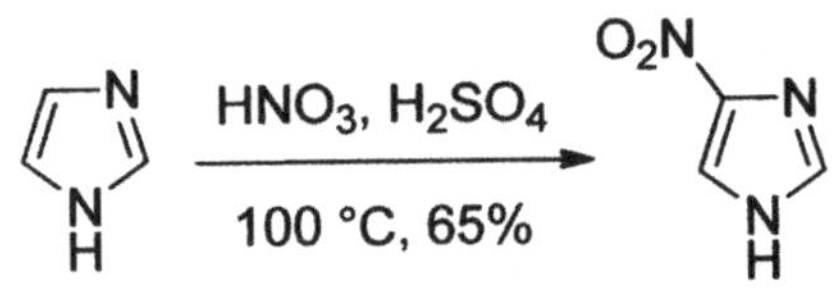

It is important to remember that substituents on the ring can and do alter the selectivity of a reaction. An example of this is the nitration of 2-thio-imidazole in which the 5-nitro product is obtained and isolated in about 30%.

Selectivity of nitration can be altered. It was recently reported that the use of surfactants increased the selectivity toward formation of 5-nitro-imidazole in 55% yield. Use of micelles is thought to operate by increasing the rate of reaction toward a particular product.[9]

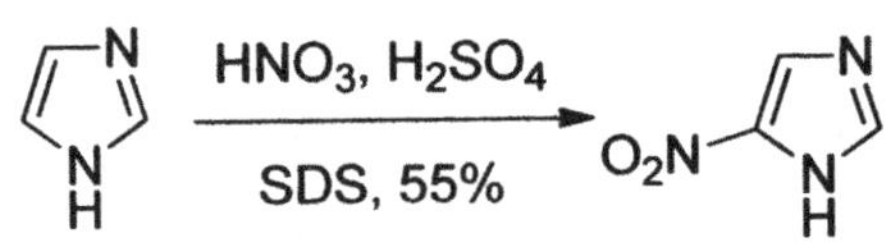

Mannich Reaction

Hydroxymethylation of *N*-substituted imidazole occurs readily at the 2 position[2] using a sealed tube. When the C2 position is substituted, reactions will take place at other sites. An example of many principles discussed, which also provides an example of the Mannich reaction and demonstrates its use in the synthesis of the imidazole portion of losartan, which is an angiotensin II receptor antagonist for the treatment of hypertension. Bromination using bromine provides a mixture of mono-4-bromo- and bis-4,5-dibromination, which on treatment of a reducing agent gives only the 4-bromo-imidazole. Subsequent Mannich reaction with base and aqueous formaldehyde at ambient temperature provides the 5-hydroxymethyl substituent. Oxidation provides the aldehyde followed by bromo displacement of the bromide with chlorine by refluxing in conc. HCl. In these examples, we see that nucleophilic addition can displace halogens on the imidazole ring system. Additionally, oxidizing agents (MnO_2) do not react directly on the imidazole ring but do react with its substituents.[10]

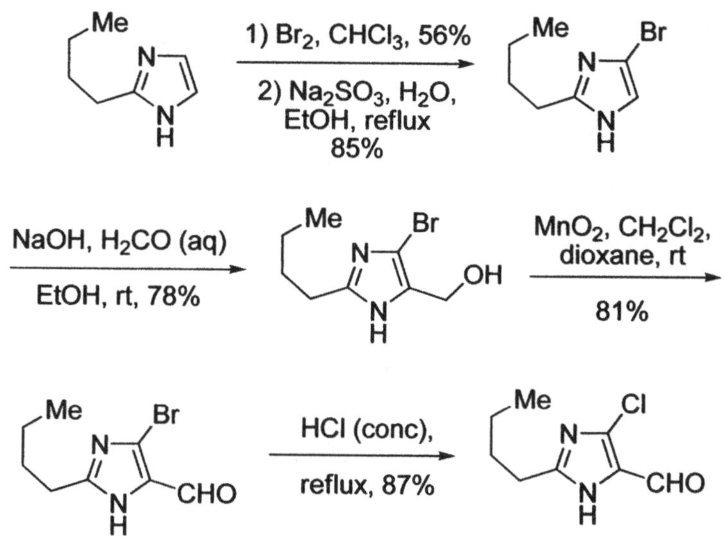

An alternative approach to this molecule has been reported by the Merck group. As is commonly done, the synthesis of the imidazole is adapted to allow introduction of appropriate substituents. The procedure prepares the imidazole by reaction of dihydroxy-acetone, ammonia, and amidine HCl. Interestingly, this reaction was reported to require high pressure, however, the Merck group found that using ammonia in MeOH allowed the condensation to take place at 45 psi and 70 °C in 69% yield. Ease of synthesis of the desired chloro compound required protection of the hydroxyl group and chlorination, which led to over chlorination. Thus, a Zn-reduction step was required to provide the desired compound. Removal of the TMS protecting group then provides the desired product.[11]

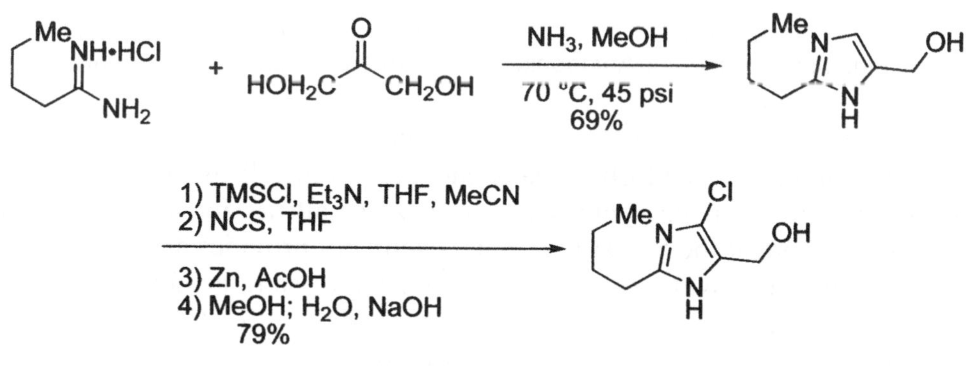

To complete the synthesis, the aldehyde is required to achieve good alkylation selectivity. The discovery group reported[12] alkylation using K_2CO_3 to provide a 97:3 ratio of *N*-selective alkylation. Sodium borohydride reduction then provides the 5-hydroxy methyl compound in 90% overall yield. This intermediate was then reacted further to form losartan, an angiotensin II receptor antagonist used to treat hypertension and coronary heart disease.

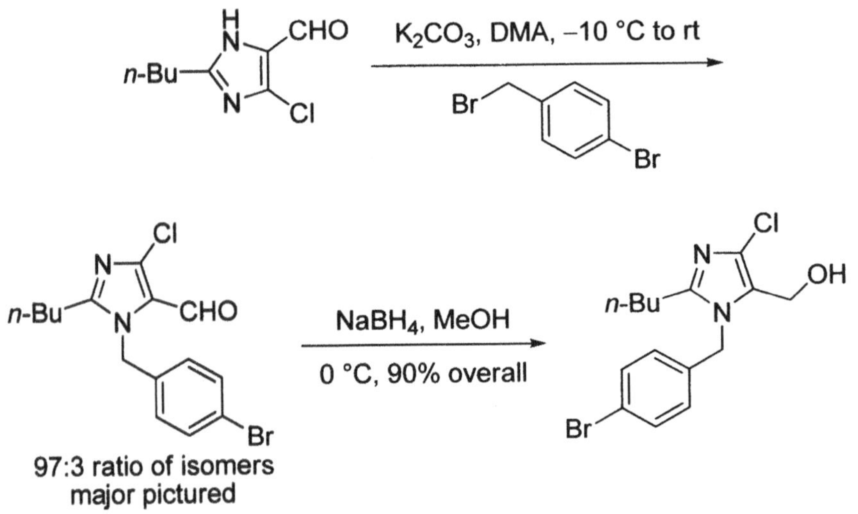

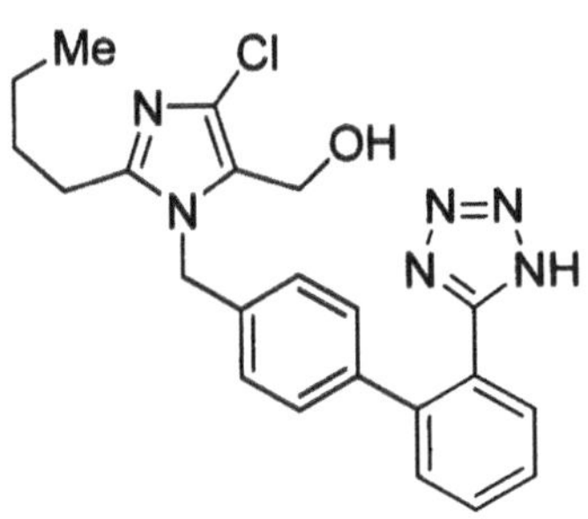

losartan
Angiotensin II Receptor Antagonist

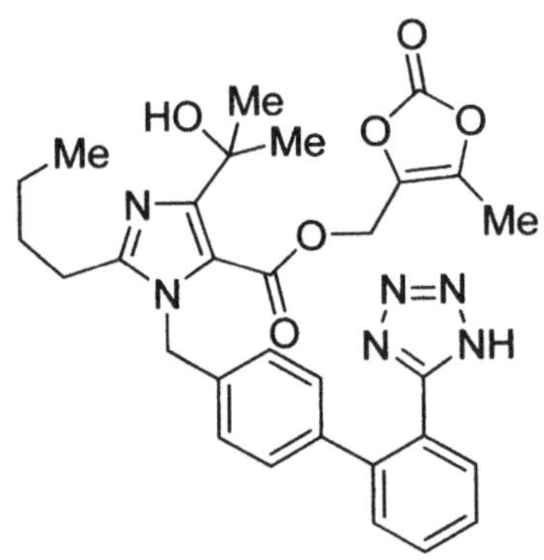

olmesartan medoxomil (a prodrug)
Angiotensin II Receptor Antagonist

Over-halogenation is a common theme in functionalization of unsubstituted imidazoles and often leads chemists to install functionality early if possible. The Wallach synthesis of imidazoles allows for the installation of a chlorine atom at the 5 position. Such chemistry provides an intermediate that is useful in the synthesis of the immunosuppressant, azathioprine. The Wallach reaction is the treatment of oxalylamides with PCl_5 or $POCl_3$ to yield the corresponding *N*-methyl-5-chloroimidazole.[13] Nitration is selective at the 4-position[14] to provide the imidazole properly functionalized to provide azathioprine.[15]

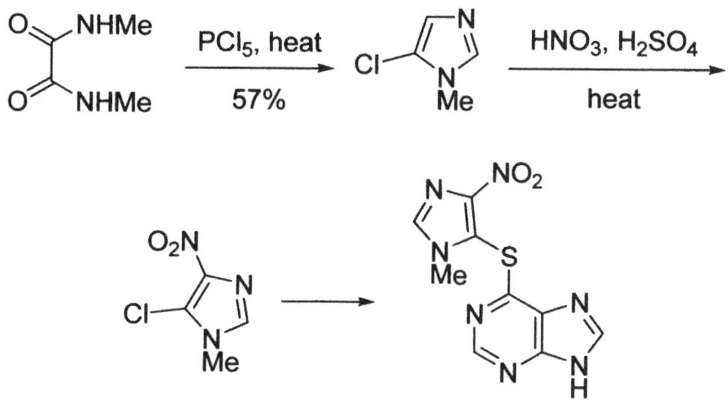

azathioprine

8.2.3 *Metalation and Direct Pd-Activation*

Metalation takes place preferentially at the C2 position in *N*-substituted imidazoles. If the 2-position is occupied, then the C5 position is metallated followed by the C4 position.[16] For the first metallation and addition of electrophile, a variety of electrophiles has been successfully used like methyl sulfate (68% yield, E= Me), Br_2 (80% yield, E = Br), or DMF (65% yield, E = CHO). LDA can also be used to deprotonate the 2-position.

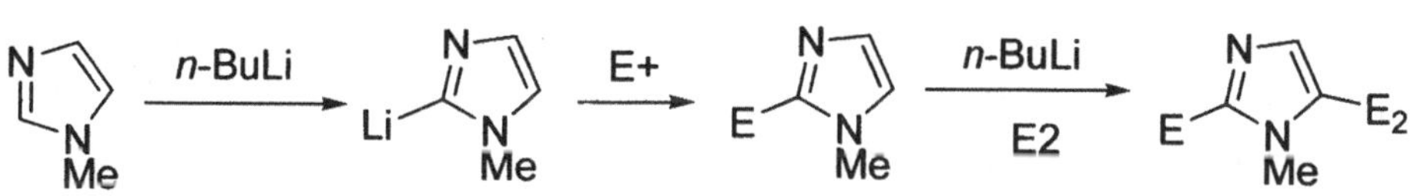

There have been examples using Pd to couple aryl iodides directly to the 5 position of the imidazole ring. In the example shown, *N*-benzyl imidazole coupled with phenyl- or methoxy-phenyliodides efficiently and was subsequently brominated at the 4-position to undergo further reaction.[17]

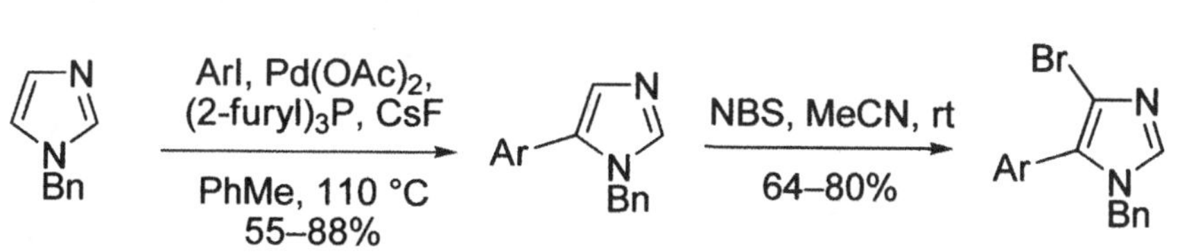

Ar = phenyl- or methoxyl-, phenyl-

8.3 ***Construction of the Imidazole Ring***

There are a variety of imidazole syntheses, some of which have names attached to them. A survey of these reactions will be overviewed followed by more modern modifications of syntheses.

8.3.1 Debus

Debus is credited for first reporting the reaction of glyoxal, NH_3, and formaldehyde to make imidazole.[2] It has been determined that primary amines and/or other salts of ammonia can be used. While it is best if a symmetrical dicarbonyl component is used or mixtures are likely to result, it was recently shown that polarizing the dicarbonyl can selectively provide one product.

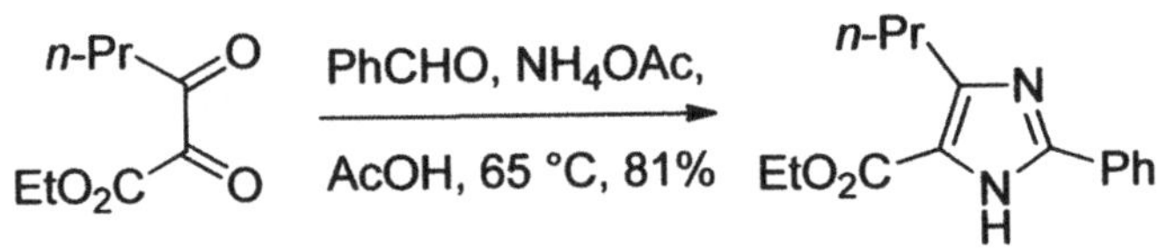

A modification to this reaction can be seen in the reaction of a dicarbonyl equivalent, acyl-vinyl-phosphonium salt.[18] Reaction of the phosphonium salt with an appropriately functionalized formamidine provides the imidazole phosphonium salt. The imidazole phosphonium salt can be further functionalized to an intermediate used to synthesize the cimetidine, an H2-receptor histamine antagonist used to treat ulcers.

A proposed mechanism for the condensation of formamidine sulfinic acid with the dicarbonyl equivalent involves attack of the amidine nitrogen atoms on both of the electrophilic centers with loss of water. The imidazole phosphonium salt does not react directly with sodium methoxide but instead generates the reactive methylene imidazole intermediate, which is trapped by MeOH. Reactivation can be promoted in protic solvents to elongate the chain to provide cimetidine.

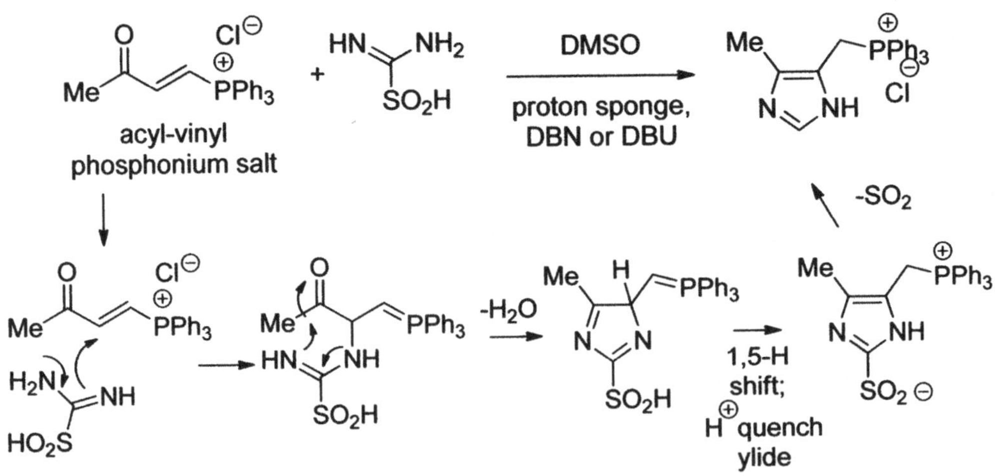

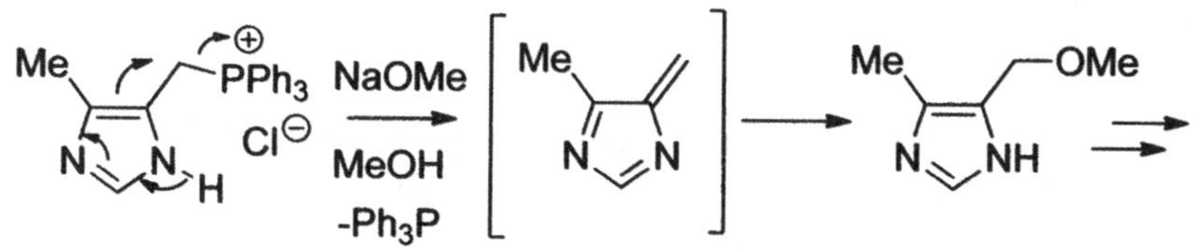

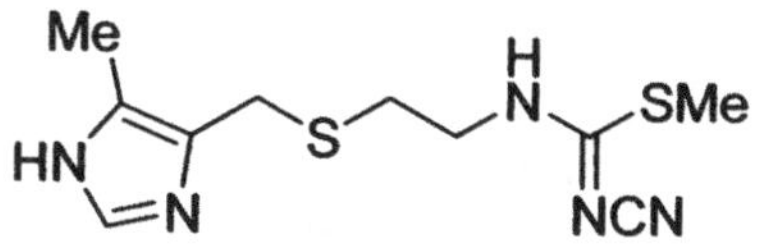

cimetidine, anti-ulcer
H_2-Receptor Histamine Antagonist

8.3.2 Weidenhagen

This is the reaction of an α-hydroxy or acetyoxy-ketone with NH_3 and an aldehyde to provide a 2,4-di-substituted imidazole.

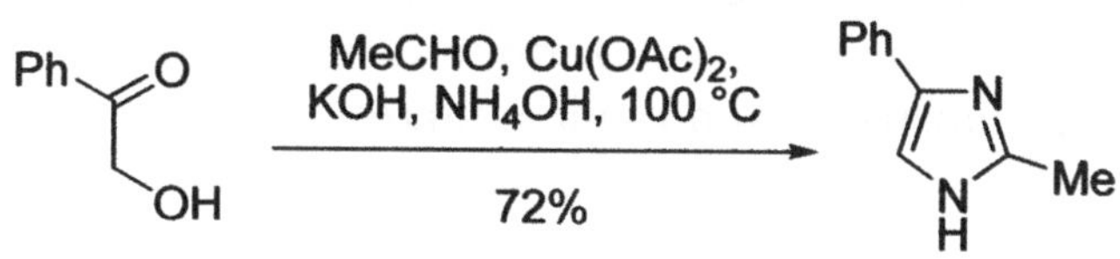

An improvement on this reaction has been described using catalytic iodine.[19] Several 1,2,4,5-tetra-substituted imidazoles were prepared in excellent yields.

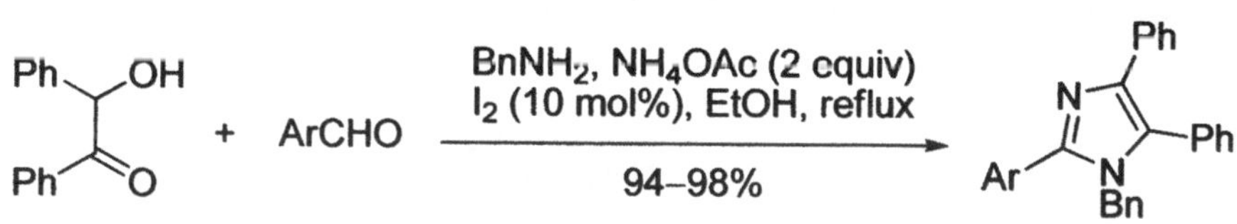

Use of thiourea for this condensation is another variation for this reaction providing 2-thiosubstituted imidazoles. Typically, the reaction is promoted thermally. Another classic method to generate 2-thio-imidazoles is the condensation of protected glycine with formate, followed by reaction with thiocyanide to form the imidazole.[20] In this instance, the 2-thio-3-carboxylate was isolated in 63% overall yield.

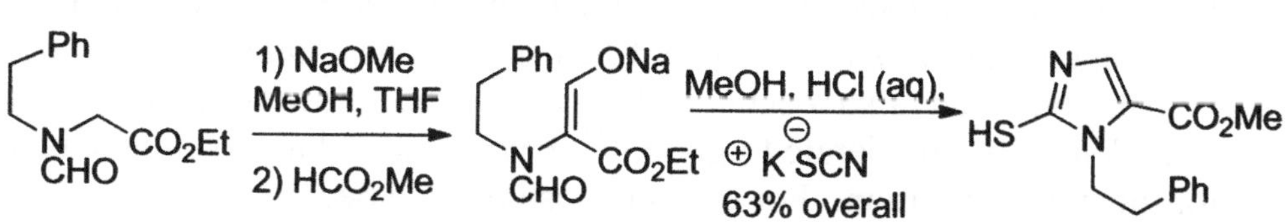

8.3.3 Bredereck

Bredereck reported the reaction of α-hydroxy, bromo, amino, or acetoxy-ketones with formamide to provide 4,5-di-substituted imidazoles. The formamide acts both as the N and C2 source.

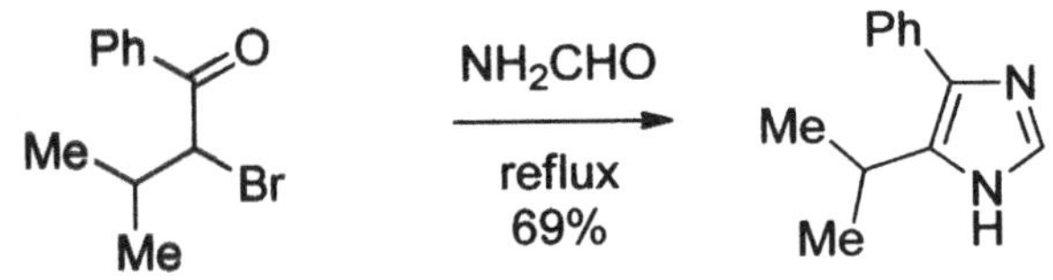

If the substituents around the ketone are too large, then yields are low. Oxazoles can form during this reaction but can oftentimes be overcome by increasing the amount of formamide or nitrogen (N) source. Many other N sources can be used with these substrates. For example, acetimidates, KSCN, guanidines, amidines, and formimidamides have been used with many of these keto substrates to provide imidazoles.

8.3.4 Radiszewski

This reaction is a variation of the conditions reported by Debus in which a dicarbonyl is reacted with an amine salt and an aldehyde in the presence of an acid or Lewis acid. If an *N*-substituted imidazole is desired, then an amine can be used. Recent Lewis acids that have been used are $ZrCl_4$[21] or $InCl_3$.[22]

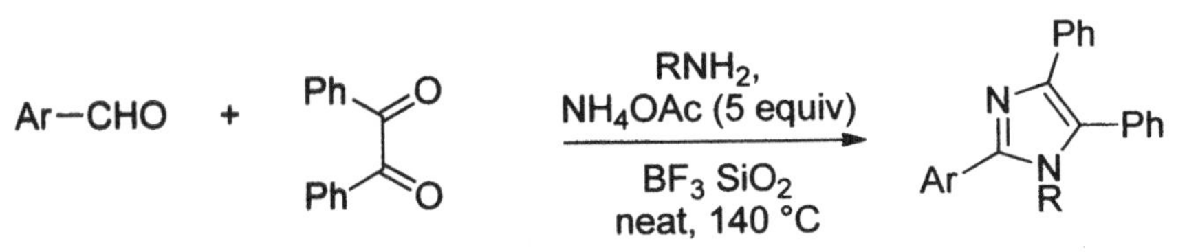

8.3.5 van Leusen

van Leusen and co-workers reported the cycloaddition of tosyl-methylisocyanide (TosMIC) with imines to afford 1, 5-di-substituted and 1, 4, 5-trisubstituted imidazoles. The procedure requires a base and the TosMIC reagent can be substituted but has limitations. Other than imines, the cycloaddition can also occur with acetimidates, nitriles, and iminoyl chlorides. Yields are moderate to good.[2,23] EWGs other than tosyl have been used for this transformation. For example, esters have been used.

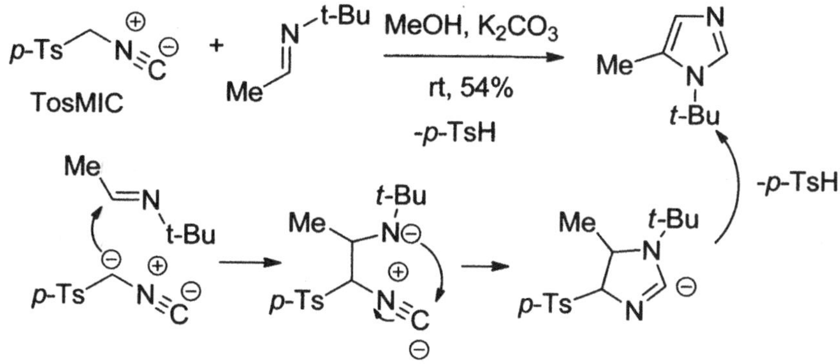

8.3.6 Use of α-Amido-Ketones

From a disconnection perspective, the use of α-amido ketones is appealing for many functionalized imidazoles. However, it sometimes suffers from many steps and poor overall yields. Much development work has gone into the preparation of the amido-ketone. For example, the Merck group has published the preparation of substrates for cyclization using a Stetter-like multicomponent coupling reaction (MCR).[24]

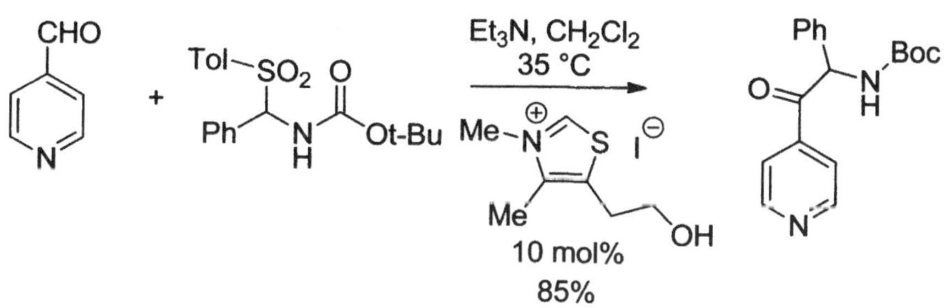

After the excellent development work on the process to prepare the α-amido-ketone, the work was further stretched into the one-pot synthesis of imidazole. This method proved somewhat general and gave good yields of many 4,5-di-substituted imidazoles.[25]

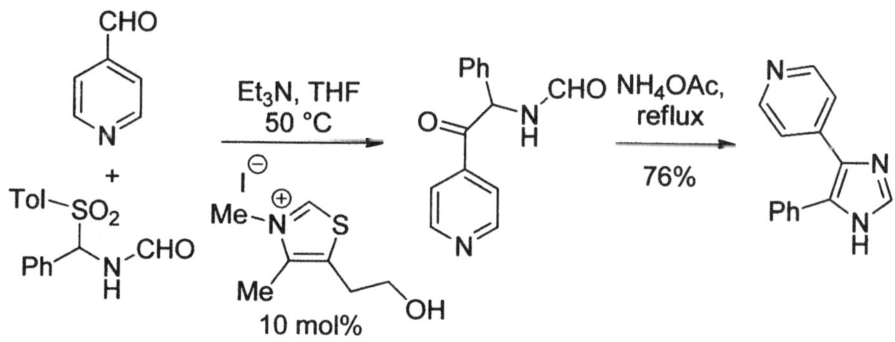

More classical preparation of imidazole using this condensation has been recently described. In this case, the starting amido-ketone was prepared by condensation of the ketimine protected glycine anion with an acid chloride. Removal of the protecting group and amide formation provided the desired starting material. Reaction of this complex α-amidoketone with a complex amine promoted by benzoic acid and heat provided the imidazole in good yield.[26] These reactions are believed to proceed via condensation to form the imine followed by cyclization to give the imidazole.

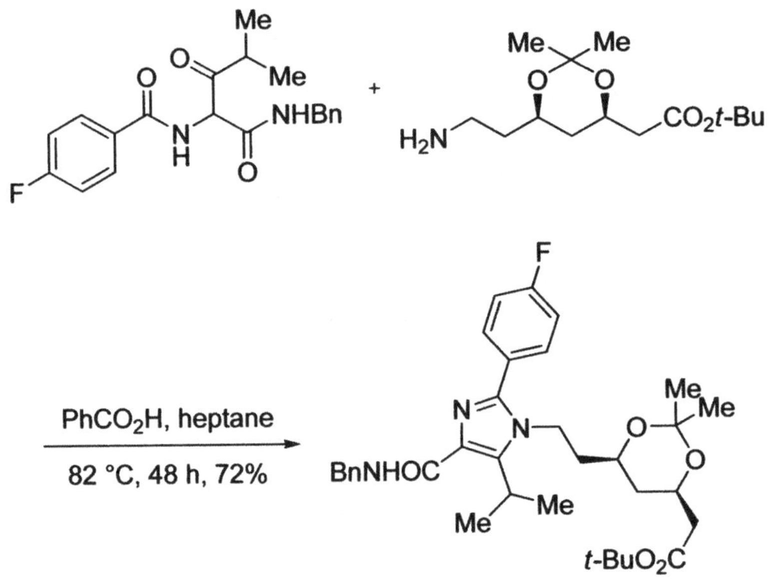

Activating groups other than acid can be used to promote cyclization of these substrates. Phosphorous reagents like PCl_5, $POCl_3$, and $Ph_3P{\bullet}Cl_2$ have been used to promote the cyclization.[2]

8.3.7 Use of DAMN Reagent

Diaminomaleonitrile (DAMN) is a reactive 2-carbon and 2 nitrogen building block excellent for forming imidazoles by reaction with an electrophilic one-carbon building block (for example the use of ortho-esters as shown below). Reaction of DAMN in refluxing MeCN with the pentanoate ortho-ester provides a good yield of the imidazole. Important to the use of DAMN is the ability to achieve mono addition to one of the two nitriles. This is achieved by hydrolysis and esterification. The two esters are then differentiated by reaction with Grignard reagents to provide the hydroxyl-isopropyl product which is a building block for the synthesis of olmesartan.[27,28] The Grignard reagent is selective for addition to the ester at the 4-position.

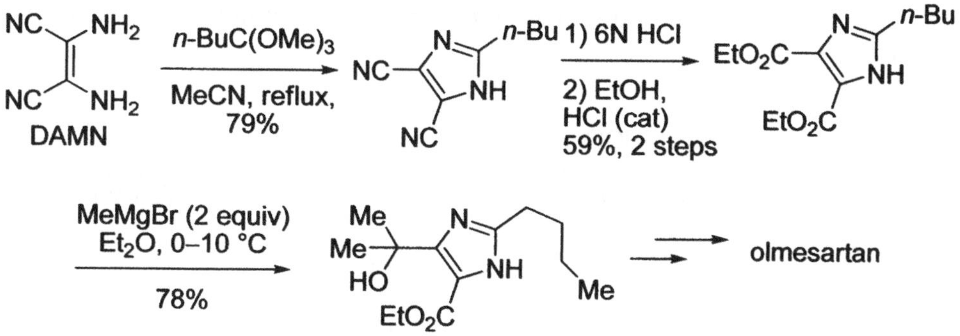

8.3.8 *Fused-Imidazole Rings*

Imidazole-fused rings have come to significant importance, particularly for the treatment of various cancers. Temozolomide generates a methylating agent *in vivo* and is used for the treatment of cancers. The drug is believed to operate *via* uncatalyzed degradation of the compound in plasma (promoted by nucleophilic addition of water followed by decarboxylation) to provide an intermediate that undergoes further degradation ("1,3-H shift" and fragmentation) to generate methyl diazonium, a potent DNA alkylator.[29]

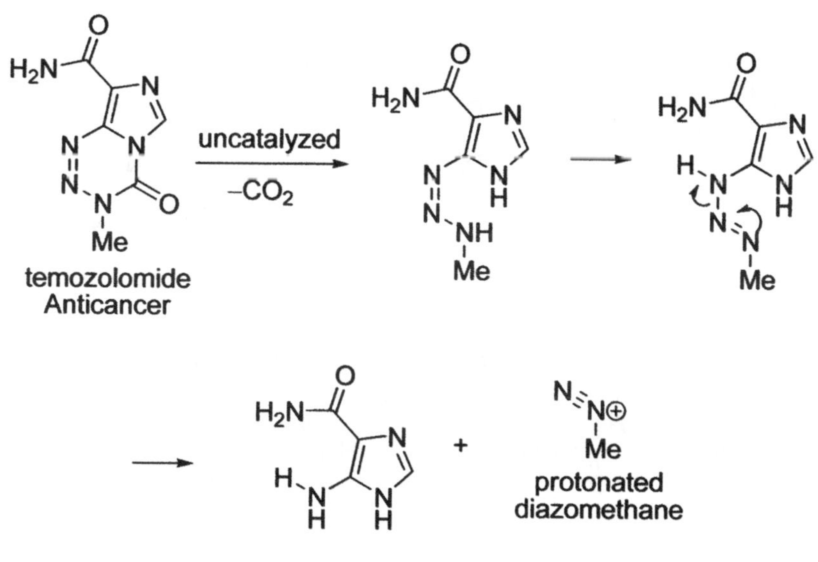

There are several reported syntheses of this 4,5-di-substituted imidazole. A published route from an academic group prepared the imidazole from 1,4-di-nitroimidazole. The proposed mechanism for this transformation is nucleophilic addition (cine addition) of cyanide to the imidazole between the two nitro groups.[30] After substitution, elimination of nitrous acid followed by *N*-protonation provided the 4-cyano-5-nitroimidazole. Subsequent transformation further elaborates the structure to temozolomide.[29]

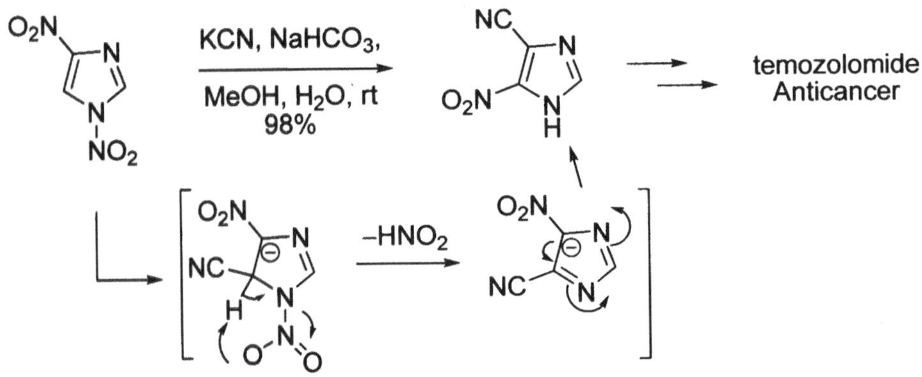

A more direct route involves the treatment of α-amino-cyano-amide with a protected formamidimide. The reaction is reported to take place at ambient temperature and provide good yields of the 1,4,5-tri-substituted and more appropriately functionalized imidazole, which is subsequently transformed into temozolomide.[31]

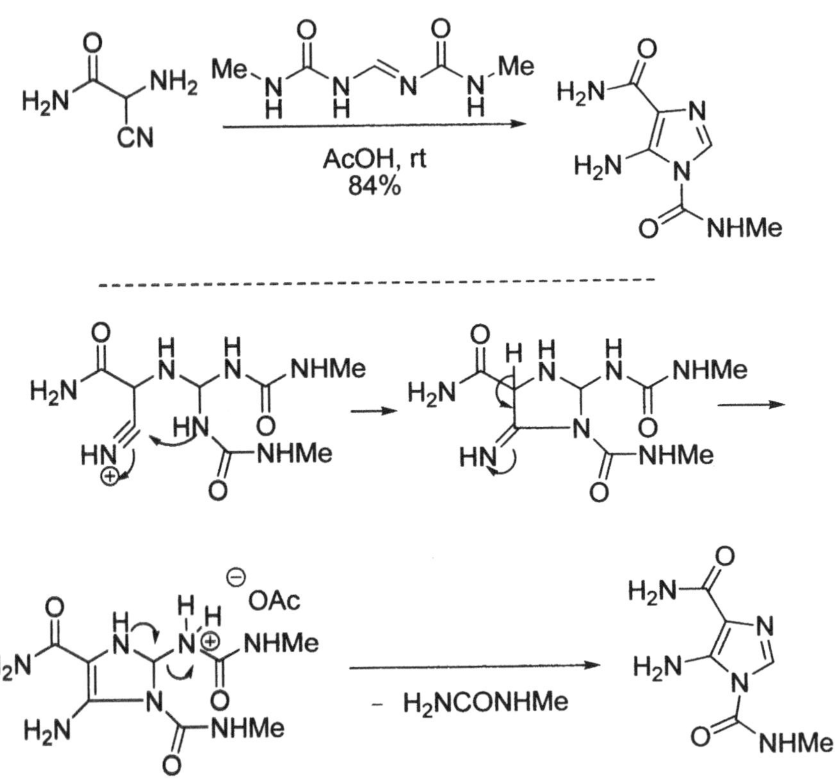

Another pharmaceutically important fused-imidazole ring system is the popular sleeping aid medication zolpidem. Bromination of 4-methylacetophenone and condensation with methylated 2-amino pyridine provides the fused-imidazole in good overall yield. Note that the ring nitrogen on the aminopyridine reaction reacts with the bromide carbon. Mannich-type alkylation at the unsubstituted 5-position provides the dimethylaminomethyl substituent in good yield. Further elaboration yields zolpidem.[32]

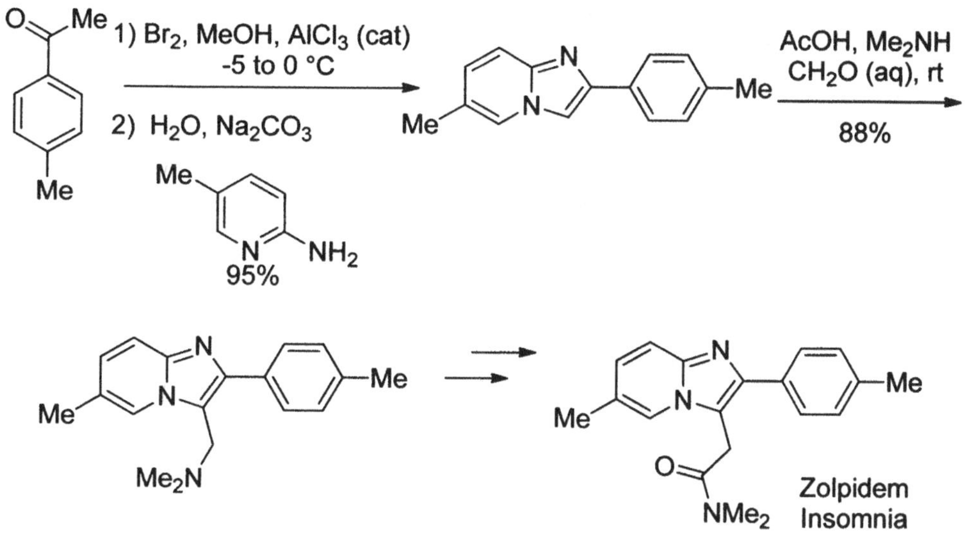

8.3.9 Miscellaneous Imidazole Ring Construction

Imidazole synthesis utilizing 2-aminopyridines as amine source

The use of microwave has recently been used to provide 5-amino-imidazoles from 2-aminopyridines. At a higher temperature and longer reaction times, dehydration occurs to form the imidazo-pyridinium salts. Reaction with hydrazine adds to the pyrimidinium salt, which then undergoes rearrangement to open the ring. The unstabilized diene undergoes hydrolysis to provide the 2-amino-4,5-di-substituted 1*H*-imidazole. Chemically, PPA can be used to form the pyrimidinium salt also.[33,34]

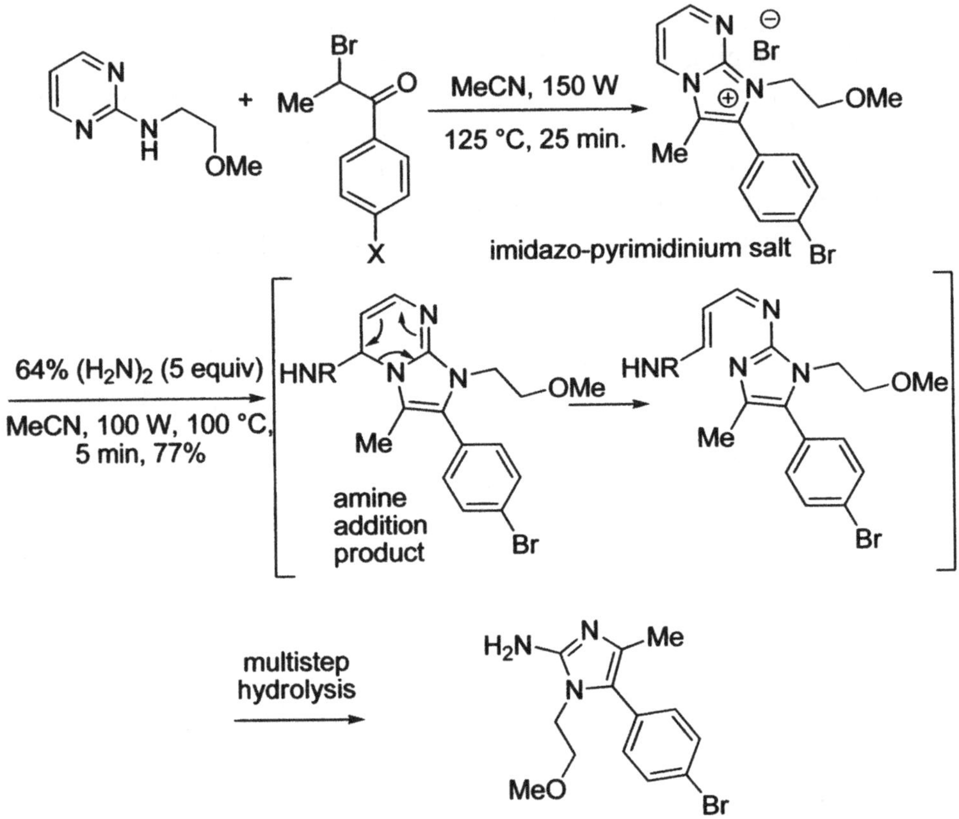

Interestingly, when lower temperature and energy microwave is used, loss of water does not occur, and the imidazolin-pyridinium salt forms. When this intermediate is subjected to the hydrolysis conditions, a Dimroth-type rearrangement is proposed to occur which leads to the alkyl substituent that was at the 1-position of the imidazole ring to now be on the 2-amino-substituent as shown.

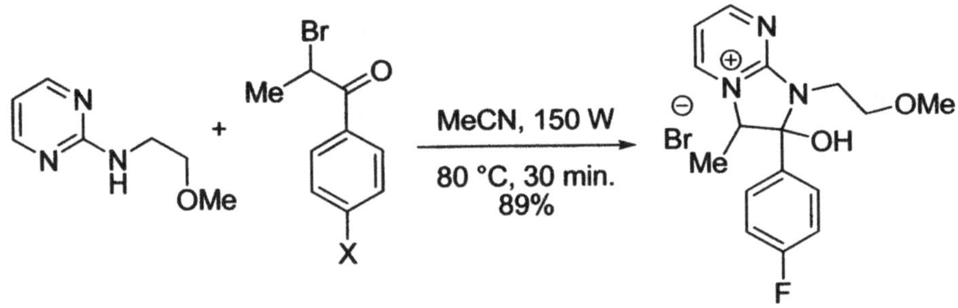

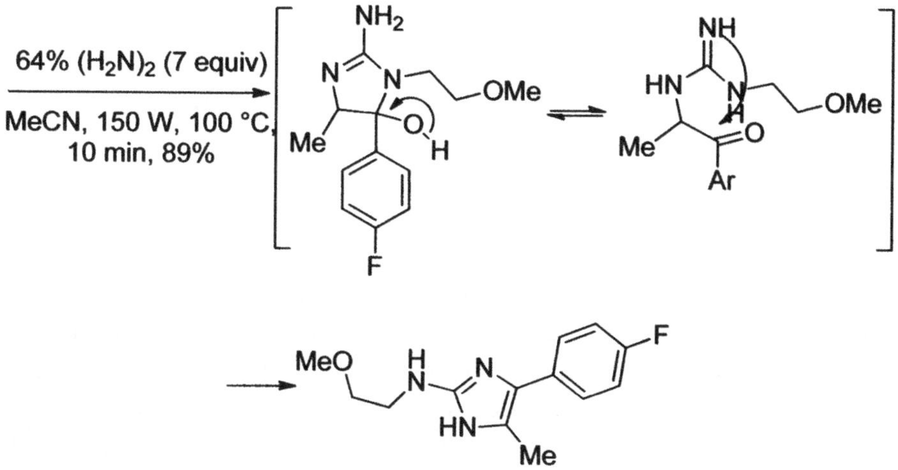

Pd- and Cu-Promoted Cyclizations to Prepare Imidazoles

Cyclization of an imino-Heck type reaction has been reported. Cyclization of a protected amidine gave reasonable yield of the 1,3,5-tri-substituted imidazole in good yield. Pd is proposed to insert into the O–N bond of the activated ester, cyclize, eliminate, and rearomatize.[21]

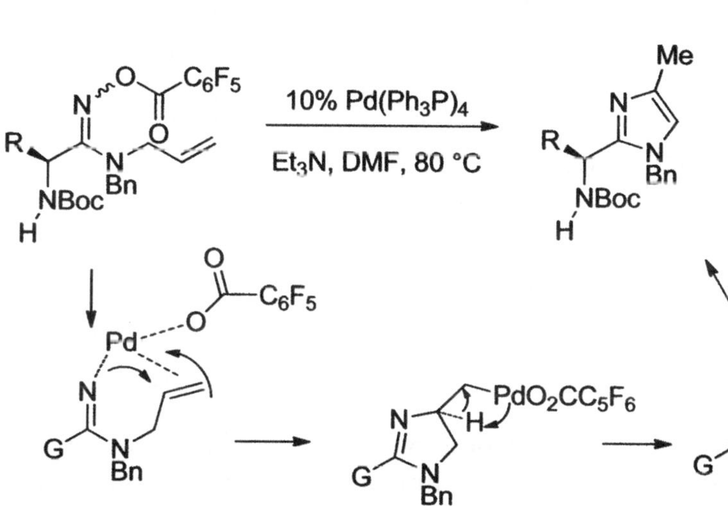

R	Yield%
Bn	70
Me	56
i-Bu	68

A Stetter-type multi-component reaction (MCR) was described earlier for the preparation of the precursor for imidazoles. Another MCR has been applied to the formation of imidazoles from simple building blocks.[21]

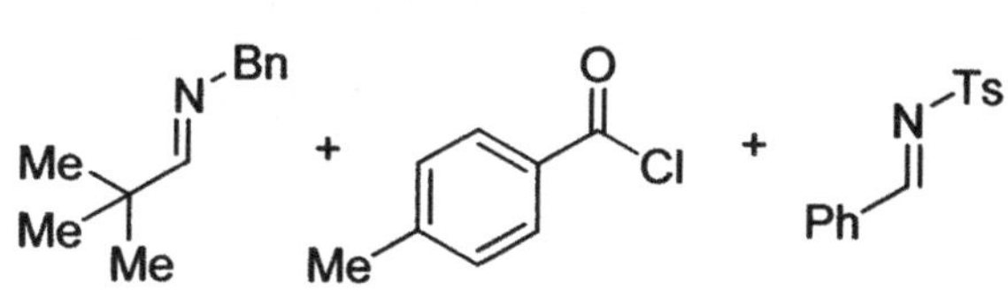

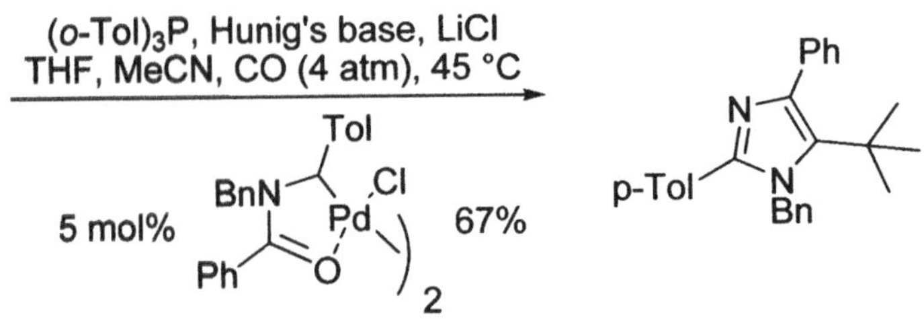

This Pd-catalyzed transformation uses two imines: an acid chloride and CO. Mechanistically, this reaction is proposed to proceed via Pd addition to the activated imine, followed by CO insertion. Reductive elimination of Pd with HCl provides the ketene, which undergoes formation of the Münchnone. Münchnones are known to undergo cycloaddition reactions with electron-deficient imines to yield cycloaddition adducts, which lose CO_2 and TsH to yield imidazole.

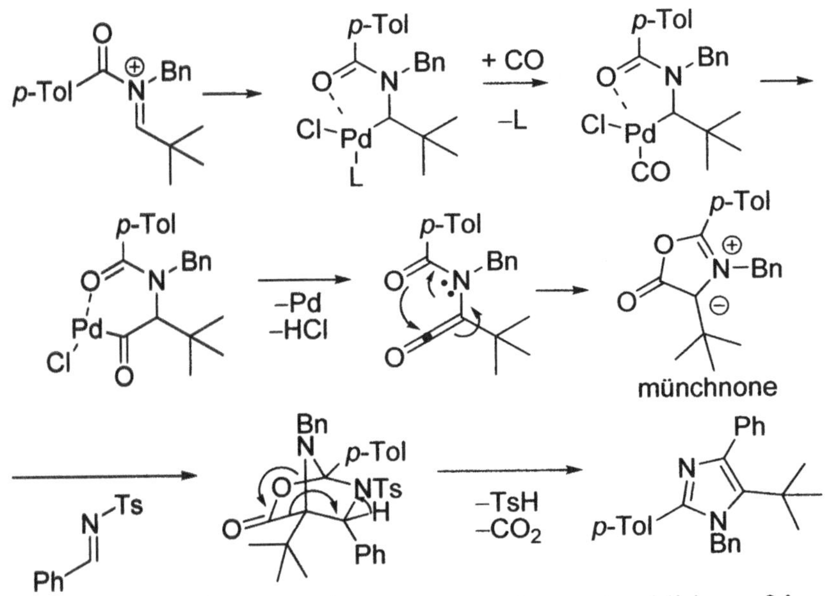

Copper(I) has been used to promote the cycloaddition of isonitriles. The method to prepare 1-aryl-4-carboxylimidazoles is general. In addition to esters, amides and phosphonates were shown to work. The ligand for the catalyst is 1,10-phenanthroline.

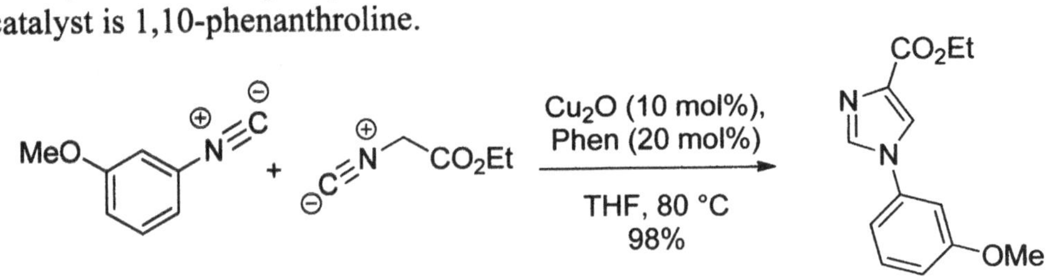

Oxidative Cyclization

SeO_2 has been used to promote cyclization of 1,3-diaza-1,3-butadienes to 5-hydroxy-1*H*-imidazoles.[35] The proposed mechanism is oxidation to the alcohol, 1,5-hydride shift to make the enol, tautomerization to the aldehyde followed by cyclization with loss of dimethyl amine to provide the product.

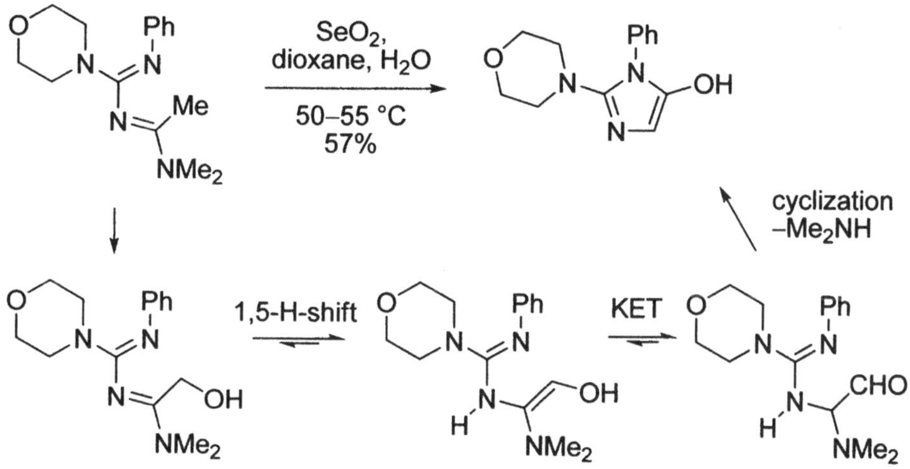

8.4 *Conversion of Imidazolines to Imidazoles*

Imidazolines can be dehydrogenated using a variety of agents to provide imidazoles. Some of the agents commonly used to perform this oxidation are: MnO_2, $BaMnO_4$, Magtrieve in PhMe, or $PhI(OAc)_2$ with K_2CO_3 in DMSO.[2,19,22] Thus, imidazolines are also a good entry into imidazoles.

8.5 *Possible Liabilities of Imidazole-Containing Drugs*

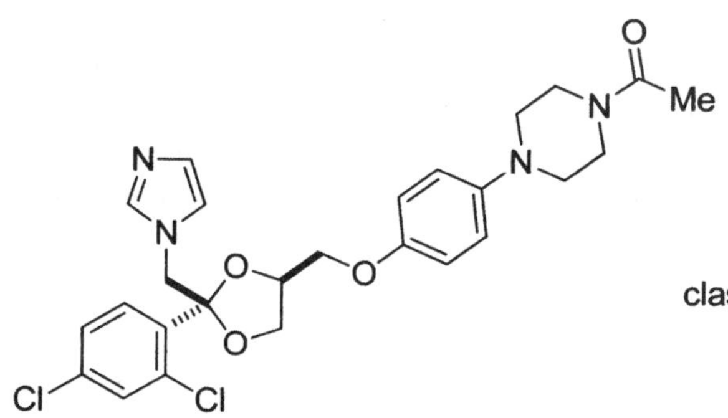

ketoconazole
anti-fungal
classical CYP inhibitor

Ketoconazole is an imidazole containing drug that is well known for its inhibition of CYPs. It is typically used in drug development in drug-drug interaction (DDI) studies. While this imidazole containing drug does inhibit

CYP oxidation, the imidazole structure may not be the only the structural moiety creating all of the inhibition. However, imidazoles can chelate to the heme Fe of the CYP450 enzyme complex and display slow dissociative kinetics. While reversible, this can disrupt the oxidation of another drug and greatly diminish the therapeutic window that was observed in animal studies.

As mentioned in the Introduction, imidazole is found in biologic systems and, as mentioned, is quite robust chemically. For example, the imidazole ring system is not readily oxidized; *N*-oxidation does not readily occur. Very powerful oxidizing agents are required to oxidize the imidazole ring. Chemically, this was recently reported by the use of trimethylsilyl fluorosulfonyldifluoroacetate (TFDA).[36] Mechanistically, this proceeds similar to the acyl transfer reaction described earlier.

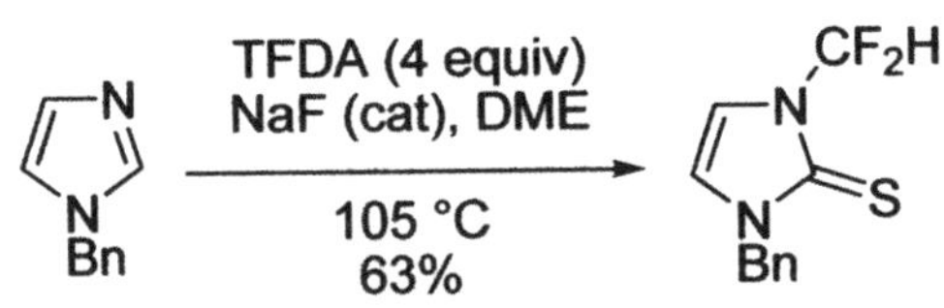

8.6 *Introduction to Benzimidazole*

Benzimidazole Numbering and Tautomerism

The parent compound, benzimidazole, is a crystalline solid with a melting point of 169–171 °C. Benzimidazoles, like imidazoles, exist as a tautomeric pair of compounds. The tautomerism of 1*H*-benzimidazoles is sufficiently rapid that the compounds are observed as a single species on the NMR timescale at room temperature.[37] 1*H*-Benzimidazoles with substituents at the 4(7) or 5(6) positions appear to be a single compound rather than a tautomeric mixture; however, replacement of the NH proton with an alkyl group or other substituent results in formation of a pair of regioisomeric compounds. Thus, regiospecificity is a consideration in the synthesis of all *N*-substituted benzimidazoles that bear any substituents at the C4–C7 positions.

^{1}H-NMR and ^{13}C-NMR spectra of the parent benzimidazole are a simple combination of the imidazole fragment and the benzene ring fragment. In the ^{1}H-NMR spectrum, the imidazole-H2 shows up as a singlet at 8.20 ppm, whereas the benzene-related H4+H7 and H5+H6 have chemical shifts at ~7.60 ppm and 7.20 ppm, respectively, both as multiplets. In the

^{13}C-NMR spectrum of benzimidazole, C2 is still the farthest downfield, with a chemical shift of 143.3 ppm. Meanwhile, C4+C7 and C5+C6 have chemical shifts at ~ 124.6 ppm and 117.0 ppm, respectively.

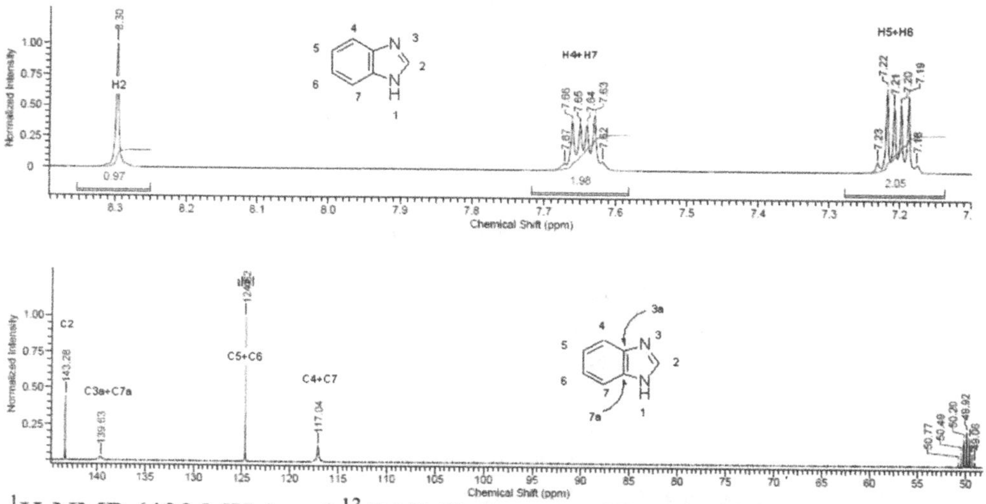

^{1}H-NMR (400 MHz) and ^{13}C-NMR Spectra of Benzimidazole in DMSO-d_6

The benzimidazole core has been incorporated into numerous pharmaceutical agents spanning a range of target classes and therapeutic indications. The proton-pump inhibitor Nexium (esomeprazole), which is prescribed for gastroesophageal reflux disease (GERD) and was the third top-selling drug in the United States in 2011 ($6.2 billion)[38] contains a benzimidazole as part of its pharmacophore. Likewise, several other drugs in this class (Prevacid, AcipHex, Protonix, and Kapidex/Dexilant) are substituted benzimidazoles. Other benzimidazole drugs include Emadine (emedastine), an H_1 receptor antagonist for allergic conjunctivitis (also sold under the brand names Daren and Remicut for allergic rhinitis and uticaria in Japan); Acardi (pimobendan), a PDE3 inhibitor used in veterinary medicine for heart failure (also approved in Japan for human use); Amias (candesartan cilexetil) and Micardis (telmisartan), both angiotensin (AT_1) receptor antagonists for chronic hypertension, heart failure, and other cardiovascular disease indications; and Albenza (albendazole), an anthelmintic for treatment of veterinary and human infections of parasitic worms.

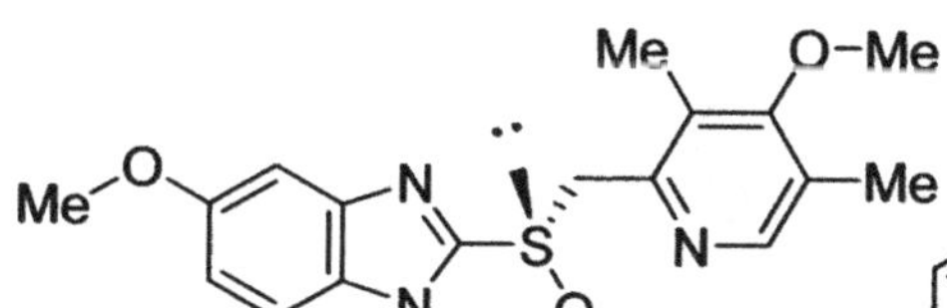

Nexium
esomeprazole

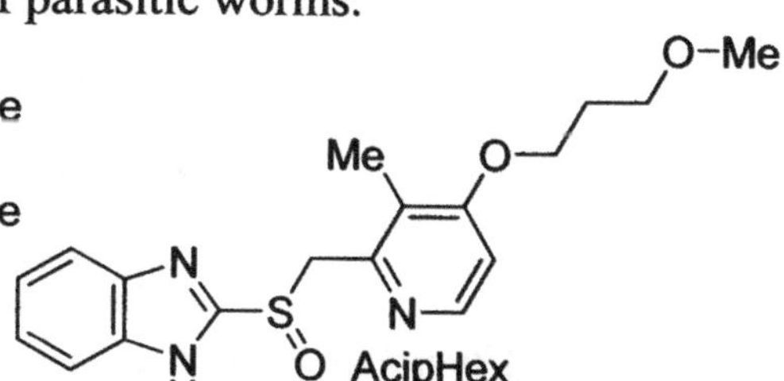

AcipHex
rabeprazole

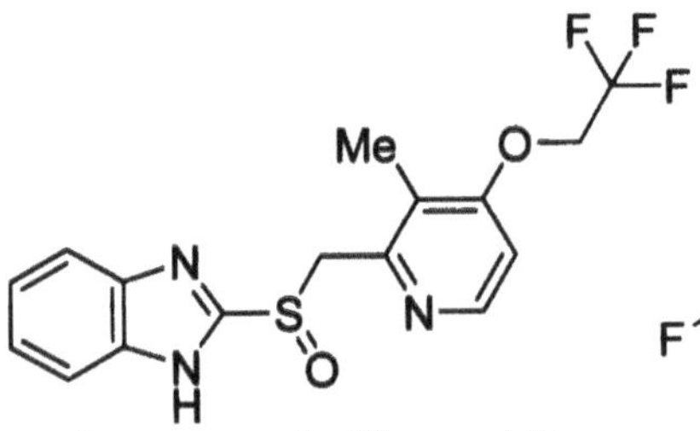

lansoprazole (Prevacid)

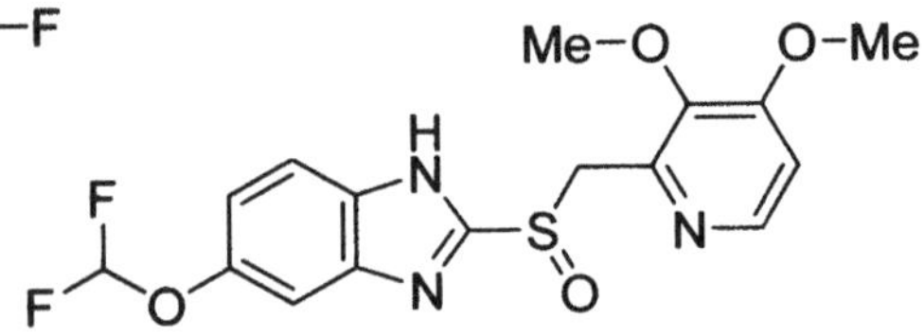

pantoprazole (Protonix)

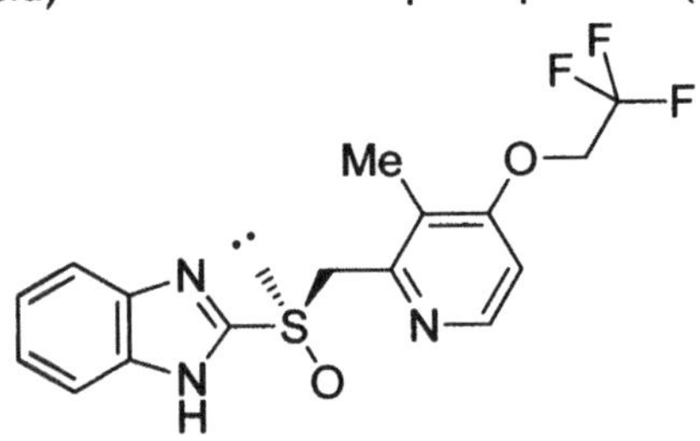

Kapidex/Dexilant
Dexlansoprazole

Proton-Pump Inhibitors for Gastroesophageal Reflux Disease

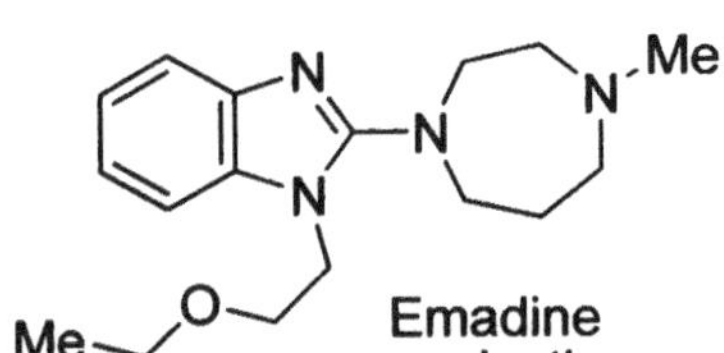

Emadine
emedastine

H_1 Receptor Antagonist for Allergies

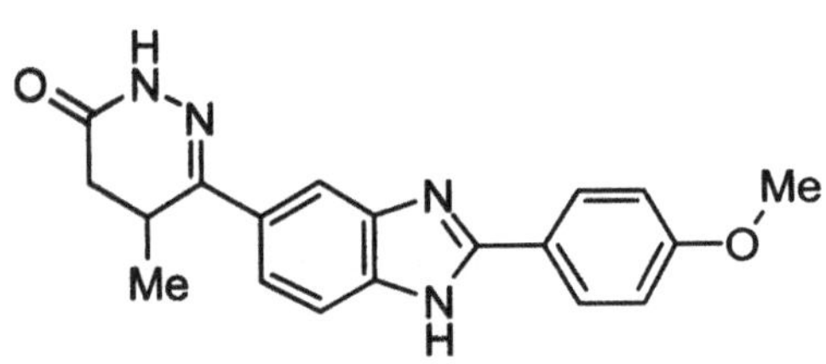

Acardi
Pimobendan

PDE3 Inhibitor for Heart Failure (Veterinary Medicine)

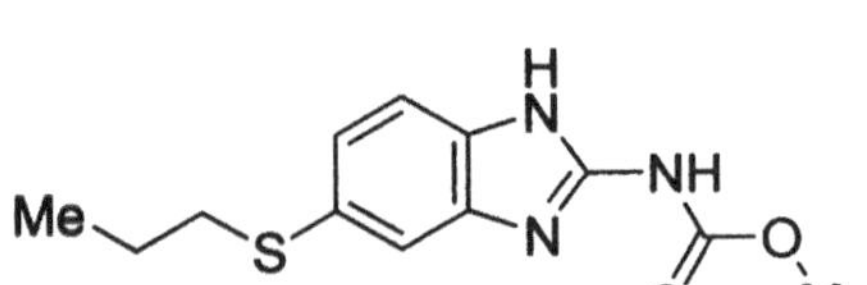

Albenza
albendazole

Albendazole for Eradication of Parasitic Worms

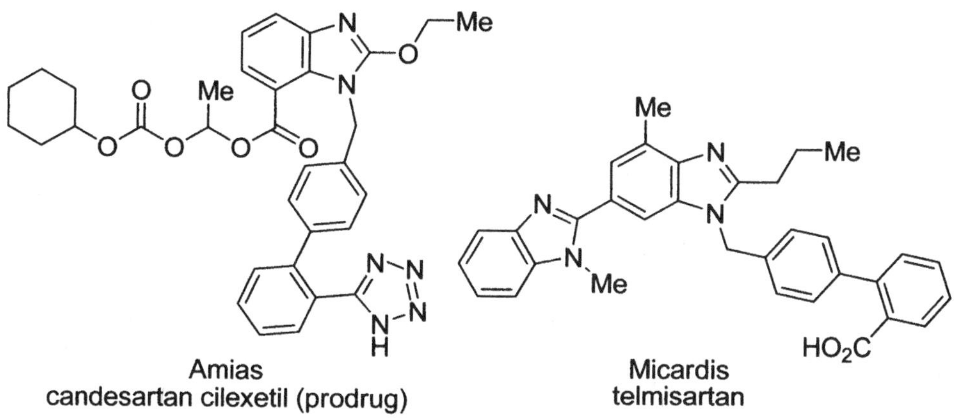

Angiotensin (AT_1) Receptor Antagonists for Hypertension and Heart Failure

8.7 Synthesis of Benzimidazoles: Classical Approaches

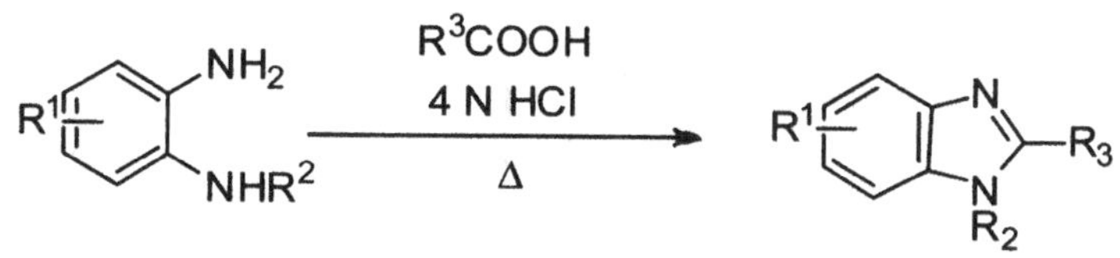

The classical synthesis of benzimidazoles (known as the Phillips Reaction), devised by Phillips,[39,40] uses a 1,2-diaminobenzene, which is condensed with a carboxylic acid in the presence of 4 N HCl with heating. In current practical applications, replacement of the carboxylic acid reagent with various carboxylic acid equivalent reagents lends broad utility and versatility to this method. Numerous modifications of this general approach are widely used in the synthesis of benzimidazole-containing drugs. For example, in a scalable synthesis of Nexium (esomeprazole magnesium), the benzimidazole core is assembled by the reaction of 3,4-diaminoanisole with potassium ethyl xanthogenate in refluxing ethanol/water.[41,42]

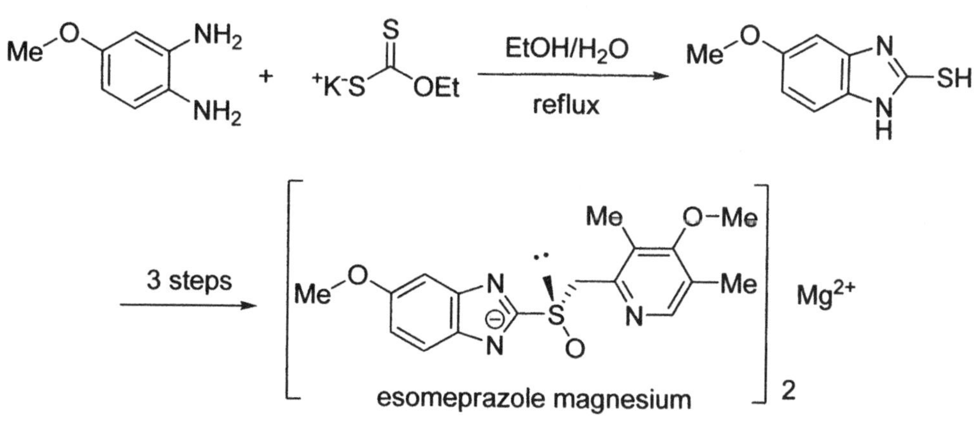

The selection of carboxylic acid surrogates for the benzimidazole-forming reaction is governed in part by the targeted functionality desired or required at the C2-position of the benzimidazole ring. Whereas use of the xanthate derivative provides efficient access to the 2-mercapto substitution for Nexium, the 2-carbamoyl substituent of Albenza (albendazole) prompts the use of reagents in which an amino or carbamoyl group could be incorporated directly. Thus, Albenza has been prepared in a two-step process from 4-(propylthio)benzene-1,2-diamine by reaction with cyanogen bromide followed by acylation with methyl chloroformate. Alternatively, the route can be truncated to a one-pot operation using either carbomethoxycyanamide or cyanamide with methylchloroformate.[43,44]

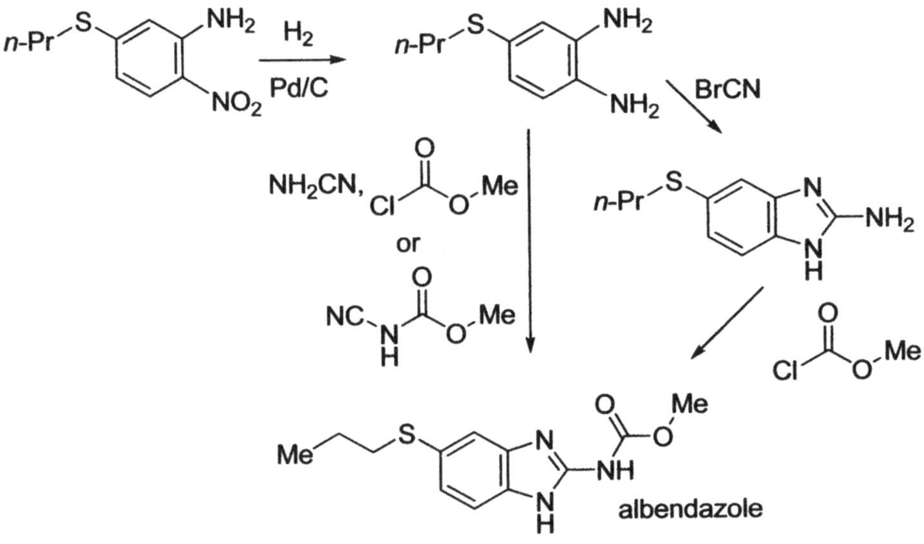

In one of the reported syntheses of pimobendan, cyclization to form the benzimidazole is achieved as the final step. In this case, dehydrative cyclization of the functionalized 2-aminoanilide intermediate in refluxing acetic acid provides pimobendan efficiently.[45]

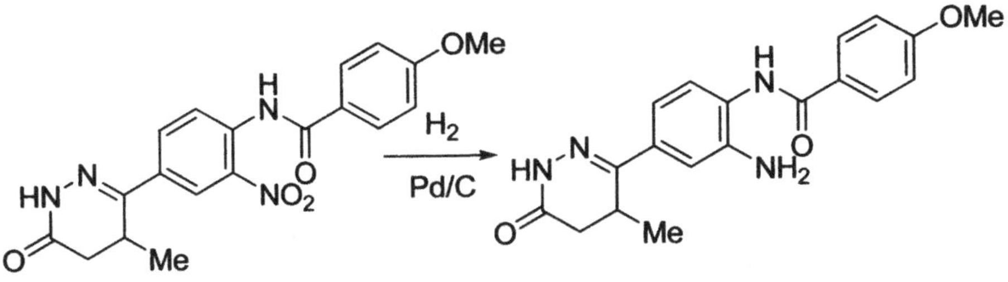

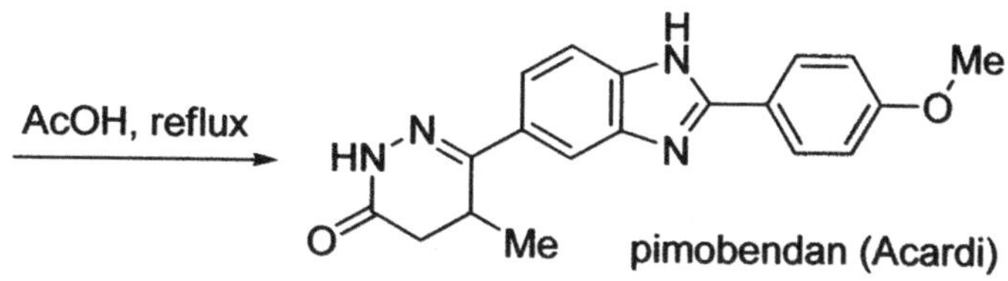

A similar approach has been used in the preparation of rivoglitazone, a PPARγ agonist which proceeded to Phase III clinical trials for the treatment of diabetes.[46] One of the reported protocols for preparation of rivoglitazone involves reductive cyclization by catalytic hydrogenation in the presence of acid to induce benzimidazole formation.

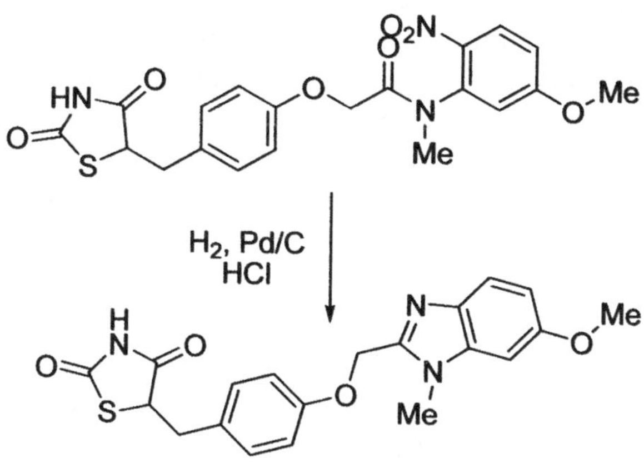

rivoglitazone
PPARγ agonist, Type 2 Diabetes

A recent patent for candesartan cilexetil describes the formation of the benzimidazole at a relatively late stage of the synthesis by cyclization of the highly functionalized diaminobenzene core under modified Phillips conditions with tetraethyl orthocarbonate at 90 °C.[47] Alternative syntheses of candesartan cilexetil use similar condensation conditions on variously elaborated benzimidazole precursors.[48,49]

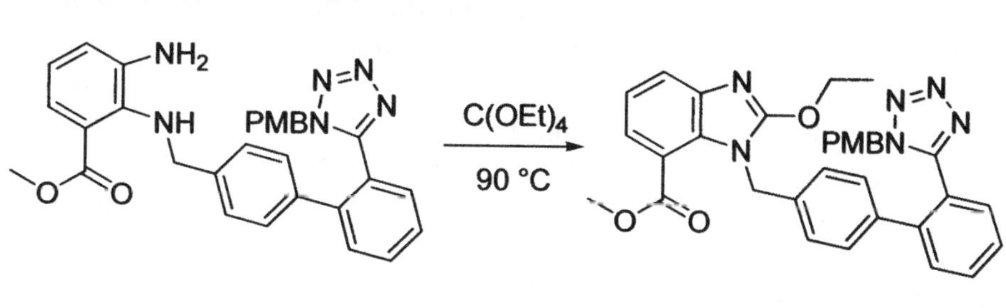

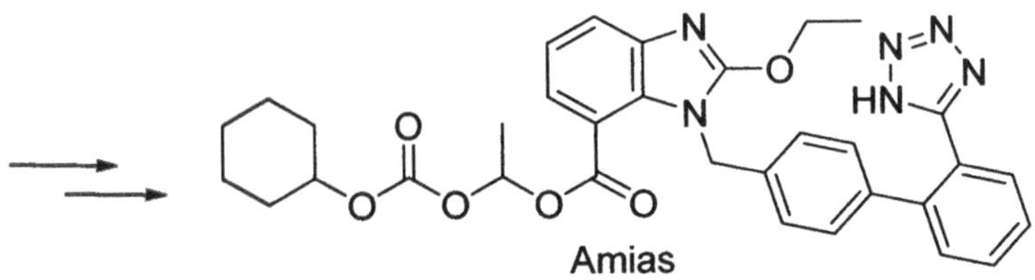

Amias
Candesartan cilexetil (prodrug)

Numerous syntheses have been reported of the AT_1 receptor antagonist telmisartan, which contains two benzimidazoles.[50] Although typical reductive cyclization conditions are used in most of the reported syntheses, one patent describes unique cyclization conditions for formation of the second benzimidazole ring system.[51] In this one-pot protocol, the 2-propyl-4-methyl-1*H*-benzimidazole-6-carboxylic acid is first activated with 2-chloro-4,6-dimethoxy-1,3,5-triazine in *N*-methyl morpholine (NMM) and methanol, then treated with *N*-methyl-*o*-phenylenediamine and heated to reflux to provide the cyclized product. This protocol allow for synthesis of the benzimidazole system under nonacidic conditions and moderate temperatures.

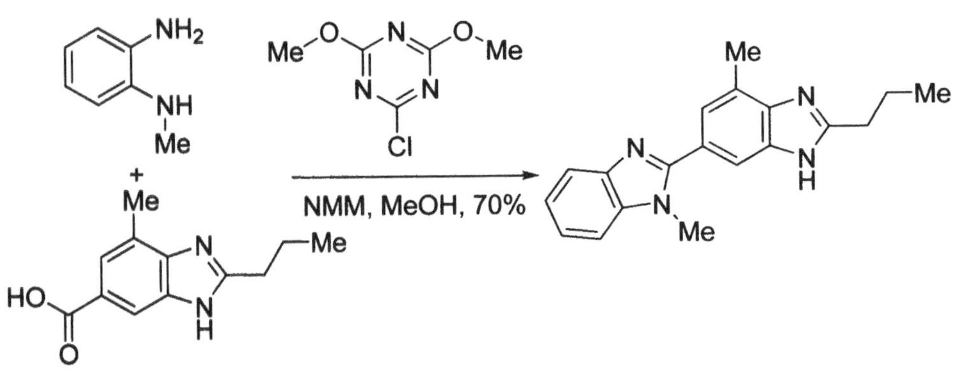

One key area of current research in the synthesis of benzimidazoles is the continued search for mild reaction conditions to avoid the harsh dehydrating conditions often required in the Phillips-type condensation/ cyclization approach. A recent report by Liu and co-workers exemplifies a mild variant of the classical orthoester condensation,[52] which may soon find utility in drug synthesis. In this report, a variety of substituted 1,2-diaminobenzenes is condensed with triethyl-orthoesters using catalytic $Ga(OTf)_3$. Under these solvent-free conditions, the reactions proceed at room temperature, with very short reaction times (5–90 min), thus providing a convenient new method for benzimidazole synthesis.

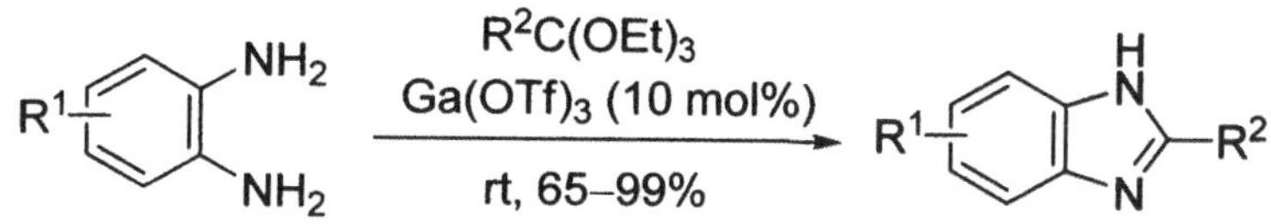

8.8 *Construction of the Benzimidazole Core Using Transition Metal-Mediated Approaches*

8.8.1 C–N Bond Formation from Aryl Halide

In recent years, significant research effort has been directed toward identification of new methods for benzimidazole synthesis.[53] Recent advances, particularly in transition metal-mediated reactions, have ushered in novel synthetic approaches for the construction and functionalization of this useful ring system. The present overview of new transition metal-mediated synthetic approaches focuses primarily on methodologies reported from 2008 to early 2012 that may soon find broader application in drug synthesis.

Brain and Brunton reported one of the early palladium-catalyzed cyclizations, which provided benzimidazoles in moderate to high yield.[54] In this study, $Pd(PPh_3)_4$ was used to catalyze the intramolecular *N*-arylation of *o*-bromoamidines to afford variously substituted 1-alkyl or 1-phenyl-benzimidazoles. This report set the stage for additional investigations of transition-metal mediated *N*-arylation as a tactic for benzimidazole synthesis.

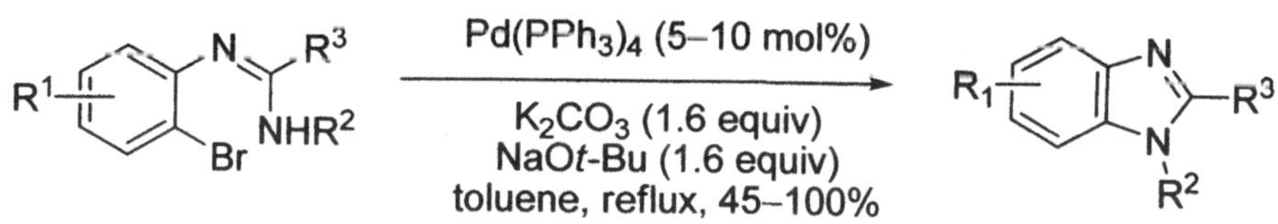

In more recent adaptations of metal-catalyzed benzimidazole construction, copper catalysts have gained preeminence for mediating either intermolecular or intramolecular C–N bond formation with aryl bromides or iodides. The utility of copper catalysis in intermolecular amidations was highlighted by Zheng and Buchwald in their report of the stepwise formation of *N*-alkyl benzimidazoles using intermolecular amidation as a key step.[55] Copper-catalyzed amidation of *o*-iodo- or *o*-bromo-*N*-alkylanilines is achieved using CuI (5 mol%), diamine ligand (L), and Cs_2CO_3 in 1,4-dioxane. The resultant anilide is then subjected to dehydrative conditions in analogy to the classical Phillips approach.

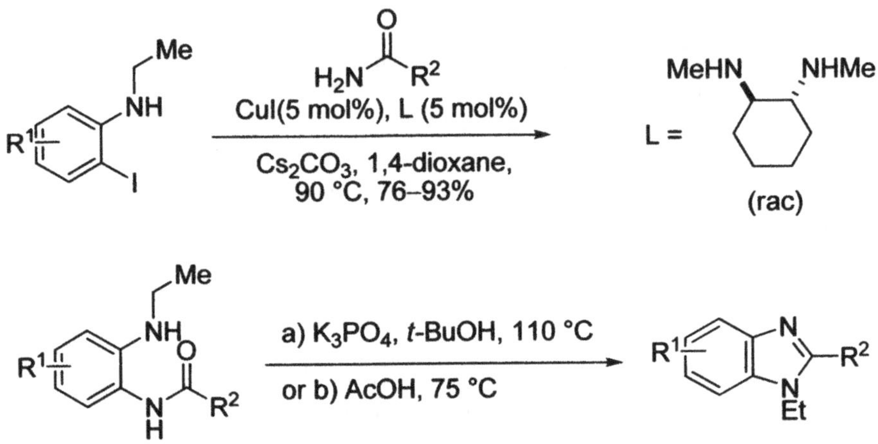

The focus on optimizing Cu-catalyzed processes and broadening their scope has continued over the past several years. Many of the recent permutations of the Cu-mediated arylation approach rely on processes in which the C–N bond formation is the last step in a multistep or cascade reaction sequence, thus avoiding acid-mediated dehydrations to complete the benzimidazole synthesis. The direct coupling of imidates (for 2-alkyl- or 2-arylimidazoles) or guanidines (for 2-aminoimidazoles) with halobenzenes provides a useful method for constructing the benzimidazole nucleus. Deng and co-workers have demonstrated a tandem diamination approach using 1,2-dihalobenzenes.[56] The preferred conditions for these couplings use CuI (15 mol%), *N,N'*-dimethylethylenediamine (DMEDA, 30 mol%), and Cs_2CO_3 in DMA or NMP at 150–170 °C to afford benzimidazole products in moderate yields.

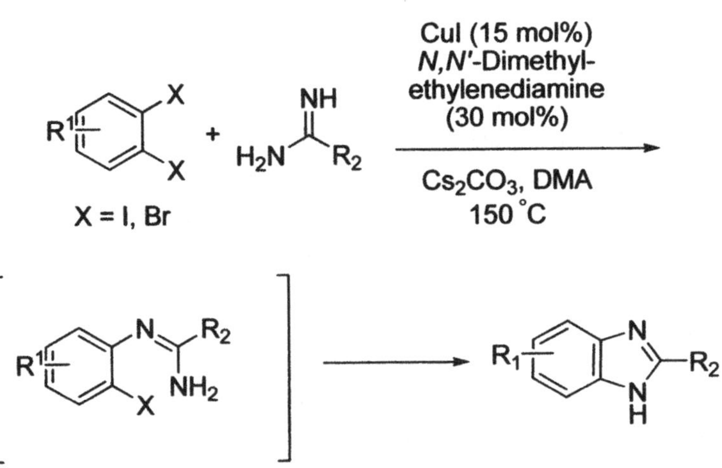

In further attempts to enhance the utility of copper-mediated *N*-arylation chemistry toward construction of 2-aminobenzimidazoles, two cascade processes employing nucleophilic addition to carbodiimides and intramolecular *N*-arylation have been reported. In the first report, Lu and Bao have reported a one-pot cascade process in which nucleophilic addition of an appropriate amine (e.g., aliphatic amine or imidazole) to an *o*-haloarylcarbodiimide produces a guanidine intermediate, which can undergo *N*-arylation to afford a 2-aminobenzimidazole.[57] More recently, a complementary cascade sequence has been reported by Xi and co-workers.[58] In this case, treatment of an *o*-haloaniline and a carbodiimide with CuI and sodium *t*-butoxide in NMP at 90–110 °C affords the corresponding 2-amino benzimidazole products in moderate to high yield. This reaction is believed to proceed via the intermediacy of a guanidine, which undergoes copper-mediated *N*-arylation as in the previously described cases. When unsymmetrically substituted carbodiimides are employed, cyclization of the guanidine intermediate occurs preferentially from an *N*-aryl rather than an *N*-alkyl group.

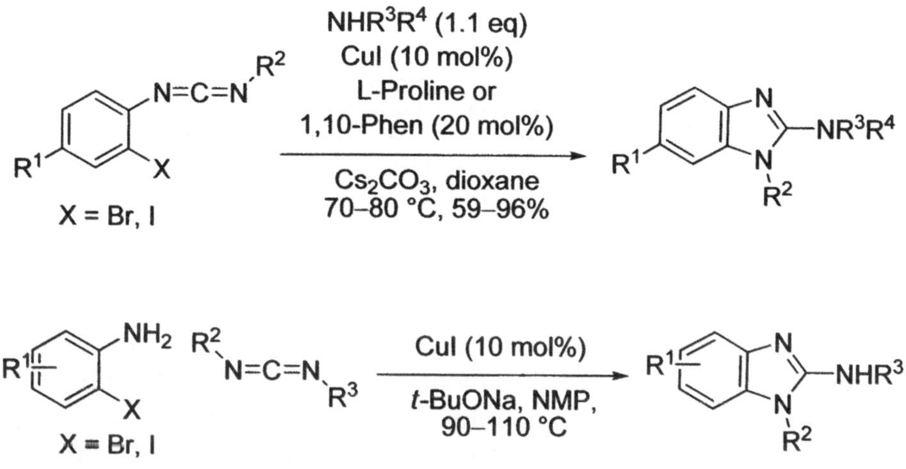

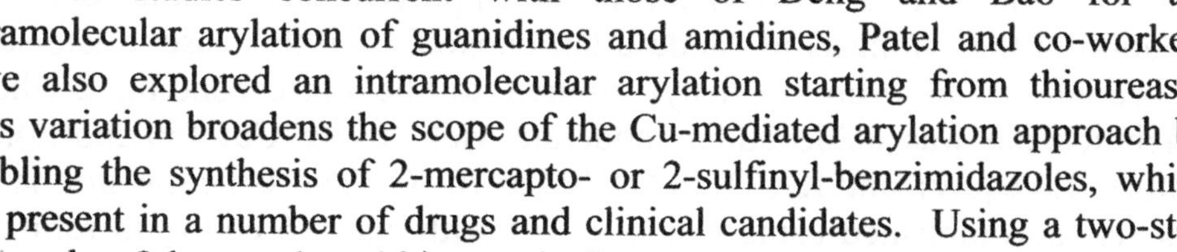

In studies concurrent with those of Deng and Bao for the intramolecular arylation of guanidines and amidines, Patel and co-workers have also explored an intramolecular arylation starting from thioureas.[59] This variation broadens the scope of the Cu-mediated arylation approach by enabling the synthesis of 2-mercapto- or 2-sulfinyl-benzimidazoles, which are present in a number of drugs and clinical candidates. Using a two-step protocol, a 2-bromophenylthiourea is first treated with triethylamine and an alkylating reagent (e.g., MeI and BnBr) in CH_3CN to provide an *S*-alkylated product. This isolated *S*-alkylated product is then treated with CuI (5 mol%), 1,10-phenanthroline (10 mol%), and K_2CO_3 in dioxane at 85 °C to afford desired 2-mercapto benzimidazole in moderate to high yield. Although this protocol has not yet been applied to synthesis of drug molecules or drug-like candidates, the method seems appropriate for the synthesis of useful 2-mercapto- or 2-sulfinyl-benzimidazoles in drug discovery applications.

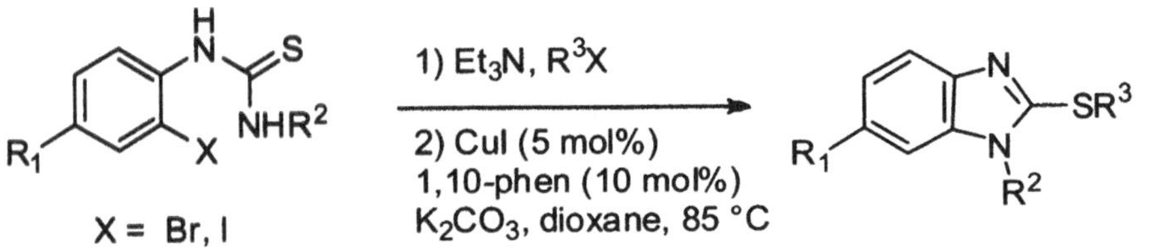

To date, most of the reported copper-mediated *N*-arylations applied toward benzimidazoles use a polar organic solvent (e.g., DMA, 1,4-dioxane, DMSO, or NMP.) Peng and co-workers have recently reported a modified intramolecular arylation of amidines to provide C2-alkyl and aryl benzimidazoles in moderate to high yield.[60] Remarkably, these arylations, which use Cu_2O (5 mol%), dimethylethylenediamine (DMEDA, 10%), and K_2CO_3, are conducted in water, lending potential environmental and economic advantage to this method.

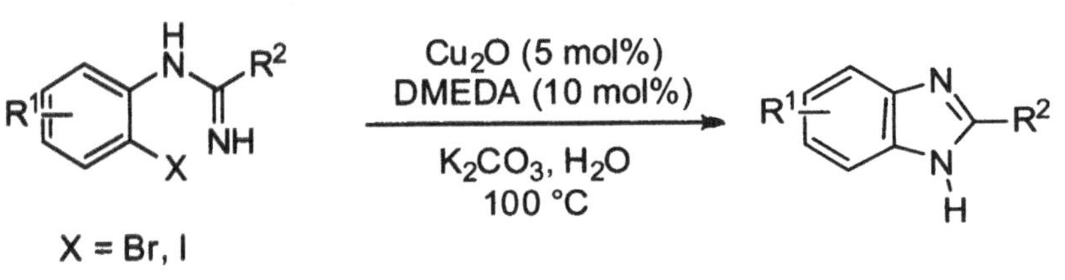

The exploration of the scope and utility of intramolecular Cu-mediated *N*-arylations for the synthesis of substituted benzimidazoles has only received significant attention among synthetic organic chemists for the past five years. Thus, this methodology has not yet been broadly applied among medicinal and process chemists; however, a few reports have begun to exemplify the utility of this approach in drug synthesis. In one recent example, Xiang and co-workers have employed an intramolecular *N*-arylation in the synthesis of candesartan cilexetil.[61] In this case, a late-stage cyclization of an *o*-bromophenyl isourea to the corresponding 2-ethoxy benzimidazole was achieved in moderate yield using CuI (50 mol%), L-proline (1.0 equiv), and K_2CO_3 in DMSO at 70 °C. The same transformation was also accomplished using catalytic CuI with NaH (6 equiv) in refluxing THF. Although further optimization of these conditions would be required to ensure a safe, cost-effective, and scalable process, these results provide preliminary exemplification of the use of this method for benzimidazole construction in drug synthesis.

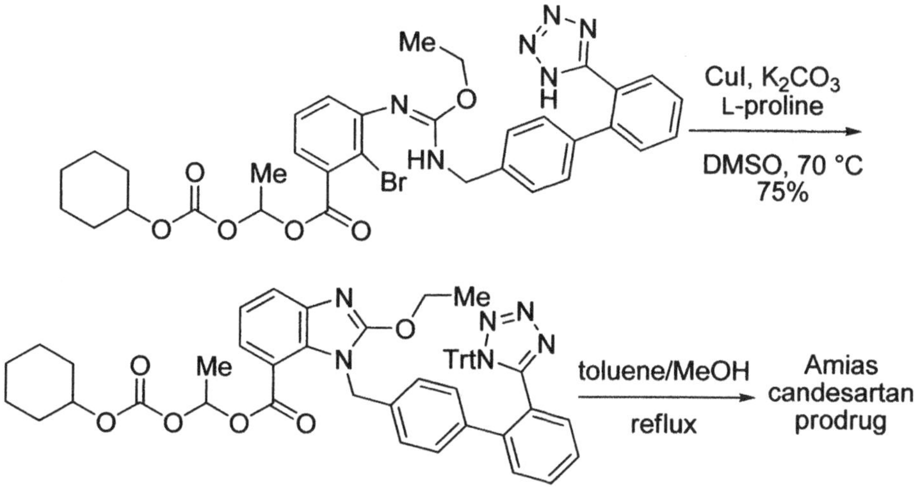

Another recent copper-mediated approach for the construction of benzimidazole drugs uses a somewhat different strategy than those previously described. Inspired in part by Driver and Shen's iron(II)-catalyzed synthesis of benzimidazoles from aryl azides,[62] Lee and co-workers have described a copper-catalyzed three-component synthesis of benzimidazoles from an *o*-haloaniline, an aldehyde, and sodium azide.[63] Under the optimized conditions, the reaction proceeds with CuCl (5 mol%), TMEDA (5 mol%) in DMSO at 120 °C to afford benzimidazoles in moderate to high yields. The mechanism for this reaction has not been fully explored, but may involve the intermediacy of a 2-iminoaryl azide. This approach has been applied toward the one-pot synthesis of tiabendazole, a fungicide, and parasiticide sold under the trade names Mintezol and Tesederm.

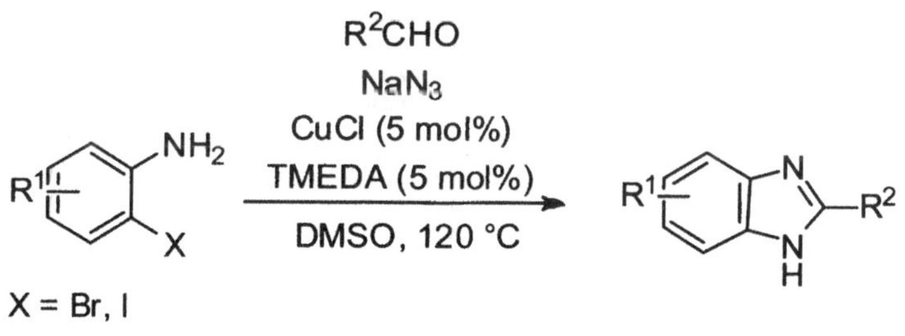

One-pot synthesis of tiabendazole

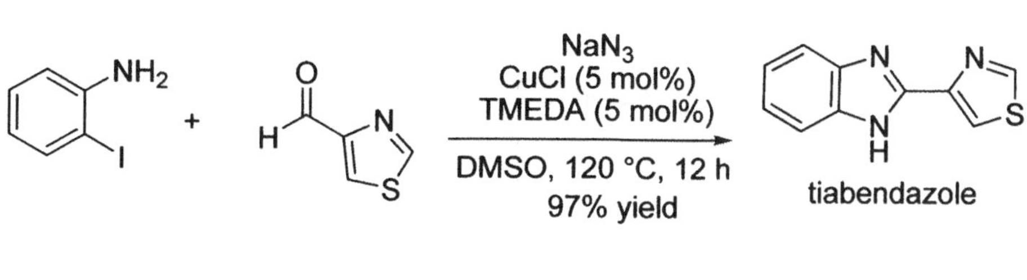

8.8.2 C–H Functionalization

The development of new methods for transition metal-mediated C–H bond functionalization has received significant interest in recent years in the context of heterocyclic synthesis.[64] The C–H functionalization approach toward C–N bond formation offers a potential advantage over the more traditional metal-mediated approaches that require more highly functionalized (e.g., aryl halide) starting materials. The synthesis of benzimidazoles *via* C–H functionalization approaches has received only limited attention to date. Nevertheless, the initial reports show that this is a promising new area for continued investigation, and further applications toward drug synthesis will likely emerge in due course. Brasche and Buchwald have described a copper-mediated C–H insertion method for intramolecular *N*-arylation of phenyl amidines.[65] The optimized conditions involve treatment of an *N*-phenylamidine with $Cu(OAc)_2$ (15 mol%) and acetic acid in DMSO at 100 °C under an oxygen atmosphere. (The authors report that catalytic activity was also observed under an air atmosphere, but product formation was considerably slower.) A variety of functionalized 2-aryl benzimidazoles was accessed using this approach.

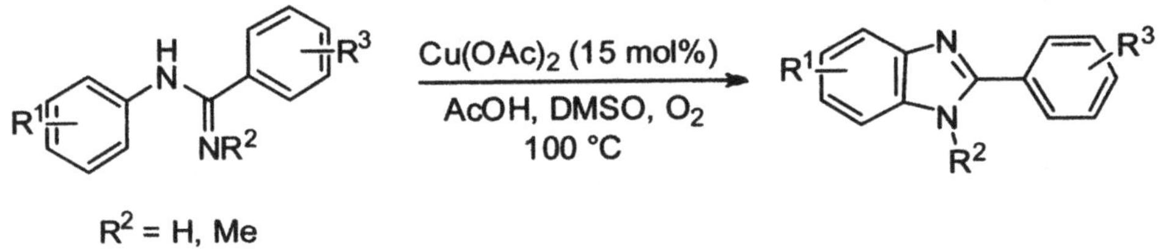

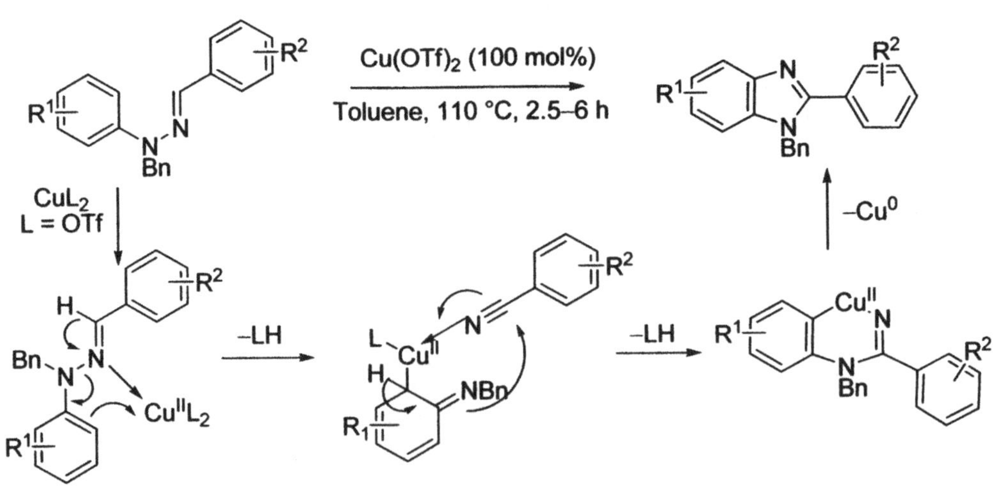

Using a somewhat different C–H functionalization approach to benzimidazoles, Punniyamurthy and co-workers have explored the conversion of *N*-benzyl bisarylhydrazones to 2-aryl-*N*-benzyl-benzimidazoles.[66] This conversion is optimally achieved using stoichiometric $Cu(OTf)_2$ in toluene at 110 °C. (This transformation was also reported using catalytic $Cu(OTf)_2$ (20 mol%) under an oxygen atmosphere, albeit in lower yield). This reaction likely proceeds via a copper-amidine intermediate analogous to the intermediate accessed by Brasche and Buchwald in their direct intramolecular arylation of *N*-aryl amidines.

8.9 Alternative Cyclization Approach Toward Benzimidazoles: Process Route Toward BYK405879

In the vast majority of cases, benzimidazole syntheses begin with a functionalized benzene intermediate, which may undergo additional functionalization prior to cyclization to the desired benzimidazole target. Webel, Palmer, and co-workers at Nycomed have recently reported a novel process route toward BYK405879, a drug candidate for treatment of "acid-related diseases."[67] Their unconventional approach of building the benzimidazole from an appropriately functionalized imidazole was designed to provide scale-up advantages over the previous, more conventional medicinal chemistry route. In this new process route, (1,2-dimethyl-1*H*-imidazol-5-yl)methanol undergoes a hypochlorite-TEMPO oxidation to the corresponding aldehyde, which is telescoped without isolation to a subsequent Stobbe condensation with diethyl succinate. This product is again telescoped without isolation to the cyclization reaction, in which acetic anhydride is employed to promote the dehydrative aromatization to the desired benzimidazole. This benzimidazole can be converted via a 4-step reaction sequence to BYK308944, a key intermediate in the synthesis of BYK405879.

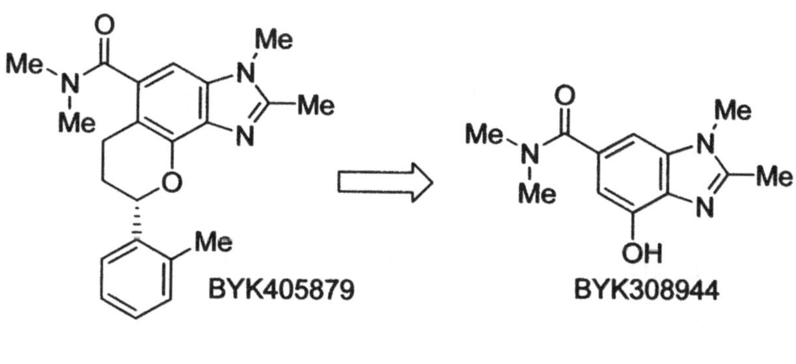

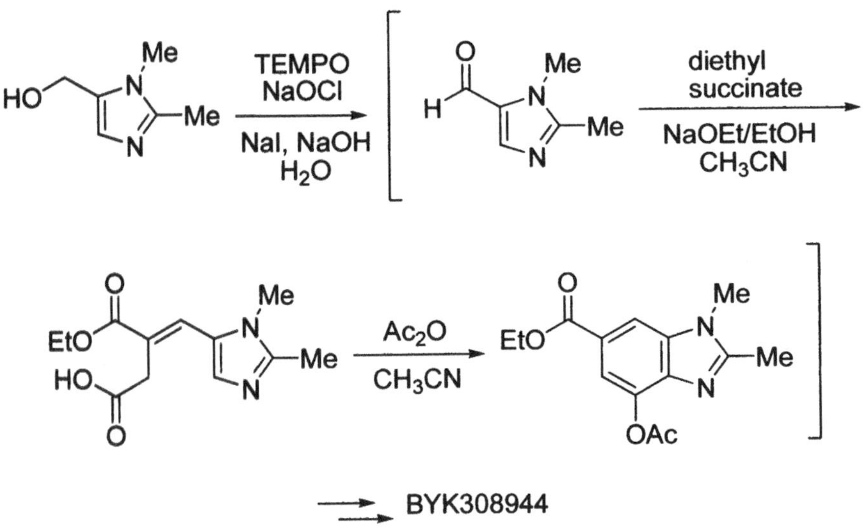

8.10 *Problems*

8.10.1 Propose a reasonable mechanism for the following transformation:

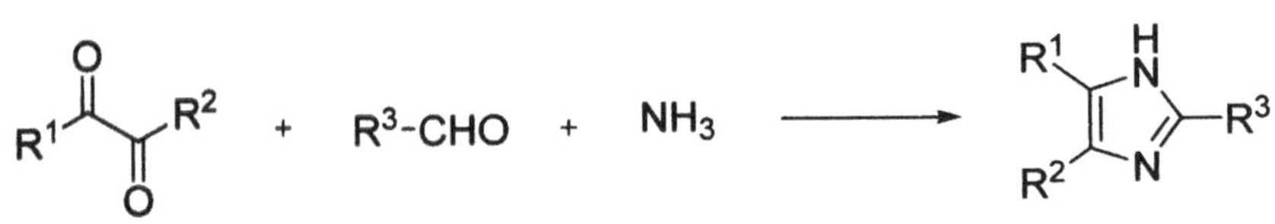

8.10.2 Propose a reasonable mechanism for the following transformation[68]:

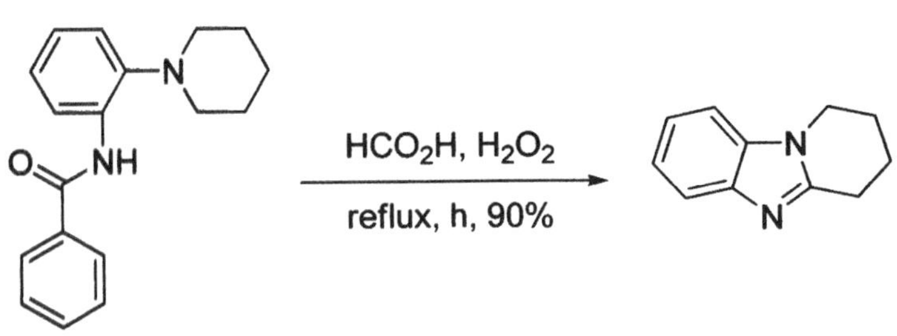

8.10.3 Propose a reasonable mechanism for the following transformation[69]:

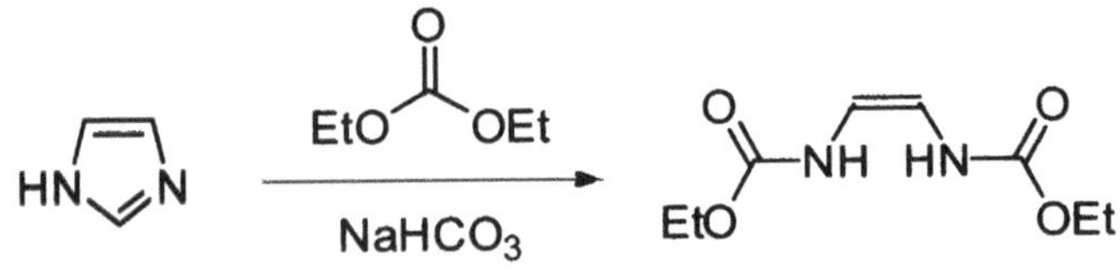

8.10.4 Propose a reasonable mechanism for the following transformation[70]:

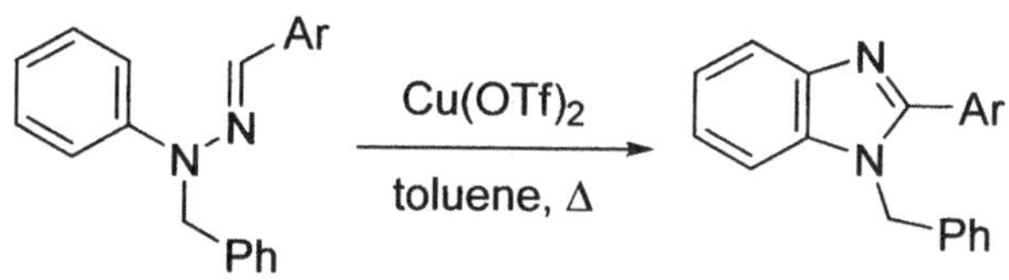

8.10.5 Propose a reasonable mechanism for the following transformation[71]:

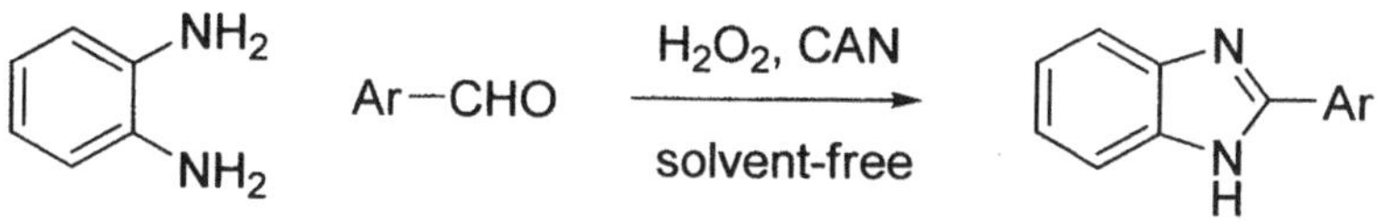

8.10.6 Suggest a mechanism for the following copper-catalyzed cross-cycloaddition between two different isocyanides to produce an imidazole as shown below.[72]

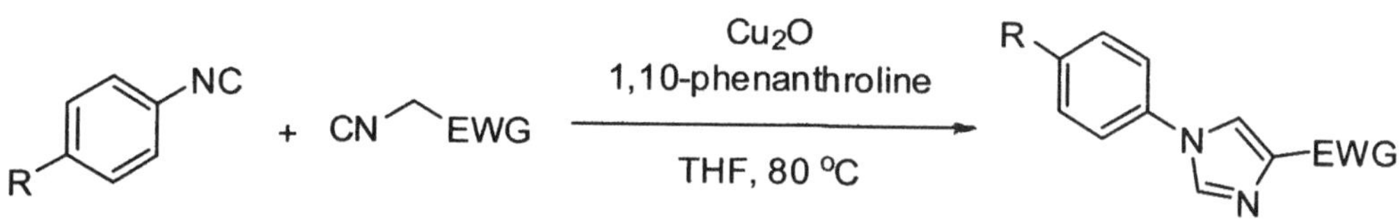

8.10.7 Propose a mechanism for the synthesis of benzimidazoles from common arylamino oximes.[73]

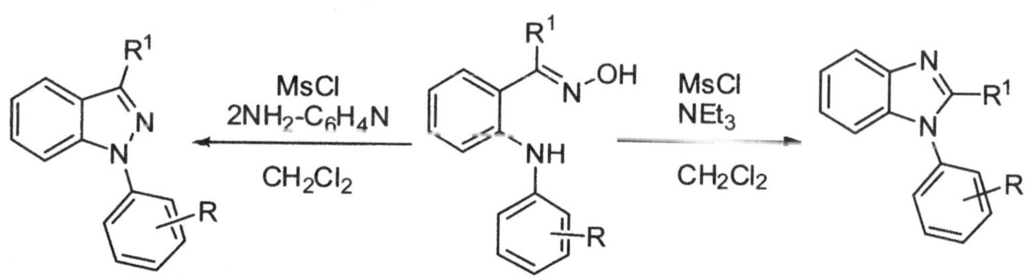

8.10.8 Propose a mechanism for the palladium-catalyzed multi-component one-step synthesis of imidazoles from imines and acid chlorides.[74]

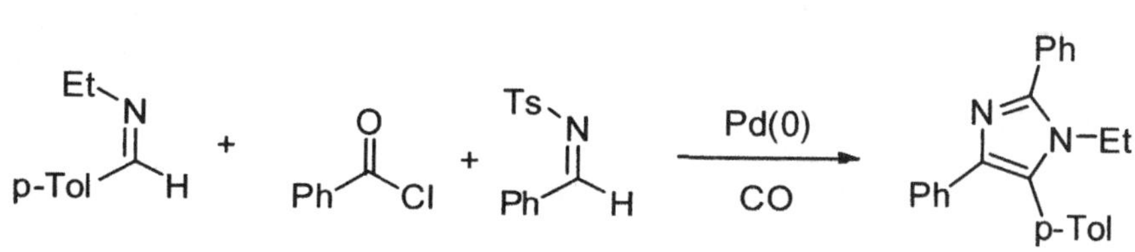

8.10.9 Alkylation of 4(5)-iodoimidazole with benzyl bromide provides a mixture of the 1-benzyl 4- and 5-idodo-1*H*-imidazoles. Provide conditions and draw the mechanism for selective conversion to 1-benzyl-4-iodo-1*H*-imidazole.[4]

8.10.10 Write a reasonable synthesis for the following compound (1-benzyl-4-bromo-2-phenyl-1*H*-imidazole-5-carbaldehyde) from 2-phenyl-1*H*-imidazole.[75]

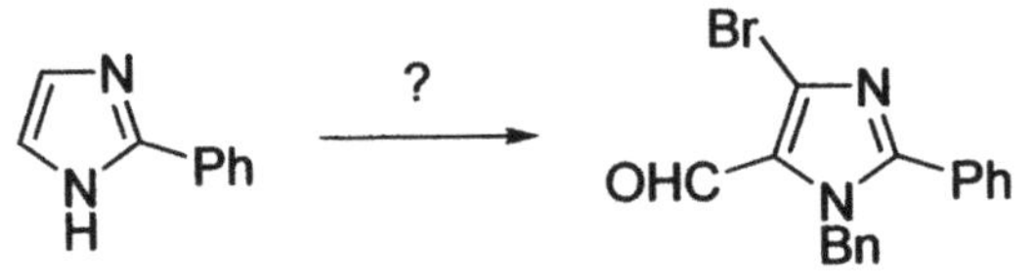

8.11 References

1. 1*H*-imidazole is used to distinguish the structure from the 2*H* and 4*H* isomers. The H denotes the atom with the sp^3 hybridization.

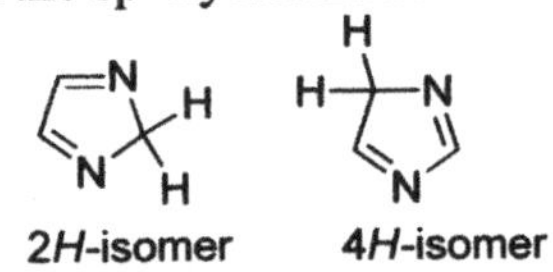

2. Grimmett, M. R. *Sci. of Synthesis*, Vol. 12, *Imidazoles*, 12.3, p 325, **2002,** Ed. Neier, R. 2002, Georg Thieme Verlag, New York.
3. Gilchrist, T. L. *Heterocyclic Chemistry* **1985,** Longman Scientific and Technical, and Wiley: New York, NY.
4. Du, H.; He, Y.; Sivappa, R.; Lovely, C. J. *Synlett* **2006,** 965–992.
5. Bhatnagar, A.; Sharma, P. K.; Kumar, N. *J. PharmTech Res*. **2011,** *3*, 28–282.
6. Liu, H.; Kerdesky, F. A.; Black, L. A.; Fitzgerald, M.; Henry, R.; Esbenshade, T. A.; Hancock, A. A.; Benanni, Y. L. *J. Org. Chem*. **2004,** *69*, 192–194.
7. Grabowski, E. J. J.; Liu, T. M. H.; Salce, L.; Schoenewaldt, E. F. *J. Med. Chem.* **1997,** *17*, 547–549.
8. Khodakovskiy, P. V.; Volochnyuk, D. M.; Panov, D. M.; Pervak, I. I.; Zarudnitskii, E. V.; Shishkin, O. V.; Yurchenko, A. A.; Shivanyuk, A.; Tolmachev, A. A. *Synthesis* **2008,** 948–956.
9. Samant, B. S.; Sukhthankar, M. G.; *Bioorg. Med. Chem. Lett.* **2011,** *21*, 1015–1018.
10. Watson, S. P. *Synth. Commun*. **1992,** *22*, 2971–2977.
11. Shi, Y.-J.; Frey, L. F.; Tschaen, D. M.; Verhoeven, T. R. *Synth. Commun*. **1993,** *23*, 2623–2630.
12. Larsen, R. D.; King, A. O.; Chen, C. Y.; Corley, E. G.; Foster, B. S.; Roberts, F. E.; Yang, C.; Lieberman, D. R.; Reamer, R. A.; Tschaen, D.; Verhoeven, T. R.; Reider, P. J.; Lo, Y. S.; Rossano, L. T.; Brookes, A. S.; Meloni, D.; Moore, J. R.; Arnett, J. F. *J. Org. Chem*. **1994,** *59*, 6391–6394.
13. Blicke, F. F.; Godt, H. C. *J. Am. Chem. Soc*. **1954,** *76*, 3653–3655.
14. Kiselyov, A. S.; Piatnitsk, E. L.; Samet, A. V.; Kisliy, V. P.; Semenov, V. V. *Bioorg. Med. Chem. Lett.* **2007,** *17*, 1369–1375.
15. Aleksandrova, E. V.; Kochergin, P. M. *Pharmaceut. Chem. J.* **2010,** *44*, 381–385.

16. Iddon, B. *Heterocycles* **1985,** *23,* 417–443.
17. Bellina, F.; Cauteruccio, S.; Rossi, R. *Tetrahedron* **2007,** *63,* 4571–4624.
18. Webb, R. L.; Lewis, J. J. *J. Heterocycl. Chem.* **1981,** 1301–1303.
19. Yet, L. *Progr. Heterocycl. Chem.* **2008,** *19,* 208–241.
20. Godefroi, E. F.; Janssen, P. A. J.; Van der Eycken, C. A.. M.; Van Heertum, A. H. M. T.; Niemegeers, C. J. E. *J. Org. Chem.* **1965,** *8,* 220–223.
21. Kamijo, S.; Yamamoto, Y. *Chem Asian J.* **2007,** *2,* 568–578.
22. Yet, L. *Progr. Heterocycl. Chem.* **2009,** *21,* 244–260.
23. van Leusen, A. M.; Wildeman, J.; Oldenziel, O. H. *J. Org. Chem.* **1977,** *42,* 1153.
24. Murry, J. A.; Frantz, D. E.; Soheili, A.; Tillyer, R.; Grabowski, E. J. J., Reider, P. J. *J. Am. Chem. Soc.* **2001,** *123,* 9696–9697.
25. Frantz, D. E.; Morency, L.; Soheili, A.; Murry, J. A.; Grabowski, E. J. J.; Tillyer, R. D. *Org. Letters*. **2004,** *6,* 843–846.
26. Bowles, D. M.; Bolton G. L.; Boyles, D. C.; Curran, T. T.; Hutchings, R. H.; Larsen, S. D.; Miller, J. M.; Park, W. K. C.; Ritsema, K. G.; Schineman, D. C.; Tamm, M. *OPRD*, **2008,** *12,* 1183–1187.
27. Yanagisawa, H.; Amemiya, Y.; Kanazaki, T.; Shimoji, Y.; Fujimoto, K.; Kitahara, Y.; Sada, T.; Mizuno, M.; Ikeda, M.; Miyamoto, S.; Furukawa, Y.; Koike, H. *J. Med. Chem.* **1996,** *39,* 323–338.
28. Babu, K. S.; Reddy, M. S.; Tagore, A. R.; Reddy, G. S.; Sebastian, S.; Varma, M. S.; Venkateswarlu, G.; Bhattachary, A.; Reddy, P. P.; Anand, R. V. *Synth. Commun.* **2009,** *39,* 291–298.
29. Wanner, M. J.; Koomen, G.-J. *J. Chem. Soc., Perkin Trans. 1* **2002,** 1877–1880.
30. Suwinski, J.; Swierczek, K. *Tetrahedron Lett.* **1998,** *39,* 3331–3332.
31. Kuo, S.-C.; Mas, J. L.; Hou, D. US 2006/0229456 (2006).
32. Sumalatha, Y.; Reddy, T. R.; Reddy, P. P.; Satyanarayana, B. *ARKIVOC* **2009,** 315–320.
33. Ermolat'ev, D. S.; Vander Eycken, E. V. *J. Org. Chem.* **2008,** *73,* 6691–6697.
34. Ermolat'ev, D. S.; Babaev, E. V.; Vander Eycken, E. V. *Org. Lett.* **2006,** *8,* 5781–5784.
35. Kumar, V.; Anand, A.; Mahajan, M. P. *Synlett* **2006,** 2199–2202.
36. Xu, W.; Abboud, K. A.; Ghiviriga, I.; Dolbier, W. R.; Rapp, M.; Wnuk, S. F. *Org. Lett.* **2006,** *8,* 5549–5551.
37. Grimmett, M. R. *Sci. of Synthesis*, Volume 12, *Benzimidazoles*, 12.4, Ed. Neier, R. **2002**, Georg Thieme Verlag, New York.
38. IMS National Sales Perspectives, **2012** (www.imshealth.com).
39. Phillips, M. A. *J. Chem. Soc.* **1928,** 172–177.
40. Phillips, M. A. *J. Chem. Soc.* **1931,** 1143–1153.
41. Graul, A.; Castaner, R.; Castaner, J. *Drugs Fut.* **1999,** *24,* 1178–1183.
42. Junggren, U. K.; Sjöstrand, S. E. EP 5129 (1979).
43. Walter, T. J. US 4152522 (1979).
44. Gyurik, R. J.; Theodorides, V. J. US 3915986 (1975).
45. Volkhard, A.; Heider, J.; Eberlein, W.; Diederen, W.; Haarmann, W. US 4361563 (1982).
46. Davies, S.; Castaner, R.; Bolos, J. *Drugs Fut.* **2008,** *33,* 926–931.
47. Seki, M. WO 2011/061996 (2011).
48. Naka, T.; Nishikawa, K.; Kato, T. US 6004989 (1999).
49. Venkataraman, S.; Reddy, P. P.; Reddy, G. S.; Reddy, S. S.; Narasimha Murthy, H. WO 2006015134 (2006).

50. Yet, L. in *Art of Drug Synthesis*, Li, J. J.; Johnson, D. S. eds., Wiley: New York, NY. **2007,** pp 129–141.
51. Heitger, H.; Rodriguez-Delhi, J. M.; Dach, R. US 7608722 (2009).
52. Liu, J.; Liu, Q.; Xu, W.; Wang, W. *Chin. J. Chem.* **2011,** *29*, 1739–1744.
53. Alamgir, M.; Black, D. S. C.; Kumar, N. *Top. Heterocycl. Chem.* **2007,** *9*, 87–118.
54. Brain, C. T.; Brunton, S. A. *Tetrahedron Lett.* **2002,** *43*, 1893–1895.
55. Zheng, N.; Buchwald, S. L. *Org. Lett.* **2007,** *9*, 4749–4751.
56. Deng, Z.; McAllister, H.; Mani, N. S. *J. Org. Chem.* **2009,** *74*, 5742–5745.
57. Lv, X.; Bao, W. *J. Org. Chem.* **2009,** *74*, 5618.
58. Wang, F.; Cai, S.; Liao, Q.; Xi, C. *J. Org. Chem.* **2011,** *76*, 3174–3180.
59. Murru, S.; Patel, B. K.; Le Bras, J.; Muzart, J. *J. Org. Chem.* **2009,** *74*, 2217–2220.
60. Peng, J.; Ye, M.; Zong, C.; Hu, F.; Feng, L.; Wang, X.; Wang, Y.; Chen, C. *J. Org. Chem.* **2011,** *76*, 716–719.
61. Wang, P.; Zheng, G.-J.; Wang, Y.-P.; Wang, X.-J.; Li, Y.; Xiang, W.-S. *Tetrahedron* **2010,** *66*, 5402–5408.
62. Shen, M.; Driver, T. G. *Org. Lett.* **2008,** *10*, 3367–3370.
63. Kim, Y.; Kumar, M. R.; Park, N.; Heo, Y.; Lee, S. *J. Org. Chem.* **2011,** *76*, 9577–9583.
64. Stokes, B. J.; Driver, T. G. *Eur. J. Org. Chem.* **2011,** 4071–4088.
65. Brasche, G.; Buchwald, S. L. *Angew. Chem. Int. Ed.* **2008,** *47*, 1932–1934.
66. Guru, M. M.; Ali, M. A.; Punniyamurthy, T. *J. Org. Chem.* **2011,** *76*, 5295–5308.
67. Webel, M.; Palmer, A. M.; Scheufler, C.; Haag, D.; Müller, B. *Org. Proc. Res. Dev.* **2010,** *14*, 142–151.
68. (a) Suleman, A.; Skibo, E. B. *J. Med. Chem.* **2002,** *45*, 1211–1220. (b) Craigo, W. A.; LeSueur, B. W.; Skibo, E. B. *J. Med. Chem.* **1999,** *42*, 3324–3333.
69. Warshawsky, A.; Altman, J.; Kahana, N.; Arad-Yellin, R.; Deshe, A.; Hasson, H.; Shoef, N.; Gottlieb, H. *Synthesis* **1989,** 825–829.
70. Guru, M. M.; Ali, M. A.; Punniyamurthy, T. *J. Org. Chem.* **2011**, *76*, 5295–5308.
71. Bahrami, K.; Khodaei, M. M.; Naali, F. *J. Org. Chem.* **2008,** *73*, 6835–6837.
72. Kanazawa, C.; Kamijo, S.; Yamamoto, Y. *J. Am. Chem. Soc.* **2006,** *128*, 10662–10663.
73. Wray, B. C.; Stambuli, J. P. *Org. Lett.* **2010,** *12,* 4576–4579.
74. Siamaki, A. R.; Arndtsen, B. A. *J. Am. Chem. Soc.* **2006,** *128*, 6050–6051.
75. (a) Bahnous, M. *et al. Tetrahedron Lett.* **2006,** *47,* 1949–1951. (b) Becher, J. *et al. Synthesis* **2006,** 530–533.

Chapter 9 Triazoles and Tetrazoles

Timothy J. Hagen and Zheng Zhang

9.1 Introduction

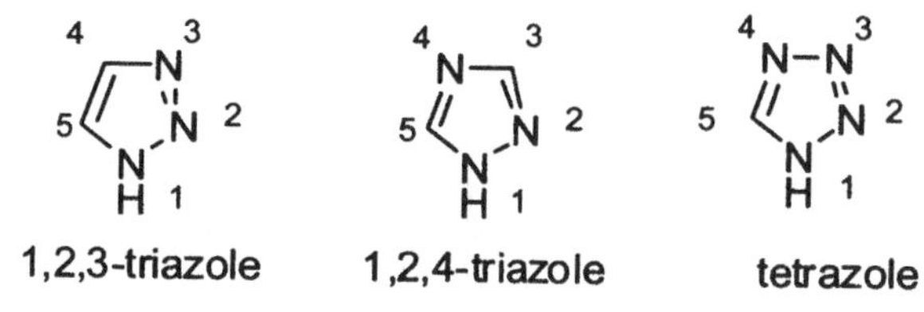

Triazole refers to either one of two isomeric five-member ring compounds with molecular formula $C_2H_3N_3$. The 1,2,3-triazole is a colorless low melting solid with a melting point of 23–25 °C and a boiling point of 203 °C/752 mmHg, while the 1,2,4-triazole is a colorless crystalline solid with a melting point of 119–121 °C. Both compounds are corrosive and an irritant to both skin and eyes. Triazole is a five-membered aromatic heterocycle with three N heteroatoms. The aromaticity of triazole tautomers was assessed by the Bird indices: 2*H*-1,2,3-triazole (88) was found to be slightly more aromatic than its 1*H*-isomer (73).[1] The small difference in Bird indices supports only a weak influence of the aromaticity and the lower stability of the 1*H*-isomer was explained by the nitrogen lone-pair repulsion that destabilizes cyclic azo-derivatives. Tetrazole has a melting point of 157–158 °C and a Bird index of 72, which is similar to pyrazole (73).

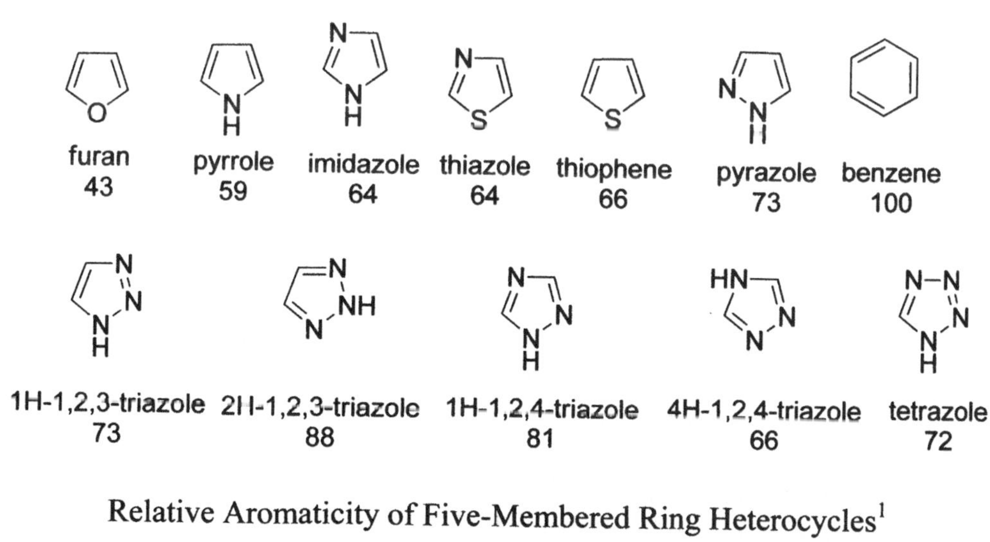

Relative Aromaticity of Five-Membered Ring Heterocycles[1]

The bond lengths for the 1*H*-1,2,4-triazole are shown below,[2] which impact the coupling constants in their ^{1}H-NMR spectra.

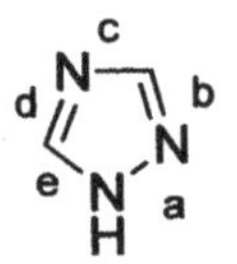

a = 1.359 Å
b = 1.323 Å
c = 1.359 Å
d = 1.324 Å
e = 1.331 Å

^{1}H-1,2,4-Triazole's Bond Lengths

In 1,2,4-triazole's ^{1}H-NMR spectrum, the chemical shift for both H3 and H5 is the same (due to the tautomerization between the nitrogen atoms) at 8.32 ppm. Like other NH-containing heterocycles, the chemical shift of triazole's NH largely depends on the solvent used to do the NMR spectrum.

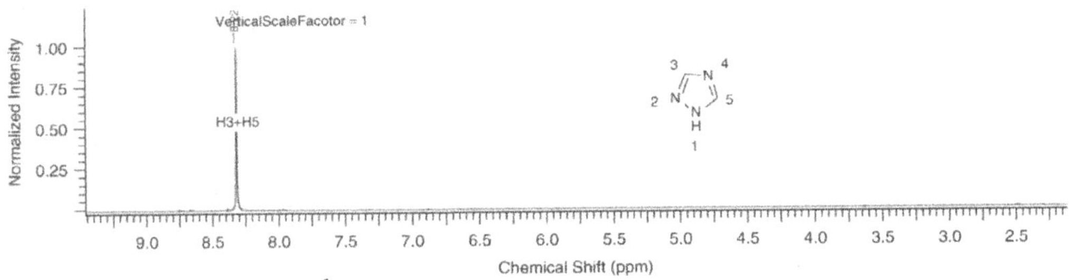

^{1}H-NMR Spectrum of Triazole

In 1,2,4-triazole's ^{13}C-NMR spectrum, the chemical shift for both C3 and C5 is the same again, due to the tautomerization between the nitrogen atoms, at 145.98 ppm.

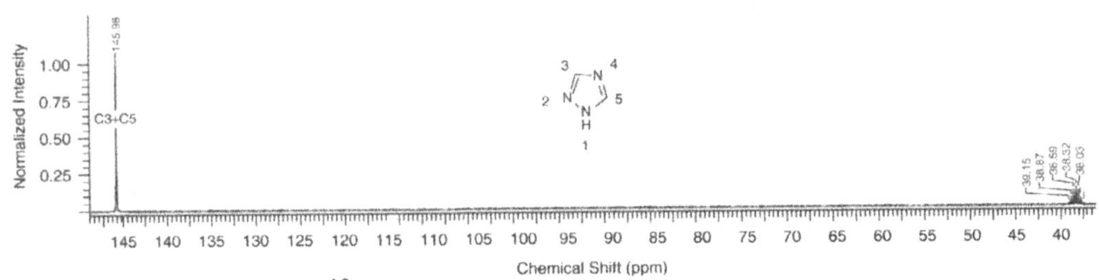

^{13}C-NMR Spectrum of Triazole

In 1,2,3-triazole's ^{1}H-NMR spectrum, the chemical shift for both H4 and H5 is the same at 7.90 ppm. Like other NH-containing heterocycles, the chemical shift of triazole's NH largely depends on the solvent used to do the NMR spectrum. In 1,2,3-triazole's ^{13}C-NMR spectrum, the chemical shift for both C4 and C5 is the same at 130.9 ppm

In tetrazole's ^{1}H-NMR spectrum, the chemical shift for the H5 is at 9.50 ppm. Like other NH-containing heterocycles, the chemical shift of triazole's NH largely depends on the solvent used to do the NMR spectrum.

In tetrazole's ^{13}C-NMR spectrum, the chemical shift for the C5 is at 143.1 ppm.

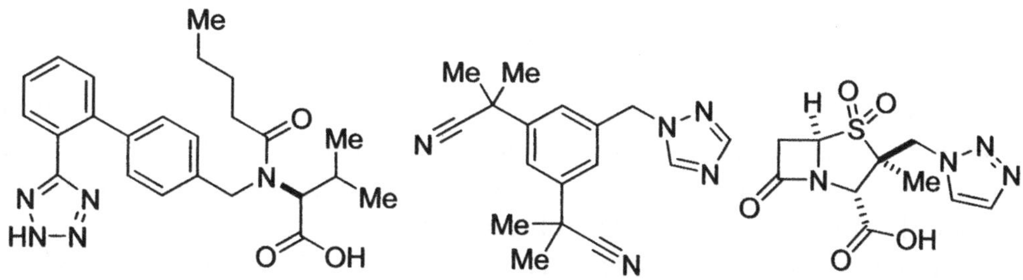

valsartan (Diovan) anastrozole (Arimidex) tazobactam (Zosyn)

Triazoles and tetrazoles do not exist in nature. However, many synthetic medicines do contain triazoles and tetrazoles. For instance, valsartan, an angiotensin receptor blocker (ARB) indicated for treatment of high blood pressure and other cardiovascular disease contains a tetrazole.[3] Anastrozole contains a 1,2,4-triazole and is an aromatase-inhibiting drug approved for treatment of breast cancer.[4] Tazobactam is a 1,2,3-triazole containing compound that inhibits the action of bacterial β-lactamases and is used to treat bacterial infection in combination with the beta-lactam antibiotic piperacillin.[5]

Five-membered ring systems with more than one N atom including triazole and tetrazoles have been recently reviewed.[6]

9.2 *Reactivity of the Triazole and Tetrazole Ring*

9.2.1 *Substitution of the 1,2,3-Triazole*

Triazole undergoes halogenation, and bromination of the triazole provides the 4,5-dibromo derivative. The 4,5-dibromotriazole was prepared by bromination of triazole with NBS.[7] Triazoles are readily alkylated on nitrogen by alkylhalides, dimethylsulfate, diazomethane, and conjugate addition. In addition, triazoles are acylated by acyl halides and anhydrides and the alkylations are not regioselective.[8]

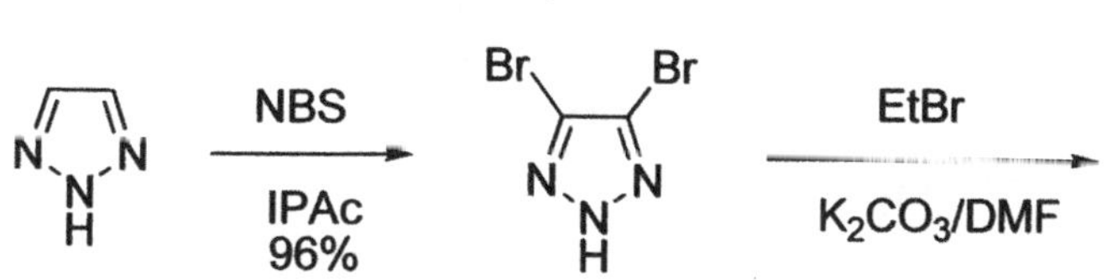

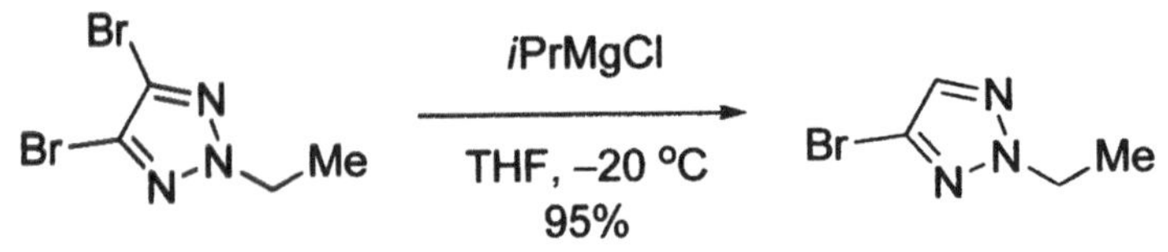

The selective formation of N-2 substituted traizoles has been recently reported.[9] The alkylation of 4,5-dibromotriazole with ethylbromide produces the *N*-2 alkylated product in an 87:13 ratio.[9] The dibromo-triazole is then readily converted into the monobromo-triazole.

Recently propargyl alcohols were reported to undergo $FeCl_3$-catalyzed triazole propargylation to yield propargylated triazoles.[10] A mixture of N-2 and N-1 isomers is generally produced. The products of the reaction were then further derivatized via click chemistry to synthesize bis-triazole systems.

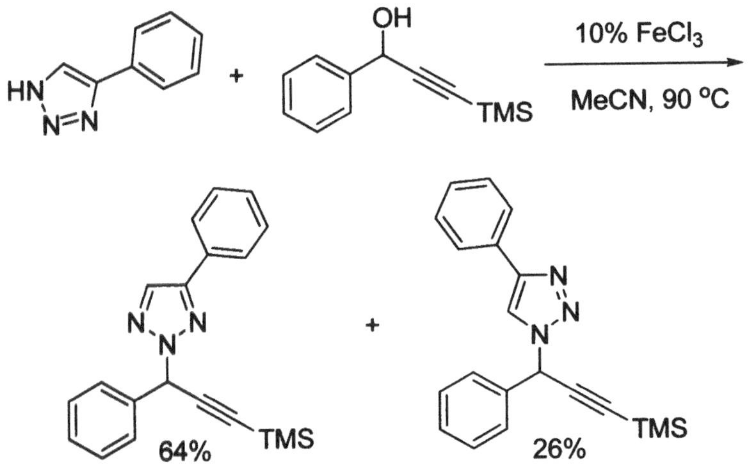

Biaryl-1,2,3-triazoles have been reported to undergo direct palladium-catalyzed arylation to yield annulated phenanthrenes.[11]

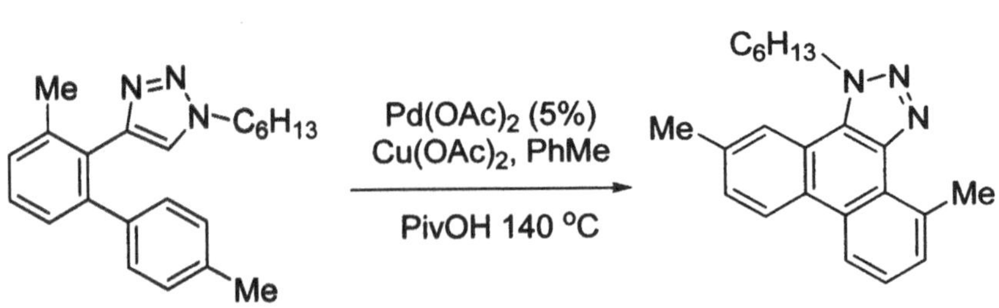

9.2.2 Substitution of the 1,2,4-Triazole

1,2,4-Triazoles are readily alkylated on nitrogen by alkyl halides although it is often difficult to predict which of the possible isomers will be formed.[8] 1,2,4-Triazole is chlorinated at C-3 through a 1-chloro-derivative that can be isolated.[8] The 3,5-dibromo-1,2,4-triazole is prepared from the parent 1,2,4-triazole in a 90% isolated yield.[12] The 3,5-dibromo-1,2,4-triazole can be reacted with stannyl derivatives in a microwave to yield 5-aryl-1,2,4-triazoles that can then be further elaboarated to 3,5-diaryl-1,2,4-triazoles.[12]

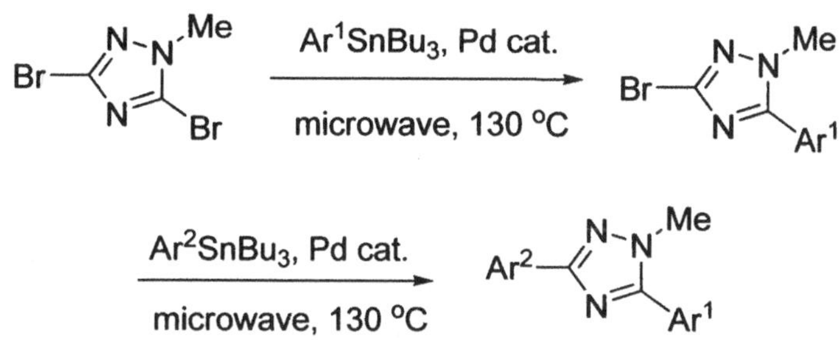

9.2.3 Alkylation of triazoles

Alkylation of 1,2,3-triazole

The 1,2,3-triazole is alkylated on nitrogen by alkyl halides, dimethyl sulfate, diazoalkanes, methyl *p*-toluenesulfonate, or through Mannich reaction. Unsubstitued 1,2,3-triazole prefers alkylation on N-1 when treated with methyl iodide and base. Alternatively, diazomethane regioselectively provides the N-2 alkylated product. The corresponding N-1 alkylated triazole tends to dialkylated at N-3, while the N-2 alkylated triazole needs a more powerful alkylating reagent such as methyl fluorosulfonate to give the 1,2-disubstitued compound.[13]

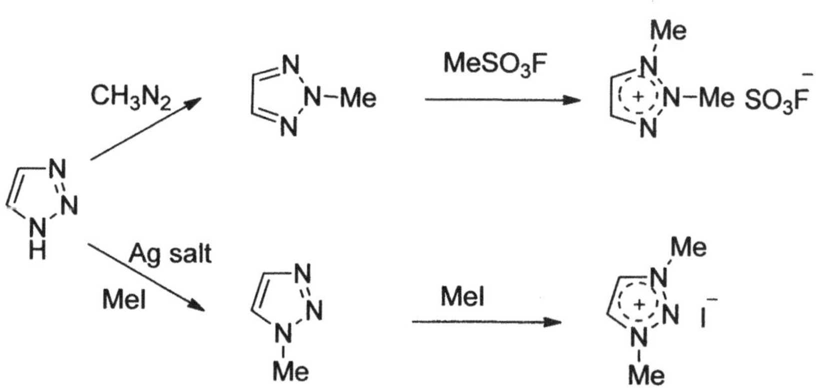

Additionally, substituents on the triazole ring will direct the orientation of alkylation. The 4-phenyl-1H-1,2,3-triazole tends to give the 1-methyl and 2-methyl isomers in 62% and 38% when treated with dimethyl sulfate. The 1-methyl-5-phenyl derivative was not found due to steric effects.[13]

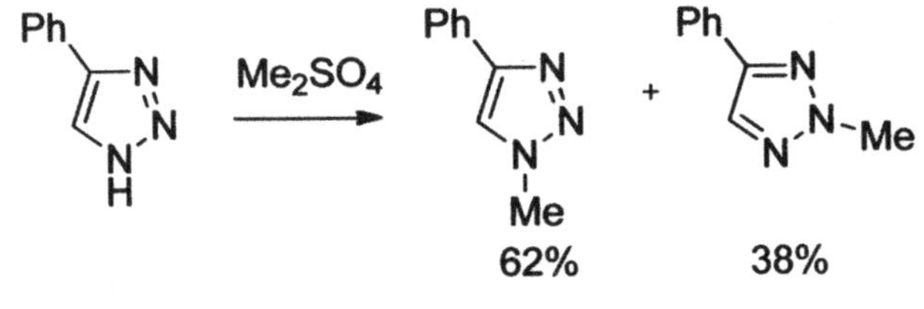

Alkylation of 1,2,4-triazole

The tautomerism of 1,2,4-triazole showed as below:

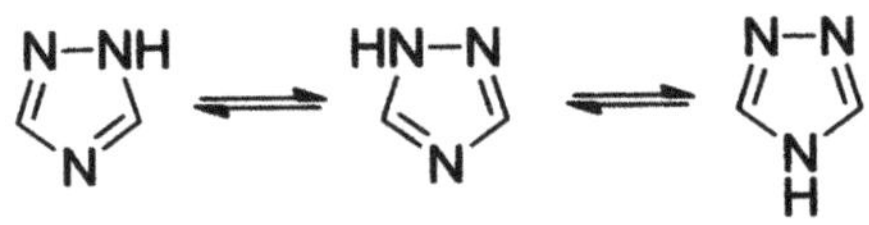

The alkylation of 1,2,4-triazole prefers to give the N-1 substitued product rather than the N-4 substituted product. However, the alkylation product of either N-1 or N-2 is difficult to predict when triazole has 3- or 5-substituents. Generally, the ratio of both products depends on the properties of the alkylating agent.[14]

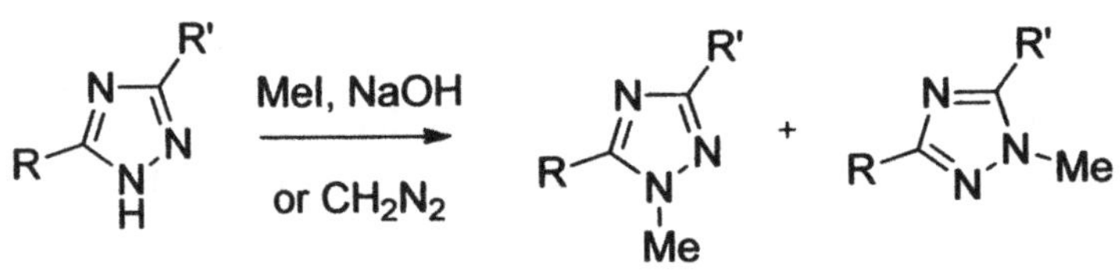

An example of alkylation of a 3-substituted triazole is shown below:

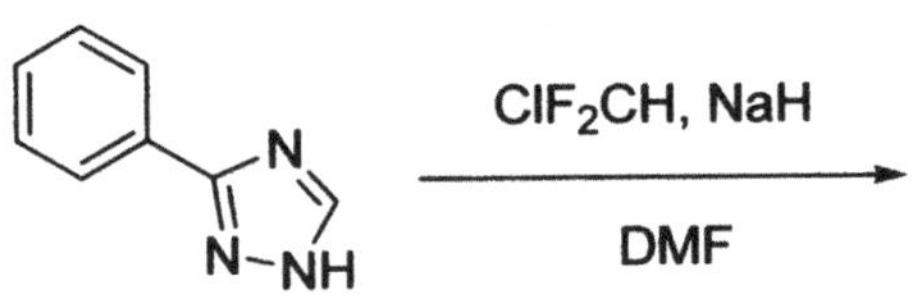

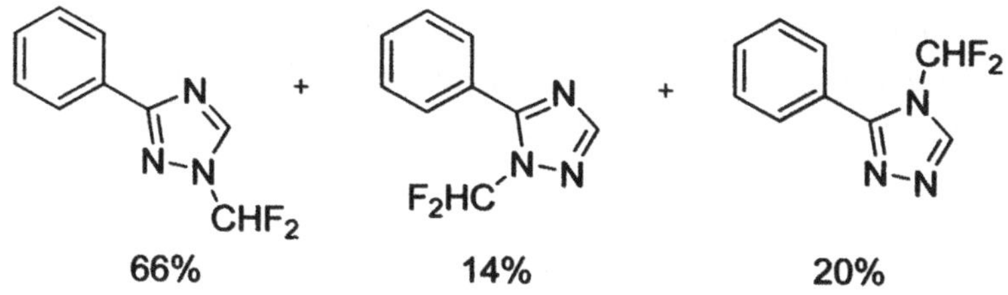

N-Difluoromethylation of 3-phenyl-1,2,4-triazole yields three products by using chlorodifluoromethane and base with N-1 substituted compound as the major product.[15] Additionally, *N*-trimethylsilylazole can be alkylated at N-1 with *n*-butylbromide in a 90% yield.[16]

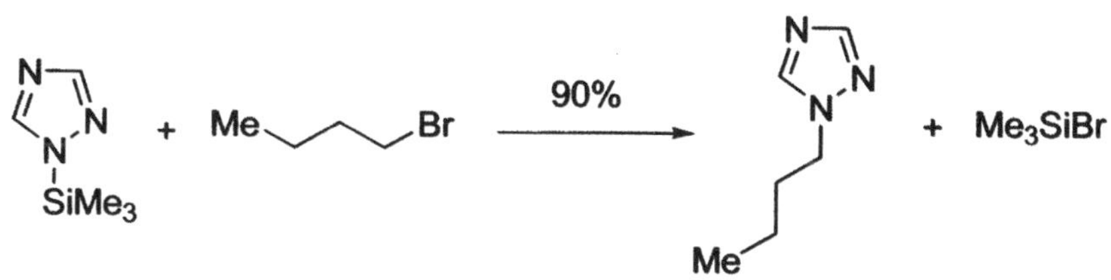

Anastrozole is an aromatase inhibitor that is used to treat breast cancer by decreasing the amount of estrogen synthesized by the human body. The synthesis of anastrozole entails a selective bromination and alkylation reaction with 1, 2, 4-triazole.[17]

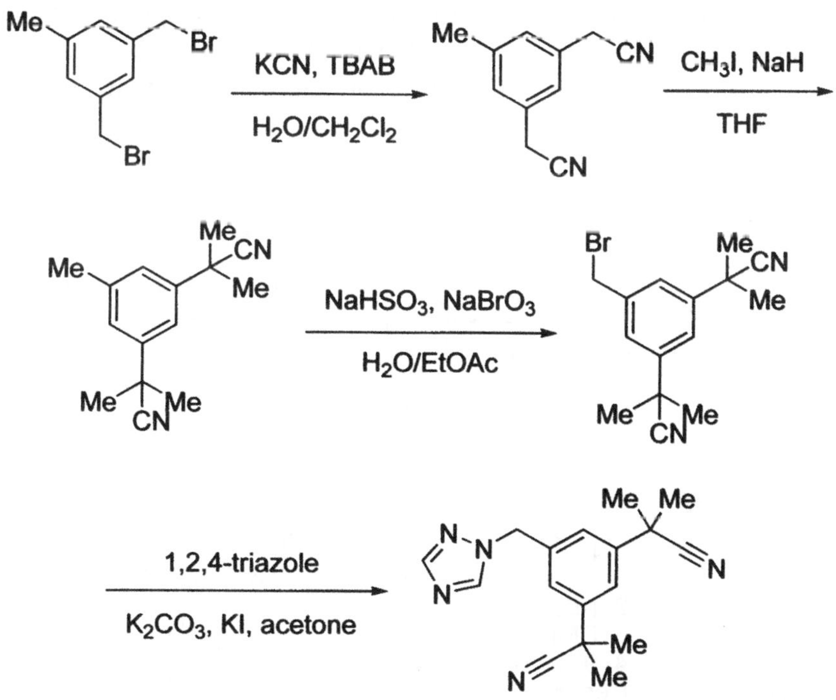

Fluconazole is used to treat *oropharyngeal candidiasis* and *cryptococcal meningitis* in AIDS as an anti-fungal agent. Fluconazole has been synthesized by S_N2 displacement using 1,2,4-triazole as illustrated below.[18]

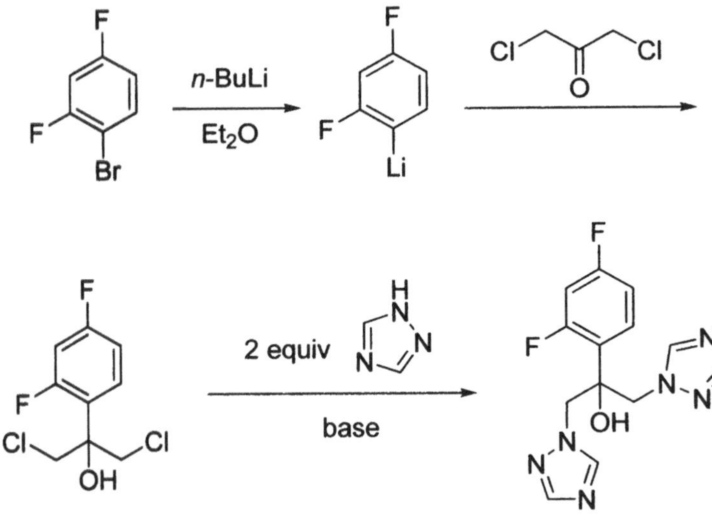

Letrozole, which reversibly binds to the heme of the cytochrome P450 unit, blocks the production of estrogens. It is an oral non-steroidal aromatase inhibitor that can be used to treat local or metastatic breast cancer. Side effects such as signs and symptoms of hypoestrogenism have been found when taking letrozole. The synthesis of letrozole involves alkylation of the 1,2,4-triazole affording a mixture of regioisomers that are separated using chromatography.[19]

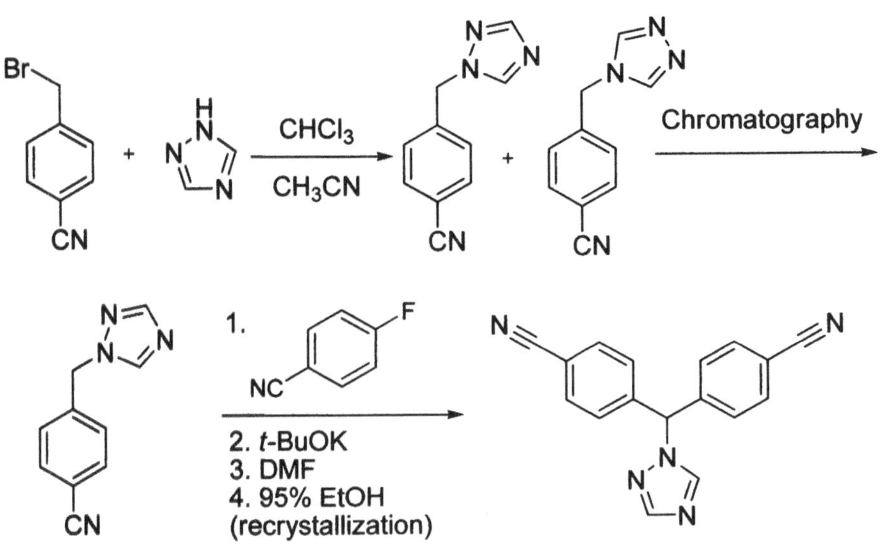

Ribavirin is an anti-viral drug that acts like nucleoside anti-metabolite to interfere with RNA metabolism. It is used for severe RSV infection, hepatitis C infection, and other viral infections but with hemolytic anemia as a major side effect. The synthesis of ribavirin can be derived from the chemoenzymatic method. The synthesis begins with simple ribose, using CRL (*Candida rugosa*) as a lipase can selectively remove the 5-acetyl group. Then the phosphorylation at the 5-hydroxyl group is taken by phosphorous oxychloride and followed by deprotection of the OH group. The last step involves enzymatic catalysis of triazole coupling with high yield.[20]

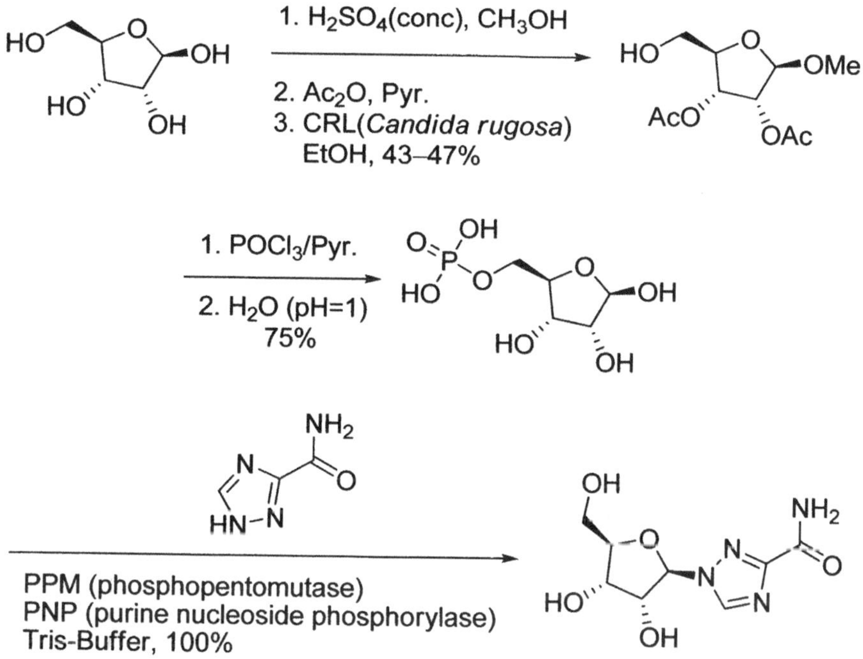

Terconazole is an anti-fungal agent used to treat vaginal fungal infections. The synthesis of terconazole starts with pre-synthesized compound **2** and then is followed by alkylation of the 1,2,4-triazole followed by ether formation.[21]

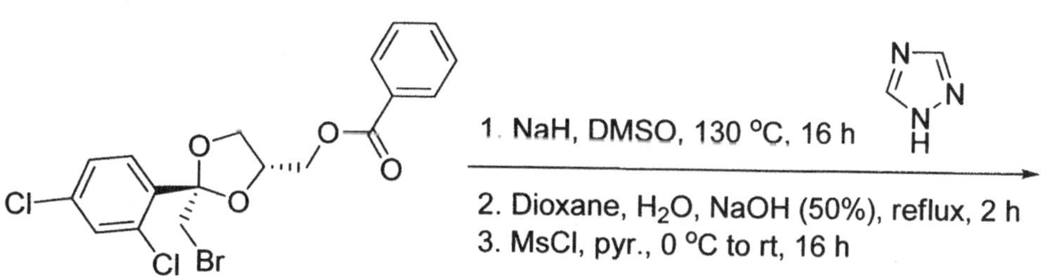

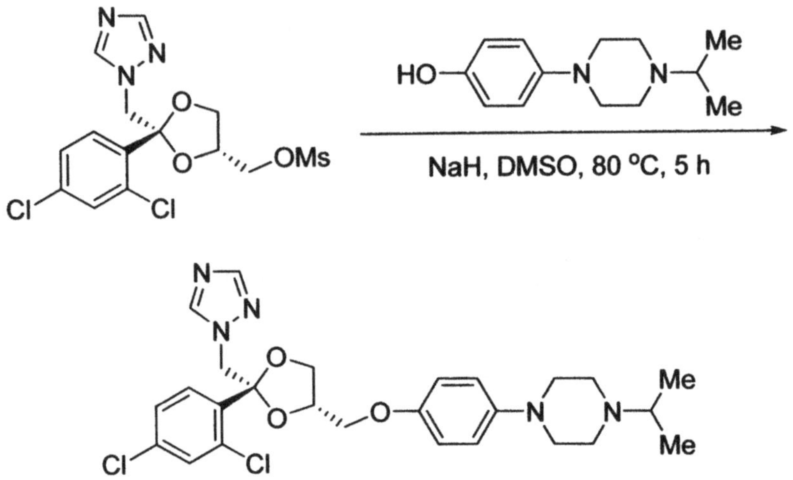

9.2.4 *Substitution of the Tetrazole*

Tetrazole can be lithiated on the 5-position; however, attempts to carry out substitution reactions sometimes fail because the intermediates are easily cleaved and eliminate nitrogen above –50 °C.[8] A series of 1,5-disubstituted tetrazoles were synthesized from a Pd/Cu catalyzed direct arylation of 1-substituted tetrazoles in the presence of tris(2-furyl)phosphine (TFP) and cesium carbonate.[22]

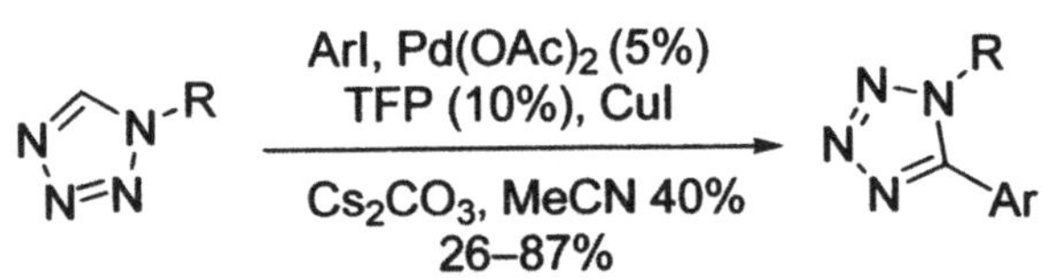

9.2.5 *Reactions of 1,2,3-Triazoles and Tetrazoles*

The Dimroth rearrangement is a rearrangement reaction taking place with certain 1,2,3-triazoles having a 5-amino substituent that is heated. During the rearrangement, the endocyclic and exocyclic nitrogen atoms switch place.[23] The position of the equilibrium in the rearrangement is dependent on the nature of the substituents and on the pH of the solvent.

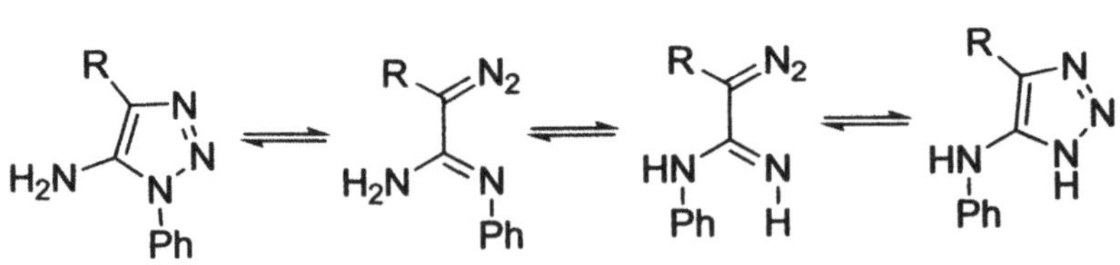

When treated with heat or light nitrogen, can be eliminated from tetrazoles and 1,2,3-triazoles. Nitrile imides can be formed by photolysis or thermolysis of 2,5-disubstituted tetrazoles.[8] The nitrile imides can then react via 1,3-dipolar cycloaddition or 1,5-electrocyclization. Photolysis of 1,5-diphenyltetrazole yields 2-phenylbenzimidazole.[24]

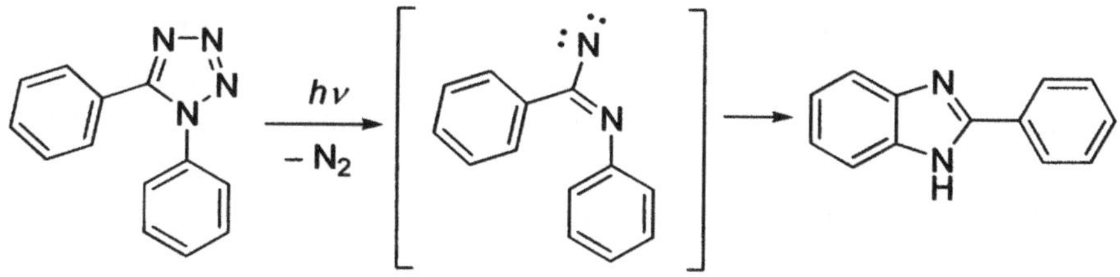

The preparation of carbazoles involving the pyrolysis of 1-phenyl-1,2,3-benzotriazole prepared from *ortho*-amino-diphenylamine and nitrous acid is known as the Graebe–Ullmann reaction. It is known that this reaction involves the cyclization of a diradical or iminocarbene intermediate, which isomerizes to carbazole via a [1,3]-hydrogen shift.

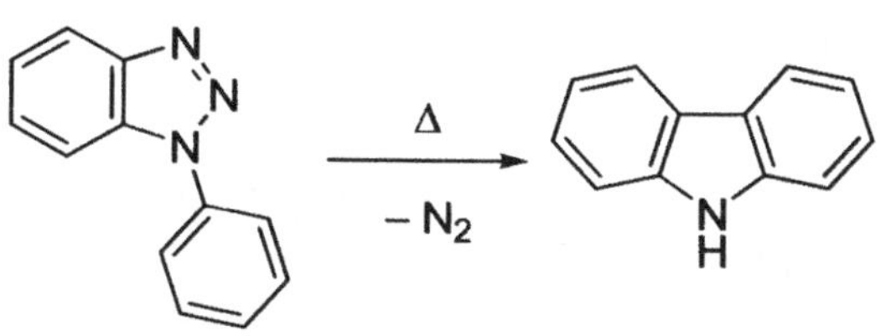

The Huisgen tetrazole rearrangement is the formation of a 2,4-disubstituted 1,3,4-oxadiazole from a *C*-substituted tetrazole by reaction with an acid chloride in the presence of a base at elevated temperature.[25] The intermediate acyl-tetrazole is prone to ring opening due to decreased aromaticity. The loss of N_2 and resultant ring closure provides the 1,3,4-oxadiazole ring. Recently the Husigen tetrazole rearrangement has been used to synthesize 1,3,4-oxadiazole containing HIV-1 non-nucleoside reverse transcriptase inhibitors.[26]

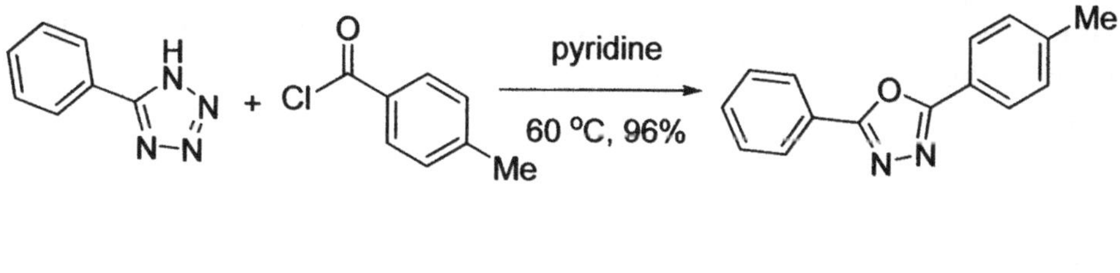

9.3 Construction of the Triazole Ring

9.3.1 Construction of the 1,2,3-Triazole Ring

Click chemistry was defined as an approach for the rapid synthesis of drug-like molecules using a few practical and reliable reactions.[27] An example of a click reaction is the Huisgen 1,3-dipolar cycloaddition of alkynes to azides to form 1,4-disubsituted-1,2,3-triazoles. The reaction is efficient, mild, and requires no protecting groups and purification in many cases.[28] The triazole is an isostere of the amide bond, however, it does not undergo cleavage like amides. They are also resistant to oxidation and reduction.

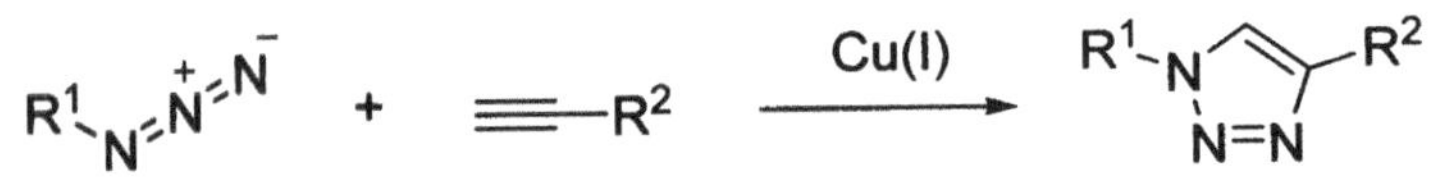

The ruthenium-catalyzed cycloaddition of azides to alkynes to form 1,5-disubstituted triazoles is complimentary to the copper catalyzed route.[29] The Cu(I)-catalyzed reaction is limited to terminal alkynes, and the Ru(II)-catalyzed reaction is active with internal alkynes.

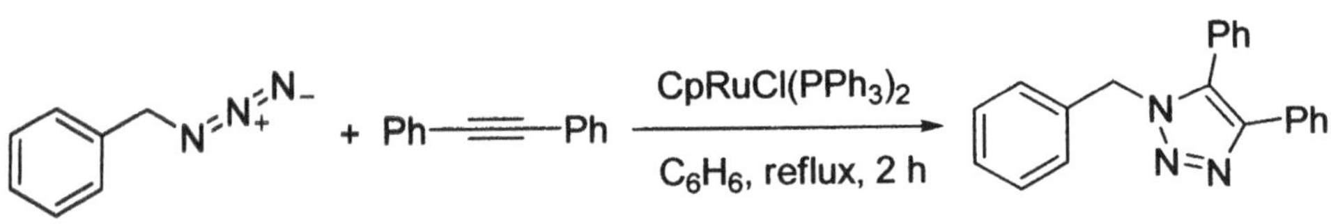

Aliphatic azides are limited in availability and can be prepared by the hydroazidation of deactivated olefins in the presence of a cobalt catalyst.[30]

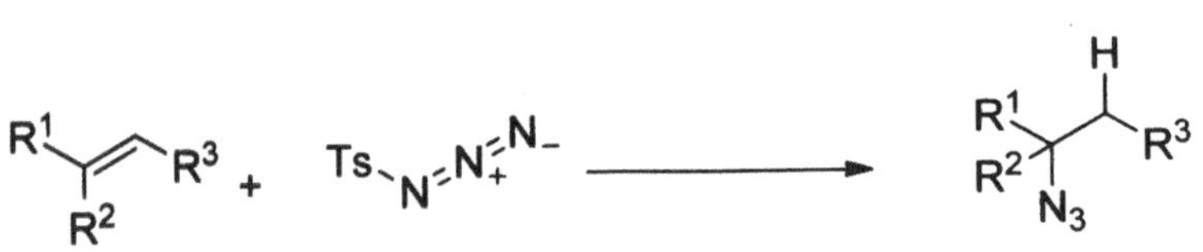

Alizapride is a dopamine antagonist with less affection on the CNS (central nervous system). It has a prokinetic and an antiemetic effect that can be used in the treatment of nausea and vomiting, including a postoperative side effect.

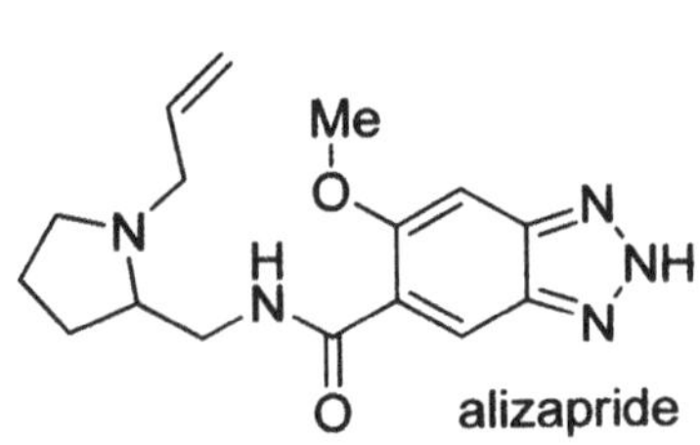

alizapride

Two traditional methods of the synthesis of alizapride involved construction of the benztriazole and then followed by coupling with amine.[31,32]

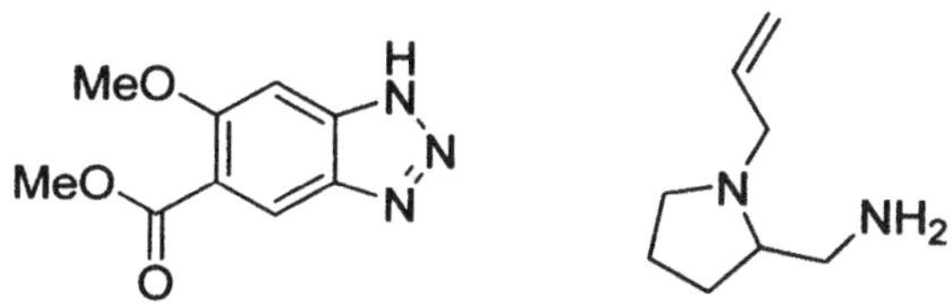

The Sandmeyer-type reaction[33] was used to form the aryl diazonium salt intermediate and followed by cyclization to obtain 1,2,3-benzotriazole.

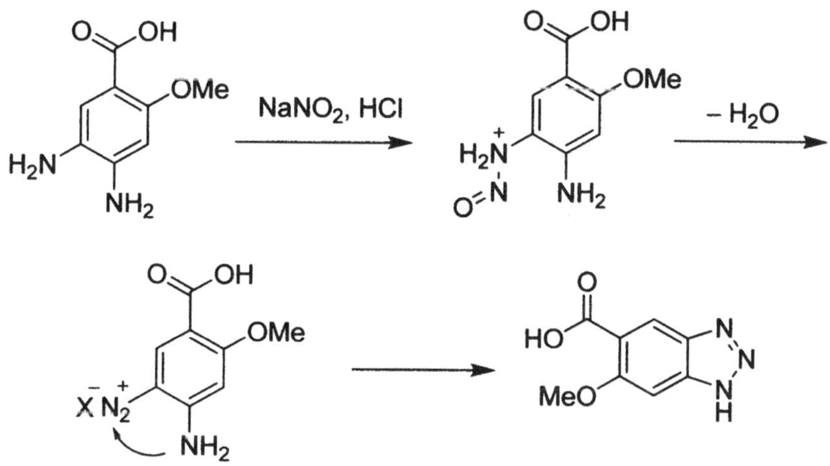

The new synthetic pathway[32] avoided using toxic intermediate dimethyl sulfate in the old methods and had a high yield. It also constructed the intermediate first by acetylation, methylation, and nitration and then by catalytic reduction, cyclization, and methylation. Final step involved amide coupling and, followed by formation of hydrochloric salt as shown below.

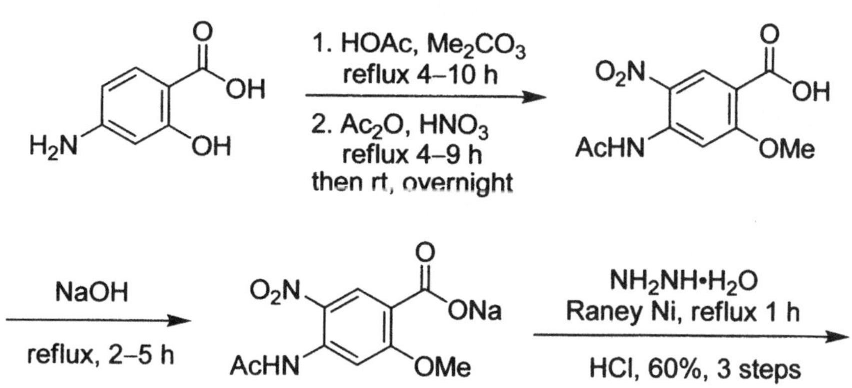

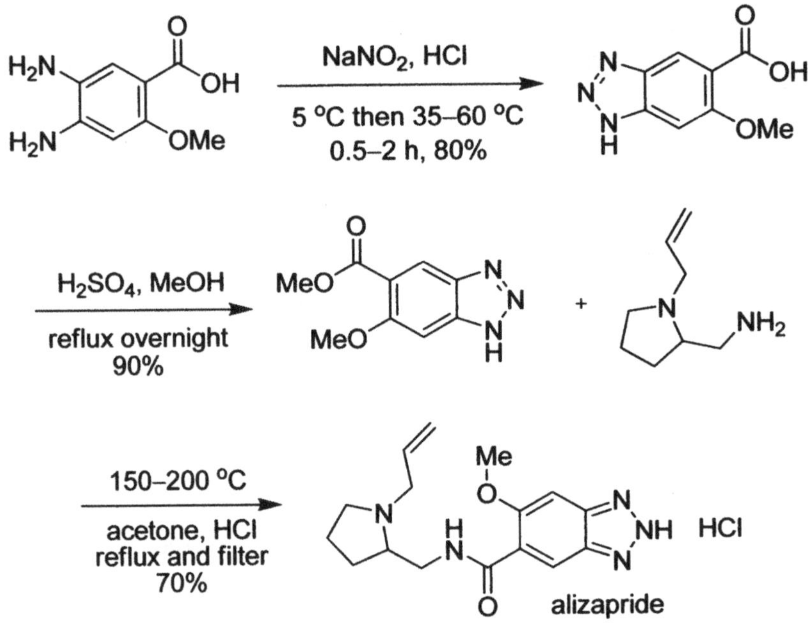

Tazobactam is a β-lactamase inhibitor that is used with piperacillin to enhance the anti-bacterial effect of piperacillin.

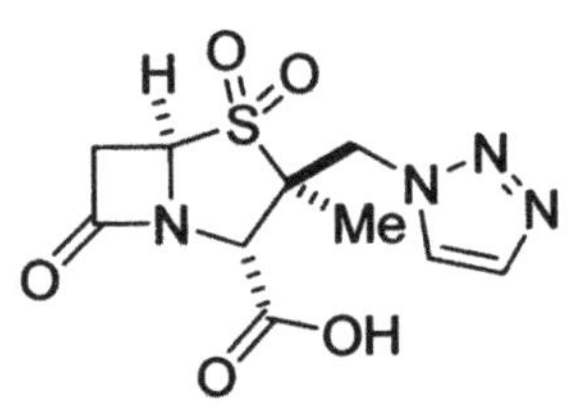

tazobactam (Zosyn)

In 1984, Micetich and co-workers reported on a method to synthesize tazobactam that involved the 1,3-dipolar addition of an azido penicillin acid derivative with TMS-acetylene.[34] The 1,3-dipolar addition is followed by deprotection of trimethylsilyl and benzyl group.[35]

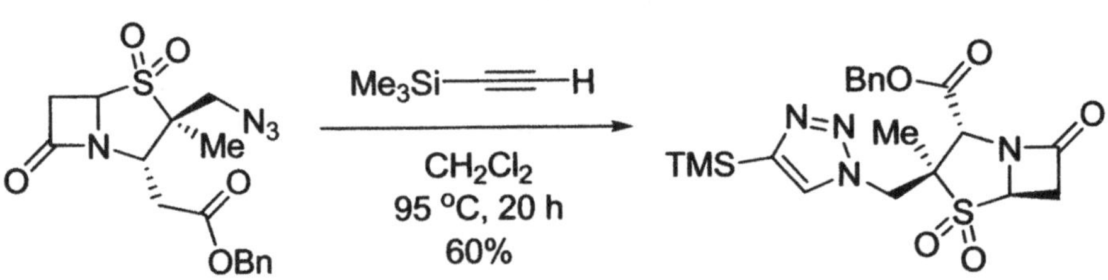

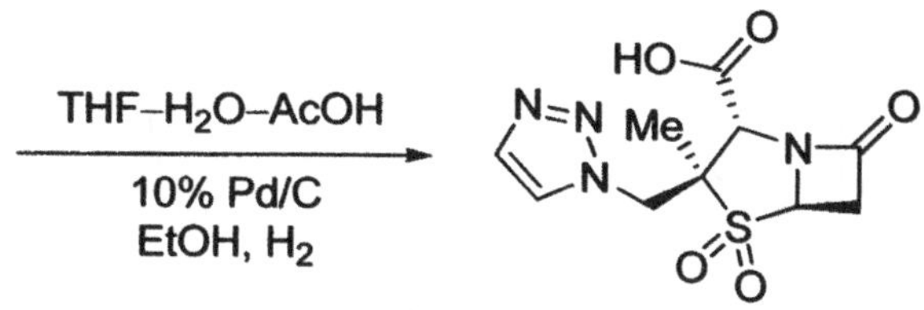

9.3.2 Construction of the 1,2,4-Triazole Ring

As a benzodiazepine-type drug, alprazolam is used to treat anxiety and panic disorders. The synthesis starts with a hydrazine benzodiazepine derivative followed by the imine formation, intramolecular cyclization, and oxidation to obtain alprazolam.[36]

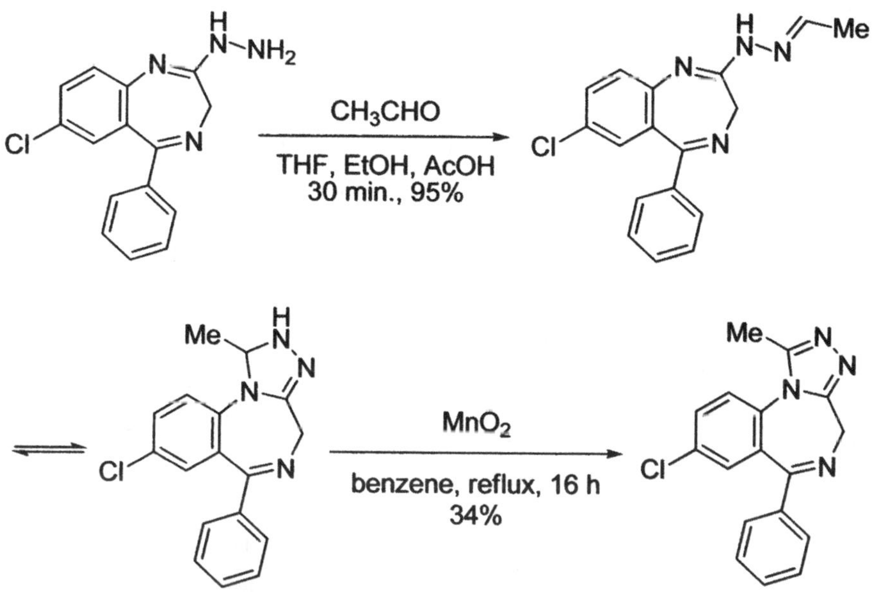

The Pellizzari reaction is the chemical reaction of an amide and a hydrazide to form a 1,2,4-triazole. Dapiprazole is an alpha-blocker that is used to reverse mydriasis after eye examination. The synthesis of dapiprazole starts with construction of a hydrazide-type structure and then follows by ring opening of piperidinone and condensation with the acylhydrazide moiety to form the final product.[37]

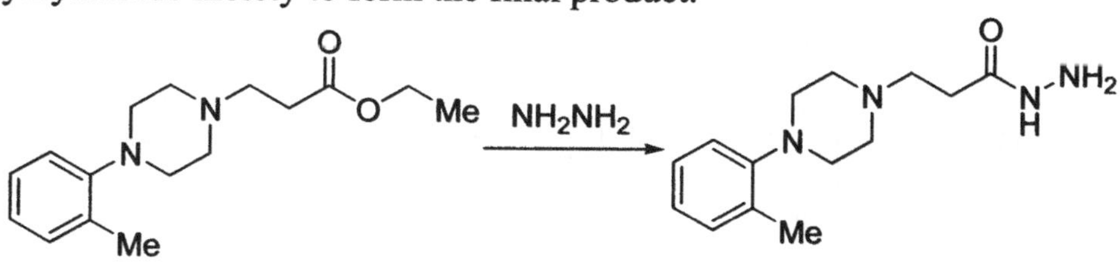

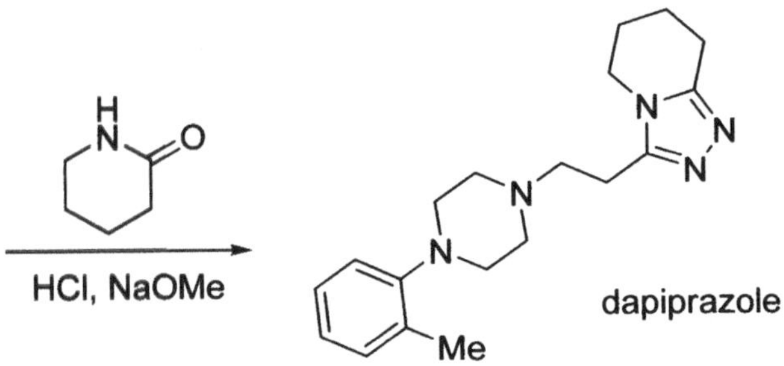

Deferasirox is used to reduce chronic iron overload in patients with long-term blood transfusion as an iron chelator. It is the first oral medication approved in the United States for such a purpose. A one-pot synthesis of deferasirox has been report that involves palladium-catalyzed coupling and triazole formation.[38]

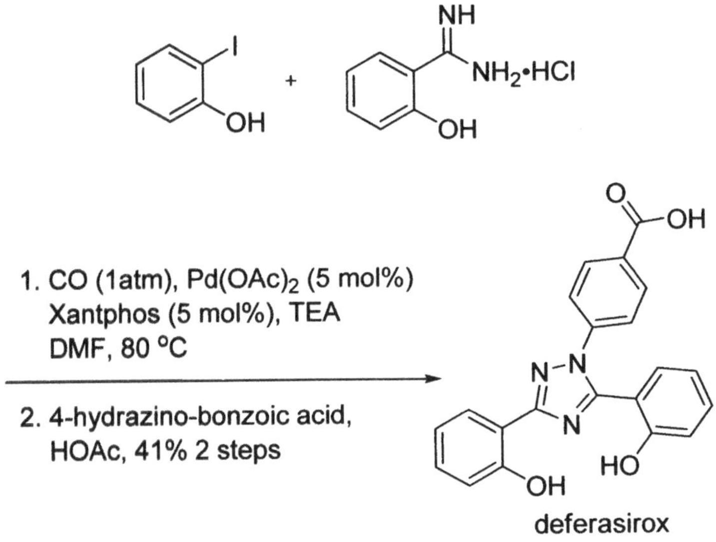

Estazolam is a benzodiazepine-type drug with anti-convulsant, hypnotic, and muscle relaxant properties. It is used in the treatment of short-term insomnia. The synthesis of estazolam involves formation of benzodiazepine followed by condensation with formic acid to yield the triazole moiety.[39]

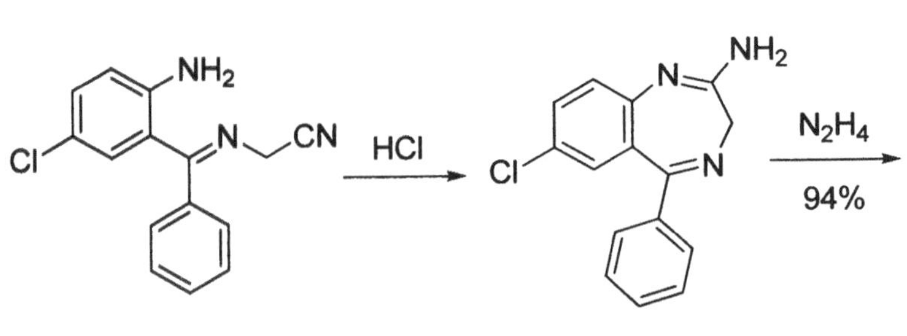

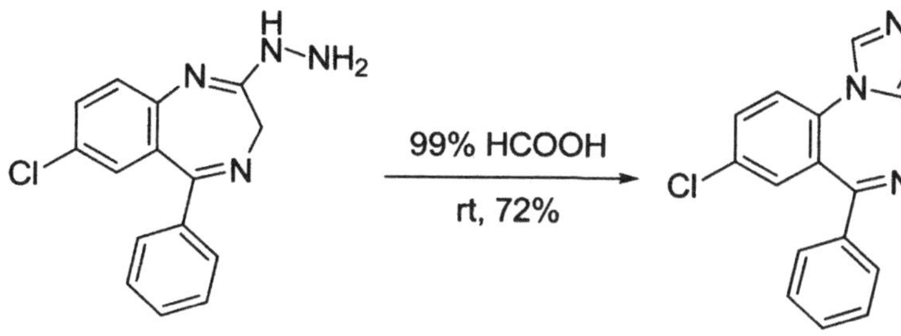

Nefazodone is an anti-depressant drug that was withdrawn from United States in 2004 due to its possibility of hepatic injury that could lead to death. The synthesis of nefazodone involves cyclization for the formation of triazole and coupling reactions with piperazine moiety.[40]

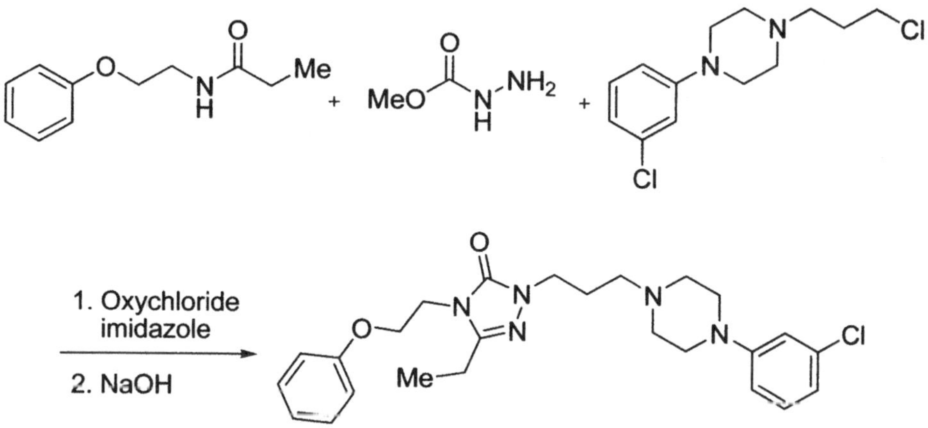

Sitagliptin is a peptidase-4 (DPP-4) inhibitor that is used as an anti-diabetic drug either alone or in combination with metformin or a thiazolidinedione for control of type 2 diabetes mellitus. The synthesis of sitagliptin contains triazole formation followed by coupling with a acid derivative.[41]

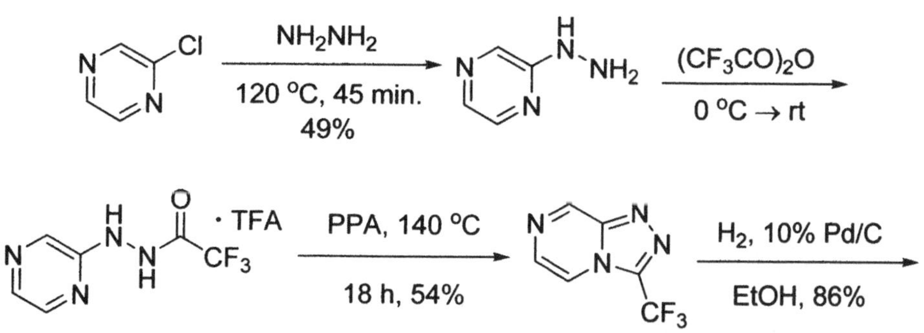

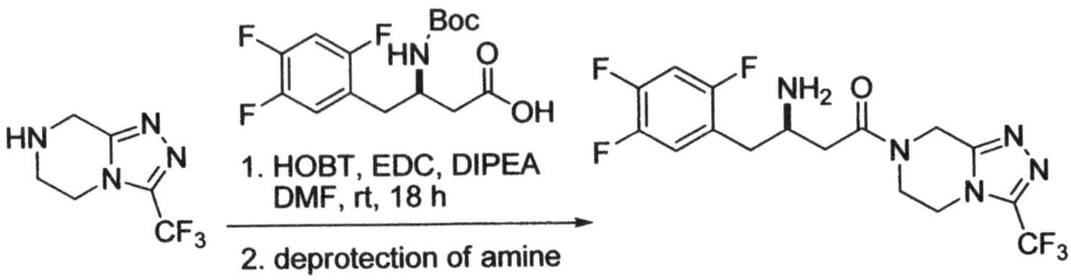

Trazodone is a serotonin uptake inhibitor that has anti-depressant activity. It has been found to be more efficient in treating depressive disorders associated with insomnia and anxiety. The synthesis of trazodone involves construction of piperazine moiety and triazole moiety and is followed by a coupling reaction between two moieties.[42] A microwave synthesis has also been an optimized method for synthesizing trazodone.[42]

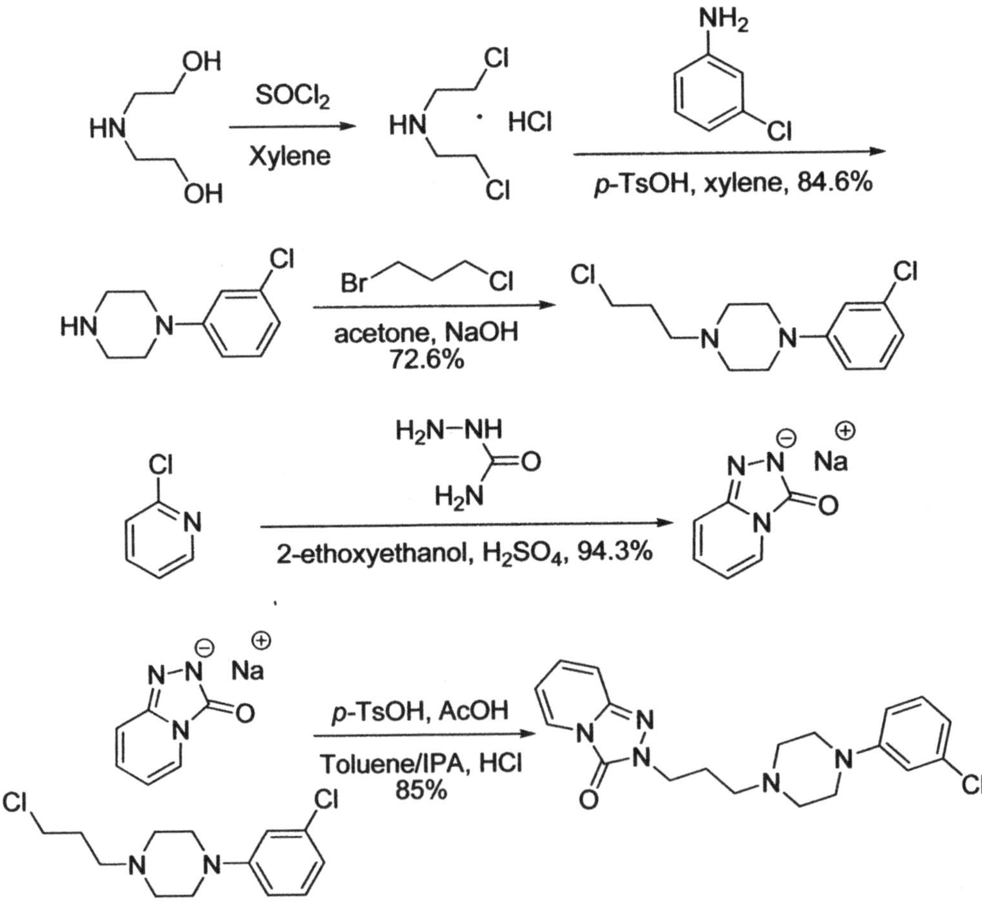

The benzodiazepine drug, triazolam, is used to treat severe insomnia. With the risk of psychiatric adverse drug reactions, it has been withdrawn from the United Kingdom, however, it is still marketed in the United States.

The synthesis of triazolam starts with a benzodiazepine derivative. After activation of the cyclic system, *N*-nitrosoamidine is treated with acetylhydrazine to yield the corresponding amidine.[43]

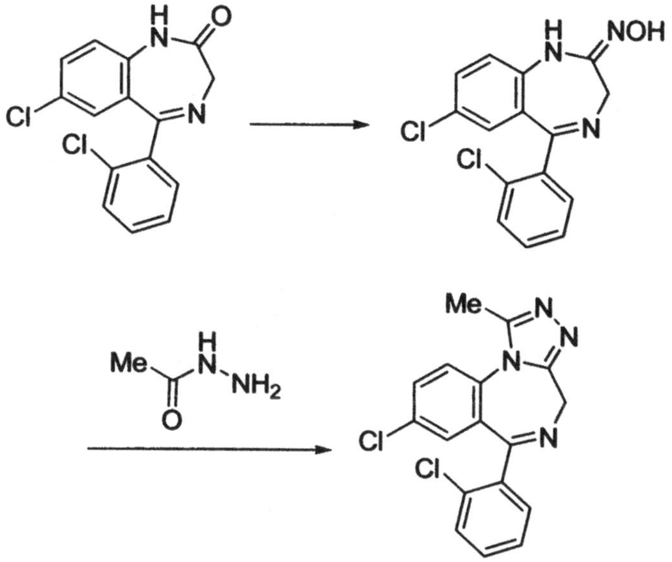

9.3.3 Construction of the Tetrazole Ring

The synthesis of tetrazole usually involves a variation of the Finnegan tetrazolc synthesis that is the addition of hydrazoic acid to a carbon-nitrogen multiple bond. Forasatan is an angiotensin I (AT1) antagonist that binds to the AT1 receptor and can be used alone or with other anti-hypertensive agents to treat hypertension. The forsartan contains an 1,2,4-triazole and a tetrazole. The tetrazole is an isostere for a carboxylic acid group. The tetrazole moiety in forasartan is synthesized by a 1,3-dipolar addition of tributyltinazide to a benzonitrile derivative to form the tetrazole moiety.[44]

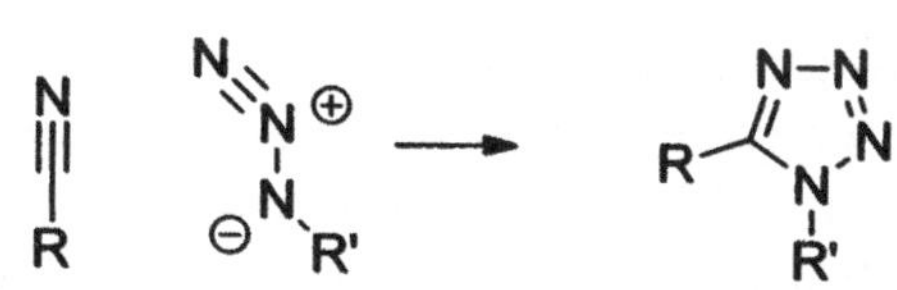

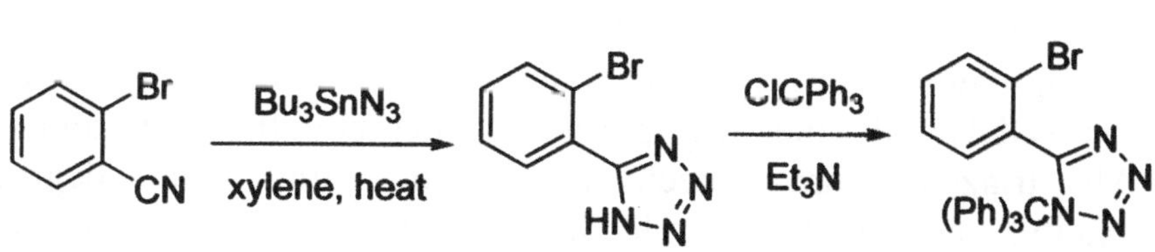

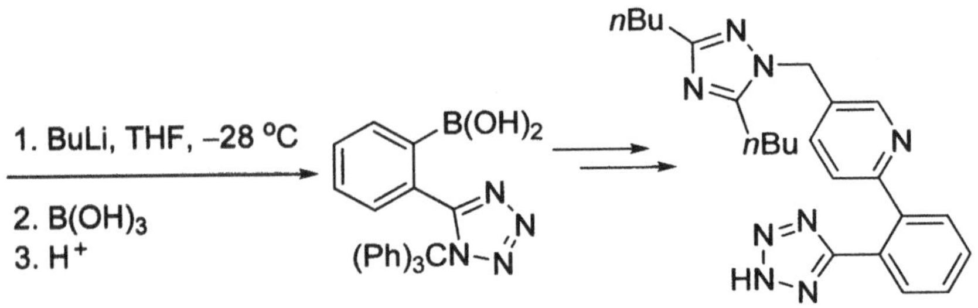

Cilostazol has been used to treat intermittent claudication in individuals with peripheral vascular disease. A similar molecule as cilostazol may have the risk of death in patients with congestive heart failure. The synthesis of cilostazol contains a 1,3-dipolar addition for the construction of the tetrazole ring, and the resulting tetrazole was coupled with 6-hydroxy-3,4-dihydroquinolin-2(1*H*)-one with the aide of potassium hydroxide.[45]

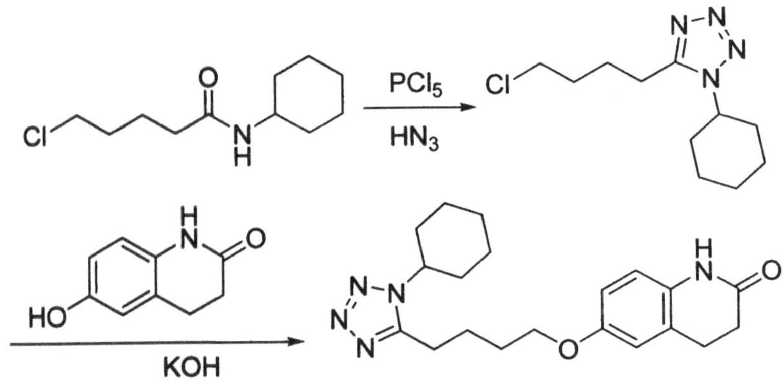

9.4 *Possible Liabilities of Triazole-containing Drugs*

Triazoles are known metal chelators, and triazoles have been associated with a number of adverse events and significant drug–drug interactions (DDIs). Triazoles are metabolized by the CYPP450 enzymes, and all triazoles exhibit some degree of drug–drug interactions. Triazoles can be substrates, inducers, and inhibitors of CYP enzymes, and, when administered with other agents that interfere with the CYP450 system, may result in significant alteration of plasma triazole levels. P-glycoprotein is a transporter protein involved in the absorption and distribution of triazoles. Triazoles can function as substrates for P-glycoprotein (Pgp) and/or inhibitors, thus, creating drug–drug interactions (DDI) with other agents that also interact with this protein.

All triazoles have been associated with some degree of hepatotoxicity, ranging from mild hepatitis to cholestasis and, rarely, fulminant hepatic failure.[46] Although not entirely clear, it seems that liver toxicity may be related to higher plasma drug levels, with most data coming from patients treated with voriconazole.[46]

9.5 Problems

9.5.1 Propose a reasonable mechanism for the following transformation[47]:

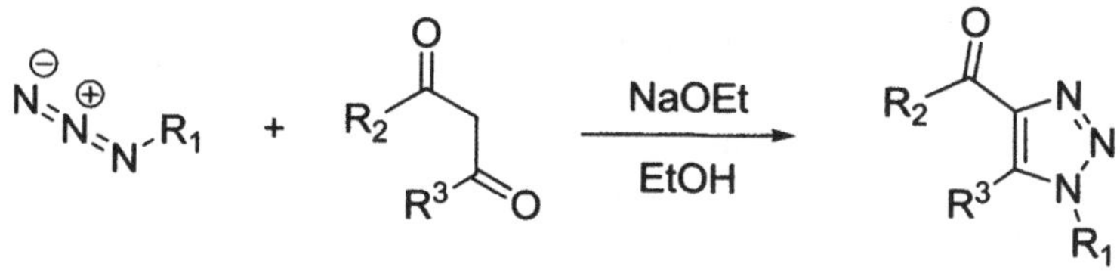

9.5.2 Propose a reasonable mechanism for the thermal transformation of triazole to carbazole[48]:

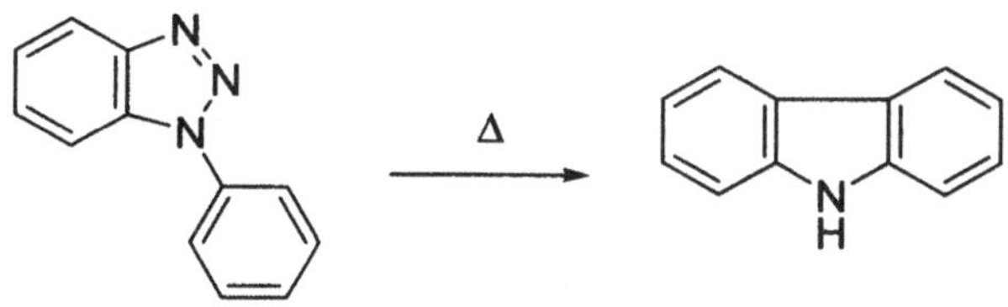

9.5.3 Propose a reasonable mechanism for the following transformation[49]:

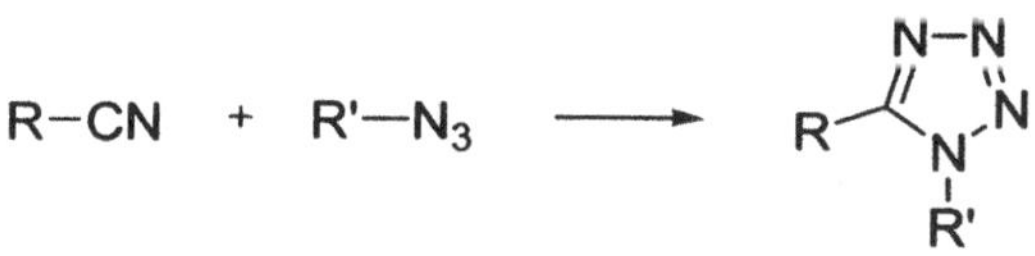

9.5.4 Propose a reasonable mechanism for the following transformation[25]:

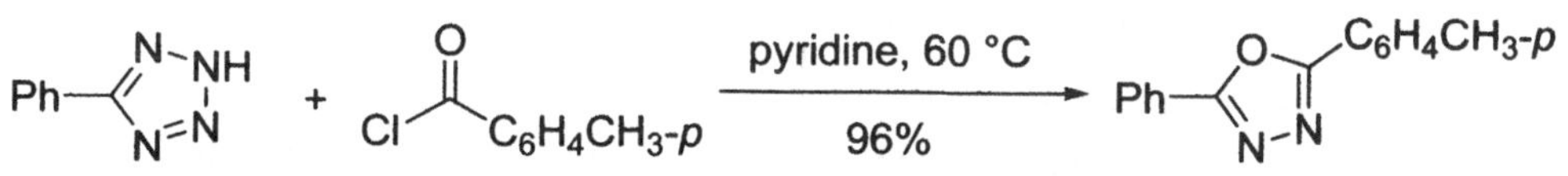

9.5.5 Determine the structures of **A**–**C** from the following transformations[50]:

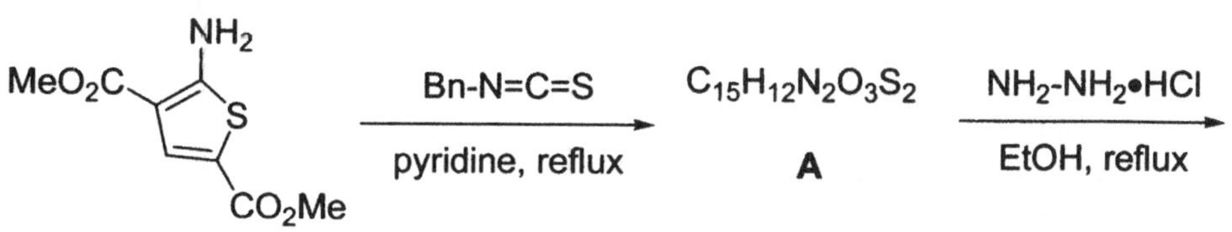

$$\underset{\textbf{B}}{C_{15}H_{16}N_4O_3S} \xrightarrow[\text{reflux}]{HC(OEt)_3} \underset{\textbf{C}}{C_{16}H_{12}N_4O_3S}$$

9.6 References

1. Bird, C. W. *Tetrahedron* **1985,** *41*, 1409–1414.
2. Buzykin, B. I.; Mironova, E. V.; Nabiullin, V. N.; Gubaidullin, A. T.; Litvinov, I. A. *Russ. J. Gen. Chem.* **2006,** *76,* 1471–1486.
3. "Diovan prescribing information" Novartis. http://www.pharma.us.novartis.com/product/pi/pdf/diovan.pdf.
4. "Arimidex prescribing information" Astrazeneca. http://www1.astrazenecaus.com/pi/arimidex.pdf
5. "Zosyn prescribing information" Wyeth http://www.4legpharma.com/Product%20Inserts/Pfizer/Zosynpowder.pdf
6. Yet, L. *Prog. Heterocycl. Chem.* **2011,** *23*, 231–266.
7. Iddon, B.; Nicholas, M. J. *J. Chem. Soc., Perkin Trans.* **1996,** 1341–1347
8. Gilchrist, T. L.; Gymer, G. E. *Adv. Heterocycl. Chem.* **1974,** *16,* 33–85
9. Wang, X. J.; Zhang, L.; Krishnamurthy, D.; Senanayake, C. H.; Wipf. P. *Org. Lett.* **2010,** *12*, 4632–4635
10. Ackermann, L.; Jeyachandran, R.; Potukuchi, H. K.; Novak, P.; Buttner, L. *Org. Lett.* **2010,** *12*, 2056–2059.
11. Albrecht, Z. *Synthesis* **1998,** *9*, 1357–1361
12. Cebrian, C.; Cozar, A.; Prieto, P.; Diaz-Ortiz, A; Hoz, A.; Carrillo, J. R.; Rodriguez, A. M.; Montillab, F. *Synlett* **2010,** 55–60.
13. Alan R. K.; Charles W. R.; Eric F. V. S. *Compr. Heterocycl. Chem.* **1996,** 4.
14. Alan, R. K.; Charles W. R.; Eric F. V. S.; Richard J. K. T. *Compr. Heterocycl. Chem.* **2008,** 5.
15. Gilchrist, T. L. *Heterocyclic Chemistry (Third Edition),* Prentice Hall: Upper saddle River, NJ., 1997.
16. Gasparini, J. P.; Gassend, R.; Maire, J. C. *J. Organometal. Chem.* **1980,** *188,* 141–150.
17. Hester, J. B. Jr.; Rudzik, A. D.; Von Voigtlander, P. F. *J. Med. Chem.* **1980,** *23*, 643–647.
18. Meguro, K.; Tawada, H.; Miyano, H.; Sato, Y.; Kuwada, Y. *Chem. Pharm. Bull.* **1973,** *21*, 2382–2390
19. Kadzimirsz, D.; Kinzl, M. EP 2210894 (2010).
20. Madding, G. D.; Smith, D. W.; Sheldon, R. I.; Lee, B. *J. Heterocycl. Chem.* **1985,** *22*, 1121–1126
21. Kim, D.; Wang, L. P.; Beconi, M.; Eiermann, G. J.; Fisher, M. H.; He, H. B.; Hickey, G. J.; Kowalchick, J. E.; Leiting, B.; Lyons, K.; Marsilio, F.; McCann, M. E.; Patel, R. A.; Petrov, A.; Scapin, G.; Patel, S. B.; Roy, R. S.; Wu, J. K.; Wyvratt, M. J.; Zhang, B. B.; Zhu, L.; Thornberry, N. A.; Weberet A. E. *J. Med. Chem.* **2005,** *48,* 141–151
22. Spulak, M.; Lubojacky, R.; Senal P.; Kunes, J.; Pour, M. *J. Org. Chem.* **2010,** *75*, 241–244
23. Snyder, N. L.; Adams, T. P. *Name Reactions in Carbocyclic Chemistry II*, Li, J. J., Ed.; Wiley: Hoboken, NJ, **2011**, 554–590.
24. Kirmse, W. *Angew. Chem.* **1959,** *71*, 537–541

25. Qiao, J. X. In *Name Reactions in Carbocyclic Chemistry-II*, Li, J. J., Ed.; Wiley: Hoboken, NJ, **2011,** 309–316.
26. Cullen, M. D.; Deng, B. L.; Hartman, T. L.; Watson, K. M.; Buckheit, R. W.;
27. Kolb, H. C.; Sharpless, K. B. *Drug Discov. Today* **2003,** *8,* 1128–1137.
28. Rostovtsev, V. V.; Green, L. G.; Fokin, V. V.; Sharpless, K. B. *Angew. Chem. Int. Ed.* **2002,** *41,* 2596–2599
29. Zhang, L.; Chen, X. G.; Peng, X.; Herman, H. Y. S.; Ian, D. W.; Sharpless, K. B.; Fokin, V. V.; Jia, G. C. *J. Am. Chem. Soc.* **2005,** *127,* 15998–15999.
30. Waser, J. *J. Am. Chem. Soc.* **2005,** *127,* 8294–8295.
31. Butler, R. N. *Chem. Rev.* **1975,** *75,* 241–257.
32. Zhang, G. M.; Sun, L.; Shen, X. W. CN 101066968 A20071107 (2007).
33. *Name Reactions in Carbocyclic Chemistry II*, Li, J. J., Ed.; Wiley: Hoboken, NJ, **2011**.
34. Micetich, R. G.; Yamabe, S.; Yamazaki, T.; Ishida, N.; Tanaka, M.; Ishizawa, T.; Kajitani, M. EP 97446 A1 19840104 (1984).
35. Micetich, R. G.; Maiti, S. N.; Spevak, P.; Hall, W. T.; Yamabe, S.; Ishida, N.; Tanaka, M.; Yamazaki, T.; Nakai, A.; Ogawas, K. *J. Med. Chem.* **1987,** *30,* 8.
36. Jackson, T.; Lawrence Woo, L.W.; Trusselle, M. N.; Chander, S. K.; Purohit, A.; Reed, M. J.; Potter, B. V. L. *Org. Biomol. Chem.* **2007,** *5,* 2940–29552.
37. Bruno, S.; Leandro, B. *Ger. Offen.* 2915318 (1979).
38. Steven, S.T.; Nicole, B. *Angew. Chem., Int. Ed.* **2010,** *49,* 325–328, S325/1–S325/25.
39. Schrawat, V. K.; et al. *Int. Appl.* **2009,** 2009069140.
40. Marisa, T. P.; Leon, A. B.; Claudio, A. P.; Montserrat, J. M.; Iribarren, A. D. *Tetrahedron Lett.* **2008,** *49,* 2642–2645.
41. Garcia, M. L.; Mauleon, D.; Carganico, G. *Tetrahedron: Asymmetry* **1995,** *6,* 2365–2368
42. Pai, N. R.; Dubhashi, D. S.; Vishwasrao, S.; Pusalkar, D. *J. Chem. Pharm. Res.* **2010,** *2,* 506–517.
43. Fustero, S.; González, J.; Del Pozo, C. *Molecules* **2006,** *11,* 583–588.
44. Reitz, D. B. EP 508445 A1 19921014 (1992).
45. Nishi, T.; Tabusa, F.; Tanaka,T.; Shimizu,T.; Kanbe, T.; Kimura, Y.; Nakagawa, K. *Chem. Pharm. Bull.* **1983,** *31,* 1151–1157.
46. Neofytos, D.; Avdic, E.; Magiorakos, A.P. *Drug, Healthc. Patient Saf.* **2010,** *2,* 27–38.
47. Hill, M. D. In *Name Reactions in Carbocyclic Chemistry-II*, Li, J. J., Ed.; Wiley: Hoboken, NJ, **2011,** 269–277.
48. Li, J.; Cook, J. M. In *Name Reactions in Heterocyclic Chemistry*, Li, J. J., Ed.; Wiley: Hoboken, NJ, 2004, pp 132–134.
49. Curran, T. T. In *Name Reactions in Carbocyclic Chemistry-II*, Li, J. J., Ed.; Wiley: Hoboken, NJ, **2011,** 278–298.
50. Li, J. J.; et al. US 6894057 (2003).

PART III SIX-MEMBERED HETEROCYCLES WITH ONE HETEROATOM 397

Chapter 10 Pyridines

Jennifer Xiaoxin Qiao

10.1 Introduction

Pyridine is a flammable, colorless, and relatively stable and unreactive liquid with a characteristic pungent, unpleasant odor. Anhydrous pyridine has a boiling point of 115 °C and a density of 0.9819 g/cm^3. Pyridine is miscible with water and virtually all organic solvents. It is weakly basic, and with hydrochloric acid it forms a crystalline hydrochloride salt, which melts at 145–147 °C. Pyridine was first isolated from bone pyrolysates. Its name was derived from the Greek for fire "pyr" and the suffix "idine" was used to designate aromatic bases. Pyridine is also used as a solvent, and its derivatives have been used as pharmaceuticals, vitamins, food flavorings, paints, dyes, rubber products, adhesives, insecticides, and herbicides, *etc.*[1] Pyridine can also be formed from the breakdown of many natural materials in the environment.

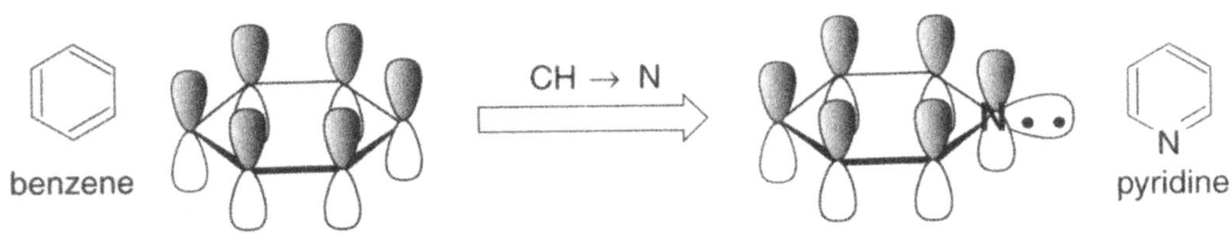

Pyridine, with 6π-electrons, is an electron-deficient aromatic heterocycle containing a ring nitrogen atom. The aromatic π-electron system does not require the participation of the lone pair of electrons on the nitrogen atom. The ring nitrogen is more electronegative than the ring carbons, making the two α-ring carbons and γ-ring carbon more electropositive than otherwise would be expected from benzene (see the natural atomic charges of pyridine and benzene calculated at RHF//6-31G**). The resonance structures and the natural atomic charges of pyridine predict its electron-deficient nature and rationalize its much higher dipole moment relative to benzene, which is mainly responsible for the higher boiling point (115 vs. 80 °C) and high water solubility of pyridine.

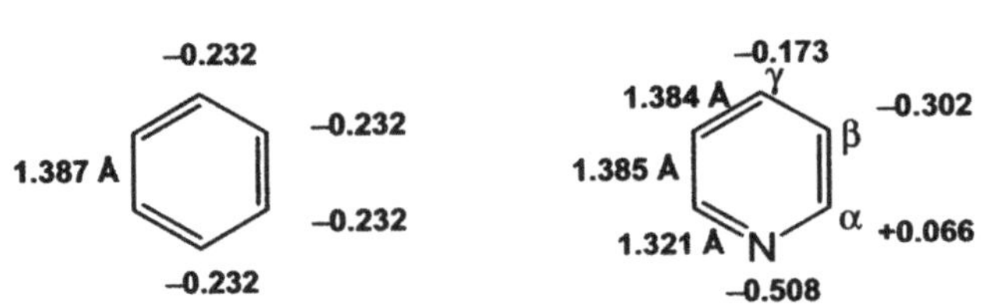

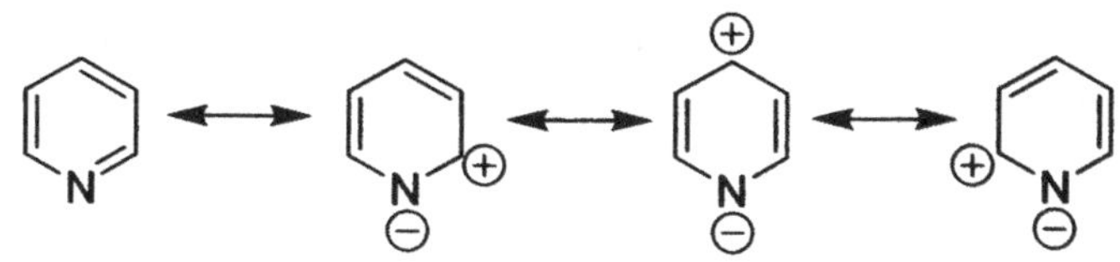

Consequently, pyridine has a reduced susceptibility to electrophilic substitution compared to benzene, while being more susceptible to nucleophilic attack. One unique aspect of pyridine is the protonation, alkylation, and acylation of its nitrogen atom. The resultant salts are still aromatic, however, and they are much more polarized. Details for reactivity of pyridine derivatives, in particular, reactions on the pyridine nitrogen and the Zincke reaction, as well as *C*-metallated pyridines, halogen pyridines, and their uses in the transition metal-catalyzed C–C and C–N cross-coupling reactions in drug synthesis, will be discussed in Section 10.2.

Pyridine is prepared commercially by the gas-phase, high-temperature reaction of crotonaldehyde, formaldehyde, steam, air, and ammonia over a silica-alumina catalyst in 60–70% yield. We will illustrate some of the common laboratory methodologies for the construction of substituted-pyridine derivatives in Section 10.3.

The ^{1}H NMR spectrum of pyridine shows three signals with the integral intensity ratio of 2:1:2, which correspond to the three chemically different protons in the molecule: α-protons (chemical shift 8.6 ppm), γ-proton (7.7 ppm), and β-protons (7.3 ppm). The α- and γ-protons are further downfield in comparison to the β-protons because of the lower electron density in the α- and γ-positions, which can be derived from the resonance structures. The coupling constants of the pyridine protons are usually similar to those of benzene, except for the coupling constant between the two α-protons being reduced from 7–8 Hz to 4–6 Hz.

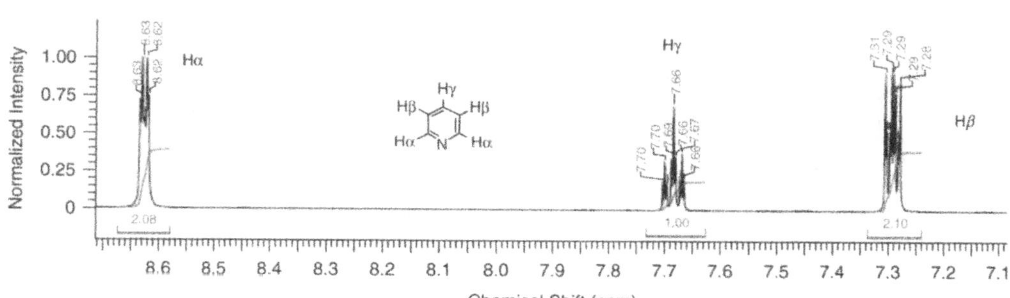

^{1}H-NMR of Pyridine

Cα and Cγ are further downfield with chemical shifts of 149.4 ppm and 135.4 ppm, respectively, in the ^{13}C-NMR spectrum of pyridine in chloroform-D. Again, Cβ has a chemical shift of 123.3 ppm similar to that of benzene (129 ppm).

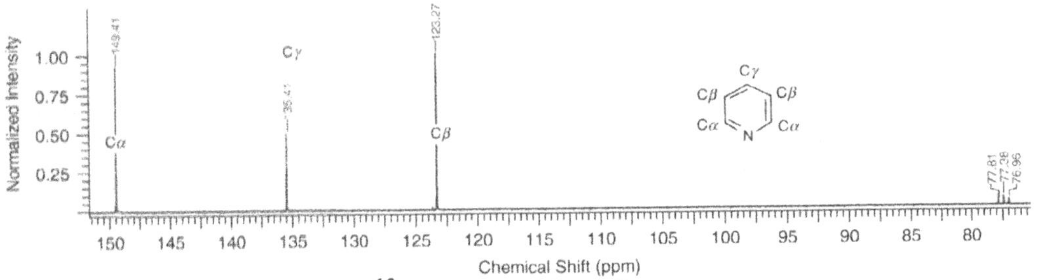

^{13}C-NMR of Pyridine

The pyridine system is found in natural products, for example, in nicotine (**1**) from tobacco, ricinine (**2**) from castor bean, vitamins such as pyridoxine or vitamin B_6 (**3**) and vitamin P (**4**), and alkaloids such as coniine and piperine. Free pyridine is present in tobacco smoke. Diploclidine (**5**) and nakinadine A (**6**) are two examples of recently isolated and structurally diverse natural products containing the pyridine core.[2,3]

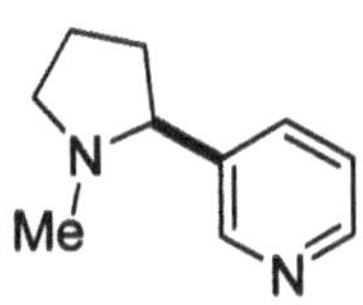
1

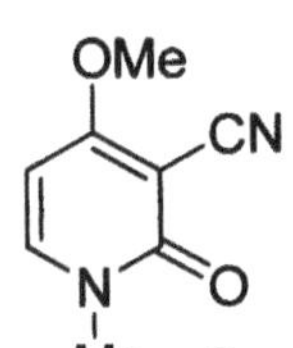
2

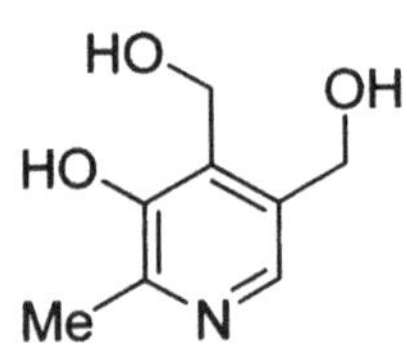
3

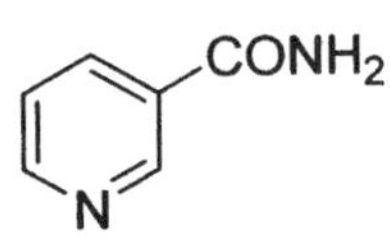
4

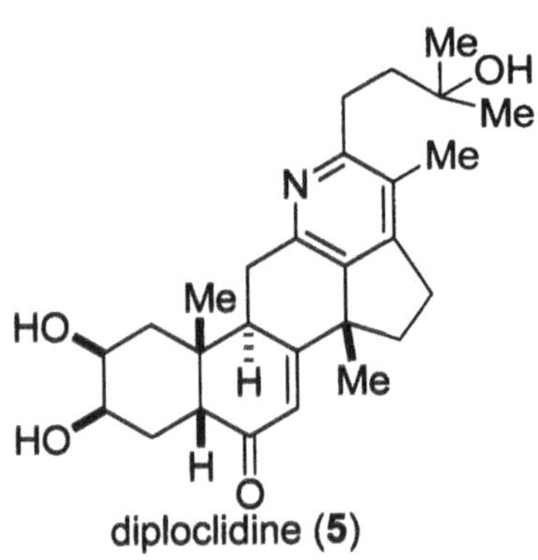
diploclidine (**5**)

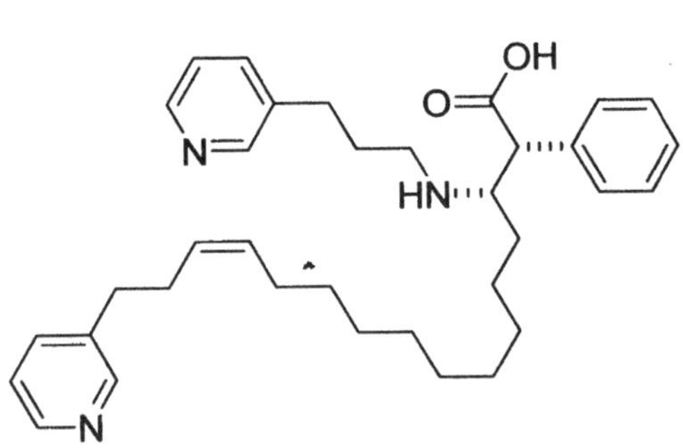
nakinadine A (**6**)

10.1.1 Pyridine-Containing Drugs

Simple substituted pyridines such as isoniazid (**7**) showed antibacterial activity. The structurally related ethionamide (**8**) is an active antibiotic prodrug against tuberculosis in humans. Nicotine (**1**) and epibatidine (**9**) are naturally occurring agonists of neural nicotinic acetylcholine receptors (nAChRs). Structural modifications of **1** and **9** led to a series of pyridine-containing compounds such as **10–13** as agonists of nAChRs with improved pharmacological properties. Among them, altinicline (**10**, SIB-1508Y) stimulates the release of dopamine and acetylcholine in the brain in both rodent and primate models,[4] and it was progressed in Phase II clinical trials for Parkinson's disease,[5] although its current status is unclear. Further introduction of a hydrophobic or hydrogen-bonding alkynyl group into the C-5 position of the pyridine ring (compound **14**) significantly increased the selectivity for α4β2 subtype over α3β4 subtype.[6]

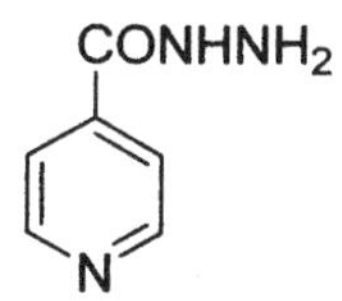

isoniazid (**7**)

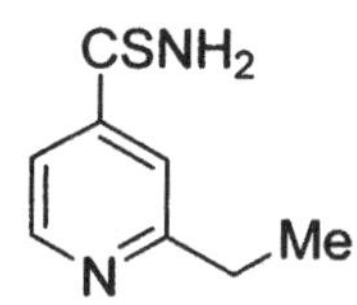

ethionamide (**8**)

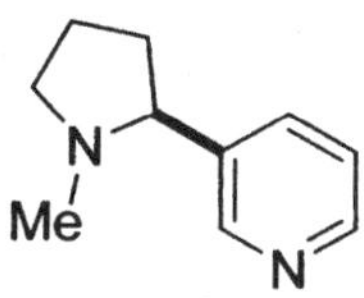

1

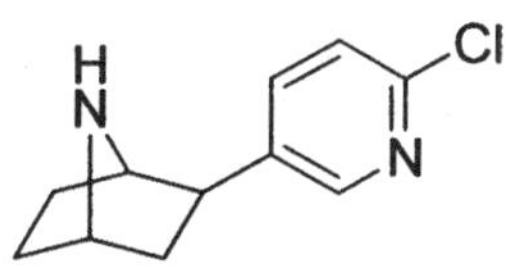

epibatidine (**9**)

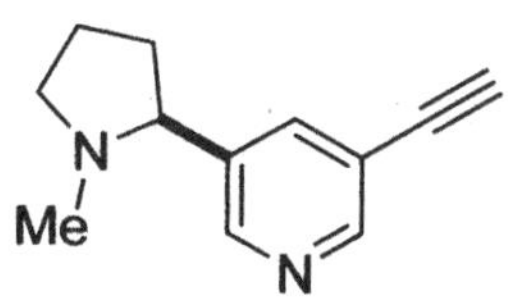

altinicline (**10**)

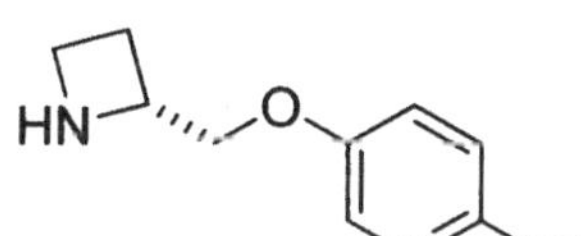

11

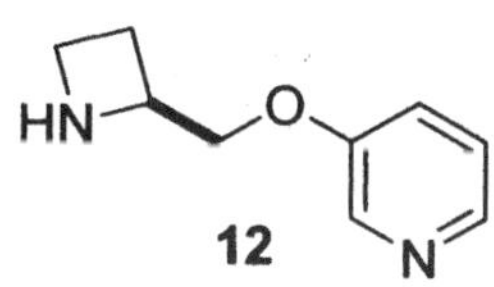

12

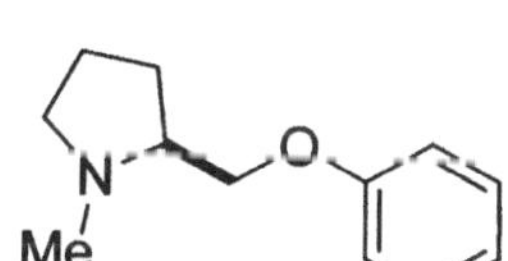

13

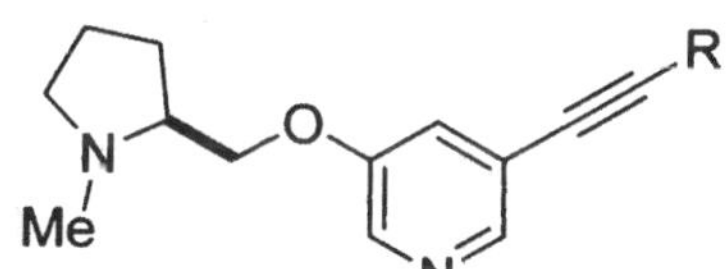

14

The pyridine ring is presented widely in biologically active compounds due to the fact that pyridine rings are generally bioisosteres of the benzene ring despite the presence of the basic nitrogen atom. Numerous marketed drugs contain pyridine ring, for example, pirbuterol (**15**, Maxair) is a β_2 adrenergic bronchodilator used in the treatment of asthma, available (as pirbuterol acetate) as a breath-activated metered-dose inhaler ("autohaler"). Mafloquine (**16**) and enpiroline (**17**) are two anti-malarial reagents with mafloquine in the (*RS,SR*)-*erythro*-form and enpiroline in the (*SS,RR*)-*threo*-form. Acrivastine (**18**), a second-generation H_1-receptor antagonist anti-histamine, is used for the treatment of allergies and hay fever. In the United

States, acrivastine (**18**) is the active ingredient in the Semprex-D brand. Imatinib mesylate (Gleevec, **19**) is a tyrosine kinase inhibitor and was approved for the treatment of ten different cancers including chronic myelogenous leukemia (CML) by 2001. Atazanavir (Reyataz, **20**) is a once-daily, anti-viral drug of the protein kinase class, and, it is prescribed for human immunodeficiency virus (HIV).

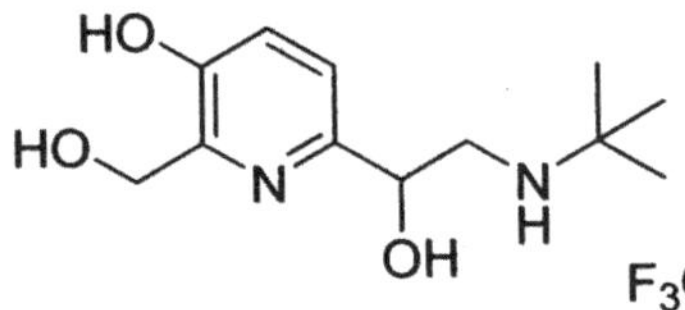

15, pirbuterol

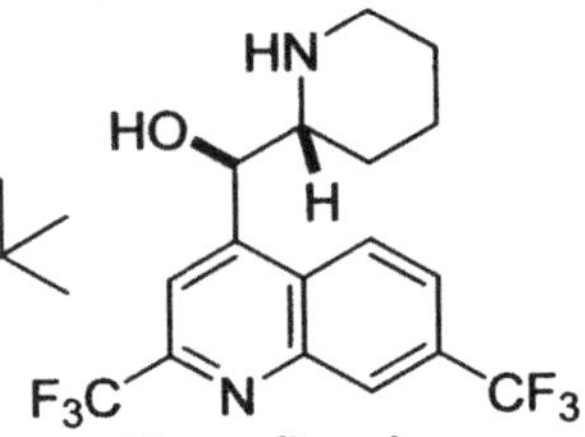

16, mafloquine

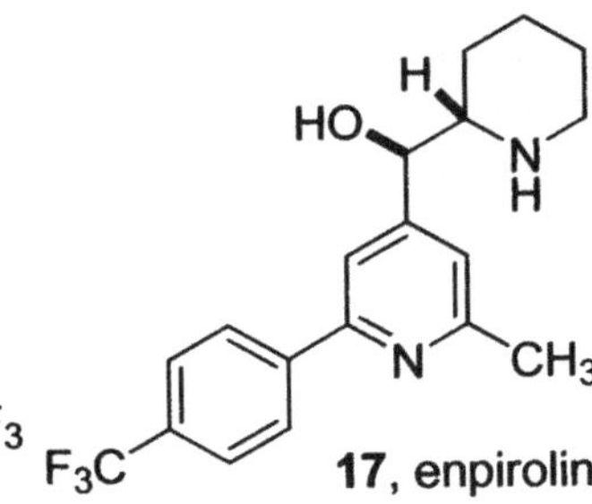

17, enpiroline

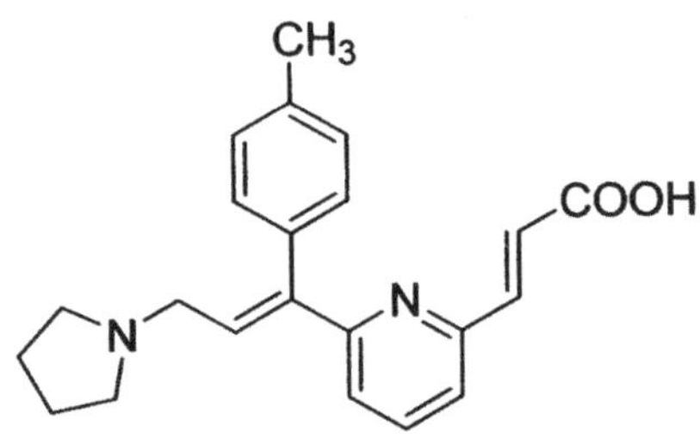

18, acrivastine

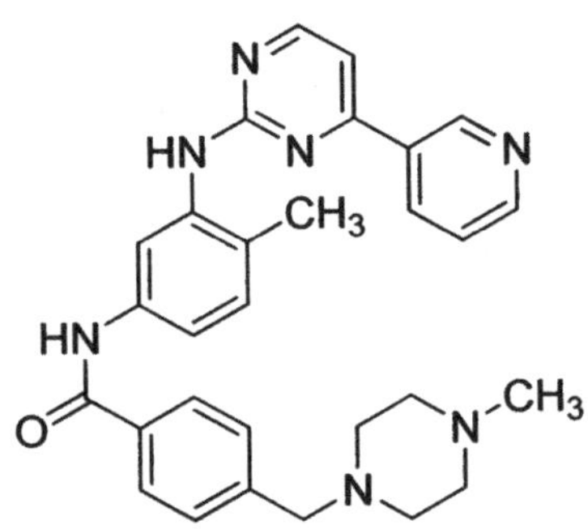

19, imatinib

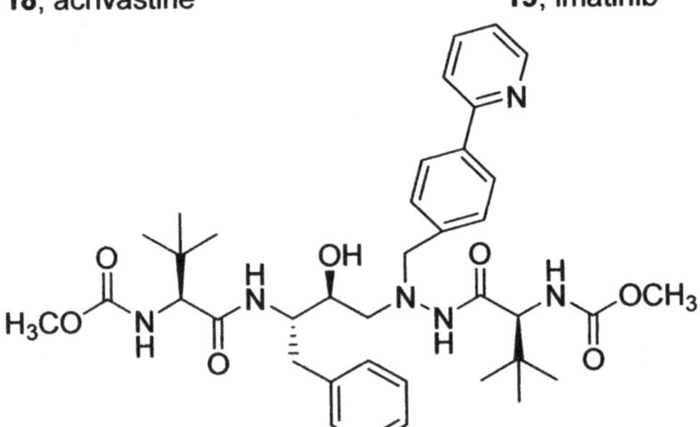

20, atazanavir

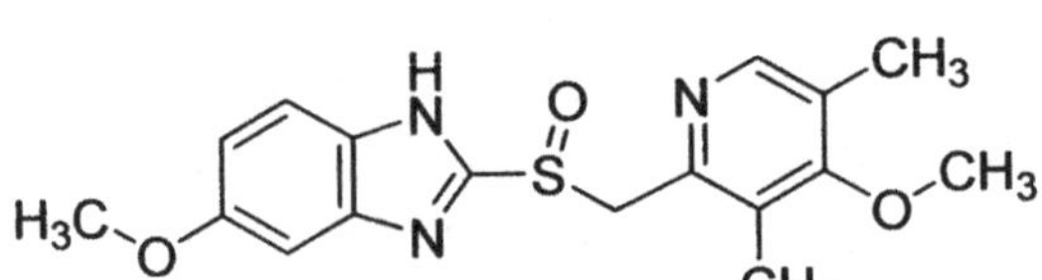

21, omeprazole

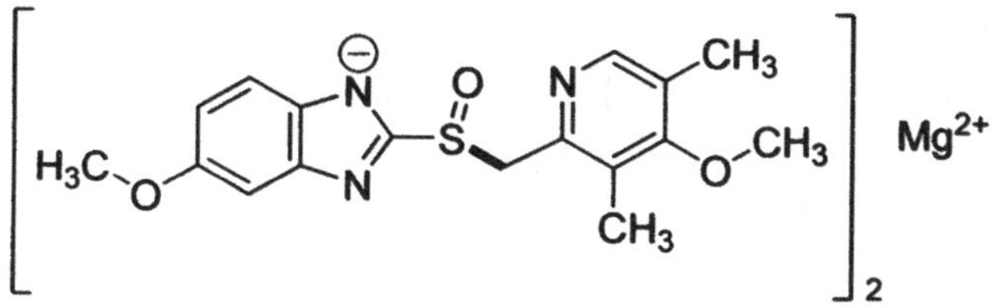

22, esomeprazole

Omeprazole (Prilosec, **21**), the first H^+/K^+-ATPase inhibitor, also known as a proton pump inhibitor (PPI), was marketed as a treatment for gastric ulcers since 1988. It functions by preventing acid production in the mucosa. Omeprazole was the best-selling drug for several years until its patent expiration in 2001, at which time, esomeprazole (Nexium, **22**), the (*S*)-enantiomer of racemic omeprazole (**21**), was launched. The mechanism of action of omeprazole (**21**), the "omeprazole cycle" was investigated.[7,8] Omeprazole (**21**) behaves more like a prodrug because pyridinium sulfydryl **23** is the actual inhibitory species in the "omeprazole cycle."

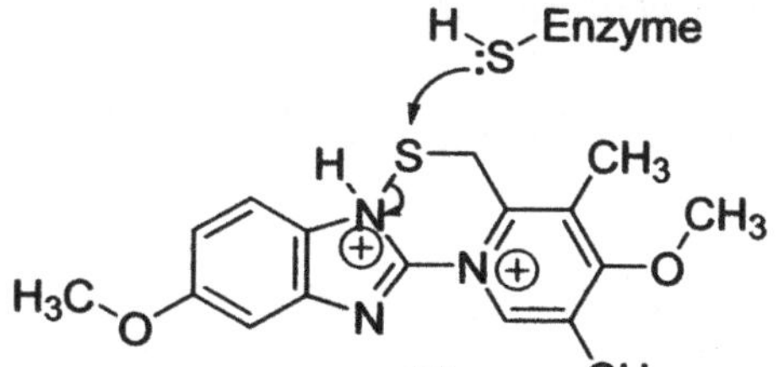

23

The 2,3,4-trisubstituted pyridine derivative **24** was designed as a potential L-Pro-L-Leu-Gly-NH_2 (PLG) tripeptidomimetic scaffold based on conformational and electrostatic comparison with the natural peptide.[9] Compound **24** exhibits higher potency with enhanced the response of the dopamine agonist *N*-propylapomorpholine (NPA) at human D_2 receptors compared to PLG in a cell-based assay.

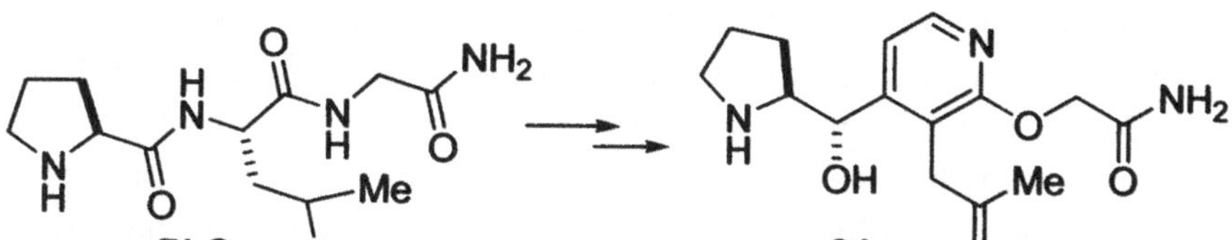

PLG **24**

10.1.2 Potential Liabilities for Pyridine-Containing Drugs

2,6-Unsubstituted pyridines bearing sterically unencumbered ring nitrogen are well known to bind tightly via chelation to the heme iron of CYP450 enzymes and display slow-offset kinetics.[10] For these reasons, binding assays for representative CYP450 enzymes are often included in the secondary pharmacology screening panels.

Several researchers investigated the structural requirements for the induction of hepatic microsomal cytochrome P450 2B1/2 and cytochrome P450 1A1/2. It was revealed that pyridine-containing compounds having lipophilic groups are inducers of hepatic P450, and compounds having aromatic groups and taking coplanar conformational structures are potent inducers of P450 1A1/2.[11]

Because of the potential liability of pyridine containing compounds as potent inhibitors or inducers of CYP450s, potential drug–drug interactions (DDI) need to be carefully considered. For instance, esomeprazole (**22**) is a competitive inhibitor of the enzymes CYP2C19 and CYP2C9, and it may therefore interact with drugs that depend on them for metabolism, such as diazepam and warfarin; *i.e.*, the concentrations of these drugs may increase if they are used concomitantly with esomeprazole.

10.2 Reactivity of the Pyridine Ring

10.2.1 Electrophilic Attack at Nitrogen of the Pyridine Ring[12]

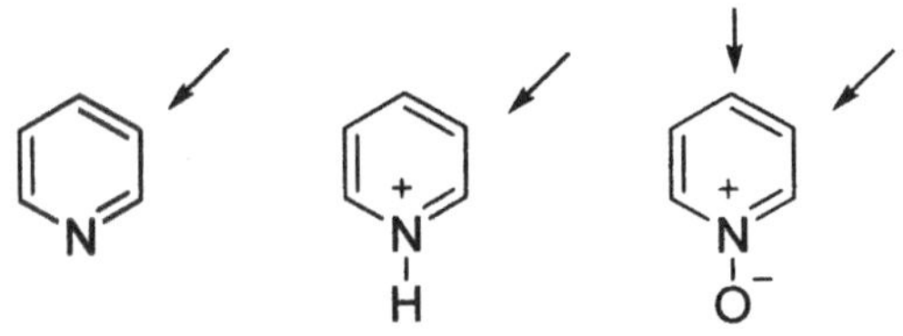

General reactivity pattern for electrophilic substitution

Most pyridines do not undergo electrophilic substitution reaction (S_EAr) at the ring carbon atoms due to the decreased electron density compared with benzene. And if the S_EAr reaction does occur, it occurs exclusively at the 3-position. Strong nucleophiles, such as amide ion, hydroxides, and organolithium compounds, can attack the ring carbon atoms of pyridines via an S_NAr reaction preferably at the 2- or 4-positions. Electron-withdrawing groups on the pyridine ring, such as nitro and nitrile, favor the S_NAr reaction. S_NAr reactions with halogen or alkoxyl substituted pyridines proceed much faster than the corresponding unsubstituted pyridines with the halogen or the alkoxy group as the leaving group.

On the other hand, pyridine is a weak base (p*K*a = 5.22) that can be protonated or can form salts with strong protonic acids. Some of the resulting pyridinium salts are commercial reagents. For instance, pyridinium perbromide is used as brominating agent, and pyridinium dichloromate

(PDC) is used as a mild oxidizing agent. Pyridine can also form coordination compounds with various Lewis acids. For example, sulfur trioxide pyridinium complex ($Py{\cdot}SO_3$) is employed as a mild sulfonating agent. The lone-pair electrons of pyridine nitrogen can react with transition metals to form complex ions such as in **25**. The complex ion pyridinium chlorochromate (PCC) is an oxidizing agent for converting alcohols to aldehydes and ketones. Under mild conditions, pyridine derivatives can be quarternized with alkyl halides by a S_N2 reaction, and the resulting *N*-alkyl pyridinium compounds are used as versatile synthetic intermediates or used as final products. Later on in this section, the Zincke reaction and its applications in the medicinal chemistry will be discussed. *N*-Acylpyridinium salts can also be generated *in situ* by a reaction of pyridine with an acyl halide. Quaternary salts of pyridines and related compounds have been reviewed.[13] Several chiral *N*-alkylpyridinium and related salts such as **26–30** have been studied as electrophiles for asymmetric nucleophilic addition reactions.[14]

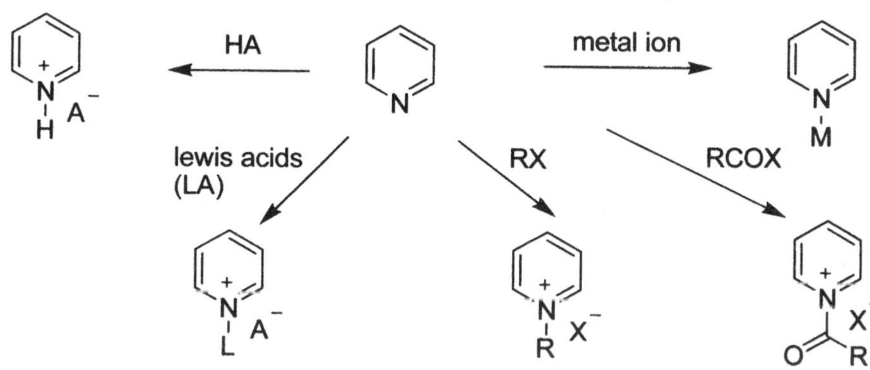

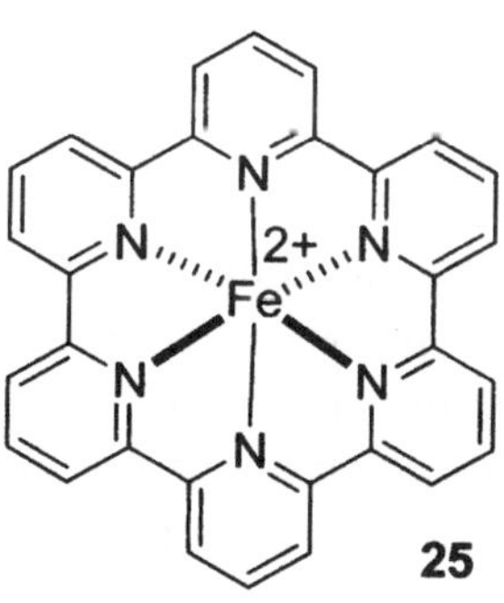

25

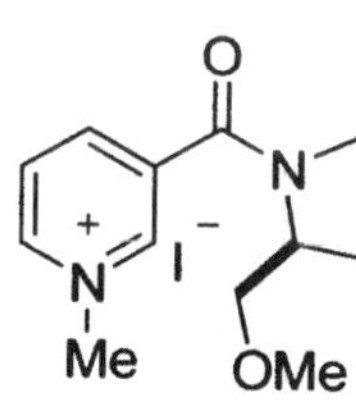

26

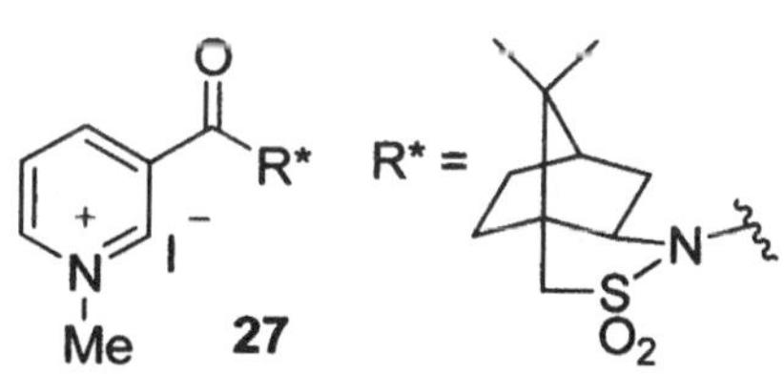

27

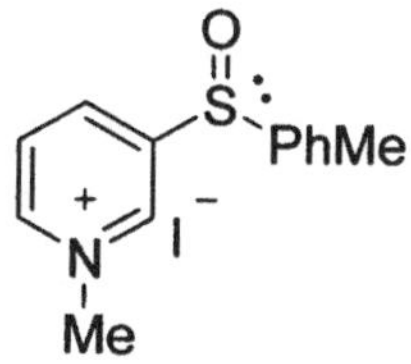

28

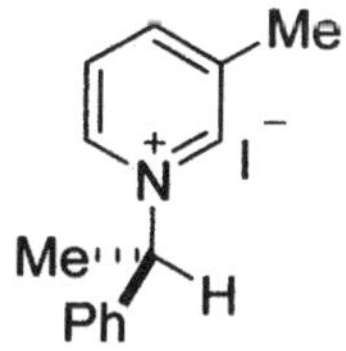

29

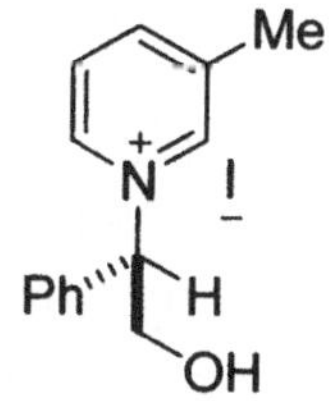

30

Pyridine *N*-oxides are also versatile intermediates in organic synthesis.[15] They are obtained by oxidation of pyridine using oxidizing agents such as peracids, H_2O_2/AcOH, H_2O_2/manganese tetrakis(2,6-dichlorophenyl)-porphyrin, H_2O_2/methyltrioxorhenium, dimethyldioxirane, bis(trimethylsilyl) peroxide, Caro's acid, oxaziridines,[16] trifluoroacetic anhydride (TFAA)/H_2O_2–urea complex,[17] O_2/ruthenium,[18] H_2O_2/molecular sieves,[19] O_2/cobalt,[20] trichloroisocyanuric acid/AcOH,[21] bromamine-T/$RuCl_3$,[22] and TBHP/$MoCl_5$.[23] Besides activating the ring for substitution reactions, the *N*-oxide moiety can also serve as an effective nitrogen-protecting group.[24]

Short-lived carbenes can react with pyridine to form the corresponding pyridinium ylide, which are far more stable than the starting carbenes. For example, pyridinium tungstate **33**, prepared from phenyl ethoxy carbene **31** and dihydropyridine **32**, serves as an effective cyclopropanation reagent to give products **34** and **35** in a 95:5 ratio and in 35% yield.[25]

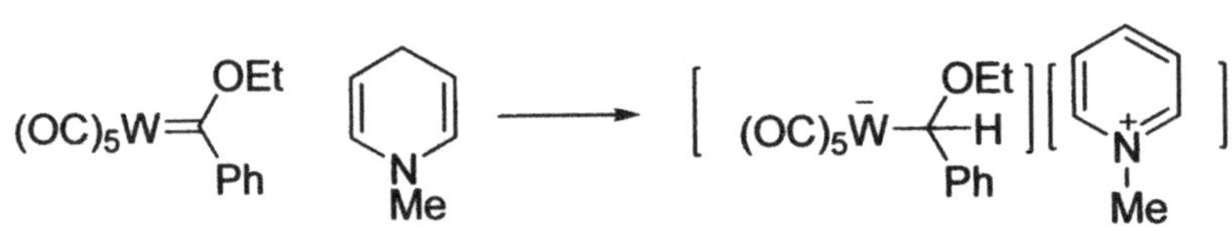

31 **32** **33**

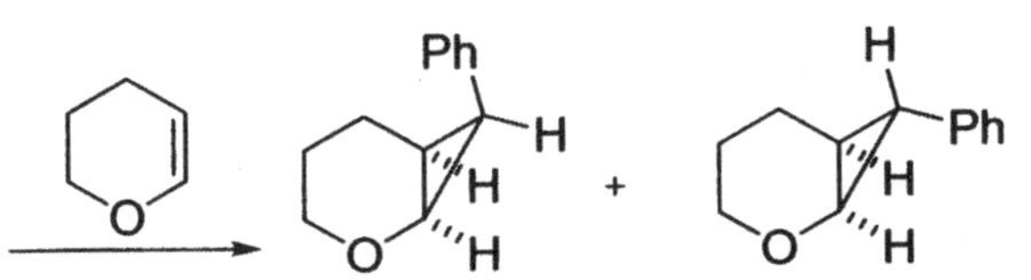

34 95:5 **35**

The Zincke reaction

The Zincke reaction[26–28] is an overall amine exchange process that converts Zincke salt, i.e., *N*-(2,4-dinitrophenyl)pyridinium salts (e.g., **36**), generated by reaction of pyridine or its derivatives with 2,4-dinitrochlorobenzene, to *N*-aryl or *N*-alkyl pyridiniums **37** upon treatment with the appropriate aniline or alkyl amine. This reaction proceeds via nucleophilic addition, ring opening, amine exchange, and electrocyclic reclosure, a sequence that also requires a series of proton transfers. The Zincke process has been applied to the preparation of a wide range of pyridinium salts, in particular, those unattainable by direct *N*-arylation or *N*-alkylation, such as electron-deficient, weakly nucleophilic pyridines and α-substituted electrophiles. In addition, α-chiral alkyl amines provide the corresponding *N*-alkyl pyridinium salts with retention of configuration in the Zincke process.

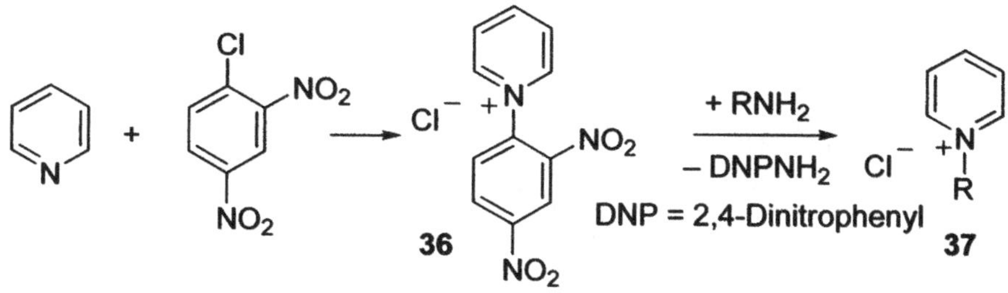

In addition to reactions with amines, Zincke salts also react with other nitrogen nucleophiles, such as hydroxylamine[29,30] and hydrazine,[29–31] providing various *N*-substituted pyridine derivatives, such as pyridine *N*-oxides **39** and *N*-aminopyridinium salts **40**.

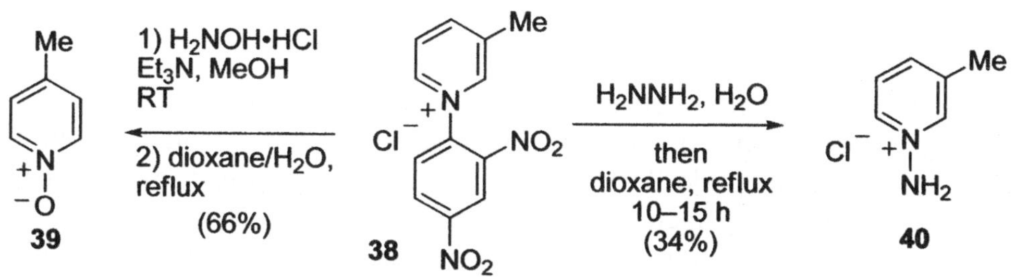

Reactions with *N*-acyl or *N*-sulfonyl hydrazines gave rise to iminopyridinium ylides and ylide precursors such as **43** and **44**.[30–33] Benzoyl hydrazines are also used in the Zincke reaction under similar conditions.[34,35]

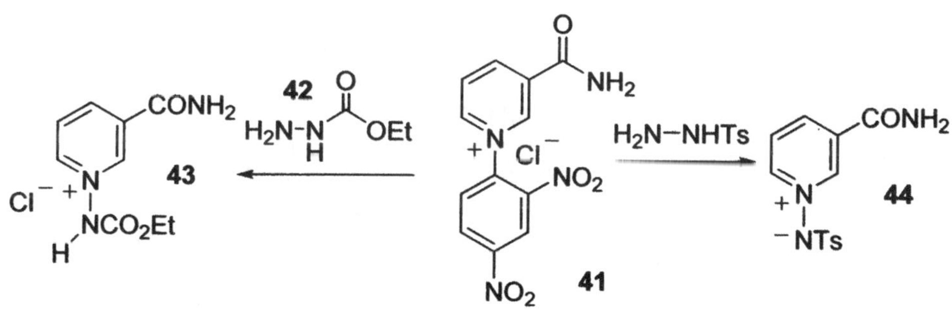

Zincke salts played an important role in the synthesis of NAD^+/NADH co-enzyme analogues. The nicotinamide-derived Zincke salt **41** has been widely used.[36–39] For instance, Zincke salt **41** was used to link with various adenine derivatives via the tether having a phosphonate functionality (**45**→**46**)[39c] to study through-space interaction between the pyridinium and base portions.

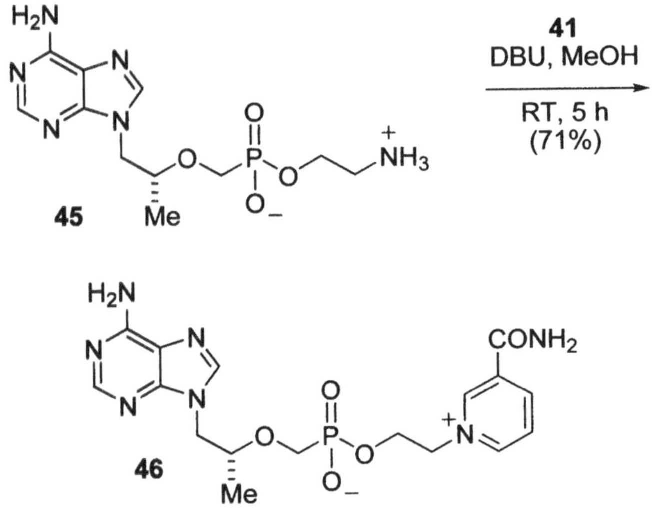

Solid-phase Zincke reaction was applied for the search of activators of the cystic fibrosis transmembrane conductance regulator protein.[40] On the other hand, the tripeptide TRH (pGlu–His–Pro–NH_2) was shown to be a hypothalamic releasing factor for the regulation of pituitary function.[41] A solid-phase Zincke reaction was used to prepare analogues of TRH having the central histidine replaced with a 1,4-dihydropyridine unit (such as **48**).[41] Compound **48** was expected to cross the hydrophobic blood–brain barrier (BBB) but to be trapped within the central nervous system upon oxidation to the hydrophilic pyridinium form.

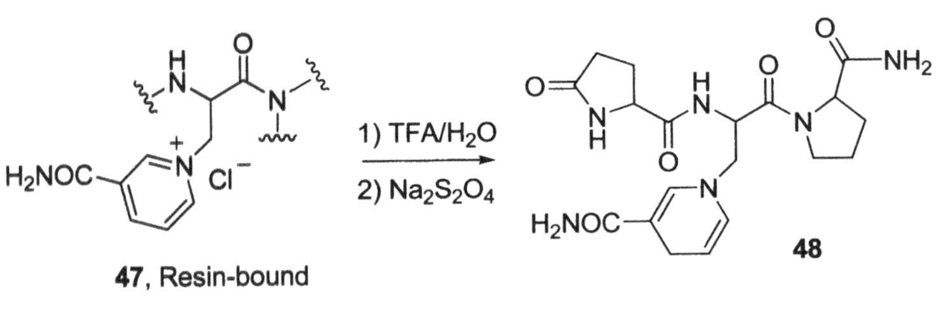

A redox system (**50**/**51**) to affect brain delivery of γ-aminobutyric acid (GABA) derivatives and analogues was also developed.[26a,42] Zincke reaction of **41** with acetal **49** followed by dithionite reduction afforded the 1,4-dihydropyridine prodrug **50**, which was hydrolyzed and oxidized *in vivo* to the active GABA analogue **51**. The neutral and lipophilic 1,4-dihydropyridine **50** can penetrate the blood–brain barrier (BBB), whereas the oxidized pyridinium salt **51** is retained in the brain for an extended period and then eliminated.

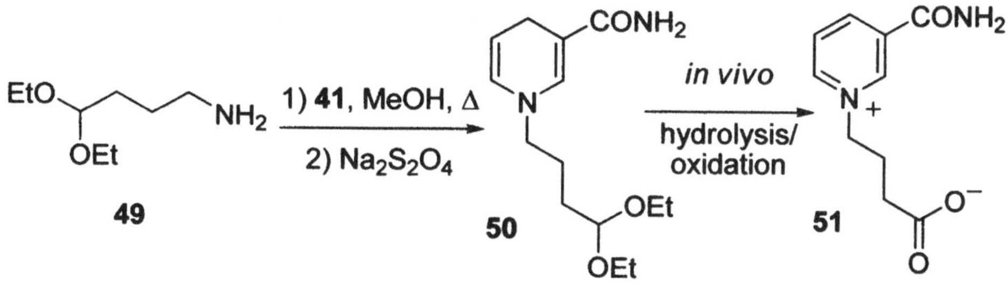

10.2.2 C–C/C–N Cross-Coupling Reactions with Organometallic Reagents[43]

Among all the methods of introducing a pyridine moiety into a drug-like molecule, metal-catalyzed carbon–carbon or carbon–nitrogen bond formation reactions of *C*-metallated pyridines or pyridine halides are the most important and widely used.

Synthesis of C-*metallated pyridines via directed lithiation*

Direct lithiation of pyridine is a common method of generating the precursors required for palladium-catalyzed cross-coupling reactions. Most methods employ the concept of directed *ortho*-metallation (DoM).[44] A directing metallation group (DMG) enables the precomplexation with the metal, resulting in metallation at adjacent to (*ortho* to) the substituent for deprotonation. Once deprotonation has been completed, the DMG assists in the stabilization of the metallated species. Known DMGs are halogens, CF_3, OH, OR, $OCONR_2$, $OSONR_2$, SO_2NR_2, SOR, NHCOR, $NHCO_2R$, CO_2H, 2-oxazolino, CONHR, $CONR_2$, COR, N–O, *etc.*

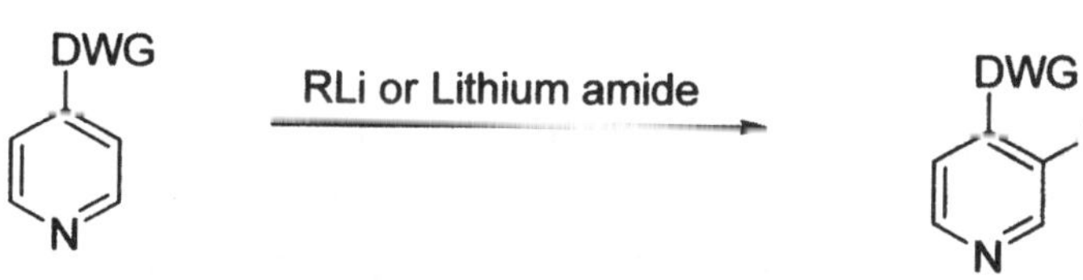

R = *n*Bu, *s*Bu, Ph, etc
lithium amide = LDA, LiTMP, etc
DMG = halogens, CF_3, OH, OR, $OCONR_2$, $OSONR_2$, SO_2NR_2, SOR, NHCOR, $NHCO_2R$, CO_2H, 2-oxazolino, CONHR, $CONR_2$, COR, etc

Regioselectivity of the reaction is affected by coordination and induction

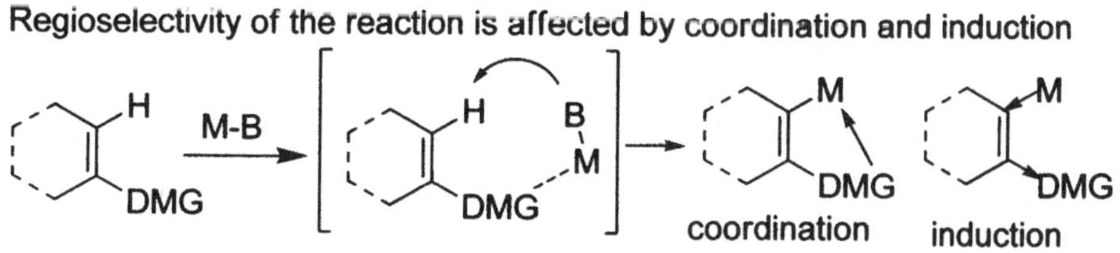

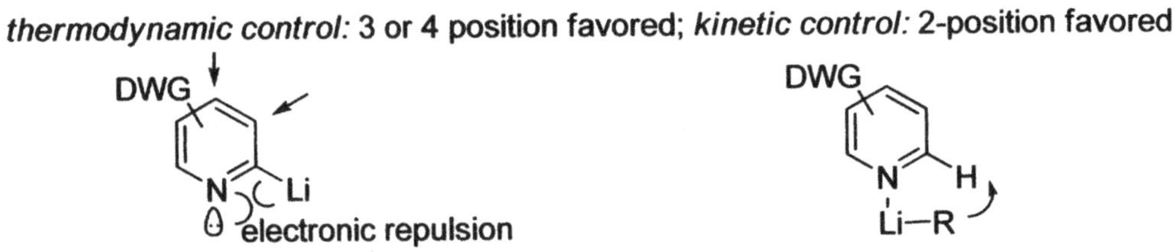

The kinetic acidity of the *ortho*-hydrogen atoms in the pyridine halides has the following order 4- > 3- > 2-position, which the same order as the kinetic acidity of the hydrogens in the pyridine. Gribble and Saulnier first reported the DoM of 2-, 3-, and 4-halo-pyridines with LDA.[45]

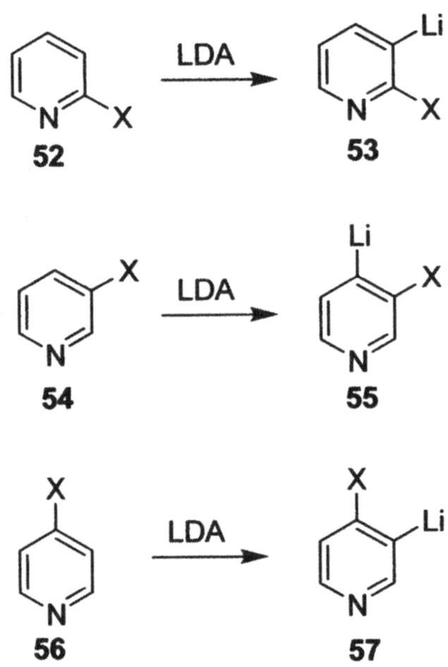

Direct lithiation at the 2-postion of the pyridine ring could be accomplished using either hexafluoroacetone,[46] a Lewis acid (such as $BF_3{\cdot}Et_2O$ complex),[47] or BuLi–LiDMAE.[48] For example, lithiation of adduct **58** was obtained from the reaction of pyridine with the highly polarized carbonyl group of hexafluoroacetone with LiTMP to give 2-lithio species **59**. Compound **59** was then trapped with electrophiles such as I_2 or Br_2 to produce the desired 2-halopyridines **60** and **61**, respectively.[46]

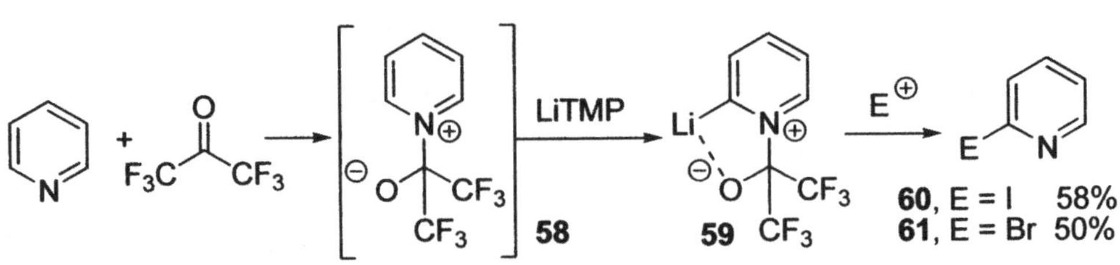

Treatment of 2-methoxypyridine with LDA afforded 3-lithio species **62**, whereas the complex base BuLi–LiDMAE led to the 6-lithio species **63**.[49]

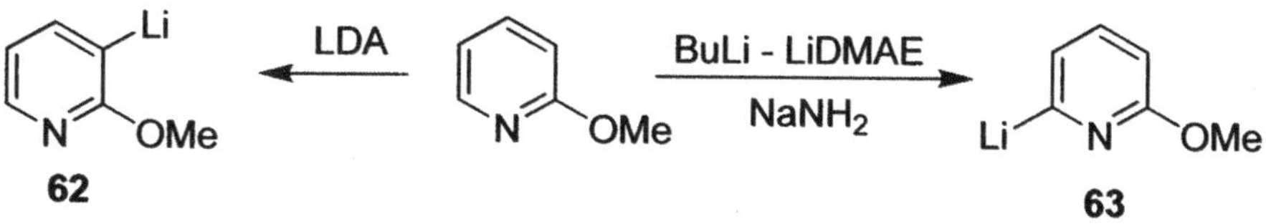

Synthesis of C-metallated pyridines via metal–halogen exchange

The widely available halogen pyridines can be used as starting materials for the preparation of *C*-metallated pyridines. For example, 2,6-dibromopyridine can be desymmetrized by mono-lithiation with *n*BuLi, followed by trapping the resultant anion **64** with an electrophile to afford compounds **65**.[50]

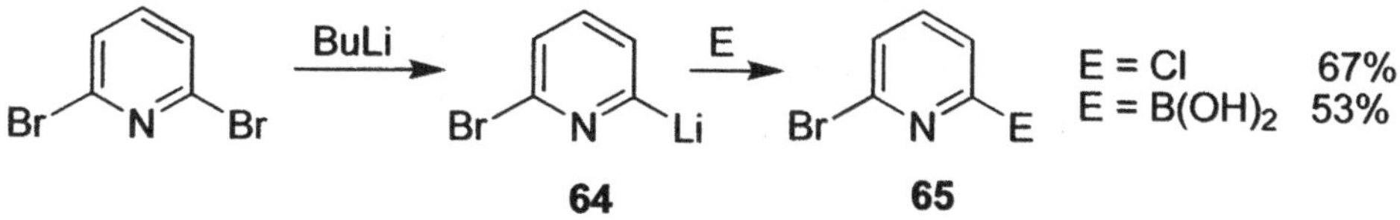

Bromine–magnesium exchange offers good chemoselectivity with multiply substituted pyridines.[51] Mono-bromo-substituted pyridines as well as dibromopyridines were easily transformed into the corresponding magnesio-derivatives followed by trapping with electrophiles. For example, 2,5-dibromopyridine produced the corresponding 2-bromo-5-substituted pyridine **66**.

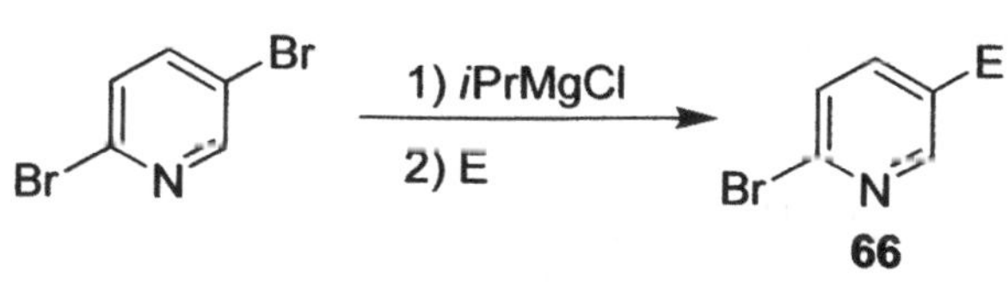

The Bu_2iPrMgLi complex, generated by the reaction of butyllithium with isopropylmagnesium chloride in a 2:1 ratio, was shown to carry out the bromine–magnesium exchange of **67** at –10 °C efficiently,[52] whereas *i*PrMgCl did not give the desired product and exposure of **67** to butyllithium at –100 °C formed the lithio-picoline species.

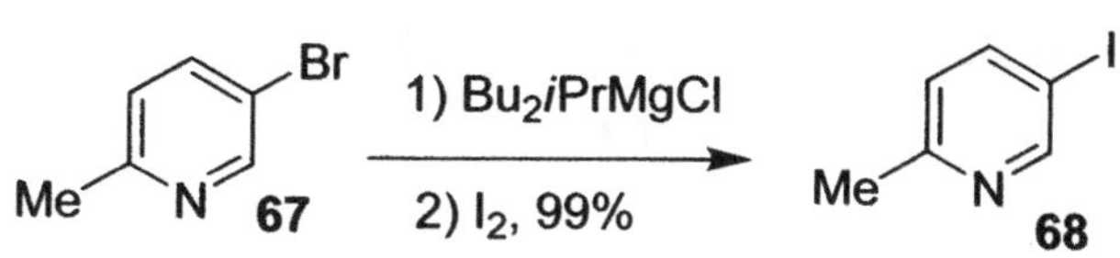

For 2,3,5-trihalo-substituted pyridine **69**, lithiation occurred at the more kinetically acidic 4-position, and then trapping with iodine afforded the 4-iodo derivative **70**. Subsequent lithiation produced 2-lithio-species **71** but underwent a "halogen dance" to the more stable intermediate 2-iodo-4-lithio derivative **72**.[53]

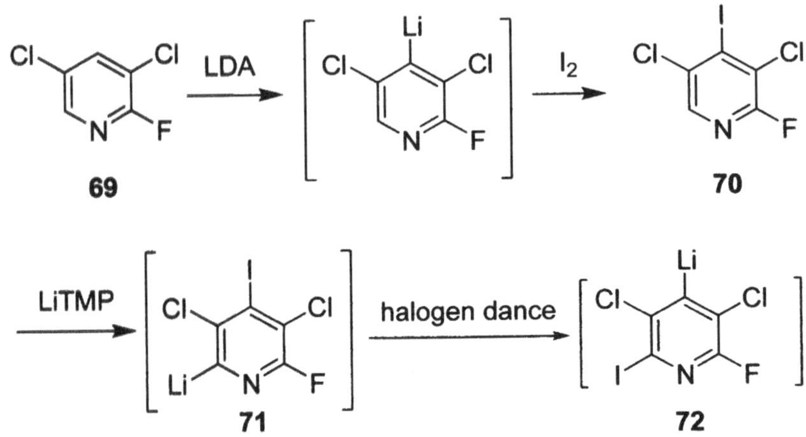

Compound **24** was designed as tri-peptide PLG mimetic as we discussed in Section 10.1.1. Lithiation of 2-fluoropyridine followed by trapping with iodine gave the iodide **73**. Halogen dancing using LDA and methallyl bromide gave the 3-substituted pyridine **74**. Iodine–magnesium exchange followed by trapping with the Boc-protected proline aldehyde afforded the alcohol **75** in an 85:15 diastereomeric mixture. Treatment of **75** with benzyl bromide in the presence of NaH and TBAI gave the benzyl ether **76**. Nucleophilic substitution of **76** with glycolamide produced **77**, which was deprotected to provide **24**.[9]

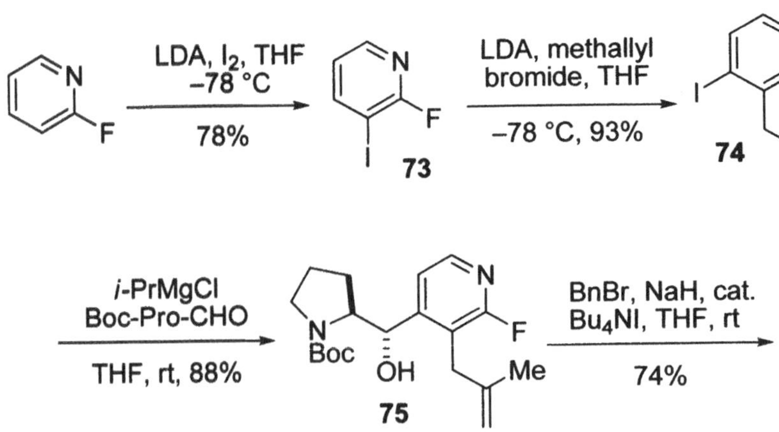

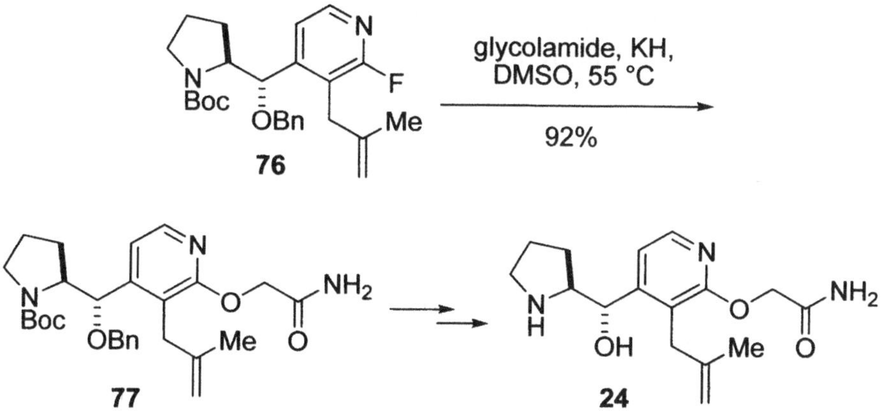

Metal-catalyzed C–C and C–N cross-coupling reactions in the synthesis of pyridine-containing drugs[54]

The *C*-metallated pyridines can undergo the following types of reactions:

(a) Lithium or Grignard pyridine derivatives nucleophilc attack to electrophiles

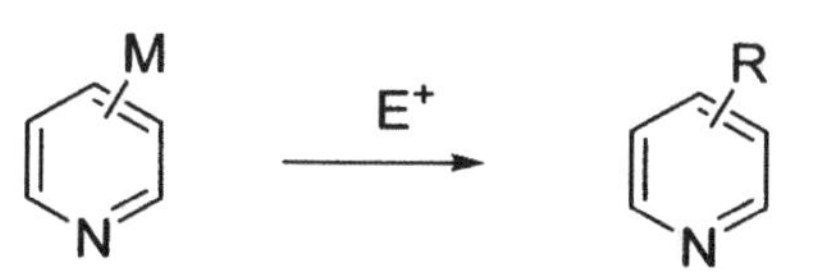

M = Li, MgBr, MgCl

E^+ = DMF, $ClSiMe_3$, CO_2, I_2, ClCOR, RCOR', etc
R = CHO, TMS, COOH, I, COR, C(OH)RR', etc

(b) Metal-catalyzed cross-coupling reactions of (i) halopyridines or related compounds with organometallic derivatives and (ii) metal-containing pyridines with halides or related compounds.

i) 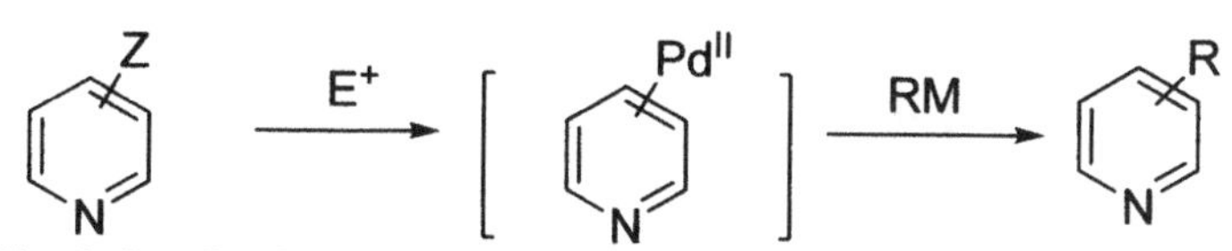

Z = I, Br, Cl, OTf

ii) 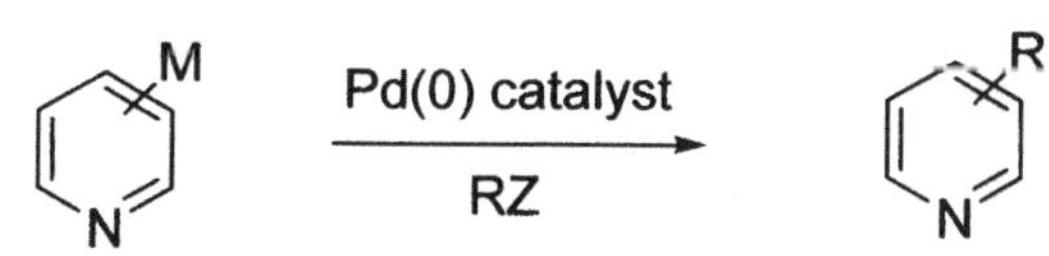

M = SnR_3, $B(OH)_2$, etc
R = aryl, alkenyl, alkynyl
Z = Br, I, Cl, OTf

Figure 1. Overview of metal-catalyzed C–C and C–N/C–O cross-coupling reactions of halopyridines or *C*-metallated pyridines

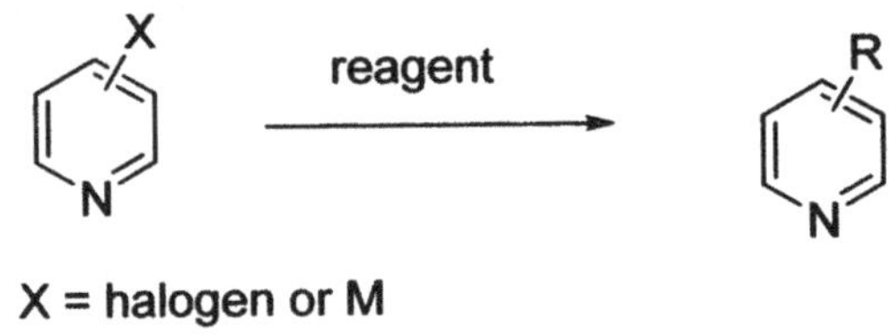

Kumada coupling[55]
RMgX, Pd(0) cat.
R = alkyl, alkenyl, alkynyl, aryl
Can access carbon hybridizations other than sp^2; Great range of reaction temperatures (–20 °C to elevated), but room temperature is the most common; Either Ni or Pd catalyst can be used; Functional groups must be compatible with the Grignard reagent.

Stille coupling[56]

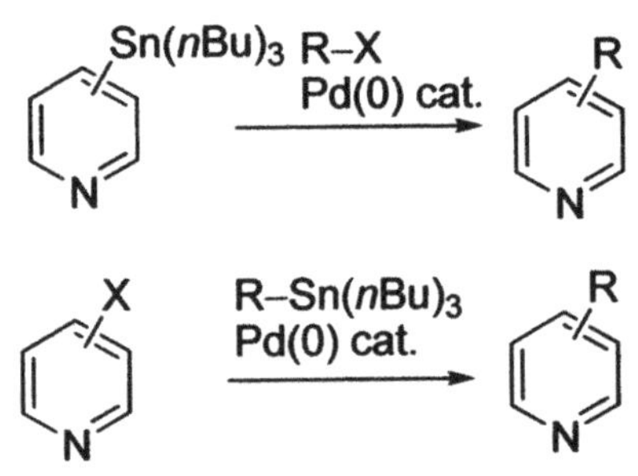

R = alkenyl, alkynyl, aryl
Typically sp^2 or sp-hybridized stannanes with organohalides or triflates, but alkyl, allyl and benzylstannanes also been used; Wide tolerance of functional groups. Major drawback: use of potentially toxic tin reagents.

Suzuki–Miyaura coupling[57]

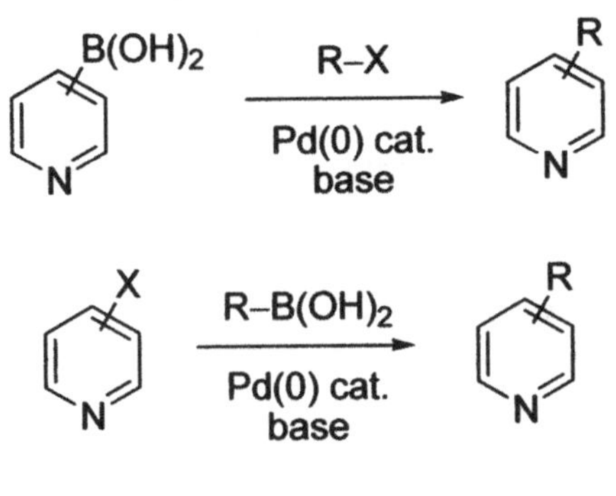

Also boronate and MIDA boronate

R = alkenyl, alkynyl, aryl
Good stability, relatively low toxicity in general, and easy access to commerically available boronic acids; Great functional group tolerability; Straightforward experimental procedures.

Sonogashira coupling[58]
Halogenpyridines (or triflates) + terminal alkynes (silylated alkynes)

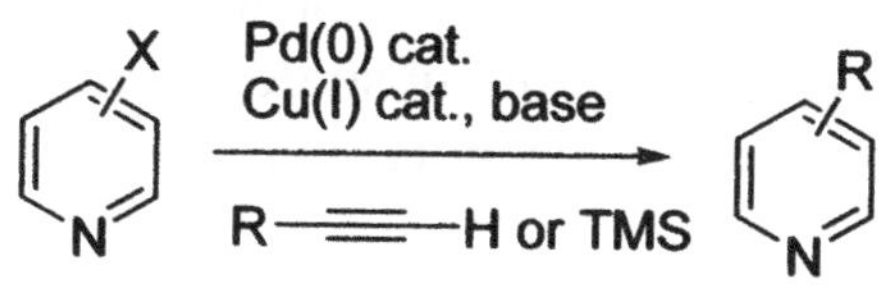

X = Br, I, OTf

R = alkenyl, alkynyl
Palladium-copper catalyzed; Comparable to the Suzuki or Stille reactions in its scope and functional group tolerability for alkenyl and alkynyl (hetero)aryls.

Heck reaction[59]

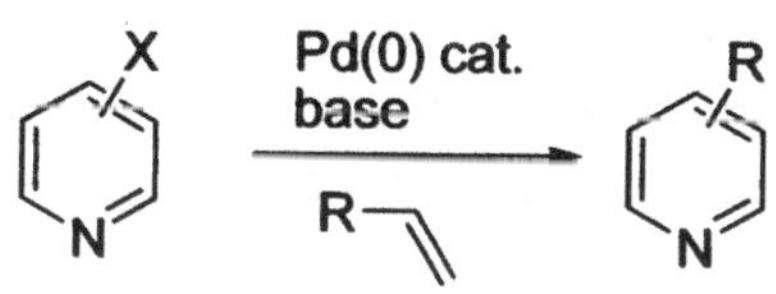

X = Br, I, OTf

R = alkenyl
Reaction can be inter-molecular or intra-molecular; Great versatility and applicable to a wide range of aryl and alkene species.

Negishi coupling[60]
RZnX, Pd (0) cat.
R = alkyl, alkenyl, alkynyl, aryl
Compatible with a wide variety of functional groups; The rate of insertion is dependent upon the nature of the substitution pattern: benzylic and allylic (0 °C), primary halides require more forcing conditions; Rieke Zinc can be paired up with less reactive organohalides.

Hiyama coupling[61]
$RSiR_3'/F^-$
Aryl and alkenyl halides or triflates with organosilyl compounds, Pd (0) cat.
R = alkenyl, alkynyl, aryl
Comparable to the Stille reaction but without the application of potential toxic tin reagents

Buchwald–Hartwig reaction[62]
R_2NH/base/Pd(0) cat.
R = R_2N
Palladium-catalyzed (catalytic); Elevated temperature and NaO*t*Bu as base under inert atmosphere; Wide range of substrates: alcohols, amides, amines, anilines, carbamates, sulfonamides, phenols, and thiols.

Chan–Lam reaction[63,62j]

$$\underset{\text{(organometalloid)}}{\text{R-M}} + \underset{\text{(Heteroatom nucleophile)}}{\text{H-YR'}} \xrightarrow[\text{CH}_2\text{Cl}_2\text{, air; Y= N, O, or S}]{\text{Cu(OAc)}_2\text{, base}} \underset{\text{(C-heteroatom cross-coupled product)}}{\text{R-YR'}} + \text{M-H}$$

R = YR
Copper-catalyzed (stoichiometric or catalytic) at room temperature and weak base under air; Wide scope of substrates: alcohols, amides, amines, anilines, azides, hydantoins, hydrazines, imides, imines, nitroso, phenols, pyrazinones, pyridones, purines, pyrimidines, sulfonamides, sulfinates, sulfoximines, thiols, thiourea, ureas, and sulfonylguanidines

Applications of palladium-catalyzed reactions in drug discovery are summarized below:

CGP 60474 (**80**)[64] is a phenylamino-pyrimidine-type protein kinase C (PKC) inhibitor with a high degree of selectivity versus other serine/threonine and tyrosine kinases. The structurally related imatinib (**19**, Gleevec) is a tyrosine kinase inhibitor currently on the market for chronic myeloid leukemia (CML). CGP 60474 was synthesized using a Negishi cross-coupling to obtain the key intermediate **79** followed by two subsequent nucleophilic substitution reactions. Halogen–lithium exchange of 2-fluoro-4-iodopyridine followed by trapping with thoroughly dried zinc chloride generated **78**, which subsequently reacted with 2,4-dichloropyrimidine regioselectively to generate pyridinyl-pyrimidine **80** in 90% yield in a one-pot reaction.[65]

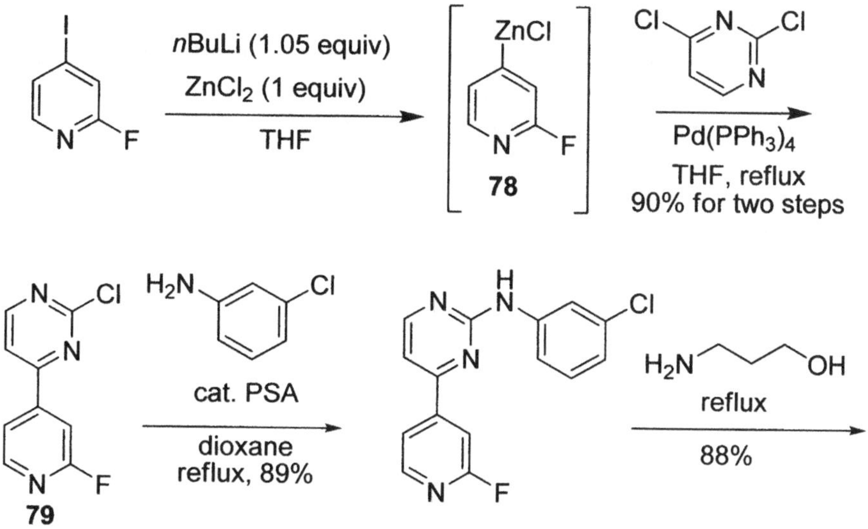

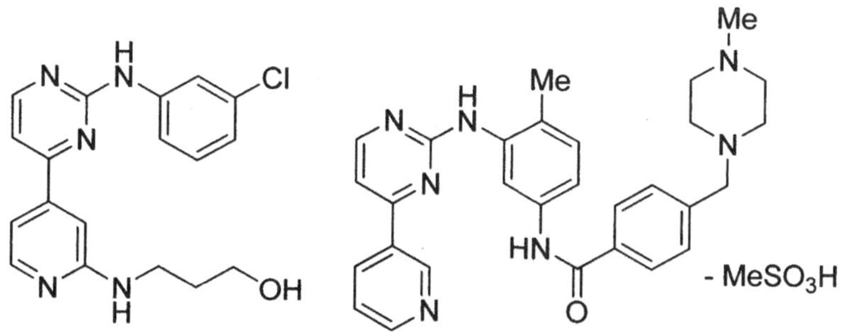

80, CGP 60474 19, imatinib (Gleevec)

On the other hand, structurally similar compounds **83** were potent vascular endothelial growth factor (VEGF) receptor inhibitors with anti-tumor activities. The key intermediate **82** was prepared using a Stille coupling reaction of stannane **81** and 2,6-dichloropyrazine.[66]

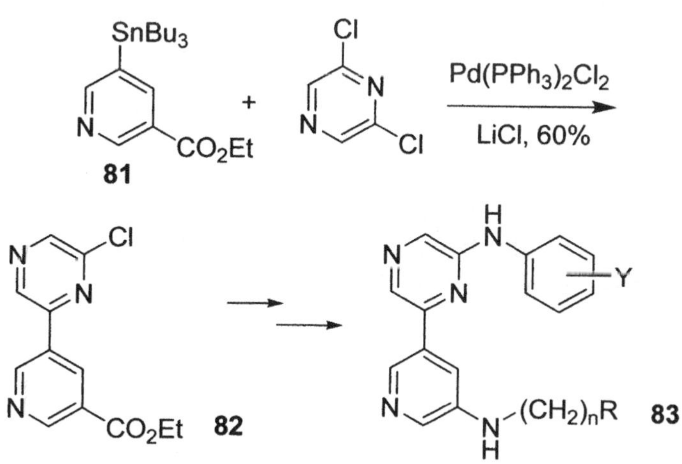

A series of imidazo[4,5-*b*]pyridines **89** are high-affinity corticotrophin-releasing factor (CRF) receptor ligands, and they were prepared from intermediates **86** and **88**, which were in turn synthesized *via* either the Suzuki or the Negishi coupling reaction.[67]

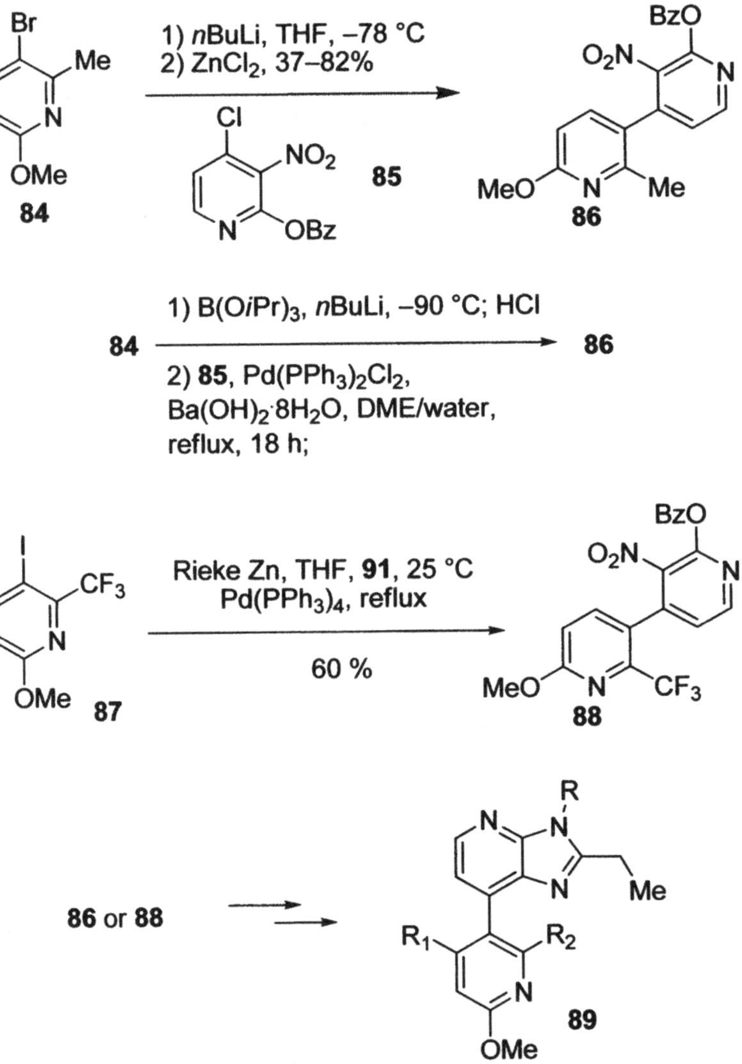

Pyridine-containing unnatural amino acids **91** and **94** were prepared by Negishi coupling of alkylzinc species such as **90**[68] or Suzuki coupling of the boron derivative generated by hydroboration of the vinyl-amino-alcohol **92**.[69]

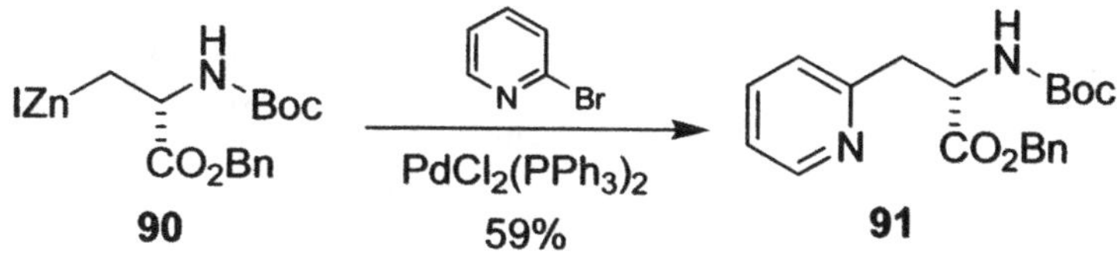

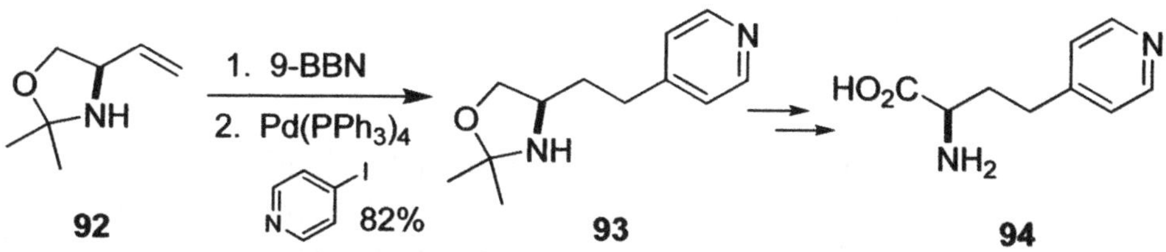

The synthesis of the nonhydrolyzable phosphor-tyrosine mimetic **98**, a potential anticancer reagent, involved a Stille reaction of **95** with **96** to form the bis-aryl intermediate **97**.[70]

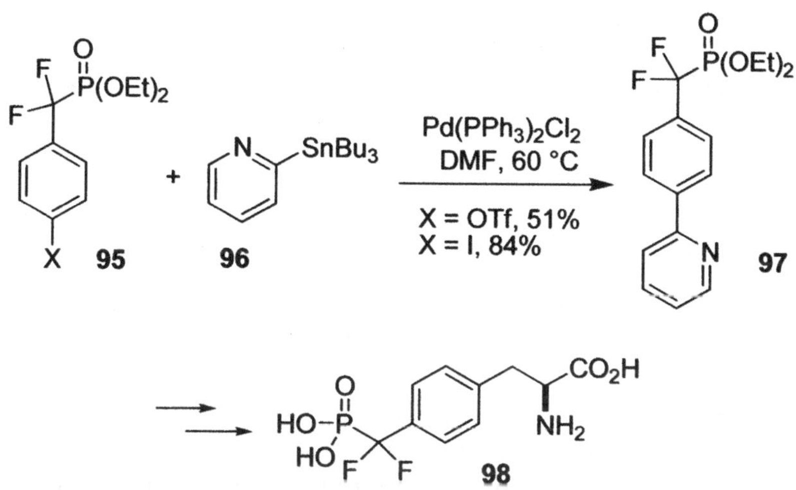

SSR182289A (**102**) is a selective and potent orally active thrombin inhibitor.[71] The key unnatural amino acid **101** was constructed using a Sonogashira reaction of acetylene **99** and 2-chloropyridine **100** followed by reduction of the nitro group and then hydrolysis of the methyl ester.

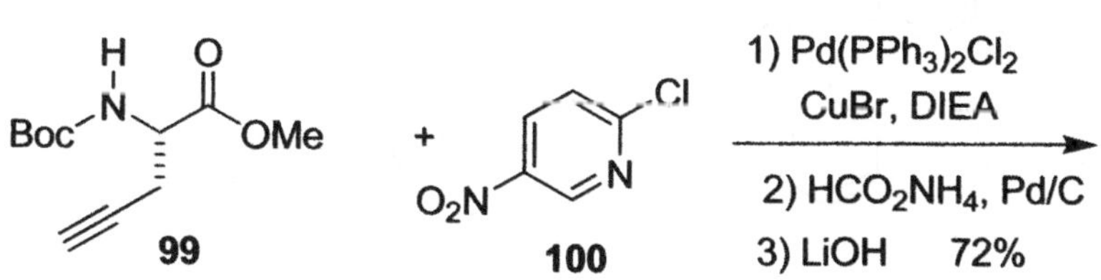

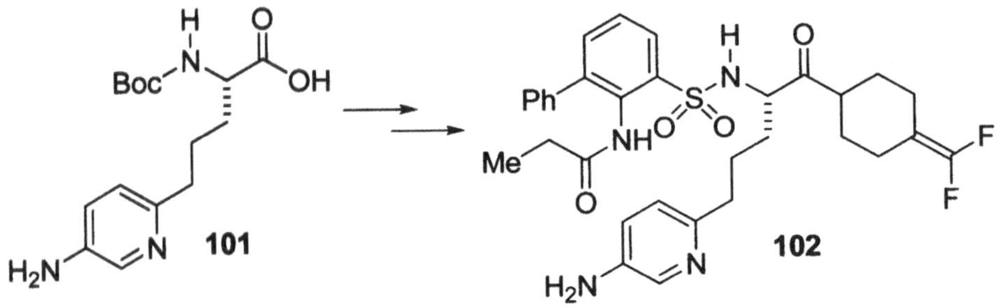

Altinicline (**10**), being structurally similar to nicotine, is an inhibitor of cholinergic nicotinic receptors for the treatment of Alzheimer's disease. It was synthesized *via* the Sonogishira coupling of iodochloro analogue **103** with terminal alkyne **104**, followed by reduction of the chloride of **105** and removal of the silyl protection group.[72]

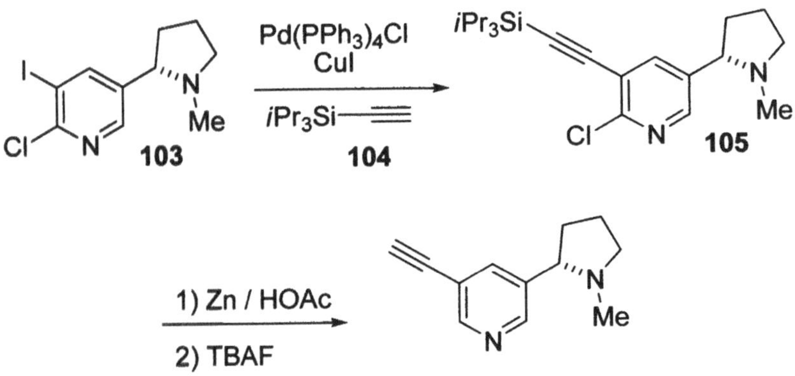

10, (altinicline)

The key step in the synthesis of CDP840 (**108**), a phosphodiesterase IV (PDE IV) inhibitor for the treatment of asthma[73] included the Liebeskind variation of the Suzuki reaction; i.e., the thiopyrimidine ether **106** was cross-coupled with pyridyl-4-boronic acid to afford **107**. Catalytic hydrogenation of **107** then afforded racemic **108**.

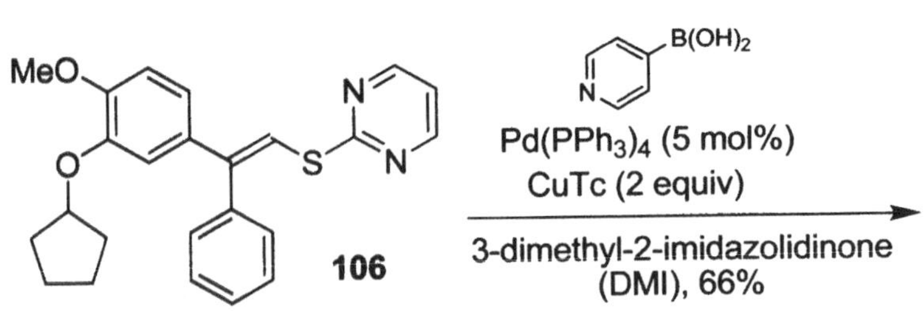

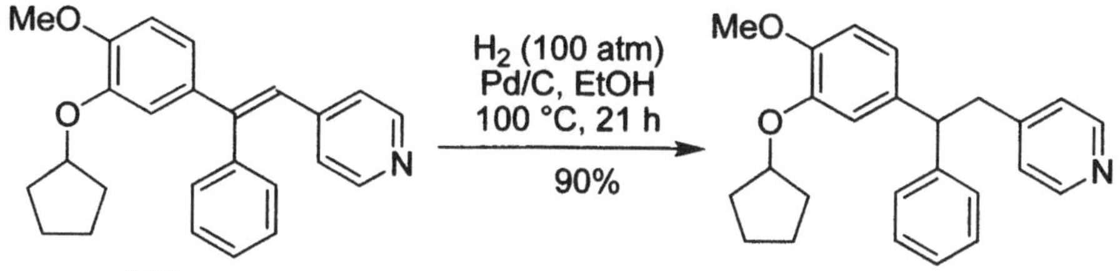

107 108

In the synthesis of PDE IV inhibitor **113**, the Suzuki–Miyaura reaction between the chloride salt **111** and the boronic acid **112** failed under typical conditions; however, desired product **113** was obtained by using Fu ligand tri-*t*-butylphosphine with palladium to ligand ratio of 1:1.[74] When the palladium-catalyzed cross-coupling was conducted under an atmosphere of carbon monoxide, the organo-palladium intermediate can be captured by carbon monoxide prior to the transmetallation step to form carbonyl-inserted derivative.[75,76] The carbonyl-inserted palladium intermediates can also be captured by other nucleophiles such as an alcohol to form an ester or an amine to form an amide.[77] Herein, the ester **109** was obtained from 2,5-dichloropyridine *via* the aforementioned selective palladium-catalyzed carbonylation reaction. Conversion of **109** to carbinol **110** with methyl Grignard followed by oxidation of pyridine with urea hydrogen peroxide (UHP)/trifluoroacetic acid anhydride (TFAA) afforded the chloride **111**.

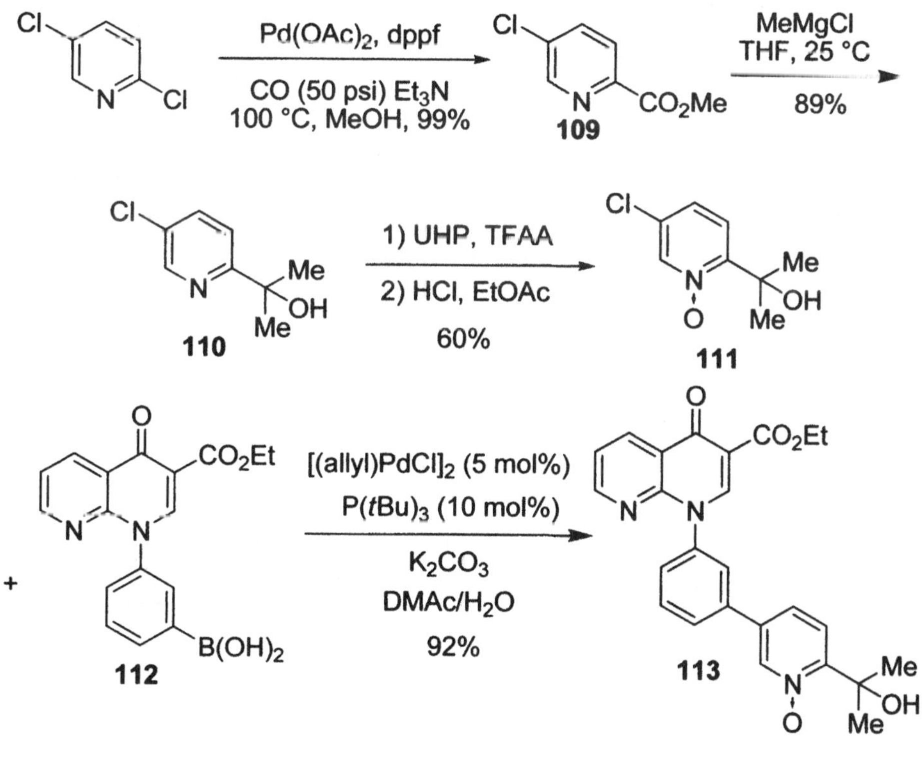

2,3,5-Tri-substituted pyridines **116** were designed as potent AKT inhibitors that were more selective against ROCK1.[78] Two sequential Suzuki–Miyaura reactions starting from pyridine **114** provided **115** and then the desired analogues **116** after an additional Suzuki coupling with $PhB(OH)_2$.

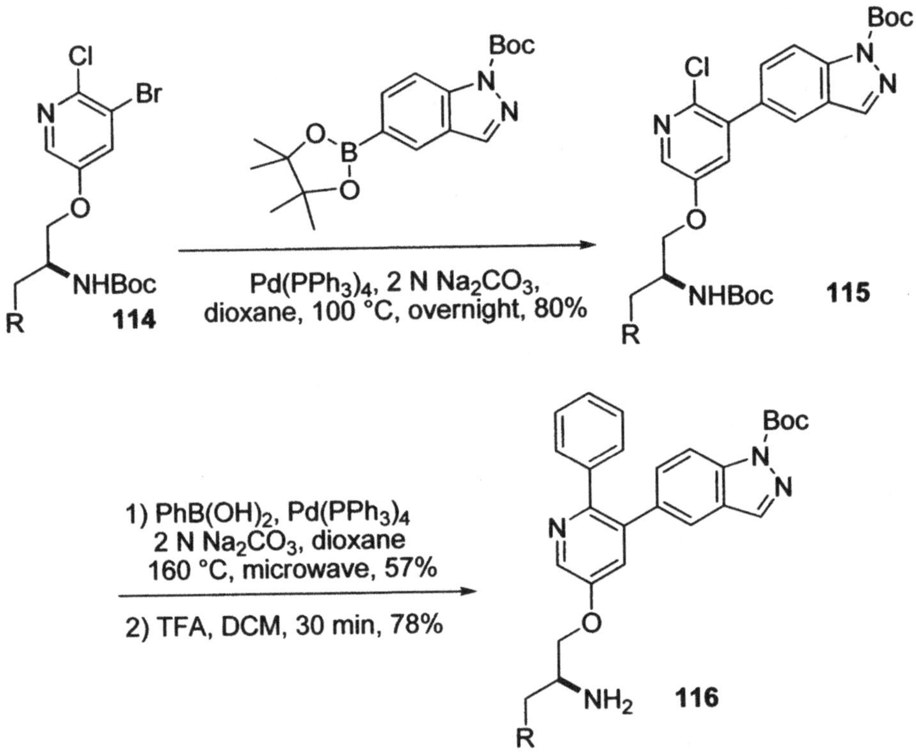

(±)-Epibatidine (**9**) is a nonopioid analgesic and nicotinic acetylcholine receptor (nAChR) antagonist, and its synthesis involved a key Stille cross-coupling reaction between pyridine stannane **117** and iodide **118** with Pd[(*o*-tolyl)$_3$P]$_2$Cl$_2$ as the catalyst to form **119**.[79]

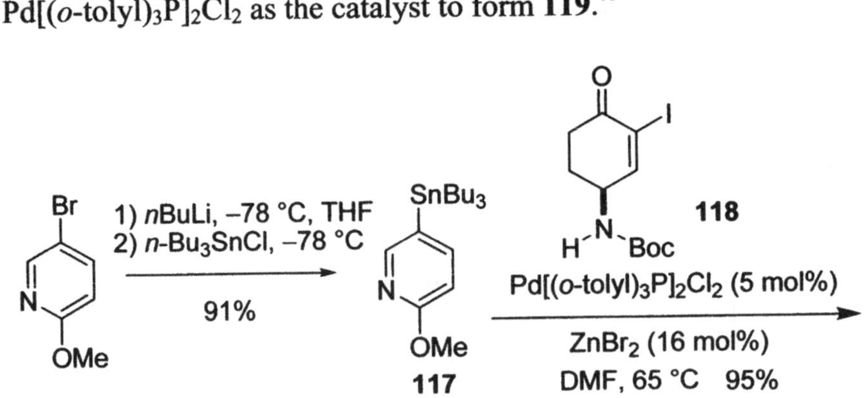

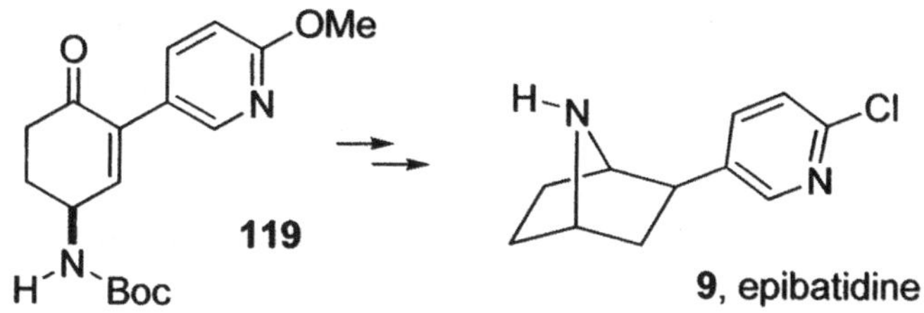

(+)-(*S*)-WS75624B (**123**) is an endothelin-converting enzyme (ECE) inhibitor that is potentially useful in the treatment of hypertension.[80] Stannane **120** reacting with iodide **121** afforded **122** which was further elaborated into **123**.

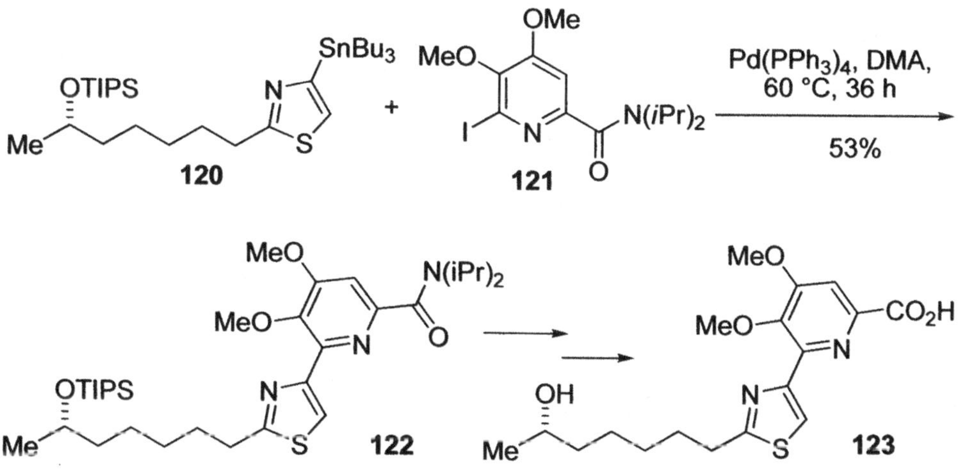

A series of indazole-pyridine analogues was developed as protein kinase B/Akt inhibitors.[81] Among them, compound **126** was the most potent and selective. It possessed an improved cardiovascular safety profile compared with other analogues in the series. It was synthesized via a Stille coupling of trimethyltin derivative **124** and the bromide **125** using $Pd_2(dba)_3$ and $(o\text{-tol})_3P$ as the catalyst.

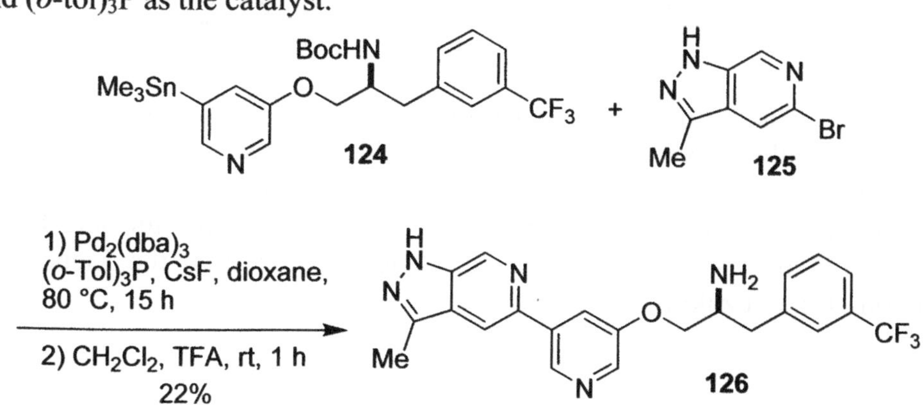

Camptothecin (**127**) is an alkaloid with potent anti-tumor activities. SAR studies showed that substitution at the 7-position of **127** led to compounds, such as SN-38 (**128**) and BNP-1350 (**129**), with improved biological activity. The Sonogashira cross-coupling of the advanced intermediate **130** with alkyne **131** afforded the common intermediate **132** for the assembly of both **128** and **129**.[82]

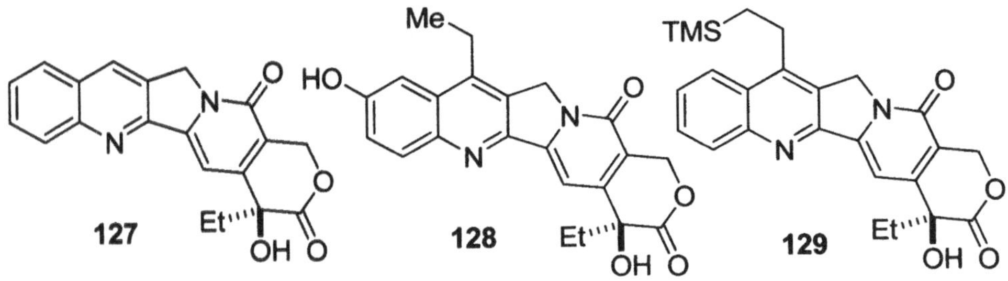

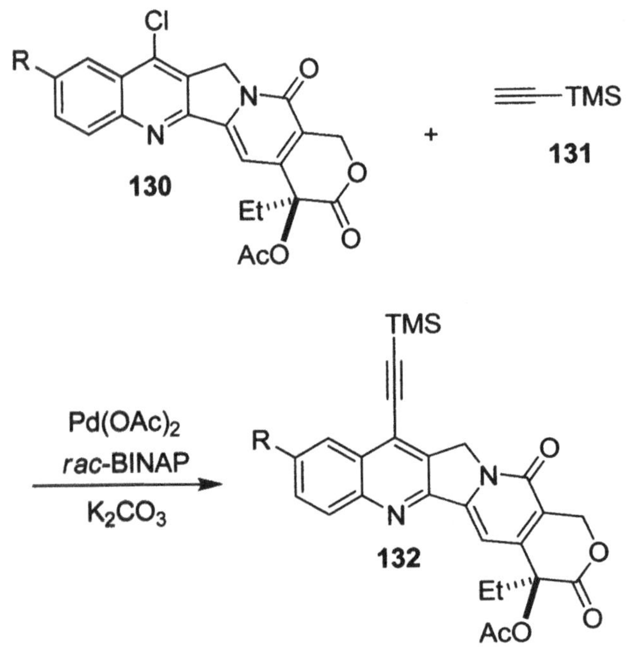

The 2-indolinone pharmacophore is present in a large number of pharmaceutically active compounds and could be constructed by a tandem Heck–Suzuki reaction sequence.[83] Intramolecular Heck reaction of **133** afforded a cyclized intermediate that, *in situ*, reacted with boronic acid **134** to produce **135** in a single-pot sequence. The reaction used the Liebeskind variation of the Suzuki reaction (CuTC).

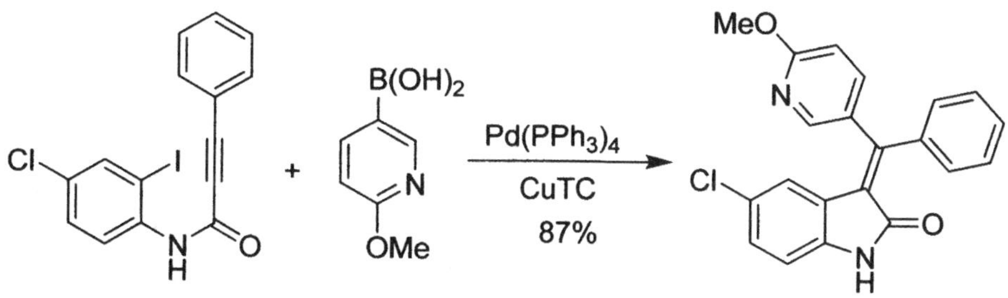

133 134 135 *E*:*Z* = 25:1

Benzodiazepines are "privileged structures" that are frequently present in biologically active compounds. Pyridobenzodiazepinone **137** was synthesized by an intramolecular Buchwald–Hartwig amination of **136**.[84]

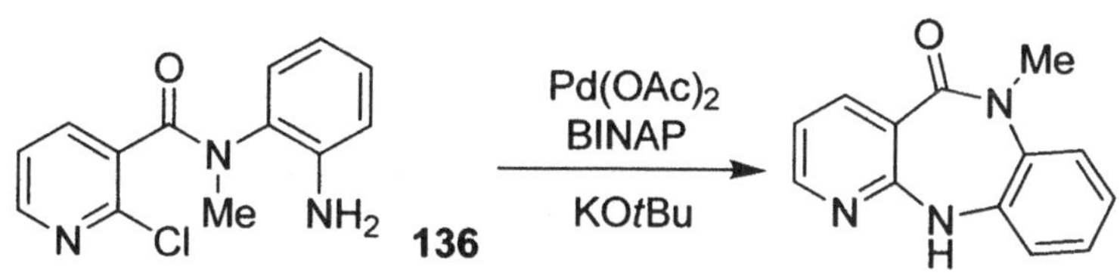

136 137

10.3 Construction of the Pyridine Ring

10.3.1 Synthesis via Condensation Reactions

Historically, the pyridine ring has been mainly constructed *via* a variety of condensation reactions[85] of smaller molecules such as amines and carbonyl compounds.

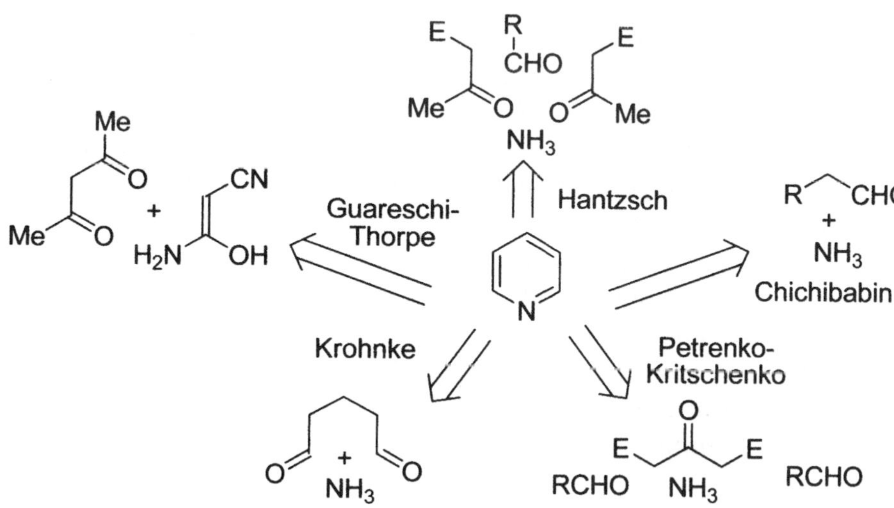

Hantzsch (dihydro)pyridine synthesis

The Hantzsch pyridine synthesis[86] is the condensation of two equivalents of a β-dicarbonyl compound such as ethylacetoacetate, one equivalent of an aldehyde, and one equivalent of a nitrogen donor such as ammonia (or ammonium acetate) in refluxing alcohol or acetic acid. The immediate resulted from this three-component coupling, 1,4-dihydropyridine **138**, is then oxidized, driven by aromatization, to substituted pyridine **139**. Saponification and decarboxylation of the 3,5-ester substituents then leads to symmetric 2,4,6-tri-substituted pyridine **140**.

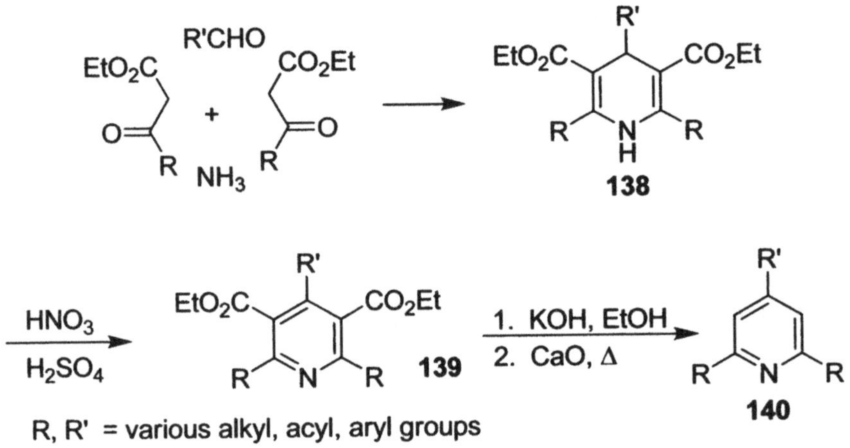

Several research laboratories investigated the mechanism of the Hantzsch dihydropyridine synthesis.[87–90] A comprehensive discussion on the mechanism is reported recently.[85g]

The Hantzsch chemistry was used to construct model systems of NADH coenzyme to understand the mechanistic details of this biological reducing agent. Later on, the 1,4-dihydropyridines were identified to pose various biological activities such as vasodilator, bronchodilator, antitumor, hepato-protective, and gero-protective activity. Some of them are well-known calcium channel blockers and commercialized as cardiovascular and antihypertensive agents such as nifedipine (**141**), amlodipine (Norvasc, **142**) or nimodipine. Amlodipine (**142**) was one of the most prescribed medicines for hypertension and one of the biggest selling drugs worldwide. Improvements and modifications of the Hantzsch pyridine synthesis have also been reported in the past several decades. Herein we only give several recent literature examples.

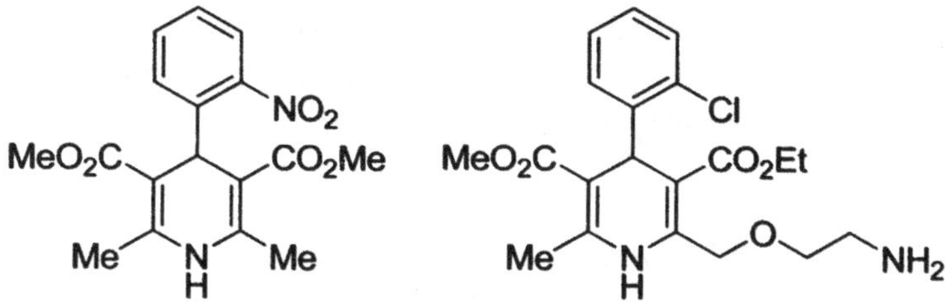

141, nifedipine **142**, amlodipine (Norvasc)

Several reagents and reaction conditions such as TMSI (generated *in situ* from TMSCl and NaI),[91] magnesium nitride (Mg_3N_2),[92] triphenylphosphine (PPh_3),[93] thiamine hydrochloride (vitamin B_1),[94] grinding under solvent-free conditions,[95] PTSA with ultrasonic irradiation[96] were recently reported to mediate efficiently the Hantzsch dihydropyridine synthesis. For instance, 1,4-dihydropyridines **143** were obtained in good yields using thiamine hydrochloride (vitamin B_1) as the catalyst under solvent-free conditions at room temperature.[94]

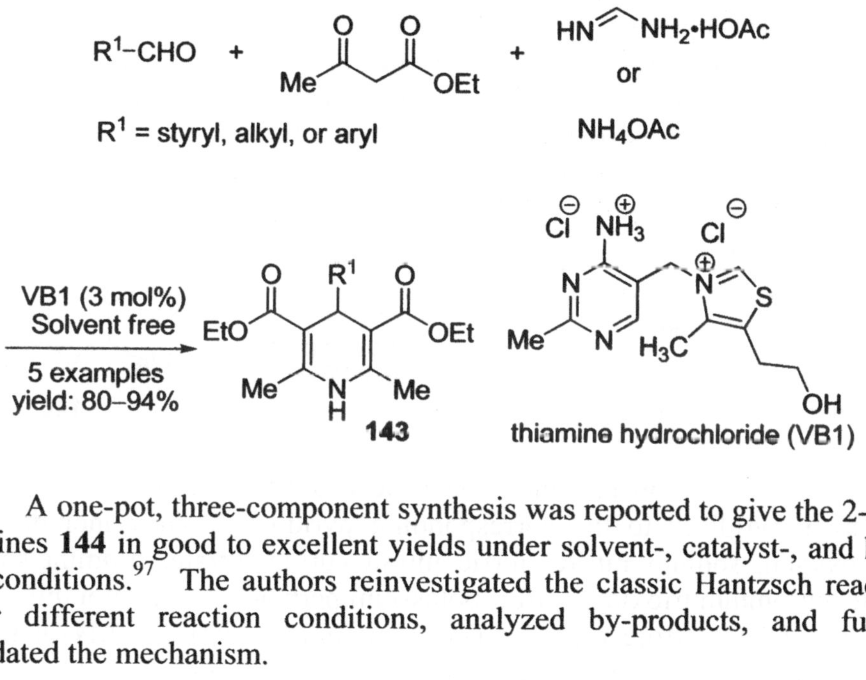

A one-pot, three-component synthesis was reported to give the 2-arylpyridines **144** in good to excellent yields under solvent-, catalyst-, and heat-free conditions.[97] The authors reinvestigated the classic Hantzsch reaction under different reaction conditions, analyzed by-products, and further elucidated the mechanism.

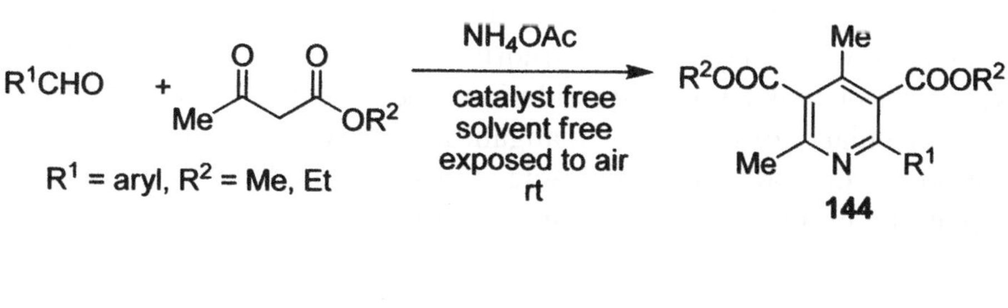

Pyridinyl analogues **145** were obtained using water as solvent and the resulting 1,4-dihydeopyridine were subsequently aromatized by ferric chloride or potassium permanganate in a one-pot synthesis.[98]

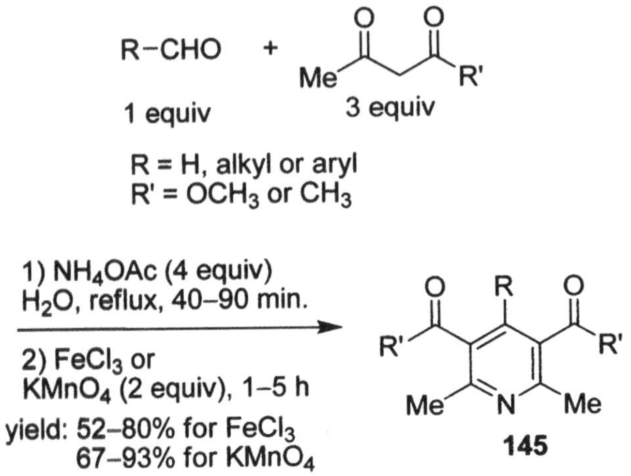

145

Hantzsch pyridine derivatives **146** were prepared by the reaction of aldehydes and β-dicarbonyls in the presence of ammonium chlorate at 80 °C under solvent-free conditions.[99] Ammonium chlorate was used both as ammonia source and as oxidizing agent source for the direct synthesis and oxidation of Hantzsch 1,4-dihydropyridines to pyridines.

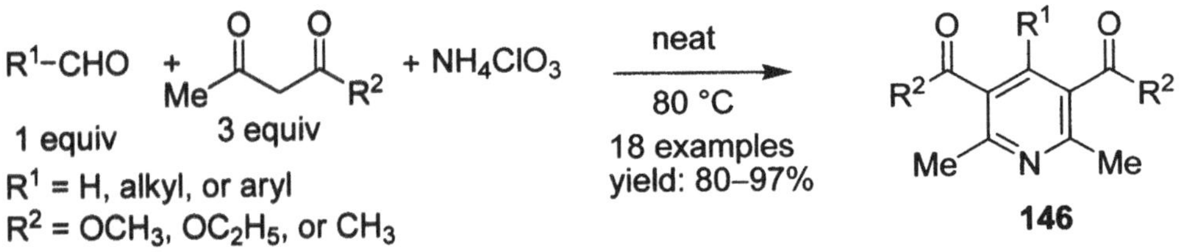

146

Many oxidation reagents[85] have been used for aromatization of the 1,4-dihydropyridines to the corresponding pyridines. They include nitric acid, oxygen, sodium nitrite, ferric nitrate/cupric nitrate, bromine/sodium acetate, chromium trioxide, sulfur, potassium permanganate, chloranil, DDQ, Pd/C, and DBU. More recently, ceric ammonium nitrate (CAN) has been found to be an efficient reagent to carry out this transformation.[100] In addition, the oxidation of the Hantzsch dihydropyridines can be accomplished with Claycop (montmorillonite K-10 clay supported cupric nitrate),[101] 9-phenyl-10-methylacridinium perchlorate,[102] activated-carbon with molecular oxygen,[103] manganese dioxide under microwave irradiation,[104] or iodoxybenzoic acid (IBX) in DMSO[105] to afford the corresponding pyridine derivatives in good to high yields.

The Hantzsch-type of pyridine synthesis was also used in the total synthesis and stereochemical assignment of the thiopeptide antibiotic micrococcin P1.[106]

A series of 1,4-dihydropyridines **149** bearing pyrazole derivatives was shown to have anti-tubercular activity against *Mycobacterium tuberculosis* $H_{37}Rv$.[107]

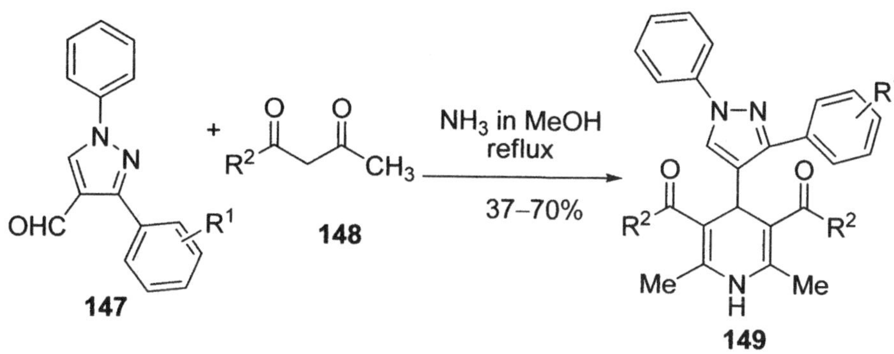

A potassium channel opener Z0947 (**152**) was used for treatment of urinary urge incontinence. The asymmetric synthesis[108] to support clinical trials involves the condensation of unsaturated ketone **150** and **151** in acetonitrile containing TMSCl. The product was not isolated but treated with aqueous ammonia and ammonium chloride. After workup, **152** was obtained in 95% *ee.*

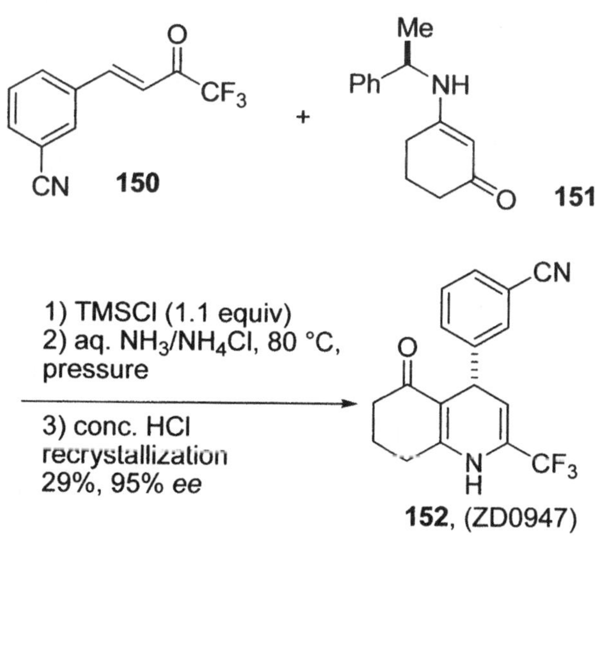

Guareschi–Thorpe pyridine synthesis

When using cyanoacetic esters instead of aldehydes, the Guareschi–Thorpe pyridine synthesis assembles pyridines **155** by the condensation of acetoacetic esters **153** with cyanoacetic esters **154** in the presence of ammonia.[109] A variation of this method involves the reaction of cyanoacetic ester **156** with β-diketone **157** in the presence of ammonia to generate 2-hydroxypyridine **158**.[110] The mechanism of this reaction has been studied, and it was initiated by an ester/amide exchange on cyanoacetic ester **156** with ammonia.[111]

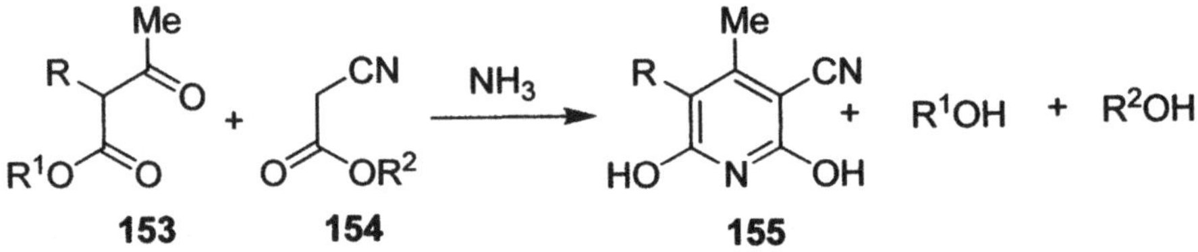

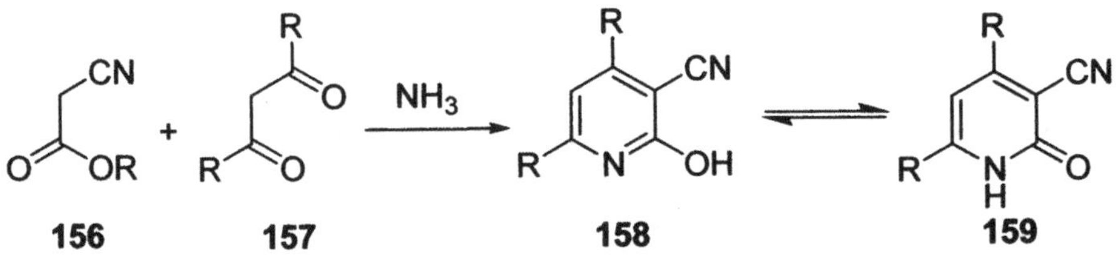

Ethionamide (2-ethylthioisonicotinamide, Trecator SC, **8**) is an antibiotic prodrug used in the treatment of tuberculosis. One synthetic pathway involves the condensation of diketo-ester **160** with cyanoacetamide **161** followed by hydrolysis of the resulting pyridone **162** into give pyridone acid **163**. Treatment of **163** with $POCl_3$ converts the lactam to imine chloride and simultaneous ester formation in ethanol to give **164**. Hydrogenation of **164** to remove the chloride, amide formation, and sequential conversion to the thioamide provided **8**.[112]

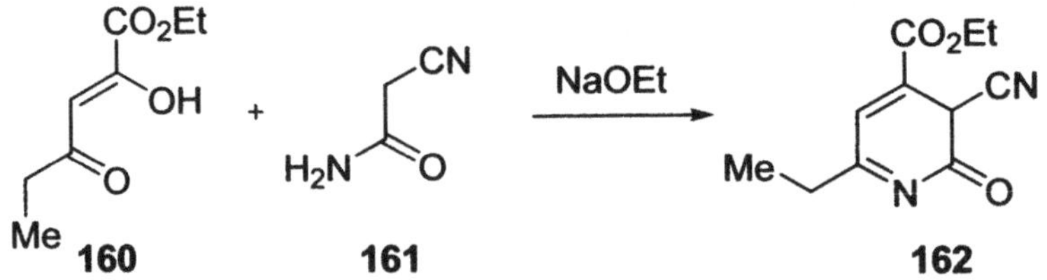

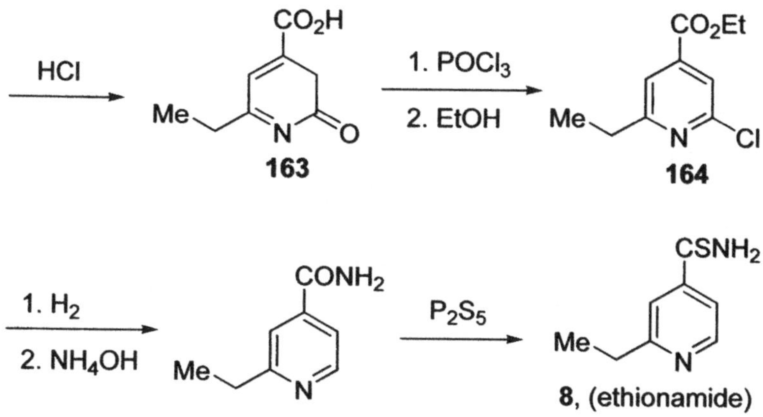

Chichibabin (Tschitschibabin) pyridine synthesis[113]

The classic Chichibabin reaction was carried out by passing vapors of aliphatic aldehyde **165** and ammonia over alumina at 300–400 °C to produce the corresponding 2,3,5-trisubstituted pyridine **166**. Ammonia serves as not only a base to catalyze an aldol reaction between two molecules of **165** but also as the source of nitrogen for the resultant pyridyl ring through the formation of an enamine with a third molecule of **165**.[114] AcOH/NH_4OAc was used to replace ammonia as the solvent/nitrogen source to improve the yield of this reaction dramatically and make the reaction easy to handle.[115]

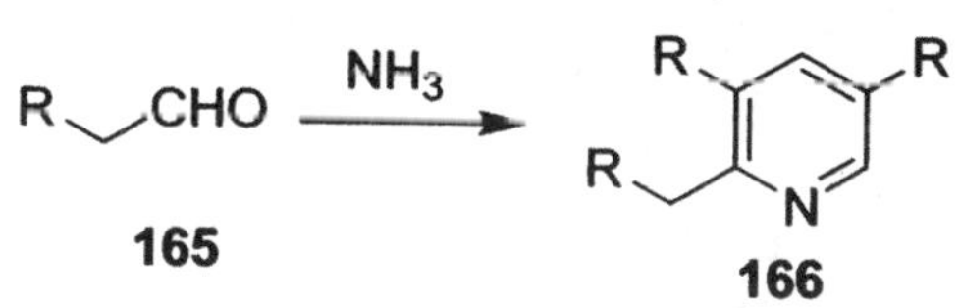

Snider reported synthesis of ficuseptine (**169**) and juliprosine (**172**) containing 2,3-dihydro-1*H*-indolizinium alkaloids *via* biomimetic intramolecular Chihibabin pyridine synthesis.[116] Two molecules of aldehyde **167** and one molecule of 4-aminobutanal dimethyl acetal **168** in acetic acid at 95 °C gave **169** in 52% yield. Meanwhile, two molecules of aldehyde **170** and one molecule of 1-pyrroline **171** in acetic acid at room temperature gave **172** in 72% yield.

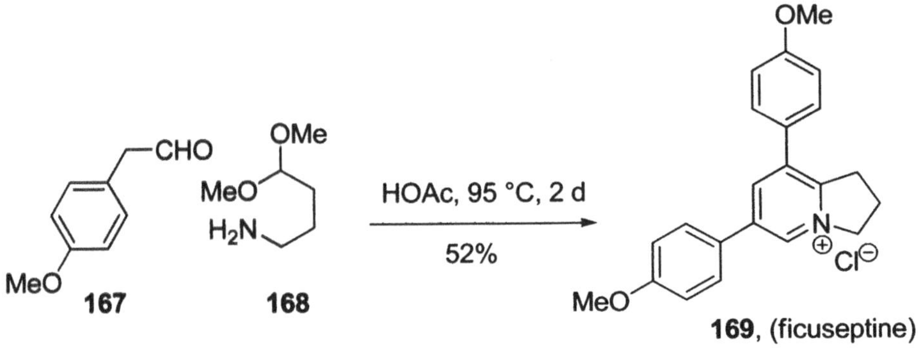

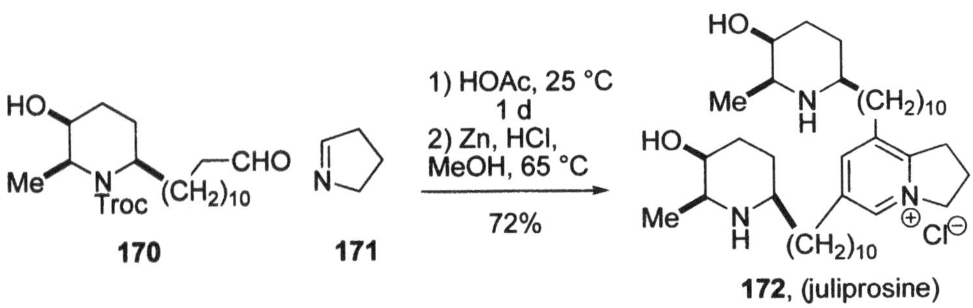

Recently, the Baran group discovered a mild "abnormal" Chichibabin pyridine synthesis using benzylammonium chloride as the amine source and $Yb(OTf)_3$ as the catalyst to form pyridinium **173** instead of **174** in their pursuit of biochemical origin of the haouamine alkaloids[117] and the total synthesis of haouamine alkaloids.[118]

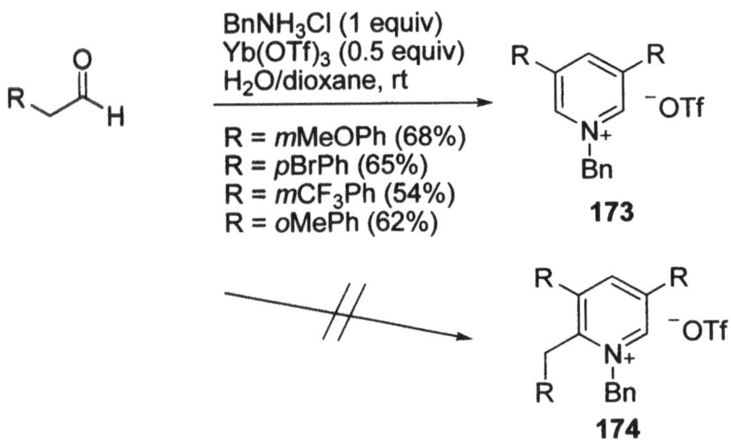

Bohlmann–Rahtz pyridine synthesis[119]

Bohlmann and Rahtz reported the preparation of 2,3,6-trisubstituted pyridines **178**[120] by Michael addition of acetylenic ketones (ynone precursors, **175**) with enamines (β-amino crotonates, **176**). The δ-aminoketones **177** can be isolated and subsequently heated at temperatures greater than 120 °C to facilitate the cyclodehydration to afford **178**.

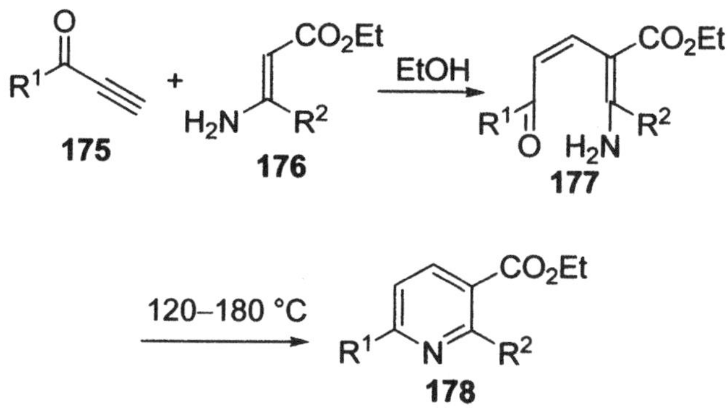

The Bagley group later developed a mild, single-step variant of the reaction, wherein acetic acid or Amberlyst 15 ion-exchange resin was used to promote cyclodehydration at a lower temperature (50 °C) to give 2,3,6-tri- and 2,3,4,6-tetra-substituted pyridines **181** with alkyl, aryl, heteroaromatic, heteroatom, and ester substituents in moderate to excellent yields.[121]

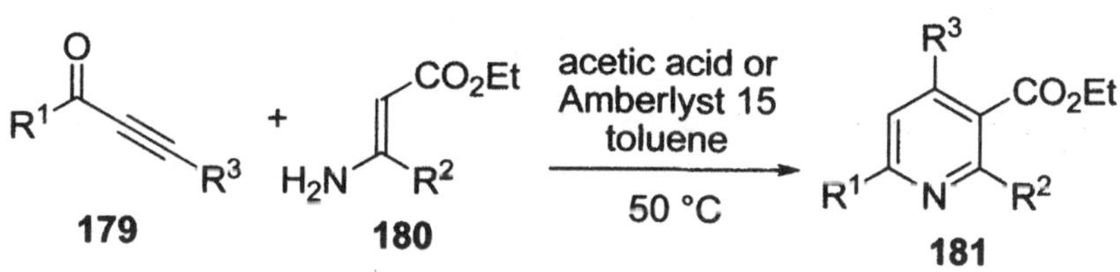

Alternatively, one could use Lewis acids[122] such as $ZnBr_2$, $Yb(OTf)_3$ or CuBr iodine in catalytic amounts, as well as catalytic amount of iodine.[123] Microwave irradiation was also applied in the reaction.[124]

The Bohlmann–Rahtz synthesis is viable for the construction of di- or tri-substituted pyridines and has been applied in the synthesis of complex target molecules. The Baldwin group[125] was one of the first to employ this method for the construction of pyridine substituted α-amino acids. Exposure of alkynyl ketone **182** to 3-aminocrotoyl ester **183** or ketone **184** in ethanol at reflux gave rise to the desired pyridyl-β-alanines **185** and **186**, respectively, which were subsequently deprotected to afford the α-amino acid L-azatyrosine analogues **187** and **188**, respectively.[125a]

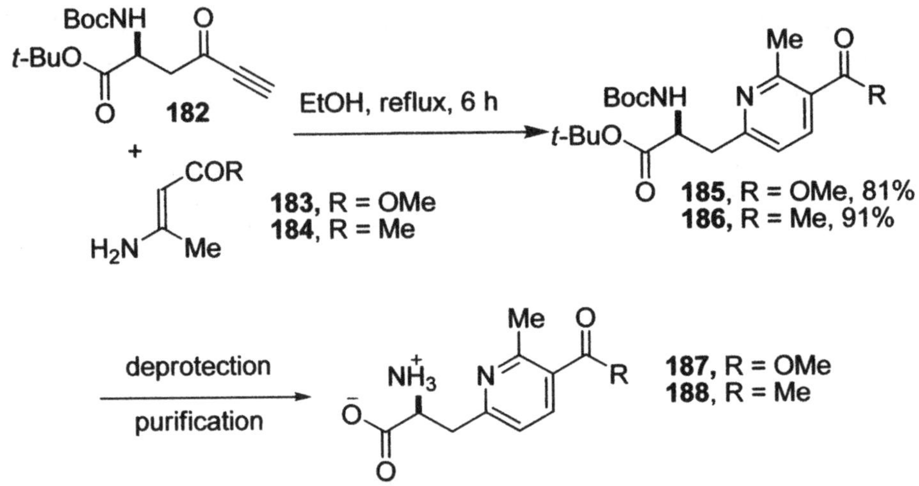

Thiopeptide antibiotics are a class of highly modified macrocyclic sulfur-containing peptides, and nearly all the thiopeptide antibiotics identified to date inhibit protein synthesis in bacteria. The Bohlmann–Rahtz pyridine synthesis is a useful methodology in the synthetic approach for thiopeptide antibiotics.[126–130] For example, one-step Bohlmann–Rahtz assembly of **189** and **190** in the presence of ammonium acetate in acetic acid at reflux afforded the corresponding pyridine-thiazole cores of thiopeptide antibiotics **191** in 63% yield.[132] The TBS protecting group was also replaced by an acetate, probably as a consequence of acid-promoted cleavage and consequent Fischer-type esterification of the liberated alcohol.

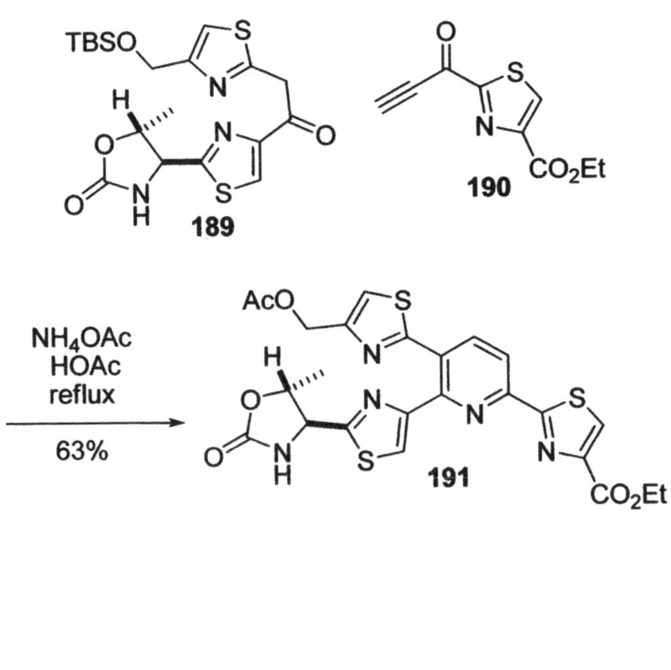

Bohlmann–Rahtz method has also been applied in the solution-phase synthesis of a library of functionalized pyridine scaffold[131] as well as the preparation of a 2,3′;6′3″-terpyridine scaffold as an α-helix mimetic.[132]

Kröhnke pyridine synthesis

Kröhnke annulation involves the condensation reaction of an α-pyridinium methyl ketone salt such as phenacylpyridinium bromide **191** to an enones such as benzalacetophenone **193** to afford 2,4,6-triphenylpyridine **194** in 90% yield.[133] Kröhnke found that glacial acetic acid and ammonium acetate were the ideal conditions to promote the desired Michael addition.

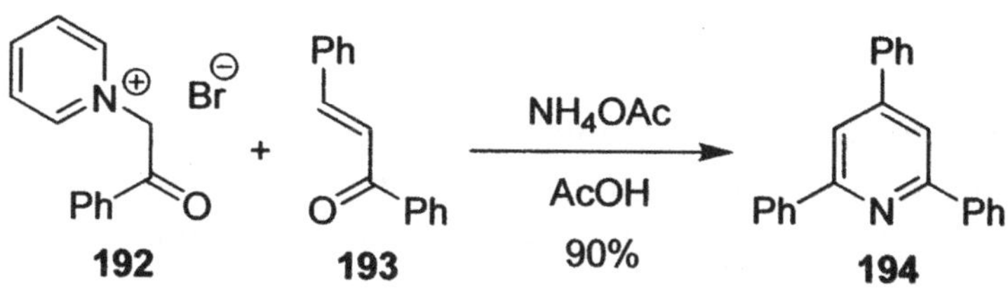

The Kröhnke synthesis was used as the key step in the preparation of several highly conjugated, chiral bridging ligands such as **198.**[134,135]

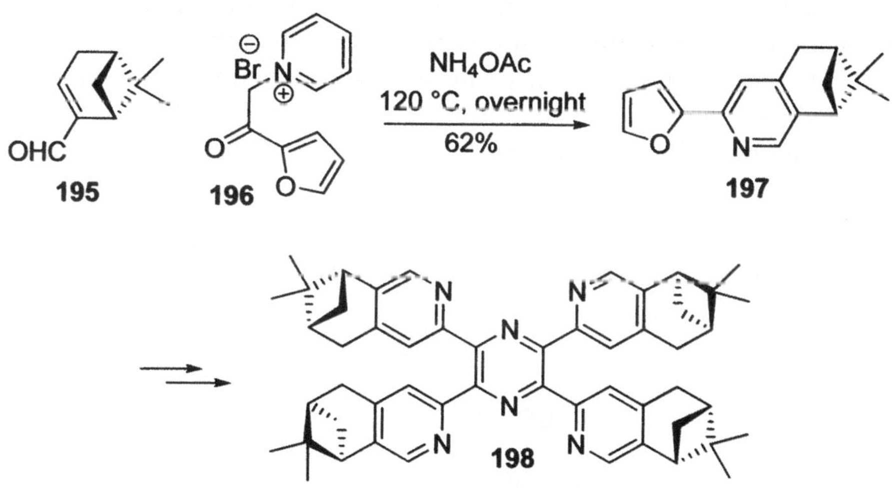

2,4,6-Tri-substituted pyridine derivatives **201** were prepared by Katritzky et al. using α-benzotriazolyl ketones **199** and α,β-unsaturated ketones **200** in the presence of ammonium acetate in refluxing acetic acid in good yields.[136] Fused 2,3,4,6-tetrasubstituted pyridines **203** were also formed from the appropriate fused bicyclic ketone substrates **202.**[137]

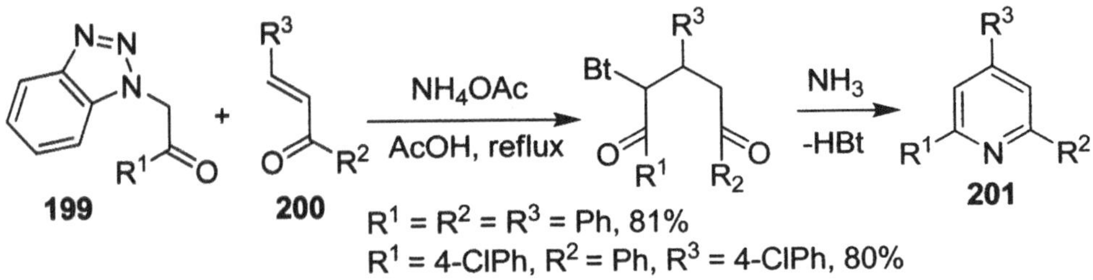

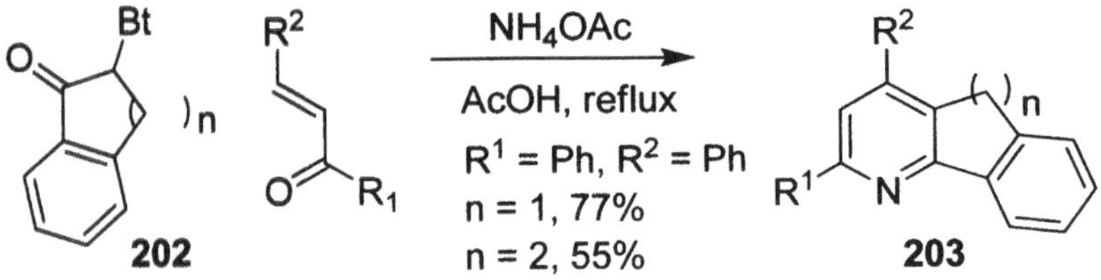

Intramolecular Kröhnke pyridine synthesis[138] as well as Kröhnke pyridine synthesis under microwave irradiation[139] have also been reported. In addition, indium trichloride can be used as a catalyst in the Kröhnke pyridine synthesis.[140]

Synthesis of anti-malarial reagent enpiroline (**17**) involved the formation of the pyridine skeleton **206** *via* the Kröhnke reaction of aroyl acrylic acid **204**, acyl pyridinium salt **205**, and ammonia acetate. Condensation of the carboxyl group in **206** with 2-lithiopyridine afforded the diaryl ketone **207**. The relative higher basicity of the terminal pyridinyl ring allowed selective reduction of this ring *via* hydrogenation in the presence of an acid and simultaneously reduction of the ketone to the alcohol. The desired isomer was then obtained by fractional crystallization to afford **17**.[141]

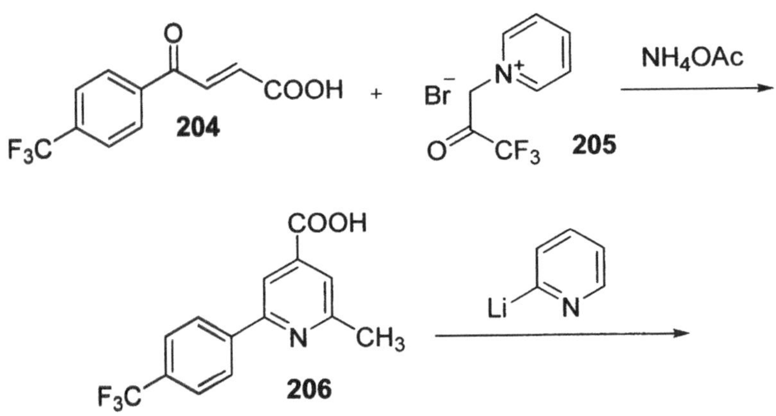

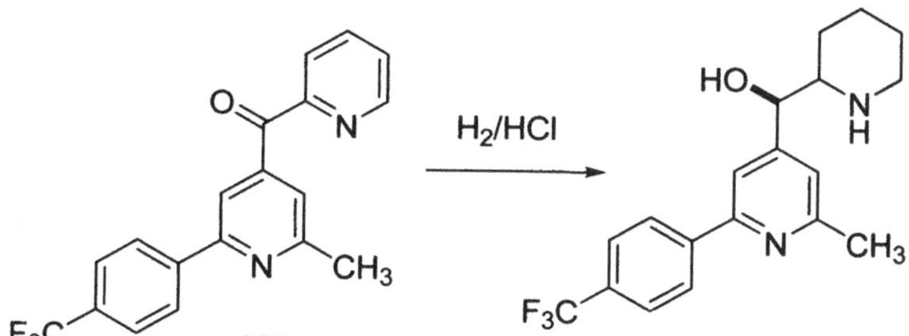

207

17, (enpiroline)

10.3.2 Synthesis via Cycloaddition Reactions

[4 + 2]-Cycloaddition reactions

Pyridines can be achieved by the [4 + 2] hetero Diels–Alder cycloaddition of i) an alkene dienophile and an 3-azadiene such as 1,2,4-triazine (the Boger reaction), oxazole (Kondrat'eva pyridine synthesis), oxazinone, pyrimidine, or ii) an alkene and a 4-azadiene; or iii) a butadiene and an azadienophile.

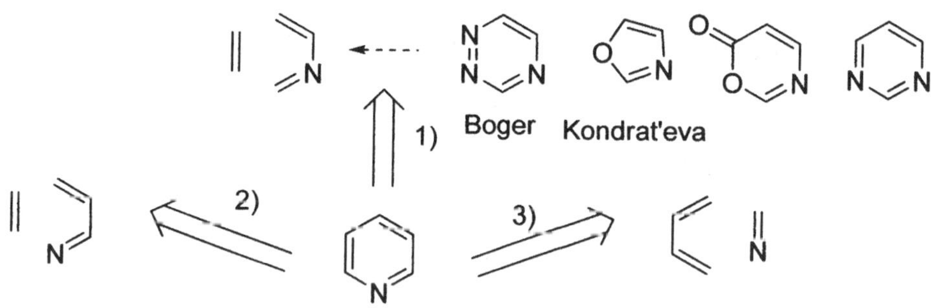

1,2,4-Triazine as the azadiene (Boger reaction)

The Boger pyridine synthesis[142] involves an inverse-electron demand[143] hetero-Diels–Alder [4 + 2] reaction of the electron-deficient 1,2,4-triazine **208** with an electron-rich dienophile such as enamine **209** followed by a facile retro-Diels–Alder of the unstable intermediate bicyclic species **210** to liberate nitrogen gas and aromatization via loss of pyrrolidine to afford the appropriately substituted pyridine derivative **211**. The pyrrolidine enamine **209** can be generated *in situ* from the corresponding ketone with pyrrolidine. There is a strong preference for the nucleophilic carbon of the dienophile to add to C-3 of the triazine.

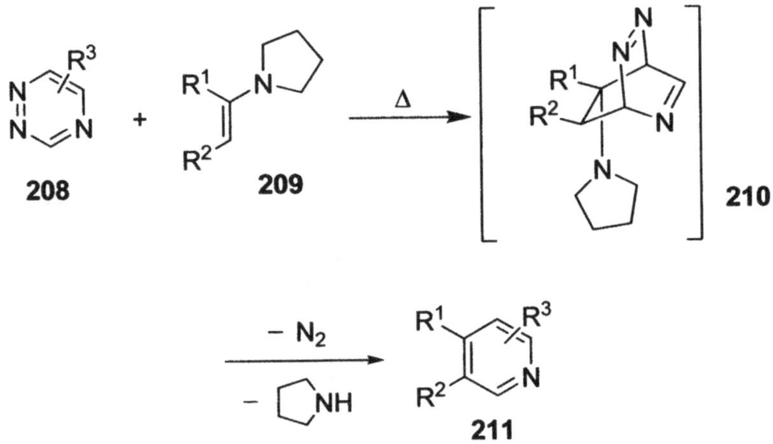

Boger initially developed the aforementioned methodology to construct penta-substituted pyridine **216** in the formal total synthesis of antitumor antibiotic streptonigrin **217**.[144] The requisite 1,2,4-triazene **214** was generated *via* a Diels–Alder/retro-Diels–Alder sequence between nitrogen-containing dienophile **212** and 1,2,4,5-tartazine **213**. Exposure of enamine **214** to **215** resulted in the formation of **216**. Similar chemistry was also applied in the synthesis of the structurally related lavendamycin **217**.[145]

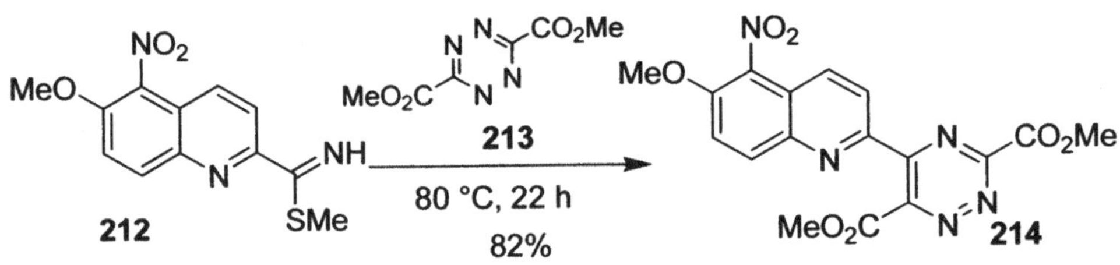

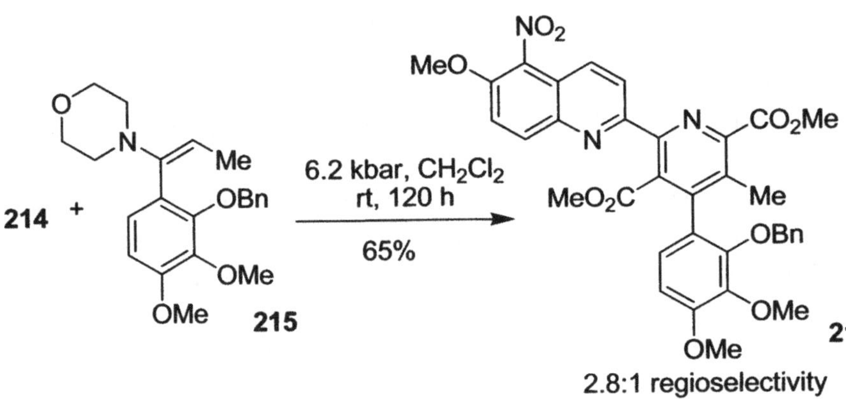

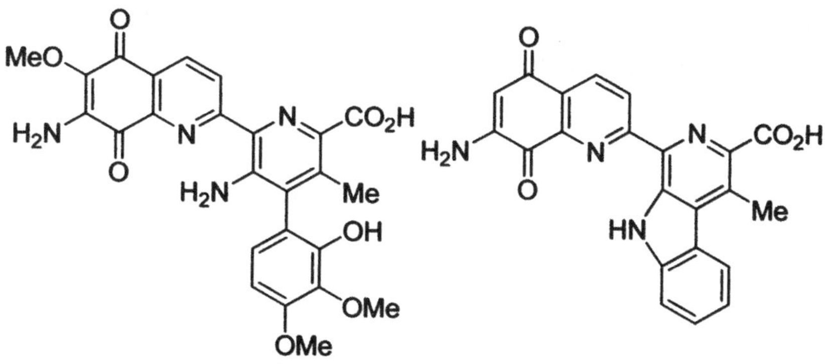

217, (streptonigrin) **218**, (lavendamycin)

Several examples of intramolecular Boger reactions toward the synthesis of pyridine-containing heterocyclic systems were reported by the Taylor group[146] and the Snyder group.[147] For instance, intramolecular cyclization of triazine **219**, after loss of nitrogen, afforded **220**. Alternatively, triazine **221** generated bicyclic systems **222**, which was then oxidized to **223**.[147]

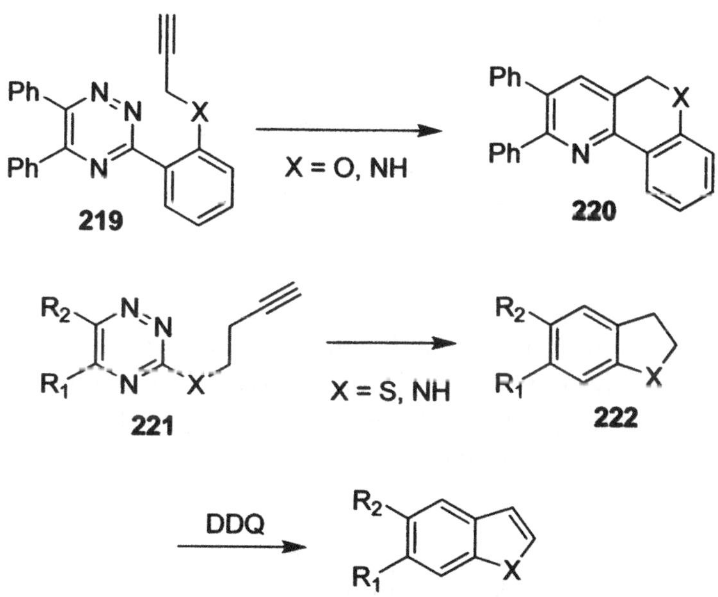

Oxazoles as azadienes (Kondrat'eva pyridine synthesis)

Kondrat'eva reported the first example of a cycloaddition of oxazoles **224** and alkenes to afford pyridine derivatives **225**.[148] However, the reaction is very sensitive to oxazole substitutions, and only certain substituted oxazoles undergo the cycloaddition reaction.

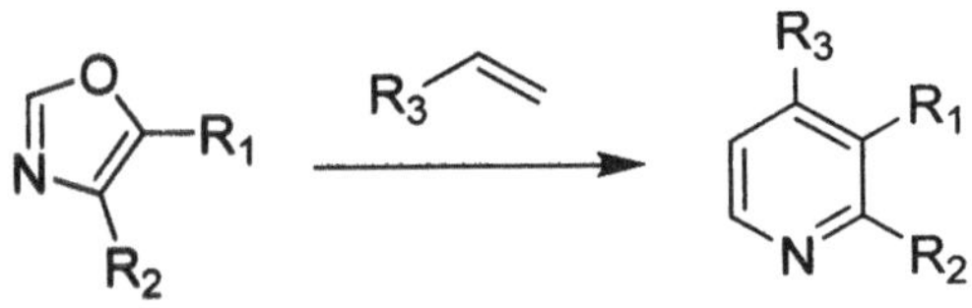

224 225

One approach to the anti-tumor compound ellipticine (228)[149] used the Kondrat'eva pyridine methodology to form pyridinyl derivative 227. Addition of methyllithium and hydrolysis of 227 then led to 228.

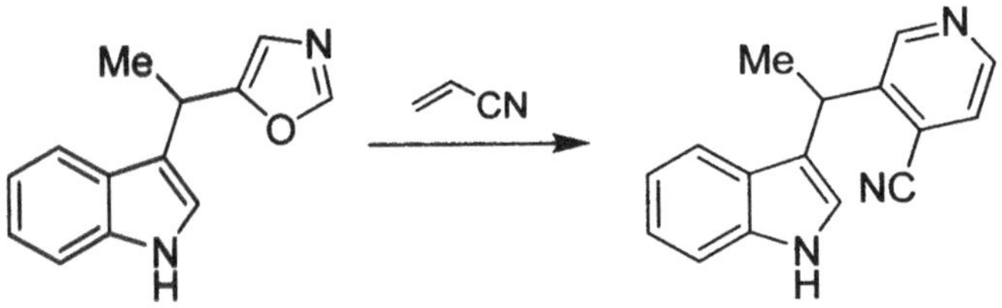

226 227

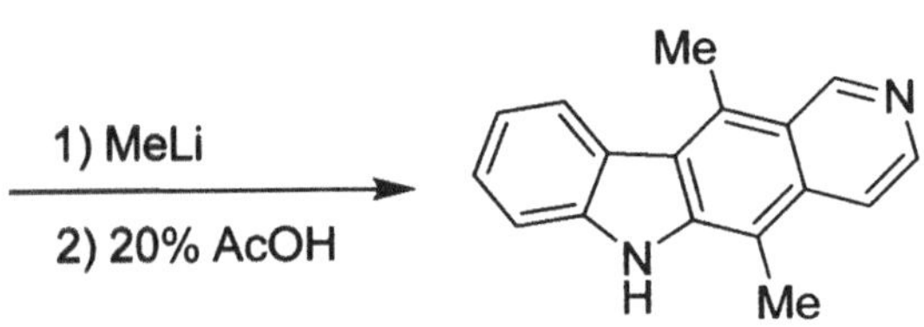

228

Cycloaddition of oxazole 229 with maleic anhydride initially gave the oxabicyclooctane system 230. Upon exposure to acidic ethanol, 230 fragmented to afford pyridine 231. Reduction of the ester 231 with $LiAlH_4$ generated vitamin B_6, pyridoxine (3).[150]

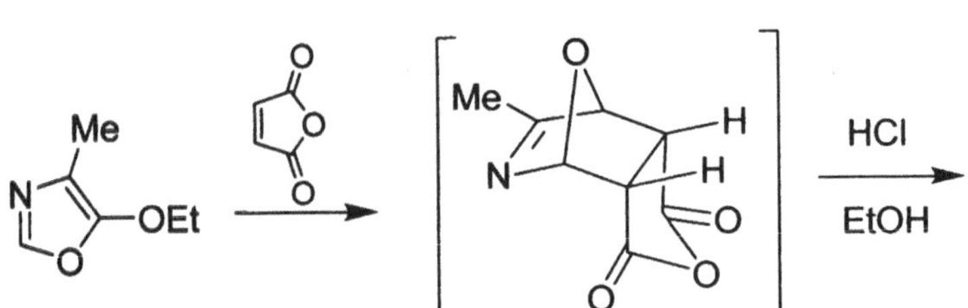

229 230

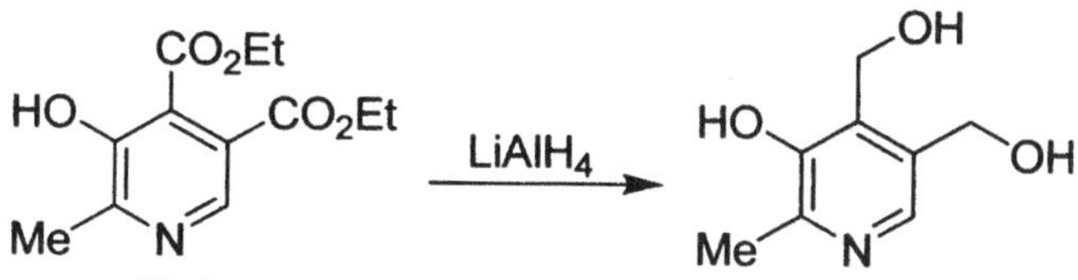

231 3

Intramolecular Kondrat'eva synthesis was also developed in the synthesis of several pyridine-containing natural products.[151,152]

Oxazinones and pyrimidines as dienes in the [4+2]-cycloaddition reactions

In an approach to the AB rings of rubrolone (**235**), Boger[153] used oxazinones as the dienes to replace triazines. Reaction of 1,3-oxazin-6-one **232** with enamines **233** produced the corresponding adduct. The adduct lost CO_2, and the pyrrolidine aromatized to give pyridine analogue **234**.

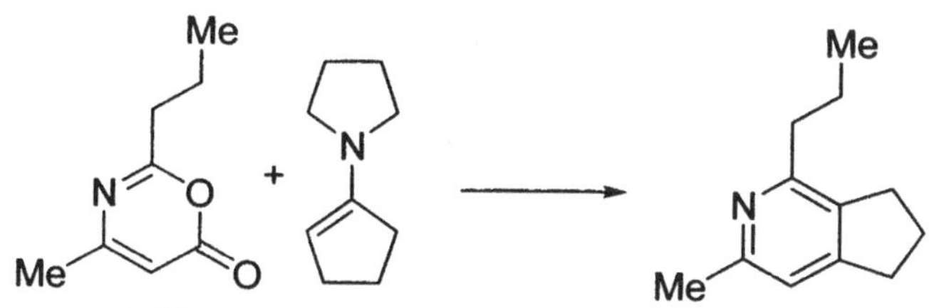

232 233 234

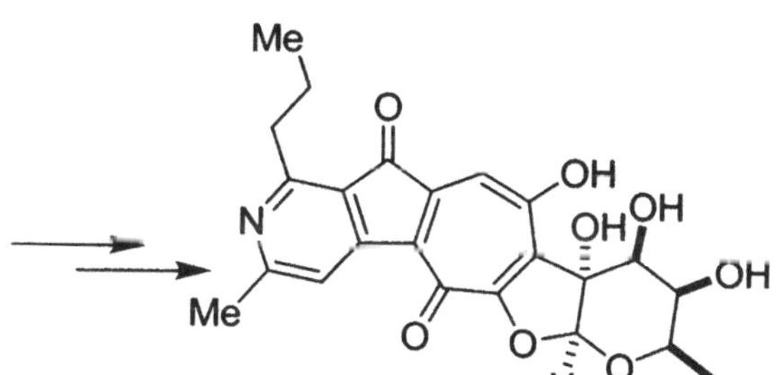

235

Pyrimidines such as **236**, **238**, and **240** can also undergo DA/retro-DA sequence. The resulting regiochemistry of the resulting pyridines **237**, **239**, and **241** is dependent upon the dienophile and the substitution pattern of the parent pyrimidines.[154]

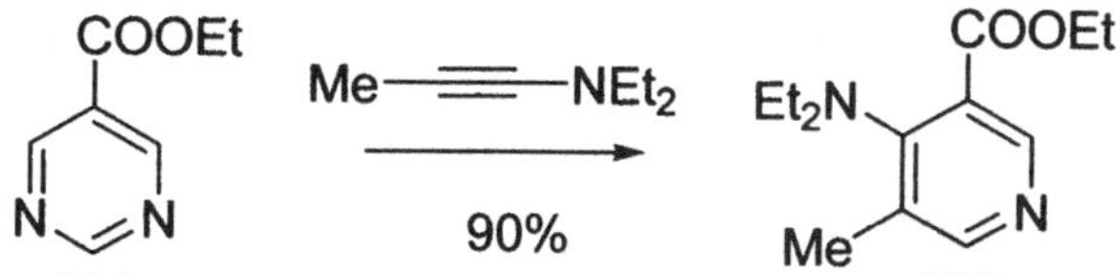

236 237

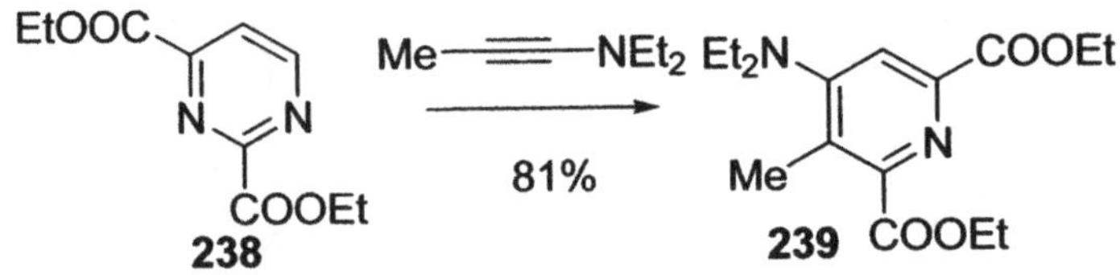

238 239

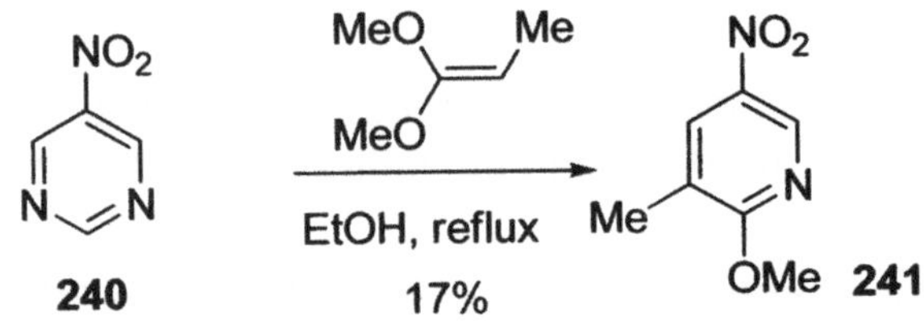

240 241

Pyrimidines with two or more complementary electron-donating groups, e.g., **242**, are capable of undergoing normal DA reaction with activated dienophiles although the yields are often only moderate.[154b]

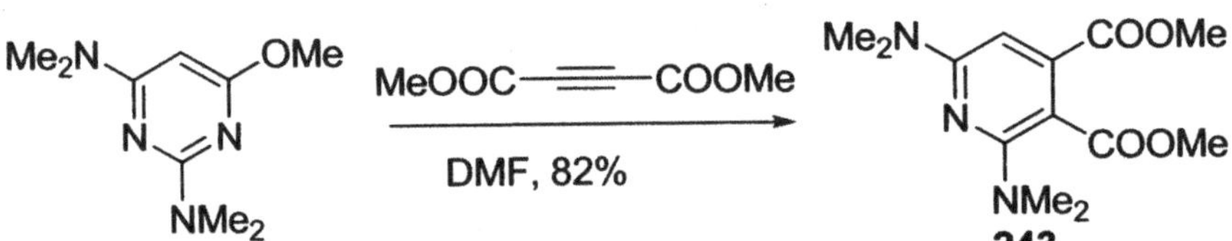

242 243

1-Azadienes as dienes in the [4+2]-cycloaddition reactions

In the total syntheses of fredericamycin **244**[155] and camptothecin **245**,[156] *N*-sulfonyl-1-aza-1,3-butadienes **246** and **249** reacted with electron-rich dienophiles **247** and **250** in an inverse-electron demand Diels–Alder reaction to afford pyridines **248** and **251** after treatment with base (DBU or NaOEt), respectively.

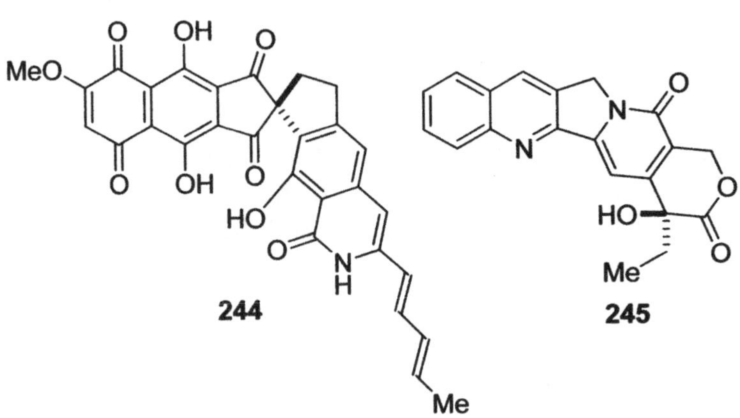

244 245

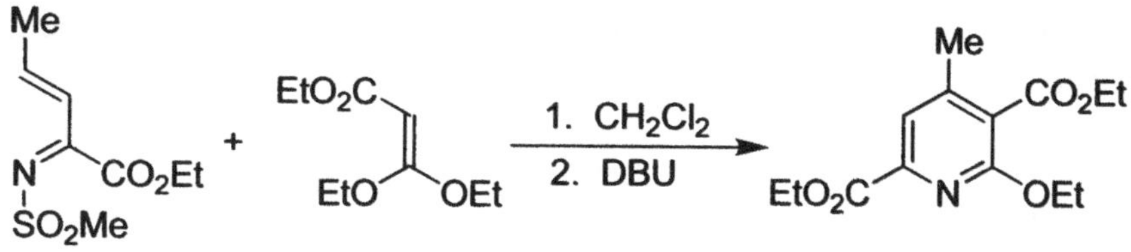

246 247 248

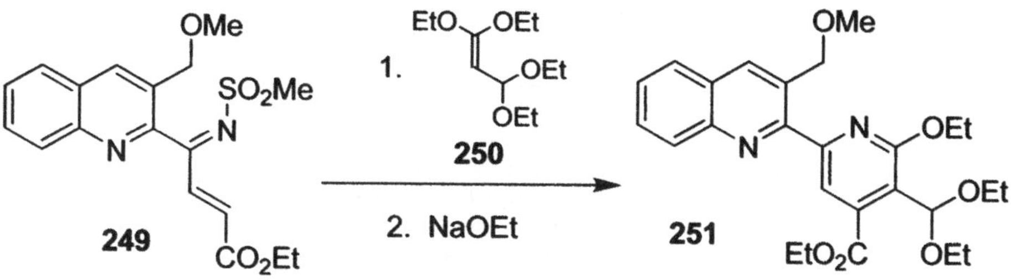

249 250 251

Ultrasound irradiation facilitated the Diels–Alder reaction of dimethylhydrazone **252** with acetylene **253** to afford pyridine **254** in shorter reaction time and increased yields in comparison to the conventional heating conditions.[157]

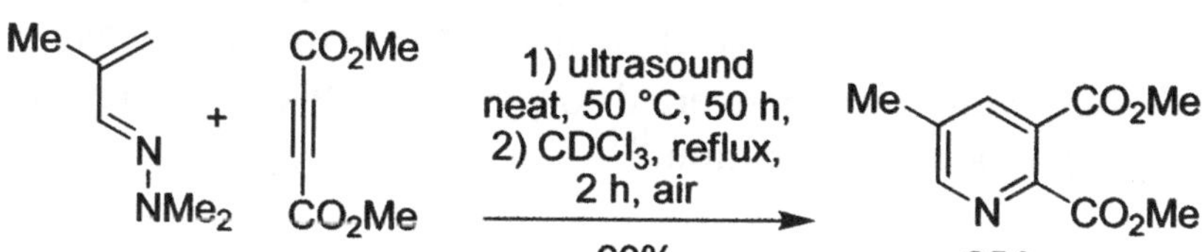

252 253 254

Heating **255** led to double intramolecular Diels–Alder reaction followed by aromatization to afford 2,2′-bipyridine **256**.[158]

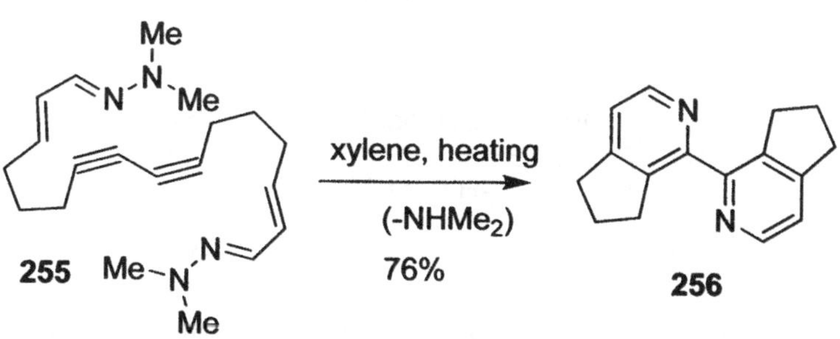

255 256

1,2,3-Triazene **257** can also serve as the 1-azadiene to undergo inverse-electron demand Diels–Alder reactions with alkynes or enamines such as **259** to give substituted pyridines **258** or **260**. Anderson and Boger[159] recently reported a systematic study of this reaction, demonstrating that the reactivity of the cycloaddition can be modulated by the C5 substituent (e.g., R = CO_2Me > Ph > H).

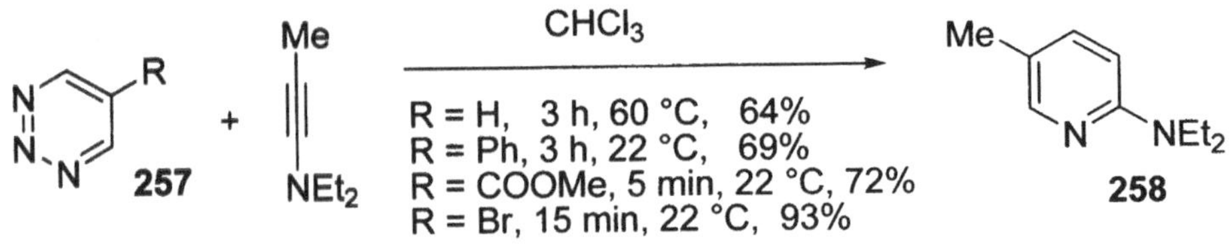

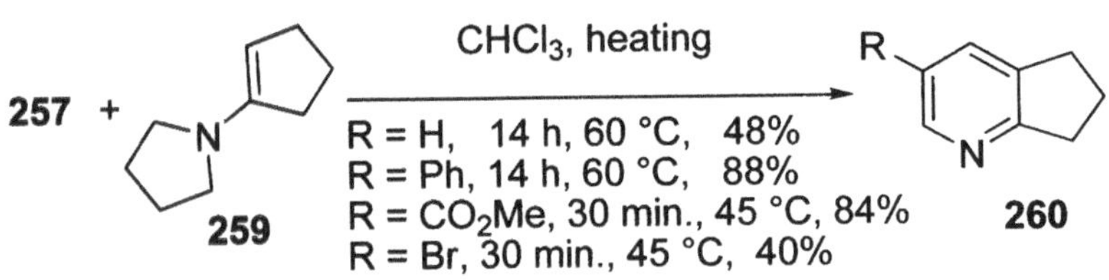

Azadienophiles and dienes in the [4+2]-cycloaddition reaction

Pyridines can also be constructed by the Diels–Alder reaction of azadienophiles, such as nitriles or imines, and dienes.[160] Imines usually need to be activated with Lewis acids such as $Yb(OTf)_3$, $ZnCl_2$, and Et_2AlCl.

Cycloaddition of oximino derivatives **261** with dienes **262** afforded the adduct **263**, which was converted to pyridines **264** by either heating in EtOH or base-promoted elimination.[161]

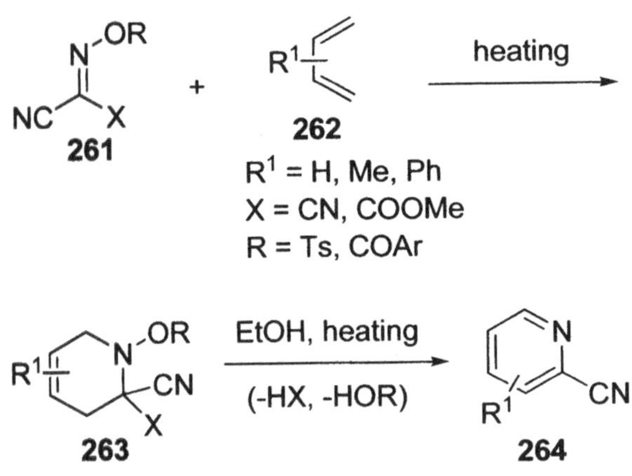

On the other hand, the Diels–Alder reaction of dienes **265** with oximinosulfonates **266** (derived from Meldrum's acid) in the presence of dimethylaluminum chloride afforded the cycloaddition adduct, which was then reacted with alkoxide and oxidizing agent (NCS) to give the desired pyridines **267**.[162]

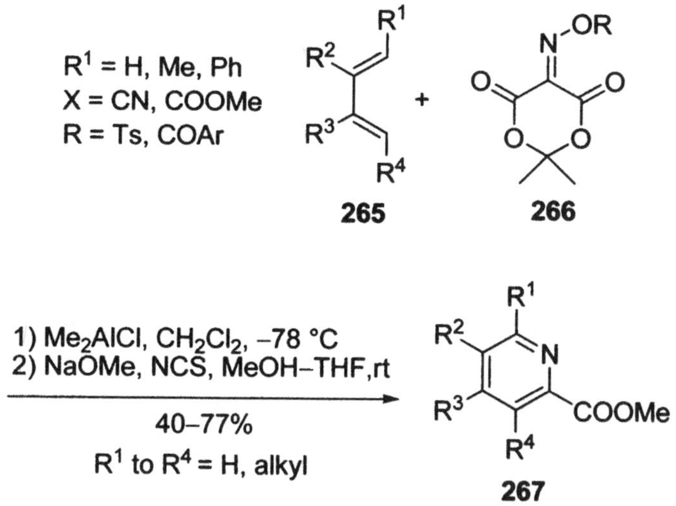

Weinreb[163] reported the intramolecular Diels–Alder reaction of oximino malonates tethered to a diene **268** under high dilution conditions. The cycloaddition adduct **269** underwent double eliminations with cesium carbonate to afford the pyridine **270**.

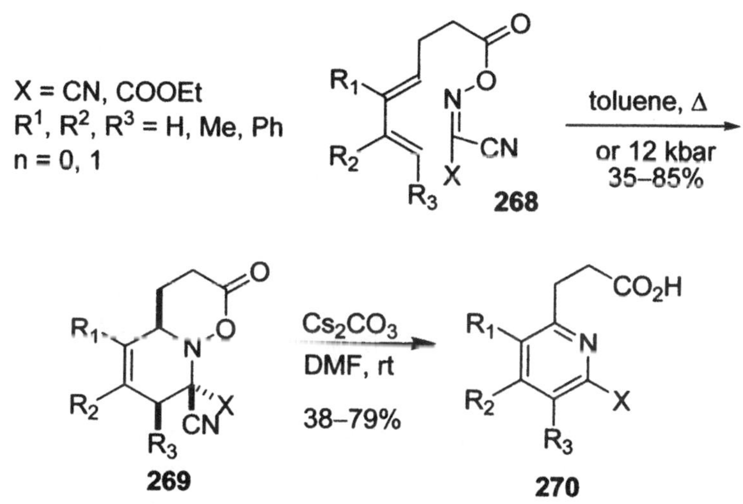

[2+2+2]-Cycloaddition reactions

Transition metal-catalyzed [2 + 2 + 2]-cycloaddition reaction of two alkynes with a nitrile is an atom-economical and powerful method to synthesize versatile and highly substituted pyridines.[164]

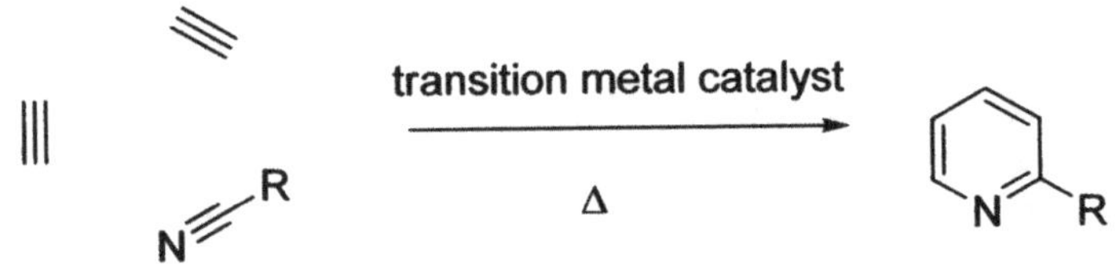

A number of transition metal complexes has been developed for the [2 + 2 + 2]-cycloaddition reaction, among which cobalt-based catalysts are most widely used. In 1973, Yamazaki and Wakatsuki reported the first synthesis of pyridines using stoichiometric and later on, catalytic cobaltacyclopentadienes.[165]

Irradiation of the reaction mixture with UV-Vis light (350–500 nm), or sunlight enabled the reaction to be carried out at room temperature.[166] The photochemistry process provided improved chemoselectivity by avoiding the homocyclotrimerization for the alkynes.[167] For example, photo-catalyzed [2 + 2 + 2]-cycloaddition of various nitriles with acetylene at room temperature in the presence of $CpCo(COD)_2$ provided the corresponding 2-pyridines such as **271** in 3–4 h in good to excellent yields.[168]

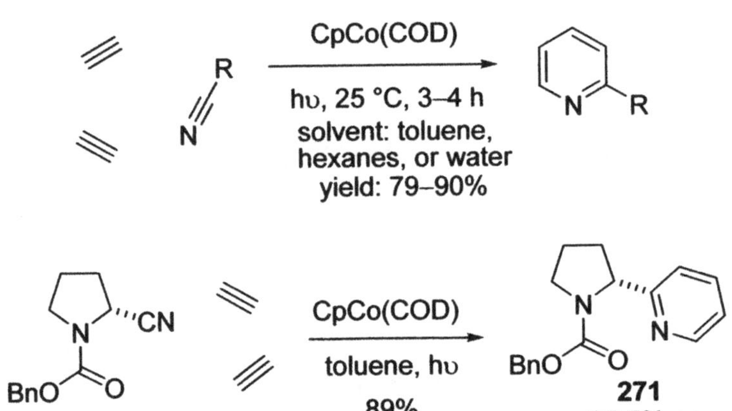

271
> 99.5% *ee*

The limitation of the reaction is obvious. When terminal alkynes reacted with nitrile, they give a mixture of 2,4,6-trisubstituted and 2,3,6-trisubstituted pyridines, with the former being the dominant product. The electron density of the metal affected the yields and regioselectivity of the reaction.[164b,169,170] In general, two approaches are employed to control regioselectivity and to enhance reactivity by tethering two of the three reaction components: i) reaction of diakynes such as α,ω-diynes, cyanoalkynes and nitriles or ii) reaction of nitrilediynes with alkynes, thus, providing bicyclic or multicyclic pyridines.

Besides Co,[171] a number of transition metal complexes such as Ru,[172] Rh,[173] Ni,[174] Ti,[175] Zr/Ni[176], and Fe[177] catalyst systems, have also been developed over the past few decades. For example, an iron-catalyzed [2 + 2 + 2]-cycloaddition of diynes **272** and unactivated nitriles **273** leading to pyridine compounds **274** at room temperature was reported recently.[177d] The catalyst is generated *in situ* from an inorganic iron salt such as FeI_2 and a diphosphine ligand such as 1,3-bis(diphenylphosphino)propane (dppp). The reaction exhibited high reactivity and regioselectivity with the optimized reaction conditions.

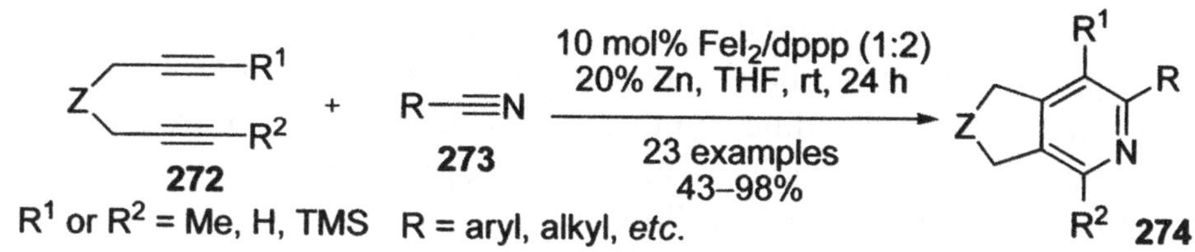

Unlike tethered diynes that are commonly seen in [2 + 2 + 2]-pyridine syntheses, nitrile-diyne substrates are less explored. Recently, Snyder and coworkers reported microwave promoted cobalt-catalyzed [2 + 2 + 2]-reaction of nitrile-diynes **275** to afford tetrahydronaphthyridines **276** in moderate to excellent yields.[178]

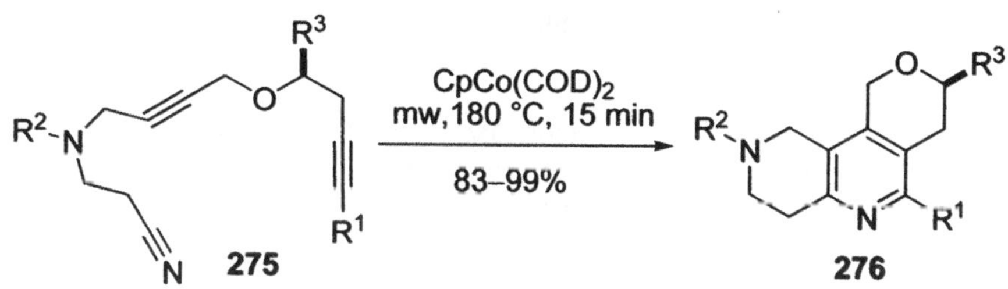

R^1 = H, Me, Et, *n*Bu, CH_2OPh, CH_2OH, TMS, Ph
R^2 = H, Me, *i*Bu, *i*Pr, CH_2CH_2OPh
R^3 = Me, $CH_2OC_6H_4OPh$

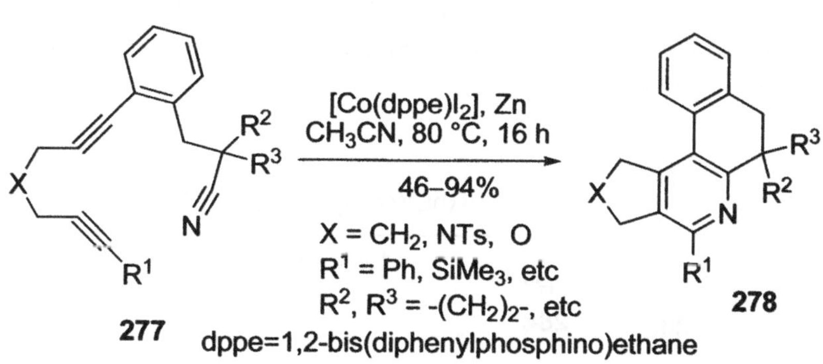

Chang and co-workers developed an intramolecular cobalt-catalyzed cyclotrimerization of both the symmetric and the asymmetric nitrile-diyne substrates to afford fused-polycyclic pyridines. A bulky substituent was required at the terminal alkyne carbon of **277** for the reaction to proceed in moderate to excellent yields to deliver pyridine derivatives **278**.[179]

Recent advances in pyridine synthesis with organomethallics

Significant amount of research work on transition metal-catalyzed pyridine synthesis have been development recently, as evidenced by recent reviews.[180] Herein, we only present several representative examples.

Application of carbon–carbon cross-coupling/6π-electrocyclization cascade reactions in pyridine synthesis is well known. Larock and co-workers developed a palladium-catalyzed coupling of vinylic imines **279** with terminal alkynes **280** followed by subsequent copper-catalyzed cyclization of **281** to give aryl, vinyl, and alkyl-substituted pyridines **282** in moderate yields.[181]

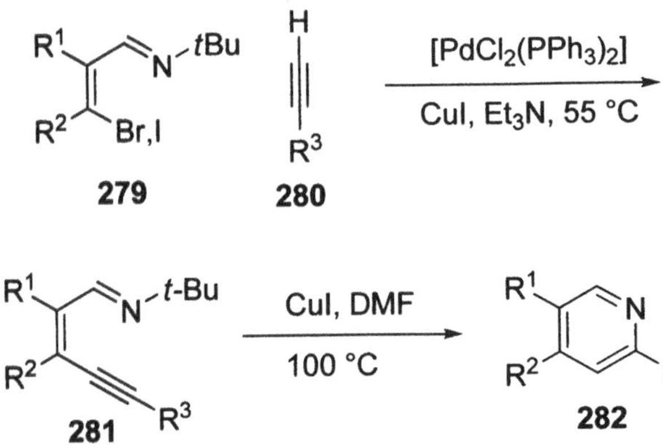

279 **280** **281** **282**

Ellman and co-workers developed a rhodium-catalyzed C–H alkenylation followed by 6π-electrocyclization to give di-, tri-, tetra-, and penta-substituted pyridines. The dihyropyridine intermediates **285** can be directly aromatized *via* hydrogenolysis to pyridine derivatives **286** in moderate to good yields.[182]

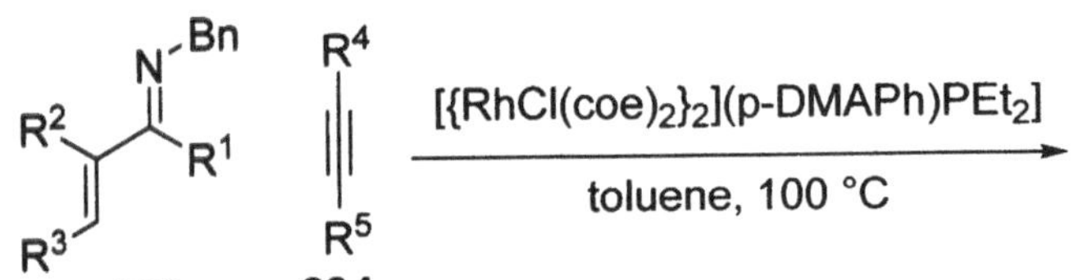

283 **284**

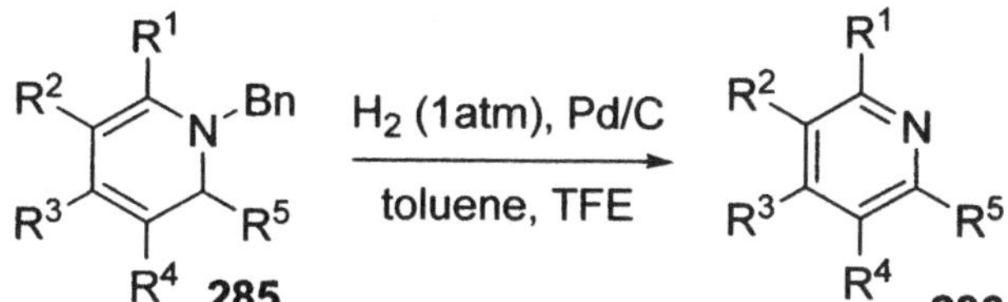

285 286

Cheng and co-workers reported a one-pot synthesis of substituted pyridines through a rhodium-catalyzed C–H alkenylation of α,β-unsaturated ketoximes **287** with symmetrical alkyne substrates **288**. 6π-Electrocyclization of the azatriene intermediates **289** and subsequent loss of water afforded the desired pyridines **290** in moderate to good yields.[183]

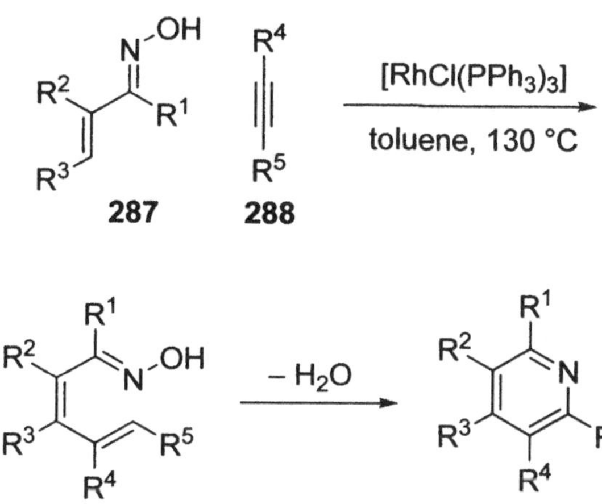

287 288 289 290

Liebeskind reported a copper-catalyzed Chan–Lam C–N cross-coupling methodology for *N*-imination of boronic acids by using oxime *O*-carboxylates **291** as iminating agents and either Cu(I)-thiophene-2-carboxylate (CuTC) or $Cu(OAc)_2$ as the catalyst under nonbasic and nonoxidizing conditions.[184] Subsequently, the *N*-alkenylated α,β-unsaturated ketoxime *O*-pentafluorobenzoates **293**, were precursors in a cascade reaction for the one-pot synthesis of tri-, tetra-, and penta-substituted pyridines **295** in moderate to excellent isolated yields.[185]

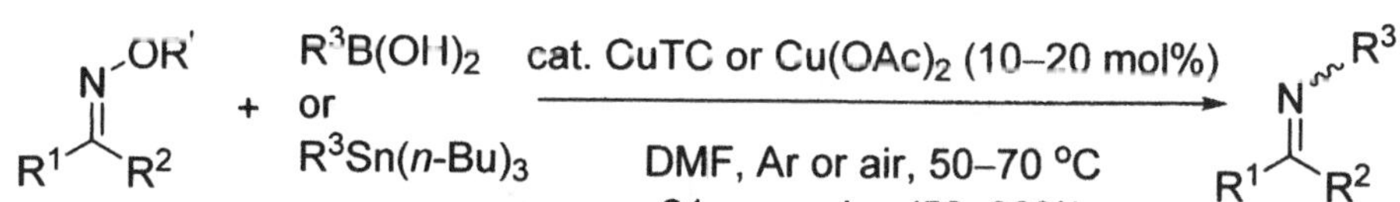

291 292

R' = Ac, COC_6F_5

R^1, R^2 = aryl, heteroaryl, alkyl

R^3 = aryl and alkenyl for B; aryl, heteroaryl, and alkenyl for Sn

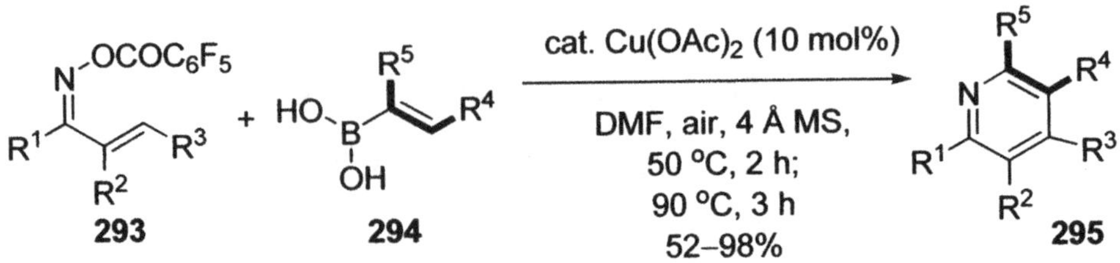

10.3.3 Synthesis via Rearrangement Reactions

Certain substituted pyridines can be obtained *via* rearrangement of pyridine *N*-oxides (Boekelheide reaction) or *via* rearrangement of an alternative heterocycle system, i.e., pyrrole (Ciamician–Dennstedt reaction).

Boekelheide reaction

The Boekelheide reaction[186–190] involves treating pyridine *N*-oxides **296** with acylating agents to afford rearranged products **297**. Traditionally, the rearrangement occurs at the α-position, but oxygen migration could also occur on the alkyl chain of substituents attached to the γ-position. This process requires the activation of pyridine *N*-oxides by electrophilic agents such as Ac_2O (in the original example), P_2O_5, TFAA, or TsCl.

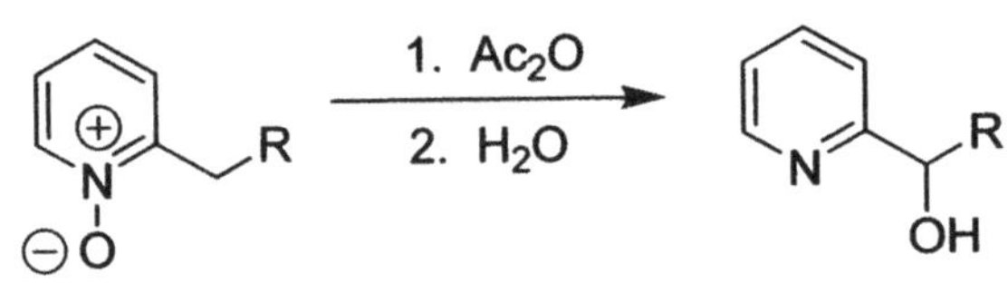

296 297

Many research laboratories looked at the mechanism of the Boekelheide rearrangement over the years.[191–199] Although still not completely understood, the ion-pair mechanism is now the generally accepted explanation for this reaction.[199b,d] And when R is a neopentyl group (**298**), the reaction gave rise to the corresponding alkene **301**, further confirming the ion-pair mechanism.

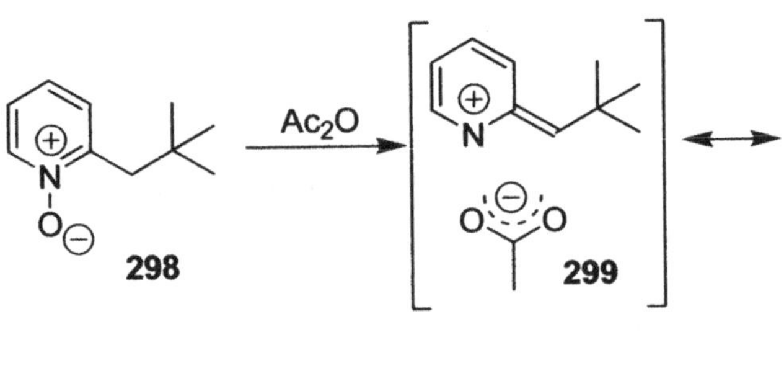

298 299

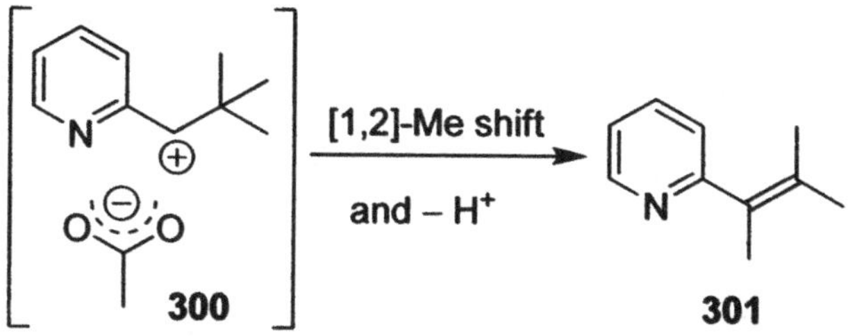

The Boekelheide reaction was applied by the Nicolaou group[200] in the synthesis of a model system of the thiopeptide antibiotic thiostrepton (**302**). The tetrahydroquinoline **303** was converted into the *N*-oxide by *m*-CPBA oxidation followed by treatment with TFAA and then hydrolysis to afford key intermediate alcohol **304** as a diastereomeric mixture.

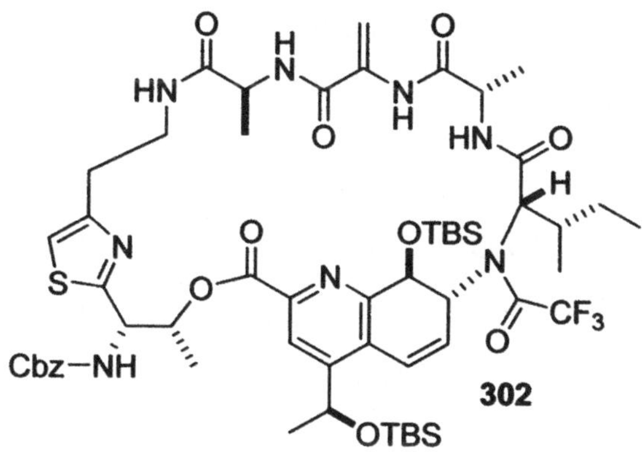

302

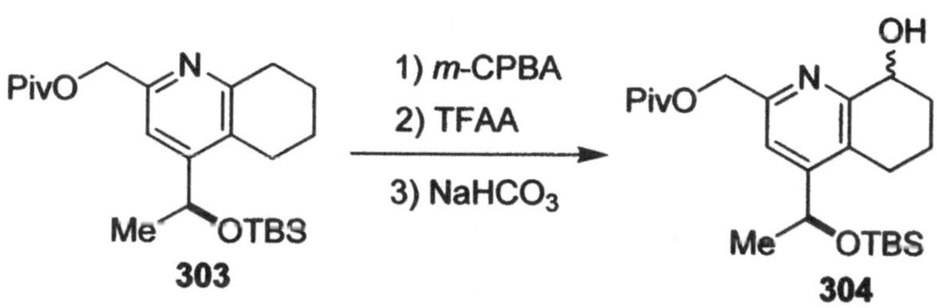

A Fe(II)-binding agent pyrimidine **307** was synthesized starting from the bis-homophenylalanine **305**[201] *via* the Boekelheide reaction with TFAA and hydrolysis to yield the advanced intermediate **306**.

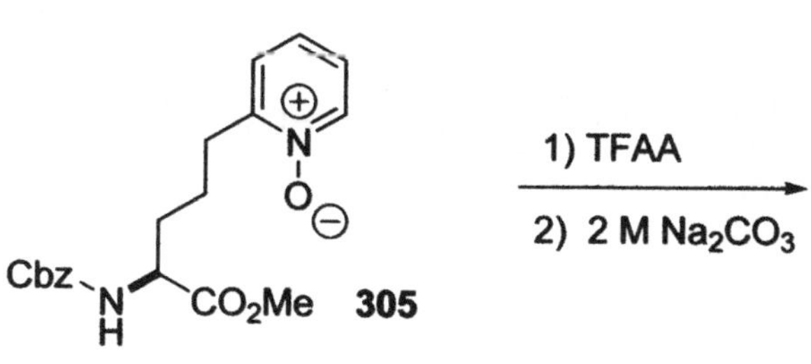

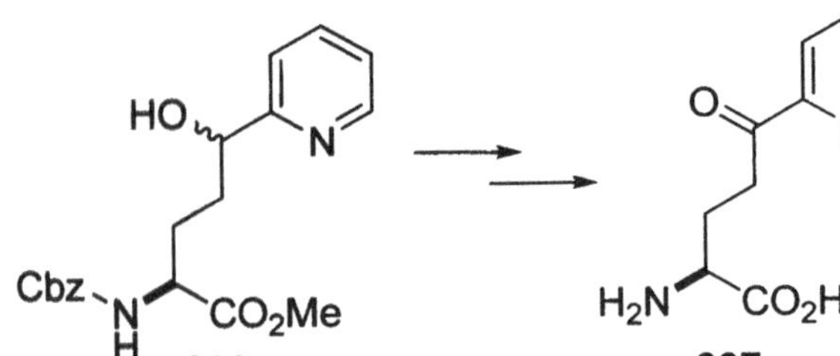

306 307

Ticolubant (**310**) is a leukotriene receptor antagonist that exhibits anti-inflammatory activities. The Wittig reaction with 3-hydroxy-6-methylpicolinaldehyde followed by a Mitsunobu reaction with 2-phenylethanol gave phenyl ether **308**. Oxidation with *m*CPBA converted the pyridine ring into the corresponding *N*-oxide, which then underwent the Boekelheide reaction with TFAA to afford the methyl alcohol **309**. Reaction with thionyl chloride and then 2,5-dichlorothiophenol followed by saponification gave **310**.[202]

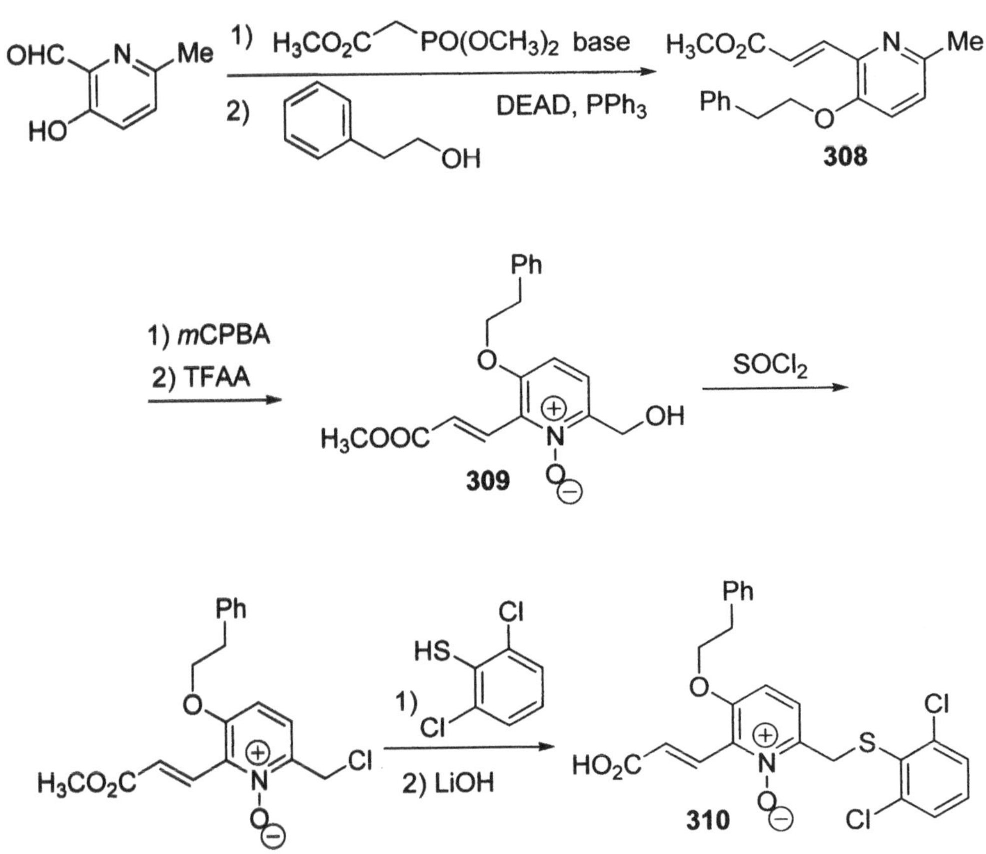

308 309 310

When the *N*-alkoxypyridinyl compound is subjected to cyanide, it undergoes the Reissert–Henze reaction[203] to afford the 2-cyano-derivative with loss of the alkoxy group. Ethylation of *N*-oxide **311** with diethylsulfate afforded intermediate **312**, which was treated with cyanide to afford the corresponding cyano-pyridine **313**, an intermediate used in the synthesis of *S*-adenosylmethionine decarboxylase inhibitor **314**.[204]

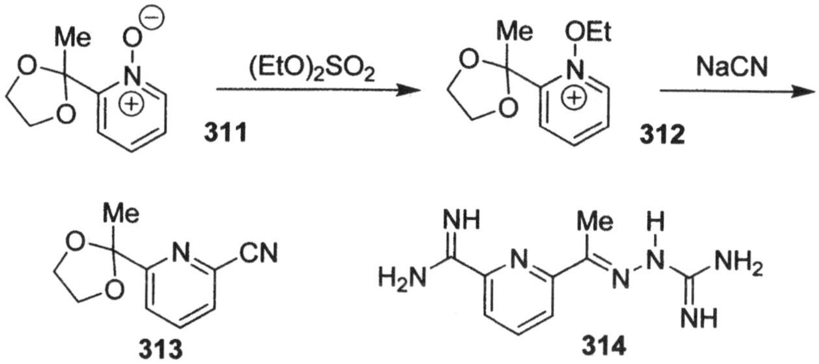

Later on, trimethylsilylcyanide[205] and diethyl phosphorocyanidate (DEPC)[206] were also found to activate the *N*-oxide in the above the Reissert–Henze reaction. In addition, lithium diethylphosphite was used as the nucleophile other than cyanide as shown in the transformation of *N*-methoxy pyridine **315** into the diethylphosphonate **316**.[207]

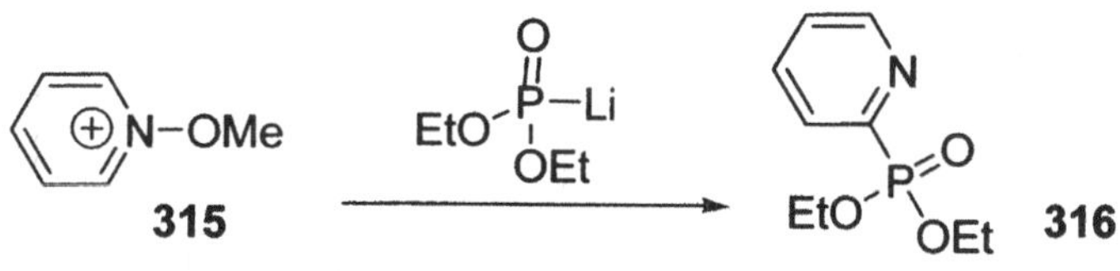

In addition, the *N*-oxide **317** was treated with the Tebbe reagent to yield the corresponding 2-methyl-isoquinoline derivative **318**.[208]

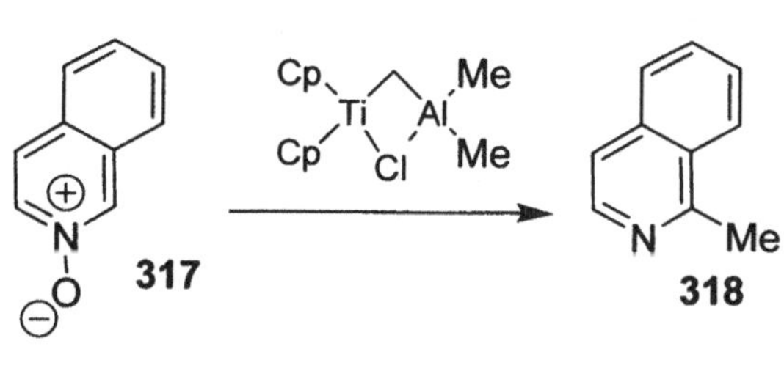

Recently, the first asymmetric Boekelheide rearrangement was reported using (*R*)-Mosher's acyl chloride [(*R*)-(–)-α-methoxy-α-(trifluoromethyl) phenylacetyl chloride, (*R*)-(–)MTPA-Cl] as the activator.[209]

Thus, treatment of **319** with (*R*)-(–)MTPA-Cl using TEA as the base at –78 °C in 2-propanol allowed the fast and complete rearrangement to the corresponding esters **320** with an 88:12 diastereomeric ratio. Saponification **320** with LiOH led to the enantiomerically enriched 1-(2-pyridinyl)alkyl alcohol derivative **321**.

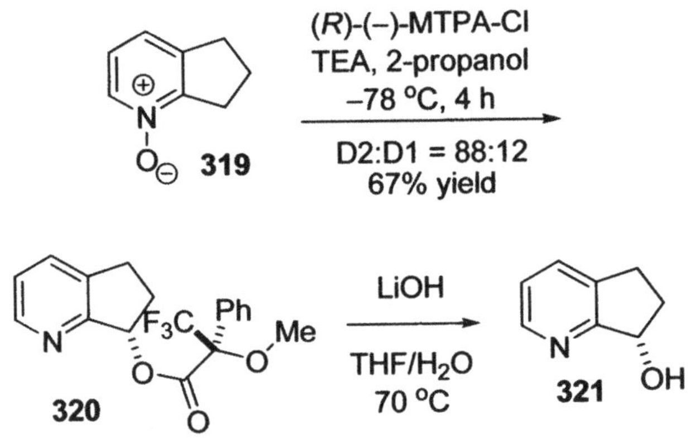

The synthesis of pyridinyl substituted pyrrolidines and piperidines from (aminoalkyl)-pyridine *N*-oxides was reported using di-*tert*-butylsilyl bis(trifluoromethanesulfonate) as an activator, which unlike commonly used Ac_2O, TFAA, *etc.* is compatible with the presence of amino groups in a Boekelheide type reaction.[210] Treatment of the pyridine *N*-oxides **322–325** with di-*tert*-butylsilyl bis(trifluoromethane-sulfonate) and triethylamine gave the corresponding pyrrolidnyl or piperidinyl compounds **326–329** in moderate to good yields. And in all the cases, microwave irradiation (50 °C, 250W, 20 min.) gave higher yields.

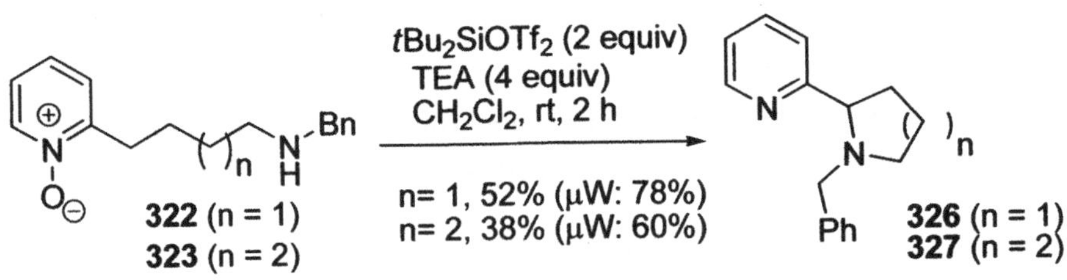

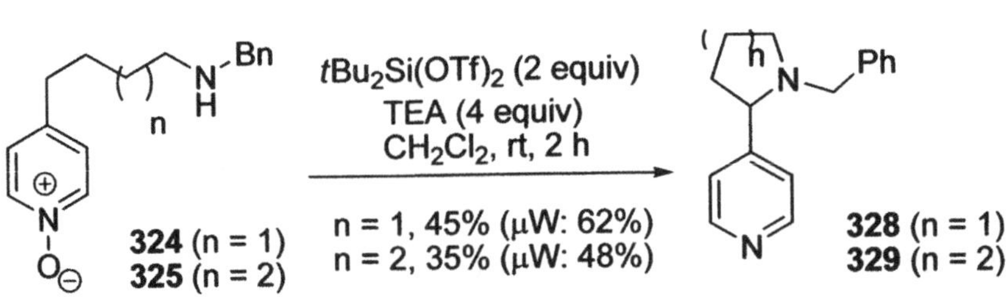

Ciamician–Dennstedt rearrangement

The Ciamician–Dennstedt reaction involves the addition of intermediate dihalocarbene, generated from haloforms (CHX_3, X = Cl, Br, or I) and a strong base, to a pyrrole (**330**) to form an unstable dihalogenocyclopropane, which rearranges to a 3-halogenopyridine such as **331**.[211] The reaction was expanded for indoles to provide 3-chloroquinolines.[212]

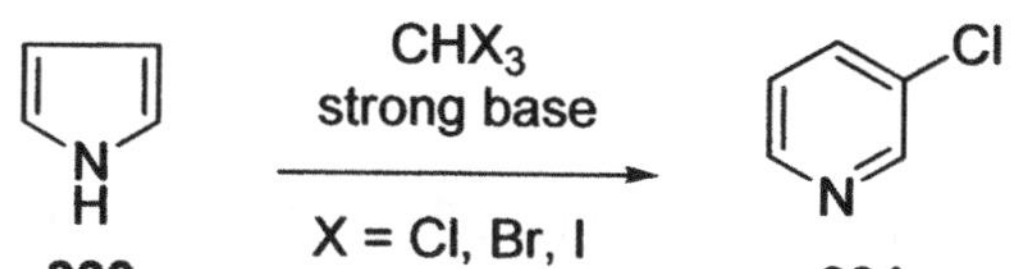

330 **331**

The yield of a typical Ciamician–Dennstedt reaction was only in 20s%. Yields can be significantly improved by using a phase-transfer catalyst such as benzyltriethylammonium chloride[213] and tetra-*n*-butylammonium hydrogen sulfate.[214] Despite its rare usage, the Ciamician–Dennstedt reaction is the only way to date to make *C*-bridged calix[4]pyridine **333**.[215] Four sequential treatments of calix[4]pyrrole **332** with sodium trichloroacetate result in all four possible geometric isomers of **333** in 26% overall yield.

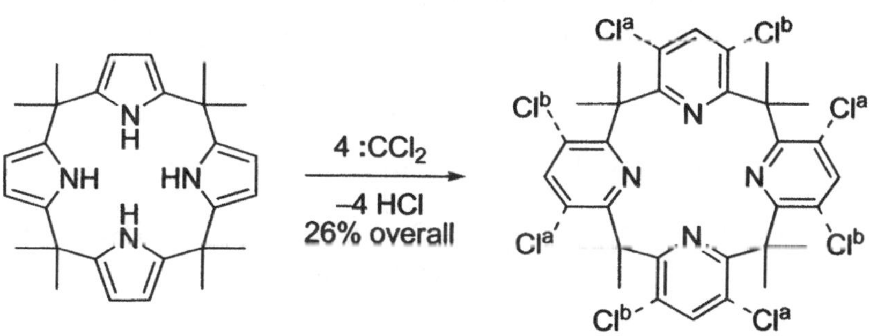

332 **333**, four geometric isomers

10.3.4. *Synthesis via Transformation of Another Heterocycle*

The furan derivatives **334** containing a 2-acyl functionality were treated with ammonia at 150 °C in a sealed-tube to provide the 3-hydroxy pyridines **335** in low yields.[216]

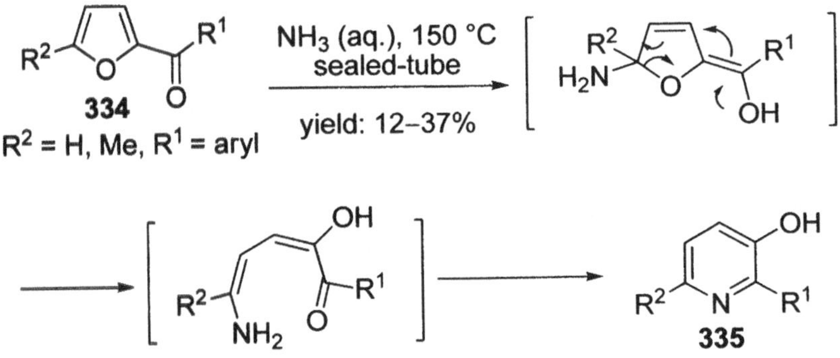

Treatment of 2*H*-pyran-2-ones **336** with urea in refluxing pyridine temperature afforded 2-aminopyridines **337** regioselectively.[217]

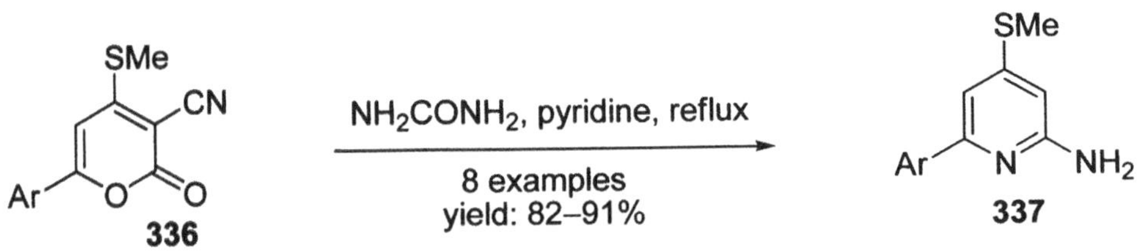

Pyrylium salts are prone to nucleophilic attack. Reaction of pyrylium salts **339** with ammonia or ammonium salts undergo ring-opening/ring-closing reaction sequences to afford the corresponding pyridine derivatives **340**. This reaction was also called the Baeyer pyridine synthesis.[218,219]

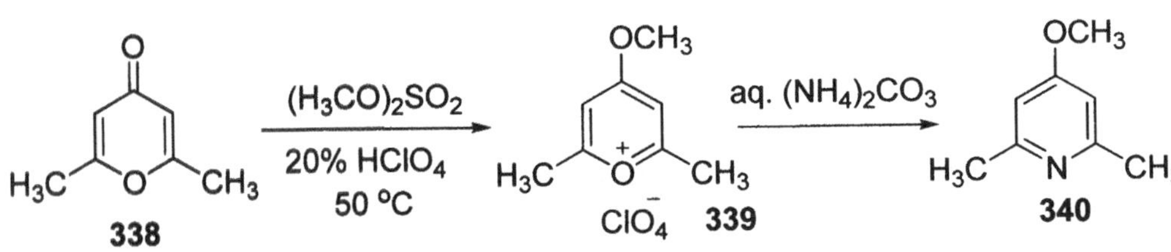

Reaction of pyrylium salts with hydroxylamines and primary amines produces pyridine *N*-oxides and pyridinium salts, respectively. Some of the pyridinium salts are biologically active. For example, compound **342** is a carbonic anhydrase activator[220] and derivatives like **344** show anti-cholinesterase activity.[221]

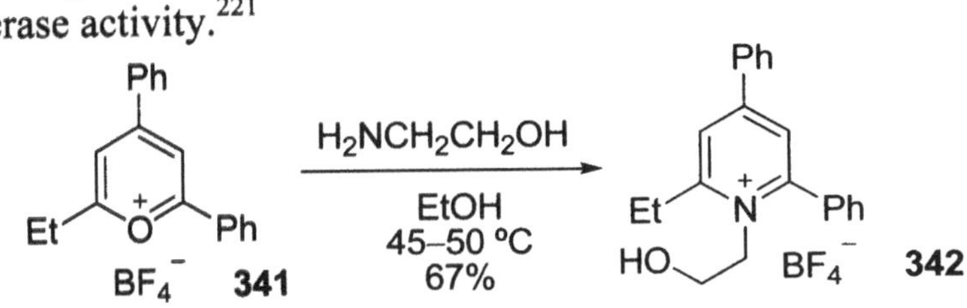

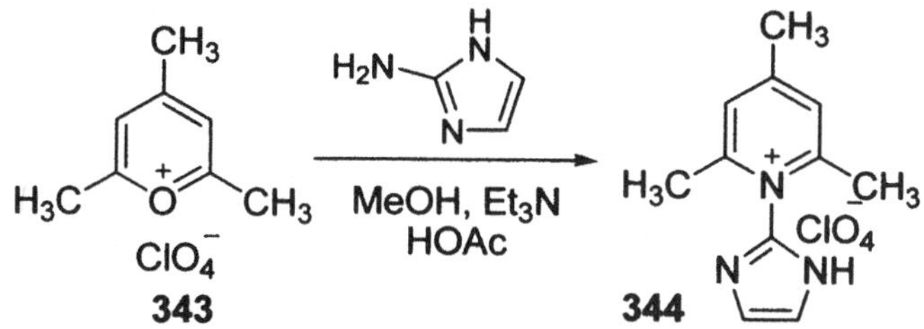

343 344

β-Carbolines are a class of indole alkaloids that are structurally similar to tryptophan, which has been shown to be useful for a variety of neuroscience applications.[222] The Baeyer pyridine synthesis was applied for the preparation of the β-carboline derivative ambocarb (**346**).[223]

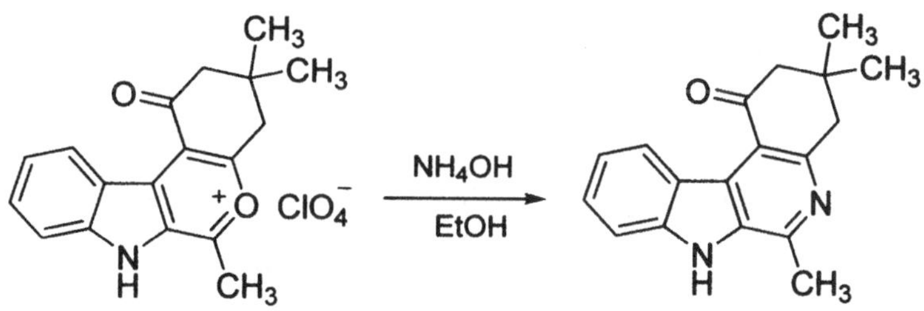

345 **346**, ambocarb

10.4 Problems

10.4.1 Propose a reasonable mechanism for the following transformation[224]:

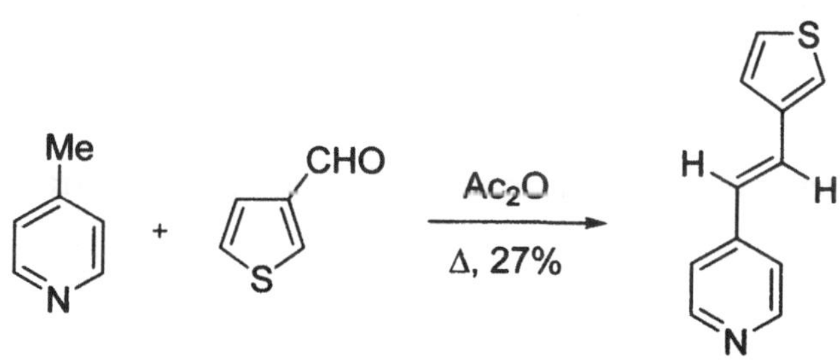

10.4.2 Propose a reasonable mechanism for the following transformation[225]:

TFAA, trifluoroacetic anhydride

10.4.3 Propose a reasonable mechanism for the following transformation[226]:

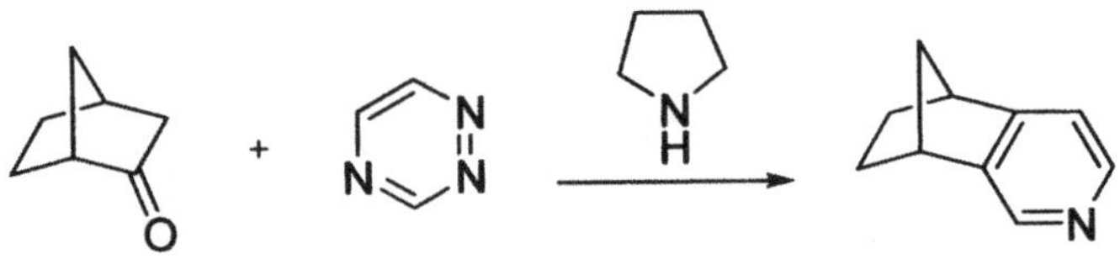

10.4.4 Propose a reasonable mechanism for the following transformation:[227]

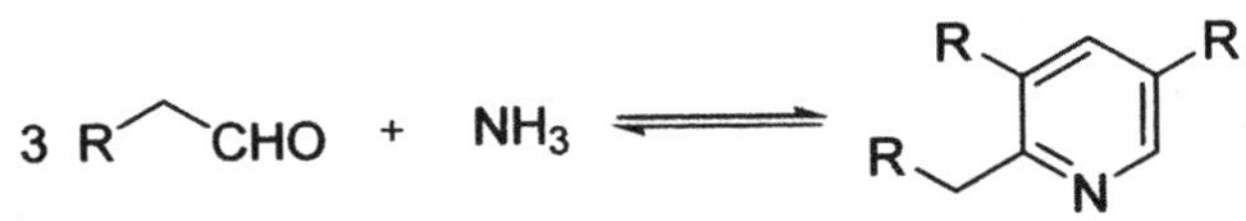

10.4.5 Propose a reasonable mechanism for the following transformation[228]:

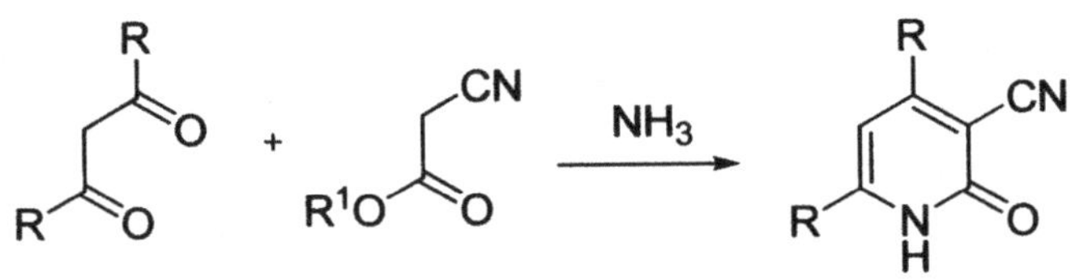

10.4.6 Propose a reasonable mechanism for the following transformation[229]:

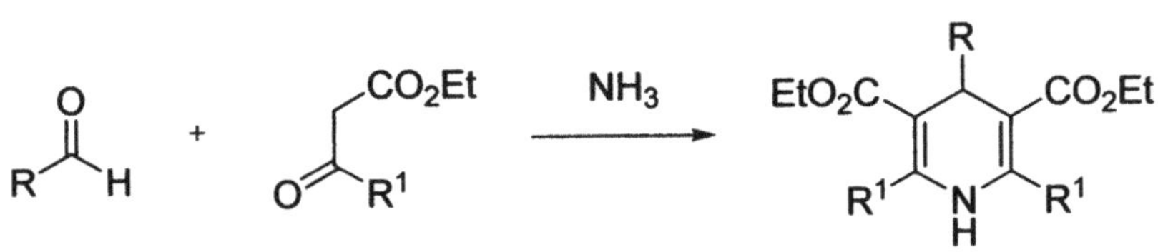

10.4.7 Propose a reasonable mechanism for the following transformation[230]:

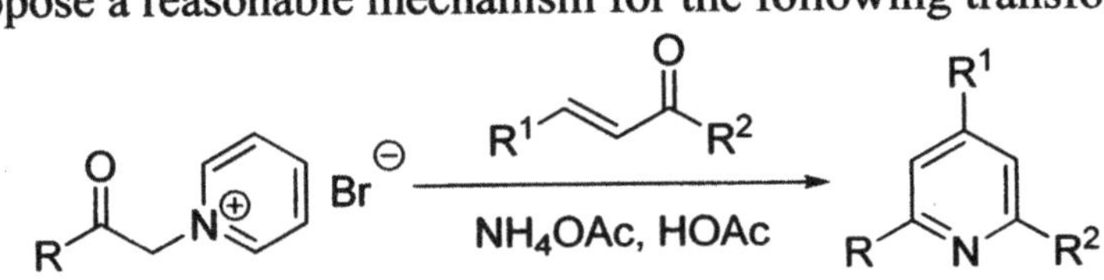

10.4.8 Propose a reasonable mechanism for the following transformation[231]:

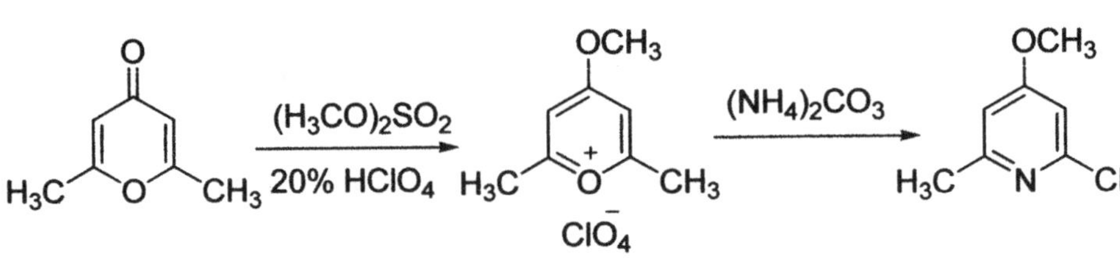

10.4.9 Propose a reasonable mechanism for the following transformation[232]:

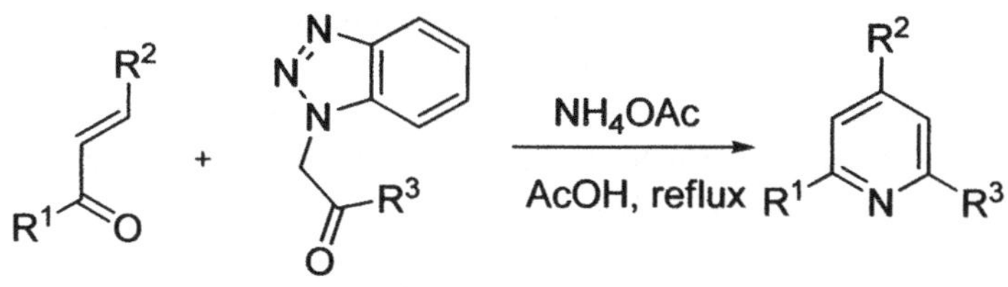

10.4.10 Propose a reasonable mechanism for the following transformation[153]:

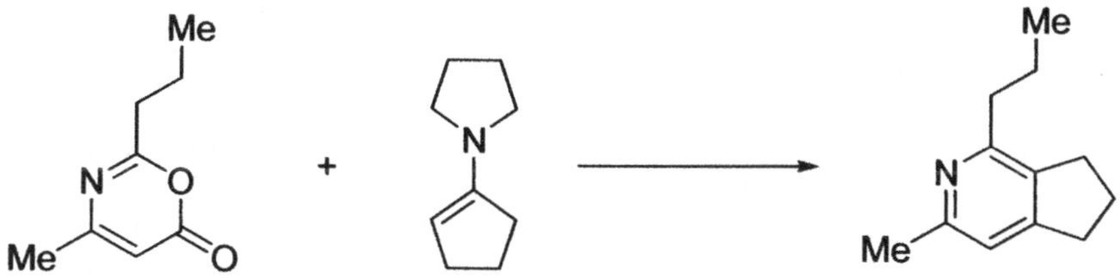

10.4.11 Propose a reasonable mechanism for the following transformation[217]:

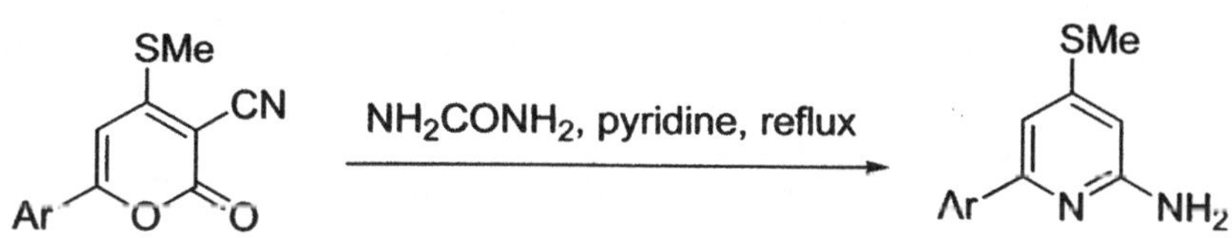

10.5 References

1. (a) Jones, G. *Comprehensive Heterocyclic Chemistry II, Vol. 5* (Eds.: Katritzky, A. R.; Rees, C. W.; Scriven, E. F. V.; McKillop A.), Pergamon, Oxford, **1996,** 167–243. (b) Henry, G. D. *Tetrahedron* **2004,** *60,* 6043–6060. (c) Joule, J. A.; Mills, K. *Heterocyclic Chemistry*, 4th ed., Blackwell Science, Cambridge, **2000,** 63–120. (d) Michael, J. P. *Nat. Prod. Rep.* **2005,** *22,* 627–646.
2. Jayasinghe, L.; Jayasooriya, C. P.; Hara, N.; Fujimoto, Y. *Tetrahedron Lett.* **2003,** *44,* 8769–8771.
3. Kubota, T.; Nishi, T.; Fukushi, E.; Kawabata, J.; Fromont, J.; Kobayashi, J. *Tetrahedron Lett.* **2007,** *48,* 4983–4985.
4. Rao, T.; Adams, P.; Correa, L.; Santori, E.; Sacaan, A.; Reid, R.; Cosford, N. *Brain Res.* **2008,** *1234,* 16–24.
5. The Parkinson Study Group. *Neurology* **2006,** *66,* 408–410.
6. Wei, Z.-L.; Xiao, Y.; Yuan, H.; Baydyuk, M; Petukhov, P. A.; Musachio, J. L.; Kellar, K. J.; Kozikowski, A. P. *J. Med. Chem.* **2005,** *48,* 1721–1724.
7. Lindberg, P.; Bränström, A.; Wallmark, B.; Mattsson, H.; Rikner, L.; Hoffman, K.-J. *Med. Res. Rev.* **1990,** *10,* 1–54.

8. Lindberg, P.; Nordberg, P.; Alminger, T.; Bränström, A.; Wallmark, B. *J. Med. Chem.* **1986,** *29*, 1327–1329.
9. Saitton, S.; Del Tredici, A. L.; Mohell, N.; Vollinga, R. C.; Boström, D.; Kihlberg, J.; Luthman K. *J. Med. Chem.* **2004,** *47,* 6595–6602.
10. Riley R. J.; Parker A. J.; Trigg S.; Manners C. N. *Pharm. Res.* **2001,** *18*, 652–655.
11. Yasuna, K.; Yoichi, M.; Eiichi, K.; Teruo, F.; Takaaki, A.; Seisho, T.; Takemi, Y.; Yukio, K. *J. Biochem.* **1993,** *114*, 697–701.
12. (a) *Comprehensive Heterocyclic Chemistry II, Vol. 5* (Eds.: Katritzky, A. R.; Rees, C. W.; Scriven, E. F. V.; McKillop A.), Pergamon: Oxford, **1996**. (b) *Comprehensive Heterocyclic Chemistry I, Vol. 2* (Eds.: Katritzky, A. R.; Rees, C. W.), Pergamon: Oxford, **1984**. (c) *Kirk-Othmer Encyclopedia of Chemical Technology* (4th Edition), V20, pp 1–33.
13. Sliwa, W. *Curr. Org. Chem.* **2003,** *7*, 995.
14. (a) Bennasar, M.-L.; Zulaica, E.; Alonso, Y.; Mata, I.; Molins, E.; Bosch, J. *J. Chem. Soc., Chem. Commun.* **2001,** 1166. (b) Amat, M.; Coll, M.-D.; Llor, N.; Escolano, C.; Molins, E.; Miravitlles, C.; Bosch, J. *Tetrahedron Asymmetry* **2003,** *14*, 1691. (c) dos Santos, D. C.; de Freitas Gil, R. P.; Gil, L.; Marazano, C. *Tetrahedron Lett.* **2001,** *42*, 6109. (d) Guilloteau-Bertin, B.; Compere, D.; Gil, L.; Marazano, C.; Das, B. C. *Eur. J. Org. Chem.* **2000,** 1391. (e) Leleu, S.; Papamicae, C.; Marsais, F.; Dupas, G.; Levacher, V. *Tetrahedron Asymmetry* **2004,** *15*, 3919. (f) Barbier, D.; Marazano, C.; Riche, C.; Das, B. C.; Potier, P. *J. Org. Chem.* **1998,** *63*, 1767.
15, (a) Marchais-Oberwinkler, S.; Nowicki, B.; Pike, V. W.; Halldin, C.; Sandell, J.; Chou, Y.-H.; Gulyas, B.; Brennum, L. T.; Farde, L.; Wikstrom, H. V. *Bioorg. Med. Chem.* **2005,** *13*, 883. (b) Chelucci, G.; Murineddu, G.; Pinna, G. A. *Tetrahedron: Asymmetry* **2004,** *15*, 1373–1389. (c) Krebs, F. C. *Tetrahedron Lett.* **2003,** *44*, 6643.
16. Youssif, S. *ARKIVOC* **2001,** 242.
17. Caron, S.; Do, N. M.; Sieser, J. E. *Tetrahedron Lett.* **2000,** *41*, 2299.
18. Jain, S. L.; Sain, B. *J. Chem. Soc., Chem. Commun.* **2002,** 1040.
19. Prasad, M.; Kamalakar, G.; Madhavi, G.; Kulkarni, S. J.; Raghavan, K. V. *J. Mol. Catal. A* **2002,** *186*, 109.
20. Jain, S. L.; Sain, B. *Angew. Chem. Int. Ed. Engl.* **2003,** *42*, 1265.
21. Zhong, P.; Guo, S.; Song, C. *Synth. Commun.* **2004,** *34*, 247.
22. Sharma, V. B.; Jain, S. L.; Sain, B. *Tetrahedron Lett.* **2004,** *45*, 4281.
23. Musina, L. A.; Shul'ts, E. E.; Kricherskii, L. A.; Adekenov, S. M.; Shakirov, M. M.; Tolstikov, G. A. *Russ. Chem. Bull.* **2006,** *55*, 331.
24. Beres, M.; Timari, G.; Hajos, G. *Tetrahedron Lett.* **2002,** *43*, 6035.
25. Martin-Vaca, B.; Rudler, H.; Audouin, M.; Nicolas, M.; Durand-Reville, T.; Vissiere, B. *J. Organomet. Chem.* **1998,** *567*, 119–126.
26. [R] (a) Cheng, W.-C.; Kurth, M. J. *Org. Prep. Proc. Int.* **2002,** *34*, 585. [R] (b) Becher, J. *Synthesis* **1980,** 589. [R] (c) Kost, A. N.; Gromov, S. P.; Sagitullin, R. S. *Tetrahedron* **1981,** *37*, 3423.
27. Zincke, Th. *Justus Liebigs Ann. Chem.* **1904,** *330*, 361.
28. König, W. *J. Prakt. Chem.* **1904,** *69*, 105.
29. Tamura, Y.; Tsujimoto, N.; Mano, M. *Chem. Pharm. Bull.* **1971,** *19*, 130.
30. Tamura, Y.; Tsujimoto, N. *Chem. Ind.* **1970,** 926.
31. Ágai, B.; Lempert, K. *Tetrahedron* **1972,** *28*, 2069.
32. (a) Knaus, E. E.; Redda, K. *J. Heterocyclic Chem.* **1976,** *13*, 1237. (b) Yeung, J. M.; Corleto, L. A.; Knaus, E. E. *J. Med. Chem.* **1982,** *25*, 191. (c) Redda, K. K.; Melles, H.; Rao, K. N. *J. Heterocyclic Chem.* **1990,** *27*, 1041.

33. Epsztajn, J.; Lunt, E.; Katritzky, A. R. *Tetrahedron* **1970,** *26*, 1665.
34. Tamura, Y.; Miki, Y.; Honda, T.; Lkeda, M. *J. Heterocyclic Chem.* **1972,** *9*, 865.
35. Rao, K. N.; Redda, K. K. *J. Heterocyclic Chem.* **1995,** *32*, 307.
36. Lettré, H.; Haede, W.; Ruhbaum, E. *Justus Liebigs Ann. Chem.* **1953,** *579*, 123.
37. Shifrin, S. *Biochim. Biophys. Acta* **1965,** *96*, 173.
38. Secrist, J. A., III; Leonard, N. J. *J. Am. Chem. Soc.* **1972,** *94*, 1702.
39. (a) Juricová, K.; Smrcková, S.; Holy, A. *Collect. Czech. Chem. Commun.* **1995,** *60*, 237. (b) Hocková, D.; Votavová, H.; Holy, A. *Tetrahedron: Asymmetry* **1995,** *6*, 2375. (c) Hocková, D.; Masojídková, M.; Holy, A. *Collect. Czech. Chem. Commun.* **1996,** *61*, 1538. (d) Hocková, D.; Holy, A. *Collect. Czech. Chem. Commun.* **1997,** *62*, 948.
40. Eda, M.; Kurth, M. J.; Nantz, M. H. *J. Org. Chem.* **2000,** *65*, 5131.
41. Tatrai, K.; Perjési, P.; Zharikova, A. D.; Li, X.; Prokai, L. *Bioorg. Med. Chem. Lett.* **2002,** *12*, 2171.
42. Woodard, P. A.; Winwood, D.; Brewster, M. E.; Estes, K. S.; Bodor, N. *Drug Des. Del.* **1990,** *6*, 15.
43. (a) Malleron, J.-L.; Fiaud, J.-C.; Legros, J.-Y. *Handbook of Palladium-Catalyzed Organic Reactions* Academic Press: London, **1997**. (b) Tolman, C.A. *Chem. Rev.* **1977,** *77*, 313–348. (c) Wolfe, J. P.; Tomori, H.; Sadighi, J. P.; Yin, J.; Buchwald, S. L. *J. Org. Chem.* **2000,** *65*, 1158–1174. (d) Li, J. J. *Prog. Heterocyclic Chem.* **2000,** *12*, 37–56. (e) Hillier, A. C.; Grasa, G. A.; Viciu, M. S.; Lee, H. M.; Yang, C.; Nolan, S. P. *J. Organometal. Chem.* **2002,** *653*, 69–82. (f) Littke, A. F.; Fu, G. C. *Angew. Chem. Int. Ed.* **2002,** *41*, 4176–4211. (g) Negishi, E-I. *Handbook of Organopalladium Chemistry for Organic Synthesis*, Wiley: NY **2002**. (h) Cardenas, D. J. *Angew. Chem. Int. Ed.* **2003,** *42*, 384–387. (i) Brase, S.; Kirchoff, J. H.; Kibberling, J. *Tetrahedron* **2003,** *59*, 885–939. (j) Chelucci, G.; Orru, G.; Pinna, G. A. *Tetrahedron* **2003,** *59*, 9471–9515. (k) Valentine, D. H., Jr.; Hillhouse, J. H. *Synthesis* **2003,** 2437–2460. (l) Schlummer, B.; Scholz, U. *Adv. Synth. Catal.* **2004,** *346*, 1599–1626. (m) Nakamura, I.; Yamamoto, Y. *Chem. Rev.* **2004,** *104*, 2127–2198. (n) Kappe, C. O. *Angew. Chem. Int. Ed.* **2004,** *43*, 6250–6284. (o) Nakamura, I.; Yamamoto, Y. *Chem. Rev.* **2004,** *104*, 2127–2198. (p) Schroter, S.; Stock, C.; Bach, T. *Tetrahedron* **2005,** *61*, 2245–2267. (q) Handy, C. J.; Manoso, A. S.; McElroy, W.T.; Seganish, W. M.; DeShong, P. *Tetrahedron* **2005,** *61*, 12201–12225. (r) Shaughnessy, K. H.; DeVasher, R. B. *Curr. Org. Chem.* **2005,** *9*, 585–604. (s) Christmann, U.; Vilar, R. *Angew. Chem. Int. Ed.* **2005,** *44*, 366–374. (t) Frieman, B. A.; Taft, B. R.; Lee, C.-T.; Butler, T.; Lipshutz, B. H. *Synthesis* **2005,** 2989–2993. (u) Nicolaou, K. C.; Bulger, P. G.; Sarlah, D. *Angew. Chem. Int. Ed.* **2005,** *44*, 4442–4489.
44. (a) Beak, P.; Meyer, A. I. *Acc. Chem. Res.* **1986,** *19*, 356–363. (b) Itami, K.; Kamei, T.; Mitsudo, K.; Nokami, T.; Yoshida, J.-I. *J. Org. Chem.* **2001,** *66*, 3970–3976. (c) Mongin, F.; Queguiner, G. *Tetrahedron* **2001,** *57*, 4059–4090. (d) Whisler, M. C.; MacNeil, S.; Snieckus, V.; Beak, P. *Angew. Chem. Int. Ed.* **2004,** *43*, 2206–2225.
45. (a) Gribble, G. W.; Saulnier, M. G. *Tetrahedron Lett.* **1980,** *21*, 4137–4140. (b) Comins, D. L.; Myoung, Y. C. *J. Org. Chem.* **1990,** *55*, 292–298.
46. Taylor, S. L.; Lee, D. Y.; Martin, J. C. *J. Org. Chem.* **1983,** *48*, 4158–4159.
47. (a) Kessar, S. V.; Singh, P.; Singh, K. N.; Dutt, M. *J. Chem. Soc., Chem. Commun.* **1991,** 570–571. (b) Kessar, S. V.; Singh, P. *Chem. Rev.* **1997,** *97*, 721–737.
48. Gros, P.; Fort, Y.; Caubere, P. *J. Chem. Soc., Perkin Trans. I* **1997,** 3597–3600.

49. Gros, P.; Fort, Y.; Caubere, P. *J. Chem. Soc., Perkin Trans. I* **1998,** 1685–1689.
50. Bouillon, A.; Lancelot, J.-C.; Santos, J. S. D. O.; Collot, V.; Bovy, P. R.; Rault, S. *Tetrahedron* **2003,** *59*, 10043–10049.
51. (a) Trecourt, F.; Breton, G.; Bonnet, V.; Mongin, F.; Marsais, F.; Queguiner, G. *Tetrahedron* **2000,** *56*, 1349–1360. (b) Ila, H.; Baron, O.; Wagner, A. J.; Knochel, P. *Chem. Commun.* **2006,** 583–593.
52. Kii, S.; Akao, A.; Iida, T.; Mase, T.; Yasuda, N. *Tetrahedron Lett.* **2006,** *47*, 1877–1879.
53. Bobbio, C.; Schlosser, M. *Eur. J. Org. Chem.* **2001,** 4533–4536.
54. (a) Chinchilla, R.; Nájera, C.; Yus, M. *Chem. Rev.* **2004**, *104*, 2667; (b) Galatsis, P. *Pyridines* in *Name Reaction in Heterocyclic Chemistry,* ed. Li, J. J., Wiley: Hoboken, NJ, 2005, pp 302–374. (c) Conzález-Bello, C.; Castedo, L. *Six-membered heterocycles: Pyridines*, in *Modern Heterocyclic Chemistry*, 1st Edition, Eds. Alvarez-Builla, J.; Jose Vaquero, J.; Barluenga, J., Wiley-VCH, 2011.
55. (a) Tamao, K.; Sumitami, K.; Kumada, M. *J. Am. Chem. Soc.* **1972,** *94*, 4374–4376. (b) Tamao, K. *J. Organometal. Chem.* **2002,** *653*, 23–26.
56. (a) Pattenden, G.; Sinclair, D. J. *J. Organometal. Chem.* **2002,** *653*, 261–268. (b) Kosugi, M.; Fugami, K. *J. Organometal. Chem.* **2002,** *653*, 50–53. (c) Espinet, P.; Echavarren, A. M. *Angew. Chem. Int. Ed.* **2004,** *43*, 470–4734. (d) Echavarren, A. M. *Angew. Chem. Int. Ed.* **2005,** *44*, 3962–3965.
57. (a) Suzuki, A. *J. Organometal. Chem.* **2002,** *653*, 83–90. (b) Kotha, S.; Lahiri, K.; Kashinath, D. *Tetrahedron* **2002,** *58*, 9633–9695. (c) Bellina, F.; Carpita, A. Rossi, R. *Synthesis* **2004,** 2419–2440. (d) Sasaki, M.; Fuwa, H. *Synlett* **2004,** 1851–1874. (e) Tyrell, E.; Brookes, P. *Synthesis* **2004** 469–483. (f) Miura, M. *Angew. Chem. Int. Ed.* **2004,** *43*, 2201–2203. (g) Leadbeater, N. E. *Chem. Commun.* **2005,** 2881–2902. (h) Baudoin, O. *Eur. J. Org. Chem.* **2005,** 4223–3229. (i) Franzen, R.; Xu, Y. *Can. J. Chem.* **2005,** *83*, 266–272. (j) Suzuki, A. *Chem. Commun.* **2005,** 4759–4763.
58. (a) Sonogashira, K.; Tohda, Y.; Hagihara, N. *Tetrahedron Lett.* **1975,** *16*, 4467–4470. (b) Castro, C. E.; Stephens, R. D. *J. Org. Chem.* **1963,** *28*, 2163. (c) Stephens, R. D.; Castro, C. E. *J. Org. Chem.* **1963,** *28*, 3313–3315. (d) Sonogashira, K. *J. Organomet. Chem.* **2002,** *653*, 46–49. (e) Tykwinski, R. R. *Angew. Chem. Int. Ed.* **2003,** *42*, 1566–1568. (f) Nagy, A.; Novak, Z.; Kotschy, A. *J. Organomet. Chem.* **2005,** *690*, 4453–4461.
59. (a) Heck, R. F. *J. Am. Chem. Soc.* **1968,** *90*, 5518–5548. (b) de Meijere, A.; Meyer, F. E. *Angew. Chem. Int. Ed.* **1994,** *33*, 2379–2411. (c) Dounay, A. B.; Overman, L. E. *Chem. Rev.* **2003,** *103*, 2945–2963. (d) Lin, B-L.; Liu, L.; Fu, Y.; Luo, S-W.; Chen, Q.; Guo, Q-X. *Organomet.* **2004,** *23*, 2114–2123. (e) Shibasaki, M.; Vogl, E. M.; Ohshima, T. *Adv. Synth. Catal.* **2004,** *346*, 1533–1552. (f) Guiry, P. J.; Kiely, D. *Curr. Org. Chem.* **2004,** *8*, 781–794. (g) Alonso, F.; Beletskaya, I. P.; Yus, M. *Tetrahedron* **2005,** *61*, 11711–11835. (h) Oestereich, M. *Eur. J. Org. Chem.* **2005,** 783–792.
60. (a) Chen, D.; Kotti, S.; Li, G. *Chemtracts* **2005,** *18*, 193–199. (b) Negishi, E.; Hu, Q.; Huang, Z.; Qian, M.; Wang, G. *Aldrichimica Acta* **2005,** *38*, 71–88. (c) Negishi, E.; Zeng, X.; Tan, Z.; Qian, M.; Hu, Q.; Huang, Z. *Palladium- or Nickel-Catalyzed Cross-Coupling With Organometals Containing Zinc, Aluminum, And Zirconium: The Negishi Coupling.* in *Metal-Catalyzed Cross-Coupling Reactions* (2nd Ed.), **2004,** *2*, 815–889. (d) Lessene, G. *Australian J Chem.* **2004,** *57*, 107.

61. (a) Yoshida, J.-I. *J. Organomet. Chem.* **2002,** *653*, 105–113. (b) *J. Organomet. Chem.* **2002,** *653*, 69.
62. (a) Guram, A. S.; Rennels, R. A.; Buchwald, S. L. *Angew. Chem. Int. Ed.* **1995,** *34*, 1348–1350. (b) Louie, J.; Hartwig, J. F. *Tetrahedron Lett.* **1995,** *36*, 3609–3612. (c) Louie, J.; Driver, M. S.; Hamann, B. C.; Hartwig, J. F. *J. Org. Chem.* **1997,** *62*, 1268–1273. (d) Hartwig, J. F. *Angew. Chem. Int. Ed.* **1998,** *37*, 2046–2067. (e) Hartwig, J. F.; Kawatsura, M.; Hauck, S. I.; Shaughnessy, K. H.; Alcazar-Roman, L. M. *J. Org. Chem.* **1999,** *64*, 5575–5580. (f) Ali, M. H.; Buchwald, S. L. *J. Org. Chem.* **2001,** *66*, 2560–2565. (g) Schlummer, B.; Scholz, U. *Adv. Synth. Catal.* **2004,** *346*, 1599–1626. (h) Loones, K. T. J.; Maes, B. U. W.; Rombouts, G.; Hostyn, S.; Diels, G. *Tetrahedron* **2005,** *61*, 10338–10348. (i) Senra, J. D.; Aguiar, L. C. S.; Simas, A. B. C. *Current Organic Synthesis* **2011,** *8*, 53–78. (j) Fischer, C.; Koenig, B. *Beilstein J. Org. Chem.* **2011,** *7*, 59–74. (k) Janey, J. M. *Buchwald-Hartwig Amination.* in *Name Reactions for Functional Group Transformations* **2007,** 564–609.
63. (a) Qiao, J. X.; Lam, P. Y. S. *Recent advances in Chan-Lam coupling reaction: copper-promoted C-heteroatom bond cross-coupling reactions with boronic acids and derivatives.* in *Boronic Acids* (2nd Edition), **2011,** *1*, 315–361. (b) Qiao, J. X.; Lam, P. Y. S. *Synthesis* **2011,** 829–856. (c) Thai, L.; Burgess, K. *Chemtracts* **1998,** *11*, 1037–1040. (d) Ley, S. V.; Thomas, A. W. *Angew. Chem., Int. Ed.* **2003,** *42*, 5400. (e) Zimmermann, J. PCT Int. Appl. **1995,** WO 9509853. (f) Zimmermann, J. Eur. Pat. Appl. **1993,** EP 564409. (g) Eberle, M.; Stierli, D.; Pillonel, C.; Ziegler, H. PCT Int. Appl. **2001,** WO 2001093682. (h) Furet, P.; Zimmermann, J.; Capraro, H. G.; Meyer, T.; Imbach, P. *J. Comput.-Aided Mol. Des.* **2000,** *14*, 403.
64. Stanetty, P.; Schnurch, M.; Mihovilovic, M. D. *Synlett* **2003,** 1862–1864.
65. Stanetty, P.; Hattinger, G.; Schnürch, M.; Mihovilovic, M. D. *J. Org. Chem.* **2005,** *70*, 5215–5220.
66. Kuo, G.-H.; Prouty, C.; Wang, A.; Emanuel, S.; DeAngelis, A.; Zhang, Y.; Song, F.; Beall, L.; Connoly, P. J.; Karnachi, P.; Chen, X.; Gruninger, R. H.; Sechler, J.; Fuentes-Pesquera, A.; Middleton, S. A.; Jolliffe, L.; Murray, W. V. *J. Med. Chem.* **2005,** *48*, 4892–4909.
67. Arvanitis, A. G.; Arnold, C. R.; Fitzgerald, L. W.; Frietze, W. E.; Olson, R. E.; Gilligan, P. J.; Roberston, D. W. *Bioorg. Med. Chem. Lett.* **2003,** *13*, 289–291.
68. Jackson, R. F.; Wishart, N.; Wood, A.; James, K.; Wythes, M. J. *J. Org. Chem.* **1992,** *57*, 3397–3404.
69. Sabat, M.; Johnson, C. R. *Org. Lett.* **2000,** *2*, 1089–1092.
70. Cockerill, G. S.; Easterfield, H. J.; Percy, J. M.; Pintat, S. *J. Chem. Soc., Perkin Trans. 1* **2000,** 2591–2599.
71. Altenburger, J.-M.; Lassalle, G. Y.; Matrougui, M.; Galtier, D.; Jetha, J.-C.; Bocskei, Z.; Berry, C. N.; Lunven, C.; Lorrain, J.; Herault, J-P.; Schaeffer, P.; O'Conner, S. E.; Herbert, J.-M. *Bioorg. Med. Chem.* **2004,** *12*, 1713–1730.
72. Wagner, F, F.; Comins, D. L. *J. Org. Chem.* **2006,** *71*, 8673.
73. Muraoka, N.; Mineno, M.; Itami, K.; Yoshida, J.-I. *J. Org. Chem.* **2005,** *70*, 6933–6936.
74. Albaneze-Walker, J.; Murry, J. A.; Soheili, A.; Ceglia, S.; Springfield, S. A.; Bazaral, C.; Dormer, P. G.; Hughes, D. L. *Tetrahedron* **2005,** *61*, 6330–6336.
75. Couve-Bonnaire, S.; Carpentier, J.-F.; Mortreux, A.; Castanet, Y. *Tetrahedron* **2003,** *59*, 2793–2799.
76. Sakamoto, T.; Yasuhara, A.; Kondo, Y.; Yamanaka, H. *Chem. Pharm. Bull.* **1992,** *40*, 1137–1139.

77. Beller, M.; Magerlein, W.; Indolese, A. F.; Fischer, C. *Synthesis* **2001,** 1098–1109.
78. Lin, H.; Yamashita, D. S.; Zeng, J.; Xie R.; Wang, W.; Nidarmarthy, S.; Luengo, J. I.; Rhodes, N.; Knick, V. B.; Choudhry, A. E.; Lai, Z.; Minthorn, E. A.; Strum, S. L.; Wood, E. R.; Elkins, P. A.; Concha, N. O.; Heerding, D. A. *Bioorg. Med. Chem. Lett.* **2010,** *20,* 673–678.
79. Sirisoma, N. S.; Johnson, C. R. *Tetrahedron Lett.* **1998,** *39,* 2059–2062.
80. Stangeland, E. L.; Sammakia, T. *J. Org. Chem.* **2004,** *69,* 2381–2385.
81. Zhu, G.-D.; Gandhi, V. B.; Gong, J.; Thomas, S.; Woods, K. W.; Song, X.; Li, T.; Diebold, R. B.; Luo, Y.; Liu, X.; Guan, R.; Klinghofer, V.; Johnson, E. F.; Bouska, J.; Olson, A.; Marsh, K. C.; Stoll, V. S.; Mamo, M.; Polakowski, J.; Campbell, T. J.; Martin, R. L.; Gintant, G. A.; Penning, T. D.; Li, Q.; Rosenberg, S. H.; Giranda, V. L. *J. Med. Chem.* **2007,** *50,* 2990–300.
82. Luo, Y.; Gao, H.; Li, Y.; Huang, W.; Lu, W.; Zhang, Z. *Tetrahedron* **2006,** *62,* 2465–2479.
83. Cheung, W. S.; Patch, R. J.; Player, M. R. *J. Org. Chem.* **2005,** *70,* 3741–3744.
84. Cheng, J.; Trudell, M. L. *Org. Lett.* **2001,** *3,* 1371–1374.
85 [R] (a) Eisner, U.; Kuthan, J. *Chem. Rev.* **1972,** *72,* 1–42. [R] (b) Stout, D. M.; Meyers, A. I. *Chem. Rev.* **1982,** *82,* 223–243. [R] (c) Sausins, A.; Duburs, G. *Heterocycles* **1988,** *27,* 269–289. [R] (d) Sausins, A.; Duburs, G. *Heterocycles* **1988,** *27,* 291–314. [R] (e) Hill, M. D. *Chem. Eur. J.* **2010,** *16,* 12052–12062. [R] (f) Lavilla, R. *J. Chem. Soc., Perkin Trans. 1* **2002,** 1141–1156. [R] (g) Li, J. J. *Palladium Chemistry in Pyridine Alkaloids* In *Progress of Heterocyclic Chemistry,* **2000,** *12,* 37–54.
86 Hantzsch, A. *Justus Liebieg Ann. Chem.* **1882,** *215,* 1–83.
87 (a) Katritzky, A. R.; Ostercamp, D. L.; Yousaf, T. I. *Tetrahedron* **1986,** *42,* 5729–5738. (b) Katritzky, A. R.; Ostercamp, D. L.; Yousaf, T. I. *Tetrahedron* **1987,** *43,* 5171–5186.
88. (a) Bottorff, E. M.; Jones, R. G.; Kornfeld, E. C.; Mann, M. J. *J. Am. Chem. Soc.* **1951,** *73,* 4380–4383. (b) Berson, J. A.; Brown, E. *J. Am. Chem. Soc.* **1955,** *77,* 444–477. (c) Marsi, K. L.; Torre, K. *J. Org. Chem.* **1964,** *29,* 3102–3103. (d) Meyers, A. I.; Sircar, J. C.; Singh, S. *J. Heterocyclic Chem.* **1967,** *4,* 461–462.
89. Singh, B.; Lesher, G. Y. *J. Heterocyclic Chem.* **1980,** *17,* 1109–1110.
90. Cocivera, M.; Effio, A.; Chen, H. E.; Vaish, S. *J. Am. Chem. Soc.* **1976,** *98,* 7362–7366.
91 Sabitha, G.; Reddy, G. S. K. K.; Reddy, C. S.; Yadav, J. S. *Tetrahedron Lett.* **2003,** *44,* 4129–4131.
92 Bridgwood, K. L.; Veitch, G. E.; Ley, S. V. *Org. Lett.* **2008,** *10,* 3627–3629.
93 Debache, A.; Ghalem, W.; Boulcina, R.; Belfaitah, A.; Rhouati, S.; Carboni, B. *Tetrahedron Lett.* **2009,** *50,* 5248–5250.
94 Lei, M.; Ma, L.; Hu, L. *Synth. Comm.* **2011,** *41,* 1969–1976.
95 Zonouz, A. M.; Moghani, D. *Synth. Comm.* **2011,** *41,* 2152–2160.
96 Kumar, A.; Maurya, R. A. *Synlett* **2008,** 883–885.
97 Shen, L.; Cao, S.; Wu, J.; Zhang, J.; Li, H.; Liu, N.; Qian, X. *Green Chem.* **2009,** *11,* 1414–1420.
98 Xia, J.-J., Wang, G.-W. *Synthesis* **2005,** 2379–2383.
99 Nasr-Esfahani, M.; Karami, B.; Behzadi, M. *J. Heterocyclic Chem.* **2009,** *46,* 931.
100 Maquestiau, A.; Mayence, A.; Vanden Eynde, J.-J. *Tetrahedron Lett.* **1991,** *32,* 3839–3840.
101 Fang, X.; Liu, Y.-C.; Li. C. *J. Org. Chem.* **2007,** *72,* 8609–8610.

102 Nakamichi, N.; Kawashita, Y.; Hayashi. M. *Synthesis* **2004,** 1015–1020.
103 Bagley, M. C.; Lubinu, M. C. *Synthesis* **2006,** 1283–1288.
104 Yadav, J. S.; Reddy, B. V. S.; Basak, A. K.; Barishya, G.; Narsaiah, A. V. *Synthesis* **2006,** 451–454.
105 Marzabadi, M. R.; Hong, X.; Gluchowski, C. *Tetrahedron Lett.* **1998,** *39,* 5293–5296.
106 Lefranc D.; Ciufolini, M. A. *Angew. Chem. Int. Ed.* **2009,** *48,* 4198–4201.
107 Trivedi, A.; Dodiya, D.; Dholariya, B.; Kataria, V.; Bhuva, V.; Shah, V. *Chem. Biol. Drug. Des.* **2011,** *78,* 881–886.
108 Ashworth, I.; Hopes, P.; Levin, D.; Patel, I.; Salloo, R. *Tetrahedron Lett.* **2002,** *43,* 4931–4933.
109 (a) Guareschi, I. *Mem. Reale Accad. Sci. Torino II* **1896,** *46,* 7. (b) Guareschi, I. *Mem. Reale Accad. Sci. Torino II* **1896,** *46,* 11. (c) Guareschi, I. *Mem. Reale Accad. Sci. Torino II* **1896,** *46,* 25. (a) Thole, F. B.; Thorpe, J. F. *J. Chem. Soc.* **1911,** *99,* 422. (d) Thole, F. B.; Thorpe, J. F. *J. Chem. Soc.* **1915,** *99,* 445. (e) Kon, G. A. R.; Thorpe, J. F. *J. Chem. Soc.* **1919,** *115,* 686. (f) Kon, G. A. R.; Thorpe, J. F. *J. Chem. Soc.* **1919,** *115,* 701. (g) Day, J. N. E.; Thorpe, J. F. *J. Chem. Soc.* **1920,** *117,* 1465. (h) Ingold, C. K.; Perren, E. A.; Thorpe, J. F. *J. Chem. Soc.* **1922,** *121,* 1765. (i) Birch, S. F.; Thorpe, J. F., *J. Chem. Soc.* **1922,** *121,* 1821. McElvain, S. M.; Lyle, R. E., Jr. *J. Am. Chem. Soc.* **1950,** *72,* 384–389.
110 Collins, D. J.; Jones, A. M. *Aust. J. Chem.* **1989,** *42,* 215–221.
111 Liberman, D.; Rist, N.; Grumbach, F.; Cals, S.; Moyeux, M.; Rouaix, A. *Bull. Soc. Chim. Fr.* **1958,** 687.
112 Sprung, M. M. *Chem. Rev.* **1940,** *40,* 297–338.
113 (a) Frank, R. L.; Seven, R. P. *J. Am. Chem. Soc.* **1949,** *71,* 2629–2635. (b) Frank, R. L.; Riener, E. F. *J. Am. Chem. Soc.* **1950,** *72,* 4182–4183. (c) Farley, C. P.; Eliel, E. L. *J. Am. Chem. Soc.* **1956,** *78,* 3477–3484.
114 Weiss, M. *J. Am. Chem. Soc.* **1952,** *74,* 200–202.
115 Snider, B. B.; Neubert, B. J. *Org. Lett.* **2005,** *7,* 2715–2718.
116 Burns, N. Z.; Baran, P. S. *Angew. Chem. Int. Ed.* **2008,** *47,* 205–208.
117 Burns, N. Z.; Jessing, M.; Baran, P. S. *Tetrahedron,* **2009,** *65,* 6600–6610.
118 [R] Bagley, M. C.; Glover, C.; Merritt, E. A. *Synlett,* **2007,** 2459–2482.
119 Bohlmann, F.; Rahtz, D. *Chem. Ber.* **1957,** *90,* 2265–2272.
120 Bagley, M. C.; Dale, J. W.; Bower, J. *Synlett* **2001,** 1149–1151.
121 (a) Bagley, M. C.; Dale, J. W.; Hughes, D. D.; Ohnesorge, M.; Phillips, N. G.; Bower, J. *Synlett* **2001,** 1523–1526. (b) Bagley, M. C.; Dale, J. W.; Bower, J. *Chem. Commun.* **2002,** 1682–1683. (c) Bagley, M. C.; Dale, J. W.; Ohnesorge, M.; Xiong, X.; Bower, J. *J. Comb. Chem.* **2003,** *5,* 41–44. (d) Cacchi, S.; Fabrizi, G.; Filisti, E. *Org. Lett.*, **2008,** *10,* 2629–2632.
122 Bagley, M. C.; Glover, C.; Chevis, D. *Synlett* **2005,** 649–651
123 Bagley, M. C.; Lunn, R.; Xiong, X. *Tetrahedron Lett.* **2002,** *43,* 8331–8334.
124 (a) Adlington, R. M.; Baldwin, J. E.; Catterick, D.; Pritchard, G. J.; Tang, L. T. *J. Chem. Soc., Perkin Trans. 1* **2000,** 303–305. (b) Adlington, R. M.; Baldwin, J. E.; Catterick, D.; Pritchard, G. J.; Tang, L. T. *J. Chem. Soc. Perkin Trans. 1* **2000,** 2311–2316.
125 Bagley, M. C.; Dale, J. W.; Merritt, E. A.; Xiong, X. *Chem. Rev.* **2005,** *105,* 685.
126 Bagley, M. C.; Bashford, K. E.; Hesketh, C. L.; Moody, C. J. *J. Am. Chem. Soc.* **2000,** *122,* 3301–3313.
127 Bagley, M. C.; Xiong, X. *Org. Lett.* **2004,** *6,* 3401.
128 Merritt, E. A.; Bagley, M. C. *Synlett* **2007,** 954–958.

129 Aulakh V. S.; Ciufolini, M. A. *J. Org. Chem.* **2009,** *74*, 5750–5753.

130 Bashford, K. E.; Burton, M. B.; Cameron, S.; Cooper, A. L.; Hogg, R. D.; Dane, P. D.; McManus, D. A.; Matrunola, C. A.; Moody, C. J.; Robertson, A. A. B.; Warne, M. R. *Tetrahedron Lett.* **2003,** *44*, 1627–1629.

131 Davis, J. M.; Truong, A.; Hamilton, A. D. *Org. Lett.* **2005,** *7*, 5405–5408.

132 (a) Kröhnke, F.; Zecher, W. *Angew. Chem. Int. Ed.* **1962,** *1*, 626–632. (b) Kröhnke, F. *Synthesis* **1976,** 1–24. (c) Zecher, W.; Kröhnke, F. *Ber.* **1961,** *94*, 690–698. (d) Zecher, W.; Kröhnke, F. *Angew. Chem. Int. Ed.* **1963,** *2*, 380.

133 Sauers, A. L.; Ho, D. M.; Bernhard, S. *J. Org. Chem.* **2004,** *69*, 8910–8915.

134 Sala, X.; Rodriguez, A. M.; Rodriguez, M. R.; Romero, I.; Parella, T.; von Zelewsky, A.; Llobet, A.; Benet-Buchholz, J. *J. Org. Chem.* **2006,** *71*, 9283–9290.

135 Katritzky, A. R.; Abdel-Fattah, A. A. A.; Tymoshenko, D. O.; Essawy, S. A. *Synthesis* **1999,** 2114–2118.

136 Katritzky, A. R.; Abdel-Fattah, A. A. A.; Belyakov, S. A.; Fahmy, A. F. M. *J. Chem. Res.* **1998,** 334–335.

137 Kelly, T. R.; Lee, Y.-J.; Mears, R. J. *J. Org. Chem.* **1997,** *62*, 2774–2781.

138 Yan, C.-G.; Wang, Q.-F.; Cai, X.-F.; Sun, J. *Cent. Eur. J. Chem.* **2008,** *6*, 188–198.

139 Saikai, P.; Prajapati, D.; Sandhu, J. S. *Tetrahedron Lett.* **2003,** *44*, 8725–8727.

140 Ash, A. B.; LaMontagne, M. P.; Markovac, A. U.S.3886167 (1975).

141 [R] (a) Boger, D. L. *Tetrahedron* **1983,** *39*, 2869–2939. [R] (b) Boger, D. L. *Chem. Rev.* **1986**, *86*, 781–793. [R] (c) Behforouz, M.; Ahmadian, M. *Tetrahedron* **2000,** *56*, 5259–5288. [R] (d) Buonora, P.; Olsen, J.-C.; Oh, T. *Tetrahedron* **2001,** *57*, 6099–6138. [R] (e) Jayakumar, S.; Ishar, M. P. S.; Mahajan, M. P. *Tetrahedron* **2002,** *58*, 379–471.

142 Woodward, R. B.; Hoffmann, R. *The Conservation of Orbital Symmetry*, VCH Publishers: New York, NY, 1970.

144 (a) Boger, D. L.; Panek, J. S. *J. Org. Chem.* **1982,** *47*, 3763–3765. (b) Boger, D. L.; Panek, J. S. *J. Org. Chem.* **1983,** *48*, 621–623. (c) Boger, D. L.; Panek, J. S. *Tetrahedron Lett.* **1983,** *24*, 4511–4514. (d) Boger, D. L.; Panek, J. S. *J. Am. Chem. Soc.* **1985,** *107*, 5745–5754. (e) Kende, A. S.; Lorah, D. P.; Boatman, R. J. *J. Am. Chem. Soc.* **1981,** *103*, 1271–1273.

145 Boger, D. L.; Panek, J. S. *Tetrahedron Lett.* **1984,** *25*, 3175–3178

146 (a) Taylor, E. C.; French, L. G. *Tetrahedron Lett.* **1986,** *27*, 1967–1970. (b) Taylor, E. C.; Macor, J. E.; Pont, J. L. *Tetrahedron* **1987,** *43*, 5145–5158. (c) Taylor, E. C.; Pont, J. L. *Tetrahedron Lett.* **1987,** *28*, 379–382. (d) Taylor, E. C.; Macor, J. E. *J. Org. Chem.* **1987**, *52*, 4280–4287. (e) Taylor, E. C.; Macor, J. E. *J. Org. Chem.* **1989,** *54*, 4984–4989.

147 (a) Benson, S. C.; Li, J.-H.; Snyder, J. K. *J. Org. Chem.* **1992,** *57*, 5285–8587. (b) Li, J.-H.; Snyder, J. K. *J. Org. Chem.* **1993,** *58*, 519–519. (c) Benson, S. C.; Lee, L.; Snyder, J. K. *Tetrahedron Lett.* **1996,** *37*, 5061–5064. (d) Wan, Z.-K.; Snyder, J. K. *Tetrahedron Lett.* **1998,** *39*, 2487–2490.

148. Turchi, I. J.; Dewar, M. J. S. *Chem. Rev.* **1975,** *75*, 389–437.

149 Kozikowski, A. P.; Hasan, N. M. *J. Org. Chem.* **1977,** *42*, 2039–2040.

150 (a) Firestone, R. A.; Harris, E. E.; Reuter, W. *Tetrahedron* **1967,** *23*, 943–955. (b) Muhlradt, P. F.; Morino, Y.; Snell, E. E. *J. Med. Chem.* **1967,** *10*, 341–344.

151 (a) Levin, J. I.; Weinreb, S. M. *J. Am. Chem. Soc.* **1983,** *105*, 1397–1398. (b) Levin, J. I.; Weinreb, S. M. *J. Org. Chem.* **1984,** *49*, 4325–4332

152 Subramanyam, C.; Noguchi, M.; Weinreb, S. M. *J. Org. Chem.* **1989,** *54*, 5580–5585

153 Boger, D. L.; Wysocki, R. J., Jr. *J. Org. Chem.* **1989,** *54*, 714–718
154 (a) Marcelis, A. T. M.; Van der Plas, H. C. *J. Org. Chem.* **2006,** *51*, 67–71. (b) Rickborn, B. *Organic Reactions* **1998,** *53*.
155 Boger, D. L. *J. Heterocyclic Chem.* **1996,** *33*, 1519–1531
156 Blagg, B. S. J.; Boger, D. L. *Tetrahedron* **2002,** *58*, 6343–6349
157 Villcampa, M.; Perez, J. M.; Avendano, C.; Menendez, J. C. *Tetrahedron* **1994,** *50*, 10047–10054
158 Bushby, N.; Moody, C. J.; Riddick, D. A.; Waldron, I. R. *J. Chem. Soc. Perkin Trans. 1* **2001,** 2183–2193
159 (a) Anderson E. D.; Boger, D. L. *J. Am. Chem. Soc.* **2011,** *133*, 12285–12292. (b) Anderson E. D.; Boger, D. L. *Org. Lett.* **2011,** 2492–2494.
160 (a) Boger, D. L.; Weinreb, S. M. **1984,** *Hetero-Diels–Alder Methodology in Organic Synthesis*, Academic Press, Orlando, Chapter 2. (b) Weinreb, S. M. **1991,** *Heterodienophile Additions To Dienes,* in *Comprehensive Organic Synthesis*, Vol. 4 (eds. Trost, B. M. and Fleming, I.), Pergamon, Oxford, p401.
161 Rüffer, U.; Breitmaier, E. J. *J. Chem. Soc.Chem. Commun.* **1989,** 623.
162 Renslo, A. R.; Danheiser, R. L. *J. Org. Chem.* **1998,** *63*, 7840.
163 Bland, D. C.; Raudenbush, B. C.; Weinreb, S. M. *Org. Lett.* **2000,** *2*, 4007–4009.
164 [R] (a) Vollhardt, K. P. C. *Angew. Chem. Int. Ed. Engl.* **1984,** *23*, 539–644. [R] (b) Bönnemen, H. *Angew. Chem. Int. Ed. Engl.* **1985,** *24*, 248–262. [R] (c) Varela, J. A.; Saà, C. *Chem. Rev.* **2003,** *103*, 3787–3801. [R] (d) Nakamura, I.; Yamamoto, Y. *Chem. Rev.* **2004,** *104*, 2127–2198. [R] (e) Yamamoto, Y. *Curr. Org. Chem.* **2005,** *9*, 503–519. [R] (f) Kotha, S.; Brahmachary, E.; Lahiri, K. *Eur. J. Org. Chem.* **2005,** 4741–4767. [R] (g) Chopade, P. R.; Louie, J. *Adv. Synth. Catal.* **2006,** *348*, 2307–2327. [R] (h) Heller, B.; Hapke, M. *Chem. Soc. Rev.* **2007,** *36*, 1085–1094.
165 (a) Wakatsuki, Y.; H. Yamazaki, H. *J. Chem. Soc., Chem. Commun.* **1973,** 280. (b) Wakatsuki, Y.; H. Yamazaki, H. *Tetrahedron Lett.* **1973,** *14*, 3383–3384. (c) Wakatsuki, Y.; H. Yamazaki, H. *J. Chem. Soc., Dalton Trans.* **1978,** 1278–1282
166 (a) Wagler, P.; Heller, B.; Ortner, J.; Funken, K.-H.; *Oehme, G. Chemie Ingenieur Technik* **1996,** *68*, 823. (b) Oehme, G.; Heller, B,; Wagler, P. *Energy* **1997,** *22*, 327.
167 Heller, B.; Heller, D.; Oehme, G. J. *Mol. Cata. A Chem.* **1998,** *136*, 219. (b) Heller, D.; Wagler, P.; Oehme, G. *Mol. Cata. A Chem.* **1996,** *110*, 211.
168 Heller, B.; Sundermann, B.; Buschmann, H.; Drexler, H.-J.; You J.; Holzgrabe, U.; Heller, E.; Oehme, G. *J. Org. Chem.* **2002,** *67*, 4414.
169 Bönnemen, H.; Brijoux, W.; Brinkmann, R.; Meurers, W.; Mynott, R.; von Philipsborn, W.; Egolf, T. *J. Organomet. Chem.* **1984,** *272*, 231.
170 Wakatsuki, Y.; Yamazaki, H. *Bull. Chem. Soc. Jap.* **1985,** *58*, 2715.
171 Selected examples for cobalt-catalyzed [2+2+2]-cycloaddition to form pyridines reported after 2007: (a) Chang, H.-T.; Jeganmohan, M. Chang, C.-H. *Org. Lett.* **2007,** *9*, 505. (b) Kase, K.; Goswami, A.; Ohtaki, K.; Tanabe, E.; Saino, N.; Okamoto, S. *Org. Lett.* **2007,** *9*, 931. (c) Goswami, A.; Ohtaki, K.; Kase, K.; Ito, T.; Okamoto, S. *Adv. Synth. Catal.* **2008,** *350*, 143. (d) Zhou, Y.; Porco, J. A.; Snyder, J. K. *Org. Lett.* **2007,** *9*, 393. (e) Young, D. D.; Deiters, A. *Angew. Chem. Int. Ed.* **2007,** *46*, 5187. (f) Geny, A.; Agenet, N.; Iannazzo, L.; Malacria, M.; Aubert, C.; Gandon, V. *Angew. Chem. Int. Ed.* **2009,** *48*, 1810. (g) Garcia, P.; Moulin, S.; Miclo, Y.; Leboeuf, D.; Gandon, V.; Aubert, C.; Malacria, M. *Chem. Eur. J.* **2009,** *15*, 2129. (h) Yuan, C.; Chang, C.-T.; Axelrod, A.; Siegel, D. *J. Am. Chem. Soc.* **2010,** *132*, 5924. (i) Hapke, M.; Weding, N.; Spannenberg, A.

Organomet. **2010,** *29*, 4298. (j) Hapke, M.; Kral, K.; Fischer, C.; Spannenberg, A.; Gutnov, A.; Redkin, D.; Heller, B. *J. Org. Chem.* **2010,** *75*, 3993

172 Ruthenium-catalyzed [2+2+2]-cycloaddition to form pyridines: (a) Yamamoto, Y.; Ogawa, R.; Itoh, K. *J. Am. Chem. Soc.* **2001,** 123, 6189. (b) Yamamoto, Y.; Okuda, S.; Itoh, K. *Chem. Commun.* **2001,** 1102. (c) Yamamoto, Y.; Kinpara, K.; Saigoku, T.; Takagishi, H.; Okuda, S.; Nishiyama, H.; Itoh, K. *J. Am. Chem. Soc.* **2005,** *127*, 605. (d) Yamamoto, Y.; Kinpara, K.; Nishiyama, H.; Itoh, K. *Adv. Synth. Catal.* **2005,** *347*, 1913. (e) Yamamoto, Y.; Kinpara, K.; Ogawa, R.; Nishiyama, H.; Itoh, K. *Chem. Eur. J.* **2006,** *12*, 5618. (f) Varela, J. A.; Castedo, L.; C. Saa, C. *J. Org. Chem.* **2003,** *68*, 8595–8598

173 Rhodium-catalyzed [2+2+2]-cycloaddition to form pyridines: (a) Tanaka, K.; Suzuki, N.; Nishida, G. *Eur. J. Org. Chem.* **2006,** 3917. (b) Tanaka, K.; Hara, H.; Nishida, G.; Hirano, M. *Org. Lett.* **2007,** *9*, 1907. (c) Wada, A.; Noguchi, K.; Hirano, M.; Tanaka, K. *Org. Lett.* **2007,** *9*, 1295. (d) Shibata, T.; Uchiyama, T.; Endo, K. *Org. Lett.* **2009,** *11*, 3906. (e) Komine, Y.; Tanaka, K. *Org. Lett.* **2010,** *12*, 1312.

174 Nickel-catalyzed [2+2+2]-cycloaddition to form pyridines: (a) McCormick, M. M.; Duong, H. A.; Zuo, G.; Louie, J. *J. Am. Chem. Soc.* **2005,** *127*, 5030. (b) Tekavec, T. N.; Zuo, G.; Simon, K.; Louie, J. *J. Org. Chem.* **2006,** *71*, 5834. (c) Stolley, R. M.; Maczka, M. T.; Louie, J. *Eur. J. Org. Chem.* **2011,** 3815–3824.

175 Titanium-catalyzed [2+2+2]-cycloaddition to form pyridines: (a) Suzuki, D.; Tanaka, R.; Urabe, H.; Sato, F. *J. Am. Chem. Soc.* **2002,** *124*, 3518. (b) Suzuki, D.; Nobe, Y.; Watai, Y.; Tanaka, R.; Takayama, Y.; Sato, F.; Urabe, H. *J. Am. Chem. Soc.* **2005,** *127*, 7474. (c) Tanaka, R.; Yuza, A.; Watai, Y.; Suzuki, D.; Takayama, Y.; Sato, F.; Urabe, H. *J. Am. Chem. Soc.* **2005,** *127*, 7774.

176 Zr/Ni bimetallic catalyst system for pyridine formation: (a) Takahashi, T.; Tsai, F.-Y.; Martin, K. *J. Am. Chem. Soc.* **2000,** *122*, 4994. (b) Takahashi, T.; Tsai, F.-Y.; Li, Y.; Wang, H.; Kondo, K.; Yamanaka, M.; Nakajima, K.; Kotora, M. *J. Am. Chem. Soc.* **2002,** *124*, 5059. (c) Zhou, L.; Yamanaka, M.; Kanno, K.-I.; Takahashi, T. *Heterocycles* **2008,** *76*, 923

177 Reports on iron-catalyzed [2 + 2 + 2]-cycloaddition to form pyridines: (a) Karine, F.; Toupet, L.; Guerchais, V. *Organomet.* **2002,** *21*, 2578–2580. (b) Schmidt, U.; Zenneck, U. *J. Organomet. Chem.* **1992,** *440*, 187. (c) Knoch, F.; Kremer, F.; Schmidt, U.; Zenneck, U.; Floch, P. L.; Mathey, F. *Organomet.* **1996,** *15*, 2713. (d) Wang, C.; Li. X.; Wu, F.; Wan, B. *Angew. Chem. Int. Ed.* **2011,** *50*, 7162–7166. (e) D'Souza, B. R.; Lane, T. K.; Louie, J. *Org. Lett.* **2011,** *13*, 2936–2939.

178 Zhou, Y.; Porco, J. A., Jr.; Snyder, J. K. *Org. Lett.* **2007,** *9*, 393.

179 Chang, H.-T.; Jeganmohan, M. Cheng, C.-H. *Org. Lett.* **2007,** *9*, 505–508.

180 Reviews on construction of pyridine cores: (a) Henry, G. D. *Tetrahedron* **2004,** *60*, 6043. (b) Hill, M. D. *Chem. Eur. J.* **2010,** *16*, 12052.

181 Roesch, K. R.; Larock, R. C. *J. Org. Chem.* **2002,** *67*, 86–94.

182 Colby, D. A.; Bergman, R. G.; Ellman, J. A. *J. Am. Chem. Soc.* **2008,** *130*, 3645–3651.

183 Parthasarathy, K.; Jeganmohan, M.; Cheng, C.-H. *Org. Lett.* **2008,** 10, 325–328.

184 Liu, S.; Yu, Y., Liebeskind L., S. *Org. Lett.* **2007,** *9*, 1947.

185 Liu, S.; Liebeskind, L. S. *J. Am. Chem. Soc.* **2008,** *130*, 6918.

186 [R] (a) Katritzky, A. R.; Lam, J. N. *Heterocycles* **1992,** *33*, 1011–1049. (b) Katritzky, A. R.; Lagowski, J. M. *Chemistry of the Heterocylic N-Oxides* Academic Press: NY, 1971.

187 (a) Kobayashi, G.; Furukawa, S. *Pharm. Bull. Jpn.* **1953,** *1*, 347–349. (b) Okuda, S. *Pharm. Bull. Jpn.* **1955,** *3*, 316–318.

188 (a) Boekelheide, V.; Linn, W. J. *J. Am. Chem. Soc.* **1954,** *76*, 1286–1291. (b) Boekelheide, V.; Harrington, D. L. *Chem. Ind.* **1955,** 1423–1424. (c) Boekelheide, V.; Lehn, W. L. *J. Org. Chem.* **1961,** *26*, 428–430.
189 Bullitt, O. H., Jr.; Maynard. J. T. *J. Am. Chem. Soc.* **1954,** *76*, 1370–1371.
190 Berson, J. A.; Cohen, T. *J. Am. Chem. Soc.* **1955,** *77*, 1281–1283.
191 Bell, S. C.; Childress, S. J. *J. Org. Chem.* **1962,** *27*, 1691–1695.
192 [R] Padwa, A.; Gunn, D. E. Jr.; Osterhout, M. H. *Synthesis* **1997,** 1353–1377.
193 Pachter, I. J. *J. Am. Chem. Soc.* **1953,** *75*, 3026–3027.
194 Muth, C. W.; R. S. Dallak, R. S.; M. L. DeMatte, M. L.; G. F. Chovenec, G. F. *J. Org. Chem.* **1968,** *33*, 2762–2767.
195 Kobayashi, G.; Furukawa, S. *Pharm. Bull.* **1953,** *1*, 347–349.
196 Okuda, S. *Pharm. Bull.* **1955,** *3*, 316–318.
197 (a) Traynelis, V. J.; Martello, R. F. *J. Am. Chem. Soc.* **1958,** *80*, 6590–6593. (b) Traynelis, V. J. Sr.; Gallagher, A. I.; Martello, R. F. *J. Org. Chem.* **1961,** *26*, 4365–4368.
198 (a) Cohen, T.; Fager, J. H. *J. Am. Chem. Soc.* **1965,** *87*, 5701–5710. (b) Koenig, T. *J. Am. Chem. Soc.* **1966,** *88*, 4045–4049. (c) Koenig, T.; Wieczorek, J. S. *J. Org. Chem.* **1968,** *33*, 1530–1532.
199 (a) Oae, S.; Kitao, T.; Kitaoka, Y. *J. Am. Chem. Soc.* **1962,** *84*, 3359–3362. (b) Bodalski, R.; Katritzky, A. R. *Tetrahedron Lett.* **1968,** 257–260. (c) Tamagaki, S.; Kozuka, S.; Oae, S. *Tetrahedron Lett.* **1968,** 4765–4768. (d) Bodalski, R.; Katritzky, A. R. *J. Chem. Soc. (B)* **1968,** 831–838. (e) Oae, S.; Kozuka, S. *Tetrahedron* **1964,** *20*, 2671–2676.
200 Nicolaou, K. C.; Safina, B. S.; Funke, C.; Zak, M.; Zecri, F. J. *Angew. Chem. Int. Ed.* **2002,** *41*, 1937–1940.
201 Collier, P. N.; Campbell, A. D.; Patel, I.; Taylor, R. J. K. *Tetrahedron* **2002,** *58*, 6117–6125.
202 Daines, R. A.; Chambers, P. A.; Folley, J. J., Griswold, D. E.; Kingsbury, W. D.; Martin, L. D.; Schmidt, D. B.; Sham, K. K.; Sarau, H. M. *J. Med. Chem.* **1992,** *35*, 2274.
203 (a) Feely, W. E.; Beavers, E. M. *J. Am. Chem. Soc.* **1959,** *81*, 4004–4007. (b) Fife, W. K.; Scriven, E. F. V. *Heterocycles* **1984,** *22*, 2375–2394
204 (a) Stanek, J.; Caravatti, G.; Frei, J.; Furet, P.; Mett, H.; Schneider, P.; Regenass, U. *J. Med. Chem.* **1993,** *36*, 2169–2171. (b) Stanck, J.; Caravatti, G.; Capraro, H.-G.; Furet, P.; Mett, H.; Schneider, P.; Regenass, U. *J. Med. Chem.* **1993,** *36*, 46–54
205 (a) Fife, W. K. *J. Org. Chem.* **1983,** *48*, 1375–1377. (b) Fife, W. K. *Heterocycles* **1984,** *22*, 1121–1124.
206 (a) Harusawa, S.; Hamada, Y.; Shioiri, T. *Heterocycles* **1981,** *15*, 981–984. (b) Susuki, H.; Iwata, C.; Sajurai, K.; Tokumoto, K.; Takahashi, H.; Hanada, M.; Yokoyama, Y.; Murakami, Y. *Tetrahedron* **1997,** *53*, 1593–1606.
207 Chen, D.; Martell, A. E.; Motekaitis, R. J.; McManus, D. *Can. J. Chem.* **1998,** *76*, 445–451.
208. Nicolaou, K. C.; Koumbis, A. E.; Snyder, S. A.; Simonsen, K. B. *Angew. Chem. Int. Ed.* **2000,** *39*, 2529–2533.
209. Andreotti, D.; Miserazzi, E.; Nalin, A.; Pozzan, A.; Profeta, R.; Spada, S. *Tetrahedron Lett.* **2010,** *51*, 6526–6530.
210. Massaro, A.; Mordini, A.; Mingardi, A.; Klein, J.; Andreotti, D. *Eur. J. Org. Chem.* **2011,** 271–279.

211. Ciamician, G. L.; Dennsted, M. *Chem. Ber.* **1881,** *14,* 1153. Corwin, A. H. *Heterocyclic Compounds* **1950,** *1,* 309. Mosher, H. S. *Heterocyclic Compounds* **1950,** *1,* 475; Skell, P. S.; Sandler, R. S. *J. Am. Chem. Soc.* **1958,** *80,* 2024. Vogel, E. *Angew. Chem.* **1960,** *72,* 8.
212. Kermack, W. O.; Perkin, W. H.; Robinson, R. *J. Chem. Soc.* **1922,** *122,* 1896.
213. (a) Kwon, S.; Nishimura, Y.; Ikeda, M.; Tamura, Y. *Synthesis* **1976,** 249. (b) De Angelis, F.; Gambacorta, A.; Nicoletti, R. *Synthesis* **1976,** 798.
214. Joshi, K. C.; Renuka, J.; Seema, J. *J. Indian Chem. Soc.* **1993,** *70,* 567.
215. Král, V.; Gale, P. A.; Anzenbacher, P. Jr.; K. Jursíková; Lynch, V.; Sessler, J. L. *J. Chem. Soc., Chem. Comm.* **1998,** 9.
216 Chubb, R. W. J.; Bryce, M. R.; Tarbit, B. *J. Chem. Soc., Perkin Trans. 1* **2001,** *16,* 1853–1854.
217 Goel, A.; Singh, F. V.; Sharon, A.; Maulik P. R. *Synlett* **2005,** 623–626.
218 Baeyer, A. *Ber.* **1910,** *43,* 2337–2343.
219 (a) Baeyer, A.; Piccard, J. *Liebigs Ann. Chem.* **1912,** *384,* 208–224. (b) Baeyer, A.; Piccard, J.; Gruber, W. *Liebigs Ann. Chem.* **1915,** *407,* 332–369.
220 Pedersen, C.; Harrit, N.; Buchardt, O. *Acta. Chem. Scand* **1970,** *24,* 3435–3443.
221 Uncuţa, C.; Căproiu, M. T.; Câmpeanu, V.; Petride, A.; Dănilă, M. G.; Plăveţi, M.; Balaban, A. T. *Tetrahedron* **1998,** *54,* 9747–9764.
222 Said, S. A; Fiksdahl, A. *Tetrahedron: Asymmetry* **2001,** *12,* 1947–1951.
223 Ilies, M.; Banciu, M. D.; Ilies, M. A.; Scozzafava, A.; Caproiu, M. T.; Supuran, C. T. *J. Med. Chem.* **2002,** *45,* 504–510.
224 Efange, S. M. N.; Michelson, R. H.; Tan, A. K.; Krueger, M. J.; Singer, T. P. *J. Med. Chem.* **1993,** *36,* 1278–1283.
225 Galatsis, P. In *Name Reactions in Heterocyclic Chemistry*; Li, J. J.; Corey, E. J., Eds.; Wiley: Hoboken, NJ, **2005,** 340–349.
226 Galatsis, P. In *Name Reactions in Heterocyclic Chemistry*; Li, J. J.; Corey, E. J., Eds.; Wiley: Hoboken, NJ, **2005,** 323–339.
227 Galatsis, P. In *Name Reactions in Heterocyclic Chemistry*; Li, J. J.; Corey, E. J., Eds.; Wiley: Hoboken, NJ, **2005,** pp 308–309.
228 Galatsis, P. In *Name Reactions in Heterocyclic Chemistry*; Li, J. J.; Corey, E. J., Eds.; Wiley: Hoboken, NJ, **2005,** pp 307–308.
229 Galatsis, P. In *Name Reactions in Heterocyclic Chemistry*; Li, J. J.; Corey, E. J., Eds.; Wiley: Hoboken, NJ, **2005,** pp 304–307.
230 Galatsis, P. In *Name Reactions in Heterocyclic Chemistry*; Li, J. J.; Corey, E. J., Eds.; Wiley: Hoboken, NJ, **2005,** 311–313.
231 Conlon, D. A. In *Name Reactions in Heterocyclic Chemistry-II*, Li, J. J., Ed.; Wiley: Hoboken, NJ, **2011,** pp 338–346.
232 Conlon, D. A. In *Name Reactions in Heterocyclic Chemistry-II*, Li, J. J.; Ed.; Wiley: Hoboken, NJ, **2011,** pp 347–350.

Chapter 11 Quinolines and Isoquinolines

Alexandros L. Zografos

11.1 Introduction

Quinoline and isoquinoline are heterocycles in which a benzene ring and a pyridine ring are fused through carbon. The isomeric heterocycles 1- and 2-azanaphthalene, better known by their trivial names quinoline and isoquinoline, have been the subject of extensive studies since their discovery in the extracts of coal tar at the beginning of nineteenth century. Since then their heterocyclic ring systems were found incorporated in several hundreds of natural products and were used as pharmacophore units in dozens of pharmaceuticals especially anti-bacterials, better known under the general name of "Quinolones".[1]

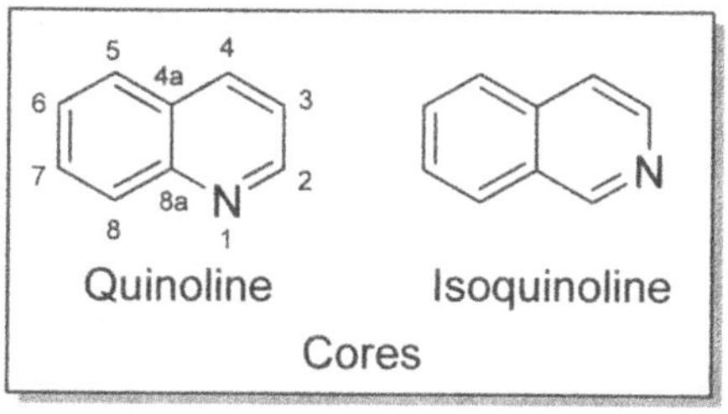

The quinolines are historically among the most important anti-malarial drugs ever used. Throughout the 20th century, the immense use of chloroquine, the most famous drug of this group, along with quinine, provided well-founded hopes for the eradication of malaria before the World War II.[2] These drugs were followed by newer and more effective anti-malarials by decorating, most commonly C2-, C4- and C8-position of the quinoline core, with appropriate groups. Among them, amodiaquine, which was introduced in 1940, piperaquine in the late 1960s, and mefloquine in 1980, still provide important knowledge for the treatment of resistant malaria parasite of the future.

The chemotherapy of malaria basically involves killing the asexual parasites (hemozoin) and providing supporting therapy to the host to boost its immune system. Although the complete mechanism of action for quinoline anti-malarials is not known, it is believed that these drugs mostly act during the blood stages of the parasite's life cycle.[3] Quinoline drugs act as inhibitors of polymerization of heme in the food vacuole of the parasite, preventing the disposal of the polymers to the cytoplasm where hemozoin is formed.[4,5] This leads to intraparasitic accumulation of free heme, which is highly toxic to the parasite. In addition to heme, several other targets are postulated to contribute to the anti-parasitic action of quinolines, such as phospholipases,[6] DNA,[7] tyrosine kinase,[8] etc.

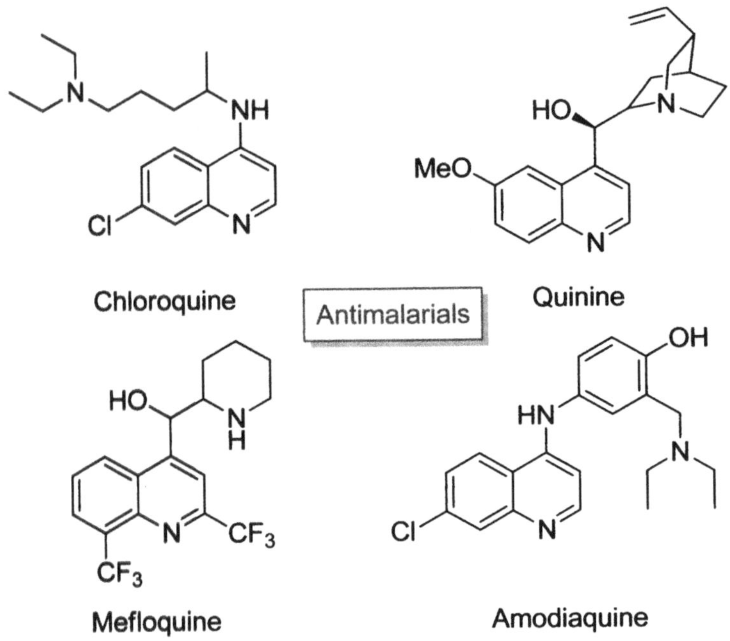

3-Carboxylic acid substituted quinolin-4-one derivatives constitute another class of biologically active compounds often known under the collective name "quinolones". The development of quinolone antibacterials since the discovery of naphthyridine agent, nalidixic acid, has progressed with periods of great innovation.

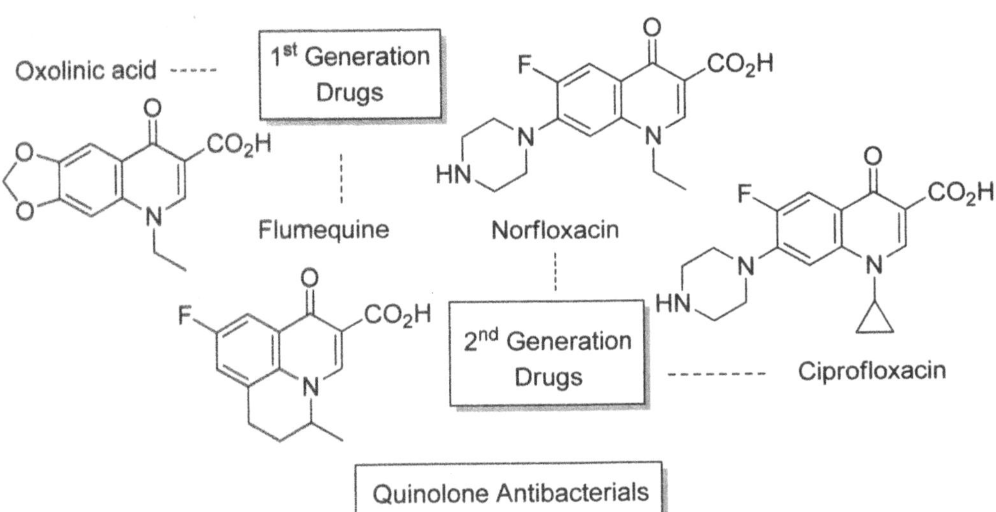

Starting with the limited range activity against Gram-negative bacteria of cinoxacin in 1960s, to the broad spectrum anti-bacterials ciprofloxacin and ofloxacin of the mid-1980s, quinolones still consist of the major group of synthetic antibiotics with activity that ranges from *Enterobacteriaceae* to Gram-positive pathogens, including *Streptococci* and *Staphylococci*.

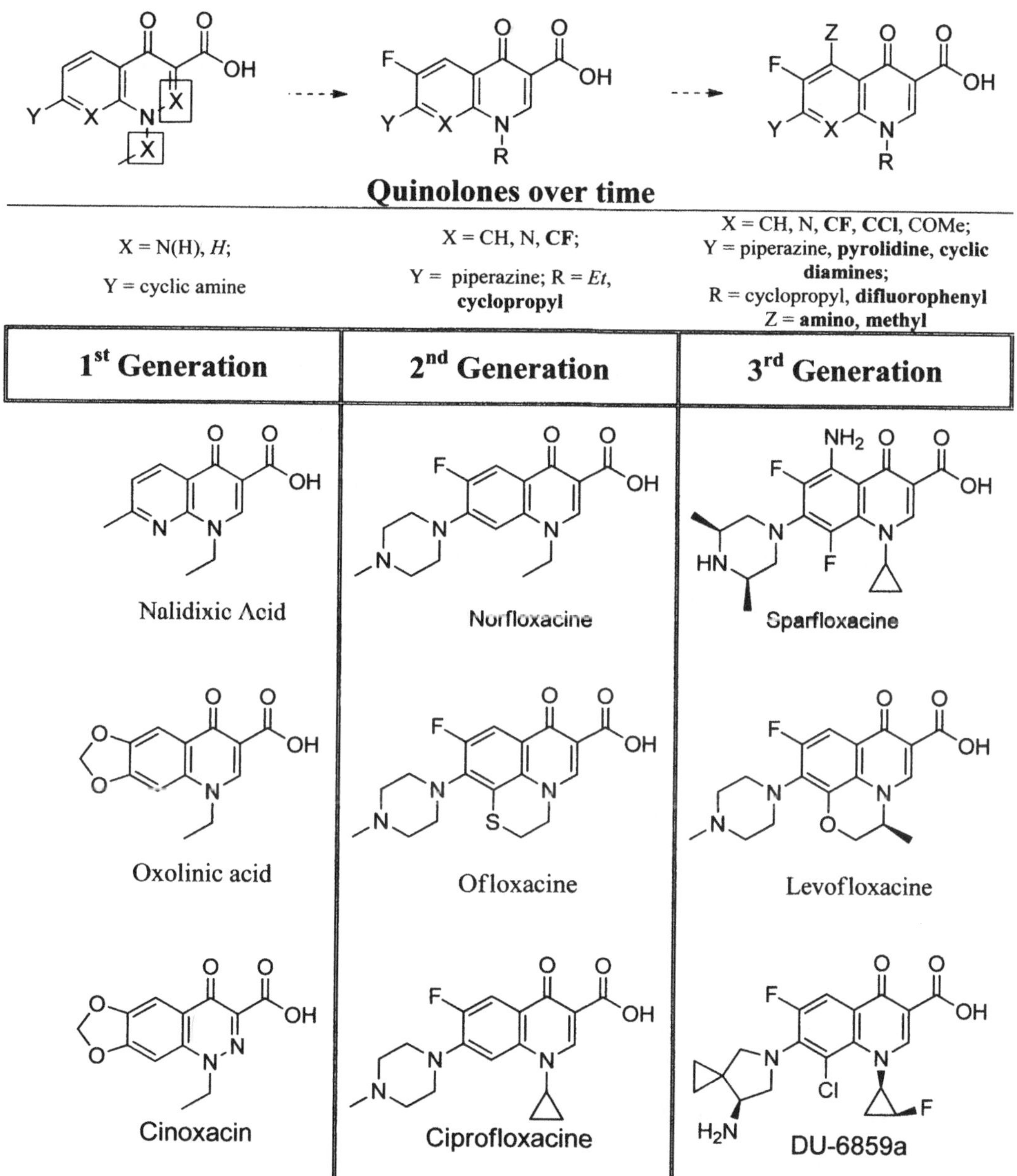

* Substituent written in italics is a choice that nowadays is abandoned. Substituent in bold is a choice used in anti-bacterials today.

Apart from the basic quinoline and isoquinoline cores, several other fused quinoline heterocycles are recognized as useful pharmacophore units. Pyrroloquinoline-based alkaloids attracted considerable interest due to their unique biological activity. The best-known example of this category is camptothecin and its analogues isolated from the stem woods of the Chinese joy tree *Camptotheca acuminate*. These derivatives were found to possess strong antitumor and retroviral activity by binding to and stabilizing a complex of DNA and enzyme topoisomerase I.[9] Recently, irinotecan, a potent anticancer drug, was marketed by Pfizer based on the modification of camptothecin structure to boost its bioavailability.[10]

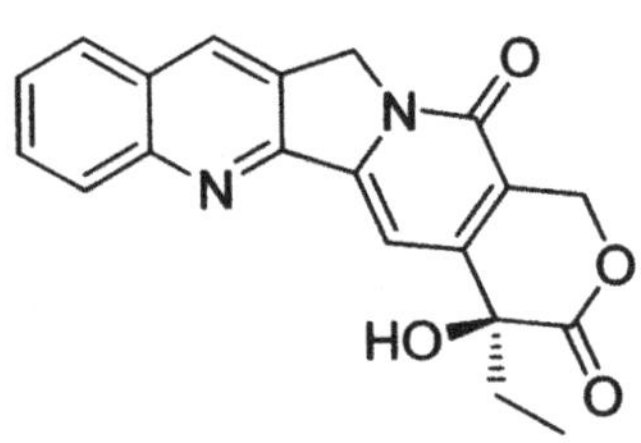

Camptothecin

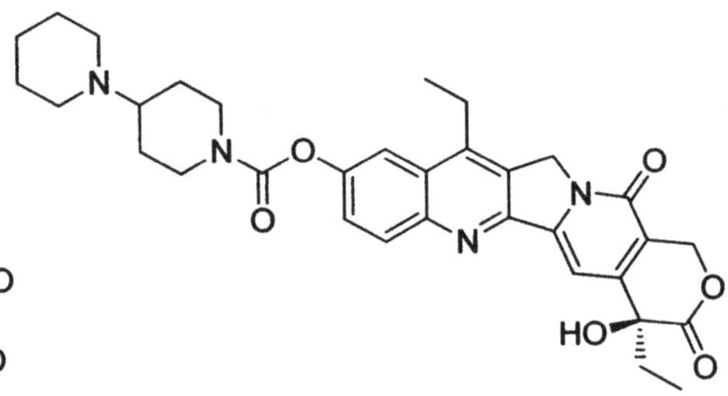

Irinotecan

Pyranoquinoline alkaloids are another important group of quinoline derivatives that possess strong biological activities, from potent inhibitors of platelet aggregation such as zanthodioline, to cytotoxic compounds acting selectively to breast cancer cells such as huajiaosimuline.[11]

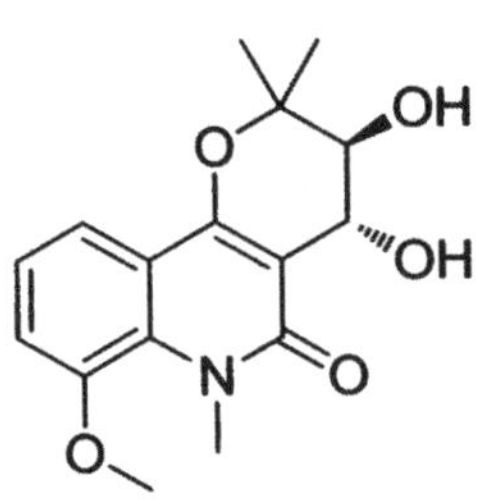

Zanthodioline

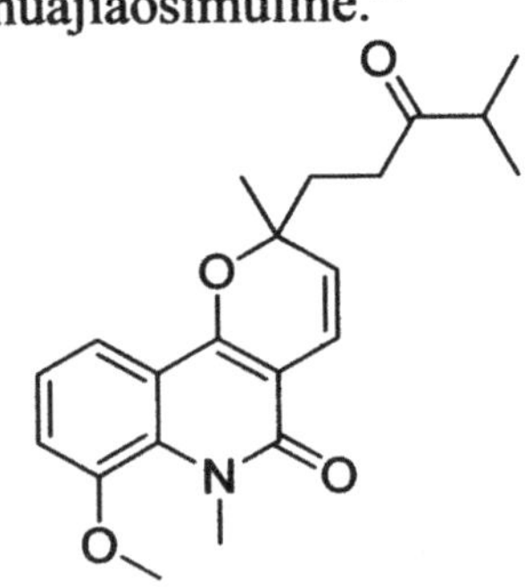

Huajiaosimuline

11.2 Reactivity of the Quinoline and Isoquinoline Ring

Introduction of heteroatoms in benzene ring results in an irregular distribution of electron density, and this strongly influences reactivity and physical properties of quinoline and isoquinoline heterocycles. The electronegative nitrogen atom draws electron density away from the ring carbon atoms, resulting in a permanent dipole moment. As a result, some carbon atoms in the ring have a partial positive charge and quinolines and isoquinolines are described as electron-poor or π-electron deficient systems.[12]

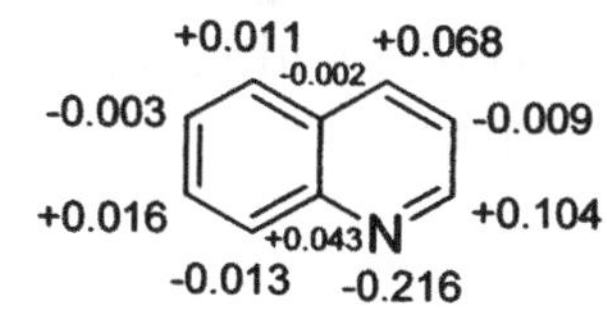

Dipole moments of quinoline

Witnessing this irregular distribution of electron density across the heterocycle, its ^{1}H-NMR indicates a doublet of doublets signal in the downfield region of 8.79 ppm for proton at C2 position. Coupling constant between H2 and H3 is 4.18 Hz whereas a coupling constant of 1.76 Hz can be measured between H2 and H4. Proton at position C4 is observed at 8.02 ppm as a doublet with a coupling constant of 8.19 Hz. H5 has a chemical shift of 7.95 ppm presented as a doublet of doublets with the higher coupling constant being 8.24 Hz. The resonance of H8 is found at 7.63 ppm doublet peak with a coupling constant of 8.57 Hz, whereas H7 as a triplet of doublets at 7.57 ppm. Finally, further highfield can be found H6 and H3 as triplet of doublets at 7.38 ppm and doublet of doublets at 7.20 ppm, respectively.

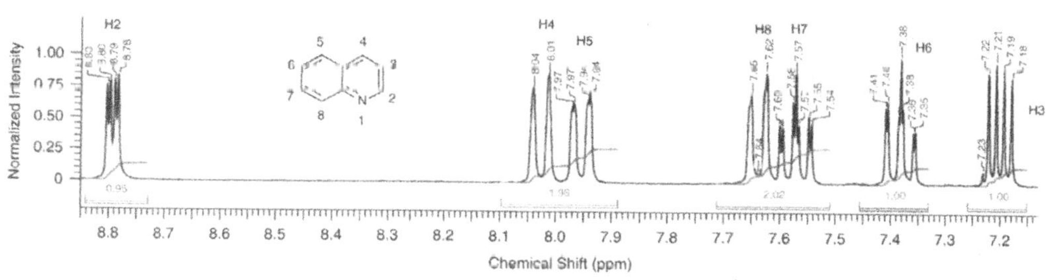

The same logic can also be witnessed for isoquinoline. In this case, the lower down-field signal is a broad singlet for H1 at 9.24 ppm representing the lower electron density possessed in this position in comparison with the doublet for H3 being at 8.50 ppm. H8, H5, H6, and H7 follow at higher field, ending with a broad singlet at 7.23 ppm for C4.

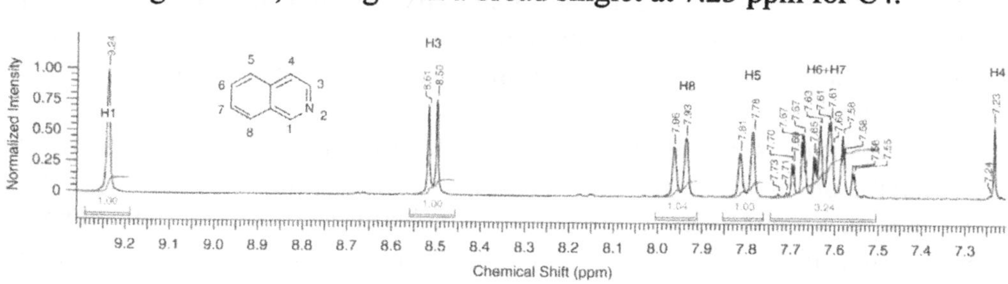

For ^{13}C-NMR, quinoline presents a further down-field signal with a chemical shift of 150.23 ppm for C2 due to the inductive effect of nitrogen atom. C8a has a chemical shift of 148.16 ppm, whereas C4 has the value of 135.81 ppm. At a higher field of 129.32 ppm and 129.26 ppm C7 and C8 can be found, followed by the values of C4a, C5 and C6 at 128.11 ppm, 127.63 ppm, and 126.35 ppm, respectively. Finally, further high field C4 is observed with a chemical shift of 120.88 ppm

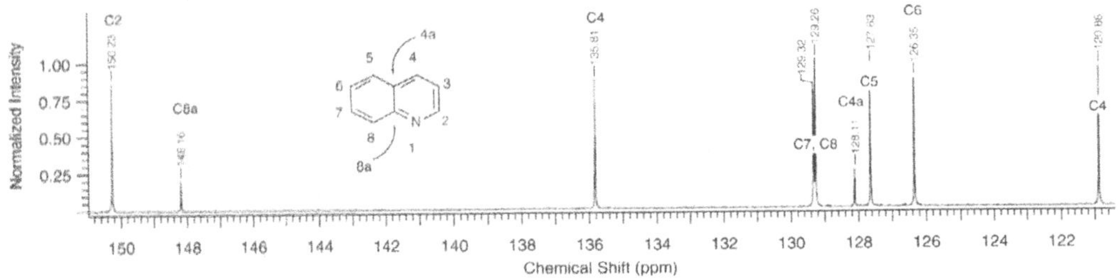

On the other hand, isoquinoline shows a peak at 152.34 ppm for C1, followed by a 142.86 ppm peak for C3 representing the two carbons, which are poor shielded due to the inductive effect from nitrogen atom. C4a gives a resonance value of 135.51 ppm, whereas C6, C-8a, C8, C7, and C5 give values between 130.06 to 126.23 ppm. Once again, C4 provides the higher field signal at 120.21 ppm.

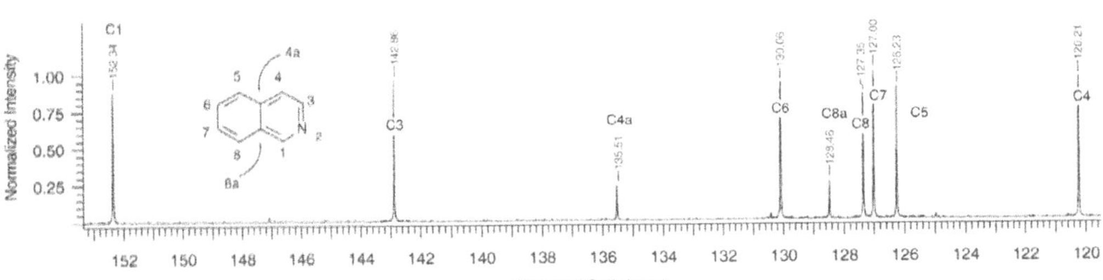

Following the dipole moments presented, quaternary alkylations on nitrogen take place readily. But unlike pyridine, both quinoline and isoquinoline heterocyclic nitrogen promote the reaction with nucleophiles. What is more, electrophilic substitution takes place much more easily than in pyridine, and the substituents are generally located in carbocyclic ring preferentially in the more activated positions of the benzene ring, with a positional selectivity in the case of quinolines in the order of C8 > C5 >>> other positions.

Substituents possessing strong electron-donating character enhance the basicity on nitrogen atom, promoting electrophilic substitution on this atom and as a consequence electrophilic substitution on either pyridine or benzene ring of the heterocycle. In contrast, electron-withdrawing groups on pyridine moiety of the quinoline or isoquinoline core accelerate the nucleophilic substitution on this side.

11.2.1 Protonation

Quinoline and isoquinolines are related to pyridine, exactly as is naphthalene related to benzene. They are 10-π-electron aromatic systems with the heterocyclic ring possessing strong polarized character. The key differences between heterocyclic and nonheterocyclic sides in quinoline and isoquinoline systems are (i) the departure of heterocyclic ring from the perfectly regular hexagonal geometry caused by the presence of the shorter carbon–nitrogen bond, (ii) the replacement of a hydrogen in the plane of heterocyclic ring with an unshared electron pair, located in an sp^2 hybrid orbital, not at all involved in the aromatic π-electron sextet, (iii) the strong permanent dipole due to the greater electronegativity of nitrogen compared with carbon atom (*vide supra*) and (iv) the presence of a polarized imine group.[13] The polarized character of the heterocycle along with the inability of the nitrogen lone-pair to involve in the aromatic sextet is responsible for the basic and nucleophilic reactivities observed by the aforementioned heterocycles.

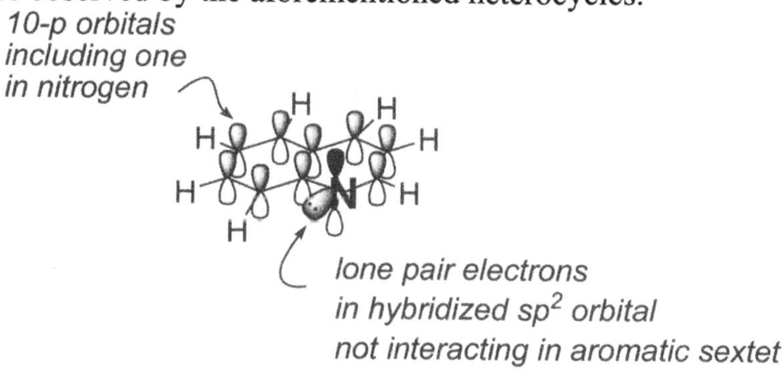

Quinoline orbitals

Some other physical properties are summarized in the Table below.[14]

Physical Properties of Quinoline and Isoquinoline

	Value	
Property	Quinoline	Isoquinoline
mp, °C	–15.6	26.5
bp, °C	238	243
ΔH_{vap}, kJ/mol	46.4	49.0
n_D^{20}	1.6268	1.6148
d^{20}, g/cm^3	1.0929	1.0986
K_a	8.9×10^{-10}	2.5×10^{-9}
Viscocity at 30 °C	2.997	3.2528
T_c	509	530

Both quinoline and isoquinoline are weak bases showing a p*K*a of 4.94 and 5.40, respectively. Treatment with strong acids, such as sulfuric acid, affords protonation in the nitrogen, providing the corresponding quinolinium and isoquinolinium salts.[15]

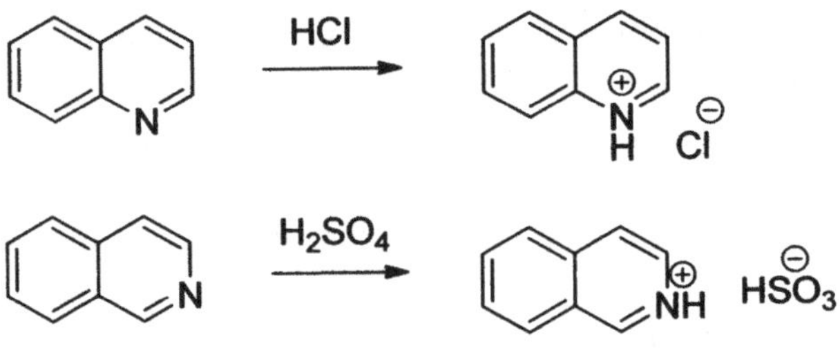

N-Protonated isoquinolines gives facile protonation of the carbocyclic ring as evidenced in its kinetic study with deuteriosulfuric acid at high temperatures. Thus, protonation favors C5 and C8 positions at higher acid strength, while at lower acid strength, a zwitterion is formed between nitrogen and position C1.[16]

11.2.2 Electrophilic Addition to the Nitrogen Atom

The lone-pair electron on the nitrogen is responsible for electrophilic reactions on its side. Thus, heterocyclic nitrogen atom of quinoline and isoquinolines react with electrophiles, as alkyl halides, alkyl sulfates, alkyl-toluene-*p*-sulfonates, etc., providing quinolinium and isoquinolinium salts.

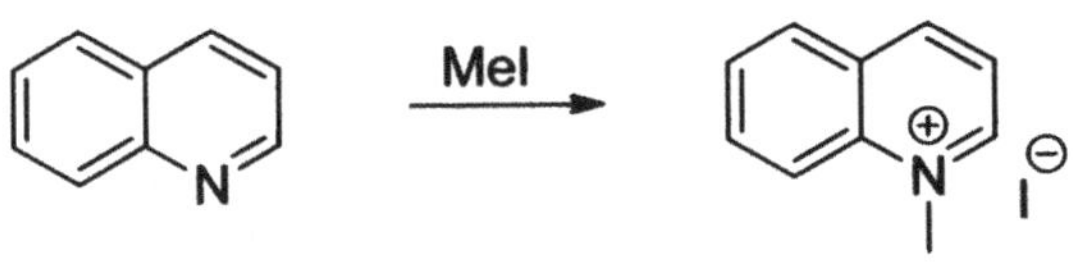

This method was used for the late introduction of ethyl group on nitrogen atom of quinoline core in the synthesis of oxolinic acid antibacterial.[17]

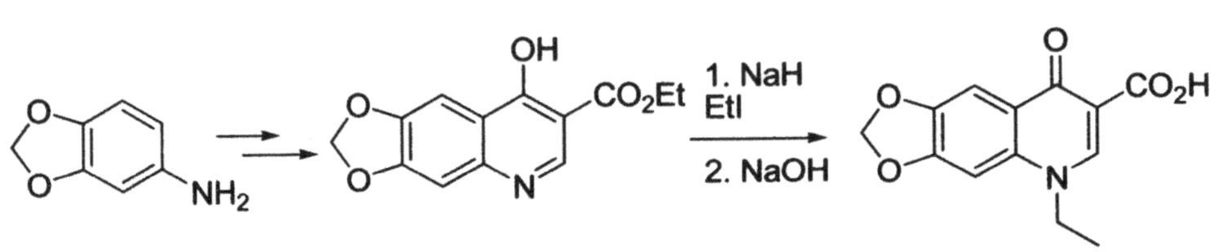

oxolinic acid

Another reaction affording the hydrazine function on quinoline or isoquinoline core is feasible by the nucleophilic attack of nitrogen atom to 2,4-dinitrophenyl *O*-hydroxylamine in the presence of potassium carbonate, as this was well presented in the synthesis of amifloxacin antibiotic.[18]

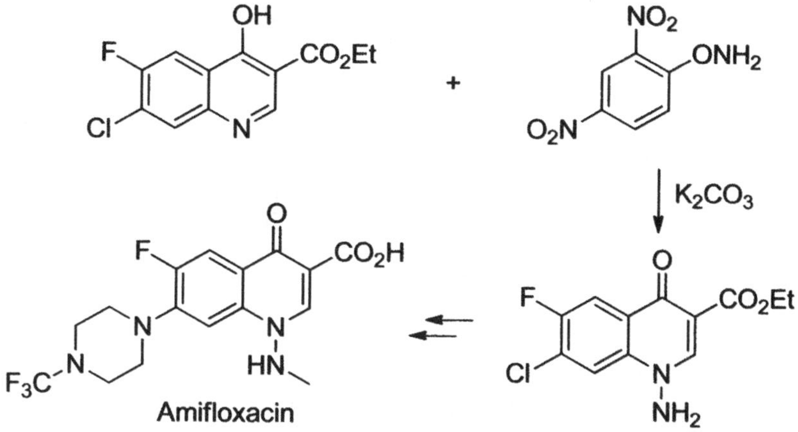

Reaction of peracids with quinolines and isoquinolines affords the corresponding *N*-oxides. Typical conditions are treatment of heterocycles with RCO_2H in the presence of H_2O_2 at −100 °C or *m*-CPBA at 0 °C. The quinoline/isoquinoline nitrogen atom reacts less readily with peracids than the tertiary amines. Large substituents and electron-withdrawing groups slow the reaction.

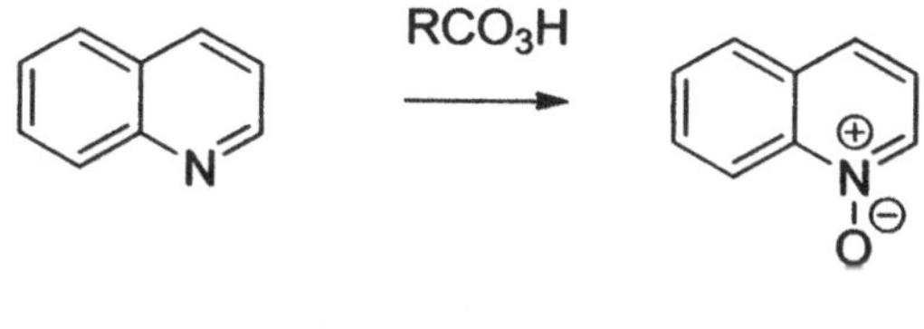

Nitrogen atom can also react with electron-deficient alkenes or alkynes, introducing the corresponding alkyl or alkene group on the heterocycles. This ability was elegantly illustrated in the synthesis of tetrahydroisoquinoline neuromuscular blocking agent cicatracurium.[19]

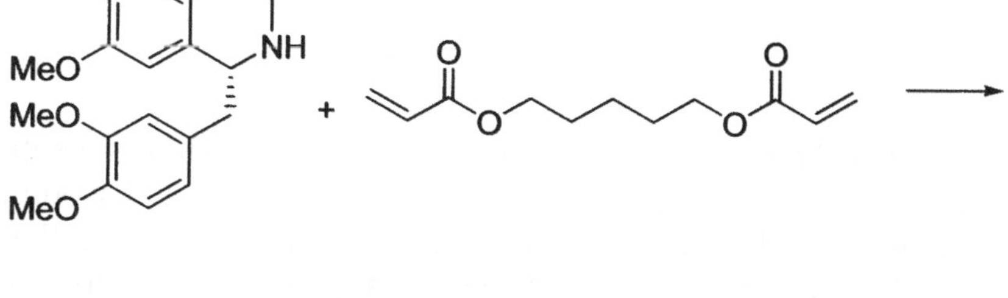

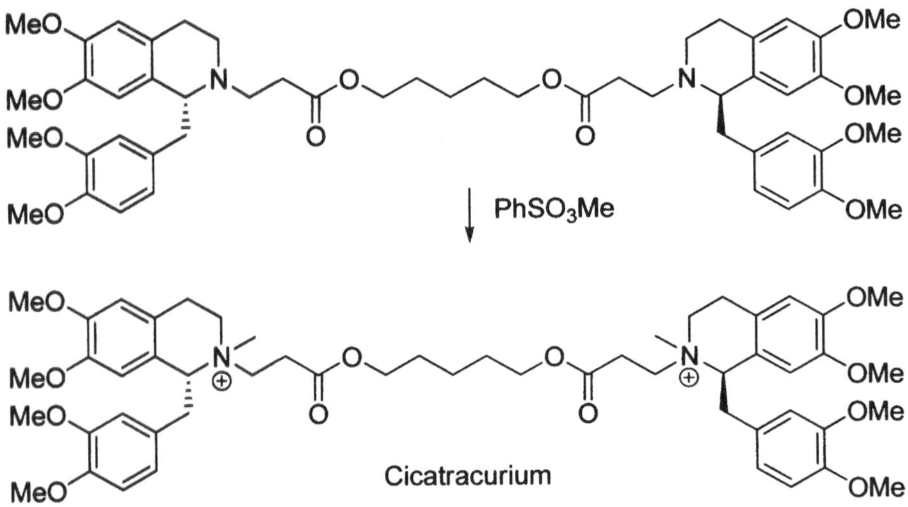

Whenever an appropriate electrophile is present in the reaction mixture along with an alkyne, the *N*-alkylated product reacts with the electrophile, producing an intermediate prone for intramolecular nucleophilic cyclization to the quinolinium or isoquinolinium salt as this is demonstrated in the synthesis of pyrrole–isoquinolines through the formation of mesoanionic intermediates.

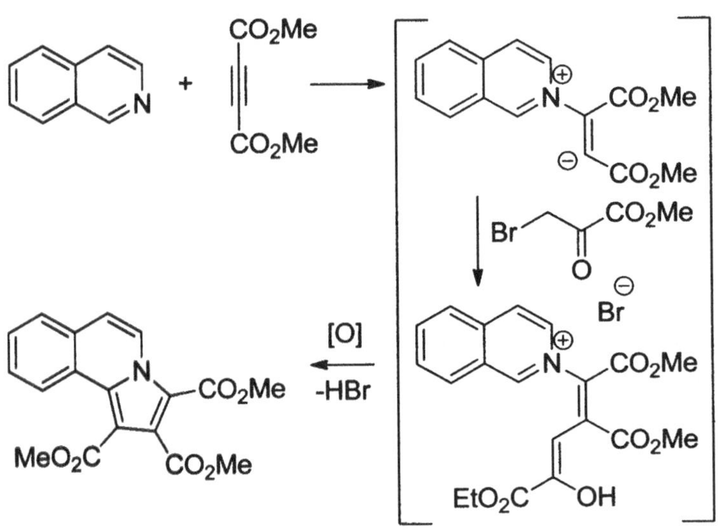

11.2.3 Electrophilic Substitution at Carbon Atom

In contrast to pyridines that are very resistant to electrophilic substitution at carbon without strong activating substituents, quinolines and isoquinolines are susceptible for substitution on the benzene ring. Following the dipole density described above, positions C5 and C8 are the only positions prone to substitution. These substitutions probably involvie attack on the preformed quinolinium or isoquinolinium cations.[13]

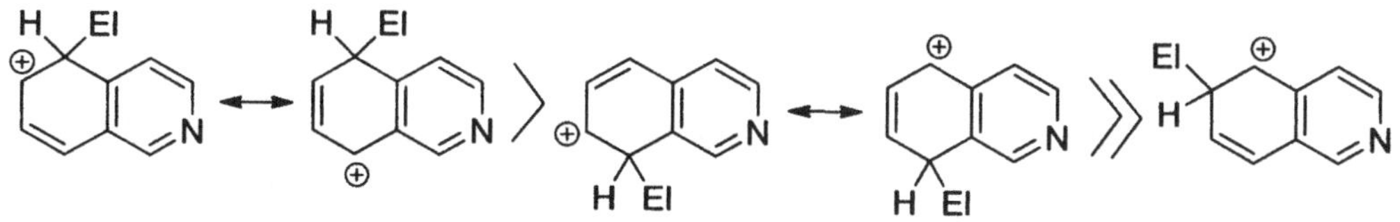

Nitration takes place in the presence of fuming acid and concentrated sulfuric acid under mild conditions and mono-nitrations are occurred exclusively at C5 and C8 positions to provide usually a mixture of products for quinolines and C5 products for isoquinolines.[20]

,

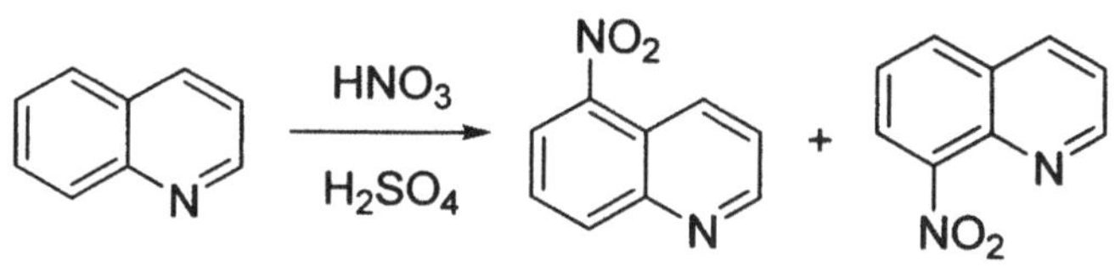

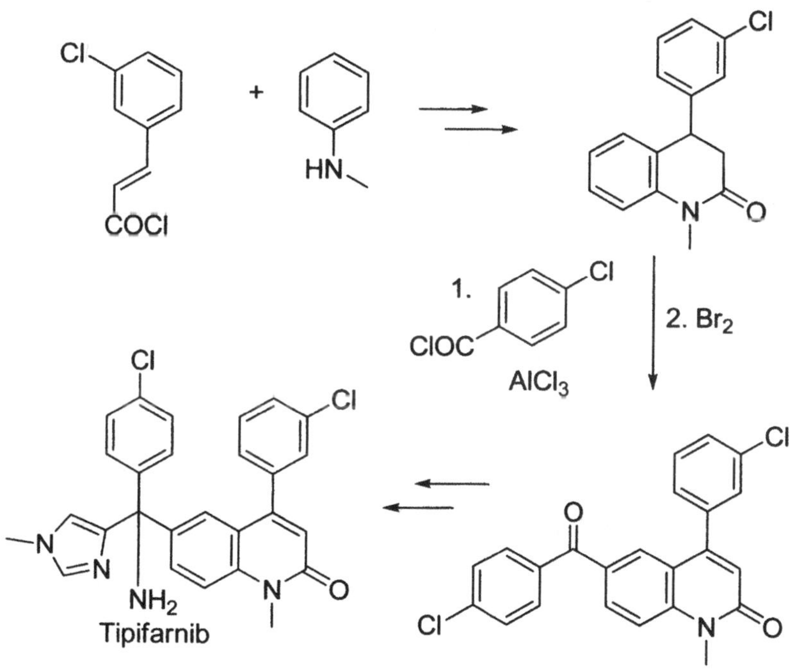

Typical electrophilic reactions as Friedel–Craft acylation and Mannich reaction are not possible due to the nucleophilic nitrogen, which rapidly reacts with the electrophile. However, methods employing prior protection of nitrogen functionality are effective tricks to induce acylation on the desired direction. The preparation of antibiotic agent tipifarnib highlights a clever way for an electrophilic substitution to occur at the C6 position by the action of a Friedel–Craft reaction on the reduced quinolinone-2 core instead followed by a subsequent oxidation by bromine to produce the 4-aryl substituted quinolone core.[21]

On the other hand, Mannich reaction of 5-chloro-8-hydroxy quinoline with formaldehyde and *N,N*-diethylpropylenediamine proceeds according to the expected 7-substitution based on the prior substitution of positions C5 and C8 in a 5,8-disubstituted quinoline to afford the anti-malarial agent clamoxyquin.[22]

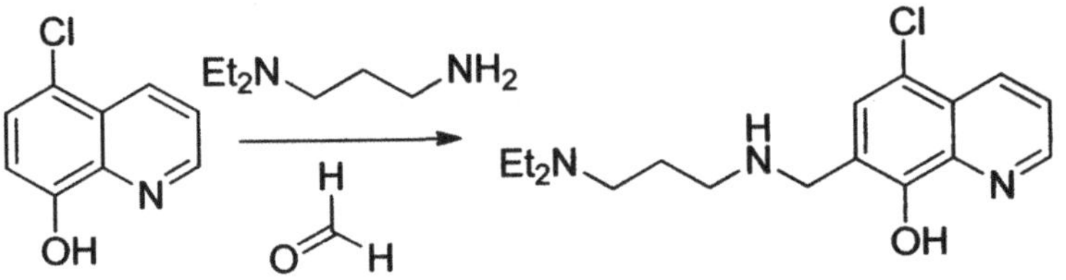

Clamoxyquin

Quinolines and isoquinolines can also react with electrophiles at the pyridine side. This can be rationalized by a different reaction mechanism involving the prior introduction of a nucleophile in the heterocyclic quinoline/isoquinoline ring followed by an electrophilic substitution involving attack on the intermediate enamine. Notable is the electrophilic bromination of isoquinoline hydrobromide in a solvent like nitrobenzene that provides 81% yield of *4*-bromoisoquinoline, in contrast to the bromination or chlorination of an isoquinoline–aluminum chloride complex that affords 78% of *5*-bromoisoquinoline. Exhaustive bromination or chlorination under Lewis acid conditions usually yields mixtures of 5,8-halogenated isoquinolines along with 5,7,8-trisubstituted derivatives.[23]

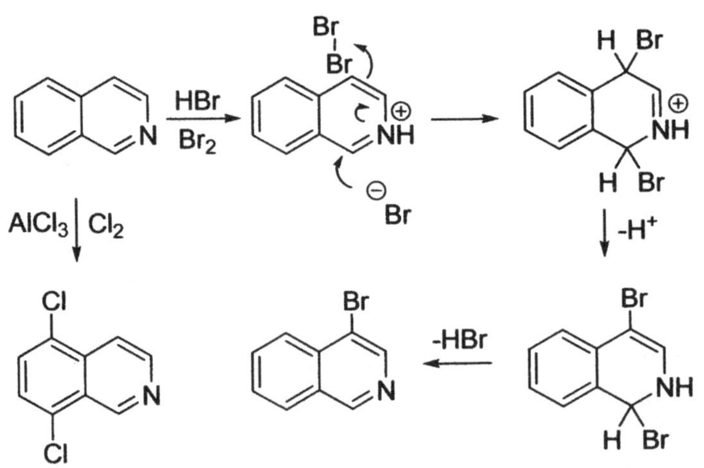

Electron-donating groups on the heteroatom promote electrophilic substitution on heterocyclic ring following the typical electrophilic enamine attack described above.[13]

11.2.4 Nucleophilic Substitution

Quinolines and isoquinolines are susceptible to addition of nucleophiles only at the heterocyclic ring. In quinoline, there is only one α- and one γ-position prone for attack, whereas in isoquinoline, two α-type positions in which 1-position is much more favored due to the inability of nitrogen atom to delocalize a negative charge. In general, the nucleophilic attack is easier in these bicyclic systems in comparison to pyridine compounds because of the associated resonance stabilization offered by the fused-benzene ring.

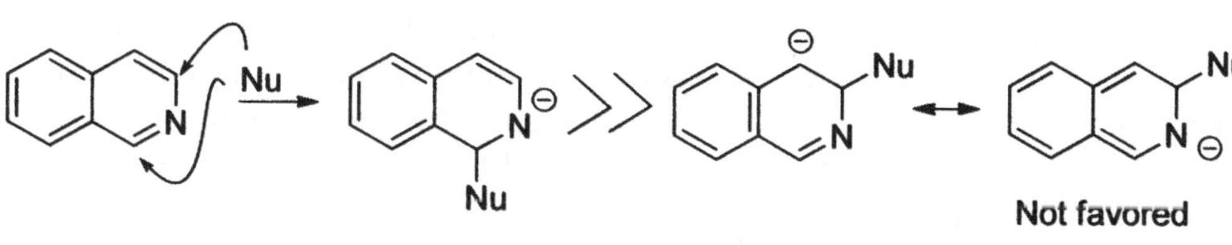

Both quinolines and isoquinolines can be hydroxylated by heating them with potassium hydroxide, forming the tautomeric quinolin-2-one and isoquinolin-1-one derivatives. The formation of these adducts can be regarded as an S_NAr process that proceeds with the evolution of hydrogen.[24]

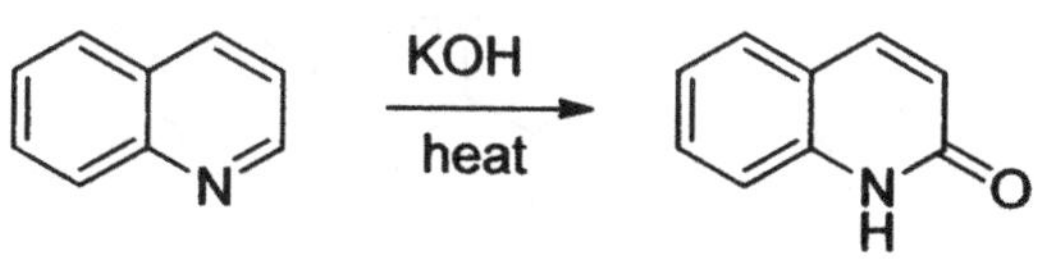

Similar to that of pyridine, the Chichibabin amination on quinoline and isoquinoline proceeds with alkali metal amides in liquid ammonia. In accordance to that, the reaction of quinoline with liquid ammonia initially forms a complex, which allows amide anion to add to the heterocyclic core of quinoline and isoquinoline bicycle, obtaining 2- or 4-aminoquinolines and 1-aminoisoquinolines, respectively, in good yields.[25]

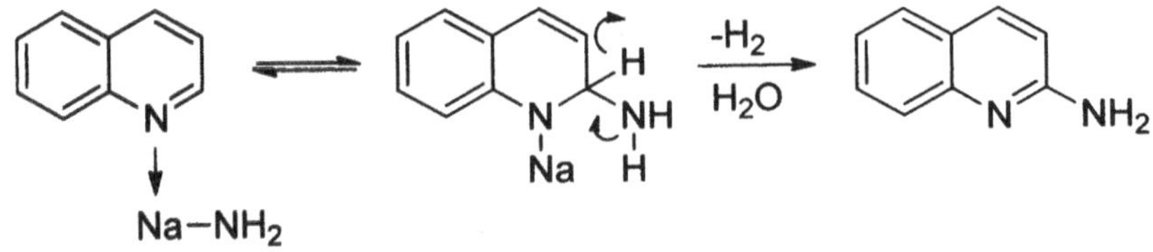

The same logic can be followed on the nucleophilic attack of alkyl or aryl groups on C2 position of quinoline and C1 position of isoquinoline cores by organometalic species (lithium or Grignard reagents). The reaction seems to proceed in two steps as this is demonstrated in the alkylation of isoquinolile below. Addition at the C1 position gives a dehydroisoquinoline-*N*-lithio derivatives, which can be hydrolyzed to furnish an isolable 1-substituted 1,2-dihydroisoquinoline. It was followed by an oxidation process to yield the full aromatized product.[26]

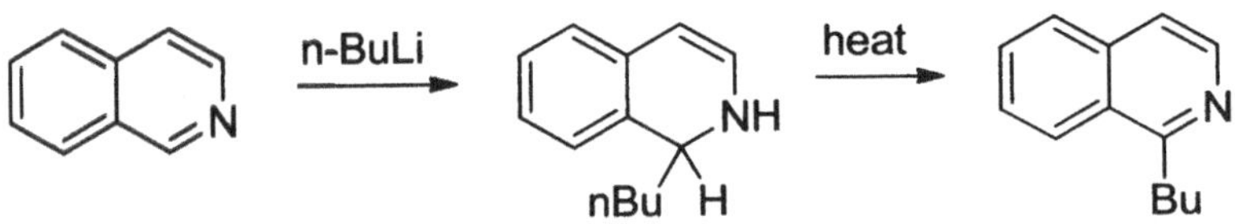

In addition to the nucleophilic substitution of hydride atom from quinolines and isoquinolines, a reaction of high value for introduction of variety of substituents is the nucleophilic displacement of leaving groups on the heterocycle. As an example, 4-chloro-8-trifluoromethylquinoline reacts in a nucleophilic displacement of chlorine atom with methyl anthranilate, to provide the precursor of NSAID antibacterial flocatfenine.[27]

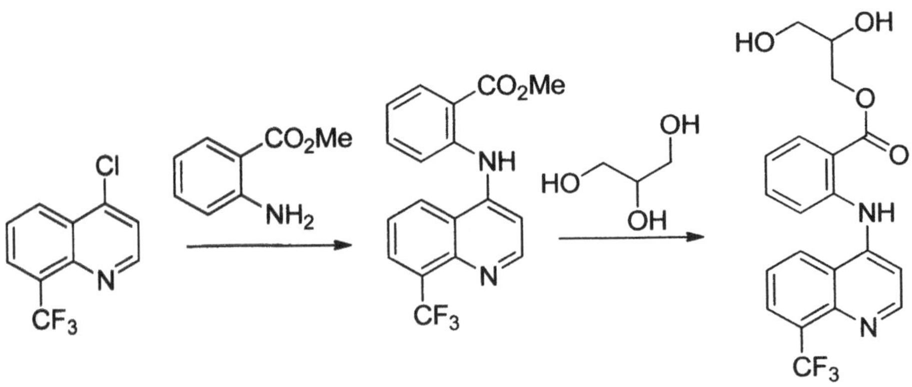

Flocatfenine

Due to the full positive charge on the nitrogen atom of quinolinium and isoquinolinium salts, these compounds are much more reactive toward nucleophilic attack in comparison to the corresponding free heterocycles. A good example for the formation of 1-alkylated isoquinoline heterocycles is demonstrated by the use of Reisert compounds. These compounds are formed when isoquinoline is treacted with an acid chloride (usually benzoyl chloride), in the presence of a cyanide. The cyanide acts as the nucleophile generating the isolable stable adducts. The acidic proton can be easily removed by strong base and used as the site of alkylation using electrophiles. The hydrolytic conditions produced the C1 alkylated product.[28]

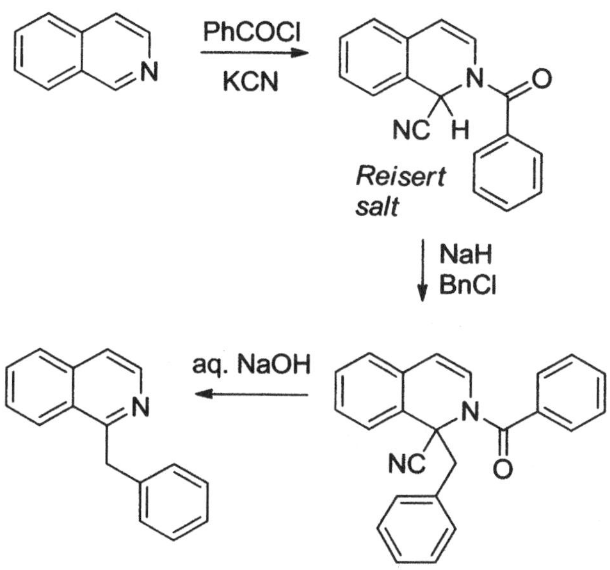

Usually under nucleophilic conditions, dihydroquinoline compounds are formed by introduction of nucleophiles at α- or γ-positions. Depending on the salt, rearomatization usually occurs. A practical way of achieving a β-chemoselective introduction of a nitro-group on quinoline ring is taking advantage of the prior formation of a nitro-quinolinium salt and nucleophilic attack of hydrogen sulphide. Transposition of the nitro group on the more electrophilic β-position followed by elimination of sulfuric acid completes the overall substitution. The same logic can also be applied to the introduction of bromine atom in 3-position of the heterocyclic ring by the action of catalytic bromide salt and excess bromine in $PhNO_2$ in accordance to the previously described bromination of isoquinolines.[23]

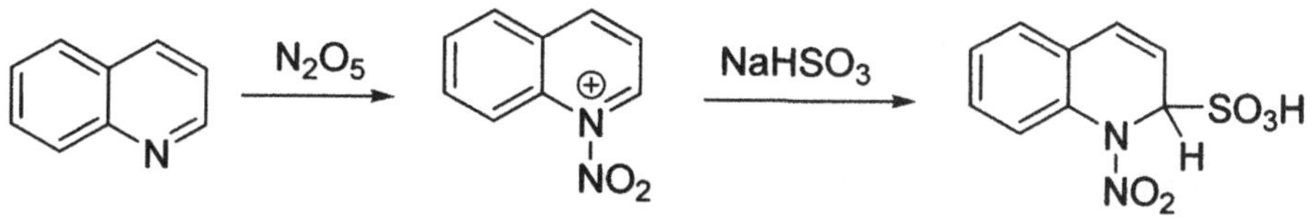

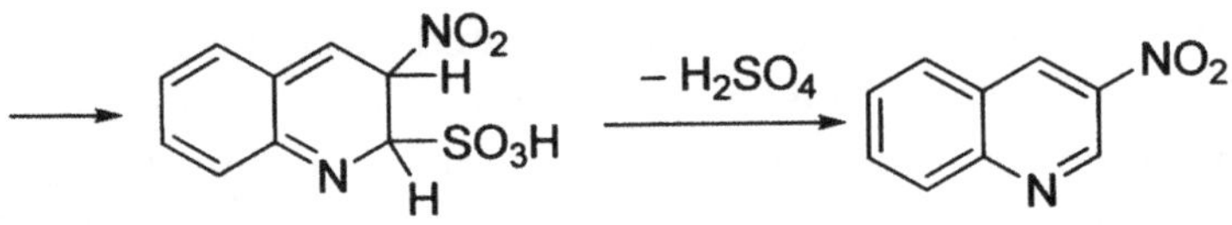

Another application of quinolinium and isoquinolinium salts is the formation of chiral salts following the use of Zincke's salt.[29] These products can be treacted with nucleophiles (Grignard, stannane reagents, or even softer nucleophiles) to afford dihydroquinolines or isoquinolines diasteroselectively.[30]

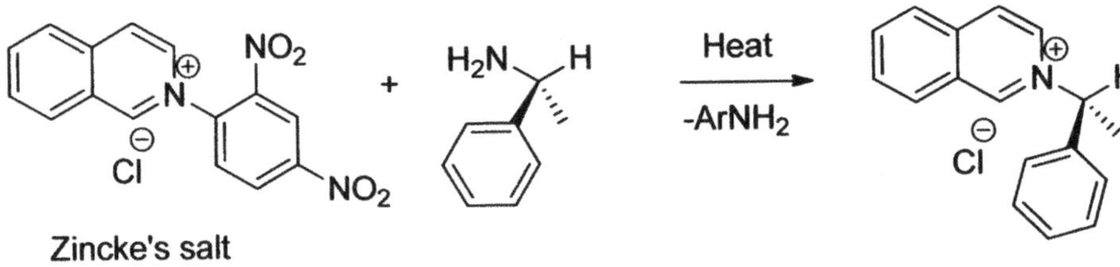

11.2.5 Amphiphilic Character of Quinoline-N-*Oxides*

Quinoline and isoquinoline *N*-oxides, formed by the action of peracids in quinoline and isoquinoline core, are useful synthetic precursors as they possess both electrophilic and nucleophilic character. The positive charge on nitrogen or the negative charge on oxygen atom can be delocalized to the α- or γ-positions, depending on demand from the reagent used.

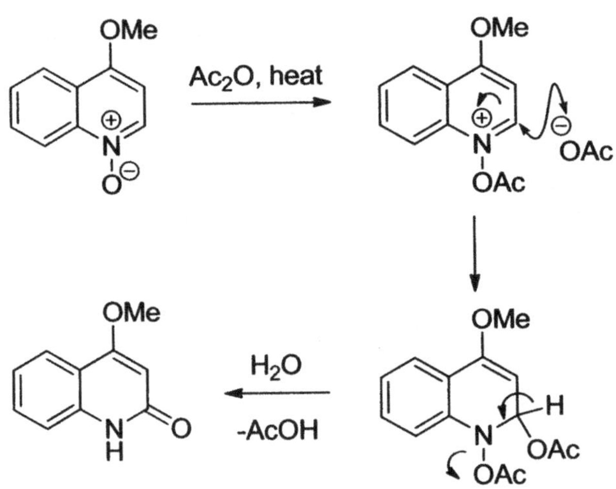

Following exactly the same logic, quinoline and isoquinoline *N*-oxides can be easily transformed to the corresponding quinolinones or isoquinolin-ones in reaction with acetic anhydride or halogen-substituted heterocycles in reaction with phosphorus oxy halogen reagents. The latter case is used for the synthesis of anti-malarial drug chloroquine from 7-chloro-4-quinolone carboxylic acid under decarboxylative heating conditions with phosphopous oxychloride.[31]

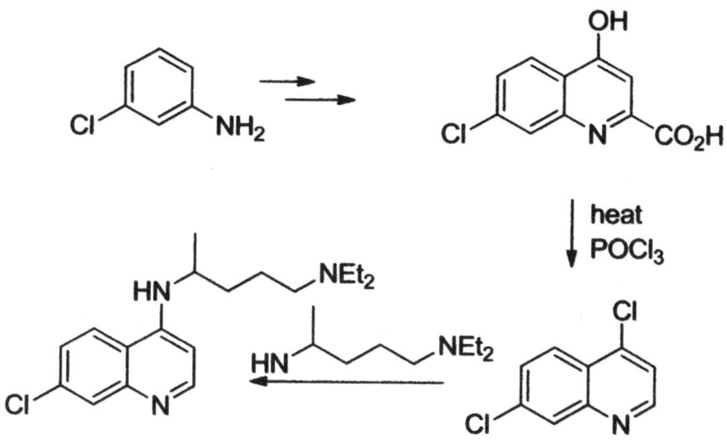

Chloroquine

N-Oxides activated with acetic anhydride are suitable for nucleophilic attack by stable carbon nucleophiles as malonates and acetoacetates at C2 or C4 position.

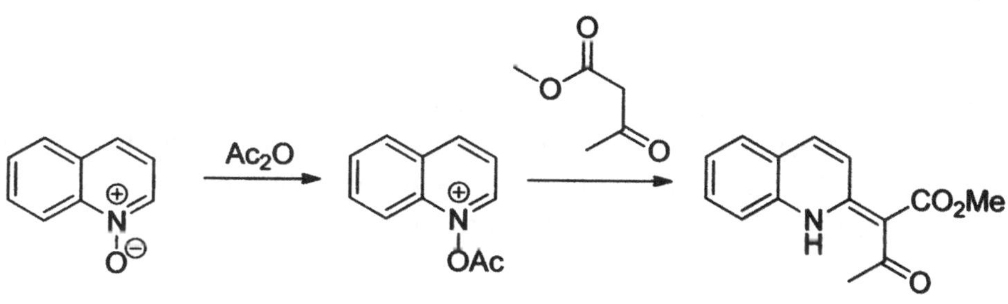

11.2.6 Metalation of Quinolines and Isoquinolines

Direct deprotonation of quinolines or isoquinolines can only be succeeded via directed *ortho*-metalation. Metal–halogen exchange can be used to produce organometallic nucleophiles from the halides located either on the benzene or the pyridine ring.[32]

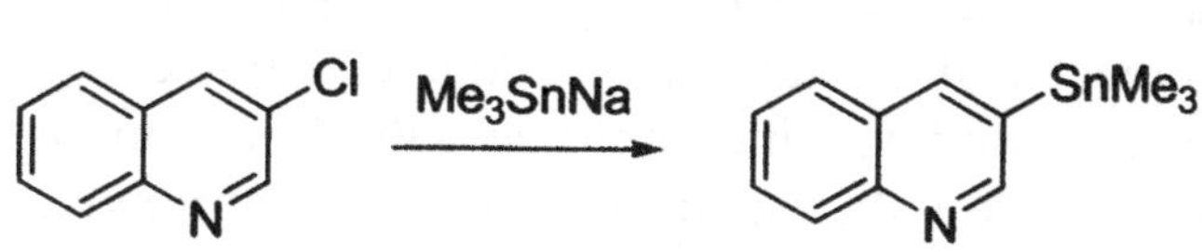

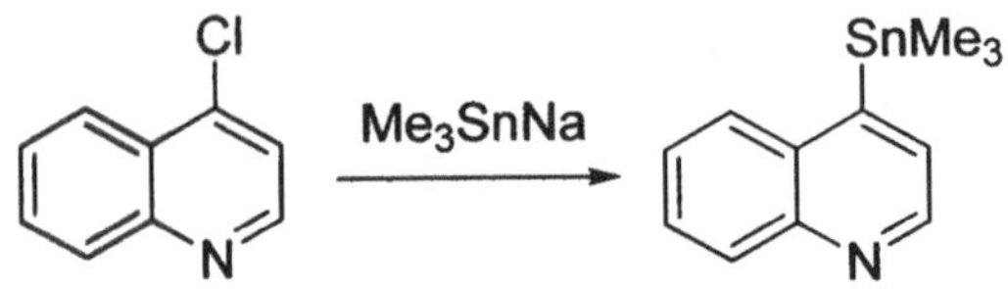

11.2.7 Palladium-Catalyzed Oxidative Coupling

Oxidative cyclizations are generally facilitated by the use of $Pd(OAc)_2$ in acetic acid under reflux. The initial step in these oxidative cyclization reactions is believed to be the electrophilic palladation of the aromatic ring. An example is presented in the preparation of anti-malarial agent quindoline, isolated from a West African plant *Cryptolepis sanguinolenta*, which was synthesized through an oxidative cyclization of the appropriately 3-substituted quinoline in the presence of two equivalents of $Pd(OAc)_2$ in trifluoroacetic acid.[33]

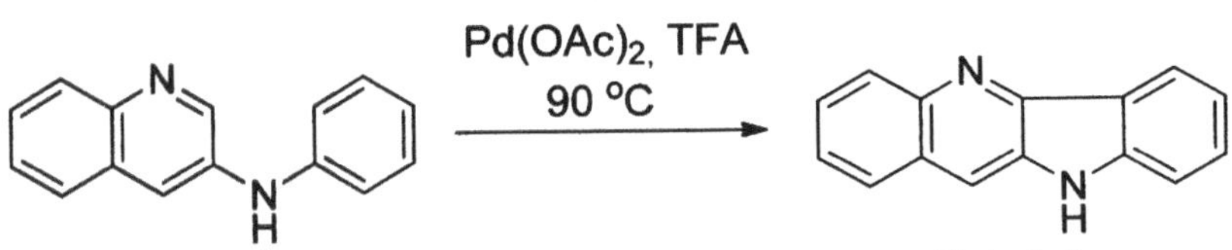

11.2.8 Cross-Coupling Reactions

Metal-catalyzed cross-coupling reactions have become, in recent years, an important tool in the organic synthetic arsenal. Among the several known cross-coupling reactions that use a quinoline fragment as a coupling partner, Suzuki and Stille are more widespread. On the other hand, only a few precedents exist for Negishi and Hiyama reactions.

Negishi coupling

Quinolinyl moiety has been applied in the Negishi reaction either as an electrophile or as nucleophile. 2- or 4-substituted quinolinyl triflates or bromides have been used extensively for introduction of aromatic rings at the C2 or C4 positions of the heterocycle. In a representative example, Murata et al. employed a Negishi reaction in his effort toward the formal synthesis of antitumor compound camptothecin. In accordance to that, 2-chloropyridine was allowed to react with lithium naphthalenide, followed by zinc chloride, to afford the corresponding zinc pyridine salt. Reaction of the resulting organozinc intermediate with 2-chloro-3-quinoline carboxylate provided the hetero biaryl core of camptothecin.[34]

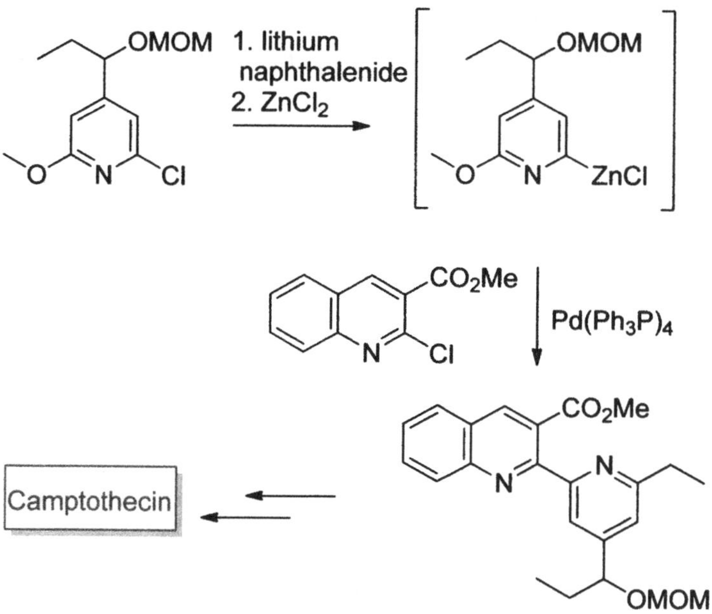

On the other hand, quinolinyl nucleophiles as the quinolinylzinc derivatives presented in the following example have been prepared by the *in situ* transmetallation of quinolinyllithium salts realized by direct halogen–metal exchange of 6-bromo quinoline derivative. The zinc salt was allowed to react with several aryl bromides, affording clean Negishi reactions for the formation of 6-substituted derivatives, which are potent inhibitors of steroid 5α reductases of types 1 and 2.[35]

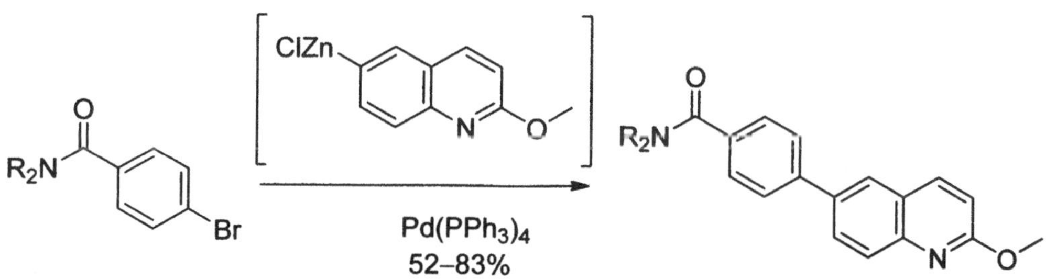

Suzuki coupling

As in the case of the Negishi reaction, quinolinyl moiety can be incorporated in the Suzuki reaction either by using substituted quinoline halides or by the use of quinolinylborane complexes. Quinolinylborane reagents can be generally prepared *in situ* by the Miyuara reaction.[36] Treatment of quinoline bromides or iodides with diboron compounds in the presence of palladium catalyst provides quinolinylboranes in good yields. Subsequent reaction with aryl bromides affords the corresponding substituted quinolines.[37] On the other hand, dialkylquinolinyl boranes can be prepared from halogen–metal exchange with *n*-BuLi, followed by quenching with $R_2B{\cdot}OEt$ or 9-BBN–Br.[38]

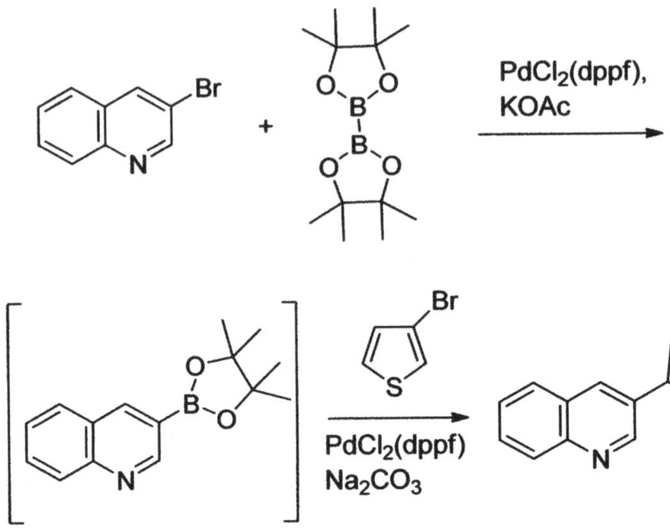

The use of quinoline bromides or iodides in reaction with alkyl boranes proceeds usually smoothly, under basic conditions in the presence of palladium catalyst. Hiyama and his group carried out a Suzuki reaction as part of their endeavor in the preparation of synthetic analogues of 3-hydroxy-3-methylglutaryl coenzyme A (HMG-CoA) reductase inhibitors.[39]

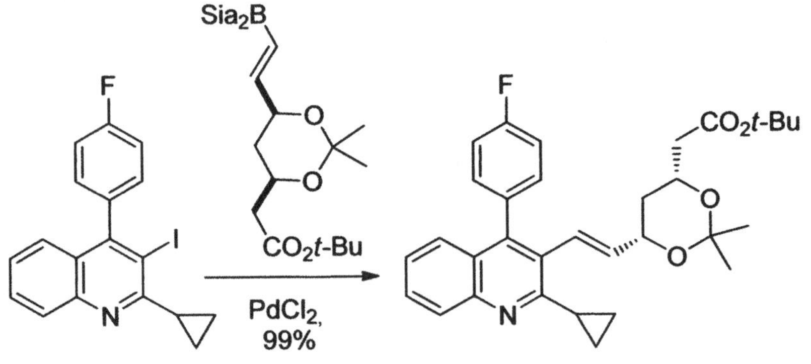

Sonogashira coupling

The Sonogashira reaction can also be applied in quinoline halogens, attaching alkynes the quinoline ring. This method has been employed as an easy way to connect the quinoline pharmacophore unit in interesting biological molecules as in the derivatization of 2-fluoro-6-*O*-propargyl-11,12-carbamate ketolide of erythromycin.[40]

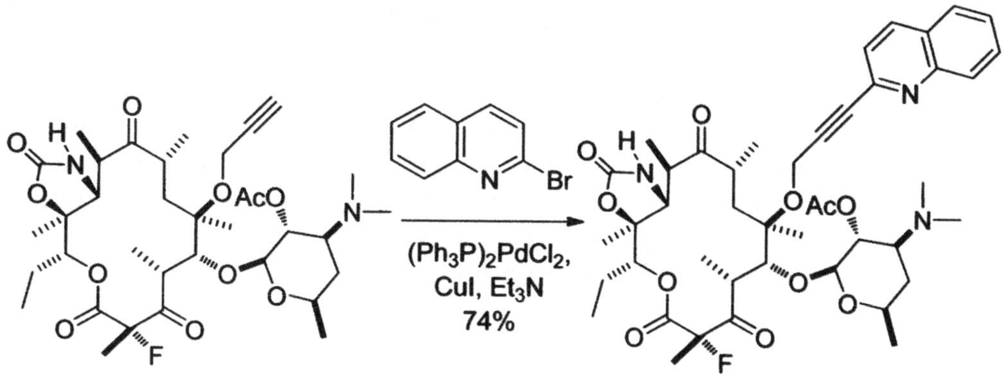

Stille coupling

In respect to the Suzuki and the Negishi couplings, quinoline stannanes provide an easy entry to highly substituted derivatives. As a highlighted example, 2-trimethylstannyl-5,8-dimethoxyquinoline reacts with appropriately substituted pyridyltriflate in the presence of palladium tetrakis-(triphenylphospine), lithium chloride, and copper bromide to produce 2-pyridylquinoline, an advance intermediate for the total syntheses of anti-tumor compounds streptonigrin and lavendamycin.[41]

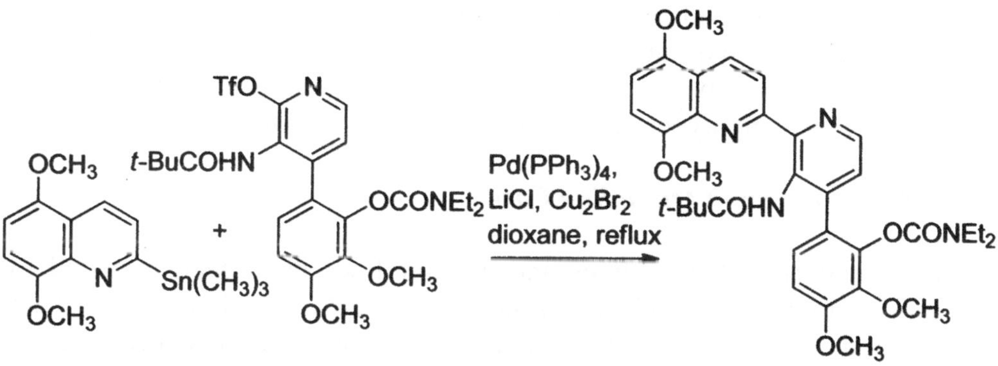

Oxidation and Reduction of Quinolines and Isoquinolines

Reduction

In the quinoline ring system, the heterocyclic ring is more easily reduced than the benzene ring, forming tetrahydro-compounds or even decahydroquinoline derivatives under hydrogenation conditions.[42] Quinolines have also been reduced to 1,2,3,4-tetrahydroquinolines by zinc borohydride and dimethylaniline under sonication conditions[43] or with indium metal in ethanol.[44] Other conditions have also been described for the selective reduction of pyridine ring working equally well on quinoline and isoquinoling heterocycles using a reducing system of nickel chloride and sodium borohydride[45,46] or *N,N*-dialkylaminoborohydrides.[47] Concerning isoquinolines, catalytic hydrogenation reduces the pyridine ring to tetrahydro-compounds when performed on cupric chromite,[48] Raney-nickel,[49] platinum oxide,[50] or palladium.[51] On the other hand, benzene can be reduced in favour of pyridine ring when isoquinolines are hydrogenated in the presence of palladium in trifluoroacetic acid.[52]

Oxidation

Probably the most common oxidation process for both quinolines and isoquinolines is their reaction with peracids forming *N*-oxides, which their uses were previously described. Apart from the nitrogen heteroatom, oxidation can affect both pyridine and carbocyclic rings of the heterocycles. Usually strong oxidizing reaction conditions, such as alkaline potassium permanganate, opens the benzene ring, affording dicarboxylic derivatives of pyridine.[53] Appropriate substitution on carbocyclic ring can provide quinoline-5,8-diones by using mild oxidizing reagents like *N*-bromosuccinimide and sulfuric acid.[54]

On the other hand, the pyridine ring portion of quinolines and isoquinolines can be oxidized to various hydroxyl-derivatives by enzymes.[55]

1,2,3,4-Tetrahydroquinoline and 1,2,3,4-tetrahydroisoquinoline derivatives are easily aromatized to quinolines and isoquinolines, respectively, by using hypervalent iodine reagents or DDQ.[56]

11.3 Construction of Quinoline Core

The great majority of quinolines and isoquinolines have been prepared by ring construction, instead of transformation of preexisting derivatives. They are obtained by variants of two main routes. The first route involves the cyclization of mono-substituted benzene rings, usually *N*-substituted anilines (Skraup, Doebner–von Miller, Knorr, Conrad–Limpach), and the second route involves the intramolecular condensation of *o*-disubstituted benzenes for the formation of the requisite pyridine ring (Friedlander, Pfitzinger reaction, etc.).

11.3.1 Camps Quinoline Synthesis[57]

Camps quinoline synthesis entails the base-catalyzed intramolecular condensation of a 2-acetamido acetophenone to substituted quinolines. Usually, 2-acetamido compounds are elaborated by the reaction of substituted benzoxazinones with the appropriate carbon nucleophiles.

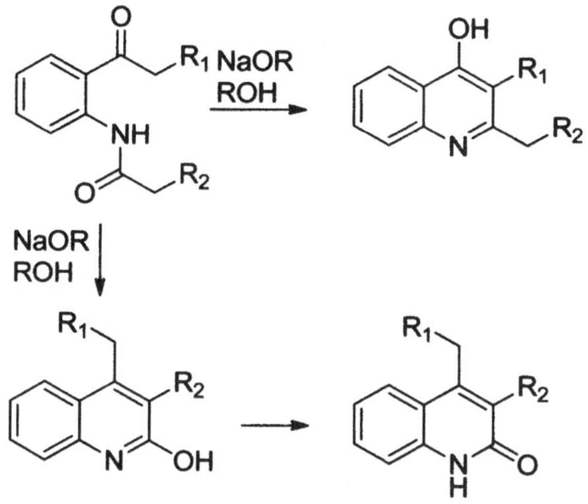

The reaction mechanism is based on the deprotonation of the amide under the basic reaction conditions and its condensation on the ketone to afford substituted 2-quinolones. In the case of 4-hydroxy-substituted products, the reaction pathway follows deprotonation of the ketone, followed by its reaction on the amide.

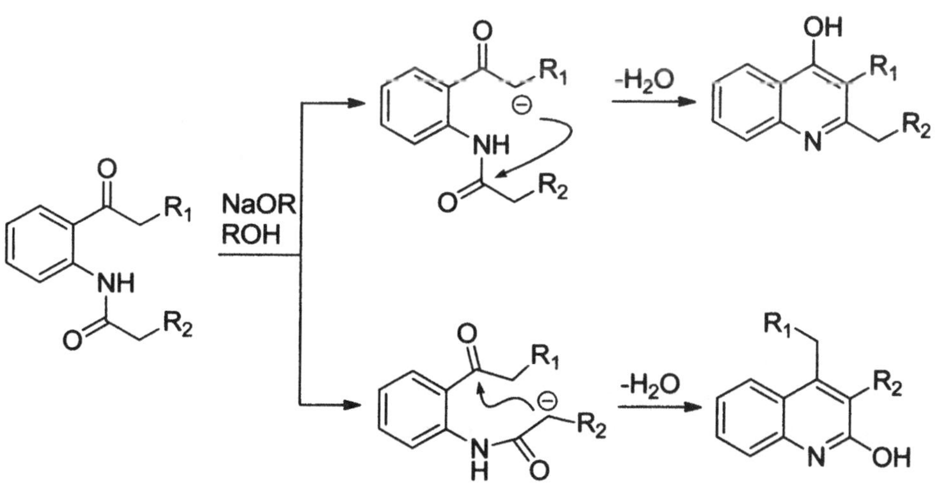

An interesting modification of Camps reaction was presented recently. It was based on the use of isatoic anhydride or benzoxazinones as active components for the *in situ* formation of intermediate amides, which were readily cyclized to the desired products. Thus, reaction of isatoic anhydride or benzoxazinones with malonates and/or ketoesters afforded libraries of quinoline compounds in high yields even at multikilogram scale.[58]

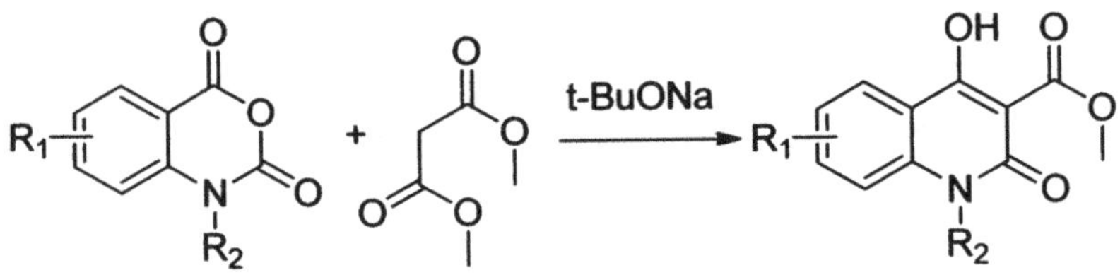

The described method was expanded to provide an easy access to quinoline alkaloid analogues by using one-step reaction between benzoxazinones and ketene silyl-acetals in the presence of titatium tetrachloride.[59]

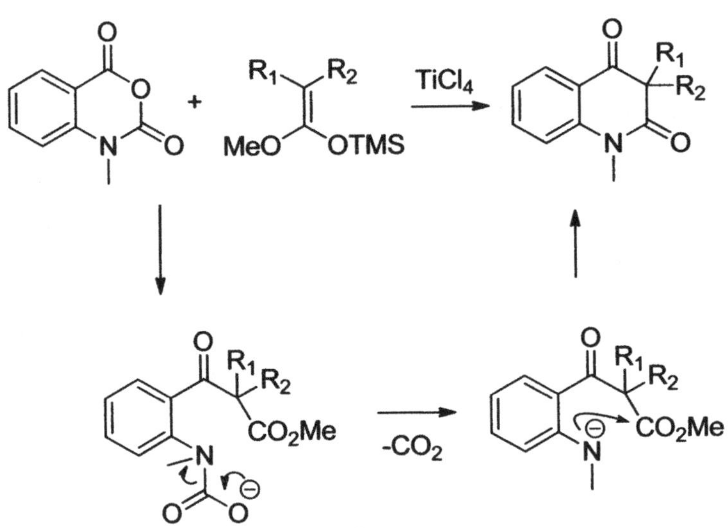

11.3.2 Combes Quinoline Synthesis[60]

Condensation of arylamine with 1,3-diketone, keto-aldehyde, or dialdehyde, followed by acidic Friedel–Craft type cyclization and dehydration, is well-known as the Combes reaction for quinoline synthesis.

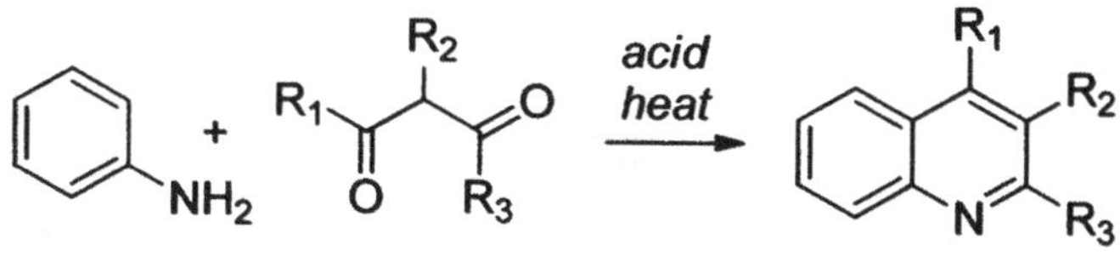

The initial condensation step proceeds usually well, even when sterically congested arylamines are used. On the other hand, the cyclization and dehydration step is affected by Friedel–Crafts-type directing substituents within the ring, following the proposed mechanism.[61] Strong electron-withdrawing groups on aromatic ring can completely prohibit the cyclodehydration step, producing only enamines.

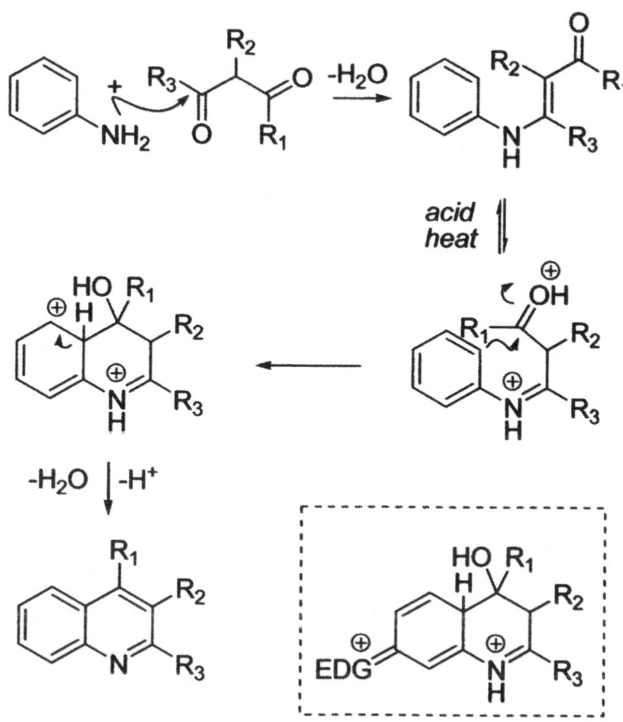

Several conditions are used for the difficult cyclization step, avoiding the acidic hydrolytic conditions in the reaction mixture that can hydrolyze the initially formed enamine. Thus, use of concentrated inorganic acids is preferential (e.g., conc. sulfuric and hydrochloric acids), as well as chloroacetic acid, *p*-toluene sulfonic acid, etc.

Unsymmetrical diones have also been used as coupling partners for the Combes reaction, providing selectively, in some cases, otherwise difficult accessible quinoline compounds.[62,63]

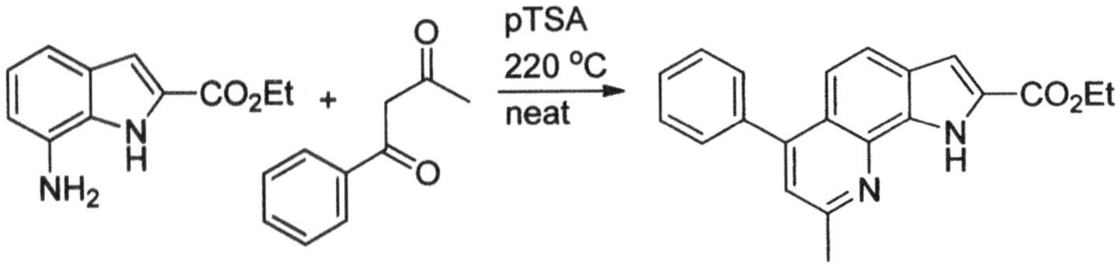

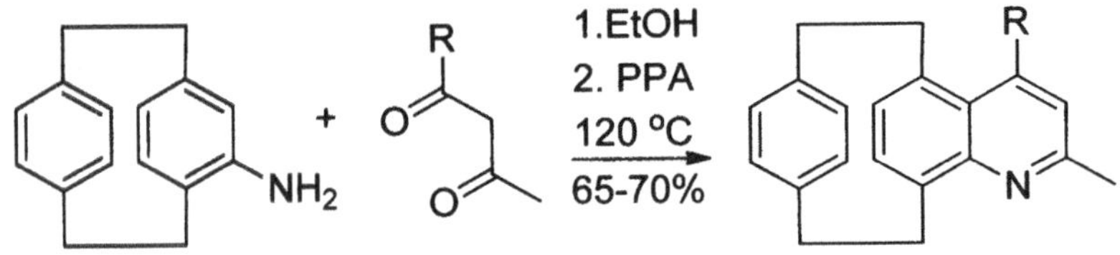

An unconventional modification of the Combes reaction was also used to access pharmacological interesting 3-formyl quinolines by treacting anilines with vinamidium salts.[64]

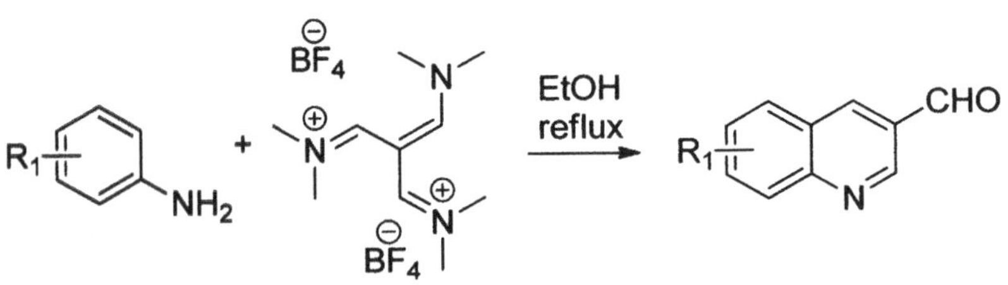

11.3.3 Conrad–Limpach[65] and Knorr[66] Reaction

Primary arylamines and β-ketoesters react in the presence of strong acid to provide 2-quinolones (Knorr reaction), whereas their thermal reaction yields 4-quinolones (Conrad–Limpach reaction). These reactions present a typical example of kinetic versus thermodynamic control, which has been employed in the synthesis of many quinolone derivatives for more than a century.

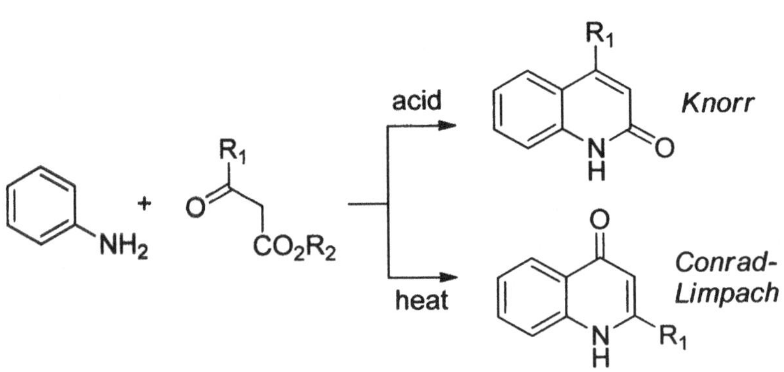

Different reaction mechanisms are postulated for the synthesis of 2- or 4-quinolone derivatives. The Knorr reaction involves a nucleophilic attack of aniline nitrogen on the ester of β-ketoester component, providing anilide, which undergoes Friedel–Crafts cyclization followed by dehydration with sulfuric acid to yield 2-quinolones. On the other hand, a completely different pathway is involved in the case of Conrad–Limpach reaction. Condensation of aniline derivative with β-ketoester provides the corresponding enamino-ester. Enolization was followed by a 6π-electrocyclization reaction to form a 4-quinolone. A postulated alternative reaction pathway invokes the formation of an iminoketene prior to the cyclization step.

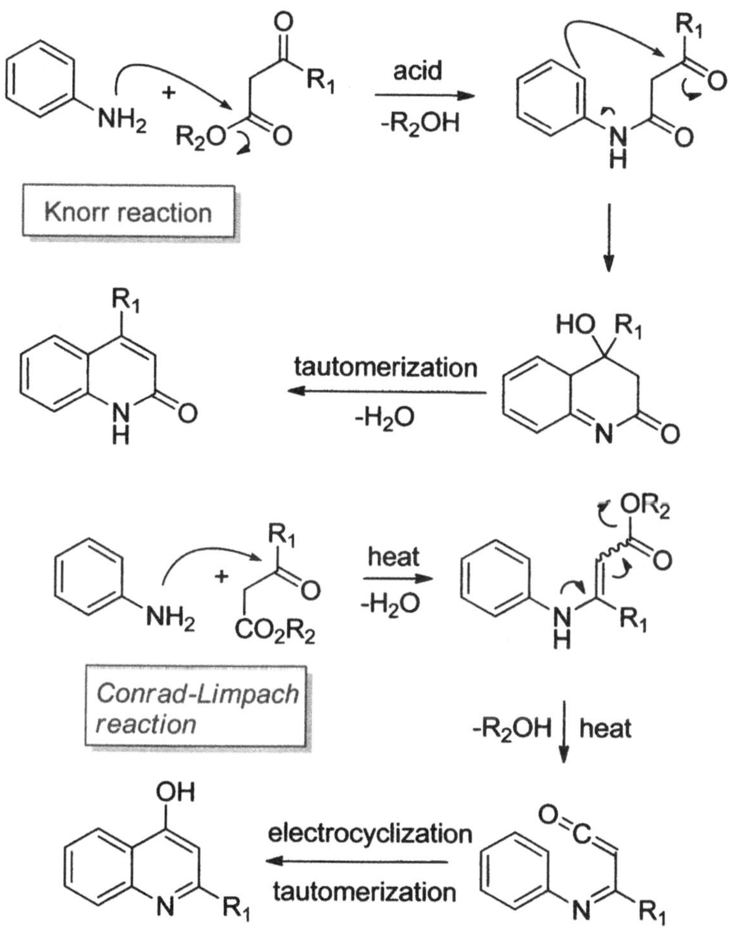

Both reactions are promoted by electron-donating groups on the aromatic ring following the mechanisms described above.

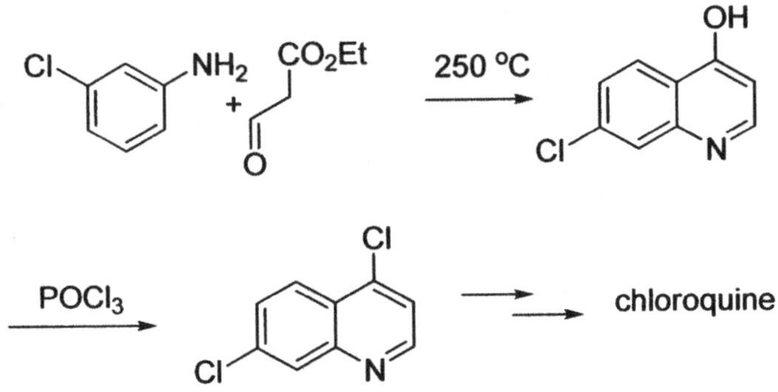

Some applications of the described reactions include the preparation of chloroquine, an important anti-malarial, which is made by reacting 4,7-dichloroquinoline with 4-diethylamino-1-methylbutylamine at 180 °C. 7-Chloro-4-hydroxyquinoline, a precursor of 4,7-dichloroquinoline, was routinely prepared by heating 3-chloroaniline with the ethyl ester of formylacetic acid, through the Conrad–Limpach quinoline synthesis.[67]

The same method was also used for the synthesis of the anti-malarial compound mefloquine. Heterocyclization of 2-trifluoromethylaniline with trifluoroacetic acid ethyl ester provides 2,8-*bis*-(trifluoromethyl)-4-hydroxyquinoline. Bromination with phosphorous tribromide, followed a carbon monoxide introduction, affords 2,7-trifluoromethyl-4-carboxylic acid quinoline. Amide formation and pyridine reduction completes the synthesis providing mefloquine in high yields.[68]

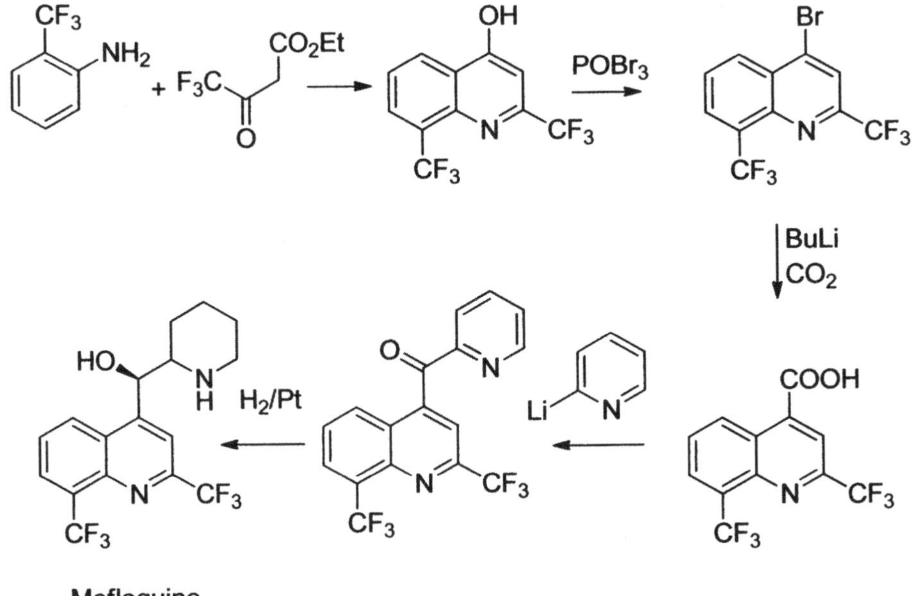

Several improvements of the Conrad–Limpach reaction exist. In one of them, Leonard and his group reported the synthesis of endochin through an alkylation of the intermediate enamino-ester by base and alkyl bromide. The formed enamino-ester was cyclized under heating conditions to produce the promising antimalarial compound.[69]

On the other hand, the high importance of Knorr reaction in drug design is highlighted in the synthesis of quinine. Reaction of *p*-anisidine and acetoacetic ester in the presence of sulfuric acid provides 6-methoxylepidine. Substitution of hydroxyl group with chloride and subsequent reduction provides the 2-unsubstituted quinoline. Condensation and oxidation affords a quininic ester, which is just three steps away from the synthesis of strong anti-malarial compound quinine.[70]

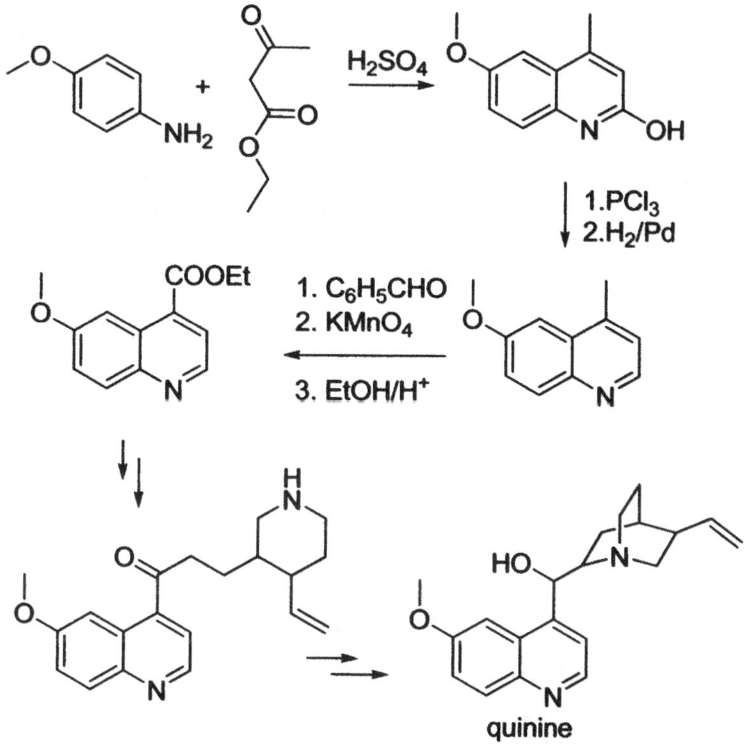

11.3.4 Friedlander[71] and Pfitzinger Syntheses

The methods cited so far for the preparation of quinoline frameworks suffer from the mixtures obtained usually when *meta*-substituted anilines are used. The Friedlander quinoline synthesis avoids this problem by using an *ortho*-acylaniline derivative in reaction with enolizable carbonyl compounds.[72]

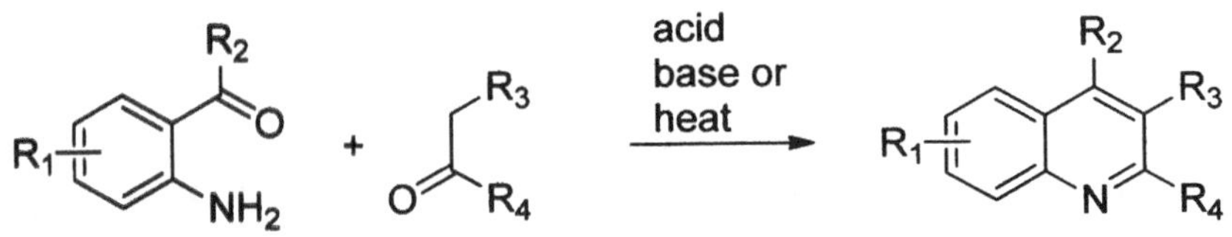

The reaction can be catalyzed either by acidic or basic conditions, or just by heating the reaction mixture. The outcome of the condensation was found to be dependent on the type of the catalyst used, with acid catalysis leading predominantly to formation of the thermodynamic product, whereas the base driving mostly to the kinetic.

Despite the fact that Friedlander synthesis of quinolines has been known for more than a century, the reaction is still not completely understood. Two possible mechanisms are still debated. The first involves preformation of a Schiff base, while the other a Claisen condensation as the initial step. In either case, the intermediates are cyclodehydrated to form the quinoline core. Evidences for both mechanisms have been reported based on the isolation of formed intermediates.[73] We have to point out that none of the described intermediates have been isolated from the exact conditions following the one-step Friedlander synthesis, providing a lot of space for doubt for the mechanism of the original reaction.

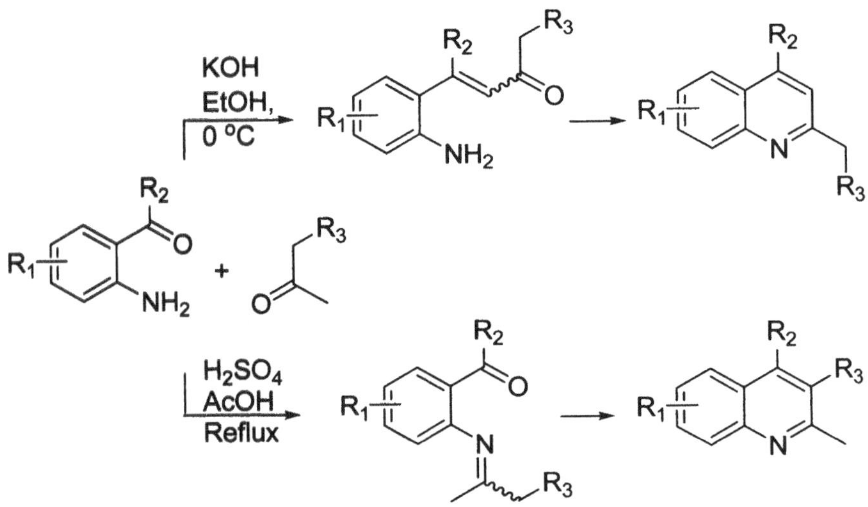

The Friedlander reaction is quite versatile and can be used either by electron-rich or electron-poor aromatic components. The only limitation is the sustainability of the *ortho*-acylaniline for self condensation. In this logic, protected anilines and imines in some cases are used to avoid the undesired self condensation of the starting anilines. On the other hand, the Friedlander condensation is applicable to the aldehydes, ketones, carboxylic acids, esters, amides, nitriles, and aldoximines that have an adjacent enolizable group. In general, the reactivity of the methylene group is the principal factor that influences the condensation. With less active methylene groups, more drastic conditions are required.

Two related reactions of quinoline construction, the Pfitzinger and Niementowski reaction, can be considered as extensions of the Friedlander synthesis. *Niementowski reaction*[74] uses anthranilic acid for the synthesis of 4-hydroxy quinolines, and *Pfitzinger reaction*[75] applies substituted isatins as precursors for *ortho*-aminophenylglyoxylic acid, which is used as the condensation component with carbonyl compounds to produce 4-carboxylic acid quinolines.

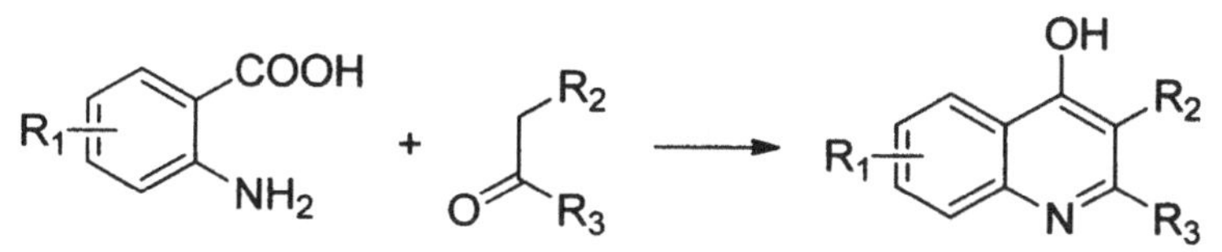

Niementowski reaction

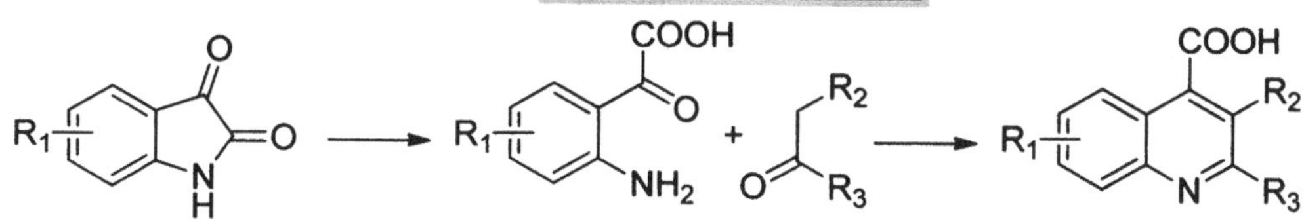

Pfitzinger reaction

Several other modifications using the logic of the Friedlander reaction exist in the literature. The use of *ortho*-nitrobenzaldehydes and ketones as starting materials in the presence of tin chloride and zinc chloride provided the products in high yields.[76]

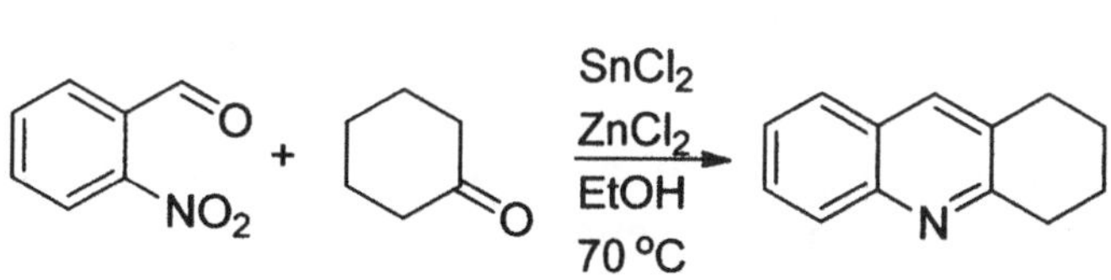

Variations employing hindered bases such as 1,3,3-trimethyl-6-azabicyclo[3.2.1]octane (TABO),[77] introduction of formyl group *in situ*[78] or microwave irradiation represent the modern alternatives of this method.[79]

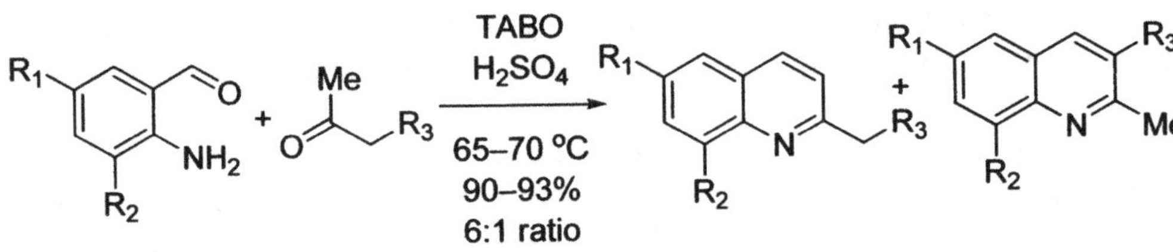

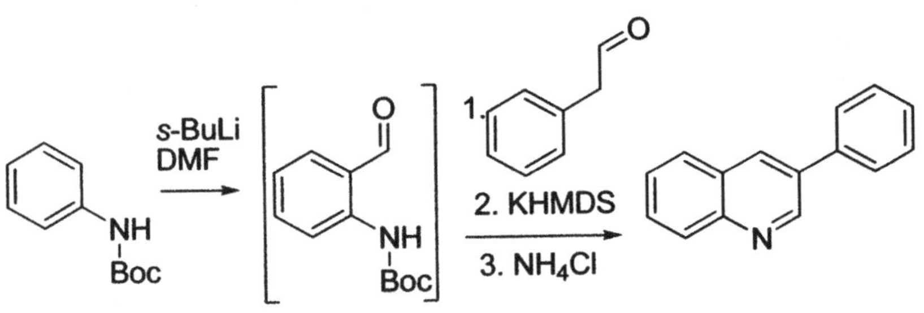

In addition, Schiff bases can also be applied as intermediates of quinoline formation. Thus, Schiff bases derived from 2-(trifluoromethyl) aniline and a methyl naphthone, mediated by lithium 2-(dimethylamino)-ethylamide, were used to furnish a series of substituted 2-(2-naphthyl)quinolines designed to target DNA.[80]

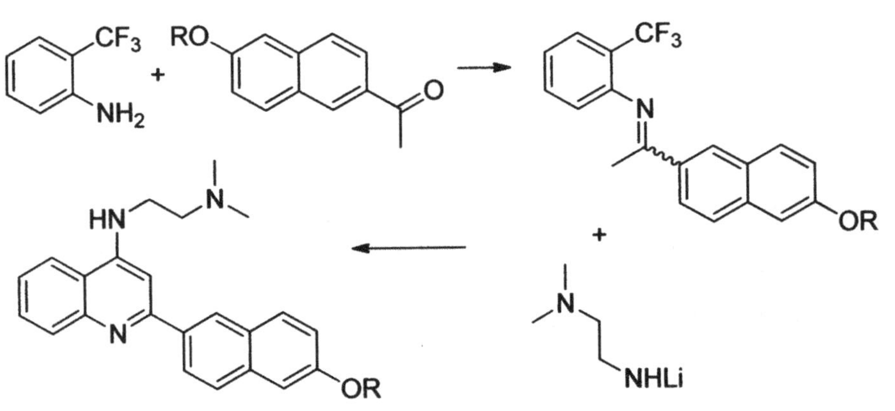

Neurokinines comprise of a group of peptides involved in nerve transmission. The nonpeptidic neurokinin antagonist talnetant has been evaluated for its effect on irritable bowel syndrome (IBS) and urinary incontinence as well as depression and schizophrenia. The quinoline portion of this compound has been prepared through a Pfitzinger condensation of isatin with methoxy acetophenone. Methoxy ether cleavage and amide formation then affords talnetant.[81]

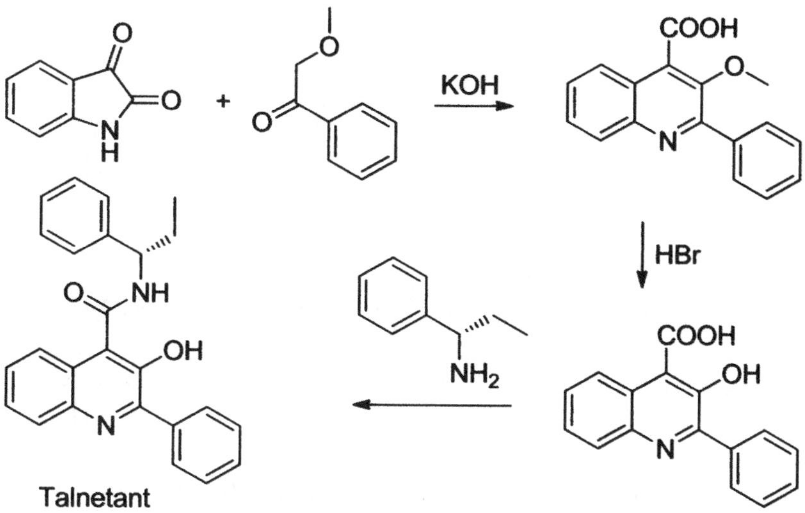

11.3.5 Gould–Jacobs Reaction[82]

The condensation reaction of aniline with alkoxy-methylenemalonic esters, followed by cyclization to 4-hydroxy-3-carbalkoxy quinolines, is described as the Gould–Jacobs reaction. The reaction can use various derivatives of methylene malonic ester such as keto-malonic esters, malonitriles, malonamides, and Meldrum acid.

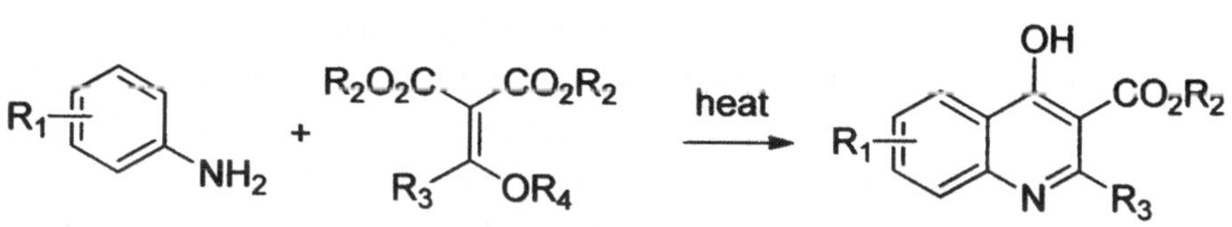

The reaction mechanism is postulated to involve initial conjugate addition of aniline derivative to methylene malonate and elimination of the alcohol providing β-anilinoacrylate. Heating the reaction mixture at 180–320 °C promotes the second alcohol elimination, producing what is believed to be an imino-ketene, which readily undergoes a 6π-electrocyclization to afford the desired 4-hydroxy-3-carbalkoxy quinoline derivative after tautomerization.[83]

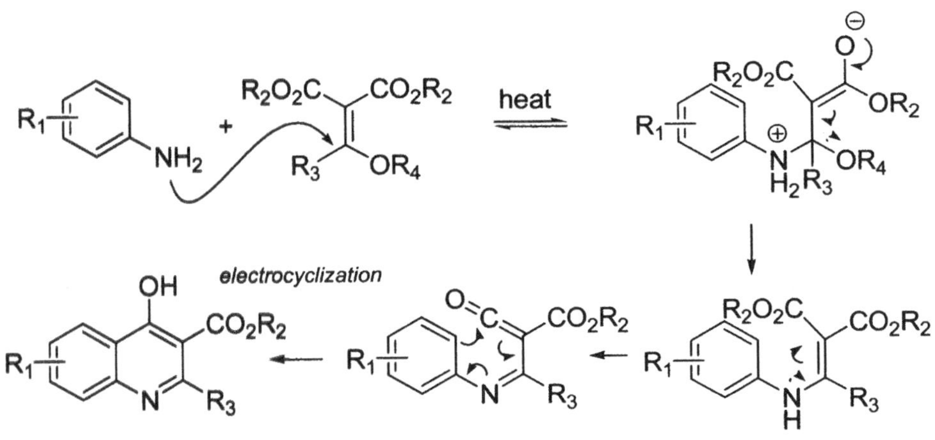

The Could–Jacobs reaction is dependable on ring substitution. Electron-donating groups accelerate the cyclization reaction. *ortho-*, *para-*Directing groups positioned *meta* to the amine usually provide only 7-substituted quinolines. The mixture of 5,7-products was observed with electron-withdrawing groups and fluoro-substituents, and it can be explained by the electron-donating ability of the carbon to cyclize.

Due to its easiness and its ability to use cheap, commercially available starting materials, the Gould–Jacobs reaction contributed significantly to the development of quinolone anti-bacterials as one of the most effective antibacterial drugs today. The synthesis of an array of 7,8-dialkoxy-4-hydroxyquinoline carboxylates, an important class of drugs that are toxic to coccidian, a protozoan that can devastate commercially poultry flocks, is a typical example involving the addition of *bis-iso*-butylalkoxyaniline to diethyl ethoxymethylenemalonate to synthesize of buquinolate[84] or the use of piperonylamine for the synthesis of the strong antibiotic oxolinic acid.[85]

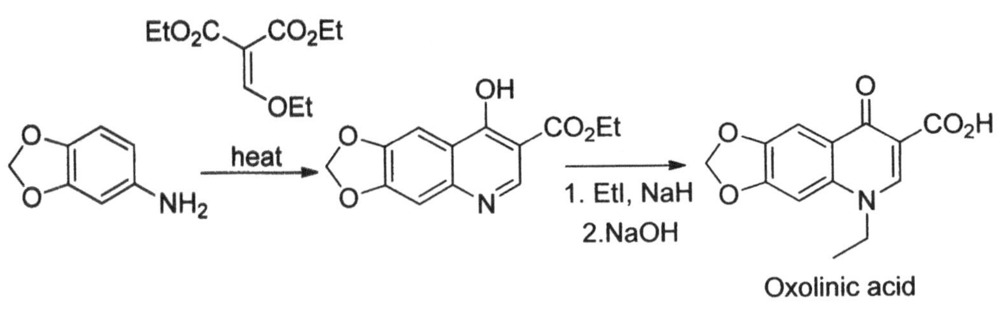

Oxolinic acid

The antiprotozoal spectum of activity is apparently changed by replacing alkoxy substituents with alkyl or amino groups. Thus, by using the appropriate substituted anilines in a Gould–Jacobs reaction, a series of antibacterial drugs can be synthesized as nequinate, amquinate, perfloxacin,[86] ibafloxacin,[87] mirisetron, etc.

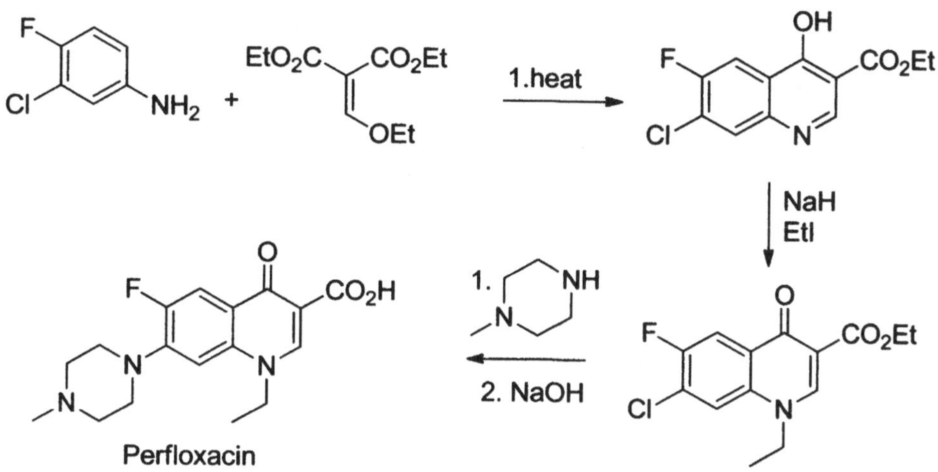

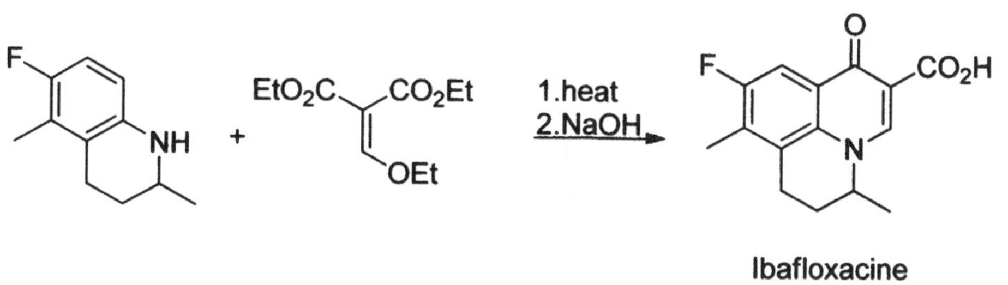

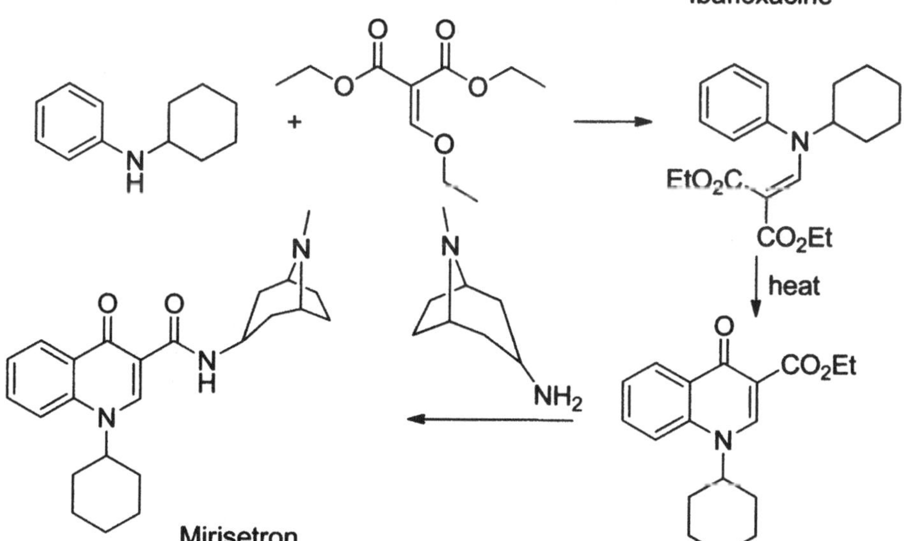

4-Hydroxy-quinoline carboxylates, prepared using the Could–Jacobs reaction, have also been frequently used as precursors for the synthesis of biologically interesting quinoline compounds. For example, the synthesis of the 4-hydroxy-7-chloroquinoline, a precursor used for the synthesis of anti-malarial chloroquine, was afforded by Could–Jacobs reaction of 3-chloroaniline with ethoxymethylenemalonic ester after hydrolysis and decarboxylation of the formed 3-carboethoxy-4-hydroxy quinoline compound.[88]

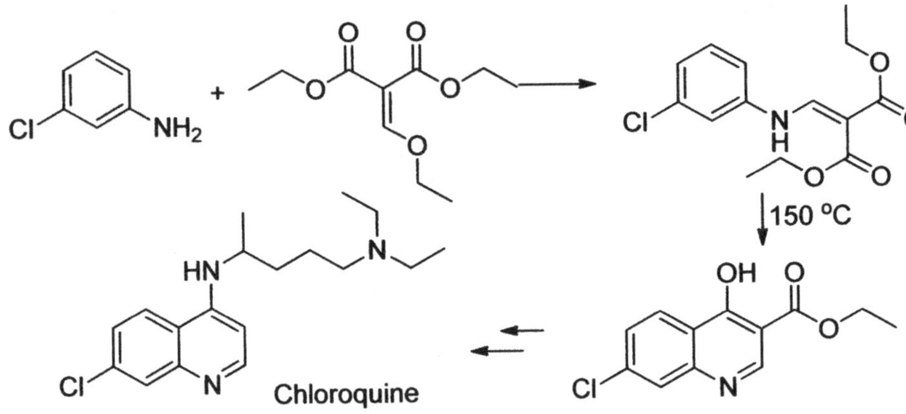

11.3.6 Meth–Cohn Quinoline Synthesis[89]

The application of Vilsmeier's reagent in acylanilides to produce 2-chloro-3-substituted quinolines is known as the Meth–Cohn quinoline synthesis.

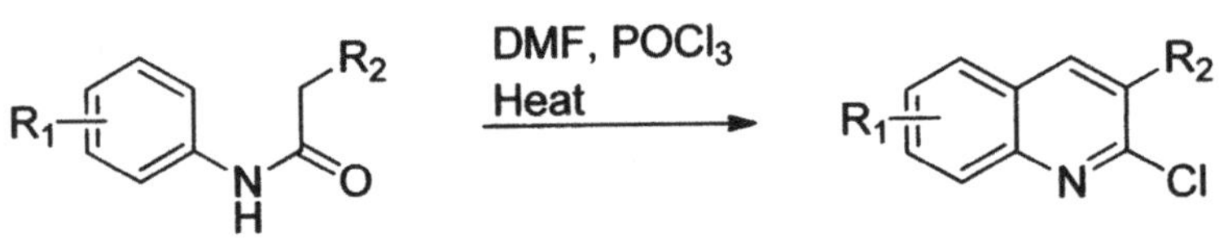

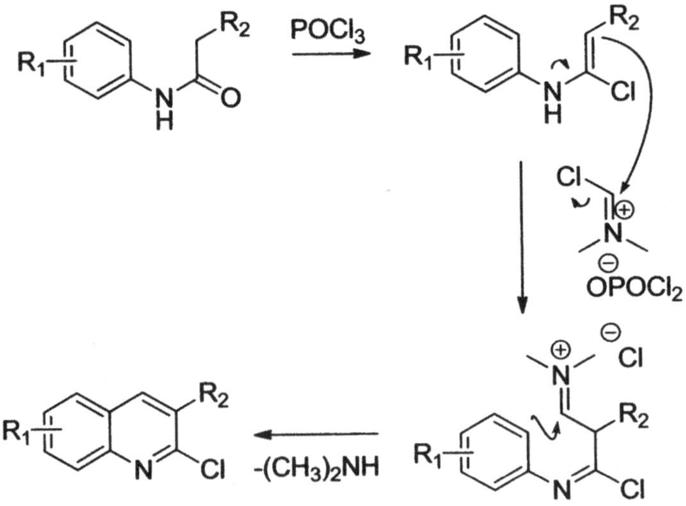

The reaction mechanism involves initial conversion of acylanilides in to α-iminochloride by the action of phosphorous oxychloride. The formed chloroenamine is treated with Vilsmeier reagent, providing the active iminium salt, which is readily cyclized to provide 2-chloro-3-alkyl quinolines.[90]

As in the case of other quinoline syntheses, which started with substituted anilines, electron-donating groups on the arene accelerate the reaction. *meta*-Substituted anilines react selectively, yielding only 7-substituted quinolines.

The Meth–Cohn quinoline synthesis provides a versatile and reliable entry in to 3-substituted 2-chloroquinolines. In several cases, the Meth–Cohn reaction plays a pivotal role for the preparation of biological important compounds as in the synthesis of E-ring modified derivatives of camptothecin[91] or for the syntheses of selected NMDA antagonists with analgesic activity.[92]

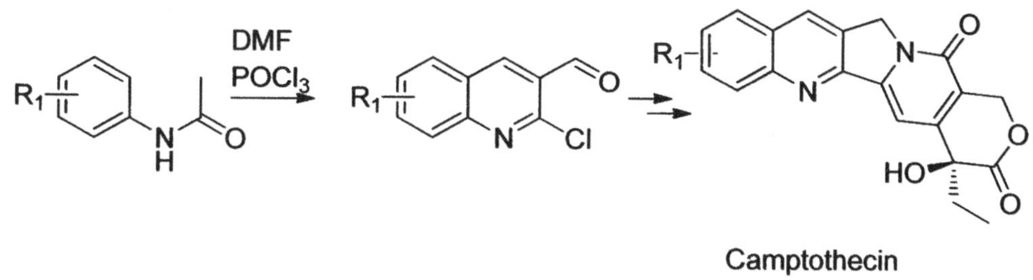

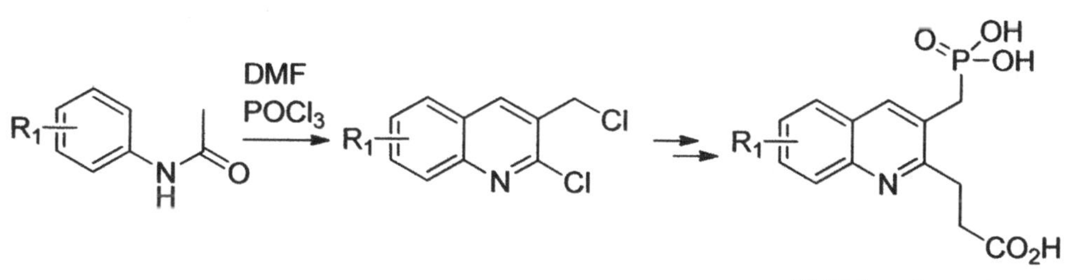

11.3.7 Skraup/Doebner–von Miller Reaction[93,94]

In 1880, Skraup synthesized quinolines by reaction of aniline and glycerol in a solution of sulfuric acid and an oxidizing reagent. What Skraup discovered was generalized one year later by Doebner and von Miller by using α,β-unsaturated aldehydes instead of glycerol in order to prepare substituted quinolines.

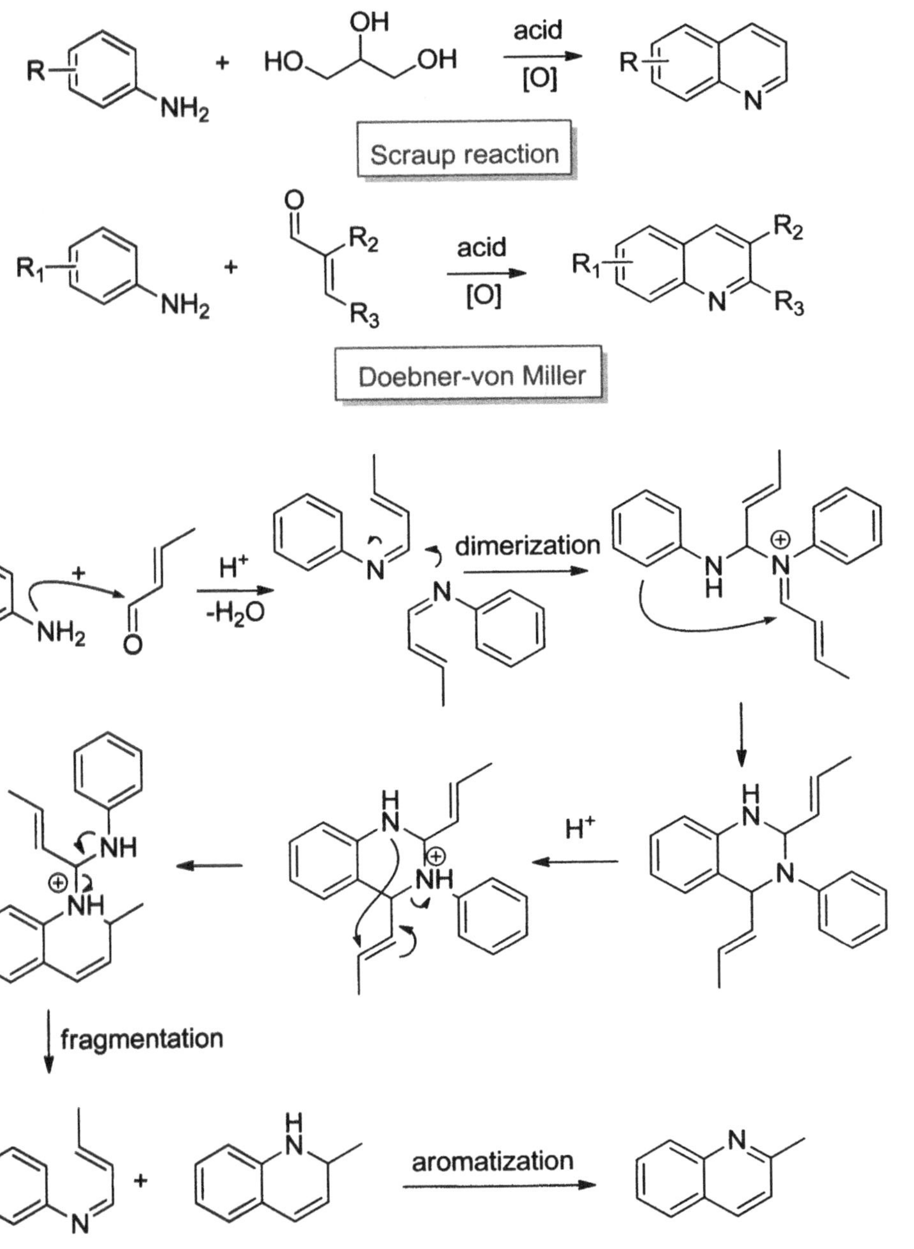

Although the Skraup/Doebner–von Miller reaction represents one of the most common reaction for the synthesis of quinoline core for more than a century, its mechanism is still dedebated.[95] To date, both of the two more popular mechanistic explanations are involving fragmentation–recombination pathways. In the first one, initially the amine reacts with the aldehyde or ketone under acidic conditions to form an imine. Dimerization and Pictet–Spengler type cyclization forms a diazetine core. Protonation and subsequent cyclization–ring cleavage reaction assembles the isoquinoline nucleus.

Fragmentation and aromatization leads to the final product, regenerating the active imine component.[96]

Usually, the reaction is low yielding depending on the substituents either on the aromatic ring or on the unsaturated aldehyde reaction partner. The low yields are also correlated with the general difficulty to work up and isolate the desired products from the reaction mixture. An improved method was reported lately from Matsugi, et al. using a biphasic system as reaction media for the Doebner–von Miller reaction. The method proved to be advantageous in terms of yield and easy work-up process.[97] Other modifications concern the use of different Lewis acids as promoters for the later cyclization process ($InCl_3$, lanthanides, etc.).[98]

The Doebner–von Miller reaction was used as a method of choice for the synthesis of several biological active quinoline compounds. The anti-malarial drug primaquine, which acts against the hepatic stage of plasmodia infection, was synthesized by a Skraup reaction from 4-methoxy-2-nitroaniline and glyceron in the presence of sulfuric acid. The nitro group was then reduced and alkylated with 4-bromo-1-phthalimidopentane to provide the protected primaquine, which was deprotected by using hydrazine.[99]

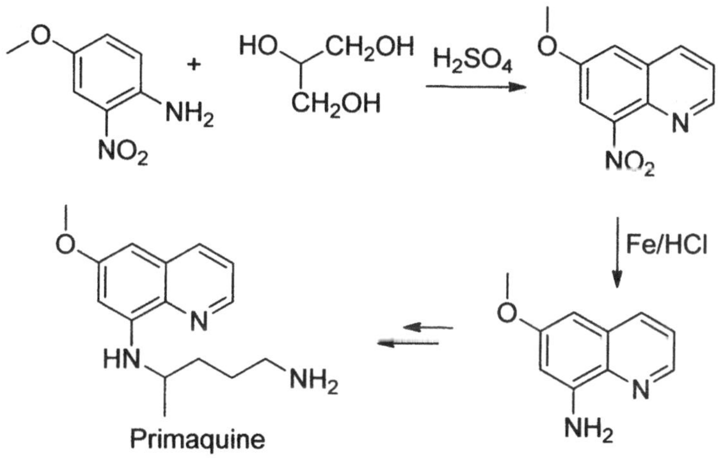

The Skraup reaction was also used for the synthesis of the amebocide drug iodoquinol. Amebiasis is an infection of the body by the protozoa *Entamoeba histolytica* with symptoms of acute ameba dysentery, which is accompanied by bloody diarrhea, vomiting, fever, and dehydration. Iodoquinol is considered as a drug of choice for treating asymptomatic or moderate forms of amebiasis. Its preparation relies on the reaction of 2-aminophenol with glycerol in the presence of sulfuric acid.[100]

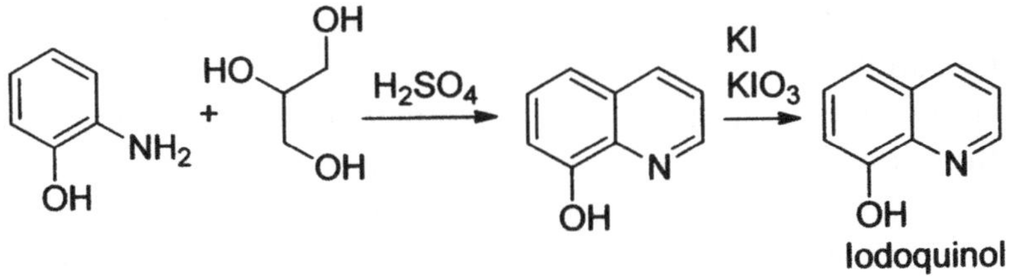

11.3.8 Modern Methods

Except the classical methods described above, modern processes for the construction of quinoline and isoquinoline cores are gaining ground in drug discovery. Although most of them have never been used for the synthesis of specific quinoline drugs, their references are expected to serve as an inspiration for designing more efficient routes for the drugs of the future. Basically these methods can be subdivided in three major categories: i. cyclization of ynones, enones, alkynes and alkenes; ii. cycloaddition and pericyclic reactions and iii. metal-metadiated cyclizations.

Cyclization of ynones and alkynes

Internal ynones and alkynes have been frequently used as starting materials for the synthesis of quinoline compounds. Ynones can act as Michael acceptors with various nucleophiles, affording intermediate enones, which are readily cyclized to quinolines.[101]

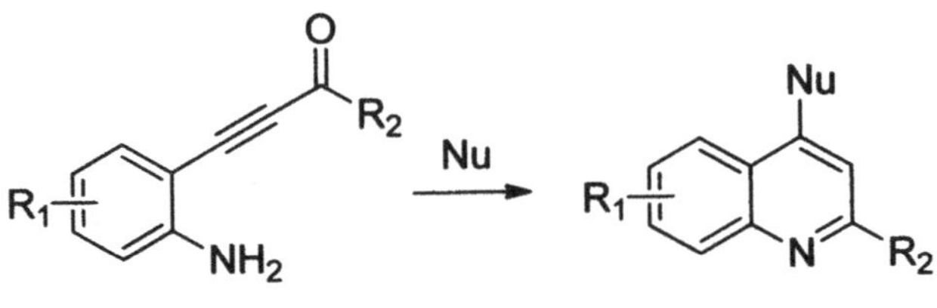

Mono-substituted anilines bearing an internal alkyne can be cyclized to quinolines either by metal-mediated hydroarylation of alkynes or through an electrophile-mediated Friedel–Craft type reaction.[102]

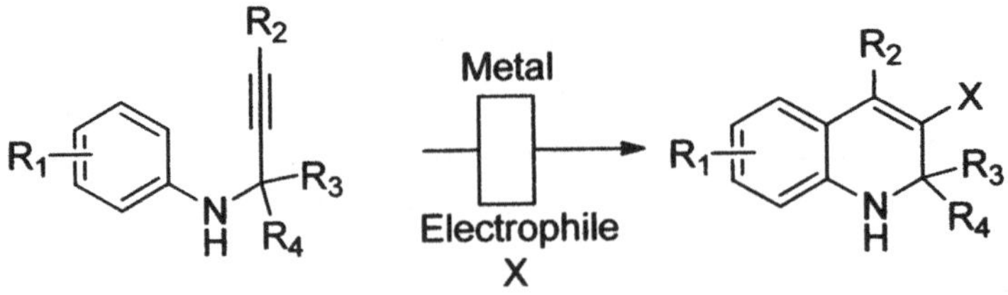

Metal-mediated or metal-catalyzed cyclizations

Several metals have been reported of being able to hydroarylate alkynes as in the examples shown below. Thus, di-substituted-propargyl anilines can be cyclized either in the presence of copper(I) chloride at reflux or by a mixture of palladium acetate and trifluoroacetic acid.[103]

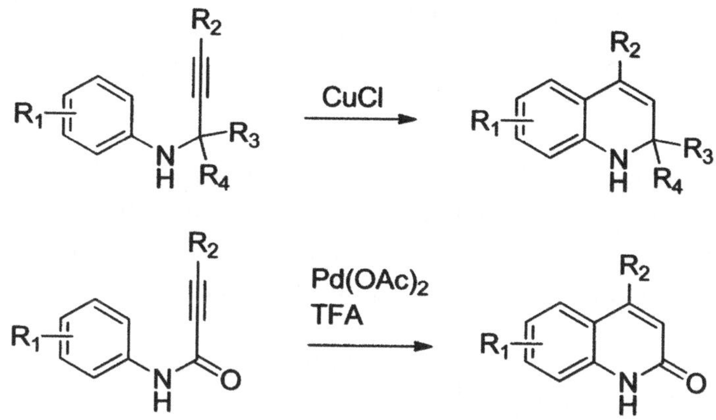

Cross-coupling reactions are essential tools for the synthesis of quinoline heterocycles not only by employing quinoline core as one of the starting components (see Section 11.2.7), but also by using them as the key strategy for the construction of quinoline rings. A Sonogashira-type multicomponent domino reaction was used for the synthesis of highly substituted quinolines from aryl iodides, primary amines and carbon monoxide.[104]

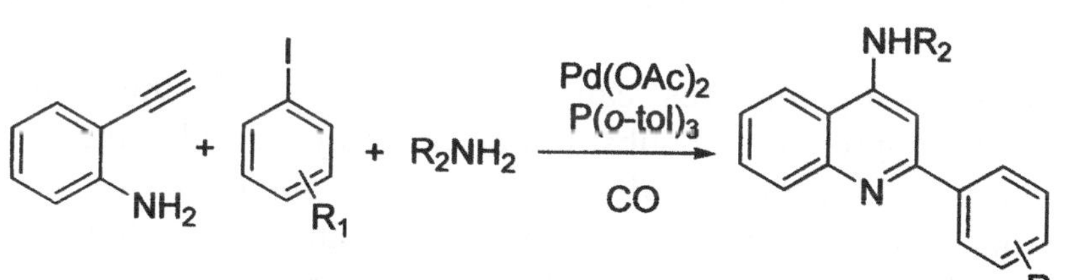

Intramolecular or intermolecular Heck-type reactions were also used in the synthesis of poly-substituted quinoline compounds. Palladium-catalyzed reaction between vinyl or aryl halides and *ortho*-allyl-substituted-*N*-tosyl-anilides produces dihydroquinolines in an intermolecular fashion, where reaction of acrylates intramolecularly affords 4-quinolones.[105,106]

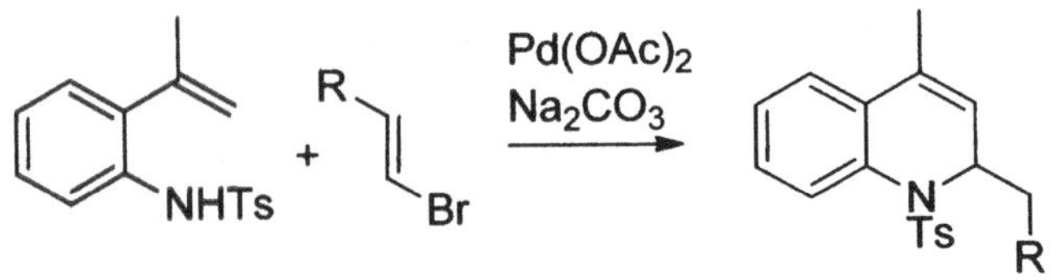

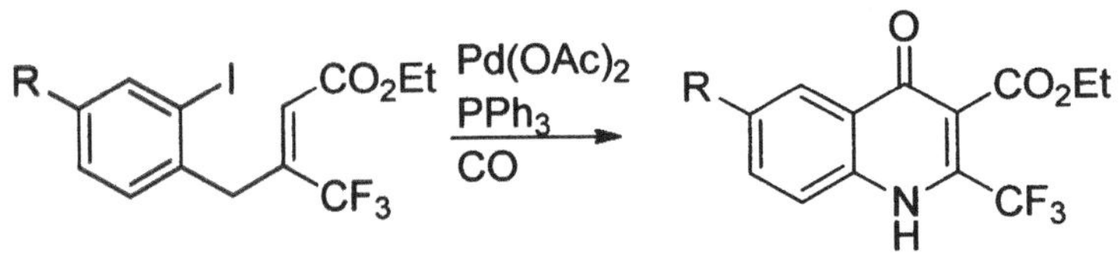

A modification of the Negishi cross-coupling has also been used for a wide range of nitrogen heterocycles. In accordance to that, the zinc reagent derived from protected iodoaniline can be treacted with α-substituted carbethoxy vinyl-triflates to produce quinolin-2-ones.[107]

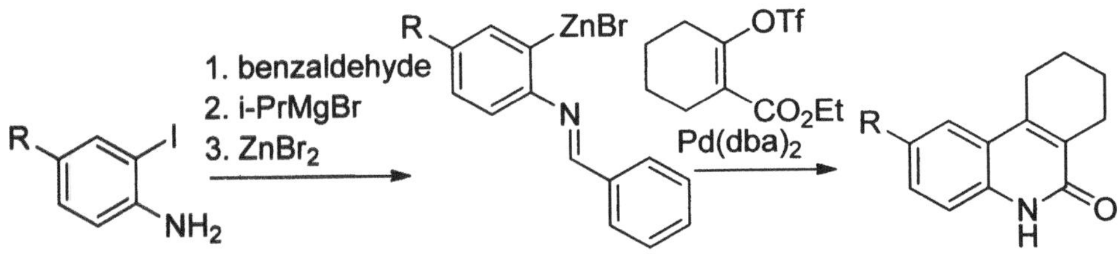

Electrocyclic ring-closing methods

The preparation of quinoline derivatives by cycloaddition and pericyclic processes is of great importance for the synthesis of otherwise difficult accessible drugs.

Pyrrole[3,4-*b*]quinolines can be formed through the coupling of anilines with *N*-propargylic-substituted heterocyclic aldehydes in the presence of a mild Lewis acid, as this was highlighted by the formal synthesis of camptothecin. The sequence initiates by imine formation, followed by a formal intramolecular Diels–Alder reaction, known as Povarov reaction.[108]

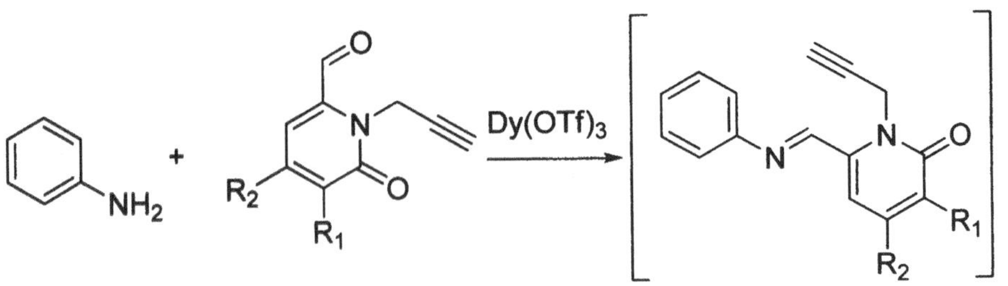

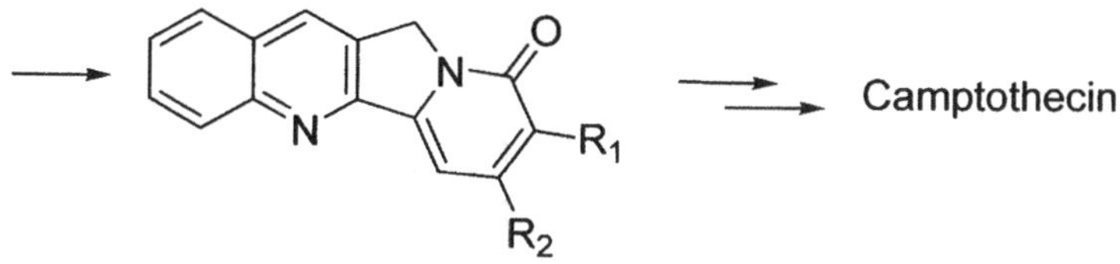

The aza-Diels–Alder reaction can also be used as a convenient method for the synthesis of substituted tetrahydroquinolines by *in situ* generation of azadiene compounds, generated from the appropriate carbonates, and its reaction with dienophiles.[109]

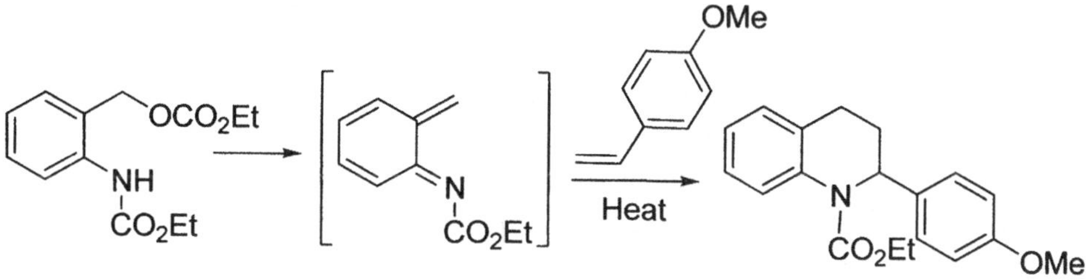

On the other hand, 6π-electrocyclization reaction of enolizable vinyl quinine mono- and di-imide substrates provides a high-yielding process to access dihydroquinolines derivatives.[110]

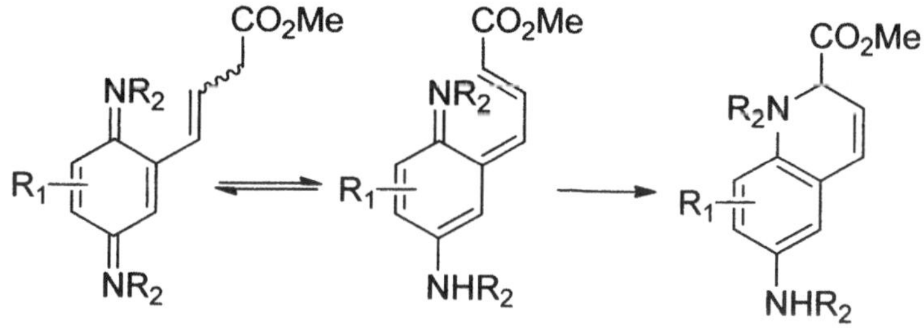

Electrocyclization processes are also postulated to be involved during the synthesis of quinolin-2-ones by treacting benzoxazinones with ketoesters.[111]

11.4 Construction of Isoquinoline Core

11.4.1 Bischler–Napieralski Reaction[112]

Reaction of phenylethyl amides with a dehydrating agent, such as P_2O_5 or $POCl_3$, was first described by August Bischler and Bernard Napieralski in 1893 to afford 3,4-dihydro-isoquinolines in good yields.

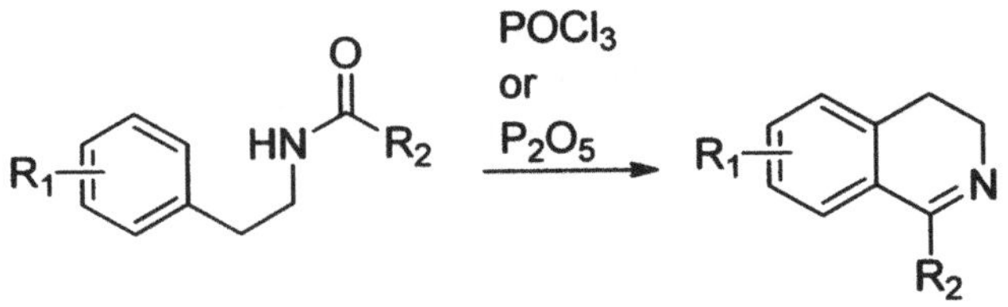

Detailed mechanistic studies by Fodor revealed that the reaction is initiated by a nucleophilic attack of phenylethylamide to the dehydrating agent to afford intermediate imidoyl chlorides. Upon heating, these compounds are converted to nitrilium salts, which were cyclized through an intramolecular electrophilic aromatic substitution, providing the desired dihydroisoquinolines.[113] Fodor prepared imidoyl chlorides using mild condition and promoted their cyclization to dihydroisoquinolines by using Lewis acids. His results supported the intermediacy of nitrilium salts as the key intermediates for the synthesis of dihydroisoquinolines.[114]

The mechanism has been further supported by the isolation and characterization of retro-Ritter by-products observed from the Bischler–Napieralski reaction. The retro-Ritter by-products are formed by decomposition of the proposed nitrilium salts.

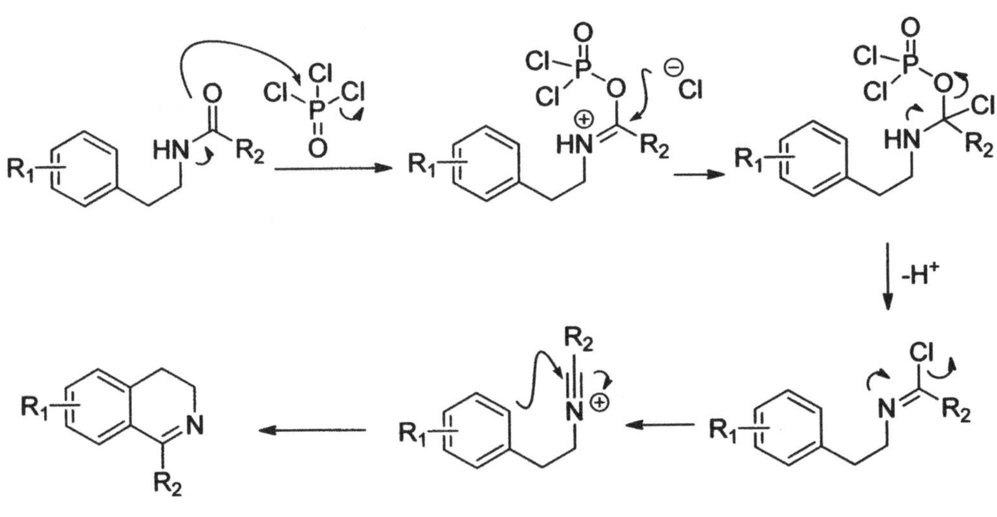

As it would be expected, the Bischler–Napieralski reaction as an intramolecular electrophilic substitution reaction is accelerated in the presence of an electron-donating group on the aromatic ring. With respect to that, electronic effects influence the regioselectivity of the reaction, leading the substitution typically to the carbon bearing the higher electron density. On the other hand, substitution on the phenylethyl side chain is usually well tolerated.[115]

Several biological potent compounds bearing a isoquinoline core in their structures relied their synthesis on a Bischler–Napieralski reaction. The isoquinoline core of anticholinergic agent solifenacin, intended as a drug for treating urinary incontinence, is prepared from benzamide of 2-phenylethylamine with phosphorous oxychloride, using a classic Bischler–Napieralski reaction. The formed imine is reduced, resolved, and transformed to the asymmetric amide of the drug.[116]

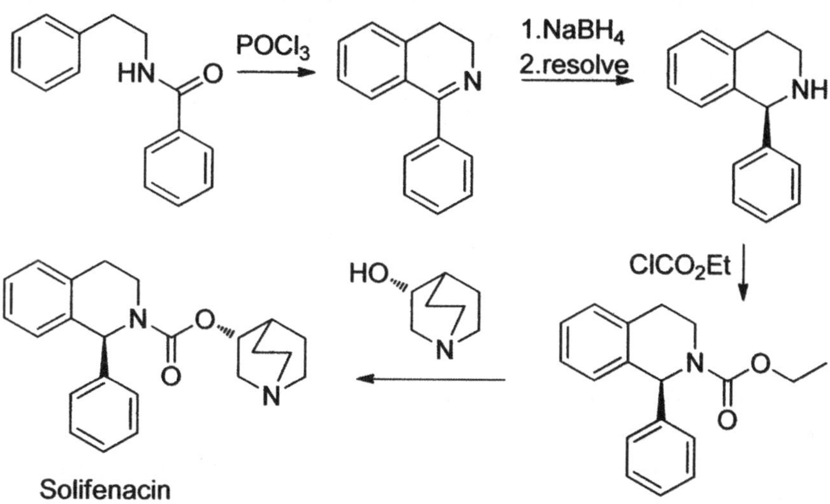

Solifenacin

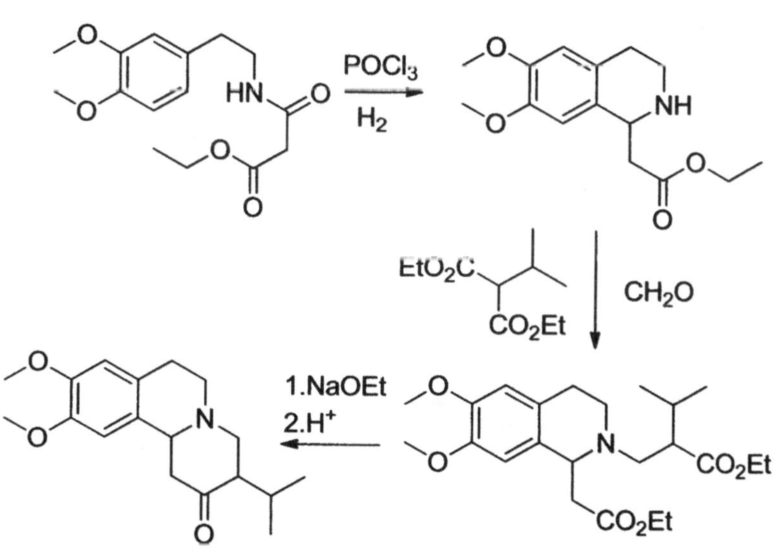

Tetrabenazine

The tricyclic isoquinoline compound tetrabenazine, a drug that has been used as an anti-psychotic agent and was also being studied for the treatment of some of the deleterious effects of the anti-psychotic dopamine antagonists, is prepared in a similar manner following the Bischler–Napieralski reaction of the appropriate substituted arylethylamine. Reduction and subsequent alkylation of the amine was followed by Dieckmann condensation to provide tetrabenazine.[117]

11.4.2 Pictet–Spengler Reaction[118]

The Pictet–Spengler reaction is a modification of the Bischler–Napieralski reaction. It is probably the most used method for the synthesis of 1,2,3,4-tetrahydro isoquinoline core. Therefore, phenylethyl amines are used in reaction with carbonyl compounds in the presence of protic or Lewis acids.

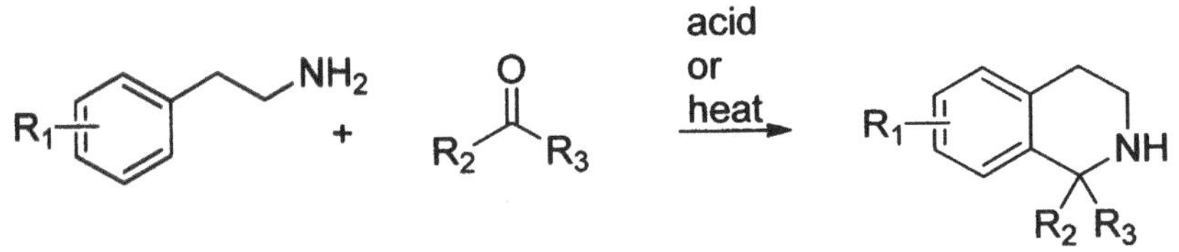

Reaction of phenylethyl amine with aldehyde or ketone forms intermediate imine, which under the acidic conditions is protonated to give the highly reactive iminium ion. Electrophilic substitution on the aromatic ring provides 1,2,3,4-tetrahydro isoquinoline after rapid loss of a proton and concomitant rearomatization.[119]

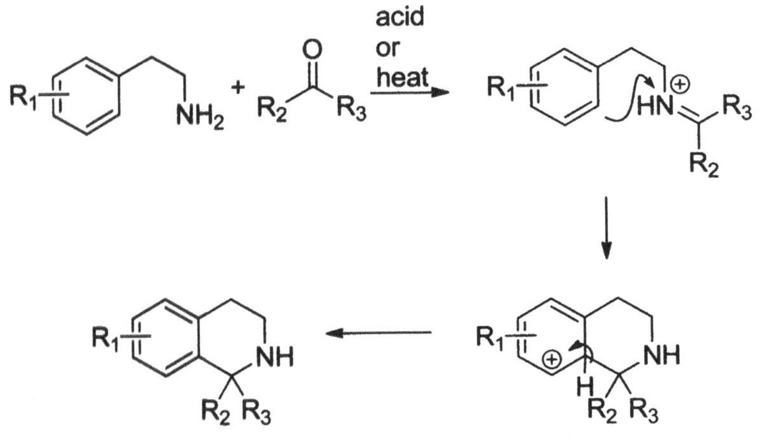

As in the case of the Bischer–Napieralski reaction, regiochemical and stereochemical issues also exist in the Pictet–Spengler reaction. Electrophilic substitution occurs on the more electron-rich carbon of the aromatic ring, whereas phenylethyl amines possess a substituent in α-position leads to a modest preference of *anti*-positioned substituent in the C1 position of the tetrahydro-isoquinolinoline core. This preference was found to be enhanced by the use of superacids as catalysts.

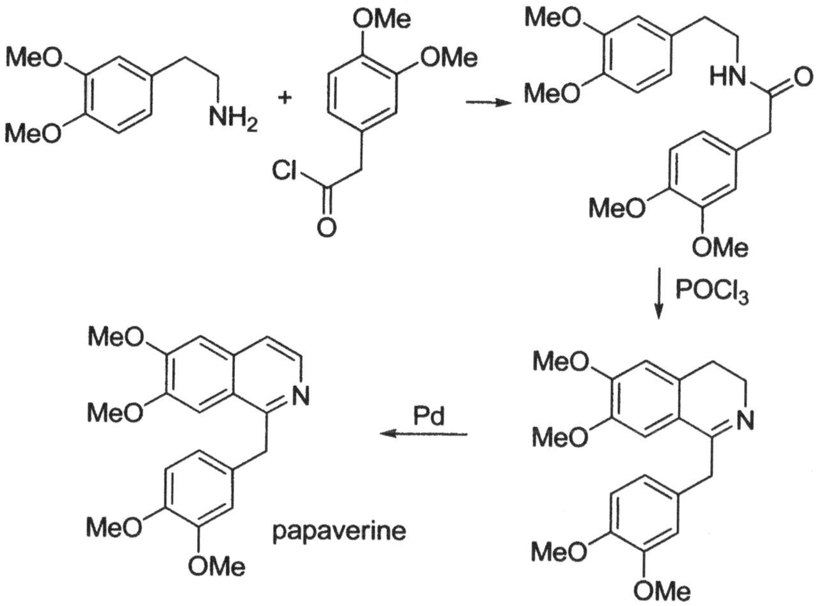

The Pictet–Spengler reaction is frequently used for the synthesis of pharmacophore isoquinoline core in the last century. Several examples well illustrate its efficiency. Papaverine is the best-known drug that incorporates the isoquinoline nucleus. This natural product that accompanies the opioids in various *papaver* species, exhibits muscle-relaxing properties and has been used as a spasmolytic drug and a vasodilator to improve cerebral blood circulation. The first synthesis of this drug was based on a Pictet–Spengler reaction of homoveratrylamine with homoveratroyl chloride to give the corresponding amide. Treatment with strong Lewis acid formed isoquinoline, which after aromatization afforded papaverine.[120]

In another example, variation in one of the starting components by using the carboxylic salt of α-epoxide in reaction with β-phenethylamine (dopamine) under the Pictet–Spengler conditions was used for the synthesis of trimethoquinol. This compound is one of a small group of compounds devoid of an aminoalcohol moiety that acts as a β-adrenergic agonist investigated as a potent bronchodilator.[121]

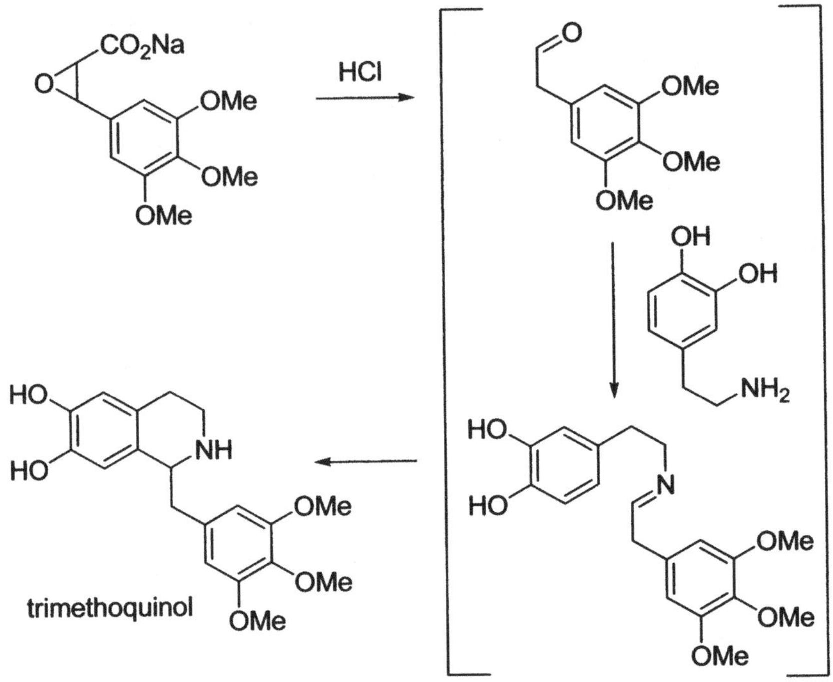

An analogous methodology was used for the synthesis of selective ACE inhibitor quinapril. Condensation of *S*-phenylalanine with formaldehyde in the presence of sulfuric acid produced an intermediate imine, which was readily cyclized to for predominantly only one enantiomer of the tetrahydroisoquinoline core. Protection of carboxylic acid moiety was followed by acylation with the appropriate acid to form quinapril after deprotection.[122]

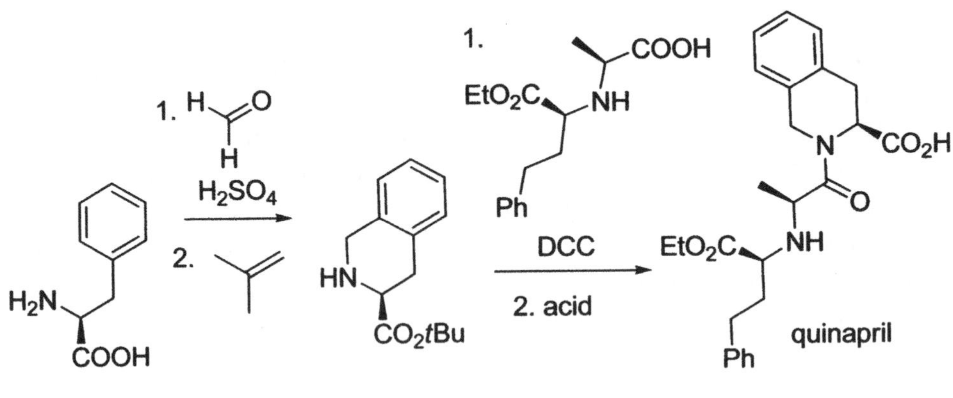

The Pictet–Spengler reaction has also been developed on solid-support materials. This advancement provided the opportunity for the synthesis of an array of structurally diverse analogues of saframycin A, an important anti-tumor antibiotic.[123]

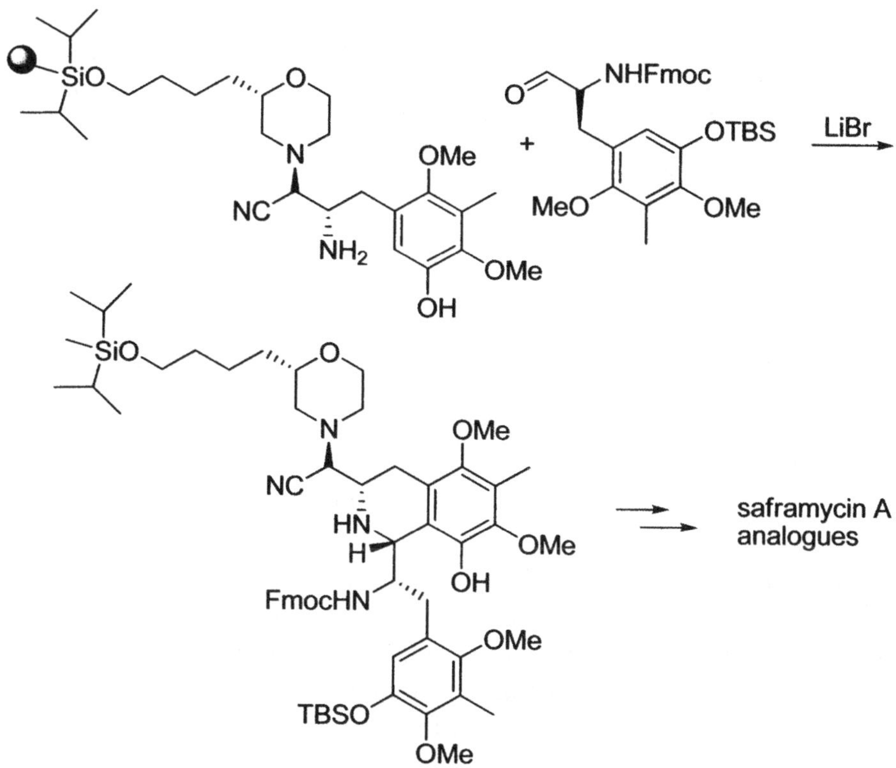

11.4.3 Pictet–Gams Isoquinoline Synthesis[124]

Another modification of the Bischler–Napielarski reaction concerning the preparation of aromatized isoquinolines instead of the 3,4-dihydro-derivative was described by Pictet and Gams in 1909.[124] From the synthetic standpoint, the Pictet–Gams isoquinoline synthesis is advantageous for the preparation of aromatized isoquinoline cores. However, the Bichler–Napieralski was the method of choice for several years due to the capricious character of the reaction based on the poor understanding of the reaction mechanism before the emergence of the Pictet–Gams reaction.

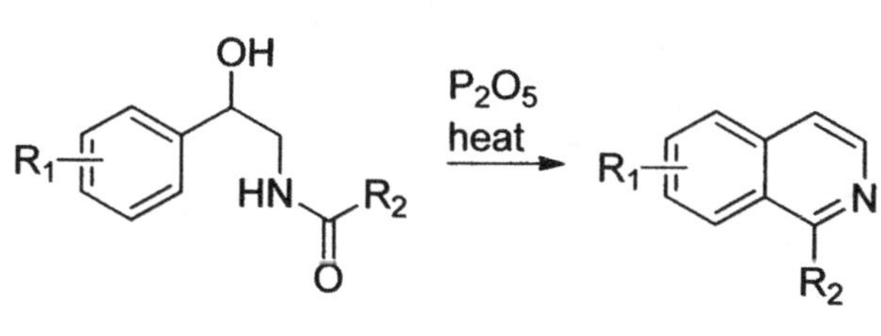

In 1977, a postulated mechanism involving the formation of an oxazoline intermediate was described by Fritton.[125] The mechanism managed to explain the unexpected products observed when different 3-positioned alkyl groups were used.

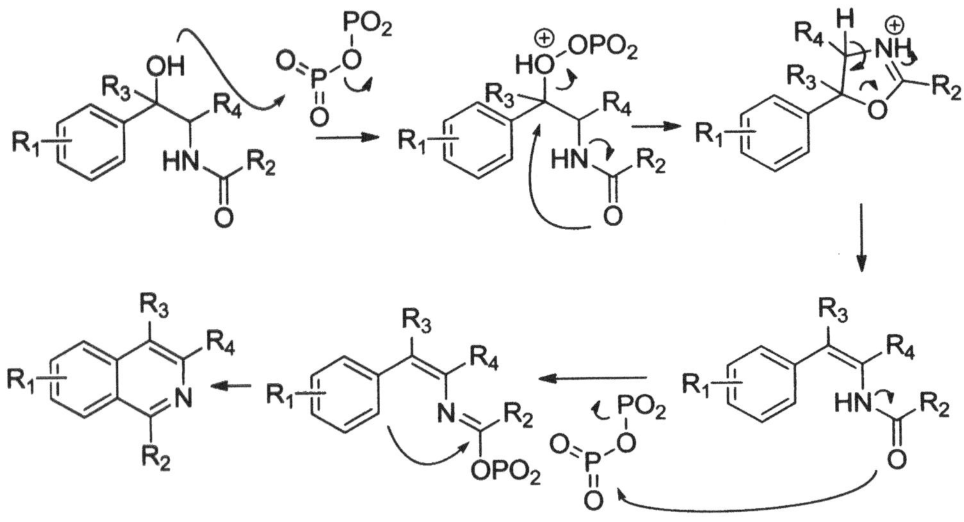

Nowadays, several modifications of the Pictet–Gams reaction exist in the literature, constituting the reaction as the method of choice for the synthesis of isoquinoline frameworks when acid labile substituents are not present in the molecule. Modifications include the use of cinnamic esters, instead of 2-phenyl-2-hydroxy ethanamines, as starting materials.[126]

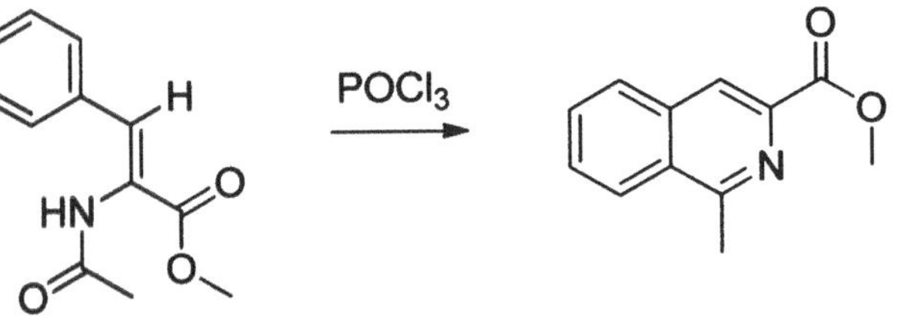

Special concern was also given in the development of milder reaction conditions using mixtures of trifluoroacetic anhydride and 2-chloropyridine or DMAP for electrophilic amide activation.[127]

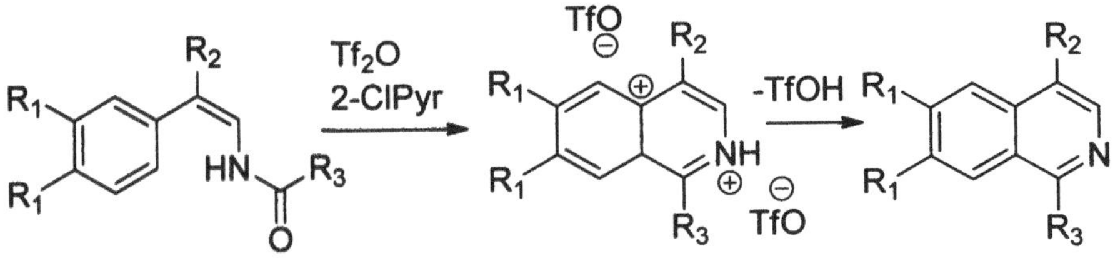

The Pictet–Gams reaction was used for the synthesis of difficult accessible isoquinoline derivatives as the tri-isoquinoline artificial receptor of resorcinol.[128]

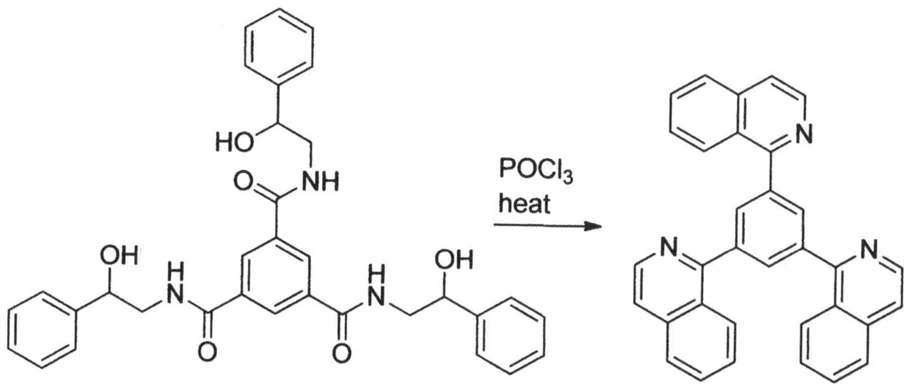

11.4.4 Pomeranz–Fritsch Reaction[129]

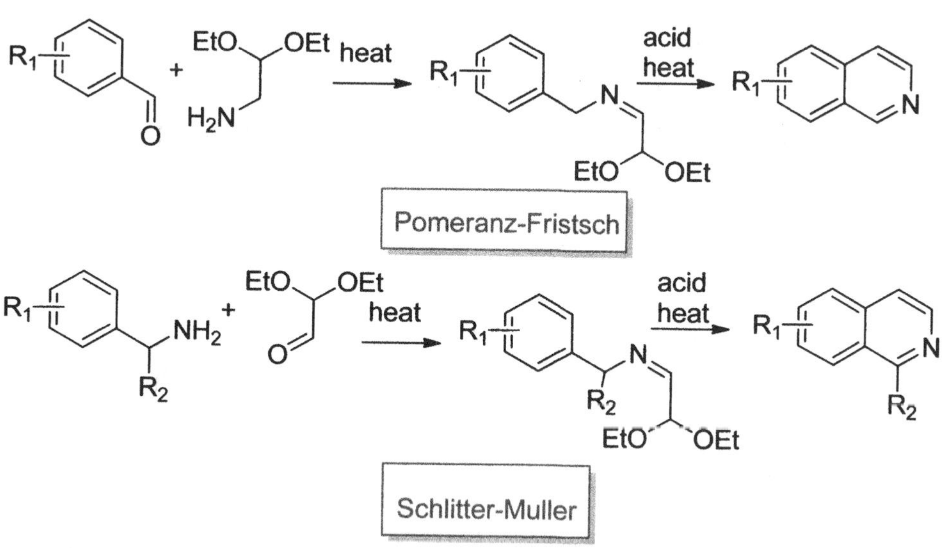

The Pomeranz–Fritsch reaction involves the initial condensation of an aryl aldehyde with 2-aminoaldehyde acetal to provide aldimine, which is then isolated and cyclized under strong acid conditions to afford the isoquinoline framework. The most known variation of this reaction is using benzylamine and glyoxal diethylacetal as the coupling pair for the initial condensation (known as the Schlitter–Muller variation), which is able to provide 1-substituted isoquinolines.[130]

The mechanism of the reaction involves intramolecular electrophilic substitution of aryl group on the acetal side as shown in the scheme. Based on that, electron-donating groups on the aromatic ring (especially *para-* to the position of ring closure) greatly enhance the ability of aromatic ring to cyclize.[131]

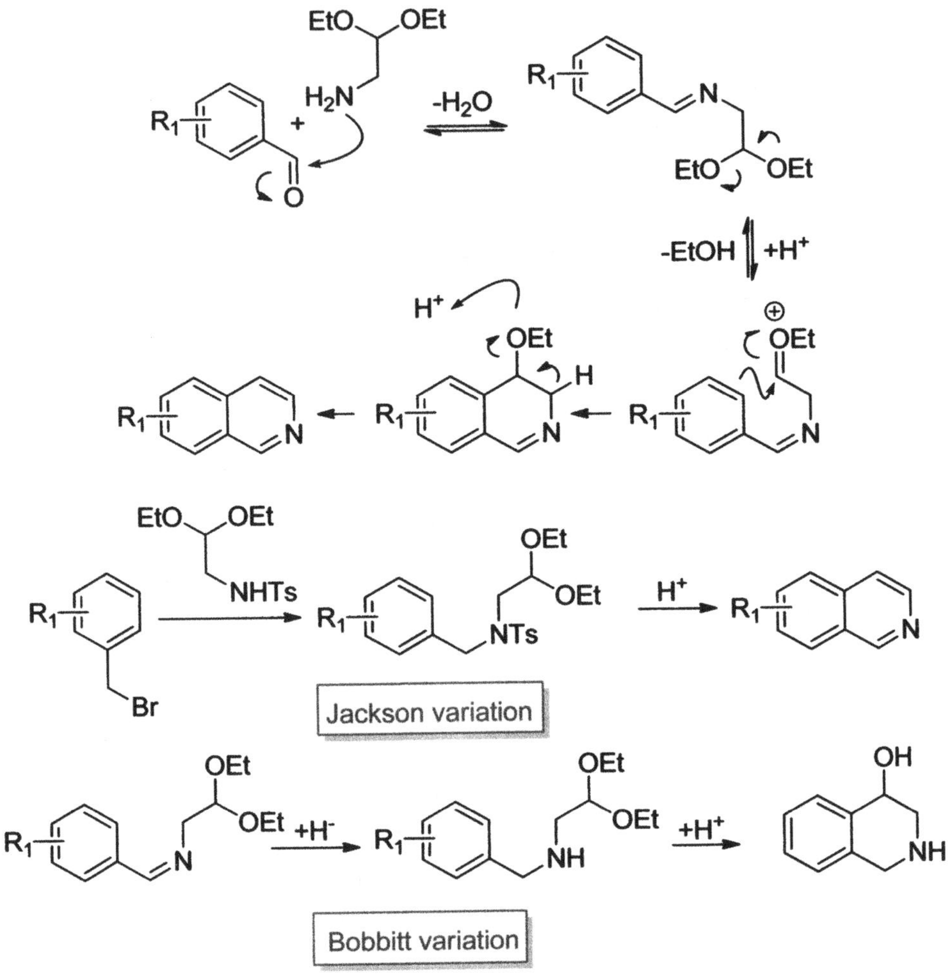

Usually, the reaction is poor in yield limited by the imine hydrolysis in the strong acidic conditions used for the cyclization. Two solutions were used to overcome this drawback: i. The use of milder acidic conditions[132] (e.g., trifluoroactic acid/boron trifluoride mixture) and ii. Carrying out the cyclization on the amine oxidation state (Bobbit and Jackson variation).[133]

11.4.5 Gabriel–Colman Rearrangement[134]

The Gabriel–Colman reaction involves rearrangement of phthalimidoyl acetates to 4-hydroxy isoquinoline cores by the use of hydrolytic basic conditions.

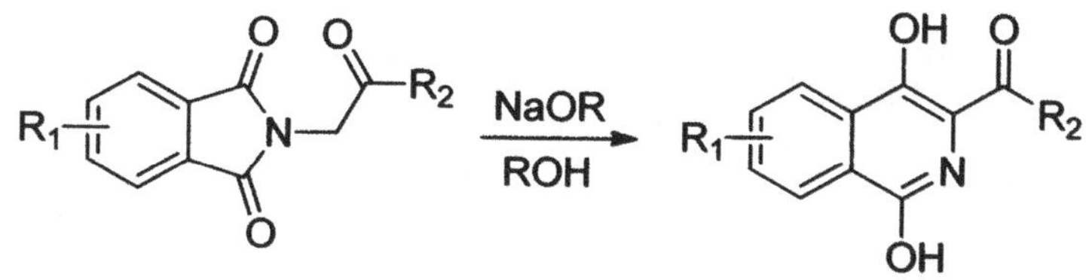

The reaction is postulated to invoke alkoxide attack on one of the amide carbonyls, furnishing amide anion. Enolate formation and Dieckmann condensation follow to provide the isoquinoline core after aromatization.[135]

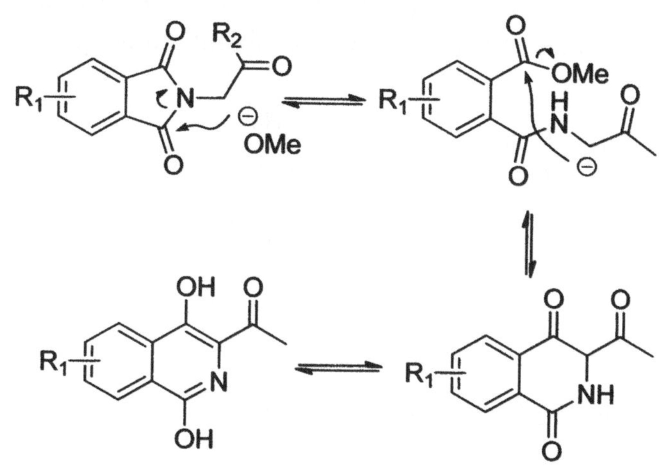

The rearrangement was found to be substrate specific, and several factors have been found to contribute in the yield of the product, such as the polarity of the solvent, the temperature, and, in some cases, the pressure.

11.4.6 Modern Methods

Cyclization of alkynes and alkenes

Methods that rely on alkynes and alkenes for the synthesis of isoquinoline core are based on two basic modes for their cyclization. The first uses electrophilic reagents as activators of alkynes or alkenes for the cyclization process, while the second method uses a nucleophilic atom usually nitrogen in reaction with the alkyne or alkene moiety.

As electrophiles I_2, ICl, PhSeCl usually are used, leading to isoquinolines bearing substitution on 4-position of the heterocycle. This can be further functionalized through cross-coupling reactions (see section 11.2.7) to provide highly substituted derivatives.

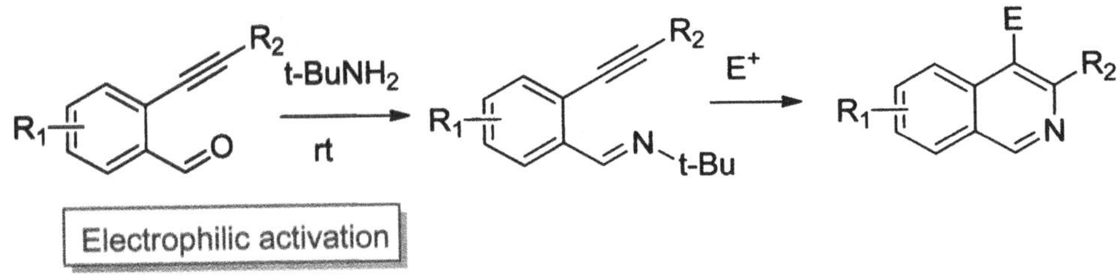

Concerning the nucleophilic cyclization, the reaction sequence usually employs the transformation of an aryl halide to the lithiated salt and its subsequent reaction with alkyl or aryl nitriles. The formed intermediate anion is readily cyclized to provide the 1,4-substituted isoquinolines.

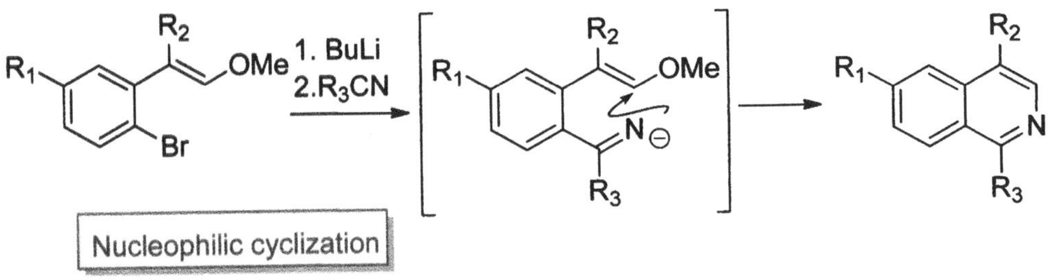

Metal-mediated or metal-catalyzed processes

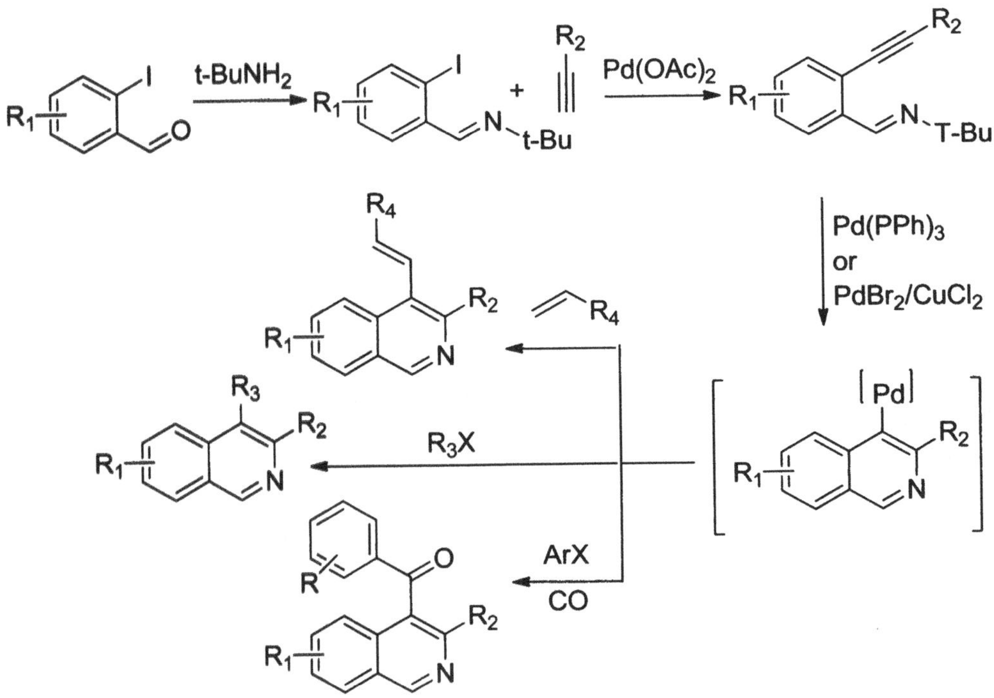

As in the case of quinolines, isoquinolines can be prepared by metal-mediated annulation processes. Reaction of *t*-butylimines of iodo-benzaldehydes with acetylenes in the presence of palladium catalyst affords 3,4-disubstituted isoquinolines. The power of this reaction relies on its ability to introduce different type of substituents on 4-position of the heterocycle by using the formed intermediate palladium complex in the cross-coupling reaction with aryl and alkyl halides or alkenes.[136,137] This methodology was successfully applied to the total synthesis of decumbenine B.[138]

Other metals like rhodium(I) have also been used for the direct cyclization of iminoaryls with alkynes without the prior existence of halogen atom on the aryl ring to produce highly substituted isoquinolines in good yields.[139]

Electrocyclic ring-closing methods

Electrocyclization or [4 + 2] processes have been reported for the construction of isoquinoline core either from aza-diene intermediates (as in the case of quinolines) or from benzyne intermediates.

In the first case, phosphazenes react with aldehydes or even ketenes to produce aza-diene, which under heating conditions, undergoes electrocyclization to afford the isoquinoline ring. This methodology has been used for the synthesis of marine alkaloid renierol.[140]

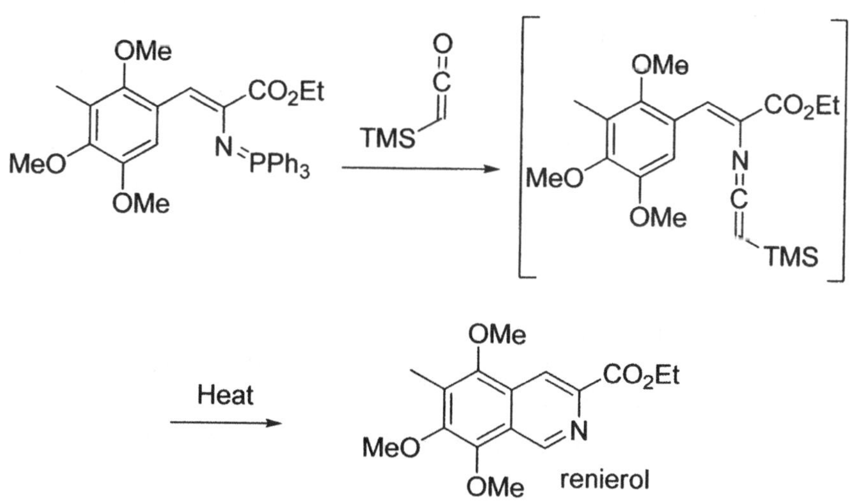

Arynes can be prepared and used as isoquinoline precursors in a formal [4 + 2] reaction with enamides or amidoacrylates. Arynes are liberated from fluoride anion reaction of silyl aryl triflates with cesium fluoride.

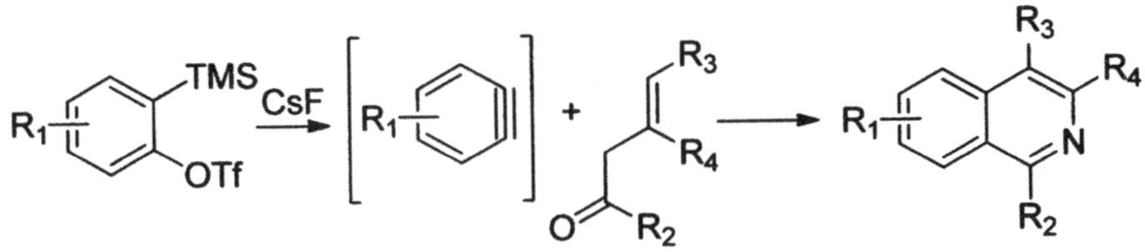

11.5 Possible Liabilities of Drugs Containing Quinoline and Isoquinoline Ring

Unsubstituted quinoline and isoquinoline are poisons when they enter the body by any of the normal routes, i.e., ingestion, or subcutaneous or intraperitoneal injection. There is also evidence that quinoline and isoquinoline are mutagenic, and long exposure can produce lung problems. Toxicity data are typically LD_{50} (rat) 330–360 mg/Kg and dermal LD_{50} (rat) 540–590 mg/Kg. Substituted quinoline and isoquinoline derivatives in general are highly toxic compounds with LD_{50} values down to 10 mg/Kg.

In contrast, quinoline-containing anti-malarial drugs such as chloroquine (launched 1934) and hydroxychloroquine have been used for many years to treat rheumatoid arthritis and systemic lupus erythematosus (SLE).[141] Although the efficacy of the antimalarial drugs are lower than methotrexate in treating RA, they remain popular choices *because of their good safety profile.* From a mechanism of action perspective, they are thought to have an inhibitory effect on Toll-like receptor (TLR) signaling which would account for their efficacy in rheumatic diseases.[142]

One of their major classes of drugs, quinolone antibiotics, is associated with serious problems such as the syndrome of hemolysis. In some cases, uremia, coagulopathy, and hyperbilirubinemia were observed for fluoroquinolone antibiotics like temafloxacin.[143] Other adverse reactions have also been reported on the central nervous system with symptoms of headaches, insomnia, and dizziness. Skeletal problems still remain theoretical for quinolones in the prenatal formation, but tendinitis and tendon rupture have occurred in small number of adult patients.[144]

Some quinolones are only slightly soluble at neutral or alkaline conditions (e.g., norfloxacin and ciprofloxacine). They induce nephrotoxicity in laboratory animals due to crystallization of the drugs in renal tubules.[145]

On the other hand, several studies on the carcinogenicity of quinolones revealed no indication of carcinogenic effect even after life-long exposure.[146]

Quinolone drugs have been described to interact significantly with other drugs. These interactions both reduce intestinal absorption after oral administration and may result in potentially serious problems via inhibition of various metabolic pathways, notably the cytochrome P450 system in the liver and γ-aminobutyric acid (GABA) neuro-inhibitory receptors in the brain.[147]

11.6 Problems

11.6.1 Propose a reasonable mechanism for the following transformation[148]:

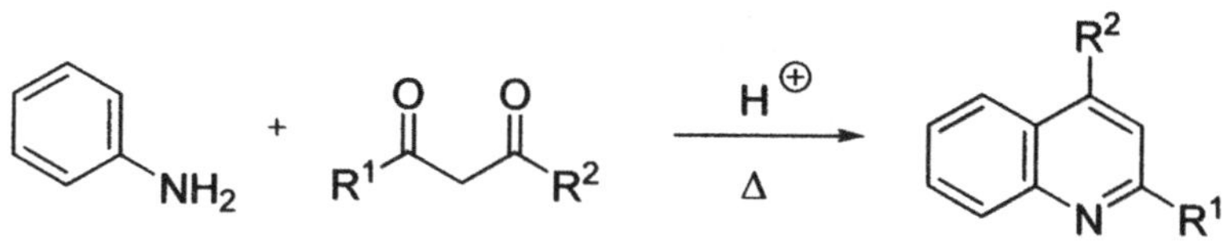

11.6.2 Propose a reasonable mechanism for the following transformation[149]:

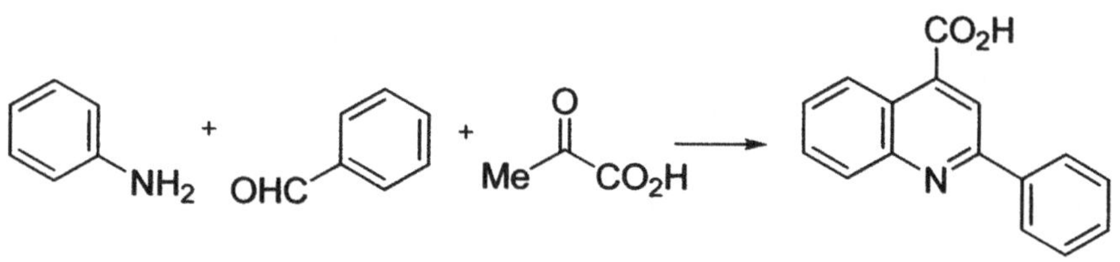

11.6.3 Propose a reasonable mechanism for the following transformation:[150]

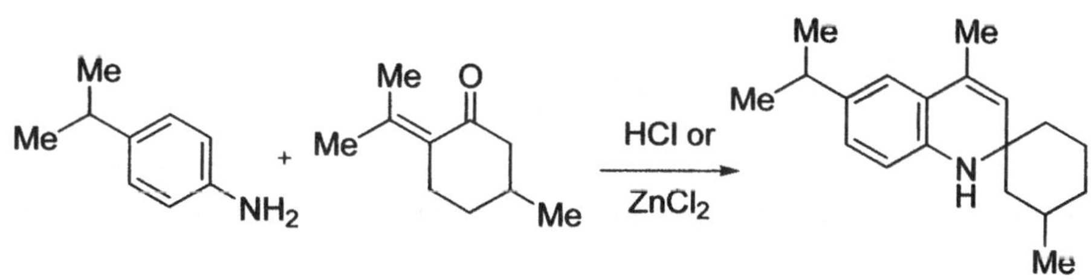

11.6.4 Propose a reasonable mechanism for the following transformation[151]:

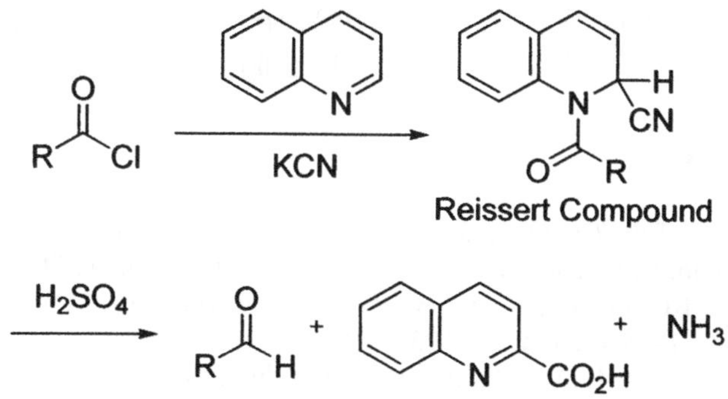

11.6.5 Propose a reasonable mechanism for the following transformation[152]:

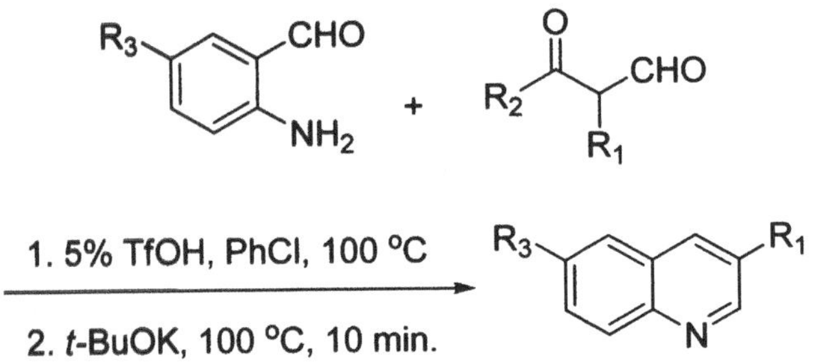

11.6.6 Propose a reasonable mechanism for the following transformation[153]:

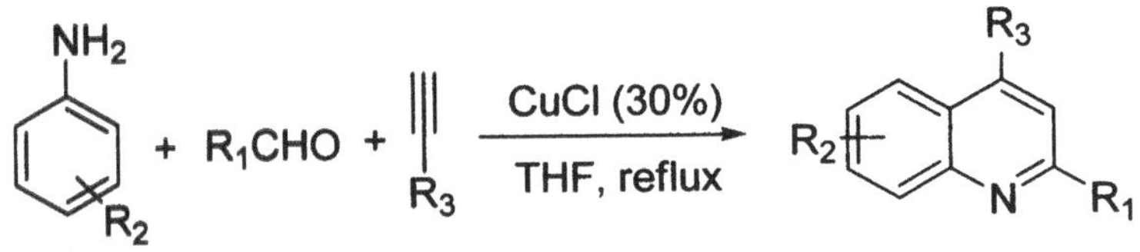

11.6.7 Propose a reasonable mechanism for the following transformation[154]:

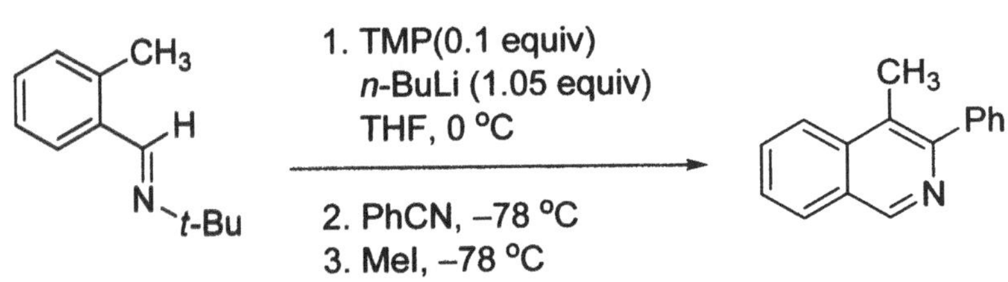

11.6.8 Propose a reasonable mechanism for the following transformation[155]:

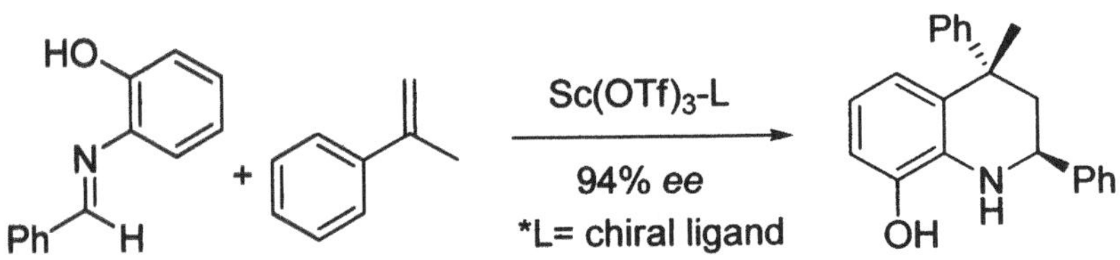

11.7 References

1. Adriole, V. T. Ed., *The Quinolones*, Academic Press: Waltham, MA, 2nd ed., **1998**.
2. Kaur, K.; Jain, M.; Reddy, R. P.; Jain, R. *Eur. J. Med. Chem.* **2010,** *45*, 3245–3264.
3. Baird, J. K.; Reickmann, K. H. *Trends Parasitol.* **2003,** *19*, 115–120.
4. Fitch, C. D. *Life Sci.* **2004,** *74*, 1957–1972.
5. Baird, J. K.; Fryauff, D. J.; Hofmann, S. L. *Clin. Infect. Dis.* **2003,** *37*, 1659–1667.

6. Vander Jagt, D. L.; Hunsaker, L. A.; Campos, N. M. *Mol. Biochem. Parasitol.* **1986,** *18*, 389–400.
7. Ciak, J.; Hahn, F. E. *Science* **1996,** *151*, 347–349.
8. Sharma, A.; Mishra, N. C. *Indian J. Biochem. Biophys.* **1999,** *36*, 299–304.
9. (a)Das, B.; Krishnaiah, M.; Venkateswarlu, K.; Das, R. *Nat. Prod. Commun.* **2006,** *55*, 263; (b) Twin, H; Batey, R. A. *Org. Lett.* **2004,** *6*, 4913.
10. Duffour, J.; Gourgou, S.; Desseigne, F.; Debrigode, C.; Mineur, L.; Pinguet, F.; Poujol, S.; Chalbos, P.; Bressole, F.; Ychou, M. *Cancer Chem. Pharmacol.* **2007,** *60*, 283.
11. (a) McLaughlin, M. J.; Hsung, R. P. *J. Org. Chem.* **2001,** *66*, 1049; (b) Michael, J. P. *Nat. Prod. Rep.* **1999,** *16*, 697; (c) Brader, G.; Bacher, M.; Greger, H.; and Hofer, O. *Phytochemistry* **1996,** *42*, 881; (d) Michael, J. P. *Nat. Prod. Rep.* **1995,** *12*, 77.
12. Finley, T. K. in *Van Nostrands Encyclopedia of Chemistry,* Chapter *Quinolines and Isoquinolines*; Wiley: NY, **2005**.
13. Joule, J. A.; Mills, K. *Heterocyclic Chemistry at a Glance*; Blackwell Publishing Group: Boston, MA, **2007**.
14. Jones, G. in *Quinolines*, Part I, **1977**; Part II, **1982**; Part III, **1990**; Grenthe, G. in *Isoquinolines*, Part I, **1981**; Kathawala, F. G.; Coppola, G. M.; Schuster, H. F. in *Isoquinolines* Part II, **1990**; Coppola G. M.; Schuster, H. F. in *Isoquinolines,* Part III, **1995**, Weissberger A.; Taylor, E. C., Eds.; *The Chemistry of Heterocyclic Compounds*, Vols. 32 and 38, Wiley-Interscience, New York. NY.
15. Krakower, E.; Reeves, L. W. *Spectrochimica Acta* **1964,** *20*, 71–77.
16. Katritzky, A. R.; Bressel, U.; Lea, J. R. *J. Chem. Soc. B* **1971,** 4.
17. Kaminsky, D.; Meltzer, R. J.; *J. Med. Chem.* **1968,** *11*, 160.
18. Wentland, M. P.; Bailey, D. M.; Cornett, J. B.; Dobson, R. A.; Powles, R. G.; Wagner, R. B. *J. Med. Chem.* **1984,** *27*, 1103.
19. Hill, D. A.; Turner, G. L. US 5,453,510 (1995).
20. Dewar, M. J. S.; Maitlis, P. M. *J. Chem. Soc.* **1957,** 2521.
21. Engelhardt, P.; Hogberg, M.; Johanssen, N.-G.; Zhou, X.-X.; Lindborg, B. *US Patent* **1999,** 5,869,493.
22. Burckhalter, J. H.; Brinigar, W. S.; Thompson, P. E. *J. Org. Chem.* **1961,** *26*, 4070.
23. Kress, T. J.; Costantino, S. M. *J. Heterocycl. Chem.* **1973,** *10*, 409.
24. Vandewalle, J .J. M.; de Ruiter, E.; Reimlinger, H.; Lenaers, R. A. *Chem. Ber.* **1975,** *108*, 3898.
25. (a) Zoltewicz, J.; Helmick, L. S.; Oestreich, T. M.; King, R. W.; Kandetzki, P. E. *J. Org. Chem.* **1973,** *38*, 1947; (b) Tondys, H.; van der Plas, H. C.; Wozniak, M. *J. Heterocycl. Chem.* **1985,** *22*, 311.
26. Geissman, T. A.; Schlatter, M. J.; Webb, I. D.; Roberts, J. D. *J. Org. Chem.* **1946,** *11*, 741.
27. Allais, A. *Chim. Ther.* **1973,** *8*, 154.
28. a) Koizumi, T.; Takeda, K.; Yoshida, K.; Yoshii, E. *Synthesis* **1977,** 497; b) Collado, D.; Perez-Inestrosa, E.; Suau, R. *J. Org. Chem.* **2003,** *68*, 3574.
29. Barbier, D.; Marazano, C.; Das, B. C.; Potier, P. *J. Org. Chem.* **1996,** *61*, 9596.
30. Barbier, D.; Marazano, C.; Riche, C.; Das, B. C.; Potier, P. *J. Org. Chem.* **1998,** *63*, 1767.
31. Surrey, A. R.; Hammer, H. F. *J. Am. Chem. Soc.* **1946,** *68*, 113.
32. (a) Yamamoto, Y.; Yanagi, A. *Heterocycles* **1981,** *16*, 1161.; (b) Echavarren, A. M.; Stille, J. M. *J. Am. Chem. Soc.* **1988,** *110*, 4051.
33. Fan, P.; Ablordeppey, S. Y. *J. Heterocycl. Chem.* **1997,** *34*, 1789.
34. Murata, N.; Sugihara, T.; Kondo, Y.; Sakamoto, T. *Synlett* **1997,** 298.
35. Baston, E.; Palusczak, A.; Hartmann, R. W. *Eur. J. Med. Chem.* **2000,** *35*, 931.

36. Ishiyama, T.; Murata, M.; Miyaura, N. *J. Org. Chem.* **1995,** *60*, 7508.
37. Giroux, A.; Han, Y.; Prasit, P. *Tetrahedron. Lett.* **1997,** *38*, 3841.
38. Ishikura, M.; Oda, I.; Terashima, M. *Heterocycles* **1985,** *23*, 2375.
39. Miyachi, N; Yanagawa, Y.; Iwasaki, H.; Ohara, Y.; Hiyama, T. *Tetrahedron Lett.* **1993,** *34*, 8267.
40. Phan, L. T.; Clark, R. F.; Rupp, M.; Or, Y. S.; Chu, D. T. W.; Ma, D. *Org. Lett.* **2000,** *2*, 2951.
41. Godard, A.; Rocca, P.; Fourquez, J.-M.; Rovera, J.-C.; Marsais, F.; Queguiner, G. *Tetrahedron Lett.* **1993,** *34*, 7919.
42. (a) Fache, F. *Synlett* **2004,** 2827; (b) Fujita, K. I.; Kitatsuji, C.; Furukawa, S.; Yamaguchi, R. *Tetrahedron Lett.* **2004,** *45*, 3215.
43. Ranu, B. C.; Jana, U.; Sarkar, A. *Synth. Commun.* **1998,** *28*, 485.
44. Moody, C. J.; Pitts, M. R. *Synlett* **1998,** 1029.
45. Alvarez, M.; Bros, M. A.; Gras, G.; Ajana, W.; Joule, J. A. *Eur. J. Org. Chem.* **1999,** 1173.
46. Nose, A.; Kudo, T. *Chem. Pharm. Bull.* **1988,** *36*, 1529.
47. Flaniken, J. M.; Collins, C. J.; Lanz, M.; Singaram, B. *Org. Lett.* **1999,** *1*, 799.
48. Leonard, N. J.; Leubner, G. W. *J. Am. Chem. Soc.* **1949,** *71*, 3408.
49. Barbier, A. M.; Rumpf, P. *Bull. Soc. Chim. Fr.* **1953,** 293.
50. Clarke, C. B.; Pinder, A. R. *J. Chem. Soc.* **1958,** 1967.
51. Neumeyer, J. L.; Boyce, C. B. *J. Org. Chem.* **1973,** *38*, 2291.
52. Meyers, A. I.; Bailey, T. R. *J. Org. Chem.* **1986,** *51*, 872.
53. Hoogerwerff, S.; van Dorp, W. A *Chem. Ber.* **1880,** *12*, 747.
54. Kim, D. W.; Choi, H. Y.; Lee, K. J.; Chi, D. Y. *Org. Lett.* **2001,** *3*, 445.
55. Boyd, D. R.,; Sharma, N. D.; Carroll, J. G.; Malone, J. F.; Mackerracher, D. G.; Allen, C. C. R. *Chem. Commun.* **1998,** 683.
56. Varma, R. S.; Kumar, D. *Tetrahedron Lett.* **1998,** *39*, 9113.
57. (a) Camps, R. *Chem. Ber.* **1899,** *32*, 3228. (b) Camps, R. *Arch. Pharm.* **1899,** *237*, 659.
58. (a) Beutner, G. L.; Kuethe, J. T.; Yasuda, N. *J. Org. Chem.* **2007,** *72*, 7058; (b) Mitsos, C. A.; Zografos, A. L.; Igglessi-Markopoulou, O. *J. Org. Chem.* **2003,** *68*, 4567.
59. Zografos, A. L.; Mitsos, C. A.; Igglessi-Markopoulou, O. *Org. Lett.* **1999,** *1,* 1953.
60. Combes, A. *Bull. Soc. Chim. Fr.* **1888,** *49*, 89.
61. Jones, G. *Heterocyclic Compounds,* Wiley, New York, **1977**, *Quinolines Vol 32*, pp 119–125.
62. Aly, A. A *Tetrahedron* **2003,** *59*, 1739.
63. El Ouar, M.; Knouzi, N.; Hamelin, J. *J. Chem. Res. (S)* **1998,** 92.
64. Tom, N. J; Ruel, E. M. *Synthesis* **2001,** 1351.
65. Reitsema, R. H. *Chem. Rev.* **1948,** *47*, 47.
66. Bergstrom, F. W. *Chem. Rev.* **1944,** *43*, 3577.
67. Northey, E. H.; Dreisbach, P. F. US 2.478.125 (1949).
68. Ohnmacht, C. J.; Patel, A. R.; Lutz, R. E. *J. Med. Chem.* **1971,** *14*, 926.
69. Leonard, N. J.; Hebrandson, H. F.; Van Heyningen, E. M. *J. Am. Chem. Soc.* **1946,** *68*, 1279.
70. Renard, J. O. G. *Fr.* 1.279.901 (1961).
71. Friedlander, P. *Chem. Ber.* **1882,** *15*, 2572.
72. a) Cheng, C.-C.; Yan, S.-J. *Org. Reactions,* **1982,** *28*, 37. b) Jones, G. *Heterocyclic Compounds, Quinolines,* **1977**, vol. *32*; Wiley: New York, p. 181–191. c) Elderfield, R. C. *Heterocyclic Compounds;* Elderfield, R. C., Ed.; Wiley: New

York, **1952;** Volume *4*, *Quinoline, Isoquinoline and Their Benzo Derivatives*, p. 4547.

73. (a) Armit, J. W.; Robinson, R. *J. Chem. Soc.* **1922,** *121*, 827; (b) Fehnel, E. A.; Deyrup, J. A; Davidson, M. B. *J. Org. Chem.* **1958,** *23*, 1996.
74. von Niementowski, S. *Chem. Ber.* **1894,** *27*, 1394.
75. Pfitzinger, W. J. *Prakr. Chem.* **1886,** *33*, 100.
76. McNaughton, B. R.; Miller, B. L. *Org. Lett.* **2003,** *5*, 4257.
77. Dormer, P. G.; Eng, K. K.; Farr, R. N.; Humphrey, G. R.; McWilliams, J. C.; Reider, P. J.; Sager, J. W.; Volante, R. P. *J. Org. Chem.* **2003,** *68*, 467.
78. Ubeda, J. I.; Villacampa, M.; Avendano, C. *Synlett* **1997,** 285.
79. Jia, C. S.; Zhang, Z.; Tu, S. J.; Wang, G. W. *Org. Biomed. Chem.* **2006,** *4*, 104.
80. Chaires, J. B.; Ren, J.; Henary, M.; Zegrocka, O.; Bishop, G. R.; Strekowski, L. *J. Amer. Chem. Soc.* **2003,** *125*, 7272.
81. Giardina, G. A. M.; Raveglia, R. F.; Grugni, M.; Sarau, H. M.; Farina, C.; Medhurst, A. D.; Graziani, D.; Schmidt, D. B.; Rigolio, R.; Luttann, M.; et al. *J. Med. Chem.* **1999,** *42*, 1053.
82. Reitsema, R. H. *Chem. Rev.* **1948,** *53*, 43.
83. Curran, T. T. in *Name Reactions in Heterocyclic Chemistry*, Li, J. J.; Corey, E. J. Eds., Wiley: Hoboken, NJ, **2005,** 423–436.
84. Watson, J. US 3,267,106 (1965).
85. Kaminsky, D.; Meltzer, R. I. *J. Med. Chem.* **1968,** *11*, 160.
86. Koga, H.; Murayama, S.; Suzue, S.; Irikura, T. H. *J. Med. Chem.* **1980,** *23*, 1538.
87. Stern, I. M. EP 1984; *Chem. Abstr.* **1984,** *101*, 110764.
88. Price, C. C.; Roberts, R. M. *J. Am. Chem. Soc.* **1946,** *68*, 1204.
89. Meth-Cohn, O. *Heterocycles* **1993,** *35,* 539.
90. Meth-Cohn, O; Narine, B.; Tamowski, B. *J. Chem. Soc., Perkin Trans. 1* **1981,** 1520.
91. (a) Toyota, M.; Komori, C.; Ihara, M. *J. Org. Chem.* **2000,** *65*, 7110; (b) Mekouar, K.; Genisson, Y.; Leue S.; Greene, A. E. *J. Org. Chem.* **2000,** *65*, 5212; (c) Lavergne, O.; Lesueur-Ginot, L.; Pla Rodas, F.; Kasprzyk, P. G.; Pommier, J.; Demarquay, D.; Prevost, G.; Ulibarri, G.; Rolland, A.; Schiano-Liberatore, A.-M.; Harnett, J.; Pons, D.; Camara, J .; Bigg, D. C. H. *J. Med. Chem.* **1988,** *41,* 5410.
92. (a) Swahn, B.-M.; Claesson, A.; Pelcman, B.; Besidski, Y.; Molin, H.; Sandberg, M. P.; Berge, O.-G. *Bioorg. Med. Chem. Lett.* **1996,** *6*, 1635; (b) Swahn, B.-M.; Anderson, F.; Pelcman, B.; Soderberg, J.; Claesson, A. *J. Labelled. Compounds Radio.* **1997,** *39*, 259.
93. Skraup, Z. H. *Ber.* **1880,** *13*, 2086.
94. Doebner, O.; von Miller, W. *Ber. Dtsch. Chem. Ges.* **1883,** *16*, 2464.
95. Denmark, S. E.; Venkatraman, S. *J. Org. Chem.* **2006,** *71*, 1668.
96. Eisch, J. J.; Dluzniewski, J. *J. Org. Chem.* **1989,** *54*, 1269.
97. Matsugi, M.; Tabusa, F.; Minamikawa, J. *Tetrahedron Lett.* **2000,** *41,* 8523.
98. (a) Ranu, B. C.; Hajra, A.; Dey, S. S.; Jana, U. *Tetrahedron* **2003,** *59*, 813; (b) Theoclitou, M. E.; Robinson, L. A. *Tetrahedron Lett.* **2002,** *43*, 3907; (c) Reddy, B. M.; Ganesh, I. *J. Mol. Cat. A-Chem.* **2000,** *151*, 289.
99. (a) Elderfield, R. C.; Mertel, H. E.; Mitch, R. T.; Wempen, I. M.; Werble, E. *J. Am. Chem. Soc.* **1955,** *77*, 4816; (b) Curd, F. H. S.; Rose, F. L. *J. Chem. Soc.* **1946,** 729.
100. Papesch, V.; Burtner, R. R. *J. Am. Chem. Soc.* **1936,** *58*, 1314.
101. Arcadi, A.; Marinelli, F.; Rossib, E. *Tetrahedron* **1999,** *55*, 13233.
102. Williamson, N. M.; March, D. R.; Ward, D. *Tetrahedron Lett.* **1995,** *42*, 7721.
103. Jia, C.; Piao, D.; Kitamura, T.; Fujiwara, Y. *J. Org. Chem.* **2000,** *65*, 7516.

104. Abbiati, G.; Arcadi, A.; Canevari, V.; Capezzuto, L.; Rossi, E. *J. Org. Chem.* **2005,** *70*, 6454.
105. (a) Larock, R. C.; Yang, H.; Pace, P. *Tetrahedron Lett.* **1998,** *39*, 1885; (b) Larock, R. C.; Pace, P.; Yang, H. *Tetrahedron Lett.* **1998,** *39*, 2515; (c) Cossy, J.; Poitevin, C.; Gomez Pardo, D.; Peglion, J. L.; Dessinges, A. *Tetrahedron Lett.* **1998,** *39*, 2965; (d) Larock, R. C.; Pace, P.; Yang, H.; Russell, C. E.; Cacchi, S.; Fabrizi, G. *Tetrahedron* **1998,** *54*, 9961; (e) Cossy, J.; Poitevin, C.; Gomez Pardo, D.; Peglion, J. L.; Dessinges, A. *J. Org. Chem.* **1998,** *63*, 4554.
106. (a) Latham, E. J; Stanforth, S. P. *Chem. Commun.* **1996,** 2253; (b) Latham, E. J; Stanforth, S. P. *J. Chem. Soc., Perkin Trans 1* **1997,** 2059.
107. Lindsay, D. M.; Dohle, W.; Jensen, A.E.; Kopp, F.; Knochel, P. *Org. Lett.* **2002,** *4*, 1819.
108. (a) Das, B.; Krishnaiah, M.;Venkateswarlu, K.; Das, R. *Nat. Prod. Commun. 1* **2006,** 255–263.
109. Nishiyama, K.; Kubo, H.; Sato, T.; Higashiyama, K.; Ohmiya, S. *Heterocycles* **1998,** *48*, 1103.
110. Parker, K. A.; Mindt, T. L. *Org. Lett.* **2002,** *4*, 4265.
111. Zografos, A. L; Mitsos, C. A.; Igglessi-Markopoulou, O. *J. Org. Chem.* **2001,** *66*, 4413.
112. Bischler, A.; Napieralski, B. *Chem. Ber.* **1893,** *26*, 1903.
113. Fodor, G.; Gal, J.; Phillips, B. A. *Angew. Chem. Int. Ed.* **1972,** *11*, 919.
114. (a) Nagubandi, S.; Fodor, G. *J. Heterocycl. Chem.* **1980,** *17*, 1457; (b) Fodor, G.; Nagubandi, S. *Tetrahedron* **1980,** *36*, 1279.
115. (a) Fowler, F. W. in *Comprehensive Heterocyclic Chemistry,* Katritzky, A. R.; Rees, C. W., Eds., Pergamon: Oxford, **1984**; *Vol. 2*, pp410–416; (b) Jones, G. in *Comprehensive Heterocyclic Chemistry II,* Katritzky, A. R.; Rees, C. W.; Scriven, D. F. V., Eds.; Elsevier: Oxford, **1996**; *Vol. 5*, pp179–181.
116. Mealy, N.; Castaner, J. *Drugs Future* **1999,** *24*, 871.
117. Brossi, A.; Lindlar, H.; Walter, M.; Schnider, O. *Helv. Chim. Acta* **1958,** *41*, 119.
118. Pictet, A.; Spengler, T. *Ber.* **1911,** *44*, 2030.
119. Cox, E. D.; Cook, J. M. *Chem. Rev.* **1995,** *95*, 1797.
120. Pictet A.; Finkelstein, M. *Chem. Ber.* **1909,** *42*, 1979.
121. Yamato, E.; Hirakura, M.; Sugasawa, S. *Tetrahedron* **1966,** *8*, 129.
122. Klutchko, S.; Blankley, C. J.; Fleming, R. W.; Hinkley, J. M.; Werner, A. E.; Nordin, I.; Holmes, A.; Hoefle, M. L.; Cohen, D. M.; Essenburg, A. D.; Kaplan, H. R. *J. Med. Chem.* **1986,** *29*, 1953.
123. Myers, A. G.; Lanman, B. A. *J. Am. Chem. Soc.* **2002,** *124*, 12969.
124. (a) Pictet, A.; Gams, A. *Ber.* **1909,** *42*, 2943; (b) Pictet, A.; Gams, A. *Ber.* **1910,** *43*, 2384; (c) Pictet, A.; Gams, A. *Ber.* **1911,** *44*, 2480.
125. (a) Fritton, A. O.; Frost, J. R.; Zakaria, M. M.; Andrew, G. J. *Chem. Soc. Chem. Commun.* **1973**, 889; (b) Fritton, A. O.; Ardabilchi, N.; Frost, J. R.; Oppong-Boachire, F. *Tetrahedron Lett.* **1977,** 4107.
126. Griesbeck, A. G.; Bondock, S.; Lex, J. *J. Org. Chem.* **2003,** *68*, 9899.
127. Movassaghi, M.; Hill, M. D. *Org. Lett.* **2008,** *10*, 3485.
128. Dyker, G.; Gabler, M.; Nouroozian, M.; Schulz, P. *Tetrahedron Lett.* **1994,** *35*, 9697.
129. (a) Pomeranz. C. *Monatsch. Chem.* **1893,** *14,* 116; (b) Fritsch, P. *Dtsch Chem. Ges Ber.* **1893,** *26*, 419.
130. Schlittler, E.; Müller, J. *Helv. Chim. Acta* **1948,** *31*, 914.
131. Brown, E. V. *J. Org. Chem.* **1977,** *42*, 3208.

132. Kucznierz, R.; Dickhaut, J.; Leinert, H.; Von Der Saal, W. *Synth. Commun.* **1999,** 29, 1617.
133. Birch, A. J.; Jackson, A. H.; Shannon, P. V. R. *J. Chem. Soc., Perkin Trans. 1* **1974,** 2185.
134. Gabriel, S.; Colman. J. *Chem. Ber.* **1900,** *33*, 980.
135. Gensler, W. J. *Heterocyclic Compounds* Vol. 4, Elderfield, R. C. Ed., Wiley, New York, N.Y. **1952,** 378.
136. (a) Roesch, K. R.; Larock, R. C. *J. Org. Chem.* **1998,** *63*, 5306; (b) Roesch, K. R.; Zhang, H.; Larock, R. C. *J. Org. Chem.* **2001,** *66*, 8042.
137. (a) Dai, G.; Larock, R. C. *Org. Lett.* **2002,** *4*, 193; (b) Huang, Q.; Larock, R. C. *Tetrahedron Lett.* **2002,** *43*, 3557; (c) Dai, G.; Larock, R. C. *Org. Lett.* **2001,** *3*, 4035.
138. Roesch, K. R.; Larock, R. C. *J. Org. Chem.* **2002**, *67*, 86.
139. Lim, S.-G.; Lee, J. H.; Moon, C. W.; Hong, J.-B.; Jun, C.-H. *Org. Lett.* **2003**, *5*, 2759.
140. Molina, P.; Vidal, A.; Tovar, F. *Synthesis* **1997,** 963.
141. Brinkmann, V.; Billich, A.; Baumruker, T.; Heining, P.; Schmouder, R.; Francis, G.; Aradhye, S.; Burtin, P. *Nat. Rev. Drug Discov.* **2010,** *9*, 883–897.
142. Manitpisitkul, W.; Wilson, N. S.; Haririan, A. *Expert Opin. Drug Saf.* **2010,** *9*, 959–969.
143. Blum, M. D.; Graham, D. J.; McCloskey, C. A. *Clin. Infect. Dis.* **1994,** *18*, 946.
144. Royer, R. J.; Pierfitte, C.; Netter, P. *Therapie* **1994,** *49*, 75.
145. Schlüter, G. *Excerpta Medica* **1986,** 61.
146. Takahama, S.; Hirohashi, M.; Kato, M.; Shimada, H. *J. Toxicol. Env. Health* **1995,** *45*, 1.
147. Akahane, K.; Sekiguchi, M.; Une, T.; Osada, Y. *Antimicrob. Agents Chemother.* **1989,** *33*, 1704.
148. Curran, T. T. In *Name Reactions in Heterocyclic Chemistry*; Li, J. J., Corey, E. J., Eds.; Wiley: Hoboken, NJ, **2005,** pp 390–397.
149. Pflum, D. A. In *Name Reactions in Heterocyclic Chemistry*; Li, J. J., Corey, E. J., Eds.; Wiley: Hoboken, NJ, **2005,** pp 407–410.
150. Moore, A. In *Name Reactions in Heterocyclic Chemistry*; Li, J. J., Corey, E. J., Eds.; Wiley: Hoboken, NJ, **2005,** pp 488–494.
151. Kanai, M.; Kato, N.; Ichikawa, E.; Shibasaki, M. *Synlett* **2005,** 1491–1508.
152. Luo, W.; Mu, Q.; Qiu, W.; Liu, T.; Yang, F.; Liu, X.; Tang, J. *Tetrahedron* **2011,** *67*, 7090.
153. Huma, H. Z. S.; Halder, R.; Kalra, S. S.; Das, J.; Iqbal, J. *Tetrahedron Lett.* **2002,** *43*, 6485.
154. Si, C.; Myers, A. G. *Angew. Chem., Int. Ed.* **2011,** *50*, 10409.
155. Xie, M.; Liu, X.; Zhu, Y.; Zhao, X.; Xia, Y.; Lin, L.; Feng, X. *Chem. Eur. J.* **2011,** *17*, 13800.

PART IV SIX-MEMBERED HETEROCYCLES WITH TWO HETEROATOMS

Chapter 12 Pyrazines and Quinoxalines

Micheal Fultz and William Rollyson

12.1 Introduction

Pyrazine is a colorless solid with a melting point of 54–55 °C and a boiling point of 115 °C. It turns to a light-amber color upon exposure to air and/or light. Pyrazine is miscible with water.

Quinoxaline is a crystalline solid with a low melting point range 29–32 °C. This diazine has a boiling point of 220 °C and has a moderate solubility in aqueous solutions. It is isomeric with quinazoline, phthalazine, and cinnoline.[1]

Pyrazines, with 6 π-electrons, are electron-*deficient* (also known as electron-*poor*) aromatic heterocycles because of the increased electro-negativity of the nitrogen atoms. Due to the presence of the electronegative nitrogen atoms, the electron density on the ring carbons is less than one. The lone-pair electrons do not take part in the delocalization so this molecule can act as a mild base.

Structurally speaking, pyrazine is a planar hexagon, similar to benzene, in both bond angles and lengths. As expected, the C–N bonds and C–N–C bond angles are shorter than their C–C and C–C–N counterparts.

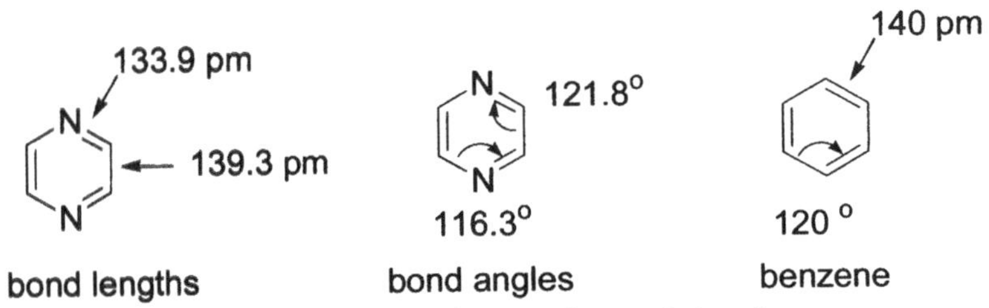

Pyrazine's Bond Lengths and Angles

Due to its symmetry, the ^{1}H-NMR of pyrazine in chloroform has a single peak at 8.87 ppm. The proton signal is further downfield than the traditional aromatic ring protons because of the increased electro-negativity of the neighbouring nitrogen atoms. As for the ^{13}C-NMR, the sole signal is present at 144.9 ppm.

Quinoxaline has a more complex proton NMR. Protons 2 and 3 appear as a singlet at 8.54 ppm. The multiplet at 7.54–7.84 comes from protons 5 and 8, while protons 6 and 7 account for the multiplet peaks between 7.42–7.45 ppm. The carbon NMR can be separated into two regions. The carbons bonded to the nitrogen are located at 144.6 and 142.6

ppm. The non-nitrogen bearing carbons are not shifted as far and are located at 129.9 and 129.6 ppm.

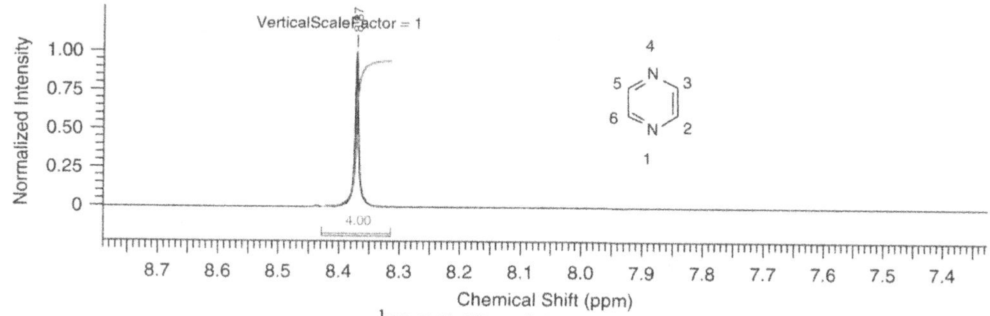

^{1}H-NMR of Pyrazine

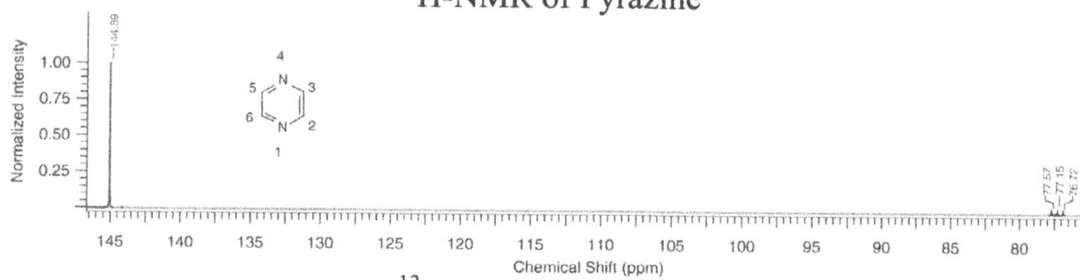

^{13}C-NMR of Pyrazine

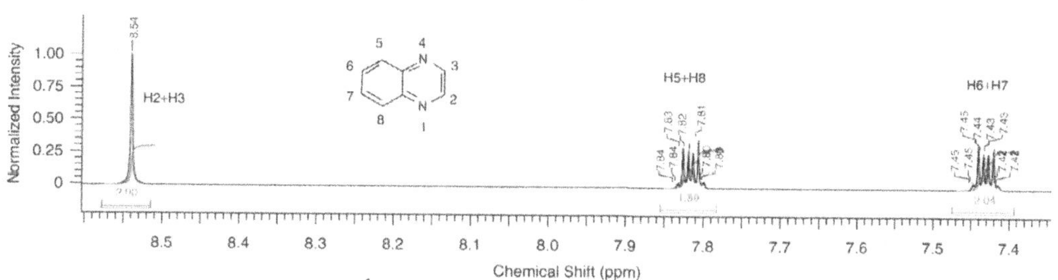

^{1}H-NMR of Quinoxaline

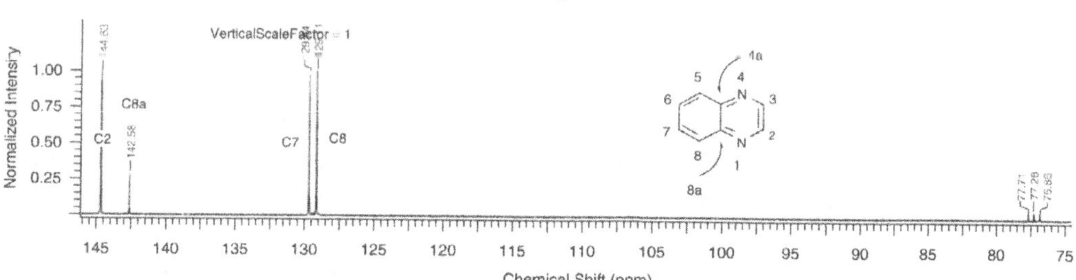

^{13}C-NMR of Quinoxaline

Although pyrazines and quinoxalines are not the most common heterocycles in nature and medicinal chemistry, they have a significant role in both fields. An example of their synthetic role can be seen in the preparation of (+)-cephalostatin 1, a complex 13-ring structure by Shair and co-workers in 2009.[2] This molecule has shown significant biological activity against cancer cells with a p16 mutation. The central pyrazine was used as a linchpin in uniting the advanced intermediates to form the carbon backbone of this complex natural product.

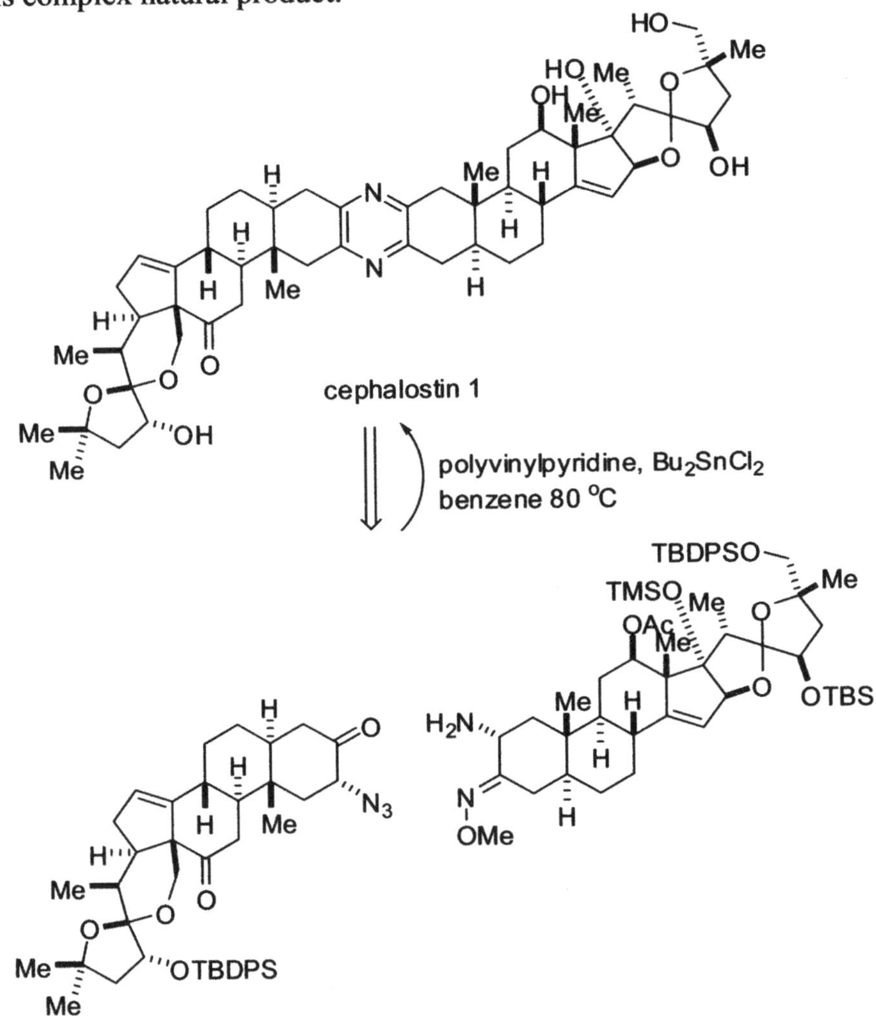

Several biologically-active natural products contain pyrazines. A variety of sources, such as the deep-water sponge *Dragmacidon*, incorporate these rings into their complex structures. This ring structure tautomerizes between the pyrazine ring and the unsaturated amide.[3]

Interest in diazine natural products by the pharmaceutical industry as building blocks as well as application within material science and supramolecular chemistry has led to extensive efforts devoted to synthetic methodology.[4] Many drug candidates with the quinoxaline and pyrazine core structures are in clinical trials in antiviral,[5] anticancer, and antibacterial areas.[6] Diversely-substituted quinoxalines and pyrazines or their derivatives are embedded with a variety of functional groups as biological agents. This has led to a significant amount of research in both synthesis and biological activity against Alzheimer's disease,[7] central nervous disorders,[8] drug abuse,[9] as well as displaying characteristics of antiviral,[10] antitumor, antibacterial, antifungal, anti-inflammatory, anti-tubercular,[11] anticonvulsant, anti-malarial, antileishanial, and trypanocidal agents.[1,12]

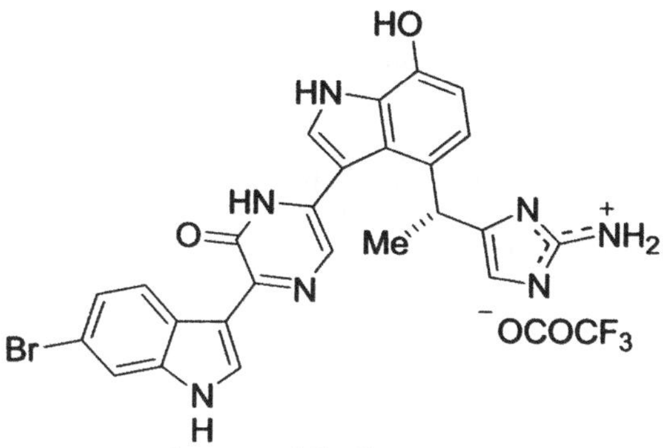

dragmacidin D

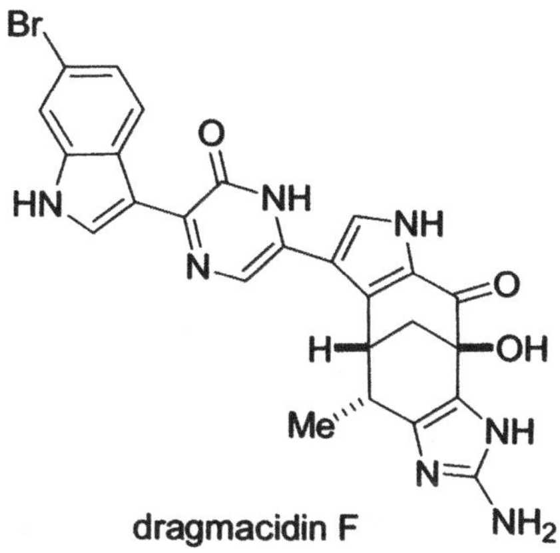

dragmacidin F

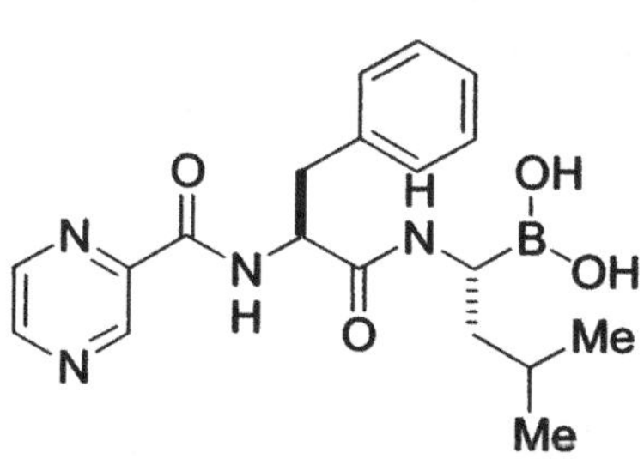

Velcade
Malignancy of bone marrow

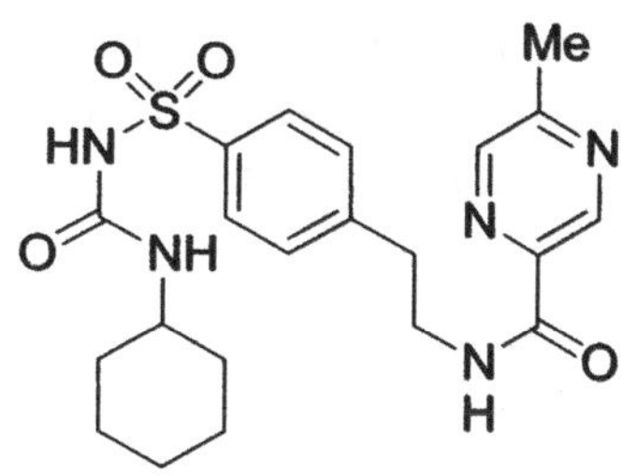

Glipizide
Antidiabetic

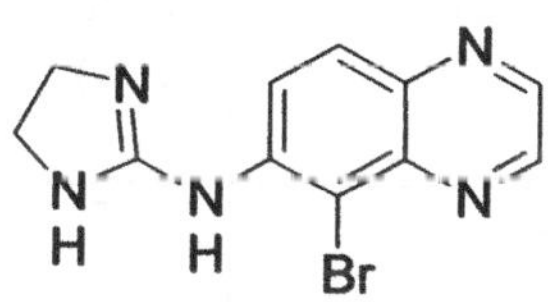

Brimonidine
Antiglucoma agent

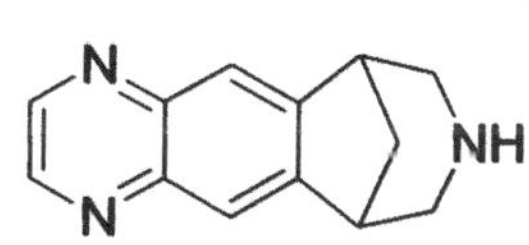

Chantix
Smoking Cessation

12.2 Formation of Diazines

Formation of pyrazines and quinoxalines has been a focal point of many years of study. The most common method to make these rings relies on the condensation of 1,2-diamines with 1,2-dicarbonyls in ethanol or acetic acid in 35–85% yields.[1] The mechanism proceeds through a double imine formation.[13] Recently improved methods have been reported, employing transition-metal catalysis[13,14] and microwaves.[15]

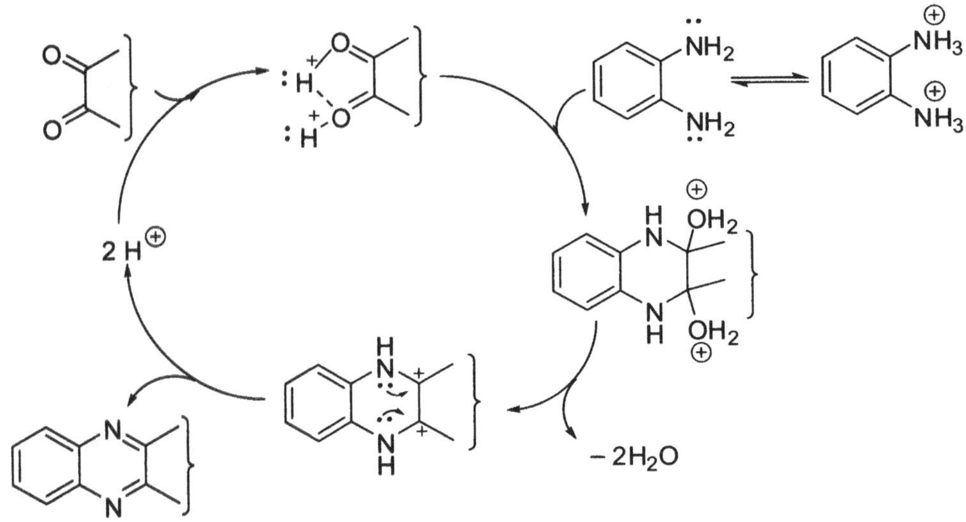

Recently, polyaniline salts have received attention as a mild polymer-based solid acid catalyst in organic chemistry.[16] When properly prepared, these salts contain 23.3% sulfuric acid. This polymeric acid is then used to catalyze the condensation of aromatic 1,2-diamines with 1,2-dicarbonyl compounds at room temperature.

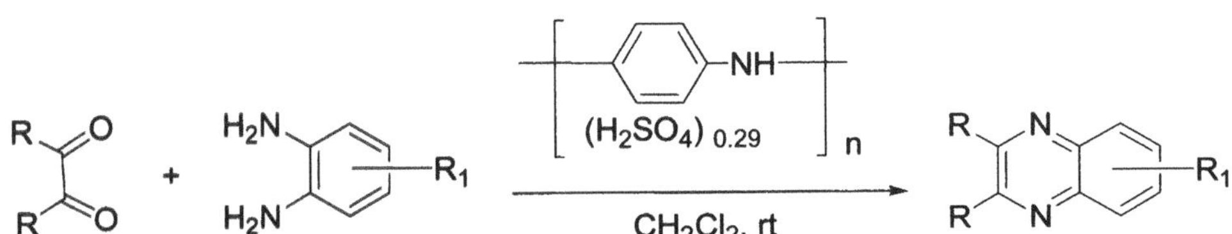

Carbonyl	Diamine	Product	Yield (%)	Reaction Time (min)
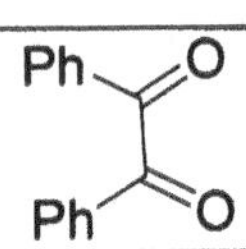	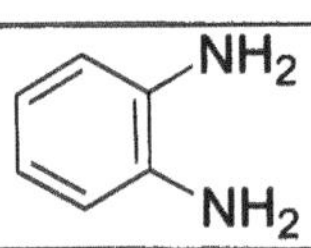	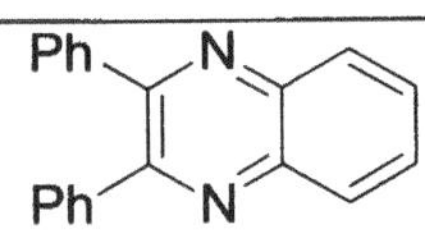	95	20
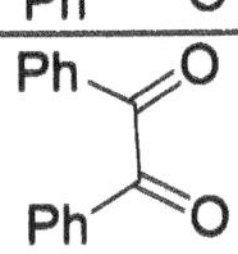	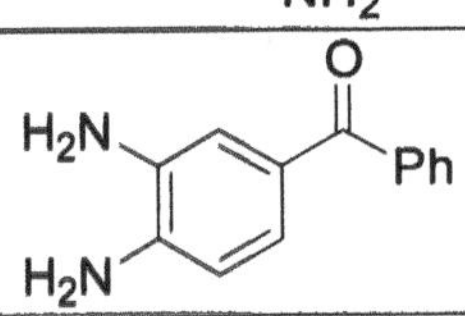	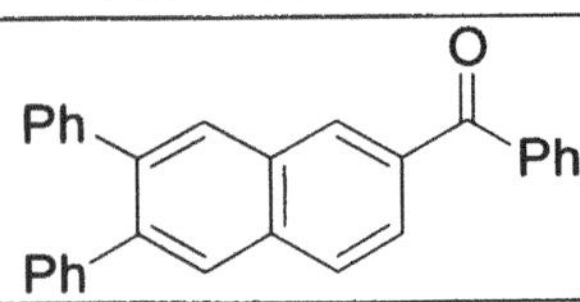	90	25
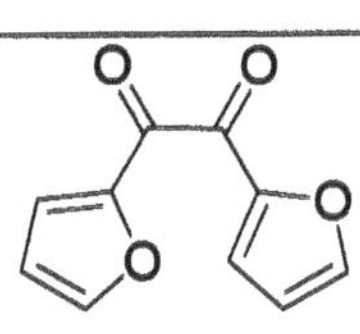	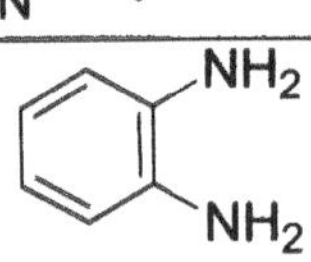	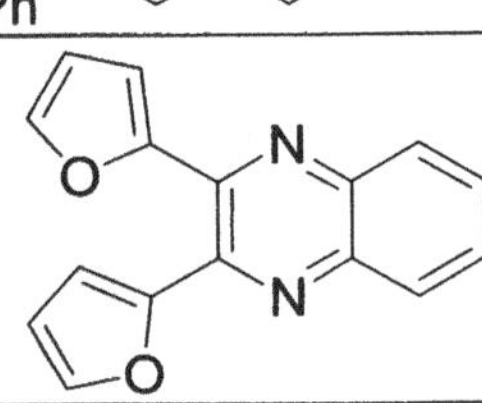	92	10

This particular method is greener than some traditional methods because it allows for the isolation and re-use of the catalyst multiple times with little or no detrimental effect to its activity. As seen the table below, the catalyst maintains reactivity through numerous cycles or reactions showing no loss of reactivity after six cycles of cyclization.

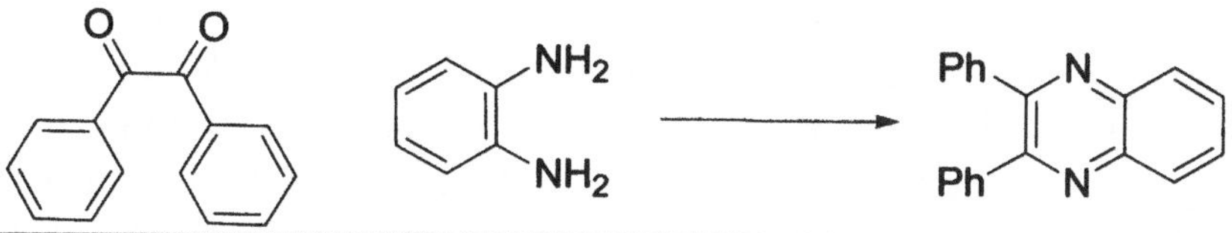

Benzil (mg)	Catalyst (mg)	Time (min)	Yield (%)
1000	50	20	95
770	38.5	20	95
680	34	20	94
560	28	20	94
330	16.5	20	94
240	12	20	93

In the effort to expand the potential chemotherapeutic drug candidates, quinoxalines and related heterocycles were examined.[16] Condensation of the 1,2-diaminobenzene with oxalic acid provides the quinoxaline tautomer in high yields. Treatment of this mixture of tautomers with phosphoryl trichloride provides the multifunctional ring. Displacing a chloride with sodium azide provides the tetrazole, which is considered as a planar acidic analogue of the carboxylic acid functionality. It can increase binding affinity,[17] potency,[18] and bioavailability.[19] Condensation of the hydrazine with benzaldehyde completes these drug candidates.

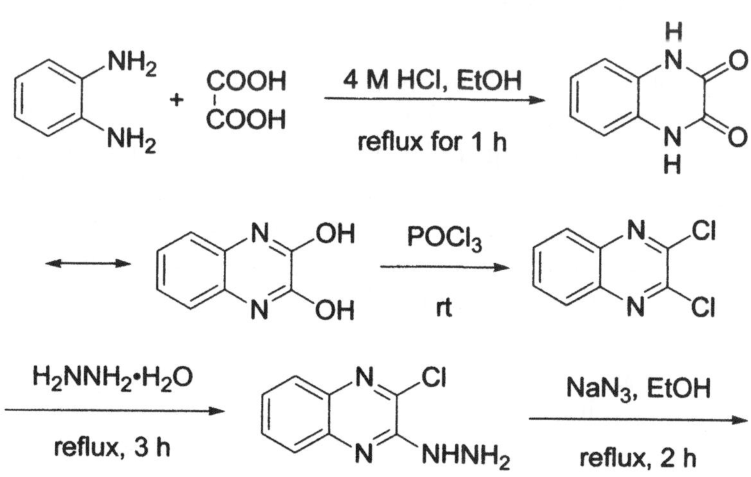

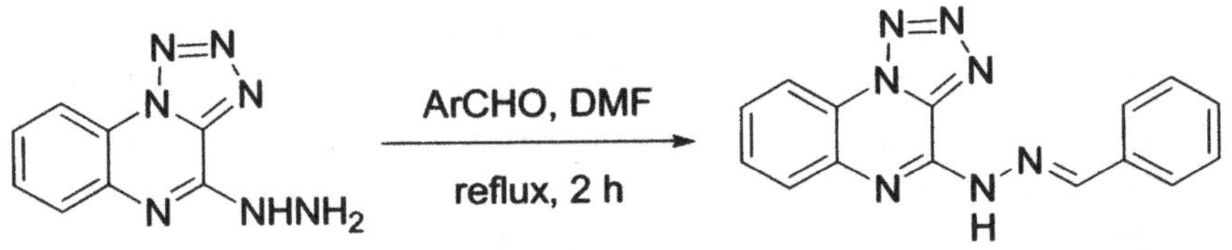

Titanium catalysis has provided another green method for the condensation of diamines with 1,2-dicarbonyls to provide quinoxalines.[13] This synthesis is done at room temperature and in less than one hour. Attaching electron-donating groups to the amino phenyl ring increased yields, while electron-withdrawing groups significantly lowered isolated yields of the desired products. When strong electron-withdrawing groups were present, even with longer reaction times, the isolated yield never passed 55%.

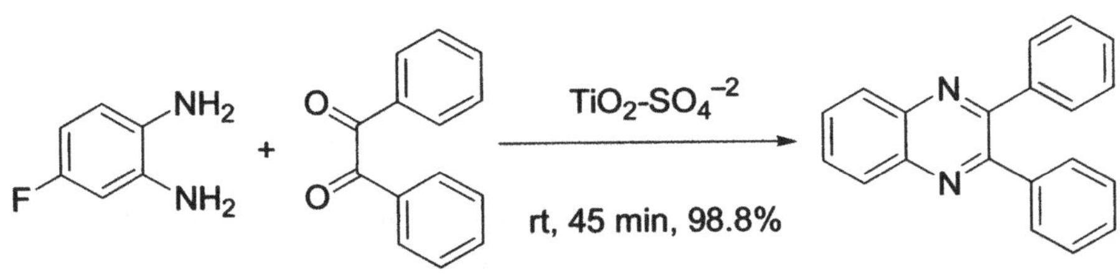

There are numerous methods of forming the pyrazine ring through the condensation of two α-amino carbonyls.[20] Some variants have been used successfully toward the synthesis of several biologically-active compounds.

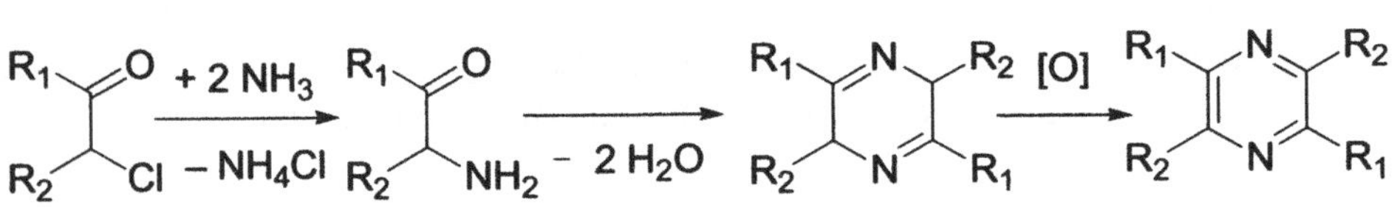

The synthesis of novel quinoxaline derivatives was key in the study of inhibitors of Pim kinases.[21] This structure-activity relationship (SAR) study found several promising molecules that demonstrated antiproliferative properties against solid cancer cell lines (PA1, PC3, and DU145). The synthesis of these compounds relied on the previously optimized condensation of diamines with α-halogenated carbonyls followed by oxidation to provide the aromatic bicycle.[21]

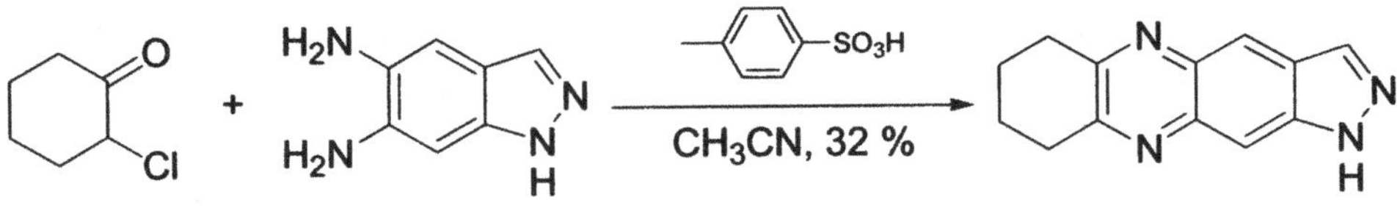

Alzheimer's disease is the most common form of dementia in older people, accounting for 50–70 % of dementia cases.[22] The exact mechanism leading to the development of this disease is not fully understood, but modern neuroimaging techniques such as positron emission tomography (PET) and single-photon emission tomography radio-labelled probes may aid in the diagnosis and monitoring of patients. To synthesize these probes, it was necessary to selectively incorporate iodine-125 into the drug candidates.

The synthesis began with the condensation of 4-bromo-1,2-diaminobenzene with 2-bromo-1-(4-(dimethylamino)phenyl)ethanone to provide two separable isomeric products. The products were crystallized and examined by X-ray crystallography to confirm the structures. The desired product was the major compound isolated at 36.7% yield. Formation of the aryl tributyltin was accomplished under a standard Stille coupling procedure, followed by conversion to the isotopically-enriched iodine under acidic oxidizing conditions to provide one of the desired isotopically-labelled drugs needed for the monitoring of the Alzeheimer's patients.

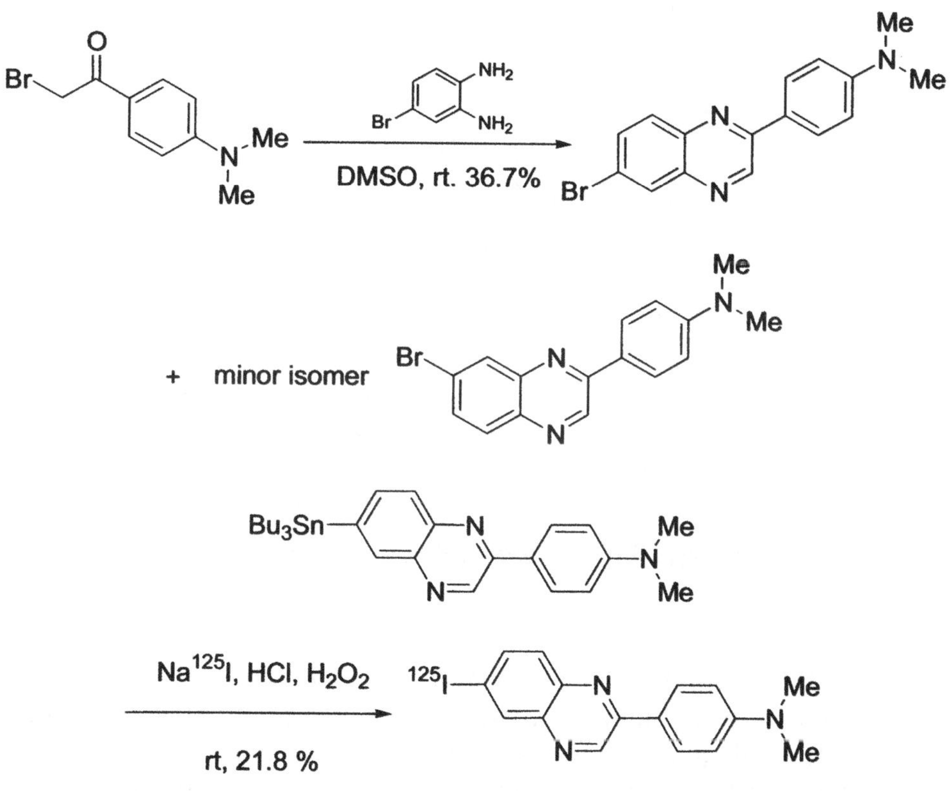

The Gutknecht pyrazine synthesis[23] has overcome the difficulty of forming the multiple regioisomers that plagued the Staedel and Rügheimer synthesis.[20] The synthesis begins with the reduction of oximino-ketones to provide the α-amino carbonyl. The amine then intermolecularly condenses with a second α-amino carbonyl to provide the cyclic di-imine. This di-imine is then oxidized in the presence of air to complete the aromatic pyrazine ring. This method is still limited in the fact it can only produce symmetrically pyrazine rings and cannot be used to provide pyrazine itself.[24] The Gutknecht reaction was used as the key step in the efforts toward synthesis of barrenazine A.[25]

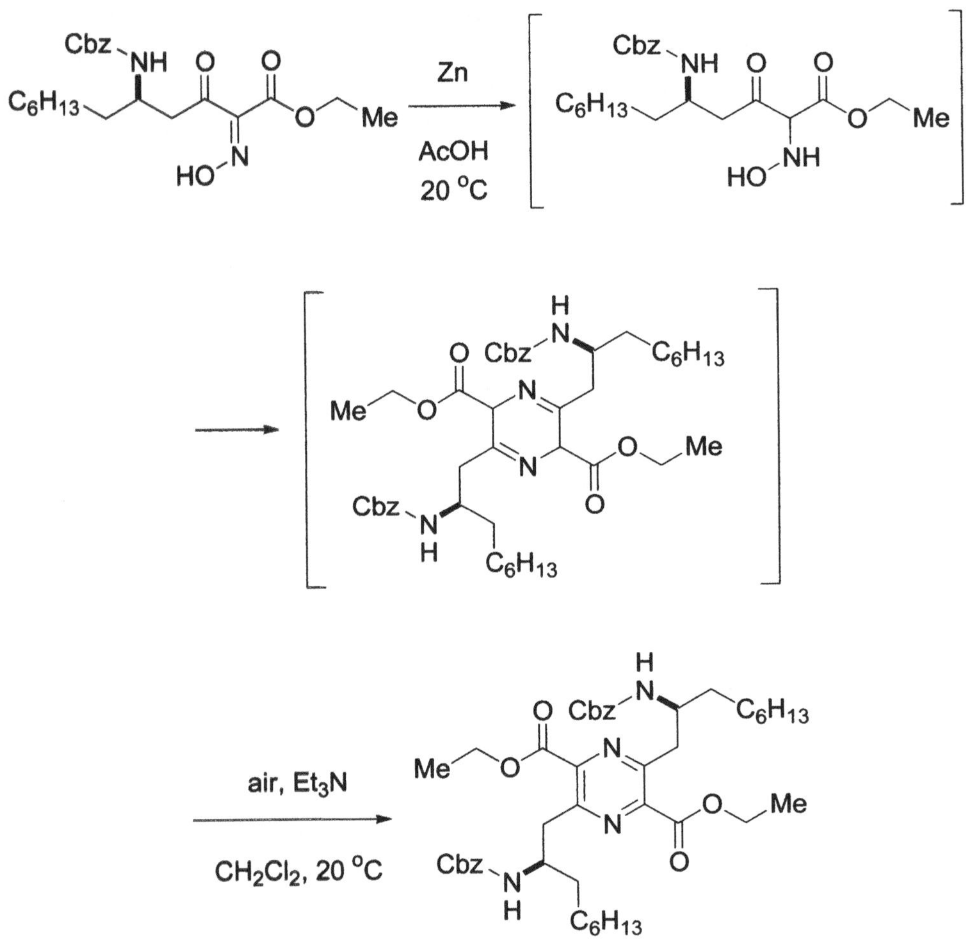

In 1965 researchers at the American University of Beirut observed that combining benzofurazan oxides with morpholine-cyclohexene in methanol afforded quinoxaline-1,4-dioxide in 48% yield.[26] There is some debate in the literature as to the exact mechanism of the Beirut reaction.[27] In the case of benzofurazan oxides reacting with an enamine, the following mechanism is generally accepted in the literature. The first step is nucleophilic addition of the enamine to the electrophilic benzofurazan oxide to form the ammonium zwitterion. Ring closure occurs via condensation of the imino-oxide onto the iminium functionality to provide the heterocycle. Finally, elimination of the amine provides the quinoxaline-1,4-dioxide.[28]

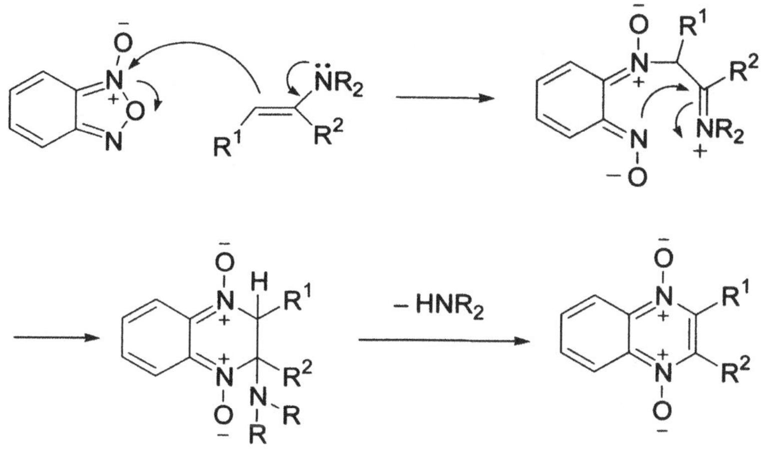

This reaction has been expanded to include not only enamines but also 1,3-dicarbonyls. The conditions used generally employ a drying agent to remove the water generated in the final step of the reaction. The variation has been used as the key step to synthesize numerous compounds, some of which have been used in clinical trials. Due to the poor effectiveness of many of the current drugs on the market, new medications to treat tuberculosis are needed. In the course of that study the benzofurazan oxide was treated with 1,3-dicarbonyl to provide possible candidates for screening.[29]

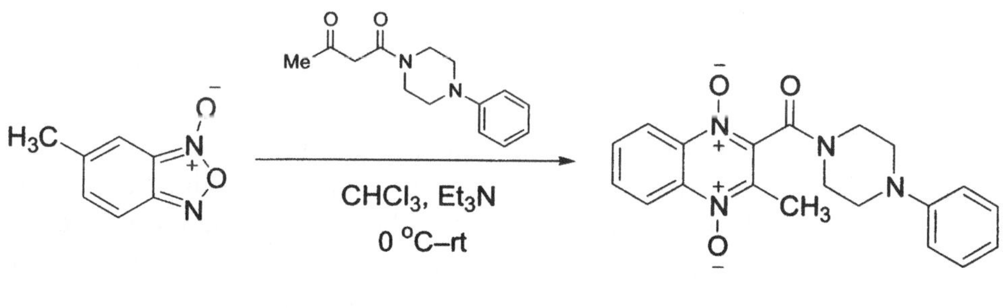

12.3 Reactivity of the Molecules

12.3.1 Reactivity of the nitrogen

The aromatic nitrogen atoms of the pyrazine and quinoxaline are only weakly basic. Since neither of the lone pair of electrons are part of the aromaticity of the molecule, either nitrogen can be protonated. Upon protonation the aromatic rings have a higher acidity value than any of the diazine isomers.[30,31]

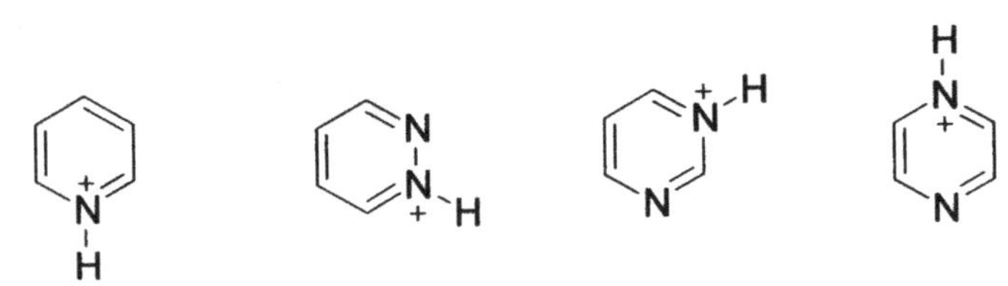

pKa = 5.14 2.24 1.78 0.65

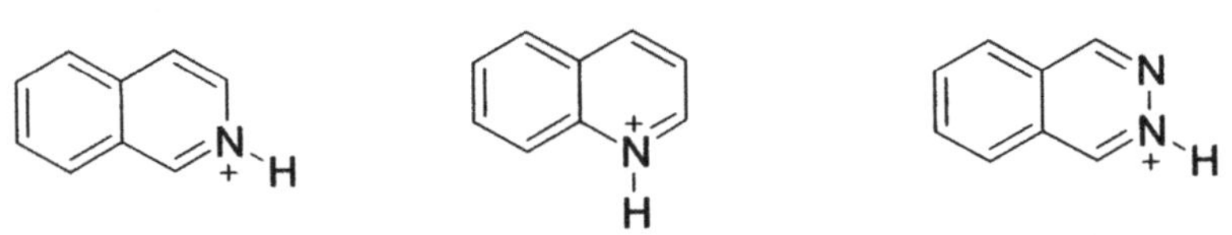

pKa = 5.14 4.97 3.39

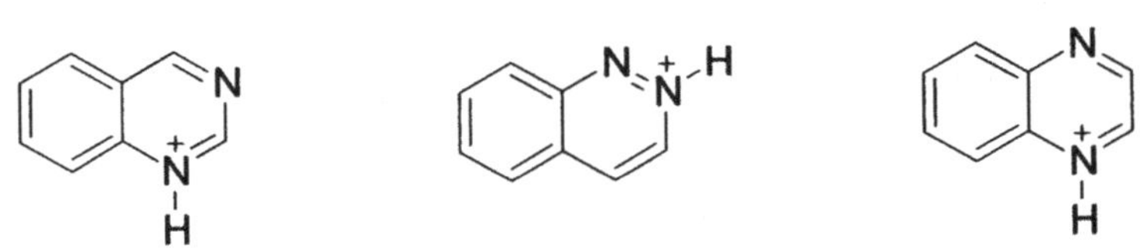

pKa = 3.31 2.67 0.60

Oxidation of the nitrogen in the diazine core has become a necessary reaction to enhance the reactivity for selected reactions. This reaction has become standardized since the late 1950s when Klein[32] demonstrated the effectiveness of hydrogen peroxide in the selectivity of this reaction. Since that time, other oxidizing agents (*m*-CPBA,[33] Oxone,[34] and PAA[35]), have been utilized.

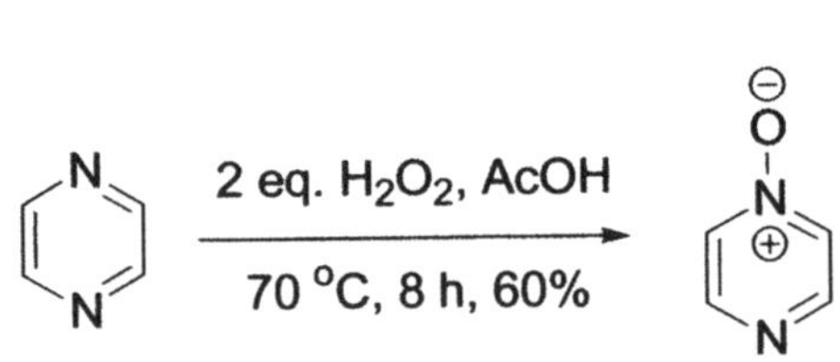

This oxidation of both nitrogen atoms in the ring can occur in high yields by adjusting the reaction conditions and equivalents of hydrogen peroxide present in the reaction.

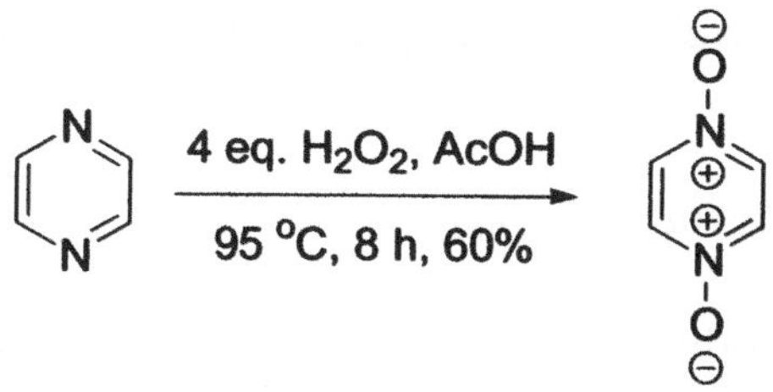

This oxidation can be exploited to region-selectively oxygenate the α positions of the ring when the oxide is dissolved in refluxing acetic anhydride followed by hydrolysis with sodium hydroxide. This alcohol can then tautomerize to form the amide.

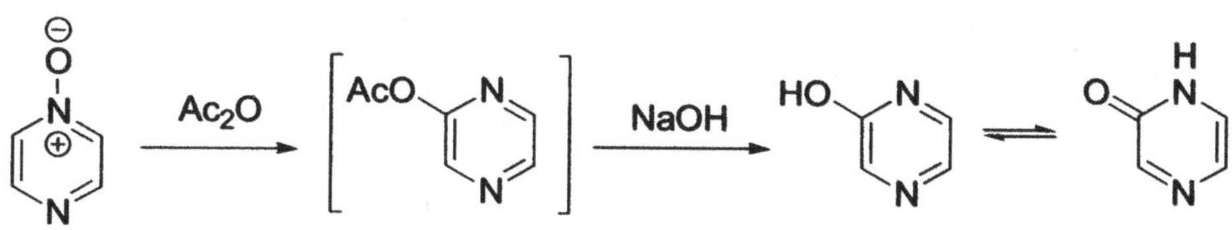

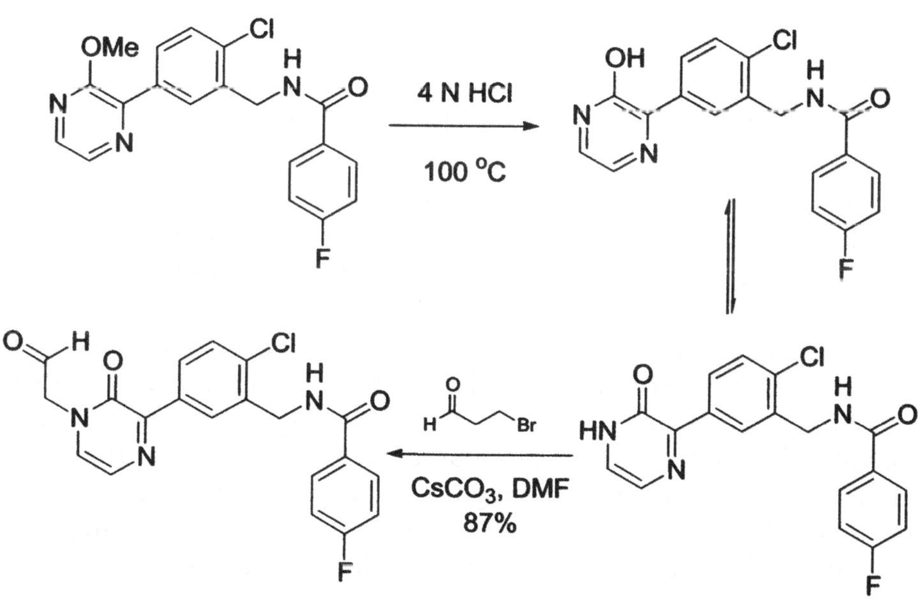

The cathepsins compromise many proteases whose function involves protein degradation, processing, and turnover. Since proteases possess numerous extended binding sites, it was anticipated that cathepsin S would be useful for fragment-based drug discovery.[36] Toward this goal, the deprotection of the pyrazine below was done with hydrochloric acid at elevated temperatures. This then provides the aromatic alcohol that then tautomerizes to provide the unsaturated amide. The equilibrium between the pyrazine and the amide allows the alkylation of the nitrogen to provide tertiary amide in high yield. This alkylation prevents the tautomerization to reform the pyrazine.

12.3.2 Reactivity of the diazine ring

Quinoxaline is a by-product of cooking some food and can inhibit the growth of some ciliate protozoa and plan-pathogenic fungi.[37] The metabolism of this ring system by *Pseudomonas putida* can provide the quinoxaline *cis*-5,6-dihydrodiol, 5-hydroxyquinoxaline, and 2(1*H*)-quinoxalinone. When quinoxaline is metabolized by *Streptomyces badius*, two different products were isolated. Both of these two products, 2(1*H*)-quinoxalinone and 3,4-dihydro-2(1*H*)-quinoxalinone, were only isolated from the treatment of *Streptomyces badius* with quinoxaline. Six other *Streptomyces* species were tested and only produced 2(1*H*)-quinoxalinone. The mechanism for the formation of 3,4-dihydro-2(1*H*)-quinoxalinone was not determined.

2(1*H*)-quinoxalinone   3,4-dihydro-2(1*H*)-quinoxalinone

The reaction of quinoxalines and pyrazines with radicals has been exploited to further develop the regioselective functionalization of these ring systems.[38] The substitution of protonated heteroaromatic bases by nucleophilic carbon radicals is a general reaction due to the variable radical sources and the high regioselectivity of the very simple experimental conditions. Using cerium ammonium nitrate (CAN), the nucleophilic carbamoyl radical was formed in a two-step process with *N*-hydroxyphthalimide (NHPI). It can reproduce many Friedel–Crafts aromatic substitutions but with the opposite regioselectivity. The high regioselectivity on the quinoxaline rings appears to be related to the low reversibility for the addition of the $^{\bullet}CONH_2$ radical to the ring under the reaction conditions.

Protonation of the heteroaromatic ring considerably increases the reactivity towards nucleophilic radicals (3–6 orders of magnitude) compared to the unprotonated base. The introduction of the carbonyl decreases the basicity of the heterocyclic ring, reducing the reactivity of the molecule and preventing disubstitution.

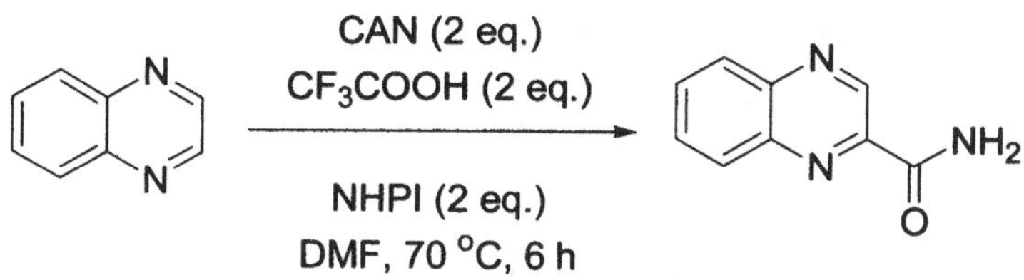

Due to the presence of two electronegative oxygen rings, the pyrazine and quinoxaline rings are poor substrates for electrophilic aromatic substrates. These nitrogen atoms do increase the heterocycles reactivity toward nucleophilic aromatic reaction. This reaction is usually done starting from the halogenated aromatic ring.[39–41] The substitution reaction proceeds in high yields to provide a single product.

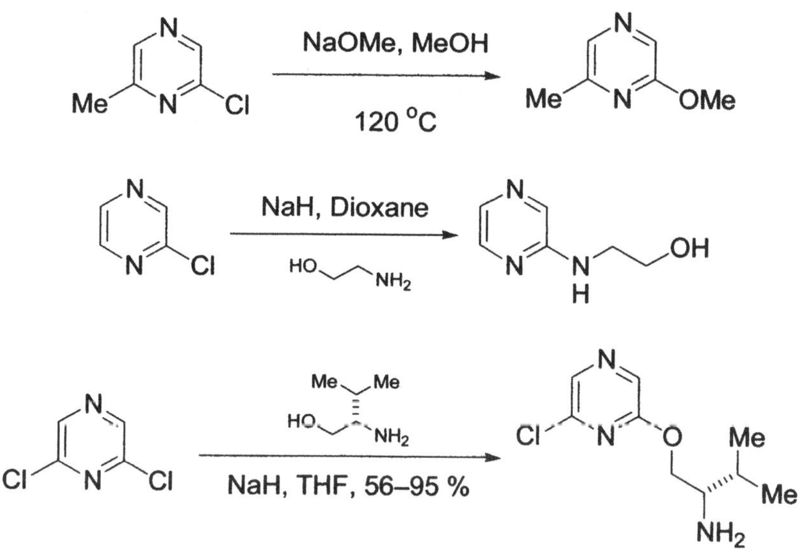

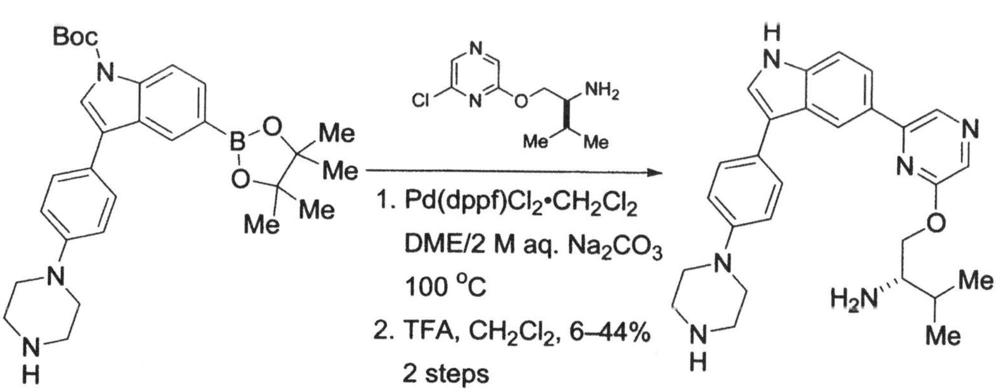

In efforts to develop selective inhibitors of the three members of the Pim kinase family, a high through-put screen (HTS) of functionalized pyrazines was needed to further study and enhance the biological activity compared to that of the base pyrazine moiety.[42] The screening was accomplished by taking 2,6-dichloropyrazine upon treatment with sodium hydride and secondary amines or alcohols displaced one chloride to provide the functionalized ring. Suzuki coupling of the remaining chloride to the aryl boronates and followed by deprotection of the Boc group provided the desired compound in 6–44% yield over two steps.

With cancer continuing to be a major health problem in both developing and industrial nations, it has become the number one killer worldwide and new drugs are needed to treat this problem.[43] Toward this effort upon the formation of the quinoxaline, it was treated under acidic conditions to condense with benzaldehyde to form the alkene. Chlorination of the oxygen produces the aromatic chloride that can undergo nucleophilic aromatic substitution to provide a diverse array of biological active drug candidates. The candidate below inhibits 97.29% of ovarian cancer cells. This study led to the conclusion that the styryl group, at position 3 led to the broadest spectrum of biological activity when compared to the substituted styrene.

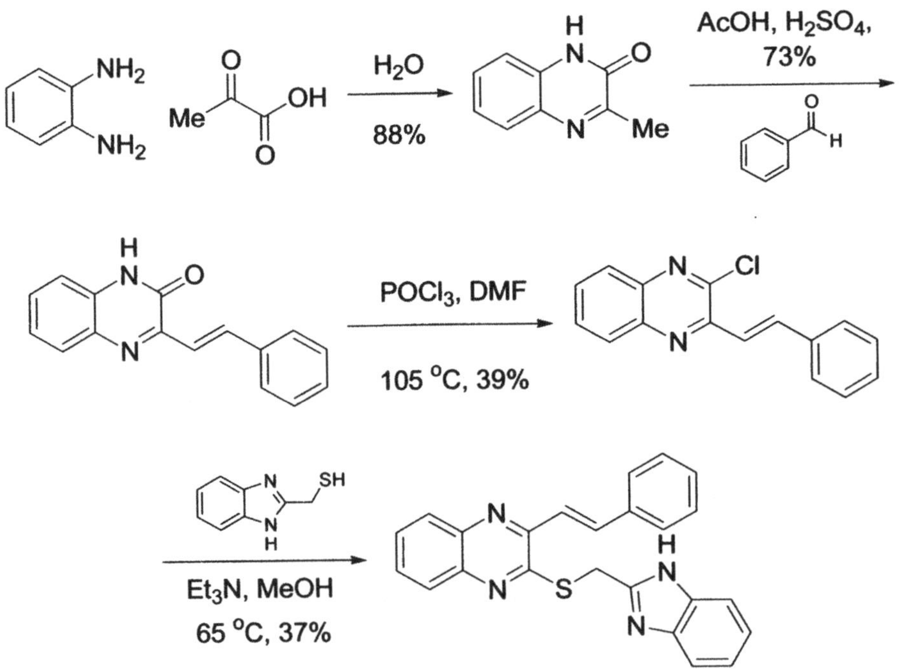

Selective fluorination of a range of heteroaromatic compounds to provide the 2-fluoro-derivatives has been achieved using an elemental fluorine–iodine mixture.[44] This fluorination of heteroaromatic rings has become of vital interest in academia and industry due to the effects the fluorine atom can have on the physical, chemical, and biological properties of the molecule.

One major limitation of this type of reaction is the inertness of the quinoxalines that contain strong electron-withdrawing groups such as 6-nitroquinoxaline and 6,7-dichloroquinoxalines, possibly due to the decreased nucleophilicity of the ring nitrogen atoms.

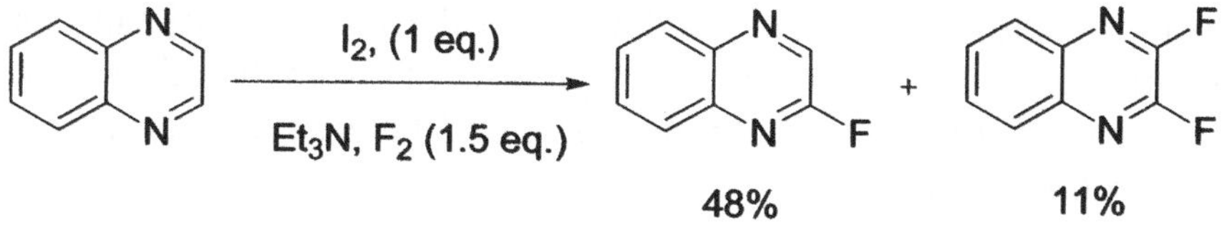

12.3.3 Metallation of the diazine ring

The metallation of diazines, in particular pyrazine and quinoxalines, is extremely difficult due to nucleophilic addition reactions related to the low LUMO energy levels of the rings.[45] The functionalization of these rings was an important synthetic goal due to the many uses of the molecules.

While the regioselective metallation of aromatic rings generally requires the presence of heteroatom substituents,[46] the inductive electron-withdrawing effect of the aromatic nitrogen activates the *ortho*-hydrogens to strong bases.

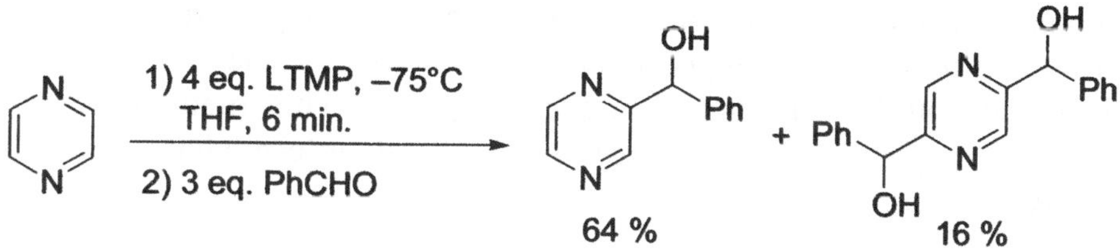

Initial attempts at this important reaction were done using lithium and magnesium amide bases. The reaction condition typically employed is at low temperatures with an excess of the base as well as the electrophile needed to capture the metallated ring.

When conditions for step 1 was longer than 15 minutes before the electrophile was added, (even at low temperatures), only dimeric compounds were recovered.

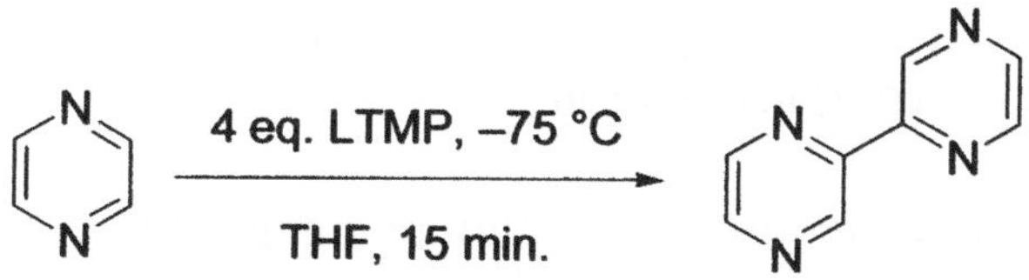

It has been demonstrated that the metallation of diazines can be challenging due to the facile competitive nucleophilic addition reaction to provide the dimer above.[43,47] Dong and co-workers speculated that the less reactive organozinc intermediate might not dimerize.[48] Upon deprotonation of the diazine ring using magnesium bases, poor yields of the desired halogenated product were isolated due to the high reactivity of the magnesium species. Dong and co-workers recently determined that stabilization of the reactive intermediate with the inclusion of zinc chloride provided the aryl iodide in excellent yields. This metallation procedure provides a pathway to form many aromatic and heterocyclic organometallic reagents at room temperature.

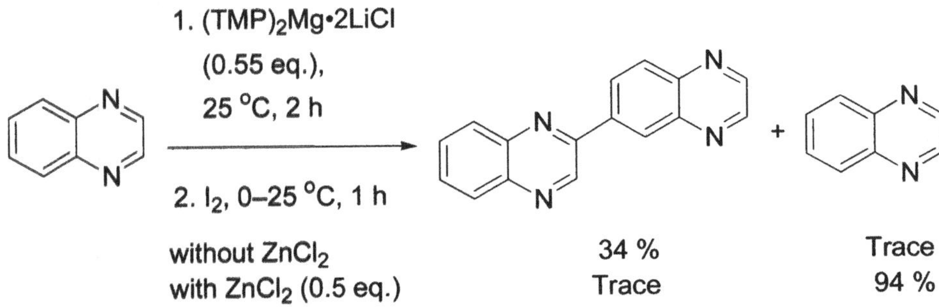

Continuing to examine the scope of the zincation of pyrazine rings, Dong[49] used $TMP_2Mg{\cdot}2LiCl$ (TMP = 2,2,6,6-tetramethyl-piperamidyl) to deprotonate pyrazine in the presence zinc chloride to provide the zincated aryl ring. Addition of copper cyanide followed by the allylic bromide provided the coupled product in moderate yield. This metallation of the quinoxaline ring provides the *ortho*-metallated intermediate with high selectivity.

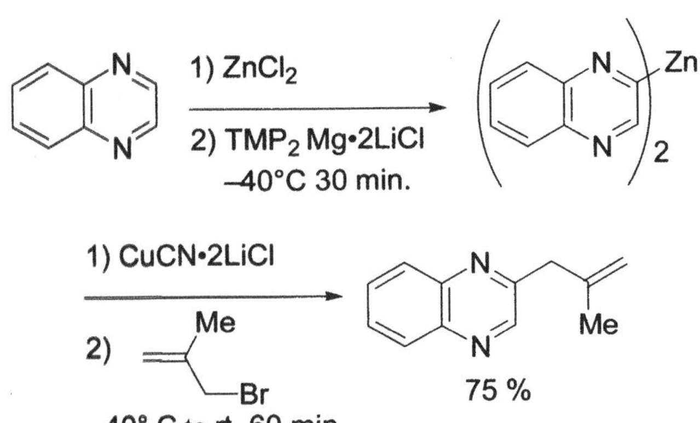

When inductively electron-withdrawing groups are located on the carbocycle of a quinoxaline, the metallation occurs selectively at the *ortho*-

position to that substituent. A second functionalization would occur at the first available site to the benzo-fused ring.

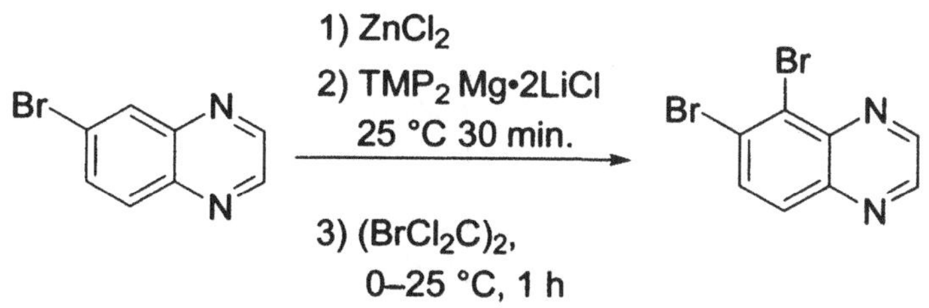

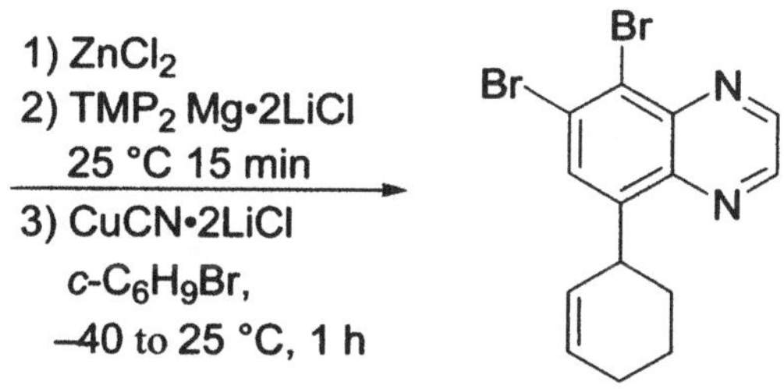

Difunctionalization of pyrazines to form 2,5-diiodopyrazine was an important synthetic goal because of the multiple applications of this type of molecule. The first dihalogenation of pyrazine was accomplished using cadmium ion salt in moderate yield.[50] This reaction is controllable to the point that modifying the equivalence of *N,N,N',N'*-tetramethylethylene-diamine (TMEDA) to provide the selective formation of either the monohalogenated or dihalogenated aromatic ring.

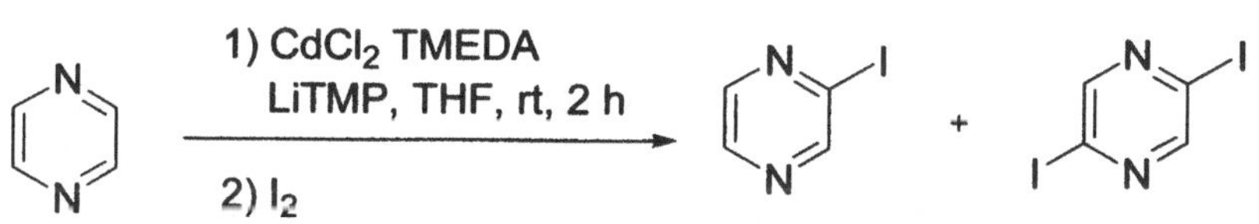

TMEDA (eq)	% Recovered	% Recovered
0.33	63%	0%
1.00	0%	58%

12.4 Coupling Reactions

12.4.1 Transition-metal coupling reactions

Direct arylation of aromatic compounds by C–H activation with aryl halides by transition metal catalysis has become an important synthetic process.[51] However until recently the reported procedures were limited to only electron-rich aromatic rings and heteroaromatic compounds. Biaryl compounds containing pyrazine were isolated in moderate to high yields when potassium *tert*-butoxide and Cy_3PAuCl were used in the reaction. This was accomplished when aryl bromides were treated with 5 equiv of pyrazine at 100 °C under the developed coupling procedure.

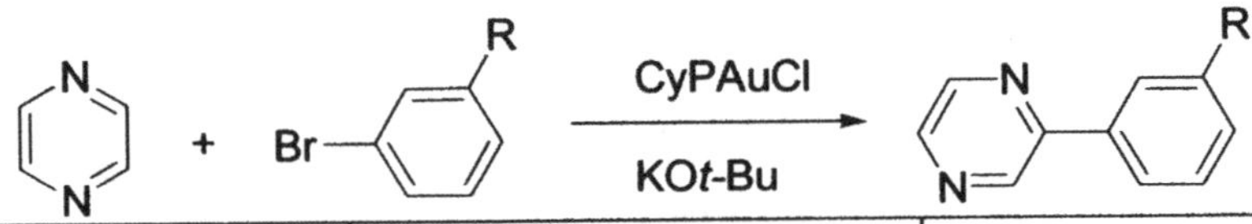

R	Reaction Time (h)	Yield (%)
H	24	82
2-methyl	12	58
2-CF_3	24	31
4-OMe	12	60

Direct C–H arylation on the pyrazine ring has also been accomplished using internal alkynes.[52] This reactivity has been attributed to the coordination of the Lewis acid to the nitrogen lone pair, increasing the acidity of the C(2)–H bond. This reaction was incredibly selective to provide the (*E*)-alkene (> 95:5). Dimethyl zinc was the most optimal catalyst used. When higher temperatures are used, an increased yield of the alkene was isolated, minimizing the amount of the dienylated product formed. When unsymmetrical alkynes are used the regioslectivity of the alkenylation is based upon the substituents off the starting alkyne. The end of the alkyne that bears the smaller substituent is the side that bonds to the aromatic ring while the larger substituent is selectively located in the *trans* position to the aromatic ring.

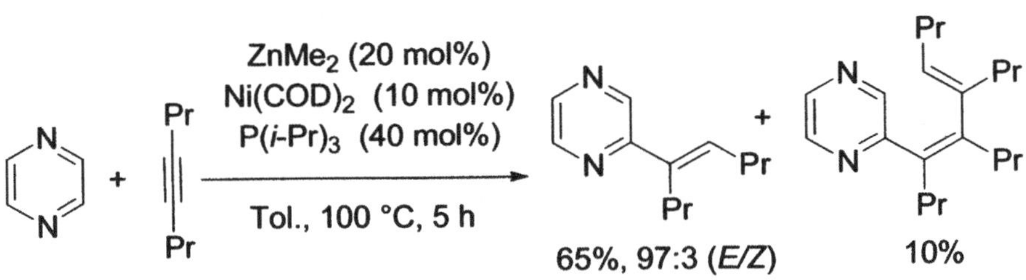

The use of trimethyl aluminum or lower reaction temperatures can slow the reductive elimination and/or promote the second insertion of additional alkynes into the carbon–nickel bond to give the dienylated product. One limitation to this reaction was that terminal alkynes were not applicable to this reaction due to rapid oligo/trimerization.

Nickel-catalyzed cross-coupling reaction like the Kumada coupling has been used to form biologically-active molecules. This coupling reaction has been described between fluoroarenes and aryl Grignard reagents with commercially available ligands to form biaryl systems at room temperature.[4] Nickel proved to be a better catalysis than palladium for this reaction. It was combined with a strongly Lewis basic, electron-donating and sterically demanding ligand to force this reaction to proceed.

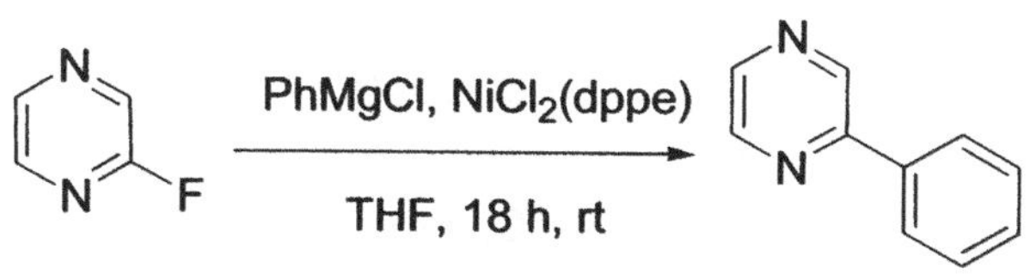

Catalytic direct arylation of aromatic C–H bonds has emerged as an atom efficient alternative to conventional cross-coupling between halides and organometallics for the synthesis of biaryl compounds.[53] Despite significant advancements in this methodology, C–H bonds in electron-deficient hetero-aromatic compounds are often not applicable. It is known aryl lithium and aryl magnesium reagents add to pyridine in a 1,2-fashion to provide the dearomitized dihydropyridine that is then converted to the desired product. This pathway could be potentially useful complement to the direct arylation method if the following criteria are met:

1) Mild arylating agent (such as aryl zinc or boron compound)
2) One-pot operation, no excess isolation of dihydroaromatics

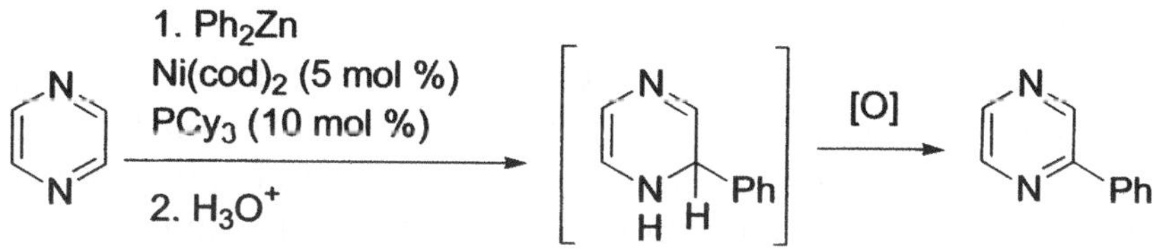

The biaryl coupling of electron-deficient aromatic rings such as pyrazines and quinoxalines can be promoted by the presence of potassium *t*-butoxide alone.[54] The evidence collected to determine the mechanism of this coupling reaction supports a radical-based process. This reaction can proceed under both microwave irradiation as well as traditional heating. However, the latter requires higher temperatures and longer reaction times. This coupling was base-dependent, showing a significant drop in yields when either the metal or anions were changed. When other electron-deficient aromatic rings such as pyridine, pyridazine, pyrimidine, or quinoxaline were used as the substrate, no significant regioselective preferences were observed. This coupling was suitable for halogenated phenyl, heteroaromatic, and alkene coupling partners.

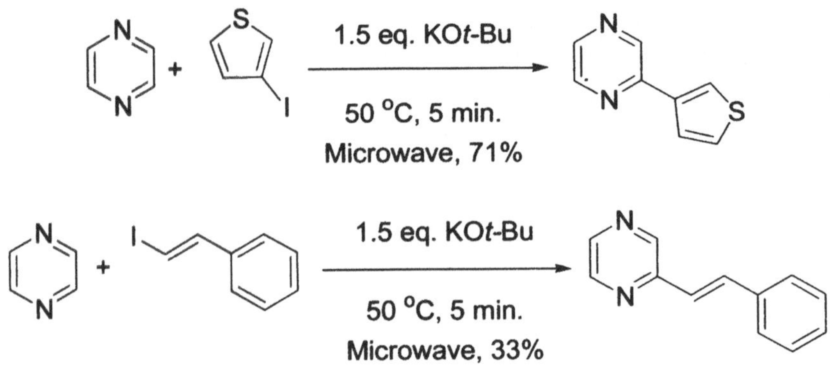

12.4.2 Palladium-catalyzed reactions

Pyrazines and quinoxalines are electron-deficient, 6π-electron heteroaromatic compounds. The inductive effects of the nitrogen atoms induce a partial positive charge on the carbon atoms. The use of palladium-catalyzed cross-coupling reactions in heterobiaryl coupling is linked to, and limited by, the synthetic and commercial availability of heteroaromatic organometallic reagents.[33] Electron-deficient nitrogen heterocycles are expensive, and difficult to prepare and use in cross-coupling reactions.

It is important to note that there are successful examples of biaryl coupling using electron-rich and sterically-hindered phosphine ligands on palladium(0).[55]

Literature research in this area yields few efficient reactions for palladium-catalyzed reactions of pyrazine and the diazine ring of quinoxaline.[56] One of the challenges associated with coupling the pyrazine ring is the free electrons on the nitrogen atoms that could bind and poison the palladium catalyst. The organometallic compounds needed for this coupling are unstable and are rarely isolated, and frequently decompose under cross-coupling reaction conditions.

The formation of the diazinylstannane can be accomplished through two separate pathways. In one pathway the pyrazine can be lithiated followed by quenching with tributyltin chloride.[57] The second pathway provides much higher yields: the chloropyrazine or quinoxaline is treated with a one equivalent of stannyl anion to provide a multifunctional aromatic ring. This equivalent can be manipulated, controlling the number of substitutions on the pyrazine ring.[57a]

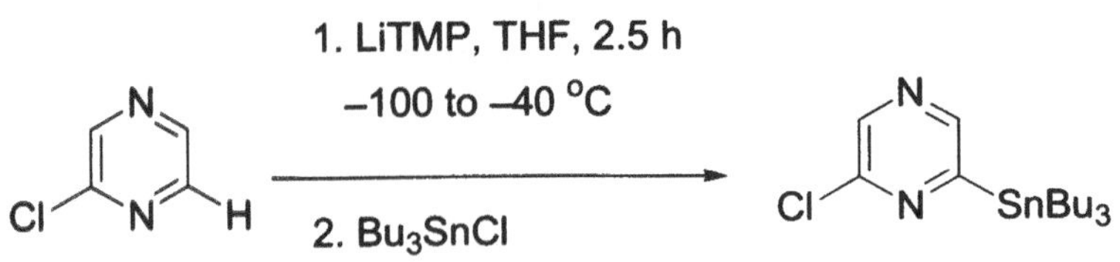

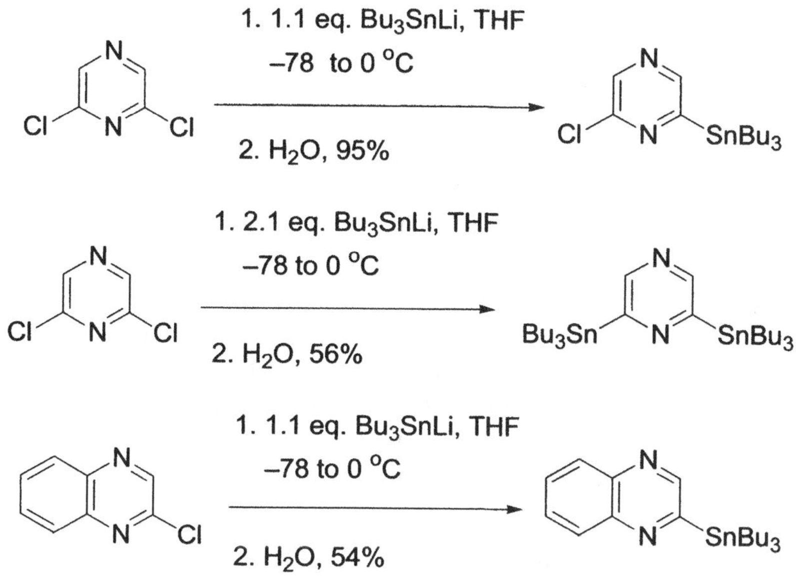

In an effort to synthesize a series of trisubstituted pyrazines to study their kinase inhibition activity, a series of regioselective substitution reaction was accomplished under two separate pathways. The first was more amenable to rapid structure activity relationships at the N_1 position. At this position a wide variety of benzylic, alkyl and amide functionalities were present.[58]

The first pathway began with a Suzuki coupling of the aryl boronate with 5-bromopyrazin-2-amine to provide the difunctional pyrazine. Halogenation α to the amino group followed by the TIPS deprotection provides a late stage intermediate that could be used for rapid screening for kinase inhibitors. Heating the diazinyl bromide in the presence of the desired amine displaces the bromide under microwave irradiation. This also causes an intramolecular nucleophile attack by the incorporated nitrogen to one of the Boc carbonyls to provide the cyclized urea in 16–54% yield.

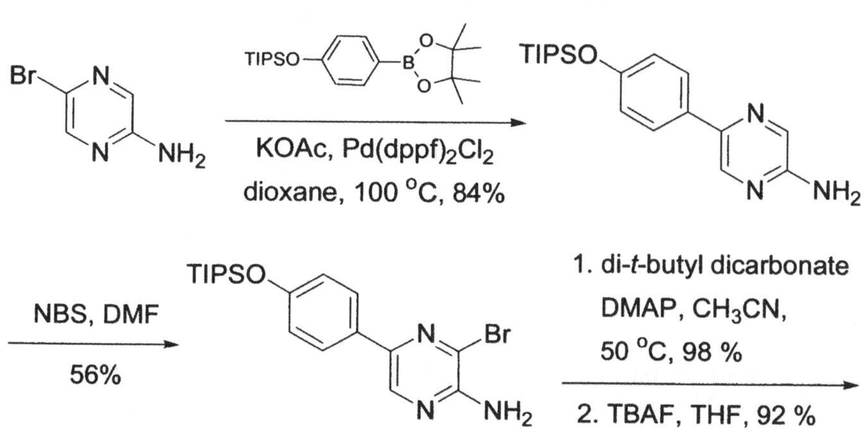

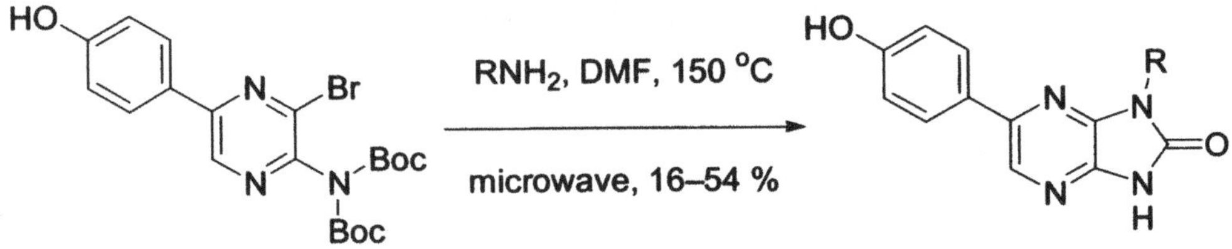

The second pathway began with 3,5-dibromopyrazin-2-amine. Substituting the bromide *ortho-* to the amine was accomplished in low yields using a variety of primary amines at elevated temperatures to provide a variety of aromatic amines. Cyclization to form the urea was accomplished using carbonyl diimidazole or urea at high temperatures. The target for study was completed with a Suzuki coupling with the boronic acid under microwave irradiation in 20–60% yields.

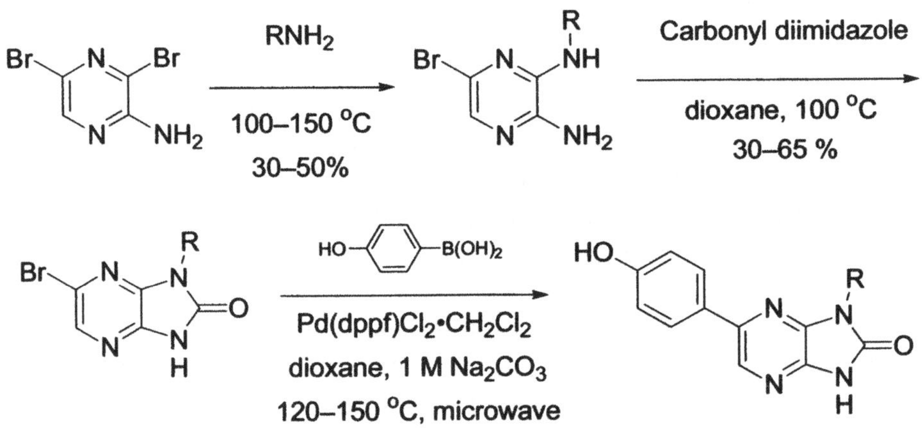

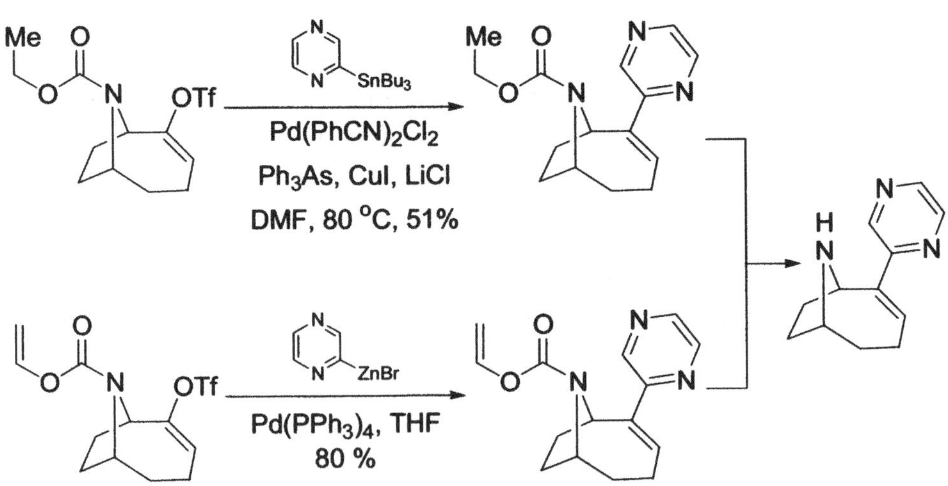

To further illustrate the broad biological activity that pyrazines can display, Sharples[59] and Seitz[60] synthesized the similar intermediates toward the same biologically-active molecule. This molecule is a biological isostere of (–)-epibatidine, which is a modulator of the nicotinic acetylcholine receptor. When comparing the two different palladium-catalyzed coupling reactions, it is interesting to see the significant increase in yield delivered by the Negeshi coupling (80%) as compared to the Stille protocol (51%).

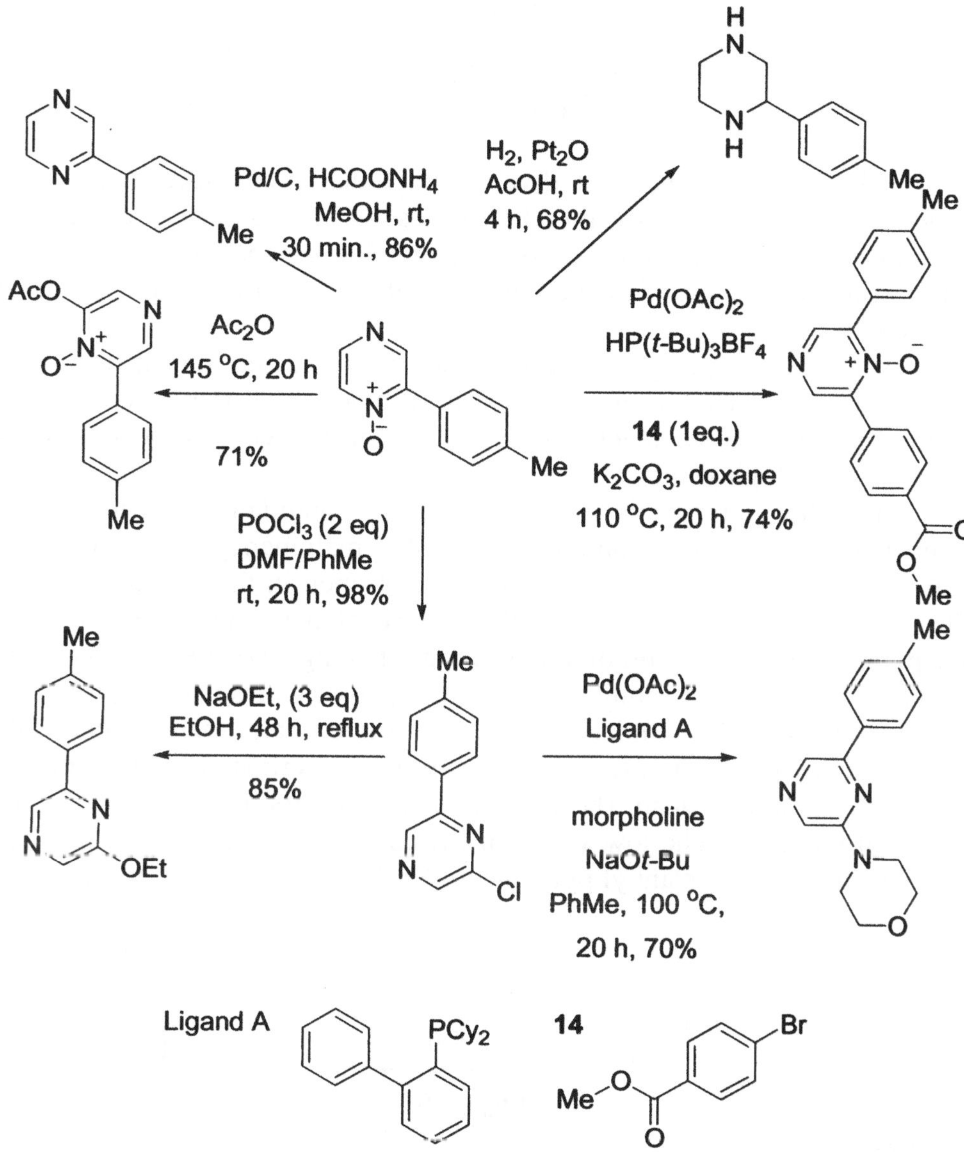

Fagnou[33] demonstrated a unique way to overcome the difficulties of palladium catalysis mentioned above by oxidizing a single nitrogen of the pyrazine to provide the *N*-oxide. Direct arylation of this ring with palladium catalysis provides the diaryl ring system in good yield and a single reported regioisomer. Coordination by the oxidized nitrogen to the palladium is no longer a possibility and, with the direct arylation, the organometallic reagent is no longer a requirement.

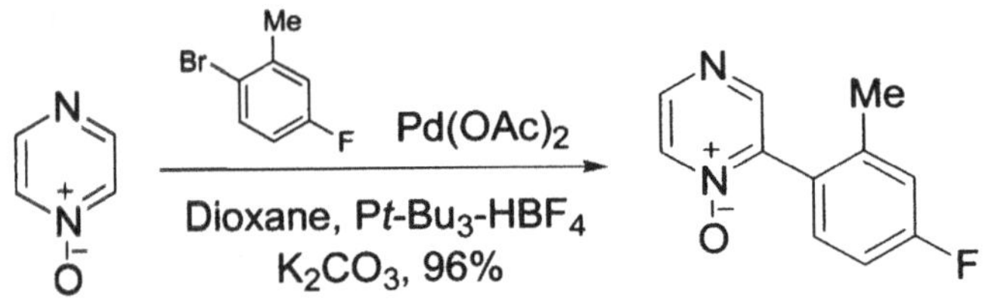

This reaction displays a diverse functional group tolerance. This reaction showed no significant differences in yields when comparing electron-donating groups and electron-withdrawing groups. The substitution pattern of the aryl halide displayed minimal effects on the isolated yields of the products.

After the coupling was completed, the oxidized nitrogen can be manipulated in many different ways to provide a variety of different structural motifs. These reactions display a high-degree of regioselectivity and moderate to high yields.

Steriodial pyrazine is a promising drug candidate to treat prostate cancer, which is the most common age-related malignancy cause of cancer death worldwide. It killed an estimated 23,000 men in the United States in 2004 and making it the second leading cause of death in American men. This pyrazine in the nanomolar concentration range competes effectively with labelled R1881 for binding with both types of androgen receptors in a dose dependent manner. Indeed it is 30-fold more potent than the clinically used anti-androgen, flutamide.[61] The electronic properties of the heterocycle coupled played a significant (however poorly understood) importance to the potency of the drug. The low yields of the coupling reactions were justified to the instability of the stannyl pyrazine under the conditions that were used.

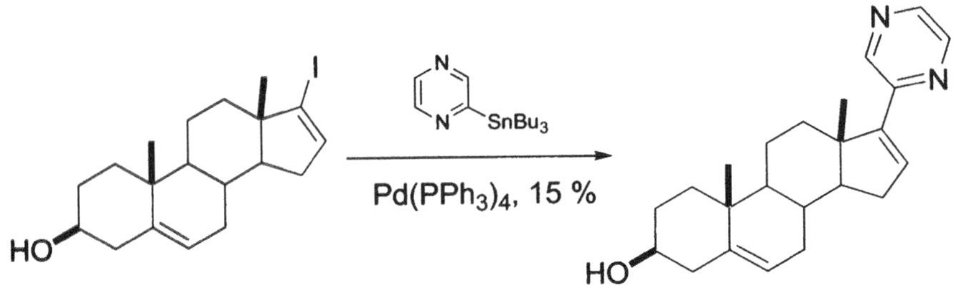

DNA-damaging agents, such as cisplatin, irinotecan, gemcitabone, and radiation, represent a historic cornerstone for the treatment of tumors.[62] However, there are certain cancers for which these agents only provide a limited benefit. These cancers are proficient in repairing the damaged DNA. Studies to prevent this repair by inhibiting two phosphoinositol 3-kinase-like kinase (PIKK) family members ATM (ataxia telangiectasia mutated) and ATR (ATM and Rad-3 related) led to the regioselective Suzuki coupling of a dibrominated pyrazine. This pyrazine first underwent coupling *ortho-* to the amino group followed by a second coupling to provide the multicyclic ring system, displaying high levels of inhibition against the targeted kinases.[62]

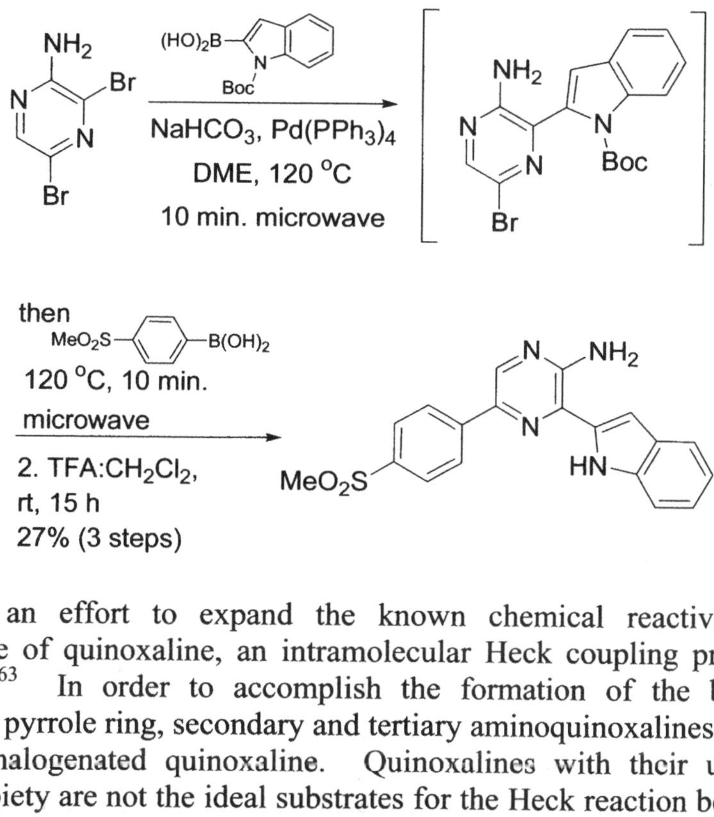

In an effort to expand the known chemical reactivity on the heterocycle of quinoxaline, an intramolecular Heck coupling protocol was optimized.[63] In order to accomplish the formation of the biologically significant pyrrole ring, secondary and tertiary aminoquinoxalines are reacted with the halogenated quinoxaline. Quinoxalines with their unique 1,4-diazine moiety are not the ideal substrates for the Heck reaction because they are not only more labile than the simple benzene counterpart but are also strong chelating agents which can coordinate to and poison the palladium catalyst. Aminoquinoxalines are especially strong chelating agents, and catalytic efficiency is difficult to achieve for such substrates. Jeffery's "ligand-free" conditions gave higher yields than the traditional coupling procedures. The enhanced reactivity and yields are presumably due to the coordination and thereby solvation of the palladium intermediates by chloride ions present in the reaction mixture.

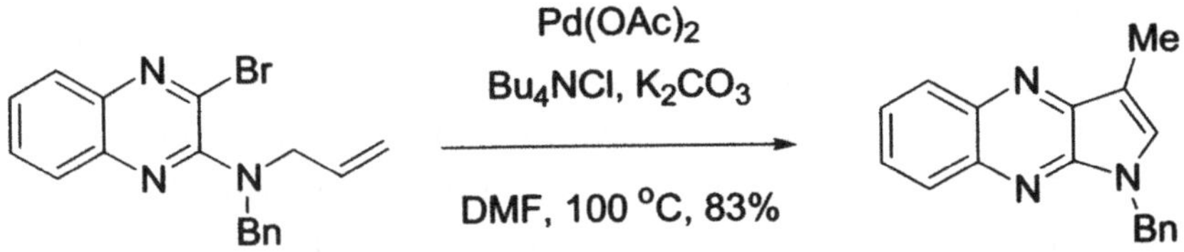

12.5 Problems

12.5.1 Radical reactions have become a powerful pathway to provide functionalization to the diazine rings. Provide a pathway to produce a carbamoyl radical using formamide, cerium ammonium nitrate, and *N*-hydroxyphthalimide and then treat that radical with quinoxaline to provide the functionalized product.[38]

12.5.2 Dihydrocephalostatin 1 was synthesized by Fuchs by forming the pyrazine core via stitching together azido-ketone and amino-methoxyimine by using dibutyltin chloride and polyvinylpyridine in benzene. What is the mechanism for the formation of this ring?[64]

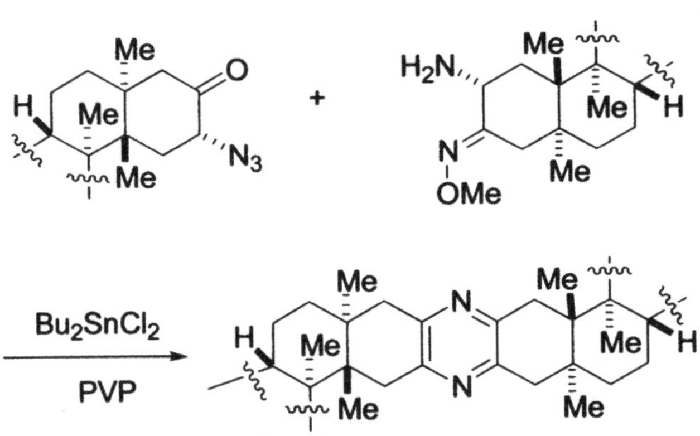

12.5.3 Selective fluorinations reactions can be an incredible asset in biological studies. Methods to selectively fluorine pyrazines and quinoxalines have been displayed using iodine. Mechanistically explain how this fluorination proceeds so selectively.[44]

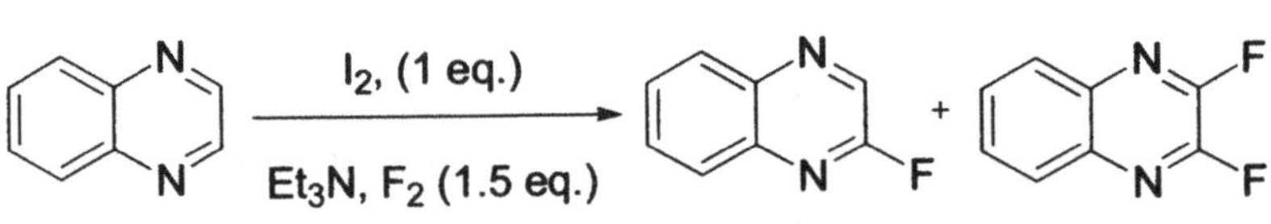

48% 11%

12.5.4 Selective alkenylation of pyrazines has been accomplished using nickel catalysis. Write out a mechanism that explains both the alkenylation product as well as the dialkenylation product in higher stereo- and regioselectivity.[52]

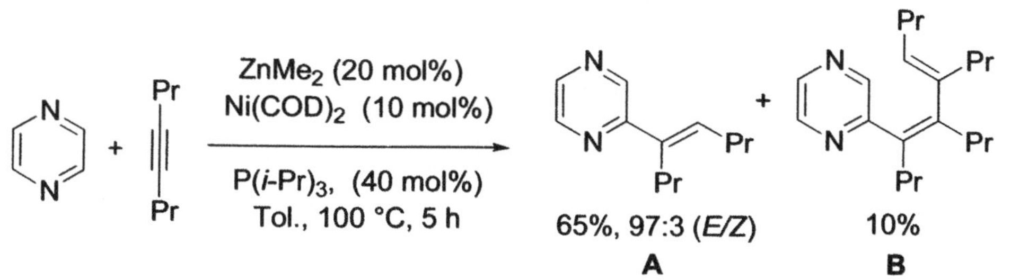

12.5.5 Regioselective reactions have become a hallmark of synthetic chemistry. Predict the product below.[41]

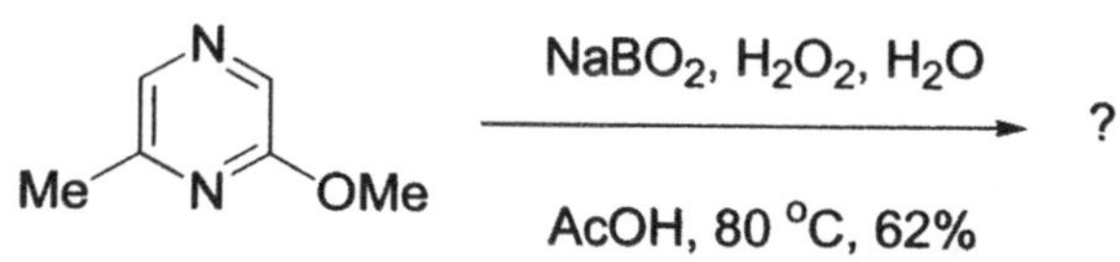

12.5.6 Provide an effective synthetic pathway to convert quinoxaline to the antibiotic pyrazinamide.[65]

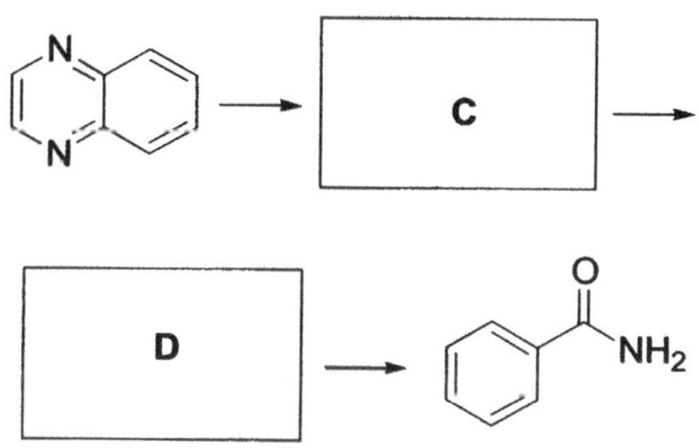

12.5.7 Palladium catalysis has been demonstrated to be broad in scope and a very powerful method to quickly build in functional groups. Provide a reasonable mechanism for the reaction below.[63]

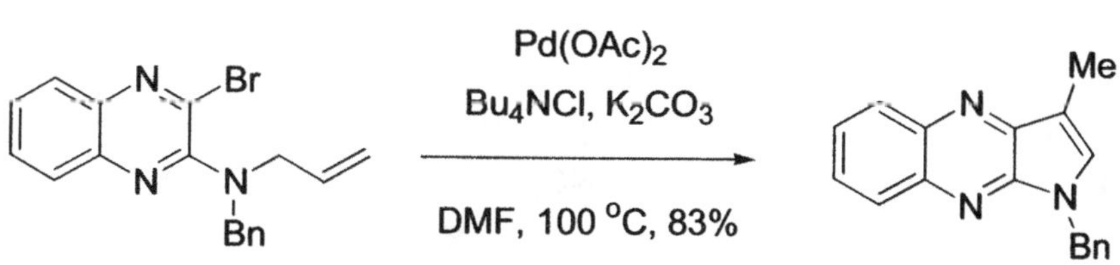

12.5.8 In an attempt to functionalize carbon-6 of 2,3-diphenylquinoxaline, it was subjected to trimethylsilyl cyanide and benzoyl chloride to provide an unexpected ring-opened product. Draw a mechanism to justify that product.[66]

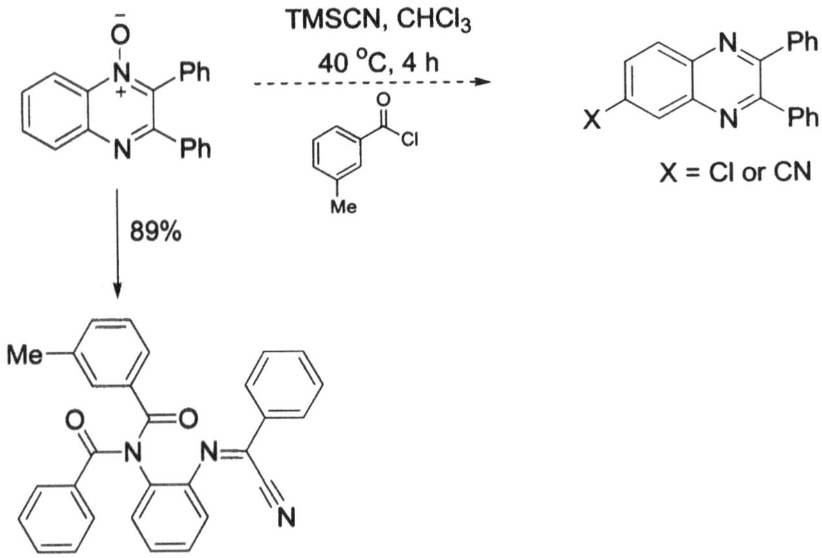

12.5.9 Propose a reasonable mechanism for the following transformation:[67]

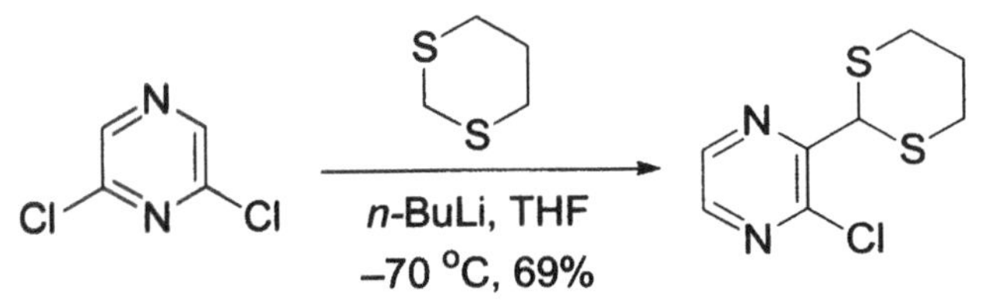

12.5.10 Predict the regiochemical outcome of the following transformation, which offers only one major product and explain the selectivity:[68]

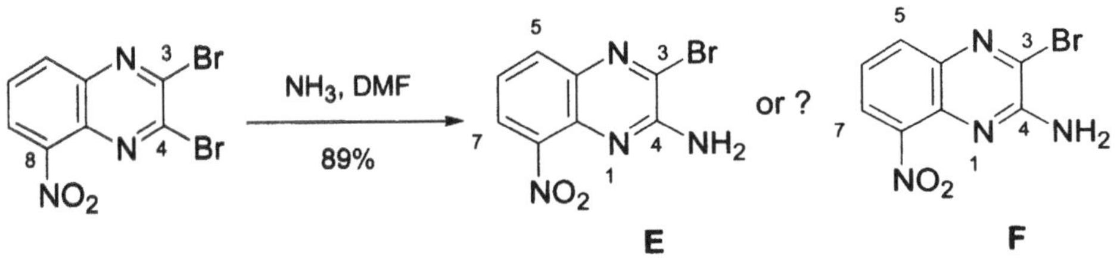

12.6 References

1. Husain, A.; Madhesia, D. *J. Pharm. Res.* **2011,** *4*, 924–929.
2. Fortner, K. C.; Kato D.; Tanaka Y.; Shair, M. D. *J. Am. Chem. Soc.* **2010,** *132*, 275–280.
3. (a) Cutignano, A.; Bifulco, G.; Bruno, I.; Casapullo, A.; Gomez-Paloma, L.; Riccio, R. *Tetrahedron* **2000**, *56*, 3743-3748. (b) Faulkner, D. J. *Nat. Prod. Rep.* **2002,** *19*, 1–48.
4. Mongin, F.; Mojovic, L.; Guillamet, B.; Trécourt, F.; Quéguiner, G. *J. Org. Chem.* **2002,** *67*, 8991–8994.
5. Harmenberg, J.; Akesson-Johansson, A.; Graslund, A.; Malmfors, T.; Bergman, J.; Wahren, B.; Akerfeldt, S.; Lundbald, L. Cox, S. *Antiviral Res.* **1991,** *15*, 193–204.
6. Naylor, M. A.; Stephen, M. A.; Nolan, J.; Sutton, B.; Tocher, J. H.; Fielden, E. M.; Adams, G. E.; Strafford, I. J. *Anticancer Drug Des.* **1993,** *8*, 439–461.
7. Huang, W.; Tang, L.; Shi, Y.; Huang, S.; Xu, L.; Sheng, R.; Wu, P.; Li, J.; Zhou, N.; Hu, Y. *Bioorg. Med. Chem.* **2011,** *19*, 7158–7167.
8. Verhoest, P. R.; Proulx-Lafrance, C.; Corman, M.; Chenard, L.; Helal, C. J.; Hou, X.; Kleiman, R.; Liu, S.; Marr, M.; Menniti, F. S.; Schmidt, C. J.; Vanase-Frawley, M.; Schmidt, A. W.; Williams, E. D.; Nelson, F. R.; Fonseca, K. R.; Liras, S. *J. Med. Chem.* **2009,** *52*, 7946–7949.
9. Béguin, C.; Duncan, K. K.; Munro, T. A.; Ho, D. M.; Xu, W.; Liu-Chen, L.-Y.; Carlezon, W. A.; Cohen, B. M. *Bioorg. Med. Chem.* **2009,** *17*, 1370–1380.
10. Shibinskaya, M. O.; Karpenko, A. S.; Lyakhov, S. A.; Andronati, S. A.; Zholobak, N. M.; Spivak, N. Y.; Samochina, N. A.; Shafran, L. V.; Zubritsky, M. J.; Galat, V. F. *Eur. J. Med. Chem.* **2011,** *46*, 794–798.
11. Vergara, F. M. F.; da S. Lima, C.; de O. Henriques, M. D. G.; Candéa, A. L. P.; Lourenço, M. C. S.; de L. Ferreira, M.; Kaiser, C. R.; de Souza, M. V. N. *Eur. J. Med. Chem.* **2009,** *44*, 4954–4959.
12. Singh, D. P.; Hashim, S. R. Singhal, R. G. *Int. J. Drug Dev. Res.* **2010,** *2*, 810–815.
13. Krishnakumar, B.; Velmurugan, R.; Jothivel, S.; Swaminathan, M. *Cat. Comm.* **2010,** *11*, 997–1002.
14. Heravi, M. M.; Bakhtiari, K.; Bamoharram, F. F.; Tehrani, M. H. *Monatsh. Chem.* **2007,** *138*, 465–467.
15. Gomez-Caro, L. C.; Sanchez-Sanchez, M.; Bocanegra-Garcia, V.; Rivera, G.; Monge, A. *Quimica Nova* **2011,** *34*, 1147–1151.
16. Srinvas, C.; Kumar, C. N. S. S. P.; Rao, V. J.; Palaniappan, S. *J. Mol. Cat. A Chem.* **2007,** *265*, 227–230.
17. Natarajan, U.; Kaliappan, I.; Singh, N. K. *Der Pharma Chemica.* **2010,** *2*, 159–167.
18. (a) Bory, P. R.; Reitz, D. B.; Collins, J. T.; Chamberlain, T. S.; Olins, G. M.; Smita, G. J.; McGraw, D. E.; Gaw, J. F. *J. Med. Chem.* **1993,** *36*, 101–110. (b) Marshall, W. S.; Goodson, T.; Gullinan, G. J.; Swanson, D.; Haisch, K. D.; Rinkema, L. E.; Leisch, J. H. *J. Med. Chem.* **1987,** *30*, 682–689.
19. Ashton, W. T.; Cantone, C. L.; Chang, L. L.; Hutchins, S. M.; Strelitz, R. A.; Malcolm, M.; Chang, S. L.; Lotti, V. J.; Farust, K. A.; Chen, T. B.; Patricia, B.; Terry, W. S.; Kivlighn, S. D.; Siegl, P. S. *J. Med. Chem.* **1993,** *36*, 591–609.
20. (a) Staedel, W.; Rügheimer L. *Berichte der deutschen chemischen Gesellschaft.* **1876**, *9*, 563–564. (b) Gastaldi, G. *Gazz. Chim. Ital.* **1921,** *51*, 233–255.
21. Gavara, L.; Saugues, E.; Alves, G.; Debiton, E.; Anizon, F.; Moreau, P. *Eur. J. Med. Chem.* **2010,** *45*, 5520–5526.
22. Cui, M.; Ono, M.; Kimura, H.; Liu, B.; Saji, H. *Bioorg. Med. Chem. Lett.* **2011,** *21*, 4193–4196.

23. Gutknecht, H. *Chem. Ber.* **1879,** *12*, 2290–2292.
24. Wright, S. W. *Gutknecht Reaction* In *Name Reactions in Heterocyclic Chemistry-II*, Li, J. J., Ed.; Wiley: Hoboken, NJ, 2011, 430–439.
25. Buron, F.; Turck, A.; Plé, N.; Bischoff, L.; Marsais, F. *Tetrahedron Lett.* **2007,** *48*, 4327–4330.
26. Haddadin, M. J.; Issidorides, C. H. *Tetrahedron Lett.* **1965,** 3253.
27. Tinsley, J. M. *Beirut Reaction*. In *Name Reactions in Heterocyclic Chemistry*, Li, J. J., Ed.; Wiley: Hoboken, NJ, 2004, 504–509.
28. Kluge, A. F.; Maddox, M. L.; Lewis, G. S. *J. Org. Chem.* **1980,** *45*, 1909.
29. Zarranz, B.; Jaso, A.; Aldana, I.; Monge, A. *Bioorg. Med. Chem.* **2003,** *11*, 2149–2153.
30. Brown, H. C.; McDaniel, D. H.; Hafliger, O. *Dissociation Constants*, In *Determination of Organic Structures by Physical Methods*, Baude, E. A. and Nachod, F. C., Eds. Academic Press: New York, 1955, 567–662.
31. Analytical Chemistry, In *Handbook of Chemistry and Physics*, 71st Ed.; Lide, D. R. Ed.; CRC Press: Boca Raton, 1991; 833–834.
32. Klein, B.; Berkowitz, J. *J. Am. Chem. Soc.* **1959,** *81,* 5160–5166.
33. Leclerc, J.-P.; Fagnou, K., *Angew. Chem., Int. Ed.* **2006,** *45*, 7781–7786.
34. McKay, S. E.; Sooter, J. A.; Bodige, S. G.; Blackstock, S. C. *Heterocyclic Comm.* **2001,** *7*, 307–312.
35. Ohta, A.; Akita, Y.; Hara, M. *Chem. Pharm. Bull.* **1979,** *27*, 2027–2041.
36. Ameriks, M. K.; Bembenek, S. D.; Burdett, M. T.; Choong, I. C.; Edwards, J. P.; Gebauer, D.; Gu, Y.; Karisson, L.; Purkey, H. E.; Staker, B. L.; Sun, S.; Thurmond, R. L.; Zhu, J. *Bioorg. Med. Chem. Lett.* **2010,** *20*, 4060–4064.
37. Sutherland, J. B.; Evans, F. E.; Freeman, J. P.; Willaims, A. J. *Lett. Appl. Microbiol.* **1996,** *22*, 199–201.
38. Minisci, F.; Recupero, F.; Punta, C.; Gambarotti, C.; Antonietti, F.; Fontana, F.; Pedulli, G. F. *Chem. Commun.* **2002,** 2496–2497.
39. Plé, N.; Turck, A.; Couture, K.; Quéguiner, G. *Synthesis* **1996,** *7*, 838–842.
40. Tarrago, G.; Zidane, I.; Marzin, C.; Tep, A. *Tetrahedron* **1988,** *44*, 91–100.
41. Dickschat, J. S.; Reichenbach, H.; Wagner-Döbler, I.; Schulz, S. *Eur. J. Org. Chem.* **2005,** 4141–4153.
42. Nishiguchi, G. A.; Atallah, G.; Bellamacina, C.; Burger, M. T.; Ding, Y.; Feucht, P. H.; Garcia, P. D.; Han, W.; Klivansky, L.; Lindvall, M. *Bioorg. Med. Chem. Lett.* **2011,** *21*, 6366–6369.
43. Noolvi, M. N.; Patel, H. M.; Bhardwaj, V.; Chauhan, A. *Eur. J. Med. Chem.* **2011,** *46*, 2327–2346.
44. Chambers, R. D.; Parsons, M.; Sandford, G.; Skinner, C. J.; Atherton, M. J.; Moilliet, J. S. *J. Chem. Soc., Perkins Trans. 1*, **1999,** 803–810.
45. Seggio, A.; Chevallier, F.; Vaultier, M.; Mongin, F. *J. Org. Chem.* **2007,** *72*, 6602–6605.
46. Plé, N.; Turck, A.; Couture, K.; Quéguiner, G. *J. Org. Chem.* **1995,** *60*, 3781–3786.
47. (a) Chevallier, F.; Mongin, F. *Chem. Soc. Rev.* **2008,** *37*, 595–609. (b) Quéguiner, G. *J. Heterocycl. Chem.* **2000,** *37*, 615–621.
48. Dong, Z.; Clososki, G.; Wunderlich, S. H.; Unsinn, A.; Li, J.; Knochel, P. *Chem. Eur. J.* **2009,** *15*, 457–468.
49. Dong, Z.-B.; Zhu, W.-H., Zhang, Z.-G.; Li, M.-Z. *J. Organomet. Chem.* **2010,** *695*, 775–780.
50. L'Helgoual'ch, J-M.; Bentabed-Ababsa, G.; Chevallier, F.; Derdour, A.; Mongin, F. *Synthesis* **2008,** *24*, 4033–4035.
51. Li, M.; Hua, R. *Tetrahedron Lett.* **2009,** *50*, 1478–1481.

52. Nakao, Y.; Kanyiva, K. S.; Hiyama, T. *J. Am. Chem. Soc.* **2008,** *130*, 2448–2449.
53. Tobisu, M.; Hyodo, I.; Chatani. N. *J. Am. Chem. Soc.* **2009,** *131*, 12070–12071.
54. Yanagisawa, S.; Ueda, K.; Taniguchi, T.; Itami, K. *Org. Lett.* **2008,** *10*, 4673–4676.
55. (a). Reetz, M. T.; Lohmer, G.; Schwickardi, R. *Angew. Chem., Int. Ed. Engl.* **1998,** *37,* 481–483. (b). Littke, A. F.; Fu, G. C. *Angew. Chem., Int. Ed. Engl.* **1998,** *37,* 3387–3388. (c) Buchwald, S. L.; Wolfe, J. P. *Angew. Chem., Int. Ed. Enl.* **1999,** *38,* 2413–2416.
56. Li, J. J. *J. Org. Chem.* **1999,** *64*, 8425–8427.
57. (a) Darabantu, M.; Boully, L.; Turck, A.; Plé, N. *Tetrahedron* **2005,** *61*, 2897–2905. (b) Buron, F.; Ple, N.; Turck, A.; Quéguiner, G. *J. Org. Chem.* **2005,** *70,* 2616–2621.
58. Mortensen, D. S.; Perrin-Ninkovic, S. M.; Harris, R.; Lee, B. G. S.; Shevlin, G.; Hickman, M.; Khambatta, G.; Bisonette, R. R.; Fultz, K. E.; Sankar, S. *Biorg. Med. Chem. Lett.* **2011,** *21*, 6793–6799.
59. Sharples, C. G. V.; Karig, G.; Simpson. G. L.; Spencer, J. A.; Wrigt, E.; Millar, N. S.; Wonnacott, S. W.; Gallagher, T. *J. Med. Chem.* **2002,** *45*, 3235–3245.
60. Gohlke, H.; Gündisch, D.; Schwarz, S.; Seitz, G.; Tilotta, M. C.; Wegge, T. *J. Med. Chem.* **2002,** *45*, 1064–1072.
61. Handratta, V. D.; Vasaitis, T. S.; Njar, V. C. O.; Gediya, L. K.; Ktaria, R.; Chopra, P.; Newman, D.; Farquhar, R.; Guo, Z.; Qiu, Y.; Brodie, A. M. H. *J. Med. Chem.* **2005,** *48*, 2972–2984.
62. Charrier, J.-D.; Durrant, S. J.; Golec, J. M. C.; Kay, D. P.; Knegtel, R. M. A.; MacCormick, S.; Mortimore, M.; O'Donnell, M. E.; Pinder, J. L.; Reaper, P. M.; Rutherford, A. P.; Wang, P. S. H.; Young, S. C.; Pollard, J. R. *J. Med. Chem.* **2011,** *54*, 2320–2330.
63. (a) Li, J. J. *J. Org. Chem.* **1999,** *64,* 8425–8427. (b) McCrindle, R.; Ferguson, G.; Arsenault, G. J.; McAlees, A. J.; Stephenson, D. K. *J. Chem. Res., Synop.* **1984,** 360–361.
64. Guo, C.; Bhandaru, S.; Fuchs, P. L. *J. Am. Chem. Soc.* **1996,** *118*, 10672–10673.
65. Zhou, Q.; Zheng, Y.-Y.; Li, B. D.; Lu, C.-X. *Jiangsu Huagong* **2005,** *33,* 124–126.
66. Nasielski, J.; Heilporn, S.; Nasielski-Hinkens, R. *Tetrahedron* **1989,** *45*, 7795–7804.
67. Torr, J. E.; Large, J. M.; Horton, P. N.; Hursthouse, M. B.; McDonald, E. *Tetrahedron Lett.* **2006,** *47*, 31–34.
68. (a) Li, J. J.; Yue, W. S. *Tetrahedron Lett.* **1999,** *40*, 4507–4510. (b) Iwata, S.; Sakajyo, M.; Tanaka, K. *J. Heterocyclic Chem.* **1994,** *31*, 1433–1438.

Chapter 13 Pyrimidines

Sha Lou and Ji Zhang

13.1 Introduction

Pyrimidine is a six-membered heterocyclic ring with a structure similar to benzene except that two C–H units are replaced with nitrogen atoms. As a result, its bond lengths and bond angles differ from benzene as shown below. For example, the benzene ring bond length is 1.40 Å with a bond angle of 120 °. The pyrimidine ring is virtually flat, however, the electron distribution is no longer distributed equally at each atom. There is considerable depletion of electron density at the 2-, 4-, and 6-positions; a moderate depletion at the 5-position, and significantly increased electron density at the nitrogen atoms. Pyrimidine is a much weaker base (pK_a 1.31) than pyridine (pK_a 5.2) because the second ring-nitrogen shares the available π-electrons with the first nitrogen. The dipole moment of pyrimidine is measured to be in the range of 2.10 to 2.40 D. Proton NMR and ^{13}C NMR are shown below.

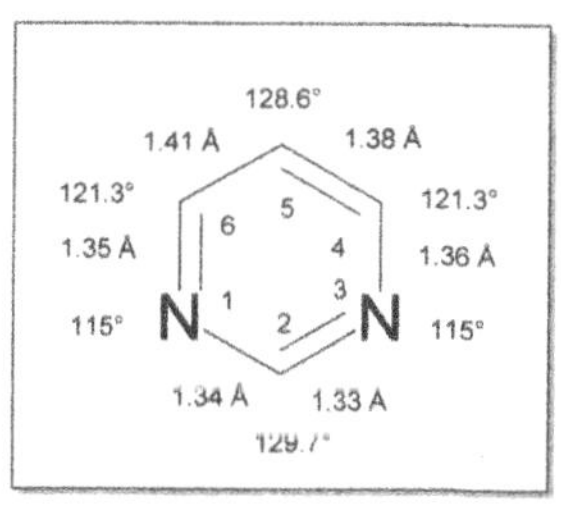

Proton NMR

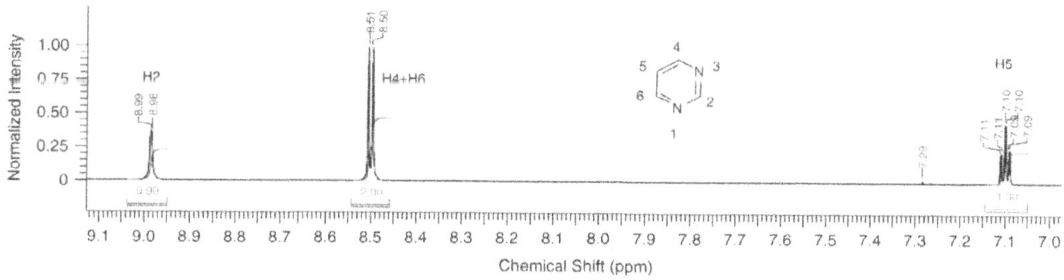

^{13}C NMR

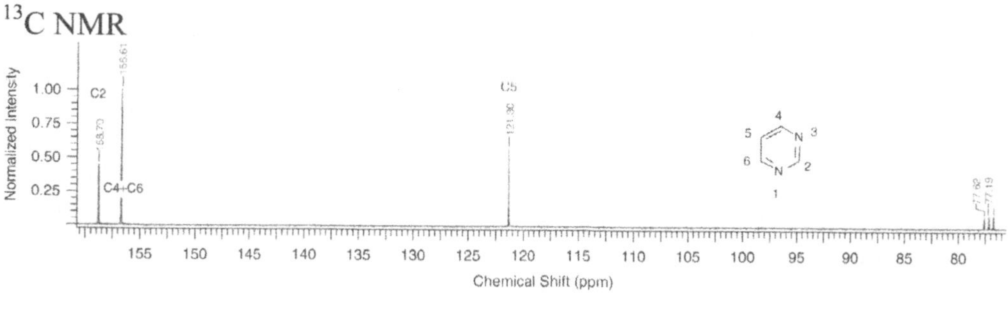

A large number of natural products and pharmaceuticals possess the pyrimidine structure. For example, sulfadiazine is a sulfonamide antibiotic that contains a 5-aminopyrimidine.[1] It eliminates bacteria that cause infections by stopping the production of folic acid inside the bacterial cell, and it is commonly used to treat urinary tract infections (UTIs). Silver sulfadiazine is a topical sulfonamide/silver antibacterial used as creams for the treatment of burns, including chemical burns. It prevents the growth of a wide array of bacteria, as well as yeast, on the damaged skin. Silver sulfadiazine is also an orally administered chemotherapeutic agent used in the treatment of metastatic breast and colorectal cancers. The chemotherapy agent 5-fluorouracil (5-FU) has been used against cancer for about 40 years. Capecitabine is a prodrug that is enzymatically converted to 5-FU in the tumor, where it inhibits DNA synthesis and slows growth of tumor tissue. 5-FU is an inhibitor that mainly works through irreversible inhibition of thymidylate synthase and blocks the synthesis of the pyrimidine thymidine, which is a nucleoside required for DNA replication.

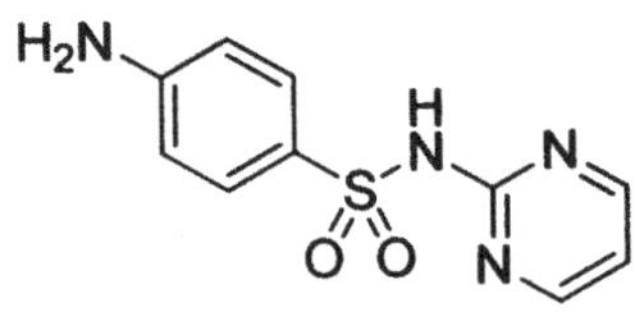

sulfadiazine

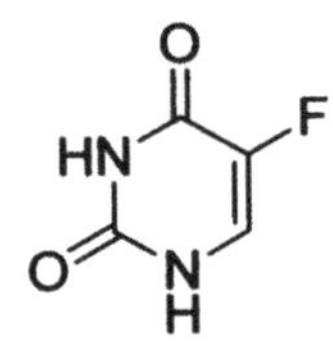

5-FU

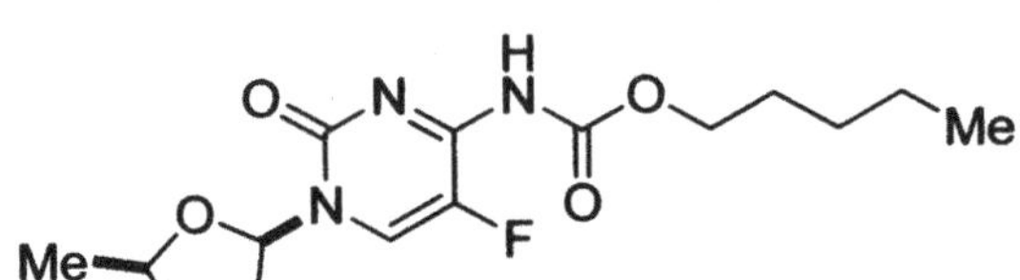

capecitabine

Pyrimidine-fused ring systems are abundant in nature and often used as the core structures in numerous pharmaceuticals. For instance, purines contain the pyrimidine moiety as part of a more complex, fused-ring heterocyclic structure. This class of heterocycles including substituted purines and their tautomers is the most widely distributed kind of nitrogen-containing heterocycles in nature. Two of the four deoxyribonucleotides and two of the four ribonucleotides, the respective building blocks of DNA and RNA, are purines. Adenine is one of the two purine nucleobases (the other being guanine) used in forming nucleotides of the nucleic acids. In DNA, adenine binds to thymine *via* two hydrogen bonds to assist in stabilizing the nucleic acid structures. In RNA, which is used for protein synthesis, adenine binds to uracil. It forms adenosine triphosphate (ATP), which transports chemical energy within cells for metabolism. Caffeine, a purine degradant, acts as a central nervous system stimulant, and it is the world's most widely consumed psychoactive drug.

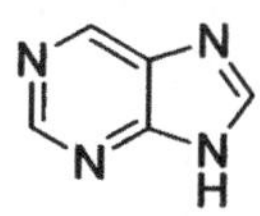

purine

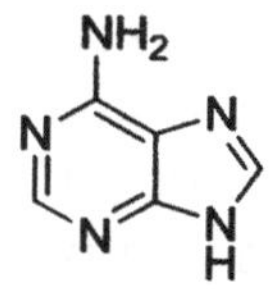

adenine

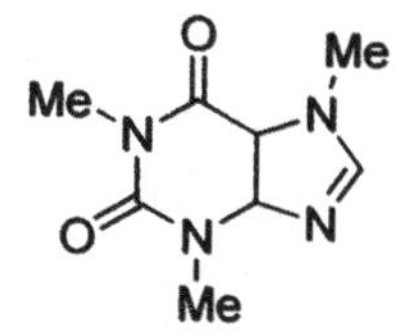

caffeine

Adenine itself was successfully incorporated into a number of pharmaceuticals. Adefovir dipivoxil, also known as bis-POM PMEA, with trade names Preveon and Hepsera, is used for the treatment of hepatitis B and herpes simplex virus infection.[2] It is an orally administered nucleotide analog as a reverse transcriptase inhibitor (NRTI). Reverse transcriptase is an enzyme crucial to viral production. A second NRTI that contains adenine is Tenofovir disoproxil fumarate (TDF), which has been used in the treatment of HIV/AIDS[3] and hepatitis B.[4] TDF is a prodrug of tenofovir (also known as PMPA) designed to improve absorption and cell permeability of the active moiety under the trade name Viread. Both adefovir dipivoxil and TDF are marketed by Gilead Sciences.

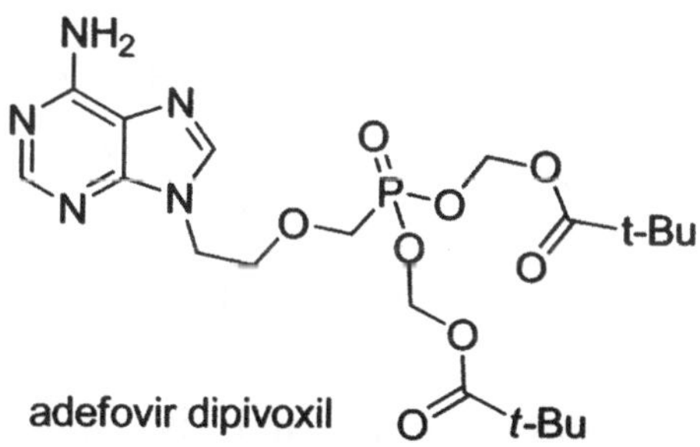

adefovir dipivoxil

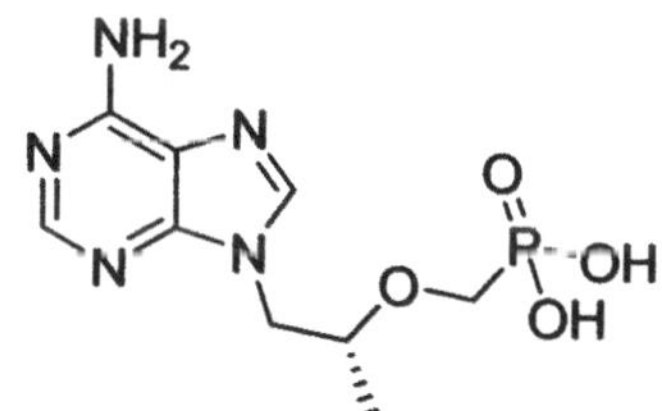

tenofovir

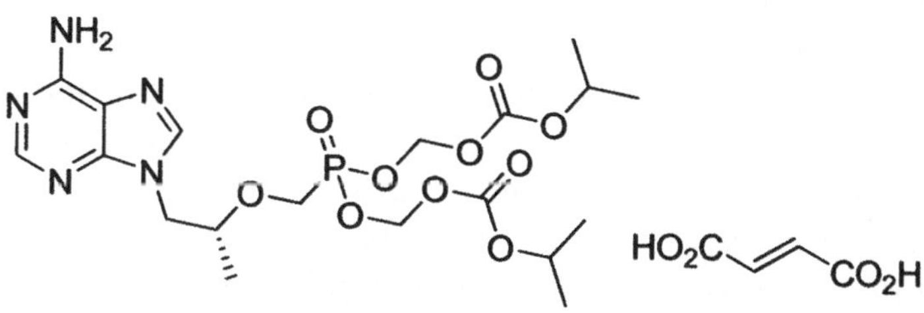

tenofovir disoproxil fumarate

Guanine, one of the four main nucleobases found in DNA and RNA, is another derivative of purine, consisting of a fused pyrimidine-imidazole ring system. Its uses in pharmaceutical development can be exemplified with the following two important drugs. Valacyclovir is an antiviral drug used in the management of herpes simplex, herpes zoster (shingles), and herpes B. It is a prodrug, being converted *in vivo* to acyclovir (9-[(2-hydroxyethoxy)-methyl]guanine). It is marketed by GlaxoSmithKline under the trade names Valtrex and Zelitrex and is currently available in generic form in the United States. Valganciclovir hydrochloride (Valcyte) is manufactured by Hoffmann–La Roche and is an antiviral medication used to treat cytomegalovirus (CMV) infections. As the L-valyl ester of ganciclovir, it is a prodrug for ganciclovir. After oral administration, it is rapidly converted to ganciclovir by intestinal and hepatic esterases. Human CMV infection is typically unnoticed in healthy people, but it can be life threatening for the immunocompromised, such as HIV-infected persons, organ transplant recipients, or newborn infants.

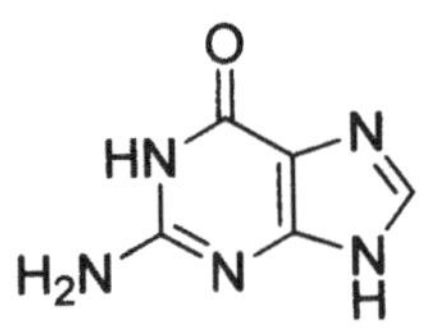

guanine

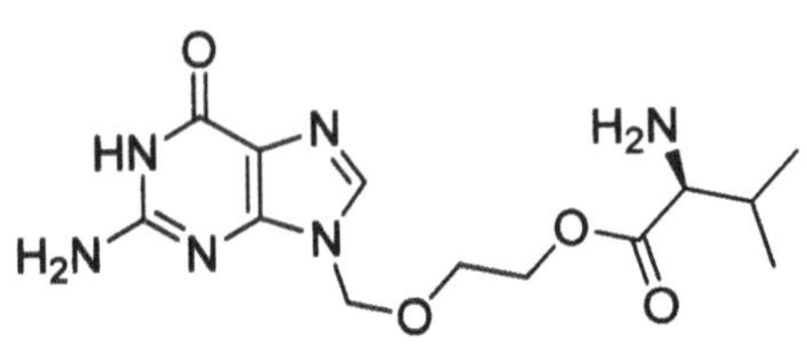

valacyclovir

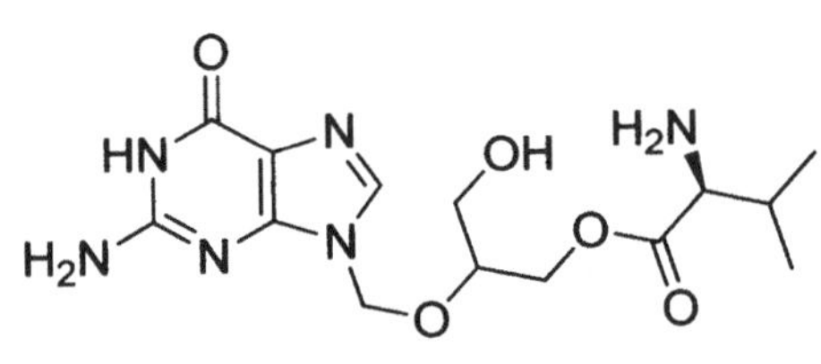

valganciclovir

Risperidone contains a pyrimidine-4-one and is an atypical antipsychotic sold under the trade name Risperdal. It is used to treat schizophrenia (including adolescent schizophrenia), schizoaffective disorder, the mixed and manic states associated with bipolar disorder, and irritability in people with autism. The drug was developed by Janssen–Cilag and first released in 1994. Pemetrexed (brand name Alimta) is a chemotherapy drug manufactured and marketed by Eli Lilly. Its indications are the treatment of pleural mesothelioma as well as non-small cell lung cancer.[5] Methotrexate (MTX) is an antimetabolite and antifolate drug. It contains a 2,4-diaminopteridine heterocyclic ring system and is used in the treatment of cancer, autoimmune diseases, and ectopic pregnancy, and for the induction of medical abortions. It acts by inhibiting the metabolism of folic acid, thereby blocking folic acid-dependent steps in the synthesis of purines and pyrimidines.[6] Methotrexate began to replace the more toxic antifolate aminopterin starting in the 1950s.

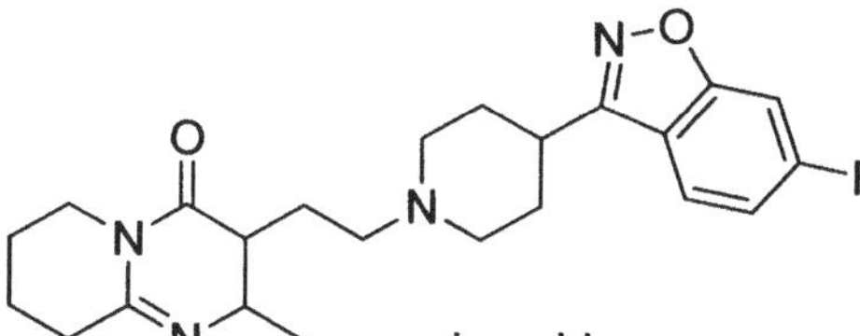

risperidone

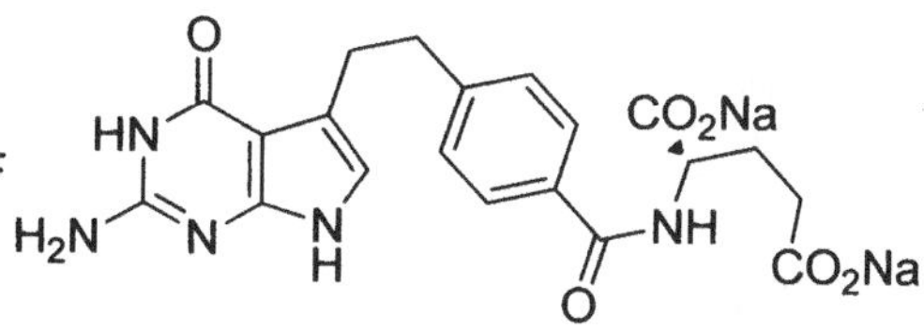

$7H_2O$ pemetrexed

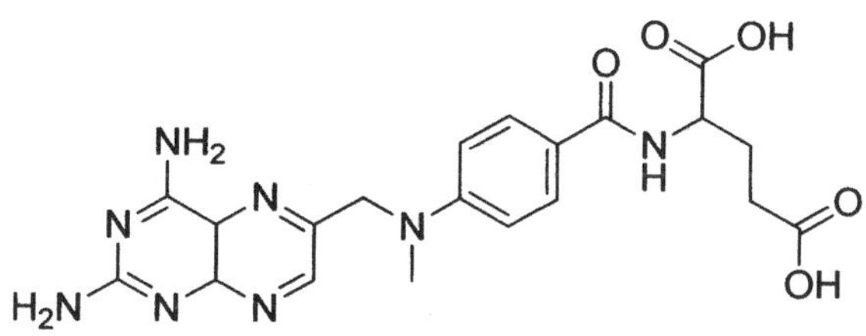

methotrexate

Given the extensive presence of the pyrimidine ring in numerous compounds of medicinal value, this chapter will primarily focus on the contemporary methods to construct pyrimidine rings and pyrimidine-fused systems. In addition, the syntheses of a few commercial drugs are discussed to showcase the preparations of pyrimidines in industrial settings.

13.2 Construction of the Pyrimidine Ring

13.2.1 Synthesis Involving Formation of Two Bonds

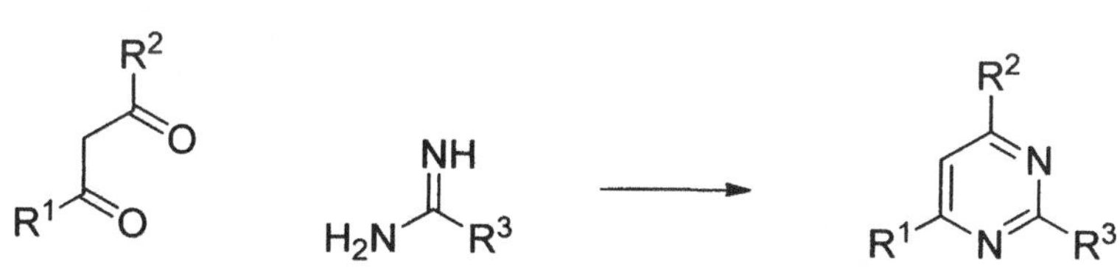

Condensation of an amidine with a 1,3-diketone is a popular method for the direct preparation of the six-membered-ring pyrimidine. Numerous methods exist for the synthesis of the prerequisite diketone and other dicarbonyl compounds, making this a very attractive strategy for the preparation of substituted pyrimidines.

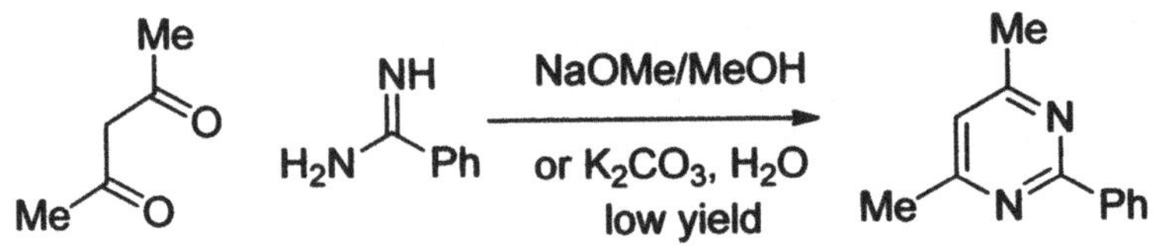

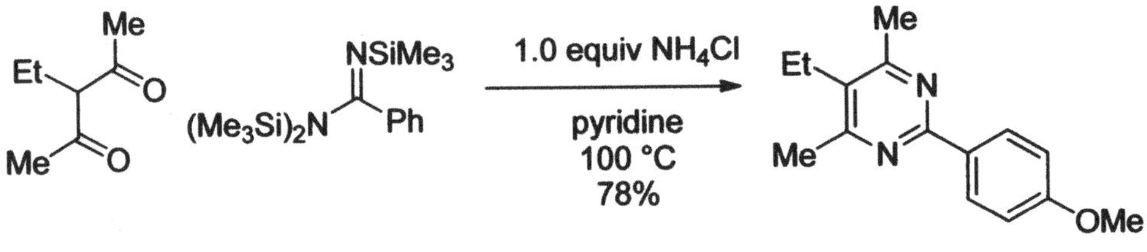

Indeed, the condensation of a three-carbon unit (typically a 1,3-dione or enone) with an N–C=N fragment (Pinner method) is widely used for the preparation of pyrimidines.[7] However, synthesis of highly substituted pyrimidines using the Pinner method generally requires long reaction times and provides the desired compounds in poor yields. The rate-limiting step in this double condensation reaction is postulated to be the dehydration of the various hydroxy-hydropyrimidine intermediates. On the other hand, *N,N,N*-tris-(trimethylsilyl)-amidines as an amidine equivalent would produce hexamethyldisiloxane rather than water. It was anticipated that the condensation reaction would be driven toward the forward direction by the silyl substitution because production of hexamethyldisiloxane is more thermodynamically favoured. NH_4Cl was identified as a general promoter for the synthesis of pyrimidines presumably by activating 1,3-dicarbonyl component.[8]

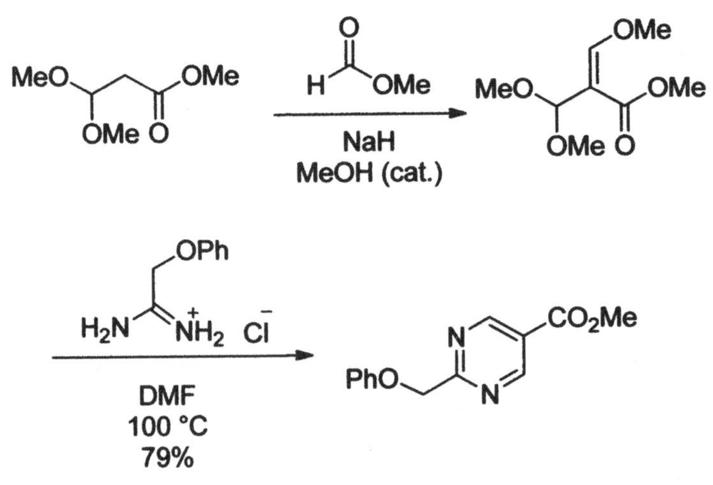

A method for the synthesis of 2-substituted pyrimidine-5-carboxylic esters is described based on the similar concept of condensation of amidine with 1,3-dicarbonyl compounds.[9] The sodium salt of 3,3-dimethoxy-2-methoxycarbonylpropen-1-ol has been found to react with a variety of amidinium salts to afford the corresponding 2-substituted pyrimidine-5-carboxylic esters.

A flow process was reported for the synthesis of β-keto esters via the $BF_3 \cdot OEt_2$-catalyzed formal C–H insertion of ethyl diazoacetate into aldehydes.[10] The β-keto esters were then condensed with a range of amidines to give a variety of 2,6-substituted pyrimidin-4-ols. The crude reaction mixture containing the β-keto esters was treacted with acetamidine hydrochloride, DBU, and EtOH to provide the corresponding pyrimidin-4-ol in excellent yields.

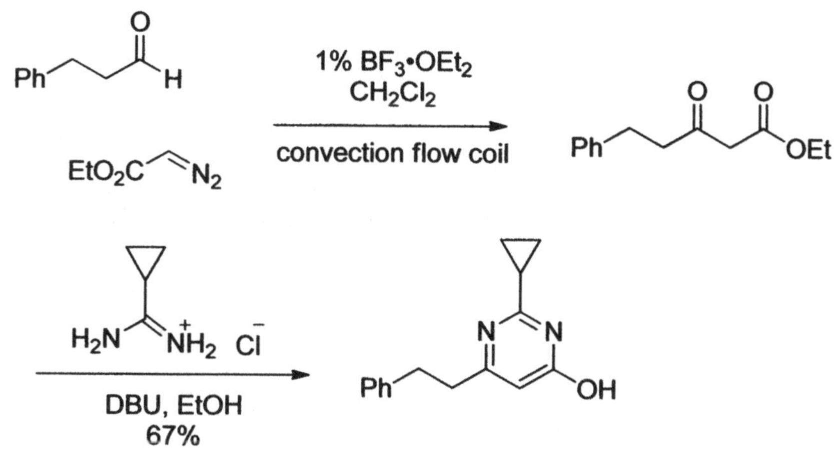

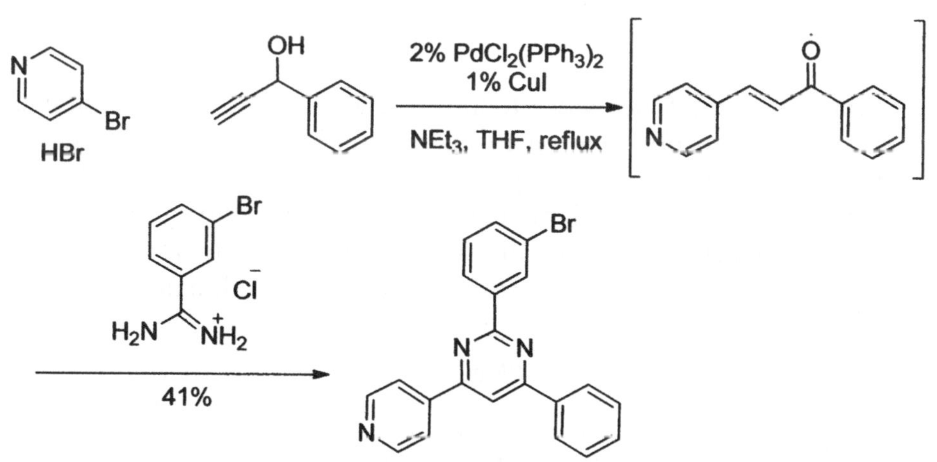

Amidine addition to α,β-unsaturated carbonyl derivatives is another popular method to construct pyrimidine rings. 2,4,6-Tri(hetero)aryl-substituted pyrimidines can be readily synthesized in a three-component, one-pot process based upon a coupling–isomerization sequence of an electron-poor (hetero)aryl halide and a terminal propargyl alcohol followed by a cyclocondensation of amidinium salts with chalcones (1,3-diaryl propenones).[11] Mechanistically, this isomerization occurring after the cross-coupling reaction is base catalyzed. The mild reaction conditions for the Sonogashira coupling reaction are compatible with many functional groups.

Polysubstituted pyrimidines were synthesized from *in situ* generated α,β-unsaturated imines and the corresponding amidine or guanidine derivatives in a convenient one-pot procedure.[12] It was proposed that the described transformations proceed *via* the initial formation of α,β-unsaturated imine that undergo nucleophilic attack by a bidentate nucleophile (amidine or guanidine). This step is then followed by elimination of ammonia and aromatization to yield the observed polysubstituted pyrimidine.

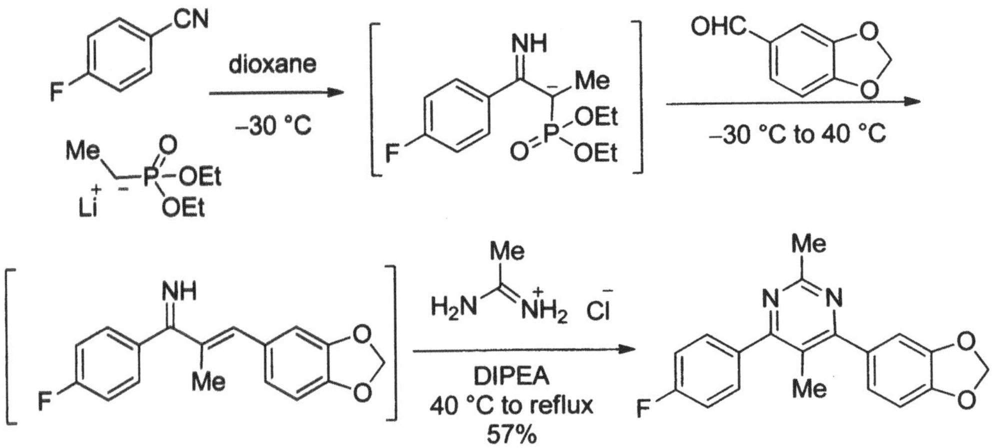

Diacetylenic ketones were found to react smoothly with amidines to yield a range of densely functionalized pyrimidines in high yields.[13] It is worth noting that the primidine product was obtained as a single regio-isomer, attributed to the acetylenic carbon bearing the ester group being the most electron deficient, making this the preferential site for nucleophilic attack of the amidines, for instance, methyl carbamimidothioate.

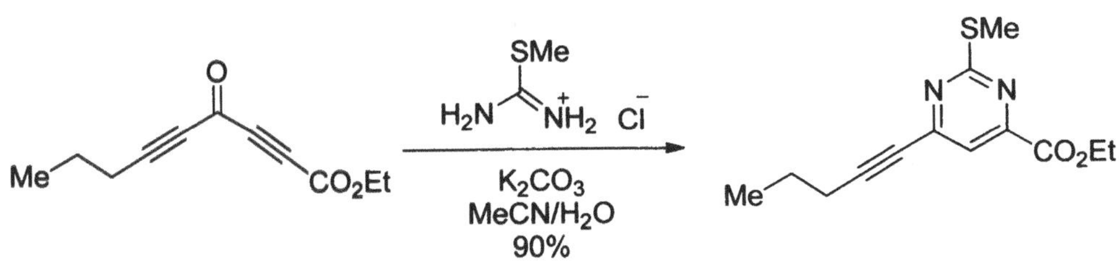

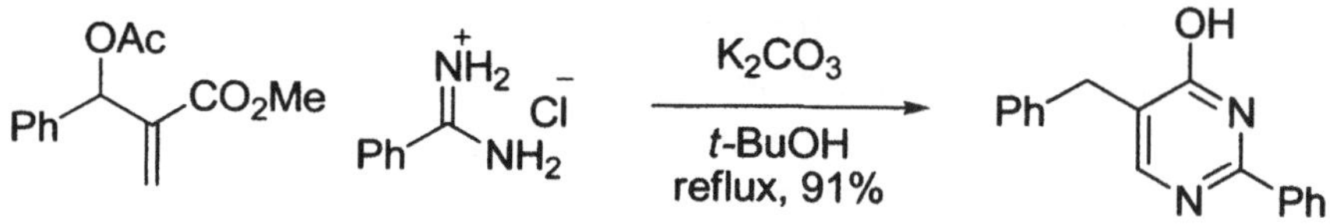

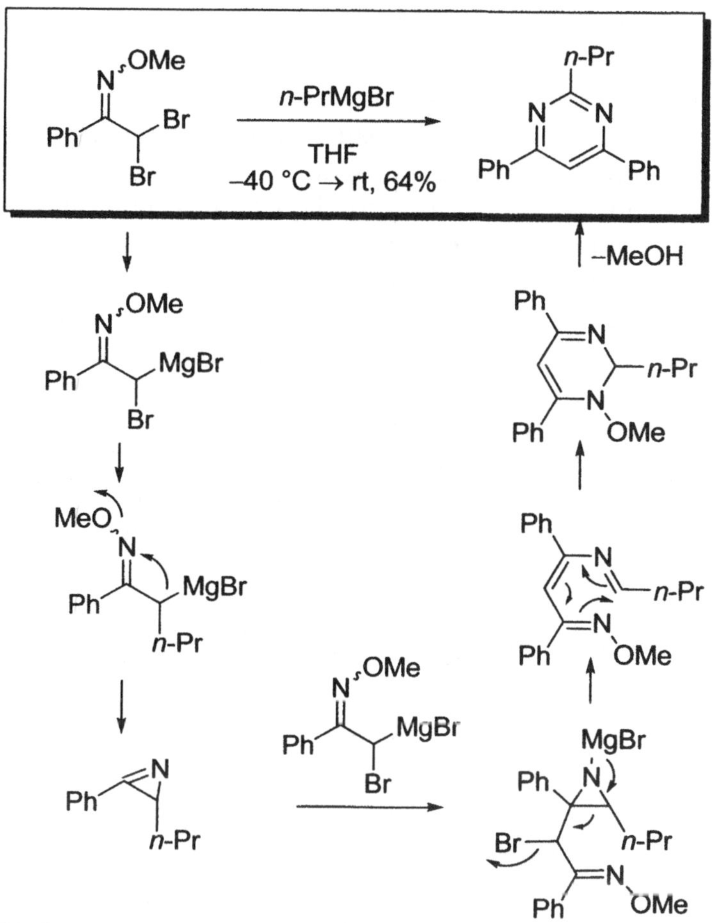

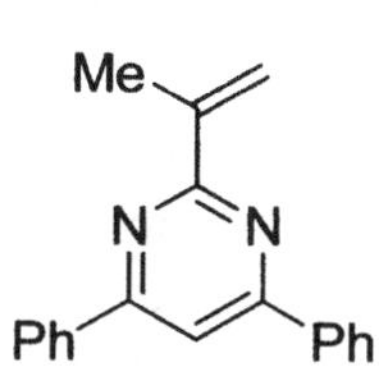

62%

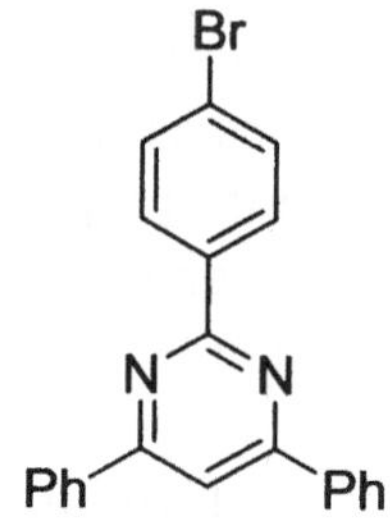

80%

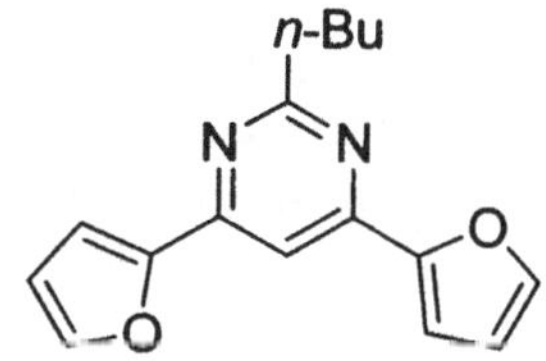

65%

A convenient method for the synthesis of substituted pyrimidines involves the reaction of amidine or guanidine derivatives with a 1,3-dielectrophilic, three-carbon unit. Synthesis of 2,4,5-trisubstituted pyrimidines from Baylis–Hillman adducts and amidines is a good example.[14] The ester group in the reactant can be replaced with acetyl and cyano groups, and the corresponding products are isolated in moderate to excellent yields.

The reaction of α,α-dibromo-oxime ethers with a variety of Grignard reagents efficiently provides 2,4,6-trisubstituted pyrimidines.[15] α,α-Dibromo-oxime ethers are easily prepared from the corresponding α,α-dibromo-ketones upon treatment with *O*-methyl hydroxylamine hydrochloride in methanol. In addition to alkyl groups, both aryl and vinyl groups can also be introduced in this manner. This protocol enables the synthesis of heteroaromatic-substituted pyrimidines. A plausible mechanism for this transformation has been proposed to involve an azirine intermediate. Bromine–magnesium exchange affords magnesium carbenoid, which is then alkylated at the α-position with the Grignard reagent. α-Magnesiated oxime ethers undergo Neber-type cyclizations to provide highly reactive azirines. The reaction of an azirine with bromine–magnesium exchange affords an aziridine intermediate, which yields a diimine intermediate via ring opening. An electrocyclization of the diimine provides a heterocyclic ring system which is converted to the pyrimidine upon elimination of methanol.

13.2.2 Synthesis Involving Formation of Three or More Bonds

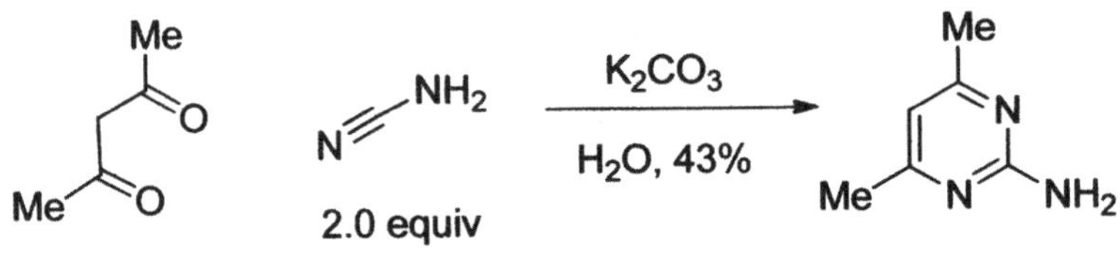

Another versatile approach to pyrimidine synthesis uses the N–C fragments. Nitriles are a common N–C source and have been used to form pyrimidines in many syntheses. Cyanamide is a particularly useful nitrile derivative in the synthesis of pyrimidines as illustrated.[16]

An improved and convenient preparation of pyrimidines and condensed pyrimidines from ketones and nitriles was reported.[17] In the presence of acetonitrile, the cation generated from methyl ethyl ketone with trifluoromethanesulfonic anhydride is trapped by the nucleophile, whereby a resonance-stabilized nitrilium species is formed. A second molecule of nitrile reacts with nitrilium intermediate, and after elimination of triflic acid, cyclization, and loss of a proton, the pyrimidine product was obtained in a useful yield. The formation of pyrimidines is accompanied by a minimum amount of vinyl triflates. This result demonstrates that the trapping of the cationic species by means of the nitrile occurs faster than the loss of a proton.

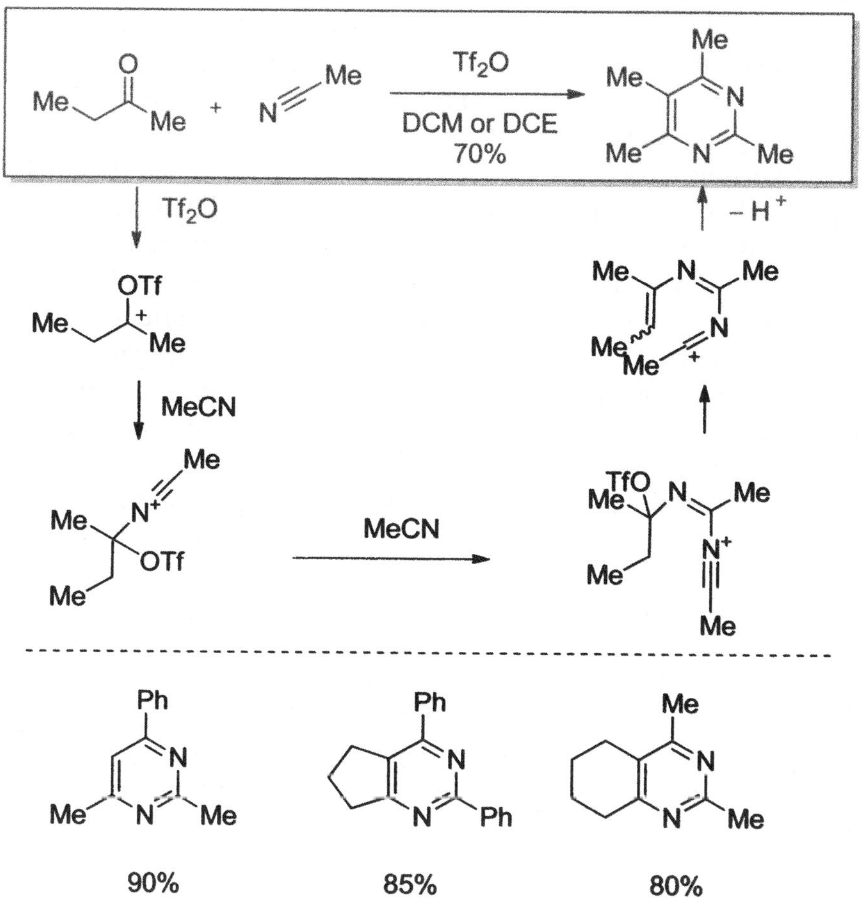

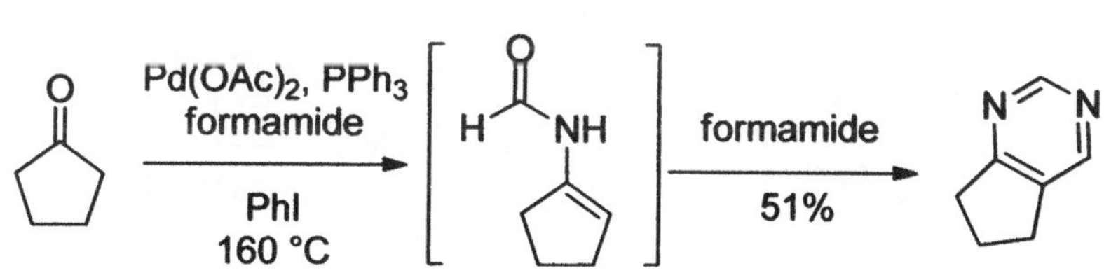

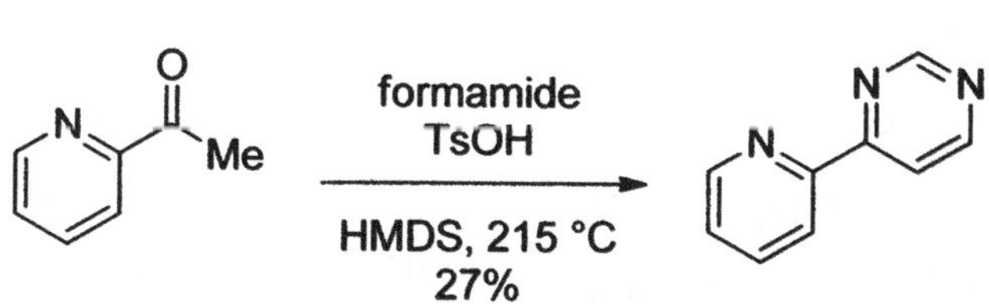

The Bredereck-type synthesis, which is well known as a conventional preparation of pyrimidine derivatives, generally requires a reaction temperature of more than 160 °C, and the product yield is moderate. A simple, high-yielding synthesis of pyrimidines from ketones in the presence of HMDS and formamide is reported.[18] Under microwave irradiation, heteroaromatic, aryl, aliphatic, and cyclic ketones cyclized to give pyrimidines in good yields.

A two-step procedure is described to convert the Biginelli 3,4-dihydropyrimidin-2(1*H*)-one to various multifunctionalized pyrimidines via the Kappe dehydrogenation[19] and a new mild PyBroP-mediated coupling with C, N, O, and S nucleophiles, which provides a readily accessible multifunctionalized pyrimidine template for diversity-oriented synthesis. Kappe et al. reported an unexpected but clean dehydrogenation of the Biginelli DHPMs by using 50–60% nitric acid. In the presence of a suitable activating reagent, 2-hydroxypyrimidine, the tautomerized pyrimidin-2(1*H*)-one, could form a highly reactive intermediate, which could be easily attacked by a nucleophile to furnish the 2-substituted pyrimidine.[20]

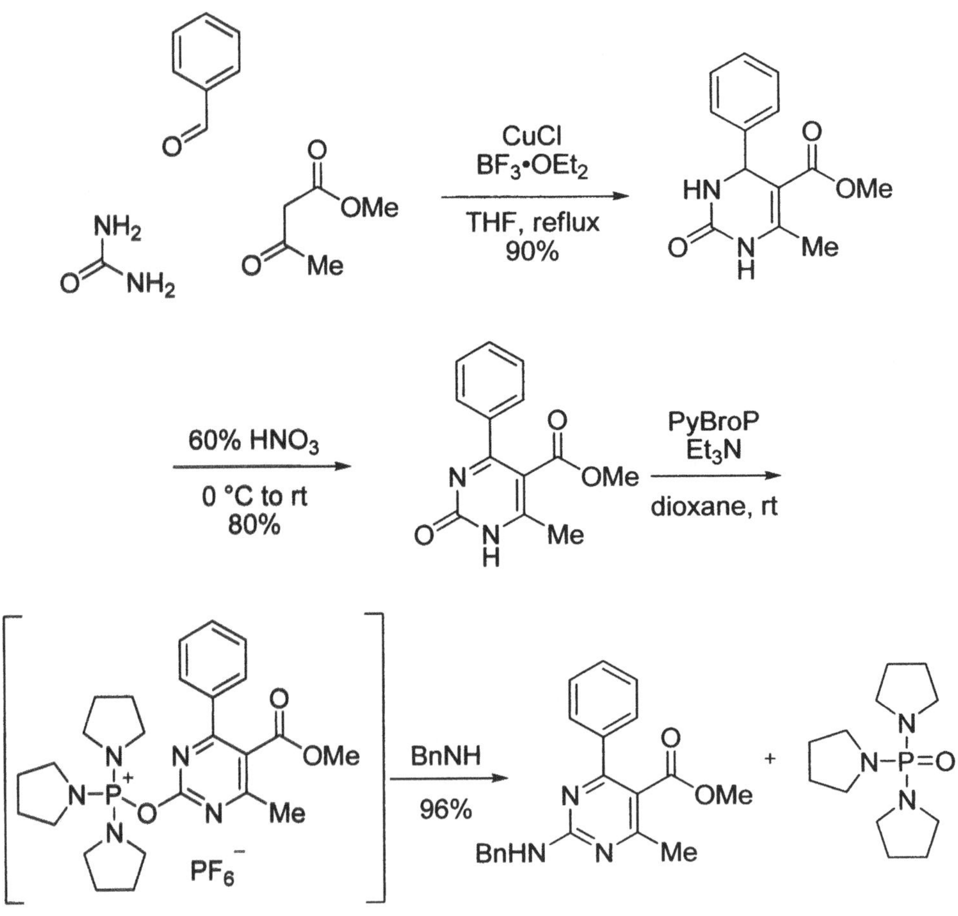

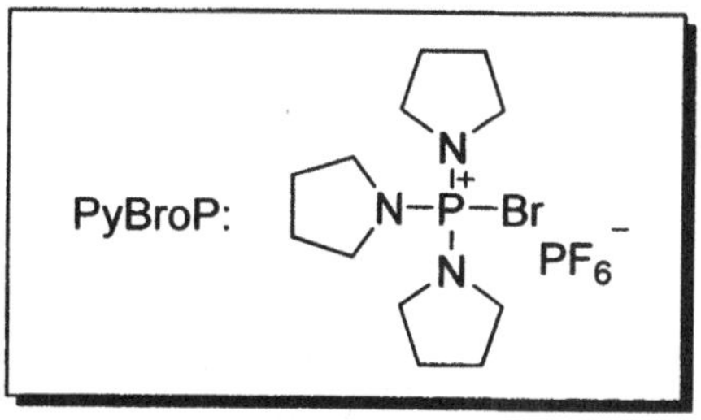

Phosphonium-based reagents PyBOP and PyBroP were found to be very effective in this transformation, affording oxytripyrrolidino-phosphonium intermediates in high yields, with PyBroP being slightly superior to PyBOP. Weak nucleophiles such as *N*-methyl benzenesulfonamide, imidazole, indole, diethyl malonate, and phenol were coupled smoothly with oxytripyrrolidino-phosphonium in high yields. Sodium *tert*-butoxide was found to be an excellent base to ensure a clean and complete reaction.

A $ZnCl_2$-catalyzed, three-component coupling reaction involving a variety of functionalized enamines, triethyl orthoformate, and ammonium acetate, leads to the production of 4,5-disubstituted pyrimidine derivatives in a single step. This type of [3+1+1+1] annulation process has not previously been reported.[21]

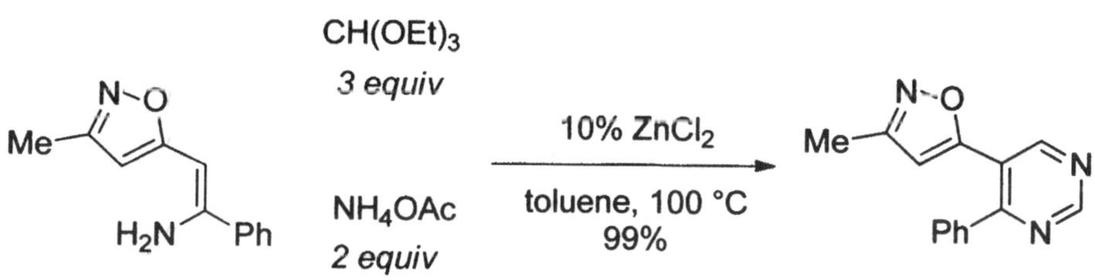

13.2.3 Synthesis of Pyrimidine-Fused Systems

Movassaghi and co-workers reported a mild, convergent, and single-step procedure for the conversion of readily available *N*-vinyl and *N*-aryl amides to the corresponding substituted pyrimidines.[22] The unique reactivity associated with electrophilic activation of amides was achieved using 2-chloropyridine (2-ClPyr) in combination with trifluoromethanesulfonic anhydride (Tf_2O). Highly activated amide derivatives were trapped with weakly nucleophilic nitriles to provide the corresponding pyrimidine derivatives directly. In the process to identify the optimum reagent combination, in contrast to pyridine, 2-ClPyr was found not to add to Tf_2O. Reversible addition of nitrile and expulsion of 2-ClPyr and TfOH provided the nitrilium ion, which cyclized to pyrimidine derivatives.

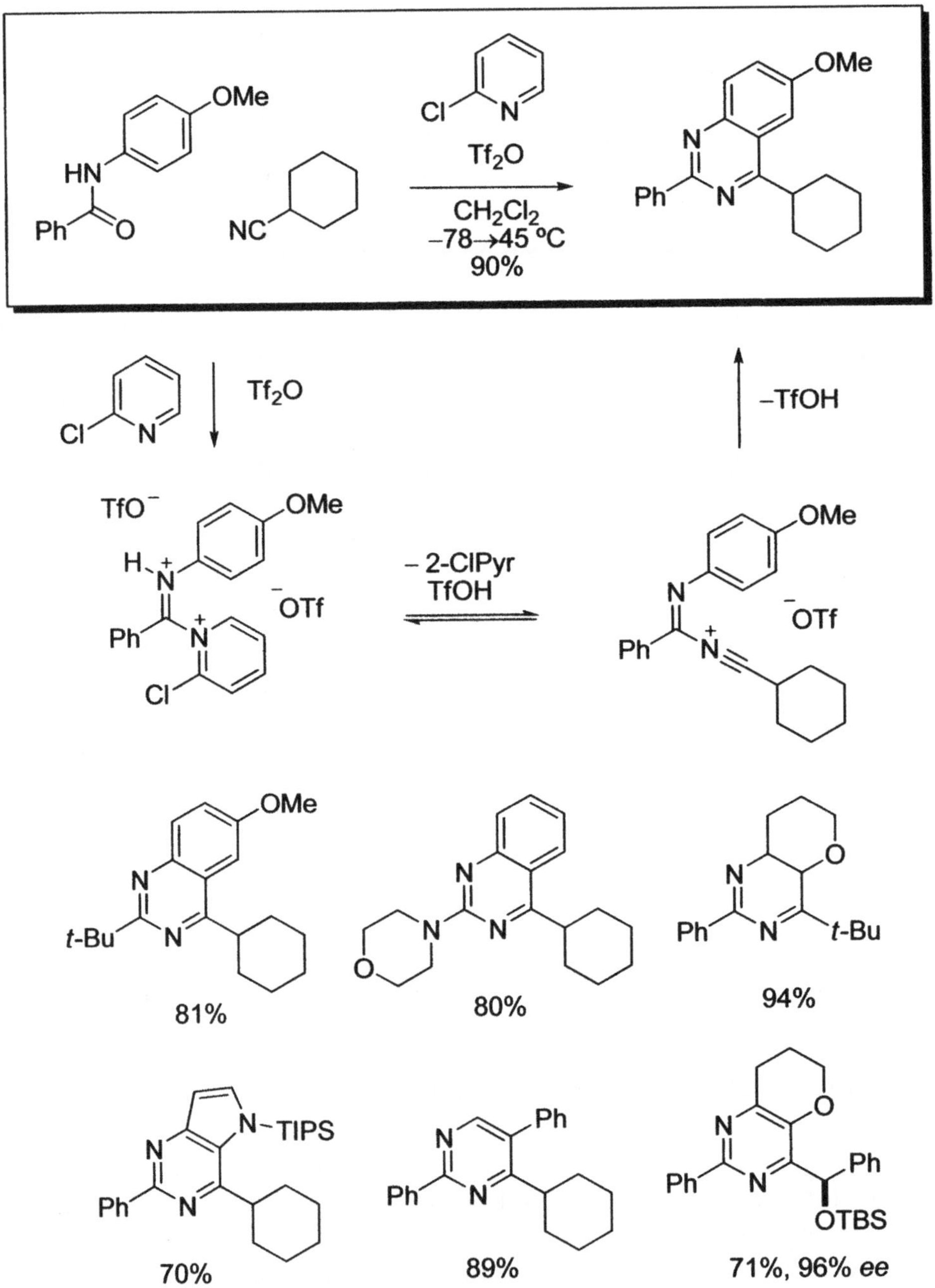

Researchers found electron-rich *N*-vinyl and *N*-aryl amides proceeded to afford the corresponding pyrimidine derivatives at the ambient temperature; however, less reactive substrates required heating for reactions to proceed. The use of epimerizable substrates provided the corresponding pyrimidine derivatives without any loss in optical activity.

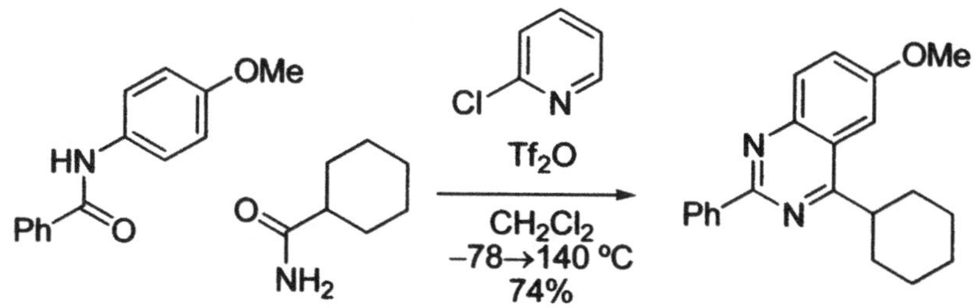

Because the dehydration of primary amides to the corresponding nitriles using Tf_2O and triethylamine has been reported, the primary amides can be used as nitrile surrogates and adds to the utility of this chemistry.[23] The direct condensation of cyanic acid derivatives with *N*-vinyl/aryl amides affords the corresponding C4-heteroatom substituted pyrimidines. The use of cyanic bromide and thiocyanatomethane in this chemistry provides versatile azaheterocycles poised for further derivatization. The synthesis of a variety of previously inaccessible C2- and C4-pyrimidine derivatives using this methodology is described.[24]

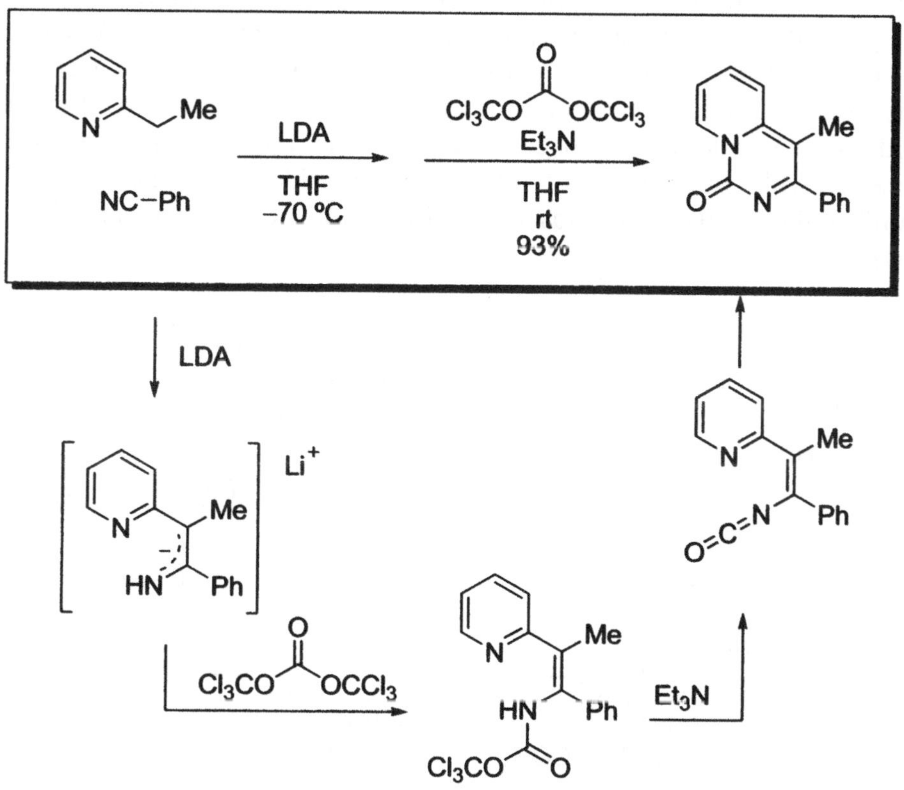

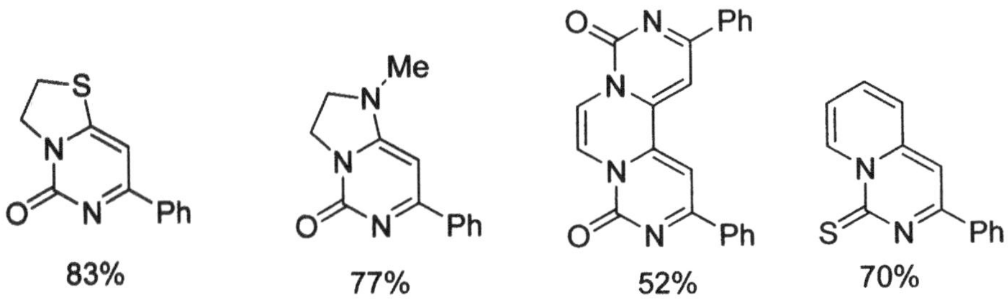

A facile and efficient [3 + 2 + 1]-annulation toward the synthesis of 3,4-fused pyrimidin-2-one and pyrimidin-2-thione derivatives was reported starting from an α-acidic imine compound, a nitrile, and triphosgene or carbon disulfide.[25] This methodology featured a range of 3,4-fused heterobicyclic pyrimidine derivatives as products using commercially available reagents 1-aza-allylic anions. Additionally, heterocycles containing more than one heteroatom, such as 2-methylthiazoline and 1,2-dimethylimidazole, afforded the desired five-membered-ring fused pyrimidines in good yields. The synthesis of pyrimidin-2-thione derivatives can be achieved by using carbon disulfide (CS_2) as a C-1 unit. 1-Azaallylic anion generated from α-acidic imine derivative in the presence of LDA reacts with and a nitrile produce the corresponding adduct, which is trapped by triphosgene to produce carbamate intermediate with concurrent elimination of lithium chloride and phosgene. Upon base addition, isocyanate was generated with liberations of hydrogen chloride and phosgene. Final intramolecular cyclization leads to the formation of the corresponding 3,4-fused pyrimidin-2-one derivative.

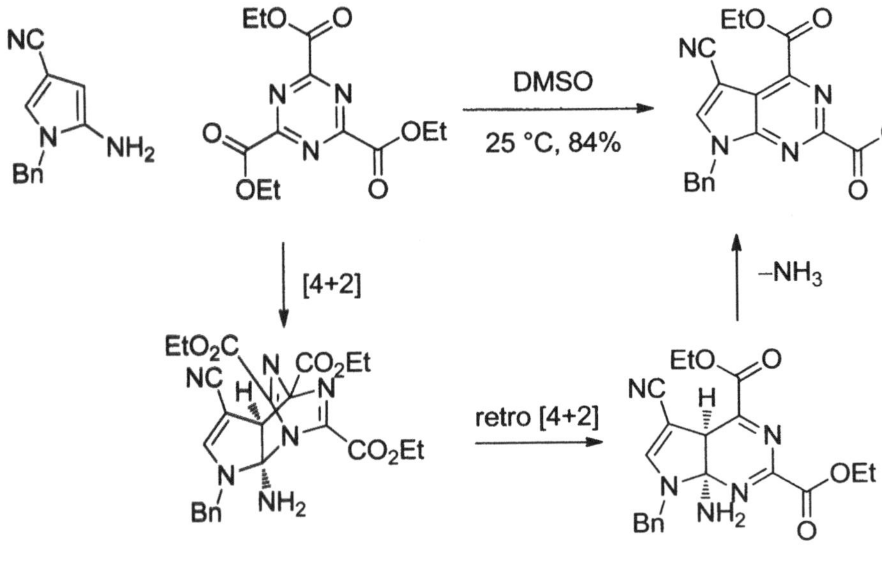

Tubercidin, toyocamycin, and sangivamycin are naturally occurring anti-bacterials, and their synthetic analogues have shown intriguing biological activities. Pyrrolo[2,3-*d*]pyrimidines are the central nucleus of these natural products. Thus, the inverse electron demand Diels–Alder reaction of 2-amino-4-cyanopyrroles with 1,3,5-triazines was reported.[26] This methodology is suitable for one-pot syntheses of highly substituted and highly functionalized pyrrolo[2,3-*d*]pyrimidines. It was suggested that the initial [4 + 2] cycloaddition reaction is a stepwise reaction, and the subsequent reactions may proceed via a retro Diels–Alder (RDA) reaction of the [4 + 2] cycloadduct with the loss of XCN followed by elimination of ammonia or ammonium chloride to produce pyrrolo[2,3-*d*]pyrimidine product in a regioselective manner.

Derivatives of pyrrolo[2,3-*d*]pyrimidines have also been extensively investigated as inhibitors of epidermal growth factor receptor (EGF-R) protein tyrosine kinase, and their potentials in the treatment for proliferative diseases involving mitogenic signaling from the EGF-R. Large amounts in the 100-kg range of this heterocyclic compound was needed to support development of pyrrolo[2,3-*d*]pyrimidine-derived anti-tumor reagent. Thus, a short and efficient synthesis of 4-(3-chlorophenylamino)-5,6-dimethyl-7*H*-pyrrolo[2,3-*d*]pyrimidine starting from cheap alanine and malononitrile was developed.[27]

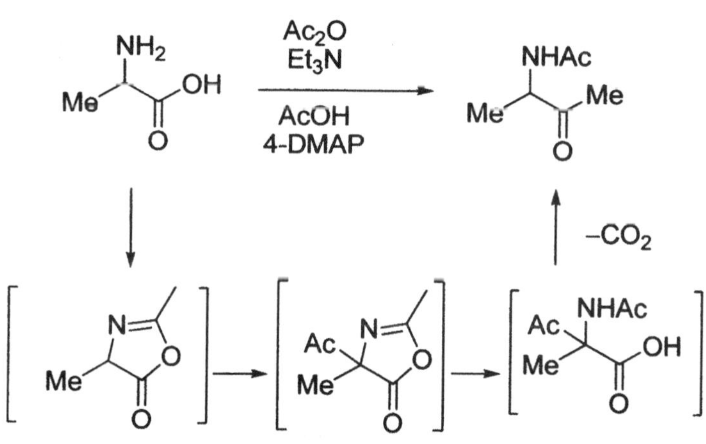

A Dakin–West reaction on plant scale is demonstrated by elaboration of a modified procedure avoiding uncontrolled release of carbon dioxide. It seems to be generally accepted that the mechanism involves dehydrative formation of an azlactone (oxazolinone), which is then *C*-acylated (in equilibria with *O*-acylation), undergoes ring-opening hydrolysis, and then decarboxylates to form the acylamino ketone. Pyrimidine ring formation is achieved in a simple one-pot reaction, which is followed by a simple isomerization. Laboratory experiments had made clear that the mechanism of this pyrimidine ring formation involves a series of ortho ester–iminoester–amidine equilibriums between aniline, pyrrole-amine, and ortho ester. It was reasoned that precipitation of the product from the reaction mixture drove these equilibriums to the product side. Chlorophenyl–pyrrolopyrimidine was converted to its isomer, which is referred to as Dimroth rearrangement, and the reaction temperature was the most important parameter. The drug substance free base was brought in its mesylate salt form by crystallization from ethanol/water in the presence of methylsulfonic acid and final recrystallization from acetone/water, which yielded a white powder with high purity.

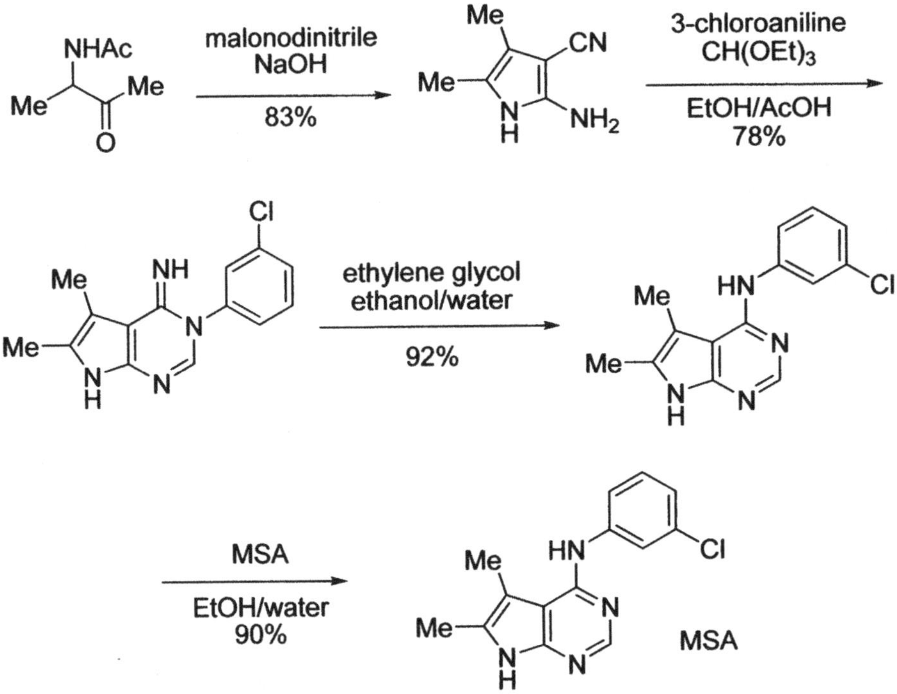

Vega developed an efficient synthesis of pyrazolo[3,4-*d*]pyrimidines based on the electrophilic properties of nitrilium salts formed by reaction of 5-aroylaminopyrazoles, nitriles and Lewis acids.[28] Treating iminolyl halides with nitriles with $SnCl_4$ gave rise of the nonisolated nitrilium salts, which underwent cyclization to generate the desired pyrazolo[3,4-*d*]pyrimidines in moderate to good yields. A similar approach was extended to the preparation of 4-amino-2-phenylquinazolines via cyclization of *N*-arylbenzamides with cyanamide catalyzed by $TiCl_4$.[29]

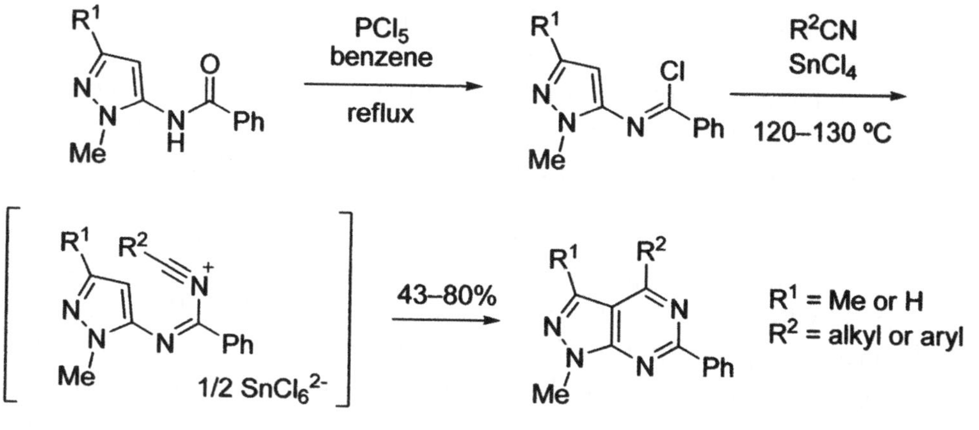

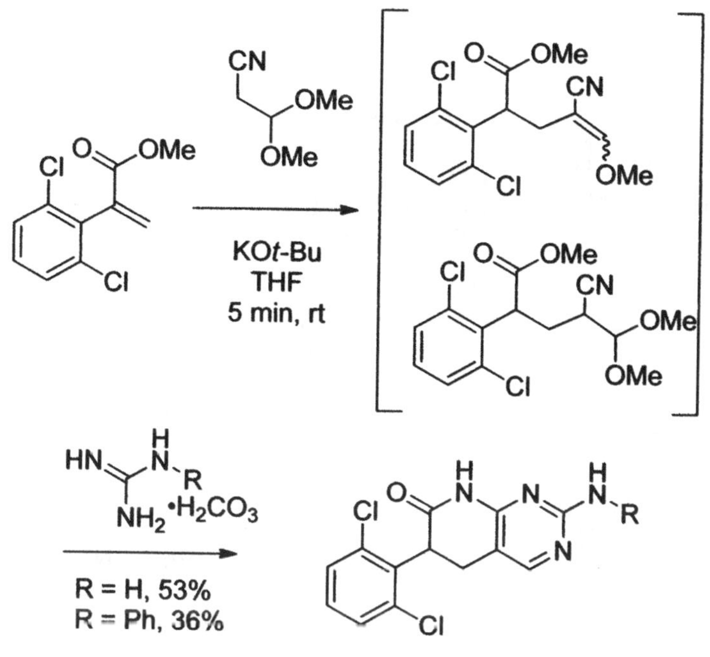

Pyrido[2,3-*d*]pyrimidines represent a heterocyclic ring system with several biological activities including inhibiting the protein kinase catalytic activity by blocking the ATP-binding site and preventing the phosphorylation of the corresponding natural substrates.[30] Particularly, 6-aryl-substituted 5,6-dihydropyrido[2,3-*d*]-pyrimidine-7(8*H*)-ones are selective inhibitors of the kinase insert domain-containing receptor and fibroblast growth factor receptor.[31] The synthesis of these compounds is usually achieved by multistep procedure in which the pyridine ring is constructed onto a preformed pyrimidine ring.[31] Thus, a convenient route to 4-unsubstituted 5,6-dihydropyrido[2,3-*d*]pyrimidines is reported.[32] An Michael addition of 3,3-dimethyoxypropanenitrile to 2-aryl-substituted acrylates leads to a mixture of Michael adduct and its derived enol ether. These intermediates can be converted to 4-substituted pyrido[2,3-*d*]-pyrimidines upon treatment with a guanidine system with microwave irradiations.

A convenient synthesis of highly functionalized dihydropyrido[2,3-*d*]pyrimidines via a double [5 + 1]-annulations strategy was developed.[33] The double annulation route starts from easily available α-alkenoyl-α-carbamoyl ketene-(*S,S*)-acetals. 2-Amino-3-carbamoyl-5,6-dihydro-4-pyridones was firstly created in excellent yield by a formal [5C + 1N] annulation reaction of ketene-(*S,S*)-acetals with ammonia. In the second step, 7,8-dihydropyrido[2,3-d]pyrimidin-4(3*H*)-ones (when R = aryl) and 7,8-dihydropyrido[2,3-d]pyrimidines (when R = H), were prepared in good yields by reacting 2-amino-3-carbamoyl-5,6-dihydro-4-pyridones with excessive Vilsmeier reagent (DMF/$POCl_3$) via a second [5 + 1] annulation step.

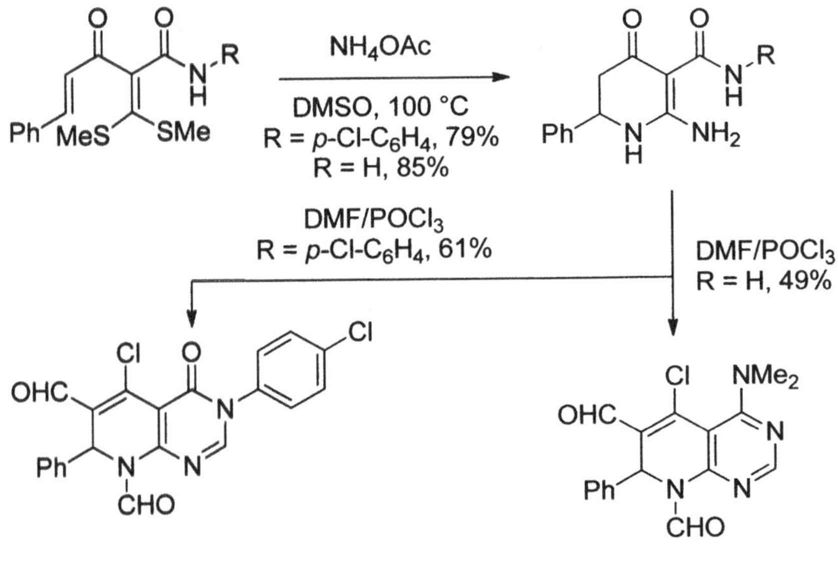

One of the reasons why many reactions were never attempted in water was because of the insolubility of many organic compounds in this media. But at a higher temperature, some reactions become more accessible. In addition to a high temperature, if higher pressures are also applied, then the solubility of organic compounds in water could be even improved in a greater extent. Therefore, the use of the microwave synthesizer, which allows for both high temperatures and higher pressures, may be very beneficial for running reactions when poor solubility is a limiting factor. A preliminary investigation indicated that the reaction of 1,3-diketones with benzamidine hydrochloride afforded pyrimidines in a good yield.[34]

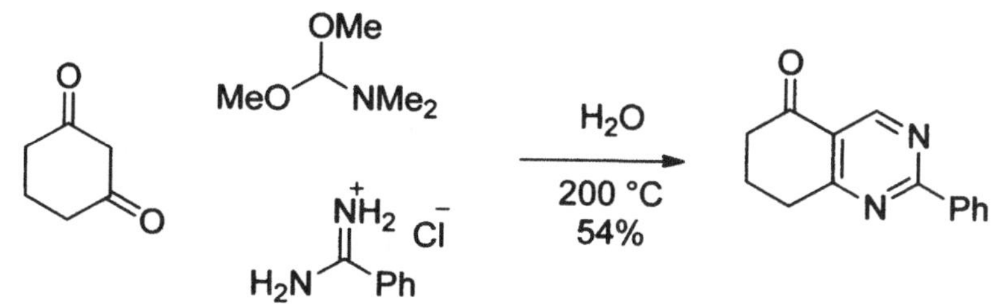

The Baran group has reported an interesting one-step synthesis of 4-monosubstituted and 4,5-disubstituted pyrimidines.[35] For example, 4,5-disubstituted pyrimidines are synthesized from the corresponding ketone in one step using inexpensive reagents, formamidine acetate, and *n*-propanol. Contrasted to other methods, this process appears quite amenable to large-scale use in industrial settings.

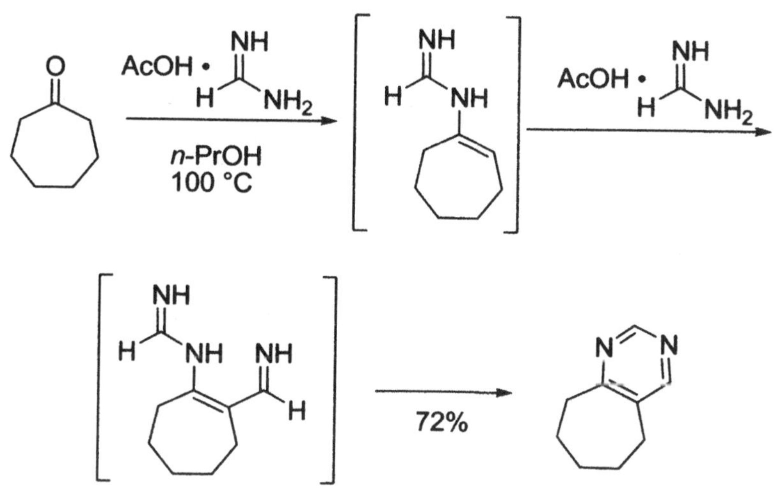

Pyrimidine derivatives are synthesized from *N*-substituted lactams and Viehes salt with a short reaction sequence, good yields of the targeted heterocyclic compounds, as well as their convenient isolations and purifications.[36] Addition of dry DMF to the reaction mixture proved to be highly beneficial in increasing the yields of the targeted heterocycles. This may be attributed to the improved solubility of amidines in the toluene/DMF (2:1) solvent mixture. Pyrimidines reacted with *N*-methylbenzylamine in dry DMF at 140 °C in a sealed-tube to furnish products of formal nucleophilic aromatic substitution of the N-Me_2 group.

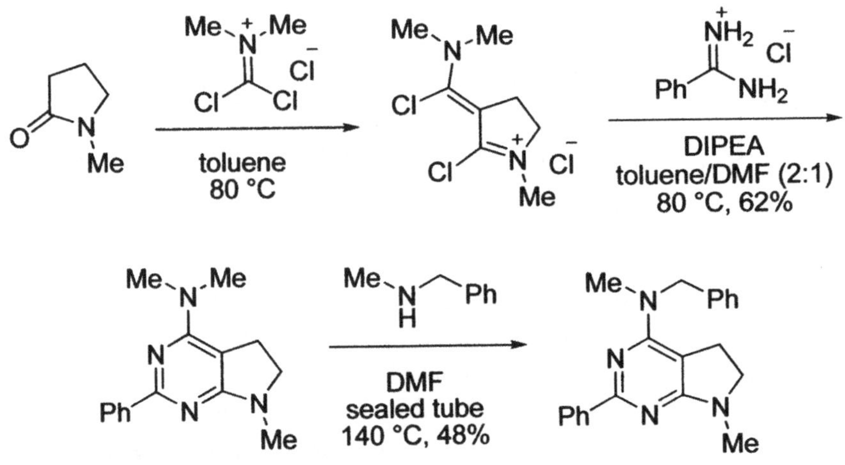

The condensation of *N*-triphenylphosphoraniliden-benzamidine with acyclic α,β-unsaturated aldehydes produces dihydropyrimidines in good to high yields.[37] The reaction mechanism probably involves an aza-Wittig reaction followed by a 6e-π-electrocyclic ring closure of the azatriene intermediate to give dihydropyrimidines, which is oxidized to the corresponding pyrimidines.

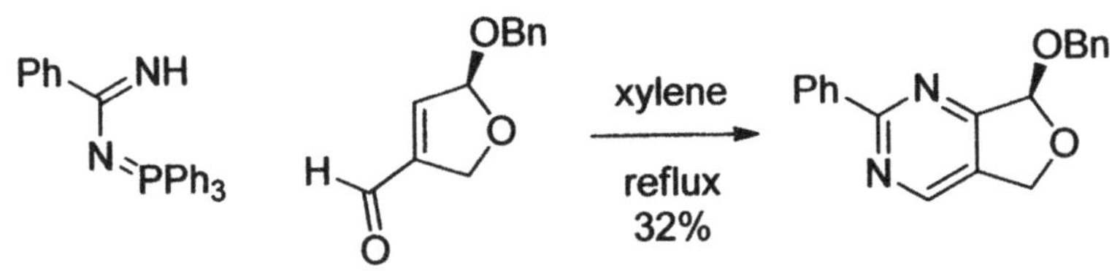

13.3 *Synthesis of Pyrimidine-Containing Drugs*

13.3.1 Allopurinol

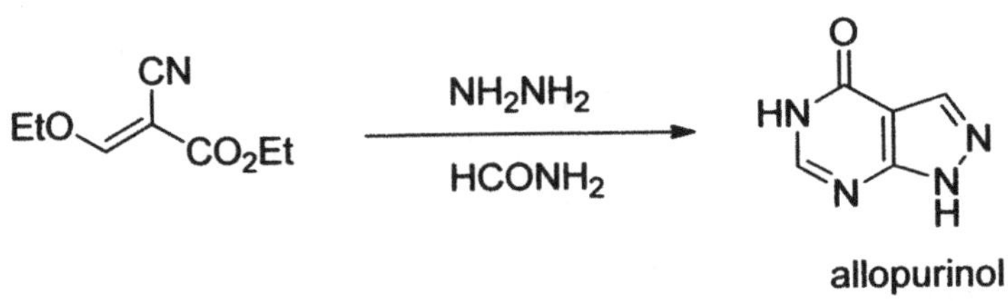

Allopurinol is a drug used primarily to treat hyperuricemia, which means excess uric acid in blood plasma and its complications, including chronic gout. It should be noted that allopurinol is not a uricosuric which means that it does not increase the excretion of uric acid in the urine so it can be used in patients with poor kidney function. Allopurinol has been marketed in the United States since 1966 under the trade name of Zyloprim. After it becomes a generic drug, a variety of brand names including Allohexal and Progout are used. Its synthesis is very simple. Condensation of hydrazine with ethyl 2-cyano-3-ethoxyacrylate in the presence of formamide leads to allopurinol.[38] Hydrzine obviously gives rise to the pair of adjacent nitrogen atoms in the pyrazole ring and ester group is the origin of the carbonyl group.

13.3.2 Trimethoprim

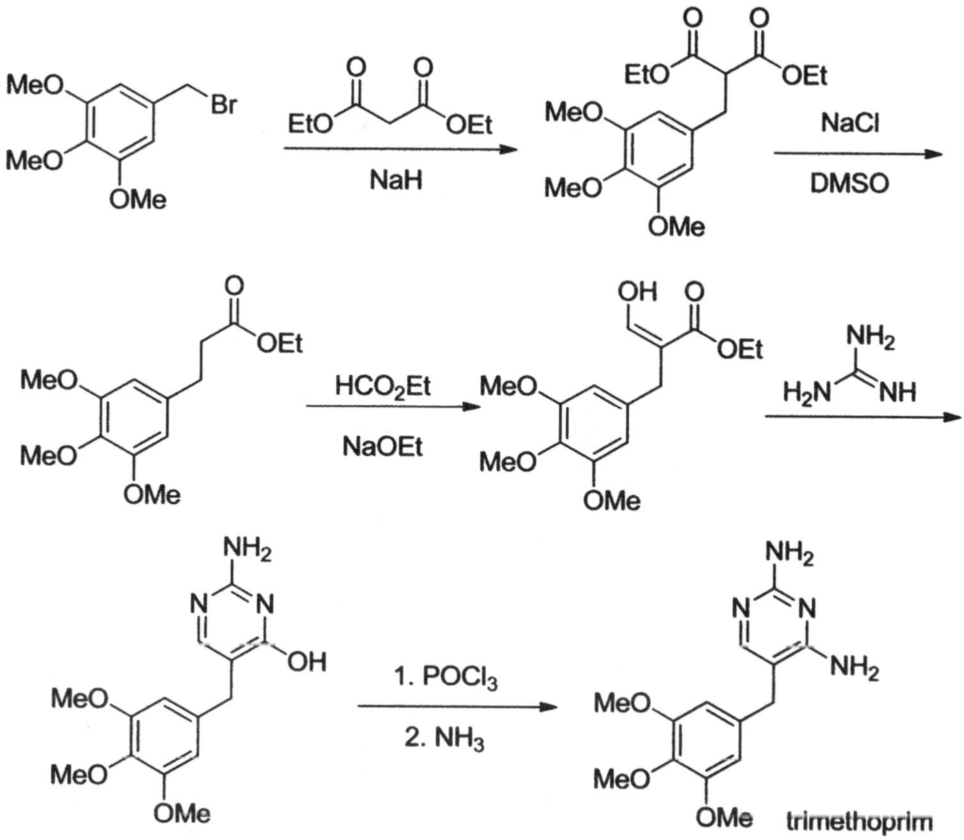

Trimethoprim is a bacteriostatic antibiotic known as dihydrofolate reductase inhibitors, mainly used in the prevention and treatment of urinary tract infections. It was formerly marketed by GlaxoSmithKline under trade names including Proloprim, Monotrim, and Triprim. One recent trimethoprim's synthesis starts from the alkylation of malonate.[39] It was proved to be higher yielding to add the aldehyde function group after decarboxylation rather than trying to reduce an ester to an aldehyde. Decarboxylation is carried out using the Krapcho conditions. Concentration with ethyl formate and cyclization with guanidine gives the pyrimidine ring system. Aromatic nucleophilic substitution in the pyrimidone then gives trimethoprim.

13.3.3 Imatinib

Imatinib mesylate (Gleevec, *N*-(4-methyl-3-(4-(pyridin-3-yl)-pyrimidin-2-ylamino) phenyl)-4-((4-methyl piperazin-1-yl)methyl) benzamide methanesulfonate, STI571) is known as an inhibitor of tyrosine kinases and is used for the treatment of chronic myeloid leukemia and gastrointestinal stromal tumors.[40] It was developed by Novartis Pharma AG and is a 2-phenylamino-pyrimidine derivative.

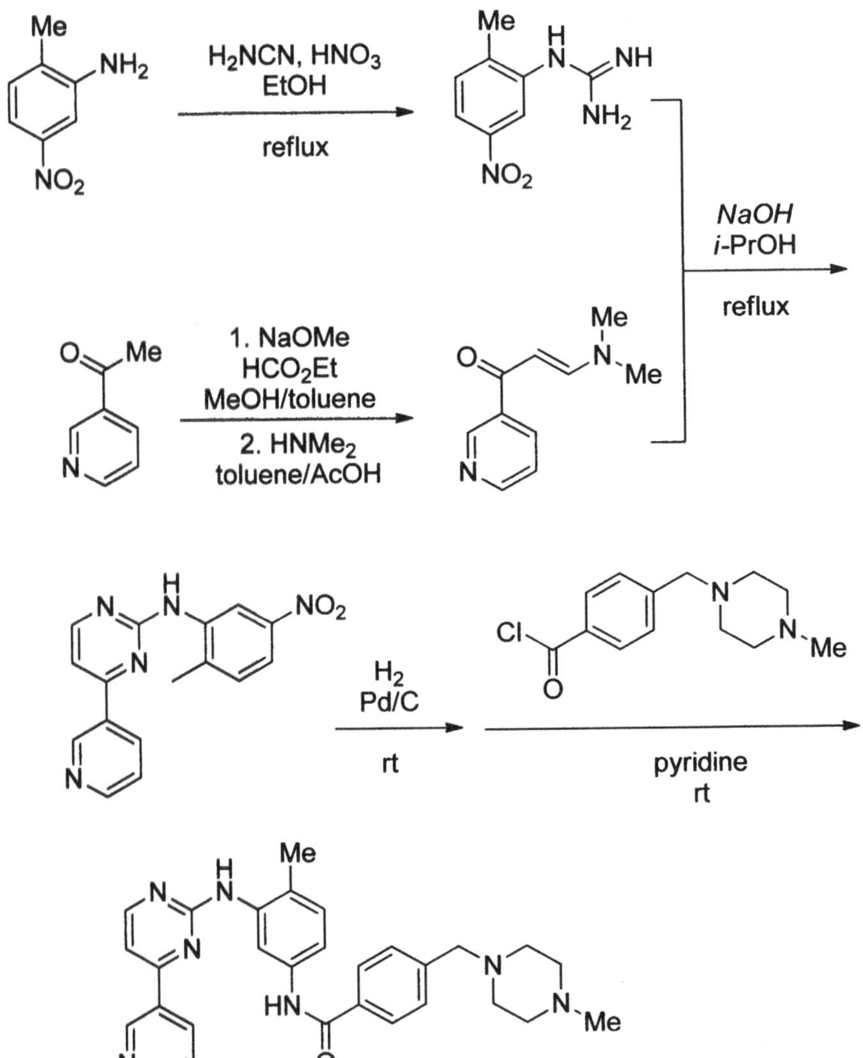

imatinib

Imatinib's first synthesis was completed by Zimmermann in 1993.[41] However, the yields of all steps were not shown in original paper and patent document. Cyanamide was used for the synthesis of phenylguanidine fragments. The synthesis of enaminone was formed by Na-enolate addition to ethyl formate followed by condensation with dimethyl amine. A pyrimidine ring was constructed by NC=N fragment addition to the enaminone part. Reduction of nitro group and acetylation led to the final product, imatinib.

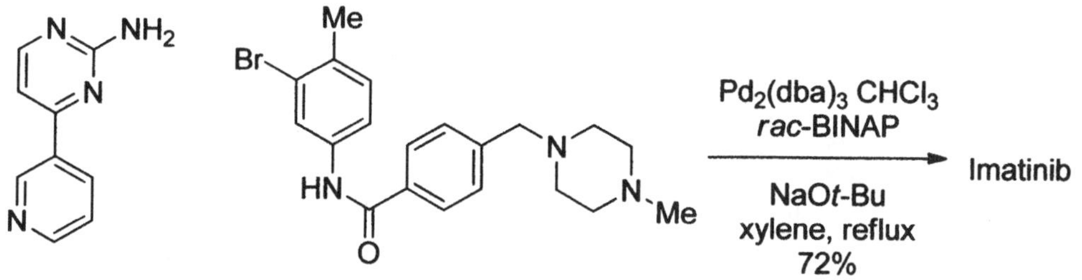

Loiseleur and co-workers developed a process for the preparation of imatinib base, in which the final step involved a $Pd_2(dba)_3CHCl_3$-catalyzed C–N coupling reaction with the of use of *rac*-BINAP as ligand.[42] Others provided an improved process based on Loiseleur's method.[43] However, the concerns of these approaches are the use of a toxic, hazardous reagent cyanamide and using relatively high-cost palladium as the catalyst for C–N coupling reaction.

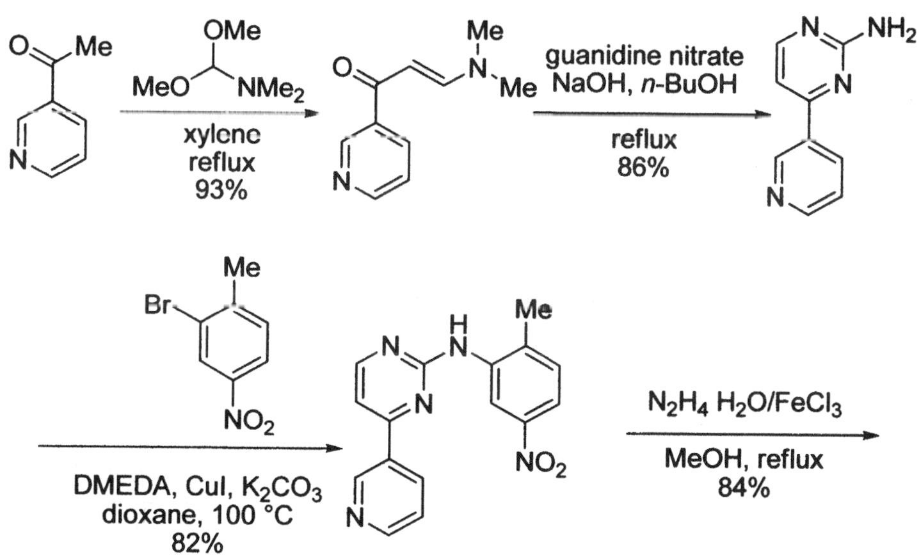

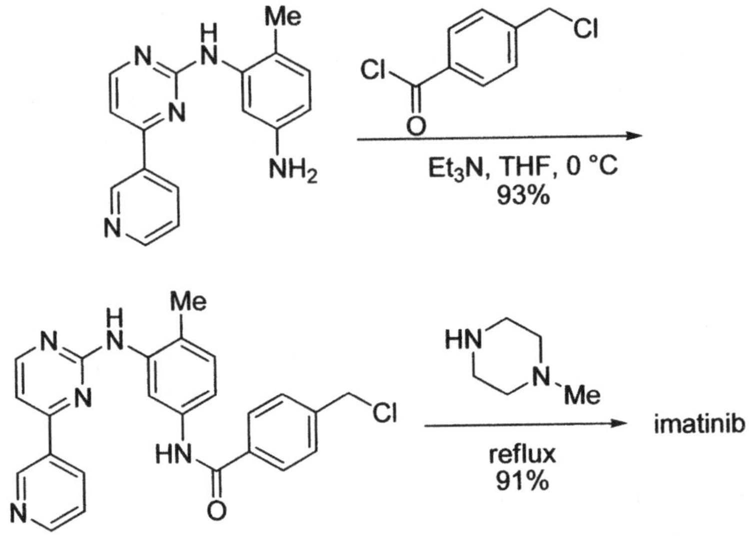

An alternative imatinib synthesis was presented with an aim to improve industrial production of imatinib.[44] 2-Acetylpyridine was alkylated with the acetal of *N,N*-dimethylformamide (DMFDMA) to enamine. A pyrimidine ring was formed with base and reagent guanidine nitrate and nitrotoluene fragment was added in an Ullmann-type reaction with CuI generating secondary amine. The nitro group was reduced by hydrazine/$FeCl_3$ to the amine, which was then converted to amide with acid chloride. The final step is the addition of piperazine to form imatinib. This route has eradicated the use of toxic cyanamide, sodium metal, and relatively expensive palladium, but has also introduced toxic hydrazine and the harmful and explosive guanidine nitrate. The final step was only demonstrated on a 0.5 gram-scale.

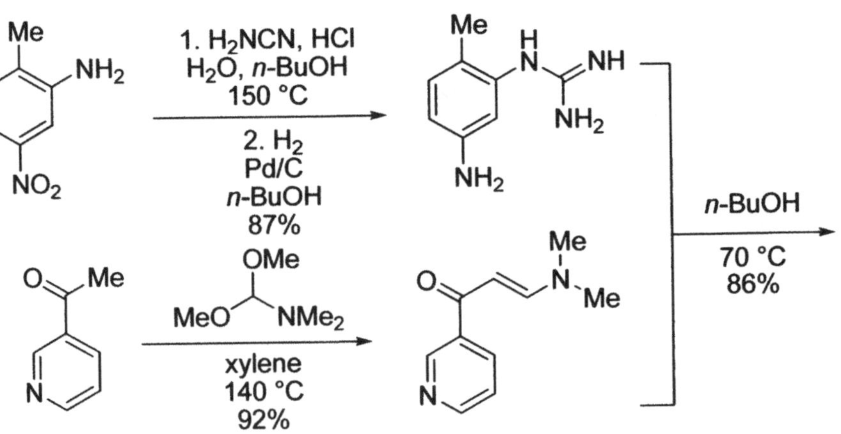

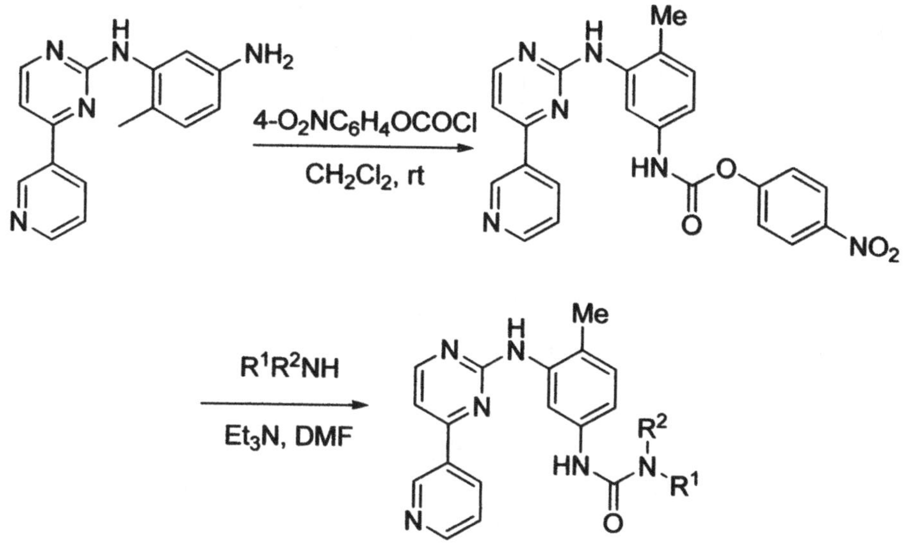

In the investigation to evaluate hypothesis that the potency of the molecule might be enhanced by replacing *N*-methylpiperazine group with an alternative binding group,[45] such as a urea moiety. The reasoning is that this group would maintain the H-bond interactions to human protein kinases. Compounds were prepared by the route in which the phenylamino-pyrimidine core structure was prepared using a variation of the principal synthesis, involving reaction of the enamine derivative of the β-aldehydoketone with guanidine. The guanidine was readily available from the commercially available aniline derivative, via reaction with cyanamid, followed by catalytic hydrogenation. Aminoformylation of 3-acetylpyridine with DMF–dimethylacetal furnished the enone intermediate. For the parallel synthesis of the urea library, the 4-nitrophenylcarbamate derivative was selected on the basis of its reactivity, together with the stability and crystallinity of the hydrochloride salt. The addition of a variety of amines to a solution of in DMF containing triethylamine provided the desired urea derivatives.

13.3.4 Bosentan

The endothelins are a family of structurally related 21-amino acid peptides that are the most potent vasoconstrictors identified so far in vascular preparations in both animals and humans. An endothelin receptor antagonist, bosentan, is a new drug for treating cardiovascular pathology, especially congestive heart failure.[46] Two generations of processes were established aiming at reducing the bulk drug manufacturing cost.

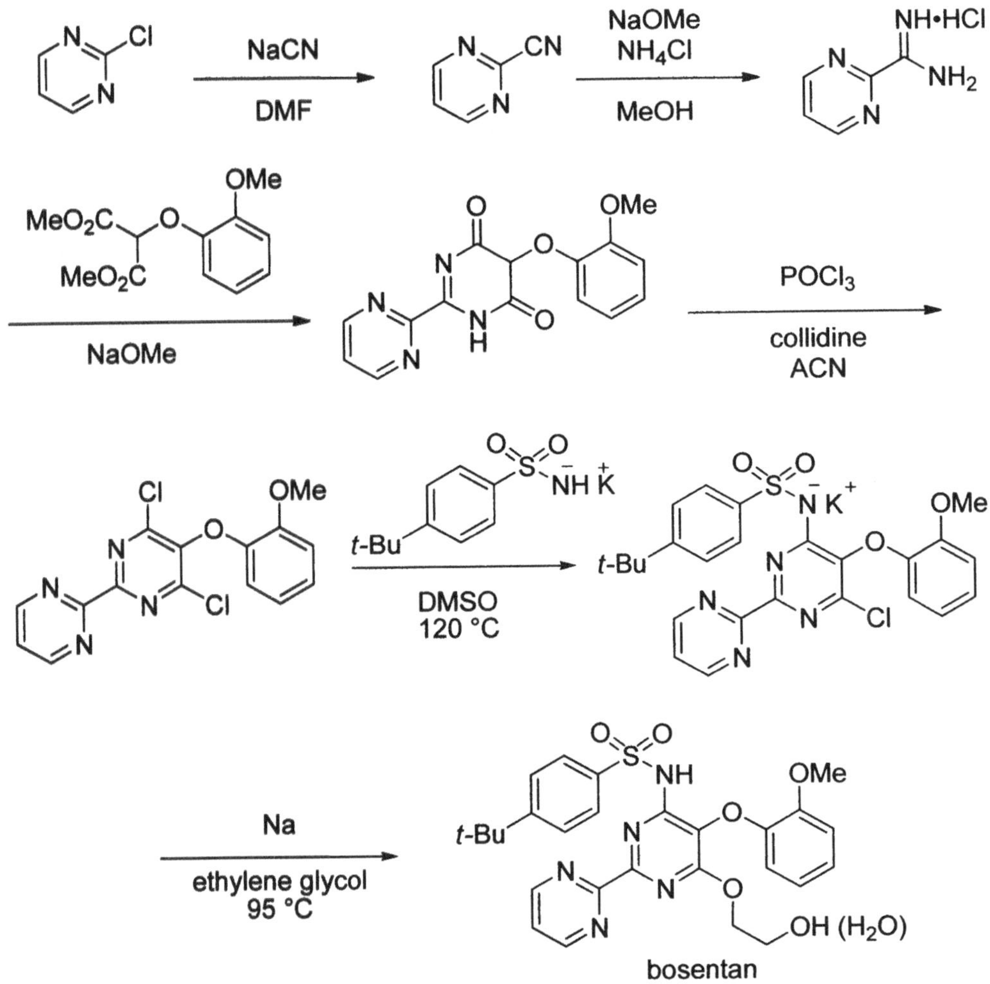

The first-generation process starts from 2-chloropyrimidine, and pyrimidine-2-carboximidamide is prepared in two steps.[47] Malonate derivative is made from dimethyl chloromalonate with guaiacol. These two fragments undergo condensation to generate pyrimidinedione, which is converted to the dichloropyrimidine with phosphorus oxychloride. Substitution of one chlorine by *tert*-butylbenzenesulfonamide led to a potassium sulfonamide salt. The other chlorine is replaced by ethylene glycol. Three crystallizations finally provide bosentan with a suitable specification grade.

In this first generation of synthesis, there are seven isolations and five drying operations including the isolation of a potent sensitizer dichloropyrimidine intermediate and a mild sensitizer-penultimate compound (reactant for the last step). A large excess (100 equiv) of ethylene glycol is required for the second chloride displacement. Dimer and pyrimidinone impurities are generated in final steps and require two recrystallizations to lower their levels.

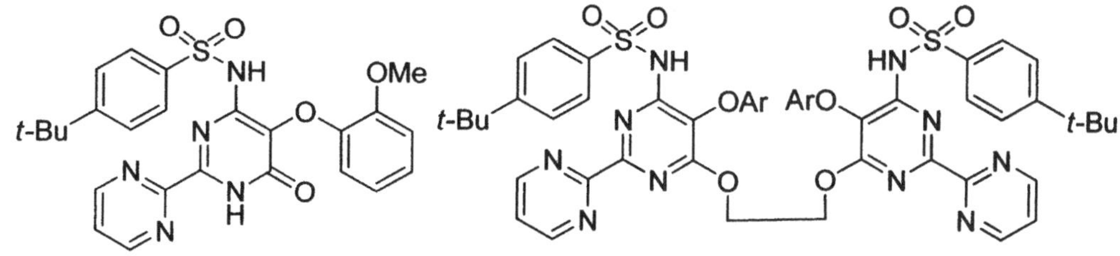

key impurities in the last step

In the second-generation process, isolation of potent sensitizer pyrimidinedichloride is removed.[48] In the S_NAr step of pyrimidinedichloride with sulfonamide, slow kinetics was observed in the absence of tetrabutylammonium bromide (TBAB). In comparison, this displacement becomes efficiently rapid when catalytic amount of TBAB is added to the anhydrous potassium carbonate suspension in toluene. To mitigate the dimer formation, the second-generation process uses monoprotected glycol, ethylene glycol mono-*tert*-butyl ether (ETB), which is commercially available and inexpensive. Deprotection of a *tert*-butyl ether with protic and Lewis acids is well precedented. However, because Lewis acid-mediated deprotection may cause metal removal in the workup, this approach is not pursued. HCl-mediated deprotection at an elevated temperature raised concerns about formation of 2-chloroethanol and the primary chloride impurity derived from bosentan. Deprotection with methanesulfonic acid also has a possibility leading to the formation of the methanesulfonate of bosentan. A clean and complete cleavage of the *tert*-butyl ether is observed using a formic acid-ether with a ratio of 2:1. Bosentan formate monoethanolate has a low solubility of in ethanol-toluene. Decant of suspension removed left starting material and *sec*-butyl ester by-product. Hydrolysis of formate in ethanol using aqueous caustic is complete in less than 1 h at 25 °C. The pyrimidinone A (<0.2%) is simultaneously hydrolyzed to pyrimidinone B. The final recrystallization provides a lowered B level to an acceptable level. In comparison to generation one, generation two improved the overall yield from pyrimidinedione to bosentan from 67% to 84%.

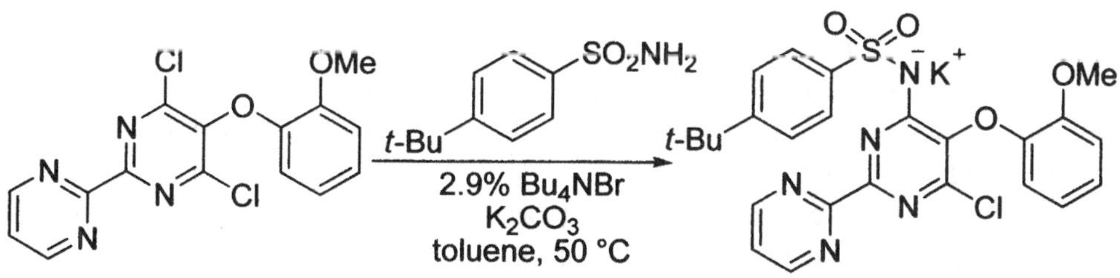

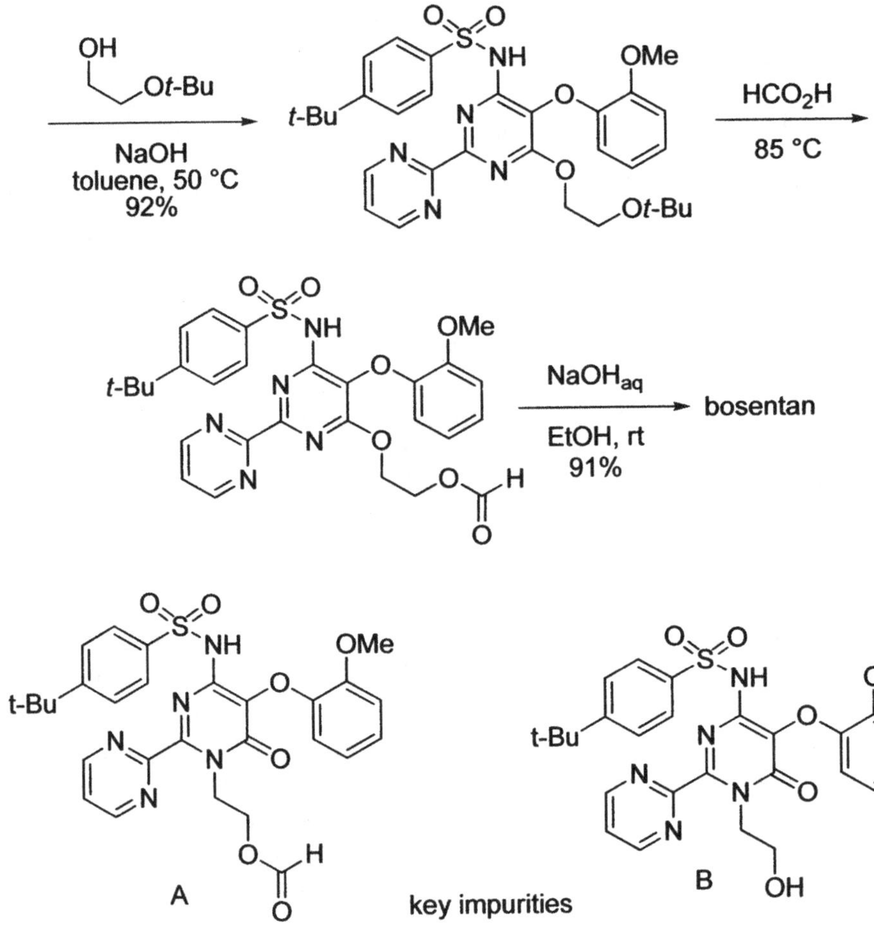

13.3.5 *Erlotinib*

Erlotinib is a drug to treat nonsmall-cells lung cancer, pancreatic cancer, and it is marketed in the United States by Genentech and OSI Pharmaceuticals and worldwide by Roche. Epidermal growth factor receptor (EGFR) is a cell membrane receptor consisting of an extracellular ligand-binding domain and an intracellular tyrosine kinase domain. Upregulation of EGFR expression is observed in many human solid tumors, such as colorectal, pancreatic, bladder, prostate, ovarian, glioma, breast, lung, renal, and head and neck cancer. Erlotinib (brand name: Tarceva) and its 4-anilinoquinazolines analogue gefitinib (brand name: Iressa) are EGFR-targeted agents. These small molecules competitively bind to the ATP binding pocket of intracellular kinase domain and block induction of downstream signaling network mediated by tyrosine kinase. And they, therefore, inhibit tyrosine kinase activity and restrict receptor catalytic activity, anti-phosphorylation and its engagement with signal transducers.

The original synthetic methods for the preparation of erlotinb involve construction of crucial intermediates of suitably substituted quinazolin-4(3*H*)-ones from 3,4-dihydroxybenzoic acid. A relatively improved synthesis of erlotinib[49] was reported in 2006. The synthesis of erlotinib starts with the *O*-alkylation of methyl 3,4-dihydroxybenzoate using either 1-chloro-2-methoxyethane or 1-bromo-2-methoxyethane with the chlorine compound requiring longer reaction time. Nitration of *O*-alkylation product gave the desired regioisomer in 92% yield and was followed by catalytic hydrogenation and cyclization with formamide performed according to the patent literature.[50] For the chlorination step it was found that the use of thionyl chloride and DMF[51] gave less reliable results than the method using phosphoryl chloride and *N,N*-diethylaniline,[52] giving chloropyrimidine in 89% yield and 96% HPLC purity. The final product was obtained as the erlotinib hydrochloride salt.

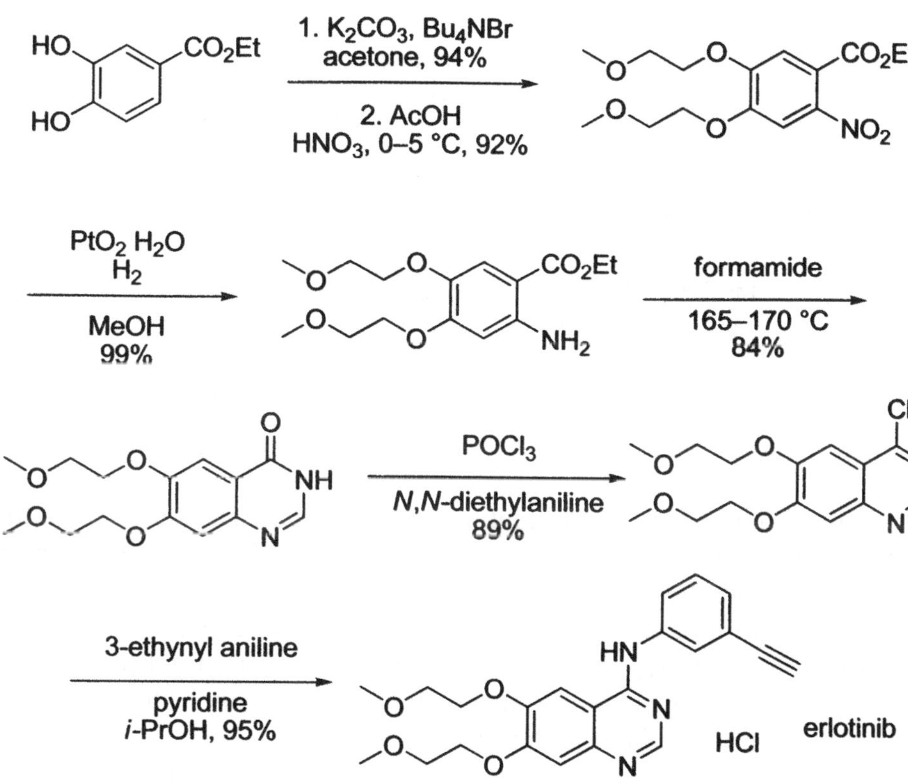

The previous synthesis involves use of corrosive chemicals such as thionyl chloride/phosphoryl chloride, costly reagents such as platinum oxide, and flammable gas such as hydrogen at high-temperature reaction conditions. To overcome these issues, an easily scalable, convergent, and versatile process for the preparation of erlotinib was reported in 56% overall yield.[53] Nitro reduction using sodium dithionite in aqueous medium gave aniline intermediate in 95% yield. 2-Amino-4,5-bis(2-methoxyethoxy)benzonitrile was treated with DMF–DMA in toluene to obtain formamidine derivative, which was reacted with 3-ethynylaniline through Dimroth rearrangement to give compound erlotinib. Upon further recrystallization, erlotinib was obtained in 70% yield with purity >99% by HPLC. The free base obtained was converted to erlotinib hydrochloride by passing HCl gas.

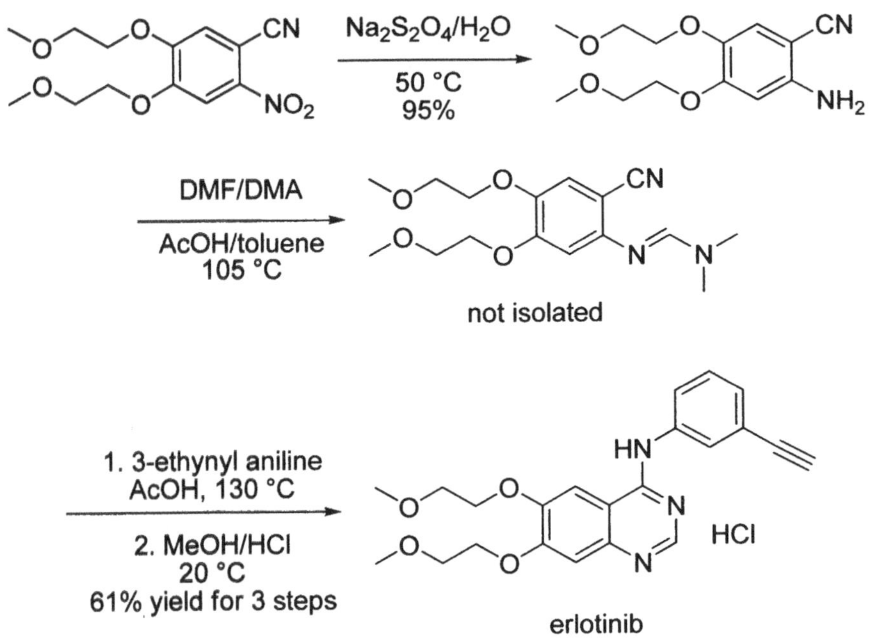

13.3.6 Rosuvastatin

Rosuvastatin is approved for the treatment of elevated LDL cholesterol, total cholesterol, and triglycerides. It is marketed by AstraZeneca as Crestor and marketed by Abbott Healthcare Pvt. Ltd. in India as "R2". Rosuvastatin[54] is a very effective inhibitor of 3-hydroxy-3-methylglutaryl coenzyme A (HMG-CoA) reductase (HMGR) and is one of the most powerful lipid-lowering agents in use as pharmaceutical agents for reducing cholesterol levels in patients at the risk of cardiovascular disease.

The process starts from a Knovengol condensation of ethyl isobutyryl acetate with 4-fluorobenzaldehyde.[55] The reaction of *S*-methyl isothiourea sulfate with α,β-unsaturated ketone followed by oxidation with DDQ gives 2-(methylthio)pyrimidine. This intermediate was further oxidized by *m*-CPBA to yield 2-(methylsulfonyl)pyrimidine. Aromatic substitution with methyl amine and sulfonylation afford methanesulfonamide derivative. The required alcohol can be synthesized by DIBAL reduction of the corresponding ester as reported by Watanabe et al.[56] Aldehyde can be prepared by chemoselective catalytic oxidation of alcohol with tetrapropylammonium perruthenate (TPAP) or NaOCl as an oxidant in the presence of catalytic amounts of 2,2,6,6-tetramethylpiperidinyloxy (TEMPO) free radical.[57] A more practical and scalable condition was reported using pyridine sulfur trioxide.[58] Rosuvastatin was easily assembled through Wittig coupling of aldehyde and ylide, which derived from a Ru-catalyzed asymmetric hydrogenation. The second stereogenic center at C-5 in rosuvastatin can later be created by diastereoselective reduction of the oxo group with Et_2BOMe and $NaBH_4$ to afford rosuvastatin ethyl ester.

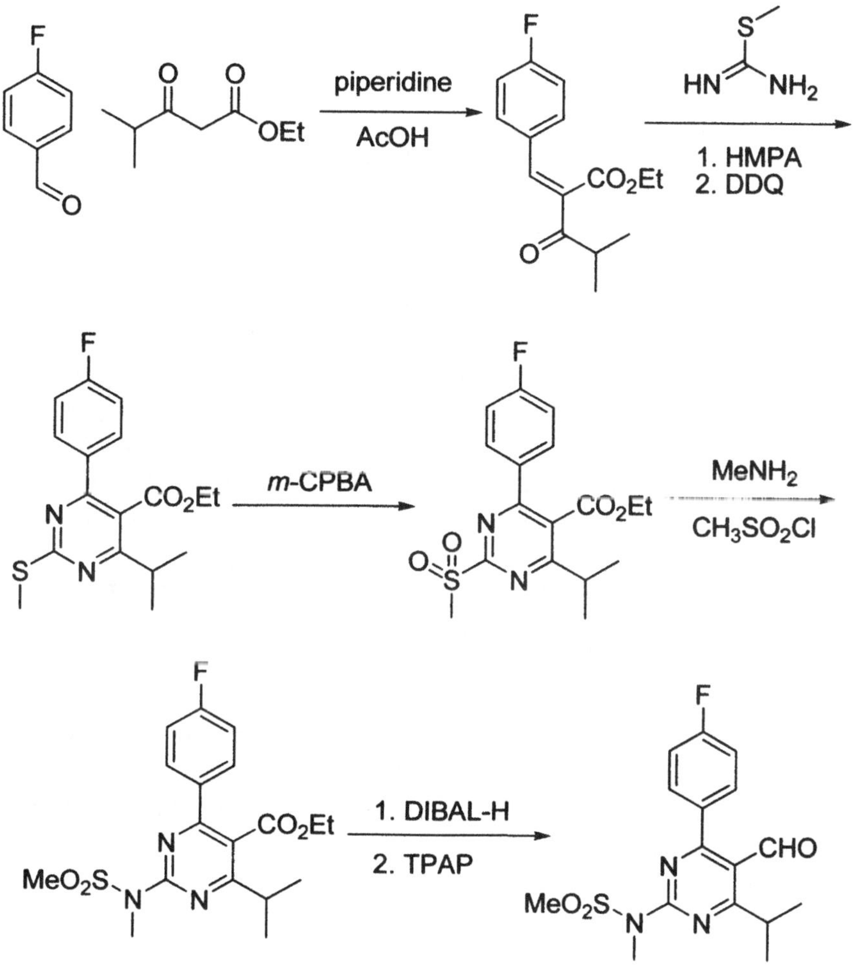

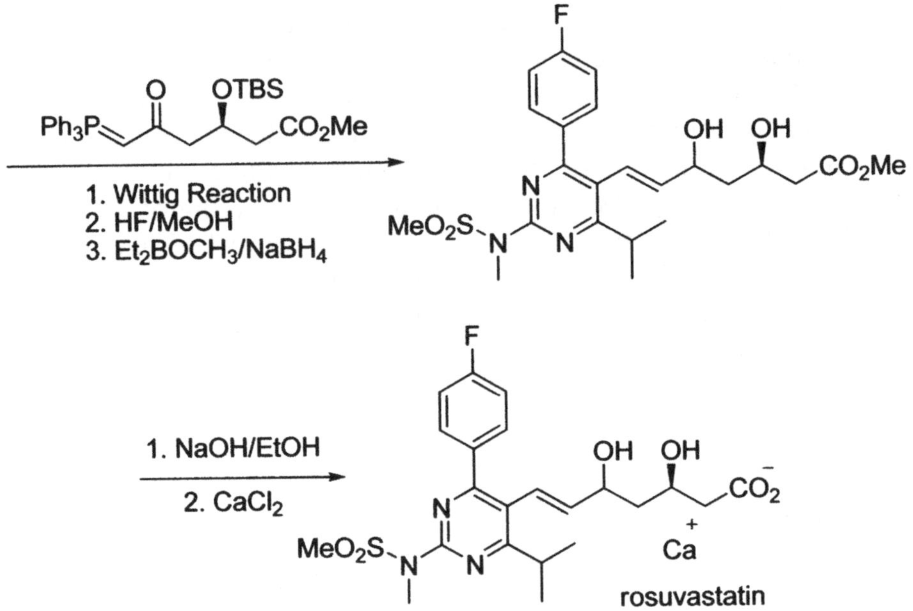

Alternatively, aldehyde intermediate could be derived from 2-(methylamino) pyrimidine derivative by introduction of the formyl group into the C-5 position of the pyrimidine ring and subsequent mesylation of the secondary amine functionality. Thus, a concise strategy was reported for the synthesis of aldehyde based on the cyclization of methylguanidine hydrochloride with 1-(4-fluorophenyl)-4-methylpentane-1,3-dione, followed by regioselective iodization of the pyrimidine and final substitution of the halogen atom by a CHO group.[59] Thus, treatment of pyrimidine intermediate with I_2 in DMSO at 100 °C led to the formation of pyrimidine iodide. Sulfonylation with $MeSO_2Cl$ gave sulfonamide. It was subjected to Pd-catalyzed formylation with CO/H_2 (1:1) at a pressure of 5 bar in toluene as suggested by Beller et al.[60] Replacement of the ligand Ph_3P by Ad_2P*n*-Bu improved the results in favor of the formation of desired aldehyde. Rosuvastatin was assembled following the reported steps: Wittig coupling of aldehyde and ylide, stereoselective reduction with Et_2BOMe and $NaBH_4$, and the final saponification reaction.

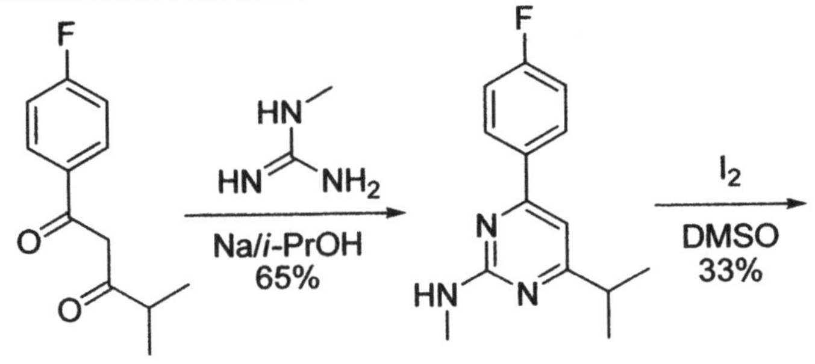

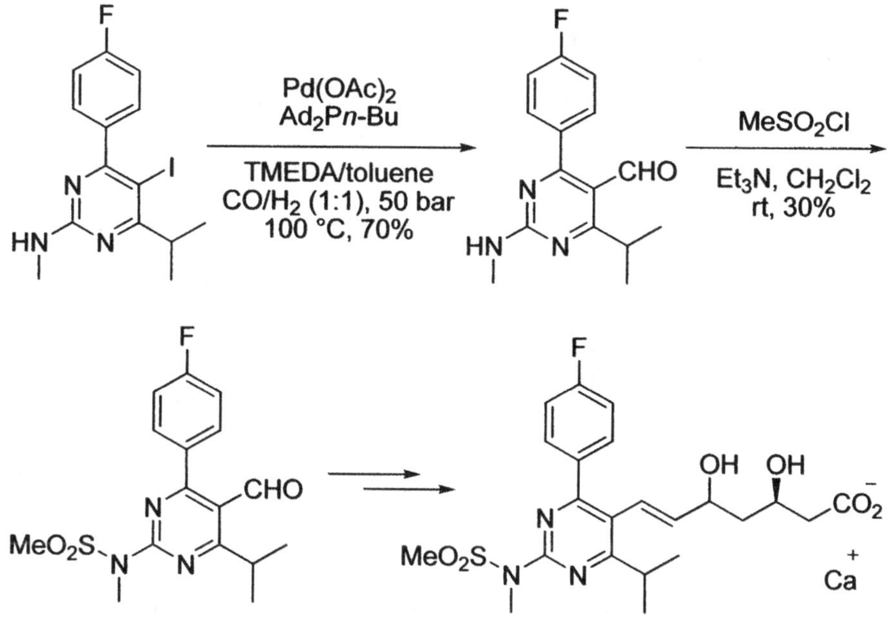

13.3.7 Sildenafil

Sildenafil citrate, sold as Viagra, is a selective inhibitor of phophodiesterase 5 (PDE5) for the treatment of male erectile dysfunction. This prescription drug was brought to market by Pfizer and was approved to use within the United States and the European Union in 1998. It has become one of the fastest-selling drugs of all time. The inhibition of PDE5 leads to higher levels of cyclic guanosine monophosphate (cGMP), which in turn leads to improved smooth muscle relaxation, increased blood flow, and hence, an improved erection.[61] It is worth to note that sildenafil is not an aphrodisiac and only helps the body's natural processes.

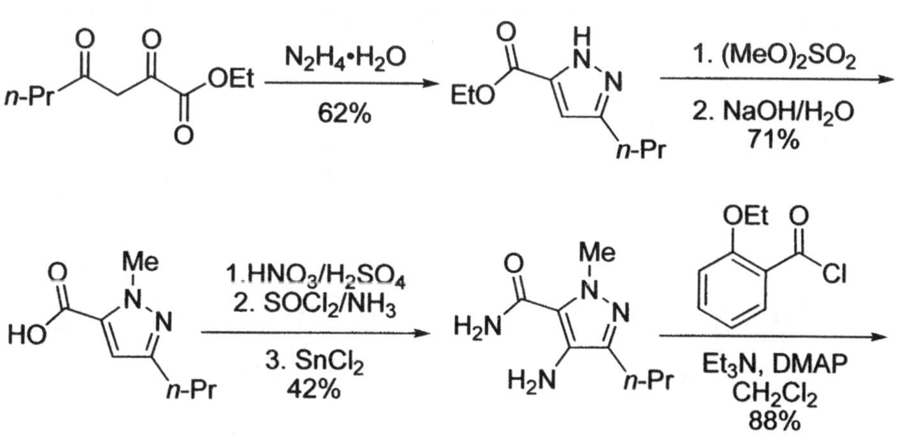

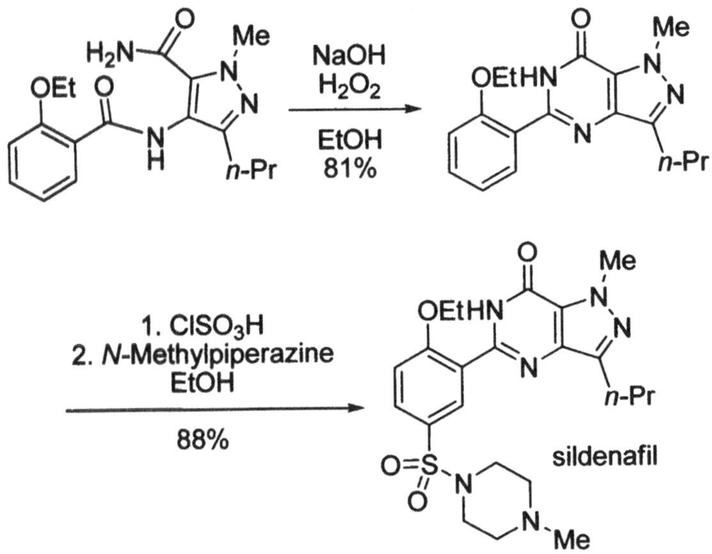

A medicinal chemistry route was used for the synthesis of early toxicity and clinical batches. The route is linear and used potentially toxic materials, such as the sulfonyl chloride, are in the final bond-forming reaction.[62] The difficulties of scaling up chlorosulfonation reactions are known due to its increased quenching time on scale-up. The optimization of the cyclization reaction to make the pyrimidinone was a rewarding process and its finding was implemented in the later scale-up campaigns. The original route used an aqueous alcoholic solution of sodium hydroxide and hydrogen peroxide to generate cyclized pyrimidinone in moderate yield (30–70%). The main side product in the presence or absence of hydrogen peroxide was the hydrolysis of the carboxamide to the corresponding acid. By conducting the cyclization under anhydrous conditions, KO*t*-Bu/*t*-BuOH, the hydrolysis side product was eliminated, and this exceptionally clean reaction proceeded in 100% isolated yield with no impurities detected.

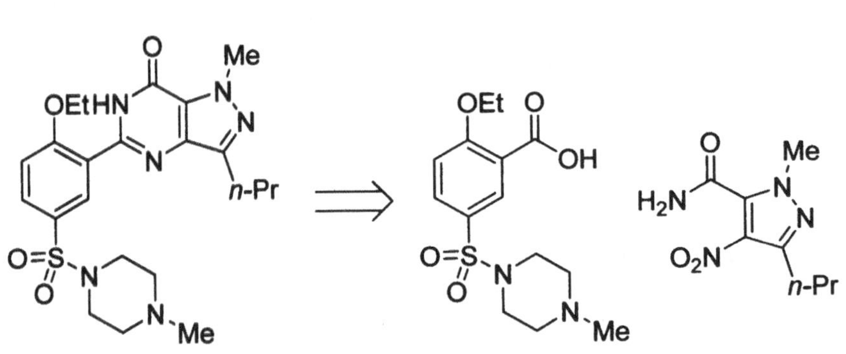

Given the scale-up issues, the development of an efficient synthesis with high throughput and process safety is highly desired. The remarkably clean cyclization reaction led to a convergent synthesis that the clean cyclization was the final bond-forming event.

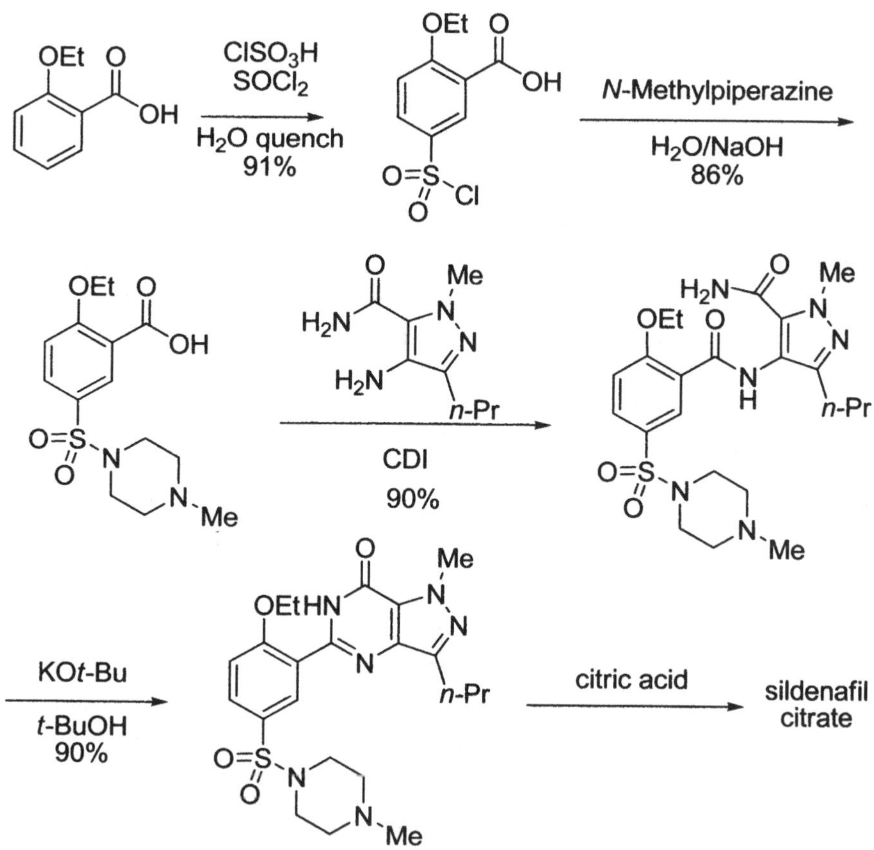

In the 2nd-generation synthesis of sildenafil citrate, chlorosulfonation of 2-ethoxybenzoic acid is straightforward. It is essential to add thionyl chloride to ensure the intermediate sulfonic acid is converted to the sulfonyl chloride. Through quite extensive investigations, crystalline form of the free amino acid was identified. Seeding is particularly important to achieve high crystallization yield with the zwitterionic intermediate. Upon completion of the sulfonylation of *N*-methylpiperazine, the product was isolated by pH adjustment to the isoelectric point and filtration of the precipitated sulphonamide after seeding. The nitro reduction was achieved via a palladium-catalyzed hydrogenation reaction to give the corresponding amine. Acid activation by *N,N*-carbonyldiimidazole (CDI) and acylation reactions were all performed in ethyl acetate as the solvent. All three reactions (nitro reduction, CDI acitivation and acylation) could be telescoped into a single step. The process step had a very low environmental impact because there was no aqueous waste stream, and the use of ethyl acetate as a sole solvent allowed for easy solvent recovery. Finally, heating the amide with 1.2 equiv of potassium *tert*-butoxide at reflux and acidified with 4 N hydrochloric acid to the isoelectric point (pH 7.5) at the end of the reaction afforded the sildenafil free base in good yield before citrate salt formation.

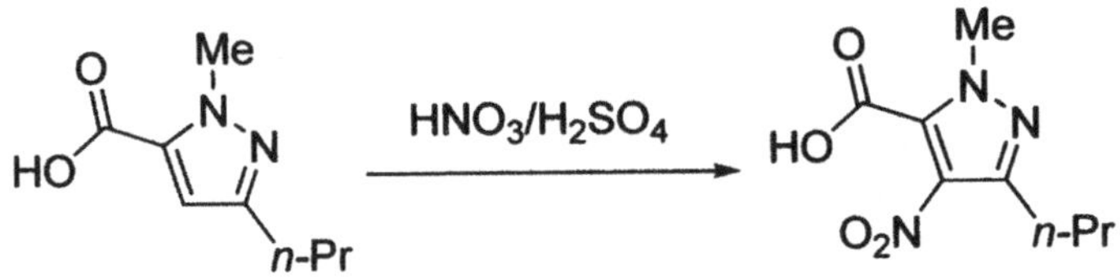

It is worth to note two exotherms were observed in the nitration step. The first exotherm was detected at 130 °C by differential scanning calorimetry (DSC) and evolved 16.2 kJ/mol. It was attributed to the decarboxylation reaction of the acid. The second, much larger exothermic event, was detected at 295 °C by DSC and evolved 294 kJ/mol. The onset temperature for the start of exothermic decarboxylation was determined to be around 100 °C. To mitigate this stability problem, the nitration procedure was broken down into three parts in order to minimize the available energy. First, the pyrazole acid was dissolved in concentrated sulfuric acid, eliminating 67 kJ/mol of energy from the reaction. Next, the fuming nitric and concentrated sulfuric acids were mixed, eliminating 44 kJ/mol from the reaction. Third, one third of the nitrating mixture was charged to the header at any one time. An HPLC analysis was then performed to ensure that the appropriate degree of reaction completion had occurred before charging the next portion of the nitrating acid mixture to the header.

Some other investigations were reported in the synthesis of pyrimidone core of sildenafil. In equation 1, the imino-ether was made from the nitrile by a Pinner reaction.[63] Zipping of two fragments (the imino ether and pyrazole amide) gave sildenafil in 85% yield. As shown in equation 2, the condensation of the aldehyde with the aminopyrazole in boiling toluene yielded dihydrosildenafil by azeoptropic distillation to remove water. Dihydrosildenafil can be oxidized using Pd/C and a small quantity of trifluoroacetic acid at high temperatures or using sodium hydrogen sulfite to give sildenafil. This route has been reported by Pfizer[64] and by Achmatowicz et al.[65] Another approach is to construct an amide bond before nitro reduction. The nitro group is then reduced with tin-(II) chloride, and the resulting amine is cyclized. Finally, the fluorine atom is displaced with the ethoxide group to give sildenafil.[66]

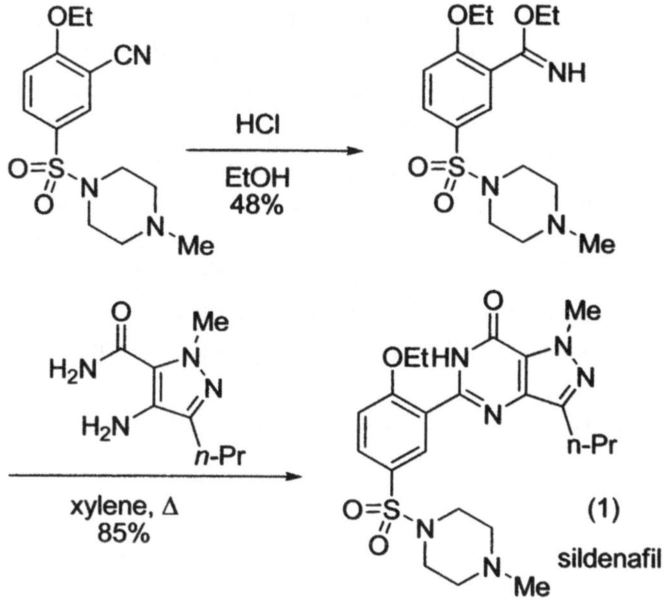
OEt
CN
HCl
EtOH
48%
OEt OEt
NH
O=S
N
N
Me
Me
O
H2N
N
H2N
n-Pr
xylene, Δ
85%
OEt HN
n-Pr
sildenafil

(1)

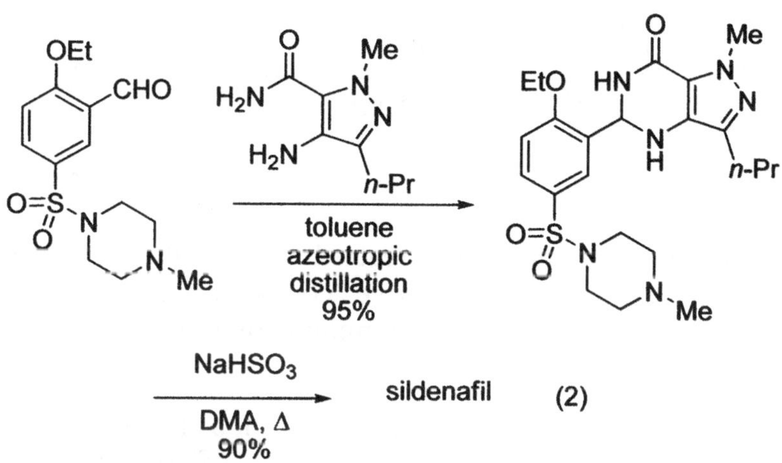
OEt
CHO
toluene
azeotropic
distillation
95%
EtO
HN
N
H
n-Pr
Me
NaHSO3
DMA, Δ
90%
sildenafil

(2)

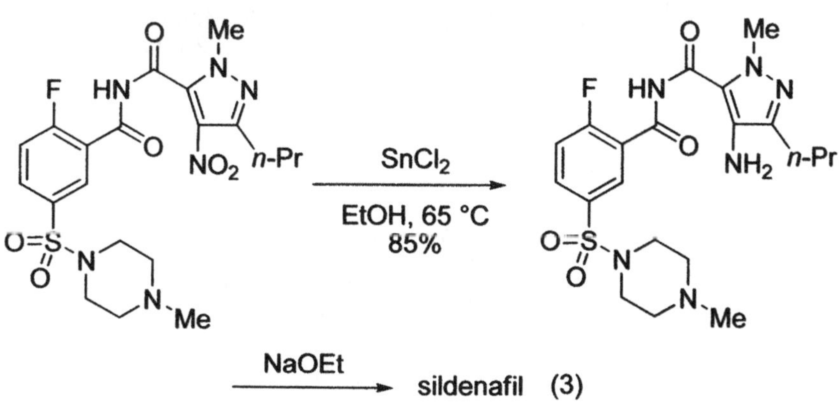
F
HN
NO2
n-Pr
SnCl2
EtOH, 65 °C
85%
NH2
Me
NaOEt
sildenafil

(3)

Development of methodologies for the synthesis of pyrimidines remains a vibrant area of chemical research. Modification of conventional strategies involving N–C=N fragment condensation with 1,3-dicarbonyl derivatives or α,β-unsaturated carbonyl compounds still represents a common and powerful approach to construct pyrimidines. Other methods, including N–C fragment condensations, offer complementary strategies for pyrimidine synthesis. Given the wide presence of pyrimidine structure and its derivatives in our medicines, research in efficient synthesis of pyrimidine derivatives and investigations of their pharmaceutical activities in a broader spectrum of diseases areas will thrive in many years to come.

Acknowledgment. We thank Dr. David Conlon for helpful discussions.

13.4 Problems

13.4.1 Work out the mechanism in the allopurinol synthesis.[67]

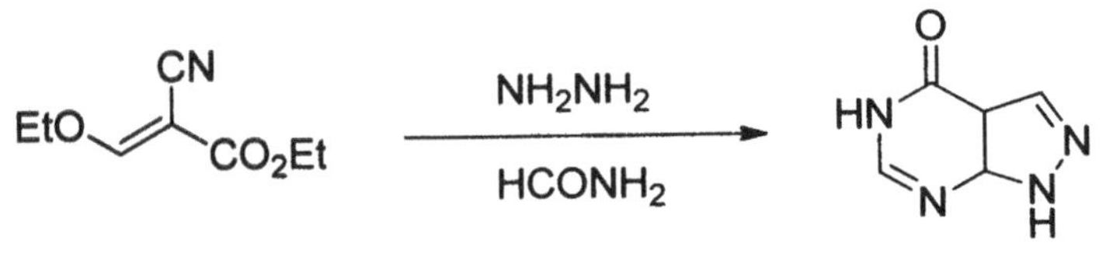

allopurinol

13.4.2 *N,N*-dimethylformimidamide reacted with 3-ethynylaniline through Dimroth rearrangement to give compound erlotinib. Suggest a reasonable mechanism for the pyrmidine ring formation.[68]

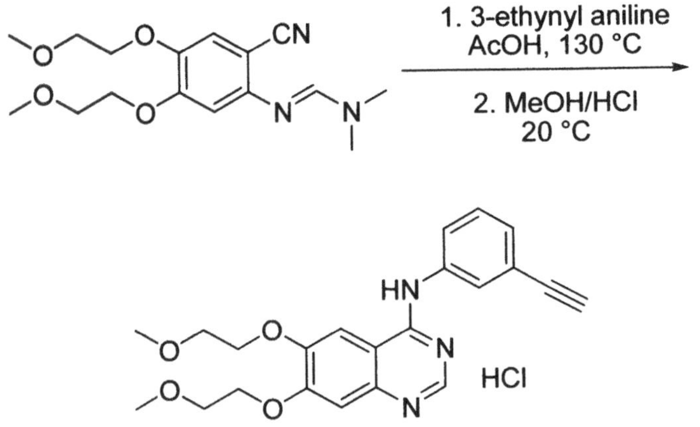

erlotinib

13.4.3 What product is possible from the following substitution?

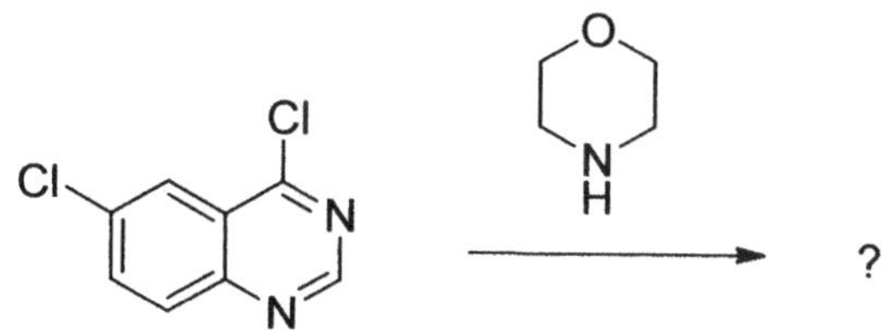

13.4.4 In the reaction of cyanamide with 1,3-diketones, it is known that urea, a possible hydrolysis product of cyanamide, did not react with diketone under the reaction conditions. Cyanoguanidine, the dimerization product of cyanamide, did not react with diketone either. Ammonia was precluded as a significant reactant. Suggest a sound mechanism for the pyrimidine formation.[69]

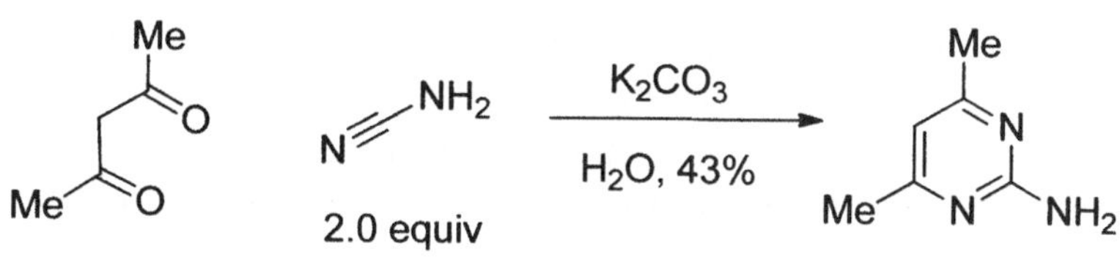

13.4.5 Suggest a plausible mechanism for the coupling reaction shown below. $ZnCl_2$ is known to coordinate with the orthoester.[70]

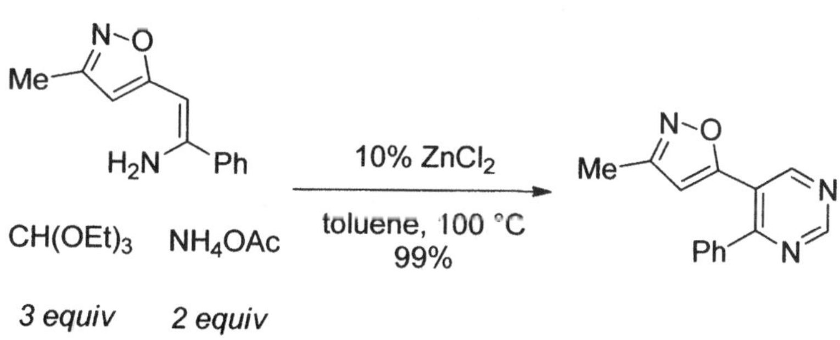

13.4.6 Proposed a desired product and a plausible mechanism.[71]

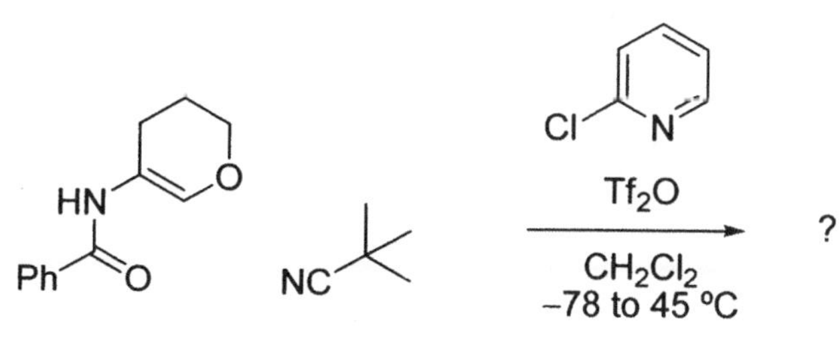

13.4.7 In the synthesis of risperidone, one fragment B started from pyridin-2-amine and 4-acetyldihydrofuran-2(3*H*)-one. Suggest intermediate **A** and **B**.[72]

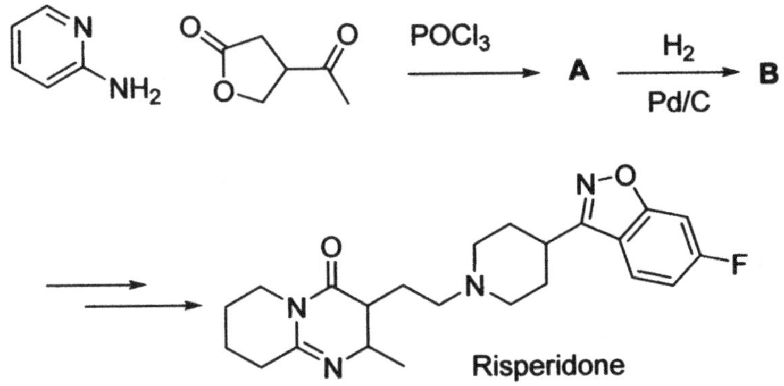

13.4.8. In the synthesis of methotrexate, 2,4,5,6-pyrimidinetetramine hydrochloride served as an intermediate. It was prepared from 5-nitroso-2,4,6-triaminopyrimidine. Propose reactants **C** and **D** to synthesize 5-nitroso-2,4,6-triaminopyrimidine and product **E** in the reaction of 2,4,5,6-Pyrimidinetetramine with 2,3-dibromopropanal.[73]

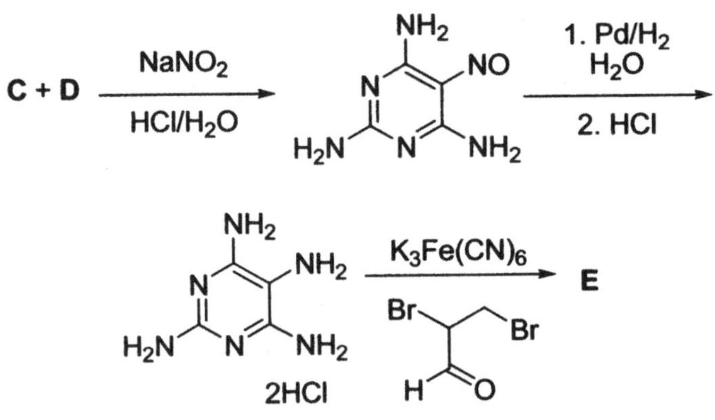

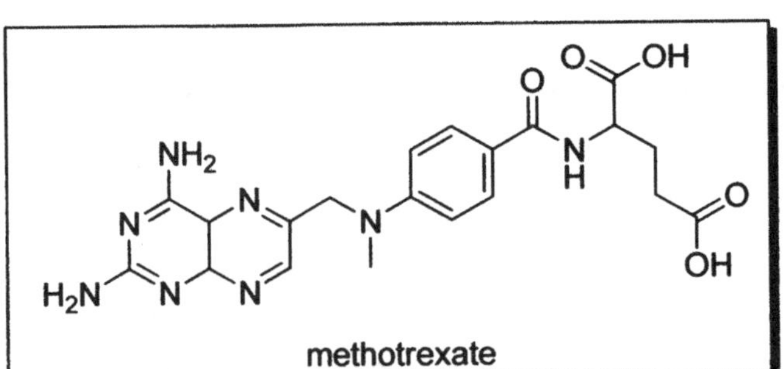

13.5 References

1 Finland, M.; Strauss, E.; Peterson, O. L. *JAMA* **1984,** *251*, 1467–1474.

2 Marcellin, P; Chang, T. T.; Lim, S. G.; Tong, M. J.; Sievert, W.; Shiffman, M. L.; Jeffers, L.; Goodman, Z.; Wulfsohn, M. S.; Xiong, S.; Fry, J.; Brosgart, C. L.; Adefovir Dipivoxil 437 Study Group *N. Engl. J. Med.* **2003,** *348*, 808–816.

3 Emau, P.; Jiang, Y.; Agy, M. B.; Tian, B.; Bekele, G.; Tsai, C. C. *AIDS Res. Ther.* **2006,** *3*, 29.

4 Arimilli, M. N.; Kim, C. U.; Dougherty, J.; Mulato, A.; Oliyai, R.; Shaw, J. P.; Cundy, K. C.; Bischofberger, N. *Antiviral Chem. Chemother.* **1997,** *8*, 557–564.

5 (a) Manegold, C. *Semin. Oncol.* **2003**, *30*, 32–36; (b) Cohen, M. H.; Justice, R.; Pazdur, R. *Oncologist* **2009,** *14*, 930–935; (c) Rossi, A.; Ricciardi, S.; Maione, P.; de Marinis, F.; Gridelli, C. *Lung Cancer* **2009,** *66*, 141–149.

6 Rajagopalan, P.T.; Zhang, Z.; McCourt, L.; Dwyer, M.; Benkovic, S. J.; Hammes, G. G. *Proc. Natl. Acad. Sci. USA* **2002,** *99*, 13481–13486.

7 (a) Pinner, A. *Ber. Dtsch. Chem. Ges.* **1884**, *17*, 2519 –2520; (b) Pinner, A. *Ber. Dtsch. Chem. Ges.* **1885**, *18*, 759–763; c) Pinner, A. *Ber. Dtsch. Chem. Ges.* **1887**, *20*, 2361–2365.

8 Ghosh, U.; Katzenellenbogen, J. A. *J. Heterocycl. Chem.* **2002,** *39*,1101–1104.

9 Zhichkin, P.; Fairfax, D. J.; Eisenbeis, S. A. *Synthesis* **2002,** 720–722.

10 Bartrum, H. E.; Blakemore, D. C.; Moody, C. J.; Hayes, C. J. *J. Org. Chem.* **2010,** *75*, 8674–8676.

11 MIller, T. J. J.; Braun, R.; Ansorge, M. *Org. Lett.* **2000,** *2*, 1967–1970.

12 Kiselyov, A. S. *Tetrahedron Lett.* **2005,** *46*, 1663–1665.

13 Adamo, M. F. A.; Adlington, R. M.; Baldwin, J. E.; Pritchard, G. J.; Rathmell, R. E. *Tetrahedron* **2003,** *59*, 2197–2205.

14 Kim, J. M.; Kim, S. H.; Kim, J. N. *Bull. Korean Chem. Soc.* **2007,** *28*, 2505–2507.

15 Kakiya, H.; Yagi, K.; Kazunari, Y.; Shinokubo, H.; Oshima, K. J. *Am. Chem. Soc.* **2002,** *124*, 9032–9033.

16 Mıller, A. *J. Org. Chem.* **1984,** *49*, 4072–4074.

17 MartÌnez, A. G.; Fernández, A. H.; Fraile, A. G.; Subramanian, L. R.; Hanack, M. *J. Org. Chem.* **1992,** *57*, 1627–1630.

18 (a) Ingebrigsten, T.; Helland, I.; Lejon, T. *Heterocycles* **2005,** *65*, 2593–2603; (b) Tyagarajan, S.; Chakravarty, P. K. *Tetrahedron Lett.* **2005,** *46*, 7889–7891.

19 Puchala, A.; Belaj, F.; Bergman, J.; Kappe, C. O. *J. Heterocycl. Chem.* **2001,** *38*, 1345–1352.

20 Kang, F.; Kodah, J.; Guan, Q.; Li, X.; Murray, W. V. *J. Org. Chem.* **2005,** *70*, 1957–1960.

21 Sasada, T.; Kobayashi, F.; Sakai, N.; Konakahara, T. *Org. Lett.* **2009,** *11*, 2161–2164.

22 Movassaghi, M.; Hill, M. D. *J. Am. Chem. Soc.* **2006,** *128*, 14254–14255

23 Movassaghi, M.; Hill, M. D. *Nat. Protoc.* **2007,** *2*, 2018-2023.

24 Ahmad, O. K.; Hill, M. D. ; Movassaghi, M. *J. Org. Chem.* **2009,** *74*, 8460–8463.

25 Sasada, T.; Moriuchi, M.; Sakai, N.; Konakahara, T. *Eur. J. Org. Chem.* **2009,** 5738–5743

26 Dang, Q.; Gomez-Galeno, J. E. *J. Org. Chem.* **2002,** *67*, 8703–8706.

27 Fischer, R. W.; Misun, M. *Org. Proc. Res. Dev.* **2001,** *5*, 581–586.

28 MadroÇero, R.; Vega, S. *Synthesis* **1987,** 628–630.

29 Zielinski, W.; Kudelko, A. *Monatsh. Chem.* **2000,** *131*, 895–899.

30 Hartmann, J. T.; Haap, M.; Kopp, H. G.; Lipp, H. P. *Curr. Drug Metab.* **2009,** *10*, 470–481.

31 Liu, J.-J.; Luk, K.-C. US 7098332 (2006).
32 Borrell, J. I. *J. Org. Chem.* **2010,** *75*, 487–490.
33 Zhao, L.; Liang, F.; Bi, X.; Sun, S.; Liu, Q. *J. Org. Chem.* **2006,** *71*, 1094–1098.
34 Molteni, V.; Hamilton, M. M.; Mao, L.; Crane, C. M.; Termin, A. P.; Wilson, D. H. *Synthesis* **2002,** 1669–1674.
35 Baran, P. S.; Shenvi, R. A.; Nguyen, S. A. *Heterocycles* **2006,** *70*, 581–586.
36 Kiselyov, A. S. *Tetrahedron Lett.* **2005,** *46*, 1177–1179.
37 Rossi, E.; Abbiati, G.; Pini, E. *Synlett* **1999,** 1265–1267.
38 Schmidt, P.; Druey, J. *Hel. Chim. Acta* **1956,** *39*, 986-991.
39 Hess, S.; Dolker, M.; Haferburg, D.; Kertscher, H. P.; Matysik, F.-M.; Ortwein, J.; Teubert, U.; Zimmermann, W.; Eger, K. *Pharmazie* **2001,** *56*, 306-310.
40 (a) Zimmermann, J.; Buchdunger, E.; Mett, H.; Meyer, T.; Lydon, N. B. *Bioorg. Med. Chem. Lett.* **1997,** *7*, 187–192; (b) Li, X. Q.; Yang, J.; Chen, X. J.; Liu, J.; Li, H. R.; Zheng, J.; He, Y.; Chen, Z.; Huang, S. *Cancer Genet. Cytogenet.* **2007,** *176*, 166–168.
41 (a) Zimmermann, J. EP564409 (1993). (b) Zimmermann, J. U.S.5521184 (1996). (c) Zimmermann, J.; Buchdunger, E.; Mett, H.; Meyer, T.; Lydon, N. B.; Traxler, P. *Bioorg. Med. Chem.Lett.* **1996,** *11*, 1221.
42 Loiseleur, O.; Kaufmann, D.; Abel, S.; Buerger, H. M.; Meisenbach,M.; Schmitz, B.; Sedelmeier, G. WO 2003066613 (2003).
43 (a) Kompella, A.; Bhujanga Rao, A. K. S.; Venkaiah Chowdary, N. W.O. 2004/108,699 (2004); (b) Szakacs, Z.; Beni, S.; Varga, Z.; Orfi, L.; Keri, G.; Noszal, B. *J. Med. Chem.* **2005,** *48*, 249.
44 Liu, Y.-F. Wang, C.-L.; Bai, Y.-J.; Han, N.; Jiao, J.-P.; Qi, X.-L. *Org. Process Res. Dev.* **2008,** *12*, 490–495.
45 Manley, P. W.; Breitenstein, W.; Brüggen, J.; Cowan-Jacob, S. W.; Furet, P.; Mestan, J.; Meyer, T. *Bioorg. Med. Chem. Lett.* **2004,** *14*, 5793–5797.
46 Neidhart, W.; Breu, V.; Burri, K.; Clozel, M.; Hirth, G.; Klinkhammer, U.; Giller, T.; Ramuz, H. *Bioorg. Med. Chem. Lett.* **1997,** *7*, 2223–2228.
47 Burri, K.; Clozel, M.; Fischli, W.; Hirth, G.; Loffler, B.-M.; Neidhart, W.; Ramuz, H. US 5292740 (1994).
48 Harrington, P. J.; Khatri, H. N.; DeHoff, B. S.; Guinn, M. R.; Boehler, M. A.; Glaser, K. A. *Org. Process Res. Dev.* **2002,** *6*, 120–124.
49 Knesl, P.; Röseling, D.; Jordis, U. *Molecules* **2006,** *11*, 286–297.
50 Schnur, R. C.; Arnold, L. D. US 5747498 (1998).
51 Kanter, J.; Pandey, A.; Robinson, J.; Scarborough, R.M. WO 2002036587 (2002).
52 Springer, R. H.; Scholten, M. B.; O'Brien, D. E.; Novinson, T.; Miller, J. P.; Robins, R. K. *J. Med. Chem.* **1982,** *25*, 235–242.
53 Chandregowda, V.; Rao, G. V.; Reddy, G. C. *Org. Process Res. Dev.* **2007,** *11*, 813–816.
54 Culhane, N. S.; Lettieri, S. L.; Skae, J. R. *Pharmacotherapy* **2005,** *25*, 990–1000.
55 Watanabe, M.; Koike, H.; Ishiba, T.; Okada, T.; Seo, S.; Hirai, K. *Bioorg. Med. Chem.* **1997,** *5*, 437–444.
56 Watanabe, M.; Koike, H.; Ishiba, T.; Okada, T.; Seo, S.; Hirai, K. *Bioorg. Med. Chem.* **1997,** *5*, 437–444.
57 Niddam-Hildesheim, V.; Chen, K. WO 2006017357 (2006).
58 Joshi, N. S.; Khile, A. S.; Kajale, Y. B.; Kamble, H. H. WO 2008059519 (2008).
59 Andrushko, N.; Andrushko, V.; König, G.; Spannenberg, A.; Börner, A. *Eur. J. Org. Chem.* **2008,** 847–853

60 Klaus, S.; Neumann, H.; Zapf, A.; Strübing, D.; Hübner, S.; Almena, J.; Riermeier, T.; Groß, P.; Sarich, M.; Krahnert, W.-R.; Rossen, K.; Beller, M. *Angew. Chem. Int. Ed.* **2006,** 45, 154–158.

61 Boolell, M.; Allen, M. J.; Ballard, S. A.; Gepi-Attee, S.; Muirhead, G. J.; Naylor, A. M.; Osterloh, I. H.; Gingell, J. C. *Int. J. Impot. Res.* **1996,** *8*, 47–52.

62 Terrett, N. K; Bell, A. S.; Brown, D.; Ellis, P. *Bioorg. Med. Chem. Lett.* **1996,** *6*, 1819–1824.

63 Dunn, P. J.; Dunne, C. WO 200198284 (2001).

64 Bunnage, M. E.; Levett, P. C.; Thomson, N. M. WO 01/98303 (2001).

65 Achmatowicz, O.; Balicki, R.; Chmielowiec, U.; Zaworska, A.; Szelejewki, W.; Magielka, S, Glowacka, A.; Wysoczynska, M.; Dzikowska, J.; Bober, L.; Landsberg, J.; Falkowski, C.; Roznerski, Z.; Marczak, B.; Kempa, A. WO 200122918 (2001).

66 Yang, D.; Wang, H.; Shi, H. Chinese Patent CN-109345.

67 Schmidt, P.; Druey, J. *Helv. Chim. Acta* **1956,** *39,* 986–991.

68 Chandregowda, V.; Rao, G. V.; Reddy, G. C. *Org. Process Res. Dev.* **2007,** *11*, 813–816.

69 Miller, A. *J. Org. Chem.* **1984,** *49*, 4072–4074.

70 Sasada, T.; Moriuchi, M.; Sakai, N.; Konakahara, T. *Eur. J. Org. Chem.* **2009,** 5738–5743

71 Movassaghi, M.; Hill, M. D. *J. Am. Chem. Soc.* **2006,** *128*, 14254–14255

72 Kim, N. D.; Lee, J. H.; Lee, M. S.; Chang, Y. K.; Lee, G. S. WO 2004035573 (2004).

73 Eyer, M.; Pfleiderer, W.; Taghavi-Moghadam, S. WO 9721711 (1997).

Chapter 14 Quinazolines and Quinazolinones

Nicole L. Snyder and Connor W. Brown

14.1 Introduction

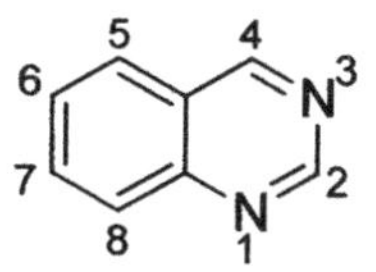

Quinazoline

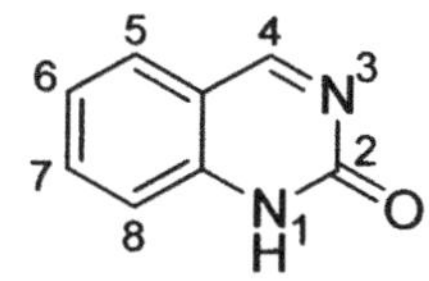

2-Quinazolinone

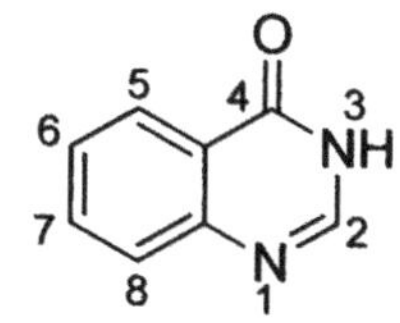

4-Quinazolinone

Quinazoline is a yellow crystalline solid with a melting point of 48 °C. Quinazoline is structurally related to 2- and 4-quinazolinones and it behaves chemically like its pyrimidine counterpart with the exception that quinazoline is more basic due the electrophilicity of the *C*4 position.

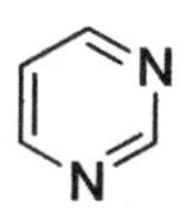

Pyrimidine
p*Ka* = 1.1

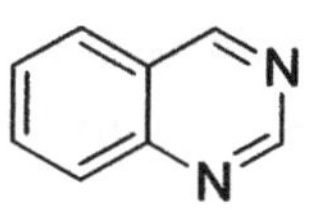

Quinazoline
p*Ka* = 3.3

p*K*a Values for pyrimidine and quinazoline

The ^{1}H-NMR spectra of quinazoline is complex due to the fused aromatic ring and six nonequivalent protons. *H*4 and *H*2, which are deshielded by the nitrogen atoms in the diazine ring, are found downfield at 9.34 ppm and 9.28 ppm with *H*4 being the most deshielded due to resonance effects. Correspondingly, *H*5–*H*8 are less shielded, with *H*8 being the least shielded (7.99 ppm) due to the electron delocalization of the heteroatom.

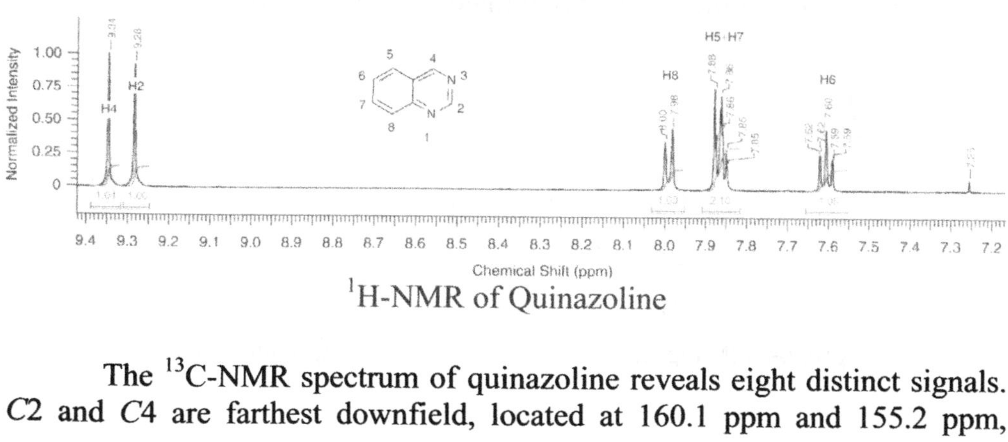

^{1}H-NMR of Quinazoline

The ^{13}C-NMR spectrum of quinazoline reveals eight distinct signals. *C*2 and *C*4 are farthest downfield, located at 160.1 ppm and 155.2 ppm, respectively. This is due to the deshielding effects of the heteroatoms in the diazine ring. Of the remaining carbon signals, *C*8a (149.9 ppm) is located farthest downfield due to proximity of *N*1 followed by *C*7, *C*8, *C*6, *C*5, and *C*4a.

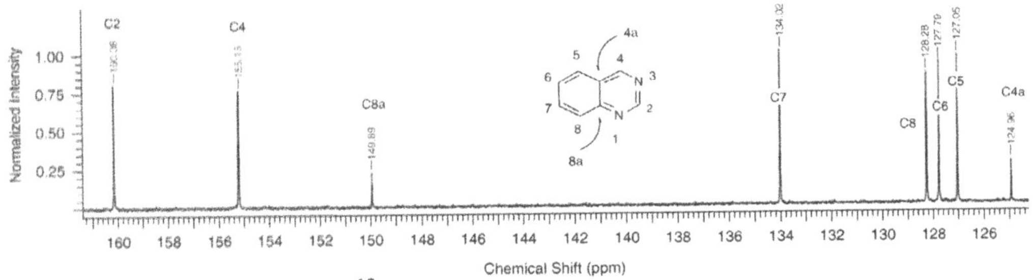

^{13}C-NMR of quinazoline

Quinazolines and quinazolinones are found in a diverse array of synthetic and naturally occurring compounds, and they have been shown to have an equally diverse number of properties including soporitic, sedative, tranquilizing, analgesic, anti-convulsant, anti-tussive, myorelacant, anti-rheumatic, anti-hypotensive, anti-allergic, bronchodilating, anti-diabetic, anti-cancer and antimicrobial.[1] Examples of compounds approved for commercial sales are shown below. The quinazoline prazosin (Minipress) is used to treat high blood pressure, and the quinazolines gefitinib (Iressa), erlotinib (Tarceva), lapatanib (Tyverb), vandetanib (Zactma) and trimetrexate are used in the treatment of various cancers. As of 2010, these compounds combined have generated billions of dollars in sales worldwide. 4-Quinazolinones such as nolatrexed and raltitrexed are also used to treat cancer, and more recently, nolatrexate, has been shown to be an effective anti-parasitic agent. Other 4-quinazolinones, such as albaconazole, are used as anti-fungals, while afloqualone (Arofuto) has been used as a muscle relaxant. Finally, the 2-quinazolinone proquazone (Biarison) has been used as an effective nonsteroidal anti-inflammatory drug (NSAID).

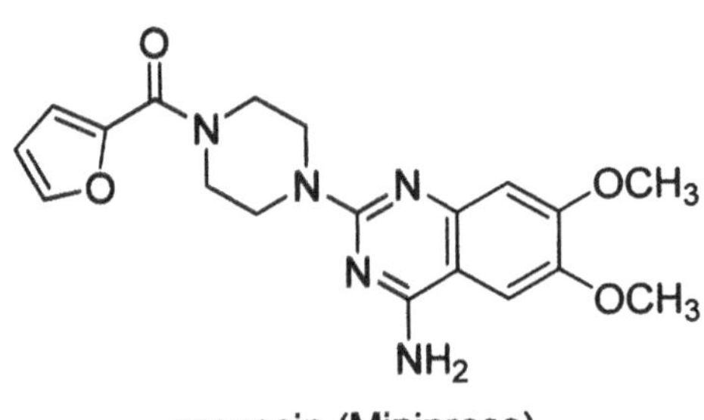

prazosin (Minipress)

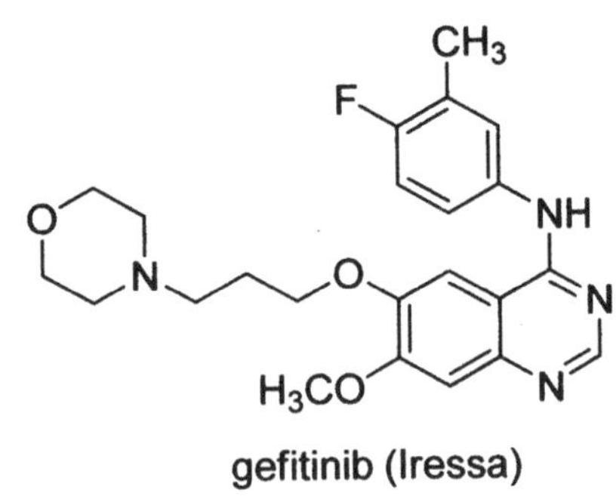

gefitinib (Iressa)

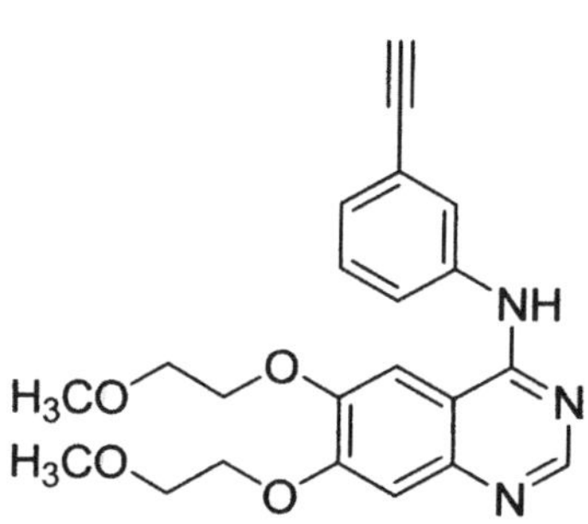

erlotinib (Tarceva)

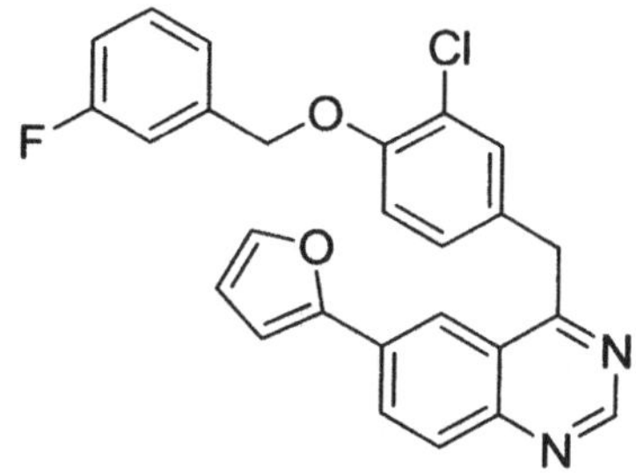

lapatinib (Tyverb)

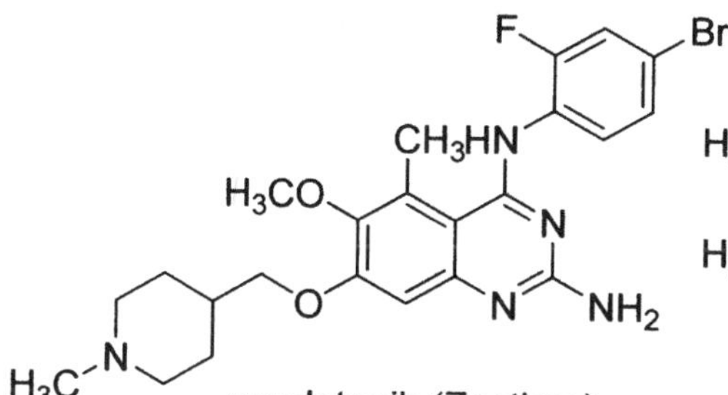

vandetanib (Zactima)

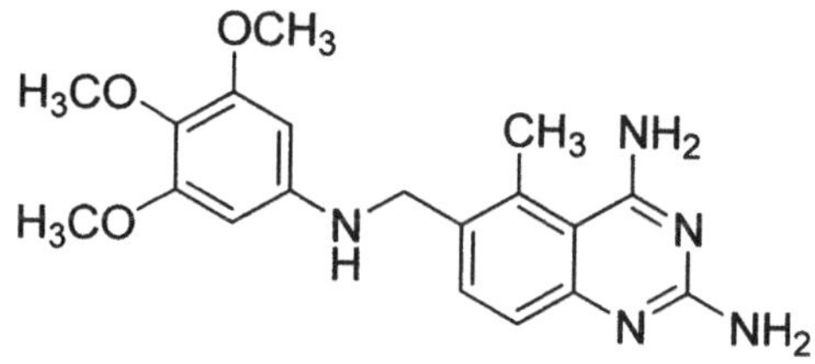

trimetrexate

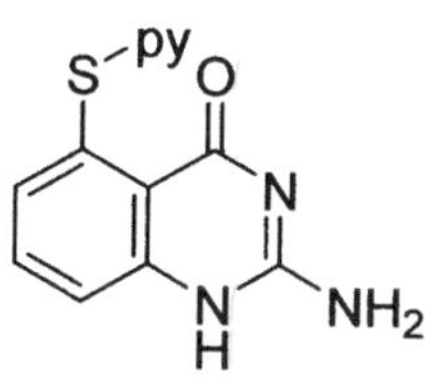

nolatrexed

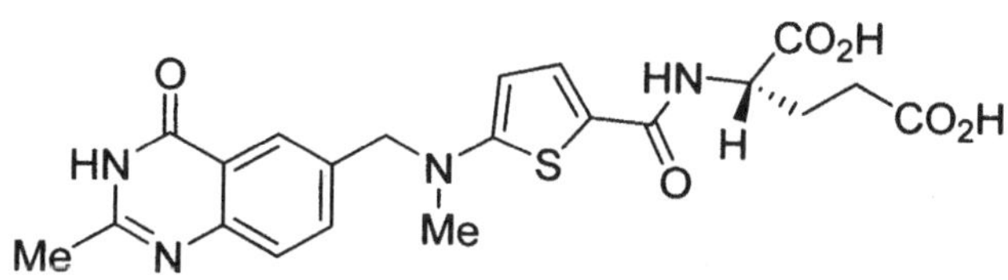

raltitrexed

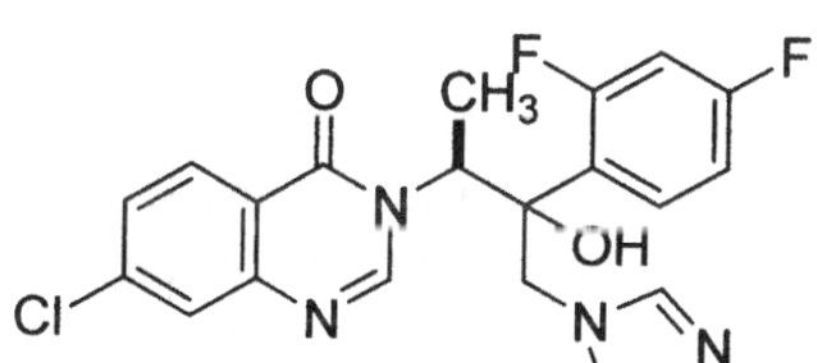

albaconazole

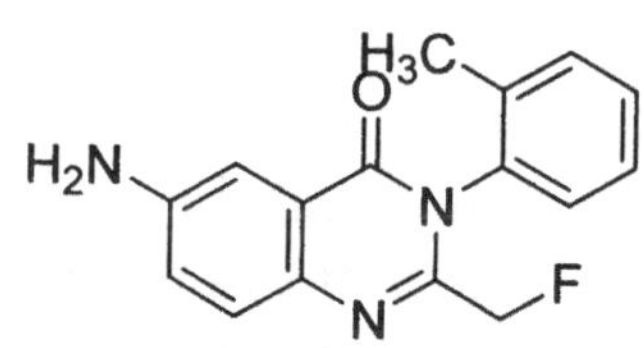

afloqualone (Arofuto)

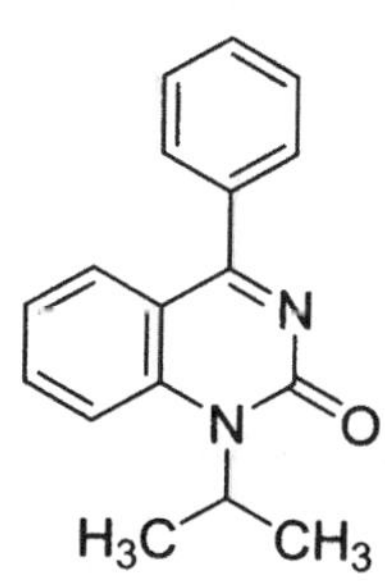

proquazone (Biarison)

14.2 Reactions of Quinazolines and Quinazolinones

14.2.1 Reactions at C4

There are few general methods for the direct functionalization of quinazolines and quinazolinones because of their deactivated nature. Therefore, these compounds are more commonly constructed from acyclic precursors. For those examples that do exist, S_EAr reactions generally take place on the benzene ring, while nucleophilic substitution reactions generally occur on the diazine ring and are selective for the *C4* position due to the electropositive nature of this position. Reactions can take place at the *C2* position, but generally only if the *C4* position is already occupied.

C4 Substitution, favored:

C2 Substitution:

The most common methods for directly functionalizing quinazoline and quinazolinone ring systems involve halogenation (usually chlorination) or thiomethylation followed by substitution with an appropriate nucleophile. Despite the limitations of this reaction sequence, a number of chemically and biologically relevant quinazoline and quinazolinone intermediates and products have been prepared in good yields. Examples where this chemistry has been employed in the synthesis of structurally significant and/or biologically relevant *C*4 substituted quinazolines and quinazolinones are highlighted in the sections below.

Halogenation

4-Chlorination of quinazolines can be accomplished using either phosphorous oxychloride ($POCl_3$),[4] thionyl chloride ($SOCl_2$),[4] or the phosphonium salt of *N*-chlorosuccinamide (NCS)[4]. The phosphonium salt of *N*-chlorosuccinamide (NBS)[4] can be used to generate the 4-bromo quinazoline derivative; however, the yield is much lower due to the instability of the brominated product.

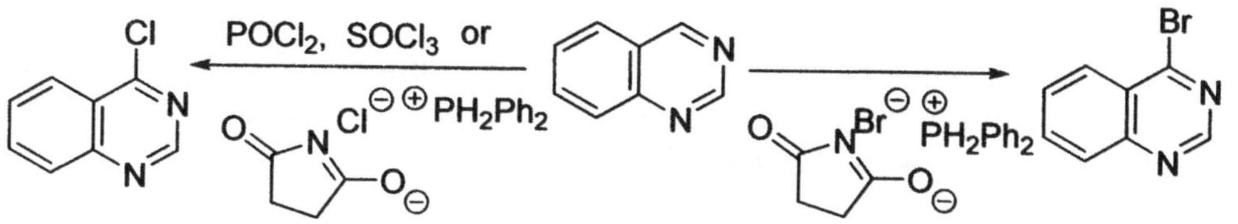

Multiple halogenations are possible when excess chlorinating reagent is used. For example, 2,4-dichloroquinazolines, which are important starting materials for synthesizing 2,4-disubstituted quinazolines, can be prepared using excess $POCl_3$.[5]

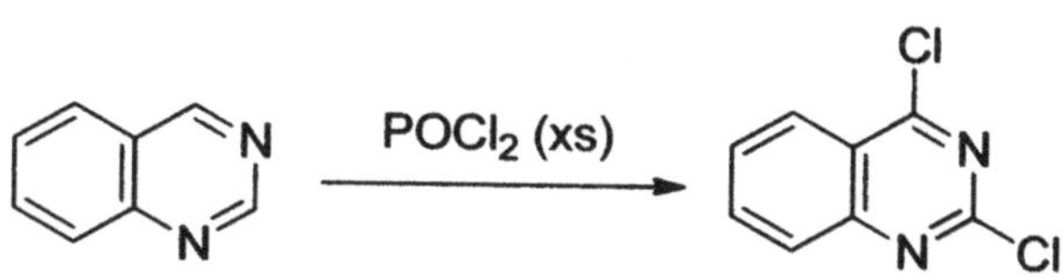

Thiomethylation

Denny and co-workers used a two-step sequence to prepare 4-thiomethyl substituted quinazolines as versatile intermediates for the synthesis diversely substituted quinazolines.[6] Treatment of 6-methyl-6*H*-pyrrolo[2,3-g]-quinazoline with Lawesson's reagent, followed by methylation using methyl iodide under basic conditions, gave the corresponding quinazoline in 32% yield over two steps.

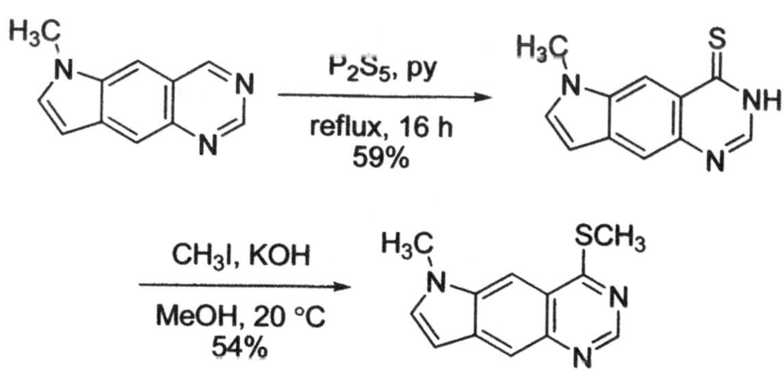

Substitution

Substitution of 4-chloro- and 4-thiomethyl quinazoline derivatives occurs readily in the presence of a nucleophile to provide the corresponding 4-

substituted quinazoline as illustrated below. This reaction formally proceeds by an addition-elimination whereby a nucleophile, Nu, attacks the *C*4 carbon pushing electrons on to *N*3. The resulting tetrahedral intermediate subsequently collapses with concomitant loss of the leaving group X to produce the *C*4 substituted product. Such ubstitution reactions have been employed in the synthesis of a number of biologically relevant quinazolines and quinazolinones as illustrated below.

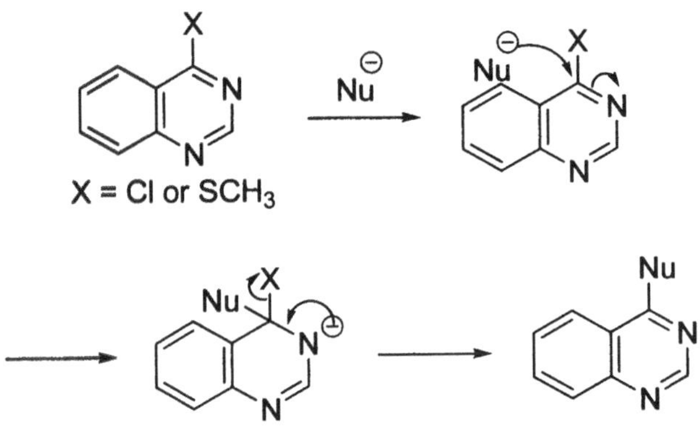

Fuller and co-workers employed a substitution reaction in their synthesis of a fluorescent photo-affinity probe, AX7593, for the detection of epidermal growth factor receptor (EGFR) signal transduction processes associated with solid tumor formation.[7] Reaction of 7-(benzyloxy)-4-chloro-6-methoxyquinazoline and 1,3-diaminobenzene in the presence of hot isopropanol generated the desired 4-quinazoline in 97% yield. Five additional steps (not shown) were required to access AX7593.

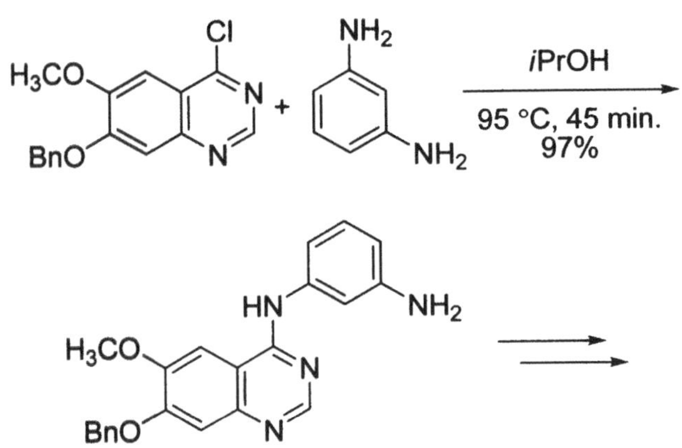

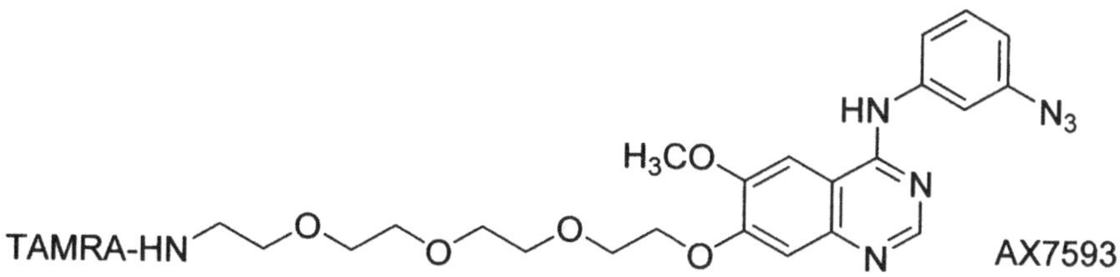

2,4-Dichloroquinazolines have also been shown to undergo substitution reactions selectively at the *C*4 position to generate the corresponding 4-substituted quinazolines. For example, reaction of 2,4-dichloroquinazoline with 2-lithiothiofuran gave the corresponding 4-thiofuran in a regioselective fashion and in 79% yield.[8] The 4-phenyl derivative was also produced in 69% yield using the same method.

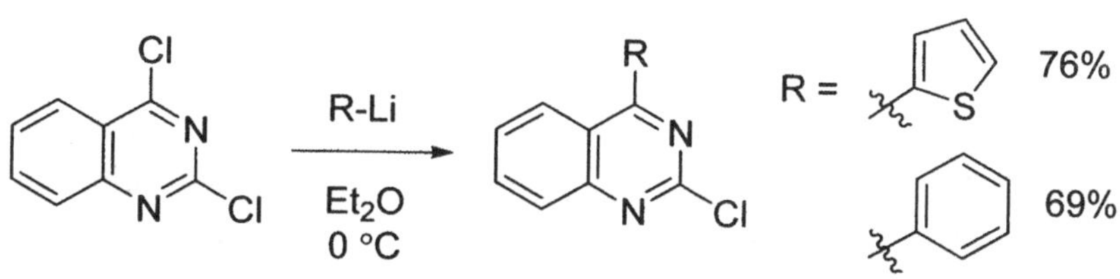

Lee and co-workers used 2,4-dichloroquinazolines in their development of 4-benzylamino phosophdiesterase (PDE) inhibitors.[9] Treatment of 2,4-dichloro-6-((triisopropylsilyl)ethynyl)quinazoline with benzyl amine under thermal conditions gave the corresponding 4-substituted product selectively. A yield was not reported for this reaction.

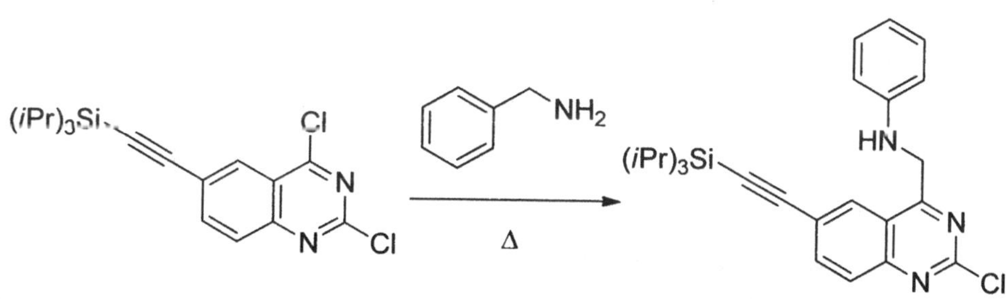

Denny and co-workers used their 4-thiomethyl substituted quinazolines to generate 4-amino substituted quinazolines as potential anti-tumor therapeutics.[6] Treatment of 6-methyl-4-(methylthio)-6*H*-pyrrolo[2,3-g]quinazoline with 4-bromoaniline under thermal conditions gave the desired 4-substituted quinazoline in 85% yield.

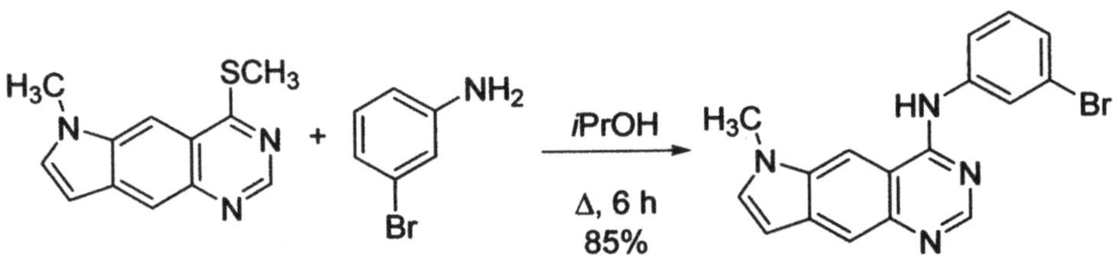

14.2.2 Reactions at C2

In general, reactions at the *C2* position of quinazolines are less favorable than reactions at the *C4* position and generally occur only when *C4* is already substituted. If *C2* selectivity is required, then quinazolinones are often used in place of quinazolines, as demonstrated in many of the examples below.

Substitution

Lee and co-workers extended their approach for the synthesis of 4-benzylamino PDE inhibitors to synthesize 2-imidazoylyl-4-benzylamino 6-substituted 2,4-dichloroquinazolinones.[9] Treatment of *N*-((2-chloro-6-((triisopropylsilyl)ethynyl)quinazolin-4-yl)methyl)aniline with imidazole under thermal conditions gave the desired product in 63% over two steps after deprotection of the silyl protecting group.

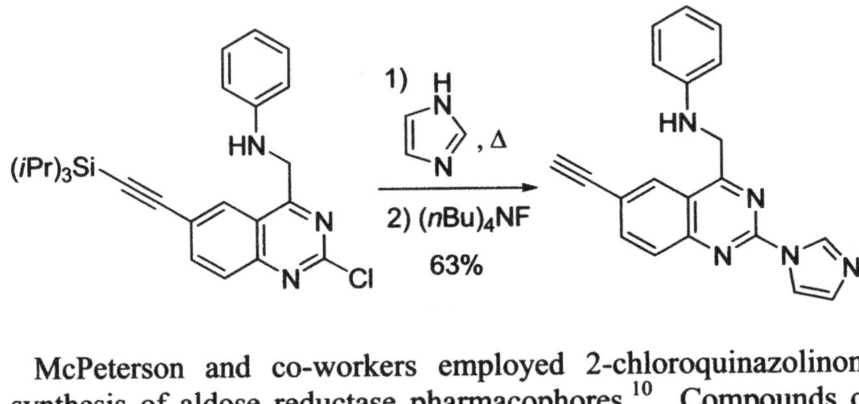

McPeterson and co-workers employed 2-chloroquinazolinones in their synthesis of aldose reductase pharmacophores.[10] Compounds of this class are useful reagents for monitoring processes such as glucose metabolism. Reaction of 2-chloro-6,7-dimethoxyquinazolin-4(3*H*)-one with benzyl amine gave the corresponding 2-substituted quinazolinone in high yield. The authors showed that this reaction worked equally well when aniline was used instead of benzyl amine (not shown) to produce the corresponding anilinoquinazolinone.

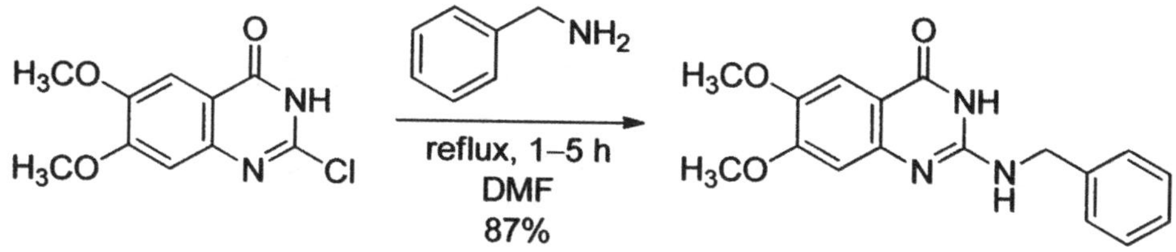

Kim and co-workers showed that the nitrile group of 3-methyl-2-cyano-4(3*H*)quinazolinones could readily exchange with a number of thio-, amino- and oxo-based nucleophiles to produce the corresponding 2-substituted quinazolinones in good to excellent yields.[11] For example, reaction of 3-methyl-2-cyano-4(3*H*)quinazolinone with sodium ethoxide in ethanol gave the desired 2-ethoxy substituted product in 96% yield.

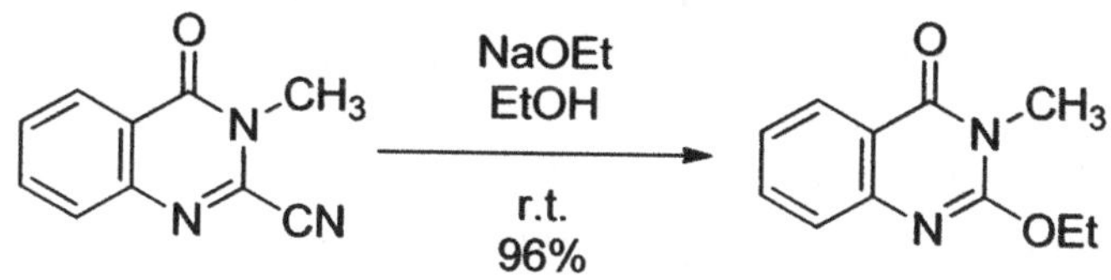

14.2.3 Metal-Mediated Substitution Reactions

Metal-mediated strategies have recently become important for the preparation of a number of synthetic compounds and natural products. In general, these processes employ mild reaction conditions, are tolerant of a number of different functional groups, and can be selectively tuned to produce the desired products in a regio- and/or stereo-selectively fashion. In recent years, metal-mediated processes have been expanded to both the construction and direct functionalization of quinazoline and quinazolinone ring systems. Several examples highlighting metal-mediated substitution reactions of biologically active compounds are illustrated below. The metal-mediated construction of quinazoline and quinazolinone ring systems is covered in the next section.

Mueller and co-workers used an iterative cross-coupling strategy to prepare a series of highly substituted quinazolines for evaluation as anti-asthmatics.[12] In one example, Stille coupling of 6,7-disubstituted 2,4-dichloroquinazolines with 3,5-dicyclopropylmethloxyphenyl tributyltin hydride gave the 4-substituted product exclusively and in good yield. A second reaction employing methyllithium, zinc chloride and palladium tetraphenylphosphine gave the 2,4-disubstituted product in high yield. This compound was shown to have excellent preliminary anti-inflammatory properties.

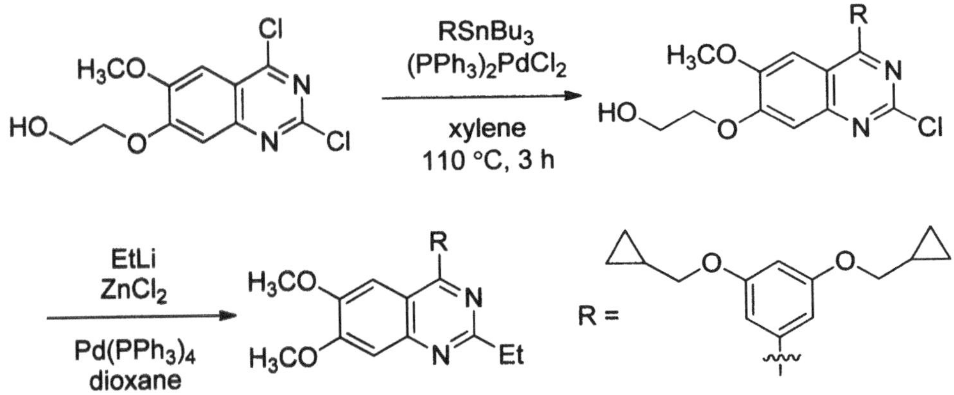

Trialkylalanes have been employed in the production of a series biologically relevant 2,4-disubstituted quinazolines.[13] In one example, reaction of 2,4-dichloroquinazoline with methylalane gave the corresponding 4-substituted product selectively in 76% yield. Treatment of the 4-substituted product with *iso*-butylalane under the same conditions gave the 2,4-disubstituted product in 61% yield over two steps.

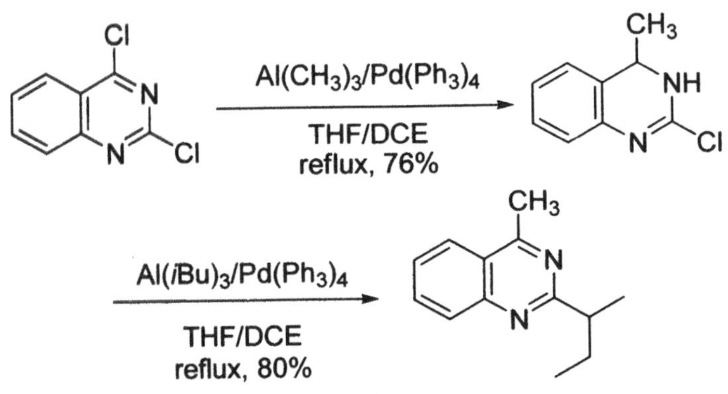

Harayma and co-workers used a palladium-assisted cyclization to synthesize rutaecarpine, a compound that has been shown to induce hypotension and vasorelaxation.[14] Treatment of an *N*-acyl-2-bromo-indole tethered quinazolinone with palladium (II) acetate in the presence of PCy_3 and potassium acetate furnished the desired product in 89% yield.

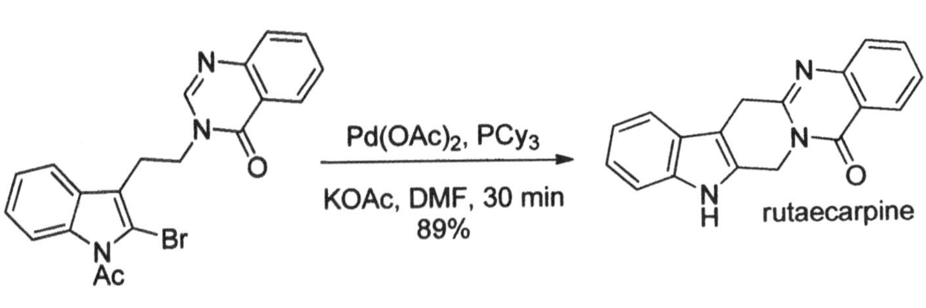

14.3 Quinazoline and Quinazolinone Synthesis

The Bischler and Niementowski syntheses are the most important methods for the synthesis of quinazolines and quinazolinones, with the latter and more well-known reaction being an improvement on the Bischler synthesis. Additional methods for the synthesis of quinazoline and quinazolinone ring systems involve various rearrangement reactions and metal-mediated processes. Examples of these reactions are illustrated below.

14.3.1 Bischler Reaction

The Bischler synthesis, first described in 1893,[15] involves the fusion of *N*-acylanthranilic acid and ammonia to generate the corresponding 2-substituted quinazolinone, which exists in equilibrium with the 4-hydroxy quinazoline tautomer. Bischler's original synthesis, shown below, involved direct conversion of ammonium-*N*-acetylanthranilate to the corresponding 2-alkyl-2,3,4-dihydro-4-quinazolinone under thermal conditions.

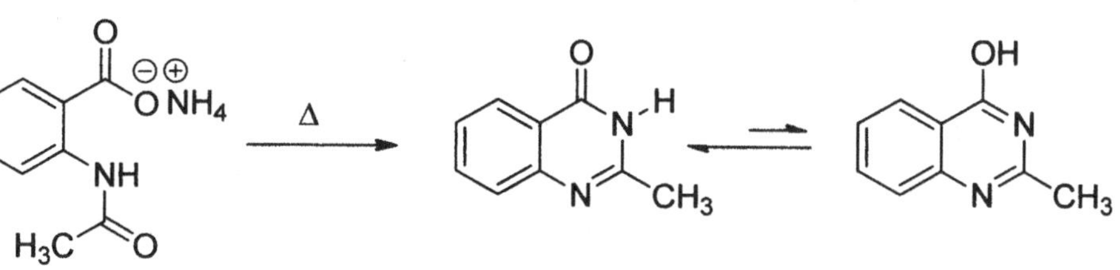

The mechanism of the Bischler synthesis is illustrated below. Fusion of ammonium-*N*-acetylanthranilate generates the diamide, which then undergoes cyclodehydration to form the corresponding quinazolinone. The quinazolinone then tautomerizes to the 4-hydroxyquinazoline.[16]

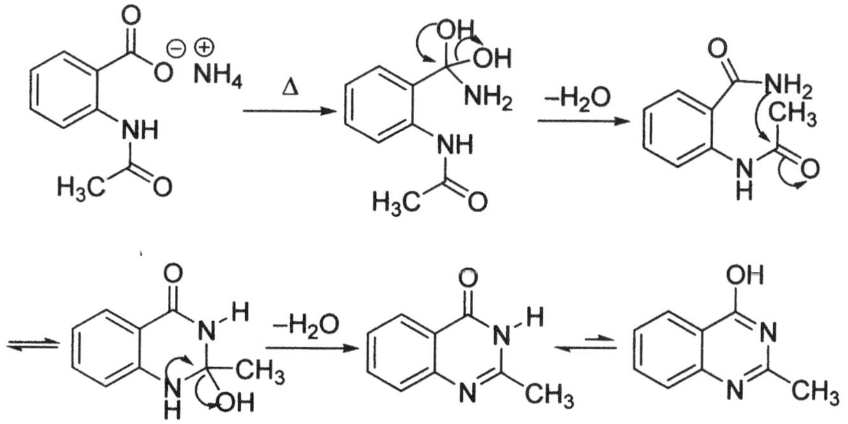

The Bischler synthesis is still practical today for the synthesis of 2-substituted quinazolines and quinazolinones. However, it has been largely replaced by the Neimentowksi reaction, which can be conducted under milder conditions and is therefore more practical for synthesis highly functionalized quinazolines and quinazolinones.[17]

14.3.2 Niementowski Reaction

The Niementowski synthesis, first described in 1895,[17] remains one of the most important methods for synthesizing quinazolines and quinazolinones. This reaction involves condensation of anthranilic acids with formamide or acetamide derivatives to form the intermediate quinazoline-4(3*H*)-ones under thermal conditions.

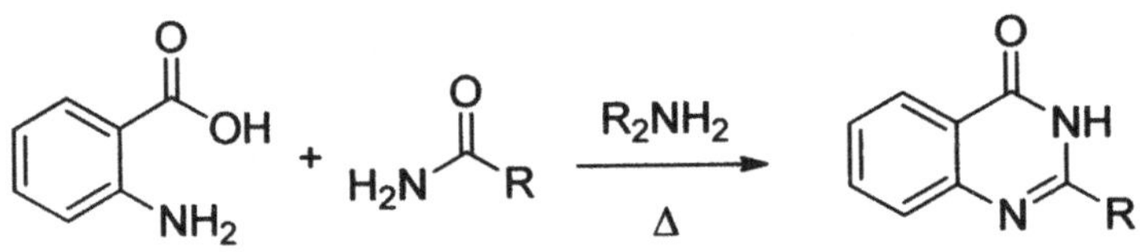

The mechanism for the Niementowski synthesis was described by Bogart and Gotthelf in 1900,[18] and is illustrated below. The reaction begins with the condensation of *o*-anthranilic acid and an acetamide derivative to form the corresponding amide with concurrent loss of ammonia. This is followed by addition of ammonia (generated *in situ*) or excess acetamide[19] to form the corresponding amide, which readily undergoes cyclodehydration under thermal conditions to form the corresponding 2-substituted quinazolinone. This quinazolinone can then tautomerize to the 4-hydroxyquinazoline when R = H.[16]

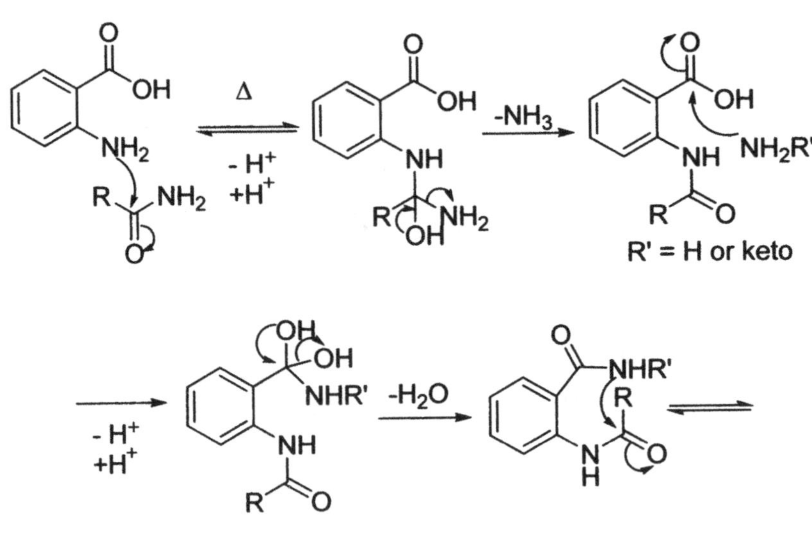

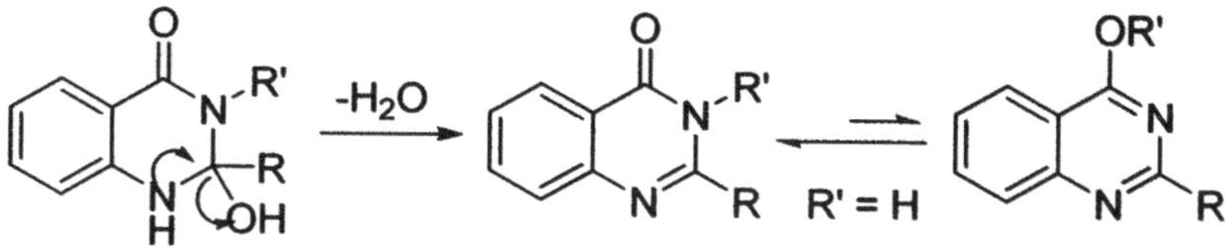

The Niementowski synthesis is still used today to prepare a host of quinazolines and quinazolinones of biological significance. For example, the original Niementowski synthesis was used as a key step in the synthesis of gefitinib (Iressa) by Richards and co-workers.[20] Fusion of 4,5-dimethyoxyanthranilic acid and formamide furnished the desired quinazolinone in 20% yield. This yield was significantly improved in later work by Örfi and co-workers using formadine acetate and formamide under microwave conditions to provide the product in quantitative yield.[21]

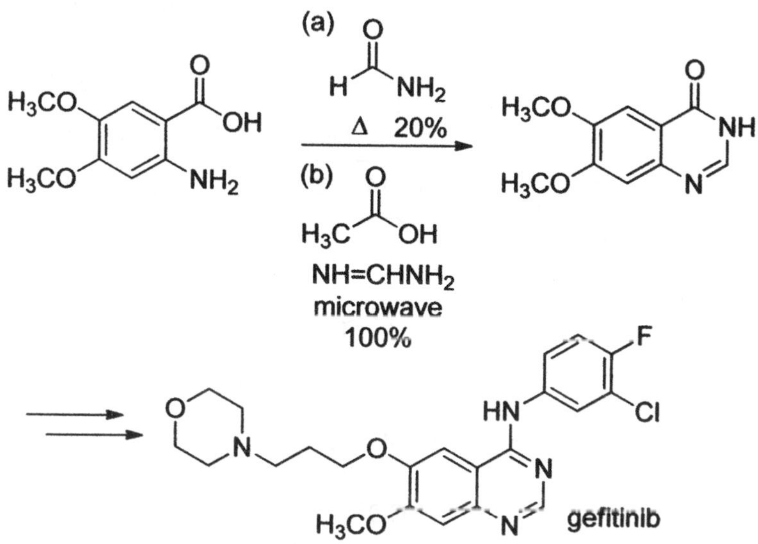

Song and co-workers used the Niementowski reaction to prepare a series of 6-fluoro-4-alkyl(aryl)thioquinazolines with anti-fungal properties.[22] Fusion of 3-fluoro anthranilic acid with formamide gave the desired quinazolinone in 54% yield. Treatment of the quinazolinone with an alkyl halide in the presence of sodium hydroxide and a phase transfer catalyst (BTEAB) gave the corresponding 4-thiosubstituted quinazoline in good to high yields. Three of the compounds generated (R = allyl, ethyl, and propyl) showed excellent anti-fungal properties against several phytopathic fungi including *Fusarium oxysproum*.

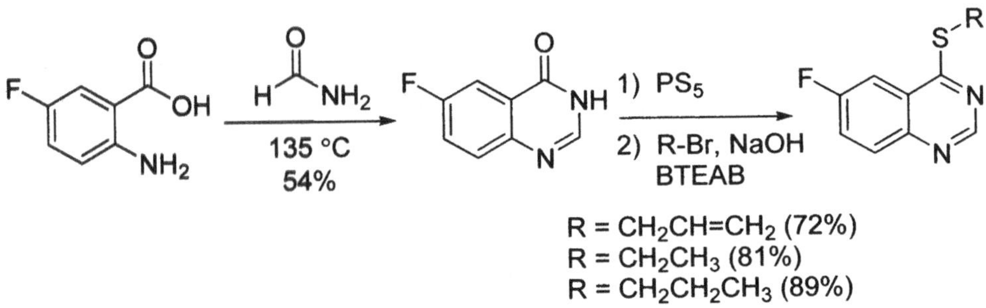

Independently, Zhou and co-workers used the same approach in their synthesis of highly functionalized quinazolines as potential anti-cancer therapeutics.[23] In one example, fusion of 2-amino-3,4,5-trimethoxybenzoic acid with formamide under thermal conditions gave the desired quinazolinone in 25% yield. A dibenzofuran quinazoline derivative produced using this process exhibited anti-proliferative properties against several tumor types, including Bcap-37, PC3, A431, and BGC823 cell lines.

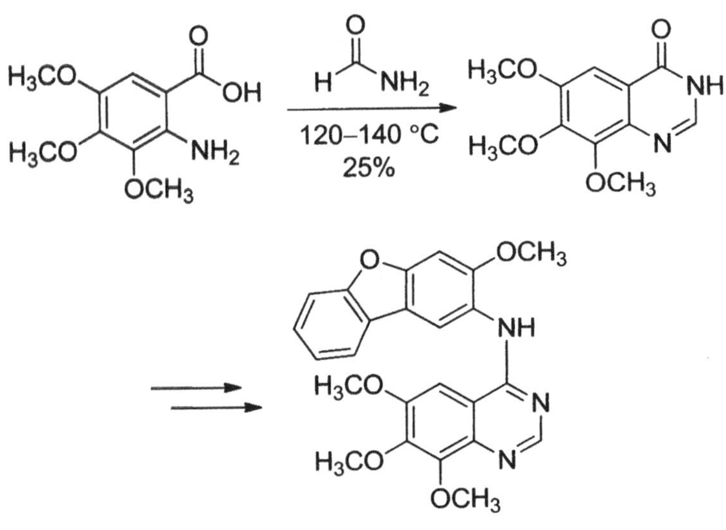

Hattori and co-workers used a Niementowski approach to synthesize a series of orally active 2-substituted-4(3*H*)quinazolinones as inhibitors of poly(ADP-ribose) polymerase.[24] Treatment of a substituted 4-fluoro amide (which was readily synthesized in three steps from commercially available 2-amino-6-fluoro-benzoic acid) with aqueous sodium hydroxide, produced the desired quinazolinone, which showed significant inhibitory activity.

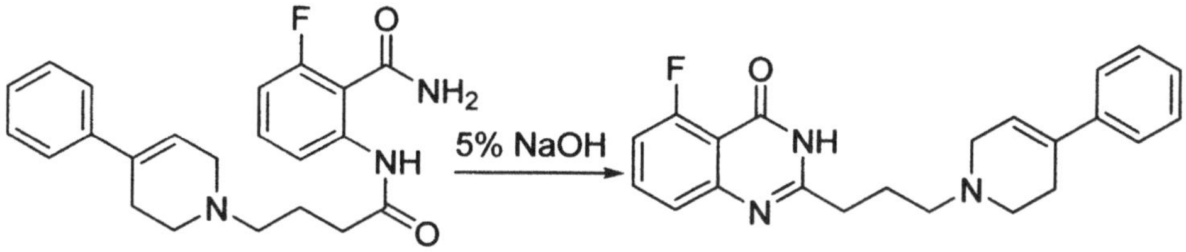

A significant number of variations on the Niementowski reaction have been developed over the years in an effort to improve reaction yields and product purity. The most common modifications involve the use of imidates, imidoylchlorides, dithiazoles, benzonitriles, or amidines. Select examples of these modifications are highlighted below.

Reid and co-workers prepared a number of napthanilic quinazolinones by condensing 3-amino-2-naphthoic acid with a furyl imidate in methanol under thermal conditions.[25] The desired product, 2-(furan-2-yl)benzo[g]quinazolin-4(3*H*)-one, was produced in 43% yield. Compounds of this class have been shown to have effective anti-cancer and antibacterial properties.

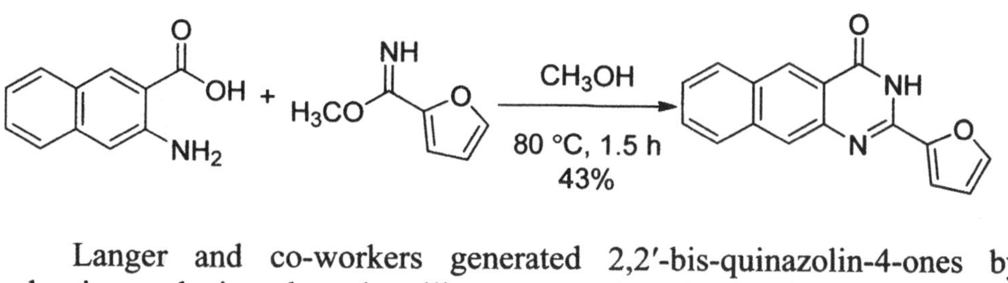

Langer and co-workers generated 2,2′-bis-quinazolin-4-ones by condensing substituted anthranilic esters with substituted bis(imidoyl)-chlorides.[26] These compounds, which are structurally similar to the quinazoline alkaloid febrifugine, have the potential to serve as anti-malarials. Treatment of methyl-2-amino-4,5-dimethoxybenzoate with bis(*p*-methoxyl-phenylimidoyl)chloride in the presence of two equivalents of TEA in refluxing toluene gave the desired product in 60% yield.

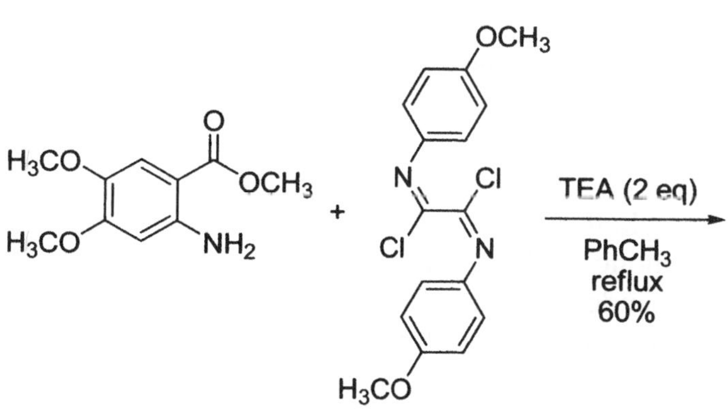

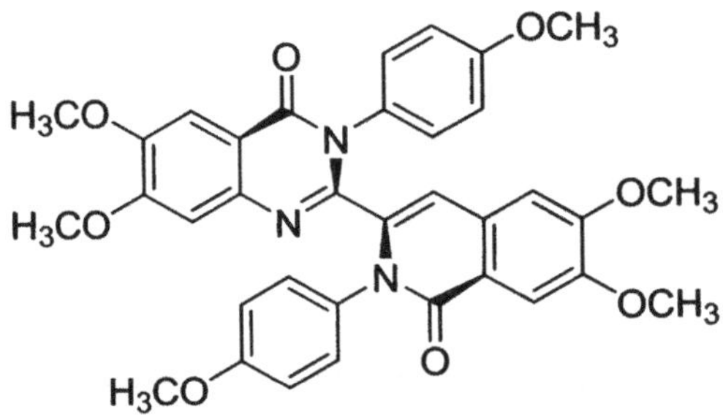

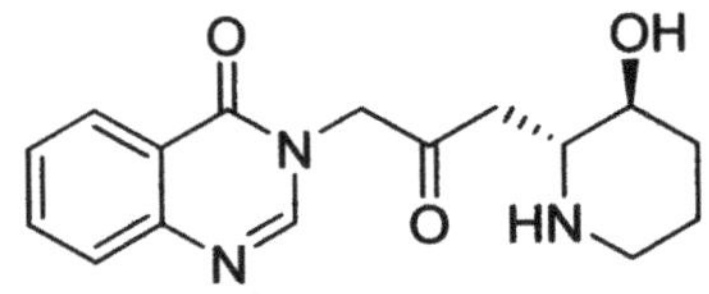

febrifugine

Kim and co-workers[27] used dithiazoles to prepare 3-substituted 2-cyanoquinazolinones. Treatment of methyl anthranilate with Appel's salt in the presence of two equivalents of pyridine furnished the corresponding methyl-*N*-(4-chloro-5*H*-1,2,3-dithiazol-5-ylidene)anthranilate (not shown). Further reaction of the ylidene with methyl amine provided the desired cyanoquinazoline in 73% yield. This work set the framework for Besson and co-workers[28] synthesis of a tricyclic congener of desoxyvasicinone, a compound with cytotoxicity against a murine leukemia P-388 cell line.

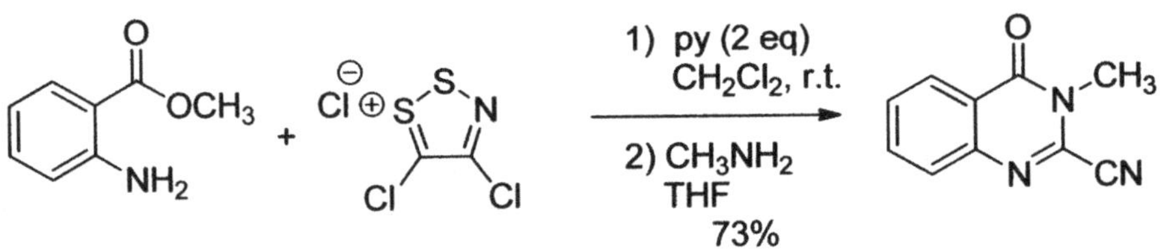

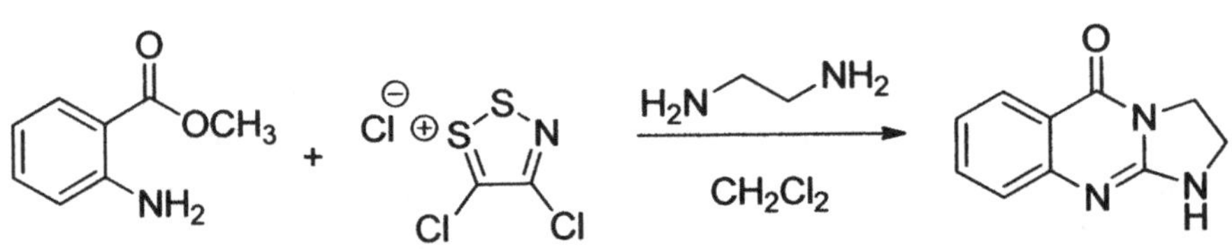

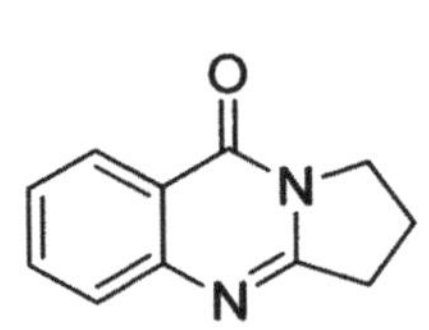

desoxyvasicinone

Bogert and Hand[29] used 2-aminobenzonitrile derivatives to generate 2-styrylquinazoline-4(3*H*)-one analogues. Treatment of 2-aminobenzonitrile with 3-phenyl-acryloyl chloride followed by oxidative ring closure, gave the desired quinazolinone in 29% yield. 6-Chloro derivatives of this class of compounds, independently prepared by Hamel and co-workers[30] (not shown), were shown to have growth inhibitory activity against leukemia cells.

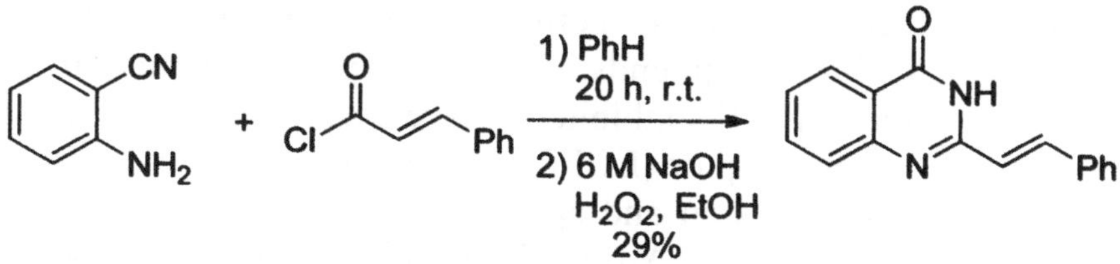

Kotsuki and co-workers used amidines to synthesize 2-substituted quinazolines.[31] Reaction of 2-fluoro-5-nitro-benzaldehyde with 4-bromo-benzyl amidine gave the corresponding quinazoline in modest yield. This approach was subsequently used by Kotsuki and co-workers to generate a number of 5-nitro- and 5-cyanoquinazolines in modest yields.

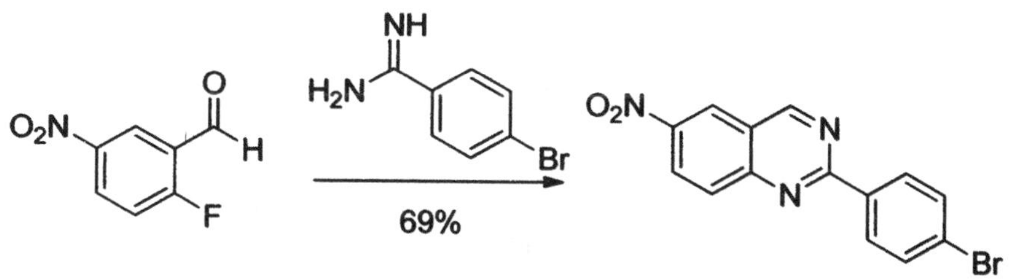

Wilson used a traceless solid-phase approach for the synthesis of the 2,4-quinazoline system of prazosin, an α-1b receptor antagonist.[32] In the final step, treatment of a highly substituted resin-bound quanidine with TFA under thermal conditions gave the desired quinazoline, prazosin, with simultaneous cleavage from the resin. Overall, prazosin was produced cleanly and in 70% yield over five steps using this method.

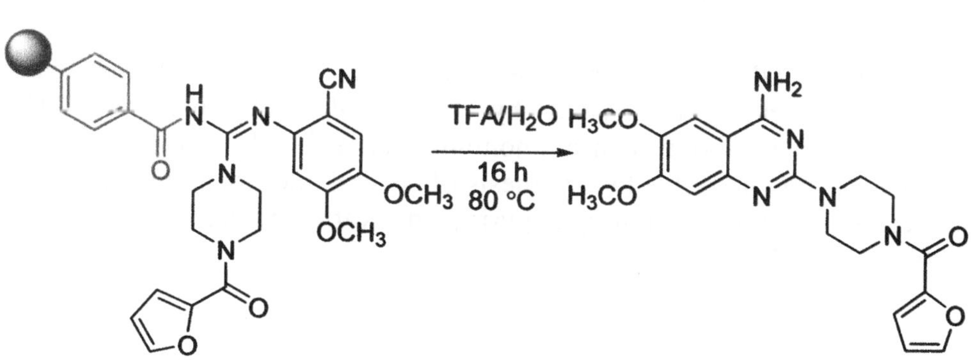

14.3.3 Cyclization and Rearrangement Approaches

Cyclization and rearrangement reactions such as the Dimroth rearrangement have played an important role in the synthesis of quinazoline-containing compounds. For example, Reddy and co-workers[33] applied the Dimroth rearrangement to their syntheses of gefitinib (Iressa) and erlotnib (Tarceva), two 4-anilinoquinazoline compounds. These compounds, which are potent protein tyrosine kinase (PTK) inhibitors, have been shown to inhibit the epidermal growth factor receptor (EGFR) which is responsible for mediating cell division, motility, adhesion and apoptosis. Reaction of *N'*-(2-cyano-4-substituted-5-methoxyphenyl)-*N,N*-dimethylformimidamide with 3-chloro-4-fluoroaniline in acetic acid at 130 °C gave gefitinib in 70% yield. Similarly, reaction of *N'*-(2-cyano-4,5-disubstituted)-*N,N*-dimethylformimidamide with 3-ethynyl aniline in refluxing acetic acid gave erlotinib in 92% yield (not shown). Besson and co-workers[34] recently used a similar, microwave-assisted Dimroth approach in their synthesis of the 4-aminoquinazoline Azixa, a homolog of Iressa and Tarceva (not shown).

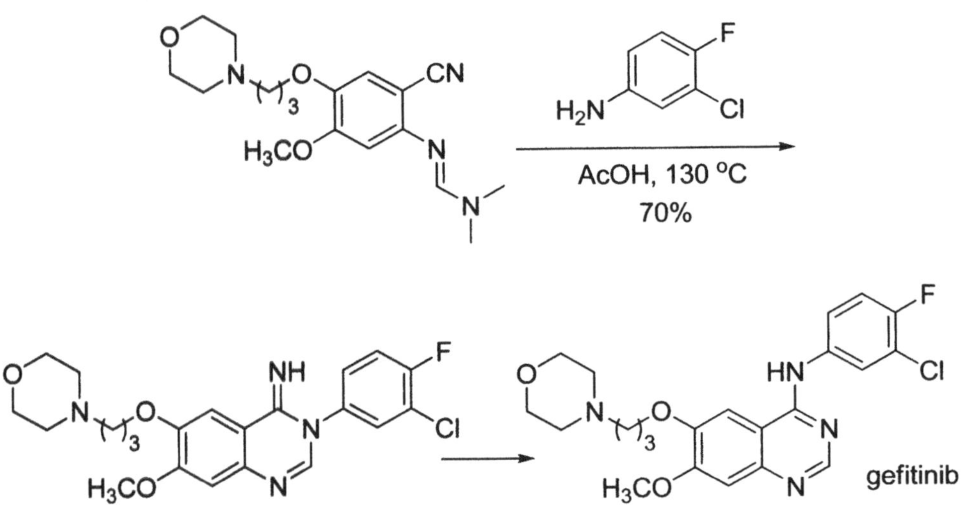

gefitinib

Hamel and co-workers used a novel approach to generate quinazolinones in modest yield by fusion of 6-chloro-2-methyl-4*H*-benzo[d][1,3]oxazin-4-one with ammonium acetate.[30] The authors went on to use this methodology for the synthesis of several additional analogues for biological evaluation, including a pyridyl derivative which showed good activity against L1210 cells.

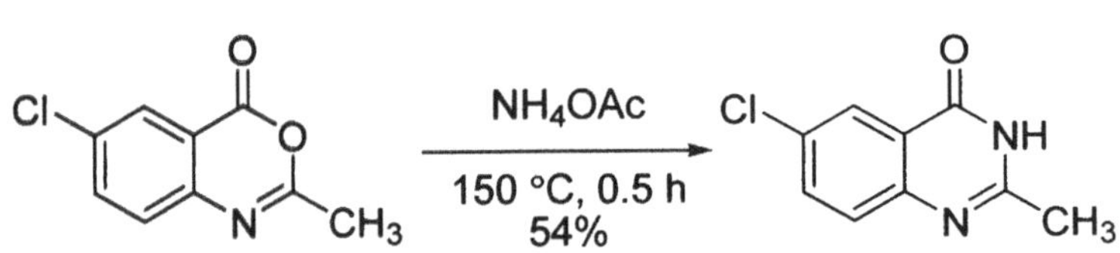

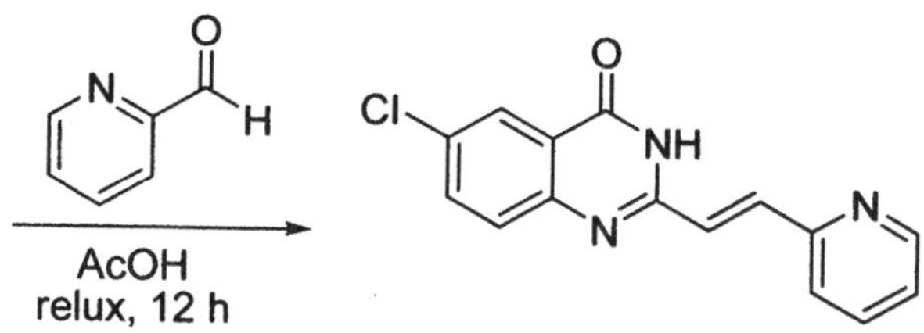

Fabis and co-workers used a base-catalyzed ring contraction to generate 2-pyrryl substituted quinazolinones.[35] Treatment of pyrrolo[2,1-*c*][1,4]benzodiazepine with hydrazine in ethanol at room temperature provided the desired quinazolinone in 60% yield.

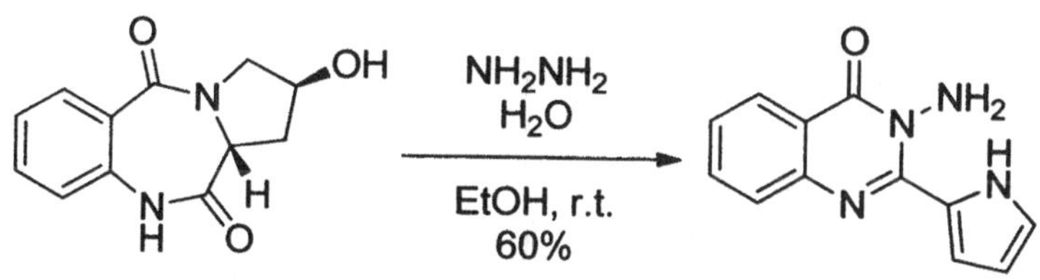

Ganesan and Wang used a base-catalyzed condensation between a tricyclic lactam and a sulfinyl benzoyl chloride derivative in the final step of their total synthesis of the cytotoxic alkaloid luotonin A.[36] The desired product was produced in 85% yield after condensing at room temperature under basic conditions for 13 h.

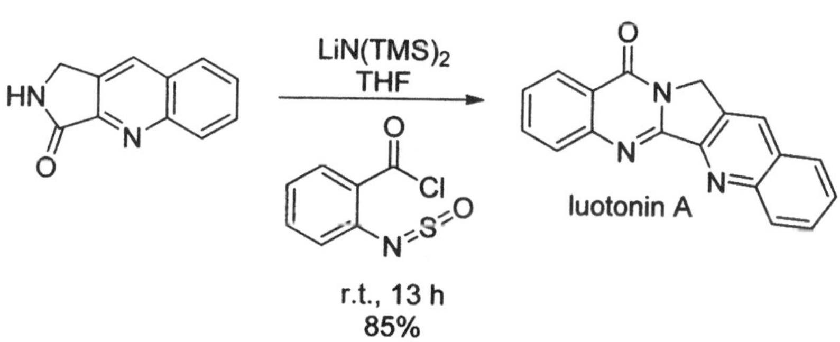

Eguchi and co-workers used an intramolecular aza-Wittig reaction in their synthesis of vasicinone,[37] an indigenous compound used as a remedy for cold, cough, bronchitis, and asthma. Treatment of a TBDMS protected chiral aza-diketone with tributylphosphine in toluene at room temperature gave the corresponding TBDMS protected *l*-vasicinone in 76% yield. Deprotection of the TBDMS group using TBAF gave the desired natural product in 52% yield (97% *ee*) over three steps starting from commercially available 2-azido benzoic acid.

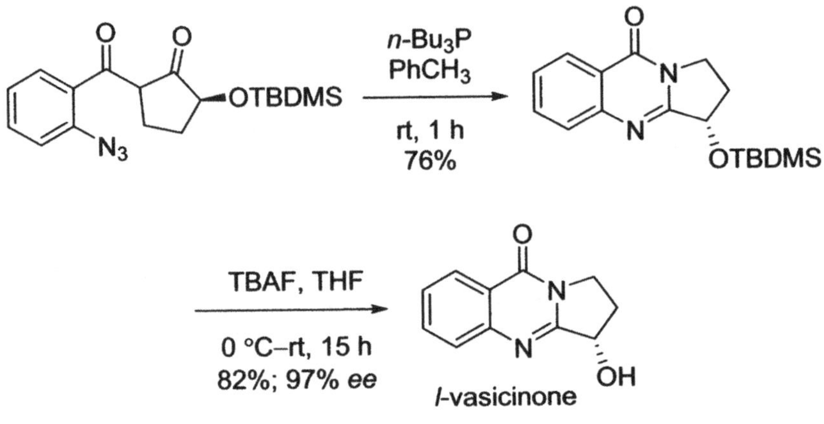

14.3.4 Transition Metal-Promoted Reactions

Transition metal-promoted reactions have recently become important in the synthesis of the basic core structures of quinazolines and quinazolinones. While many of the examples below are not necessarily biologically relevant in and of themselves, they highlight important areas of development in the synthesis of these ring classes that can be employed in the future synthesis of compounds with greater biological significance.

Wantanabe and co-workers used an intermolecular reductive *N*-heterocyclization approach to generate quinazoline.[38] Treatment of 2-nitro benzaldehyde with palladium(II) and molybdenum(V) complexes as catalysts in the presence of carbon monoxide under high pressure generated quinazoline in 29% yield.

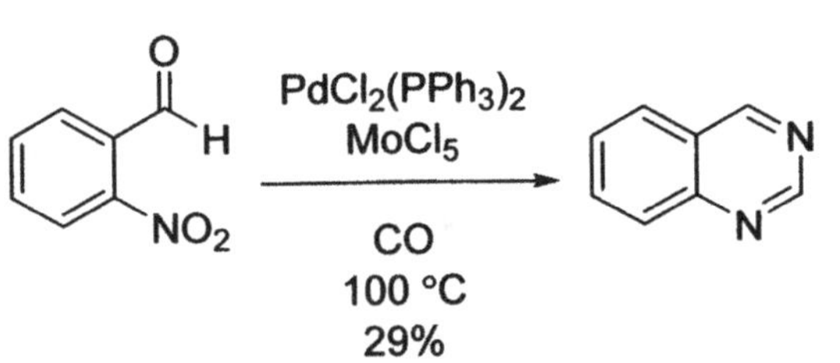

Quinazoline derivatives were prepared by Fu and co-workers using an Ullmann-type coupling.[39] Treatment of (2-bromophenyl)methylamine with benzamide in the presence of potassium carbonate and a copper(I) catalyst under thermal conditions in isopropanol gave the corresponding quinazoline in 87% yield. The authors also reported modest to good yields using several substituted (2-bromophenyl)methylamine and benzamide derivatives (not shown).

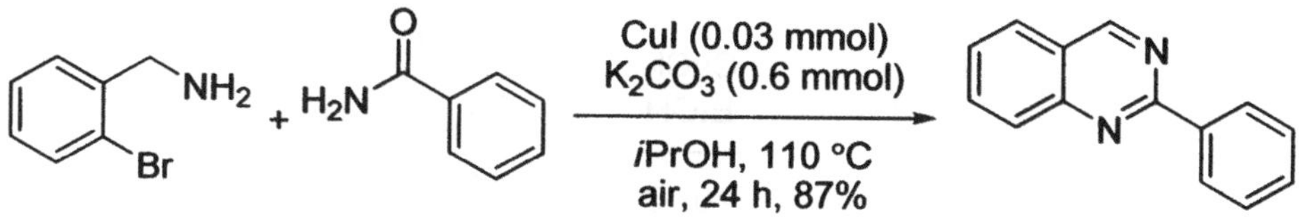

Larksarp and Alper used a palladium(II) catalyzed cyclocarbonylation reaction to generate a series of more complex and biologically relevant 4(3*H*)-quinazolinone derivatives.[40] Compounds of this class, such as the mold metabolite crysogine (*Penicillium chrysogenum*), have been shown to exhibit PTK inhibition and cholecystokinin inhibition as well as antimicrobial, anti-convulsant, anti-depressant and anti-inflammatory properties. Reaction of *o*-iodoaniline, a substituted ketenimine, and carbon monoxide with palladium(II) acetate and 1,1'-bis-diphenylphosphino-ferrocene (dppf) under thermal conditions gave the desired quinazolinone in near quantitative yield. In the same report, the authors showed that similar reactions could also be conducted with isocyanates and carbodiimides (not shown).

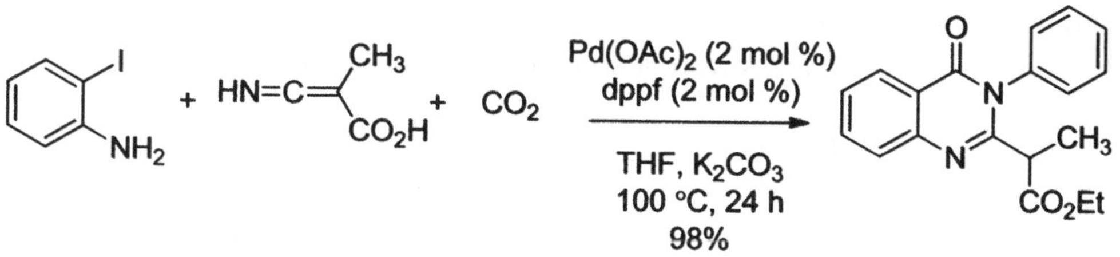

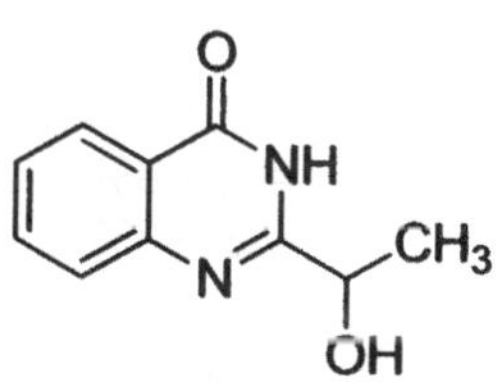

crysogine

Abdel-Jalil and co-workers used a novel copper(II)-catalyzed condensation reaction between anthranilamides and aryl, alkyl, and heteroalkyl aldehydes to generate the corresponding 4(3*H*)-quinazolinones in excellent yields.[41] In one example, treatment of anthranilamide with 2-furaldehyde gave the corresponding 2-substituted quinazolinone in 85% yield. The resulting compound could be used to readily access the antibiotic nitrofurquinazol in just a few short steps.

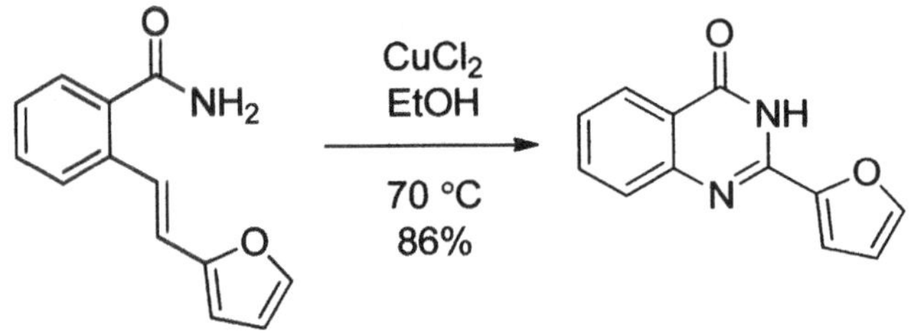

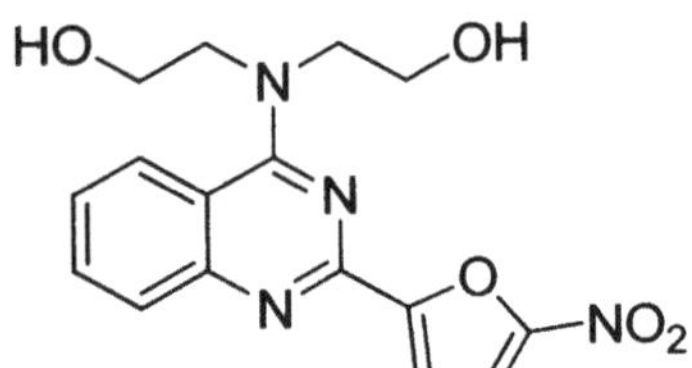

nifurquinazol

Kamal and co-workers used a transition metal-catalyzed approach in their synthesis of desoxyvasicinone.[42] Treatment of an aromatic azide with iron(III) chloride in the presence of sodium iodide and acetonitrile gave the desired product in high yield. Derivatives containing 6,6,6- (n = 2) and 6,6,7-fused (n = 3) systems were also reported in high yield.

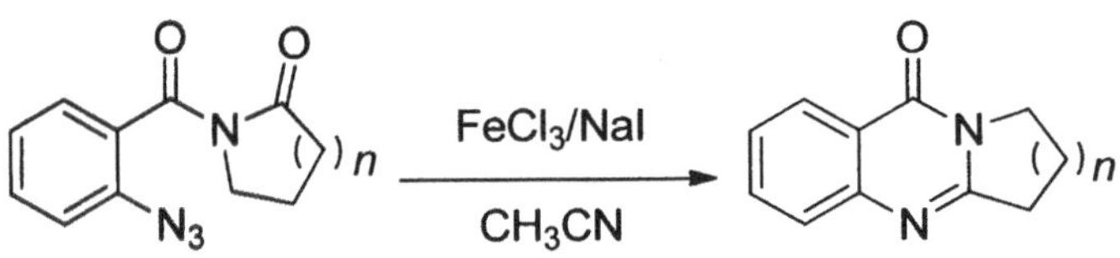

n = 1, 97% (desoxyvasicinone)
n = 2, 88%
n = 3, 95%

14.4 Synthesis of Quinazoline- and Quinazolinone-Containing Drugs

The synthesis of several commercially available and biologically significant quinazolines and quinazolinones are illustrated below. In most cases, the quinazoline or quinazolinone ring system is constructed from acyclic precursors and then functionalized accordingly.

14.4.1 Quinazoline-containing Drugs

Protein tyrosine kinases (PTKs) are vital components of many signal transduction pathways that regulate various cell functions such as cell division, motility, and apoptosis. As such, PTKs are common targets for cancer therapies. Quinazoline-based compounds such as erlotinib have been shown to be effective PTK inhibitors. Erlotinib specifically inhibits epidermal growth factor receptor (EGFR), a PTK that is expressed in high levels in tumor cells of numerous types of cancers, including pancreatic, breast, and lung cancer.

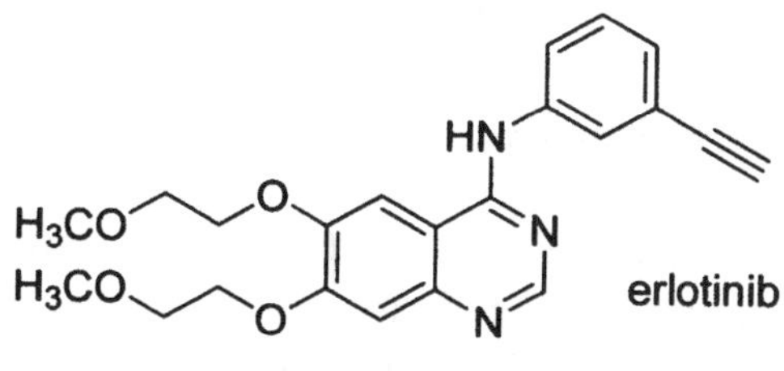

erlotinib

A facile synthesis of erlotinib, patented by Schnur and Arnold, began with the condensation of readily available ethyl-2-amino-4,5-bis-(2-methoxy-ethoxy)benzoate and formamide to produce the corresponding quinazolinone, 6,7-bis(2-methoxyethoxy)quinazolin-4(3*H*)-one, in high yield. Reduction of 6,7-bis(2-methoxyethoxy)quinazolin-4(3*H*)-one with oxalylchloride furnished the corresponding 4-chloroquinazoline in 92% yield.[43] Finally, reaction of the 4-chloroquinazolilne with 3-ethynylaniline under basic conditions afforded erlotinib in 71% yield (56% over three steps).

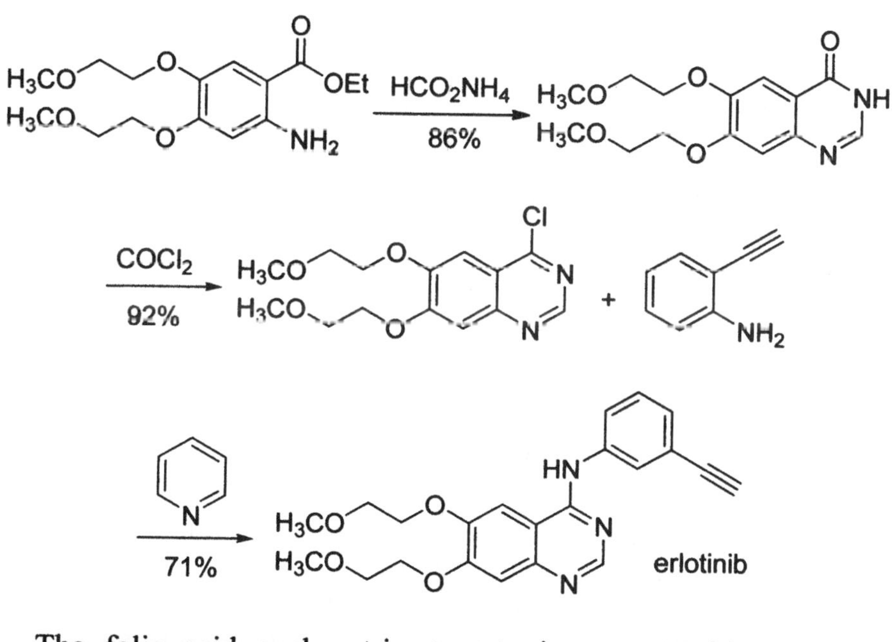

erlotinib

The folic acid analog trimetrexate is a competitive inhibitor of bacterial and protozoan dihydrofolate reductase. This compound is currently used as a treatment for *Pneumocystis carinii* pneumonia (PCP) in patients with immunocompromising conditions.

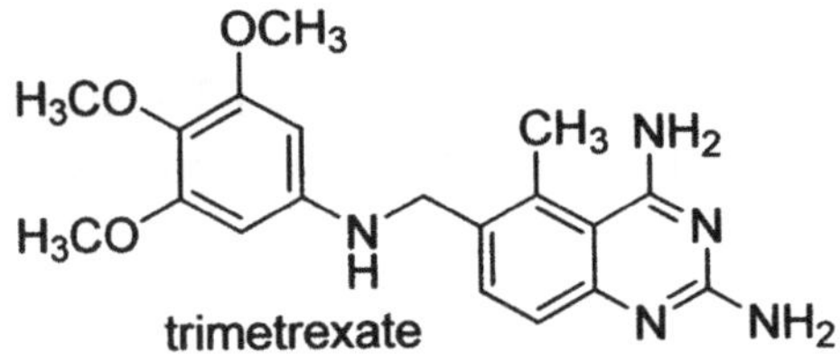

The synthesis of trimetrexate, reported by Davoll and Johnson,[44] began with an acid-catalyzed condensation reaction between 1-nitro-4-chloro-*o*-toluonitrile and guanidine to produce 2,4-diamino-5-methyl-6-nitro-quinazoline as an intermediate for further functionalization. Hydrogenation of 2,4-diamino-5-methyl-6-nitro-quinazoline produced the corresponding 2,4,6-triamino-quinazoline, which was readily converted the corresponding nitrile upon treatment with hydrochloric acid and cuprocyanide. Condensation of the nitrile with 2,3,4-trimethoxyaniline produced trimetrexate in 26% overall yield.

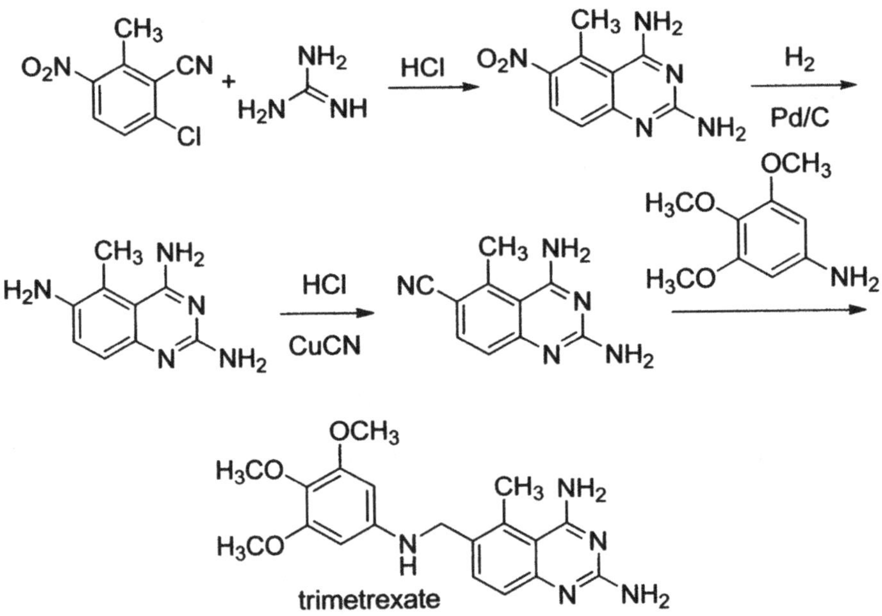

Although never commercially marketed, nifurquinazole has been shown to be a powerful nitrofuran-based bactericidal. Nitrofuran-based antibiotics function by a complex set of mechanisms that result in the degradation of bacterial macromolecules. Rapid reduction of nitrofurans via flavoproteins, specifically nitrofuran reductase, occurs inside the bacterial cell. The highly reactive species that are generated in this process are ultimately responsible for damage to ribosomal proteins and DNA, and they have also been shown to negatively impact bacterial respiration and pyruvate metabolism negatively.

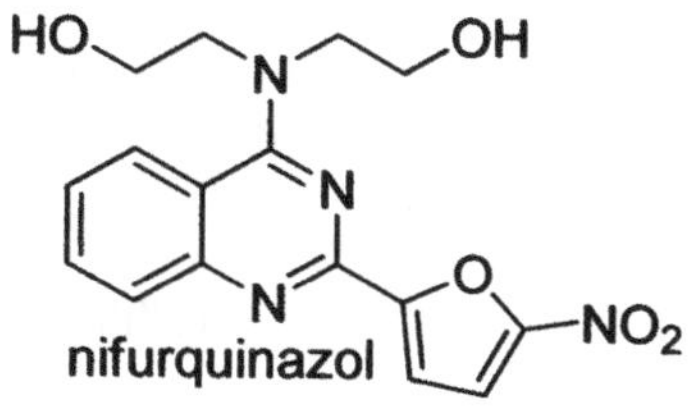

Many antibacterial nitrofurans contain heterocyclic ring substituents attached to the nitrofuran core. Nifurquinazol incorporates a quinazoline derivative attached at the 5-position of a 2-nitrofuran ring. The original synthesis of nifurquinazole formed the quinazoline via the analogous quinazolinone.[45] Condensation of anthranilic acid with ethyl-5-nitro-2-furimidate hydrochloride under basic conditions gave the quinazolinone in 89% yield. This compound was then converted to the corresponding 4-chloroquinazoline by treatment with phosphorous oxychloride. Finally, substitution with diethanolamine afforded the desired compound, nifurquinazol, in 58% yield (42% over three steps).

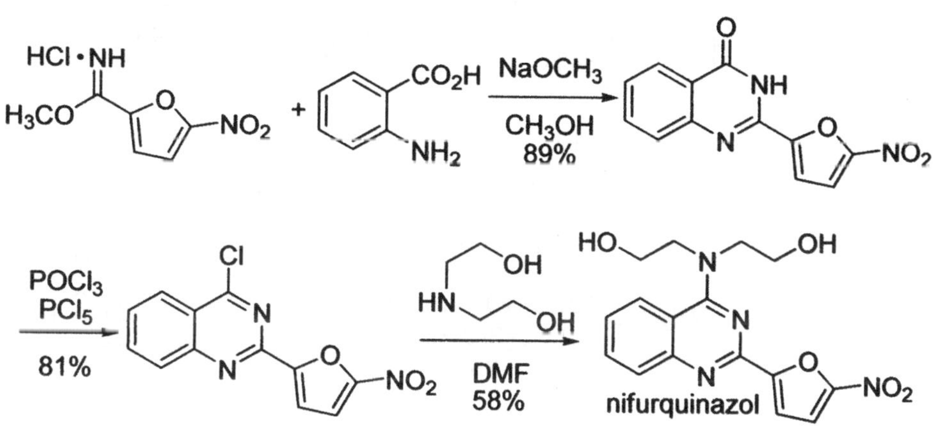

14.4.2 Quinazolinone-Containing Drugs

Many quinazolinone-based compounds function as thymidylate synthase inhibitors by directly binding to the active site of thymidylate synthase. This results in the inhibition of DNA replication as well as DNA damage, S-phase cell arrest, and caspase-dependent apoptosis. Many compounds of this class have been successfully employed as anti-cancer pharmaceutics. Two examples, nolatrexed, a compound used to treat unresectable hepatocellular carcinoma (HCC), and raltitrexed, a first-line treatment against advanced colorectal cancer, are highlighted below.

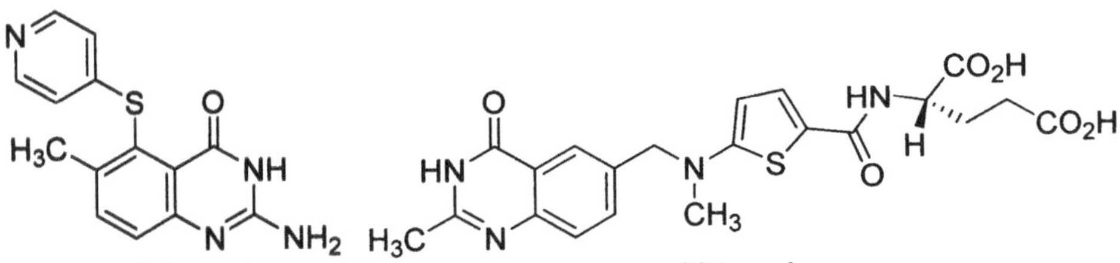

nolatrexed raltitrexed

The original synthesis of nolatrexed, reported by Webber and co-workers,[46] was recently improved by Fang and co-workers.[47] Base-catalyzed hydrolysis of 4-bromo-5-methylisatin proceeded in 84% yield to produce the corresponding methyl ester. Niementowski reaction of the ester with chloroformamidine hydrochloride in diglyme gave the corresponding quinazolinone-hydrochoride in 87% yield. Finally, Ullmann reaction of the hydrochloride with 4-mercaptopyridine produced nolatrexed in 85% yield (62% over five steps).

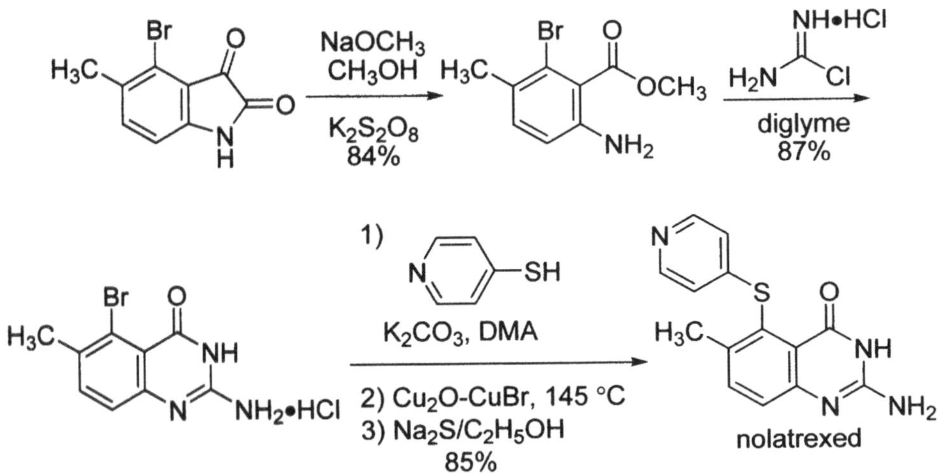

Cao and co-workers[48] reported a facile synthesis to raltitrexed that helped to overcome many of the unwanted *N,N*-dialkylation side products reported in the previous syntheses of this compound.[49] Their synthesis began with the coupling of commercially available 2,5-thiophenedicarboxylic acid and diethyl L-glutamate in the presence of *N,N'*-dicyclohexylcarbodiimide (DCC) to produce the desired product in 60% yield.[48] Conversion of the carboxylic acid using diphenylphosphoryl azide (DPPA) in the presence of *t*-butanol gave the Boc protected amine in 56% yield, which was subsequently converted under standard alkylating conditions using TBAB as a phase transfer catalyst to the methyl amine in 90% yield. Removal of the Boc protecting group using TFA, followed by coupling of the amine with 6-

bromomethyl-2-methyl-4-quinazolinone and hydrolysis of the remaining protecting groups, resulted in the production of raltitrexed in 66% yield (18% over seven steps).

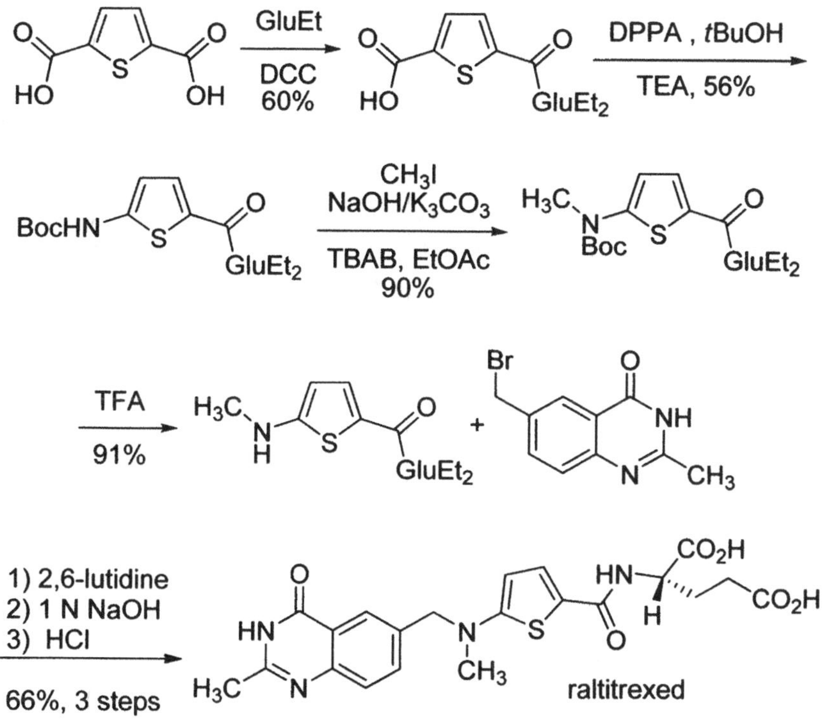

14.5 Problems

14.5.1 Propose a reasonable mechanism for the following transformation:

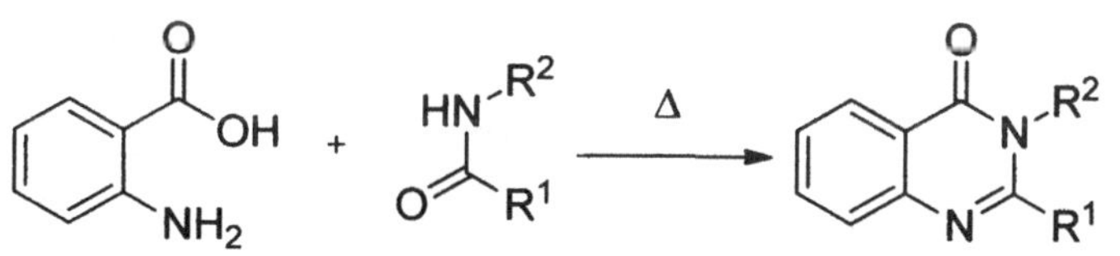

14.5.2 Provide a mechanism for the following one-pot transformation[50]:

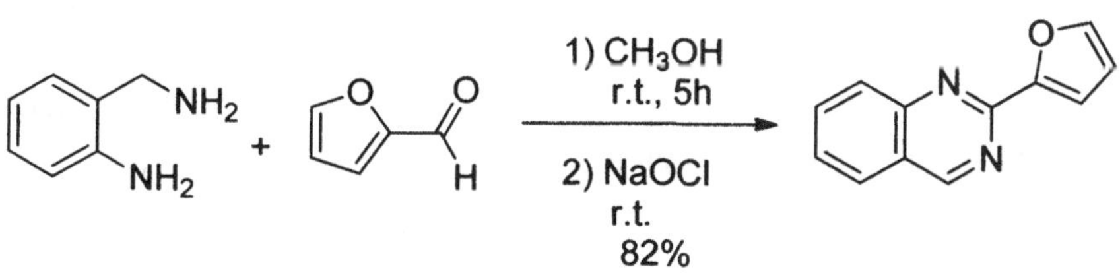

14.5.3 Provide a mechanism for the following transformation[26]:

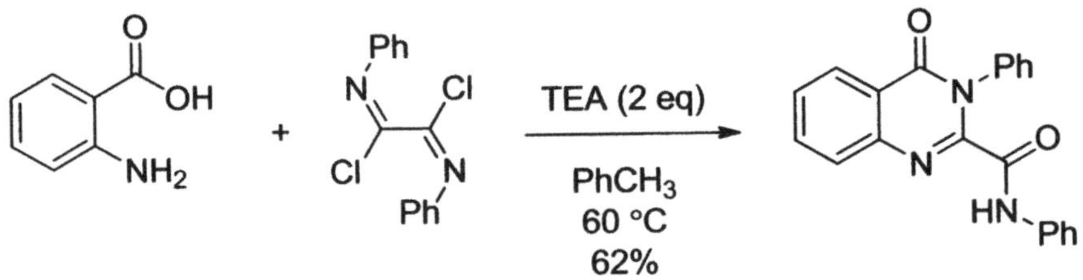

14.5.4 Benzotriazine-4-ones have been used to synthesize quinazolinones. Provide a mechanism for the transformation of **A** to **B** and predict the product of [4 + 2] addition between **B** and benzonitrile[10]:

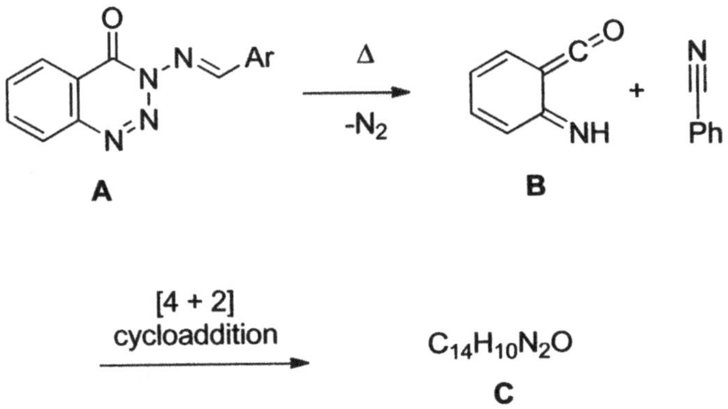

14.5.5 Predict the product of the following reaction[51]:

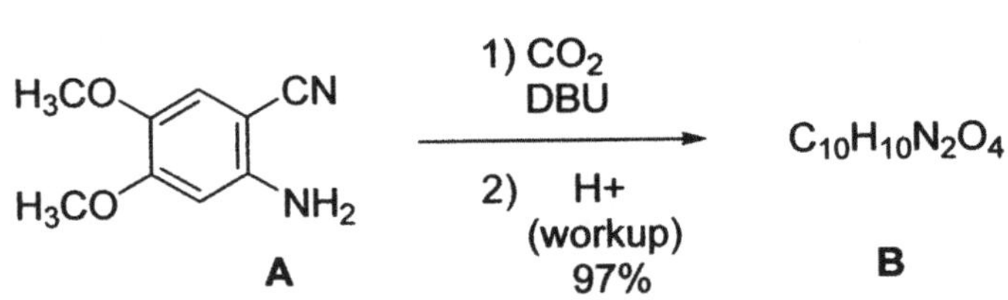

14.5.6 Predict the product of the following reaction:[52]

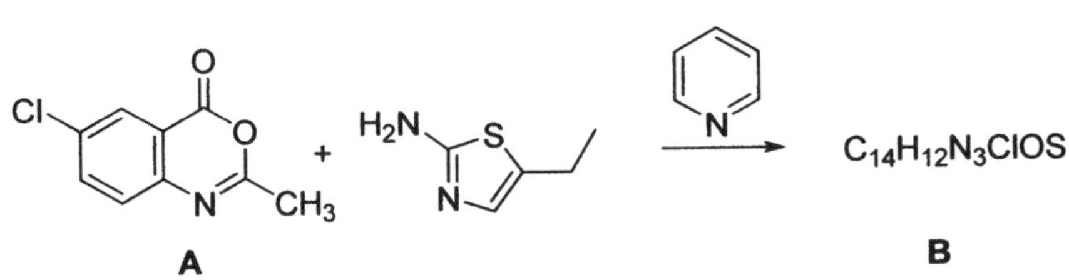

4.5.7 Predict the major products of the following reactions[53]:

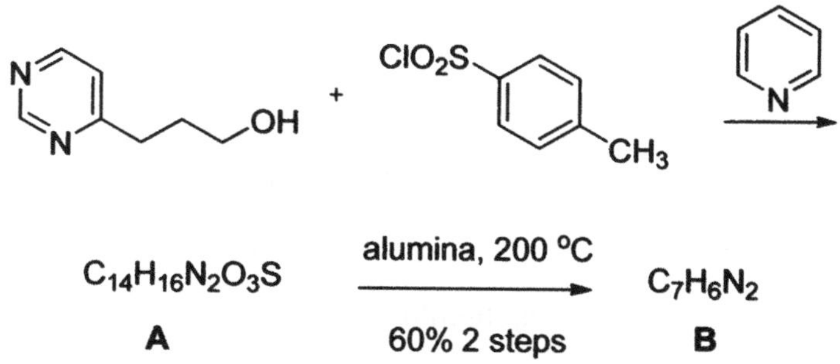

14.5.8 Predict the major products of the following reaction sequence[54]:

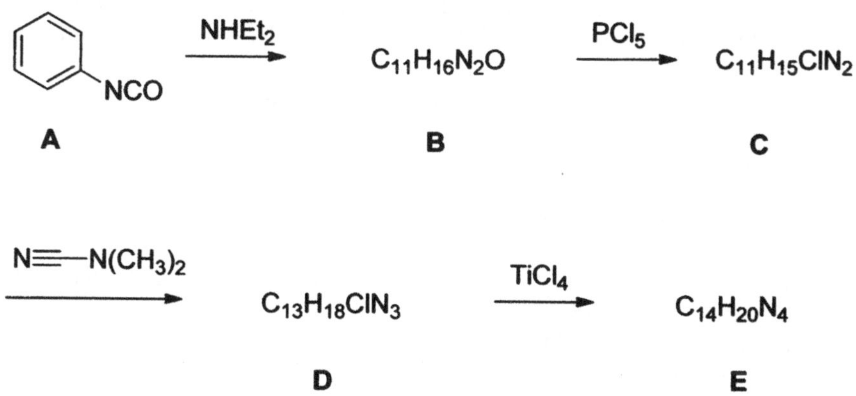

14.5.9 Quinazoline 3-oxide has been shown to be an important precursor for the synthesis chlorodiazepoxide (Librium). Propose a reasonable mechanism for this transformation:

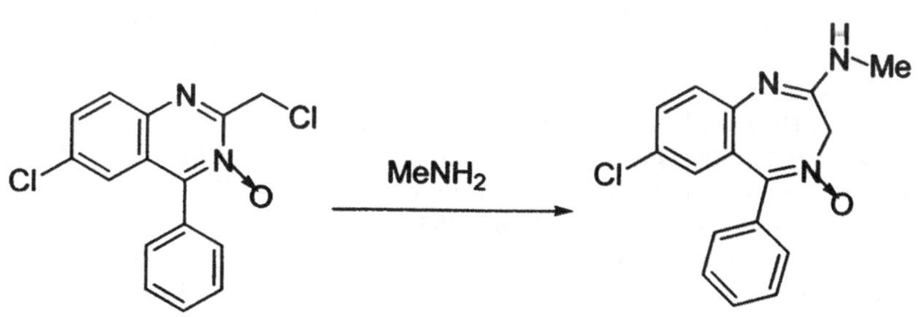

quinazoline 3-oxide chlordiazepoxide (Librium)

14.5.10 Provide a concise total synthesis for the antimalarial isofebrifugine using commercially available reagents containing no more than six carbon atoms:

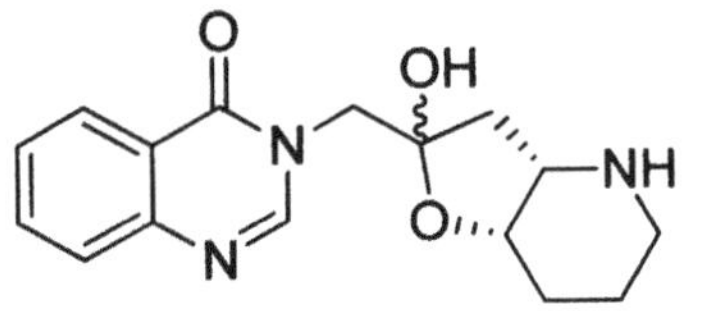

isofebrifugine

14.6 References

1 For a recent review see: Selvam, T. P.; Kumar, P. V. *Res. Pharm.* **2011,** *1,* 1–21.

2 Alexandre, F.-R.; Berecibar, A.; Wrigglesworth, R.; Bessen, T. *Tetrahedron* **2003,** *59,* 1413–1419.

3 Rewcastle, G. W.; Oalmer, B. D.; Bridges, A. J. *J. Med. Chem.* **1996,** *39,* 918–928.

4 Sugimoto, O.; Mori, M.; Moriya, K.; Tanji, K. *Helv. Chem. Acta* **2001,** *84,* 112.

5 Crenshaw, R. R.; Luke, G. M.; Partyka, R. A. U.S.4,102,855 (1978).

6 Rewcastle, G. W.; Palmer, B. D.; Bridges, A. J.; Hollis Showalter, H. D.; Sun, L.; Nelson J.; McMichael, A.; Kraker, A. J.; Fry, D. W.; Denny, W. A. *J. Med. Chem.* **1996,** *39,* 918–928.

7 Shredder, K. R.; Wong, M. S.; Nomanbhoy, T.; Leventhal, P. S.; Fuller, S. R. *Org. Lett.* **2004,** *6,* 3715–3718.

8 Harden, D. B.; Mokrosz, M. J.; Strekowsky, L. *J. Org. Chem.* **1988,** *53,* 4137–4140.

9 Lee, S. J.; Konishi, Y.; Yu, D. T.; Miskowski, T. A.; Riviello, C. M.; Macina, O. T.; Frierson, M. R.; Kondo, K.; Sugitani, M.; Sircar, J. C.; Blazejewski, K. M. *J. Med. Chem.* **1995,** *38,* 3547–3557.

10 Mc Paterson, T.; Smalley, R. K.; Suschitzky, H. *Synthesis* **1975,** 187–189.

11 Lee, H.-S.; Chang, Y.-G.; Kim, K. *J. Heterocycl. Chem.* **1998,** *35,* 659–668.

12 Charpiot, B.; Brun, J.: Donze, I.; Naef, R.; Stefani, M.; Mueller, T. *Bioorg. Med. Chem. Lett.* **1998,** *8,* 2891–2896.

13 Mangalagiu, I.; Benneche, T.; Undheim, K. *Tetrahedron Lett.* **1996,** *37,* 1309–1312.

14 Harayama, T.; Hori, A.; Serban, G.; Morikami, Y.; Matsumoto, T.; Abe, H.; Takeuchi, Y. *Tetrahedron* **2004,** *60,* 10645–10650.

15 Bischler, A.; Burkart, E. *Ber.* **1893,** *26,* 1394–1353.

16 (a) Culbertson, H.; Decisus, J. C.; Christensen, B. E. *J. Am. Chem. Soc.* **1953,** *74,* 4834–4838. (b) Mason, S. G. *J. Chem Soc.* **1957,** 4874–4880.

17 Von Niementowski, S. *J. Prakt. Chem.* **1895,** *51,* 564–572.

18 Bogart, M. T.; Gotthelf, A. H.; *J. Am. Chem. Soc.* **1900,** *22,* 522–535.

19 Patel, V. S.; Patel, S. R.; *J. Indian Chem. Soc.* **1965,** *42,* 531–535.

20 Barker, A. J.; Gibson, K. H.; Grundy, W.; Godfrey, A. A.; Barlow, J. J.; Healy, M. P.; Woodburn, J. R.; Ashton, S. E.; Curry, B J.; Scarlett, L.; Henthorn, L.; Richards, L. *Bioorg. Med. Chem. Lett.* **2001,** *11,* 1191–1914.

21 Őrfi, L.; Wáczek, F.; Pató, J; Varga, I.; Hegymegi-Barakonyi, B.; Houghten, R.A.; Kerí, G. *Curr. Med. Chem.* **2004,** *11,* 2549–2553.

22 Xu, G.-F.; Song, B.-A.; Bhadury, P. S.; Yang, S.; Zhang, P.-Q.; Jin, L.-H.; Zue, W.; Hu, D.-Y.; Lu, P. *Bioorg. Med. Chem.* **2007,** *15,* 3768–3774.

23 Liu, G.; Hu, D.-Y.; Jin, L.-H,; Song, B.A.; Yang, S.; Liu, P. S.; Bhadury, P. S.; Ma, Y.; Luo, H.; Zhou, X. *Bioorg. Med. Chem.* **2007,** *15*, 6608–6617.

24 Hattori, K.; Kido, Y.; Yamamoto, H.; Ishida, J.; Kamijo, K.; Murano, K.; Ohkubo, M.; Kinoshita, T.; Iwashita, A.; Mihara, K.; Yamazaki, S.; Matsuoka, N.; Teramura, Y.; Miyake, H. *J. Med. Chem.* **2004,** 4145–4154.

25 (a) Reid, W.; Stephan, W. *Chem. Ber.* **1963,** *96*, 1218. (b) Reid, W.; Sinharray. A. *Chem. Ber.* **1963,** *96*, 3306. (c) Reid, W.; Stephan, W. *Chem. Ber.* **1962,** *95*, 3042.

26 Wuckelt, J.; Doering, M.; Beckert, R.; Langer, P. *Synlett* **1999,** *7*, 1100–1102.

27 Lee, H.-S.; Chang, Y.-G.; Kim, K. *J. Heterocycl. Chem.* **1998,** *35*, 659–668.

28 Pereira, M.-F; Alexandre, R.; Thiéry, V.; Besson, T. *Tetraherdron Lett.* **2004,** *45*, 3097–3099.

29 (a) Bogert, M. T.; Hand, W. F. *J. Am. Chem. Soc.* **1902,** *24*, 1031–1050. (b) Bogert, M. T.; Hand, W. F. *J. Am. Chem. Soc.* **1903,** *25*, 935–947.

30 Jiang, J. B.; Hesson, D. P.; Dusak, B. A.; Dexter, D. L.; Kang, G. L.; Hamel, E. *J. Med. Chem.* **1990,** *33*, 1721–1728.

31 Kotsuki, H.; Sakai, H.; Morimoto, H.; Suenaga, H. *Synlett.* **1999,** 1993–1995.

32 Wilson, L. J. *Org. Lett.* **2001,** *3*, 585–588.

33 Chandregowda, V.; Rao, G. V.; Reddy, G. C. *Org. Process Rev. Dev.* **2007,** *11*, 813–816.

34 Foucourt, A.; Dubouilh-Benard, C.; Chosson, E.; Corbiére, C.; Buquet, C.; Iannelli, M.; Leblond, B.; Marsais, F.; Besson, T. *Tetrahedron* **2010,** *66*, 4495–4502.

35 Fabis, F.; Santos, J. S. O.; Fouchet-Jolivet, S.; Rault, S. *Tetrahedron Lett.* **2001,** *42*, 5183–5185.

36 Wang, H.; Ganesan, A. *Tetrahedron Lett.* **1998,** *39*, 9097–9098.

37 Eguchi, S.; Suzuki, T.; Okawa, T.; Matsushita, Y. *J. Org. Chem.* **1996,** *61*, 7316–7319.

38 Akazome, M.; Yamamoto, J.; Kondo, T. Watanabe, Y. *J. Organomet. Chem.* **1995,** 494, 229–233.

39 Wang, C.; Li, S.; Liu, H.; Jiang, Y.; Fu, H. *J. Org. Chem.* **2010,** *75*, 7963–7938.

40 Larksarp, C.; Alper, H. *J. Org. Chem.* **2000,** *65*, 2773–2777.

41 Abdel-Jalil, R. J.; Voelter, W.; Saeed, M. *Tetrahedron Lett.* **2004,** *45*, 3475–3476.

42 Kamal, A.; Ramana, V.; Ankati, H. B.; Ramana, A. *Tetrahedron Lett.* **2002,** *43*, 6861–6863.

43 Schnur, R. C.; Arnold, L. D. U.S. 5,747,498 (1998).

44 Davoll, J.; Johnson, A. M. *J. Chem. Soc., C* **1970,** 997–1002.

45 Burch, H. A. *J. Med. Chem.* **1966,** *9*, 408–410.

46 Webber, S. E.; Bleckman, T. M.; Attard, J.; Deal, J. G.; Kathardekar, B.; Welsh, K. M.; Webber, S.; Janson, C. A.; Matthews, D. A.; Smith, W. W.; Freer, S. T.; Jordan, S. R.; Bacquet, R. J.; Howland, E. F.; Booth, C. L. J.; Ward, R. W.; Hermann, S. M.; White, J.; Morse, C. A.; Hilliard, J. A.; Bartlett, C. A. *J. Med. Chem.* **1993,** *36*, 733–746.

47 Zhao, X.; Li, F.; Zhuang, W.; Xue, X.; Lian, Y.; Fan, J.; Fang, D. *Org. Process Res. Dev.* **2010,** *14*, 346–350.

48 Cao, S.-L.; Wan, R.; Feng, Y.-P. *Synth. Comm.* **2003,** *33*, 3510–3526.

49 Marsham, P. R.; Hughes, L. R.; Jackman, A. L.; Hayter, A. J.; Oldfield, J.; Wardleworth, J. M.; Bishop, J. A. M.; O'Connor, B. M.; Calvert, A. H. *J. Med. Chem.* **1991,** *34*, 1594–1605.

50 Maheswari, C. U.; Kumar, G. S.; Venkateshwar, M.; Kumar, R. A.; Kantam, M. L.; Reddy, K. R. *Adv. Synth. Catal.* **2010,** *352*, 341–346.

51 Mizuno, T.; Okamoto, N.; Ito, T.; Miyata, T. *Tetrahedron Lett.* **2000,** *41*, 1051–1053.

52 Párkányi, C. J.; Schmidt, D. S.; *J. Heterocycl. Chem.* **2000,** *37*, 725–729.
53 Wong, J. L.; Brown, M. S.; Rapoport, H. *J. Org. Chem.* **1965,** *30,* 2398–2404.
54 Zielinsky, W.; Kudelko, A.; Holt, E. M. *Heterocycles* **1998,** *48*, 319–328.

Index

Printed and bound by CPI Group (UK) Ltd, Croydon, CR0 4YY
18/09/2026
14965004-0002